Reference Series in Phytochemistry

This series provides a platform for essential information on plant metabolites and phytochemicals, their chemistry, properties, applications, and methods.

By the strictest definition, phytochemicals are chemicals derived from plants. However, the term is often also used to describe the large number of secondary metabolic compounds found in and derived from plants. These metabolites exhibit a number of nutritional and protective functions for human wellbeing and are used e.g. as colorants, fragrances and flavorings, amino acids, pharmaceuticals, hormones, vitamins and agrochemicals.

The series offers extensive information on various topics and aspects of phytochemicals, including their potential use in natural medicine, their ecological role, role as chemo-preventers and, in the context of plant defense, their importance for pathogen adaptation and disease resistance. The respective volumes also provide information on methods, e.g. for metabolomics, genetic engineering of pathways, molecular farming, and obtaining metabolites from lower organisms and marine organisms besides higher plants. Accordingly, they will be of great interest to readers in various fields, from chemistry, biology and biotechnology, to pharmacognosy, pharmacology, botany and medicine.

The Reference Series in Phytochemistry is indexed in Scopus.

Hosakatte Niranjana Murthy
Editor

Bioactive Compounds in Bryophytes and Pteridophytes

With 110 Figures and 67 Tables

Springer

Editor
Hosakatte Niranjana Murthy
Department of Botany
Karnatak University
Dharwad, Karnataka, India

ISSN 2511-834X ISSN 2511-8358 (electronic)
Reference Series in Phytochemistry
ISBN 978-3-031-23242-8 ISBN 978-3-031-23243-5 (eBook)
https://doi.org/10.1007/978-3-031-23243-5

This Springer imprint is published by the registered company Springer Nature Switzerland AG.
The registered company address is: Gewerbestrasse 11, 6330 Cham, Switzerland

Preface

Bryophytes are non-vascular plants which are possessing a dominant gametophytic phase in their life cycle. A large diversity of bryophytes is existing on the earth and they have been classified into hornworts, liverworts, and mosses. Bryophytes are reported to possess a wide variety of phytochemicals including sugars, lipids, nitrogen, and sulfur-containing compounds in addition to phenolics and terpenoids. In contrast, pteridophytes are sporophytes which are possessing roots, stems and leaves. These are flowerless, seedless, vascular plants that show true alternation of generations. These plants are referred to as vascular cryptogams and reproduce by spores that developed in sporangia and are classified as homosporous and heterosporous types. Phytochemically pteridophytes are rich in phenolics, flavonoids, alkaloids, and terpenoids. Both bryophytes and pteridophytes are used in traditional medicines and are reported to have diverse biological activities.

Against this backdrop, this book encompasses research work on the bioactive compounds of bryophytes and pteridophytes. The chapters presented in this volume focus on several research subjects that have provided extensive information on bioactive compounds and their biological activities. Each chapter provides specific groups containing information in the form of tables and illustrations. Abundant useful references are provided in each chapter which will be useful for future research and experimentation.

I would like to thank and express my deepest gratitude to the contributors who helped me to complete this book. I am thankful to Professor Jean-Michel Merillon and Professor Kishan Gopal Ramawat, Series Editors, for their constant encouragement. I thank Dr. Sylvia Blago and Veronika Mang for their constant help and support. Finally, I am thankful to the Springer editorial team and production team for completing this assignment successfully.

Dharwad, India
June 2023

Hosakatte Niranjana Murthy

Contents

About the Editor

Hosakatte Niranjana Murthy, Professor at Post-Graduate Department of Botany, Karnatak University, Dharwad, India, has obtained his Ph.D. degree from Karnatak University, India. He has a tremendous passion for research and academics. Since 1986, he has served in various positions at the Post-Graduate Department of Botany, Karnatak University, Dharwad, India. Apart from his teaching experience of 35 years, he possesses extensive research experience in the area of plant biotechnology. He has post-doctoral and collaborative research experience in many foreign research institutes. He worked at Biotechnology Division, Tata Energy Research Institute, New Delhi, India (1992); Crop Science Department, University of Guelph, Guelph, Canada (1993); Research Centre for the Development of Horticultural Technology, Chungbuk National University, Cheongju, South Korea (2000–2001, 2002, 2004, 2006–2007, 2013–2014); and Department of Biological Sciences, University of Nottingham, Nottingham, United Kingdom (2005–2006) as a post-doctoral fellow/visiting scientist. He is the recipient of various prestigious fellowships including Biotechnology National Associate, Biotechnology Overseas Associate (awarded by Department of Biotechnology, Ministry of Science and Technology, Government of India), Brain Pool Fellowship (awarded by Korean Society of Science and Technology, South Korea), Visiting Fellowship (awarded by Korea Science and Engineering Foundation, South Korea), and Commonwealth Post-doctoral Fellowship (awarded by Association of Commonwealth Universities, UK). He has completed more than 15 research projects funded by various agencies and guided several Ph.D. students. He has published more than 225 research articles in international peer-reviewed journals with high impact factors. His research work has been cited more than 4800 times by fellow researchers and has an H-index (Hirsch index) of 38 as recorded by Scopus.

Professor Hosakatte Niranjana Murthy has developed biotechnological methods for the production of pharmaceutically important secondary metabolites from cell and organ cultures of Ginseng, Siberian ginseng, Echinacea, St. John's wort using large-scale bioreactors along with South Korean collaborators. His experimental investigations on the use of adventitious root cultures and bioreactor technologies for the production of biomass and secondary metabolites have paved the way for the commercialization of plant secondary metabolites. Various ginseng-based commercial products have been released and are currently available in the market.

Contributors

Ivana Márcia Alves Diniz Departamento de Odontologia Restauradora, Universidade Federal de Minas Gerais, Minas Gerais, Brazil

O. Roger Anderson Biology and Paleo Environment, Lamont-Doherty Earth Observatory, Columbia University, Palisades, NY, USA

Shivananthini Balasundaram Department of Botany, Centre for Plant Biotechnology, St. Xavier's College (Autonomous), Palayamkottai, India

Anda Maria Baroi National Institute for Research & Development in Chemistry and Petrochemistry – ICECHIM Bucharest, Bucharest, Romania

University of Agronomic Sciences and Veterinary Medicine of Bucharest, Bucharest, Romania

Jayanta Barukial Debraj Roy College, Golaghat, Assam, India

Ángel Benítez Departamento de Ciencias Biológicas y Agropecuarias, Universidad Técnica Particular de Loja, Loja, Ecuador

Medha A. Bhat Department of Botany, Karnatak University, Dharwad, India

Protha Biswas Department of Life Sciences, Presidency University, Kolkata, West Bengal, India

Kaitlyn L. Blatt-Janmaat Bioinformatics and Scientific Data, Leibniz Institute of Plant Biochemistry, Halle (Saale), Germany

Department of Chemistry, University of New Brunswick, Fredericton, NB, Canada

Frédéric Bourgaud Plant Advanced Technologies, Vandœuvre-lès-Nancy, France

Cellengo, Vandœuvre-lès-Nancy, France

Bojan Božić Institute of Physiology and Biochemistry "Ivan Djaja", Faculty of Biology, University of Belgrade, Belgrade, Serbia

Roxana Ioana Brazdis National Institute for Research & Development in Chemistry and Petrochemistry – ICECHIM Bucharest, Bucharest, Romania

University "Politehnica" of Bucharest, Bucharest, Romania

Tsun-Thai Chai Center for Agriculture and Food Research, Universiti Tunku Abdul Rahman, Kampar, Malaysia

Department of Chemical Science, Faculty of Science, Universiti Tunku Abdul Rahman, Jalan Universiti, Kampar, Malaysia

Pankaj Chetia Department of Life Sciences, Dibrugarh University, Dibrugarh, Assam, India

Department of Life Science & Bioinformatics, Assam University, Silchar, Assam, India

Yit-Lai Chow Center for Agriculture and Food Research, Universiti Tunku Abdul Rahman, Kampar, Malaysia

Department of Biological Science, Faculty of Science, Universiti Tunku Abdul Rahman, Jalan Universiti, Kampar, Malaysia

Rosangela Maria Ferreira da Costa e Silva Universidade Estadual do Mato Grosso do Sul, Campus Dourados-MS, Mato Grosso do Sul, Brazil

José Maria da Fonte Ferreira CICECO -Aveiro Institute of Materials, Department of Materials and Ceramic Engineering, University of Aveiro, Aveiro, Portugal

Kakoli Das Department of Life Science and Bioinformatics, Assam University, Silchar, India

Subrata Das Department of Life Science and Bioinformatics, Assam University, Silchar, Assam, India

Tuyelee Das Department of Life Sciences, Presidency University, Kolkata, India

Anupam Das Talukdar Department of Life Science and Bioinformatics, Assam University, Silchar, Assam, India

Jamilly Bignon de Souza Programa de Pós-Graduação em Ciências Aplicadas a Produtos para a Saúde, Universidade Federal Fluminense, Niterói, Brazil

Abhijit Dey Department of Life Sciences, Presidency University, Kolkata, West Bengal, India

Damini Dey Department of Life Sciences, Dibrugarh University, Dibrugarh, Assam, India

Manabendra Dutta Choudhury Department of Life Science & Bioinformatics, Assam University, Silchar, Assam, India

Raphaël E. Duval Université de Lorraine, CNRS, Nancy, France

Faculté de Pharmacie, ABC Platform®, Vandœuvre-lès-Nancy, France

Kasper Enemark-Rasmussen Department of Chemistry, Technical University of Denmark, Lyngby, Denmark

Adrià Farràs Department of Pharmacy, Faculty of Health Sciences, Universidad San Jorge, Villanueva de Gállego (Zaragoza), Spain

Department of Biochemistry and Physiology, Faculty of Pharmacy and Food Sciences, Universitat de Barcelona, Barcelona, Spain

Irina Fierascu National Institute for Research & Development in Chemistry and Petrochemistry – ICECHIM Bucharest, Bucharest, Romania

University of Agronomic Sciences and Veterinary Medicine of Bucharest, Bucharest, Romania

Radu Claudiu Fierascu National Institute for Research & Development in Chemistry and Petrochemistry – ICECHIM Bucharest, Bucharest, Romania

University "Politehnica" of Bucharest, Bucharest, Romania

Toma Fistos National Institute for Research & Development in Chemistry and Petrochemistry – ICECHIM Bucharest, Bucharest, Romania

University "Politehnica" of Bucharest, Bucharest, Romania

Vidyarani George Department of Botany, Centre for Plant Biotechnology, St. Xavier's College (Autonomous), Palayamkottai, India

Mimosa Ghorai Department of Life Sciences, Presidency University, Kolkata, West Bengal, India

Charlotte H. Gotfredsen Department of Chemistry, Technical University of Denmark, Lyngby, Denmark

Porismita Hazarika Dibrugarh University, Dibrugarh, Assam, India

Nicolas Hocquigny Université de Lorraine, CNRS, Nancy, France

Faculté de Pharmacie, ABC Platform®, Vandœuvre-lès-Nancy, France

Silvia Juliet Iruthayamani Department of Botany, Centre for Plant Biotechnology, St. Xavier's College (Autonomous), Palayamkottai, India

Vartika Jain Department of Botany, Government Meera Girls College, Udaipur, Rajasthan, India

Ximena Jaramillo-Fierro Departamento de Química, Universidad Técnica Particular de Loja, Loja, Ecuador

Luciana de Jesus Jatoba Federal Institute of Education, Science, and Technology of São Paulo – IFSP, Hortolândia, Brazil

Sibashish Kityania Department of Life Science and Bioinformatics, Assam University, Silchar, Assam, India

Gabriela Pereira Lima Programa de Pós-Graduação em Ciências Aplicadas a Produtos para a Saúde, Universidade Federal Fluminense, Niterói, Brazil

Víctor López Department of Pharmacy, Faculty of Health Sciences, Universidad San Jorge, Villanueva de Gállego (Zaragoza), Spain

Yi Lu Department of Biotechnology and Biomedicine, Technical University of Denmark, Kongens Lyngby, Denmark

Tanja Lunić Institute of Physiology and Biochemistry "Ivan Djaja", Faculty of Biology, University of Belgrade, Belgrade, Serbia

Johnson Marimuthu alias Antonysamy Department of Botany, Centre for Plant Biotechnology, St. Xavier's College (Autonomous), Palayamkottai, India

Raíssa Volpatto Marques Department of Biotechnology and Biomedicine, Technical University of Denmark, Kongens Lyngby, Denmark

Miguel Angel Meneses Departamento de Química, Universidad Técnica Particular de Loja, Loja, Ecuador

Sissi Miguel Cellengo, Vandœuvre-lès-Nancy, France

Milica Milutinović Institute for Biological Research "Siniša Stanković"-National Institute of Republic of Serbia, University of Belgrade, Belgrade, Serbia

Montserrat Mitjans Department of Biochemistry and Physiology, Faculty of Pharmacy and Food Sciences, Universitat de Barcelona, Barcelona, Spain

Hosakatte Niranjana Murthy Department of Botany, Karnatak University, Dharwad, India

Deepa Nath Department of Botany, Gurucharan College, Silchar, Assam, India

Rajat Nath Department of Life Science and Bioinformatics, Assam University, Silchar, Assam, India

Biljana Božić Nedeljković Institute of Physiology and Biochemistry "Ivan Djaja", Faculty of Biology, University of Belgrade, Belgrade, Serbia

Eduard V. Nekrasov Amur Branch, Botanical Garden-Institute of the Far Eastern Branch of the Russian Academy of Sciences, Blagoveshchensk, Russia

Selma Ribeiro Paiva Programa de Pós-Graduação em Ciências Aplicadas a Produtos para a Saúde, Universidade Federal Fluminense, Niterói, Brazil

Laboratório de Botânica Estrutural e Funcional, Departamento de Biologia Geral, Instituto de Biologia, Universidade Federal Fluminense, Niterói, Brazil

Kristian Peters German Centre for Integrative Biodiversity Research (iDiv) Halle-Jena-Leipzig, Leipzig, Germany

Institute of Biology/Geobotany and Botanical Garden, Martin Luther University Halle-Wittenberg, Halle (Saale), Germany

Bioinformatics and Scientific Data, Leibniz Institute of Plant Biochemistry, Halle (Saale), Germany

Yvonne Poeschl German Centre for Integrative Biodiversity Research (iDiv) Halle-Jena-Leipzig, Leipzig, Germany

Institute of Biodiversity, Friedrich Schiller University, Jena, Germany

Aline Possamai Della Department of Botany, Institute of Biosciences, Laboratory of Systematics and Biogeography of Vascular Plants, University of São Paulo, São Paulo, Brazil

Minakshi Puzari Department of Life Sciences, Dibrugarh University, Dibrugarh, Assam, India

Yixian Quah Developmental and Reproductive Toxicology Research Group, Korea Institute of Toxicology, Daejeon, Republic of Korea

Matteo Radice Departamento Ciencias de la Tierra, Universidad Estatal Amazónica, Puyo, Ecuador

Arnaud Risler Université de Lorraine, CNRS, Nancy, France

Aleksander Salwinski Plant Advanced Technologies, Vandœuvre-lès-Nancy, France

Marcelo Guerra Santos Laboratório de Biodiversidade, Departamento de Ciências, Faculdade de Formação de Professores, Universidade do Estado do Rio de Janeiro, São Gonçalo, Brazil

Ionela Daniela Sardarescu (Toma) University "Politehnica" of Bucharest, Bucharest, Romania

National Research and Development Institute for Biotechnology in Horticulture, Ştefăneşti, Argeş, Romania

Henrik Toft Simonsen Department of Biotechnology and Biomedicine, Technical University of Denmark, Kongens Lyngby, Denmark

Shreeta Singha Department of Life Science and Bioinformatics, Assam University, Silchar, Assam, India

Marijana Skorić Institute for Biological Research "Siniša Stanković"-National Institute of Republic of Serbia, University of Belgrade, Belgrade, Serbia

Kandikere Ramaiah Sridhar Department of Biosciences, Mangalore University, Mangalagangotri, Mangalore, India

Wojciech J. Szypuła Department of Pharmaceutical Biology, Faculty of Pharmacy, Medical University of Warsaw, Warsaw, Poland

Sheri-Ann Tan Department of Bioscience, Faculty of Applied Sciences, Tunku Abdul Rahman University of Management and Technology, Kuala Lumpur, Malaysia

Shi-Ruo Tong Department of Physical Science, Faculty of Applied Sciences, Tunku Abdul Rahman University of Management and Technology, Kuala Lumpur, Malaysia

Henriette Uthe German Centre for Integrative Biodiversity Research (iDiv) Halle-Jena-Leipzig, Leipzig, Germany

Institute of Biodiversity, Friedrich Schiller University, Jena, Germany

Eduardo Valarezo Departamento de Química, Universidad Técnica Particular de Loja, Loja, Ecuador

Govardhana G. Yadav Department of Botany, Karnatak University, Dharwad, India

Alam Zeb Department of Biochemistry, University of Malakand, Chakdara, Pakistan

Suzana Živković Institute for Biological Research "Siniša Stanković"-National Institute of Republic of Serbia, University of Belgrade, Belgrade, Serbia

Part I
Bryophytes

Bioactive Compounds from Bryophytes

1

Kakoli Das, Sibashish Kityania, Rajat Nath, Subrata Das, Deepa Nath, and Anupam Das Talukdar

Contents

Abstract

Bryophytes are "nonvascular plants" that are also termed "amphibious plants." There are around 24,000 species of bryophytes present in the world; the major groups are Hornworts (300 species), liverworts (600 species), and mosses (14,000 species). Bryophytes are restricted to shady and moist places, such as damp trees and rocks, by the side of streams or pools. Various bioactive compounds are present in bryophytes including polyphenols, steroids, organic acids, terpenoids, and phenyl quinone. Some species of liverworts including *Bazzania*, *Conocephalum conicum*, *Riccia gangetica*, and *Radula* species have vital metabolites with significant antifungal action against *Aspergillus*, *Fusarium*, and *Penicillium*, such as Marchantin and lunularin. In earlier times, tribal people

K. Das
Department of Life Science and Bioinformatics, Assam University, Silchar, India

S. Kityania · R. Nath · S. Das · A. D. Talukdar (✉)
Department of Life Science and Bioinformatics, Assam University, Silchar, Assam, India
e-mail: anupam@bioinfoaus.ac.in

D. Nath
Department of Botany, Gurucharan College, Silchar, Assam, India

H. N. Murthy (ed.), *Bioactive Compounds in Bryophytes and Pteridophytes*,
Reference Series in Phytochemistry, https://doi.org/10.1007/978-3-031-23243-5_2

used many bryophytes to cure various diseases, such as skin disorders, hepatic disorders, cardiovascular diseases, antimicrobial, wound healing, and antipyretic. Tribal people belonging to Africa, Europe, America, Australia, China, Taiwan, Pakistan, and various regions of eastern, northern, and southern India used various bryophytes as food and ethnomedicine. Some common ethnomedicinal bryophytes used by some of the tribal communities are, namely, *Marchantia polymorpha* against inflammation, *Plagiochasma appendiculata* against skin infections, *Plytrichum* species for growth of hair, and species of *Riccia* against ringworms in children. Secondary metabolites derived from bryophytes have a wide range of pharmacological effects, including antimicrobial, antioxidant, anti-inflammatory, and anti-cancer properties.

Keywords

Bioactive compounds · Bryophytes · Ethnomedicinal · Pharmacological · Secondary metabolites

Abbreviation

GC –MS	Gas chromatography-mass spectrometry
HepG2	Hepatoma G2
MCF-7	Michigan Cancer Foundation-7
ROS	Reactive oxygen species
TEM	Transmission electron microscope

1 Introduction

Bryophytes, or nonvascular terrestrial plants, are classified into three kinds: hornworts (*Anthocerophyta*), liverworts (*Marchantiophyta*), and mosses (*Bryophyta*); they are mainly thalloid or foliose in structure. The plant body of bryophyte is not distinguished into leaf, stem, and root; morphologically they are very simple but chemically they are very complex [1]. Traditionally, various sorts of illnesses were treated using bryophytes by the tribal communities. Except oceans, bryophytes can be found in many habitats, and they are a very much important source of some bioactive compounds. In India, a total of 2504 species of bryophytes are available, which is 17.27% of the world's bryophytes. Bryophytes are always found in damp rocks, wet areas, and tree trunks in forests. Bryophytes play an important buffer system for other plants, they can be used as medicines, animal food, and cosmetics, and some species of bryophytes are also used in horticulture and fiber industries. In countries like Ireland, Poland, and Finland, some species of moss and liverworts are used as fuel. Some bryophyte species like *Marchantia polymorpha*, *Riccia* sp., and *Anthoceros* sp. have great medicinal values, for example, *Marchantia polymorpha* cures diseases like liver infection and pulmonary Tuberculosis. All over the world, 65 species of *Marchantia* are being identified and out of 65 species, 11 species are reported in India, including *Marchantia gemminata*, *Marchantia emarginata*, *Marchantia hartlessiana*, *Marchantia assamica*, *Marchantia papillata*, *Marchantia paleacea*, *Marchantia linearis*, *Marchantia polymorpha*, *Marchantia robusta*, and *Marchantia subintegra* [2]. Some mosses and liverworts are an indicator of environmental conditions, and some of the moss species help to prevent soil erosion by forming a mat-like structure as their rhizoids bind soil particles which reduces the loss of water. Some economically important bryophytes are *Leucobryum glaucum*, used for decoration and cushion making, *Floribundaria* species, used for making shepherds' homes, and *Papillaria* species, used to make repellants against insects [3]. In bryophytes, more than 2200 chemical constituents are present, such as terpenoids (monoterpenoid, sesquiterpenoid, and diterpenoids), lipids, (bis) and bibenzyls are some natural products isolated from bryophytes [4]. Lipophilic mono-, sesqui-, and diterpenoids are also identified and isolated from liverworts; the terpenoids are more than 1600 in number, but in moss, few mono- and diterpenoids and around 100 sesquiterpenoids are identified today [5]. *Pseudoscleropodium purum*, *Eurhynchium angustirete*, and *Eurhynchium striatum* were found to have essential oil that has antimicrobial activity [6]. Marchantin A, discovered in *Marchantia polymorpha* and *Marchantia emarginata* subsp. *tosana*, Palagiochin A, and Perrottein F, all are bisbibenzyls. *Marchantia polymorpha* has antiplasmodial property [7] and *Marchantia emarginata* has antioxidant and anticancerous properties. Compounds like β-phellandrene and β-caryophyllene are terpenoids found in species *Porella cordaeana* that are showing antimicrobial activities [8]. Some other compounds from *Dicranum scoparium* and *Polytrichastrum formosum* have anti-insect [9] antifeedant [10] properties.

2 Ethnobotanical Importance of Bryophytes

A lower group of plants like bryophytes are considered to have some medicinal values as they were used by tribal communities for a long to cure various diseases. Some ancient tribal communities used bryophytes as their traditional medicine due to their beneficial chemical constituents [11]. Ethnobotanical research is an important subject of study, including advances in medicines, biodiversity protection, and resource management. Tribal communities use bryophytes to cure many diseases and this knowledge helps researchers a lot in the discovery of active phytochemicals that have medicinal potential. Melghat forest (India) is rich in biodiversity, and tribal people of the locality use some of the bryophytes for diseases like respiratory disease, skin disease, cold, and fever. *Marchantia polymorpha* thalli are used against inflammation, and for skin disease they use thalli of species *Plagiochasma appendiculatum* externally and for hair growth *Polytrichum* sp., and some species of *Riccia* are also used against the treatment of ringworms, the paste of thallus is mixed with jaggery and given it to the affected children [12]. In ancient China, *Marchantia* species has been used as a medicinal plant. Bryophytes protect skin and recover skin wounds and skin diseases [13]. It was known that traditionally they were used as medicine to treat skin diseases and also they have antibacterial, antifungal, and antiviral capacities [14]. The term "ethnobryology" was given by Sevile Flowers [11], where different uses of bryophytes were discussed [15]. Treatment for diseases like ringworm by using different species of *Riccia* was also reported [16]. It was reported that tea made from *Polytrichum commune* can liquefy gall bladder stones and stones present in the kidney. Some mosses show various ethnobotanical practices, *Philonotis* sp. is used to treat broken bones [14], and extract of *Rhodobryum giganteum* can cure angina disease [17]. *Polytrichum commune* Hedw. was traditionally used to cure pulmonary tuberculosis [1]. *Polytrichum commune* is used as a traditional remedy in China to cure uterine prolapse, lymphocytic leukemia, and fever [18]. *Polytrichum commune* was found to have cytotoxic [19], antineuroinflammatory [20], and anticancerous potentials. *Polytrichum commune* is also a laxative, which means they cure constipation and diuretic problems, also known as water pills [21]. *Marchantia polymorpha* was also traditionally used to heal cuts and burns [22], *Marchantia polymorpha* is showing antiviral and antifungal activities [23]. *Reboulia hemisphaerica* (L.) Raddi was formerly employed to treat exterior wounds, blotches, and hemostasis [1], and it also has antiplatelet [24] and antimicrobial [25] properties. *Polytrichum juniperinum* Hedw. used in the treatment of skin conditions and urinary issues [22] and shows antibacterial action against *Enterococcus faecalis, Bacillus cereus*, and *Streptococcus pyogenes* [26]. It also shows anticancerous action against human breast cancer (A549), mouse Sarcoma 37 cells, and intestinal cancer (CaCo-2) [27]. *Fontinalis antipyretica* Hedw. was traditionally used to cure fever and microbial diseases and to detoxification and it was also found that *Fontinalis antipyretica* Hedw. is antiproliferative [18], antimicrobial, and antiproliferative against neoplastic cell lines [28].

3 Phytochemicals from Bryophytes and Their Bioactivity

Trease and Evans were done the phytochemical analysis of *Bryum cellulare* extract to see the presence of bioactive compounds [29]. Deora et al. performed various experiments that show a variety of antibiotic substances are present in bryophytes. *Bryum cellulare* shows various antifungal activities against *Cochliobolus lunatus* and *Diplodia maydis*, in a biochemical assay using mycoherbicide [30]. *Marchantia*, a common liverwort genus famous all over the world, possesses properties that prevent inflammation and infection. Major chemical constituents of *Marchantia* including steroids, triterpenoids, and flavonoids, such as luteolin, apigenin, and quercetin. In a study, it was found that ether extract of *Marchantia polymorpha* has isoprenoid compounds including acoradiene, thujopsene, and α- chamignen-9-one, which were examined using gas chromatography-mass spectrometry (GC-MS). Marchantin A and plagiochin E isolated from *Marchantia emarginata* and *Marchantia polymorpha,* respectively, also have anticancer and antifungal properties [31]. In the genus *Marchantia*, marchantin A, a cyclic bis (bib enzyl ether) that is present in the species *Marchantia emarginata* subsp. *tosana* induces apoptosis in human MCF-7 breast cancer cells. Marchantin A exhibits activity against several kinds of bacteria and fungi [32]. Marchantin A after observation through transmission electron microscopy (TEM) found active against *Pseudomonas aeruginosa* and Gram-negative *Pasteurella multocida* [31]. Plagiochin E is antifungal macrocyclic bis (bib benzyl) that show antifungal activity [33]. From various studies, it was found that the secondary metabolite of some bryophytes possesses potential antifungal activity. *Marchantia* thallus methanolic extract and phenolic chemicals were obtained, such as protocatechol, cinnamate, chlorogenate, vanillate, and gallate. After the phytochemical screening of bryophytes, namely, *Targionia hyphophylla*, *Anthoceros erectus*, *Plagiochasma articulata*, *Cyathodium tuberosum*, and *Asterella angusta*, it was found that they also have some bioactive compounds. Ethanol extracts of *Targionia hyphophylla* and *Plagiochasma articulata* possess more antibacterial activity than the methanol extract of *Asterella angusta.* Secondary metabolites, such as alkaloids, steroids, phenols, flavonoids, tannins, and coumarins are present in these bryophytes [4]. The phytochemical study of these bryophytes revealed the presence of phytochemical constituents like flavonoids, coumarins, phenols, sugars, and steroids [34]. The genera of *Chiloscyphus*, *Plagiochila*, and *Scapania* possess cholesterol and simple lipids (triglycerides and waxes) and complex lipid (glyco- and phospholipids) are often present. Organic acids (cis-aconitic acid, malic acid, malonic acid, shikimic acid, fumaric acid) and polyacetylenes are some other substances also isolated from bryophytes [4]. Phytochemical analysis of *Anthoceros erectus* and *Plagiochasma articulata* reveals they have a few secondary metabolites, including alkaloids, flavonoids, phenols, coumarins, steroids, and tannin [35]. Ethanol extract of *Conocephalum conicum* shows the presence of lunularin which has an anticancer property and shows cytotoxicity against the HepG2 cell in humans [36]. Phytochemical analysis of *Philonotis* sp. shows that some active compounds like triterpenoid-saponins were found, which are also antipyretic and antidotal. Methanol and ethanol extract of *Plagiochila* sp. have active phytochemicals, namely, bicyclohumslenone, plagiochilin A, plagiochilin B, and menthane monoterpenoids [37]. Methanol extract of *Radula sumatrana* showed the presence of compounds rasumatranins A-D, M, and N, which are found effective against cancer [38]. Pallidisetin A and pallidisetin B are present in *Polytrichum pallidisetum,* which have cytotoxic and anti-inflammatory action against many human tumor cell lines [39]. Phytochemicals jungermannenone A and jungermannenone B are found in *Jungermannia fauriana* and both have anticancer property [40]. The phytochemicals found in bryophytes along with their therapeutic activity are listed below in Table 1. Some important phytochemical structures (flavonoids, terpenoids, polyketides, bisbenzyls, and bisbibenzyls) obtained from various bryophyte species are shown in Figs. 1 and 2.

4 The Pharmacological Activity of Bryophytes

4.1 Antimicrobial Activity

The methanolic extract of the *Plagiochila* sp. exhibits plagiochilin A and plagiochilin B, which shows antimicrobial activity. Apigenin, kaempferol, luteolin, and glycosides have been found to show antimicrobial activity; these chemicals were derived from *Sphagnum* sp. Methanolic extract of *Marchantia* also showed antibacterial action.

Table 1 Summary of different phytochemicals and therapeutic activities of bryophytes

Scientific name	Solvent used	Phytochemicals	Therapeutic activity	References
Bryum celluare	Ethanolic extract, methanolic extract, aqueous extract	Flavonoids Terpenoids Sterols	Antifungal	[30]
Marchantia convolunta	Ethyl acetate	β-Caryophyllene, diepi-alpha-cedrene epoxide, ledene oxide, 9-cedranone, tetra de canoic acid, methyl ester 1,2,4-tripropylbenzene phytol	Cytotoxic against liver and lung carcinoma (anticancer)	[41]
Marchantia polymorpha	Methanol extract	Marchantin A; MB-G (35 a) and marchantin E	Anticancer, antipyretic, and antibacterial, decrease the cell viability of T47D and A 256 cell lines.	[42] [43]
Marchantia papillate	Methanol extract, ethanol extract, acetone extract	Riccardin C Fatty acids (31.77%), steroids (11.52%), bibenzyl (2.46%)	Anti-inflammatory, treat inflammation caused by fire, antibacterial, treat boils.	[44]
Marchantia linearis	Ethanol extract	Flavonoid	Antifungal	[45]
Marchantia convoluta	Methanol extract	Flavonoid	Antiviral (hepatitis B), anti-inflammatory, cytotoxic (HepG2 and H1299 cell lines)	[46]
Marchantia paleacea	Methanol extract	Flavonoids, saponins	Antimicrobial	[47]
Marchantia polymorpha	Tryptone, agar, glucose, and yeast extract	Plagiochin E	Antifungal, macrocyclic bis (bibenzyl) against Candida al bicans)	[48]
Asterella angusta	Methanol extract	Alkaloids, coumarins, sugar, flavonoids, phenols	Antibacterial, antifungal	[49]
Targionia hyphophylla	Ethanol extract	Alkaloids, coumarins, sugar, steroids, tannins	Antibacterial	[37]

(continued)

Table 1 (continued)

Scientific name	Solvent used	Phytochemicals	Therapeutic activity	References
Plagiochasma articulata	Ethanol extract	Coumarins, sugar, steroids, tanins	Antifungal, antibacterial	[50]
Riccardia sp.	Methanol extract	Riccardin A and B, sacullatal	Antileukemic activity	[51]
Philonotis sp.	Methanol extract	Triterpenoid-saponins	Antidotal, antipyretic	[52]
Plagiochila sp.	Methanol extract	Bicyclohumslenone, plagiochilin A, plagiochilin B; menthanemono terpenoids	Antimicrobial and antileukemic activity.	[37]
Pallavicinia sp.	Methanol extract	Sacullatal	Antimicrobial	[53]
Reboulia hemisphaerica	Ethanol extract	Benzoquinone	Hemostasis cures wounds	[1]
Frullania tamarisci	Methanol extract	Tamariscol, frullanolide	Antiseptic activity	[1]
Cratoneuron filicinum	Methanol extract	Terpenoids, acetogenins	Cures heart disease	[3]
Philonotis Fontana	Methanol extract	Alcohol, amides, aromatic amines	Cure heat burns	[11]
Philonotis sp.	Ethanol extract	Alkaloid, steroids, glycosides	Cure burns, used as an antidote	[1]
Oreas martina	Methanol extract	Glycosides, fatty acids, terpenoids	Cures epilepsy, hemostasis, nervous disorders	[1]
Ditrichum pallidum	Ethanol extract	Phenols, fatty acids	Treats convulsions	[3]
Weisia viridula	Methanol extract	Phenols	Antimicrobial	[1]
Plagiochasma appendiculantum	Methanol extract	Alkaloids, flavonoids, carbohydrates, saponins	It cures skin diseases	[12]
Dumortiera hirsuta	Ethanol extract	Riccardin D	Antibiotics, anticancer	[29]
Leptodictyum riparium	Methanol extract	Phenolic compounds	Used in the treatment of uropathy	[3]
Rhodobryum roseum	Ethanol extract	Flavonoids and phenols	Used in the treatment of nervous disorders and cardiovascular diseases.	[3]
Fissidens nobilis	Methanol extract	Terpenoids	In the treatment of hair fall	[29]

(continued)

Table 1 (continued)

Scientific name	Solvent used	Phytochemicals	Therapeutic activity	References
Taxiphyllum taxirameum	Methanol extract	Phenols	Hemostasis, treatment of external wounds	[1]
Mnium cuspidatum	Methanol extract	Saponarin, flavonoids	Used to cure nose bleeding	[3]
Radula sumatrana	Methanol extract	Rasumatranin A-D, M and N	Anticancer	[38]
Conocephalum conicum	Ethanol extract	Lunularin	Anticancer, cytotoxicity	[36]
Plagiochasma intermedium	Methanol extract	Pakyonol	It is effective against prostate cancer	[54]
Bazzanianovae-zelandiae	Methanol extract	Naviculyl caffeate	Cytotoxic	[55]
Polytrichum pallidisetum	Methanol extract	Pallidisetin A	Anti-inflammatory	[56]
Polytrichum pallidisetum	Methanol extract	Pallidisetin B	It is cytotoxic against U-251MG and RPM 1–7951 human tumor cell lines	[39]
Jungermannia fauriana	Methanol extract	Jungermannenone A	Anticancer	[57]
Jungermannia fauriana	Methanol extract	Jungermannenone B	Anticancer	[40]
Sphagnum palustre	Ethanol extract	Fulvic acid	Anticancer	[58]

4.2 Antifungal Activity

Some species of bryophytes, such as *Diplophyllum albicans, Pogonatum aloides, Plagiothecium denticulatum,* and Cinnamolid show antifungal action against the spore germination of *Uromyces fabae*, *Septoria nodorum*, *Botrytis cinerea*, and *Alternaria brassicola*. Methanol extract of *Odontoschisma denudatum* and *Herberta adunca* prevents the growth of pathogenic fungi like *Pythium debaryanum, Pizoctonia solani,* and *Botrytis cinerea,* as they possess antifungal substances including(−) α herebertenol, and (−) - β herbertenol [59]. Macrocyclic bis (bibenzyl) and Plagiochin E were extracted; *Marchantia polymorpha* L. shows antifungal properties. Ethanol extract of *Marchantia linearis* shows antifungal properties and the thallus of *M. linearis* is utilized in the in vivo organic control of pathogen growth [45].

4.2.1 Antibacterial Activity

Marchantin A shows antibacterial activities against some Gram-positive bacteria, such as *Streptococcus pyogenes*, *Streptococcus viridans*, *Staphylococcus aureus*,

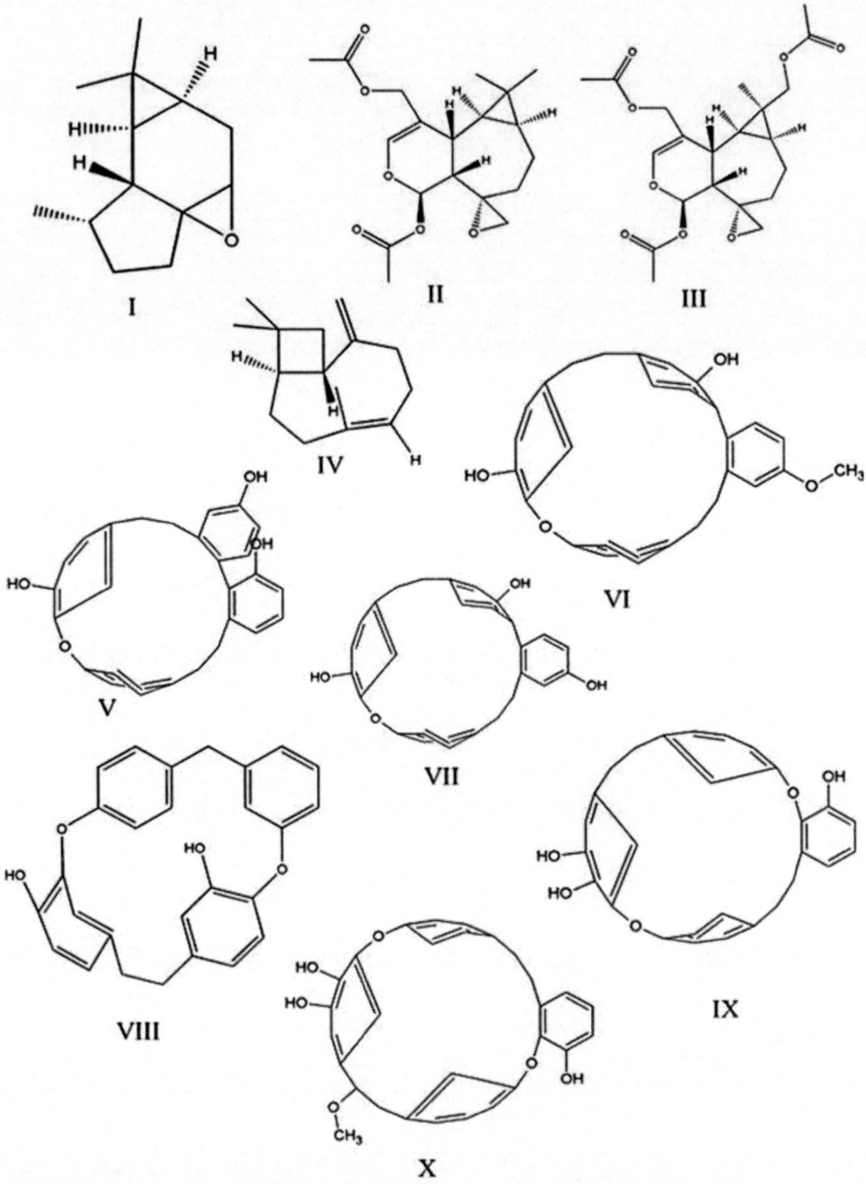

Fig. 1 Structure of some bioactive phytochemicals from bryophytes. (I- ledene oxide, II- plagiochilin A, III- plagiochilin B, IV- β-caryophyllene, V- plagiochilin E, VI- riccardin A, VII- riccardin C, VIII- riccardin B, IX- marchantin A, X- marchantin E)

and Gram-negative, such as *Pseudomonas aeruginosa*, *Escherichia coli*, and *Pasteurella multocida.* Marchantin A helps cure diseases caused by *Streptococcus pyogenes* and *Staphylococcus aureus*.

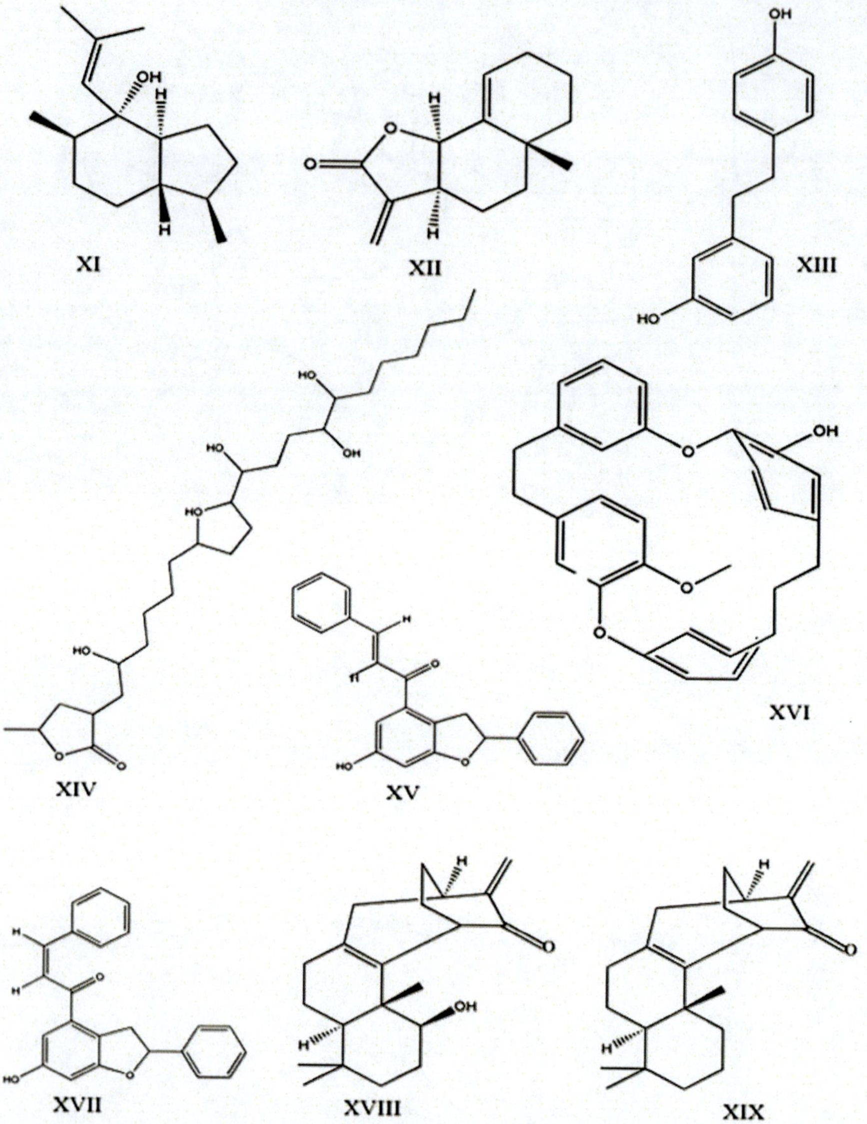

Fig. 2 Structure of some bioactive phytochemicals from bryophytes. (XI- tamariscol, XII- frullanolide, XIII- lunularin, XIV- acetogenins, XV- pakyonol, XVI- pallidisetin A, XVII- pallidisetin B, XVIII- jungermannenone A, XIX- jungermannenone B)

4.2.2 Antiviral Activity

Bryophytes like *Marchantia convoluta* are antiviral against hepatitis B, humic acid present in *Sphagnum* fought with some viruses, and peat humic acid exhibits antiviral action against types 1 and 2 of the herpes simplex virus [60].

4.3 Cytotoxic Activity

Marchantin A, marchantin D, and marchantin E isolated from the *Marchantia paleacea* and *Marchantia tosana* showed cytotoxic activity, resulting in apoptotic cell death of a tumorous cell. The extracts of *Polytrichum juniperinum* that possess anticancer activity against mice Sarcoma was also reported [61]. Liverworts *Diplophyllum taxifolium* and *Diplophyllum albicans* have diplophyllin, which shows anticancer action against epidermoid carcinoma in humans [62]. Sesquiterpenoids isolated from *Frullania monocular*, *Frullania tamarisci*, *Porella japonica*, and *Conocephalum supradecompositum* show inhibitory activity against carcinoma [62]. α-Methylene γ-lactone *Diplophyllum taxifolium* and *Diplophyllum albicans* shows anticancer activity against epidermoid carcinoma in humans. *Plagiochila fasciculata* is also found to exhibit properties to inhibit leukemia.

4.4 Antioxidant Activity

Ethanolic extracts of *Polytrichastrum alpinum* and *Saniona uncinata* possess phenolic substances with free radical scavenging potential thus balancing the number of reactive oxygen species (ROS) development in cells. *Hypnum mammillatum* and *Brachythecium rutabulum* also possess antioxidant properties [63]. Natural antioxidant compounds like flavonoids, phenolics, and tannins found in some species, such as, *Marchantia* show antioxidant properties playing the role of free radical scavenging agents, luteolin from methanolic extract of *Marchantia* thallus, that is, vanilate, gallate, einnamate, protocatechol, caffeate, and sinapic, are responsible for major antioxidant activities [63]. Ethyl acetate and methanolic extract of *Marchantia polymorpha* also have some antioxidant properties. Some mosses like *Polytrichasteum alpinum* and *Sanionia uncinate* are used as antioxidants in cosmetics industries and for medicinal purposes. Methanolic extracts of *Leucobryum bowringii*, *Plagiochilla beddomei*, and *Octoblepharum* show antioxidant properties [64].

5 Conclusions

There is an increased interest in phytochemical analysis by modern researchers. Extensive literature studies and article analysis have revealed that bryophytes are very much important due to their chemical constituents with potential biological activities. This may be the basis for a new source of therapeutic compounds to develop health supplements. Bryophytes that are crucial for medicine are used in biotechnological methods to produce secondary metabolites. Though a lot of underexplored species of bryophytes still exist, which are believed to be having medicinally significant phytochemicals. Various commercial products like perfumes, cosmetics, and insect-repellant are also produced from these bryophytes. Various eco-friendly products can be obtained from bryophytes and proper study of underexplored bryophytes species can generate various commercially important products along with therapeutics. Bryophytes embrace several therapeutically important phytochemicals, which can give rise to a potential drug lead near future against various diseases. The in vivo culture of bryophytes helps in the production and isolation of phytochemicals at a larger scale. Secondary metabolites obtain from bryophytes show numerous pharmacological effects, such as anti-inflammatory, antioxidant, and anti-microbial. Bryophytes are ethnobotanically important plants that are extensively used by tribal communities throughout the world in the treatment of various diseases and they have great medicinal values. The phytochemicals of bryophytes have various biological activities and can be used in the development of various biotechnological as well as pharmaceutical products. Thus, bryophytes represent a therapeutically potential group that may welfare health of the society.

Acknowledgments We are thankful to Bioinformatics Centre, Assam University, Silchar, for software support through DBT-Bioinformatics Infrastructure Facility and the e-journal access facility through the DBT e-Library Consortium (DeLCON).

References

1. Asakawa Y, Ludwiczuk A, Nagashima F (2013) Phytochemical and biological studies of bryophytes. Phytochemistry 91:52–80. https://doi.org/10.1016/j.phytochem.2012.04.012
2. Udar R (1976) Recent advances in Hepaticae (liverworts) in India. Rec Adv in Bot 1:24–54
3. Pant G, Tewari SD (1989) Various human uses of bryophytes in the Kumaun region of northwest Himalaya. Bryologist 92:120–122. https://doi.org/10.2307/3244026
4. Marko S, Aneta B, Dragoljub G (2001) Bryophytes as a potential source of medicinal compounds. Pregl Rev 21(1):17–29
5. Ludwiczuk A, Asakawa Y (2019) Bryophytes as a source of bioactive volatile terpenoids–a review. Food Chem Toxicol 132:110649. https://doi.org/10.1016/j.fct.2019.110649
6. Tosun G, Yayli B, Özdemir T, Batan N, Bozdeveci A, Yayli N (2015) Volatiles and antimicrobial activity of the essential oils of the mosses *Pseudoscleropodium purum*, *Eurhynchium striatum*, and *Eurhynchium angustirete* grown in Turkey. Rec Nat Prod 9(2):237–242
7. Jensen S, Omarsdottir S, Bwalya AG, Nielsen MA, Tasdemir D, Olafsdottir ES (2012) Marchantin a, a macrocyclic bisbibenzyl ether, isolated from the liverwort *Marchantia polymorpha*, inhibits protozoal growth in vitro. Phytomedicine 19(13):1191–1195. https://doi.org/10.1016/j.phymed.2012.07.011
8. Bukvicki D, Gottardi D, Tyagi AK, Veljic M, Marin PD, Vujisic L, Vannini L (2014) *Scapania nemorea* liverwort extracts: investigation on volatile compounds, in vitro antimicrobial activity and control of *Saccharomyces cerevisiae* in fruit juice. LWT-Food Sci Technol 55(2):452–458. https://doi.org/10.1016/j.lwt.2013.09.029
9. Abay G, Altun M, Karakoc OC, Gul F, Demirtas I (2013) Insecticidal activity of fatty acid-rich Turkish bryophyte extracts against *Sitophilus granarius* (Coleoptera: Curculionidae). Comb Chem High Throughput Screen 16(10):806–816
10. Rempt M, Pohnert G (2010) Novel acetylenic oxylipins from the moss *Dicranum scoparium* with antifeeding activity against herbivorous slugs. Angew Chem Int Ed 49(28):4755–4758. https://doi.org/10.1002/anie.201000825
11. Flowers S (1957) Ethnobryology of the Gosuite Indians of Utah. Bryologist 60(1):11–14. https://doi.org/10.2307/3240044

12. Shirsat RP (2008) Ethnomedicinal uses of some common bryophytes and pteridophytes used by tribals of Melghat region (Ms), India. Ethnobog leafl 2008(1):92. https://opensiuc.lib.siu.edu/ebl/vol2008/iss1/92
13. Ando H (1983) Use of bryophytes in China 2. Mosses indispensable to the production of Chinese gallnuts. Proc Bryol Soc Jap 3:124–125
14. Saxena DK (2004) Uses of bryophytes. Resonance 9(6):56–65
15. Harris ES (2008) Ethnobryology: traditional uses and folk classification of bryophytes. Bryologist 169-217. https://doi.org/10.1639/0007-2745(2008)111
16. Glime JM (2007) Economic and ethnic uses of bryophytes. Flora North America 27:14–41
17. Wu PC (1977) *Rhodobryum giganteum* (Schwaegr.) par can be used for curing cardiovascular disease. Acta Phytotax Sin 15(93)
18. Cheng X, Xiao Y, Wang X, Wang P, Li H, Yan H, Liu Q (2012) Anti-tumor and pro-apoptotic activity of ethanolic extract and its various fractions from *Polytrichum commune* L. ex Hedw in L1210 cells. J Ethnopharmacol 143(1):49–56. https://doi.org/10.1016/j.jep.2012.05.054
19. Fu P, Lin S, Shan L, Lu M, Shen YH, Tang J, Zhang WD (2009) Constituents of the moss *Polytrichum commune*. J Nat Prod 72(7):1335–1337. https://doi.org/10.1021/np800830v
20. Guo ZF, Bi GM, Zhang YH, Li JH, Meng DL (2020) Rare benzonaphthoxanthenones from Chinese folk herbal medicine *Polytrichum commune* and their anti-neuroinflammatory activities in vitro. Bioorg Chem 102:104087. https://doi.org/10.1016/j.bioorg.2020.104087
21. Hu RL (1987) Bryology. Higher Education Press, Beijing, p 465
22. Ozturk M, Gökler İ, Altay V (2018) Medicinal bryophytes distributed in Turkey, Springer. In: Plant and human health, vol 1, Cham, pp 323–348. https://doi.org/10.1007/978-3-319-93997-1_8
23. Dey A, De JN (2011) Antifungal bryophytes: a possible role against human pathogens and in plant protection. Res J Bot 6(4):129
24. Wei HC, Ma SJ, Wu CL (1995) Sesquiterpenoids and cyclic bisbibenzyls from the liverwort *Reboulia hemisphaerica*. Phytochemistry 39(1):91–97. https://doi.org/10.1016/0031-9422(94)00860-V
25. Sharma C (2013) Comparative evaluation of antimicrobial activity of methanolic extract and phenolic compounds of a liverwort, *Reboulia hemispherica*. Arch Bryol 192:1–6
26. Karpinski TM, Adamczak A (2017) Antibacterial activity of ethanolic extracts of some moss species. Herba Pol 63:3. https://doi.org/10.1515/hepo-2017-0014
27. Kļaviņa L, Bikovens O, Šteinberga I, Maksimova V, Eglīte L (2012) Characterization of chemical composition of some bryophytes common in Latvia. Environ Exp Biol 10:27–34
28. Drobnik J, Stebel A (2021) Four centuries of medicinal mosses and liverworts in European Ethnopharmacy and scientific pharmacy: a review. Plan Theory 10(7):1296. https://doi.org/10.3390/plants10071296
29. Azuelo AG, Sariana LG, Pabualan MP (2011) Some medicinal bryophytes: their ethnobotanical uses and morphology. Asian J Biodives 2:49–80. https://doi.org/10.7828/ajob.v2i1.92
30. Deora GS, Guhil N (2015) Phytochemical analysis and antifungal activity of moss *Bryum cellulare* against some Phytopathogenic fungi. Int J Pharm Sci 6(2):688–691. https://doi.org/10.13040/IJPSR.0975-8232.6(2).688-91
31. Jantwal A, Rana M, Joshi Rana A, Upadhyay J, Durgapal S (2019) Pharmacological potential of genus *Marchantia*: a review. J Pharmacogn Phytochem 8(2):641–645
32. Huang WJ, Wu CL, Lin CW, Chi LL, Chen PY, Chiu CJ, Chen CN (2010) Marchantin a, a cyclic bis (bibenzyl ether), isolated from the liverwort *Marchantia emarginata* subsp. *tosana* induces apoptosis in human MCF-7 breast cancer cells. Cancer Lett 291(1):108–119. https://doi.org/10.1016/j.canlet.2009.10.006
33. Oiso Y, Toyota M, Asakawa Y (1999) Occurrence of a bis-bibenzyl derivative in the Japanese fern *Hymenophyllum barbatum*: first isolation and identification of perrottetin H from the pteridophytes. Chem Pharm Bull 47(2):297–298

34. Horn A, Pascal A, Lončarević I, Volpatto Marques R, Lu Y, Miguel S, Simonsen HT (2021) Natural products from bryophytes: from basic biology to biotechnological applications. Crit Rev Plant Sci 40(3):191–217. https://doi.org/10.1080/07352689.2021.1911034
35. Chavhan A (2017) Phytochemical screening and antibacterial activity of bryophytes. Int J Life Sci 5(3):405–408
36. Lu ZQ, Fan PH, Ji M, Lou HX (2006) Terpenoids and bisbibenzyls from Chinese liverworts *Conocephalum conicum* and *Dumortiera hirsuta*. J Asian Nat Prod Res 8(1–2):187–192. https://doi.org/10.1080/1028602042000325537
37. Rycroft DS, Heinrichs J, Cole WJ, Anton H (2001) A phytochemical and morphological study of the liverwort *Plagiochila retrorsa* Gottsche, new to Europe. J Bryol 23(1):23–34. https://doi.org/10.1179/jbr.2001.23.1.23
38. Wang X, Cao J, Dai X, Xiao J, Wu Y, Wang Q (2017) Total flavonoid concentrations of bryophytes from Tianmu Mountain, Zhejiang Province (China): phylogeny and ecological factors. PLoS One 12(3):e0173003. https://doi.org/10.1371/journal.pone.0179837
39. Ivanova V, Kolarova M, Aleksieva K, Dornberger KJ, Haertl A, Moellmann U, Chipev N (2007) Sanionins: anti-inflammatory and antibacterial agents with weak cytotoxicity from the antarctic moss *Sanionia georgico-uncinata*. Prep Biochem Biotechnol 37(4):343–352. https://doi.org/10.1080/10826060701593241
40. Liu W, Li H, Cai PJ, Wang Z, Yu ZX, Lei X (2016) Scalable Total synthesis of rac-Jungermannenones B and C. Angew Chem Int Ed 55(9):3112–3116. https://doi.org/10.1002/anie.201511659
41. Xiao JB, Chen XQ, Zhang YW, Jiang XY, Xu M (2006) Cytotoxicity of *Marchantia convoluta* leaf extracts to human liver and lung cancer cells. Braz J Med Biol Res 39(6):731–738. https://doi.org/10.1590/S0100-879X2006000600005
42. Suire C, Bouvier F, Backhaus RA, Bégu D, Bonneu M, Camara B (2000) Cellular localization of isoprenoid biosynthetic enzymes in *Marchantia polymorpha*. Uncovering a new role of oil bodies. Plant Physiol 124(3):971–978. https://doi.org/10.1104/pp.124.3.971
43. Beike AK, Decker EL, Frank W, Lang D, Vervliet-Scheebaum M, Zimmer AD, Reski R (2010) Applied bryology-bryotechnology. Bryophyte Divers Evol 31(1):22–32. https://doi.org/10.11646/bde.31.1.7
44. Negi K, Tewari SD, Chaturvedi P (2018) Antibacterial activity of marchantia papillata raddi subsp. Grossibarba (Steph) Bischl against *Staphylococcus aureus*. Indian J Tradit Knowl 17(4): 763–769
45. Krishnan R, Kannan KV, Murugan K (2014) Antifungal activity of the ethanolic extracts of *Marchantia linearis* Lehm and Lindenb. Against some pathogenic fungi. J Aquat Biol Fish 2: 556–563
46. Chen XQ, Xiao JB (2010) RP-HPLC-DAD detrmination of flavonoids: separation of quercetin, luteolin and apigenin in *Marchantia convoluta*. Iran J Pharm Res 3:175–181. https://doi.org/10.22037/IJPR.2010.634
47. Siregar ES, Pasaribu N, Sofyan MZ (2021) Antioxidant activity of liverworts *Marchantia paleacea* Bertol. From North Sumatra Indonesia. In IOP conference series: earth and environmental. Science 713(2021):012061. https://doi.org/10.1088/1755-1315/713/1/012061
48. Wu XZ, Cheng AX, Sun LM, Lou HX (2008) Effect of plagiochin E, an antifungal macrocyclic bis (bibenzyl), on cell wall chitin synthesis in *Candida albican*s 1. Acta Pharmacol Sin 29(12): 1478–1485. https://doi.org/10.1111/j.1745-7254.2008.00900.x
49. Khurram M, Hameed A, Amin MU, Gul A, Ullah N, Hassan M, Manzoor W (2011) Phytochemical screening and in vitro evaluation of anticandidal activity of *Dodonaea viscosa* (L.) Jaeq(Sapindaceae). Afr J Pharm Pharmacol 5(11):1422–1426. https://doi.org/10.5897/AJPP11.404
50. Bodade RG, Borkar PS, SAIFUL AM, Khobragade CN (2008) In vitro screening of bryophytes for antimicrobial activity. J Med Plant Res 7(4):23–28. DOR: 20.1001.1.2717204.2008.7.25.16.2

51. Asakawa Y, Toyota M, Tori M, Hashimoto T (2000) Chemical structures of macrocyclic bis (bibenzyls) isolated from liverworts (Hepaticae). Spectroscopy 14(4):149–175. https://doi.org/10.1155/2000/570265
52. Asakawa Y (2007) Biologically active compounds from bryophytes. Pure Appl Chem 79(4): 557–580. https://doi.org/10.1351/pac200779040557
53. Subhisha S, Subramoniam A (2005) Antifungal activities of a steroid from *Pallavicinia lyellii*, a liverwort. Indian J Pharm 37(5):304. https://doi.org/10.4103/0253-7613.16854
54. Ji M, Shi Y, Lou H (2011) Overcoming of P-glycoprotein-mediated multidrug resistance in K562/A02 cells using riccardin F and pakyonol, bisbibenzyl derivatives from liverworts. Biosci Trends 5(5):192–197. https://doi.org/10.5582/bst.2011.v5.5.192
55. Burgess EJ, Larsen L, Perry NB (2000) A cytotoxic sesquiterpene caffeate from the liverwort *Bazzanian ovae-zelandiae*. J Nat Prod 63(4):537–539. https://doi.org/10.1021/np990492x
56. Zheng GQ, Ho DK, Elder PJ, Stephens RE, Cottrell CE, Cassady JM (1994) Ohioensins and pallidisetins: novel cytotoxic agents from the moss *Polytrichum pallidisetum*. J Nat Prod 57(1): 32–41. https://doi.org/10.1021/np50103a005
57. Guo YX, Lin ZM, Wang MJ, Dong YW, Niu HM, Young CY, Yuan HQ (2016) Jungermannenone a and B induce ROS-and cell cycle-dependent apoptosis in prostate cancer cells in vitro. Acta Pharmacol Sin 37(6):814–824. https://doi.org/10.1038/aps.2016.26
58. Dey A, Mukherjee A (2015) Therapeutic potential of bryophytes and derived compounds against cancer. J Acute Dis 4(3):236–248. https://doi.org/10.1016/j.joad.2015.04.011
59. Matsuo A, Yuki S, Higashi R, Nakayama M, Hayashi S (1982) Structure and biological activity of several sesquiterpenoids having a novel herbertane skeleton from the liverwort *Herberta adunca*. In Proceedings of the 25th Symposium on chemistry of natural products. Symposium papers 242–249
60. Klöcking R, Thiel KD, Sprössig M (1976) Antiviral activity of humic acids from peat water. Proc. 5th Internat. Peat Congr., Poznabn, Poland 1:446–455
61. Paliwal AJ, Arjun M, Madhav NV, Murthy E, Aruna M (2014) Endangered treatment of traditional medicinal amphibian plants (bryophytes). In Proceedings of national seminar on traditional medicine & health practice, pp 61–75
62. Ohta Y, Andersen NH, Liu CB (1977) Sesquiterpene constituents of two liverworts of genus *Diplophyllum*: novel eudesmanolides and cytotoxicity studies for enantiomeric methylene lactones. Tetrahedron 33(6):617–628. https://doi.org/10.1016/0040-4020(77)80301-3
63. Gökbulut A, SatilmiŞ B, Batçioğlu K, Cetin B, Şarer E (2012) Antioxidant activity and luteolin content of *Marchantia polymorpha* L. Turk J Biol 36(4):381–385. https://doi.org/10.3906/biy-1106-15
64. Manoj GS, Murugan K (2012) Phenolic profiles, antimicrobial and antioxidant potentiality of methanolic extract of a liverwort, *Plagiochila beddomei* Steph. Indian J Nat Prod Resour 3(2): 173–183. http://nopr.niscpr.res.in/handle/123456789/14417

Therapeutic Potential of Bryophytes and Its Future Perspective

2

Jayanta Barukial and Porismita Hazarika

Contents

Abstract

Plants have many therapeutic potential for curing various ailments of human beings. However, the study of bryophytes considering their therapeutic potential toward human beings is still in its infancy. Recent public demand for plant-based medicine, as well as the emergence of antibiotic-resistant microorganisms, has prompted biologists to seek out novel plant-based natural medicines. Furthermore, bryophyte's potential antibacterial capabilities can be utilized for medicinal purposes against the relevant infection. Current studies on bioactive compounds in bryophytes have revealed significant multiple numbers of secondary metabolites in this elegant group of plants. Hence, this study aims to review the recent research on their clinical activities with respect to antidiabetic, anti-inflammatory,

J. Barukial (✉)
Debraj Roy College, Golaghat, Assam, India

P. Hazarika
Dibrugarh University, Dibrugarh, Assam, India
e-mail: rs_porismitahazarika@dibru.ac.in

H. N. Murthy (ed.), *Bioactive Compounds in Bryophytes and Pteridophytes*,
Reference Series in Phytochemistry, https://doi.org/10.1007/978-3-031-23243-5_7

antimicrobial, antitumor, and antioxidant properties corresponding to the ethnomedicinal reports throughout the different communities of the world. It has immense value in near future considering the bioprocesses of their genome, as well as innovative drug discovery by using genetic engineering and biotechnology. Some of the important bioactive compounds in different bryophytes are benzyl benzoate, p-hydroxycinnamic acid, 7,8-dihydroxycoumarin, marchantins, riccardins, triterpenoid saponin, tetracyclic diterpene, sesquiterpenes, diplophyllin, plagiochiline, plagiochin E, and perrotetin E.

Keywords

Antidiabetic · Anti-inflammatory · Antimicrobial · Antitumor · Bioactive compounds · Biopharmaceuticals · Traditional uses

Abbreviations

Akt	Protein kinase B
COX	Cyclooxygenase
H1N1	Hemagglutinin type 1 and neuraminidase type 1
HL	Human leukemia
KB	Ubiquitous KERATIN-forming tumor cell line
LOX	Lipoxygenase
LPS	Lipopolysaccharide
LXR	The liver-X-receptor
MB	Metastatic breast
MDA	Malondialdehyde
MDR	Multidrug-resistant
MRSA	Methicillin-resistant *Staphylococcus aureus*
NF-kB	Nuclear factor kappa B
NO	Nitric oxide
PC	Prostate cancer

1 Introduction

Bryophytes are the second biggest type of terrestrial plant after flowering plants. Bryophytes are divided into three categories: hornworts (Anthocerotopsida), liverworts (Marchantiopsida), and mosses (Bryopsida) [37]. These plants' chemical components can be employed as biologically active agents. Predation, UV radiation, high temperature, and microbial decomposition are all known to cause bryophytes to create a variety of secondary metabolites to counteract biotic and abiotic stress. Bryophytes have emerged as a possible biopharming tool for the synthesis of sophisticated biopharmaceuticals in recent years. Even though bryophytes have the potential to be employed in medicine, their application in practical research with consequences for human health has yet to be completely explored [43]. They produce a significant number of secondary metabolites [43]. Many bryophyte compounds have demonstrated intriguing biological activity, particularly with regard to their application in medicine and agriculture for the overall benefit of living beings [37], exhibiting a variety of activities, such as antimicrobial, antifungal, cytotoxic, antitumor, and insecticidal properties [7]. Chemical studies have been conducted on around 3.2% of mosses and 8.8% of liverworts. Current studies on the medicinally active ingredients of bryophytes are being conducted in order to cure diseases, such as skin diseases, cardiovascular diseases, hepatic problems, and a variety of other maladies [80].

1.1 Traditional Uses of Bryophytes

Bryophytes are being used by diverse tribal populations in South, North, and Eastern India, Poland, Argentina, Australia, New Zealand, Turkey, Japan, Taiwan, and Pakistan, to alleviate hepatic abnormalities, skin issues, chronic diseases, such as cardiovascular, as fever reducers, antimicrobial compounds, wound healing, and many other afflictions [13] (Table 1). Various moss species, such as *Philonotis, Bryum,* and *Mnium*, have been mashed into a paste and administered as a poultice by the Chinese and Native Americans. In the Himalayan area of India, burnt moss ash combined with fat and honey is used as an ointment for cuts, burns, and wounds [20, 60]. *Marchantia polymorpha* Linn. possesses antipyretic, antihepatic, antidotal, and diuretic effects and is used to heal open wounds. *M. polymorpha* Linn. has been used to treat inflammation-related disorders by the Khampti people in Arunachal Pradesh, India. *M. palmata* Nees. is a medicinal plant that is used to heal boils. To minimize edema and pus development, a thick paste of thalli is applied to the skin [26, 47]. *Polytrichum* species are generally used as a diuretic or to halt bleeding, but they can also be used to grow long, black hair [22]. *Plagiochasma appendiculatum* and *Targionia hypophylla* L. are used to treat skin problems [7]. *Frullania ericoides* (Nees ex Mart.) Mont is a plant that is used to treat head lice and nourish the hair [57]. Scabies, itches, and other skin problems are treated by *Targionia hypophylla* L. [57]. In the Himalayas, *Riccia* species were used to cure ringworm [20]. Cystitis, bronchitis, tonsillitis, and tympanitis can all be treated with *Haplocladium microphyllum* (Hedw.) Broth. [20]. Hemorrhage is treated with *Sphagnum* sp. *Sphagnum teres* (Schimp.) Ångström are used to alleviate eye conditions [20], snakebites, gallstones, cuts, burns, scalds, fractures, and distended tissue, which are all treated with *Conocephalum conicum.* Hair growth and burns are treated with *Fissidens japonicum* [4] (Fig. 1).

2 Therapeutic Potential of Bryophytes

The existence of specific new chemicals in bryophytes is critical in today's medical science environment. It has been discovered that there are several therapeutic applications for various human illnesses. Terpenoids, phenols, glycosides, and fatty acids are among the many physiologically active substances found in bryophytes. Other

Table 1 Some bryophytes with their traditional uses and mode of uses

S. no.	Bryophyte species	Family	Traditional uses	Mode of use	References
1	*Marchantia palmate* Nees.	Marchantiaceae	Treatment of boils to minimize swelling and pus development	The leaf paste is applied instantly after it has been freshly prepared	[26, 47]
2	*Marchantia polymorpha* Linn.	Marchantiaceae	Cuts, fractures, venomous snakebites, burns, scalds, open wounds, and many other inflammation-related disorders are all treated with this plant	The skin is treated with a hot water decoction and a thick thalli paste	[26, 47]
3	*Philonotis* sp.	Bartramiaceae	Used to put damaged bones back together	Crushed into a paste and used as a poultice	[6, 7]
4	*Rhodobryum giganteum* (Schwägr.) Paris	Bryaceae	In the treatment of cardiovascular disorders and neurological prostration	Not found	[7, 20]
5	*Rhodobryum roseum* (Hedw.) Limpr.	Bryaceae	In the treatment of cardiovascular disorders and neurological prostration	Not found	[7, 20]
6	*Polytrichum commune* Hedw.	Polytrichaceae	Mostly used to halt bleeding, but it is also used to make hair grow long and black	Boiling as a tea to relieve a cold Calyptras oil extract is used in hair	[20, 22]
7	*Targionia hypophylla* L.	Targioniaceae	Scabies, itches, and other skin problems are treated with this plant	The entire thallus, combined with the leaves of Mayilsikkai (*Actiniopteris radiata,* Pteridaceae), is made for purpose of treating children with scabies, itches, and dermatitis, a paste was developed and infused with two teaspoons of coconut oil	[7, 57]
8	*Frullania ericoides* (Nees ex Mart.) Mont.	Jubulaceae	Used to treat head lice and nourish hair	Approximately 50 g of the whole plant is ground into a paste that is used on the hair every other day after being roasted in coconut oil	[57]
9	*Riccia* sp.	Ricciaceae	Used to get rid of ringworms	Not found	[20]

10	*Haplocladium microphyllum* (Hedw.) broth.	Thuidiaceae	To treat cystitis, bronchitis, tonsillitis, and tympanitis	Not found	[20]
11	*Sphagnum teres* (Schimp.) Ångström	Sphagnaceae	Cure eye disorders	Not found	[20]
12	*Bryum* sp.	Bryaceae	Used to treat cuts, burns, and wounds	Crushed to form a paste and used as a poultice As an ointment, moss ash combined with fat and honey is utilized	[20, 60]
13	*Mnium* sp.	Mniaceae	Used as a bandage for cuts, burns, and wounds	Crushed into a paste and used as a poultice As an ointment, moss ash is combined with fat and honey.	[20, 60]
14	*Conocephalum conicum* (L.) Underw.	Conocephalaceae	Used to treat wounds, burns, scalds, fractures, swollen tissue, dangerous snake bites, and gallstones	Not found	[4]
15	*Fissidens japonicum*	Fissidentaceae	Used for hair growth, burns, and choloplania	Not found	[4]

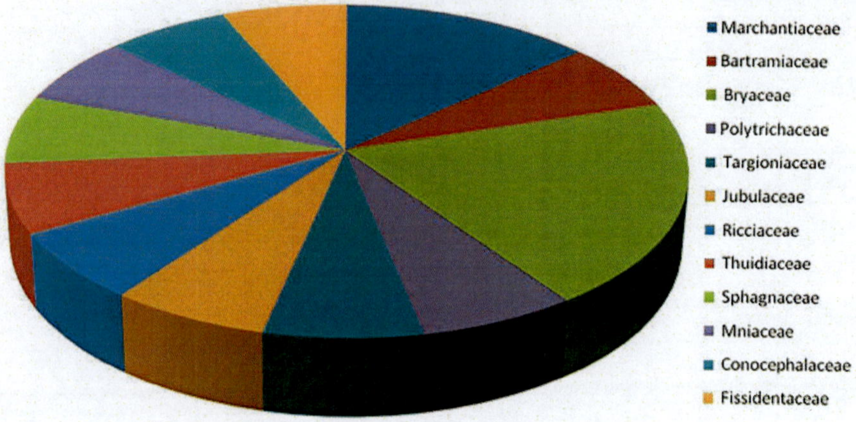

Fig. 1 Family of Bryophytes that possesses traditional value

components found in bryophytes include aliphatic chemicals, phenyl quinines, polyols, amino acids, essential fats, oligo- and polysaccharides, as well as heterocyclic and phenolic compounds, although little research has been conducted to correlate therapeutic benefits with specific bryophytes species [11]. This tiny, slow-growing group of plants, which includes liverworts, hornworts, and mosses, is a natural resource that has been studied for antibacterial, antioxidant, anti-inflammatory, antivenomous, antileukemic, and anticancer action. Modern phytochemists and biochemists have extracted a large number of physiologically active chemical compounds from bryophytes that might be used in the pharmaceutical sector (Table 2). Certain bryophyte compounds have been shown to hinder the development of bacteria. Three *Radula* spp. suppress the development of *Staphylococcus aureus*. Many bryophyte species have been proven to have anticancer action. Diplophylline, the first anticancer-active chemical, was discovered in liverworts. This chemical has substantial anticancer action in humans. Certain mosses contain polyunsaturated fatty acids, which are previously recognized to have essential medical applications, such as decreasing atherosclerosis and cardiovascular disease, reducing collagen-induced thrombocyte aggregation, and lowering triacylglycerols and cholesterol in plasma [56]. Many liverworts have also been shown to contain antitumor sesquiterpenoids. Furthermore, they are widely used in surgical dressings, diapers, and other human medical uses. Their application is not limited to Asia [18], but is also known in Brazil [55], England [77], North America [53, 54], Germany [18], as well as in China [17, 78].

2.1 Some of the Active Therapeutic Compounds from Bryophytes and Their Biological Activities

Some of the significant bioactive compounds with their biological activities (Fig. 2) and their chemical structures are shown below (Figs. 3, 4, 5, 6, 7, 8, 9, 10, 11, 12, and 13).

Table 2 Several bryophytes with their major bioactive constituents

S. no.	Name of bryophytes	Major bioactive constituents	References
1	*Isotachis japonica* Stephani	Benzyl benzoate, benzyl cinnamate, and B-phenylethyl cinnamate	[41]
2	*Riccardia* sp.	Riccardins A and B, Sacullata	[6]
3	*Marchantia polymorpha* L.	Marchantin A, MB-G Marchantin D and E	[8]
4	*Diplophyllum albicans* (L.) Dumort.	Diplophylline	[61]
5	*Diplophyllum taxifolium* (Wahlenb.) Dumort.	Diplophylline	[61]
6	*Pallavicinia* sp.	Bicyclohumulenone, Plagiochilinea, Plagiochilide, Plagiochilal B	[6]
7	*Plagiochila* sp.	Sacullatal	[6]
8	*Plagiochila rutilans* Lindenb.	Menthane monoterpenoids	[23]
9	*Philonotis* sp.	Triterpenoid saponins	[4]
10	*Rhodobryum giganteum* (Schwägr.) Paris	p-Hydroxycinnamic acid, 7–8-dihydroxycoumarin	[4]
11	*Bryum* sp.	Triterpenoid saponins	[8]
12	*Plagiomnium* sp.	Triterpenoid saponins	[6]
13	*Mnium* sp.	Triterpenoid saponins	[6]
14	*Physcomitrella patens* (Hedw.) Bruch & Schimp.	Tetracyclic diterpene, namely, 16α–hydroxykaurane (16α-hydroxy-ent-kaurane, Kaurenol, $C_{20}H_{34}O$)	[58, 59]
15	*Porella cordaeana* (Huebener) Moore	Sesquiterpene hydrocarbons and monoterpene	[12]
16	*Herbertus aduncus* (Dicks.) Gray	(−)-Alpha-herbertenol;(−)-beta-herbertenol, and(−)-alpha-formylherbertenol	[42, 39, 40]
17	*Plagiochasma appendiculatum* Lehm. & Lindenb.	Plagiochin E, 13,13′-oisoproylidenericcardin D, and neomarchantin A	[49, 72]
18	*Conocephalum conicum* (L.) Underw.	Plagiochin E, 13,13′-oisoproylidenericcardin D, and neomarchantin A	[49, 72]
19	*Marchantia polymorpha* L.	Plagiochin E, 13,13′-oisoproylidenericcardin D, and neomarchantin A, marchantin A	[2, 49, 72]
20	*Marchantia paleacea* Bertol.	Marchantin A	[1, 2]
21	*Marchantia emarginata* subsp. *tosana*	Marchantin A	[1, 2]
22	*Riccardia multifida* (L.) Gray	Riccardins	[1, 2]
23	*Radula perrottetii* Gottsche ex Stephani	Perrottetin E	[1, 2, 71]
24	*Herbertus borealis* Crundw.	Herbertane sesquiterpenoids	[10]

Fig. 2 Major Bioactive Compounds in Bryophytes

Fig. 3 Benzyl benzoate

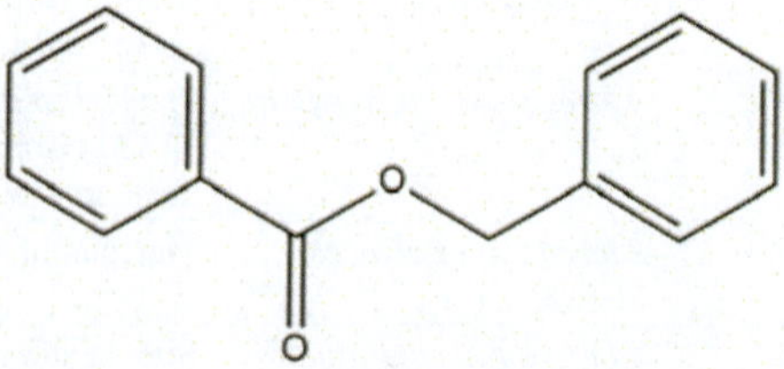

2.1.1 Benzyl Benzoate

The ester of benzyl alcohol with benzoic acid is benzyl benzoate (Fig. 3). It comes in the form of a white solid or a transparent oily liquid with a subtle fragrant odor. Benzyl benzoate is entirely insoluble in water and glycerol but miscible in alcohol, chloroform, ether, and oils. Benzyl benzoate (BB) is one of the oldest medications

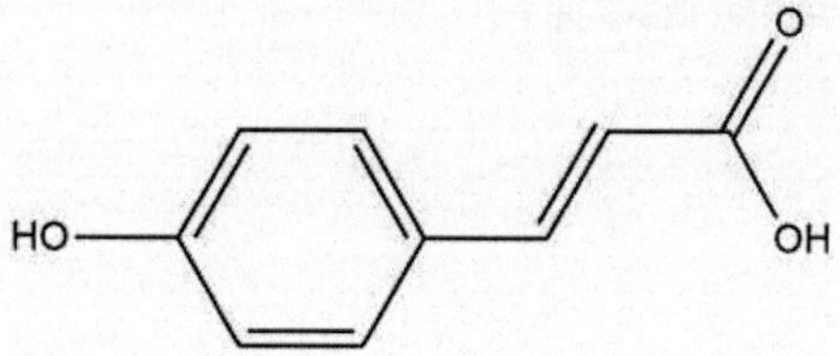

Fig. 4 p-Hydroxycinnamic Acid

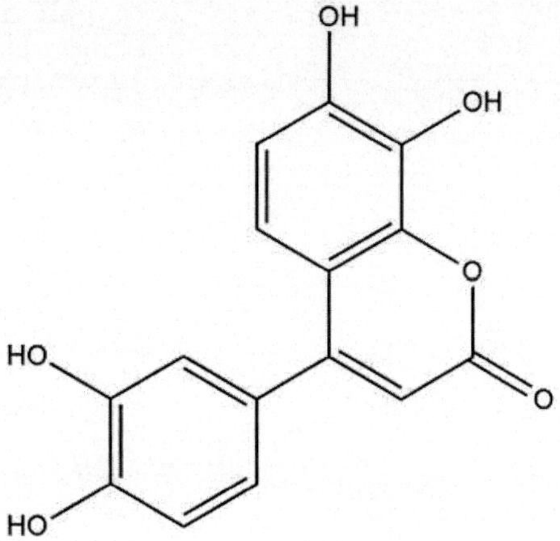

Fig. 5 7, 8-Dihydroxycoumarin

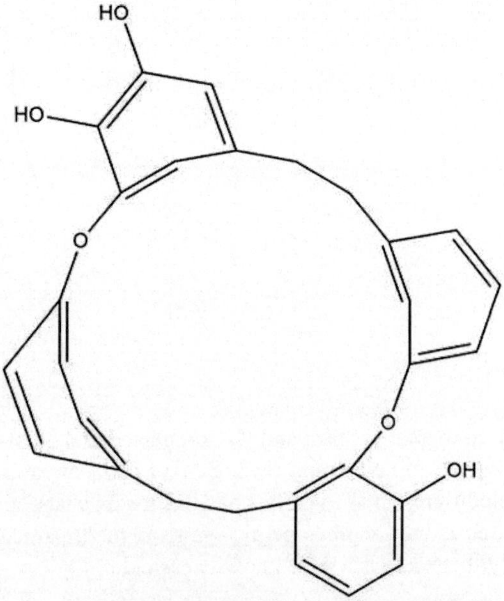

Fig. 6 Marchantin A

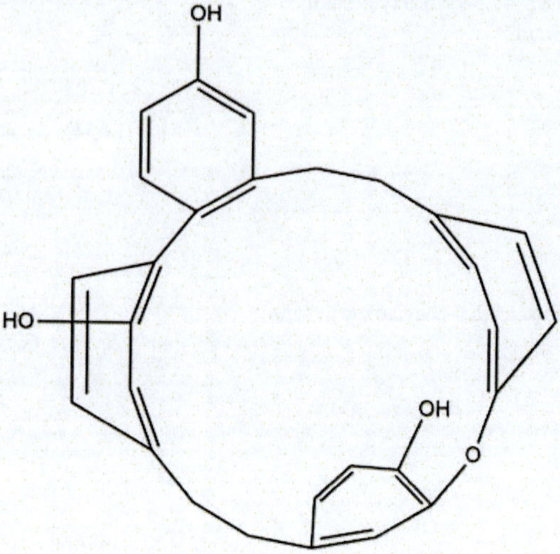

Fig. 7 Riccardin C

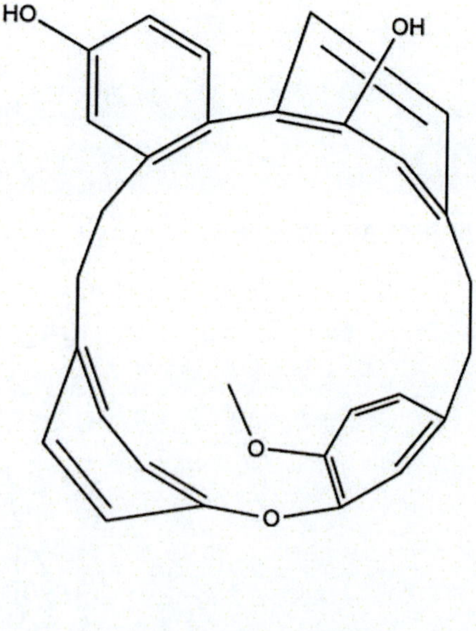

Fig. 8 Riccardin F

used to treat scabies, and it is suggested as a "first-line intervention" for the disease's cost-effective therapy [62]. Benzyl benzoate and its derivatives may help to lower blood pressure [51]. By boosting macrophage activity, benzyl benzoates might be used as immunotherapeutic agents in the treatment of infectious illnesses [14]. Benzyl benzoate is a well-known acaricide [44].

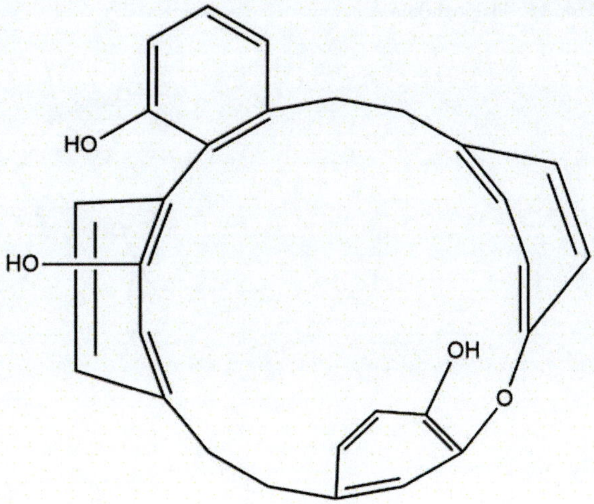

Fig. 9 Riccardin D

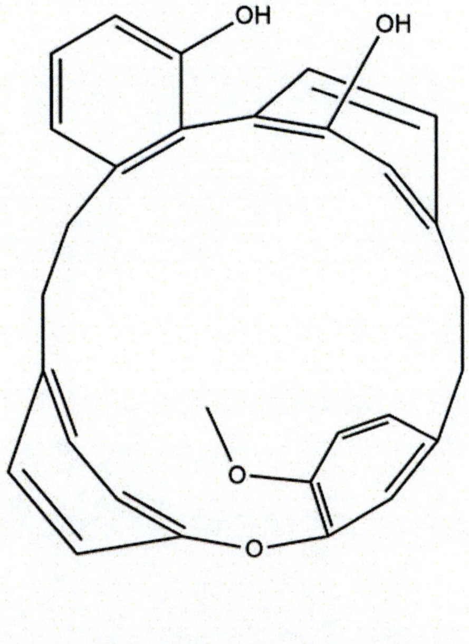

Fig. 10 Riccardin G

2.1.2 p-Hydroxycinnamic Acid

Coumarin, commonly known as p-hydroxy-cinnamic acid (Fig. 4), is a cinnamic acid derivative. It aids in the reduction of high blood pressure and stroke. Because p-hydroxy-cinnamic acid has anticoagulant effects, it increases blood circulation [32, 76]. It has antiviral efficacy against H1N1 viruses and antitrypanosomal activity against *Trypanosoma brucei* [31, 48].

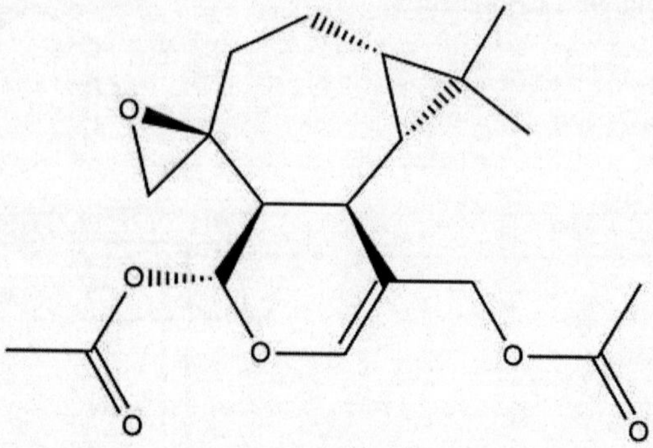

Fig. 11 Plagiochiline A

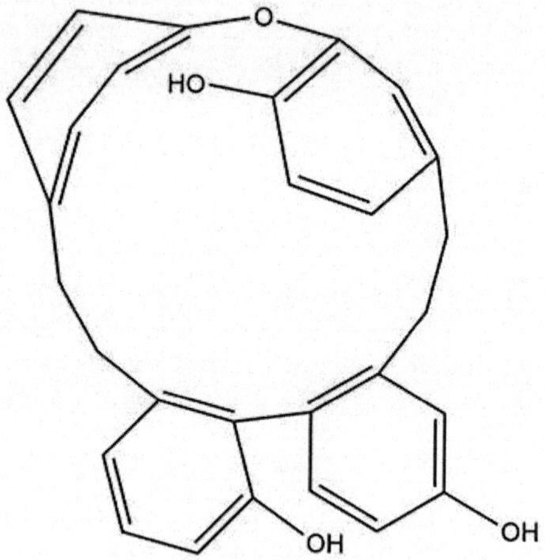

Fig. 12 Plagiochin E

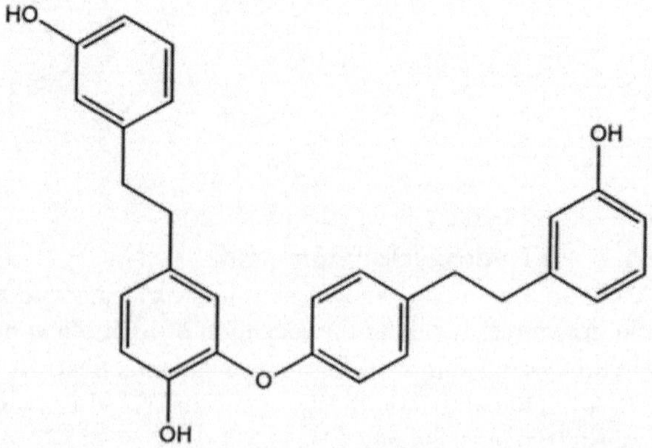

Fig. 13 Perrottetin E

2.1.3 7,8-Dihydroxycoumarin

It is a naturally occurring coumarin (Fig. 5) molecule with antioxidant and anti-inflammatory properties. Because of its insulin-stimulating properties and subsequent control of the apoptotic pathway, it might be utilized to treat diabetes [74]. In A549 human lung adenocarcinoma cells, 7,8-dihydroxycoumarin performs a concentration-dependent function in the activation of apoptosis via reduction of Akt/NF-kB signaling. As a result, 7,8-dihydroxycoumarin might be a natural option for the treatment and prevention of lung adenocarcinoma [75].

2.1.4 Marchantins

Marchantin A (Fig. 6) has been discovered to have antibacterial, antitumor, and antileukemia properties, as well as antioxidant, antifungal, and 5-LOX, COX, and calmodulin inhibitory properties [25, 28, 81]. Chemoresistant prostate cancer PC3 cells were found to be susceptible to marchantin M [16].

2.1.5 Riccardins

Riccardins have been discovered to have a wide range of pharmacological effects, including antifungal, liver X receptor-modulating, anticancer, and NOS-inhibiting properties [81]. Riccardin A and B were discovered to have cytotoxic action [16]. Riccardin C showed cytotoxicity and anti-MRSA action in a prostate cancer cell line. LXRa agonist/LXRb antagonist riccardins C (Fig. 7) and F (Fig. 8) were discovered to be an LXRa agonist/LXRb antagonist and an LXRa antagonist, respectively. Riccardin D (Fig. 9), a demethylated derivative of riccardin G (Fig. 10), has been demonstrated to inhibit a variety of human cancer cells [50]. Riccardin C was discovered to be effective against PC3 cells from chemoresistant prostate cancer. Riccardin D was discovered to have a strong antiproliferative impact on the HL-60, K562, and MDR K562/A02 human leukemia cell lines [16].

2.1.6 Triterpenoid Saponins

Triterpenoid saponins are naturally occurring sugar conjugates of triterpenes that, when shaken with water, generate a stable froth [35]. Due to their indispensable roles as potent antimicrobials, antioxidants, health restoratives, antiaging, improved cognitive function, memory, nervine stimulators, and most importantly as potent antineoplastic molecules, these important plant-derived secondary metabolites have a huge pharmaceutical demand [9, 15, 35].

2.1.7 Tetracyclic Diterpene

Tetracyclic diterpenes are diterpenoid natural products that are formed by the cyclization of appropriately orientated pimaradienes [21]. It inhibited the growth of multidrug-resistant (MDR) and methicillin-resistant *Staphylococcus aureus* (MRSA) strains of *Staphylococcus aureus* [63]. It has antibacterial, insecticidal, and anthelmintic qualities, as well as inhibitory effects on important enzymes, analgesic, and cytotoxic properties [73].

2.1.8 Sesquiterpenes

C15-Terpenoids made up of three isoprene units are known as sesquiterpenes. Lactones, alcohols, acids, aldehydes, and ketones are examples of naturally occurring hydrocarbons or oxygenated forms. Essential oils and aromatic components in sesquiterpenes have a variety of pharmacological properties, including antiplasmodial, cytotoxic, antifungal, antiviral, and antibacterial action [5, 19, 64, 69, 79].

2.1.9 Diplophyllin

Diplophyllin is an ent-eudesmanolide that has been shown to be effective in the treatment of human epidermoid cancer [16]. KB cell lines are toxic to diplophyllin [30].

2.1.10 Plagiochiline

Plagiochiline A (Fig. 11) slows cell division by preventing cytokinesis from being completed, especially during the abscission stage. It also lowers the survival of DU145 cells in clonogenic experiments and causes significant cell death in these cells [66].

2.1.11 Plagiochin E

Plagiochin E (Fig. 12) is a phenolic macrocyclic bisbibenzyl chemical that is new to science. It has potent antifungal and anticancer properties [65].

2.1.12 Perrottetin E

Perrottetin E (Fig. 13) is a cytotoxic bis(bibenzy1) ether [71]. It is cytotoxic [24, 50].

3 Major Therapeutic Activities of Bryophytes

3.1 Antitumor Activities

Cytotoxic 8,9-secokaurane diterpenes from the New Zealand liverwort *Lepidolaena taylorii* (Gottsche) Trevis. were shown to be potent against human tumor cell lines. Furthermore, two 8,9-secokauranes from the New Zealand liverwort *Lepidolaena palpebrifolia* were shown to be cytotoxic. Tumor growth inhibitors costunolide and tulipinolide were identified from the liverworts *Conocephalum supradecompositum* Stephani, *Frullania monocera* (Taylor) Gottsche, Lindenb. & Nees, *Frullania tamarisci* (L.) Dumort., *Marchantia polymorpha* Linn., *Porella japonica* (Sande Lac.) Mitt., and *Wiesnerella denudata* (Mitt.) Stephani. The cytotoxicity of several secondary metabolites isolated from the liverwort *Ptilidium pulcherrimum* (Weber) Hampe against the PC3, MDA-MB-231, and HeLa cell lines has been described, with ursane triterpenoids showing considerable cytotoxicity against PC3 cells [16]. *Tetraphis pellucida, Plagiomnium cuspidatum, Metaneckera menziesii, Bryoandersonia illecebra, Dicranella heteromalla, Racomitrium sudeticum, Polytrichum ohioense,* and *Anomodon attenuatus* (Hedw.) Huebenerare the most promising cytotoxic moss species. *Dumortiera hirsuta* and *Bazzania trilobata* were the most active liverworts [33].

3.2 Antidiabetic Activities

In alloxan-induced diabetes, *Taxithelium nepalense* has strong antidiabetic action [67]. Investigation of three species of liverworts, *P. striatus, P. epiphylla,* and *B. oshimensis*, revealed that they had antidiabetic properties [29]. *Lunularia cruciata* has a high inhibitory action against α-glucosidase and α-amylase [46]. *Octoblepharum albidum* also has the ability to combat diabetes by interacting with digestive enzymes, such as α-glucosidase and α-amylase [68]. *Marchantia subintegra*, *Marchantia emarginata*, and *Plagiochasma cordatum* have antidiabetic properties [45]. *Hedwigia ciliata* P. Beauv. extracts had a considerable antidiabetic activity, which was mediated by the inhibition of α-glucosidase [36].

3.3 Anti-Inflammatory Activities

The ability of *Dicranum majus* and *Thuidium delicatulum* to block the LPS-induced NO pathway demonstrates their efficacy in reducing the inflammatory response [38]. Anti-inflammatory properties of *Porella densifolia* were investigated [34]. *Philonotishastate* has anti-inflammatory properties that are moderately strong, effective, and noticeable [52]. *Corsinia coriandrina* (Spreng.) Lindb. (Corsiniaceae), *Mannia androgyna* (L.) A. Evans (Aytoniaceae), *Plagiochasma rupestre* (J.R. Forst et G. Forst) Steph. (Aytoniaceae), *Porella cordaeana* (Huebener) Moore (Porellaceae), *Porella platyphylla* (L.) Pfeiff. (Porellaceae), *Reboulia hemisphaerica* (L.) Raddi (Aytoniaceae), *Riccia fluitans* L. (Ricciaceae), and *Targionia hypophylla* L. (Targioniaceae) shows anti-inflammatory activities in vitro [70].

3.4 Antimicrobial Activities

Bacillus cereus, Bacillus megaterium, Bacillus subtilis, Cryptococcus neoformans, Acinetobacter calcoaceticus, and *Staphylococcus aureus* are all sensitive to marchantin A, making it a powerful antibacterial [3]. Antifungal effects of riccardin D and B, as well as dihydroptychantol, were detected against *Candida albicans*, with minimum inhibitory quantities ranging from 0.25 to 0.8 μg and minimum inhibitory concentrations ranging from 16.64 μg mL^{-1} [7, 27]. The growth of *B. subtilis* and *S. aureus* was suppressed by ether and methanol extracts from the liverwort *Mastigophora diclados* (Brid.) Nees [30]. Some pathogenic fungi (*Botrytis cinerea, Rhizoctonia solani,* and *Pythium debaryanum*) are inhibited by a methanolic extract of bryophytes, *Herberta adunca*, and *Odontoschisma denudatum*. (−) – Herbertenol, (−) – herebrtenol, (−) – formylherbertenol, and (+) –acetoxyodontoschismenol are antifungal compounds [37].

4 Future Perspective

Bryophytes, which contain a wide range of secondary metabolites, might be a viable source of bioactive chemicals with enormous medicinal potential. They might be the source of numerous developed metabolic pathways that could be sensibly regulated for the production of various innovative medicinal compounds due to their presence in various niches and inhabiting the most diversified group of the plant kingdom. Furthermore, genetic engineering appears to be the most potential application of bryophytes in medicine. Bryophytes are already being employed to manufacture human blood-clotting proteins, while others have been shown to inhibit thromb activity. Recent public demand for plant-based medicine, as well as the emergence of antibiotic-resistant microorganisms, has prompted biologists to seek out novel plant-based natural medicines. Furthermore, bryophyte's potential antibacterial capabilities can be utilized for medicinal purposes against the relevant infection. They might be the source of numerous developed metabolic pathways that could be sensibly handled for the production of various innovative medicinal chemicals since they are present in various niches and occupy the most diversified group of the plant kingdom [43].

5 Conclusion

Future drug development programs will use phytochemical data-mining tools to identify, quantify, evaluate, conformational analysis, clinical assessment, monitoring, bioactivity analysis, and create novel medications based on these unique bioactive compounds present in bryophytes. Considering all of these aspects, bryophytes are prospective sources of herbal medicines and components for a wide range of active medicinal goods [7].

References

1. Asakawa Y (1981) Biologically active substances obtained from bryophytes. J Hattori Bot Lab 50:123–142
2. Asakawa Y (1982) Chemical constituents of the Hepaticae. In: Fortschritte der Chemie Organischer Naturstoffe/Progress in the chemistry of organic natural products. Springer, Vienna, pp 1–285
3. Asakawa Y (1990) Biologically active substances from bryophytes. In: Bryophyte development: physiology and biochemistry. CRC Press, London, pp 259–287
4. Asakawa Y (2007) Biologically active compounds from bryophytes. Pure Appl Chem 79:557–580
5. Awouafack MD, Tane P, Kuete V, Eloff JN (2013) Chapter 2: Sesquiterpenes from the medicinal plants of Africa. In: Kuete V (ed) Medicinal plant research in Africa. Elsevier, Oxford, pp 33–103

6. Azuelo AG, Sariana LG, Pabualan MP (2011) Some medicinal bryophytes: their ethnobotanical uses and morphology. Asian J Biodiversity 2(1):49–80
7. Bandyopadhyay A, Dey A (2022) The ethno-medicinal and pharmaceutical attributes of bryophytes: a review. Phytomed Plus:100–255
8. Beike AK, Decker EL, Frank W, Lang D, Vervliet-Scheebaum M, Zimmer AD et al (2010) Applied bryology-bryotechnology. Bryophyte Diversity Evol 31:22–32
9. Biswas T, Dwivedi UN (2019) Plant triterpenoid saponins: biosynthesis, in vitro production, and pharmacological relevance. Protoplasma 256(6):1463–1486
10. Buchanan MS, Connolly JD, Rycroft DS (1996) Herbertane sesquiterpenoids from the liverworts *Herbertus aduncus* and *H. borealis*. Phytochemistry 43:1245–1248
11. Buck WR, Allen BH, Pursell RA (2005) Recent literature on bryophytes. Bryologist 108:158–175
12. Bukvicki D, Gottardi D, Veljic M, Marin PD, Vannini L, Guerzoni ME (2012) Identification of volatile components of liverwort (*Porella cordaeana*) extracts using GC/MS-SPME and their antimicrobial activity. Molecules 17:6982–6995
13. Chandra S, Chandra D, Barh A, Pankaj, Pandey RK, Sharma IP (2016) Bryophytes: hoard of remedies, an ethno-medicinal review. J Tradit Complement Med 7:94–98
14. Choi SZ, Choi SU, Bae SY, Pyo SN, Lee KR (2005) Immunobioloical activity of a new benzyl benzoate from the aerial parts of *Solidago virgaaurea* var.*gigantea*. Arch Pharm Res 28:49
15. Desai SD, Desai DG, Kaur H (2009) Saponins and their biological activities. Pharma Times 41: 13–16
16. Dey A, Mukherjee A (2015) Therapeutic potential of bryophytes and derived compounds against cancer. J Acute Dis 4:236–248
17. Ding H (1982) Medicinal spore-bearing plants of China. Shanghai Science and Technology Press, Shanghai, pp 1–409
18. Frahm JP (2004) Recent developments of commercial products from bryophytes. Bryologist 107:277–283
19. Geetha B, Nair MS, Latha PG, Remani P (2012) Sesquiterpene lactones isolated from *Elephantopus scaber* L. inhibits human lymphocyte proliferation and the growth of tumour cell lines and induces apoptosis in vitro. J Biotechnol Biomed 2012:721285
20. Glime JM (2007) Economic and ethnic uses of bryophytes. Flora North Am 27:14–41
21. Hanson JR (2013) The tetracyclic diterpenes. Elsevier
22. Harris ES (2008) Ethnobryology: traditional uses and folk classification of bryophytes. Bryologist:169–217
23. Heinrichs J, Groth H, Gradstein SR, Rycroft DS, Cole WJ, Anton H (2001) *Plagiochilarutilans* (Hepaticae): a poorly known species from tropical America. Bryologist 104:350–361
24. Ivković I, Novaković M, Veljić M, Mojsin M, Stevanović M, Marin PD, Bukvički D (2021) Bis-bibenzyls from the liverwort *Pellia endiviifolia* and their biological activity. Plan Theory 10:1063
25. Iwai Y, Murakami K, Gomi Y, Hashimoto T, Asakawa Y, Okuno Y, Ishikawa T, Hatakeyama D, Echigo N, Kuzuhara T (2011) Anti-influenza activity of marchantins, macrocyclic bisbibenzyls contained in liverworts. PLoS One 6:e19825
26. Jantwal A, Rana M, Joshi Rana A, Upadhyay J, Durgapal S (2019) Pharmacological potential of genus *Marchantia*: a review. J Pharmacogn Phytochem 8:641–645
27. Jensen JSR, Omarsdottir S, Thorsteinsdottir JB, Ogmundsdottir HM, Olafsdottir ES (2012) Synergistic cytotoxic effect of the microtubule inhibitor marchantin A from *Marchantia polymorpha* and the Aurora kinase inhibitor MLN8237 on breast cancer cells in vitro. Planta Med 78:448–454
28. Jensen S, Omarsdottir S, Bwalya AG, Nielsen MA, Tasdemir D, Olafsdottir ES (2012) Marchantin A, a macrocyclic bisbibenzyl ether, isolated from the liverwort Marchantia polymorpha, inhibits protozoal growth in vitro. Phytomedicine 19:1191–1195

29. Kocazorbaz EK, Kerem T, Moulahoum H, Ün RN (2021) Phytochemical and bioactivity analysis of several methanolic extracts of nine bryophytes species. Sakarya Univ J Sci 25: 938–94931
30. Komala I, Ito T, Nagashima F, Yagi Y, Asakawa Y (2010) Cytotoxic, radical scavenging and antimicrobial activities of sesquiterpenoids from the Tahitian liverwort *Mastigophora diclados* (Brid.) Nees (Mastigophoraceae). J Nat Med 64:417–422
31. Li HB (2014) Research on antiviral constituents in Re-Du-Ning injection (I). Chin Tradit Herb Drugs 45:1682–1688
32. Lone R, Shuab R, Koul K (2014) Role of cinnamate and cinnamate derivatives in pharmacology. Glob J Pharmacol 8:328–335
33. Ludwiczuk A, Asakawa Y (2019) Bryophytes as a source of bioactive volatile terpenoids – a review. Food Chem Toxicol 132:110649
34. Ludwiczuk A, Asakawa Y (2020) Terpenoids and aromatic compounds from bryophytes and their central nervous system activity. Curr Org Chem 24:113–128
35. Mahato SB, Sarkar SK, Poddar G (1988) Triterpenoid saponins. Phytochemistry 27:3037–3067
36. Mandić MR, Oalđe MM, Lunić TM, Sabovljević AD, Sabovljević MS, Gašić UM et al (2021) Chemical characterization and in vitro immunomodulatory effects of different extracts of moss *Hedwigia ciliata* (Hedw.) P. Beauv. from the Vršačke Planine Mts., Serbia. PLoS One 16(2): e0246810
37. Marko S, Aneta B, Dragoljub G (2001) Bryophytes as a potential source of medicinal compounds. Pregl Rev 21:17–29
38. Marques RV, Sestito SE, Bourgaud F, Miguel S, Cailotto F, Reboul P, Jouzeau J-Y, Rahuel-Clermont S, Boschi-Muller S, Simonsen HT (2022) Anti-inflammatory activity of bryophytes extracts in LPS-stimulated RAW264. 7 murine macrophages. Molecules 27:1940
39. Matsuo A, Yuki S, Higashi R, Nakayama M, Hayashi S (1982) Structure and biological activity of several sesquiterpenoids having a novel herbertane skeleton from the liverwort *Herberta adunca*. In: Book structure and biological activity of several sesquiterpenoids having a novel herbertane skeleton from the liverwort *Herberta adunca*. Papers Tokyo, pp 242–249
40. Matsuo A, Yuki S, Nakayama M, Hayashi S (1982) Three new sesquiterpene phenols of the ent-herbertane class from the liverwort Herberta adunca. Chem Lett 11:463–466
41. Matsuo A, Yuki S, Nakayama M (1983)(−)-Herbertenediol and(−)-herbertenolide, two new sesquiterpenoids of the ent-herbertane class from the liverwort Herberta adunca. Chem Lett 12: 1041–104243
42. Matsuo A, Nozaki H, Kubota N, Uto S, Nakayama M (1984) Structures and conformations of (–)-isobicyclogermacrenal and(–)-lepidozenal, two key sesquiterpenoids of the cis-and trans-10, 3-bicyclic ring systems, from the liverwort Lepidozia vitrea: X-ray crystal structure analysis of the hydroxy derivative of(–)-isobicyclogermacrenal. J Chem Soc Perkin Trans 1:203–214 41 41
43. Mishra R, Pandey VK, Chandra R (2014) Potentialofbryophytes as therapeutics. Int J Pharm Sci Res 5:3584–3593
44. Monteiro IN, Monteiro ODS, Costa-Junior LM, da Silva Lima A, Andrade EHDA, Maia JGS, Mouchrek Filho VE (2017) Chemical composition and acaricide activity of an essential oil from a rare chemotype of *Cinnamomum verum* Presl on *Rhipicephalus microplus* (Acari: Ixodidae). Vet Parasitol 238:54–57
45. Mukhia S, Mandal P, Singh D, Singh D (2017) Study of bioactive phytoconstituents and in-vitro pharmacological properties of Thallose Liverworts of Darjeeling Himalaya. J Pharm Res 11: 490–501
46. Mukhia S, Mandal P, Singh D, Singh D (2019) Comparison of pharmacological properties and phytochemical constituents of in vitro propagated and naturally occurring liverwort *Lunularia cruciata*. BMC Complement Altern Med 19:1–16
47. Namsa ND, Tag H, Mandal M, Kalita P, Das AK (2009) An ethnobotanical study of traditional anti-inflammatory plants used by the Lohit community of Arunachal Pradesh, India. J Ethnopharmacol 125:234–245

48. Naß J, Efferth T (2018) The activity of *Artemisia* spp. and their constituents against trypanosomiasis. Phytomedicine 47:184–191
49. Niu C, Qu JB, Lou HX (2006) Antifungal bis [bibenzyls] from the Chinese liverwort*Marchantia polymorpha* L. Chem Biodivers 3:34–40
50. Novakovic M, Bukvicki D, Andjelkovic B, Ilic-Tomic T, Veljic M, Tesevic V, Asakawa Y (2019) Cytotoxic activity of riccardin and perrottetin derivatives from the liverwort *Lunularia cruciata*. J Nat Prod 82:694–701
51. Ohno O, Ye M, Koyama T, Yazawa K, Mura E, Matsumoto H, Ichino T, Yamada K, Nakamura K, Ohno T, Yamaguchi K, Ishida J, Fukamizu A, Uemura D (2008) Inhibitory effects of benzyl benzoate and its derivatives on angiotensin II-induced hypertension. Bioorg Med Chem 16:7843–7852
52. Oyedapo OO, Makinde AM, Ilesanmi GM, Abimbola EO, Akinwunmi KF, Akinpelu BA (2015) Biological activities (anti-inflammatory and anti-oxidant) of fractions and methanolic extract of *Philonotis hastate* (Duby Wijk & MargaDant). Afr J Tradit Complement Altern Med 12:50–55
53. Pejin B, Newmaster S, Sabovljevic M, Miloradovic Z, Grujic-Milanovic J, Ivanov M et al (2011) Antihypertensive effect of the moss *Rhodobryum ontariense* in vivo. J Hypertens 29: e315–e316
54. Pejin B, Vujisic L, Sabovljevic M, Tesevic V, Vajs V (2011) Preliminary data on essential oil composition of the moss *Rhodobryum ontariense* (Kindb.) Kindb. Cryptogam Bryol 32:113–117
55. Pinheiro MdFdS, Lisboa RCL, Brazão RdV (1989) Contribuição ao estudo de briófitas como fontes de antibióticos. Acta Amazon 19: 139–145
56. Radwan SS (1991) Sources of C20-polyunsaturated fatty acids for biotechnological use. Appl Microbiol Biotechnol 35:421–430
57. Remesh M, Manju C (2009) Ethnobryological notes from Western Ghats, India. Bryologist 111: 532–537
58. Sabovljevi A, Sokovi M, Pejin B, Sabovljevi M (2010) Comparison of extract bioactivities of in-situ and vitro grown selected bryophyte species. Afr J Microbiol Res 4:808–812
59. Sabovljevi A, Sokovi M, Pejin B, Sabovljevi M (2011) Bio-activities of extracts from some axenically farmed and naturally grown bryophytes. J Med Plant Res 5:565–571
60. Saxena D (2004) Uses of bryophytes. Resonance 9:56–65
61. Saxena DK, Harinder (2004) Uses of bryophytes. Resonance 9:56–65
62. Sharma G, Dhankar G, Thakur K, Raza K, Katare OP (2016) Benzyl benzoate-loaded microemulsion for topical applications: enhanced dermatokinetic profile and better delivery promises. AAPS PharmSciTech 17:1221–1231
63. Smyrniotopoulos V, Vagias C, Rahman MM, Gibbons S, Roussis V (2010) Ioniols I and II, tetracyclic diterpenes with antibacterial activity, from *Sphaerococcus coronopifolius*. Chem Biodivers 7:666–676
64. Sotanaphun U, Lipipun V, Suttisri R, Bavovada R (1999) A new antiviral and antimicrobial sesquiterpene from *Glyptopetalum sclerocarpum*. Planta Med 65:257–258
65. Speicher A, Groh M, Zapp J, Schaumlöffel A, Knauer M, Bringmann G (2009) A synthesis-driven structure revision of 'plagiochin E', a highly bioactive bisbibenzyl. Synlett 2009: 1852–1858
66. Stivers NS, Islam A, Reyes-Reyes EM, Casson LK, Aponte JC, Vaisberg AJ, Hammond GB, Bates PJ (2018) Plagiochiline A inhibits cytokinetic abscission and induces cell death. Molecules 23:1418
67. Tatipamula VB, Killari KN, Ketha A, Sastry VG (2017) *Taxithelium napalense* acts against free radicals and diabetes mellitus. Bangladesh J Pharmacol 12:197–203
68. Tatipamula VB, Ketha A, Nallapaty S, Kottana H, Koneru ST (2021) Moss *Octoblepharum albidum* Hedw.: isolation, characterization, in vitro and in vivo antidiabetic activities. Adv Tradit Med 21:351–360

69. Tchuendem MHK, Mbah JA, Tsopmo A, Foyere Ayafor J, Sterner O, Okunjic CC, Iwu MM, Schuster BM (1999) Anti-plasmodial sesquiterpenoids from the African *Reneilmia cincinnata*. Phytochemistry 52:1095–1099
70. Tosun A, Küpeli Akkol E, Süntar I, Özenoğlu Kiremit H, Asakawa Y (2013) Phytochemical investigations and bioactivity evaluation of liverworts as a function of anti-inflammatory and antinociceptive properties in animal models. Pharm Biol 51:1008–1013
71. Toyota M, Tori M, Takikawa K, Shiobara Y, Kodama M, Asakawa Y (1985) Perrottetins E, F, and G from *Radula perrottetii* (liverwort) – isolation, structure determination, and synthesis of perrottetine. Tetrahedron Lett 26:6097–6100
72. Vashistha H, Dubey R, Pandey N (2007) Antimicrobial activity of three bryophytes against human pathogens. In: Current trends in bryology. Bishen Singh Mahendra Pal Singh, Dehra Dun, pp 47–59
73. Villa-Ruano N, Lozoya-Gloria E, Pacheco-Hernández Y (2016) Kaurenoic acid: a diterpene with a wide range of biological activities. Stud Nat Prod Chem 51:151–174
74. Vinayagam R, Xu B (2017) 7, 8-Dihydroxycoumarin (daphnetin) protects INS-1 pancreatic β-cells against streptozotocin-induced apoptosis. Phytomedicine 24:119–126
75. Wang Y, Li C-F, Pan L-M, Gao Z-L (2013) 7, 8-Dihydroxycoumarin inhibits A549 human lung adenocarcinoma cell proliferation by inducing apoptosis via suppression of Akt/NF-κB signaling. Exp Ther Med 5:1770–1774
76. Wang K-H, Li S-F, Zhao Y, Li H-X, Zhang L-W (2018) In vitro anticoagulant activity and active components of safflower injection. Molecules 23:170
77. Wren RC (1956) Potter's new cyclopaedia of botanical drugs and preparations. Pitmann, London
78. Wu P (1982) Some uses of mosses in China. Bryol Times 13:5–7
79. Wu C-L, Chien S-C, Wang S-Y, Kuo Y-H, Chang S-T (2005) Structure-activity relationships of cadinane-type sesquiterpene derivatives against wood-decay fungi. Wood Res Technol 59:620–627
80. Yayintas OT, Irkin LC (2018) Bryophytes as hidden treasure. J Sci Perspect 2:71–82
81. Zhao P, Song C (2018) Macrocyclic bisbibenzyls: properties and synthesis. Stud Nat Prod Chem 55:73–110

Volatile Compounds and Oils from Mosses and Liverworts

3

Eduardo Valarezo, Miguel Angel Meneses, Ximena Jaramillo-Fierro, Matteo Radice, and Ángel Benítez

Contents

E. Valarezo (✉) · M. A. Meneses · X. Jaramillo-Fierro
Departamento de Química, Universidad Técnica Particular de Loja, Loja, Ecuador
e-mail: bevalarezo@utpl.edu.ec; mameneses@utpl.edu.ec; xvjaramillo@utpl.edu.ec

M. Radice
Departamento Ciencias de la Tierra, Universidad Estatal Amazónica, Puyo, Ecuador
e-mail: mradice@uea.edu.ec

Á. Benítez
Departamento de Ciencias Biológicas y Agropecuarias, Universidad Técnica Particular de Loja, Loja, Ecuador
e-mail: arbenitez@utpl.edu.ec

H. N. Murthy (ed.), *Bioactive Compounds in Bryophytes and Pteridophytes*, Reference Series in Phytochemistry, https://doi.org/10.1007/978-3-031-23243-5_8

Abstract

Bryophytes are small, non-vascular plants, such as mosses, liverworts, and hornworts. They play a vital role in regulating ecosystems; some bryophyte species are among the first to colonize open ground. Bryophytes do not have seeds or flowers; instead, they reproduce via spores, and these species are also very good indicators of habitat quality. Volatile compounds and essential oils have been extracted by hydrodistillation and by solvent from some species of bryophytes. The chemical composition of the isolated oils and the nature of the volatile compounds have been determined by gas chromatography coupled to mass spectrometry and gas chromatography coupled to the flame initiation detector. This chapter presents the chemical composition of 76 essential oils from mosses and liverworts, identifying a significant variety of compositions. These findings open new potential research trends to investigate the biological activity of these essential oils. The growing demand for natural products by pharmaceutical, cosmetic, and food markets points to new opportunities for the protection and sustainable use of natural resources.

Keywords

Bryophytes · Bioactive compounds · Essential oil · Liverworts · Mosses · Volatile compounds

Abbreviations

a.s.l.	above sea level
CF	Chemical Formula
CN	Compound number
DH	Diterpene hydrocarbons
EAE	Ethyl acetate extract
EE	Ethanol extract
EO	Essential oil
FID	Flame ionization detector
GC	Gas chromatography
HR	High-resolution
IR	Infrared spectroscopy
ME	Methanol extract
MH	Monoterpenes hydrocarbons
MM	Monoisotopic mass
MS	Mass spectrometry

NMR	Nuclear magnetic resonance
OD	Oxygenated diterpenes
OM	Oxygenated monoterpenes
OS	Oxygenated sesquiterpenes
OT	Other compounds
RI	Retention indices
RIf	Reference Retention indices
SPME	Solid-phase microextraction
TLC	Thin-layer chromatography
tr	trace (< 0.05%)
UV	Ultraviolet-visible spectroscopy

1 Introduction

Bryophytes or bryophytes in sensu lato (broad sense, to differentiate them from bryophytes in sensu stricto or mosses) are non-vascular land plants [1]. The term bryophyte comes from Ancient Greek *βρύον* (brúon, bryon) meaning tree moss, liverwort, and "φυτόν" (phutón, phyton) meaning plant. It is believed that they are the descendants of green algae and were the first to evolve 500 million years ago after colonizing terrestrial spaces [2]. In this traditional division, we have the Bryophyta in the strict sense (mosses), Hepatophyta (liverworts), and Anthocerotophyta (hornworts) [3]. Bryophytes do not have seeds or flowers. Instead, they reproduce via spores. Two main parts are recognized in a bryophyte: 1. the gametophyte, which produces gametes and is photosynthetic (the rhizoids, unicellular in liverworts and hornworts or multicellular in most mosses) that fix the gametophyte to the substrate, and 2. a sporophyte, which is ephemeral or short-lived and dependent on the gametophyte [4]. The bryophytes present an alternation of generations between the dependent sporophyte generation, which produces the spores, and the independent gametophyte generation, which produces the sex organs and sperm and eggs [5, 6].

The appearance of the gametophyte can be thallose or foliose, and in all cases, the cortical cells are responsible for photosynthesis. The said gametophyte is formed by the activity of the meiospore that forms or not, depending on the taxonomic groups, a previous protonema that in some mosses is green and branched. In all bryophytes, the sexual organs are usually found at the termination of special branches, more or less protected [7]. Only in certain thallose liverworts and hornworts are the antheridia and archegonia embedded in the thallus [8]. The archegonium generally has several cells in the neck canal, a character that separates bryophytes from pteridophytes, which do not show more than a single cell, and an egg cell in the belly. The antheridium, usually sticky, forms numerous biflagellate spermatozoa. Once the egg cell is fertilized, the zygote is formed which, without a resting state, divides numerous times and finally forms the capsule or sporophyte inside which, after meiosis of the spore stem cells, the haploid meiospores originate again.

Bryophytes have poorly differentiated tissues and do not have conduction vessels, that is, they do not have xylem or phloem, and they do not have roots, stems, or true leaves, but rather a vegetative body with very primitive structures, with cells that do not constitute a tissue [6]. Hence, the "roots" that they present are called rhizoids, the "stem" cauloid, and the "little leaves" phyllodes. It is worth noting that due to the low level of organization that these organisms possess, none of these structures just mentioned are true organs as they appear in tracheophytes or vascular plants.

Bryophytes are very numerous and are found all over the world, from polar and alpine regions to the tropics. The bryophytes belong to 18,000–25,000 species distributed in 1473 plant genera within 165 plant families [9]. With regards terrestrial life, they prefer, however, for their development humid places. Bryophytes do not live in extremely arid sites or seawater, they do not have special tissues for taking in water or mineral salts [10], nor do they have internal tissues for the circulation of sap; however, some tracheiform conductor-type cells may be present. The intake of water is carried out, directly through the leaves, never through the rhizoids, since these serve only for fixing to the substrate. For all these reasons, a level of organization is assigned to thallophytes. Bryophytes provide an important buffer system for other plants living nearby and are very good indicators of habitat quality [11]. Furthermore, bryophytes contain numerous potentially useful compounds, including oligosaccharides, polysaccharides, sugar alcohols, amino acids, fatty acids, aliphatic compounds, phenylquinones, and aromatic and phenolic substances [12].

On the other hand, essential oils (EOs), volatile oils, or simply essences are the natural aromatic substances responsible for the fragrances of leaves, flowers, and other plant organs [13]. The most common method for extracting EOs is steam extraction (distillation or hydrodistillation) [14]. Essential oils are synthesized and secreted by glandular hairs, oil cells, or secretory ducts or cavities [15]. Essential oil is especially abundant in botanical families Apiaceae, Asteraceae, Conifers, Lamiaceae, Myrtaceae, and Rutaceae, [16]. In general, EOs constitute values <0.01% up to values >3% of the dry weight of the plant [17]. Essential oils are complex mixtures of low molecular weight organic compounds, especially compounds of a terpene nature, which can be monoterpenes (10 carbons), sesquiterpenes (15 carbons), and diterpenes (20 carbons). These monoterpenes and sesquiterpenes can be, in turn, acyclic, monocyclic, and bicyclic and also oxygenated and non-oxygenated [18]. The main biosynthetic pathways are the mevalonate pathway, the methyl-erithrytol-pathway, and finally the shikimic acid pathway; mentioned biosynthetic pathways lead to sesquiterpenes, monoterpenes, and diterpenes and phenylpropenes, respectively [19].

The compounds present in essential oils mostly contain from 8 to 20 carbons and molecular masses from 80 to 350 Da. The low molecular weight of the EO compounds and the lack of some intermolecular forces between them, such as the dipole-dipole force (hydrogen bonds), make these compounds volatile, hence their name and characteristic odor. Gas chromatography (GC), either coupled to mass spectrometry (MS) or coupled to the flame initiation detector (FID), is the technique used to determine the volatile compounds quantitatively and qualitatively in essential oil.

Some oils are almost monomolecular since they have almost exclusively a single component; others are rich in 2–3 molecules. However, most are polymolecular, containing 3–4 major molecules, a number of minor molecules, and sometimes hundreds of different molecules that are only present in trace amounts. Essential oils are used in the chemical, food, cosmetic, and pharmaceutical industries, due to their chemical structure, smell, and active ingredient. From the pharmacological point of view, the properties of essential oils are highly variable due to the heterogeneity of their components [20].

The chemical composition of the essential oil of various species of vascular plants has demonstrated great chemical diversity in several countries [21]. In contrast studies on bryophytes, chemistry is still poorly known [22], for example, about 5% of bryophytes have been chemically studied [23].

2 Mosses

The mosses class, the most numerous of the bryophytes, includes 11,000–13,000 species of small nonvascular spore-bearing land distributed in more than 600 genera that are grouped into three orders: Bryales, Sphagnales, and Andraeales [24]. They live all over the world, and many of them are pioneers on rocky substrates where life is very unfavorable for vascular plants. Mosses are distributed throughout the world except in salt water and are commonly found in moist shady locations. They are best known for those species that carpet woodland and forest floors. Ecologically, mosses break down exposed substrata, releasing nutrients for the use of more complex plants that succeed them [25]. They also aid in soil erosion control by providing surface cover and absorbing water, and they are important in the nutrient and water economy of some vegetation types. Economically important species are those in the genus *Sphagnum* that form peat. Sphagnum acidifies its environment, which retards the growth of bacteria and fungi, making the decomposition of organisms in the medium very slow [26].

Of all the bryophytes, it is the group with the most resistant species to drought and the cold climate of the poles. It is the group that has been most useful to man. Archaeological studies in Europe indicate that they served as a bandage and tinder for the Romans who lived in the north of England, and the Vikings used them to fill their sandals. Also, they have been widely used in folk medicine to cure various ailments, especially in China [27]. In the northern hemisphere, sphagnum moss peat bogs are the largest reservoirs of carbon on earth, far larger than tropical rain forests [28]. This moss was used as a bandage for its antiseptic and absorbent properties (much greater than those of cotton). All mosses have leaves, stems, and rhizoids [29]. There are erect, creeping, and hanging mosses. The sporophyte of mosses is the most complex of all bryophytes; it is made up of the foot (which joins it to the gametophyte), the seta that raises the capsule above the gametophyte, and the capsule that contains the spores. On the rim of the urn, there are filamentous structures that contribute to the dispersal of spores.

Mosses are green plants, generally small, measuring from a few millimeters to 20–30 cm, although in some erect forms or with hanging stems they sometimes reach almost 1 meter [30]. They live on the ground, on rocks or as epiphytes, and in humid places and rooted or floating in bodies of fresh water; they are not marine. The most conspicuous part of any moss is the gametophyte which consists of a stem with 3 to 5 rows of leaves in a radial arrangement. By their anatomy, stems and leaves have simple structures; the former may have a central axis or cord of smaller cells surrounded by several layers of larger parenchymal cells which in turn are surrounded by one or several layers of thick-walled epidermal cells [31]. In certain cases, the epidermis is made up of large, thin-walled cells known as hyalodermis.

The stems may be more or less covered by uniseriate or branched filaments called paraphylls or carry other special structures, pseudoparaphylls, to protect the meristematic zones. For their part, the leaves are often sheeted one cell thick, except in the middle part where a nerve with supporting and conduction cells can be found. Leaf cells can exhibit various shapes and sizes; their ornamentations or thickenings give them special mechanical or physiological properties, particularly related to water economy or photosynthesis. The stems bear smooth or papillose rhizoids at the base or along their body, sometimes in such abundance as to form a tomentum [32]. The rhizoids are multicellular filaments, with oblique transverse walls. The sexual organs – archegonia and antheridia – are protected by modified leaves and by intermingled paraphyses that help maintain moisture. The archegonia form in an apical or lateral position, while the antheridia vary in position on monoecious or dioecious stems [33].

The biological cycle of moss can be summarized as follows: a meiospore germinates, giving rise to a protonema which, in most mosses, has a filamentous shape. In the upper part of the protonema, that is, chlorophyllous, a pyramidal cell stands out due to its oblique transverse partitions, which will generate, through repeated segmentation in three directions, the gametophyte or moss itself. Each of these cells originated from the activity of the apical and subsequently divides repeatedly. The inner part of each of them will originate from the stem and the outer part from the leaves [34]. The size reached by the gametophyte is generally small, and only in some tropical mosses does it reaches 50 cm. Despite being the stem of a very primitive structure, some conductive and mechanical elements can be recognized in it. The sexual organs are located at the apices of the ramifications, and usually, the male ones are close to the female ones, although in certain dioecious species there is sexual dimorphism. Antheridia are normally formed from a superficial cell by segmentation. The archegonia, sometimes protected by special sheets, have numerous cells in the neck canal [4].

Once fertilization has been carried out, thanks to the action of water as a vector and the sucrose secreted by the archegonium as a chemotactically orienting substance, the zygote divides transversely. The apical cell will form, after numerous divisions, the capsule, while the basal one will form a sterile tissue, also diploid (see chromosome), which will constitute the peduncle of the capsule or seta. In Andraeales and Bryales, the seta is of gametophytic origin. In the formation of the capsule, two groups of cells or tissues stand out, one external or amphithecium and the other internal or endothecium [35]. In Bryales and Andraeales, the fertile tissue, which must give rise to meiospores, derives from the outer layer of the endothecium, while in Sphagnales it does so from its inner layer. The column is formed in all of them from the inner layer of the endothecium. Unlike liverworts and hornworts, mosses do not form disintegrating elaters. Capsule dehiscence varies with orders. In Bryales, it is made by means of an operculum, and they have teeth or peristomes, originating in the outer layer of the capsule, which collaborates in the release of the spores. The Sphagnales and Andraeales lack a peristome, and in the latter order, the capsule is opened by four longitudinal fissures.

The studies on the volatile composition and essential oil of mosses are few. Recent (last twenty years) research demonstrated the presence of essential oil and volatile compounds in mosses, the compounds found are of a natural monoterpene hydrocarbon, oxygenated monoterpenes, sesquiterpene hydrocarbons, oxygenated sesquiterpenes, diterpene hydrocarbons, oxygenated diterpenes, and other compounds (oxygenated or not). The volatile compounds found in mosses have a highly variable chemical formula and structure; are aliphatic, aromatic, terpenoid, non-terpenoid, oxygenated, non-oxygenated, etc.; and have been found in species belonging to various genera of the Class mosses [36–40].

2.1 *Antitrichia curtipendula* (Hedw.) Brid

The essential oils obtained by hydrodistillation from *A. curtipendula* collected in Turkey were analyzed by GC-MS and GC-FID. In this essential oil, 15 components were characterized, representing almost 85.61% of the total. The major component was tetradecanal (20.23%), a compound with the chemical formula (CF) $C_{14}H_{28}O$ and monoisotopic mass (MM) 212.2 Da; the other main compounds were nonanal (19.96%), hexahydrofarnesyl acetone (14.26%), and β-ionone (10.43%) [41].

2.2 *Brachythecium albicans* (Hedw.) Schimp

The main compounds found in the essential oil of *B. albicans* were nonanal (41.0%) and 4,4-dimethyl-E-2-pentene (6.6%). This oil was rich in aldehydes (51.3%), hydrocarbons (13.5%), and alcohols (4.3%). The amounts of terpenoids present in this moss are generally less than non-terpenoid compounds. *B. albicans* was collected in water from Sebinkarahisar, Kinik, Gümüshane, Turkey (at a height of ca. 1370 m) [42].

2.3 *Brachythecium salebrosum* (F. Weber & D. Mohr) Schimp

A total of 39 compounds were identified, constituting over 85.2% of the total oil composition of *B. salebrosum*. The main constituents of *B. salebrosum* essential oil were n-nonanal (66.3%), n-octanol (3.1%), hexahydrofarnesyl acetone (3.1%), n-heptanal (2.7%), and n-octanal (1.7%). In addition, n-nonanal (66.3%), n-undecanal (0.5%), and hexahydrofarnesyl acetone (3.1%) were present in this essential oil. Aliphatic aldehydes were the major constituents of *B. salebrosum* with a ratio of 73.3%. *B. salebrosum* was collected on the stony place in Corylus sp. communities from Akçaabat-Yidizli Village, Trabzon, Turkey (at a height of ca. 450 m) [43].

2.4 *Breutelia tomentosa* (Sw. ex Brid.) A. Jaeger

An acrocarpous moss, *B. tomentosa* is a widespread species that is distinguished by the robust habit, in dense or loose, glossy, yellowish-green, or green tufts and strongly plicate leaves (Fig. 1). Its habitat is on soil, humus occasionally epiphytic, or in rocks. In Ecuador, it is most frequently encountered in the montane forests and paramo between 1500–4600 m a.s.l. [44, 45].

The chemical composition of the essential oil of *B. tomentosa* collected in Ecuador was analyzed by GC-MS and GC-FID; the data of the qualitative and quantitative composition are shown in Table 1. Table 1 also shows the type of compound, the chemical formula (CF), and the monoisotopic mass (MM). Twenty-six components were determined in essential oil, representing 87.29% of the total oil; the principal constituents (> 5%) are found to be sesquiterpene hydrocarbons (epizonarene (CN: 11, 8.68%, CF: $C_{15}H_{24}$, MM: 204.2), α-selinene (6.69%), (Z)-thujopsene (6.15%), β-selinene (5.66%), viridiflorene (5.12%)) and oxygenated sesquiterpene (geranyl isovalerate (6.80%)). In essential oils of *B. tomentosa*, the principal groups were sesquiterpene hydrocarbons (42.27%), oxygenated sesquiterpenes (12.66%), oxygenated monoterpenes (6.80%),

Fig. 1 *B. tomentosa*

Table 1 Chemical composition of the essential oil from *Breutelia tomentosa*

CN	Compound	RI	Rif	%	Type	CF	MM (Da)
1	4-Octen-3-one	940	948	0.16	OT	$C_8H_{14}O$	126.1
2	9,10-DehydroIsolongifolene	1356	1361	1.38	SH	$C_{15}H_{20}$	200.2
3	α-Isocomene	1384	1387	0.54	SH	$C_{15}H_{24}$	204.2
4	β-Elemene	1390	1389	1.62	SH	$C_{15}H_{24}$	204.2
5	(Z)-Thujopsene	1423	1429	6.15	SH	$C_{15}H_{24}$	204.2
6	(E)-α-Bergamotene	1429	1432	1.98	SH	$C_{15}H_{24}$	204.2
7	Germacrene D	1486	1484	3.60	SH	$C_{15}H_{24}$	204.2
8	β-Selinene	1488	1489	5.66	SH	$C_{15}H_{24}$	204.2
9	Viridiflorene	1492	1496	5.12	SH	$C_{15}H_{24}$	204.2
10	α-Selinene	1495	1498	6.69	SH	$C_{15}H_{24}$	204.2
11	Epizonarene	1506	1501	8.68	SH	$C_{15}H_{24}$	204.2
12	γ-Selinene	1528	1532	0.85	SH	$C_{15}H_{24}$	204.2
13	Zierone	1568	1574	3.35	OS	$C_{15}H_{22}O$	218.2
14	(Z)-α-Copaene-8-ol	1605	1595	3.11	OS	$C_{15}H_{24}O$	220.2
15	Khusimone	1617	1604	3.77	OT	$C_{14}H_{20}O$	204.2
16	Geranyl isovalerate	1619	1606	6.80	OS	$C_{15}H_{26}O_2$	238.2
17	Selina-3,11-dien-6-α-ol	1648	1644	3.08	OS	$C_{15}H_{24}O$	220.2
18	α-Cadinol	1661	1652	3.12	OS	$C_{15}H_{26}O$	222.2
19	Cyclotetradecane	1675	1669	3.71	OT	$C_{14}H_{28}$	196.2
20	1-Heptadecene	1690	1696	1.72	OT	$C_{17}H_{34}$	238.3
21	1-Pentadecanal	1709	1713	1.60	OT	$C_{15}H_{30}O$	226.2
22	Hexadecanal	1783	1792	2.53	OT	$C_{16}H_{32}O$	240.2
23	Hexahydrofarnesyl acetone	1833	1847	3.32	OT	$C_{17}H_{26}O_2$	262.2
24	1-Hexadecanol	1883	1874	3.57	OT	$C_{16}H_{34}O$	242.3
25	Rimuene	1904	1896	2.14	DH	$C_{20}H_{32}$	272.5
26	Phytol	1931	1942	3.04	OD	$C_{20}H_{40}O$	296.3
Monoterpene hydrocarbons (MH)				–			
Oxygenated monoterpenes (OM)				–			
Sesquiterpene hydrocarbons (SH)				42.27			
Oxygenated sesquiterpenes (OS)				19.46			
Diterpene hydrocarbons (DH)				2.14			
Oxygenated diterpenes (OD)				3.04			
Other compounds (OT)				20.38			
Total identified				87.29			

–: not detected

oxygenated diterpenes (3.04%), diterpene hydrocarbons (2.14%), as well as other compounds (20.38%) including khusimone, cyclotetradecane, 1-hexadecanol, hexahydrofarnesyl acetone, hexadecanal, 1-heptadecene, and 1-pentadecanal.

2.5 *Bryum pallescens* Schleich. ex Schwagr

The main compounds found in the essential oil of *B. pallescens* were nonanal (29.3%) and Z-phytol (8.9%). This oil was rich in aldehydes (41.7%), hydrocarbons (7.9%), and alcohols (9.2%). The amounts of terpenoids present in this moss are generally less than non-terpenoid compounds. *B. pallescens* was collected in water from Sebinkarahisar, Temeltepe, Gümüshane, Turkey (at a height of ca. 1243 m) [42].

2.6 *Campylopus richardii* Brid

An acrocarpous moss, *C. richardii* is a widespread species that is distinguished by the robust habit-forming dense tufts; blackish–brown, foliate stems; and gradually large leaves ending in a comal tuft (Fig. 2). Its habitat is on soil, logs, rocks, and occasionally epiphytic. In Ecuador, it is most frequently encountered in the semi-dry lowland, montane forests, and paramo between 1000–5000 m a.s.l. [44, 45].

In the essential oil from Ecuadorian species *C. richardii*, 33 components were identified, representing 90.90% of the total essential oil, and epi-α-muurolol (CN: 22, 15.13%, CF: $C_{15}H_{26}O$, MM: 222.2), α-cadinol (12.51%), cadalene (6.67%), and β-cadinene (6.02%) were the main constituents (Table 2). In essential oils of *C. richardii*, the principal groups were oxygenated sesquiterpene (39.43%), sesquiterpene hydrocarbons (37.75%), diterpene hydrocarbons (2.75%), and monoterpene hydrocarbons (0.83%). Likewise, other compounds (10.14%) were found, such as hexahydrofarnesyl acetone, 1-hexadecanol, methyl linolenate, and hexadecanal.

Fig. 2 *C. richardii*

Table 2 Chemical composition of the essential oil from *Campylopus richardii*

CN	Compound	RI	Rif	%	Type	CF	MM (Da)
1	α-Phellandrene	1001	1002	0.16	MH	$C_{10}H_{16}$	136.1
2	Limonene	1022	1024	0.67	MH	$C_{10}H_{16}$	136.1
3	Bicycloelemene	1321	1331	0.18	SH	$C_{15}H_{24}$	204.2
4	α-Copaene	1373	1374	0.11	SH	$C_{15}H_{24}$	204.2
5	β-Elemene	1390	1389	0.26	SH	$C_{15}H_{24}$	204.2
6	β-Funebrene	1399	1413	2.16	SH	$C_{15}H_{24}$	204.2
7	(Z)-Thujopsene	1423	1429	1.02	SH	$C_{15}H_{24}$	204.2
8	γ-Maaliene	1432	1435	0.36	SH	$C_{15}H_{24}$	204.2
9	α-Patchoulene	1448	1454	4.19	SH	$C_{15}H_{24}$	204.2
10	Alloaromadendrene	1453	1458	3.20	SH	$C_{15}H_{24}$	204.2
11	Germacrene D	1486	1484	1.11	SH	$C_{15}H_{24}$	204.2
12	β-Selinene	1488	1489	1.03	SH	$C_{15}H_{24}$	204.2
13	Bicyclogermacrene	1498	1500	4.28	SH	$C_{15}H_{24}$	204.2
14	Cuparene	1508	1504	3.86	SH	$C_{15}H_{22}$	202.2
15	β-Cadinene	1516	1520	6.02	SH	$C_{15}H_{24}$	204.2
16	(Z)-Calamenene	1525	1528	2.27	SH	$C_{15}H_{22}$	202.2
17	γ-Selinene	1528	1532	1.03	SH	$C_{15}H_{24}$	204.2
18	Elemol	1550	1548	3.72	OS	$C_{15}H_{26}O$	222.2
19	Globulol	1591	1590	3.25	OS	$C_{15}H_{26}O$	222.2
20	Viridiflorol	1599	1592	2.22	OS	$C_{15}H_{26}O$	222.2
21	epi-α-Muurolol	1641	1640	15.13	OS	$C_{15}H_{26}O$	222.2
22	Cubenol	1652	1645	2.60	OS	$C_{15}H_{26}O$	222.2
23	α-Cadinol	1661	1652	12.51	OS	$C_{15}H_{26}O$	222.2
24	Cadalene	1684	1675	6.67	SH	$C_{15}H_{18}$	198.1
25	1-Pentadecanal	1709	1713	0.97	OT	$C_{15}H_{30}O$	226.2
26	Hexadecanal	1783	1792	1.04	OT	$C_{16}H_{32}O$	240.2
27	Hexahydrofarnesyl acetone	1833	1847	3.10	OT	$C_{17}H_{26}O_2$	262.2
28	1-Hexadecanol	1883	1874	2.49	OT	$C_{16}H_{34}O$	242.3
29	Rimuene	1904	1896	1.62	DH	$C_{20}H_{32}$	272.5
30	Hexadecanoic acid	1963	1959	0.78	OT	$C_{16}H_{32}O_2$	256.2
31	Methyl linolenate	2045	2047	1.40	OT	$C_{19}H_{32}O_2$	292.2
32	Kaurene	2050	2042	1.13	SH	$C_{15}H_{24}$	204.2
33	Bis(2-ethylhexyl) phthalate	2529	2550	0.36	OT	$C_{24}H_{38}O_4$	390.3
Monoterpene hydrocarbons (MH)				0.83			
Oxygenated monoterpenes (OM)				–			
Sesquiterpene hydrocarbons (SH)				38.88			
Oxygenated sesquiterpenes (OS)				39.43			
Diterpene hydrocarbons (DH)				1.62			
Oxygenated diterpenes (OD)				–			
Other compounds (OT)				10.14			
Total identified				90.9			

–: not detected

2.7 *Eurhynchium angustirete* (Broth.) T.J. Kop

Seven volatile compounds were identified in the essential oils from *E. angustirete* collected in Turkey. The volatile components identified accounted for 99.9% of the oil composition. *E. angustirete* oil was characterized by high amounts of hydrocarbons (81.6%). The remaining components in the essential oil were monoterpenes (18.3%). The main constituents were eicosane (28.6%), tetracosane (19.8%), tricosane (17.2%), pentacosane (16.0%), and α-pinene (11.6%) [46].

2.8 *Eurhynchium pulchellum* (Hedw.) Jenn

A total of 39 compounds were identified, constituting over 80.9% of the total oil composition of *E. pulchellum*. The major components of *E. pulchellum* were n-nonanal (36.2%), n-pentadecanal (12.0%), hexahydrofarnesyl acetone (10.0%), n-decane (8.1%), and n-undecanal (5.1%). In addition, n-nonanal (36%), n-undecanal (5.1%), and hexahydrofarnesyl acetone (10.0%) were present in this essential oil. Aliphatic aldehydes were the major constituents of *E. pulchellum* with a ratio of 57.9%. *Eurhynchium pulchellum* (Hedw.) Jenn. was collected on the edge of the forest under the *Carpinus betulus*, *Alnus glutinosa*, and *Rhododendron ponticum* from Akçaabat-Yidizli Village, Trabzon, Turkey (at a height of ca. 500 m) [43].

2.9 *Eurhynchium striatum* (Schreb. ex Hedw.) Schimp

A total of 34 volatile compounds were identified in the essential oils of *E. striatum* collected in Turkey. The volatile components identified accounted for 97.3% of the oil composition. The essential oil of *E. striatum* had a high content of ketone (48.1%) and aldehydes (23.3%) with 3-octanone (48.1%, CF: $C_8H_{16}O$, MM: 128.12), nonanal (13.7%), and tetradecanol (5.7%) being the major constituents [46].

2.10 *Fontinalis antipyretica* Hedw

At least 27 steam volatile compounds have been isolated from Sweden *F. antipyretica*. Ten (ethanal, ethyl formate, ethyl acetate, ethanol, hexanal, 2-heptanone, ethyl hexanoate 1-hexyl acetate, 2-octanone, and ethyl heptanoate) of these have been identified, the main compound being hexanal. Tetracosanoic acid has also been isolated from this moss [47].

2.11 *Hylocomium splendens* (Hedw.) Schimp

Fifty-eight components were identified from the oil of *H. splendens*, representing 75.4% of the total oil. The essential oil of *H. splendens* was rich in monoterpenes

(30.8%). The major compounds of this oil were β-pinene (11.6%) and α-pinene (8.9%), as well as limonene, camphene, and heptadecene. This moss was collected at different locations in Artvin, Turkey [48].

2.12 *Hypnum cupressiforme* Hedw

The major components of *H. cupressiforme* were nonanal (12.5%) and 2 E-tetradecen-1-ol (6.9%). This essential oil was rich in non-terpenoid components, such as aldehydes (15.6%) and terpenoid components, such as sesquiterpene hydrocarbons (12.7%). *H. cupressiforme* was collected growing on tree bodies from Turkey (at a height of ~1627 m) [40].

2.13 *Homalia trichomanoides* (Hedw.) Brid

The qualitative chemical composition of *H. trichomanoides* EO collected in Germany was determined by GC-MS. The identified compounds are shown in Table 3. In the essential oil of this species were found 56 compounds. The main group of compounds present in this species was sesquiterpene hydrocarbons (42.86%), followed by oxygenated monoterpenes (12.50%), as well as monoterpene hydrocarbons (7.14%), oxygenated sesquiterpenes (5.36%), diterpene hydrocarbons (5.36%), and oxygenated diterpenes (1.79%). Other compounds (25.00%) were also found, including benzaldehyde, phenylacetaldehyde, bornyl acetate, geosmin, and others [37].

2.14 *Homalothecium lutescens* (Hedw.) H. Rob

The major components of *H. lutescens* were nonanal (36.8%) and tricosane (6.5%). This essential oil was rich in non-terpenoid components, such as aldehydes (50.9%) and terpenoid components, such as sesquiterpene hydrocarbons (11.0%). *H. lutescens* was collected growing on stones near streams from Turkey (at a height of ~1348 m) [40].

2.15 *Leptodontium viticulosoides* (P. Beauv.) Wijk & Margad

An acrocarpous moss, *L. viticulosoides* is a widespread species that is distinguished by the robust habit, absence of a central stand, and strongly recurved leaf margins (Fig. 3). Its habitat is on soil and rocks, occasionally epiphytic. In Ecuador, it is a common specie of montane forests and páramo between 1500–4700 m a.s.l. [44, 45].

The qualitative and quantitative composition of essential oil *from* *L. viticulosoides* collected in Ecuador is shown in Table 4. Twenty-nine components were identified, representing 89.39% of the total oil; the principal group was sesquiterpene hydrocarbons, such as β-selinene (CN: 7, 13.52%, CF: $C_{15}H_{24}$, MM: 204,2), α-selinene (10.50%), β-bisabolene (9.12), cadalene (7.26), selina-3,7(11)-

Table 3 Chemical composition of the essential oil from *Homalia trichomanoides*

CN	Compound	*Homalia trichomanoides*	Type	CF	MM (Da)
1	2-Heptanone	x	OT	$C_7H_{14}O$	114.10
2	*n*-Heptanal	x	OT	$C_7H_{14}O$	114.10
3	Benzaldehyde	x	OT	C_7H_6O	106.04
4	α-Pinene	x	MH	$C_{10}H_{16}$	136.13
5	Camphene	x	MH	$C_{10}H_{16}$	136.13
6	3-Octanone	x	OT	$C_8H_{16}O$	128.12
7	β-Pinene	x	MH	$C_{10}H_{16}$	136.13
8	Phenylacetaldehyde	x	OT	C_8H_8O	120.05
9	Limonene	x	MH	$C_{10}H_{16}$	136.13
10	E-2-Octenal	x	OT	$C_8H_{14}O$	126.10
11	*n*-Nonanal	x	OT	$C_9H_{18}O$	142.14
12	*trans*-Pinocarveol	x	OM	$C_{10}H_{16}O$	152.12
13	*m*-Dimethoxybenzene	x	OT	$C_8H_{10}O_2$	138.07
14	Camphor	x	OM	$C_{10}H_{16}O$	152.12
15	Borneol	x	OM	$C_{10}H_{18}O$	154.14
16	Terpinen-4-ol	x	OM	$C_{10}H_{18}O$	154.14
17	Myrtenol	x	OM	$C_{10}H_{16}O$	152.12
18	β-Cyclocitral	x	OM	$C_{10}H_{16}O$	152.12
19	Bornyl acetate	x	OT	$C_{12}H_{20}O_2$	196.15
20	2E,4E-Decadienal	x	OM	$C_{10}H_{16}O$	152.12
21	Pentalenene	x	SH	$C_{15}H_{24}$	204.19
22	Maali-1,3-diene	x	SH	$C_{15}H_{22}$	202.17
23	Anastreptene	x	SH	$C_{15}H_{22}$	202.17
24	α-Copaene	x	SH	$C_{15}H_{24}$	204.19
25	Geosmin	x	OT	$C_{12}H_{22}O$	182.17
26	α-Ionone	x	OT	$C_{13}H_{20}O$	192.15
27	Longifolene	x	SH	$C_{15}H_{24}$	204.19
28	α-Barbatene	x	SH	$C_{15}H_{24}$	204.19
29	Peculiar oxide	x	OS	$C_{15}H_{26}O$	222.20
30	α-Cedrene	x	SH	$C_{15}H_{24}$	204.19
31	Aristolene	x	SH	$C_{15}H_{24}$	204.19
32	*trans*-α-Bergamotene	x	SH	$C_{15}H_{24}$	204.19
33	Aromadendrene	x	SH	$C_{15}H_{24}$	204.19
34	epi-β-Santalene	x	SH	$C_{15}H_{24}$	204.19
35	β-Barbatene	x	SH	$C_{15}H_{24}$	204.19
36	E-β-Farnesene	x	SH	$C_{15}H_{24}$	204.19
37	α-Humulene	x	SH	$C_{15}H_{24}$	204.19
38	β-Santalene	x	SH	$C_{15}H_{24}$	204.19
39	β-Acoradiene	x	SH	$C_{15}H_{24}$	204.19
40	β-Ionone	x	OT	$C_{13}H_{20}O$	192.15
41	Germacrene D	x	SH	$C_{15}H_{24}$	204.19
42	Bicyclogermacrene	x	SH	$C_{15}H_{24}$	204.19

(continued)

Table 3 (continued)

CN	Compound	*Homalia trichomanoides*	Type	CF	MM (Da)
43	2-Tridecanone	x	OT	$C_{13}H_{26}O$	198.19
44	γ-Cadinene	x	SH	$C_{15}H_{24}$	204.19
45	*cis/trans*-Calamenen	x	SH	$C_{15}H_{22}$	202.17
46	β-Sesquiphellandrene	x	SH	$C_{15}H_{24}$	204.19
47	Zonarene	x	SH	$C_{15}H_{24}$	204.19
48	δ-Cuprenene	x	SH	$C_{15}H_{24}$	204.19
49	1-*epi*-Cubenol	x	OS	$C_{15}H_{26}O$	222.20
50	α-Cadinol	x	OS	$C_{15}H_{26}O$	222.20
51	Mintsulfide	x	SH	$C_{15}H_{24}S$	236.16
52	Isopimara-8(14),15-diene	x	DH	$C_{20}H_{32}$	272.47
53	Heneicosa-6,9,12,15-tetraene	x	OT	$C_{21}H_{36}$	288.28
54	Abietatriene	x	DH	$C_{20}H_{30}$	270.23
55	16-Kaurene	x	DH	$C_{20}H_{32}$	272.25
56	Arachidonic acid	x	OD	$C_{20}H_{32}O_2$	304.24
Monoterpene hydrocarbons (MH)		4			
Oxygenated monoterpenes (OM)		7			
Sesquiterpene hydrocarbons (SH)		24			
Oxygenated sesquiterpenes (OS)		3			
Diterpene hydrocarbons (DH)		3			
Oxygenated diterpenes (OD)		1			
Other compounds (OT)		14			
Total identified		56			

Fig. 3 *L. viticulosoides*

Table 4 Chemical composition of the essential oil from *Leptodontium viticulosoides*

CN	Compound	RI	Rif	%	Type	CF	MM (Da)
1	α-Cubebene	1343	1345	0,45	SH	$C_{15}H_{24}$	204.2
2	β-Elemene	1390	1389	1.26	SH	$C_{15}H_{24}$	204.2
3	α-Gurjunene	1392	1409	0.25	SH	$C_{15}H_{24}$	204.2
4	(E)-β-Caryophyllene	1410	1417	5.32	SH	$C_{15}H_{24}$	204.2
5	(E)-α-Bergamotene	1429	1432	1.31	SH	$C_{15}H_{24}$	204.2
6	β-Santalene	1449	1457	1.58	SH	$C_{15}H_{24}$	204.2
7	β-Selinene	1488	1489	13.52	SH	$C_{15}H_{24}$	204.2
8	α-Selinene	1495	1498	10.5	SH	$C_{15}H_{24}$	204.2
9	α-Muurolene	1504	1500	1.43	SH	$C_{15}H_{24}$	204.2
10	β-Bisabolene	1510	1505	9.12	SH	$C_{15}H_{24}$	204.2
11	d-Cadinene	1520	1522	3.91	SH	$C_{15}H_{24}$	204.2
12	Selina-3,7(11)-diene	1535	1545	5.59	SH	$C_{15}H_{24}$	204.2
13	Germacrene B	1556	1559	0.89	SH	$C_{15}H_{24}$	204.2
14	Nerolidol	1565	1562	2.53	OS	$C_{15}H_{26}O$	222.2
15	Caryophyllene oxide	1578	1582	1.24	OS	$C_{15}H_{24}O$	220.2
16	Globulol	1591	1590	1.09	OS	$C_{15}H_{26}O$	222.2
17	Viridiflorol	1599	1592	1.00	OS	$C_{15}H_{26}O$	222.2
18	Khusimone	1617	1604	2.85	OT	$C_{14}H_{20}O$	204.2
19	epi-α-Muurolol	1641	1640	3.73	OS	$C_{15}H_{26}O$	222.2
20	α-Cadinol	1661	1652	2.73	OS	$C_{15}H_{26}O$	222.2
21	Isospathulenol	1673	1666	1.62	OS	$C_{15}H_{24}O$	220.2
22	Cadalene	1684	1675	7.26	SH	$C_{15}H_{18}$	198.1
23	1-Pentadecanal	1709	1713	1.60	OT	$C_{15}H_{30}O$	226.2
24	(E, E)-Farnesol	1718	1713	3.74	OS	$C_{15}H_{26}O$	222.2
25	Hexadecanal	1783	1792	0.97	OT	$C_{16}H_{32}O$	240.2
26	Hexahydrofarnesyl acetone	1833	1847	2.14	OT	$C_{17}H_{26}O_2$	262.2
27	Sandaracopimaradiene	1935	1922	0.36	DH	$C_{20}H_{32}$	272.5
28	Hexadecanoic acid	1963	1959	0.80	OT	$C_{16}H_{32}O_2$	256.2
29	Octadecanal	2011	2017	0.60	OT	$C_{18}H_{36}O$	268.3
Monoterpene hydrocarbons (MH)				–			
Oxygenated monoterpenes (OM)				–			
Sesquiterpene hydrocarbons (SH)				62.39			
Oxygenated sesquiterpenes (OS)				17.68			
Diterpene hydrocarbons (DH)				0.36			
Oxygenated diterpenes (OD)				–			
Other compounds (OT)				8.96			
Total identified				89.39			

–: not detected

diene (5.59%), and (E)-β-caryophyllene (5.32). In essential oils of *L. viticulosoides*, the principal groups were sesquiterpene hydrocarbons (62.39%), oxygenated sesquiterpenes (17.68%), and diterpene hydrocarbons (0.36%). Likewise, other compounds (8.96%) were found, including khusimone, hexahydrofarnesyl acetone, and 1-pentadecanal.

2.16 *Macromitrium perreflexum* Steere

This species is a rare species that is distinguished by its robust habit and remarkably reflexed leaves (Fig. 4). *M. perreflexum* is an epiphytic, common on trunks and branches of shrubs (canopy), and occasionally on rocks [44, 45]. It is a rare species in Ecuador and distributed in montane forests and paramo between 2200–4700 m a.s.l.

Twenty-five components were identified in *M. perreflexum*, EO from Ecuador, representing 90.21% of the total oil, and the major compounds were selina-3,11-dien-6-α-ol (CN: 20, 19.71%, CF: $C_{15}H_{24}O$, MM: 220.2), curcuphenol (10.60%), bicyclogermacrene (9.68%), γ-eudesmol (7.32%), and viridiflorene (5.12%) (Table 5). In essential oils of *M. perreflexum*, the principal groups were oxygenated sesquiterpene (49.55%), sesquiterpene hydrocarbons (39.38%), diterpene hydrocarbons (0.68%), and other compounds (0.60%), such as hexahydrofarnesyl acetone.

Fig. 4 *M. perreflexum*

Table 5 Chemical composition of the essential oil from *Macromitrium perreflexum*

Peak #	Compound	RI	Rif	%	Type	CF	MM (Da)
1	Bicycloelemene	1321	1331	0.09	SH	$C_{15}H_{24}$	204.2
2	Longicyclene	1369	1371	2.33	SH	$C_{15}H_{24}$	204.2
3	β-Gurjunene	1428	1431	0.51	SH	$C_{15}H_{24}$	204.2
4	β-Barbatene	1441	1440	0.84	SH	$C_{15}H_{24}$	204.2
5	Alloaromadendrene	1453	1458	1.17	SH	$C_{15}H_{24}$	204.2
6	α-Amorphene	1475	1483	2.73	SH	$C_{15}H_{24}$	204.2
7	Germacrene D	1486	1484	3.95	SH	$C_{15}H_{24}$	204.2
8	β-Selinene	1488	1489	2.15	SH	$C_{15}H_{24}$	204.2
9	Viridiflorene	1492	1496	5.12	SH	$C_{15}H_{24}$	204.2
10	bicyclogermacrene	1498	1500	9.68	SH	$C_{15}H_{24}$	204.2
11	α-Muurolene	1504	1500	1.26	SH	$C_{15}H_{24}$	204.2
12	β-Bisabolene	1510	1505	3.78	SH	$C_{15}H_{24}$	204.2
13	β-Cadinene	1516	1520	2.43	SH	$C_{15}H_{24}$	204.2
14	γ-Selinene	1528	1532	1.00	SH	$C_{15}H_{24}$	204.2
15	Cubenene	1553	1552	2.34	SH	$C_{15}H_{24}$	204.2
16	Spathulenol	1572	1577	1.87	OS	$C_{15}H_{24}O$	220.2
17	Viridiflorol	1599	1592	4.36	OS	$C_{15}H_{26}O$	222.2
18	Ledol	1611	1602	1.59	OS	$C_{15}H_{26}O$	222.2
19	γ-Eudesmol	1640	1630	7.32	OS	$C_{15}H_{26}O$	222.2
20	Selina-3,11-dien-6-α-ol	1648	1644	19.71	OS	$C_{15}H_{24}O$	220.2
21	Cedrenol	1657	1647	2.52	OS	$C_{15}H_{24}O$	220.2
22	Curcuphenol	1724	1717	10.60	OS	$C_{15}H_{22}O$	218.2
23	Xanthorrhizol	1754	1751	1.58	OS	$C_{15}H_{22}O$	218.2
24	Hexahydrofarnesyl acetone	1833	1847	0.60	OT	$C_{17}H_{26}O_2$	262.2
25	Rimuene	1904	1896	0.68	DH	$C_{20}H_{32}$	272.5
Monoterpene hydrocarbons (MH)				–			
Oxygenated monoterpenes (OM)				–			
Sesquiterpene hydrocarbons (SH)				39.38			
Oxygenated sesquiterpenes (OS)				49.55			
Diterpene hydrocarbons (DH)				0.68			
Oxygenated diterpenes (OD)				–			
Other compounds (OT)				0.6			
Total identified				90.21			

–: not detected

2.17 *Leucodon sciuroides* (Hedw.) Schwägr

Forty-one compounds were identified in *L. sciuroides* EO, representing 87.6%. This moss was collected at different locations in Artvin, Turkey. The aldehydes (49.9%) were the major constituents in the oil of *L. sciuroides*. n-Nonanal (26.8%) and heptanal (13.7%) were the main compounds of this oil, as well as tetradecanol, eicosane, and octanal [48].

2.18 *Mnium hornum* Hedw

Thirty-two compounds were found in the essential oil of the German species *M. hornum* [37]. The main group of compounds present in this species was sesquiterpene hydrocarbons (26.79%), followed by oxygenated sesquiterpenes (8.93%), as well as monoterpene hydrocarbons (5.36%), oxygenated monoterpenes (5.36%), and diterpene hydrocarbons (5.36%) (Table 6). Other compounds (8.93%) were also found, including n-heptanal, bornyl acetate, α-terpinyl acetate, epoxydecalin, and geosmin.

2.19 *Mnium marginatum* (Dicks.) P. Beauv

Table 7 shows the 37 compounds found in the essential oil of the species *M. marginatum* collected in Germany. The main group of compounds present in this species was sesquiterpene hydrocarbons (33.93%), followed by oxygenated sesquiterpenes (10.71%), as well as monoterpene hydrocarbons (5.36%), oxygenated monoterpenes (1.79%), and oxygenated diterpenes (1.79%). Other compounds (12.50%) were also found, including n-heptanal, benzaldehyde, 3-octanone, bornyl acetate, geosmin, and others [37].

2.20 *Mnium stellare* Hedw

Twenty compounds were found in the essential oil of *M. stellare* collected in Germany [37]. The main group of compounds present in this species was sesquiterpene hydrocarbons (19.64%), followed by oxygenated monoterpenes (1.79%) and oxygenated sesquiterpenes (1.79%) (Table 8). Other compounds (12.50%) were also found, including n-heptanal, 3-octanol, phenylacetaldehyde, n-nonanal, and others.

2.21 *Neckera complanata* (Hedw.) Huebener

The essential oils obtained by hydrodistillation from *N. complanata*, collected in Turkey, were analyzed by GC-FID and GC-MS. Twenty-one compounds in the oil representing 71.61% were identified. 3-Octanone (22.26%, CF: $C_8H_{16}O$, MM:1281) were the major constituents [49].

Table 6 Chemical composition of the essential oil from *Mnium hornum*

CN	Compound	*Mnium hornum*	Type	CF	MM (Da)
1	*n*-Heptanal	x	OT	$C_7H_{14}O$	114.10
2	α-Pinene	x	MH	$C_{10}H_{16}$	136.13
3	Camphene	x	MH	$C_{10}H_{16}$	136.13
4	Limonene	x	MH	$C_{10}H_{16}$	136.13
5	Camphor	x	OM	$C_{10}H_{16}O$	152.12
6	β-Cyclocitral	x	OM	$C_{10}H_{16}O$	152.12
7	Bornyl acetate	x	OT	$C_{12}H_{20}O_2$	196.15
8	2E,4E-Decadienal	x	OM	$C_{10}H_{16}O$	152.12
9	α-Terpinyl acetate	x	OT	$C_{12}H_{20}O_2$	196.15
10	Epoxydecalin	x	OT	$C_{12}H_{20}O$	180.15
11	α-Copaene	x	SH	$C_{15}H_{24}$	204.19
12	Geosmin	x	OT	$C_{12}H_{22}O$	182.17
13	Longifolene	x	SH	$C_{15}H_{24}$	204.19
14	α-Cedrene	x	SH	$C_{15}H_{24}$	204.19
15	β-Barbatene	x	SH	$C_{15}H_{24}$	204.19
16	10-*epi*-Muurola-4,11-diene	x	SH	$C_{15}H_{24}$	204.19
17	*ar*-Curcumene	x	SH	$C_{15}H_{22}$	202.17
18	γ-Curcumene	x	SH	$C_{15}H_{22}$	202.17
19	β-Selinene	x	SH	$C_{15}H_{24}$	204.19
20	α-Selinene	x	SH	$C_{15}H_{24}$	204.19
21	β-Curcumene	x	SH	$C_{15}H_{22}$	202.17
22	δ-Cadinene	x	SH	$C_{15}H_{24}$	204.19
23	β-Bazzanene	x	SH	$C_{15}H_{24}$	204.19
24	γ-Cuprene	x	SH	$C_{15}H_{24}$	204.19
25	*trans*-α-Bisabolene	x	SH	$C_{15}H_{24}$	204.19
26	α-Calacorene	x	SH	$C_{15}H_{20}$	200.16
27	Deoxopinguisone	x	OS	$C_{15}H_{22}O$	218.17
28	Maalian-5-ol	x	OS	$C_{15}H_{26}O$	222.20
29	α-Cadinol	x	OS	$C_{15}H_{26}O$	222.20
30	10,11-Dihydro-α-cuparenone	x	OS	$C_{15}H_{22}O$	218.17
31	Fukinanolide	x	OS	$C_{15}H_{22}O_2$	234.33
32	Abietatriene	x	DH	$C_{20}H_{30}$	270.23
Monoterpene hydrocarbons (MH)		3			
Oxygenated monoterpenes (OM)		3			
Sesquiterpene hydrocarbons (SH)		15			
Oxygenated sesquiterpenes (OS)		5			
Diterpene hydrocarbons (DH)		1			
Oxygenated diterpenes (OD)		–			
Other compounds (OT)		5			
Total identified		32			

Table 7 Chemical composition of the essential oil from *Mnium marginatum*

CN	Compound	*Mnium marginatum*	Type	CF	MM (Da)
1	*n*-Heptanal	x	OT	$C_7H_{14}O$	114.1
2	Benzaldehyde	x	OT	C_7H_6O	106.0
3	α-Pinene	x	MH	$C_{10}H_{16}$	136.1
4	Camphene	x	MH	$C_{10}H_{16}$	136.1
5	3-Octanone	x	OT	$C_8H_{16}O$	128.1
6	Limonene	x	MH	$C_{10}H_{16}$	136.1
7	*n*-Nonanal	x	OT	$C_9H_{18}O$	142.1
8	n-Decanal	x	OM	$C_{10}H_{20}O$	156.2
9	Bornyl acetate	x	OT	$C_{12}H_{20}O_2$	196.2
10	Maali-1,3-diene	x	SH	$C_{15}H_{22}$	202.2
11	1-epi-α-Pinguisene	x	SH	$C_{15}H_{24}$	204.2
12	Anastreptene	x	SH	$C_{15}H_{22}$	202.2
13	α-Copaene	x	SH	$C_{15}H_{24}$	204.2
14	β-Funebrene	x	SH	$C_{15}H_{24}$	204.2
15	Geosmin	x	OT	$C_{12}H_{22}O$	182.17
16	α-Barbatene	x	SH	$C_{15}H_{24}$	204.2
17	Aristolene	x	SH	$C_{15}H_{24}$	204.2
18	β-Cedrene	x	SH	$C_{15}H_{24}$	204.2
19	γ-Maaliene	x	SH	$C_{15}H_{24}$	204.2
20	Isobazzanene	x	SH	$C_{15}H_{24}$	204.2
21	β-Barbatene	x	SH	$C_{15}H_{24}$	204.2
22	α-Humulene	x	SH	$C_{15}H_{24}$	204.2
23	β-Acoradiene	x	SH	$C_{15}H_{24}$	204.2
24	Germacrene D	x	SH	$C_{15}H_{24}$	204.2
25	Bicyclogermacrene	x	SH	$C_{15}H_{24}$	204.2
26	Cuparene	x	SH	$C_{15}H_{22}$	202.2
27	β-Bisabolene	x	SH	$C_{15}H_{24}$	204.2
28	α-Chamigrene	x	SH	$C_{15}H_{24}$	204.2
29	*trans*-α-Bisabolene	x	SH	$C_{15}H_{24}$	204.2
30	Palustrol	x	OS	$C_{15}H_{26}O$	222.2
31	Germacrene-D-4-ol	x	OS	$C_{15}H_{26}O$	222.2
32	T-Cadinol	x	OS	$C_{15}H_{26}O$	222.2
33	α-Cadinol	x	OS	$C_{15}H_{26}O$	222.2
34	α-Cuparenone	x	OS	$C_{15}H_{20}O$	216.2
35	Diplophyllolide	x	OS	$C_{15}H_{20}O_2$	232.2
36	Heneicosa-6,9,12,15-tetraene	x	OT	$C_{21}H_{36}$	288.3
37	Manool	x	OD	$C_{20}H_{34}O$	290.3
Monoterpene hydrocarbons (MH)		3			
Oxygenated monoterpenes (OM)		1			
Sesquiterpene hydrocarbons (SH)		19			
Oxygenated sesquiterpenes (OS)		6			
Diterpene hydrocarbons (DH)		–			
Oxygenated diterpenes (OD)		1			
Other compounds (OT)		7			
Total identified		37			

Table 8 Chemical composition of the essential oil from *Mnium stellare*

CN	Compound	*Mnium stellare*	Type	CF	MM (Da)
1	*n*-Heptanal	x	OT	$C_7H_{14}O$	114.1
2	3-Octanone	x	OT	$C_8H_{16}O$	128.1
3	3-Octanol	x	OT	$C_8H_{18}O$	130.1
4	Phenylacetaldehyde	x	OT	C_8H_8O	120.1
5	*n*-Nonanal	x	OT	$C_9H_{18}O$	142.1
6	*m*-Dimethoxybenzene	x	OT	$C_8H_{10}O_2$	138.1
7	*n*-Nonanal	x	OT	C_9H_{20}	128.2
8	β-Cyclocitral	x	OM	$C_{10}H_{16}O$	152.1
9	α-Copaene	x	SH	$C_{15}H_{24}$	204.2
10	β-Barbatene	x	SH	$C_{15}H_{24}$	204.2
11	α-Humulene	x	SH	$C_{15}H_{24}$	204.2
12	β-Acoradiene	x	SH	$C_{15}H_{24}$	204.2
13	β-Chamigrene	x	SH	$C_{15}H_{24}$	204.2
14	Bicyclogermacrene	x	SH	$C_{15}H_{24}$	204.2
15	α-Cuprenene	x	SH	$C_{15}H_{24}$	204.2
16	Cuparene	x	SH	$C_{15}H_{22}$	202.2
17	β-Bisabolene	x	SH	$C_{15}H_{24}$	204.2
18	β-Bazzanene	x	SH	$C_{15}H_{24}$	204.2
19	α-Cadinene	x	SH	$C_{15}H_{24}$	204.2
20	T-Cadinol	x	OS	$C_{15}H_{26}O$	222.2
Monoterpene hydrocarbons (MH)		–			
Oxygenated monoterpenes (OM)		1			
Sesquiterpene hydrocarbons (SH)		11			
Oxygenated sesquiterpenes (OS)		1			
Diterpene hydrocarbons (DH)		–			
Oxygenated diterpenes (OD)		–			
Other compounds (OT)		7			
Total identified		20			

2.22 *Neckera crispa* Hedw

Essential oil of Turkish *Neckera crispa* was analyzed by GC-FID and GC-MS. Forty-two compounds representing 82.12% were identified. The essential oil of *N. crispa* was rich in β-phellandrene (20.00%, MH, CF: C10H16, MM: 136.1), camphene (10.36%), and γ-bisabolene-E (5.51%) [49].

2.23 *Phyllogonium viride* Brid

Twenty-seven compounds were identified in essential oil obtained from *P. viride*, whose samples were collected in southern Brazil. The compounds majorly found were β-bazzanene (20.30%, SH, CF: $C_{15}H_{24}$, MM: 204.19), β-caryophyllene

(17.06%), β-chamigrene (14.02), and germacrene B (11.72%), and β-barbatene (6,10%) [50].

2.24 *Plagiomnium acutum* (Lindb.) T.J. Kop

The ether extract of the Japanese moss *Plagiomnium acutum* was chromatographed on silica gel and Sephadex LH-20 to give α-cedrene (SH, CF: $C_{15}H_{24}$, MM: 204.19); β-cedrene, α-acoradiene, and ent-β-cedrene (ent-sesquiterpene); and 3, 7-dolabelladiene-18-ol (dolabellane diterpenoid) [51].

2.25 *Plagiomnium undulatum* (Hedw.) T.J. Kop

The EO from *P. undulatum* collected in Bavaria and Baden-Württemberg (Germany), Chamonix (France), and Sicily (Italy) was analyzed by GC-MS [37]. The results are shown in Table 9. Twenty-seven compounds were found in the essential oil of *P. undulatum*. The main group of compounds present in this species was monoterpene hydrocarbons (8.93%), followed by oxygenated monoterpenes (7.14%) and sesquiterpene hydrocarbons (7.14%), as well as oxygenated sesquiterpenes (3.57%), diterpene hydrocarbons (1.79%), and oxygenated diterpenes (1.79%). Other compounds (17.86%) were also found, including *n*-heptanal, 3-octananone, 1-octanol, bornyl acetate, geosmin, and others.

For *P. undulatum* collected on the soil of the forest under the *Carpinus betulus* and *Alnus glutinosa* from Akçaabat-Yidizli Village, Trabzon, Turkey (at a height of ca. 550 m), a total of 39 volatile compounds were identified, constituting over 88.8% of total oil composition. The main components of *P. undulatum* was γ-elemene (24.1%), δ-cadinene (11.7%), α-cadinol (9.5%), τ-muurolol (7.3%), and n-nonanal (6.1%). In addition, n-nonanal (6.1%), n-undecanal (1.3%), and hexahydrofarnesyl acetone (2.6%) were present in this essential oil. Sesquiterpene hydrocarbons (51.7%) were shown to be the main group in EO from *P. undulatum* [43].

2.26 *Plagiothecium undulatum* (Hedw.) Schimp

Forty-three compounds were found in the essential oil of *P. undulatum* collected in Austria [37]. The main group of compounds present in this species was sesquiterpene hydrocarbons (37.50%), followed by monoterpene hydrocarbons (10.71%) and oxygenated monoterpenes (10.71%), as well as diterpene hydrocarbons (1.79%) and oxygenated diterpenes (1.79%) (Table 10). Other compounds (14.29%) were also found, including *n*-heptanal, E-2-nonenal, E, E-2,4-nonadienal, bornyl acetate, epoxydecalin, and others.

Table 9 Chemical composition of the essential oil from *Plagiomnium undulatum*

CN	Compound	*Plagiomnium undulatum*	Type	CF	MM (Da)
1	*n*-Heptanal	x	OT	$C_7H_{14}O$	114.10
2	α-Pinene	x	MH	$C_{10}H_{16}$	136.13
3	Camphene	x	MH	$C_{10}H_{16}$	136.13
4	3-Octanone	x	OT	$C_8H_{16}O$	128.12
5	β-Pinene	x	MH	$C_{10}H_{16}$	136.13
6	2-Pentylfuran	x	OT	$C_9H_{14}O$	138.10
7	Δ-3-Carene	x	MH	$C_{10}H_{16}$	136.13
8	Limonene	x	MH	$C_{10}H_{16}$	136.13
9	E-2-Octenal	x	OT	$C_8H_{14}O$	126.10
10	1-Octen-3-ol	x	OT	$C_8H_{16}O$	128.12
11	1-Octanol	x	OT	$C_8H_{18}O$	130.14
12	*n*-Nonanal	x	OT	$C_9H_{18}O$	142.14
13	Pinocarvone	x	OM	$C_{10}H_{14}O$	150.10
14	Borneol	x	OM	$C_{10}H_{18}O$	154.14
15	β-Cyclocitral	x	OM	$C_{10}H_{16}O$	152.12
16	Bornyl acetate	x	OT	$C_{12}H_{20}O_2$	196.15
17	2E,4E-Decadienal	x	OM	$C_{10}H_{16}O$	152.12
18	Geosmin	x	OT	$C_{12}H_{22}O$	182.17
19	β-Cedrene	x	SH	$C_{15}H_{24}$	204.19
20	Amphora-4,11-diene	x	SH	$C_{15}H_{24}$	204.19
21	β-Ionone	x	OT	$C_{13}H_{20}O$	192.15
22	β-Bisabolene	x	SH	$C_{15}H_{24}$	204.19
23	β-Sesquiphellandrene	x	SH	$C_{15}H_{24}$	204.19
24	1-*epi*-Cubenol	x	OS	$C_{15}H_{26}O$	222.20
25	α-Cadinol	x	OS	$C_{15}H_{26}O$	222.20
26	Abietatriene	x	DH	$C_{20}H_{30}$	270.23
27	Manool	x	OD	$C_{20}H_{34}O$	290.26
Monoterpene hydrocarbons (MH)		5			
Oxygenated monoterpenes (OM)		4			
Sesquiterpene hydrocarbons (SH)		4			
Oxygenated sesquiterpenes (OS)		2			
Diterpene hydrocarbons (DH)		1			
Oxygenated diterpenes (OD)		1			
Other compounds (OT)		10			
Total identified		27			

2.27 *Pleurochaete squarrosa* (Brid.) Lindb

Forty components were identified from the oil of *P. squarrosa* collected in Turkey, representing 88.6% of the total oil. The major compounds were nonanal (24.6%), heptanal (12.2%), eicosane (7.7%), and octanal (3.8%) [52].

Table 10 Chemical composition of the essential oil from *Plagiothecium undulatum*

CN	Compound	*Plagiothecium undulatum*	Type	CF	MM (Da)
1	*n*-Heptanal	x	OT	$C_7H_{14}O$	114.1
2	α-Pinene	x	MH	$C_{10}H_{16}$	136.1
3	Camphene	x	MH	$C_{10}H_{16}$	136.1
4	β-Pinene	x	MH	$C_{10}H_{16}$	136.1
5	Myrcene	x	MH	$C_{10}H_{16}$	136.1
6	Δ-3-Carene	x	MH	$C_{10}H_{16}$	136.1
7	Limonene	x	MH	$C_{10}H_{16}$	136.1
8	2-Nonanone	x	OT	$C_9H_{18}O$	142.1
9	*E*-2-Nonenal	x	OT	$C_9H_{16}O$	140.1
10	Camphor	x	OM	$C_{10}H_{16}O$	152.1
11	Myrtenal	x	OM	$C_{10}H_{14}O$	150.1
12	E, E-2,4-Nonadienal	x	OT	$C_9H_{14}O$	138.1
13	β-Cyclocitral	x	OM	$C_{10}H_{16}O$	152.1
14	Carvone	x	OM	$C_{10}H_{14}O$	150.1
15	4,8a-Dimethyloctalin	x	OM	$C_{12}H_{17}$	161.1
16	Bornyl acetate	x	OT	$C_{12}H_{20}O_2$	196.2
17	2E,4E-Decadienal	x	OM	$C_{10}H_{16}O$	152.1
18	α-Terpinyl acetate	x	OT	$C_{12}H_{20}O_2$	196.2
19	Epoxydecalin	x	OT	$C_{12}H_{20}O$	180.2
20	α-Longipinene	x	SH	$C_{15}H_{24}$	204.2
21	Longicyclene	x	SH	$C_{15}H_{24}$	204.2
22	β-Bourbonene	x	SH	$C_{15}H_{24}$	204.2
23	Sativene	x	SH	$C_{15}H_{24}$	204.2
24	β-Longipinene	x	SH	$C_{15}H_{24}$	204.2
25	Longifolene	x	SH	$C_{15}H_{24}$	204.2
26	α-Cedrene	x	SH	$C_{15}H_{24}$	204.2
27	β-Caryophyllene	x	SH	$C_{15}H_{24}$	204.2
28	*trans*-α-Bergamotene	x	SH	$C_{15}H_{24}$	204.2
29	Sesquisabinene	x	SH	$C_{15}H_{24}$	204.2
30	β-Barbatene	x	SH	$C_{15}H_{24}$	204.2
31	α-Humulene	x	SH	$C_{15}H_{24}$	204.2
32	β-Ionone	x	OT	$C_{13}H_{20}O$	192.2
33	*ar*-Curcumene	x	SH	$C_{15}H_{22}$	202.2
34	γ-Muurolene	x	SH	$C_{15}H_{24}$	204.2
35	Germacrene D	x	SH	$C_{15}H_{24}$	204.2
36	α-Muurolene	x	SH	$C_{15}H_{24}$	204.2
37	β-Curcumene	x	SH	$C_{15}H_{22}$	202.2
38	γ-Cadinene	x	SH	$C_{15}H_{24}$	204.2
39	*cis/trans*-Calamenen	x	SH	$C_{15}H_{22}$	202.2
40	β-Sesquiphellandrene	x	SH	$C_{15}H_{24}$	204.2
41	*trans*-α-Bisabolene	x	SH	$C_{15}H_{24}$	204.2

(continued)

Table 10 (continued)

CN	Compound	*Plagiothecium undulatum*	Type	CF	MM (Da)
42	Abietatriene	x	DH	$C_{20}H_{30}$	270.2
43	Arachidonic acid	x	OD	$C_{20}H_{32}O_2$	304.2
Monoterpene hydrocarbons (MH)		6			
Oxygenated monoterpenes (OM)		6			
Sesquiterpene hydrocarbons (SH)		21			
Oxygenated sesquiterpenes (OS)		–			
Diterpene hydrocarbons (DH)		1			
Oxygenated diterpenes (OD)		1			
Other compounds (OT)		8			
Total identified		43			

2.28 *Pohlia nutans* (Hedw.) Lindb

The major components of *P. nutans* were nonanal (7.8%) and E-2-tetradecen-1-ol (7.1%). This essential oil was rich in non-terpenoid components, such as aldehydes (33.4%) and terpenoid components, such as sesquiterpene hydrocarbons (15.3%). *P. nutans* (Hedw.) Lindb. was collected growing on rocks from Turkey (at a height of ~1560 m a.s.l.) [40].

2.29 *Polytrichum commune* Hedw

In Turkish *P. commune* essential oil, 25 components were identified, representing almost 95.48% of total oils. The main components were biformene (13.06%, DH, CF: $C_{20}H_{32}$, MM: 272.3), hexahydrofarnesyl acetone (9.99%), (9Z,12Z)-octadecadienoic acid (9.51%), bornyl acetate (8.10%), and α-pinene (6.53%), E-β-ocimene (6.48%), and camphene (6.31%), respectively [41].

2.30 *Pseudoscleropodium purum* (Hedw.) M. Fleisch

A total of 65 volatile compounds were identified in the essential oils of *P. purum* collected in Turkey. The volatile components identified accounted for 97.7% of the oil composition. *P. purum* oil contained high amounts of mono and sesquiterpenes (28.2% and 26.9%). α-Pinene (16.1%), β-longipinene (8.7%), and heptanal (8.1) were the most abundant compounds [46].

2.31 *Rhacocarpus purpurascens* (Brid.) Paris

A pleurocarpous moss, *R. purpurascens* is a widespread species that is distinguished by its medium size, pale olive-green to pale golden color, irregularly to regularly pinnately branch, stem and branches tips attenuate, and leaves oblong-lanceolate with gradual and too abrupt long piliferous tips (Fig. 5). Its habitat is on soil, rock, and humus, occasionally epiphytic. In Ecuador, it is most frequently encountered in the montane forests and paramo between 1500–4600 m a.s.l. [44, 45].

Thirty-one compounds were identified from the essential oil of Ecuadorian *R. purpurascens*, representing 93.84% of the total essential oil, and α-cadinol (CN: 20, 36.84%, CF: $C_{15}H_{26}O$, MM: 222.2), α-santalene (8.35%), nerolidol (5.08%), and (Z)-α-copaene-8-ol (4.40%) were the major components (Table 11). In essential oils of *R. purpurascens*, the principal groups were oxygenated sesquiterpene (58.55%), sesquiterpene hydrocarbons (29.16%), oxygenated diterpenes (2.32%), diterpene hydrocarbons (2.06%), and other compounds (1.75%), including hexahydrofarnesyl acetone and 2-α-acetoxy-amorpha-4-7(11)-diene.

2.32 *Rhodobryum ontariense* (Kindb.) Paris

The essential oil the moss *R. ontariense* collected in Serbia was obtained by hydrodistillation and analyzed by GC-MS. Thirteen compounds were identified, representing 86.41% of the total oil. The main chemical constituents were OD phytol (31.95%, CF: $C_{20}H_{40}O$, MM: 296.3), 1-octen-3-ol (15.44%), α-pinene (11.55%), and n-nonanal (7.28%) [53].

Fig. 5 *R. purpurascens*

Table 11 Chemical composition of the essential oil from *Rhacocarpus purpurascens*

CN	Compound	RI	Rif	%	Type	CF	MM (Da)
1	α-Copaene	1373	1374	0.15	SH	$C_{15}H_{24}$	204.2
2	α-Bourbonene	1378	1376	tr	SH	$C_{15}H_{24}$	204.2
3	β-Patchoulene	1380	1379	0.6	SH	$C_{15}H_{24}$	204.2
4	β-Cubebene	1386	1387	0.24	SH	$C_{15}H_{24}$	204.2
5	β-Elemene	1390	1389	0.04	SH	$C_{15}H_{24}$	204.2
6	α-Santalene	1406	1416	8.35	SH	$C_{15}H_{24}$	204.2
7	(E)-α-Bergamotene	1429	1432	2.57	SH	$C_{15}H_{24}$	204.2
8	Epi-β-Santalene	1444	1445	3.30	SH	$C_{15}H_{24}$	204.2
9	(E)-β-Farnesene	1447	1454	3.44	SH	$C_{15}H_{24}$	204.2
10	Germacrene D	1486	1484	2.31	SH	$C_{15}H_{24}$	204.2
11	bicyclogermacrene	1498	1500	3.63	SH	$C_{15}H_{24}$	204.2
12	β-bisabolene	1510	1505	2.40	SH	$C_{15}H_{24}$	204.2
13	β-Sesquiphellandrene	1518	1521	1.89	SH	$C_{15}H_{24}$	204.2
14	Germacrene B	1556	1559	0.24	SH	$C_{15}H_{24}$	204.2
15	Nerolidol	1565	1562	5.08	OS	$C_{15}H_{26}O$	222.2
16	Salvial-4(14)-en-1-one	1603	1594	3.26	OS	$C_{15}H_{24}O$	220.2
17	(Z)-α-Copaene-8-ol	1605	1595	4.40	OS	$C_{15}H_{24}O$	220.2
18	epi-α-Muurolol	1641	1640	1.84	OS	$C_{15}H_{26}O$	222.2
19	β-Eudesmol	1660	1649	0.63	OS	$C_{15}H_{26}O$	222.2
20	α-Cadinol	1661	1652	36.84	OS	$C_{15}H_{26}O$	222.2
21	Isospathulenol	1673	1666	0.84	OS	$C_{15}H_{24}O$	220.2
22	α-Bisabolol	1688	1685	1.95	OS	$C_{15}H_{26}O$	222.2
23	(E,E)-Farnesol	1718	1713	0.58	OS	$C_{15}H_{26}O$	222.2
24	Aristolone	1757	1762	2.46	OS	$C_{15}H_{22}O$	218.2
25	2-α-acetoxy-Amorpha-4-7 (11)-diene	1792	1805	0.45	SH	$C_{15}H_{24}$	204.2
26	Hexahydrofarnesyl acetone	1833	1847	1.30	OT	$C_{17}H_{26}O_2$	262.2
27	Rimuene	1904	1896	0.46	DH	$C_{20}H_{32}$	272.5
28	Callitrisin	1921	1941	0.67	OS	$C_{15}H_{20}O_2$	232.2
29	Phytol	1931	1942	2.32	OD	$C_{20}H_{40}O$	296.3
30	Bifloratriene	1973	1977	1.19	DH	$C_{20}H_{32}$	272.3
31	Kaurene	2050	2042	0.41	SH	$C_{15}H_{24}$	204.2
Monoterpene hydrocarbons (MH)				–			
Oxygenated monoterpenes (OM)				–			
Sesquiterpene hydrocarbons (SH)				30.02			
Oxygenated sesquiterpenes (OS)				58.55			
Diterpene hydrocarbons (DH)				1.65			
Oxygenated diterpenes (OD)				2.32			
Other compounds (OT)				1.30			
Total identified				93.84			

–: not detected

2.33 *Sphagnum auriculatum* Schimp

The essential oils from *S. auriculatum* collected in Portugal were isolated by hydrodistillation and analyzed by GC-MS. Six compounds were determined, of which three were identified. The main compounds were as yet unidentified sesquiterpene (20.3%) and n-heneicosane (8.2%) [54].

2.34 *Sphagnum subnitens* Russow & Warnst

Ten compounds were determined, of which eight were identified in essential oils from *S. subnitens* collected in Portugal. The essential oil was isolated by hydrodistillation and analyzed by GC-MS. The main compounds were zierene (29.1%, SH, CF: $C_{15}H_{24}$, MM: 204.2) and phyllocladene (11.2%) [54].

2.35 *Syntrichia intermedia* Brid

The main compounds found in the essential oil of *S. intermedia* Brid. were E-2-tetradecen-1-ol (9.9%) and nonanal (8.3%). This oil was rich in aldehydes (18.0%), hydrocarbons (24.1%), and alcohol (13.5%). The amounts of terpenoids present in this moss are generally less than non-terpenoid compounds. *S. intermedia* was collected on soil from Sebinkarahisar, Ekecek, Gümüshane, Turkey (at a height of ca. 1360 m) [42].

2.36 *Taxiphyllum wisgrillii* (Garov.) Wijk & Margad

Twenty-one compounds were found in the essential oil of *T. wisgrillii* collected in Bavaria (Germany) [37]. The main group of compounds present in this species was sesquiterpene hydrocarbons (17.86%), followed by oxygenated monoterpenes (5.36%) and oxygenated sesquiterpenes (1.79%) (Table 12). Other compounds (12.50%) were also found, including *n*-heptanal, 3-octanone, 1-octen-3-ol, *n*-nonanal, and others.

2.37 *Thuidium peruvianum* Mitt

A pleurocarpous moss, *T. peruvianum* is a widespread and common component of tropical Andes, species that is distinguished by the medium to large size, olive dark green to yellowish-brown or golden color, strongly differentiated stem, and branch leaves (Fig. 6). Its habitat is on soil, rocks and tree bases, lowland rainforest, montane forests, and paramo, occasionally epiphytic [44, 45]. Ecuador is most frequently encountered in the montane forests and paramo between 1000–4600 m a.s.l.

Phytol (CN: 26, CF: $C_{20}H_{40}O$, MM: 296,3) with 21.72% was the main compound in EO from Ecuadorian *T. peruvianum*. Valerenol (10.07%), β-selinene (9.26%), and

Table 12 Chemical composition of the essential oil from *Taxiphyllum wisgrillii*

CN	Compound	*Taxiphyllum wisgrillii*	Type	CF	MM (Da)
1	*n*-Heptanal	x	OT	$C_7H_{14}O$	114.1
2	Benzaldehyde	x	OT	C_7H_6O	106.0
3	3-Octanone	x	OT	$C_8H_{16}O$	128.1
4	E-2-Octenal	x	OT	$C_8H_{14}O$	126.1
5	1-Octen-3-ol	x	OT	$C_8H_{16}O$	128.1
6	*n*-Nonanal	x	OT	C_9H_{20}	128.2
7	α-Terpineol	x	OM	$C_{10}H_{18}O$	154.1
8	β-Cyclocitral	x	OM	$C_{10}H_{16}O$	152.1
9	2E,4E-Decadienal	x	OM	$C_{10}H_{16}O$	152.1
10	α-Terpinyl acetate	x	OT	$C_{12}H_{20}O_2$	196.2
11	β-Caryophyllene	x	SH	$C_{15}H_{24}$	204.2
12	β-Barbatene	x	SH	$C_{15}H_{24}$	204.2
13	α-Humulene	x	SH	$C_{15}H_{24}$	204.2
14	β-Acoradiene	x	SH	$C_{15}H_{24}$	204.2
15	β-Chamigrene	x	SH	$C_{15}H_{24}$	204.2
16	Z, E-α-Farnesene	x	SH	$C_{15}H_{24}$	204.2
17	Bicyclogermacrene	x	SH	$C_{15}H_{24}$	204.2
18	β-Bisabolene	x	SH	$C_{15}H_{24}$	204.2
19	β-Bazzanene	x	SH	$C_{15}H_{24}$	204.2
20	α-Cadinene	x	SH	$C_{15}H_{24}$	204.2
21	Deoxopinguisone	x	OS	$C_{15}H_{22}O$	218.2
Monoterpene hydrocarbons (MH)		–			
Oxygenated monoterpenes (OM)		3			
Sesquiterpene hydrocarbons (SH)		10			
Oxygenated sesquiterpenes (OS)		1			
Diterpene hydrocarbons (DH)		–			
Oxygenated diterpenes (OD)		–			
Other compounds (OT)		7			
Total identified		21			

α-bisabolol (7.60%) were the other main compounds of a total of 28 components (Table 13). In this oil, the compounds that represent 86.31% were identified. The principal groups were oxygen sesquiterpenes (28.85%), oxygenated diterpenes (21.72%), sesquiterpene hydrocarbons (18.84%), diterpene hydrocarbons (1.84%), and monoterpene hydrocarbons (0.45%), as well as other compounds (14.61%), including hexahydrofarnesyl acetone, 1-hexadecanol, 1-pentadecanal, hexadecanoic acid, hexadecanal, and octadecanal.

2.38 *Tortella inclinata* var. *densa* (Lorentz & Molendo) Limpr

Thirteen compounds were identified from the essential oil of *T. inclinata* var. *densa*, collected in Turkey, representing 93.8% of the total essential oil. Eicosane (27.2%), nonanal (14.8%), and undecanal (7.7%) were the major components [52].

Fig. 6 *T. peruvianum*

2.39 *Tortella tortuosa* (Schrad. ex Hedw.) Limpr

In the essential oil of *T. tortuosa*, 33 components were identified, representing 99.4% of the essential oil. Eicosane (15.7%, SH, CF: $C_{20}H_{42}$, MM: 282.3), nonanal (9.1%), heptanal (8.3%), and α-pinene (4.4%) were the main constituent [52].

2.40 *Tortula muralis* Hedw

The major components of *T. muralis* were nonanal (18.3%) and tetradecanol (4.3%). This essential oil was rich in non-terpenoid components, such as aldehydes (26.9%) and terpenoid components, such as sesquiterpene hydrocarbons (6.7%). *T. muralis Hedw.* was collected growing on soil from Turkey (at a height of ~1520 m a.s.l) [40].

3 Liverworts

The class Liverworts groups some 7000–9000 species that are found all over the world, but preferably in the tropical American regions [55]. They live mainly in shady and cool areas during the summer. Very few are truly aquatic. In the fossil

Table 13 Chemical composition of the essential oil from *Thuidium peruvianum*

CN	Compound	RI	Rif	%	Type	CF	MM (Da)
1	α-Phellandrene	1001	1002	0.10	MH	$C_{10}H_{16}$	136.1
2	Limonene	1022	1024	0.35	MH	$C_{10}H_{16}$	136.1
3	α-Cubebene	1343	1345	0.21	SH	$C_{15}H_{24}$	204.2
4	α-Bourbonene	1378	1376	0.09	SH	$C_{15}H_{24}$	204.2
5	β-Elemene	1390	1389	0.42	SH	$C_{15}H_{24}$	204.2
6	(E)-β-Caryophyllene	1410	1417	1.22	SH	$C_{15}H_{24}$	204.2
7	γ-Elemene	1430	1434	0.66	SH	$C_{15}H_{24}$	204.2
8	β-Selinene	1488	1489	9.26	SH	$C_{15}H_{24}$	204.2
9	γ-Cadinene	1513	1513	2.65	SH	$C_{15}H_{24}$	204.2
10	β-Cadinene	1516	1520	3.83	SH	$C_{15}H_{24}$	204.2
11	Germacrene B	1556	1559	0.50	SH	$C_{15}H_{24}$	204.2
12	Nerolidol	1565	1562	2.66	OS	$C_{15}H_{26}O$	222.2
13	Viridiflorol	1599	1592	0.44	OS	$C_{15}H_{26}O$	222.2
14	Cedrol	1607	1600	2.51	OS	$C_{15}H_{26}O$	222.2
15	Ledol	1611	1602	1.21	OS	$C_{15}H_{26}O$	222.2
16	epi-α-Muurolol	1641	1640	1.66	OS	$C_{15}H_{26}O$	222.2
17	α-Cadinol	1661	1652	1.63	OS	$C_{15}H_{26}O$	222.2
18	Valerenol	1669	1655	10.07	OS	$C_{15}H_{24}O$	220.2
19	(Z)-α-Bisabolene epoxide	1681	1675	1.07	OS	$C_{15}H_{24}O$	220.2
20	α-Bisabolol	1688	1685	7.60	OS	$C_{15}H_{26}O$	222.2
21	1-Pentadecanal	1709	1713	2.35	OT	$C_{15}H_{30}O$	226.2
22	Hexadecanal	1783	1792	1.80	OT	$C_{16}H_{32}O$	240.2
23	Hexahydrofarnesyl acetone	1833	1847	3.92	OT	$C_{17}H_{26}O_2$	262.2
24	1-Hexadecanol	1883	1874	3.31	OT	$C_{16}H_{34}O$	242.3
25	Rimuene	1904	1896	1.84	DH	$C_{20}H_{32}$	272.5
26	Phytol	1931	1942	21.72	OD	$C_{20}H_{40}O$	296.3
27	Hexadecanoic acid	1963	1959	1.94	OT	$C_{16}H_{32}O_2$	256.2
28	Octadecanal	2011	2017	1.29	OT	$C_{18}H_{36}O$	268.3
Monoterpene hydrocarbons (MH)				0.45			
Oxygenated monoterpenes (OM)				–			
Sesquiterpene hydrocarbons (SH)				18.84			
Oxygenated sesquiterpenes (OS)				28.85			
Diterpene hydrocarbons (DH)				1.84			
Oxygenated diterpenes (OD)				21.72			
Other compounds (OT)				14.61			
Total identified				86.31			

–: not detected

state, they are known for sure since the Jurassic, although it seems that they already existed in the Carboniferous. The plants are not economically important to humans but do provide food for animals, facilitate the decay of logs, and aid in the disintegration of rocks by their ability to retain moisture [56].

Liverworts (from the Greek hepar meaning liver) derive their name from the shape of some of them that resemble the lobes of the liver. In the past, many were used to cure ailments of that organ [26]. Some extracts of these plants have shown activity against fungi and bacteria and have been used successfully in laboratory experiments to control pathogenic fungal pests in tomato, pepper, and wheat crops [57]. Liverworts have two types of bodies: one has a stem and leaves (foliose liverworts) and the other is ribbon-shaped or flattened green sheets (thallose liverworts). Both have delicate hair-like structures, called rhizoids, that help them attach to and absorb substances from the substrate.

The liverwort sporophyte is the simplest of all bryophytes; it has three parts: the foot that joins it to the gametophyte; the seta, a pedicel that raises the capsule above the gametophyte; and the capsule, which contains the spores and structures as springs that contribute to their dispersion. The capsule opens into four valves [26]. Thallose liverworts, which are branching and ribbon-like, grow commonly on moist soil or damp rocks, while leafy liverworts are found in similar habitats as well as on tree trunks in damp woods. Foliose liverworts are more numerous and diverse than thalloses.

Liverworts have apical growth, and their size, very small, reaches 30 cm in some species. In leafy liverworts, the gametophyte is formed by a more or less underground lower part devoid of chlorophyll, on which, thanks to the activity of a meristematic cell (of growth tissue) with two or three faces, a series of leaves develop: distichous or tristic disposition. Given the dorsiventral disposition that most of the foliose liverworts show, the activity of the lower face of the meristematic cell usually originates from the underleaf or small leaves located on the ventral face of the stem.

After fertilization, the basal cells of the young sporophyte constitute a haustorial (sucking) tissue that takes water and nutrients from the gametophyte. The upper cells give rise to the foot, and the wall of the capsule is formed from the amphithecium, which in this case has only one layer of cells. In thallous liverworts, such as Marchantia, the structure of the flattened thallus is complete, and it shows aeriferous pores in the upper epidermis as stomata. On the underside, unicellular rhizoids and multicellular scales are visible, which are interpreted as amphigastrium. The sexual organs are situated on top of special branches, antheridiophores, and archegoniophores [58].

The antheridiophore consists of an elongated peduncle topped by a wide octolobed disk where the antheridia are embedded in crypts. The archegoniophore is analogous, but the lobes are deeply indented at maturity. The archegonia, which originally formed on the upper face, due to a subsequent folding, are located on the lower face upon reaching maturity. When the water drops hit the archegoniophore disc, they drag the biflagellate spermatozoa that can hit the archegoniophore and fertilize the egg cell. The zygote develops the embryo, and the foot or haustorial tissue and the sporiferous tissue derived from the endothecium can be seen. After meiosis, numerous elongated meiospores and elaters (sterile cells of the sporogonium) are formed. The capsule opens once mature by four or six valves. Within the liverworts, several orders can be separated, from the most primitive to the most modern, are Calobryales, Jungermanniales, Metzgeriales, Sphaerocarpales, and Marchantiales.

Liverworts produce secondary metabolites in abundance [59] with more than 1500 terpenoids and 350 aromatic compounds flavonoids [60–62]. In this context, liverworts produce terpenoids and aromatic compounds, many of which exhibit diverse interesting biological properties related to antitumor, antimicrobial, antifungal, antioxidative, and insecticidal activities and cytotoxic and insect antifeedant [60]. Several studies have shown abundant secondary metabolites in foliose liverworts.

3.1 *Asterella marginata* (Nees) S.W. Arnell

The volatile constituents of liverworts *A. marginata* from South Africa were determined by GC-MS and GC-FID. Forty-one constituents were determined, representing 96.32% of the total. The main component was elaidic acid methyl ester with 28.61%, and other main compounds were β-caryophyllene oxide 19.03% and α-barbatene 8.75% [59].

3.2 *Dumortiera hirsuta* (Sw.) Nees

Volatile compounds of South African liverwort *D. hirsuta* were obtained with cyclohexane as solvent. In a volatile fraction 43, chemical components were identified, representing 86.87% of the total. The main constituents were β-caryophyllene oxide (23.80%), alloaromadendrene oxide (9.83%), n-hexadecanoic acid (6.89%), cuparene (6.80%), and β-barbatene (5.25%) [59].

3.3 *Fossombronia swziensis* Perold

The analysis of volatile components of *F. swziensis*, collected in South Africa, showed the presence of 81 constituents which represents 89.92% of the total. The main constituents were b-caryophyllene oxide (15.75%), p-cymene (5.35%), and allo-aromadendra-4(15),10(14)-diene (5.3%) [59].

3.4 *Frullania brasiliensis* Raddi

A foliose liverwort *F. brasiliensis* is the most common Neotropical species. It is recognized by recurved apiculate leaf apices and the terete perianth. Underleaves bifid to 1/5–1/3 with margins always recurved (Fig. 7). Its habitat is on the bark of trees, soil, and rock in lowland and montane rainforests and paramo and semi-deciduous and dry forests, 100–3900 m a.s.l. [63, 64].

Fig. 7 *F. brasiliensis*

In the essential oil of *F. brasiliensis* collected in Ecuador, 25 components were determined, as shown in Table 14. The identified components represent 80.12% of the total oil. The components were grouped as oxygenated sesquiterpenes (50.81%), sesquiterpene hydrocarbons (27.47%), and oxygenated monoterpenes (0.45%), diterpene hydrocarbons (0.23%), as well as other compounds (1.16%). The main components were OS τ-muurolol (CN: 21, 32.14%, CF: $C_{15}H_{26}O$, MM: 222,2), germacrene-D (11.98%), rosifoliol (5.08%), τ-cadinol (4.71%), and elemol (4.19%).

3.5 *Herbertus juniperoideus* (Sw.) Grolle

A foliose liverwort *H. juniperoideus* is a widespread species that is distinguished by the leaf lobes not overlapping at the sinus, the short leaf tips, and the broad vitta bifurcating very high up the lamina, just below the sinus. Underleaves are similar to leaves but more symmetrical (Fig. 8). Its habitat is on bark, rock, and soil in montane forests and paramo, 1000–3750 m a.s.l. [45, 63, 64].

Twenty-seven components were identified in the essential oil of Ecuadorian *H. juniperoideus* (Table 15) representing 88.21% of total constituents in the oil. The components were grouped as sesquiterpene hydrocarbons (46.71%), oxygenated sesquiterpenes (39.08%), and diterpene hydrocarbons (0.87%), oxygenated monoterpenes (0.79%), as well as other compounds (0.76%). The main components

Table 14 Chemical composition of the essential oil from *Frullania brasiliensis*

CN	Compound	RI	RIf	%	Type	CF	MM (Da)
1	1-Octen-3-ol, acetate	1107	1110	0.45	OM	$C_{10}H_{18}O_2$	170.1
2	Thymol methyl ether	1217	1232	1.16	OT	$C_{11}H_{16}O$	164.1
3	δ-Elemene	1324	1335	0.10	SH	$C_{15}H_{24}$	204.2
4	α-Cubebene	1335	1345	0.42	SH	$C_{15}H_{24}$	204.2
5	Silphiperfola-5,7(14)-diene	1355	1363	0.25	SH	$C_{15}H_{22}$	202.2
6	α-Copaene	1363	1374	0.44	SH	$C_{15}H_{24}$	204.2
7	β-Bourbonene	1370	1387	0.74	SH	$C_{15}H_{24}$	204.2
8	β-Elemene	1383	1389	3.03	SH	$C_{15}H_{24}$	204.2
9	Longifolene	1392	1407	1.77	SH	$C_{15}H_{24}$	204.2
10	β-Cubebene	1393	1387	0.63	SH	$C_{15}H_{24}$	204.2
11	Aromadendrene	1423	1439	1.07	SH	$C_{15}H_{24}$	204.2
12	β-Gurjunene	1448	1431	1.54	SH	$C_{15}H_{24}$	204.2
13	Germacrene-D	1466	1484	11.98	SH	$C_{15}H_{24}$	204.2
14	Viridiflorene	1476	1496	2.17	SH	$C_{15}H_{24}$	204.2
15	Bicyclogermacrene	1480	1500	tr	SH	$C_{15}H_{24}$	204.2
16	α-Calacorene	1527	1544	0.80	SH	$C_{15}H_{20}$	200.2
17	γ-Selinene	1528	1522	2.53	SH	$C_{15}H_{24}$	204.2
18	Elemol	1570	1548	4.19	OS	$C_{15}H_{26}O$	222.2
19	Viridiflorol	1579	1592	3.17	OS	$C_{15}H_{26}O$	222.2
20	Rosifoliol	1592	1600	5.08	OS	$C_{15}H_{26}O$	222.2
21	τ-muurolol	1642	1640	32.14	OS	$C_{15}H_{26}O$	222.2
22	τ-cadinol	1648	1638	4.71	OS	$C_{15}H_{26}O$	222.2
23	Torreyol	1657	1656	0.66	OS	$C_{15}H_{26}O$	222.2
24	Acorenone	1673	1692	0.86	OS	$C_{15}H_{24}O$	220.2
25	Sandaracopimaradiene	1935	1935	0.23	DH	$C_{20}H_{32}$	272.3
Monoterpene hydrocarbons (MH)				–			
Oxygenated monoterpenes (OM)				0.45			
Sesquiterpene hydrocarbons (SH)				27.47			
Oxygenated sesquiterpenes (OS)				50.81			
Diterpene hydrocarbons (DH)				0.23			
Other compounds (OT)				1.16			
Total identified				80.12			

–: not detected

were sesquiterpene hydrocarbons, bicyclogermacrene (CN: 14, 18.23%) and germacrene D (4.67%), and oxygenated sesquiterpenes, caryophyllene oxide (15.29%), spathulenol (11.90%), and viridiflorol (8.93%).

3.6 *Leptoscyphus hexagonus* (Nees) Grolle

A foliose liverwort *L. hexagonus* is distinguished by plant robust and glossy brown color; leaves are strongly convex, erect, and appressed and somewhat expanded

Fig. 8 *H. juniperoideus*

ventrally; apex is rounded; ventral leaf base is auriculate and with a tooth, occasionally entire; underleaves are large, bifid to 1/4–1/3 (Fig. 9). Its habitat is on bark in the upper montane forest and shrubby paramo, 2500–4000 m a.s.l. [63, 64].

Twenty-one chemical components were identified in the essential oil of Ecuadorian *L. hexagonus* representing 85.85% of the total components oil. The components are grouped as oxygenated sesquiterpenes (66.24%) and sesquiterpene hydrocarbons (19.61%) (Table 16). The main components were oxygenated sesquiterpenes, cabreuva oxide D (33.77%), elemol (18.55%), and viridiflorol (8.03%), and sesquiterpene hydrocarbons, bicyclogermacrene (6.70%).

3.7 *Leptolejeunea elliptica* (Lehm. & Lindenb.) Besch

The volatile compounds of *Leptolejeunea elliptica* collected in Tokushima (Japan) were isolated using headspace solid-phase microextraction and analyzed by GC-MS. Twenty-six components were identified, and the two main compounds were 1-ethyl-4-methoxybenzene (51.71%) and 1-ethyl-4-hydroxybenzene (13.66%); other compounds were present in concentrations minor than 5% [65].

Table 15 Chemical composition of the essential oil from *Herbertus juniperoideus*

CN	Compound	RI	RIf	%	Type	CF	MM (Da)
1	1-Octen-3-ol, acetate	1107	1110	0.79	OM	$C_{10}H_{18}O_2$	170.1
2	Bicycloelemene	1321	1331	0.33	SH	$C_{15}H_{24}$	204.2
3	α-Cubebene	1335	1345	0.15	SH	$C_{15}H_{24}$	204.2
4	α-Copaene	1363	1374	0.27	SH	$C_{15}H_{24}$	204.2
5	β-Bourbonene	1370	1387	0.32	SH	$C_{15}H_{24}$	204.2
6	β-Elemene	1383	1389	0.78	SH	$C_{15}H_{24}$	204.2
7	Longifolene	1392	1407	2.58	SH	$C_{15}H_{24}$	204.2
8	α-Gurjunene	1412	1409	1.15	SH	$C_{15}H_{24}$	204.2
9	Aromadendrene	1423	1439	1.61	SH	$C_{15}H_{24}$	204.2
10	(E)-β-Farnesene	1447	1454	3.20	SH	$C_{15}H_{24}$	204.2
11	Germacrene-D	1466	1484	4.67	SH	$C_{15}H_{24}$	204.2
12	Alloaromadendrene	1474	1458	3.30	SH	$C_{15}H_{24}$	204.2
13	Viridiflorene	1476	1496	3.69	SH	$C_{15}H_{24}$	204.2
14	Bicyclogermacrene	1480	1500	18.23	SH	$C_{15}H_{24}$	204.2
15	δ-Cadinene	1505	1522	3.38	SH	$C_{15}H_{24}$	204.2
16	Sesquiphellandrene	1513	1521	3.05	SH	$C_{15}H_{24}$	204.2
17	Caryophyllene oxide	1562	1582	15.29	OS	$C_{15}H_{24}O$	220.2
18	Spathulenol	1562	1577	11.90	OS	$C_{15}H_{24}O$	220.2
19	Viridiflorol	1579	1592	8.93	OS	$C_{15}H_{26}O$	222.2
20	τ-Muurolol	1642	1640	2.14	OS	$C_{15}H_{26}O$	222.2
21	Selin-11-en-4-α-ol	1667	1658	0.82	OS	$C_{15}H_{26}O$	222.2
22	3-Oxo-7.8-dihydro-β-ionol	1711	1695	0.48	OT	$C_{13}H_{20}O_2$	208.2
23	Hexahydrofarnesyl acetone	1833	1843	0.28	OT	$C_{18}H_{36}O$	268.3
24	5,15-Rosadiene	1904	1896	0.39	DH	$C_{20}H_{32}$	272.3
25	Sandaracopimaradiene	1935	1935	0.17	DH	$C_{20}H_{32}$	272.3
26	Sclarene	1985	1974	0.16	DH	$C_{20}H_{32}$	272.3
27	Kaurene	2060	2042	0.15	DH	$C_{20}H_{32}$	272.3
Monoterpene hydrocarbons (MH)				–			
Oxygenated monoterpenes (OM)				0.79			
Sesquiterpene hydrocarbons (SH)				46.71			
Oxygenated sesquiterpenes (OS)				39.08			
Diterpene hydrocarbons (DH)				0.87			
Other compounds (OT)				0.76			
Total identified				88.21			

–: not detected

3.8 *Lophozia ventricosa* (Dicks.) Dumort

The species *Lophozia ventricosa* was collected in Altenau (Germany), and the essential oil was isolated by hydrodistillation. The chemical composition was analyzed by GC and GC-MS. 93.3% of the total oil was formed by 28 constituents. The main components were maaloxide (45.6%), eudesma-4(15),7(11)-dien-8-one (28.9%), 1(10)-spirovetivene-7-β-ol (5.7%), and eudesma-4(15),11-dien-8-one (5,7%) [66].

Fig. 9 *L. hexagonus*

3.9 *Marchantia pappeana* Lehm

Fifty-eight volatile components were identified in a crude extract of South African liverwort *M. pappeana*, representing 88.44% of the total volatile compounds. The main components were (Z)-α-bisabolene (5.98%), 11-nordrim-8-en-12-al (6.81%), and (Z)-biformene (11.60%) [59].

3.10 *Marchantia polymorpha.* subsp. *ruderalis* Bischl. & Boissel.-Dub

In the crude extract of South African liverwort *M. polymorpha* subsp. *ruderalis*, 54 volatile constituents were identified, representing 99.86% of the volatile compounds. The main components were β-chamigrene (23.93%), thujopsene (16.10%), β-acoradiene (11.76%), and α-barbatene (6.29%) [59].

3.11 *Marchesinia brachiata* (Sw.) Schiffn

Aromatic compounds 3,4-dimethoxy-1-vinylbenzene and 2,4,5-trimethoxy-1-vinylbenzene were isolated from the Ecuadorian liverwort *M. brachiata*, together with a known flavone, apigenin-7,4′-dimethylether. Their structures were confirmed by extensive NMR spectroscopic analysis [67].

3.12 *Marsupella aquatica* (Lindenb.) Schiffn

The species *Marsupella aquatica* was collected near Gaschurn/Montafon (Austria). The volatile compounds were isolated by hydrodistillation and then separated in

Table 16 Chemical composition of the essential oil from *Leptoscyphus hexagonus*

CN	Compound	RI	RIF	%	Type	CF	MM (Da)
1	α-Copaene	1363	1374	0.68	SH	$C_{15}H_{24}$	204.2
2	Longifolene	1392	1407	0.33	SH	$C_{15}H_{24}$	204.2
3	α-Longipinene	1399	1350	3.62	SH	$C_{15}H_{24}$	204.2
4	Caryophyllene	1412	1417	0.22	SH	$C_{15}H_{24}$	204.2
5	Aromadendrene	1423	1439	0.34	SH	$C_{15}H_{24}$	204.2
6	Aristolediene	1436	1435	1.49	SH	$C_{15}H_{22}$	202.2
7	cis-Thujopsene	1443	1429	0.18	SH	$C_{15}H_{24}$	204.2
8	Dehydroaromadendrene	1445	1460	1.29	SH	$C_{15}H_{24}$	204.2
9	α-Patchoulene	1470	1454	1.16	SH	$C_{15}H_{24}$	204.2
10	Eremophilene	1473	1489	0.39	SH	$C_{15}H_{24}$	204.2
11	Viridiflorene	1476	1496	1.33	SH	$C_{15}H_{24}$	204.2
12	Bicyclogermacrene	1480	1500	6.70	SH	$C_{15}H_{24}$	204.2
13	α-Selinene	1485	1498	1.30	SH	$C_{15}H_{24}$	204.2
14	Cabreuva oxide D	1492	1479	33.77	OS	$C_{15}H_{24}O$	220.2
15	Valencene	1508	1496	0.58	SH	$C_{15}H_{24}$	204.2
16	trans-Cycloisolongifol-5-ol	1515	1513	2.57	OS	$C_{15}H_{24}O$	220.2
17	Spathulenol	1562	1577	0.57	OS	$C_{15}H_{24}O$	220.2
18	Elemol	1570	1548	18.55	OS	$C_{15}H_{26}O$	222.2
19	Viridiflorol	1579	1592	8.03	OS	$C_{15}H_{26}O$	222.2
20	Ledol	1591	1602	2.57	OS	$C_{15}H_{26}O$	222.2
21	Drimenol	1752	1757	0.18	OS	$C_{15}H_{26}O$	222.2
Monoterpene hydrocarbons (MH)				–			
Oxygenated monoterpenes (OM)				–			
Sesquiterpene hydrocarbons (SH)				19.61			
Oxygenated sesquiterpenes (OS)				66.24			
Diterpene hydrocarbons (DH)				–			
Other compounds (OT)				–			
Total identified				85.85			

–: not detected

flash column chromatography. The chemical components were identified by GC and GC-MS. The main compounds were (+)-amorpha-4,11-diene 1 (9.6%) and (−)-amorpha-4,7(11)-diene 2 (25.2%) [68].

3.13 *Mylia nuda* Inoue & B.Y. Yang

The species *M. nuda* was collected in Ilan (Taiwan). The essential oil was isolated by hydrodistillation. The chemical components were identified by GC and GC-MS. In *M. nuda*, the samples investigated were found to lack barbatene-, isobazzanene-, chamigrene-, cuprenene-, myltaylene-, and cyclomyltaylane-type compounds [69].

3.14 *Plagiochila asplenioides* (L.) Dumort

The species *P. asplenioides* was collected in Hamburg (Germany), and the essential oil was isolated by hydrodistillation. The chemical composition was performed by GC and GC-MS. The components identified were reported as maali-1,3-diene, anastreptene, italicene, α-barbatene, β-funebrene, γ-maaliene, amaaliene, β-barbatene, β-acoradiene, β-chamigrene, γ-curcumene, (+)-δ-selinene,(−)-bicyclogermacrene, α-cuprenene, α-chamigrene, and β-bazzanene from the non-oxygenated fraction and in the oxygenated fraction (+)-maalian-5-ol, rosifoliol, gymnomitr-3(15)-en-4β-ol, plagiochilide and 3α-acetoxybicyclogermacrene [70].

3.15 *Pallavicinia lyellii* (Hook.) Carruth

The crude extract of *P. lyellii*, collected in South Africa, showed 74 constituents, representing 90.12% of the total volatile constituents. The main components were ε-cuprenene (14.21%), cuparene (5.51%), and muurola-4-en-3, 8-dione (5.17%) [59].

3.16 *Plagiochila bifaria* (Sw.) Lindenb.

The essential oils from *P. bifaria* collected in Madeira (Portugal) were analyzed by GC and GC-MS. The main components were methyl everninate (1–35%), peculiar oxide (13–16%), and *ent*-eudesm-4(15)-ene-6-one (9–19%) and ent-7-hydroxy-eudesm-4-en-6-one (11%) [71].

3.17 *Plagiochila maderensis* Gottsche ex Steph.

The essential oils from *P. maderensis* collected in Madeira (Portugal) were analyzed by GC and GC-MS. The main components were terpinolene (34–60%) and bicyclogermacrene (5.3%) [71].

3.18 *Plagiochila retrorsa* Gottsche

The essential oils from *P. retrorsa* collected in Madeira (Portugal) were analyzed by GC and GC-MS. The main components were β-phellandrene (0.5–46.0%), allo-ocimene (4.9–15.0%), neo-allo-ocimene (4.2–9.6%), peculiar oxide (8.9–11.9%), bicyclogermacrene (3.6–6.0%), ent-eudem-4(15)-ene-6-one (1.7–12.7%), and ent-7-hydroxy-eudesm-4-en-6-one (0.5–8.5%) [71].

3.19 *Plagiochila stricta* Lindenb.

The essential oils from *P. stricta* collected in Madeira (Portugal) were analyzed by GC and GC-MS. The main components were terpinolene (0.3–8.1%), allo-ocimene (6.7–19.1%), neo-allo-ocimene (3.7–11.4%), peculiar oxide (11.1–21.0%), bicyclogermacrene (3.6–17.3%), and spathulenol (2.1–14.2%) [71].

3.20 *Plagiochasma rupestre* (J.R. Forst. & G. Forst.) Steph

South African liverwort *P. rupestre* presents 76 volatile constituents, which represent 91.54% of the volatile compounds. The main components were isocembrene (12.61%), alloaromadendrene oxide (7.72%), cuparene (6.92%), allo-aromadendra-4(15), 10(14)-diene (6.66%), and cis-β-elemene (5.55%) [59].

3.21 *Plicanthus hirtellus* (F. Weber) R.M. Schust.

Samples of *P. hirtellus* collected in Sao Tome and Principe afforded strong-smelling light-yellow oils. *P. hirtellus* volatiles were dominated by anastreptene (13%) and spathulenol (14%). Carvone (4%) and 1,8-cineole (2%) were also detected in this species volatiles [72].

3.22 *Radula boryana* (F. Weber) Nees ex, Mont.

The volatile compound of *R. boryana* collected in Sao Tome and Principe was characterized by high amounts of p-cymene (17%), along with other monoterpene hydrocarbons, and by the unusual presence of the oxygen-containing monoterpenes, thymol, and carvacrol [72].

3.23 *Radula aquilegia* (Hook. f. & Taylor) Gottsche, Lindenb. & Nees

The volatile compounds of *R. aquilegia* collected in Madeira (Portugal) were isolated by distillation-extraction and analyzed by GC and GC-MS. The main compounds were trans-β-farnesene (63.7%) and eremophilene (7.7%). On the other hand, the main compounds identified in *R. aquilegia* collected in Azores (Switzerland) were pentalenene (5.8%), β-acoradiene (31.3%), α-acoradiene (13%), drima-7,9(11)-diene (11.1%), β-selinene (16.6%), eremophilene (14.6%), valencene (39.3%), and α-helmiscapene (10.7%) [73].

3.24 *Radula carringtonii* J.B. Jack

The main compounds of *R. carringtonii* collected in Madeira (Portugal) were valencene (65.4%), α-selinene (25.9%), UI A (8.1%), xanthorrhizol (5.2%), and 2,2-dimethyl-5-hydroxy-7-(2-phenylethyl)-chromene (7.3%). Among the volatile compounds of *R. carringtonii* collected in Azores (Portugal), valencene (57.6%), α-selinene (16.2%), 7-epi-α-Selinene, and 2,2-dimethyl-5-hydroxy-7-(2-phenylethyl)-chromene were identified (10.4%) [73].

3.25 *Radula complanata* (L.) Dumort.

Among the identified volatile compounds of *R. complanata* collected in Swiss, isolated by distillation-extraction, and analyzed by GC and GC-MS were 3-methoxy bibenzyl (52.2%) and valencene (5.2%).

3.26 *Radula holtii* Spruce

The volatile compounds of *R. holtii* collected in Madeira (Portugal) were isolated by distillation-extraction and analyzed by GC and GC-MS. The main compounds were α-pinene (5.3%), n-decane (5.0%), n-undecane (6.5%), pentalenene (11.4%), β-bisabolene (15.3%), bisabola-1,3,5,7(14),10-pentaene (12.7%), and myli-4(15)-ene (11.3%). On the other hand, the main compounds identified in *R. holtii* collected in Portugal (mainland) were α-phellandrene (7.5%), β-bisabolene (9.9%), cis--γ-Bisabolene (22.1%), Bisabola-1,3,5,7(14),10-pentaene (8.6%), and myli-4(15)-ene (14,5%) [73].

3.27 *Radula jonesii* Bouman, Dirkse & K. Yamada

Among the identified volatile compounds of *R. jonesii* collected in Madeira (Portugal), isolated by distillation-extraction, and analyzed by GC and GC-MS were *trans*-β-farnesene (42,2%) and *cis*-β-farnesene (5,1%) [73].

3.28 *Radula lindenbergiana* Gottsche ex C. Hartm.

The main identified volatile compounds present in *R. lindenbergiana* collected in Madeira (Portugal) were petasitene (5,8%), *cis*-γ-bisabolene (5,6%), and 3-methoxy bibenzyl (56,1%) and in *R. lindenbergiana* collected in Portugal (mainland) were 3-methoxy bibenzyl (63,8%) and cis-γ-bisabolene (8,5%) [73].

3.29 *Radula nudicaulis* Steph

The main identified volatile compounds present in *R. nudicaulis* collected in Madeira (Portugal) were β-helmiscapene (36.0%), *cis*-γ-bisabolene (25.2%), and 2,2-dimethyl-5-hydroxy-7-(2-phenylethyl)-chromene (7.2%) [73].

3.30 *Radula perrottetii* Gottsche ex Steph.

The species *R. perrottetii* was collected in Tokushima (Japan). The essential oil was isolated by hydrodistillation, and the chemical composition was analyzed by GC and GC-MS. The components identified were reported as Δ-3-carene, α-terpinene, *p*-cymene, (Z)-β-ocimene, γ-terpinene, terpinolene, β-elemene, 7-epi-α-cedrene, α-gurjunene, α-cedrene, aristolene, γ-maaliene, eremophila-1(10),6-diene, calarene, valerena-4,7(11)-diene, selina-3,7-diene, β-acoradiene, allo-aromadendrene, 4,5-di-epi-aristolochene, selina- 4,7-diene, β-chamigrene, eremophila-1(10),7-diene, eremophilene, hinesene, cuparene, α-chamigrene, (E)- γ-bisabolene, γ-cuprenene, and bicyclohumulenone [74].

3.31 *Radula wichurae* Steph.

The main identified volatile compounds present in *R. wichurae* collected in Madeira (Portugal) were pentalenene (5.3%), β-santalene (5.0%), drima-7,9(11)-diene (6.8%), eremophilene (8.2%) and valencene (18.3%) and in *R. wichurae* collected in Azores (Portugal) were pentalenene (8.2%), β-santalene (6.2%), drima-7,9(11)-diene (11.7%), eremophilene (7.0%), and valencene (30.7%) [73].

3.32 *Riccia albolimbata* S.W. Arnell

The analysis of the crude extract of South African liverwort *R. albolimbata* shows 37 constituents, representing 86.20% of the total volatile compounds. The main components were behenic acid methyl ester (5.35%), linoleic acid methyl ester (5.72%), and 8a-hydroxyeudema-3,11-diene (6.85%), pentadecanoic acid (12.23%), erucic acid methyl ester (12.36%), and n-hexadecanoic acid (25.30%) [59].

3.33 *Scapania nemorea* (L.) Grolle

The volatile compounds of *S. nemorea* (L.) collected in Jastrebac (Serbia) were isolated using three different solvents, methanol, ethanol, and ethyl acetate. The identification was performed by SPME-GC-MS. Sixty-two compounds were identified in the methanol extract (ME, 90.0% of the total), 49 in the ethanol extract (EE, 99.2% of the total), and 48 in the ethyl acetate extract (EAE, 98.1% of the total). The main compounds in the three extracts were β-bazzanene (11.0% ME, 17.9% EE, and 14,6% EAE), isobazzanene (10,2% ME, 15.8% EE, and 11.7% EAE), aromadendrene (8.8% ME, 12,9% EE, and 10.6% EAE), bicyclogermacrene (8.8% ME), γ-muurolene (8.1% EE, and 5.7% EAE), cis-α-Bisabolene (5.5% EE) and β-barbatene (6.2% EAE) [75].

3.34 *Symphyogyna podophylla* (Thunb.) Mont. & Nees

In the crude extract of *S. podophylla*, collected in South Africa, 31 constituents were identified, representing 89.1% of the total. The main components were 8(17), 14-labdadiene-6,13-diol (20.4%), β-barbatene (16.62%), trans-linalool oxide (7.87%), β-caryophyllene (7.73%), and 2,15-valparadiene (5.02%) [59].

3.35 *Syzygiella anomala* (Lindenb. & Gottsche) Steph

A foliose liverwort *S. anomala* is a widespread species that is distinguished by robust habit and reddish-purple in color or violet-colored; leaves are ovate-triangular, with recurved dorsal leaf margin and decurrent dorsal base with opposite leaf bases connected both dorsally and ventrally; and leaf apex is subacute to slightly bifid (Fig. 10). Its habitat is on bark, moist rock, and soil in montane forests and paramo, 1500–3600 m a.s.l. [45, 63, 64].

Fig. 10 *S. anomala*

In the essential oil of *S. anomala collected* in Ecuador, 27 chemical components were identified representing 90.17% of the total oil. The components are grouped as sesquiterpene hydrocarbons (65.39%), oxygenated sesquiterpenes (23.48%), oxygenated monoterpenes (1.16%), and monoterpene hydrocarbons (0.14%) (Table 17). The main constituents in the EO of *S. anomala* were sesquiterpene hydrocarbons (silphiperfola-5,7(14)-diene (CN:4, 25.22%, CF $C_{15}H_{22}$, MM: 202.2),

Table 17 Chemical composition of the essential oil from *Syzygiella anomala*

CN	Compound	RI	RIf	%	Type	CF	MM (Da)
1	β-Phellandrene	1023	1025	0.14	MH	$C_{10}H_{16}$	136.1
2	1-Octen-3-ol, acetate	1107	1110	1.16	OM	$C_{10}H_{18}O_2$	170.1
3	Bicycloelemene	1321	1331	0.23	SH	$C_{15}H_{24}$	204.2
4	Silphiperfola-5,7(14)-diene	1355	1363	25.22	SH	$C_{15}H_{22}$	202.2
5	Isoledene	1358	1374	0.10	SH	$C_{15}H_{24}$	204.2
6	α-Gurjunene	1412	1409	0.73	SH	$C_{15}H_{24}$	204.2
7	Caryophyllene	1412	1417	0.10	SH	$C_{15}H_{24}$	204.2
8	β-Barbatene	1431	1440	3.99	SH	$C_{15}H_{24}$	204.2
9	cis-Thujopsadiene	1437	1465	7.00	SH	$C_{15}H_{22}$	202.2
10	Dehydroaromadendrene	1445	1460	1.61	SH	$C_{15}H_{24}$	204.2
11	(E)-β-Farnesene	1447	1454	1.95	SH	$C_{15}H_{24}$	204.2
12	Viridiflorene	1476	1496	6.51	SH	$C_{15}H_{24}$	204.2
13	Bicyclogermacrene	1480	1500	8.42	SH	$C_{15}H_{24}$	204.2
14	Cuparene	1488	1504	0.55	SH	$C_{15}H_{22}$	202.2
15	Trichodiene	1501	1533	0.18	SH	$C_{15}H_{24}$	204.2
16	β-Vetispirene	1517	1493	8.01	SH	$C_{15}H_{22}$	202.2
17	γ-Dehydro-Ar-himachalene	1545	1530	0.79	SH	$C_{15}H_{20}$	200.2
18	Maaliol	1553	1566	0.84	OS	$C_{15}H_{26}O$	222.2
19	Caryophyllene oxide	1562	1582	8.98	OS	$C_{15}H_{24}O$	220.2
20	β-Oplopenone	1564	1575	6.40	OS	$C_{15}H_{24}O$	220.2
21	Globulol	1571	1590	1.83	OS	$C_{15}H_{26}O$	222.2
22	Viridiflorol	1579	1592	1.01	OS	$C_{15}H_{26}O$	222.2
23	Cubeban-11-ol	1582	1595	0.38	OS	$C_{15}H_{26}O$	222.2
24	Rosifoliol	1592	1600	1.04	OS	$C_{15}H_{26}O$	222.2
25	Muurola-4,10(14)-dien-1β-ol	1613	1630	0.36	OS	$C_{15}H_{24}O$	220.2
26	Valerenal	1650	1668	0.18	OS	$C_{15}H_{22}O$	218.2
27	Aristolone	1757	1762	2.46	OS	$C_{15}H_{22}O$	218.2
Monoterpene hydrocarbons (MH)				0.14			
Oxygenated monoterpenes (OM)				1.16			
Sesquiterpene hydrocarbons (SH)				65.39			
Oxygenated sesquiterpenes (OS)				23.48			
Diterpene hydrocarbons (DH)				–			
Other compounds (OT)				–			
Total identified				90.17			

–: not detected

bicyclogermacrene (8.42%), β-vetispirene (8.01%), cis-thujopsadiene (7.0%), and viridiflorene (6.51%)) and oxygenated sesquiterpenes (caryophyllene oxide (8.98%) and β-oplopenone (6.4%)).

3.36 *Tritomaria polita* (Nees) Jørg.

The essential oil of *T. polita*, collected in Tyrol (Austria), was extracted by hydrodistillation and analyzed by GC and GC-MS. The known sesquiterpene hydrocarbons β-elemene, aromadendrene, allo-aromadendrene (1.5%), 4,5-di-epi-aristolochene, α-amorphene, eremophilene, (+)-α-selinene (5%), δ-amorphene, and selina-3,7 (11)-diene were identified [76].

4 Conclusions

As reported in the present chapter, mosses and liverworts represent a very important source of essential oils which are just partially investigated. The wide variety of compounds reported opens up new perspectives for further research trends concerning the biological activity of essential oils; these studies could identify new potential applications in the pharmaceutical, cosmetic, and food sectors, as is the case for essential oils obtained from tree and herb species that have already been extensively studied. The ever-increasing demand for natural products by various market sectors represents an important opportunity for the protection and sustainable use of natural resources, also enhancing the traditional ethnobotanical knowledge of the cultures and territories of origin of the species mentioned.

References

1. Frego KA (2007) Bryophytes as potential indicators of forest integrity. For Ecol Manag 242:65–75. https://doi.org/10.1016/j.foreco.2007.01.030
2. Smithsonian Tropical Research Institute (2021) Bryophytes tiny plants in a big changing world. https://stri.si.edu/story/bryophytes. Accessed 23 Jan 2022
3. Glime JM, Chavoutier L (2017) Glossary. Bryophyte ecology. Michigan Technological University and the International Association of Bryologists, Houghton
4. Morales Baquero CP, Ospino Cerpa JD, Jiménez Vásquez JA, Berbén Henriquez AM, Negritto MA (2017) Briófitos: un mundo en miniatura. INFOFLORA Bol Bot 1:12
5. Burrows BA, McCubbin AG (2018) Reproduction, overview by phylogeny: plant. In: Skinner MK (ed) Encyclopedia of reproduction, 2nd edn. Academic, Oxford
6. Schofield WB (2019) Bryophyte. https://www.britannica.com/plant/bryophyte. Accessed 05 Jan 2022
7. León-Martínez G, Vielle-Calzada J-P (2019) Chapter Twenty: Apomixis in flowering plants: developmental and evolutionary considerations. In: Grossniklaus U (ed) Current topics in developmental biology. Academic
8. Frangedakis E, Shimamura M, Villarreal JC, Li F-W, Tomaselli M, Waller M et al (2021) The hornworts: morphology, evolution and development. New Phytol 229:735–754. https://doi.org/10.1111/nph.16874

9. Sabovljević MS, Ćosić MV, Jadranin BZ, Pantović JP, Giba ZS, Vujičić MM et al (2022) The conservation physiology of bryophytes. Plan Theory 11:1282–1294. https://doi.org/10.3390/plants11101282
10. Ćosić M, Vujičić MM, Sabovljević MS, Sabovljević AD (2019) What do we know about salt stress in bryophytes? Plant Biosyst 153:478–489. https://doi.org/10.1080/11263504.2018.1508091
11. Benítez Á, Prieto M, Aragón G (2015) Large trees and dense canopies: key factors for maintaining high epiphytic diversity on trunk bases (bryophytes and lichens) in tropical montane forests. Forestry 88:521–527. https://doi.org/10.1093/forestry/cpv022
12. Glime JM (2017) Household and personal uses. Bryophyte ecology, vol 1-1. Michigan Technological University and the International Association of Bryologists, Houghton
13. Ali B, Al-Wabel NA, Shams S, Ahamad A, Khan SA, Anwar F (2015) Essential oils used in aromatherapy: a systemic review. Asian Pac J Trop Biomed 5:601–611. https://doi.org/10.1016/j.apjtb.2015.05.007
14. Aramrueang N, Asavasanti S, Khanunthong A (2019) Chapter 10: Leafy vegetables. In: Pan Z, Zhang R, Zicari S (eds) Integrated processing technologies for food and agricultural by-products. Academic, London
15. Sharifi-Rad J, Sureda A, Tenore GC, Daglia M, Sharifi-Rad M, Valussi M et al (2017) Biological activities of essential oils: from plant chemoecology to traditional healing systems. Molecules 22:70–102. https://doi.org/10.3390/molecules16121047110.3390/molecules22010070
16. Ibrahim TA, El-Hela AA, El-Hefnawy HM, Al-Taweel AM, Perveen S (2017) Chemical composition and antimicrobial activities of essential oils of some coniferous plants cultivated in Egypt. Iran J Pharm Sci 16:328–337
17. Valarezo E, Ojeda-Riascos S, Cartuche L, Andrade-González N, González-Sánchez I, Meneses MA (2020) Extraction and study of the essential oil of copal (*Dacryodes peruviana*), an amazonian fruit with the highest yield worldwide. Plan Theory 9:1658–1671. https://doi.org/10.3390/plants9121658
18. Perveen S (2018) Introductory chapter: terpenes and terpenoids. In: Perveen S, Al-Taweel A (eds) Terpenes and terpenoids. IntechOpen, London
19. Başer KHC, Buchbauer G (2016) Handbook of essential oils. Science, technology and applications, Chromatographia, 2nd edn. Springer, Berlin
20. Micić D, Ostojić S, Pezo L, Blagojević S, Pavlić B, Zeković Z et al (2019) Essential oils of coriander and sage: investigation of chemical profile, thermal properties and QSRR analysis. Ind Crop Prod 138:111438. https://doi.org/10.1016/j.indcrop.2019.06.001
21. Ramawat KG, Dass S, Mathur M (2009) The chemical diversity of bioactive molecules and therapeutic potential of medicinal plants. In: Ramawat KG (ed) Herbal drugs: ethnomedicine to modern medicine. Springer, Berlin/Heidelberg
22. Asakawa Y (2007) Biologically active compounds from bryophytes. Pure Appl Chem 79:557–580. https://doi.org/10.1351/pac200779040557
23. Sabovljevic A, Sabovljevic M (2008) Bryophytes, a source of bioactive and new compounds. In: Govil JN, Singh VK (eds) Phytopharmacology and therapeutic values IV. Studium Press LLC, Houston
24. Magill RE (2010) Moss diversity: new look at old numbers. Phytotaxa 9:167–174. https://doi.org/10.11646/phytotaxa.9.1.9
25. Britannica, The Editors of Encyclopaedia (2022) Moss. https://www.britannica.com/plant/moss-plant. Accessed 03 May 2022
26. Salazar Allen N (2011) El mundo de las plantas pequeñas: Las Briofitas. Impreso por Editora Novo Art, S.A., Pánama
27. Harris ESJ (2008) Ethnobryology: traditional uses and folk classification of bryophytes. Bryologist 111:169–217
28. Vanderpoorten A, Goffinet B (2009) Introduction to bryophytes. Cambridge University Press, Cambridge

29. New Hampshire PBS, Squam Lakes Natural Science Center (2022) NatureWorks: bryophytes. https://nhpbs.org/natureworks/. Accessed 12 Feb 2022
30. Ávila-Pérez P, Longoria-Gándara LC, García-Rosales G, Zarazua G, López-Reyes C (2018) Monitoring of elements in mosses by instrumental neutron activation analysis and total X-ray fluorescence spectrometry. J Radioanal Nucl Chem 317:367–380. https://doi.org/10.1007/s10967-018-5896-z
31. Ray S, Bhattacharya S (2022) Manual for bryophytes: morphotaxonomy, diversity, spore germination, conservation. CRC Press, Boca Raton
32. Crandall-Stotler BJ, Bartholomew-Began SE (2022) Morphology of mosses (Phylum Bryophyta). http://www.efloras.org/index.aspx. Accessed 02 Apr 2022
33. Delgadillo-Moya C (2014) Biodiversity of Bryophyta (mosses) in Mexico. Rev Mex Biodivers 85:S100–S1S5. https://doi.org/10.7550/rmb.30953
34. The Bryophyte Groups (2015) What is a moss? https://www.anbg.gov.au/bryophyte/what-is-moss.html. Accessed 05 Apr 2022
35. Shaw J, Anderson LE (1988) Peristome development in mosses in relation to systematics and evolution. II. *Tetraphis pellucida* (Tetraphidaceae). Am J Bot 75:1019–1032. https://doi.org/10.1002/j.1537-2197.1988.tb08809.x
36. Pejin B, Vujisic L, Sabovljevic M, Tesevic V, Vajs V (2011) Preliminary data on essential oil composition of the moss *Rhodobryum ontariense* (Kindb.) Kindb. Cryptogam Bryol 32:113–117. https://doi.org/10.7872/cryb.v32.iss1.2011.113
37. Saritas Y, Sonwa MM, Iznaguen H, König WA, Muhle H, Mues R (2001) Volatile constituents in mosses (Musci). Phytochemistry 57:443–457. https://doi.org/10.1016/S0031-9422(01)00069-3
38. Li L, Zhao J (2009) Determination of the volatile composition of *Rhodobryum giganteum* (Schwaegr.) Par. (Bryaceae) using solid-phase microextraction and gas chromatography/mass spectrometry (GC/MS). Molecules 14. https://doi.org/10.3390/molecules14062195
39. Kahriman N, Ozdemir T, BeyzaCansu T, Volga C, Yaylı N (2009) Essential oils in mosses (*Brachythecium salebrosum*, *Eurhynchium pulchellum* and *Plagiomnium undulatum*) grown in Turkey. Asian J Chem 21:5505–5509
40. Üçüncü O, Cansu TB, Ozdemir T, Alpay Karaoğlu Ş, Yayli N (2010) Chemical composition and antimicrobial activity of the essential oils of mosses (*Tortula muralis* Hedw., *Homalothecium lutescens* (Hedw.) H. Rob., *Hypnum cupressiforme* Hedw., and *Pohlia nutans* (Hedw.) Lindb.) from Turkey. Turk J Chem 34:825–834. https://doi.org/10.3906/che-1002-62
41. Yucel TB (2021) Chemical composition and antimicrobial and antioxidant activities of essential oils of *Polytrichum commune* (Hedw.) and *Antitrichia curtipendula* (Hedw.) Brid. grown in Turkey. Int J Second Metab 8:272–283. https://doi.org/10.21448/ijsm.945405
42. Ozdemir T, Ucuncu O, Cansu TB, Kahriman N, Yayli N (2010) Volatile constituents in mosses (*Brachythecium albicans* (Hedw.) Schimp., *Bryum pallescens* Schleich. ex Schwagr and *Syntrichia intermedia* Brid.) grown in Turkey. Asian J Chem 22:7285–7290
43. Özdemir T, Yayli N, Cansu TB, Volga C, Yayli N (2009) Essential oils in mosses (*Bracizythecium salebrosum*, *Eurhynchium puichellum* and *Plagiomnium undulatum*) grown in Turkey. Asian J Chem 21:5505–5509
44. Churchill SP, Linares C. EL (1995) Prodromus Bryologiae Novo-Granatensis: introducción a la flora de musgos de Colombia. vol Parte 2. Universidad Nacional de Colombia-Instituto de Ciencias Naturales, Museo de Historia Natural, Bogotá
45. Gradstein S, Churchill S, Salazar Allen N (2001) Guide to the bryophytes of tropical America. New York Botanical Garden Press, Bronx
46. Tosun G, Yaylı B, Özdemir T, Batan N, Bozdeveci A, Yaylı N (2015) Volatiles and antimicrobial activity of the essential oils of the mosses *Pseudoscleropodium purum*, *Eurhynchium striatum*, and *Eurhynchium angustirete* grown in Turkey. Rec Nat Prod 9:237–242
47. Bendz G, Svensson L (1971) Volatile compounds from *Fontinalis antipyretica*. Phytochemistry 10:3283–3285. https://doi.org/10.1016/S0031-9422(00)97391-6

48. Cansu TB, Yayli B, Özdemir T, Batan N, Karaoğlu SA, Yayli N (2013) Antimicrobial activity and chemical composition of the essential oils of mosses (*Hylocomium splendens* (Hedw.) Schimp. and *Leucodon sciuroides* (Hedw.) Schwägr.) growing in Turkey. Turk J Chem 37:213–219. https://doi.org/10.3906/kim-1204-72
49. Cansu TB, Özdemir T, Batan N, Yayli B, Karaoglu SA, Yayli N (2014) Essential oil analysis and antimicrobial activity of *Neckera complanata* (Hedw.) huebener and *Neckera crispa* Hedw. (Neckeraceae) grown in Turkey. Asian J Chem 26:2005–2008. https://doi.org/10.14233/ajchem.2014.15618A
50. Klegin C, de Moura NF, Oliveira de Sousa MH, Frassini R, Roesch-Ely M, Bruno AN et al (2021) Chemical composition and cytotoxic evaluation of the essential oil of *Phyllogonium viride* (Phyllogoniaceae, Bryophyta). Chem Biodivers 18:e2000794. https://doi.org/10.1002/cbdv.202000794
51. Toyota M, Kimura K, Asakawa Y (1998) Occurrence of ent-sesquiterpene in the Japanese moss-*Plagiomnium acutum* : first isolation and identification of the ent-sesqui-and dolabellane-type diterpenoids from the musci. Chem Pharm Bull 46:1488–1489. https://doi.org/10.1248/cpb.46.1488
52. Tosun G, Yayli B, Özdemir T, Batan N, Yayli N, Karaoglu SA (2014) Chemical composition and antimicrobial activity of essential oils from *Tortella inclinat*a var. Densa, *T. tortusa* and *Pleurochaete squarrosa*. Asian J Chem 26:2001–2004. https://doi.org/10.14233/ajchem.2014.15618
53. Boris P, Ljubodrag V, Marko S, Vele T, Vlatka V (2011) Preliminary data on essential oil composition of the moss *Rhodobryum ontariense* (Kindb.) Kindb. Cryptogam Bryol 32:113–117. https://doi.org/10.7872/cryb.v32.iss1.2011.113
54. Sim-Sim M, Abreu M, Garcia C, Sérgio C, Figueiredo AC (2017) Essential oil composition of two *Sphagnum* species grown in Portugal and their in vitro culture establishment. Nat Prod Commun 12. https://doi.org/10.1177/1934578X1701200839
55. Von Konrat M, Soderstrom L, Renner MAM, Hagborg A, Briscoe L, Engel JJ (2010) Early land plants today (ELPT): how many liverwort species are there? Phytotaxa 9:22–40. https://doi.org/10.11646/phytotaxa.9.1.5
56. Britannica, The Editors of Encyclopaedia (2019) Liverwort. https://www.britannica.com/plant/liverwort. Accessed 22 Jan 2022
57. Jan-Peter F (2004) Recent developments of commercial products from bryophytes. Bryologist 107:277–283. https://doi.org/10.1639/0007-2745(2004)107[0277:RDOCPF]2.0.CO;2
58. Simpson MG (2010) Chapter 3: Evolution and diversity of green and land plants. In: Simpson MG (ed) Plant systematics, 2nd edn. Academic, San Diego
59. Linde J, Combrinck S, Vuuren SV, Rooy JV, Ludwiczuk A, Mokgalaka N (2016) Volatile constituents and antimicrobial activities of nine South African liverwort species. Phytochem Lett 16:61–69. https://doi.org/10.1016/j.phytol.2016.03.003
60. Asakawa Y, Ludwiczuk A, Nagashima F (2013) Phytochemical and biological studies of bryophytes. Phytochemistry 91:52–80. https://doi.org/10.1016/j.phytochem.2012.04.012
61. Asakawa Y, Ludwiczuk A, Nagashima F (2013) Chemical constituents of bryophytes: bio- and chemical diversity, biological activity, and chemosystematics. Progress in the chemistry of organic natural products. Springer, New York
62. Nagashima F, Asakawa Y (2011) Terpenoids and bibenzyls from three Argentine liverworts. Molecules 16:10471–10478. https://doi.org/10.3390/molecules161210471
63. Gradstein SR (2020) Checklist of the liverworts and hornworts of Ecuador. Frahmia 17:1–40
64. Gradstein SR (2021) The liverworts and hornworts of Colombia and Ecuador. Memoirs of The New York Botanical Garden. Springer Cham, Cham
65. Sakurai K, Tomiyama K, Yaguchi Y, Asakawa Y (2020) Characteristic odor of the Japanese liverwort (*Leptolejeunea elliptica*). J Oleo Sci 69:767–770. https://doi.org/10.5650/jos.ess19262
66. Song C, Zhu T, Lu R, König WA (2007) Essential oil composition of liverwort *Lophozia ventricosa*. Chin J Appl Environ Biol 13:458–460

67. Nagashima F, Murakami Y, Asakawa Y (1999) Aromatic compounds from the Ecuadorian liverwort *Marchesinia brachiata*: a revision. Phytochemistry 51:1101–1104. https://doi.org/10.1016/S0031-9422(99)00146-6
68. Adio AM, von Reuß SH, Paul C, Muhle H, König WA (2007) Sesquiterpenoid constituents of the liverwort *Marsupella aquatica*. Tetrahedron Asymmetry 18:1245–1253. https://doi.org/10.1016/j.tetasy.2007.05.020
69. von Reuß SH, Wu C-L, Muhle H, König WA (2004) Sesquiterpene constituents from the essential oils of the liverworts *Mylia taylorii* and *Mylia nuda*. Phytochemistry 65:2277–2291. https://doi.org/10.1016/j.phytochem.2004.04.039
70. Adio AM, König WA (2005) Sesquiterpene constituents from the essential oil of the liverwort P*lagiochila asplenioides*. Phytochemistry 66:599–609. https://doi.org/10.1016/j.phytochem.2005.01.015
71. Figueiredo AC, Sim-Sim M, Costa MM, Barroso JG, Pedro LG, Esquível MG et al (2005) Comparison of the essential oil composition of four *Plagiochila species*: *P. bifaria*, *P. maderensis*, *P. retrorsa* and *P. stricta*. Flavour Fragr J 20:703–709. https://doi.org/10.1002/ffj.1627
72. Figueiredo AC, Garcia C, Sim-Sim M, Sérgio C, Pedro LG, Barroso JG (2010) Volatiles from *Plicanthus hirtellus* (F. Weber) R.M. Schust. and Radula boryana (F. Weber) Nees (Hepaticae) grown in São Tomé e Príncipe Archipelago. Flavour Fragr J 25:219–222. https://doi.org/10.1002/ffj.1998
73. Figueiredo AC, Sim-Sim M, Barroso JG, Pedro LG, Esquível MG, Fontinha S et al (2009) Liverwort *Radula* species from Portugal: chemotaxonomical evaluation of volatiles composition. Flavour Fragr J 24:316–325. https://doi.org/10.1002/ffj.1943
74. Tesso H, König WA, Asakawa Y (2005) Composition of the essential oil of the liverwort *Radula perrottetii* of Japanese origin. Phytochemistry 66:941–949. https://doi.org/10.1016/j.phytochem.2005.03.003
75. Bukvicki D, Gottardi D, Tyagi AK, Veljic M, Marin PD, Vujisic L et al (2014) *Scapania nemorea* liverwort extracts: investigation on volatile compounds, in vitro antimicrobial activity and control of *Saccharomyces cerevisiae* in fruit juice. LWT Food Sci Technol 55:452–458. https://doi.org/10.1016/j.lwt.2013.09.029
76. Adio AM, Paul C, König WA, Muhle H (2003) Volatile constituents in the liverwort *Tritomaria polita*. Phytochemistry 64:637–644. https://doi.org/10.1016/S0031-9422(03)00298-X

Anticancerous Compounds from Bryophytes: Recent Advances with Special Emphasis on Bis(bi)benzyls

4

Vartika Jain, Mimosa Ghorai, Tuyelee Das, and Abhijit Dey

Contents

Abstract

Bryophytes are small-sized, spore-forming, non-vascular plants considered lower plants in the Plant Kingdom. Liverworts, hornworts, and mosses are three major classes of bryophytes that usually grow in moist habitats. Bryophyte flora has

V. Jain
Department of Botany, Government Meera Girls College, Udaipur, Rajasthan, India
e-mail: vartikabotany@gmail.com

M. Ghorai · A. Dey (✉)
Department of Life Sciences, Presidency University, Kolkata, West Bengal, India
e-mail: mimosaghorai@gmail.com; abhijit.dbs@presiuniv.ac.in

T. Das
Department of Life Sciences, Presidency University, Kolkata, India
e-mail: tuyeleedas25@gmail.com

H. N. Murthy (ed.), *Bioactive Compounds in Bryophytes and Pteridophytes*,
Reference Series in Phytochemistry, https://doi.org/10.1007/978-3-031-23243-5_3

been used by ethnic communities for various purposes and most importantly for the treatment of diseases. Though bryophytes gained less attention from scientific communities, still, many of the bryophyte species have been subjected to evaluation of pharmacological activities and bioactive constituents. The prominent bioactivity of many bryophytes obtained is against cancer cells. Many anticancer compounds have been isolated among which aromatic bibenzyl compounds and terpenoids have shown strong anticancer potential in vitro and/or in vivo studies. Liverworts contain a high amount of cytotoxic bibenzyl compounds. The present chapter summarizes various anticancer bioactive molecules isolated from bryophytes and their structure along with a special emphasis on the cytotoxic properties of bibenzyl compounds and its derivatives.

Keywords

Liverworts · Marchantin · Perrottetin · Plagiochin · Riccardin

Abbreviations

A-172	Human glioblastoma cell line
APC	Adenomatous Polyposis Coli
Bax	Bcl-2 associated X Protein
BSC	Monkey kidney cells
CDC	Cyclin-Dependent Kinase
DHA	6.7 Dihydroptychantol A
DNA	Deoxyribonucleic acid
ED_{50}	Median Effective Dose
HCT116	Human Colorectal Carcinoma
HeLa	Henrietta Lacks
HepG2	Hepatocellular Carcinoma
HIV	Human Immunodeficiency Virus
HL-60	Human promyelocytic leukemia
IC_{50}	Half maximal inhibitory concentration
ID_{50}	Infectious Dose
K562/A02	Human myelogenous leukemia
KB	Human epithelial carcinoma cell
LOVO	Colon adenocarcinoma
LXRα	Liver X Receptor alpha
M	Mitosis
MCF-7	Human breast adenocarcinoma
MDA-MB-435	Human breast ductal carcinoma
MDR	Multi-Drug Resistance
Min	Multiple Intestinal Neoplasia
MTT	(3-(4, 5-dimethylthiazolyl-2)-2, 5-diphenyltetrazolium bromide)
NT2/D1	Human embryonal teratocarcinoma cell line
P388	Leukemia cell line
PARP	Poly Adenosine diphosphate-Ribose Polymerase
PC3	Prostrate Cancer cell line

P-gp	Permeability glycoprotein
UV	Ultra Violet
VCR	Vincristine

1 Introduction

Bryophytes are popularly known as "Amphibians of Plant Kingdom" along with pteridophytes. These are considered among the lower group of plants having characteristics, such as no vascular system, jacketed sex organs, and heteromorphic alternation of generation in which the gametophytic phase is dominant. There are approximately 24,000 species in this group of terrestrial plants, making it the second largest [1, 2]. Being small in size and surviving in some special habitat conditions make them difficult for collection and identification. Therefore, they remain neglected for the assessment of the phytopharmaceutical potential for a long time. Recent studies have identified a number of bioactive molecules and pharmacological activities associated with bryophytes.

Some of the major pharmaceutical activities discovered so far are anti-inflammatory, antioxidant, antifungal, antibacterial and antiviral, antiarthritic, cytotoxic, neuroprotective, nitrous oxide and acetylcholinesterase inhibitory, muscle relaxant, etc. [3–11]. The major bioactive compounds isolated from various bryophytes belong to different secondary metabolites, such as terpenoids, highly unsaturated fatty acids, alkanones, phytosterols, flavonoids, and flavonoid glycosides, phenylpropanoids, benzenoids, and bibenzyl derivatives besides others [10, 12–15].

Cancer is a dreadful disease spreading rapidly in all countries of the world. Although synthetic anticancer drugs are available, yet they are very costly along with serious side effects. Plant-derived drugs are considered safer than synthetic medicine; hence, the search for anticancer compounds from plants is rampant [16]. Various bryophyte species have also been screened for anticancer activities and the results are overwhelming [17]. The present chapter deals with the cytotoxic/antiproliferative/multidrug resistance (MDR) activities of bryophytes and bis(bi)-benzyls as anticancerous compounds in particular.

2 Anticancer Activity of Bryophytes

Several species of bryophyte flora have been screened for their anticancer potential [17], and many promising anticancer compounds have also been isolated (Table 1). The cytotoxicity of five species of mosses, *Tortula muralis*, *Dryptodon pulvinatus*, *Hypnum cupressiforme*, *Ceratodon purpureus*, and *Rhytidiadelphus squarrosus*, was evaluated by Wolski et al. [18] using the 3-(4,5-dimethylthiazol-2-yl)-2,5-diphenyltetrazolium bromide (MTT) assay after 24 h of exposure of L929 cells. The significant ($P < 0.05$) effect of *R. squarrosus* extract on cell viability ($68.6 \pm 9.2\%$) was observed at a concentration of 125 μg/mL. A sharp decrease in

Table 1 Anticancer activity of compounds of interest from bryophytes

S. No.	Compound	Bryophyte	Type of activity	Activity against cell line/organisms	References
1.	Muscicolone	*Frullania muscicola*	Cytotoxicity	Human tumor cells	[43]
2.	8,9-secokaurane diterpenes	*Lepidolaena taylorii* and *Lepidolaena palpebrifolia*	Cytotoxicity	Human tumor cell lines	[41, 42]
3.	Marsupellone and Acetoxymarsupellone	*Marsupella emarginata*	Cytotoxicity	P388	[66]
4.	Diplophyllin- an ent-eudesmanolide	*Diplophyllum albicans* and *Diplophyllum taxifolium*	Cytotoxicity	Human epidermoid carcinoma	[68]
5.	Naviculyl caffeate	*Bazzania novae-zelandiae*	Cytotoxicity	Human tumor cell lines	[79]
6.	Glaucescenolide	*Schistochila glaucescens*	Cytotoxicity	P388	[80]
7.	Plagiochiline-A-15-yl Octanoate, 14-hydroxyplagiochi line-A-15-yl	*Plagiochila ovalifolia*	Cytotoxicity	P388	[78]
8.	12-hydroxychiloscyphone	*Chiloscyphus rivularis*	Cytotoxicity	Human lung carcinoma cells	[77]
9.	Jungermannenones A,B,C and D	*Jungermannia* species	Tumor-inhibiting activity	–	[50]
10.	(−)-Alpha-herbertenol, (−)-herbertenediol9, (−)-mastigophorene C, (−)-mastigophorene D and (−)-diplophyllolide A	*Mastigophora diclados*	Cytotoxicity	HL-60 and KB	[81]
11.	Ohioensin A	*Polytrichum ohioense*	Cytotoxicity	PS, MCF-7	[17, 118]
12.	Ohioensin B	*Polytrichum ohioense*	Cytotoxicity	Mouse leukemia, HT-29, human colon adenocarcinoma	[17, 118]
13.	Ohioensin H	*Polytrichum commune*	Cytotoxicity	MDA-MB-435, human T-cell leukemia (6 T-CEM), A549, LOVO, HepG2	[17, 118]

14.	Pallidisetin A and pallidisetin B	*Polytrichum pallidisetum*	Cytotoxicity	RPMI-7951, U-251	[119]
15.	Ansamitocin P-3	*Claopodium crispifolium* and *Anomodon attenuatus*	Cytotoxicity	A-549, HT-29	[122]
16.	Marchantin A	*Marchantia emarginata* subsp. *tosana* *Marchantia polymorpha* and *Marchantia tosana* *Reboulia hemisphaerica*	Anticancer activity Cytotoxicity Cytotoxicity Microtubule depolymerization activity	MCF-7 CC50 L6 cell KB cells Human cervical carcinoma cell line -	[92] [94] [70] [131]
17.	Marchantin C	*Schistochila glaucescens* *Reboulia hemisphaerica*	Cytotoxicity Apoptotic activity Microtubule depolymerization Matrix metallopeptidase reduction Angiogenic inhibition	P388 Human cervical carcinoma Human cervical carcinoma cell line T98G, U87 gl ioma cells T98G glioma cells	[80] [130] [131] [132]
18.	Marchantin M	*Asterella angusta*	Cytotoxicity and apoptosis	Chemoresistant PC3	[106]
19.	Neomarchantins A and B	*Schistochila glaucescens*	Cytotoxicity	P388	[80]
20.	Plagiochin E	*Marchantia polymorpha*	Proapoptotic activity MDR reversal	*Candida albicans* adriamycin-resistant K562/A02 cells	[135] [96]

(continued)

Table 1 (continued)

S. No.	Compound	Bryophyte	Type of activity	Activity against cell line/organisms	References
21.	Isoplagiochins A and B	*Plagiochila fruticosa*	Inhibition of in vitro tubulin polymerization	–	[136]
22.	Perrottetin E Perrottetin E, 10′-hydroxyperrottetin E, and 10,10′-dihydroxyperrottetin E	*Radula perrottetii* *Pellia endiviifolia*	Cytotoxicity Cytotoxicity	KB NT2/D1, A-172	[105] [90]
23.	Perrottetin F	*Lunularia cruciata*	Cytotoxicity	(NT2/D1) and (A-172)	[90]
24.	Riccardin A and Riccardin B	*Riccardia multifida*	Cytotoxicity		[108]
25.	Riccardin C	*Plagiochasma intermedium* and *Reboulia hemisphaerica* *Asterella angusta*	Apoptotic activity Antiproliferative activity	PC3 –	[106] [106]
26.	Riccardin D	*Monoclea forsteri* *Dumortiera hirsuta*	Antiproliferative activity, induced apoptosis Inhibition of hyphal growth Antiproliferative activity Antiproliferative activity Prevention of intestinal polyposis	A172 cells *Candida albicans* H460 HL-60, K562, MDR K562/A02 cells APC Min/+ mice	[96] [3] [137] [138] [140]
27.	Riccardin F	*Plagiochasma intermedium*	Alteration of P-gp-mediated drug resistance	Adriamycin-resistant K562/A02	[141]

28.	Dihydroptychantol A	*Asterella angusta*	Chemotherapeutic MDR reversal activity	Adriamycin-resistant K562/A02, vincristine-resistant KB/VCR cells Chemoresistant human U87 cell	[83, 142] [39]
29.	Lunularin 14-hydroxylunularin	*Dumortiera hirsuta* *Ricciocarpus natans*	Cytotoxicity Cytotoxicity	HepG2 NCTC-clone 929 fibroblast J 774 murine macrophage, peritoneal macrophage (BALB/c mice)	[39] [145]
30.	Pakyonol	*Plagiochasma intermedium*	P-gp	Adriamycin-resistant K562/A02	[102]
31.	3,3,4,4-Tetramethoxybibenzyl	*Frullania inouei*	Cytotoxicity	KB, KB/VCR, K562 or K562/A02, vincristine-resistant KB/VCR, adriamycin-resistant K562/A02	[87]
32.	4-Hydroxy-3-methoxybibenzyl	*Plagiochila fasciculata*	Cytotoxicity	BSC	[99]
33.	3,5-Dihydroxybibenzyl	*Radula amoena*	Cytotoxicity	HepG-2, A549	[146]
34.	Paleatin B	*Marchantia paleacea* var. *diptera*	Cytotoxicity	KB, P-388	[93]
35.	Pusilatin B and C	–	Cytotoxicity	KB	[85]
36.	Brittonin A and B	*Frullania inouei*	Cytotoxicity and proapoptotic activity	KB, KB/VCR, K562/A02	[87]
37.	Chrysotobibenzyl	*Frullania inouei*	Cytotoxicity	KB, KB/VCR K562, K562/A02	[87]

the viability of the L929 cells at higher concentrations of both *R. squarrosus* and *D. pulvinatus* extracts was also observed. A significant antiproliferative activity (~50%) of the extracts of a moss, *Hedwigia ciliata*, was observed against the human breast adenocarcinoma MDA-MB-231 cell line [19].

Antiproliferative and cytotoxic potential of a moss *Bryum* was tested against three cell lines, MCF-12A (human breast epithelial cells), SKBR 3 (human breast cancer cells), and HeLa (human cervix cancer cells). When treated with 500 μg/mL and 1000 μg/mL concentrations, a notable anticancer effect was observed against human breast cancer cells (SKBR 3), which decreased survival rates to 69% and 40%, respectively. Anticancer activity (76 $\pm$ 5%) was also observed against HeLa cells. However, low cytotoxicity (18 $\pm$ 5%) was observed against MCF-12A cells [20].

Yayıntaş et al. [21] have shown the antitumoral potential of butanol fraction of a liverwort *Marchantia polymorpha* against HeLa and lung carcinoma (A549) cells in MTT assay. Ether extracts of eight liverworts, namely, *Riccia fluitans* L. (Ricciaceae), *Porella cordaeana* (Huebener) Moore (Porellaceae), *Porella platyphylla* (L.) Pfeiff. (Porellaceae), *Corsinia coriandrina* (Spreng.) Lindb. (Corsiniaceae), *Mannia androgyna* (L.) A. Evans (Aytoniaceae), *Plagiochasma rupestre* (J.R. Forst et G. Forst) Steph. (Aytoniaceae), *Reboulia hemisphaerica* (L.) Raddi (Aytoniaceae), and *Targionia hypophylla* L. (Targioniaceae), were cytotoxic against Sp2/0 and YAC-1 cell lines without any effect on HeLa cells. When used at 1 mg/mL concentration, *Riccia fluitans*, *Porella cordaeana*, and *Targionia hypophylla* showed strong cytotoxicity against YAC-1 cells. Whereas *T. hypophylla* and *P. cordaeana* were the most active species against Sp2/0 cells with an inhibition rate of 86% [22].

Vollar et al. [23] screened 168 aqueous and organic extracts of 42 bryophyte species for in vitro antiproliferative activity against a panel of human gynecological cancer cell lines by using the MTT assay. Inhibition of proliferation ($\geq$25%) was observed for 41 species against at least one of the cancer cell lines at 10 μg/mL. Vollar et al. reported some promising antiproliferative bryophyte species [23]. Several extracts of *Brachythecium rutabulum*, *Climacium dendroides*, *Encalypta streptocarpa*, *Neckera besseri*, *Pleurozium schreberi*, and *Pseudoleskeella nervosa* were active in the antiproliferative assay.

Yağlıoğlu et al. [24] showed that hexane and ethyl acetate extracts of two mosses, *Rhytidiadelphus triquetrus* and *Tortella tortuosa*, had antiproliferative and cytotoxic activities against HeLa and C6 cell lines at higher concentrations (100, 75, and 50 mg/mL) than 5-fluorouracil as standard.

Anticancer activities of dichloromethane extract of *Dicranum scoparium* were shown by Abay et al. [25] against HeLa cell lines with a strong antiproliferative activity of the fraction-9 at concentrations of 100 and 50 μg/ml. Liktor-Busa et al. [26] have shown antiproliferative activity of *Abietinella abietina, Climacium dendroides, Pseudoscleropodium purum, Rhytidiadelphus squarrosus, Syntrichia ruralis,* and *Plagiomnium cuspidatum* against the HeLa, A2780, and T47D human cancer cell lines with more than 50% activity were observed at the concentrations of 10 or 30 μg/ml. Using the MTT assay, an 80% methanolic extract of *Lepidozia borneensis* showed cytotoxicity against human breast cancer (MCF-7) with an IC_{50}

value of 47.33 ± 7.37 μg/mL. Apoptosis occurred during the first 24 h of treatment and significantly increased to 30.8% after 72 h of treatment [27]. A high inhibitory activity (0.9–5 μg/mL) of some extracts of mosses, *Sphagnum magellanicum, Dicranum polysetum,* and *Pleurozium schreberi*, was also observed against the rat glioma cells [28]. 80% methanol extracts of *Pogonatum cirratum* subsp. *fuscatum* and *Sphagnum cuspidatum* were shown to inhibit the proliferation of ovarian carcinoma (Caov-3) and hepatocellular carcinoma (HepG2) cells, respectively [29]. The selected cancer cell lines could not be inhibited by *Sphagnum cuspidatum* subsp. *subrecurvum*, *Sphagnum junghuhniannum*, and *Pogonatum cirratum* subsp. *macrophyllum.* Using concentrations of 85 and 170 μg/mL, Oztopcu-Vatan et al. observed antiproliferative activities of the acetone extract C of *Homalothecium sericeum* (Hedw.) Schimp. (Brachytheciaceae) against rat glioma (C6) cell line [30]. An ethyl acetate extract C of an aquatic moss *Fontinalis antipyretica* Hedw. demonstrated in vitro antiproliferative properties against C6 cells *in* concentrations of 80 and 160 μg/mL.

3 Liverworts

Liverworts are the most prominent group of bryophytes and possess different secondary metabolites. More than 3000 compounds have been isolated from liverworts so far. Various liverwort species contain monoterpenes, diterpenoids, triterpenoids, and sesquiterpenoids. Liverworts also contain flavonoids, highly unsaturated fatty acids, lignans, steroids, and volatile aromatic compounds, such as phenolic bibenzyls, benzylphthalides, and phenanthrenes, and their dihydro analogs or dihydrostelbenes are also found in liverworts [15, 31–33]. Besides, some rare compounds, such as *seco*-africanes, *seco*-cuparanes, noraristolanes, 1,10-*seco*-aromadendranes, 2,3-*seco*-aromadendranes, neotrifaranes, tridensanes, ricciocarpanes, modified pacifigorgianes, pinguisanes, chenopodanes, and riccardiphanes have been isolated from liverworts [31, 34, 35].

In liverworts, monoterpenes, such as α-pinene, β-pinene, and limonene are most frequently observed [31, 32, 36]. One of the cytotoxic monoterpenes, found in *Trichocolea* species, is isoprenyl phenyl ether. For example, a major cytotoxic compound, namely, Methyl 4-[(5-oxogeranyl)oxy]-3-methoxybenzoate, was isolated from *Trichocolea mollissima* [37]. From the *Trichocolea*, geranyl phenyl ethers based on cytotoxic monoterpenoids were synthesized [38]. A moderately cytotoxic monoterpene ester, 2 alpha, 5 beta-dihydroxybornane-2-cinnamate have been isolated from Chinese *Conocephalum conicum* which was cytotoxic for human HepG2 cells [39]. Monoterpenes are also reported from *Jungermannia vulcanicola* [40]. A variety of diterpenoids, such as 5,10-*seco*-clerodane, 9,10-*seco*-clerodane, infuscane, *seco*-infuscane, spiroclerodane, *epi*homoverrucosane, *abeo*-labdane, sacculatane, cyathanes, fusicoccanes, cembranes, dolabellanes, vibsanes, neodenudatanes, verticillanes, viscidanes, and prenylguaianes are also found in liverworts [31, 35]. Among these, many compounds have shown cytotoxic activity. For example, 8,9-secokaurane diterpenes, isolated from *Lepidolaena taylorii* and

Lepidolaena palpebrifolia, were found to be cytotoxic against human tumor cell lines [41, 42]. Muscicolone, an ent-labdane type diterpene, isolated from *Frullania muscicola,* has also demonstrated cytotoxic activity against human tumor cells [43].

Entkaurene-type diterpenoids isolated from liverworts have also shown cytotoxic action against human leukemia cell lines [44]. It has been shown that *Jungermannia truncata* contains *ent*-11alpha-hydroxy-16-kauren-15-one which has apoptosis-inducing properties against HL-60 (acute promyelocytic leukemia) cells in a caspase-dependent manner [45, 46] as well as promoting apoptosis by tumor necrosis factor in human leukemia cells [47]. In HL-60 cells, the same compound induced apoptosis via p38 mitogen-activated protein kinase p38 [48]. A New Zealand liverwort *Jungermannia* species have also shown to possess cytotoxic kaurene- and entkaurene-type diterpenoids [49]. According to Kondoh et al. [50], Jungermannenones A, B, C, and D from *Jungermannia* species inhibit tumor growth through caspase-dependent mechanisms. Liu et al. [51] have isolated cis-clerodane diterpenoids from *Gottschelia schizopleura* which have been screened for cytotoxic activity against A549, liver hepatoblastoma (HepG2), colon adenocarcinoma (LOVO), and MDA-MB-435 cell lines. Cembrane-type diterpenoids and anadensin from *Chandonanthus hirtellus* have shown weak cytotoxicity against HL-60 cells, whereas fusicoccane-type diterpenoids and fusicoauritone 6 alpha-methyl ether have shown weak cytotoxicity against human epithelial carcinoma (KB) cell lines [52]. Weak cytotoxicity (30 mg/disk) of *Ent*-kaurene clavigerins A–D obtained from *Lepidolaena clavigera* was observed against BSC cells [53]. A weak inhibitory activity of a new atisane-2 derivative isolated from *Lepidolaena clavigera* against mouse lymphocytic leukemia cells (P-388) was observed with an IC_{50} value of 16 mg/mL [54]. Cytotoxic activity of α-Zeorin ($C_{30}H_{52}O_2$) has also been demonstrated against P-388 cells with an IC_{50} of 1.1 mg/ml [55, 56].

Triterpenoids have been shown to possess cytotoxicity and anticancer potential in various in vitro, in vivo, as well as preclinical studies [57]. Interestingly, plant-derived triterpenoids possess immense cytotoxic potential [58]. *Ptilidium pulcherrimum* secondary metabolites are cytotoxic against PC3 (prostate cancer cell line), MDA-MB-231, and HeLa cell lines; particularly, ursane triterpenoids were found to be cytotoxic for PC3 cells [59]. Triterpenes have also been isolated from other liverworts, to name a few *Fossombronia alaskana Fossombronia pusilla, Conocephalum japonicum, Nardia scalaris,* and *Blepharidophyllum densifolium* [60–63].

Sesquiterpenes have 15 carbon atoms with three isoprene units as a backbone. These compounds have shown immense therapeutic potential as anti-inflammatory, antimicrobial, antiviral, and anticancer agents [64, 65]. Marchantiophyta contains more than 900 sesquiterpenoids, eudesmane, and aromadendrane. In addition, cuparane, pinguisane, and barbatane (= gymnomitrane) also have been found. Many bryophytes have been shown to possess sesquiterpenoid compounds having cytotoxicity [17]. Sesquiterpenoids, marsupellone ($C_{15}H_{52}O$), and acetoxymarsupellone ($C_{17}H_{24}O_3$) isolated from *Marsupella emarginata* have shown cytotoxic activity (ID_{50}-1 mg/mL) against leukemia cells (P388) [66]. Some pinguisanoids, porellacetals A–D isolated from *Porella cordaeana*, have also shown anticancer potential [67]. It has been shown that diplophyllin, an ent-eudesmanolide isolated from *Diplophyllum albicans* and *Diplophyllum taxifolium,* is significantly cytotoxic to human epidermoid carcinoma [68]. Sesquiterpenoid compounds, costunolide and tulipinolide, have been isolated from *Conocephalum supradecompositum*, *Frullania monocera*, *Frullania tamarisci*, *Marchantia polymorpha*, *Porella japonica*, *Wiesnerella denudata, Lepidozia vitrea*, *Plagiochila semidecurrens*, and *Plagiochila ovalifolia* which are tumor growth inhibitors [69–75]. DNA-damaging sesquiterpenes were also found in *Porella cordeana*, *Frullania nisquallensis*, and *Chiloscyphus rivularis* [76]. Yeast-based DNA-damaging assays showed that 12-hydroxychiloscyphone, isolated from methyl ethyl ketone extract of *Chiloscyphus rivularis*, was cytotoxic and active against human lung carcinoma cells [77]. Plagiochiline-A-15-yl octanoate and 14-hydroxyplagiochiline-A-15-yl which are 2,3-secoaromadendrane-type sesquiterpenoids isolated from ether extract of *Plagiochila ovalifolia* have shown cytotoxicity against murine leukemia tumor cells (P-388) [78]. Naviculyl caffeate obtained from *Bazzania novae-zelandiae* has shown cytotoxic potential against human tumor cell lines [79]. Glaucescenolide, a sesquiterpene lactone isolated from *Schistochila glaucescens*, has also shown cytotoxicity against P388 cells [80].

Some herbertane-type sesquiterpenoids, such as (−)-alpha-herbertenol, (−)-diplophyllolide A, (−)-herbertenediol, (−)-mastigophorene C, and (−)-mastigophorene D, isolated from *Mastigophora diclados*, have shown cytotoxic potential against HL-60 and KB cell lines [81]. A weak cytotoxic action of chandolide-a zierane sesquiterpene gamma-lactone, isolated from *Chandonanthus hirtellus*, was observed against the HL-60 cell line [52]. Pinguisane- and germacrane-type sesquiterpenoids isolated from *Frullania* sp. and *Porella perrottetiana* have shown cytotoxicity against HL-60 and KB cell lines [82].

Bibenzyls are steroidal ethane derivatives and several bibenzyls, and/or their derivatives have been isolated from many liverworts, for example, *Asterella angusta* [83], *Bazzania trilobata* [84], *Blasia pusilla* [85], *Cavicularia densa* [86], *Dumortiera hirsuta* [39], *Frullania inouei* [87], *Jubula japonica* [88], *Lepidozia incurvata* [89], *Lunularia cruciata* [90], *Marchantia paleacea* [91], *Marchantia emarginata* subsp. *tosana* [92], *Marchantia paleacea* var. *diptera* [93], *Marchantia polymorpha* [94], *M. tosana* [70], *Marsupidium epiphytum* [95], *Monoclea forsteri* [96], *Pellia endiviifolia* [90], *Plagiochila* sp. [97, 98], *Plagiochila fasciculata* [99], *Plagiochila fruticosa* [100], *Plagiochila diversifolia P. permista* var. *intergerrima* [101], *Plagiochasma intermedium* [102], *Porella perrottetiana* [82], *Ptychanthus striatus* [103], *Radula marginata* [104], *Radula perrottetii* [105], *Reboulia hemisphaerica* [106], *Ricciocarpos natans* [107], *Riccardia multifida* [108], *Riccardia multifida* subsp. *decrescens* [109], and *Schistochila glaucescens* [80].

Bibenzyl and their dimeric form bis(bi)benzyls have shown various bioactivities including antimicrobial, anticancer, antioxidant, cyclooxygenase, calmodulin modulation, LXRα activating, HIV preventive, lipoxygenase, tyrosinase, and microtubule polymerization. Some bibenzyl cannabinoids (tetrahydrocannabinol type), perrottetinene, and perrottetineic acid may also have neurological effects

[10, 11]. Anticancer activity of various bibenzyl compounds isolated from liverworts has been discussed separately.

4 Hornworts

Some mono-, sesqui-, and diterpenoids have been isolated from *Anthoceros caucasicus*, *A. punctatus*, and *A. agrestis*, for example, aristolene, anastreptene, α- and β-pinene, β-myrcene, β-barbatene, β-bazzanene, δ-cuprenene, camphene, diplophyllolide, limonene, maaliol, terpinolene, and veticadinoxide [110, 111]. Additionally, *Anthoceros laevis* and *A. punctatus* have been found to contain methyl *p*-coumarate. Rosmarinic acid has also been isolated from *Anthoceros punctatus*, *A. agrestis*, and *Megaceros flagellaris*. An alkaloid anthocerodiazonin has also been isolated from *Anthoceros agrestis* grown in tissue cultures [112]. Though some of these compounds have shown anticancer potential [113, 114], however, the anticancer activity of hornworts has not yet been reported. As compared with liverworts and mosses, hornworts have been less researched for phytoconstituents and need attention from the scientific fraternity.

5 Mosses

Mosses are rich in terpenoids, flavonoids, benzoic, and cinnamic acid derivatives, such as chlorogenic, ellagic, and ferulic acids, coumarins, and benzonaphthoxanthenones [10, 115]. Some of these compounds have shown anticancerous potential as described below.

Ohioensins are benzonaphthoxanthenones, a class of flavonoids, which have been isolated from a variety of moss species, including *Polytrichum ohioense* Renauld & Cardot. (ohioensin A–E), *Polytrichastrum alpinum* (Hedw.) G.L. Sm. (ohioensin F and G), and *Polytrichum commune* Hedw. (Ohioensin H). Furthermore, 1-O-methylohioensin B, 1-O-methyldihydroohioensin B, and 1,14-di-O-methyldihydroohioensin B were isolated from *Polytrichum pallidisetum* Funck.

These compounds have shown cytotoxic potential against various cell lines. Ohioensin A has been found to be cytotoxic for murine leukemia (PS) and MCF-7; ohioensin B against mouse leukemia (HT-29) and human colon adenocarcinoma; ohioensin C, D, and E against 9PS and P388; and ohioensin H against human breast adenocarcinoma (MDA-MB-435), human T-cell leukemia (6 T-CEM), A549, LOVO, and HepG2. 1-O-Methylohoensin B has shown cytotoxicity against HT-29, human colon adenocarcinoma, human melanoma (RPMI-7951), and human glioblastoma multiforme (U-251 MG) cell lines, and l-O-methyldihydroohioensin B against U-251 MG and 1,14-di-O-methyldihydroohioensin B have shown cytotoxic potential against A549 and RPMI-7951 cell lines [17, 116–118].

The cinnamoyl bibenzyls pallidisetin A and pallidisetin B, isolated from *Polytrichum pallidisetum*, exhibit cytotoxicity against RPMI-7951 and U-251 cells

[119]. *Polytrichum juniperinum* extracts showed cytotoxic activity against sarcoma 37 in mice [120] as well as photoprotective effects against UV-induced DNA damage in hamster lung fibroblasts V79 cells [121].

Significant cytotoxicity by a compound ansamitocin P-3 ($C_{32}H_{43}ClN_2O_9$), isolated from *Claopodium crispifolium* and *Anomodon attenuatus*, has been shown against A-549 and HT-29 cell lines [122]. Besides, many other compounds, such as 15-methoxyansamitocin P-3 ($C_{33}H_{45}ClN_2O_{10}$), maytanbutine ($C_{36}H_{50}ClN_3O_{10}$), and trewiasine ($C_{37}H_{52}ClN_3O_{11}$) isolated from *Anomodon attenuatus*, *Claopodium crispifolium*, *Isothecium subdiversiforme*, and *Thamnobryum sandei* have shown antitumor potential [122–124].

A recent study by Klegin et al. [125] has identified β-bazzanene, β-caryophyllene, β-chamigrene, and germacrene B from the essential oil of *Phyllogonium viride* Brid without any cytotoxic potential in breast and colorectal tumor cells (MCF-7 and HCT-116). Similarly, cytotoxic potential of ethyl alcohol extracts of mosses, namely, *Abietinella abietina*, *Homalothecium sericeum*, *Tortella tortuosa*, *Syntrichia ruralis*, and *Bryoerythrophyllum rubrum*, on 5-fluorouracil-resistant colorectal cancer HCT116 and HT29 cell lines has been demonstrated [126]. These moss species could be further explored for the isolation of anticancerous molecules.

6 Anticancer Activity of Bis(bi) Benzyl Compounds Isolated from Liverworts

There are 103 characterized Bis(bi) benzyl compounds that have been isolated from different liverworts [127]. The cytotoxic/antiproliferative potential of some of the major compounds is discussed in brief. Chemical structures of some of the bioactive bibenzyl molecules are given in Fig. 1.

6.1 Marchantin

Bis-bibenzyls, such as of marchantin type have shown several bioactivities including cytotoxic action. Marchantins are the first cyclic bis(bibenzyls) molecules characterized from liverwort *Marchantia polymorpha*. Syntheses of marchantin A from liverworts are also reported [115, 128].

An anticancer compound, marchantin A ($C_{28}H_{24}O_5$), isolated from *Marchantia emarginata* subsp. *tosana*, inhibited the growth of MCF-7 cells by inducing apoptosis, upregulating the expression of p21 and p27 genes, and decreasing cyclin D1 and B1 expression with an IC_{50} of 4.0 μg/ml [92].

It has also been reported that marchantin A isolated from *M. polymorpha* exhibits cytotoxic activity in rat myeloblast CC50 L6 cells with an IC_{50} value of 6.64 μM [94]. Marchantin A isolated from *M. polymorpha* and *M. tosana* has also been reported to be cytotoxic against KB cells [70]. In addition, Jensen et al. observed a decrease in cell viability of A256 (human breast cancer) cell line [94]. Marchantin C

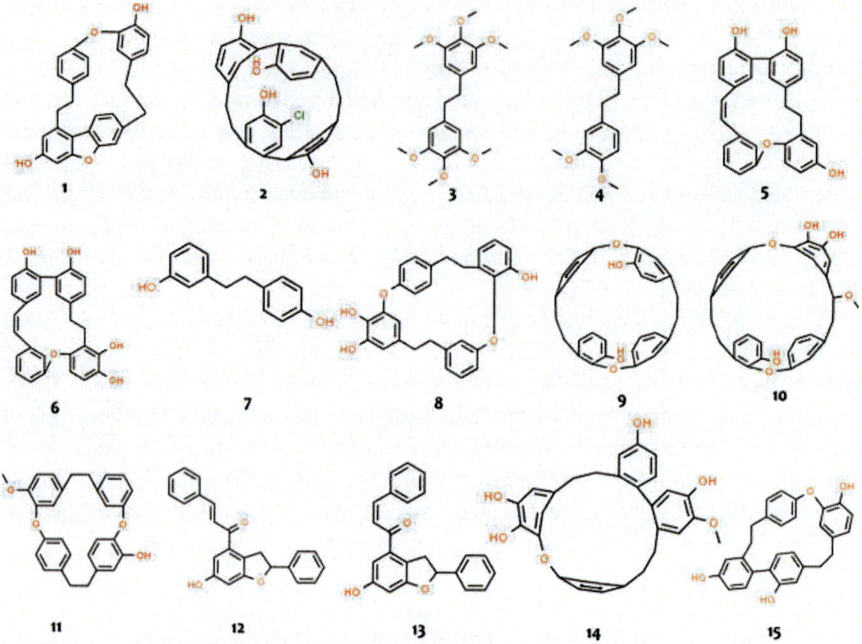

Fig. 1 Structures of bioactive bibenzyls: 1. Asterelin A ($C_{28}H_{22}O_4$) 2.Bazzanin S ($C_{28}H_{23}ClO_4$) 3.Brittonin A ($C_{20}H_{26}O_6$) 4.Chrysotobibenzyl ($C_{19}H_{24}O_5$) 5.Isoplagiochin A ($C_{28}H_{22}O_4$) 6.Isoplagiochin B ($C_{28}H_{22}O_5$) 7.Lunularin ($C_{14}H_{14}O_2$) 8.Marchantin A ($C_{28}H_{24}O_5$) 9.Marchantin C ($C_{28}H_{24}O_4$) 10.Marchantin E ($C_{29}H_{26}O_6$) 11.Pakyonol ($C_{29}H_{26}O_4$) 12.Pallidisetin A 13.Pallidisetin B ($C_{23}H_{18}O_3$) 14.Plagiochin A ($C_{29}H_{26}O_6$) 15.Riccardin C ($C_{28}H_{24}O_4$) (from https://www.chemspider.com/)

($C_{28}H_{24}O_4$) is another well-known bis-bibenzyl isolated from liverworts. Isolated from a liverwort *Schistochila glaucescens,* it has shown cytotoxicity against P388 [80]. Marchantin C also induced a dose-dependent proapoptotic effect in human glioma A172 cells through regulation of Bax-Bcl-2 proteins [129]. Decreased microtubule quantity along with in vivo and in vitro antitumor activity by arresting cell cycle at G(2)/M phase in A172 and HeLa cells is also demonstrated by marchantin C. Human cervical carcinoma xenografts treated with marchantin C have shown increased apoptosis with cyclin B1, Bax, and caspase-3 [130]. Marchantins A and C, isolated from *Reboulia hemisphaerica*, have shown strong microtubule depolymerization activities in human cervical carcinoma cell line HeLa [131]. Reduction in matrix metallopeptidase in marchantin C treated T98G and U87 glioma cells observed to inhibit the migration of cancer cells [132]. Vincristine resistance was also found to be altered by marchantin C and its synthetic dimethyl ether derivative, 7,8-dehydromarchantin C in KB/VCR cells by retardation of P-glycoprotein (P-gp) activity [133]. Angiogenic inhibition of T98G glioma cells by marchantin C has also been reported [134].

Marchantin M, isolated from *Asterella angusta*, is a cyclic bis-bibenzyl compound and has shown cytotoxicity against chemoresistant PC3 cells by elicitation of apoptosis along with upregulation of Bax expression, PARP cleavage, and caspase-3 activity [106].

6.2 Neomarchantins

Neomarchantins A and B, isolated from *Schistochila glaucescens*, were found to induce apoptosis against P388 cells with an IC_{50} values of 18 and 7.6 μg/ml, respectively [80].

6.3 Plagiochin

Plagiochin E, a macrocyclic bis-bibenzyl, isolated from *M. polymorpha* had shown proapoptotic activity in *Candida albicans* by performing chromatin condensation and nuclear fragmentation, metacaspase activation, cytochrome c release, and down-regulation of CDC28, CLB2, and CLB4 expression leading to G2/M cell cycle arrest [135]. Plagiochin E has also shown a reversal effect on MDR in adriamycin-resistant K562/A02 cells [96].

6.4 Isoplagiochin

In vitro tubulin polymerization inhibition was inhibited by macrocyclic bis (bibenzyls) and isoplagiochins A ($C_{28}H_{22}O_4$) and B ($C_{28}H_{22}O_5$) and isolated from *Plagiochila fruticosa* with an IC_{50} of 50 and 25 μM, respectively. These compounds were also isolated from *P. diversifolia* Lindenb & Gottsche and *P. permista* var. *intergerrima* Herzog [101, 136].

6.5 Perrottetin

Perrottetin E is a prenyl bibenzyl, isolated from *Radula perrottetii*, and has shown cytotoxicity against the human KB cells with an ID_{50} of 12.5 μg/ml [105]. Recently, the modest cytotoxic activity of perrottetin E, 10′-hydroxyperrottetin E, and 10,10′-dihydroxyperrottetin E obtained from methylene-chloride/methanol extract of *Pellia endiviifolia* against U-937 (acute monocytic leukemia cells), K-562 (human chronic myelogenous leukemia cells), and HL-60 and significant cytotoxicity against human embryonal teratocarcinoma cell line (NT2/D1) and human glioblastoma cell line (A-172) were observed. The perrottetin F phenanthrene derivative isolated from *Lunularia cruciata* also showed similar results [90].

6.6 Riccardin

Riccardins are well-known bioactive bis-bibenzyl compounds isolated from various liverworts. Cytotoxicity of riccardin A ($C_{29}H_{26}O_4$) and riccardin B ($C_{28}H_{24}O_4$), isolated from *Riccardia multifida*, has been reported [108]. Riccardin C ($C_{28}H_{24}O_4$), a cyclic bis-bibenzyl, is reported from *P. intermedium* and *Reboulia hemisphaerica*. It has shown cytotoxic potential via inducing apoptosis by modulating the expression levels of Bcl-2, Bax, and PARP against PC3 cells. Antiproliferative activity of riccardin C, isolated from *A. angusta*, was also observed [106].

Riccardin D is a macrocyclic bis(bi)benzyl which was detected in *Monoclea forsteri* and demonstrated antiproliferative activity on human glioma A172 cells and induced apoptosis [96]. It was also isolated from *D. hirsuta* and significantly inhibited hyphal growth of *Candida albicans* [3]. Antiproliferative activity of riccardin D was also observed against human umbilical vascular endothelial cells along with angiogenic reduction in H460 (human lung cancer carcinoma) cell line [137], and antiproliferative effect on human leukemia cell lines, HL-60, K562, and MDR K562/A02 cells was also observed with dependence on DNA topoisomerase-II [138]. Moreover, its brominated and aminomethylated derivatives have also shown antiproliferative activity against KB, MCF-7, and PC3 cell lines [139]. Riccardin D, isolated from *D. hirsuta*, has also led to the prevention of intestinal polyposis in APC Min/+ mice [140].

Riccardin F, isolated from *P. intermedium*, has shown alteration of P-gp-mediated drug resistance in adriamycin-treated cancer cell line K562/A02 [141].

6.7 Dihydroptychantol A (DHA)

A macrocyclic bis-bibenzyl, dihydroptychantol A (DHA) derived from *Asterella angusta* and its thiazole derivatives have shown chemotherapeutic MDR reversal activity by evaluating against adriamycin-resistant K562/A02 cells, vincristine-resistant KB/VCR cells, and their parental cells by MTT assay [83, 142]. Cell lines KB/VCR and K562 were found to be the most resistant and sensitive, respectively. A sharp decline in cell viability of both the cell lines was also observed [84]. PP-gp mediated remarkable MDR reversal and adriamycin cytotoxicity towards K562/A02 of DHA and its derivatives was also detected by MTT assays [87, 143].

DHA which was synthesized chemically has been shown to induce autophagy with IC_{50} values of 29.6 μM (24 hr) and 24.7 μM (48 hr) in human osteosarcoma U2OS cells. Cell cycle arrest at G2/M-phase with upregulation of cyclin B1 and enhanced expression of nuclear p53 p21Wafl/Cip1 (p53 target gene) was also observed along with decreased expression of cytoplasmic p53 in the treated cells [144]. DHA also reversed chemoresistant human glioblastoma U87 cell line with IC_{50} values of 21.2 μM (24 hr) and 23.7 μM (48 hr) [39].

6.8 Lunularin

Moderate cytotoxic activity of a monomeric bibenzyl compound lunularin ($C_{14}H_{14}O_2$) isolated from *Dumortiera hirsuta* was observed in human HepG2 cells with an IC_{50} value of 7.4 mg/ml [39]. A growth inhibitory effect was observed with 14-hydroxylunularin isolated from *Ricciocarpus natans* in NCTC-Clone 929 fibroblast J 774 murine macrophage and peritoneal macrophage (BALB/c mice) [145].

6.9 Other Cytotoxic Bis(Bibenzyls)

Aside from the above-mentioned well-known bis-bibenzyls, cyclic bis-bibenzyl compounds, such as pakyonol ($C_{29}H_{26}O_4$) and plagiochin E ($C_{28}H_{24}O_4$) have been shown to decrease Bcl-2 (anti-apoptotic protein) and increase Bax expression (pro-apoptotic protein). Furthermore, PARP cleavage and caspase-3 activity were observed by MTT assays and Western blots in chemoresistant PC3 cells after exposure to these cyclic bis-bibenzyl compounds [106].

Pakyonol (Fig. 1), isolated from *Plagiochasma intermedium*, was also alleviated by P-gp-mediated MDR in adriamycin-induced tumor K562/A02 cell line after treatment for 48 h at a concentration of 3 μg/ml [102]. The methylated bibenzyl 3,3,4,4-tetramethoxybibenzyl obtained from *Frullania inouei* has shown cytotoxicity against several human cancer cell lines K562, KB, K562/A02, or KB/VCR along with MDR reversal activity in vincristine-resistant KB/VCR and adriamycin-resistant K562/A02 cells [87]. Similarly, 4-hydroxy-3-methoxybibenzyl, isolated from *Plagiochila fasciculata*, has shown cytotoxicity at a 60 μg/well concentration against the monkey kidney cells (BSC), but did not significantly inhibit the growth of P-388 leukemia cells [99]. Some prenylated bibenzyl compounds isolated from *Radula* spp. have also shown cytotoxic action. For example, 2-carbomethoxy-3,5dihydroxystilbene and 3,5-dihydroxybibenzyl isolated from *Radula amoena* have exhibited moderate cytotoxicity in human cancer cell lines [146]. Similarly, methyl 2,4-dihydroxy-3-(3-methyl-2-butenyl)-6-phenethylbenzoate isolated from *Radula constricta* exhibited cytotoxicity against human lung cancer cells (A549 and NCI-H1299) with IC_{50} values of 6.0 and 5.1 μM, respectively [147].

Paleatin B, an acyclic bis-bibenzyl, was obtained from *Marchantia paleacea* var. *diptera* (Nees & Mont.) S. Hatt. Paleatin B showed cytotoxicity against KB and P-388 cell lines [93].

Pusilatins (A–D) are bis(bibenzyl) dimers that have been isolated from liverwort *Blasia pusilla*. Pusilatin B and C were both cytotoxic against KB cells with ED_{50} of 13.1 μg/ml and 13 μg/ml, respectively. It also inhibited DNA polymerase β activity with IC_{50} values of 13 and 5.16 μM, respectively [85].

Cytotoxic and proapoptotic activity of brittonin A ($C_{20}H_{26}O_6$) and B isolated from *Frullania inouei* was demonstrated against K562/A02 KB and KB/VCR. Another compound chrysotobibenzyl ($C_{19}H_{24}O_5$) isolated from *F. inouei* has also shown cytotoxic potential against human cell lines K562, K562/A02, KB, and KB/VCR

[87]. Extraction of bibenzyl compounds from *Frullania* sp. and *Porella perrottetiana* and determination of their cytotoxic action using water-soluble tetrazolium-8 colorimetric assay have shown their positive role against HL-60 and KB cell lines [82].

7 Conclusion

Bryophytes are small plants that remained ignored over a very long period. However, in the recent past, several phytochemical and pharmacological investigations have been carried out on various bryophytes species, and some intriguing observations were received.

Liverworts and mosses are found to be rich in some unique phytoconstituents, for example, marchantin, plagiochin, pakyonol, ricciocarpanes, pinguisanes, etc., and also shown anticancer activities in various models. Several bibenzyl and bis-bibenzyl compounds have been isolated from bryophytes among which many have shown chemotherapeutic potential against various cancer cell lines. In order to establish their efficacy against many types of cancer, more research is needed, especially clinical trials. The present chapter briefly describes anticancerous compounds isolated from liverworts, mosses, and hornworts and throws light on the bibenzyl compounds having multifarious health-beneficial activities. Moreover, it also emphasizes the promising bryophyte species which could be explored for isolation of novel chemotherapeutic molecules using recent drug discovery technology.

References

1. Sabovljevic MS, Sabovljević AD, Ikram NKK, Peramuna AV, Bae H, Simonsen HT (2016) Bryophytes – an emerging source for herbal remedies and chemical production. Plant Genet Resour Characterisation Util 14:314–327
2. Martínez-Abaigar J, Núñez-Olivera E (2021) Novel biotechnological substances from bryophytes. In: Sinha RP, Häder D-P (eds) Natural bioactive compounds. Academic Press, Cambridge, MA
3. Cheng A, Sun L, Wu X, Lou H (2009) The inhibitory effect of a macrocyclic bisbibenzyl riccardin D on the biofilms of *Candida albicans*. Biol Pharm Bull 32:1417–1421
4. van Hoof LD, Vanden Berghe DA, Petit E, Vlietnick AJ (1981) Antimicrobial and antiviral screening of bryophyta. Fitoterapia 52:223–229
5. Wu XZ, Cheng AX, Sun LM, Lou HX (2008) Effect of plagiochin E, an antifungal macrocyclic bis(bibenzyl), on cell wall chitin synthesis in *Candida albicans*. Acta Pharmacol Sin 29: 1478–1485
6. Scher JM, Schinkovitz A, Zapp J, Wang Y, Franzblau SG, Becker H et al (2010) Structure and anti-TB activity of trachylobanes from the liverwort *Jungermannia exsertifolia* ssp. *cordifolia*. J Nat Prod 73:656–663
7. Singh M, Govindarajan R, Nath V, Rawat AK, Mehrotra S (2006) Antimicrobial, wound healing and antioxidant activity of *Plagiochasma appendiculatum* Lehm. et Lind. J Ethnopharmacol 107:67–72

8. Ivanova V, Kolarova M, Aleksieva K, Dornberger KJ, Haertl A, Moellmann U et al (2007) Sanionins: anti-inflammatory and antibacterial agents with weak cytotoxicity from the Antarctic moss *Sanionia georgico-uncinata*. Prep Biochem Biotechnol 37:343–352
9. Dey A, De JN (2012) Antioxidative potential of bryophytes: stress tolerance and commercial perspectives: a review. Pharmacol 3:151–159
10. Ludwiczuk A, Asakawa Y (2020) Terpenoids and aromatic compounds from bryophytes and their central nervous system activity. Curr Org Chem 24:113–128
11. Nandya S, Dey A (2020) Bibenzyls and bisbybenzyls of bryophytic origin as promising source of novel therapeutics: pharmacology, synthesis and structure-activity. Daru J Pharm Sci 28:701–734
12. Asakawa Y (2001) Recent advances in phytochemistry of bryophytes-acetogenins, terpenoids and Bis(Bibenzyl)s from selected Japanese, Taiwanese, New Zealand, Argentinean and European liverworts. Phytochemistry 56:297–312
13. Asakawa Y (2010) ChemInform abstract: chemical constituents of the bryophytes. Chemin 27: 1–562
14. Chen F, Ludwiczuk A, Wei G, Chen X, Crandall-Stotler B, Bowman JL (2018) Terpenoid secondary metabolites in bryophytes: chemical diversity, biosynthesis and biological functions. Crit Rev Plant Sci 37:210–231
15. Xie C-F, Lou H-X (2009) Secondary metabolites in bryophytes: an ecological aspect. Chem Biodivers 6:303–312
16. Sagbo IJ, Otang-Mbeng W (2021) Plants used for the traditional management of cancer in the Eastern Cape Province of South Africa: a review of ethnobotanical surveys, ethnopharmacological studies and active phytochemicals. Molecules 26:4639. https://doi.org/10.3390/molecules26154639
17. Dey A, Mukherjee A (2015) Therapeutic potential of bryophytes and derived compounds against cancer. J Acute Dis 4:236–248
18. Wolski GJ, Sadowska B, Fol M, Podsędek A, Kajszczak D, Kobylińska A (2021) Cytotoxicity, antimicrobial and antioxidant activities of mosses obtained from open habitats. PLoS One 16(9):e0257479
19. Mandić MR, Oalđe MM, Lunić TM, Sabovljević AD, Sabovljević MS, Gasˇić UM et al (2021) Chemical characterization and *in vitro* immunomodulatory effects of different extracts of moss *Hedwigia ciliata* (Hedw.) P. Beauv. from the Vrsˇačke Planine Mts., Serbia. PLoS One 16(2): e0246810. https://doi.org/10.1371/journal.pone.0246810
20. Onbasli D, Yuvali G (2021) *In vitro* medicinal potentials of *Bryum capillare*, a moss sample, from Turkey. Saudi J Biol Sci 28:478–483. https://doi.org/10.1016/j.sjbs.2020.10.031
21. Yayıntaş OT, Yılmaz S, Sökmen M (2019) Determination of antioxidant, antimicrobial and antitumor activity of bryophytes from Mount Ida (Canakkale, Turkey). Indian J Tradit Knowl 18:395–401
22. Önder A, Özenoğlu H (2019) Evaluation of cytotoxic effects on ethereal extracts of some selected liverworts. FABAD J Pharm Sci 44:119–125
23. Vollár M, Gyovai A, Szűcs P, Zupkó I, Marschall M, Csupor-Löffler B, Bérdi P, Vecsernyés A, Csorba A, Liktor-Busa E, Urbán E, Csupor D (2018) Antiproliferative and antimicrobial activities of selected bryophytes. Molecules 23:1520
24. Yağlioğlu MŞ, Gökhan ABAY, Demirtaş İ, Yağlioğlu AŞ (2017) Phytochemical screening, antiproliferative and cytotoxic activities of the mosses *Rhytidiadelphus triquetrus* (Hedw.) Warnst. and *Tortella tortuosa* (Hedw.) Limpr. Anatolian Bryol 3:31–42
25. Abay G, Altun M, Koldas S, Riza Tufekci A, Demirtas I (2015) Determination of antiproliferative activities of volatile contents and HPLC profiles of *Dicranum scoparium* (Dicranaceae, Bryophyta). Comb Chem High Throughput Screen 18:453–463
26. Liktor-Busa E, Urbán E, Zupkó I, Szűcs P, Csupor D (2015) In vitro antibacterial and antiproliferative screening of Hungarian bryophytes. Planta Med 81(16):66
27. Abu Bakar MF, Abdul Karim F, Suleiman M, Isha A, Rahmat A (2015) Phytochemical constituents, antioxidant and antiproliferative properties of a liverwort, *Lepidozia borneensis* Stephani from Mount Kinabalu, Sabah, Malaysia. Evidence-based Complement Altern Med 2015:1–9

28. Klavina L, Springe G, Nikolajeva V, Martsinkevich I, Nakurte I, Dzabijeva D, Steinberga I (2015) Chemical composition analysis, antimicrobial activity and cytotoxicity screening of moss extracts (Moss Phytochemistry). Molecules 20:17221–17243. https://doi.org/10.3390/molecules200917221
29. Karim AF, Suleiman M, Rahmat A, Abu Bakar MF (2014) Phytochemicals, antioxidant and antiproliferative properties of five moss species from Sabah, Malaysia. Int J Pharma Pharmaceut Sci 6:292–297
30. Oztopcu-Vatan P, Savaroglu F, Filik-Iscen C, Kabadere S, Ilhan S, Uyar R (2011) Antimicrobial and antiproliferative activities of *Homalothecium sericeum* (Hedw.) Schimp. extracts. Fresenius Environ Bull 20:461–466
31. Asakawa Y (1995) Chemical constituents of the bryophytes. In: Herz W, Kirby GW, Moore RE, Steglich W, Tamm C (eds) Progress in the chemistry of organic natural products. Springer, Vienna
32. Asakawa Y, Ludwiczuk A, Nagashima F (2013) Chemical constituents of bryophytes: bio- and chemical diversity, biological activity, and chemosystematics. In: Kinghorn AD, Falk H, Kobayashi J (eds) Progress in the chemistry of organic natural products. Springer, Vienna
33. Stafford HA (2000) The evolution of phenolics in plants. In: Ibrahim R, Varin L, De Luca V, Romeo J (eds) Evolution of metabolic pathways, vol 34. Elsevier Science, Oxford
34. Asakawa Y (1982) Chemical constituents of the Hepaticae. In: Herz W, Grisebach H, Kirby GW (eds) Progress in the chemistry of organic natural products. Springer, Vienna
35. Ludwiczuk A, Asakawa Y (2014) Fingerprinting of secondary metabolites of liverworts: chemosystematic approach. J AOAC Int 97:1234–1243
36. Asakawa Y, Ludwiczuk A, Nagashima F (2013) Phytochemical and biological studies of bryophytes. Phytochemistry 91:52–80
37. Perry NB, Foster LM, Lorimer SD, May BC, Weavers RT (1996) Isoprenyl phenyl ethers from liverworts of the genus *Trichocolea*: cytotoxic activity, structural corrections, and synthesis. J Nat Prod 59:729–733
38. Baek SH, Perry NB, Weavers RT (1998) Synthesis of geranyl phenyl ethers based on the cytotoxic monoterpenoids from the liverwort genus *Trichocolea*. J Nat Prod 61:1143–1145
39. Lu ZQ, Fan PH, Ji M, Lou HX (2006) Terpenoids and bisbibenzyls from Chinese liverworts *Conocephalum conicum* and *Dumortiera hirsuta*. J Asian Nat Prod Res 8:187–192
40. Yokouchi Y, Satake K, Ambe Y (1984) Monoterpene composition of the essential oil of the aquatic liverwort *Jungermannia vulcanicola* Steph. Bryologist 87:323–326
41. Perry NB, Burgess EJ, Tangney RS (1996) Cytotoxic 8,9-secokaurane diterpenes from a New Zealand liverworts, *Lepidolaena taylorii*. Tetrahedron Lett 37:9387–9390
42. Perry NB, Burgess EJ, Baek SH, Weavers RT, Geis W, Mauger AB (1999) 11-oxygenated cytotoxic 8,9-secokauranes from a New Zealand liverwort, *Lepidolaena taylorii*. Phytochemistry 50:423–433
43. Lou HX, Li GY, Wang FQ (2002) A cytotoxic diterpenoid and antifungal phenolic compounds from *Frullania muscicola* Steph. J Asian Nat Prod Res 4:87–94
44. Nagashima F, Kasai W, Kondoh M, Fujii M, Watanabe Y, Braggins JE (2003) New ent-kaurene-type diterpenoids possessing cytotoxicity from the New Zealand liverwort *Jungermannia* species. Chem Pharm Bull 51:1189–1192
45. Nagashima F, Kondoh M, Kawase M, Simizu S, Osada H, Fujii M et al (2003) Apoptosis-inducing properties of ent-kaurene-type diterpenoids from the liverwort *Jungermannia truncata*. Planta Med 69:377–379
46. Suzuki I, Kondoh M, Nagashima F, Fujii M, Asakawa Y, Watanabe Y (2004) A comparison of apoptosis and necrosis induced by ent-kaurene-type diterpenoids in HL-60 cells. Planta Med 70:401–406
47. Suzuki I, Kondoh M, Harada M, Koizumi N, Fujii M, Nagashima F et al (2004) An ent-kaurene diterpene enhances apoptosis induced by tumor necrosis factor in human leukemia cells. Planta Med 70:723–727

48. Kondoh M, Suzuki I, Harada M, Nagashima F, Fujii M, Asakawa Y et al (2005) Activation of p38 mitogen-activated protein kinase during ent-11alpha-hydroxy-16-kauren-15-one-induced apoptosis in human leukemia HL-60 cells. Planta Med 71:275–277
49. Nagashima F, Kondoh M, Fujii M, Takaoka S, Watanabe Y, Asakawa Y (2005) Novel cytotoxic kaurane-type diterpenoids from the New Zealand liverwort *Jungermannia* species. Tetrahedron 61:4531–4544
50. Kondoh M, Nagashima F, Suzuki I, Harada M, Fujii M, Asakawa Y et al (2005) Induction of apoptosis by new ent-kaurene diterpenoids isolated from type-the New Zealand liverworts *Jungermannia* species. Planta Med 71:1005–1009
51. Liu CM, Zhu RL, Liu RH, Li HL, Shan L, Xu XK et al (2009) cis-Clerodane diterpenoids from the liverwort *Gottschelia schizopleura* and their cytotoxic activity. Planta Med 75:1597–1601
52. Komala I, Ito T, Nagashima F, Yagi Y, Kawahata M, Yamaguchi K et al (2010) Zierane sesquiterpene lactone, cembrane and fusicoccane diterpenoids, from the Tahitian liverwort *Chandonanthus hirtellus*. Phytochemistry 71:1387–1394
53. Perry NB, Burgess EJ, Foster LM, Gerard PJ, Toyota M, Asakawa Y (2008) Insect antifeedant sesquiterpene acetals from the liverwort *Lepidolaena clavigera*. 2. Structures, artifacts and activity. J Nat Prod 71:258–261
54. Perry NB, Burgess EJ, Baek SH, Weavers RT (2001) The first atisane diterpenoids from a liverwort: polyols from *Lepidolaena clavigera*. Org Lett 3:4243–4245
55. Neves M, Morais R, Gafner S, Hostettmann K (1998) Three triterpenoids and one flavonoid from the liverwort *Asterella blumeana* grown *in vitro*. Phytother Res 12:21–24
56. Wong SM, Oshima Y, Pezzuto J, Fong H, Farnsworth N (1986) Plant anticancer agents XXXIX. Triterpenes from *Isis missouriensis* (Iridaceae). J Pharm Sci 75:317–320
57. Bishayee A, Ahmed S, Brankov N, Perloff M (2011) Triterpenoids as potential agents for the chemoprevention and therapy of breast cancer. Front Biosci 16:980–996
58. Setzer WN, Setzer MC (2003) Plant-derived triterpenoids as potential antineoplastic agents. Mini Rev Med Chem 3:540–556
59. Guo DX, Du Y, Wang YY, Sun LM, Qu JB, Wang XN et al (2009) Secondary metabolites from the liverwort *Ptilidium pulcherrimum*. Nat Prod Commun 4:1319–1322
60. Grammes C, Burkhardt G, Becker H (1994) Triterpenes from *Fossombronia* liverworts. Phytochemistry 35:1293–1296
61. Toyota M, Asakawa Y (1993) Sesqui- and triterpenoids of the liverwort *Conocephalum japonicum*. Phytochemistry 32:1235–1237
62. Benes I, Vaněk T, Buděšínský M, Herout V (1981) A triterpenoid of the serratane type from the liverwort *Nardia scalaris*. Phytochemistry 20:2591–2252
63. Flegel M, Becker H (1999) Di- and triterpenoids from the liverwort *Blepharidophyllum densifolium*. Z Naturforsch C 54:481–487
64. Chadwick M, Trewin H, Gawthrop F, Wagstaff C (2013) Sesquiterpenoids lactones: benefits to plants and people. Int J Molecular Sci 14(6):12780–12805
65. Abu-Izneid T, Rauf A, Shariati MA, Khalil AA, Imran M, Rebezov M, Uddin MS, Mahomoodally MF, Rengasamy K (2020) Sesquiterpenes and their derivatives-natural anticancer compounds: an update. Pharma Res 161:105165
66. Asakawa Y (2008) Liverworts-potential source of medicinal compounds. Curr Pharm Des 14: 3067–3088
67. Tan CY, Inagaki M, Chai HB, Lambrechts MK, Onder A, Kiremit HO, Rakotondraibe LH (2017) Phytochemical and cytotoxic investigations of pinguisanoids from liverwort *Porella cordaeana*. Phytochem Lett 19:77–82. https://doi.org/10.1016/j.phytol.2016.12.003
68. Ohta Y, Andersen NH, Liu CB (1977) Sesquiterpene constituents of two liverworts of genus *Diplophyllum*: novel eudesmanolides and cytotoxicity studies for enantiomeric methylene lactones. Tetrahedron 33(6):617–628
69. Asakawa Y (1981) Biologically active substances obtained from bryophytes. J Hattori Bot Lab 50:123–142

70. Asakawa Y, Toyota M, Taira Z, Takemoto T (1982) Biologically active cyclic bisbenzyls and terpenoids isolated from liverworts. In: Proceeding of the 25th symposium on chemistry of natural products. Tokyo, p. 337–344
71. Matsuo A, Atsumi K, Nadaya K, Nakayama M, Hayashi S (1981) 13C NMR chemical shifts of ovalifoliene and related compounds with 2,3-seco-alloaromadendrane skeleton: structure of (+)-9alphaacetoxyovalifoliene, a plant growth inhibitor. Phytochemistry 20(5):1065–1068
72. Matsuo A, Atsumi K, Nakayama M, Hayashi S (1981) Structure of ent-2,3-secoalloaromadendrane sesquiterpenoids having plant growth inhibitory activity from *Plagiochila semidecurrens* (liverwort). J Chem Soc Perkin Trans 1:2816–2824
73. Matsuo A, Nadaya K, Nakayama M, Hayashi S (1981) Plant growth inhibitors isolated from the liverwort, *Plagiochila ovalifolia*. Nippon Kagaku Kaishi 1981(5):665–670
74. Matsuo A, Kubota N, Nakayama M, Hayashi S (1981) (−)-Lepidozenal, a sesquiterpenoid with a novel trans-fused bicycle [8.1.0] undecane system from the liverwort *Lepidozia vitrea*. Chem Lett 10(8):1097–1000
75. Matsuo A, Nozaki A, Kubota N, Uto S, Nakayama M (1984) Structures and conformation of (−)-isobicyclogermacrenal and 9(−)-lepidozenal, two key sesquiterpenoids of the cis- and trans-10,3-bicyclic ring system, from the liverwort *Lepidozia vitrea*. J Chem Soc Perkin Trans 1:203–214
76. Gunatilakaa AAL, Kingston DGI (1997) DNA-damaging natural products with potential anticancer activity. Stud Nat Prod Chem 20(Part F):457–505
77. Wu C, Gunatilaka AA, McCabe FL, Johnson RK, Spjut RW, Kingston DG (1997) Bioactive and other sesquiterpenes from *Chiloscyphus rivularis*. J Nat Prod 60(12):1281–1286
78. Toyota M, Tanimura K, Asakawa Y (1998) Cytotoxic 2,3-secoaromadendrane-type sesquiterpenoids from the liverwort *Plagiochila ovalifolia*. Planta Med 64(5):462–464
79. Burgess EJ, Larsen L, Perry NB (2000) A cytotoxic sesquiterpene caffeate from the liverwort *Bazzania novae-zelandiae*. J Nat Prod 63(4):537–539
80. Scher JM, Burgess EJ, Lorimer SD, Perry NB (2002) Cytotoxic sesquiterpene and unprecedented sesquiterpene-bisbibenzyl compounds from the liverwort *Schistochila glaucescens*. Tetrahedron 58(39):7875–7882
81. Komala I, Ito T, Nagashima F, Yagi Y, Asakawa Y (2010) Cytotoxic, radical scavenging and antimicrobial activities of sesquiterpenoids from the Tahitian liverwort *Mastigophora diclados* (Brid.) Nees (Mastigophoraceae). J Nat Med 64(4):417–422
82. Komala I, Ito T, Nagashima F, Yagi Y, Asakawa Y (2011) Cytotoxic bibenzyls, and germacrane- and pinguisane-type sesquiterpenoids from Indonesian, Tahitian and Japanese liverworts. Nat Prod Commun 6(3):303–309
83. Qu J, Xie C, Guo H, Yu W, Lou H (2007) Antifungal dibenzofuran bis (bibenzyl)s from the liverwort *Asterella angusta*. Phytochemistry 68:1767–1774
84. Martini U, Zapp J, Becker H (1998) Chlorinated macrocyclic bisbibenzyls from the liverwort *Bazzania trilobata*. Phytochemistry 47:89–96
85. Yoshida T, Hashimoto T, Takaoka S, Kan Y, Tori M, Asakawa Y et al (1996) Phenolic constituents of the liverwort: four novel cyclic bisbibenzyl dimers from *Blasia pusilla* L. Tetrahedron Lett 52:14487–14500
86. Toyota M, Yoshida T, Kan Y, Takaoka S, Asakawa Y (1996) (+)-Cavicularin: a novel optically active cyclic bibenzyldihydrophenanthrene derivative from the liverwort *Cavicularia densa* Steph. Tetrahedron Lett 37:4745–4748
87. Guo DX, Xiang F, Wang XN, Yuan HQ, Xi GM, Wang YY et al (2010) Labdane diterpenoids and highly methoxylated bibenzyls from the liverwort *Frullania inouei*. Phytochemistry 71: 1573–1578
88. Toyota M, Asakawa Y (1993) Bibenzyl and sesquiterpenoids from the liverwort *Jubula japonica*. Phytochemistry 34(4):1135–1137
89. Scher JM, Zapp J, Schmidt A, Becker H, Bazzanins L-R (2003) Chlorinated macrocyclic bisbibenzyls from the liverwort *Lepidozia incurvata*. Phytochemistry 64:791–796

90. Ivković I, Novaković M, Veljić M, Mojsin M, Stevanović M, Marin PD, Bukvički D (2021) Bis-Bibenzyls from the liverwort *Pellia endiviifolia* and their biological activity. Plan Theory 10:1063
91. So ML, Chan WH, Xia PF, Cui Y (2002) Two new cyclic bis(bibenzyl)s, isoriccardinquinone A and B from the liverwort *Marchantia paleacea*. Nat Prod Lett 16(3):167–171
92. Huang WJ, Wu CL, Lin CW, Chi LL, Chen PY, Chiu CJ et al (2010) Marchantin A, a cyclic bis (bibenzyl ether), isolated from the liverwort *Marchantia emarginata* subsp. *tosana* induces apoptosis in humanMCF-7 breast cancer cells. Cancer Lett 291:108–119
93. Schwartner C, Bors W, Michel C, Franck U, Müller-Jakic B, Nenninger A et al (1995) Effect of marchantins and related compounds on 5-lipoxygenase and cyclooxygenase and their antioxidant properties: a structure activity relationship study. Phytomedicine 2:113–117
94. Jensen S, Omarsdottir S, Bwalya AG, Nielsen MA, Tasdemir D, Olafsdottir ES (2012) Marchantin A, a macrocyclic bisbibenzyl ether, isolated from the liverwort *Marchantia polymorpha*, inhibits protozoal growth *in vitro*. Phytomedicine 19:1191–1195
95. Toyota M, Omatsu I, Braggins J, Asakawa Y (2011) Novel prenylbibenzyls from the New Zealand liverwort *Marsupidium epiphytum*. Chem Pharm Bull 59:480–483
96. Shi YQ, Qu XJ, Liao YX, Xie CF, Cheng YN, Li S et al (2008) Reversal effect of a macrocyclic bisbibenzylplagiochin E on multidrug resistance in adriamycin-resistant K562/A02 cells. Eur J Pharmacol 584:66–71
97. Anton H, Kraut L, Mues R, Maria IZM (1997) Phenanthrenes and bibenzyls from a *Plagiochila* species. Phytochemistry 46(6):1069–1075
98. Anton H, Schoeneborn R, Mues R (1999) Bibenzyls and bisbibenzyls from a neotropical *Plagiochila* species. Phytochemistry 52(8):1639–1645
99. Lorimer SD, Perry NB (1994) Antifungal hydroxy-acetophenones from the New Zealand liverwort, *Plagiochila fasciculata*. Planta Med 60:386–387
100. Hashimoto T, Kanayama S, Kan Y, Tori M, Asakawa Y (1996) Isoplagiochins C and D, new type of macrocyclic bis(bibenzyls), having two biphenyl linkages from the liverwort *Plagiochila fruticosa*. Chem Lett 25(9):741–742
101. Asakawa Y (2016) Polyphenols in bryophytes: structures, biological activities, and bio-and total syntheses. Recent Adv Polyphenol Res 5:36–66
102. Xie CF, Qu JB, Wu XZ, Liu N, Ji M, Lou HX (2010) Antifungal macrocyclic bis(bibenzyls) from the Chinese liverwort *Plagiochasma intermedium* L. Nat Prod Res 24:515–520
103. Hashimoto T, Ikeda H, Takaoka S, Tanaka M, Asakawa Y (1999) Ptychantols A-C, macrocyclic bis(bibenzyls), possessing a transstilbene structure from the liverwort *Ptychantus striatus*. Phytochemistry 52(3):501–509
104. Toyota M, Shimamura T, Ishii H, Renner M, Braggins J, Asakawa Y (2002) New bibenzyl cannabinoid from the New Zealand liverwort *Radula marginata*. Chem Pharm Bull 50: 1390–1392
105. Asakawa Y (2008) Liverworts – potential source of medicinal compounds. Curr Pharm Design 14:3067–3088
106. Xu AH, Hu ZM, Qu JB, Liu SM, Syed AKA, Yuan HQ et al (2010) Cyclic bisbibenzyls induce growth arrest and apoptosis of human prostate cancer PC3 cells. Acta Pharmacol Sin 31:609
107. Kunz S, Becker H (1994) Bibenzyl derivatives from the liverwort *Ricciocarpos natans*. Phytochemistry 36:675–677
108. Asakawa Y, Toyota M, Taira Z, Takemoto T, Kido M (1983) Riccardin A and riccardin B, two novel cyclic bibenzyls possessing cytotoxicity from the liverwort *Riccardia multifida* (L.) S. Gray. J Org Chem 48:2164–2167
109. Yoshida T, Toyota M, Asakawa Y (1997) Isolation, structure elucidation, and chemical derivatization of a new cyclic bisbibenzyl dimer, pusilatin E, from the liverwort *Riccardia multifida* subsp *decrescens*. J Nat Prod 60:145–147
110. Mekem Sonwa M, Konig WA (2003) Chemical constituents of the essential oil of the hornwort *Anthoceros caucasicus*. Flavour Frag J 18:286–289

111. Xiong W, Fu J, Kollner TG, Chen X, Jia Q, Guo H, Qian P, Guo H, Wu G, Chen F (2018) Biochemical characterization of microbial type terpene synthases in two closely related species of hornworts, *Anthoceros punctatus* and *Anthoceros agrestis*. Phytochemistry 149:116–122
112. Trennheuser F, Burkhardt G, Becker H (1994) Anthocerodiazonin, an alkaloid from *Anthoceros agrestis*. Phytochemistry 37:899–903
113. Mukhtar YM, Adu-Frimpong M, Xu X, Yu J (2018) Biochemical significance of limonene and its metabolites: future prospects for designing and developing highly potent anticancer drugs. Biosci Rep 38(6):BSR20181253
114. Anwar S, Shamsi A, Shahbaaz M et al (2020) Rosmarinic acid exhibits anticancer effects via MARK4 inhibition. Sci Rep 10:10300. https://doi.org/10.1038/s41598-020-65648-z
115. Commisso M, Guarino F, Marchi L, Muto A, Piro A, Degola F (2021) Bryo-activities: a review on how bryophytes are contributing to the arsenal of natural bioactive compounds against fungi. Plan Theory 10:203. https://doi.org/10.3390/plants10020203
116. Zheng GQ, Chang CJ, Stout TJ, Clardy J, Cassady JM (1989) Ohioensin-A: a novel benzonaphthoxanthenone from *Polytrichum ohioense*. J Am Chem Soc 111:5500–5501
117. Cassady JM, Baird WM, Chang CJ (1990) Natural products as a source of potential cancer chemotherapeutic and chemopreventive agents. J Nat Prod 53:23–41
118. Fu P, Lin S, Shan L, Lu M, Shen YH, Tang J et al (2009) Constituents of the moss *Polytrichum commune*. J NatProd 72(7):1335–1337
119. Zheng GQ, Ho DK, Elder PJ, Stephens RE, Cottrell CE, Cassady JM (1994) Ohioensins and pallidisetins: novel cytotoxic agents from the moss *Polytrichum pallidisetum*. J Nat Prod 57: 32–41
120. Belkin M, Fitzgerald DB, Felix MD (1952) Tumor-damaging capacity of plant materials. II. Plants used as diuretics. J Natl Canc Inst 13:741–744
121. Pereira BK, Rosa RM, da Silva J, Guecheva TN, Oliveira IM, Ianistcki M et al (2009) Protective effects of three extracts from Antarctic plants against ultraviolet radiation in several biological models. J Photochem Photobiol B 96:117–129
122. Suwanborirux K, Chang CJ, Spjut RW, Cassady JM (1990) Ansamitocin P-3, a maytansinoid, from *Claopodium crispifolium* and *Anomodon attenuatus* or associated actinomycetes. Exp Dermatol 46:117–120
123. Spjut RW, Cassady JM, McCloud T, Norris DH, Suffness M, Cragg GM et al (1988) Variation in cytotoxicity and antitumor activity among samples of a moss, *Claopodium crispifolium* (Hook.) Ren. & Card. (Thuidiaceae). Econ Bot 42:62–72
124. Sakai K, Ichikawa T, Yamada K, Yamashita M, Tanimoto M, Hikita A et al (1988) Antitumor principles in mosses: the first isolation and identification of maytansinoids, including a novel 15- methoxyansamitocin P-3. J Nat Prod 51:845–850
125. Klegin C, de Moura NF, Oliveira de Sousa MH, Frassini R, Roesch-Ely M, Bruno AN, Bitencourt TC, Flach A, Bordin J (2021) Chemical composition and cytotoxic evaluation of the essential oil of *Phyllogonium viride* (Phyllogoniaceae, Bryophyta). Chem Biodivers 18: e2000794
126. Özerkan D, Erol A, Altuner EM, Canlı K, Kuruca DS (2021) Some bryophytes trigger cytotoxicity of stem cell-like population in 5-fluorouracil resistant colon cancer cells. Nutr Cancer:1–11. https://doi.org/10.1080/01635581.2021.1933098
127. Yoshida K, Cheynier V, Quideau S (2016) Recent advances in polyphenol research. Wiley, Chichester. https://doi.org/10.1002/9781118883303
128. Kodama M, Shiobara Y, Sumitomo H, Matsumura K, Tsukamoto M, Harada C et al (1988) Total syntheses of marchantin A and riccardin B, cytotoxic bis(bibenzyls) from liverworts. J Org Chem 53:2–77
129. Shi YQ, Liao YX, Qu XJ, Yuan HQ, Li S, Qu JB et al (2008) Marchantin C, a macrocyclic bisbibenzyl, induces apoptosis of human glioma A172 cells. Cancer Lett 262:173–182
130. Shi YQ, Zhu CJ, Yuan HQ, Li BQ, Gao J, Qu XJ et al (2009) Marchantin C, a novel microtubule inhibitor from liverwort with anti-tumor activity both in vivo and in vitro. Cancer Lett 276:160–170

131. Gao J, Li X, Lv BB, Sun B, Zhu CJ, Lou HX et al (2009) LC-DAD/MS/MS detection of Macrocyclic Bisbibenzyls from the liverwort *Reboulia hemisphaerica* and the cell-based screening of their microtubule inhibitory effects. Chin J Nat Med 7:123–128
132. Shen J, Li G, Liu Q, He Q, Gu J, Shi Y et al (2010) Marchantin C: a potential anti-invasion agent in glioma cells. Cancer Biol Ther 9:33–39
133. Xi GM, Sun B, Jiang HH, Kong F, Yuan HQ, Lou HX (2010) Bisbibenzyl derivatives sensitize vincristine-resistant KB/VCR cells to chemotherapeutic agents by retarding P-gp activity. Bioorg Med Chem 18:6725–6733
134. Lv Y, Song Q, Shao Q et al (2012) Comparison of the effects of marchantin C and fucoidan on sFlt-1 and angiogenesis in glioma microenvironment. J Pharm Pharmacol 64:604–609
135. Wu XZ, Chang WQ, Cheng AX, Sun LM, Lou HX (2010) Plagiochin E, an antifungal active macrocyclic bis (bibenzyl), induced apoptosis in *Candida albicans* through a metacaspase-dependent apoptotic pathway. Biochim Biophys Acta Gen Subj 1800:439–447
136. Morita H, Tomizawa Y, Tsuchiya T, Hirasawa Y, Hashimoto T, Asakawa Y (2009) Antimitotic activity of two macrocyclic bis(bibenzyls), isoplagiochins A and B from the liverwort *Plagiochila fruticosa*. Bioorg Med Chem Lett 19:493–496
137. Sun CC, Zhang YS, Xue X, Cheng YN, Liu HP, Zhao CR et al (2011) Inhibition of angiogenesis involves in anticancer activity of riccardin D, a macrocyclic bisbibenzyl, in human lung carcinoma. Eur J Pharmacol 667:136–143
138. Xue X, Qu XJ, Gao ZH, Sun CC, Liu HP, Zhao CR et al (2012) Riccardin D, a novel macrocyclic bisbibenzyl, induces apoptosis of human leukemia cells by targeting DNA topoisomerase II. Investig New Drugs 30:212–222
139. Li Y, Ma Y, Zhang L, Guo F, Ren L, Yang R et al (2012) *In vivo* inhibitory effect on the biofilm formation of *Candida albicans* by liverwort derived riccardin D. PLoS One 7:e35543
140. Liu HP, Gao ZH, Cui SX, Sun DF, Wang Y, Zhao CR et al (2012) Inhibition of intestinal adenoma formation in APCmin/+ mice by riccardin D, a natural product derived from liverwort plant *Dumortiera hirsuta*. PLoS One 7:e33243
141. Ji M, Shi Y, Lou H (2011) Overcoming of P-glycoprotein-mediated multidrug resistance in K562/A02 cells using riccardin F and pakyonol, bisbibenzyl derivatives from liverworts. Biosci Trends 5:192–197
142. Li X, Sun B, Zhu CJ, Yuan HQ, Shi YQ, Gao J et al (2009) Reversal of p-glycoprotein-mediated multidrug resistance by macrocyclic bisbibenzyl derivatives in adriamycin-resistant human myelogenous leukemia (K562/A02) cells. Toxicol In Vitro 23(1):29–36
143. Sun B, Yuan HQ, Xi GM, Ma YD, Lou HX (2009) Synthesis and multidrug resistance reversal activity of dihydroptychantol A and its novel derivatives. Bioorg Med Chem 17:4981–4989
144. Li X, Wu WK, Sun B, Cui M, Liu S, Gao J et al (2011) Dihydroptychantol A, a macrocyclic bisbibenzyl derivative, induces autophagy and following apoptosis associated with p53 pathway in human osteosarcoma U2OS cells. Toxicol Appl Pharmacol 251:146–154
145. Roldos V, Nakayama H, Rolón M, Montero-Torres A, Trucco F, Torres S et al (2008) Activity of a hydroxybibenzyl bryophyte constituent against *Leishmania* spp. and *Trypanosoma cruzi*: in silico, in vitro and in vivo activity studies. Eur J Med Chem 43:1797–1807
146. Fan S, Zhu R, Zhou J, Li Y, Qiao Y, Zhang C, Zhang J, Gao Y, Chen W, Lou H (2019) Prenyl bibenzyls isolated from Chinese liverwort *Radula amoena* and their cytotoxic activities. Phytochem Lett 31:53–57. https://doi.org/10.1016/j.phytol.2019.03.006
147. Zhang CY, Gao Y, Zhu RX, Qiao YN, Zhou JC, Zhang JZ, Li Y, Li SW, Fan SH, Lou HX (2019) Prenylated bibenzyls from the chinese liverwort *Radula constricta* and their mitochondria-derived paraptotic cytotoxic activities. J Nat Prod 82:1741–1751. https://doi.org/10.1021/acs.jnatprod.8b00897

Immunomodulatory Potential of *Hedwigia ciliata* and *Hypnum cupressiforme*

5

Tanja Lunić, Bojan Božić, and Biljana Božić Nedeljković

Contents

Abstract

Hedwigia ciliata (Hedw.) P. Beauv. and *Hypnum cupressiforme* Hedw. are two widespread moss representatives, which belong to the second largest group of plants in terms of species number – bryophytes. Despite their abundance and well-known usage in traditional medicine, these mosses have been overlooked for a long time when it comes to their biological activities and potential application. However, with the identification of novel, highly interesting, and diverse secondary metabolites in different extracts of *H. ciliata* and *H. cupressiforme*, the perception of these species has rather changed. Their extracts contain a plethora of polyphenols, flavonoids, and terpenoids, plenty of which have shown interesting immunomodulatory activities, such as antioxidant, antitumor, antiproliferative, anti-inflammatory, neuroprotective, antibacterial, and antifungal. The exact mechanisms by which moss extracts exert their effects are still being investigated and the research interest on this topic is continuously growing. The present chapter provides the first detailed overview of the research related to the

T. Lunić · B. Božić · B. B. Nedeljković (✉)
Institute of Physiology and Biochemistry "Ivan Djaja", Faculty of Biology, University of Belgrade, Belgrade, Serbia
e-mail: tanja.lunic@bio.bg.ac.rs; bbozic@bio.bg.ac.rs; biljana@bio.bg.ac.rs

H. N. Murthy (ed.), *Bioactive Compounds in Bryophytes and Pteridophytes*, Reference Series in Phytochemistry, https://doi.org/10.1007/978-3-031-23243-5_5

chemical composition and biological activities of mosses *H. ciliata* and *H. cupressiforme*, putting together future perspectives and strategies for the improvement of knowledge about bryophytes and their biological potential. In the following decades, species, such as *H. ciliata* and *H. cupressiforme*, and mosses in general, will certainly represent attractive candidates and sources for the production of new, sustainable, and bryophyte-derived chemicals for diverse therapeutic purposes.

Keywords

Antibacterial activity · Antifungal activity · Anti-inflammatory activities · Antioxidant activity · Antitumor activity · Bryophytes · Immunomodulation

Abbreviations

ABTS	2,2′-azinobis-3-ethylbenzthiazoline-6-sulfonic acid
AChE	Acetylcholinesterase
AD	Alzheimer's Disease
Aβ	Amyloid Beta
CNS	Central Nervous System
COX-2	Cyclooxygenase 2
DNA	Deoxyribonucleic Acid
DPPH	2,2-diphenyl-1-picrylhydrazyl
EGFR	Epidermal Growth Factor Receptor
FRAP	Ferric Reducing Antioxidant Power
HO-1	Heme Oxygenase 1
IL-1β	Interleukin 1β
IL-6	Interleukin 6
iNOS	inducible Nitric Oxide Synthase
JNK	c-Jun N-terminal kinase
LDL	Low-Density Lipoprotein
LPS	Lipopolysaccharide
MAPK	Mitogen-Activated Protein Kinase
MBC	Minimal Bactericidal Concentration
MFC	Minimal Fungicidal Concentration
MIC	Minimal Inhibitory Concentration
MMP	Matrix MetalloPeptidase
NADH	Nicotinamide Adenine Dinucleotide Hydrogen
NF-κB	Nuclear Factor Kappa B
NMR	Nuclear Magnetic Resonance
NQO1	NADH Quinone Oxidoreductase 1
NRF2	Nuclear factor erythroid 2-Related Factor 2
ORAC	Oxygen Radical Absorbance Capacity
PD	Parkinson's Disease
RB	Retinoblastoma
ROS	Reactive Oxygen Species

SOD	Superoxide Dismutase
TNF-α	Tumor Necrosis Factor-Alpha
Tyr	Tyrosinase
UV	Ultraviolet

1 Introduction

The immune system has evolved with the aim to maintain homeostasis in the body, thus protecting it from a wide range of pathogens from the outside as well as from the invasion of altered cells from the inside. The immune system is particularly important in tissue repair, a process that is significant for the overall robustness of an organism (the capability of the body to maintain its functions and performances despite deviations from homeostasis). Due to such demands, the immune system evolved into a very complex and sophisticated structure. Therefore, the immune system is a dynamic, integrative, multicomponent, plastic, and well-functioning system that is responsible for the maintenance of homeostasis in the body. In carrying out its functions (fight against pathogens, eliminate damaged and aberrant cells, and tissue repair), the immune system continuously modulates its mechanisms, either to activate them when the function is activated or to suppress them after the function has been performed (Fig. 1). Unfortunately, in some cases, the immune response is extremely pronounced and persistent, which may lead to the development of autoimmune diseases and hypersensitivity reactions. On the other hand, if

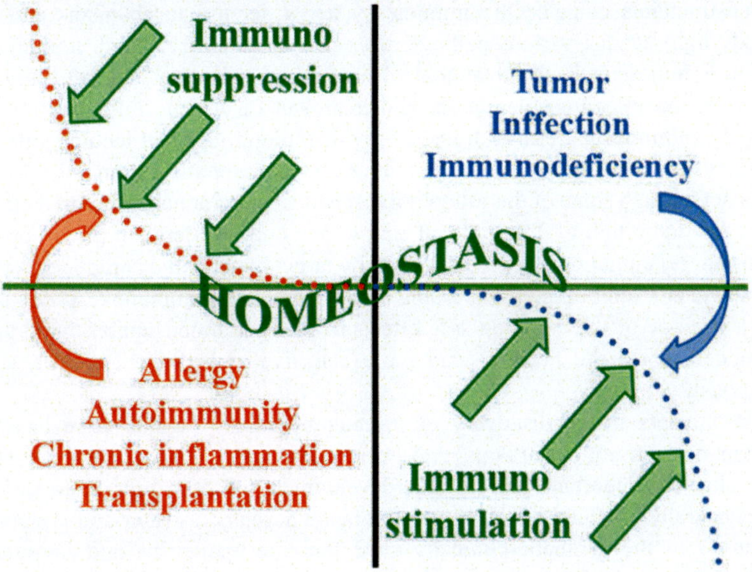

Fig. 1 Schematic presentation of immunostimulation and immunosuppression and their influence on immune homeostasis

the immune response is not strong enough, it might lead to the development of tumors. Therefore, the ability to modulate the immune response by external agents is crucial in a wide array of immunopathologies, including the prevention of infection, treatment of tumors, and the suppression of autoimmune diseases/hypersensitivity reactions. This process of immune response modulation is called immunomodulation, and in terms of immunotherapy, immunomodulation is referred to as an intervention in which the immune response is altered to the desired level (either stimulated or suppressed) [1]. Substances that have an influence on immune functions are called immunomodulators. Based on their effects on the immune system, immunomodulators are usually classified into two categories: immunostimulants and immunosuppressants [2]. These molecules represent a diverse group of recombinant, synthetic, semi-synthetic, and natural compounds. The interest in immunomodulators has increased significantly over the past few decades due to a wide range of their applications, either for stimulation or for suppression of the immune system. Immunomodulators are even being used as prophylactic agents to maintain immune homeostasis. Besides, the modification of the immune response by various pharmacological agents has been demonstrated as an efficient therapeutic strategy for many disorders. For example, the stimulation of the immune response is highly desirable in conditions, such as different tumors, infections, or immunodeficiency [1, 3]. Immunostimulants usually induce nonspecific activation of the immune system, unless they are associated with antigens (like adjuvants in vaccines). They activate different effectors' mechanisms of the immune response, including phagocytosis and intracellular killing of organisms, antigen presentation, cytotoxic and antiviral activity, cytokine release, and antibody production [4]. Therefore, immunostimulants enhance the immune system's defense mechanisms and help the body fight against various pathogens/tumors. On the other hand, immunosuppression is important in the treatment of autoimmune diseases, the prevention of organ rejection after transplantation, and treatment of chronic inflammatory processes [3]. Immunosuppressive drugs are divided into groups that include glucocorticoids, cytostatics, antibodies, drugs acting on immunophilins, and others. They usually act through some of the following mechanisms: immunodepletion of effector cells and/or inhibition of T and B cell stimulation and proliferation [5]. In optimal circumstances, immunosuppressants primarily target hyperactive components of the immune system, inhibiting or decreasing the intensity of the immune response in the body. Since one of the common side effects of many immunosuppressive drugs is immunodeficiency, the need to find more selective and efficient alternatives is continuously growing.

First attempts in the discovery of immunomodulatory agents were based on investigation of traditional plants, and even today compounds of natural (herbal) origin play an important role in the development of new immunomodulators [6, 7]. Naturally obtained immunomodulators are accepted as safer and sustainable alternatives to the synthetic-clinically used immunosuppressive and immunostimulatory drugs, which usually possess severe side effects. Among other plants, bryophytes and their extracts show significant immunomodulatory potential, which is why they have been increasingly investigated in the recent period [8–16] and represent a perspective field for future investigations.

Bryophytes represent a group of terrestrial plants that are taxonomically placed between the algae and pteridophytes and are divided into three classes: *liverworts*, *hornworts*, and *mosses* [17]. Although bryophytes are among the oldest land plants, their role and significance as sources of immunomodulatory molecules have been relatively unknown and rarely explored. However, nowadays this trend is changing and bryophytes are being utilized for different purposes ranging from the pharmaceutical industry, horticulture, and household uses to ecology [17]. Representatives of bryophytes have been used in traditional medicine around the world, especially in China, Europe, North America, and India [7]. Their antimicrobial activity was well known and they were used to treat various pathological conditions, especially those caused by bacterial infections. They have also been used traditionally to treat cardiovascular diseases, bronchitis, skin infections, wounds and burns, different types of tumors, and other conditions associated with inflammation [7, 17]. Significant immunomodulatory effects of bryophyte extracts have been related to their rich content of various phytochemicals, such as polyphenols, flavonoids, terpenoids, and carbohydrates – which have been previously reported [8, 9, 11, 18]. These constituents act through diverse mechanisms of the immune system modulation, on multiple molecular targets (see later Sect. 3, 'Moss Extract Biological Activities'). Although chemical profiles of most bryophyte species are still relatively unknown, the development of analytical techniques has facilitated these analyses and led to growing interest in bryophyte chemistry, revealing the great potential of these plants for immunomodulation.

The present chapter is focused on the immunomodulatory potential of two widespread moss species – *Hedwigia ciliata* (Hedw.) P. Beauv. and *Hypnum cupressiforme* Hedw. These mosses belong to the important subclass of Bryopsida called Bryidae, which constitute the vast majority of all moss species [19]. Mosses *H. ciliata* and *H. cupressiforme* can be found on nearly all continents, in a wide range of habitats and climate zones [20, 21]. Despite the prevalence and well-known traditional usage of both species, studies regarding their chemical composition and biological activities are scarce, while studies about their biological activities usually involve the analysis of the antioxidant, antimicrobial, or antitumor potential of their extracts [8, 9, 22]. Based on all previously presented, the present paper was prepared to review the potential of *H. ciliata* and *H. cupressiforme* extracts as well as their bioactive metabolites to modulate the immune response in different conditions.

2 Moss Chemical Composition

Although it is estimated that there are over 14,000 species of mosses around the world, only a small number of these species have been chemically analyzed, due to several following reasons [11]. Namely, mosses are morphologically quite small, which makes them difficult to collect in large quantities. Additionally, the identification of mosses is very challenging, even under the microscope, making their collection harder in comparison with higher plants. Despite these limitations, with the development of analytical methods (NMR, liquid chromatography, mass spectrometry, etc.) the interest in moss chemistry has increased.

As one of the first terrestrial plants, mosses were often exposed to unfavorable environmental conditions, such as pathogens, insects, animal attacks, drought, and UV radiation. Since they do not possess any mechanical protection like higher plants, mosses had to develop various chemical and biological mechanisms as part of their survival strategy, including the production of diverse secondary metabolites. While plant primary metabolites are directly involved in their growth and metabolism, secondary metabolites are generally synthesized through numerous biosynthetic pathways in plants with the main aim to protect them against biotic or abiotic stresses. A plethora of plant secondary metabolites exhibit promising biological activities and have the potential to be used in drug discovery and immunotherapy. In addition to being biologically active themselves, secondary metabolites may also serve as the starting components for the synthesis of novel biologically active compounds. Common moss secondary metabolites include polyphenols, phenolic acids, flavonoids, different alkaloids, saponins, terpenes, lipids, and carbohydrates [13].

Phenolic compounds together with terpenes represent some of the most interesting and well-studied secondary metabolite groups in mosses and generally in plants. Phenolic compounds are widespread throughout the plant world and show an immense structural diversity (from simple phenolic acids to different polymeric structures). They are characterized by one or more aromatic rings with one or more hydroxyl groups (polyphenols). They mainly exist in plant tissue as conjugates of mono- or polysaccharides or as esters [23]. Terpenes, also known as terpenoids, represent a wide group of natural compounds with great structural diversity and in plants are often present in a glycosylated form (saponins). Depending on the number of isoprene units, terpenes can be divided into mono, di, tri, tetra, and sesquiterpenes [24].

Chemical compositions of two moss species, *H. cupressiforme* and *H. ciliata*, which are the main focus of the present review, have been examined in a couple of studies [8, 9, 15, 25–29]. In a study that examined the chemical composition of different moss species, a qualitative analysis has been performed in extracts of *H. cupressiforme*, confirming the presence of certain classes of compounds (anthraquinones, terpenoids, flavonoids, alkaloids) [30]. Total contents of certain classes of secondary metabolites (phenols, phenolic acids, flavonoids, flavonols, terpenes) were also determined in the extracts of *H. cupressiforme* and *H. ciliata* [8, 9, 28]. A study has revealed the presence of different mono-, sesqui-, diterpenes, aldehydes, and hydrocarbons (29 compounds in total) in the essential oil of moss *H. cupressiforme* [18], while from gametophytes of the *H. cupressiforme* two biflavonoids (hypnogenol B1 and hypnum biflavonoid A), two phenyl-substituted aromadendrin derivatives (hypnum acid and hypnum acid methylester), and kaempferol have been isolated [29]. In another study, the chemical composition of *H. ciliata* ethanolic extract was found to be as follows: acetic acid, triethoxymethylsilane, tetraethyl silicate, 1,1-diethoxypentane, hepta-2,4-dienal,

R-limonene, 1,2,4,5-tetramethylbenzene, 1-ethyl-2,3-dimethylbenzene, benzoic acid, ethyl ester, 4-methyl benzoic acid, hexose, maltol, orcinaldehide, diethyltoluamide, tetradecanoic acid (myristic acid), 2-phenylmethyl-1,3-cyclohexanedione, dihydrophytol, *γ*-palmitolactone, farnesol, *n*-hexadecanoic acid, hexadecanoic acid, manoyl oxide, doconexent, kaurene, γ-stearolactone, transphytol, linoleic acid, octadecanoic acid, linoleic acid ethyl ester, monopalmitin, monostearin, squalene, *α*-tocopherol, campesterol, stigmasterol, obtusifoliol, *γ*-sitosterol, isofucosterol, cycloartenol, vitamin-E acetate, sitostenone, dihydroxyacetone (oxetane), delta-3-carene, phytol, and stearic acid [15]. Additionally, six flavonoid *O*-glycosides and one *C'*-glycoside have been identified in the extract of *H. ciliata* [26], as well as luteolin tetraglycoside7-*O*-neohesperidoside-4'-*O*-sophoroside [25].

Moreover, two recent studies have revealed the presence of 14 phenolic acids and flavonoids in the ethanolic, ethyl acetate, water/ethanolic, and pure water extracts of *H. cupressiforme* and *H. ciliata*, providing detailed in vitro biological evaluation of the mentioned extracts. This was the first time that a majority of the following compounds were identified in the extracts of *H. cupressiforme* and *H. ciliata*: gallic acid, protocatechuic acid, 5-*O*-caffeoylquinic acid, *p*-hydroxybenzoic acid, caffeic acid, quercetin 3-*O*-rutinoside, *p*-coumaric acid, quercetin 3-*O*-glucoside, isorhamnetin 3-*O*-glucoside, eriodictyol, apigenin, naringenin, kaempferol, and acacetin (Fig. 2). It is well known that the compounds identified in the extracts of

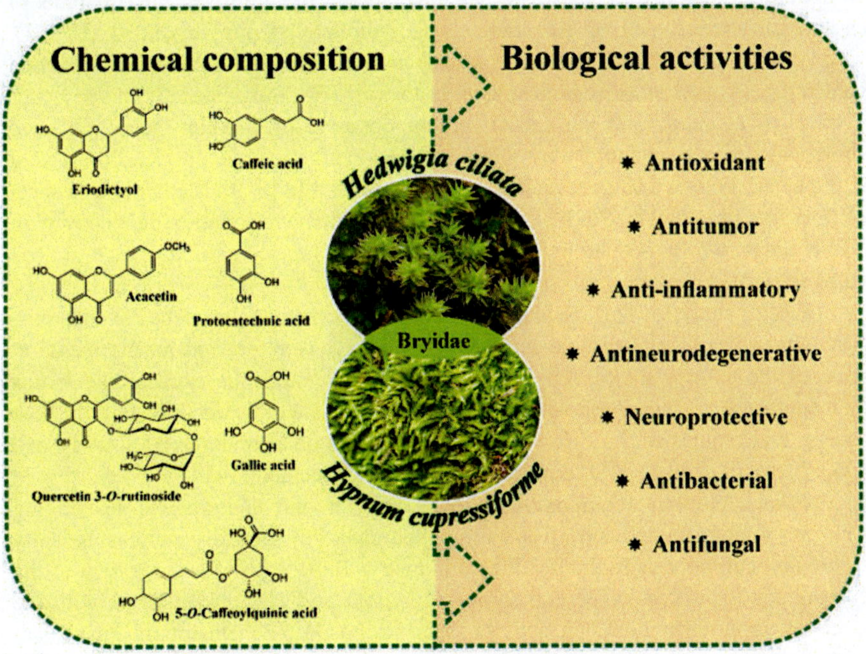

Fig. 2 Selected representatives of the secondary metabolites identified in different extracts of *H. ciliata* and *H. cupressiforme* and their biological activities are discussed in the following text

these two mosses exhibit significant biological activities, such as antitumor, antioxidant, anti-inflammatory, and antimicrobial. Therefore, each of these activities will be further depicted and discussed in the following text, with special reference to the immunomodulatory potential of the compounds identified in the corresponding extracts.

3 Moss Extracts Biological Activities

3.1 Antioxidant Activity

Oxidative stress – a phenomenon caused by an imbalance between the production of reactive oxygen species (ROS) and the antioxidant capacity of the organism – has been associated with the development of numerous chronic diseases, such as cardiovascular diseases, diabetes, neurodegenerative diseases, and cancer [8]. Since plants possess an innate ability to synthesize a wide range of antioxidants, they have been exploited for their ability to treat or prevent several human pathologies in which oxidative stress seems to be one of the causes. Among other plants, bryophytes produce a plethora of secondary metabolites (polyphenols, flavonoids, terpenes), which allows these plants to cope with biotic and abiotic stress [16].

Results from the in vitro assays for antioxidant activity of bryophytes and compounds isolated from these species have shown the ability of their extracts to capture different types of free radicals (2,2-diphenyl-1-picrylhydrazyl (DPPH), superoxide anion radical, hydroxyl radical) and the ability to inhibit beta-carotene bleaching, as well as to reduce iron (III) to iron (II). Ethyl acetate and/or water extracts of *H. cupressiforme* and *H. ciliata* obtained by Soxhlet extraction have shown an antioxidant activity comparable to the synthetic antioxidant ascorbic acid in the beta-carotene test, as revealed in two recent studies [8, 9]. Methanolic extracts of *H. ciliata* obtained by Soxhlet extraction have shown weak antioxidant activity in DPPH assay, while methanolic extracts of *H. cupressiforme* exhibited high antioxidant activity [31].

In another study, the ferric reducing antioxidant power (FRAP) and the 2,2′-azinobis-3-ethylbenzthiazoline-6-sulfonic acid ($ABTS^{+}$) assays were applied for the measurement of the antioxidant capacity of *H. cupressiforme* ethanolic extract. The results indicate that *H. cupressiforme* extract exhibited moderate antioxidant activity in both assays, with values of 7.59 ± 0.63 μM of trolox equivalent for $ABTS^{+\cdot}$ assay and 475.67 ± 6.38 μM of $FeSO_4$ equivalent in the FRAP method [14].

The large content of antioxidants in bryophytes and their radical scavenging activities qualify them as promising future sources of medicinally and cosmetically significant compounds in the healthcare industry. Table 1 gives an overview of the compounds identified in the extracts of *H. ciliata* and *H. cupressiforme* with the literature reported antioxidant activities, the proposed mechanisms of antioxidant protection, and suggested cellular targets.

Table 1 Reported antioxidant activities of the compounds identified in the extracts of *H. ciliata* and *H. cupressiforme*

Gallic acid
Exhibited 43.9% and 60% scavenging effects on DPPH and hydrogen peroxide (H_2O_2), respectively [32] At low concentrations managed to scavenge another reactive species, hypochlorous acid, HOCl [33] Restored the activities of antioxidant enzymes catalase and glutathione peroxidase, notably decreased lipid peroxidation and reduced level of malondialdehyde in the brain, kidney, and liver in mice [34]
Protocatechuic acid
Exhibited better antioxidant activity in vitro compared to trolox in both lipid and aqueous media, through chelating metal transition ions as well as through scavenging free radicals in DPPH, $ABTS^{+\cdot}$, reducing power (Fe^{3+}), reducing power (Cu^{2+}), superoxide anion radical-scavenging, hydroxyl radical-scavenging, chelating ability (Fe^{2+}), and chelating ability (Cu^{2+}) assays [35]
5-*O*-Caffeoylquinic acid
Exhibited antioxidant activity in DPPH (0.49 ± 0.01 μmol Trolox equivalents/μmol compound), $ABTS^{+\cdot}$ (0.58 ± 0.03 μmol Trolox equivalents/μmol compound), and ORAC assays (2.21 ± 0.04 μmol Trolox equivalents/μmol compound) [36]
p-Hydroxybenzoic acid
Increased antioxidant enzyme activities under heat stress reduced lipid peroxidation and enhanced heat tolerance of cucumber seedlings [37]
Caffeic acid
Exhibited 51.5% scavenging effects on DPPH radical [38] Exerted 68.2% (at concentration 10 μg mL^{-1}) and 75.8% (at concentration 30 μg mL^{-1}) lipid peroxidation inhibition, as well as potent $ABTS^{+\cdot}$ radical scavenging, DPPH radical scavenging, superoxide anion radical scavenging, total reducing power, and metal chelating activities [39]
Quercetin 3-*O*-rutinoside
Good antioxidant activity in the Trolox equivalent antioxidant capacity assay and an efficient inhibitor of lipid peroxidation [40]
p-Coumaric acid
At concentration 45 μg mL^{-1} inhibited 71.2% lipid peroxidation of linoleic acid emulsion, exhibited potent DPPH radical scavenging, $ABTS^{+\cdot}$ radical scavenging, superoxide anion radical scavenging, H_2O_2 scavenging, ferric ion (Fe^{3+}) reducing power, and ferrous ion (Fe^{2+}) chelating activities [41] Effectively scavenged hydroxyl radical, significantly inhibited low-density lipoprotein (LDL) oxidation when administered orally (317 mg per day) for 30 days, and also reduced LDL cholesterol levels in serum of Sprague-Dawley male rats [42]
Quercetin 3-*O*-glucoside
Showed high antioxidant activity in the DPPH assay with an RC_{50} value of 22 μg mL^{-1} [43]
Isorhamnetin 3-*O*-glucoside
Potent inhibitor of lipid peroxidation, good antioxidant activity assessed in vitro by DPPH and FRAP assays [44]
Eriodictyol
Reduced lipid peroxidation in isoproterenol-induced myocardial infarcted male Albino Wistar rats after 45 days of treatment [45] Reduced oxidative damage in human retinal pigment epithelial cells ARPE-19 via regulation of nuclear factor erythroid 2-related factor 2 (NRF2), heme oxygenase 1 (HO-1) activation, and has also increased the levels of intracellular glutathione [46]

(continued)

Table 1 (continued)

Apigenin
Reduced lipid peroxidation and protected antioxidant system in N-nitrosodiethylamine induced and phenobarbital promoted hepatocellular carcinogenesis in Albino Wistar rats [47] Exhibited high DNA protective effect in the presence of H_2O_2 and Fe^{2+} [48]
Naringenin
Exhibited high antioxidant capacity and hydroxyl and superoxide radical scavenger, reduced oxidative damage of lipids as well as DNA damage [49]
Kaempferol
Exhibited good DPPH and $ABTS^{+\cdot}$ radical scavenging activities [50] Exhibited good scavenging activity against peroxynitrite and hydroxyl radicals at a lower concentration, while at higher concentrations increased the expression of antioxidant enzymes [51]
Acacetin
Reduced *tert*-butyl peroxide-induced ROS generation and increased the expression of antioxidant proteins, such as HO-1, superoxide dismutase (SOD), and nicotinamide adenine dinucleotide hydrogen (NADH) quinone oxidoreductase 1 (NQO1) [52]

3.2 Antitumor Activity

Tumors represent some of the most common and dangerous diseases of today, claiming millions of lives each year worldwide [53]. Tumors are characterized by uncontrolled cell growth and division, invasion of normal tissues, and often spreading throughout the entire body. According to the degree of aggressiveness or their ability to metastasize, tumors can be divided into malignant and benign ones. In addition to well-known classical therapies (surgery, radiation, or chemotherapy), there is a constant need to develop alternative, effective, and more affordable antitumor agents which exhibit fewer side effects [54]. Natural products have received much more attention over the past decades, as there is increasing evidence for their potential to inhibition of various stages of tumorigenesis, as well as associated inflammatory processes. Approximately 60% of drugs currently used in therapies for various tumors have been isolated from natural products, whereby the plant kingdom has been the most significant source of biologically active molecules [55]. Due to the diverse content of secondary metabolites, mosses and generally bryophytes are excellent candidates for finding new, less toxic, and more selective therapeutic agents in the fight against various tumors [8, 9, 11].

Although the mechanisms of antitumor activity of bryophytes have not been fully elucidated, it has been established that their extracts can activate various biochemical pathways and cause apoptosis and/or necrosis of tumor cells [12]. Numerous studies have reported that phenolic and flavonoid compounds, together with terpenoids, represent some of the major secondary metabolites present in bryophytes responsible for their antiproliferative properties [8, 50, 56, 57]. The capability of phenolic and phenolic-like compounds to prevent and/or slow the progression of tumor cells by their interaction with the basic cellular processes associated with cell proliferation, differentiation, inflammation, apoptosis, and angiogenesis has been shown in many studies [8, 12, 58, 59]. One of the mechanisms included in the inhibition of tumor cell growth is the induction of different apoptotic pathways. For example, it has been shown that apoptosis of some tumor cells is promoted by the generation of ROS and nitric oxide (NO) [60, 61]. ROS and NO act as second messengers in cell signaling and are essential for various biological processes in normal cells, but can also be involved in the development of different pathologies. Although the roles of ROS and NO in tumors are open to question, these cells usually exhibit a consistent increase in the generation of ROS and NO, which in turn makes them more sensitive to further oxidative stress, which is one of the strategies employed in the fight against tumors [62, 63].

Water, water/ethanol, and ethyl acetate extracts of moss *H. cupressiforme* obtained by Soxhlet extraction have exhibited significant antiproliferative potential against human breast cancer cells (MDA-MB-231), causing the decrease in the tumor cell viability by approximately 50% at the extract concentration of 10 μg mL^{-1} [8]. Similar results (~50% cell viability decrease) were reported for the water/ethanol and ethyl acetate extracts of moss *H. ciliata* obtained by the same procedure of extraction, at the same concentration of extract, on MDA-MB-231 cells [9]. Extracts of both species significantly increased the production of both ROS and NO by MDA-MB-231 cells, which leads to a conclusion that this might mediate and correlate with the significant antiproliferative potential of *H. ciliata* and *H. cupressiforme* extracts. The results indicated that the antiproliferative effects of these extracts toward MDA-MB-231 cells may be caused by the increased production of ROS and NO in cancer cells; however, the exact mechanism remains to be further explored. On the other hand, the same extracts of *H. cupressiforme* and *H. ciliata* under the previously mentioned conditions did not exhibit a significant antiproliferative effect on the human colon cancer HCT-116 cell line [8, 9].

In another study, the methanolic extract of *H. cupressiforme* obtained by Soxhlet extraction inhibited the proliferation of cervical cancer HeLa cells by 11.92% at the dose of 25 μg mL^{-1}, 18.73% at the dose of 50 μg mL^{-1}, 38.01% at the dose of 100 μg mL^{-1}, and 54.47% at the dose of 200 μg mL^{-1}, thus exhibiting a strong antiproliferative effect. In the same study, only moderate antiproliferative effects were noticed in lung carcinoma epithelial cells (A549) [31].

Table 2 provides an overview of the secondary metabolites identified in the extracts of *H. ciliata* and *H. cupressiforme* and the antitumor activities with the proposed action mechanisms described for these metabolites in the literature.

3.3 Anti-inflammatory and Neuroprotective Activities

Inflammation is a complex physiological response of the immune system to tissue damage caused by physical injury, pathogen infection, toxins, and many other agents [103]. Various cells of the immune system and numerous soluble mediators participate in the process of inflammation, maintaining the host's response designed to repair the damaged tissue. A central role in the inflammatory response is played by leukocytes, cells that migrate to the site of infection/damage and release numerous

Table 2 Reported antitumor activities of the compounds identified in the extracts of *H. ciliata* and *H. cupressiforme*

Gallic acid
Exhibited antitumor properties in human lung adenocarcinoma cell line A549 by inducing cell apoptosis, elevating ROS, disrupting mitochondrial membrane potential, and activating caspase-3 [58] Inhibition of proliferation and induction of apoptosis in MCF-7 human breast carcinoma cells by activating Fas/FasL as well as the caspase-8 system [59] Inhibition of proliferation of HepG2 and SMMC 7721 human hepatocellular carcinoma cell lines and induction of apoptosis in SMMC 7721 cells [60]
Protocatechuic acid
Induced cell death in HepG2 hepatocellular carcinoma cells by stimulating the c-Jun N-terminal kinase (JNK) and p38 subgroup of the mitogen-activated protein kinase (MAPK) family [64] Exhibited apoptotic and antiproliferative effects in HL-60 leukemia cells by increasing DNA fragmentation and Bax protein expression, reducing the expression of Bcl-2 protein and retinoblastoma (RB) phosphorylation [65] Inhibited cell proliferation and induced cell cycle arrest at a sub-G1 phase in primary cultured human uterine leiomyoma cells [66] Induced apoptosis in human gastric adenocarcinoma cells via the JNK/p38 MAPK pathway activated Fas/FasL pathway, increased the translocation of Bax, and reduced the expression of Bcl-2 protein [67] Increased the apoptosis and/or slowed down the invasion and metastasis of human breast cancer MCF7 cell, lung cancer A549 cell, HepG2 cell, cervix HeLa cell, and prostate cancer LNCaP cell [68]
5-*O*-Caffeoylquinic acid
Reduced HT-29 (human colon adenocarcinoma) cell viability, increasing the apoptotic rate in the cells [69] Exhibited anti-invasive activity against non-small cell lung cancer through p53-dependent regulation of signaling pathways [70]
p-Hydroxybenzoic acid
Enhanced the sensitivity of MCF-7 human breast cancer cells to a specific HDAC6 inhibitor via promotion of the HIPK2/p53 pathway [71]
Caffeic acid
Reduced SK-Mel-28 human melanoma cancer cell viability, induced apoptosis, inhibited colony formation, modulated cell cycle, and altered caspases gene expression [57] Exerted antitumor effect through its pro-oxidant activity; elevated reactive oxygen species levels and altered mitochondrial membrane potential in HeLa and ME-180 cancer cells [72]
Quercetin 3-*O*-rutinoside
Exhibited cytotoxic and apoptotic activity against human colon carcinoma (Caco-2) and HepG2 cell line [73] Exerted anti-migratory potential by inhibiting epidermal growth factor receptor signaling in several human pancreatic cancer cell lines [74]
p-Coumaric acid
Inhibited cell proliferation of human A375 and mouse melanoma B16 cells promoted the apoptosis of these cells, notably upregulated the levels of Apafl and Bax and downregulated the levels of Bcl-2 [59] Inhibited the proliferation of human colorectal carcinoma (HCT15 and HT29) and increased the apoptosis of these cells [75]

(continued)

Table 2 (continued)

Quercetin 3-*O*-glucoside
Induced DNA topoisomerase II inhibition, cell cycle arrest, and apoptosis in hepatocellular carcinoma cells (HepG2) [76] Showed significant cytotoxicity against the Caco-2 cell line and moderate cytotoxicity against the HepG2 cells [73]
Isorhamnetin 3-*O*-glucoside
Aglycone of isorhamnetin showed antitumor activity against human hepatocellular carcinoma cells (BEL-7402) [58] Isorhamnetin inhibited NOZ and GBC-SD human gallbladder cancer cell proliferation and metastasis by inactivation of the PI3K/AKT signaling pathway [77]
Eriodictyol
Exerted strong anticancer activity against the human lung cancer cell line A549, through induction of mitochondrial apoptosis, G2/M cell cycle arrest, as well as inhibition of the m-TOR/PI3K/Akt signaling pathway [78] Inhibited proliferation, metastasis, and induced apoptosis of brain tumor cells (glioma) through the blockade of the PI3K/Akt/NF-κB signaling pathway [79] Effectively inhibited proliferation, migration, and invasion and also induced apoptosis in retinoblastoma RB Y79 cell line, by blocking the PI3K/Akt signaling pathway [80] Exhibited anticancer and apoptotic potential in human hepatocellular carcinoma HepG2 cells through cell cycle arrest and modulation of apoptosis-related proteins [81]
Apigenin
Inhibited proliferation, invasion, and migration of colorectal cancer SW480, HCT-116, DLD1, and LS174T cell lines through interaction with different signaling pathways [82–85] Inhibited cell proliferation, enhanced the immune response, and/or induced apoptosis and cell cycle arrest in breast cancer cell lines (BT-474, MDA-MD-231, T47D, MDA-MB-468, SKBR3, and MDA-MB-453) [86–90] Inhibited cell proliferation, migration, and induced apoptosis in lung cancer H1299, H460, and A549 cell lines [91, 92] Inhibited cell proliferation, migration, and activation and also induced apoptosis and cell cycle arrest in prostate cancer cells LNCaP, PC-3, 22Rv1, and DU145 [93–95]
Naringenin
Showed anticancer effects through induction of tumor cell death and inhibition of angiogenesis in B16F10 murine and SK-MEL-28 human malignant melanoma cells [96] Inhibited human lung cancer proliferation, migration, and metastasis, inducing apoptosis and arrest of tumor progression in vitro [56] Inhibited cell proliferation in HCT116 and SW480 human colorectal cancer cell lines through p38-dependent cyclin D1 downregulation and cell growth [97]
Kaempferol
Inhibited Miapaca-2, Panc-1, and SNU-213 human pancreatic cancer cell growth and migration via blockade of epidermal growth factor receptor (EGFR) related Src, ERK1/2, and AKT pathways [98] Inhibited the growth of human breast cancer (MDA-MB-231) cells, arrested cell cycle, induced apoptosis, as well as DNA damage [99] Inhibited both growth and migration of glioma cells and induced cell death through ERK and Akt-dependent pathways [100]
Acacetin
Inhibited cell growth and cell cycle progression, induced apoptosis in human prostate cancer (PCA), LNCaP, and DU145 cells [101] Induced apoptosis and altered the nuclear and cell morphology in K562 (human T cell leukemia cells) [102]

growth factors, cytokines, and reactive oxygen and nitrogen species. From the initial injury to the final recovery, progressive changes take place in the damaged tissue, aimed at removing the cause of the inflammation and healing the tissue. Inflammatory processes are necessary for adequate immune system functioning and surveillance, as well as optimal repair and regeneration following the injury. However, there are cases when the immune response becomes inadequate and uncontrolled, leading to a state of chronic inflammation, which is involved in the pathogenesis of numerous diseases, including atherosclerosis, tumors, and asthma, and some neurological disorders, such as Alzheimer's disease (AD) and Parkinson's disease (PD) [8, 104].

Extracts obtained from plants, including bryophytes, have been used to alleviate and treat different inflammation-related disorders in the traditional medicine of almost all civilizations [7]. Numerous compounds, including polysaccharides, flavonoids, fatty acids, aliphatic compounds, and aromatic and phenolic compounds, identified in bryophytes are responsible for the effects that these plants exhibit [105]. Bryophytes have been used in traditional medicine to treat cuts, burns, wounds, uropathy, inflammatory diseases (inflammation of the throat, pharynx, pneumonia), and fever [7, 105]. Although bryophytes have been used in traditional medicine for a long time, more intensive research regarding their biological potential and deeper mechanisms of action have started only recently. In particular, when it comes to the anti-inflammatory and neuroprotective potential of these plants, studies have confirmed that bryophytes have the ability to reduce the production of ROS and NO by microglia cells, to inhibit various inflammation-related enzymes (inducible nitric oxide synthase (iNOS), acetylcholinesterase (AChE), tyrosinase (Tyr)), and to provide neuroprotection of SH-SY5Y neurons induced by microglia-mediated lipopolysaccharide (LPS) neurotoxicity [8, 11, 104], which is discussed in more details in the following text.

Microglia are the resident macrophage-like cells of the central nervous system (CNS) which are closely involved in the maintaining of brain homeostasis [106]. These cells play fundamental roles in the immune response in CNS through phagocytic debris removal, as well as brain protection and repair. When microglial cells are activated, in response to infection or physical trauma, they can induce neuroinflammation through the secretion of various pro-inflammatory mediators, including NO and ROS. Large amounts of NO are produced in the organism by the enzyme iNOS, after stimulation of cells with endotoxins (e.g., LPS) and cytokines involved in pathological processes. These molecules have been related to the higher risk of developing neurodegenerative diseases, such as AD and PD, multiple sclerosis, and cerebral ischemia. For the treatment of such conditions, it is important to find new, natural compounds that inhibit the production of NO and thus reduce inflammation [8].

Moreover, the inhibition of enzymes, such as AChE and Tyr, associated with the development of neurodegenerative disorders mediated by inflammation, represents a promising pathway in finding new treatment strategies. AChE is a cholinergic enzyme whose primary function is to catalyze and promote the breakdown of the neurotransmitter acetylcholine, thus regulating its amount in synapses. Since one of the main features of AD is memory loss caused by a reduced amount of acetylcholine, the inhibition of AChE represents an effective therapeutic approach in the treatment of AD [107]. On the other hand, Tyr is a key enzyme involved in the production of melanin in the skin and hair but also contributes to the production of neuromelanin in the CNS. Neuromelanin production and accumulation and consequential neuronal damage have been linked to PD. Tyrosinase inhibition is therefore an important target in the development of drugs for PD [108]. Due to the high cost and frequent side effects that occur when using commercial drugs for neurodegenerative and neuroinflammatory diseases, there is an increased need to find new, less toxic, and more specific treatments.

Ethanolic, ethyl acetate, water/ethanolic, and water extracts of *H. cupressiforme* and *H. ciliata* obtained by Soxhlet extraction have shown promising anti-inflammatory activities by significantly reducing the production of NO in LPS-stimulated BV2 microglial cells and increasing viability/metabolic activity of these cells [8, 9]. The same extracts of these two moss species have exhibited high inhibitory activities toward AChE and Tyr enzymes when compared to standard substances galantamine and kojic acid, respectively [8, 9].

Moreover, in a recently published study, it has been demonstrated that ethyl acetate extract of *H. cupressiforme* obtained by Soxhlet extraction reduced the levels of cytokines, interleukin (IL)-6, and tumor necrosis factor (TNF)-α, in the BV2 cell supernatants compared to their levels in supernatants of only LPS-treated control cells [104]. Investigated extracts also significantly diminished the production of ROS and NO by microglial cells almost to the levels of non-stimulated control cells, thus alleviating inflammation. The supernatant transfer model system revealed that LPS-activated and moss-treated BV2 cells increased the viability of SH-SY5Y neurons, providing them with neuroprotection [104].

Table 3 encompasses the secondary metabolites identified in the extracts of *H. ciliata* and *H. cupressiforme* with their reported anti-inflammatory/neuroprotective activities, proposed mechanisms, and suggested cellular targets from the literature.

3.4 Antimicrobial Activity

In recent years, due to the growing development of pathogen resistance against commonly used antibiotics, as well as increased mortality rates from bacterial and fungal infections, there is an urgent demand for finding new, more efficient antimicrobial agents. A special place in the search for novel antimicrobial agents is occupied by plants and their extracts, whereby bryophytes have shown a promising and significant antimicrobial potential [22, 27, 127]. Although bryophytes normally grow in humid habitats, it has been noted that they are less susceptible to fungal diseases and are relatively free from microbial invasion. The absence of diseases in these species indicated that bryophytes are able to produce some constitutive or inducible broad-range antimicrobials [27]. These compounds are synthesized as a defense mechanism against different pathogens to protect otherwise delicate plants not only from bacteria and fungi but also from insects and slugs [22]. Terpenes,

Table 3 Reported anti-inflammatory/neuroprotective activities of the compounds identified in the extracts of *H. ciliata* and *H. cupressiforme*

Gallic acid
Inhibited the acetylation of the nuclear factor kappa B (NF-kB), reduced the production of cytokine in microglia cells, protected neurons from amyloid-beta (Aβ)-induced neurotoxicity, and efficiently blocked neuronal cell death [109] Significantly reduced LPS-induced increase in glial fibrillary acidic protein (a biomarker of activated astrocytes) and ED-1 (a biomarker of activated microglia), as well as iNOS and IL-1β (a pro-inflammatory cytokine) in the LPS-infused substantia nigra of rat brain after the systemic administration (100 mg kg^{-1}), thus attenuating LPS-induced neuroinflammation [110]
Protocatechuic acid
Reduced oxidative stress in cerebellar granule neurons induced by hydrogen peroxide reduced NO production in microglial cells stimulated with LPS [111] Inhibited inflammatory response of LPS-activated BV2 microglia by regulating the SIRT1/NF-κB pathway and thereby attenuated microglial activation-induced PC12 cell apoptosis [112]
5-*O*-Caffeoylquinic acid
Inhibited LPS-induced iNOS and cyclooxygenase (COX)-2 expression, as well as production of NO and pro-inflammatory mediators, TNF-α and IL-1β by BV2 microglial cells, blocking the activation of p38 MAPK and phosphorylated NF-κB p65 [113]
p-Hydroxybenzoic acid
Reduced oxidative stress in cerebellar granule neurons induced by hydrogen peroxide and protected these neurons from glutamate-induced excitotoxicity [111]
Caffeic acid
Effective 5-lipooxygenase inhibitor that downregulated NF-κBp65 in the inflammatory response in rats [114] Significantly reduced mRNA and protein levels of TNF-α, IL-6, and IL-1β at the application site of caffeic acid as well as in human keratinocytes in vitro and ameliorated skin edema in an acute and chronic model of cutaneous inflammation in mice [115]
Quercetin 3-*O*-rutinoside
Enhanced the reduced levels of brain-derived neurotrophic factor, nerve growth factor, and glutathione, thus exhibiting neuroprotection in the diabetic retina [116]
p-Coumaric acid
Exhibited neuroprotective effect on cerebral ischemia through modulation of the apoptosis mechanism – markedly decreased caspase-3 and caspase-9 immunoreactivity [117] Alleviated LPS-induced brain damage through oxidative stress reduction, significantly increasing levels of superoxide dismutase and glutathione, while decreasing AChE activity, levels of TNF-α and IL-6, and suppressing neuronal apoptosis [118]
Quercetin 3-*O*-glucoside
Increased SH-SY5Y cell viability in cells treated with Aβ downregulated the expression of apoptosis-related proteins, such as Bcl-2-associated X protein, and cleaved caspase-9 [119]
Isorhamnetin 3-*O*-glucoside
Decreased the production of inflammatory mediators derived from arachidonic acid metabolism, namely12(S)-hydroxy(5Z,8E,10E)-heptadecatrienoic acid, thromboxane B_2, prostaglandin E_2, and 12(S)-hydroxy-(5Z,8Z,10E,14Z)-eicosatetraenoic acid [44]
Eriodictyol
Reduced NO production by LPS-stimulated RAW 264.7 cells, suppressed the phagocytic activity of such activated macrophages, and reduced the expression of mRNA and the secretion of pro-inflammatory cytokines, through the blockage of NF-κB activation and phosphorylation of p38 MAPK, extracellular signal-regulated kinases 1 and 2 (ERK1/2) and JNK [120]

(continued)

Table 3 (continued)

Attenuated LPS-induced acute lung injury in a mouse model by regulating the Nrf2 pathway and inhibiting the expression of inflammatory cytokines in macrophages [121]
Apigenin
Decreased the production of pro-inflammatory cytokines through the inhibition of COX-2 and NF-kB activation [122] Protected human pluripotent stem cell-derived neurons in a model of AD by promoting a global downregulation of pro-inflammatory cytokines and NO release, reducing the frequency of spontaneous Ca^{2+} signals, and significantly reducing caspase-3/7 mediated apoptosis [123]
Naringenin
Reduced the pro-inflammatory cytokine response induced by LPS in both macrophages and cells of whole blood, inhibited the phosphorylation of macrophage kinases in LPS-stimulated macrophages [124] Attenuated the apoptosis and neurotoxicity in Aβ-stimulated AD via inhibition of caspase3, activation of PI3K/AKT, and modulation of GSK-3β signaling pathways [125]
Kaempferol
Showed inhibitory effects on activated T cell proliferation and significantly inhibited LPS-induced ROS and NO release by RAW 264.7 cells [50] Prevented ischemic brain injury and neuroinflammation by inhibiting STAT3 and NF-κB activation, exhibiting therapeutic potential for neuroinflammation-related diseases [126]
Acacetin
Alleviated *tert*-butyl peroxide-induced generation of important inflammatory mediators (COX-2, iNOS) as well as degradation of the extracellular matrix (aggrecan, collagen II, matrix metallopeptidase (MMP)13, MMP9, and MMP3) [52]

bis-bibenzyls, polyphenols, and flavonoids identified in bryophytes have been reported as antimicrobials effective against a wide range of microorganisms [127]. For instance, biflavonoids (hypnogenol B1 and hipnumflavonoid A) identified in *H. cupressiforme* have been previously reported to possess antibacterial activity against several microorganisms [29]. Bryophyte species have been used as antimicrobial agents since ancient times, in traditional Chinese and Indian medicine for surgical dressings, diapers, and other human medicinal applications. Nowadays, an increasing number of studies are examining the antibacterial, antifungal, and antiviral potential of these species. Several studies have examined the antibacterial and antifungal activities of *H. cupressiforme* and *H. ciliata* moss extracts, wherefore these are the activities that are generally the most studied when it comes to these species.

The antimicrobial activity of the *H. cupressiforme* essential oil (at a concentration of 27 mg mL^{-1} in hexane) was tested against the following bacteria: *Escherichia coli* ATCC 35218, *Yersinia pseudotuberculosis* ATCC 911, *Pseudomonas aeruginosa* ATCC 43288, *Staphylococcus aureus* ATCC 25923, *Enterococcus faecalis* ATCC 29212, and *Bacillus cereus* 709 Roma as well as fungi *Candida albicans* ATCC 60193 and *Saccharomyces cerevisiae* RSKK 251. Essential oils from *H. cupressiforme* showed antifungal activities with minimal inhibitory concentrations (MICs) of 337 and 675 μg mL^{-1} against *S. cerevisiae* and *C. albicans*, respectively. However, no antimicrobial activity was observed against the tested bacteria [18].

Ethanolic extracts of *H. cupressiforme* (10 mg mL^{-1}) exhibited antibacterial activity against *E. coli* ATCC 25922, *Klebsiella pneumoniae*, *Salmonella infantis*, and *Salmonella kentucky* showing a 7 mm inhibition zone in the disk diffusion test in all cases [22]. Ethanolic extract of *H. cupressiforme* exhibited antimicrobial effects against *E. coli* ATCC 35218 (6.9 ± 0.7 mm inhibition zone), and methanolic extract against *B. cereus* 863, *Salmonella sp.* 213, and *S. cerevisiae* TP (3–2) with inhibition zones 7.3 ± 0.9, 6.8 ± 0.6, and 7.7 ± 0.3 mm, respectively. Acetone extract has shown an antimicrobial effect against *C. albicans* ATCC 16231, with a corresponding inhibition zone of 7.0 ± 0.7, while chloroform extract of *H. cupressiforme* showed the highest antimicrobial effect with 8.4 ± 1.2 mm inhibition zone against *B. cereus* 863, while it also inhibited *Bacillus subtilis* RSKK 244, *Salmonella sp.* 213, *S. cerevisiae* TP (3–2), and *P. aeruginosa* ATCC 27853 with corresponding zones 7.6 ± 0.4, 7.2 ± 1.0, 7.8 ± 0.1, and 8.0 ± 0.0 mm, respectively. All investigated extracts of *H. cupressiforme* were inactive against *S. aureus*. All extracts were prepared at concentration 10 mg mL^{-1} [127].

Methanolic extract of *H. cupressiforme* (30 mg mL^{-1}) has shown good to moderate activity against several Gram +ve and Gram -ve bacteria, namely *B. subtilis* ATCC 6633, *E. coli* ATCC 11230, *K. pneumonia* (isolated from the animal specimen), *P. aeruginosa* ATCC 27853, *Salmonella typhimurium* CCM 5445, *S. aureus* ATCC 6538, *Streptococcus pyogenes* (isolated from the animal specimen), and *Mycobacterium smegmatis* DSM 43465, with corresponding inhibition zones of 14.2, 12.2, 10.4, 11.6, 12.4, 12.6, 9.2, and 6.0 mm, respectively. Additionally, antifungal activity of the same *H. cupressiforme* methanolic extract exhibited antifungal activity against *C. albicans* ATCC 10231, *Rhodotorula rubra* DSM 70403, and *Kluyveromyces fragilis* ATCC 8608 with corresponding inhibition zones of 9.8, 11.2, and 11.4 mm [128].

In a different study, methanolic extracts of *H. cupressiforme* (20, 10, and 5 mg mL^{-1}) were tested against the following bacteria: *Staphylococcus epidermidis* ATCC 12228, *Micrococcus flavus* ATCC 10240, *B. subtilis* ATCC 10707, *E. coli* ATCC 25922, and *Salmonella enteritidis* ATCC 13076. Antifungal activity was tested using the following species: *Aspergillus flavus* ATCC 9170, *Aspergillus fumigatus* (human isolate), *Aspergillus niger* ATCC 6275, *Penicillium funiculosum* ATCC 10509, *Penicillium ochrochloron* ATCC 9112, *Trichoderma viride* ATCC IAM 5061, and *C. albicans* (isolated directly from patients). Antifungal activities of the investigated methanolic *H. cupressiforme* extracts were higher than their antibacterial activity. The minimal fungicidal concentration (MFC) of the extract against all investigated fungi species was 5 mg mL^{-1}, while MICs were also 5 mg mL^{-1} for all investigated fungi, except for *P. funiculosum* with an MIC of 2.5 mg mL^{-1}. On the other hand, MICs for the investigated bacteria species were 10 mg mL^{-1} for *E. coli* and *S. enteritidis*, while for the other species MIC was 20 mg mL^{-1}. The minimal bactericidal concentration (MBC) of the extract against investigated bacteria was 20 mg mL^{-1} in all cases, except for *S. enteritidis* with an MBC of 10 mg mL^{-1} [10].

Ethanolic extract of *H. cupressiforme* was tested against 13 different microorganisms: 4 Gram +ve bacteria (*B. cereus* ATCC 10876, *L. monocytogenes* ATCC 7677, *Clostridium perfringens* ATCC 313124, and *S. aureus* ATCC 25923), 6 Gram -ve bacteria (*P. aeruginosa* ATCC 27853, ATCC 10876, *E. coli* ATCC 25922, *S. typhimurium* ATCC 14028, *K. pneumonia* ATCC 13883, *Shigella sonnei* ATCC 25931, *Yersinia enterocolitica* ATCC 27729), and 3 fungi (*A. niger* ATCC 9642, *C. albicans* ATCC 10231, and *S. cerevisiae* ATCC 976). Extracts exhibited good antibacterial and antifungal activities (6–14 mm inhibition zone) against the test organisms. The highest antibacterial activity (15.33 mm inhibition zone) was obtained against *K. pneumoniae*, while the weakest activity was observed against *S. sonnei* (6.00 mm inhibition zone) [129].

Regarding the antimicrobial potential of *H. ciliata*, ethanolic extract of this moss at a concentration of 9 mg mL^{-1} exhibited good activity against several Gram +ve and Gram -ve microorganisms, namely *B. subtilis* ATCC 6633, *Enterobacter aerogenes* ATCC 13048, *E. faecalis* ATCC 29212, *Enterococcus faecium*, *S. typhimurium* SL 1344, *Staphylococcus carnosus* MC1.B, *S. epidermidis* DSMZ 20044, and *Streptococcus agalactiae* DSMZ 6784, with corresponding MICs of 62.5, 500, 125, 500, 1000, 125, 125, and 250 μg mL^{-1}, respectively [15].

All the above-mentioned results indicate that extracts of *H. cupressiforme* and *H. ciliata* mosses should find a practical application in the prevention and protection of plants, animals, and/or humans against bacterial and fungal infections since these plants represent natural and nontoxic sources of wide-spectrum antibiotics that can serve as selective agents against infectious diseases.

Table 4 contains the selected secondary metabolites identified in the extracts of *H. ciliata* and *H. cupressiforme* with the antimicrobial activities and proposed action mechanisms described for these metabolites in the literature.

4 Conclusion

In the past few decades, a significant advance has been made to identify and isolate various compounds from different plant sources for therapeutic application. Being present in almost all parts of the world and occupying the second largest group of the plant kingdom, bryophytes possess immense potential for the discovery and development of novel drugs. Two widespread moss species, *H. cupressiforme* and *H. ciliata*, represent a rich source of numerous important biologically active compounds. Thus, their extracts can be used in the treatment of different conditions related to inflammation, infection, and cancer. Additional studies are needed to investigate the detailed mechanisms underlying the activities shown by these species, as well as more in vivo studies. Most importantly, extracts of *H. cupressiforme* and *H. ciliata* represent an entirely new source in the development of more effective, sustainable, and less toxic drugs for the prevention and/or adjuvant treatment of various human pathologies.

Table 4 Reported antimicrobial activities of the compounds identified in the extracts of *H. ciliata* and *H. cupressiforme*

Gallic acid
Effective against *Mannheimia haemolytica* and *Pasteurella multocida*, two crucial bovine respiratory disease-associated pathogens, with MICs of 250 and 500 μg mL^{-1}, respectively [130] Exhibited antimicrobial activity against investigated bacteria with MIC of 500 μg mL^{-1} for *P. aeruginosa*, 1500 μg mL^{-1} for *E. coli*, 1750 μg mL^{-1} for *S. aureus*, and 2000 μg mL^{-1} for *Listeria monocytogenes*; induced irreversible changes in properties of the membrane by changing hydrophobicity, decreased the negative surface charge, leading to leakage of essential intracellular constituents [131]
Protocatechuic acid
Exhibited antimicrobial effect against Gram (+) (*B. cereus* – MIC 39 μg mL^{-1}; *S. aureus* – MIC 156 μg mL^{-1}; and *S. faecalis* – 39 μg mL^{-1}) and Gram (–) (*Citrobacter freundii* – MIC 312 μg mL^{-1}; *E. coli* – MIC 78 μg mL^{-1}; and *P. aeruginosa* – MIC 156 μg mL^{-1}) bacteria and fungi (*C. albicans*, MIC 156 μg mL^{-1}, and *Microsporum audouinii*, MIC 10 μg mL^{-1}) [132]
5-*O*-Caffeoylquinic acid
Exhibited antimicrobial activity against *S. aureus* PCM 1932, *E. faecium* PCM 1859, *E. coli* PCM 2561, *Proteus vulgaris* PCM 542, *P. aeruginosa* PCM 2563, *K. pneumoniae* PCM 65542, and *C. albicans* ATCC 10231 with the same MIC_{80} of 10 mg mL^{-1} for all cases except *K. pneumoniae* where MIC_{80} was 5 mg mL^{-1} [133]
p-Hydroxybenzoic acid
Exhibited antibacterial activity against most of the investigated Gram (+) and some Gram(–) bacteria with IC_{50} of 160 and 100–170 μg mL^{-1}, respectively [134]
Caffeic acid
Exerted antibacterial activity against *S. aureus* clinical strains isolated from infected wounds, both alone and in combination with antibiotics [135]
Quercetin 3-*O*-rutinoside
Exhibited the highest antibacterial activity among the investigated compounds, where the most affected bacteria were *S. epidermidis* ATCC 10875 and *E. faecalis* ATCC 14428 with MICs of 8 μg mL^{-1} [136]
p-Coumaric acid
Exerted potent antimicrobial activity against *Shigella dysenteriae* 51302 (MIC 10 mg mL^{-1}) by irreversibly changing the cell membrane permeability, causing the loss of cells' ability to maintain macromolecules in the cytoplasm, and binding to DNA, thus inhibiting important cellular functions [137]
Quercetin 3-*O*-glucoside
Exhibited the highest antibacterial activity among investigated flavonoids against *E. coli*, *S. aureus*, *B. cereus*, *P. aeruginosa*, and *B. subtilis* [138]
Isorhamnetin 3-*O*-glucoside
Exhibited antibacterial activity, where the most affected bacteria were *S. epidermidis* ATCC 10875, *S. aureus* ATCC 13709, and *K. pneumoniae* ATCC 27736 with MICs of 32 μg mL^{-1} [136]
Eriodictyol
Significantly inhibited the growth of *S. aureus* ATCC 12600 at 50 μM (18.2 ± 6.1%) and 70 μM (19.8 ± 8.2%) compared to control [139]
Apigenin
Exerted antibacterial activity against five pathogenic bacterial strains: *P. aeruginosa*, *K. pneumoniae*, *S. typhimurium*, *Proteus mirabilis*, and *E. aerogenes* with corresponding inhibition zones of 12.24 ± 0.41, 10.52 ± 0.38, 17.36 ± 0.18, 19.12 ± 0.01, and 14.02 ± 0.03 mm, respectively [140]

(continued)

Table 4 (continued)

Naringenin
Exhibited antibacterial activity on methicillin-resistant *S. aureus* by decreasing biofilm formation and reducing the secretion of fatty acid [141]
Kaempferol
Displayed antibacterial activity and protective effect on *Helicobacter pylori* infection [142] Exhibited antimicrobial activity against *Candida parapsilosis* complex with MIC ranging from 32 to 128 μg mL^{-1}, decreasing the metabolic activity and biomass of growing biofilms of the *C. parapsilosis* complex [143]
Acacetin
Inhibited pore-forming activity of pneumolysin, the major virulence factor that contributes to the interaction between *S. pneumoniae* and the host and also reduced the virulence of *S. pneumoniae* both in vivo and in vitro [144] Exhibited antimicrobial activities against *Actinomyces naeslundii*, *Actinomyces israelii*, *Streptococcus mutans*, *Prevotella intermedia*, *Porphyromonas gingivalis*, and *Aggregatibacter actinomycetemcomitans*, with MIC values ranging from 0.25 to 1.0 mg mL^{-1} [145]

Acknowledgments This work was supported by the Grant of the Ministry of Education, Science and Technological Development of the Republic of Serbia [Contract number: 451-03-68/2022-14/200178].

References

1. Radulović NS, Filipović SI, Zlatković DB, Đorđević MR, Stojanović NM, Randjelović PJ, Mitic KV, Jevtovic-Stoimenov TM, Randelovic (2016) Immunomodulatory pinguisane-type sesquiterpenes from the liverwort *Porella cordaeana* (Porellaceae): the "new old" furanopinguisanol and its oxidation product exert mutually different effects on rat splenocytes. RSC Adv 6(48):41847–41860
2. Zhao S, Gao Q, Rong C, Wang S, Zhao Z, Liu Y, Xu J (2020) Immunomodulatory effects of edible and medicinal mushrooms and their bioactive immunoregulatory products. J Fungi 6(4): 269
3. Fontana F, Figueiredo P, Bauleth-Ramos T, Correia A, Santos HA (2018) Immunostimulation and immunosuppression: nanotechnology on the brink. Small Method 2(5):1700347
4. Sellon DC, Long MT (2013) Equine infectious diseases, 2nd edn. Elsevier Health Sciences
5. Rathee P, Chaudhary H, Rathee S, Rathee D, Kumar V (2013) Immunosuppressants: a review. Pharm Innov J 1(12):90–103
6. Shi J, Weng J-H, Mitchison TJ (2021) Immunomodulatory drug discovery from herbal medicines: insights from organ-specific activity and xenobiotic defenses. elife 10:e73673
7. Glime J (2017) Medical uses: medical conditions. In: Glime J (ed) Bryophyte ecology, vol 5. International Association of Bryologists, Seattle
8. Lunić TM, Oalđe MM, Mandić MR, Sabovljević AD, Sabovljević MS, Gašić UM, Duletic-Lausevic SN, Bozic BD, Nedeljkovic BDB (2020) Extracts characterization and In vitro evaluation of potential immunomodulatory activities of the Moss *Hypnum cupressiforme* Hedw. Molecules 25(15):3343
9. Mandić MR, Oalđe MM, Lunić TM, Sabovljević AD, Sabovljević MS, Gašić UM, Duletic-Lausevic SN, Bozic BD, Nedeljkovic BDB (2021) Chemical characterization and in vitro immunomodulatory effects of different extracts of moss *Hedwigia ciliata* (Hedw.) P. Beauv. from the Vršačke Planine Mts., Serbia. PLoS One 16(2):e0246810

10. Veljić M, Đurić A, Soković M, Ćirić A, Glamočlija J, Marin P (2009) Antimicrobial activity of methanol extracts of *Fontinalis antipyretica*, *Hypnum cupressiforme* and *Ctenidium molluscum*. Arch Biol Sci 61:225–229
11. Asakawa Y, Ludwiczuk A (2015) Chemical constituents of bryophytes: structures and biological activity. J Nat Prod 81(3):641–660
12. Dey A, Mukherjee A (2015) Therapeutic potential of bryophytes and derived compounds against cancer. J Acute Dis 4(3):236–248
13. Kļaviņa L (2018) Composition of mosses, their metabolites and environmental stress impacts. Doctoral thesis, University of Latvia, Latvia
14. Wolski GJ, Sadowska B, Fol M, Podsędek A, Kajszczak D, Kobylińska A (2021) Cytotoxicity, antimicrobial and antioxidant activities of mosses obtained from open habitats. PLoS One 16(9):e0257479
15. Canlı K, Cetin B, Altuner EM, Türkmen Y, Üzek U, Dursun H (2014) In vitro antimicrobial screening of *Hedwigia ciliata* var. *leucophaea* and determination of the ethanol extract composition by gas chromatography/mass spectrometry (GC/MS). J Pure Appl Microbiol 8: 2987–2998
16. Gahtori D, Chaturvedi P (2019) Bryophytes: a potential source of antioxidants. In: Bryophytes. IntechOpen, London
17. Rachna P, Vashishtha B (2015) Bryophytes-boon to mankind-A review. Agric Rev 36(1): 77–79
18. Üçüncü O, Cansu TB, Özdemir T, Karaoglu SA, Yayli N (2010) Chemical composition and antimicrobial activity of the essential oils of mosses (*Tortula muralis* Hedw., *Homalothecium lutescens* (Hedw.) H. Rob., *Hypnum cupressiforme* Hedw., and *Pohlia nutans* (Hedw.) Lindb.) from Turkey. Turk J Chem 34(5):825–834
19. Gao B, Chen MX, Li XS, Liang YQ, Zhang DY, Wood AJ, Oliver MJ, Zhang JH (2020) Ancestral gene duplications in mosses characterized by integrated phylogenomic analyses. J Syst Evol 60(1):144–159
20. Vujičić M, Sabovljević A, Sabovljević M (2011) Axenically culturing the bryophytes: establishment and propagation of the moss *Hypnum cupressiforme* Hedw. (Bryophyta, Hypnaceae) in in vitro conditions. Bot Serb 35(1):71–77
21. Pantović J, Veljić M, Grdović S, Sabovljevic MS (2021) An annotated list of moss species of Serbia. Phytotaxa 479(3):207–249
22. Altuner EM, Canli K, Akata I (2014) Antimicrobial screening of Calliergonella cuspidata, Dicranum polysetum and Hypnum cupressiforme. J Pure Appl Microbiol 8(1):539–545
23. Balasundram N, Sundram K, Samman S (2006) Phenolic compounds in plants and agri-industrial by-products: antioxidant activity, occurrence, and potential uses. Food Chem 99 (1):191–203
24. Cox-Georgian D, Ramadoss N, Dona C, Basu C (2019) Therapeutic and medicinal uses of terpenes. Med Plants 12:333–359
25. Osterdahl BG, Lindberg G (1977) Chemical studies on bryophytes. 18. Luteolin 7-0-Neohesperi-doside-4-O-sophorosidc, another new tetraglycoside from *Hedwigia ciliata*. Acta Chem Scand B 31:293–296
26. Osterdahl BG (1979) Chemical studies on bryophytes. 22. Flavonoid C-Glycosides of *Mnium undulatum*. Acta Chem Scand B 33:400–404
27. Filiz S, Semra I, Cansu F-I (2011) An evaluation of the antimicrobial activity of some Turkish mosses. J Med Plant Res 5(14):3286–3292
28. Erturk O, Sahin H, Erturk E, Hotaman HE, Koz B, Ozdemir O (2015) The antimicrobial and antioxidant activities of extracts obtained from some moss species in Turkey. Herba Polonica 61(4):52–65
29. Sievers H, Burkhardt G, Becker H, Zinsmeister HD (1994) Further biflavonoids and 3-′-phenylflavonoids from Hypnum cupressiforme. Phytochemistry 35(3):795–798
30. Filiz S, Semra I, Cansu F-I (2011) An evaluation of the antimicrobial activity of some Turkish mosses. J Med Plants Res 5(14):3286–3292

31. Yayıntaş OT, Yılmaz S, Sökmen M (2019) Determination of antioxidant, antimicrobial and antitumor activity of bryophytes from Mount Ida (Canakkale, Turkey). Indian J Tradit Knowl 18:395–401
32. Yen G-C, Duh P-D, Tsai H-L (2002) Antioxidant and pro-oxidant properties of ascorbic acid and gallic acid. Food Chem 79(3):307–313
33. Rosso R, Vieira TO, Leal PC, Nunes RJ, Yunes RA, Creczynski-Pasa TB (2006) Relationship between the lipophilicity of gallic acid n-alquil esters' derivatives and both myeloperoxidase activity and HOCl scavenging. Bioorg Med Chem 14(18):6409–6413
34. Li L, Ng T, Gao W, Li W, Fu M, Niu S, Zhao L, Chen RR, Liu F (2005) Antioxidant activity of gallic acid from rose flowers in senescence accelerated mice. Life Sci 77(2):230–240
35. Li X, Wang X, Chen D, Chen S (2011) Antioxidant activity and mechanism of protocatechuic acid in vitro. Funct Foods Health Dis 1(7):232–244
36. Zuo G, Kim H-Y, Guillen Quispe YN, Wang Z, Kim K-H, Gonzales Arce PH, Lim SS (2021) *Valeriana rigida* Ruiz & Pav. Root extract: a new source of Caffeoylquinic acids with antioxidant and aldose reductase inhibitory activities. Foods 10(5):1079
37. Zhang J, Li D-M, Sun W-J, Wang X-J, Bai J-G (2012) Exogenous p-hydroxybenzoic acid regulates antioxidant enzyme activity and mitigates heat stress of cucumber leaves. Sci Hortic 148:235–245
38. Chen JH, Ho C-T (1997) Antioxidant activities of caffeic acid and its related hydroxycinnamic acid compounds. J Agric Food Chem 45(7):2374–2378
39. Gülçin İ (2006) Antioxidant activity of caffeic acid (3, 4-dihydroxycinnamic acid). Toxicology 217(2–3):213–220
40. Plumb GW, Price KR, Williamson G (1999) Antioxidant properties of flavonol glycosides from green beans. Redox Rep 4(3):123–127
41. Kiliç I, Yeşiloğlu Y (2013) Spectroscopic studies on the antioxidant activity of p-coumaric acid. Spectrochim Acta A Mol Biomol Spectrosc 115:719–724
42. Zang L-Y, Cosma G, Gardner H, Shi X, Castranova V, Vallyathan V (2000) Effect of antioxidant protection by p-coumaric acid on low-density lipoprotein cholesterol oxidation. Am J Physiol Cell Physiol 279(4):C954–C960
43. Razavi SM, Zahri S, Zarrini G, Nazemiyeh H, Mohammadi S (2009) Biological activity of quercetin-3-O-glucoside, a known plant flavonoid. Russ J Bioorg Chem 5(3):376–378
44. Lesjak M, Beara I, Simin N, Pintać D, Majkić T, Bekvalac K, Orcic D, Mimica-Dukic N (2018) Antioxidant and anti-inflammatory activities of quercetin and its derivatives. J Funct Foods 40:68–75
45. Hariharan P, Palani M, Vaiyapuri M (2017) An evaluation of antioxidant potential of flavonoid eriodictyol in isoproterenol-induced myocardial infarction in rats. Indian J Pharm Educ Res 51 (4):603–612
46. Johnson J, Maher P, Hanneken A (2009) The flavonoid, eriodictyol, induces long-term protection in ARPE-19 cells through its effects on Nrf2 activation and phase 2 gene expression. Investig Ophthalmol Vis Sci 50(5):2398–2406
47. Singh JPV, Selvendiran K, Banu SM, Padmavathi R, Sakthisekaran D (2004) Protective role of Apigenin on the status of lipid peroxidation and antioxidant defense against hepatocarcinogenesis in Wistar albino rats. Phytomedicine 11(4):309–314
48. Romanova D, Vachalkova A, Cipak L, Ovesna Z, Rauko P (2001) Study of antioxidant effect of apigenin, luteolin and quercetin by DNA protective method. Neoplasma 48(2):104–107
49. Cavia-Saiz M, Busto MD, Pilar-Izquierdo MC, Ortega N, Perez-Mateos M, Muñiz P (2010) Antioxidant properties, radical scavenging activity and biomolecule protection capacity of flavonoid naringenin and its glycoside naringin: a comparative study. J Sci Food Agric 90(7): 1238–1244
50. Wang J, Fang X, Ge L, Cao F, Zhao L, Wang Z, Xiao W (2018) Antitumor, antioxidant and anti-inflammatory activities of kaempferol and its corresponding glycosides and the enzymatic preparation of kaempferol. PLoS One 13(5):e0197563

51. Kashyap D, Sharma A, Tuli HS, Sak K, Punia S, Mukherjee TK (2017) Kaempferol–A dietary anticancer molecule with multiple mechanisms of action: recent trends and advancements. J Funct Foods 30:203–219
52. Wang H, Jiang Z, Pang Z, Zhou T, Gu Y (2020) Acacetin alleviates inflammation and matrix degradation in nucleus Pulposus cells and ameliorates intervertebral disc degeneration in vivo. Drug Des Devel Ther 14:4801
53. Ma X, Yu H (2006) Cancer issue: global burden of cancer. Yale J Biol Med 79(3–4):85
54. Compton C (2020) Cancer initiation, promotion, and progression and the acquisition of key behavioral traits. In: Cancer: the enemy from within. Springer, Cham, pp 25–48
55. Shu Y-Z (1998) Recent natural products based drug development: a pharmaceutical industry perspective. J Nat Prod 61(8):1053–1071
56. Shi X, Luo X, Chen T, Guo W, Liang C, Tang S, Mo J (2021) Naringenin inhibits migration, invasion, induces apoptosis in human lung cancer cells and arrests tumour progression in vitro. J Cell Mol Med 25(5):2563–2571
57. Pelinson LP, Assmann CE, Palma TV, da Cruz IBM, Pillat MM, Mânica A, Stefanell N, Weis GCC, Allves ADO, Andrade CMD, Ulrich H, Morsch VMM, Schetinger MRC, Bagatini MD (2019) Antiproliferative and apoptotic effects of caffeic acid on SK-Mel-28 human melanoma cancer cells. Mol Biol Rep 46(2):2085–2092
58. Teng B-s, Lu Y-H, Wang Z-T, Tao X-Y, Wei D-Z (2006) In vitro anti-tumor activity of isorhamnetin isolated from *Hippophae rhamnoides* L. against BEL-7402 cells. Pharmacol Res 54(3):186–194
59. Hu X, Yang Z, Liu W, Pan Z, Zhang X, Li M, Liu X, Zheng Q, Li D (2020) The anti-tumor effects of p-coumaric acid on melanoma A375 and B16 cells. Front Oncol 10:558414
60. Kim MY, Choi EO, HwangBo H, Kwon DH, Ahn KI, Kim HJ, Ji SY, Hong SH, Jeong JW, GY KI, Park C, Choi YH (2018) Reactive oxygen species-dependent apoptosis induction by water extract of Citrus unshiu peel in MDA-MB-231 human breast carcinoma cells. Nutr Res Pract 12(2):129–134
61. Brüne B, von Knethen A, Sandau KB (1998) Nitric oxide and its role in apoptosis. Eur J Pharmacol 351(3):261–272
62. Wang J, Yi J (2008) Cancer cell killing via ROS: to increase or decrease, that is the question. Cancer Biol Ther 7(12):1875–1884
63. Wink DA, Vodovotz Y, Laval J, Laval F, Dewhirst MW, Mitchell JB (1998) The multifaceted roles of nitric oxide in cancer. Carcinogenesis 19(5):711–721
64. Yip E, Chan A, Pang H, Tam Y, Wong YH (2006) Protocatechuic acid induces cell death in HepG2 hepatocellular carcinoma cells through a c-Jun N-terminal kinase-dependent mechanism. Cell Biol Toxicol 22(4):293–302
65. Tseng T-H, Kao T-W, Chu C-Y, Chou F-P, Lin W-L, Wang C-J (2000) Induction of apoptosis by hibiscus protocatechuic acid in human leukemia cells via reduction of retinoblastoma (RB) phosphorylation and Bcl-2 expression. Biochem Pharmacol 60(3):307–315
66. Shon W-K, Cho C-H, Ramachandran S, Song D-K, Shin S-J, Kwon S-H, Cja SD (2008) Induction of apoptosis by *Hibiscus* protocatechuic acid in human uterine leiomyoma cells. Korean J Gynecol Oncol 19(1):48–56
67. Lin HH, Chen JH, Huang CC, Wang CJ (2007) Apoptotic effect of 3, 4-dihydroxybenzoic acid on human gastric carcinoma cells involving JNK/p38 MAPK signaling activation. Int J Cancer 120(11):2306–2316
68. Yin M-C, Lin C-C, Wu H-C, Tsao S-M, Hsu C-K (2009) Apoptotic effects of protocatechuic acid in human breast, lung, liver, cervix, and prostate cancer cells: potential mechanisms of action. J Agric Food Chem 57(14):6468–6473
69. Murad LD, Soares NCP, Brand C, Monteiro MC, Teodoro AJ (2015) Effects of caffeic and 5-caffeoylquinic acids on cell viability and cellular uptake in human colon adenocarcinoma cells. Nutr Cancer 67(3):532–542

70. In J-K, Kim J-K, Oh JS, Seo D-W (2016) 5-Caffeoylquinic acid inhibits invasion of non-small cell lung cancer cells through the inactivation of p70S6K and Akt activity: involvement of p53 in differential regulation of signaling pathways. Int J Oncol 48(5):1907–1912
71. Wang X-N, Wang K-Y, Zhang X-S, Yang C, Li X-Y (2018) 4-Hydroxybenzoic acid (4-HBA) enhances the sensitivity of human breast cancer cells to adriamycin as a specific HDAC6 inhibitor by promoting HIPK2/p53 pathway. Biochem Biophys Res Commun 504(4):812–819
72. Kanimozhi G, Prasad NR (2015) Anticancer effect of caffeic acid on human cervical cancer cells. In: Coffee in health and disease prevention. Elsevier, In, pp 655–661
73. Maiyo FC, Moodley R, Singh M (2016) Cytotoxicity, antioxidant and apoptosis studies of quercetin-3-O glucoside and 4-(β-D-glucopyranosyl-1→ 4-α-L-rhamnopyranosyloxy)-benzyl isothiocyanate from Moringa oleifera. Anti Cancer Agents Med Chem 16(5):648–656
74. Lee J, Han S-I, Yun J-H, Kim JH (2013) Quercetin 3-O-glucoside suppresses epidermal growth factor–induced migration by inhibiting EGFR signaling in pancreatic cancer cells. Tumor Biol 36(12):9385–9393
75. Jaganathan SK, Supriyanto E, Mandal M (2013) Events associated with apoptotic effect of p-Coumaric acid in HCT-15 colon cancer cells. World J Gastroenterol 19(43):7726
76. Sudan S, Rupasinghe HV (2014) Quercetin-3-O-glucoside induces human DNA topoisomerase II inhibition, cell cycle arrest and apoptosis in hepatocellular carcinoma cells. Anticancer Res 34(4):1691–1699
77. Zhai T, Zhang X, Hei Z, Jin L, Han C, Ko AT, Yu X, Wang J (2021) Isorhamnetin inhibits human gallbladder cancer cell proliferation and metastasis via PI3K/AKT signaling pathway inactivation. Front Pharmacol 12:628621
78. Zhang Y, Zhang R, Ni H (2020) Eriodictyol exerts potent anticancer activity against A549 human lung cancer cell line by inducing mitochondrial-mediated apoptosis, G2/M cell cycle arrest and inhibition of m-TOR/PI3K/Akt signalling pathway. Arch Med Sci 16(2):446
79. Li W, Du Q, Li X, Zheng X, Lv F, Xi X, Huang G, Yang J, Liu S (2020) Eriodictyol inhibits proliferation, metastasis and induces apoptosis of glioma cells via PI3K/Akt/NF-κB signaling pathway. Front Pharmacol 11:114
80. Wen S, Hu M, Xiong Y (2021) Effect of Eriodictyol on retinoblastoma via the PI3K/Akt pathway. J Healthc Eng 2021:6091585
81. Wang F, Wang Y-H, Wang J-J, Xu H-L, Wang C-M (2016) Eriodictyol-induced anti-cancer and apoptotic effects in human hepatocellular carcinoma cells are associated with cell cycle arrest and modulation of apoptosis-related proteins. Bangladesh J Pharmacol 11(2):285–291
82. Xu M, Wang S, Song Y, Yao J, Huang K, Zhu X (2016) Apigenin suppresses colorectal cancer cell proliferation, migration and invasion via inhibition of the Wnt/β-catenin signaling pathway. Oncol Lett 11(5):3075–3080
83. Lee Y, Sung B, Kang YJ, Kim DH, Jang J-Y, Hwang SY, Kim M, Lim HS, Yoon JH, Chung HY, Kim ND (2014) Apigenin-induced apoptosis is enhanced by inhibition of autophagy formation in HCT116 human colon cancer cells. Int J Oncol 44(5):1599–1606
84. Dai J, Van Wie PG, Fai LY, Kim D, Wang L, Poyil P, Luo J, Zhang Z (2016) Downregulation of NEDD9 by apigenin suppresses migration, invasion, and metastasis of colorectal cancer cells. Toxicol Appl Pharmacol 311:106–112
85. Chunhua L, Donglan L, Xiuqiong F, Lihua Z, Qin F, Yawei L, Liang Z, Ge W, Linin J, Ping Z, Kun L, Xuegng S (2013) Apigenin up-regulates transgelin and inhibits invasion and migration of colorectal cancer through decreased phosphorylation of AKT. J Nutr Biochem 24(10): 1766–1775
86. Seo H-S, Jo JK, Ku JM, Choi H-S, Choi YK, Woo J-K, Kim HI, Kang SY, Lee KM, Nam KW, Park N, Jang BH, Shin YC, Ko SG (2015) Induction of caspase-dependent extrinsic apoptosis by apigenin through inhibition of signal transducer and activator of transcription 3 (STAT3) signalling in HER2-overexpressing BT-474 breast cancer cells. Biosci Rep 35(6):e00276
87. Tseng TH, Chien MH, Lin WL, Wen YC, Chow JM, Chen CK, Kuo TC, Lee WJ (2017) Inhibition of MDA-MB-231 breast cancer cell proliferation and tumor growth by apigenin

through induction of G2/M arrest and histone H3 acetylation-mediated p21WAF1/CIP1 expression. Environ Toxicol 32(2):434–444
88. Cao X, Liu B, Cao W, Zhang W, Zhang F, Zhao H, Meng R, Zhang L, Niu R, Hao X, Zhang B (2013) Autophagy inhibition enhances apigenin-induced apoptosis in human breast cancer cells. Chin J Cancer Res 25(2):212–222
89. Coombs MRP, Harrison ME, Hoskin DW (2016) Apigenin inhibits the inducible expression of programmed death ligand 1 by human and mouse mammary carcinoma cells. Cancer Lett 380 (2):424–433
90. Seo HS, Ku JM, Choi HS, Woo JK, Jang BH, Go H, Shin YC, Ko SG (2015) Apigenin induces caspase-dependent apoptosis by inhibiting signal transducer and activator of transcription 3 signaling in HER2-overexpressing SKBR3 breast cancer cells. Mol Med Rep 2 (2):2977–2984
91. Lee Y-M, Lee G, Oh T-I, Kim BM, Shim D-W, Lee K-H, Kim YJ, Lim BO, Lim JH (2016) Inhibition of glutamine utilization sensitizes lung cancer cells to apigenin-induced apoptosis resulting from metabolic and oxidative stress. Int J Oncol 48(1):399–408
92. Zhou Z, Tang M, Liu Y, Zhang Z, Lu R, Lu J (2017) Apigenin inhibits cell proliferation, migration, and invasion by targeting Akt in the A549 human lung cancer cell line. Anti-Cancer Drugs 28(4):446–456
93. Gupta S, Afaq F, Mukhtar H (2002) Involvement of nuclear factor-kappa B, Bax and Bcl-2 in induction of cell cycle arrest and apoptosis by apigenin in human prostate carcinoma cells. Oncogene 21(23):3727–3738
94. Shukla S, Fu P, Gupta S (2014) Apigenin induces apoptosis by targeting inhibitor of apoptosis proteins and Ku70–Bax interaction in prostate cancer. Apoptosis 19(5):883–894
95. Shukla S, Kanwal R, Shankar E, Datt M, Chance MR, Fu P, MacLennan GT, Gupta S (2015) Apigenin blocks IKKα activation and suppresses prostate cancer progression. Oncotarget 6 (31):31216–31232
96. Choi J, Lee D-H, Jang H, Park S-Y, Seol J-W (2020) Naringenin exerts anticancer effects by inducing tumor cell death and inhibiting angiogenesis in malignant melanoma. Int J Med Sci 17(18):3049
97. Song HM, Park GH, Eo HJ, Lee JW, Kim MK, Lee JR, Lee MH, Koo JS, Jeong JB (2015) Anti-proliferative effect of naringenin through p38-dependent downregulation of cyclin D1 in human colorectal cancer cells. Biomol Ther 23(4):339–344
98. Lee J, Kim JH (2016) Kaempferol inhibits pancreatic cancer cell growth and migration through the blockade of EGFR-related pathway in vitro. PLoS One 11(5):e0155264
99. Zhu L, Xue L (2019) Kaempferol suppresses proliferation and induces cell cycle arrest, apoptosis, and DNA damage in breast cancer cells. Oncol Res 27(6):629
100. Jeong JC, Kim MS, Kim TH, Kim YK (2009) Kaempferol induces cell death through ERK and Akt-dependent down-regulation of XIAP and survivin in human glioma cells. Neurochem Res 34(5):991–1001
101. Singh RP, Agrawal P, Yim D, Agarwal C, Agarwal R (2005) Acacetin inhibits cell growth and cell cycle progression, and induces apoptosis in human prostate cancer cells: structure–activity relationship with linarin and linarin acetate. Carcinogenesis 26(4):845–854
102. Watanabe K, Kanno S-I, Tomizawa A, Yomogida S, Ishikawa M (2012) Acacetin induces apoptosis in human T cell leukemia Jurkat cells via activation of a caspase cascade. Oncol Rep 27(1):204–209
103. Chen L, Deng H, Cui H, Fang J, Zuo Z, Deng J, Li Y, Wang X, Zhao L (2018) Inflammatory responses and inflammation-associated diseases in organs. Oncotarget 9(6):7204
104. Lunić TM, Mandić MR, Oalđe Pavlović MM, Sabovljević AD, Sabovljević MS, Božić Nedeljković BĐ, Bozic BD (2022) The influence of seasonality on secondary metabolite profiles and neuroprotective activities of moss *Hypnum cupressiforme* extracts: in vitro and in silico study. Plan Theory 11(1):123
105. Glime J (2017) Medical uses: biologically active substances. In: Glime J (ed) Bryophyte ecology, vol 2. International Association of Bryologists, Seattle

106. Ginhoux F, Prinz M (2015) Origin of microglia: current concepts and past controversies. Cold Spring Harb Perspect Biol 7(8):a020537
107. Ahmed F, Ghalib RM, Sasikala P, Ahmed KM (2013) Cholinesterase inhibitors from botanicals. Phcog Rev 7(14):121
108. Carballo-Carbajal I, Laguna A, Romero-Giménez J, Cuadros T, Bové J, Martinez-Vicente M, Parent A, Gonzalez-Sepulveda M, Penuelas N, Torra A, Rodriguez-Galvan B, Ballabio A, Hasegawa T, Bortolozzi A, Gelpi E, Vila M (2019) Brain tyrosinase overexpression implicates age-dependent neuromelanin production in Parkinson's disease pathogenesis. Nat Commun 10 (1):1–19
109. Kim MJ, Seong AR, Yoo JY, Jin CH, Lee YH, Kim YJ, Lee J, Jun WJ, Yoon HG (2011) Gallic acid, a histone acetyltransferase inhibitor, suppresses β-amyloid neurotoxicity by inhibiting microglial-mediated neuroinflammation. Mol Nutr Food Res 55(12):1798–1808
110. Liu Y-L, Hsu C-C, Huang H-J, Chang C-J, Sun S-H, Lin AM-Y (2020) Gallic acid attenuated LPS-induced neuroinflammation: protein aggregation and necroptosis. Mol Neurobiol 57(1): 96–104
111. Winter AN, Brenner MC, Punessen N, Snodgrass M, Byars C, Arora Y, Linseman DA (2017) Comparison of the neuroprotective and anti-inflammatory effects of the anthocyanin metabolites, protocatechuic acid and 4-hydroxybenzoic acid. Oxidative Med Cell Longev 2017: 6297080
112. Kaewmool C, Kongtawelert P, Phitak T, Pothacharoen P, Udomruk S (2020) Protocatechuic acid inhibits inflammatory responses in LPS-activated BV2 microglia via regulating SIRT1/ NF-κB pathway contributed to the suppression of microglial activation-induced PC12 cell apoptosis. J Neuroimmunol 341:577164
113. Kim M, Choi S-Y, Lee P, Hur J (2015) Neochlorogenic acid inhibits lipopolysaccharide-induced activation and pro-inflammatory responses in BV2 microglial cells. Neurochem Res 40(9):1792–1798
114. Liang G, Shi B, Luo W, Yang J (2015) The protective effect of caffeic acid on global cerebral ischemia-reperfusion injury in rats. Behav Brain Funct 11(1):1–10
115. Zhang M, Zhou J, Wang L, Li B, Guo J, Guan X, Han Q, Zhang H (2014) Caffeic acid reduces cutaneous tumor necrosis factor alpha (TNF-α), IL-6 and IL-1β levels and ameliorates skin edema in acute and chronic model of cutaneous inflammation in mice. Biol Pharm Bull 37(3): 347–354
116. Ola MS, Ahmed MM, Ahmad R, Abuohashish HM, Al-Rejaie SS, Alhomida AS (2015) Neuroprotective effects of rutin in streptozotocin-induced diabetic rat retina. J Mol Neurosci 56(2):440–448
117. Guven M, Aras AB, Akman T, Sen HM, Ozkan A, Salis O, Sehitogiu I, Kaikan Y, Silan C, Deniz M, Cosar M (2015) Neuroprotective effect of p-coumaric acid in rat model of embolic cerebral ischemia. Iran J Basic Med Sci 18(4):356–383
118. Daroi PA, Dhage SN, Juvekar AR (2022) p-Coumaric acid mitigates lipopolysaccharide induced brain damage via alleviating oxidative stress, inflammation and apoptosis. J Pharm Pharmacol 74(4):556–564
119. Kim JH, Lee S, Cho EJ, Kim HY (2019) Neuroprotective effects of kaempferol, quercetin, and its glycosides by regulation of apoptosis. J Korea Acad-Ind Coop Soc 20(2):286–293
120. Lee JK (2011) Anti-inflammatory effects of eriodictyol in lipopolysaccharidestimulated raw 264.7 murine macrophages. Arch Pharm Res 34(4):671–679
121. Zhu GF, Guo HJ, Huang Y, Wu CT, Zhang XF (2015) Eriodictyol, a plant flavonoid, attenuates LPS-induced acute lung injury through its antioxidative and anti-inflammatory activity. Exp Ther Med 10(6):2259–2266
122. Wang J, Liu Y-T, Xiao L, Zhu L, Wang Q, Yan T (2014) Anti-inflammatory effects of apigenin in lipopolysaccharide-induced inflammatory in acute lung injury by suppressing COX-2 and NF-kB pathway. Inflammation 37(6):2085–2090
123. Balez R, Steiner N, Engel M, Muñoz SS, Lum JS, Wu Y, Wang D, Vallotton P, Sachdev P, O'Connor M, Sidhu K, Munch G, Ooi L (2016) Neuroprotective effects of apigenin against

inflammation, neuronal excitability and apoptosis in an induced pluripotent stem cell model of Alzheimer's disease. Sci Rep 6(1):1–16
124. Bodet C, La V, Epifano F, Grenier D (2008) Naringenin has anti-inflammatory properties in macrophage and ex vivo human whole-blood models. J Periodontal Res 43(4): 400–407
125. Zhang N, Hu Z, Zhang Z, Liu G, Wang Y, Ren Y, Wu Y, Geng F (2018) Protective role of naringenin against Aβ25-35-caused damage via ER and PI3K/Akt-mediated pathways. Cell Mol Neurobiol 8(2):549–557
126. Yu L, Chen C, Wang L-F, Kuang X, Liu K, Zhang H, Du JR (2013) Neuroprotective effect of kaempferol glycosides against brain injury and neuroinflammation by inhibiting the activation of NF-κB and STAT3 in transient focal stroke. PLoS One 8(2):e55839
127. Çolak E, Kara R, Ezer T, Çelik GY, Elibol B (2011) Investigation of antimicrobial activity of some Turkish pleurocarpic mosses. Afr J Biotechnol 10(60):12905–12908
128. Dulger B, Yayintas OT, Gonuz A (2005) Antimicrobial activity of some mosses from Turkey. Fitoterapia 76(7–8):730–732
129. Erturk O, Sahin H, Erturk E, Hotaman HE, Koz B, Ozdemir O (2015) The antimicrobial and antioxidant activities of extracts obtained from some moss species in Turkey. Herba Pol From Bot to Med Res 61(4):52–65
130. Rajamanickam K, Yang J, Sakharkar MK (2019) Gallic acid potentiates the antimicrobial activity of tulathromycin against two key bovine respiratory disease (BRD) causing-pathogens. Front Pharmacol 9:1486
131. Borges A, Ferreira C, Saavedra MJ, Simões M (2013) Antibacterial activity and mode of action of ferulic and gallic acids against pathogenic bacteria. Microb Drug Resist 19(4): 256–265
132. Kuete V, Nana F, Ngameni B, Mbaveng AT, Keumedjio F, Ngadjui BT (2009) Antimicrobial activity of the crude extract, fractions and compounds from stem bark of *Ficus ovata* (Moraceae). J Ethnopharmacol 124(3):556–561
133. Bajko E, Kalinowska M, Borowski P, Siergiejczyk L, Lewandowski W (2016) 5-O-Caffeoylquinic acid: a spectroscopic study and biological screening for antimicrobial activity. LWT – Food Sci Technol 65:471–479
134. Cho J-Y, Moon J-H, Seong K-Y, Park K-H (1998) Antimicrobial activity of 4-hydroxybenzoic acid and trans 4-hydroxycinnamic acid isolated and identified from rice hull. Biosci Biotechnol Biochem 62(11):2273–2276
135. Kępa M, Miklasińska-Majdanik M, Wojtyczka RD, Idzik D, Korzeniowski K, Smoleń-Dzirba J, Wasik TJ (2018) Antimicrobial potential of caffeic acid against *Staphylococcus aureus* clinical strains. Biomed Res Int 2018:Article ID 7413504
136. Rigano D, Formisano C, Basile A, Lavitola A, Senatore F, Rosselli S, Bruno M (2007) Antibacterial activity of flavonoids and phenylpropanoids from *Marrubium globosum* ssp. *libanoticum*. Phytother Res 21(4):395–397
137. Lou Z, Wang H, Rao S, Sun J, Ma C, Li J (2012) p-Coumaric acid kills bacteria through dual damage mechanisms. Food Control 25(2):550–554
138. Akroum S, Bendjeddou D, Satta D, Lalaoui K (2009) Antibacterial activity and acute toxicity effect of flavonoids extracted from Mentha longifolia. Am Eurasian J Sci Res 4(2): 93–96
139. Du Toit K, Buthelezi S, Bodenstein J (2009) Anti-inflammatory and antibacterial profiles of selected compounds found in South African propolis. S Afr J Sci 105(11):470–472
140. Nayaka HB, Londonkar RL, Umesh MK, Tukappa A (2014) Antibacterial attributes of apigenin, isolated from *Portulaca oleracea* L. Int J Bacteriol 2014:Article ID 175851
141. Song H-S, Bhatia SK, Gurav R, Choi T-R, Kim HJ, Park Y-L, Han YH, Park JY, Mi S, Park SL, Lee HS, Kim W, Kim YG, Yang YH (2020) Naringenin as an antibacterial reagent controlling of biofilm formation and fatty acid metabolism in MRSA. bioRxiv. https://doi.org/10.1101/2020.03.08.983049

142. Escandón RA, Del Campo M, López-Solis R, Obreque-Slier E, Toledo H (2016) Antibacterial effect of kaempferol and(−)-epicatechin on *Helicobacter pylori*. Eur Food Res Technol 242 (9):1495–1502
143. Rocha MFG, Sales JA, da Rocha MG, Galdino LM, de Aguiar L, Pereira-Neto WDA, Cardeiro RDA Castelo-Branco DDSCM, Sidrim JJC, Brihante RSN (2019) Antifungal effects of the flavonoids kaempferol and quercetin: a possible alternative for the control of fungal biofilms. Biofouling 35(3):320–328
144. Li S, Lv Q, Sun X, Tang T, Deng X, Yin Y, Li L (2020) Acacetin inhibits Streptococcus pneumoniae virulence by targeting pneumolysin. J Pharm Pharmacol 72(8):1092–1100
145. More G, Lall N, Hussein A, Tshikalange TE (2012) Antimicrobial constituents of Artemisia afra Jacq. ex Willd. Against periodontal pathogens. Evid Based Complement Alternat Med 2012:Article ID 252758

Extracts from the Liverwort *Bazzania trilobata* with Potential Dermo-cosmetic Properties

6

Raíssa Volpatto Marques, Aleksander Salwinski, Kasper Enemark-Rasmussen, Charlotte H. Gotfredsen, Yi Lu, Nicolas Hocquigny, Arnaud Risler, Raphaël E. Duval, Sissi Miguel, Frédéric Bourgaud, and Henrik Toft Simonsen

Contents

Supplementary Information: The online version contains supplementary material available at https://doi.org/10.1007/978-3-031-23243-5_9.

R. V. Marques · Y. Lu · H. T. Simonsen (✉)
Department of Biotechnology and Biomedicine, Technical University of Denmark, Kongens Lyngby, Denmark
e-mail: raivol@dtu.dk; yilu@dtu.dk; henrik.toft.simonsen@univ-st-etienne.fr

A. Salwinski
Plant Advanced Technologies, Vandœuvre-lès-Nancy, France
e-mail: alek-sander.salwinski@plantadvanced.com

K. Enemark-Rasmussen · C. H. Gotfredsen
Department of Chemistry, Technical University of Denmark, Lyngby, Denmark
e-mail: keras@kemi.dtu.dk; chg@kemi.dtu.dk

N. Hocquigny · R. E. Duval
Université de Lorraine, CNRS, Nancy, France

Faculté de Pharmacie, ABC Platform®, Vandœuvre-lès-Nancy, France
e-mail: raphael.duval@univ-lorraine.fr

A. Risler
Université de Lorraine, CNRS, Nancy, France
e-mail: ar-naud.risler@univ-lorraine.fr

S. Miguel
Cellengo, Vandœuvre-lès-Nancy, France
e-mail: sissi.miguel@cellengo.com

H. N. Murthy (ed.), *Bioactive Compounds in Bryophytes and Pteridophytes*, Reference Series in Phytochemistry, https://doi.org/10.1007/978-3-031-23243-5_9

Abstract

Bazzania trilobata (L.) Gray is a leafy liverwort from the family of Lepidoziaceae, well known for its antifungal properties. In this study, the 70% ethanol and methanol extracts of *B. trilobata* were investigated for new in vitro biological activities of cosmetic interest. The results showed that the total phenol content, the DPPH (1,1-diphenyl-2-picrylhydrazyl) free radical scavenging activity, and the anti-collagenase activity of the 70% ethanol extract were higher than for methanol. The methanol extract showed mild tyrosinase inhibitory activity and antimicrobial properties towards the Gram-positive bacteria *Enterococcus faecalis*. Lignans, coumarins, and bis-bibenzyls were the major classes of phenolic constituents tentatively identified in both extracts. In addition, a known drimenyl caffeate was identified in *B. trilobata* and its structure was confirmed by NMR spectroscopy. These results suggest that extracts from *B. trilobata* could be exploited as an interesting new source of natural active ingredients for cosmetic applications.

Keywords

Antioxidant · Antimicrobial · *Bazzania trilobata* · Collagenase inhibitory activity · Drimenyl caffeate · Tyrosinase inhibitory activity

Abbreviations

COSY	Homonuclear Correlation Spectroscopy
DPPH	1-Diphenyl-2-picrylhydrazyl
GAE	Gallic Acid Equivalents

F. Bourgaud
Plant Advanced Technologies, Vandœuvre-lès-Nancy, France

Cellengo, Vandœuvre-lès-Nancy, France
e-mail: frederic.bourgaud@plantadvanced.com

HMBC	Heteronuclear Multiple Bond Correlation
IC50	Half Maximal Inhibitory Concentration
J	Coupling Constant
MIC	Minimum Inhibitory Concentration
MS/MS	Tandem Mass Spectrometry
NMR	Nuclear Magnetic Resonance
TPC	Total Phenolic Content
UHPLC-HRMS	Ultrahigh Performance Liquid Chromatography-High Resolution Mass Spectrometry

1 Introduction

Bazzania trilobata (L.) Gray (Lepidoziaceae) is a leafy liverwort with a circumboreal distribution, including western Europe, eastern and western USA, and Japan, which grows in extensive gametophyte mats [1]. *B. trilobata* has been described for its antitumor [2] and antifungal properties [3, 4]. There is already a commercial antifungal and antibacterial product in Germany based on an ethanol extract of *B. trilobata* [5, 6]. Sesquiterpenes and bis-bibenzyls have been reported as antifungal constituents from *B. trilobata* [4]. Extracts and isolated compounds from other species of *Bazzania* have shown therapeutic potential with antitumor [7, 8], antimicrobial [9, 10], and inhibitory effects on nitric oxide production [11, 12]. Thus, *Bazzania* spp. is a source of valuable bioactive compounds. However, the knowledge of biological activities available from *Bazzania* spp. and other bryophytes is little compared to that of higher plants [13–15]. Therefore, this study provides additional knowledge on the new potential biological properties of extracts from *B. trilobata*.

Bioactive plant extracts have found valuable applications, especially in cosmetics and herbal remedies, including that of bryophytes [14, 16, 17]. Plant extracts rich in polyphenols are an important source of natural antioxidant ingredients for the protection of the skin against free radicals [18]. Plant metabolites are also applied as anti-wrinkle and skin-lightening agents. One of the key targets in the cosmetic industry is the discovery of inhibitors of aging-related enzymes, such as collagenase and elastase. These enzymes, when overexpressed, can lead to accelerated proteolytic degradation of collagen and elastin fibers in the extracellular matrix that impacts the integrity and elasticity of the skin [19, 20]. Another important target is tyrosinase, the main enzyme in the melanin synthetic pathway. Inhibition of its activity is one of the ways of preventing skin hyperpigmentation disorders [21]. Furthermore, it is an advantage to obtain extracts of cosmetic interest with additional antimicrobial activity. These ingredients are called preservative boosters and can contribute to lowering the concentration of synthetic preservatives in final cosmetic formulations [22]. In this study, the inhibitory effects of 70% ethanol and methanol extract from *B. trilobata* on skin aging and pigmentation-related enzymes, as well as their antioxidant and antimicrobial properties, were investigated. The phytochemical constituents of both extracts were tentatively identified.

2 Methods

2.1 Plant Material

Bazzania trilobata (L.) Gray was collected in the Black Forest, Germany (Lat. 47.911223/Long. 8.092431) in April 2018 and identified by Professor Dr. Nils Cronberg (Department of Biology, Lund University, Lund, Sweden). The specimen is identical to the voucher specimen with ID no MTRaMa13 sent for deposition at the Lund University Botanical Museum (LD). In this study, the whole plant was used for analysis [20].

2.2 Extraction Preparation

B. trilobata was dried at room temperature and ground to a fine powder using a bead mill. The dried powder was homogenized in 70% ethanol (v/v) in water and methanol for the extraction of small molecules. The solution (1:10 g/mL of dry weight to solvent ratio) was macerated for 30 min by a rotating mixer at room temperature. After centrifugation, the supernatant was collected and used for analysis. The 70% ethanol and methanol extracts were tested at the final concentrations as indicated in each experiment. For the antimicrobial analysis, the methanol extract was evaporated and the dry extract was dissolved in dimethyl sulfoxide (DMSO; Carlo Erba) at a concentration of 20.5 mg/mL [20].

2.3 Determination of Total Phenolic Content

The total phenolic content (TPC) was determined by the Folin–Ciocalteu's method [23]. Briefly, 20 μL of plant extracts (1/4 diluted), water (blank), and diluted gallic acid standard solutions (Sigma-Aldrich, ref. G7384; 0.4, 0.2, 0.1, 0.05, 0.025, 0.0125, 0.00625, 0.003125 mg/mL) were added to a microplate. Next, 100 μL of 10% Folin–Ciocalteu (Sigma-Aldrich; ref. F9252) and 80 μL of 7.5% sodium carbonate (Merck; ref. 1.06392.0500) were added to the samples, and the absorption was measured by spectrophotometer (Synergy HT, BioTek®) at 760 nm for 30 min at 25 °C. The TPC was estimated from a standard curve of gallic acid [23]. The results were expressed in terms of milligrams of gallic acid equivalent per 100 mg of dry plant material. The assay was conducted in triplicate.

2.4 DPPH Free Radical Scavenging Assay

The DPPH (1,1-diphenyl-2-picrylhydrazyl; Sigma-Aldrich) free radical scavenging activity of the extracts was determined based on the methods previously described [24]. The samples were prepared at eight different concentrations, then 70 μL of each dilution was mixed with 140 μL of methanolic DPPH solution (0.6×10^{-4} M). The same procedure was realized for the positive (ascorbic acid, Sigma-Aldrich; ref. A7506-100G) and negative (methanol) controls. In separated wells, the extracts and ascorbic acid dilutions were also mixed to 140 μL of 100% methanol for the sample's absorbance corrections. The samples were incubated for 30 min at 25 °C and the absorbance was measured at 517 nm. The IC_{50} values were estimated by the linear regression method.

$$\text{DPPH radical scavenging activity}(\%) = \frac{\text{DO}(100\%) - (\text{DO} - \text{DO}(\text{blank}))}{\text{DO}(100\%)} \times 100$$

where DO: extraction solution + DPPH solution; DO (blank): extraction solution + methanol and DO (100%): methanol + DPHH solution.

2.5 In Vitro Collagenase Inhibition Assay

Collagenase inhibition activity was measured by following the enzymatic conversion of the synthetic substrate FALGPA (*N*-[3-(2-Furyl)acryloyl]-Leu-Gly-Pro-Ala) (Bachem; ref. 4006713.0025) to FAL (N- (3[2-Furyl]acryloyl)-Leu) + Gly-Pro-Ala (GPA). The collagenase activity from *Clostridium histolyticum* (type IA, Sigma-Aldrich, ref. C9891, specific activity ≥125 CDU/mg solid) was determined by the procedure previously described by [25]. Ethylenediaminetetraacetic acid (EDTA) disodium salt dihydrate (purity ≥99%, Alfa Aesar; ref. A15161) was used as the control [20].

2.6 In Vitro Elastase Inhibition Assay

The elastase inhibitory activity was determined by a spectrophotometric method using a microplate reader, Synergy HT (Biotek). The assay is based on the detection of enzymatic-driven conversion of *N*-succinyl-Ala-Ala-Ala-p-nitroanilide (SAAApNA; Sigma-Aldrich, ref. S4760) to p-nitroanilide (pNA) that strongly absorbs at 420 nm. The reaction mixture contained 170 μL of elastase synthetic substrate: SAAApNA (1.5 mM in 50 mM Tris buffer containing, 10 mM $CaCl_2$ and 400 mM NaCl, pH 7.5) and 20 μL of plant extract (test sample) or pure solvent of the sample (blank, control). The enzymatic conversion was initiated by the addition of 10 μL of 0.05 mg/mL of porcine pancreatic elastase (Sigma-Aldrich, ref. E7885-5MG) in the same buffer as its substrate SAAApNA.

Elastase-driven conversion of SAAApNA to pNA was followed for 25 min at 25 °C by measuring an increase of the sample's absorption at 420 nm, proportional to pNA concentration. 3,4-dichloroisocoumarin (3,4-DCIC) (Sigma Aldrich, D7910, purity ≥98%) was used as a positive control. The points in the linear range of the absorbance versus time plots were applied to calculate the slopes, directly proportional to elastase activity [20]. Then, the values of elastase inhibition expressed as the percent of the activity of the test samples versus the control experiment (pure solvent) were calculated for all samples according to the following equation:

$$\text{Elastase activity(EA\%)} = \frac{\text{Slope of sample}}{\text{Slope of blank}} \times 100$$

$$\text{Elastase inhibition activity(\%)} = 100\% - \text{EA\%}$$

2.7 In Vitro Tyrosinase Inhibition Assay

The mushroom tyrosinase inhibitory activity was determined by a spectrophotometric method using a microplate reader, Synergy HT (Biotek), based on [26] with modifications. Tyrosinase-driven conversion of L-Tyr to dopachrome was followed for 25 min at 25 °C by measuring an increase of the sample's absorption at 475 nm, proportional to dopachrome concentration. Kojic acid (purity 99%, Alfa Aesar) was used as a positive control. The tyrosinase activity was determined by the procedure described by [20].

2.8 Antibacterial Assay

Antibacterial activities were screened with the concentration of the methanol extract at 512 μg/mL. The antibacterial activity was determined by the broth microdilution method based on ISO 20776-1:2006 standard [27], in accordance with CLSI [28] and EUCAST [29] guidelines. The method was previously described [30]. The following bacteria have been used in this work: *Escherichia coli* ABC5 (ATCC 25922), *Staphylococcus aureus* ABC1 (ATCC 29213), *Pseudomonas aeruginosa* ABC4 (ATCC 27853), *Klebsiella pneumoniae* ABC12 (ATCC 700603), *Staphylococcus epidermidis* ABC91 (clinical origin), *Enterococcus faecalis* ABC 3 (ATCC 29212), *Acinetobacter baumannii* ABC 14 (ATCC 19606), *Enterobacter cloacae* ABC 45 (clinical origin).

Briefly, the screening test conditions were performed as follows, positive growth control with 75 μL MHB-CA (Mueller-Hinton Broth, Cations-Adjusted) with bacteria, + 25 μL H_2O. 5 [2–8].10^5 CFU/mL per well. There were eight replicates/bacteria/microplate. A negative control without bacteria was also tested both with and without the test sample, the latter annotated as sample control. For the test, 25 μL of the sample was added. All were incubated for 24 h at 35 °C.

2.9 Antifungal Assay

The antifungal activities were screened with the concentration of the methanol extract at 512 μg/mL. *Candida albicans* ABC F1 (clinical origin) and *Aspergillus brasiliensis* ABC F16 (ATCC 16404) were used for this study. To investigate the antifungal activities of the extract, the antifungal activity was determined by the broth microdilution method according to EUCAST guidelines [31]. Again a positive growth control was tested with 100 μL 2X RPMI 1640 + 50 μL fungi suspension +50 μL H_2O. For *Candida*, there was [1–5].10^5 UFC/mL per well; and for *Aspergillus* it was [0.5–2.5].105 spores/mL per well. There is 8 replicates/fungi/

microplate. A negative control without bacteria was also tested both with and without the test sample, the latter annotated as sample control. For the test, 25 μL of the sample was added. All were incubated for 24 h at 35 °C for *Candida*, and 48 h at 35 °C for *Aspergillus*.

2.10 UHPLC-HRMS Analysis

The extracts were diluted at 10 mg/mL in ethanol absolute and a volume of 1 μl of samples was injected for analysis. Ultrahigh performance liquid chromatography-high resolution mass spectrometry (UHPLC-HRMS) was realized on Agilent 1290 Infinity II UHPLC (Agilent Technologies) with diode array detector (DAD) coupled to an Agilent 6545 QTOF with Agilent Dual Jet Stream electrospray ion source with a drying gas temperature of 325 °C, a gas flow of 8 L/min, and a sheath gas temperature of 300 °C and flow of 12 L/min. The capillary voltage was set to 4000 V and a nozzle voltage to 500 V. Analyses were performed in negative ion mode.

Mass spectra were recorded at centroid mode for m/z 100–1700 in MS mode and m/z 30–1700 in MS/MS mode, with an acquisition rate of 10 spectra/s using fixed collision energies of 10, 20, and, 40 eV and a maximum of three selected precursor ions per cycle.

The separation was performed on a reversed-phase column Agilent Poroshell 120 Phenyl Hexyl column (150 × 2.1 mm, 1.9 μm), using water/acetonitrile mobile phase, both containing 20 mM formic acid (phase A/B respectively). Phase B increased from 10% to 100% in 10 min, then held at 100% B for 2 min, returned to 10% in 0.1 min, and equilibrated for 2 min at a flow rate of 350 μL/min, and column temperature of 40 °C [20].

The LC-MS/MS raw data were processed by the open-source software MS-DIAL (version 4.60), enabling ion chromatogram extraction and peak deconvolution [32]. The processed data (mass spectrometry and spectral data) were used to tentatively identify by matching the mass spectral data of the compounds 1–9 (Fig. S1) against the records of the MS-FINDER databases (Version 3.50) [33] (http://prime.psc.riken.jp/).

2.11 Purification by Preparative Liquid Chromatography

Compound **10** was purified from the commercial Lebermooser extract (Niem-Handel, Gernsheim, Germany). The dry crude extract (3 g) was partitioned in distilled water and ethyl acetate. The ethyl acetate phase was evaporated and 226 mg of dry extract was dissolved in 4 mL of ethanol absolute and 1 mL distilled water. The resulting solution was used to separate compound 10 by preparative liquid chromatography (LC) Armen Spot Prep II (Armen) with a C18 column (250 mm × 50 mm, 10 μm, Vydac Denali; Grace). The fractions were purified using water containing 0.1% vol. of formic acid (A) and pure ACN (B) with the gradient mobile phase of B of 70% (0–20 min), 80–100% (20–21 min), 100% (21 min–25 min) at a flow rate of 120 mL/min and a UV detection at 238 and 324 nm (Fig. S2).

The fractions containing the purified compound were combined, evaporated under vacuum and 6.28 mg (purity (average UV-vis between 210 and 600 nm) >95%) of the compound was obtained. The isolate was analyzed using the HPLC Agilent 1200 system (Agilent) with an Agilent 1260 Infinity Diode array Detector (applied range: 210–600 nm) coupled to a mass spectrometer Agilent 6120 Quadrupole LC/MS (electrospray ionization and atmospheric pressure chemical ionization in negative or positive ion mode, m/z 100–1000), using a Vydac Denali C18 reverse-phase column (250 mm × 4,6 mm, 10 μm; Grace) maintained at 25 °C during all analyses. The mobile phase was composed of water containing 0.1% vol. of formic acid (A) and pure ACN (B), delivered at 1.5 mL/min with the gradient of the B phase as follows: 70% (0–20 min), 80–100% (20–21 min), 100% (21–25 min).

2.12 NMR Measurement

The presented NMR spectra were recorded on an 800 MHz Avance III HD spectrometer equipped with a 5 mm TCI CryoProbe (Bruker Biospin). ^{1}H and ^{13}C chemical shifts are reported relative to TMS (δ (1H) = 0.0 ppm, δ (13C) = 0.0 ppm) using the solvent signals as secondary reference (MeOD: δ (1H) = 3.31 ppm and δ (13C) = 49.0 ppm). The HSQC spectra were acquired using a data matrix of 4096 × 1024 complex points with acquisition times of 200 and 15 ms in F2 and F1, respectively. Adiabatic bilevel 1H decoupling was employed during acquisition. The HMBC spectra were acquired using a data matrix of 4096 × 512 complex points with acquisition times of 220 and 6 ms in F2 and F1, respectively. The DQF-COSY spectra were acquired using a data matrix of 4096 × 1024 complex points with acquisition times of 220 and 53 ms in F2 and F1, respectively [20].

3 Phytochemicals and Biological Activities of *Bazzania trilobata*

3.1 Determination of Phenolic Content of *B. trilobata* Extracts

Plants extracts containing polyphenols have shown significant redox properties with antioxidants and health benefits for humans [34]. Polyphenolic extracts have found valuable applications as active ingredients in cosmetic formulations due to their range of properties, such as antioxidants, antimicrobial, anti-inflammatory, and anti-aging activities [35]. Thus, the total phenolic content (TPC) of the 70% ethanol and methanol extracts of *B. trilobata* was determined based on the colorimetric Folin-Ciocalteu method (plant extraction and determination of TPC are given in the supplementary material). The TPC was expressed as gallic acid equivalents (Table 1).

The TPC of 70% ethanol-based was shown to be higher by 38% than the methanol-based equivalent, which is possible due to the difference in the solvent polarity that provides a better phenol extraction efficiency (Table 1). Polyphenolic compounds have

Table 1 Total phenolic content (TPC) and DPPH free radical-scavenging activity

Extracts	TPC (mg GAE/100 mg)[a]	DPPH radical scavenging activity, IC_{50} (μg/mL)
70% Ethanol	1.30	82
Methanol	0.95	122
Ascorbic acid	–	2

[a]TPC value were expressed as gallic acid equivalents (GAE) in mg per 100 mg of dry plant material

Table 2 Predicted compounds from *Bazzania trilobata* extracts

Peak	RT (min)	Molecular Formula	Experimental (m/z) [M-H]⁻	Theoretical (*m/z*) [M-H]⁻	Error (ppm)	Tentative identification
1	2.09	$C_{18}H_{12}O_{10}$	387.0348	387.0358	−2.58	Jamesopyrone [40]
2	2.43	$C_{26}H_{26}O_{14}$	561.1252	561.1250	0.36	Trilobatin A [39]
3	2.67	$C_{15}H_{14}O_{10}$	353.0515	353.0514	0.28	7,8-Dihydroxy-7-*O*-β-D-glucuronide [4]
4	2.73	$C_{35}H_{28}O_{17}$	719.1249	719.1254	−0.70	Trilobatin K [40]
5	3.24	$C_{27}H_{20}O_{12}$	535.0870	535.0882	−2.24	Trilobatin C [39]
6	6.99	$C_{28}H_{22}O_4$	421.1437	421.1445	−1.90	Isoplagiochin C [38]
7	8.01	$C_{28}H_{20}Cl_2O_4$	489.0658	489.0666	−1.64	Bazzanin B [37]
8	8.55	$C_{28}H_{19}Cl_3O_4$	523.0269	523.0276	−1.34	Bazzanin C or D [37]
9	8.57	$C_{29}H_{20}Cl_2O_4$	501.0650	501.0666	−3.19	Bazzanin K [37]
10	8.89	$C_{24}H_{32}O_4$	383.2231	383.2228	0.78	Drimenyl caffeate[a] [52]

[a]Chemical structure confirmed by NMR

shown to be abundantly present in liverworts [36]. *B. trilobata* was described as a source of rare cyclic bis-bibenzyls and chlorinated bis-bibenzyls, e.g., isoplagiochin C and bazzanins (Table 2) [4, 37, 38]. Other polyphenolic constituents, such as lignans are also highly present in *B. trilobata*, e.g., trilobatins (Table 2) [39, 40]. Moreover, coumarins were already reported from *B. trilobata* extracts, e.g., 7,8-dihydroxy-7-*O*-β-D-glucuronide (Table 2) [4, 41].

3.2 Antioxidant Activity of *B. trilobata* Extracts

Bryophytes have developed efficient antioxidant machinery to overcome biotic and abiotic stresses; this leads to a promising alternative source of antioxidant compounds [42]. Antioxidants are molecules that neutralize free radicals, which play an important role in the prevention of various diseases and skin aging [43]. In the antioxidant screening, 70% ethanol and methanol extracts of *B. trilobata* were investigated by the DPPH (1-diphenyl-2-picrylhydrazyl) scavenging assay (determination of DPPH activity is given in the supplementary material). Both extracts revealed a reducing power; however, the DPPH radical scavenging ability of the 70% ethanol (IC_{50} 82 μg/mL) and methanol (IC_{50} 122 μg/mL) extracts were lower than ascorbic acid used as a positive control [44] (Table 1). These results agree with the TPC of the extracts indicating that the 70% ethanol extract richer in phenols has stronger antioxidant properties. Phenolic compounds have a key role as antioxidants and their activity is mainly related to the number and arrangement of hydroxyl groups in their molecular structure [18].

3.3 Collagenase, Elastase, and Tyrosinase Inhibitory Activity of *B. trilobata* Extracts

To expand the knowledge of the biological activities of the extracts from *B. trilobata*, we attempted to investigate their potential as skin anti-aging and anti-pigmentation ingredients in cosmetic formulations. Therefore, the ability to inhibit the activity of three target enzymes of cosmetic interest was investigated (determination of in vitro enzymatic activities are given in the supplementary material). The results showed that the 70% ethanol extract inhibited 40% of collagenase activity at the final concentration of 8.33 mg/mL whereas the methanol extract inhibited 20% at the final concentration of 6.66 mg/mL. The extracts exhibited limited anti-collagenase activity compared to that of the positive control EDTA (94% at 1.49 mg/mL) (Fig. 1a).

Both extracts were tested for tyrosinase activity inhibition at the final concentration of 5.33 mg/mL together with the positive control kojic acid at 0.04 mg/mL. Only the methanol extract showed moderate tyrosinase inhibition of 43%, however, lower than kojic acid, which showed 99% (Fig. 1b). Furthermore, both extracts showed no elastase inhibitory activity at the final concentration of 2.66 mg/mL.

To our knowledge, this study is the first report on the effects of *B. trilobata* extracts on collagenase and tyrosinase activities. Indeed, few studies have reported the activity of extracts or isolated metabolites from bryophytes on these target enzymes. Recently, the n-hexane and chloroform extracts at 2 mg/mL of the in vitro culture of the liverwort *Marchantia polymorpha* L. were reported to inhibit tyrosinase activity (69.54% and 69.10%, respectively) [45]. Likewise, the ethanol and methanol extract of the moss *Polytrichum formosum* was also found to have collagenase and tyrosinase inhibitory activity, respectively [20]. This confirms that bryophytes are a useful source for novel compounds with anti-tyrosinase and anti-collagenase activity.

3.4 Antimicrobial Activity of *B. trilobata*

Within the cosmetic market, there is a growing demand for skin care products containing natural antimicrobial ingredients as an alternative source to the standard synthetic preservatives. In the literature, antimicrobial activity has been reported in various species of bryophytes, particularly, in liverworts [46–50]. Bryophytes are reported not to be infected by microorganisms due to their ability to produce specialized protective molecules [5, 51].

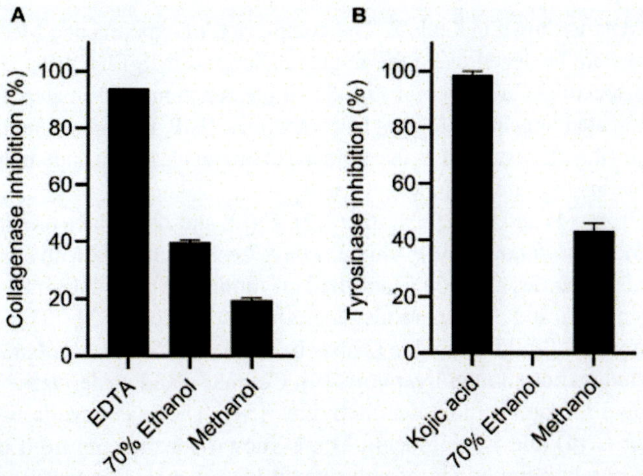

Fig. 1 Inhibitory effect of the 70% ethanol and methanol extracts on (**a**) collagenase activity and (**b**) tyrosinase activity. For collagenase activity, the 70% ethanol extract was tested at the final concentration of 8.33 mg/mL and the methanol extract at a final concentration of 6.66 mg/mL. For tyrosinase activity, both extracts were tested at a final concentration of 5.33 mg/mL. The positive control experiments were conducted using 1.49 mg/mL for EDTA (collagenase) and 0.04 mg/mL for kojic acid (tyrosinase). The results are expressed as the mean ± standard deviation ($n = 2$–3)

Thus, the antibacterial activity of the methanol extract was evaluated against Gram-positive (*Staphylococcus aureus*, *Enterococcus faecalis*, *Staphylococcus epidermidis*) and Gram-negative (*Pseudomonas aeruginosa*, *Escherichia coli*, *Klebsiella pneumoniae*, *Acinetobacter baumannii*, *Enterobacter cloacae*) bacteria of pathogenic interest (determination of the antimicrobial activity is given in the supplementary material). Antifungal activity of *B. trilobata* is well established towards phytopathogenic fungi [4]. Then, the antifungal potential against *Candida albicans* and *Aspergillus brasiliensis* was also tested. In this work, the antimicrobial activity of the methanol extract was only detected in *E. faecalis* which completely inhibited the bacterial growth at 512 μg/mL.

In other investigated species of *Bazzania*, the ethanol extract of *Bazzania tridens* evaluated with a different method showed intermediate (125–500 μg/mL) minimum inhibitory concentration (MIC) values towards *S. aureus*, *P. aeruginosa*, and *E. coli* [9]. Moreover, the sesquiterpenoid, chiloscyphenol A, isolated from the Chinese *Bazzania albifolia* showed antifungal activity against *Candida* species with MIC values of 8–32 μg/mL [10].

3.5 Chemical Constituents of *B. trilobata* Extracts

The composition of specialized metabolites of the 70% ethanol and methanol extracts was analyzed by UHPLC-HRMS (UHPLC-HRMS analysis is given in the supplementary material). The annotation of ten known compounds, based on mass spectrometry and spectral data is shown in Table 2. The comparison of experimental MS/MS and in silico spectra were analyzed using the fragmentation tool MS-FINDER [33] (MS/MS and in silico spectrums are given in Fig. S1 of the supplementary material). Figure 2 shows the extracted ion chromatogram of compounds **1–10**. Phenolic compounds were the major constituents identified in the extracts; the main classes include lignans (**1,2,4,** and **5**), coumarins (**3**), and bis-bibenzyls (**6–9**).

Jamesopyrone (**1**) and the trilobatins A (**2**), C (**5**), and K (**4**) have previously been isolated from *B. trilobata* [39, 40]. Lignans have been associated with a broad range of biological properties including antioxidant, antimicrobial, antiviral, antitumor, anti-inflammatory, and anti-neurodegenerative activities [53]. The coumarin 7,8-dihydroxy-7-*O*-β-D-glucuronide (**3**) has been identified in *B. trilobata*, although coumarins are less common in liverworts [4]. Coumarin and its derivatives are well-known to have important biological activities [54]. The macrocyclic bis-bibenzyl isoplagiochin C (**6**) and isoplagiochin D are known constituents from bryophytes proposed as parent compounds of chlorinated bis-bibenzyls of the bazzanin type [38]. Bazzanin B (**7**) and bazzanin S are bioactive chlorinated cyclic bis-bibenzyls from *B. trilobata* along with the bis-bibenzyl isoplagiochin D, and they have shown antifungal activities towards phytopathogenic fungi [4]. Several other chlorinated bis-bibenzyls, such as bazzanins C/D (**8**) and K (**9**) are biosynthesized in *B. trilobata* [37, 38, 55], and bibenzyls and bis-bibenzyls from liverworts exhibit a variety of therapeutic properties like anti-cancer, antioxidant, antimicrobial, and nitric oxide inhibitory activities [56].

We also identified in *B. trilobata* a sesquiterpene caffeate, drimenyl caffeate (**10**), which was first isolated from the liverwort *Bazzania fauriana* [52]. Compound **10** was isolated by preparative liquid chromatography and its structure was determined by 1D and 2D NMR spectra (Fig. 3 and Table 3; purification and NMR details for compound **10** are given in the supplementary material).

The trans-caffeate part of compound **10** was evident from the large ^{3}JHH coupling (roughly 16.5 Hz) between H2' and H3', the characteristic *meta*-coupling

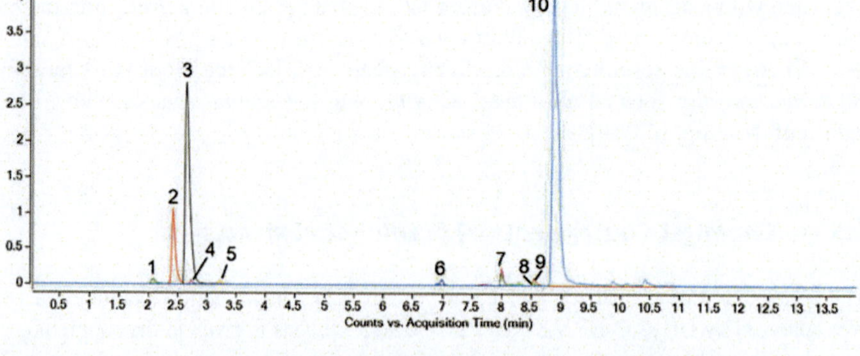

Fig. 2 Extracted ion chromatogram of compounds **1–10** from *B. trilobata* extracts

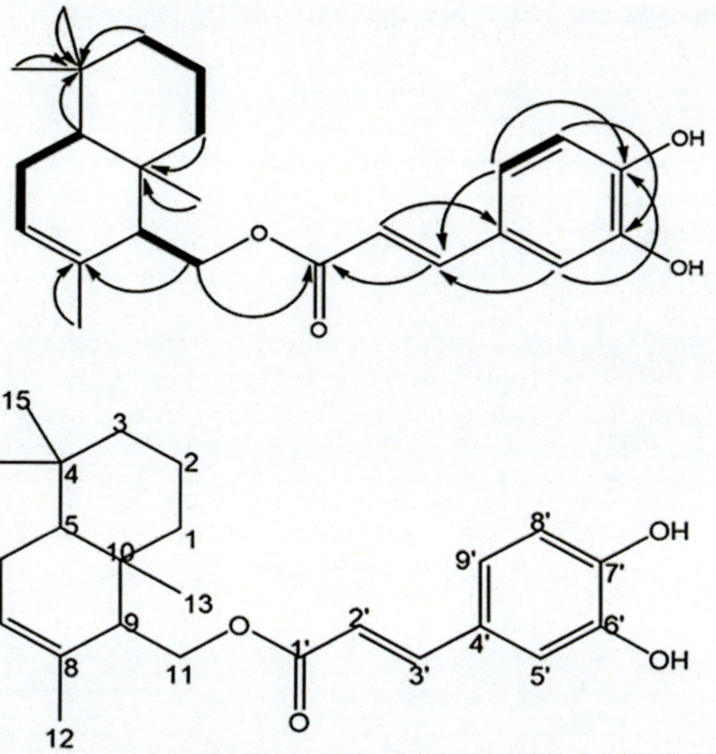

Fig. 3 HMBC (arrows) and COSY (bold bonds) correlations of drimenyl caffeate (**10**)

pattern for H5' and H9', and ortho-coupling between H8' and H9', and observed ^{1}H-^{13}C HMBC correlation peaks between H2'-C1', H8'-C6' and H9'-C7'. The HMBC correlation peak between H11 and C1' and the H11-H9 COSY correlation peak then established the other side of the ester bridge. HMBC correlation peaks between H11-C8 and H11-C10 confirmed the positioning of C8 and C10, while HMBC correlation peaks between H13-C10 and H12-C8 identified the position of these two methyl groups. The double-ring system was then further assigned using COSY correlation peaks to identify the segments C5-C6-C7 and C3-C2-C1. Lastly, shared HMBC correlation peaks to C4 for H3, and the methyl groups H14 and H15 together with an observed correlation peak between the methyl H13 and C10 completed the structure assignment. The observed ^{1}H and ^{13}C chemical shifts are in good agreement with previously published data [52].

Sesquiterpene caffeates in liverworts have only been identified in *Bazzania* spp. [41]. The sesquiterpenoid cyclomyltaylyl-3-caffeate isolated from *Bazzania japonica* showed superoxide anion release inhibitory activity [57] and the myltaylane caffeate from *Bazzania nitida* showed potent inhibition of nitric oxide production [11]. Also, naviculyl caffeate was reported as a cytotoxic sesquiterpenoid isolated from the liverwort *Bazzania novae-zelandiae* [7].

Table 3 NMR spectroscopic data (800 MHz in MeOD-d_4) for drimenyl caffeate (**10**)

Annotation	^{13}C (ppm)	^{1}H (ppm)	Multiplicity, J-couplings (Hz)	^{1}H-^{1}H COSY	(key) ^{1}H-^{13}C HMBC
1	40.5	1.99 1.14	m m	2	10
2	19.5	1.58 1.47	m m	1, 3	
3	42.9	1.43 1.20	m m	2	4
4	33.6				
5	50.9	1.24	m	6	4, 10
6	24.4	2.01 1.90	m m	5, 7	
7	124.5	5.53	s, br		
8	133.2				
9	54.6	2.1	s, br		8
10	36.8				
11	63.8	4.36 4.19	dd, 11.8; 3.1 dd, 11.8; 5.8	9	1', 8, 9, 10
12	22.1	1.68	s		7, 8, 9
13	15.0	0.87	s		9, 10
14	33.7	0.87	s		4, 15
15	22.4	0.91	s		4, 14
1'	169.1				
2'	115.1	6.19	d, 16.5	3'	1'
3'	146.3	7.49	d, 16.5	2'	1'
4'	127.4				
5'	114.9	7.01	d, 1.6	9'	7', 9'
6'	146.4				
7'	149.1				
8'	116.2	6.77	d, 8.1	9'	4', 6'
9'	122.7	6.91	dd, 8.1; 1.6	5', 8'	5', 7'

4 Conclusions

This study shows that the extracts from the liverwort *B. trilobata* have antioxidant, antimicrobial, collagenase and tyrosinase inhibitory activities. In addition, a sesquiterpene caffeate was identified in *B. trilobata*. The extracts are rich in phenolic constituents and contain a sesquiterpenoid, which possibly explains most of the biological activities. We demonstrate that *B. trilobata* has, besides its already known antifungal activities, the potential for new biotechnological applications. These results contribute to the knowledge of the medicinal properties of liverworts and in special the inhibitory effect on aging-related enzymes.

Acknowledgments This research was supported by Marie Sklodowska-Curie Actions Innovative Training Networks under the Horizon 2020 program under grant agreement n° 765115 – MossTech. The authors thank Professor Nils Cronberg, Lund University, Sweden, for support in plant identification and collection. The NMR Center • DTU and the Villum Foundation are acknowledged for access to the 800 MHz spectrometers.

References

1. Jackson MR, Bellemare J (2018) The potential for indirect negative effects of exotic insect species on a liverwort, *Bazzania trilobata* (Lepidoziaceae), mediated by the decline of a foundation tree species, *Tsuga canadensis* (Pinaceae) 1. J Torrey Bot Soc 145:183–194. https://doi.org/10.3159/TORREY-D-17-00040.1
2. Spjut RW, Suffness M, Cragg GM, Norris DH (1986) Mosses, liverworts, and hornworts screened for antitumor agents. Econ Bot 40:310–338. https://doi.org/10.1007/BF02858989
3. Tadesse M, Steiner U, Hindorf H, Dehne H-W (2004) Bryophyte extracts with activity against plant pathogenic fungi. SINET Ethiop J Sci 26:55–62. https://doi.org/10.4314/sinet.v26i1.18200
4. Scher JM, Speakman J-B, Zapp J, Becker H (2004) Bioactivity guided isolation of antifungal compounds from the liverwort *Bazzania trilobata* (L.) S.F. Gray. Phytochemistry 65: 2583–2888. https://doi.org/10.1016/j.phytochem.2004.05.013
5. Frahn J-P, Frahm J-PP (2004) Recent developments of commercial products from bryophytes. Bryologist 107:277–283. https://doi.org/10.1639/0007-2745(2004)107[0277:RDOCPF]2.0.CO;2
6. (2021) Lebermoosextrakt by Jean Pütz. https://www.jean-puetz-produkte.de/lebermoosextrakt-pflanzenstaerkungsmittel-100-ml-p-103.html. Accessed 12 Oct 2021
7. Burgess EJ, Larsen L, Perry NB (2000) A cytotoxic Sesquiterpene Caffeate from the liverwort *Bazzania novae-zelandiae*. J Nat Prod 63:537–539. https://doi.org/10.1021/np990492x
8. Liu N, Guo D-X, Wang S-Q et al (2012) Bioactive Sesquiterpenoids and Diterpenoids from the liverwort *Bazzania albifolia*. Chem Biodivers 9:2254–2261. https://doi.org/10.1002/cbdv.201100408
9. Zhu R-L, Wang' D, Xu' L, Shi' R-P (2006) Antibacterial activity in extracts of some bryophytes from China and Mongolia. J Hattori Bot Lab 100:603–615. https://doi.org/10.18968/jhbl.100.0_603
10. Zheng S, Chang W, Zhang M et al (2018) Chiloscyphenol A derived from Chinese liverworts exerts fungicidal action by eliciting both mitochondrial dysfunction and plasma membrane destruction. Sci Rep 8:326. https://doi.org/10.1038/s41598-017-18717-9
11. Harinantenaina L, Asakawa Y (2007) Chemical constituents of Malagasy liverworts. 6. A Myltaylane Caffeate with nitric oxide inhibitory activity from *Bazzania nitida*. J Nat Prod 70: 856–858. https://doi.org/10.1021/np060549x
12. Harinantenaina L, Quang DN, Nishizawa T et al (2007) Bioactive compounds from liverworts: inhibition of lipopolysaccharide-induced inducible NOS mRNA in RAW 264.7 cells by Herbertenoids and Cuparenoids. Phytomedicine 14:486–491. https://doi.org/10.1016/j.phymed.2006.09.011
13. Simonsen HT, Drew DP, Lunde C (2009) Perspectives on using *Physcomitrella patens* as an alternative production platform for thapsigargin and other terpenoid drug candidates. Perspect Med Chem 2009:1–6. https://doi.org/10.4137/PMC.S2220
14. Sabovljević MS, Sabovljević AD, Ikram NKK et al (2016) Bryophytes – an emerging source for herbal remedies and chemical production. Plant Genet Resour 14:314–327. https://doi.org/10.1017/S1479262116000320

15. Commisso M, Guarino F, Marchi L, Muto A, Piro A, Degola F (2021) Bryo-activities: a review on how bryophytes are contributing to the arsenal of natural bioactive compounds against fungi. Plants 10(2):203. https://doi.org/10.3390/plants10020203
16. Horn A, Pascal A, Lončarević I et al (2021) Natural products from bryophytes: from basic biology to biotechnological applications. CRC Crit Rev Plant Sci 40:191–217. https://doi.org/10.1080/07352689.2021.1911034
17. Marques RV, Sestito SE, Bourgaud F et al (2022) Anti-inflammatory activity of bryophytes extracts in LPS-stimulated RAW264.7 murine macrophages. Molecules 27:1940. https://doi.org/10.3390/MOLECULES27061940
18. Kusumawati I, Indrayanto G (2013) Natural antioxidants in cosmetics. In: Studies in natural products chemistry. Elsevier B.V., Oxford, UK, pp 485–505
19. Freitas-Rodríguez S, Folgueras AR, López-Otín C (2017) The role of matrix metalloproteinases in aging: tissue remodeling and beyond. Biochim Biophys Acta Mol Cell Res 1864:2015–2025. https://doi.org/10.1016/j.bbamcr.2017.05.007
20. Marques RV, Guillaumin A, Abdelwahab AB et al (2021) Collagenase and Tyrosinase inhibitory effect of isolated constituents from the Moss *Polytrichum formosum*. Plan Theory 10: 1271. https://doi.org/10.3390/plants10071271. https://creativecommons.org/licenses/by/4.0/
21. Pillaiyar T, Manickam M, Namasivayam V (2017) Skin whitening agents: medicinal chemistry perspective of tyrosinase inhibitors. J Enzyme Inhib Med Chem 32:403–425. https://doi.org/10.1080/14756366.2016.1256882
22. Herman A (2019) Antimicrobial ingredients as preservative booster and components of self-preserving cosmetic products. Curr Microbiol 76:744–754. https://doi.org/10.1007/s00284-018-1492-2
23. Singleton VL, Orthofer R, Lamuela-Raventós RM (1999) [14] Analysis of total phenols and other oxidation substrates and antioxidants by means of folin-ciocalteu reagent. In: Methods in enzymology. Academic, United States, pp 152–178
24. Sharma OP, Bhat TK (2009) DPPH antioxidant assay revisited. Food Chem 113:1202–1205. https://doi.org/10.1016/j.foodchem.2008.08.008
25. Chajra H, Salwinski A, Guillaumin A et al (2020) Plant milking technology: an innovative and sustainable process to produce highly active extracts from plant roots. Molecules 25:4162. https://doi.org/10.3390/molecules25184162
26. Kamkaen N, Mulsri N, Treesak C (2007) Screening of some tropical vegetables for anti-tyrosinase activity. Thai Pharm Health Sci J 2:15–19
27. (2006) ISO 20776–1:2006 Clinical laboratory testing and in vitro diagnostic test systems – Susceptibility testing of infectious agents and evaluation of performance of antimicrobial susceptibility test devices – Part 1: Reference method for testing the in vitro. https://www.iso.org/standard/41630.html
28. Cockerill F (2013) Performance standards for antimicrobial susceptibility testing: twenty-third informational supplement. Clinical and Laboratory Standards Institute, Wayne
29. (2021) EUCAST: MIC determination. https://eucast.org/ast_of_bacteria/mic_determination/. Accessed 17 Mar 2021
30. Elmi A, Spina R, Risler A et al (2020) Evaluation of antioxidant and antibacterial activities, cytotoxicity of *Acacia seyal* Del Bark extracts and isolated compounds. Molecules 25:2392. https://doi.org/10.3390/molecules25102392
31. (2021) EUCAST: Antifungal susceptibility testing. https://www.eucast.org/astoffungi/. Accessed 17 Mar 2021
32. Tsugawa H, Cajka T, Kind T et al (2015) MS-DIAL: data-independent MS/MS deconvolution for comprehensive metabolome analysis. Nat Methods 12:523–526. https://doi.org/10.1038/nmeth.3393
33. Tsugawa H, Kind T, Nakabayashi R et al (2016) Hydrogen rearrangement rules: computational MS/MS fragmentation and structure elucidation using MS-FINDER software. Anal Chem 88: 7946–7958. https://doi.org/10.1021/acs.analchem.6b00770

34. Olszowy M (2019) What is responsible for antioxidant properties of polyphenolic compounds from plants? Plant Physiol Biochem 144:135–143. https://doi.org/10.1016/j.plaphy.2019.09.039
35. Cherubim DJ, Martins CV, Fariña L, Lucca RA (2020) Polyphenols as natural antioxidants in cosmetics applications. J Cosmet Dermatol 19:33–37. https://doi.org/10.1111/jocd.13093
36. Asakawa Y (2016) Polyphenols in bryophytes. In: Recent advances in polyphenol research. Wiley, Chichester, pp 36–66
37. Martini U, Zapp J, Becker H (1998) Chlorinated macrocyclic bisbibenzyls from the liverwort *Bazzania trilobata*. Phytochemistry 47:89–96. https://doi.org/10.1016/S0031-9422(97)00495-0
38. Speicher A, Hollemeyer K, Heinzle E (2001) Rapid detection of chlorinated bisbibenzyls in *Bazzania trilobata* using MALDI-TOF mass spectrometry. Phytochemistry 57:303–306. https://doi.org/10.1016/S0031-9422(01)00010-3
39. Martini U, Zapp J, Becker H (1998) Lignans from the liverwort *Bazzania trilobata*. Phytochemistry 49:1139–1146. https://doi.org/10.1016/S0031-9422(97)01076-5
40. Scher JM, Zapp J, Becker H (2003) Lignan derivatives from the liverwort *Bazzania trilobata*. Phytochemistry 62:769–777. https://doi.org/10.1016/S0031-9422(02)00626-X
41. Asakawa Y, Ludwiczuk A, Nagashima F (2013) Chemical constituents of bryophytes. Bio- and chemical diversity, biological activity, and chemosystematics. Springer Vienna, Vienna
42. Gahtori D, Chaturvedi P (2020) Bryophytes: a potential source of antioxidants. In: Bryophytes. IntechOpen, United Kingdom
43. Neha K, Haider MR, Pathak A, Yar MS (2019) Medicinal prospects of antioxidants: a review. Eur J Med Chem 178:687–704. https://doi.org/10.1016/j.ejmech.2019.06.010
44. Njus D, Kelley PM, Tu Y-J, Schlegel HB (2020) Ascorbic acid: the chemistry underlying its antioxidant properties. Free Radic Biol Med 159:37–43. https://doi.org/10.1016/j.freeradbiomed.2020.07.013
45. Tran TQ, Phan HN, Bui AL, Quach PND (2020) Biological activities of in vitro liverwort *Marchantia polymorpha* L. extracts. Not Bot Hortic Agrobot Cluj-Napoca 48:826–838. https://doi.org/10.15835/nbha48211884
46. Banerjee RD, Sen SP (1979) Antibiotic activity of bryophytes. Bryologist 82:141. https://doi.org/10.2307/3242073
47. Sawant UJ, Karadge BA (2010) Antimicrobial activity of some bryophytes (liverworts and a hornwort) from Kolhapur district. Pharm J 2:29–32. https://doi.org/10.1016/S0975-3575(10)80046-X
48. Nikolajeva V, Liepina L, Petrina Z et al (2012) Antibacterial activity of extracts from some bryophytes. Adv Microbiol 02:345–353. https://doi.org/10.4236/aim.2012.23042
49. Kumar Tyagi A, Bukvicki D, Gottardi D et al (2013) Antimicrobial potential and chemical characterization of Serbian liverwort (*Porella arboris-vitae*): SEM and TEM observations. Evid-Based Complement Altern Med 2013:1–7. https://doi.org/10.1155/2013/382927
50. Negi K, Asthana AK, Chaturvedi P (2020) GC–MS analysis and antifungal activity of acetone extract of *Conocephalum conicum* (L) Underw (liverwort) against aflatoxins producing fungi. S Afr J Bot 131:384–390. https://doi.org/10.1016/j.sajb.2020.02.035
51. Ludwiczuk A, Asakawa Y (2019) Bryophytes as a source of bioactive volatile terpenoids – a review. Food Chem Toxicol 132:110649. https://doi.org/10.1016/j.fct.2019.110649
52. Toyota M, Asakawa Y (1988) Sesquiterpenoids from the liverwort *Bazzania fauriana*. Phytochemistry 27:2155–2159. https://doi.org/10.1016/0031-9422(88)80116-X
53. Zálešák F, Bon DJ-YD, Pospíšil J (2019) Lignans and Neolignans: plant secondary metabolites as a reservoir of biologically active substances. Pharmacol Res 146:104284. https://doi.org/10.1016/j.phrs.2019.104284

54. Carneiro A, Matos MJ, Uriarte E, Santana L (2021) Trending topics on Coumarin and its derivatives in 2020. Molecules 26:501. https://doi.org/10.3390/molecules26020501
55. Asakawa Y, Ludwiczuk A, Novakovic M et al (2022) Bis-bibenzyls, Bibenzyls, and Terpenoids in 33 genera of the marchantiophyta (liverworts): structures, synthesis, and bioactivity. J Nat Prod 85:729–762. https://doi.org/10.1021/ACS.JNATPROD.1C00302/ASSET/IMAGES/LARGE/NP1C00302_0026.JPEG
56. Nandy S, Dey A (2020) Bibenzyls and bisbybenzyls of bryophytic origin as promising source of novel therapeutics: pharmacology, synthesis and structure-activity. DARU J Pharm Sci 28:701–734. https://doi.org/10.1007/s40199-020-00341-0
57. Asakawa Y, Toyota M, Ueda A et al (1991) Sesquiterpenoids from the liverwort *Bazzania japonica*. Phytochemistry 30:3037–3040. https://doi.org/10.1016/S0031-9422(00)98247-5

7 Bryophytes as an Accumulator of Toxic Elements from the Environment: Recent Advances

Jayanta Barukial and Porismita Hazarika

Contents

Abstract

Toxic elements cause a serious threat to both the terrestrial and aquatic ecosystems. They are released into the environment by anthropogenic activities like the discharge of wastewaters, viz. industrial effluents, home sewage, use of chemical fertilizers, burning of fossil fuel, mining of different ores, use of radioactive elements, and nuclear reactors which contribute to heavy metal influx into the environment. Bryophytes include liverworts, hornworts, and mosses which have a significant potential to absorb heavy metals, making them useful biomonitoring tools. Because of the lack of an efficient vascular system, heavy metals deposition has been seen in bryophytes. Bryophyte tissue is a potent ion exchanger with the environment; hence, they accumulate heavy metals from the sources. Metal

J. Barukial (✉)
Debraj Roy College, Golaghat, Assam, India

P. Hazarika
Dibrugarh University, Dibrugarh, Assam, India
e-mail: rs_porismitahazarika@dibru.ac.in

H. N. Murthy (ed.), *Bioactive Compounds in Bryophytes and Pteridophytes*,
Reference Series in Phytochemistry, https://doi.org/10.1007/978-3-031-23243-5_6

absorption is extremely noticeable in bryophytes, especially in samples from contaminated streams. Mosses are the most important of the three groups of bryophytes in terms of bioaccumulation of hazardous substances from the environment. Moss species are more effective than vascular plant leaves for monitoring air pollution produced by heavy metals in urban areas. Hence, bryophytes are regarded as the best biomonitoring agent of environmental pollution. Currently, Moss bag techniques have been used to give a low-cost, flexible, and dense monitoring design that can show spatial and temporal trends but also vertical and horizontal gradients for a number of inorganic and organic pollutants. The moss bag approach will successfully overcome the issue of a lack of naturally grown mosses, allowing homogeneous biomonitoring of gaseous pollutants across all anthropogenically devastated areas. It has been utilized successfully for biomonitoring of potentially hazardous elements, such as rare earth elements and persistent organic chemicals, primarily polycyclic aromatic hydrocarbons. In this context, a more in-depth research is necessary from the forthcoming researchers in this field.

Keywords

Bioaccumulation · Biomonitoring · Bryophytes · Ecosystem · Heavy metals · Moss bag · Toxic elements

Abbreviations

Ag^{-}	Silver
Al^{-}	Aluminum
Ca^{-}	Calcium
Cd^{-}	Cadmium
CEC^{-}	Cation Exchange Capacity
CF^{-}	Contamination Factor
Co^{-}	Cobalt
Cr^{-}	Chromium
Cu^{-}	Copper
Fe^{-}	Iron
Hg^{-}	Mercury
HM^{-}	Heavy Metal
K^{-}	Potassium
Mg^{-}	Magnesium
Mn^{-}	Manganese
Mo^{-}	Molybdenum
Na^{-}	Sodium
Ni^{-}	Nickel
PAH^{-}	Polycyclic Aromatic Hydrocarbon
Pb^{-}	Lead
PTE^{-}	Potentially Toxic Trace Elements

Se^-	Selenium
Sn^-	Tin
Zn^-	Zinc

1 Introduction

People are becoming more aware of the term "pollution" and are collecting harmful substances from their surroundings. They have since learned about the dangerous substances' negative effects on their health and the health of other living beings. Copper (Cu), Iron (Fe), Molybdenum (Mo), Zinc (Zn), and, in some cases, Aluminium (Al), Nickel (Ni), and Selenium (Se) are all trace metals that organisms need as micronutrients. However, in certain circumstances, these same components may accumulate in high concentrations in species, causing ecological devastation. Cadmium (Cd), Chromium (Cr), Cobalt (Co), Copper (Cu), Iron (Fe), Lead (Pb), Mercury (Hg), Nickel (Ni), Silver (Ag), Tin (Sn) and Zinc (Zn), as well as the lighter elements Aluminium (Al), Arsenic (As), and Selenium (Se) are the most usually linked to environmental toxicity [22].

1.1 Sources of Hazardous and Toxic Materials in the Environment

Both aquatic and terrestrial ecosystems are threatened by toxic trace metals. Heavy metals from various sources have poisoned both ecosystems. Heavy metals are one of the most studied contaminants in the environment. Depending on the dose and length of exposure, almost any heavy metal or metalloid could be hazardous to biota [1]. Metals with a specific density of more than 5 gcm^{-3} are classified as heavy metals [39]. Heavy metals discharged into the atmosphere through mining, smelting, and other industrial activities eventually find their way back to the soil via dry and wet deposition. Heavy metals are released into the environment by the discharge of wastewaters, such as industrial effluents and home sewage. Chemical fertilizers and fossil fuel burning both contribute to anthropogenic heavy metal influx into the environment. Phosphate fertilizers are particularly hazardous when it comes to heavy metal levels in commercial chemical fertilizers [1]. Heavy metals harm to water and soil because they are dumped into the water, moved down streams, and eventually trapped in the water's underlying bed; or they are washed away by overflow onto the water surface [21]. The toxic effects of these metals are an issue for ecological, evolutionary, nutritional, and environmental reasons [56]. The toxicity of persons exposed is influenced by the dose, manner of exposure, chemical species, as well as their age, gender, genetics, and nutritional status. Heavy metals are persistent in the environment and can bioaccumulate in food systems. Cadmium, lead, and mercury are examples of common air pollutants emitted mostly as a result of industrial activities. They contribute to soil deposition and build-up despite the low air levels. Cadmium has also been identified as a probable human carcinogen, capable of causing lung cancer. Lead poisoning impairs the growth and neurobehavioral development of fetuses, newborns, and toddlers as well as raises blood pressure in adults.

Mercury is harmful in both its elemental and inorganic forms, but the organic molecules, particularly methyl mercury, that accumulate in the food chain, that is, in predatory fish in lakes and oceans, are the primary routes of human exposure. Long-range transboundary, air pollution is only one source of exposure to these metals, but due to their persistence and potential for global atmospheric transmission, atmospheric emissions have an impact on even the most remote places [40].

1.2 Most Hazardous Toxic Elements with Environmental Impact

Many elements are regarded as heavy metals; however, some are significant in terms of the environment. Cr, Ni, Cu, Zn, Cd, Pb, Hg, and As are among the most ecologically hazardous heavy metals and metalloids [1, 5]. Cr, Mn, Ni, Cu, Zn, Cd, and Pb are the most prevalent heavy metal contaminants found in the environment [1, 42]. In 2009, China outlined four metals, that is, Cr, Cd, Pb, and Hg, and the metalloids, as the highest priority pollutants for monitoring in the "12th 5 year plan for comprehensive prevention and control of heavy metals in the environment" [1, 23]. These metallic elements are considered systemic toxicants that can trigger organ damage even at low levels of exposure. They are also described as human carcinogens by the US Environmental Protection Agency and the International Agency for Research on Cancer [84].

2 Bryophytes and Toxic Elements

Because of their dispersion powers, bryophytes are more distributed widely than other plants [57, 80]. From an evolutionary standpoint, these are represented by the second most species-rich cluster of land-dwelling plants [37]. The mosses contain approximately 8000 species, liverworts 6000 species, and hornworts 200 species [31]. Bryophytes allocate important buffer structures for other groups of plants and thus play a crucial part in perpetuating ecosystems [36]. Bryophytes are widely used as touchstone species in air pollution, water pollution, and soil pollution. Besides, they are also used in various fields, such as material for seed beds, fuel, medicines, food, pesticides, moss gardening, treatment of waste, construction, genetic engineering, culturing, and soil conditioning [27, 65].

2.1 Sequestration of Toxic Elements by Bryophytes

A tolerant plant has a specific physiological system that enables it to work efficiently even when exposed to excessive heavy metal concentrations [83]. Bryophytes have a significant potential to absorb heavy metals, making them useful biomonitoring tools. However, depending on the element and bryophyte species employed, this capability may vary [4, 58]. Because of the lack of an efficient vascular system, heavy metals deposition has been seen in mosses and other bryophytes. This is owing to the relatively unrestricted exchange of solutes between active plant tissues and the atmosphere [30, 41]. Several investigations in the Goujiang Karst bauxite in South Western China found that gemmiferous bryophyte communities tolerate high-heavy metal substrates better than nongemmiferous communities, making gemmiferous bryophyte communities valuable in heavy metal pollution monitoring [80]. Pleurocarpous bryophytes are more susceptible to toxins than acrocarpous bryophytes. This sensitivity might be due to variations in the growth forms' water conducting systems and soluble metal absorption [48, 64]. Elevated pollution and poor water quality are likely to be problematic for aquatic bryophytes. The composition of bryophyte species is indicative of river hydromorphology in the assessment of surface water quality, while the abundance of elements in bryophyte tissue depicts water chemistry [25, 77].

Role of direct involvement of photochemical of bryophytes in the accumulation of toxic element is not established yet, but it has been revealed by some studies that the metal chelating properties involve in the sequestration of toxic elements by bryophytes. For assessing the antioxidant capacity that retains metals that induce lipid peroxidation, metal chelating activity is crucial. Chelating substances bind transition metals in the body for this reason, which prevents radical production [45]. Metal ion sequestration in the cell wall, vacuoles, and cytoplasm vesicles are all known to be involved in heavy metal tolerance in bryophytes. Heavy metal toxicity can be minimized by bryophytes by trapping toxic ions in internal and external spaces. One of the most well-known sites of metal detoxification is the cell wall [7, 8, 47]. According to some interpretations, heavy metal transport and deposition in metal-treated pollen grains may be facilitated via cell membrane pits, cytoplasm vesicles and multivesicular aggregates [20]. Herbarium moss samples might have been useful in predicting patterns in Pb and Cu deposition [67].

2.2 Ion Exchange Characteristics of Bryophytes

Bryophyte tissue is a potent ion exchanger, which has been recognized for decades [10, 74]. Metal tolerance in bryophyte is species-specific, although the mechanisms for the diverse levels of tolerance are unclear. In the mosses, the data indicate a hypothetic correlation between lamina cell shape and metal tolerance. Species with long, thin lamina cells may withstand high metal levels better than those with isodiametric cells [58]. *In comparison to tracheophyte roots, bryophyte tissues exhibit greater cell wall cation exchange capacities (CEC), which may be crucial in the sequestration and protoplasmic absorption of crucial cations like Mg*. The CEC of epilithic and wooded soil bryophytes reduces when the preferred substratum's Ca concentration and pH decline. It is likely that a lower CEC avoids excessive adsorption of the phytotoxicant AI, which becomes more readily accessible under acidic environments, although this concept is still not validated [9]. Cu^{2+}, $Pb^{2+} > Ni^{2+} > Co^{2+} > Zn^{2+}, Mn^{2+}$ is the persistence ability order for heavy metal ions in *Hylocomium splendens* [63, 73]. At ambient levels, retention efficiency in *Sphagnum* falls in the order $Fe^{3+} > Mg^{2+}$, $Ca^{2+} > K^{+}$, Na^{+}, as well as with cation exchange resins [10]. Unesterified polyuronic acids make up around 25% of the tissue dry weight in *Sphagnum acutifolium*, and there is a strong link between the quantity of these acids and the cation exchange capacity (CEC) of several *Sphagnum* species [19, 73]. The tissue abundance of pectic compounds, particularly uronic acid, is intrinsically linked to the cation exchange capacity (CEC) of *Sphagnum* [44].

3 Role of Bryophytes in Sequestration of Toxic Elements: Recent Advances

As a method for determining the levels of environmental health and assessing the harmful contaminants in the biosphere, bryo-monitoring is progressively gaining popularity [50]. Tyler and his colleagues came up with the notion of using mosses to quantify atmospheric heavy metal deposition in the late 1960s. The moss analysis approach provides a proxy, time-integrated estimate of heavy metal deposition patterns from the atmosphere to terrestrial systems [33]. As a result, many regions of the world employ these agents in the current situation to monitor the different kind of pollution [50] (listed in Table 1).

The European moss survey has been conducted every 5 years since 1990 [35]. The survey was conducted from 2000 to 2001 to investigate patterns of variation in heavy metal concentrations in mosses across Europe, identify the most contaminated places, create regional maps, and improve knowledge of long-range transboundary contamination [33]. The European moss study collects data on 10 heavy metal concentrations (As, Cd, Cr, Cu, Fe, Hg, Ni, Pb, V, and Zn) in naturally grown mosses, as well as the metals Al and Sb and nitrogen since 2005 [32, 34, 35, 71]. For 27 archival and native bryophyte specimens collected in Guangzhou from 1932 to 2018, five heavy metals (As, Cd, Cu, Pb, and Zn) were analyzed [81].

The Republic of Moldova's deposition of potentially harmful substances was assessed using the moss biomonitoring approach. The research was carried out under the auspices of the International Cooperative Program on Effects of Air Pollution on Natural Vegetation and Crops. In May 2020, samples of the moss *Hypnum cupressiforme* Hedw. were gathered from 41 sampling locations spread over the whole nation. Neutron activation analysis and atomic absorption spectrometry were used to estimate the mass fractions of 35 elements, including Na, Mg, Al, Cl, K, Ca, Sc, Ti, V, Cr, Mn, Fe, Co, Ni, Cu, Zn, As, Br, Se, Rb, Sr, Sb, Cs, Ba, Cd, La, Ce, Sm, Eu, Tb, Hf, Ta, Th, Pb, and U [85].

The total contents of eight elements (Cu, Zn, Fe, Mn, Ni, Pb, Cd, and Cr) as determined by ICP-AES and Atomic Absorption Spectrophotometry (AAS) methods were compared in four types of indigenous mosses (*Brachythecium plumosum, Eurhynchium laxirete, Taxiphyllum taxirameum,* and *Haplocladium strictulum*), which were collected from various sampling sites in the Chengdu city, China. According to the study, *T. taxirameum* had a larger potential for metal

Table 1 Name of the bryophytes that are involved in the accumulation of toxic elements

Sl. No	Name	Family	Habitat
Liverworts			
1	*Riccia fluitans* L.	Ricciaceae	Aquatic
2	*Aneura pinguis* (L.) Dumort.	*Aneuraceae*	Terricolous
3	*Pellia endiviifolia* (Dicks.) Dumort.	Peliaceae	Terricolous
4	*Solenostoma crenulatum* Mitt.	Jungermanniaceae	Corticolous, saxicolous
Mosses			
5	*Pleurochaete squarrosa* (Brid.) Lindb.	Pottiaceae	Aquatic
6	*Hypnum cupressiforme* Hedw.	Hypnaceae	Terricolous/ Saxicolous
7	*Pseudoscleropodium purum* (Hedw.) M. Fleisch.	Brachytheciaceae	Terricolous
8	*Hylocomium splendens* (Hedw.) Schimp.	Hylocomiaceae	Saxicolous
9	*Bryum pseudotriquetrum* (Hedw.) Schwaegr.	Bryaceae	Saxicolous
10	*Bryum turbinatum* (Hedw.) Turner	Bryaceae	Saxicolous
11	*Chorisodontium aciphyllum* (Hook. f. & Wilson) Broth.	Dicranaceae	Saxicolous
12	*Racomitrium lanuginosum* (Hedw.) Brid.	*Grimmiaceae*	Saxicolous
13	*Rhizomnium punctatum* (Hedw.) T.J. Kop.	Mniaceae	Terricolous
14	*Taxiphyllum barbieri* (Cardot & Copp.) Z. Iwats.	Hypnaceae	Aquatic
15	*Pohlia nutans* (Hedw.) Lindb.	Bryaceae	Terricolous
16	*Leskea angustata* Taylor	Leskeaceae	Terricolous
17	*Fabronia ciliaris* (Brid.) Brid.	Fabroniaceae	Saxicolous / Corticolous
18	*Polytrichastrum formosum* (Hedw.) G.L. Sm.	Polytrichaceae	Terricolous / saxicolous
19	*Pleurozium schreberi* (Willd. ex Brid.) Mitt.	Hylocomiaceae	Terricolous
20	*Fontinalis antipyretica* L. ex. Hedw.	Fontinalaceae	Aquatic
21	*Philonotis fontana* (Hedw.) Brid.	*Bartramiaceae*	Terricolous / saxicolous
22	*Pohlia flexuosa* Harv.	Bryaceae	Saxicolous
23	*Cinclidotus fontinaloides* (Hedw.) P. Beauv.	*Cinclidotaceae*	Saxicolous
24	*Dialytrichia mucronata* (Brid.) Broth.	*Pottiaceae*	Corticolous/ saxicolous
25	*Hygroamblystegium fluviatile* (Hedw.) Loeske	Amblystegiaceae	Aquatic / saxicolous
26	*Hygroamblystegium tenax* (Hedw.) Jenn.	Amblystegiaceae	Aquatic / saxicolous
27	*Platyhypnidium riparioides* (Hedw.) Dixon	Brachytheciaceae	Aquatic
28	*Leptodictyum riparium* (Hedw.) Warnst.	Amblystegiaceae	Aquatic
29	*Scorpiurium circinatum* (Brid.) M. Fleisch. & Loeske	Brachytheciaceae	Saxicolous

(continued)

Table 1 (continued)

Sl. No	Name	Family	Habitat
30	*Fontinalis hygrometrica* (Hedw.) P. Syd.	Fontinalaceae	Aquatic
31	*Fissidens bryoides* Hedw.	Fissidentaceae	Terricolous
32	*Cinclidotus aquaticus* (Hedw.) Bruch & Schimp.	Cinclidotaceae	Aquatic
33	*Cratoneuron filicinum* (Hedw.) Spruce	Amblystegiaceae	Terricolous
34	*Palustriella commutata* (Hedw.) Ochyra	Amblystegiaceae	Aquatic
35	*Ceratodon purpureus* (Hedw.) Brid.	Ditrichaceae	Saxicolous
36	*Sphagnum palustre* L.	Sphagnaceae	Aquatic
37	*Funaria hygrometrica* Hedw.	Funariaceae	Terricolous
38	*Brachythecium* species	Brachytheciaceae	Terricolous
39	*Eurhynchium* species	Brachytheciaceae	Terricolous

accumulation than other species, and there were substantial inter- and intraspecies variances in heavy metal concentrations [17].

In order to determine the effects of growth substrates, geographic elevation, and moss species type on the accumulation characteristics of heavy metals as well as to pinpoint heavy metal sources, concentrations of Cr, Co, Ni, Zn, Sr, Cd, Ba, and Pb in various moss species from Mountain Gongga, China, were analyzed. The findings revealed substantial differences in both the composition and geographical distribution of these components. The findings demonstrated that elevation has an impact on the variance of heavy metals in moss. The kind of moss and growth substrate had less of an impact on the metal concentration of the mosses studied for this investigation. The PMF model's findings showed that the majority of the Co, Cr, and Ni in the mosses on Mountain Gongga came from substrate sources, while other elements were predominantly linked to human activities, Pb and Cd might be ascribed to atmospheric deposition [82].

After mineral exploitation, the restoration of natural vegetation in manganese mining regions has become a crucial task. In mining regions of South western China, bryophytes have a priceless impact on ecological restoration. The findings indicated that *Bryum atrovirens* obtained from two different types of regions had a considerable capacity to accumulate Mn, with cumulants of 5588.00 μg/g and 4283.41 μg/g, respectively. All mosses demonstrated a high capacity for Cd enrichment. It demonstrated that mosses were very resistant to heavy metals [60].

In both polluted and uncontaminated sites in Villavicencio (Colombia) and its surrounds, the presence and distribution of the bioaccumulation of lead in bryophytes have been assessed. Fifty-two samples of bryophytes in total were gathered, of which 43 came from locations spread around the city's urban areas (homes, businesses, and highways), while the remaining nine came from clean regions located outside the city. Nitric and hydrochloric acids were used to treat the samples, and the results were then analyzed using atomic absorption spectrometry. Pb concentrations were found to be between 1 and 6 times higher in the commercial sector than in the residential and highway sectors. The regional variations in lead deposition are reflected in the spatial patterns of lead concentrations in bryophytes. According to this study, mosses and liverworts can be used to detect pollution hotspots in a city [72].

Over the course of 5 years, a research was conducted using *Pleurozium schreberi* next to a national highway that crosses Poland from north to east in the vicinity of Natura 2000 regions. Three places that were noticeably different were used to harvest moss samples. The amount of Zn, Ni, Pb, Co, and Cd in moss was examined in this study in relation to the effects of road transportation [59].

As markers of metal contamination, two moss species – *Physcomitrium cyathicarpum* and *Barbula constricta* – growing in various parts of Delhi, India, have been utilized. Using atomic absorption spectroscopy, the levels of significant heavy metals including Cr, Co, Cd, Cu, Fe, Hg, Ni, and Pb have been estimated in the tissues of both moss species, with Fe, Ni, Cu, and Cr having the greatest levels followed by Co, Cd, Pb, and Hg. Fe, Co, Cu, and Cr concentrations were found to be high in both species growing in the North Delhi zone, followed by South and West Delhi, indicating that areas with an industrial belt, heavy traffic, and companies that produce chemical effluents [76].

In order to assess the capability for heavy metal accumulation in mosses at several sites in the Idukki District of Kerala, India, eighteen moss species and their soil substrata were examined. Statistics revealed substantial interspecies variations in metal concentrations ($p = 5\%$), where *Campylopodium khasianum* had a greater capability for metal accumulation. In all five of the chosen sampling locations, the substratum had the greatest Cr level, followed by Ni and Pb. Regardless of the sample sites, all the mosses exhibited substantial Cr (III) accumulation relative to other metals (Cd, Cu, Pb, and Ni). *Campylopodium khasianum*, one of 18 mosses, was shown to gather the most Cr, Cd, Ni, and Pb, indicating that it may be used to clean up soil polluted with these metals [75].

4 Several Bryophytes in the Deposition of Toxic Substances from the Environment

Metal absorption is extremely noticeable in bryophytes, especially in samples from contaminated streams [14]. Mosses are the most important of the three groups of bryophytes in terms of bioaccumulation of hazardous substances from the environment. Moss species are more effective than vascular plant leaves for monitoring air pollution produced by heavy metals in urban areas [18]. Mosses that can acquire large levels of heavy metals from the environment have evolved a natural response to these circumstances [13]. Moss proves to be a promising bioindicator for elements notably Al, Cr, Sc, Th, Pb, Cd, Cu, V, and partially Zn deposition [15].

4.1 Liverworts

Pellia endiviifolia (Dicks.) Dumort. and *Aneura pinguis* (L.) Dumort. can be utilized as reliable bioindicators of water quality [16]. *Marchantia polymorpha* L. may be utilized as an adequate air pollution indicator. A combination of indices, such as chlorophyll, sugar, protein, catalase, and peroxidase in this species exposed for a short amount of time can reliably reveal pollution levels in the air. It may be regarded as a hyperaccumulator for lead since it demonstrated high levels of absorption. As a result, it can be utilized as a bioindicator or bioaccumulator species and can be used for indication or accumulation in various contaminated locations [66].

4.2 Mosses

Fontinalis antipyretica Hedw. can be utilized to detect zinc pollution in aquatic systems [53]. It acquires pollutants like heavy metals and other trace elements, making it a good indication of urban pollution in terms of the ecological threats posed. It may also be used to collect inorganic and organic contaminants [24–26, 77–79]. *Pohlia flexuosa* Harv. can tolerate large levels of hazardous metals without showing any signs of harm in its growth and development. For these metals, it possesses a tolerance and exclusion mechanism, notably for the nonessential elements As and Pb. As a result, its luxuriant and spontaneous development might be exploited as a phytostabilization pioneer plant in the black shale outcrop, where vascular plants are uncommon. Its ability to tolerate Cd toxicity may be due to the control of K and Zn uptake. *P. flexuosa* Harv., in particular, can grow and function properly in severely polluted soil (up to 486.0 mg kg^{-1} Cd and 2220 mg kg^{-1} Cu) [83]. In a biomonitoring study in remote areas of Italy and Northern Victoria, 15 chemical elements were discovered in five epigean moss species: *Hypnum cupressiforme* Hedw., *Pseudoscleropodium purum* (Hedw.) M. Fleisch., *Hylocomium splendens* (Hedw.)Schimp., *Bryum pseudotriquetrum* (Hedw.) Schwaegr., and *Chorisodontium aciphyllum* (Hook. f. & Wilson) Broth [6]. In a study conducted in Trieste, transplants of the mosses *Hypnum cupressiforme* Hedw. and *Pseudoscleropodium purum* (Hedw.) M. Fleisch. were compared as active biomonitors of some airborne trace elements (As, Cd, Cr, Cu, Fe, Hg, Mn, Pb, Ti, V, Zn). *Pseudoscleropodium purum* (Hedw.) M. Fleisch. has a strong resistance to heavy metals in the atmosphere, accumulating and losing practically all elements at equal or greater rates, especially those connected to particulate, dry depositions. The physical absorption of the coarse component of the dust by the *P. purum* (Hedw.) M. Fleisch. transplants was the predominant mechanism of heavy metal accumulation [51, 86]. Largescale patterns connected to moist depositions might be detected using *Hypnum cupressiforme* Hedw. This species was shown to be capable of removing metal ions (Co, Ni, Zn, Cd, Pb, and Cu) from aqueous solutions based on biosorption studies. These two carpet-forming moss species were also used to investigate the atmospheric deposition of these components in Kosovo [13, 28, 54, 55]. *Racomitrium lanuginosum* (Hedw.) Brid. seems to have a wider potential for monitoring long-range atmospheric transit for these harmful substances [61]. Instrumental neutron activation analysis was used to evaluate the content of a total of 36 elements in *Brachythecium* sp. and *Eurhynchium* sp. [2]. *Taxiphyllum barbieri* (Cardot & Copp.) Z. Iwats., an aquatic moss, appears to be a good indicator species for metal toxicity because it showed clear sensitivity at the microscopic level [46]. Because it acquired significant levels of Cu and Ni, *Pohlia nutans* (Hedw.) Lindb. was thought to be a pollutant-resistant species. It not only survives extreme pollution, but also conquers severely polluted (barren) places in the absence of nonferrous smelters. It produces asexual reproductive structures that are highly specialized [68], which helps these species thrive in extremely polluted environments [48, 64, 73, 86]. *Leskea angustata* Taylor and *Fabronia ciliaris* (Brid.) Brid. are two epiphytic moss species that may be used to assess environmental pollution [49]. *Bryum turbinatum* (Hedw.) Turner and *Rhizomnium punctatum* (Hedw.) T.J. Kop. have high CFs for a variety of heavy metals and can be used to analyzed chemical contamination patterns [26]. *Polytrichastrum formosum* (Hedw.) G.L. Sm. is a suitable bioindicator for a variety of chemical components [52]. *Pleurozium schreberi* (Willd. ex Brid.) Mitt. is a sensitive bioindicator of heavy metal contamination in the environment. In Poland, this species is suggested for biomonitoring. They allow you to determine the degree of contamination, the source of contamination, and the direction of contamination spread [29, 43]. Metal accumulation by aquatic bryophytes from polluted mine streams is highly recorded. *Platyhypnidium riparioides* (Hedw.) Dixon, *Dialytrichia mucronata* (Brid.) Broth, *Hygroamblystegium fluviatile* (Hedw.) Loeske, *Hygroamblystegium tenax* (Hedw.) Jenn., and *Cinclidotus fontinaloides* (Hedw.) P. Beauv. could be used as trustworthy water quality bioindicators. Metal levels are high in *Philonotis fontana* (Hedw.) Brid. and *Solenostoma crenulatum* Mitt. (burton) [16, 77, 79]. *Leptodictyum riparium* (Hedw.) Warnst. is capable of retaining large levels of trace elements and has a high tolerance for human contamination [25]. Bioaccumulation in *Scorpiurum circinatum* (Brid.) Fleisch. & Loeske revealed that moss cells resisted heavy metal toxicity and immobilizing most harmful ions extracellularly, most likely in cell wall binding sites, which are the primary site of metal detoxification [8]. *Plagiomnium affine* (Blandow ex Funck) T.J. Kop. has been discovered to have a limited capacity to collect specific elements, such as Zn, Cl, and others [58]. Pb and Zn accumulation is highest in the gametophyte and placenta of *Fontinalis hygrometrica* (Hedw.) P. Syd [7] specially in their cell walls, vacuoles, nuclei, and plastids [8]. *Sphagnum palustre* L. has proven to be a reliable bioindicator that may be used in biomonitoring research [70]. *Ceratodon purpureus* is a pollution-tolerant species that has been related to human influence [62]. *Pleurochaete squarrosa* (Brid.) Lindb. and *Hypnum cupressiforme* Hedw. were used in several biomonitoring assessments of heavy metal, nitrogen deposition, and δ 15 N signatures in a Mediterranean environment. In comparison to other pleurocarpous mosses, it is a viable biomonitor [38].

5 Perspective of Using Bryophytes in Accumulation of Toxic Elements

Moss analysis is a time-integrated surrogate measurement of metal deposition from the atmosphere to terrestrial systems. It is simpler and less expensive than traditional precipitation analysis because it eliminates the need for large numbers of precipitation collectors and a long-term program of sample collection and analysis. Because mosses have larger trace element concentrations than rainwater, analysis is easier and less prone to contamination. Although moss concentration measurements do not give a direct quantitative assessment of deposition, they may be calculated using one of many regression models that link moss survey findings to precipitation monitoring data [12, 33, 34]. Metal(loid) detoxification processes in bryophytes are probably worthy of further investigation. It is also worth noting that phytochelatin synthase (PCS) and phytochelatins (PCn) have recently been discovered in various bryophytes, indicating that PCn's involvement in metal detoxification and homeostasis in these plants might be important [11]. The moss bag approach successfully overcomes the issue of a lack of naturally grown mosses, allowing homogeneous biomonitoring of gaseous pollutants across all anthropogenically devastated areas. It has been utilized successfully for biomonitoring of potentially hazardous elements, such as rare earth elements (PTEs) and persistent organic chemicals, primarily polycyclic aromatic hydrocarbons (PAHs). Moss bag techniques will be able to give a low-cost, flexible, and dense monitoring design that can show spatial and temporal trends but also vertical and horizontal gradients for a number of inorganic and organic pollutants. It might be used to monitor heavy metals in the air for a long time [3, 69].

6 Conclusion

It has been revealed that the mosses are more relevant to accumulating toxic elements than the other groups of bryophytes (Fig. 1). In terms of habitat, saxicolous and corticolous mosses are more relevant (Fig. 2). The family Amblystegiaceae has the potential in the tolerance of toxic metals from the environment followed by the family Brachytheciaceae and Bryaceae (Fig. 3). More in-depth research on these species, which play a significant role in phylogenesis, might uncover the presence of additional critical detoxifying systems that have been lost through time and/or better define the molecular processes underlying these plants' remarkable resistance to metal(loid)s. Understanding the present and developing successful solutions to meet future difficulties can be aided by a closer look into the past [11].

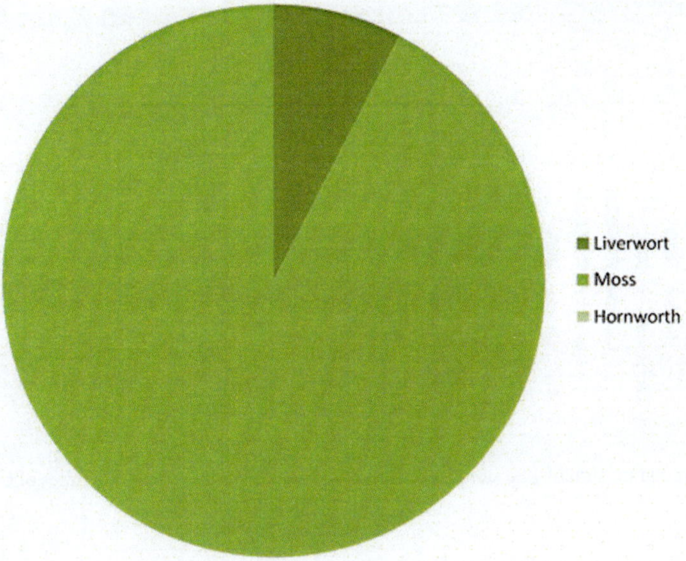

Fig. 1 Pie diagram showing rate of involvement of different groups of bryophytes in term of toxic elements accumlation

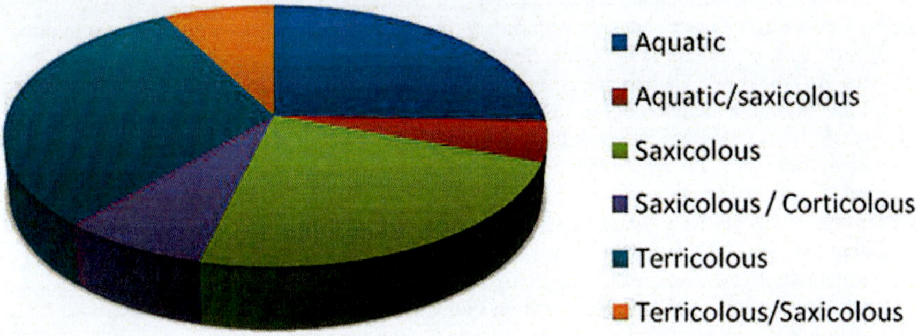

Fig. 2 Pie diagram showing habitat-wise rate of involvement of bryophytes in context of toxic elements

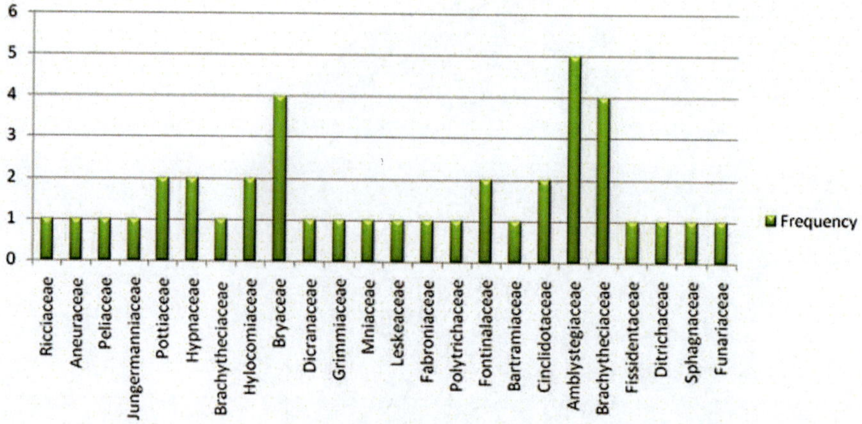

Fig. 3 Bar digram showing accumulation rate of toxic elements by different families of bryophytes

References

1. Ali H, Khan E, Ilahi I (2019) Environmental chemistry and ecotoxicology of hazardous heavy metals: environmental persistence, toxicity, and bioaccumulation. J Chem 2019:6730305
2. Aničić M, Tasić M, Frontasyeva MV, Tomašević M, Rajšić S, Mijić Z et al (2009) Active moss biomonitoring of trace elements with *Sphagnum girgensohnii* moss bags in relation to atmospheric bulk deposition in Belgrade, Serbia. Environ Pollut 157(2):673–679
3. Ares A, Aboal JR, Carballeira A, Giordano S, Adamo P, Fernández JA (2012) Moss bag biomonitoring: a methodological review. Sci Total Environ 432:143–158
4. Ares Á, Itouga M, Kato Y, Sakakibara H (2018) Differential metal tolerance and accumulation patterns of Cd, Cu, Pb and Zn in the liverwort *Marchantia polymorpha* L. Bull Environ Contam Toxicol 100(3):444–450
5. Barakat M (2011) New trends in removing heavy metals from industrial wastewater. Arab J Chem 4(4):361–377
6. Bargagli R, Brown DH, Nelli L (1995) Metal biomonitoring with mosses: procedures for correcting for soil contamination. Environ Pollut 89(2):169–175
7. Basile A, Cogoni AE, Bassi P, Fabrizi E, Sorbo S, Giordano S et al (2001) Accumulation of Pb and Zn in Gametophytes and Sporophytes of the Moss Funaria hygrometrica (Funariales). Ann Bot 87(4):537–543
8. Basile A, Sorbo S, Pisani T, Paoli L, Munzi S, Loppi S (2012) Bioacumulation and ultrastructural effects of Cd, Cu, Pb and Zn in the moss *Scorpiurum circinatum* (Brid.) Fleisch. & Loeske. Environ Pollut 166:208–211
9. Bates JW (1992) Mineral nutrient acquisition and retention by bryophytes. J Bryol 17(2):223–240
10. Bell P (1959) The ability of Sphagnum to absorb cations preferentially from dilute solutions resembling natural waters. J Ecol 47(2):351–355
11. Bellini E, Betti C, Sanità di Toppi L (2021) Responses to cadmium in early-diverging Streptophytes (Charophytes and bryophytes): current views and potential applications. Plan Theory 10(4):770
12. Berg T, Steinnes E (1997) Recent trends in atmospheric deposition of trace elements in Norway as evident from the 1995 moss survey. Sci Total Environ 208(3):197–206

13. Boquete MT, Fernández JÁ, Carballeira A, Aboal JR (2013) Assessing the tolerance of the terrestrial moss *Pseudoscleropodium purum* to high levels of atmospheric heavy metals: a reciprocal transplant study. Sci Total Environ 461–462:552–559
14. Burton M, Peterson P (1979) Metal accumulation by aquatic bryophytes from polluted mine streams. Environ Pollut (1970) 19(1):39–46
15. Čeburnis D, Steinnes E, Kvietkus K (1999) Estimation of metal uptake efficiencies from precipitation in mosses in Lithuania. Chemosphere 38(2):445–455
16. Ceschin S, Aleffi M, Bisceglie S, Savo V, Zuccarello V (2012) Aquatic bryophytes as Ecol. Indic. of the water quality status in the Tiber River basin (Italy). Ecol Indic 14(1):74–81
17. Chen Y, Yuan S, Su Y, Wang L (2010) Comparison of heavy metal accumulation capacity of some indigenous mosses in Southwest China cities: a case study in Chengdu city. Plant Soil Environ 56(2):60–66
18. Chen Y, Jiang Y, Huang H, Mou L, Ru J, Zhao J et al (2018) Long-term and high-concentration heavy-metal contamination strongly influences the microbiome and functional genes in Yellow River sediments. Sci Total Environ 637–638:1400–1412
19. Clymo R (1963) Ion exchange in Sphagnum and its relation to bog ecology. Ann Bot 27(2): 309–324
20. Dinis AM, Mesquita JF (2004) Uptake of heavy metal salt solutions by pollen grains of Magnolia× soulangeana (Magnoliaceae). Canada J Bot 82(12):1758–1767
21. Duruibe JO, Ogwuegbu M, Egwurugwu J (2007) Heavy metal pollution and human biotoxic effects. Int J Phys Sci 2(5):112–118
22. Freedman B (2018) Toxic elements. Environ Sci.Available via DIALOG.https://ecampusontario.pressbooks.pub/environmentalscience/chapter/chapter-18-toxic-elements/.
23. Fu Z, Guo W, Dang Z, Hu Q, Wu F, Feng C et al (2017) Refocusing on nonpriority toxic metals in the aquatic environment in China. ACS Publications, Washington, DC
24. Gecheva G, Yancheva V, Velcheva I, Georgieva E, Stoyanova S, Arnaudova D et al (2020) Integrated monitoring with moss-bag and mussel transplants in reservoirs. Water 12(6):1800
25. Gecheva G, Yurukova L, Ganeva A (2011) Assessment of pollution with aquatic bryophytes in Maritsa River (Bulgaria). Bull Environ Contam Toxicol 87(4):480–485
26. Gecheva G, Yurukova L, Cesa M, Cheshmedjiev S (2015) Monitoring of aquatic mosses and sediments: a case study in contaminated rivers, Bulgaria. Plant Biosyst 149(3):527–536
27. Glime JM (2007) Economic and ethnic uses of bryophytes. Flora N Am 27:14–41
28. González-Miqueo L, Elustondo D, Lasheras E, Santamaría J (2010) Use of native mosses as biomonitors of heavy metals and nitrogen deposition in the surroundings of two steel works. Chemosphere 78(8):965–971
29. Grodzinska K (1999) The environmental situation in the upper Vistula Basin (Southern Poland). In: Peakall DB, Walker CH, Migula P (eds) Biomarkers: a pragmatic basis for remediation of severe pollution in Eastern Europe. Springer Netherlands, Dordrecht, pp 29–48
30. Gupta A (1995) Heavy metal accumulation by three species of mosses in Shillong, North-Eastern India. Water Air Soil Pollut 82(3):751–756
31. Hallingback T, Hodgetts N (2000) Mosses, liverworts and hornworts. Status survey and conservation action plan for bryophytes. IUCN/SSC Bryophyte Specialist Group, Gland
32. Harmens H, Norris D (2008) Spatial and temporal trends in heavy metal accumulation in mosses in Europe (1990–2005): Programme Coordination centre for the ICP Vegetation, Centre for Ecology and Hydrology
33. Harmens H, Buse A, Büker P, Norris D, Mills G, Williams B et al (2004) Heavy metal concentrations in European mosses: 2000/2001 survey. J Atmos Chem 49(1):425–436
34. Harmens H, Norris DA, Koerber GR, Buse A, Steinnes E, Rühling Å (2007) Temporal trends in the concentration of arsenic, chromium, copper, iron, nickel, vanadium and zinc in mosses across Europe between 1990 and 2000. Atmos Environ 41(31):6673–6687
35. Harmens H, Norris D, Steinnes E, Kubin E, Piispanen J, Alber R et al (2010) Mosses as biomonitors of atmospheric heavy metal deposition: spatial patterns and temporal trends in Europe. Environ Pollut 158(10):3144–3156

36. Harris ES (2008) Ethnobryology: traditional uses and folk classification of bryophytes. Bryologist 111:169–217
37. Hassel K, Segreto R, Ekrem T (2013) Restricted variation in plant barcoding markers limits identification in closely related bryophyte species. Mol Ecol Resour 13(6):1047–1057
38. Izquieta-Rojano S, Elustondo D, Ederra A, Lasheras E, Santamaría C, Santamaría JM (2016) *Pleurochaete squarrosa* (Brid.) Lindb. as an alternative moss species for biomonitoring surveys of heavy metal, nitrogen deposition and δ15N signatures in a Mediterranean area. Ecol Indic 60: 1221–1228
39. Järup L (2003) Hazards of heavy metal contamination. Br Med Bull 68(1):167–182
40. Joint W, Organization WH (2007) Health risks of heavy metals from long-range transboundary air pollution. World Health Organization. Regional Office for Europe. Report No.: 9289071796. Available via https://apps.who.int/iris/handle/10665/107872
41. Kapfer J, Birks H, Astrup Felde V, Klanderud K, Martinessen TC, Ross L et al (2013) Long-term vegetation stability in northern Europe as assessed by changes in species co-occurrences. Plant Ecol Divers 6:289–302
42. Khan T, Muhammad S, Khan B, Khan H (2011) Investigating the levels of selected heavy metals in surface water of Shah Alam River (A tributary of River Kabul, Khyber Pakhtunkhwa). J Himal Earth Sci 44(2):71–79
43. Kłos A, Ziembik Z, Rajfur M, Dołhańczuk-Śródka A, Bochenek Z, Bjerke JW et al (2018) Using moss and lichens in biomonitoring of heavy-metal contamination of forest areas in southern and north-eastern Poland. Sci Total Environ 627:438–449
44. Knight A, Crooke W, Inkson R (1961) Cation-exchange capacities of tissues of higher and lower plants and their related uronic acid contents. Nature 192(4798):142–143
45. Kocazorbaz EK, Kerem T, Moulahoum H, Ün RN (2021) Phytochemical and bioactivity analysis of several methanolic extracts of nine bryophytes species. Sakarya Univ J Sci 25(4): 938–949
46. Kováčik J, Babula P, Hedbavny J (2017) Comparison of vascular and non-vascular aquatic plant as indicators of cadmium toxicity. Chemosphere 180:86–92
47. Krzesłowska M, Lenartowska M, Mellerowicz EJ, Samardakiewicz S, Woźny A (2009) Pectinous cell wall thickenings formation – a response of moss protonemata cells to lead. Environ Exp Bot 65(1):119–131
48. Lepp NW, Salmon D (1999) A field study of the ecotoxicology of copper to bryophytes. Environ Pollut 106(2):153–156
49. Macedo-Miranda G, Avila-Pérez P, Gil-Vargas P, Zarazúa G, Sánchez-Meza JC, Zepeda-Gómez C et al (2016) Accumulation of heavy metals in mosses: a biomonitoring study. Springer Plus 5(1):715
50. Mahapatra B, Dhal NK, Dash AK, Panda BP, Panigrahi KCS, Pradhan A (2019) Perspective of mitigating atmospheric heavy metal pollution: using mosses as biomonitoring and indicator organism. Environ Sci Pollut Res 26(29):29620–29638
51. Mariet C, Gaudry A, Ayrault S, Moskura M, Denayer F, Bernard N (2011) Heavy metal bioaccumulation by the bryophyte Scleropodium purum at three French sites under various influences: Rural conditions, traffic, and industry. Environ Monit Assess 174(1):107–118
52. Markert B, Weckert V (1993) Time-and-site integrated long-term biomonitoring of chemical elements by means of mosses. Toxicol Environ Chem 40(1–4):43–56
53. Martins RJE, Boaventura RAR (2002) Uptake and release of zinc by aquatic bryophytes (Fontinalis antipyretica L. ex. Hedw.). Water Res 36(20):5005–5012
54. Maxhuni A, Lazo P, Kane S, Qarri F, Marku E, Harmens H (2016) First survey of atmospheric heavy metal deposition in Kosovo using moss biomonitoring. Environ Sci Pollut Res 23(1): 744–755
55. Mentese S, Yayintas O, Bas B, İrkin L, Yilmaz S (2021) Heavy metal and mineral composition of soil, atmospheric deposition, and mosses with regard to integrated pollution assessment approach. Environ Manag 67:1–19

56. Nagajyoti PC, Lee KD, Sreekanth T (2010) Heavy metals, occurrence and toxicity for plants: a review. Environ Chem Lett 8(3):199–216
57. Patiño J, Vanderpoorten A (2018) Bryophyte biogeography. CRC Crit Rev Plant Sci 37(2–3): 175–209
58. Petschinger K, Adlassnig W, Sabovljevic M, Lang I (2021) Lamina cell shape and cell wall thickness are useful indicators for metal tolerance – an example in bryophytes. Plan Theory 10:274
59. Radziemska M, Mazur Z, Bes A, Majewski G, Gusiatin ZM, Brtnicky M (2019) Using mosses as bioindicators of potentially toxic element contamination in ecologically valuable areas located in the vicinity of a road: a case study. Int J Environ Res Public Health 16(20):3963
60. Ren J, Liu F, Luo Y, Zhu J, Luo X, Liu R (2021) The pioneering role of bryophytes in ecological restoration of manganese waste residue areas, Southwestern China. J Chem
61. Riget F, Asmund G, Aastrup P (2000) The use of lichen (*Cetraria nivalis*) and moss (*Rhacomitrium lanuginosum*) as monitors for atmospheric deposition in Greenland. Sci Total Environ 245:137–148
62. Rola K, Osyczka P (2018) Cryptogamic communities as a useful bioindication tool for estimating the degree of soil pollution with heavy metals. Ecol Indic 88:454–464
63. Rühling Å, Tyler G (1970) Sorption and retention of heavy metals in the woodland moss Hylocomium splendens (Hedw.). Br Sch Oikos 21:92–97
64. Salemaa M, Derome J, Helmisaari H-S, Nieminen T, Vanha-Majamaa I (2004) Element accumulation in boreal bryophytes, lichens and vascular plants exposed to heavy metal and sulfur deposition in Finland. Sci Total Environ 324(1–3):141–160
65. Saxena D (2004) Uses of bryophytes. Resonance 9(6):56–65
66. Sharma S (2007) Marchantia polymorpha L.: a bioaccumulator. Aerobiologia 23(3):181–187
67. Shotbolt L, Büker P, Ashmore M (2007) Reconstructing temporal trends in heavy metal deposition: assessing the value of herbarium moss samples. Environ Pollut 147(1):120–130
68. Smith AJE (1978) Provisional atlas of the bryophytes of the British Isles. 2nd ed. Biological Records Centre
69. Ştefănuţ S, Öllerer K, Manole A, Ion MC, Constantin M, Banciu C et al (2019) National environmental quality assessment and monitoring of atmospheric heavy metal pollution-a moss bag approach. J Environ Manag 248:109224
70. Szczepaniak K, Astel A, Bode P, Sârbu C, Biziuk M, Raińska E et al (2006) Assessment of atmospheric inorganic pollution in the urban region of Gdańsk, Northern Poland. J Radioanal Nucl Chem 270(1):35–42
71. Thöni L, Yurukova L, Bergamini A, Ilyin I, Matthaei D (2011) Temporal trends and spatial patterns of heavy metal concentrations in mosses in Bulgaria and Switzerland: 1990–2005. Atmos Environ 45(11):1899–1912
72. Trujillo-González JM, Zapata-Muñoz YL, Torres-Mora MA, García-Navarro FJ, Jiménez-Ballesta R (2020) Assessment of urban environmental quality through the measurement of lead in bryophytes: case study in a medium-sized city. Environ Geochem Health 42(10): 3131–3139
73. Tyler G (1990) Bryophytes and heavy metals: a literature review. Bot J Linn Soc 104(1–3): 231–253
74. Tyler G (2008) Bryophytes and heavy metals: a literature review. Bot J Linn Soc 104(1–3): 231–253
75. Uniyal P, Singh A, Sood A, Sharma P (2017) Assessment of accumulation of some heavy metals in mosses of Idukki District, Kerala (Western Ghats, India). Int J Plant Environ 3(01):15–19
76. Vats SK, Singh A, Koul M, Uniyal PL (2010) Study on the metal absorption by two mosses in Delhi Region (India). J Am Sci 6:176–181
77. Vázquez MD, Villares R, Carballeira A (2013) Biomonitoring urban fluvial contamination on the basis of physiological stress induced in transplants of the aquatic moss Fontinalis antipyretica Hedw. Hydrobiologia 707(1):97–108

78. Vázquez MD, Real C, Villares R (2020) Optimization of the biomonitoring technique with the aquatic Moss Fontinalis antipyretica Hedw: selection of shoot segment length for determining trace element concentrations. Water 12(9):2389
79. Vuori KM, Helisten H (2010) The use of aquatic mosses in assessment of metal pollution: appraisal of type-specific background concentrations and inter-specific differences in metal accumulation. Hydrobiologia 656(1):99–106
80. Wang S, Zhang Z, Wang Z (2015) Bryophyte communities as biomonitors of environmental factors in the Goujiang karst bauxite, southwestern China. Sci Total Environ 538:270–278
81. Wu L, Fu S, Wang X, Chang X (2020) Mapping of atmospheric heavy metal deposition in Guangzhou city, southern China using archived bryophytes. Environ Pollut 265(Pt B):114998
82. Xiao J, Han X, Sun S, Wang L, Rinklebe J (2021) Heavy metals in different moss species in alpine ecosystems of Mountain Gongga, China: geochemical characteristics and controlling factors. Environ Pollut 272:115991
83. Xu Y, Yang R, Zhang J, Gao L, Ni X (2021) Distribution and dispersion of heavy metals in the rock–soil–moss system in areas covered by black shale in the Southeast of Guizhou Province, China. Environ. Sci. Pollut. Res.(29)1:1-14.
84. Yedjou CG, Patlolla AK, Sutton DJ (2012) Heavy metals toxicity and the environment Paul B Tchounwou. Published in final edited form as: EXS.101: 133–164
85. Zinicovscaia I, Hramco C, Chaligava O, Yushin N, Grozdov D, Vergel K et al (2021) Accumulation of potentially toxic elements in mosses collected in the Republic of Moldova. Plan Theory 10(3):471
86. Zvereva E, Kozlov M (2011) Impacts of industrial polluters on bryophytes: a meta-analysis of observational studies. Water Air Soil Pollut 218:573–586

Part II

Pteridophytes

Bioactive Compounds of Pteridophytes

8

Hosakatte Niranjana Murthy, Govardhana G. Yadav, and
Medha A. Bhat

Contents

H. N. Murthy (✉) · G. G. Yadav · M. A. Bhat
Department of Botany, Karnatak University, Dharwad, India

H. N. Murthy (ed.), *Bioactive Compounds in Bryophytes and Pteridophytes*,
Reference Series in Phytochemistry, https://doi.org/10.1007/978-3-031-23243-5_10

Abstract

Pteridophytes are non-flowering plants that are possessing horticultural and medicinal value. Vegetative parts or even entire plants, fiddleheads, and rhizomes of pteridophytes are edible and rich in nutritional composition. They are also possessing plentiful phytochemicals including flavonoids, phenolic acids, lignans, coumarins, chromones, phenylpropanoids, quinones, xanthones, terpenoids, alkaloids, and glycosides. These phytochemicals are demonstrated to have several biological activities including antioxidant, anti-cancer, anti-diabetic, anti-inflammatory, anti-microbial, and neuroprotective effects. This review presents an overview of nutritional value, and phytochemicals present in pteridophytes. The biological activities of phytochemicals present in pteridophytes are also presented.

Keywords

Adiantum · Bioactive compounds · *Dryopteris* · *Equisetum* · *Lycopodium* · Phytochemicals · Pteridophytes · *Selaginella*

1 Introduction

The pteridophytes are non-flowering, vascular, and spore-bearing plants including ferns and fern-allies. Ferns and fern-allies comprise over 568 genera and about 13,000 species found in temperate and tropical regions of the world in different ecological niches, as hydrophytes, mesophytes, lithophytes, and epiphytes [1]. Pteridophytes are generally classified into two major groups, lycophytes including club mosses, spike mosses, and quillworts, and filicophytes which include ferns and horsetails [2]. Pteridophytes are having economic importance as horticultural, food, and medicinal plants. Several species of pteridophytes, such as *Lycopodium, Selaginella,* and ferns including *Angiopteris*, *Asplenium*, *Marattia*, *Nephrolepis,* and others are having aesthetic value and are used as horticultural plants [3]. The rhizome of many ferns, such as *Pteris* is rich in starch and they are used as food. The young fronds or leaf tips which are popularly called fiddleheads and fiddleheads of many ferns are used as a vegetable [4–6]. Pteridophytes have been used in the preparation of medicine in Indian, Chinese, and Oriental systems of medicine to cure many human ailments [7–9]. Pteridophytes are abundant with phytochemicals including polyphenols, alkaloids, and terpenoids [9, 10]. Many of the phytochemicals present in pteridophytes possess health-promoting activities and are useful against cancer, coronary heart disease, diabetes, high blood pressure, inflammation, infections of microbes, viruses, and parasites, psychotic diseases, spasmodic conditions, and ulcers. This contribution aims to review the nutritional benefits of pteridophytes. Further, we are presenting the phytochemicals that are present in selected pteridophytes and exploring their biological activities and health benefits.

2 Nutritional Benefits

Underground rhizomes, fiddleheads, and fronds of ferns are used as food by different communities in China, Korea, Japan, India, Philippines, Malaysia, North America, Europe, and sub-Saharan Africa (Table 1) [5, 6, 11, 12]. Rhizomes of *Angiopteris evecta, Blechnum orientale,* and underground tubers of *Nephrolepis cordifolia* are eaten in India and sub-Saharan Africa and they are a good source of starchy food [5, 6]. Leaves of *Christella dentata, Diplazium sammatii, Huperzia phlegmaria, H. squarrosa, Nephrolepis cordifolia, Ophioglossum lusoafricanum, O. reticulatum, O. ovatum, O. vulgatum* and entire plant of *Isoetes debii, Marsilea minuta, Salvinia cucullata,* and *S. natans* are consumed in various parts of the world as a vegetable (Table 1).

The fiddleheads (young fronds) are consumed in different regions of the world as salad, used in the preparation of soups, cakes, noodles, and other dishes, and the dried powder is used as herbal tea/liquor [5, 6, 12, 13]. The boiled and dried fiddleheads are stored along with salt for up to 2–3 years and used subsequently [13]. Several researchers carried out nutritional analysis of ferns which are used as food material (Table 2). The fiddleheads/edible pteridophytes contain good amounts of carbohydrates, protein, and fat. The amount of carbohydrate varies from 21.5 $g.kg^{-1}$ in *Nephrolepis cordifolia* [14] to 108.7 $g.kg^{-1}$ in *Nephrolepis biserrata* [15]. Protein content ranges from 10.3 $g.kg^{-1}$ in *Nephrolepis cordifolia* [14] to 61.3 $g.kg^{-1}$ in *Nephrolepis biserrata* [16]. Fat values vary from 1.6 $g.kg^{-1}$ in *Diplazium maximum* [16] to 11.8 $g.kg^{-1}$ in *Diplazium sammatii* [15]. The fiber content was 0.4–38.3 $g.kg^{-1}$ in different fern fiddleheads (Table 2). The energy values of fern fiddleheads vary from 319.5 to 408.5 kcal kg^{-1} and were higher than the amaranth and spinach (Table 2). Dietary fibers are an important component of plant-derived food with several health benefits, such as improving the intestinal flora and therapeutic effects against diabetes and dyslipidemia [17]. The fiber content of edible pteridophytes was comparable to leafy vegetables, viz., spinach and amaranth (Table 2). Edible pteridophytes are also rich in minerals including calcium, phosphorous, potassium, sodium, manganese, copper, zinc, and iron (Table 2). For example, iron values range from 5.5 $mg.kg^{-1}$ in *Diplazium sammatii* to 11,280 $mg.kg^{-1}$ in *Nephrolepis biserrata* (Table 2). Calcium levels vary from 0.55 $mg.kg^{-1}$ in *Ophioglossum polyphyllum* to 1900 $mg.kg^{-1}$ in *Diplazium maximum* (Table 2).

Sareen et al. [16] carried out the analysis of the amino acid composition of *Diplazium maximum* and reported the presence of alanine, arginine, aspartic acid, cysteine, glutamic acid, glycine, histidine, isoleucine, leucine, lysine, methionine, phenylalanine, proline, serine, threonine, tyrosine, valine, and tryptophan, accounting for a total amino acid content of 23.75 g.100 g^{-1} of biomass. They also reported the fat content of 1.62 g.100 g^{-1} in *Diplazium maximum* [16]. Fatty acid profiling of the fat obtained from *D. maximum* revealed palmitic acid, linoleic acid, dihomo-γ-linolenic acid, and α-linolenic acid (omega 6 fatty acid) as the major components [16]. The presence of dihomo-γ-linolenic acid is accountable for the therapeutic benefits of *D. maximum*. Similarly, the presence of omega-3 and omega-6 polyunsaturated fatty acids was also recently reported in the fiddleheads of

Table 1 Some of the edible pteridophytes

Species	Parts used	Countries	References
Allantodia dilatata (Blume) Ching	Tender leaves	China	[12]
Allantodia gigantea (Bak.) Ching	Tender leaves	China	[12]
Allantodia himalayensis Ching	Tender leaves	China	[12]
Allantodia spectabilis (Wall. ex Mett.) Ching	Tender leaves	China	[12]
Allantodia viridissima (H. Christ) Ching	Tender leaves	China	[12]
Alsophila spinulosa (Wall. ex Hook.) R. M.Tryon	Stems	China	[12]
Ampelopteris prolifera (Retz.) Copel	Fiddle heads	China, India	[6, 12]
Angiopteris esculenta Ching	Rhizome	China	[12]
Angiopteris evecta (G. Forst.) Hoffm.	Rhizome	India	[6]
Angiopteris fokiensis Hieron.	Rhizome	China	[12]
Asplenium scolopendrium L.	Fiddle heads	Europe	[4]
Asplenium unilaterale Lam.	Rhizome	China	[12]
Athyrium acutipinnulum Kodama ex Nakai	Fiddle heads	Korea	[13]
Athyrium brevifrons Tagawa	Fiddle heads; Rhizome	China, Korea	[12, 13]
Athyrium distentifolium Tausch ex Opiz	Fiddle heads	Korea, Europe	[4, 13]
Athyrium esculentum (Retz.) Sw.	Fiddle heads	Korea	[13]
Athyrium filix-femina (L.) Roth.	Fiddle heads	Europe	[4]
Athyriopsis japonica (Thunb.) Ching *var. oshimensis* (Christ) Ching	Young leaves	China	[12]
Athyrium multidentatum (Doll) Ching	Fronds	China	[12]
Athyrium pachyphyllum Ching	Fronds	China	[12]
Athyrium yokoscense (Franch. et Sav.) Chrsit	Fronds	China	[12]
Azolla pinnata R. Br.	Entire plant	India	[6]
Blotiella glabra (Bory). R. M. Tryon	Fiddle heads	Democratic Republic of Congo	[5]
Blechnum orientale Linn.	Fiddle heads, rhizome	China, India	[6, 12]
Botryhchium lanuginosum Well. Ex Hook et Giev.	Fiddle heads	Nigeria	[5]
Callipteris esculenta (Retz.) J. Sm. ex Moore et Houlst	Fronds	China	[12]
Callipteris esculenta (Retz.) J. Sm. ex Moore et Houlst. var. pubescens (Link) Ching	Fronds	China	[12]
Ceratopteris cornuta (P. Beauv.) Lepr.	Fiddle heads, leaves	Liberia	[5]

(continued)

Table 1 (continued)

Species	Parts used	Countries	References
Christella dentata (Forsk.) Brownsey & Jemmy	Leaves	Democratic Republic of Congo	[5]
Ceratopteris thalictroides (L.) Brongniart	Fiddle heads	China, India, Madagascar	[5, 6, 12]
Cibotium barometz (L.) J. Sm.	Rhizome	China	[12]
Coniogramme emeiensis Ching et Shing	Fronds, rhizome	China	[12]
Coniogramme intermedia Hieron.	Fronds, rhizome	China	[12]
Coniogramme intermedia Hieron. *var. glabra* Ching	Fronds, rhizome	China	[12]
Coniogramme japonica (Thunb.) Diels	Fronds, rhizome	China	[12]
Coniogramme jingangshanensis Ching et Shing	Fronds, rhizome	China	[12]
Coniogramme robusta Christ	Fronds, rhizome	China	[12]
Coniogramme rosthornii Hieorn.	Fronds, rhizome	China	[12]
Coniogramme simillima Ching ex Shing	Fronds, rhizome	China	[12]
Coniogramme taipaishanensis Ching et Y. T. Hsieh	Fronds, rhizome	China	[12]
Coniogramme wilsoni Hieron.	Fronds, rhizome	China	[12]
Cornopteris decurrenti-alata (Hook.) Nakai	Fronds	China	[12]
Cyrtomium fortunei J. Sm.	Fronds		[12]
Dicranopteris linearis (Burm. f.) Underw.	Fiddle heads	India	[6]
Diplazium cochleata (D. Don) C. Chr.	Fiddle heads	India	[18]
Diplazium dilatatum Blume	Fiddle heads	India	[18]
Diplazium esculatum (Rotz.) Sw.[a]	Fiddle heads	India	[6]
Diplazium maximum (D. Don) C. Chr.	Fiddle heads	India	[16, 18]
Diplazium proliferum (Lam.) Thouars	Fiddle heads	Madagascar	[5]
Diplazium sammatii (Kuhn) C. Chr.	Fiddle heads, leaves	Democratic Republic of Congo	[5]
Diplazium spectabile (Wall. ex Mett.) Ching	Fiddle heads	India	[18]
Diplazium squamigerum (Mett.) Matsum.	Fiddle heads	Japan	[13]
Drynaria baronii (Christ) Diels	Rhizomes	China	[12]
Drynaria fortunei (Kunze) J. Sm.	Rhizomes	China	[12]
Dryopteris aemula (Aiton) Kuntze	Fiddle heads	Europe	[4]

(continued)

Table 1 (continued)

Species	Parts used	Countries	References
Dryopteris affinis (Lowe) Fraser-Jenk.	Fiddle heads	Europe	[4]
Dryopteris borreri (Newman) Oberh. & Tavel	Fiddle heads	Europe	[4]
Dryopteris cambrensis Roger.	Fiddle heads	Europe	[4]
Dryopteris carthusiana (Vill.) H.P. Fuchs	Fiddle heads	Europe	[4]
Dryopteris caucasica (A. Braun) Fraser-Jenk. & Corley	Fiddle heads	Europe	[4]
Dryopteris cochleata (D. Don) C. Chr.	Fiddle heads	India	[18]
Dryopteris dilatata (Hoffm.) A. Gray	Fiddle heads	Europe	[4]
Dryopteris expansa (C. Presl) Fraser-Jenk. & Jermy	Fiddle heads	Europe	[4]
Dryopteris filix-mas (L.) Schott	Fiddle heads	Europe	[4]
Dryopteris oreades Fomin	Fiddle heads	Europe	[4]
Dryopteris remota (hybrid of *D. affinis* and *D. expansa*)	Fiddle heads	Europe	[4]
Huperzia phlegmaria (L.) Rothm. (=*Phlegmariurus phlegmaria* L. Holub)[a]	Sporophylls	India	[6]
Huperizia squarrosus (G.Forst.) Rothm., comb. superfl. [=*Phlegmariurus squarrosus*) (G.Forst.) Á.Löve & D.Löve] [a]	Sporophylls	India	[6]
Isoetes debii Sinha (=*Isoetes coromandelina* L. fil. *Isoetes sahyadriensis* Mahab.)	Entire plant	India	[6]
Lastrea limbosperma (All.) Holub	Fiddle heads	Europe	[4]
Lunathyrium acrostichoides Ching	Fronds	China	[12]
Lunathyrium coreanum (Christ) Ching	Fronds	China	[12]
Lygodium japonicum (Thanb.) Sw.	Fiddle heads	China, India	[6, 12]
Marsilea minuta L.	Entire plant	Gambia, India, Senegal	[5, 6]
Marsilea quadrifolia L.	Young leaves	China	[12]
Microsorum punctatum (L.) Copel.	Fiddle heads	India	[6]
Matteuccia intermedia C. Chr.	Fronds	China	[12]
*Matteuccia orientalis (*Hook.) Trev.	Fronds	China	[12]
Matteuccia struthiopteris (L.) Tod.	Fiddle heads	Canada, China, Europe, India, Japan, Malaysia, USA, Europe	[4, 13]
Nephrolepis auriculata (L.) Trimen.	Young leaves, tubers	China	[12]
Nephrolepis cordifolia (L.) C. Presl.[a]	Underground tubers	India, Nigeria	[5, 6]

(continued)

Table 1 (continued)

Species	Parts used	Countries	References
Nephrolepis biserrata (L.) C. Presl	Fiddle heads	India; Nigeria	[6, 19]
Nephrolepis cardifolia (Sw.) Schott.	Leaves	Democratic Republic of Congo	[5]
Neottopteris nidus (L.) J. Sm.	Young leaves	China	[12]
Onoclea sensibilis L.	Fiddle heads	Europe	[4]
Ophioglossum lusoafricanum Prantl.	Leaves	Swaziland	[5]
Ophioglassum ovatum Bory.	Leaves	Madagascar	[5]
Ophioglossum polyphyllum A. Braun	Fronds	China	[12]
Ophioglossum reticulatum L.	Leaves	India, South Africa, Swaziland, Tanzania, Zanzibar	[5, 6]
Ophiglossum vulgatum L.	Sporophylls	India, Nigeria	[5, 6]
Osmunda cinnamomea L.	Fiddle heads	China, Japan, Korea	[13]
Osmunda japonica Thunb.	Fiddle heads	China, Japan, Korea	[12, 13]
Osmunda regalis L.	Fiddle heads	India, Korea, Japan, Europe	[4, 6, 13]
Osmundastrum cinnamomeum (L.) Presl.	Fronds	China	[12]
Phegopteris connectilis (Michx.) Watt	Fiddle heads	Europe	[4]
Polypodiodes niponica (Mett.) Ching	Tender leaves	China	[12]
Polypodium vulgare L.	Fiddle heads	Europe	[4]
Polystichum aculeatum (L.) Roth	Fiddle heads	Europe	[4]
Polystichum setiferum (Forssk.) Woynar	Fiddle heads	Europe	[4]
Pteridium aquilinum L. Kuhn.	Fiddle heads	Angola, Cameroon, China, Democratic Republic of Congo, India, Madagascar, Nigeria, South Africa, Europe	[4–6, 12, 13]
Pteridium revolutum (Bl.) Nakai	Fronds, rhizome	China	[12]
Pteris wallichiana Agardh	Tender leaves	China	[12]
Salvinia cucullata Roxb.	Entire plant	India	[6]
Salvinia natans (L.) All.	Entire plant	India	[6]
Sphenomeris chinensis L.	Fiddle heads	India	[6]
Stenochlaena palustries (Burm. f.) Bedd.	Fiddle heads	China, India, Japan, Korea	[6, 13]
Tectaria coadunata (J. Sm.) C. Chr.	Fiddle heads	India	[18]
Thelypteris palustris Schott	Fiddle heads	Europe	[4]
Woodwardia japonica (L. f.) J. Sm.	Rhizome	China	[12]
Woodwardia unigemmata (Makino) Nakai	Rhizome	China	[12]

[a]Plants are also in cultivations

European ferns [4]. Consequently, edible pteridophytes are rich in carbohydrates, protein, fat, and essential mineral elements. The nutrient composition of edible pteridophytes will suffice for essential nutrition in humans.

3 Bioactive Compounds

In the current review, certain representative genera have been selected to present bioactive compounds of the pteridophytes, viz., *Lycopodium*, *Selaginella*, *Equisetum*, *Adiantum,* and *Dryopteris*. Chemical investigations of pteridophytes have revealed the occurrence of major groups, including flavonoids, phenolic acids, lignans, coumarins, chromones, phenylpropanoids, quinones, xanthones, terpenoids, alkaloids, and glycosides.

3.1 Bioactive Compounds of *Lycopodium* Species

Phytochemical reports revealed that varied species of *Lycopodium* are rich in alkaloids, glycosides, and terpenoids (Fig. 1).

Lycopodium alkaloids have been classified into four structural classes, namely, lycopodine class, lycodine class, fawcettimine class, and miscellaneous group (Table 3) [23].

All the *Lycopodium* alkaloids are made up of polycyclic carbon skeletons with varying levels of oxidation and they are kind of nitrogen heterocyclic compounds with a novel skeleton. They are tricyclic or tetracyclic compounds composed of basic skeleton $C_{16}N$ and $C_{16}N_2$ and a few of them might be $C_{14}N$, $C_{15}N_2$, $C_{22}N_2$, $C_{22}N_2$, and $C_{27}N_3$ alkaloids [24]. More than 150 lycopodine classes of alkaloids have been isolated from *Lycopodium* spp. and these are the most widely reported *Lycopodium* alkaloids. This class of compounds possesses six-membered rings, with A and C being in *cis* quinolizidine ring system and most of the ring B have carbonyl groups at C-5 and a few at C-6. Lycopodine **(2)** is one of the most common representatives of this class which was isolated from *Lycopodium alopecuroides*, *L. alpinum*, *L. clavatum*, *L. densum*, *L. japonicum*, *L. lucidulum*, *L. magellanicum*, *L. obscurum*, *L. paniculatum*, *L. serratum*, and *L. volubile* [25–44]. Huperzine A **(1)** is a lycodine class of alkaloid that was isolated from *L. phlegmaria* [45], L. *selago* [46], and *L. serratum* [32]. These compounds are different from lycopodine class in having opened A ring, and the C ring is converted into a separate hexahydropyridine ring. More than 65 alkaloids belonging to lycodine class have been isolated from different *Lycopodium* species [47]. Lycojapodine A **(3)** is a fawcettimine class of alkaloid that was extracted from *L. japonicum* [30, 44]. This class of compound can be regarded as the result of C_4–C_{13} bond breaking and forming C_4–C_{12} bond in lycopodine class. Another class is considered a miscellaneous group that does not have a uniform skeleton. Phlegmarine types are the major miscellaneous alkaloids present in *Lycopodium* species, in them, the C-4 is unconnected to C-12 or C-13. Several triterpenoids have been isolated from

Table 2 Nutritional value of some edible pteridophytes in comparison with popular leafy vegetables

Species/crop	Energy (kcal kg^{-1})	Carbohydrates (g kg^{-1})	Protein (g kg^{-1})	Fat (g kg^{-1})	Fibre (g kg^{-1})	Minerals (mg kg^{-1})								Reference
						Ca	P	K	Na	Mn	Cu	Zn	Fe	
Diplazium maximum	319.40	61.36	25.30	1.60	38.30	277.20	1315	1783	–	–		46.10	200.50	[16]
Diplazium sammatii	408.50	65.50	10.30	11.80	0.40	1900	7.00	1600	520	–	3.00	4.10	5.50	[15]
Nephrolepis biserrata	–	108.70	61.30	3.30	8.70	2232	–	5000	2500	1008	1587.5	–	11,280	[20]
Nephrolepis cordifolia	–	21.50	10.30	–	1.60	26.90	–	141.20	45.50	0.70	7.20	1.10	7.50	[14]
Ophioglossum polyphyllum	–	–	24.50	–	–	0.55	4.90	41.80	0.46	27	170	29	888	[21]
Amaranth	30.59	22.80	81.00	13.50	44.10	3300	732	5720	168	12.40	2.10	8.60	46.40	[22]
Spinach	28.37	20.50	21.40	0.60	23.80	822.90	35.50	6250	425	11.20	1.70	9.30	29.50	[22]

Lycopodium species and these are serratene type pentacyclic triterpenoids. Serratenediol (**9**) is one such triterpenoid isolated from *L. serratum* which has exhibited several biological activities [48]. Zhang et al. [49] have isolated apigenin-4′-*O*-(2″,6″-di-*O*-*p*-coumaroyl)-*β*-D-glucopyranoside (**7**), a floavone glycoside from *L. cernuum*.

3.2 Bioactive Compounds of *Selaginella* Species

A large array of compounds including alkaloids, caffeoylquinic acids, chromones, coumarins, flavonoids, lignans, phenolics, pigments, quinones, saponins, sterols, and terpenoids have been isolated from varied *Selaginella* species (Table 4; Figs. 2, 3, 4, 5, 6, and 7).

Delicatuline A (**11**), delicatuline B (**12**), 6-amino-9-purine methyl caproate (**10**), paucine (**13**), and paucine 3′-O-β-D-glucopyranoside (**14**) are some of the major alkaloids which were reported from *Selaginella delicatula* and *S. moellendorffii* [146, 147]. 5-Carboxymethyl-7-hydroxychromone (**15**), uncinoside A (**16**), and uncinoside B (**17**) are the chromone group of compounds extracted from *S. moellendorffii* and *S. uncinata* [148, 149]. 3-(4-Hydroxyphenyl)-6,7-dihydroxy coumarin (**18**), Isopimpinellin (**19**), and Umbelliferone (**20**) were certain coumarins sequestered from *S. moellendorffii* and *S. tamariscina* [150–152]. Major flavonoids which were isolated from *Selaginella* species were amentoflavone (**22**), hinokiflavone (**23**), involvenflavone A (**24**), robustaflavone (**26**), seladoeflavone A (**27**), seladoeflavone B (**28**), and uncinataflavone A (**29**) [151, 153–168]. Caffeic acid (**30**), ferulic acid (**31**), syringic acid (**32**), and vanillic acid (**33**) are some of the phenolics obtained from *Selaginella tamariscina* [150]. Major lignans which were reported from *Selaginella* are burseneolignan (**37**), pictalignan A (**39**), selaginellol (**40**), selamoellenin B (41), sinensiol B (**42**), syringaresinol (**43**), and tamariscinol U (**44**) [146, 147, 150, 169–172]. Selaginellins are a small group of pigments exclusively found in the ancient genus *Selaginella*. Selaginellin (**46**) was the first compound that was isolated in 2007 [173]. Subsequently, many more compounds have been reported including selagenellin A (**47**), selaginellin B (**48**), selaginone A (**49**), selaginpulvilin A (**50**), and selariscinin A (**51**) (Table 4) [174–185]. (1α, 3β, 25R)-Spirost-5-ene-2-diol-3-O-α-L-rhamnopyranosyl-(1 → 2)-O-[α-L-rhamnopyranosyl (1 → 4)]-O-β-D-glucopyranoside (**52**) and (2α, 3β, 12β, 25R)-spirost-5-ene-2, 3, 12-triol-3-O-α-L-rhamnopyranosyl-(1 → 2)-O-[α-L-rhamnopyranosyl-(1 → 4)]-O-β-D-glucopyranoside (**53**) were the two major saponins extracted from *S. uncinata* [186]. Several sterols were obtained from *S. delicatula* and *S. tamariscina* including stigmasterol (**54**), β-sitosterol (**55**), and β-sitostenone (**56**) [151, 187]. Selaginedorffone A (**57**) and selaginedorffone B (**58**) are a few of the terpenoids reported from *Selaginella moellendorffii* [188].

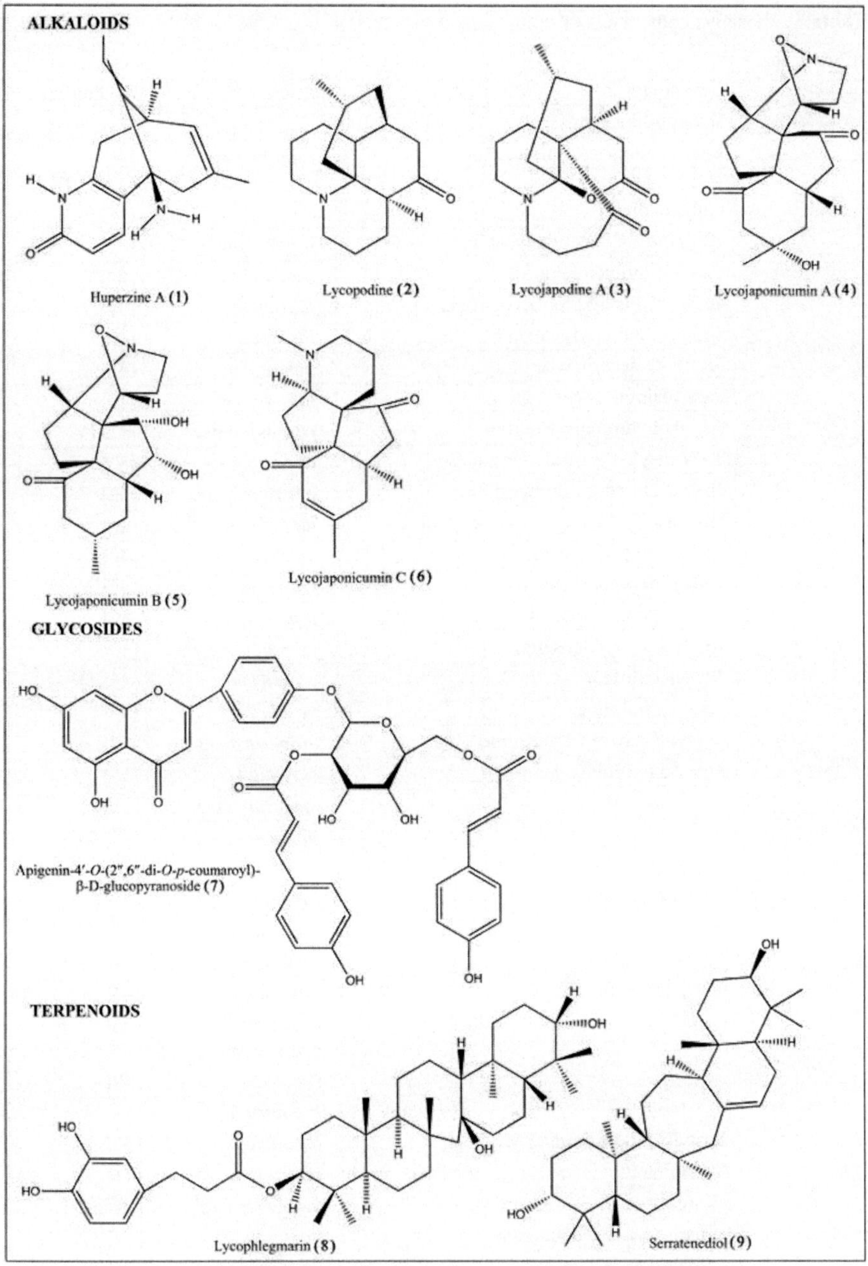

Fig. 1 Major Alkaloids, glycosides, and terpenoids isolated from *Lycopodium* spp.

Table 3 Bioactive compounds of *Lycopodium* spp.

Chemical group	Compound	Species	References
Alkaloids	**Lycopodine type**		
	11α-Hydroxy-acetylfawcettine	*L. japonicum*	[40]
	11β-Hydroxy-12-epilycodoline	*L. japonicum*	[40, 42]
	12β-Hydroxy-acetylfawcettiine N-oxide	*L. japonicum*	[42]
	17α-Methyllycoflexine	*L. japonicum*	[50]
	4α,8β,12β-trihydroxylycopodine	*L. japonicum*	[40]
	4α,8β-dihydroxylycopodine	*L. japonicum*	[40]
	5R,8R-O-acetylfawcettiine	*L. obscurum*	[41]
	6α,8β-Dihydroxylycopodine	*L. japonicum*	[40]
	8,15-Dihydrohuperzinine	*L. casuarinoides*	[51]
	8β-(Acetyloxy) obscurumine A	*L. obscurum*	[52]
	8β,11α-Dihydroxylycopodine	*L. clavatum*	[53]
	8β-Acetoxy-12β-hydroxy-lycopodine	*L. japonicum*	[40]
	8β-Hydroxy-11α-acetoxylycopodine	*L. japonicum*	[40]
	8β-Hydroxyhuperzine E	*L. japonicum*	[40]
	8β-Hydroxylycodoline	*L. japonicum*	[40]
	8β-Hydroxylycoposerramine K	*L. japonicum*	[40]
	Acetylacrifoline	*L. obscurum*	[29, 41]
	Acetylannofoline	*L. obscurum*	[29]
	Acetyldebenzoylalopecurine	*L. alopecuroides*	[54]
	Acetyldihydrolycopodine	*L. clavatum*	[26]
		L. magellanicum	[35]
		L. obscurum	[29, 43]
		L. paniculatum	[28]
	Acetylfawcettiine	*L. japonicum*	[40, 44]
		L. magellanicum	[35]
		L. obscurum	[29, 33, 43]
	Acetylfawcettine N-oxide	*L. japonicum*	[44]
		L. clavatum var. *megastachyon*	[55]
		L. serratum	[39]
		L. japonicum	[40]
	Acetyllycoposerramine M	*L. japonicum*	[42]
	Acrifoline	*L. obscurum*	[41]
	Alopecurine	*L. alopecuroides*	[25, 54]
	Anhydrodeacetylpaniculine	*L. paniculatum*	[56]
	Anhydrolycodoline	*L. alopecuroides*	[25]
		L. inundatum	[57]
		L. japonicum	[40]
		L. obscurum	[41, 43]
	Anhydrolycopodoline	*L. japonicum*	[44]

(continued)

Table 3 (continued)

Chemical group	Compound	Species	References
	Annofoline	*L. annotinum*	[58]
	Annotinine	*L. annotinum;* *L. annotinum* var. *acrifolium*	[59]
	Annotinolide A	*L. annotinum*	[60]
	Annotinolide B	*L. annotinum*	[60]
	Annotinolide C	*L. annotinum*	[60]
	Clavolonine	*L. alopecuroides*	[25]
		L. alpinum	[38]
		L. clavatum	[26]
		L. japonicum	[40, 44]
		L. magellanicum	[27]
		L. obscurum	[41, 52]
	Complanadine B	*L. complanatum*	[61]
	Deacetylfawcettiine	*L. japonicum*	[40]
		L. magellanicum	[27]
		L. obscurum	[33, 41, 62]
	Deacetyllycoclavine	*L. paniculatum*	[56]
		L. serratum	[39]
	Deacetyllycofawcine	*L. obscurum*	[52]
	Deacetylpaniculine	*L. paniculatum*	[56]
	Debenzoylalopecurine	*L. alopecuroides*	[25]
		L. inundatum	[57]
	Dehydroisofawcettiine	*L. obscurum*	[62]
	Diacetyllycofoline	*L. japonicum*	[44]
	Dihydrolycopodine	*L. clavatum*	[26]
		L. obscurum	[41]
		L. paniculatum	[28]
		L. volubile	[34]
	Diphaladine A	*L. japonicum*	[42]
	Fawcettiine	*L. clavatum*	[26]
		L. fawcettii	[63]
		L. japonicum	[40, 42, 44]
		L. magellanicum	[27]
		L. obscurum	[29, 41]
		L. serratum	[64]
	Fawcettine N-oxide	*L. japonicum*	[44]
	Flabelliformine	*L. clavatum* var. *megastachyon*	[65]
		L. lucidulum	[65]
		L. obscurum	[29, 33, 41]
	Huperzine E	*L. serratum* var. *longipetiolatum*	[32]

(continued)

Table 3 (continued)

Chemical group	Compound	Species	References
	Huperzinine N-oxide	*L. casuarinoides*	[51]
	Inundatine	*L. inundatum*	[57]
	Isofawcettiine	*L. obscurum*	[41, 52]
	Lannotinidine A	*L. annotinum; L. annotinum* var. *acrifolium*	[59]
	Lannotinidine C	*L. annotinum*	[59]
	Lannotinidine D	*L. annotinum*	[59]
	Lannotinidine E	*L. annotinum* var. *acrifolium*	[59]
	Lannotinidine F	*L. annotinum*	[59]
	Lannotinidine G	*L. annotinum*	[59]
	Lucidioline	*L. japonicum*	[66]
		L. lucidulum	[67]
		L. serratum	[39]
	Lycoclavine	*L. alpinum*	[38]
		L. clavatum var. *megastachyon*	[55]
		L. japonicum	[42]
		L. paniculatum	[56]
	Lycodoline	*L. alopecuroides*	[25]
		L. annotinum	[59, 68]
		L. chinense	[69, 70]
		L. japonicum	[66]
		L. obscurum	[29, 33, 41]
		L. serratum	[39]
	Lycofawcine	*L. japonicum*	[40]
	Lycofoline	*L. annotinum*	[58]
		L. obscurum	[41]
	Lyconesidine C	*L. chinense*	[69, 70]
	Lyconnotine	*L. annotinum*	[59, 71]
	Lycopecurine	*L. alopecuroides*	[54, 72]
	Lycoplanine B	*L. complanatum*	[73]
	Lycoplanine C	*L. complanatum*	[73]
	Lycopocarinamine A	*L. carinatum*	[74]
	Lycopocarinamine B	*L. carinatum*	[74]
	Lycopocarinamine C	*L. carinatum*	[74]
	Lycopocarinamine D	*L. carinatum*	[74]
	Lycopocarinamine E	*L. carinatum*	[74]
	Lycopocarinamine F	*L. carinatum*	[74]
	Lycopodatine A	*L. inundatum*	[57]

(continued)

Table 3 (continued)

Chemical group	Compound	Species	References
	Lycopodatine B	*L. inundatum*	[57]
	Lycopodatine C	*L. inundatum*	[57]
	Lycopodine	*L. alopecuroides*	[25]
		L. alpinum	[38]
		L. clavatum	[26, 31, 36, 75]
		L. densum	[37]
		L. japonicum	[30, 40, 42, 44]
		L. magellanicum	[27, 35]
		L. obscurum	[29, 33, 41, 43]
		L. paniculatum	[28]
		L. serratum	[39]
		L. serratum var. *longipetiolatum*	[32]
		L. volubile	[34]
	Lycoposerramine F	*L. serratum*	[39]
	Lycoposerramine G	*L. japonicum*	[40]
		L. serratum	[39]
	Lycoposerramine H	*L. serratum*	[39]
	Lycoposerramine I	*L. serratum*	[39]
	Lycoposerramine J	*L. serratum*	[39]
	Lycoposerramine K	*L. serratum*	[39]
		L. japonicum	[42]
	Lycoposerramine L	*L. serratum*	[39]
	Lycoposerramine M	*L. japonicum*	[40]
		L. serratum	[39]
	Lycoposerramine N	*L. serratum*	[39]
	Lycoposerramine O	*L. serratum*	[39]
	Malycorin B	*L. phlegmaria*	[45]
	Malycorin C	*L. phlegmaria*	[45]
	Miyoshianine A	*L. japonicum*	[66]
	Miyoshianine C	*L. japonicum*	[66]
	Obscurumine A	*L. obscurum*	[61]
	Obscurumine B	*L. obscurum*	[29, 33, 41, 61]
	Obscurumine C	*L. obscurum*	[43]
	Obscurumine O	*L. obscurum*	[33]
	Obscurumine P	*L. obscurum*	[33]
	Paniculine	*L. paniculatum*	[56]
	Serratezomine C	*L. serratum*	[39, 76]
	Serratidine	*L. serratum*	[39]

(continued)

Table 3 (continued)

Chemical group	Compound	Species	References
		L. serratum var. *longipetiolatum*	[32]
	Strictumine A	*L. obscurum*	[41]
	Strictumine B	*L. obscurum*	[41]
	α-Lofoline	*L. annotinum*	[58]
		L. japonicum	[40]
	β-Lofoline	*L. annotinum*	[58]
		L. fawcettii	[63]
	Lycodine type		
	11-Hydroxy lycodine	*L. complanatum*	[77]
	5-Acetyllycofoline	*L. obscurum*	[41]
	Carinatumin A	*L. carinatum*	[78]
	Carinatumin B	*L. carinatum*	[78]
	Casuarine A	*L. casuarinoides*	[79]
	Casuarine B	*L. casuarinoides*	[79]
	Complanadine A	*L. complanatum*	[80]
	Complanadine C	*L. complanatum*	[81]
	Complanadine D	*L. complanatum*	[81]
	Complanadine E	*L. complanatum*	[82]
	Des-N-Methylfastigiatine	*L. fastigiatum*	[83]
	Des-N-methyl-α-obscurine	*L. alpinum*	[38]
		L. fawcettii	[84]
		L. japonicum	[50]
		L. obscurum	[29, 41]
	Fastigiatine	*L. fastigiatum*	[83]
	Flabellidine	*L. paniculatum*	[56]
	Himeradine A	*L. chinense*	[85]
	Huperzine A	*L. phlegmaria*	[45]
		L. selago	[46]
		L. serratum var. *longipetiolatum*	[32]
	Huperzine B	*L. casuarinoides*	[51]
		L. serratum var. *longipetiolatum*	[32]
	Huperzine D	*L. casuarinoides*	[51]
	Huperzinine	*L. casuarinoides*	[51]
		L. platyrhizoma	[86]
	Lycodine	*L. annotinum*	[87]
		L. fawcettii	[84]
		L. japonicum	[40, 50]
		L. lucidulum	[65]
		L. clavatum var. *megastachyon*	[65]

(continued)

Table 3 (continued)

Chemical group	Compound	Species	References
		L. magellanicum	[35]
		L. obscurum	[52]
		L. platyrhizoma	[86]
		L. serratum	[64]
	Lycoparin A	*L. casuarinoides*	[88]
	Lycoparin B	*L. casuarinoides*	[88]
	Lycoparin C	*L. casuarinoides*	[88]
	Lycopladine F	*L. complanatum*	[89]
	Lycopladine G	*L. complanatum*	[89]
	Lycoplanine D	*L. complanatum*	[73]
		L. platyrhizoma	[86]
	Lycoplatyrine A	*L. platyrhizoma*	[86]
	Lycoplatyrine B	*L. platyrhizoma*	[86]
	N-demethylhuperzinine	*L. casuarinoides*	[51, 90]
	N-demethyl-α-obscurine	*L. obscurum*	[33, 43]
	N-demethyl-β-obscurine	*L. obscurum*	[33, 43]
		L. serratum	[64]
	N-Methyl-β-obscurine	*L. obscurum*	[62]
	N-Methyl huperzine B	*L. serratum* var. *longipetiolatum*	[32]
	N-Methyllycodine	*L. magellanicum*	[35]
	Sauroxine	*L. saururus*	[91]
	Selagine	*L. selago*	[46]
	α-Obscurine	*L. annotinum*	[92]
		L. japonicum	[50, 66]
		L. magellanicum	[27]
	β-Obscurine	*L. annotinum*	[92]
		L. casuarinoides	[51]
	Fawcettimine type		
	(15R)-14,15-Dihydroepilobscurinol	*L. japonicum*	[50, 93]
	11α-Hydroxyfawcettidine	*L. serratum*	[94]
	14,15-Dehydrolycoflexine	*L. japonicum*	[50, 93]
	15-epi-6-hydroxy-6,7-dehydro-8-deoxy-13 dehydroserratinine	*L. japonicum*	[44]
	2α,11α Dihydroxyfawcettidine	*L. serratum*	[94]
	2β-Hydroxylycothunine	*L. serratum*	[94]
	5-Dehydromagellanine	*L. magellanicum*	[35]
	5α-Hydroxy-6-oxodihydrophlegmariurine A	*L. japonicum*	[44]
	6-Hydroxy-6,7-Dehydrolycoflexine	*L. japonicum*	[50]
	6-Hydroxyl-6,7-dehydrolycoflexine	*L. japonicum*	[93]
		L. japonicum	[44, 93]

(continued)

Table 3 (continued)

Chemical group	Compound	Species	References
	6-Hydroxyl-6,7-dehydro-8-deoxy-13-dehydroserratinine		
	8-Deoxy-13-dehydroserratinine	*L. japonicum*	[93]
	8α-Acetoxy-fawcettimine	*L. squarrosum*	[95]
	8α-Hydroxylycothunine	*L. serratum*	[94]
	8α,11α-Dihydroxyfawcettidine	*L. serratum*	[94]
	8β-acetoxy-fawcettimine	*L. squarrosum*	[95]
	8β-Hydroxyfawcettimine	*L. squarrosum*	[95]
	Acetyllycoposerramine-U	*L. squarrosum*	[95]
	Alolycopine	*L. alopecuroides*	[54, 96]
	Alopecuridine	*L. alopecuroides*	[25]
	Dihydrolycopoclavamine A	*L. serratum*	[95]
	Fawcettidine	*L. alopecuroides*	[54]
	Fawcettimine	*L. clavatum*	[26, 95]
		L. complanatum	[97]
		L. japonicum	[30, 44]
		L. serratum	[64]
	Fewcettidine	*L. complanatum*	[97]
	Isoobscurinine	*L. japonicum*	[44]
		L. obscurum	[33]
	Isopalhinine A	*L. japonicum*	[50]
	Lycobscurine A	*L. obscurum*	[62]
	Lycobscurine B	*L. obscurum*	[62]
	Lycobscurine C	*L. obscurum*	[62]
	Lycoclavatumide	*L. clavatum*	[53]
	Lycoflexine	*L. japonicum*	[30, 50, 93]
		L. obscurum	[29, 43]
		L. squarrosum	[95]
	Lycoflexine N-oxide	*L. squarrosum*	[95]
	Lycogladine A	*L. complanatum*	[97]
	Lycogladine B	*L. complanatum*	[97]
	Lycogladine C	*L. complanatum*	[97]
	Lycogladine D	*L. complanatum*	[97]
	Lycogladine E	*L. complanatum*	[97]
	Lycogladine F	*L. complanatum*	[97]
	Lycogladine G	*L. complanatum*	[97]
	Lycogladine H	*L. complanatum*	[97]
		L. platyrhizoma	[86]
	Lycojapodine A	*L. japonicum*	[30, 44]
	Lycojaponicumin A	*L. japonicum*	[98]
	Lycojaponicumin B	*L. japonicum*	[98]

(continued)

Table 3 (continued)

Chemical group	Compound	Species	References
	Lycojaponicumin C	*L. japonicum*	[44, 98]
	Lyconesidine A	*L. chinense*	[69, 70]
	Lyconesidine B	*L. chinense*	[69, 70]
	Lycopladine A	*L. complanatum*	[99]
	Lycopladine B	*L. complanatum*	[97, 100]
	Lycopladine C	*L. complanatum*	[100]
	Lycopladine D	*L. complanatum*	[100]
	Lycoplatyrine C	*L. platyrhizoma*	[86]
	Lycopoclavamine A	*L. clavatum*	[95]
		L. japonicum	[50]
	Lycopoclavamine B	*L. clavatum*	[95]
		L. complanatum	[97]
	Lycoposerramine A	*L. serratum*	[101]
	Lycoposerramine B	*L. serratum*	[102]
	Lycoposerramine C	*L. serratum*	[103]
	Lycoposerramine D	*L. serratum*	[103]
	Lycoposerramine E	*L. serratum*	[103]
	Lycoposerramine P	*L. serratum*	[103]
	Lycoposerramine Q	*L. serratum*	[103]
	Lycoposerramine S	*L. serratum*	[103]
	Lycoposerramine U	*L. serratum*	[103]
		L. squarrosum	[95]
	Lycoposquarrosamine A	*L. squarrosum*	[95]
	Lycothunine	*L. serratum*	[94]
	Macleanine	*L. serrata*	[104]
	Magellanine	*L. magellanicum*	[27]
	Malycorin A	*L. phlegmaria*	[45]
	Megastachine	*L. megastachyum*	[105]
	N-Formyl- Lycoposerramine T	*L. serratum*	[64]
	N-Methyl- Lycoposerramine T	*L. serratum*	[64]
	Obscurinine	*L. japonicum*	[44]
		L. obscurum	[33, 43, 106]
	Obscurinine B	*L. obscurum*	[43]
	Obscurumine D	*L. obscurum*	[43]
	Obscurumine E	*L. obscurum*	[43]
	Obscurumine F	*L. obscurum*	[29, 41]
	Obscurumine G	*L. obscurum*	[29, 41]
	Obscurumine H	*L. obscurum*	[33]
	Obscurumine I	*L. obscurum*	[33]
	Obscurumine J	*L. obscurum*	[33]
	Obscurumine K	*L. obscurum*	[33]

(continued)

Table 3 (continued)

Chemical group	Compound	Species	References
		L. platyrhizoma	[86]
	Obscurumine L	*L. obscurum*	[33]
	Obscurumine M	*L. obscurum*	[33]
	Obscurumine N	*L. obscurum*	[33]
	Palhinine A	*L. japonicum*	[44, 50, 93]
	Palhinine B	*L. japonicum*	[93]
	Paniculatine	*L. magellanicum*	[35]
		L. paniculatum	[28, 56]
	Phlegmariurine B	*L. japonicum*	[44]
	Serratanidine	*L. serratum*	[94]
	Serratezomine A	*L. serratum* var. *serratum*	[76]
	Serratezomine B	*L. serratum* var. *serratum*	[76]
	Serratine	*L. serratum*	[94, 107]
	Serratinine	*L. serratum*	[107]
	Miscellaneous type		
	Anabasine	*L. platyrhizoma*	[86]
	Anatabine	*L. platyrhizoma*	[86]
	Carinatumin C	*L. carinatum*	[78]
	Casuarinin H	*L. platyrhizoma*	[86]
	Cermizine A	*L. cernuum*	[70]
	Cermizine B	*L. cernuum*	[70]
	Cermizine C	*L. cernuum*	[70]
	Cermizine D	*L. cernuum*	[70]
		L. obscurum	[52]
	Cermizine D N-oxide	*L. obscurum*	[52]
	Cernuine	*L. cernuum*	[70, 108–110]
		L. chinense	[111]
	Cernuine N-oxide	*L. cernuum*	[70]
	Cryptadine A	*L. cryptomerinum*	[112]
	Cryptadine B	*L. cryptomerinum*	[112]
	Dihydrodeoxyserralongamine B	*L. serratum* var. *longipetiolatum*	[32]
	Dihydrolycolucine	*L. lucidulum*	[113]
	Huperine E	*L. japonicum*	[30]
	Lannotinidine B	*L. annotinum*	[59]
	Lucidine A	*L. lucidulum*	[114]
	Lucidine B	*L. chinense*	[70]
		L. lucidulum	[113, 114]
	Luciduline	*L. lucidulum*	[115]
	Lucidulinone	*L. lucidulum*	[114]

(continued)

Table 3 (continued)

Chemical group	Compound	Species	References
	Lycocernuine	*L. cernuum*	[70, 108, 109]
		L. japonicum	[50]
	Lycocernuine N-oxide	*L. cernuum*	[70]
	Lycochinine A	*L. chinense*	[116]
	Lycochinine B	*L. chinense*	[116]
	Lycochinine C	*L. chinense*	[116]
	Lycojaponicumin E	*L. japonicum*	[50]
	Lyconadin A	*L. complanatum*	[77]
	Lyconadin B	*L. complanatum*	[100]
	Lyconadin D	*L. complanatum*	[82]
	Lyconadin E	*L. complanatum*	[82]
	Lycolucine	*L. lucidulum*	[113]
	Lycoperine A	*L. hamiltonii*	[117]
	Lycopladine H	*L. complanatum*	[118]
		L. japonicum	[50]
	Lycoposerramine R	*L. serratum*	[64]
	Lycoposerramine T	*L. serratum*	[64]
	Lycoposerramine V	*L. serratum*	[119]
	Lycoposerramine W	*L. serratum*	[119]
	Lycoposerramine X	*L. serratum*	[120]
	Lycoposerramine Y	*L. serratum*	[120]
	Lycoposerramine Z	*L. serratum*	[120]
	Lycospidine A	*L. complanatum*	[121]
	Nankakurine A	*L. hamiltonii*	[24]
	Nicotine	*L. cernuum*	[110]
	Oxolucidine A	*L. lucidulum*	[114]
	Palcernine A	*L. japonicum*	[50]
	Senepodine A	*L. chinense*	[70, 111, 116]
	Senepodine B	*L. chinense*	[70]
	Senepodine C	*L. chinense*	[70]
	Senepodine D	*L. chinense*	[70]
	Senepodine E	*L. chinense*	[70, 116]
	Senepodine G	*L. chinense*	[70]
	Senepodine H	*L. chinense*	[70]
	Serralongamine A	*L. serratum* var. *longipetiolatum*	[122]
	Serralongamine B	*L. serratum* var. *longipetiolatum*	[32]
	Serralongamine C	*L. serratum* var. *longipetiolatum*	[32]
	Serralongamine D	*L. serratum* var. *longipetiolatum*	[32]

(continued)

Table 3 (continued)

Chemical group	Compound	Species	References
	Serratezomine D	*L. serratum*	[123]
	Serratezomine E	*L. serratum*	[123]
	Spirolucidine	*L. lucidulum*	[124]
	Tetrahydrodeoxylucidine B	*L. serratum* var. *longipetiolatum*	[32]
Glycosides	Acylated apigenin 4′-O-β-D-glucoside	L. clavatum	[125]
	Apigenin-4′-*O*-(2″,6″-di-*O*-*p*-coumaroyl)-*β*-D-glucopyranoside	L. cernuum	[49]
Terpenoids	(3α, 8β,14α, 21β)-26, 27-Dinoroncocerane-3, 8, 14, 21-tetrol	*L. japonicum*	[126]
		L. obscurum	[127]
	(3β,8β,14α,21α)-26,27- Dinoronocerane-3,8,14,21-tetrol	*L. japonicum*	[128]
	(3β,8β,14α,21β) 26,27-dinoronocerane-3,8,14,21-tetrol	*L. japonicum*	[128]
	16-oxo-21β,24-dihydroxyserrat-14-en-3α-yl acetate	*L. complanatum*	[129]
	16-oxo-21β-hydroxyserrat-14-en-3α-yl acetate	*L. complanatum*	[129]
	21-*epi*-Serratenediol	*L. japonicum*	[130]
		L. serratum	[131]
		L. megastachyum	[132]
		L. serratum	[48]
	21-*epi*-Serratenediol-3-acetate	*L. megastachyum*	[132]
		L. serratum	[48]
	21-Episerratriol	*L. clavatum*	[133]
	21α-hydroxyserrat-14-en-3β-ol	*L. phlegmaria*	[134]
	21α-hydroxyserrat-14-en-3β-yl acetate	*L. complanatum*	[129]
		L. phlegmaria	[134]
	21β,24-dihydroxyserrat-14-en-3α-yl acetate	*L. complanatum*	[129]
	21β-hydroxyserrat-14-en-3α-yl acetate	*L. complanatum*	[129]
	21β-hydroxyserrat-14-en-3α-ol	*L. phlegmaria*	[134]
	21β-hydroxyserrat-14-en-3β-ol	*L. phlegmaria*	[134]
	21β-hydroxyserrat-14-en-3β-yl acetate	*L. complanatum*	[129]
		L. japonicum	[130]
	21β-hydroxyserrat-14-en-3β-yl-formate	*L. japonicum*	[130]
	26-Nor-8-oxo-α-onocerin	*L. japonicum*	[126]
	26-Nor-8β-hydroxy-α-onocerin	*L. obscurum*	[127]
	3, 20β, 21β, 24-Tetrahydroxyserrat-14-ene	*L. japonicum*	[126]
	3-Epilycoclavanol	*L. japonicum*	[128]
	3α,21α-Dihydroxy-16-oxoserrat-14-en-24-yl p-coumarate	*L. japonicum*	[135]

(continued)

Table 3 (continued)

Chemical group	Compound	Species	References
	3α,21β,24-trihydroxyserrat-14-en-16-one	*L. complanatum*	[136]
	3α,21β,24-Trihydroxyserrat-14-ene	*L. japonicum*	[126]
		L. clavatum	[137]
	3α,21β,29-Trihydroxyserrat-14-en-16-one	*L. complanatum*	[136]
	3α,21β-dihydroxyserrat-14-en-16-one	*L. complanatum*	[129]
	3β,21β,24-Trihydroxyserrat-14-ene	*L. japonicum*	[126]
	3β,21α-Dihydroxy-26-nor-8,14-sekogammaser- 14(27)-en-8-one	*L. clavatum*	[137]
	3β,21α-Dihydroxy-8,14-sekogammasera-8(26),14(27)-di-ene	*L. clavatum*	[137]
	3β,21β,24-trihydroxyserrat-14-en-16-one	*L. complanatum*	[136]
	3β,21β-Dihydroxyserrat-14-en-16-one	*L. complanatum*	[136]
	3β-hydroxyserrat-14-en-21β-yl-formate	*L. japonicum*	[130]
	Clavatol	*L. clavaum*	[138]
	Hydroxyserratenone	*L. phlegmaria*	[139]
	Japonicumin A	*L. japonicum*	[140]
	Japonicumin B	*L. japonicum*	[140]
	Japonicumin C	*L. japonicum*	[140]
	Japonicumin D	*L. japonicum*	[140]
	Lycernuic A	*L. complanatum*	[136]
	Lycernuic acid A	*L. cernuum*	[49]
		L. japonicum	[126]
	Lycernuic acid B	*L. cernuum*	[49]
	Lycernuic acid C	*L. cernuum*	[49]
	Lycernuic acid D	*L. cernuum*	[49]
	Lycernuic acid E	*L. cernuum*	[49]
	Lycernuic ketone	*L. japonicum*	[135]
	Lycernuic ketone A	*L. cernuum*	[49]
	Lycernuic ketone B	*L. cernuum*	[49]
	Lycernuic ketone C	*L. cernuum*	[49]
	Lycoclaninol	*L. japonicum*	[128]
	Lycoclavanin	*L. clavatum*	[31, 141]
	Lycoclavanol	*L. clavatum*	[31]
		L. complanatum	[129]
		L. japonicum	[128, 130, 140]
		L. megastachyum	[132]
	Lycojaponicuminol A	*L. japonicum*	[126]
	Lycojaponicuminol B	*L. japonicum*	[126]
	Lycojaponicuminol C	*L. japonicum*	[126]
	Lycojaponicuminol D	*L. japonicum*	[126]

(continued)

Table 3 (continued)

Chemical group	Compound	Species	References
	Lycojaponicuminol E	*L. japonicum*	[126]
	Lycojaponicuminol F	*L. japonicum*	[126]
	Lycomplanatum A	*L. complanatum*	[136]
	Lycomplanatum B	*L. complanatum*	[136]
	Lycomplanatum C	*L. complanatum*	[136]
	Lycomplanatum D	*L. complanatum*	[136]
	Lycomplanatum E	*L. complanatum*	[136]
	Lycomplanatum F	*L. complanatum*	[136]
	Lycomplanatum G	*L. complanatum*	[136]
	Lycomplanatum H	*L. complanatum*	[136]
	Lycophlegmarin	*L. phlegmaria*	[134]
	Lycopodiin A	*L. japonicum*	[128]
	Lycoxanthol	*L. lucidulum*	[142]
	Phlegmanol A	*L. phlegmaria*	[139]
	Phlegmanol B	*L. phlegmaria*	[139]
	Phlegmanol C	*L. phlegmaria*	[139]
	Phlegmanol D	*L. phlegmaria*	[139]
		L. megastachyum	[132]
	Phlegmanol E	*L. phlegmaria*	[139]
	Phlegmaric acid	*L. japonicum*	[126]
		L. phlegmaria	[139]
	Serrat-14-en-3β,21α-diol	*L. cernuum*	[49]
		L. complanatum	[129]
	Serrat-14-en-3β,21β-diol	*L. cernuum*	[49]
		L. complanatum	[129]
	Serrat-14-en-3β-yl-acetate	*L. japonicum*	[130]
	Serrat-14-ene-3β, 21β-diol	*L. japonicum*	[126]
	Serrat-14-ene-3β,21α-diol	*L. japonicum*	[126]
	Serrate-14-en-3,21-dione	*L. japonicum*	[130]
	Serratenediol	*L. serratum*	[107]
		L. complatanum	[143]
		L. japonicum	[130]
		L. phlegmaria	[139]
		L. megastachyum	[132]
		L. serratum	[48, 131]
	Serratenediol-3-acetate	*L. megastachyum*	[132]
		L. phlegmaria	[139]
		L. serratum	[131]
	serratenonediol diacetate	*L. megastachyum*	[132]
	Serratriol	*L. phlegmaria*	[139]
		L. serratum	[131]
	Tohogeninol	*L. serratum*	[144]

(continued)

Table 3 (continued)

Chemical group	Compound	Species	References
	Tohogenol	*L. complatanum*	[143]
		L. japonicum	[135]
		L. phlegmaria	[139]
		L. serratum	[48, 131, 144]
	Tohogenol diacetate	*L. megastachyum*	[132]
	α-Onoceradienedione	*L. japonicum*	[130]
	α-Onocerin	*L. clavatum*	[31, 145]
		L. japonicum	[128, 130]

3.3 Bioactive Compounds of *Equisetum* Species

Alkaloids, polyphenolics, terpenes, and sterols have been isolated from several species of *Equisetum* (Table 5).

Apigenin (**59**), kaempferol (**60**), quercetin (**62**), onitin (**68**), and their glycosides are the major polyphenols reported from *E. arvense*, *E. fluviatile*, *E. hyemale*, *E. palustre*, and *E. telmateia* (Fig. 8) [210–213].

Cholesterol (**69**), campesterol (**70**), and β-sitosterol (**55**) are some of the phytosterols which were isolated from *E. arvense* and *E. myriochaetum* [214, 215]. Terpenes, such as α-amyrin (**71**), β-amyrin (**72**), taraxerol (**73**), and germanicol (**74**) have been also sequestered from *E. arvense* (Fig. 9) [214].

3.4 Bioactive Compounds of *Adiantum* Species

Chemical investigations of *Adiantum* species revealed that triterpenoids, flavonoids, phenyl propanoids, phenolics, coumarins, and phytosterols are the major phytochemicals present in them (Table 6).

Hopane, neohopane, norphopane, fernane, adiane, and filicane are different categories of triterpenoids reported from *Adiantum* species. Hopan-22-ol (**80**) is a hopane series of triterpnoid which has been reported from *A. tetraphyllum* [232]. Neophop-13(18)-ene (**81**) is a neophane series of tritepenoid (Fig. 10) that is sequestered from *A. caudatum*, *A. cuneatum*, *A. monochlamys,* and *A. pedatum* [233–237].

21-Hydroxyadiantone (**76**) is norphopane series triterpenoid (Fig. 10) obtained from *A. venustum* [238]. Fern-9-(11)-en-28-ol (**78**) is fernane series of triterpenoid (Fig. 10) extracted from *A. capillus-veneris* and *A. lunulatum* [239, 240]. Adian-5-en-3α-of (**77**) is an adiane series triterpenoid (Fig. 10) isolated from *A. capillus-veneris* [239]. Filic-3-ene (**79**) is filicane triterpenoid (Fig. 10) obtained from

Table 4 Bioactive compounds of *Selaginella* spp.

Compound group	Compound	Species	References
Alkaloids	5-Hydroxy-N_8,N_8-dimethylpseudophrynaminol	*S. moellendorffii*	[189]
	5-Hydroxyselaginellic acid	*S. moellendorffii*	[189]
	6-Amino-9-purine methyl caproate	*S. delicatula*	[146]
	Delicatuline A	*S. delicatula*	[146]
	Delicatuline B	*S. delicatula*	[146]
	Indole-3-carboxaldehyde	*S. delicatula*	[146]
	Medioresinol-4″-*β*-glucoside	*S. delicatula*	[146]
	N-(5-Hydroxyneoselaginelloyl)-L-phenylalanine	*S. moellendorffii*	[189]
	N-(5-Hydroxyselaginelloyl)-L-phenylalanine	*S. moellendorffii*	[189]
	Neoselaginellic acid	*S. moellendorffii*	[189]
	N-Neoselaginelloyl-L-phenylalanine	*S. moellendorffii*	[189]
	N-Selaginelloyl-L-phenylalanine	*S. moellendorffii*	[189]
	Paucine	*S. moellendorffii*	[147]
	Paucine 3′-O-β-D-glucopyranoside	*S. moellendorffii*	[147]
	Selaginellic acid	*S. moellendorfii*	[189]
	β-Carboline	*S. delicatula*	[146]
Caffeoylquinic acids	3,5-Di-O-caffeoylquinic acid,	*S. delicatula*	[159]
	3,4-Di-O-caffeoylquinic acid	*S. delicatula*	[159]
	4,5-Di-O-caffeoylquinic acid	*S. delicatula*	[159]
Chromones	5-Carboxymethyl-7-hydroxychromone	*S. moellendorffii*	[149]
	Uncinoside A	*S. uncinata*	[148]
	Uncinoside B	*S. uncinata*	[148]
Coumarins	3-(4-Hydroxyphenyl)-6,7-dihydroxy coumarin	*S. tamariscina*	[151]
	Isopimpinellin	*S. moellendorffii*	[152]
	Umbelliferone	*S. tamariscina*	[150]

Flavonoids	(2″S) Chrysocauloflavone I	*S. uncinata*	[165]
	(2″*S*)-2″,3″-Dihydroamentoflavone-4′-methyl ether	*S. uncinata*	[166]
	(2″S)-2″,3″-Dihydroamentoflavone	*S. bryopteris*	[163]
		S. uncinata	[166]
	(2″S)-2″,3″-Dihydrohinokiflavone	*S. bryopteris*	[163]
	(2R) 2, 3-Dihydroamentoflavone	*S. uncinata*	[165]
	(2R)-5-Carboxymethyl-3′, 4′, 7-trihydroxyflavonone	*S. moellendorffii*	[149]
	(2S) 2,3-Dihydro-5,5″,7,7″,4′-pentahydroxy-6,6″-dimethyl-[3′-O-4‴]-biflavone	*S. uncinata*	[165]
	(2S)-5-Carboxymethyl-3′, 4′, 7-trihydroxyflavonone	*S. moellendorffii*	[149]
	(2S)-2,3-Dihydroamentoflavone	*S. bryopteris*	[163]
		S. uncinata	[166]
	(2*S*)-2,3-Dihydroamentoflavone-4′-methyl ether	*S. uncinata*	[166]
	(2S)-2,3-Dihydrohinokiflavone	*S. bryopteris*	[163]
	(2S)-5-Carboxymethyl-4′, 7-dihydroxyflavonone	*S. moellendorffii*	[149]
	(2*S*,2″*S*)-2,3,2″,3″-Tetrahydroamentoflavone-4′-methyl ether	*S. uncinata*	[166]
	(2*S*,2″*S*)-Tetrahydroamentoflavone	*S. uncinata*	[166]
	(2S,2″S)-2,3,2″,3″-Tetrahydroamentoflavone	*S. bryopteris*	[163]
	(2S,2″S)-2,3,2″,3″-Tetrahydrohinokiflavone	*S. bryopteris*	[163]
	(2S,2″S)-Tetrahydrorobustaflavone	*S. uncinata*	[164]
	2,2″,3,3" Tetrahydrorobustaflavone 7,4′,7″-trimethyl ether	*S. doederleinii*	[190]
	2,3-Dihydro-5 methylether-robustaflavone	*S. lepidophylla*	[153]
	2,3-Dihydro-5,5″,7,7″,4′-pentahydroxy-6,6″-dimethyl-(3′-*O*- 4‴)-biflavone,	*S. labordei*	[191]
	2,3-dihydroamentoflavone 7,4′,7″-trimethyl ether	*S. delicatula*	[187]
	2,3-dihydroamentoflavone 7,4′-dimethyl ether	*S. delicatula*	[187]
	2″,3″-dihydroisocryptomerin 7-methyl ether	*S. delicatula*	[187]
	2,3-Dihydrorobustaflavone	*S. lepidophylla*	[153]

(continued)

Table 4 (continued)

Compound group	Compound	Species	References
	2′,8″-Biapigenin	*S. tamariscina*	[158]
	2″,3″-Dihydro-3″,3‴-biapigenin	*S. doederleinii*	[156]
	2″,3″-dihydrorobustaflavone 7,4′, 7″-trimethyl ether	*S. delicatula*	[159]
	2″,3″-dihydrorobustaflavone 7,4′, –dimethyl ether	*S. delicatula*	[159]
	2″, 3″ Dihydroochnaflavone	*S. labordei*	[191]
	2″,3″-Dihydro-3′,3‴-biapigenin,	*S. labordei*	[191]
	3,3′-Binaringenin	*S. chrysocaulos*	[163]
		S. doederleinii	[156]
	3′-Phenol-apigenin	*S. doederleinii*	[168]
	4′-Methylamentoflavone	*S. bryopteris*	[163]
	5- Carbomethoxymethyl-4′, 7-dihydroxyflavone	*S. moellendorffii*	[149]
	5-Carboxymethyl-3′, 4′, 7-trihydroxyflavone	*S. moellendorffii*	[149]
	5-Carboxymethyl-4′,7-dihydroxyflavone	*S. moellendorffii*	[149, 155]
	5-Carboxymethyl-4′,7-dihydroxyflavone butyl ester	*S. moellendorffii*	[155]
	5-Carboxymethyl-4′,7-dihydroxyflavone ethyl ester	*S. moellendorffii*	[155]
	5-Carboxymethyl-4′-hydroxyflavone-7-O-β-D-glucopyranoside	*S. moellendorffii*	[149]
	5-Carboxymethyl-7,4′-dihydroxyflavanone 7-O-β-D-glucopyranoside	*S. moellendorffii*	[147]
	5-Carboxymethyl-7,4′-dihydroxyflavonone	*S. moellendorffii*	[192]
	5-Carboxymethyl-7,4′-dihydroxyflavonone-7-O-β-D-glucopyranoside	*S. moellendorffii*	[192]
	5-Carboxymethyl-7-hydroxychromone	*S. moellendorffii*	[149]
	6-(2-Hydroxy-5-carboxyphenyl)-apigenin	*S. tamariscina*	[151]
	6-(5-Carboxyl-2-methoxyphenyl)-apigenin	*S. uncinata*	[167]
	7,4′,7″,4‴-Tetra-O-methylamentoflavone	*S. denticulata*	[161]
	7,7″,4‴-Tri-O-methylamentoflavone	*S. bryopteris*	[163]
	7-Methylamentoflavone	*S. bryopteris*	[163]
	7′-O-Methylhinokiflavone	*S. bryopteris*	[163]

		S. tamariscina	[162]
	Amentoflavone	*S. bryopteris*	[163]
		S. delicatula	[159]
		S. doederleinii	[156]
		S. moellendorffii	[155]
		S. pulvinata	[154]
		S. tamariscina	[151, 157, 158, 193]
		S. uncinata	[165, 166]
	Amentoflavone 7,4,7,4 tetramethyl ether	*S. moellendorffii*	[155]
	Apigenin	*S. pulvinata*	[154]
	Bilobetin	*S. moellendorffii*	[155]
		S. uncinata	[165]
	Chrysocauloflavone-I	*S. chrysocaulos*	[163]
	Chrysocauloflavone-II	*S. chrysocaulos*	[163]
	Chrysocauloflavone-III	*S. chrysocaulos*	[163]
	Chrysoeriol	*S. moellendorffii*	[155]
	Cryptomerin B	*S. denticulata*	[161]
	Delicaflavone	*S. doederleinii*	[156, 194]
		S. uncinata	[165]
	Ginkgetin	*S. moellendorffii*	[155]
		S. pulvinata	[154]
	Hinokiflavone	*S. denticulata*	[161]
		S. moellendorffii	[155]
		S. pulvinata	[154]
		S. tamariscina	[162]
	Involvenflavone A	*S. involvens*	[160]
	Involvenflavone B	*S. involvens*	[160]

(continued)

Table 4 (continued)

Compound group	Compound	Species	References
	Involvenflavone C	*S. involvens*	[160]
	Involvenflavone D	*S. involvens*	[160]
	Involvenflavone E	*S. involvens*	[160]
	Involvenflavone F	*S. involvens*	[160]
	Isocryptomerin	*S. denticulata*	[161]
		S. pulvinata	[154]
		S. tamariscina	[195]
	Isoginkgetin	*S. moellendorffii*	[155]
		S. pulvinata	[154]
	Kayaflavone	*S. moellendorffii*	[155]
	Podocarpusflavone A	*S. moellendorffii*	[155]
	Robustaflavone	*S. delicatula*	[159]
		S. denticulata	[161]
		S. doederleinii	[156]
		S. lepidophylla	[153]
		S. tamariscina	[158]
		S. uncinata	[164]
	Robustaflavone 7,4′,7″-trimethyl ether	*S. doederleinii*	[190]
	Robustaflavone 4′,4‴-dimethyl ether	*S. delicatula*	[187]
	Robustaflavone 4′-methyl ether	*S. delicatula*	[159]
		S. moellendorffii	[155]
		S. uncinata	[164, 165]
	Robustaflavone 7,4′,4‴-trimethyl ether	*S. delicatula*	[187]
	Robustaflavone 7,4′-dimethyl ether	*S. delicatula*	[159, 187]
	Robustaflavone 7-methyl ether	*S. uncinata*	[165]

	Seladoeflavone A	*S. doederleinii*	[168]
	Seladoeflavone B	*S. doederleinii*	[168]
	Seladoeflavone C	*S. doederleinii*	[168]
	Seladoeflavone D	*S. doederleinii*	[168]
	Seladoeflavone E	*S. doederleinii*	[168]
	Seladoeflavone F	*S. doederleinii*	[168]
	Selagintriflavonoid A	*S. doederleinii*	[196]
	Selagintriflavonoid B	*S. doederleinii*	[196]
	Selagintriflavonoid C	*S. doederleinii*	[196]
	Selagintriflavonoid D	*S. doederleinii*	[196]
	Selagintriflavonoid E	*S. doederleinii*	[196]
	Selagintriflavonoid F	*S. doederleinii*	[196]
	Selagintriflavonoid G	*S. doederleinii*	[196]
	Selagintriflavonoid H	*S. doederleinii*	[196]
	Sotetsuflavone	*S. denticulata*	[161]
	Sumaflavone	*S. tamariscina*	[158]
	Taiwaniaflavone	*S. tamariscina*	[158]
	Uncinatabiflavone A	*S. uncinata*	[164]
	Uncinatabiflavone B	*S. uncinata*	[164]
	Uncinatabiflavone C	*S. uncinata*	[164]
	Uncinatabiflavone C 7-methyl ether	*S. uncinata*	[165]
	Uncinatabiflavone D	*S. uncinata*	[164]
	Uncinataflavone A	*S. uncinata*	[167]
	Uncinataflavone B	*S. uncinata*	[167]
Lignans	(−)-(7′*S*,8*S*,8′*R*)-4,4′-Dihydroxy-3,3′,5,5′-tetramethoxy-7′,9-epoxylignan-9′-ol-7-one	*S. moellendorffii*	[172]
	(−)-8,8′-Bisdihydrosiringenin	*S. moellendorffii*	[147]
	(−)-Lariciresinol	*S. moellendorffii*	[147]
	(7′*S*,8′*R*,8*R*)-Lyoniresinol	*S. moellendorffii*	[172]

(continued)

Table 4 (continued)

Compound group	Compound	Species	References
	(7R,8S) Dehydrodiconiferyl alcohol	*S. sinensis*	[169]
	(7R,8S)-Dihydrodehydrodiconiferyl alcohol	*S. sinensis*	[169]
	(7*S*,8*R*)-3,3′,5-Trimethoxy-4′,7-epoxy-8,5′-neolignan-4,9,9′-triol	*S. moellendorffii*	[147, 172]
	(7S,8R)-4,9-Dihydroxy-3,3′,5-trimethoxy-4′,7-epoxy-8,5′-neolignan-9′-oic acid methyl ester	*S. moellendorffii*	[147]
	(7*S*,8*S*)-3,3′,5-Trimethoxy-4′,7-epoxy-8,5′-neolignan-4,9,9′-triol	*S. moellendorffii*	[172]
		S. picta	[170]
	(8*R*)-3,5′-Dimethoxy-8, 3′-neoligna-4,4′,9,9′-tetraol	*S. moellendorffii*	[172]
	1-(4′-Hydroxy-3′-methoxyphenyl)-2-[4″-(3-hydroxy-propyl)-2″,6″-dimethoxyphenoxy] propane-1,3-dio]	*S. moellendorffii*	[172]
	3,3′,5-Trimethoxy-4′,7-epoxy-8,5′-neolign-7-ene-4,9,9′-triol 9-O-β-D-glucopyranoside	*S. moellendorffii*	[147]
	4,9-Dihydroxy-4′,7-epoxy-8′,9′-dinor-8,5′-neolignan-7′-oic acid	*S. picta*	[170]
	4-*O*-Methylcedrusin	*S. moellendorffii*	[172]
	7S,7′S,8R,8′R-Icariol A_2	*S. moellendorffii*	[147]
	Burseneolignan	*S. moellendorffii*	[172]
	Dihydrobuddlenol B	*S. moellendorffii*	[172]
	Dihydrosinapyl alcohol	*S. moellendorffii*	[147]
	Lyoniside	*S. moellendorffii*	[147]
	Pictalignan A	*S. picta*	[170]
	Pictalignan B	*S. picta*	[170]
	Pictalignan C	*S. picta*	[170]
	rel-(7R,8S)-3,3′,5-Trimethoxy-4′,7-epoxy-8,5′-neoligna-4,9,9′-triol 4-O-β-D-glucopyranoside	*S. moellendorffii*	[147]
	Selaginellol	*S. moellendorffii*	[147, 172]
	Selaginellol 4′-O-β-D-glucopyranoside	*S. moellendorffii*	[147]

	Selamoellenin A	*S. moellendorffii*	[197]
	Selamoellenin B	*S. moellendorffii*	[172]
	Selamoellenin C	*S. moellendorffii*	[172]
	Selamoellenin D	*S. moellendorffii*	[172]
	Selariscinin D	*S. tamariscina*	[171]
	Sinensiol A	*S. sinensis*	[169, 178, 199]
	Sinensiol B	*S. sinensis*	[169]
	Sinensiol C	*S. sinensis*	[169]
	Sinensiol D	*S. sinensis*	[169]
	Sinensiol E	*S. sinensis*	[169]
	Sinensiol F	*S. sinensis*	[169]
	Sinensiol G	*S. sinensis*	[169]
	Syringaresinol	*S. delicatula*	[146]
		S. moellendorffii	[147, 172]
		S. picta	[170]
		S. tamariscina	[150]
	Tamariscinol U	*S. tamariscina*	[171]
	Tamariscinol V	*S. tamariscina*	[171]
	Tamariscinol W	*S. tamariscina*	[171]
	Tamariscinoside C	*S. tamariscina*	[199]
Phenolics	Caffeic acid	*S. tamariscina*	[150]
	Ferulic acid	*S. tamariscina*	[150]
	Syringic acid	*S. tamariscina*	[150]
	Vanillic acid	*S. tamariscina*	[150]
Pigments	10-Methoxylated selaginellin M	*S. tamariscina*	[181]
	Diselaginellin A	*S. pulvinata*	[175]
	Diselaginellin B	*S. pulvinata*	[175]
	Isoselagintamarlin A	*S. tamariscina*	[200]

(continued)

Table 4 (continued)

Compound group	Compound	Species	References
	Selagibenzophenone A	*S. pulvinata*	[179]
	Selagibenzophenone B	*S. tamariscina*	[180]
	Selaginellin	*S. pulvinata*	[174, 175, 178, 182]
		S. sinensis	[173]
		S. tamariscina	[181, 183, 185]
	Selaginellin A	*S. pulvinata*	[174, 178, 182]
		S. tamariscina	[176, 183–185]
	Selaginellin B	*S. pulvinata*	[174, 175, 179]
		S. tamariscina	[176, 184]
	Selaginellin C	*S. pulvinata*	[182]
		S. tamariscina	[185]
	Selaginellin D	*S. pulvinata*	[174]
	Selaginellin E	*S. pulvinata*	[174]
		S. tamariscina	[184]
	Selaginellin F	*S. pulvinata*	[174]
	Selaginellin G	*S. pulvinata*	[154, 179]
	Selaginellin H	*S. pulvinata*	[154, 178]
	Selaginellin I	*S. tamariscina*	[201]
	Selaginellin J	*S. tamariscina*	[201]
	Selaginellin K	*S. tamariscina*	[202]
	Selaginellin L	*S. pulvinata*	[179]
		S. tamariscina	[202]
	Selaginellin M	*S. tamariscina*	[183, 185]
	Selaginellin N	*S. pulvinata*	[178]
		S. tamariscina	[185]

	Selaginellin O	*S. tamariscina*	[183, 184]
	Selaginellin P	*S. pulvinata*	[203]
	Selaginellin P	*S. tamariscina*	[204]
	Selaginellin Q	*S. pulvinata*	[203]
	Selaginellin R	*S. pulvinata*	[203]
	Selaginellin S	*S. moellendorffii*	[205]
		S. pulvinata	[179]
	Selaginellin T	*S. pulvinata*	[179]
	Selaginellin U	*S. tamariscina*	[177]
	Selaginellin V	*S. tamariscina*	[177]
	Selaginellin W	*S. tamariscina*	[177]
	selaginisoquinoline A	*S. pulvinata*	[203]
	Selaginone A	*S. tamariscina*	[180]
	Selaginone B	*S. tamariscina*	[180]
	Selaginopulvin	*S. pulvinata*	[178, 184, 206, 207]
	Selaginpulvilin A	*S. pulvinata*	[178]
		S. tamariscina	[184]
	Selaginpulvilin B	*S. pulvinata*	[178]
		S. tamariscina	[184]
	Selaginpulvilin C	*S. pulvinata*	[178]
		S. tamariscina	[184]
	Selaginpulvilin D	*S. pulvinata*	[178]
		S. tamariscina	[184]
	Selaginpulvilin E	*S. pulvinata*	[207]
		S. tamariscina	[184]
	Selaginpulvilin F	*S. pulvinata*	[207]
	Selaginpulvilin G	*S. pulvinata*	[207]

(continued)

Table 4 (continued)

Compound group	Compound	Species	References
	Selaginpulvilin H	*S. pulvinata*	[207]
	Selaginpulvilin I	*S. pulvinata*	[207]
	Selaginpulvilin J	*S. pulvinata*	[207]
	Selaginpulvilin K	*S. pulvinata*	[184]
	Selaginpulvilin L	*S. pulvinata*	[184]
	Selaginpulvilin M	*S. tamariscina*	[206]
	Selaginpulvilin N	*S. tamariscina*	[206]
	Selaginpulvilin O	*S. tamariscina*	[206]
	Selaginpulvilin P	*S. tamariscina*	[206]
	Selaginpulvilin Q	*S. tamariscina*	[206]
	Selaginpulvilin R	*S. tamariscina*	[206]
	Selaginpulvilin S	*S. tamariscina*	[206]
	Selaginpulvilin T	*S. tamariscina*	[206]
	Selagintamarlin A	*S. tamariscina*	[184]
	Selariscinin A	*S. tamariscina*	[177, 181]
	Selariscinin B	*S. tamariscina*	[181]
	Selariscinin C	*S. tamariscina*	[181]
	Selariscinin D	*S. tamariscina*	[177, 208]
	Selariscinin E	*S. tamariscina*	[208]
Quinones	1-Methoxy-3-methylanthraquinone	*S. tamariscina*	[151]
	α-Tocopheryl quinone	*S. delicatula*	[187]
Saponins	(1α, 3β, 25R)-Spirost-5-ene-2-diol-3-O-α-L-rhamnopyranosyl-(1 → 2)-O-[α-L-rhamnopyranosyl(1 → 4)]-O-β-D-glucopyranoside	*S. uncinata*	[186]
	(2α, 3β, 12β, 25R)-Spirost-5-ene-2, 3, 12-triol-3-O-α-L-rhamnopyranosyl-(1 → 2)-O-[α-L-rhamnopyranosyl-(1 → 4)]-O-β-D-glucopyranoside	*S. uncinata*	[186]
		S. uncinata	[186]

	(3β, 12β,25R)-Spirost-5-ene-3,12-diol-3-O-α-L-rhamnopyranosyl-(1 → 2)-O-[α-L-rhamnopyranosyl-(1 → 4)]-O-β-D-glucopyranoside		
	(3β, 7β, 12β, 25R)-Spirost-5-ene-3, 7, 12-triol-3-O-α-L-rhamnopyranosyl-(1 → 2)-O-[α-L-rhamnopyranosyl-(1 → 4)]-O-β-D-glucopyranoside	*S. uncinata*	[186]
Sterols	3β-(3-Hydroxybutyroxy)-16α-hydroxy-5α,17β-cholestan-21-carboxylic acid	*S. tamariscina*	[209]
	3β,16α-Dihydroxy-5α,17β-cholestan-21-carboxylic acid	*S. tamariscina*	[209]
	3β-Acetoxy-16α-hydroxy-5α,17β-cholestan-21-carboxylic acid	*S. tamariscina*	[209]
	Stigmasta-4,22-dien-3-one	*S. delicatula*	[187]
	Stigmasterol	*S. delicatula*	[187]
	β-Sitostenone	*S. delicatula*	[187]
	β-Sitosterol	*S. delicatula*	[187]
		S. tamariscina	[151]
Terpenoids	(3S,4S,5R,10S)-18(4 → 3)-*abeo*-3,4,12,18-Tetrahydroxy-8,11,13-abietatrien-7-one	*S. moellendorffii*	[188]
	(4Z,6E)-2,7-Dimethyl-8-hydroxyocta-4,6-dienoic acid 8-O-β-D-glucopyranoside	*S. moellendorffii*	[192]
	Selaginedorffone A	*S. moellendorffii*	[188]
	Selaginedorffone B	*S. moellendorffii*	[188]

A. capillus-veneris, A. caudatum, A. cuneatum, A. edgeworthii, A. monochlamys, and *A. pedatum* [234–237, 239, 241, 242]. Caffeic acid (**30**) and ferulic acid (**31**) are some of the phenolic compounds obtained from *A. tetraphyllum* (Fig. 10) [232]. Psoralen (**75**) is a coumarin extracted from *A. thalictroides* var. *hirsutum* [243]. Campesterol (**70**), stigmasterol (**54**), and β-sitosterol (**55**) are certain phytosterols reported from *Adiantum* species [232, 243, 244].

3.5 Bioactive Compounds of *Dryopteris* Species

The phytochemicals isolated from *Dryopteris* include flavonoids, phenolics, phenolic glycosides, phenylpropanoids, phloroglucinols, terpenoids, and steroids (Table 7; Figs. 11, 12, and 13).

Quercetin (**62**), rutin (**65**), quercitrin (**94**), and sutchuenoside A (**95**) are some of the flavonoids reported from several species of *Dryopteris* (Figs. 11, 12) [273–281]. (E)-4-(3,4-Dimethoxyphenyl)but-3-en-1-ol (**86**), caffeic acid (**30**), dryofracoumarin A (**82**), dryofracoumarin B (**83**), esculetin (**84**), and isoscopoletin (**85**) are some of the major phenolics and coumarins obtained from *D. crassirhizoma* and *D. fragrans* (Fig. 11) [282–285]. Monomeric, dimeric, trimeric, tetrameric, pentameric, and hexameric phloroglucinols were extracted from *Dryopteris* species. Aspidin BB (**100**), aspidinol (**101**), dryocrassin ABBA (**103**), flavaspidic acid AB (**105**), and flavaspidic acid PB (**106**) are some of the major phloroglucinols obtained from *Dryopteris* species (Fig. 13). [284–305]. Sesquiterpenes, triterpenes, nor-triterpenes, and other terpenoids were found in the genus *Dryopteris*, which were isolated from the *D. fragrans, D. championii,* and *D. crassirhizoma* (Fig. 12). Some of the steroid compounds, such as β-sitosterol (**55**) and β-sitosterol 3-O-β-D-glucopyranoside (**97**) were isolated from *D. championii, D. cycadina, and D. fragrans* [278, 291, 293, 306].

4 Biological Activities

Pteridophytes have been used in traditional systems of medicine, such as Traditional Chinese Medicine (TCM), Ayurveda, Unani, Siddha, and Oriental Medicines to cure several human ailments [7]. Pteridophytes are also utilized in the preparation of ethnomedicine to treat various diseases including neurological diseases, inflammation, hepatitis, arthritis, rheumatism, dermatosis, and cancer [338, 339]. Plant extracts and secondary metabolites isolated from pteridophytes have proven to possess antioxidant, anti-cancer, anti-diabetic, anti-inflammatory, anti-microbial, and neuroprotective effects (Fig. 14) [9, 10, 340]. Following are some of the specific examples of biological activities reported from pteridophytes in general.

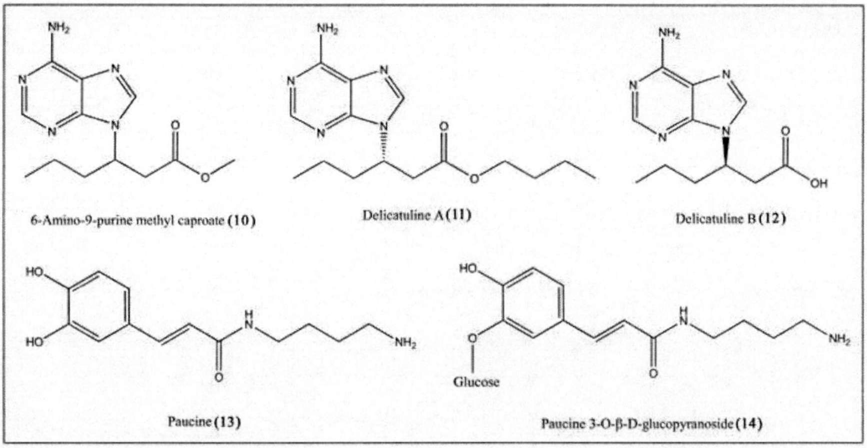

Fig. 2 Major alkaloids isolated from *Selaginella* spp.

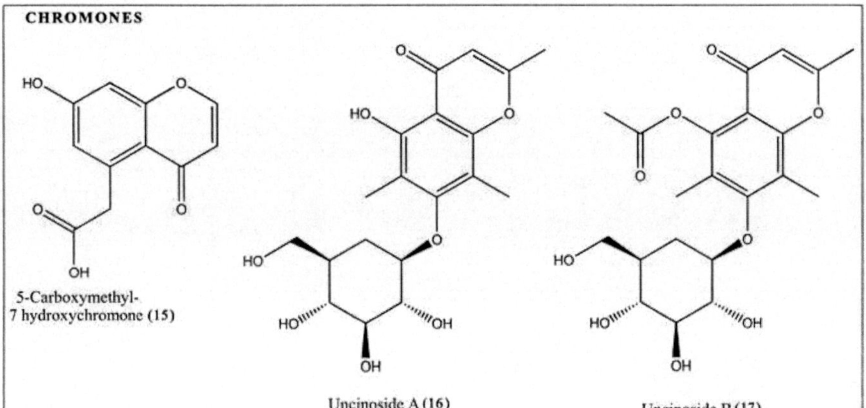

Fig. 3 Major chromones and coumarins isolated from *Selaginella* spp.

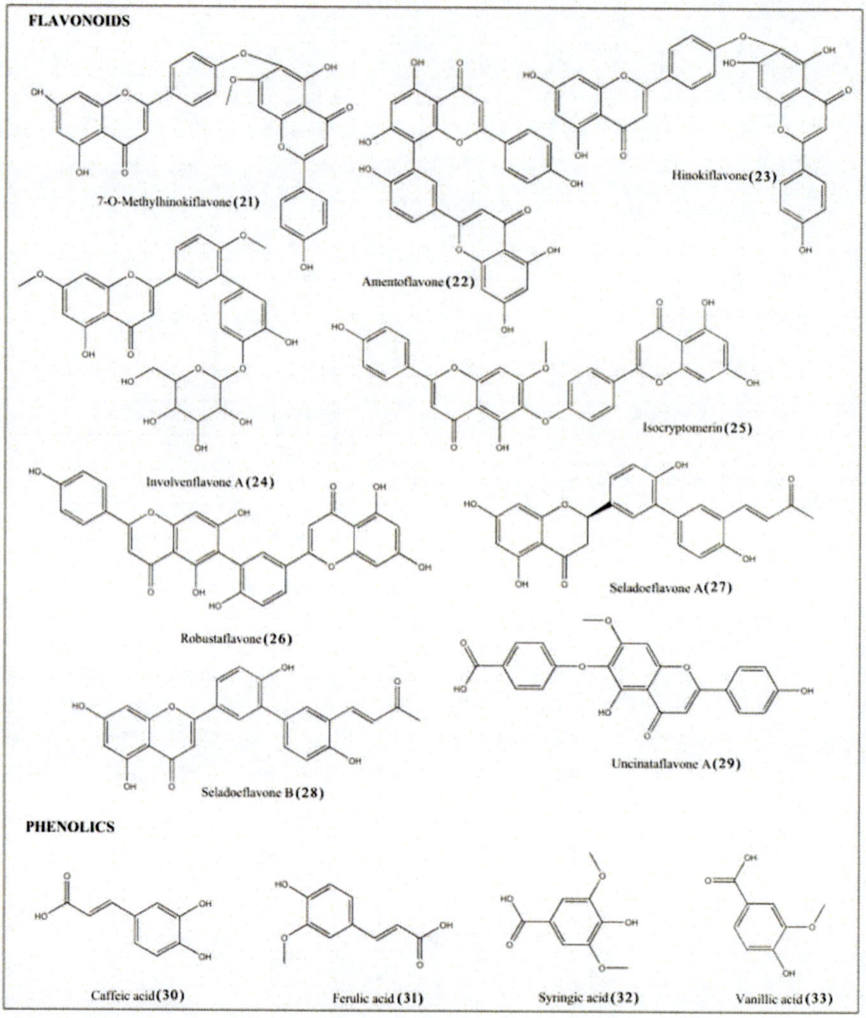

Fig. 4 Major flavonoids and phenolics isolated from *Selaginella* spp.

4.1 Anti-Alzheimer's Disease Activity

Lycopodium is useful in the treatment of neurological disorders like Alzheimer's disease (AD), anxiety, and memory loss. AD is a neurodegenerative disease, where, cholinergic neurotransmission of the central nervous system is disrupted [341]. To treat AD, it has been suggested that boosting cholinergic neurotransmission be used. Huperzine-A (Hup-A), a substance produced from multiple

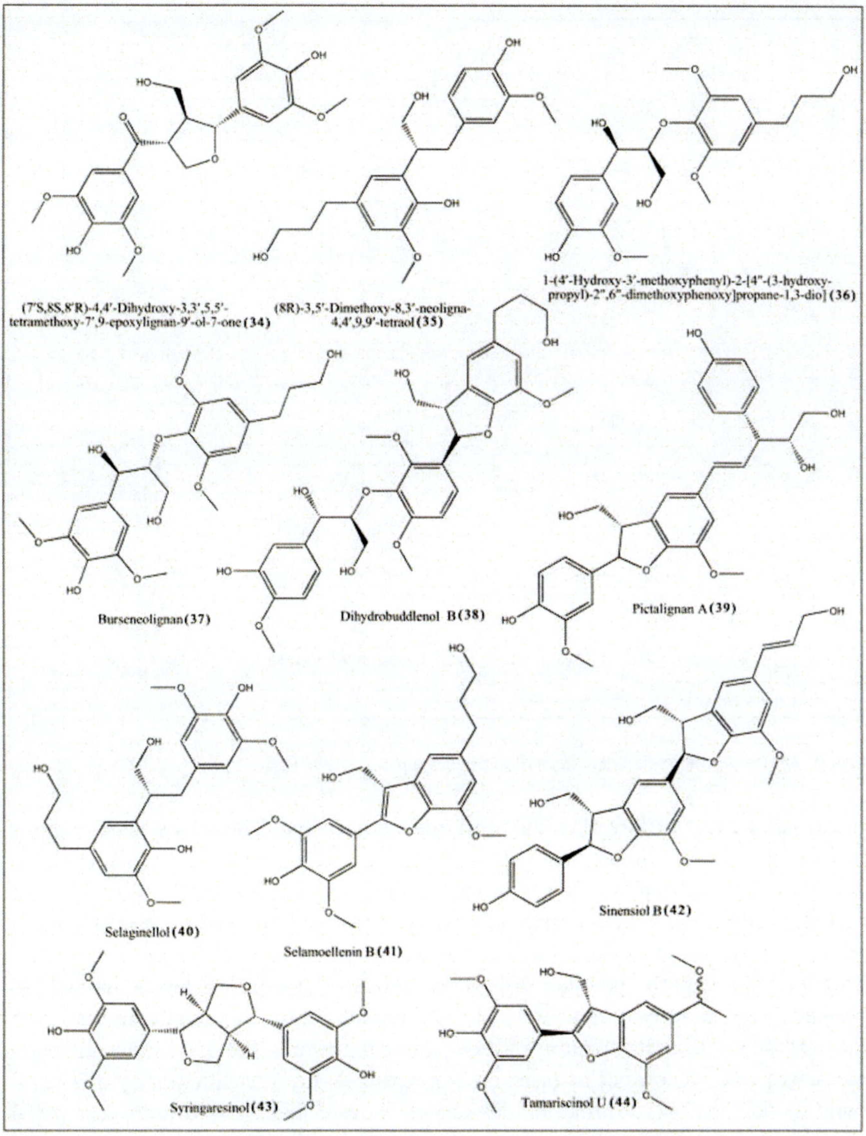

Fig. 5 Major lignans isolated from *Selaginella* spp.

Lycopodium species, has been shown in numerous studies to be a highly potent, precise, and reversible inhibitor of acetylcholinesterase (AChE), assisting in the reduction of some mild to severe AD symptoms. For instance, Cheng et al. [342] examined the effects of Hup-A in vitro on scopolamine-induced memory impairment in rats and compared those results with those of E2020 and tacrine (E2020 and tacrine are common drugs that inhibit the AChE). They proved that Hup-A is the most effective acetylcholinesterase inhibitor and that it significantly outperformed

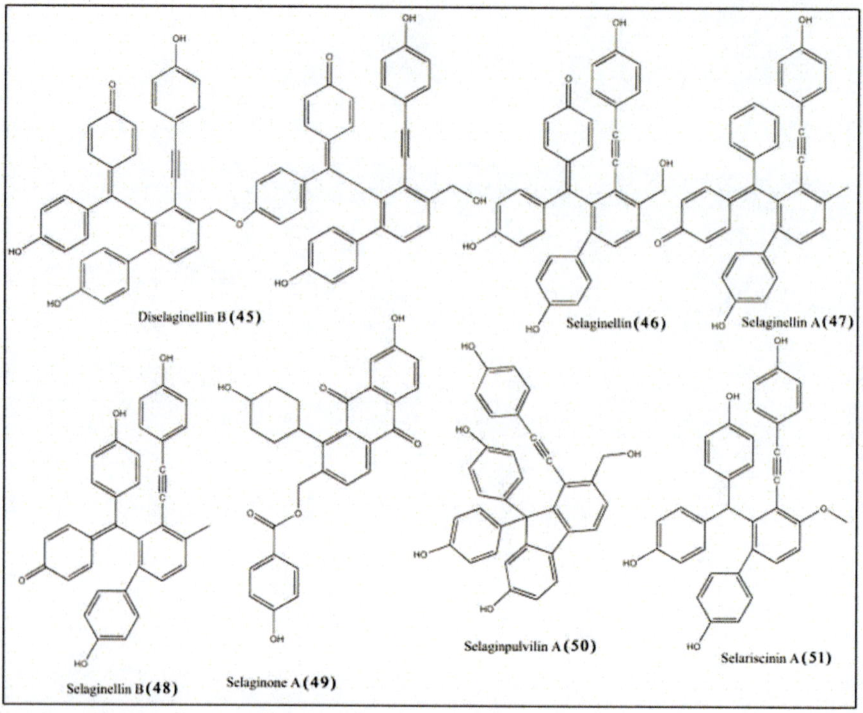

Fig. 6 Major pigments isolated from *Selaginella* spp.

tacrine or E2020 in lowering the working memory loss caused by scopolamine, proving that it is a suitable medication for the treatment of cognitive impairment in AD patients. Furthermore, it was found that Hup-A was 8-fold and twofold more potent in molar terms than donepezil and rivastigmine, respectively, for increasing cortical acetylcholine levels when donepezil, rivastigmine, and Hup-A (standard drugs) were compared for their effects on cortical acetylcholine levels and acetylcholinesterase activity in rats [344]. Hup-A inhibits AChE by directly binding with the enzyme's active site opening and preventing the regular substrate from accessing the active side, according to research employing X-ray crystallography and computer modeling [344]. Animal model studies showed that Hup-A improved spatial working memory in monkeys with experimental cognitive impairment via an adrenergic mechanism [345]. Hup-A dramatically improves memory deficiencies in elderly and AD patients, according to clinical research. For example, 202 patients with suspected or probable AD took part in a multicentre, randomized, placebo-controlled study by Zhang et al. [346]. For 12 weeks, one group of 100 patients received 400 mg of Hup-A every day, while the other 102 patients received a placebo. According to the AD Assessment Scale (ADAS-Cog), the therapy group showed improvements in cognition as well as behavior, mood, and ADL performance (ADAS non-Cog).

Table 5 Bioactive compounds of *Equisetum* spp.

Compound group	Compound	Species	References
Alkaloids	18-Deoxypalustrine	*E. palustre*	[216]
	Equisetumine	*E. debile*	[217]
	Myricoidine	*E. palustre*	[216]
	N^5-Acetylpalustrine	*E. palustre*	[216]
	N^5-Formylpalustridiene	*E. palustre*	[216]
	N^5-Formylpalustrine palustridine	*E. palustre*	[216]
	Nicotine	*E. bogotense*	[218]
		E. palustre	[218]
	Palustridiene	*E. palustre*	[216]
	Palustridine	*E. bogotense*	[218]
		E. giganteum	[218]
		E. palustre	[218]
	Palustrine	*E. bogotense*	[218]
		E. palustre	[216, 218]
	Spermidine	*E. palustre*	[216]
Flavonoids	6-Chloroapigenin	*E. arvense*	[219]
	Apigenin	*E. arvense*	[212]
		E. fluviatile	[211]
		E. palustre	[213]
	Catechin	*E. telmateia*	[220]
	Chrysin	*E.myriochaetum*	[215]
	Dichlorokaempferol	*E. arvense*	[219]
		E. x litorale	[219]
	Genkwanin	*E. palustre*	[213]
	Kaempferol	*E. palustre*	[211]
		E. telmateia	[211]
	Luteolin	*E. palustre*	[213]
		E. arvense	[212]
	Pinocembrin	*E. myriochaetum*	[215]
	Protocatechuic acid	*E. telmateia*	[220, 221]
	Quercetin	*E. arvense*	[211]
		E. hyemale	[210]
Glycosides	**Aliphatic glycosides**		
	(*Z*)-3-Hexenyl *O*-β -D-glucopyranoside	*E. debile*	[222]
	Flavonoid glycosides		
	Apigenin 4′-O-glucoside	*E. fluviatile*	[211]
	Apigenin 5-*O*-glucoside	*E. arvense*	[223]
	Apigenin-4′-glucoside	*E. arvense*	[219]
		E. x litorale	[219]
	Apigenin-5-glucoside	*E. arvense*	[219]
		E. x litorale	[219]

(continued)

Table 5 (continued)

Compound group	Compound	Species	References
	Apigenin-5-O-β-D-glucopyranoside	*E. palustre*	[213]
	Genkwanin-5-glucoside	*E. arvense*	[219]
		E. x litorale	[219]
	Genkwanin-5-O-β-D-glucopyranoside	*E. palustre*	[213]
	Kaempferol 3,7-*O*-β -D-diglucopyranoside	*E. debile*	[222]
	Kaempferol-3-sophoroside-7-O-β-D-glucopyranoside	*E. hyemale*	[210]
	Kaempferol 3,7-O-diglucoside	*E. sylvaticum*	[211]
		E. fluviatile	[211]
		E. telmateia	[211, 220]
	Kaempferol 3′-O-rutinoside	*E. palustre*	[211]
		E. sylvaticum	[211]
	Kaempferol 3-O-(6”-O-acetylglucoside)-7-O-rhamnoside	*E. telmateia*	[220]
	Kaempferol 3-O-(6″-O-acetylglucoside)-7-O-glucoside	*E. telmateia*	[211]
	Kaempferol 3-O-(6″-O-malonylglucoside)-glucoside	*E. fluviatile*	[211]
	Kaempferol 3-O-(6″-O-acetylglucoside)	*E. telmateia*	[211, 220]
	Kaempferol 3-O-(6″-O-acetylglucoside)-7-O-glucoside	*E. telmateia*	[220]
	Kaempferol 3-O-(6″-O-acetylglucoside)-7-O-rhamnoside	*E. telmateia*	[211]
	Kaempferol 3-O-acetylglucoside	*E. telmateia*	[221]
	Kaempferol 3-O-glucoside	*E. sylvaticum*	[211]
		E. fluviatile	[211]
		E. palustre	[211]
		E. telmateia	[211, 220, 221]
	Kaempferol 3-O-glucoside-7-O-rhamnoside	*E. sylvaticum*	[211]
		E. telmateia	[211]
	Kaempferol 3-O-glucoside-7-O-rhamnoside	*E. telmateia*	[220]
	Kaempferol 3-*O*-glycoside	*E. arvense*	[223]
	Kaempferol 3-O-rutinoside-7-O-glucoside	*E. palustre*	[211]
		E. sylvaticum	[211]
		E. telmateia	[211]
	Kaempferol 3-O-rutinoside-7-O-sophoroside	*E. palustre*	[211]
	Kaempferol 3-*O*-sophoroside	*E. debile*	[222]
	Kaempferol 3-*O*-sophoroside-7-*O*-β -D-glucopyranoside	*E. debile*	[222]

(continued)

Table 5 (continued)

Compound group	Compound	Species	References
	Kaempferol 3-O-β-sophoroside-7-O-β-D-glucopyranoside	*E. arvense*	[224]
	Kaempferol acetyl-dihexose	*E. telmateia*	[221]
	Kaempferol acetyl glucoside-rhamnoside	*E. telmateia*	[221]
	Kaempferol glucoside-rhamnoside	*E. telmateia*	[221]
	Kaempferol-3,7-di-O-β-D-glucopyranoside	*E. arvense*	[219]
		E. giganteum	[225]
		E. hyemale	[210]
		E. myriochaetum	[226]
		E. x litorale	[219]
	Kaempferol-3-glucoside	*E. arvense*	[212, 219]
		E. giganteum	[225]
		E. x litorale	[219]
	Kaempferol-3-O-1″-β-D-glucopyranosyl-3-O-1‴-β-D-glucopyranoside	*E. palustre*	[227]
	Kaempferol-3-O-sophoroside	*E. arvense*	[219]
		E. giganteum	[225]
		E. myriochaetum	[226]
		E. x litorale	[219]
	kaempferol-3-O-sophoroside-4′-O-β-glucoside	*E. myriochaetum*	[226]
	Kaempferol-3-O-sophoroside-7-O-glucoside	*E. giganteum*	[225]
	Kaempferol-3-O-β-D-glucopyranoside-7-O-β-D-glucopyranoside	*E. palustre*	[213]
	Kaempferol-7-O-α-L-rhamnoside-4′-O-β-D-glycopyranoside	*E. hyemale*	[210]
	Kaempferol-7-O-β-D-glucopyranoside	*E. hyemale*	[210]
	Luteolin-5-glucoside	*E. arvense*	[219]
		E. x litorale	[219]
	Luteolin-7-O-β-D-glucopyranoside	*E. palustre*	[213]
	Quercetin 3-O-(6″-O-malonylglucoside)	*E. arvense*	[211]
	Quercetin 3-O-glucoside	*E. arvense*	[211, 212, 223]
		E. palustre	[213]
		E. sylvaticum	[211]
	Quercetin-3,7-di-O-glucoside	*E. giganteum*	[225]
	Quercetin-3-O-(caffeoyl)-glucoside	*E. giganteum*	[225]
	Quercetin-tri-O-hexoside	*E. giganteum*	[225]

(continued)

Table 5 (continued)

Compound group	Compound	Species	References
	Rutin	*E. hyemale*	[228]
	Lignan glycosides		
	Isolariciresinol 3a-O-β-D-glucopyranoside	*E. debile*	[217]
	Lariciresinol 9-O-β-D-glucopyranoside	*E. debile*	[217]
	Megastigmane glycosides		
	(3S,5R,6S,7E,9S)-Megastigman-7-ene-5,6-epoxy-3,9-diol 3,9-O-β -D-diglucopyranoside	*E. debile*	[222]
	(6R,9S)-3- Oxo-a -ionol 9-O-b -D glucopyranoside	*E. debile*	[222]
	Debiloside A	*E. debile*	[229]
	Debiloside B	*E. debile*	[222, 229]
	Debiloside C	*E. debile*	[229]
	Macarangioside D	*E. debile*	[222]
	Sammangaoside A	*E. debile*	[222]
	Neolignan glycosides		
	(7*S*,8*R*)-Dehydrodiconiferyl 4-*O*-β -D-glucopyranoside	*E. debile*	[222]
	Phenolic glycosides		
	2-(Sophorosyl)-1-(4-hydroxyphenyl) ethanone	*E. hyemale*	[210]
	3-Hydroxyhispidin-3,4′-di-O-glucoside	*E. giganteum*	[225]
	Coniferin	*E. arvense*	[224]
	Equisetumoside A	*E. arvense*	[224]
	Equisetumoside B	*E. arvense*	[224]
	Equisetumoside C	*E. arvense*	[224]
	Equisetumoside B	*E. debile*	[217]
	Phenylethanoid glycosides		
	Phenyethyl *O*-β-D-glucopyranoside	*E. debile*	[222]
	Phenolic sesquiterpene glycosides		
	Onitin-9-O-glucoside	*E. arvense*	[212]
	Styrylpyrone glycosides		
	3,4-dihydroxy-6-(3′,4′-dihydroxy-E-styryl)-2-pyrone 3-O-β-D-glucopyranoside	*E. arvense*	[230]
	3′deoxyequisetumpyrone [3,4-hydroxy-6-(4′-hydroxy-E-styryl)-2-pyron-3-O-β-D-glucopyranoside]	*E. arvense*	[231]
	4′-O-methylequisetumpyrone [3,4-hydroxy-6-(3′-hydroxy-	*E. arvense*	[231]

(continued)

Table 5 (continued)

Compound group	Compound	Species	References
	4″-methoxy-E-styryl)-2-pyron-3-O-β-D-glucopyranoside]		
	Equisetumpyrone	*E. arvense*	[230, 231]
		E. fluviatile	[230]
		E. giganteum	[225]
		E. palustre	[230]
Phenolic sesquiterpenes	Onitin	*E. arvense*	[212]
Lignin	Guaiacylglycerol-β-coniferyl ether	*E. debile*	[217]
Neolignan	Debilignanoside	*E. debile*	[217]
Norisoprenoid	(3S,5R,6R,7E,9S)-9-[(β-D-Glucopyranosyl)oxy]megastigm-7-ene-3,5,6-triol	*E. debile*	[229]
	Blumenol A	*E. debile*	[229]
	Corchoinoside C	*E. debile*	[229]
	Sammangaoside A	*E. debile*	[229]
Phenolics	4-O-(*p*-Coumaroyl)shikimic acid	*E. palustre*	[213]
	5-O-Caffeoyl shikimic acid	*E. arvense*	[211]
		E. fluviatile	[211]
		E. sylvaticum	[211]
		E. telmateia	[211, 220]
	Caffeic acid	*E. hyemale*	[228]
	Caffeoyl-methylate-4-β-glucopuranoside	*E. myriochaetum*	[226]
	Chlorogenic acid	*E. hyemale*	[228]
	Coumaric acid	*E. debile*	[217]
	Dicaffeoyl meso-tartaric acid (check if these both are same)	*E. arvense*	[211]
	di-*E*-Caffeoyl-*meso*-tartaric acid	*E. arvense*	[223]
	Ferulic acid	*E. debile*	[217]
	Monocaffeoyl meso-tartaric acid	*E. arvense*	[211]
		E. fluviatile	[211]
		E. palustre	[211]
		E. sylvaticum	[211]
		E. telmateia	[211, 220]
	p-Hydroxybenzoic acid	*E. debile*	[217]
		E. telmateia	[220, 221]
Phenylhexanes	Debilitriol	*E. debile*	[217]
Terpenoids	**Sterols**		
	28-Isofucosterol	*E. arvense*	[214]
	Campesterol	*E. arvense*	[214]
	Cholesterol	*E. arvense*	[214]

(continued)

Table 5 (continued)

Compound group	Compound	Species	References
	β-Sitosterol	*E. myriochaetum*	[215]
	Epicholesterol	*E. arvense*	[214]
	Sitosterol	*E. arvense*	[214]
	Triterpenoids		
	Germanicol	*E. arvense*	[214]
	Isobauerenol	*E. arvense*	[214]
	Taraxerol	*E. arvense*	[214]
	α-Amyrin	*E. arvense*	[214]
	β-Amyrin	*E. arvense*	[214]
Triamines	Spermidine	*E. debile*	[217]
Miscellaneous	5-Hydroxymethyl-2-furfuraldehyde	*E. debile*	[217]
	5-Hydroxymethylfurfural	*E. hyemale*	[210]
	α-D-Fructofuranose	*E. hyemale*	[210]
	β-D-Glucosylsitosterol	*E. myriochaetum*	[215]
	γ-Hydroxycaprylic acid	*E. debile*	[217]

4.2 Cytotoxic Activity

Zhao et al. [285] identified coumarins from *Dryopteris fragrans* (L.) Schott, such as dryofracoumarin A (82), esculetin (84), and isoscopoletin (85) along with the compounds of other group including (E)-4-(3,4-dimethoxyphenyl)but-3-en-1-ol (86), cis-3-(3,4-dimethoxyphenyl)-4-[(E)-3,4-dimethoxystyryl] cyclohex-1-en (87), trans-3-(3,4-dimethoxyphenyl) (3,4-dimethoxyphenyl) -4-[(E)-3,4-dimethoxystyryl] methylphlorobutyrophenone (110), aspidinol (101), albicanol (98), and cyclohex-1-en (89) (107). By using the MTT [3-(4,5-dimethylthiazol-2-yl)-2,5-diphenyl tetrazolium bromide] test, all substances were assessed for their cytotoxic activities. Esculetin (84), isoscopoletin (85), and cis-3-[(E)-3,4-dimethoxystyryl]-4-(3,4-dimethoxyphenyl)] Trans-3-(3,4-dimethoxyphenyl)-4-[(E)-3,4-dimethoxystyryl] and cyclohex-1-en (87) cyclohex-1-en (89) demonstrated substantial cytotoxic effects against three cancer cell lines A549, MCF7, and HepG2. In a different investigation, Liu et al. [283] discovered a phenylpropanoid molecule called (E)-caffeic acid-9-O-D-xylpyranosyl-(12)-D-glucopyranosyl ester (96) and evaluated its anticancer activities using MTT assay against breast cancer cells (MCF-7). The chemical had significant activity against MCF-7 cells. Dryofraterpene A (99), a sesquiterpene that Zhong et al. [336] isolated from *Dryopteris fragrans* Schott, was tested for its anti-proliferative properties against five human cancer cell lines, including A549, MCF7, HepG2, HeLa, and PC-3. These effects were assessed by CCK-8 and lactate dehydrogenase (LDH) assay. Below a

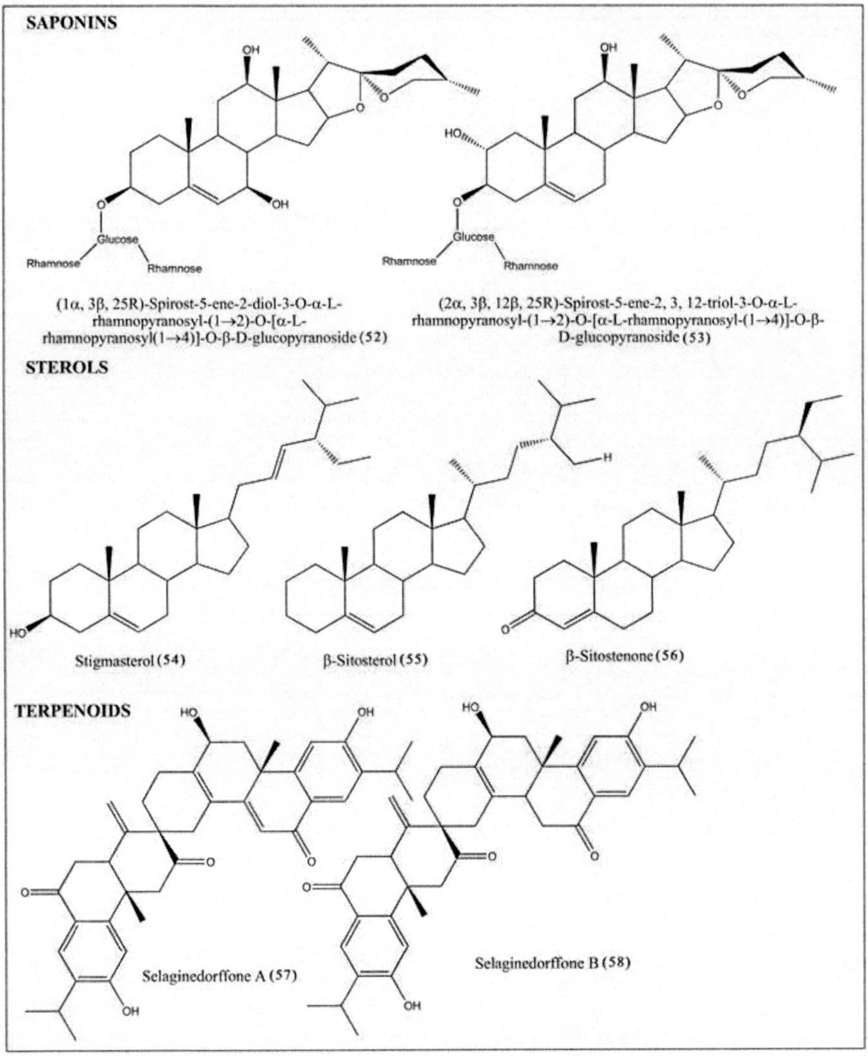

Fig. 7 Major saponins, sterols, and terpenoids isolated from *Selaginella* spp.

10 μM concentration, dryofraterpene A (99) dramatically reduced cancer cell proliferation without causing any visible necrosis.

4.3 Antitumor Activity

Alkaloids, such as lycopodine (**2**), lycojaponicumin A (**4**), lycojaponicumin B (**5**), and lycojaponicumin C (**6**) and a terpenoid serratenediol (**9**) were isolated from

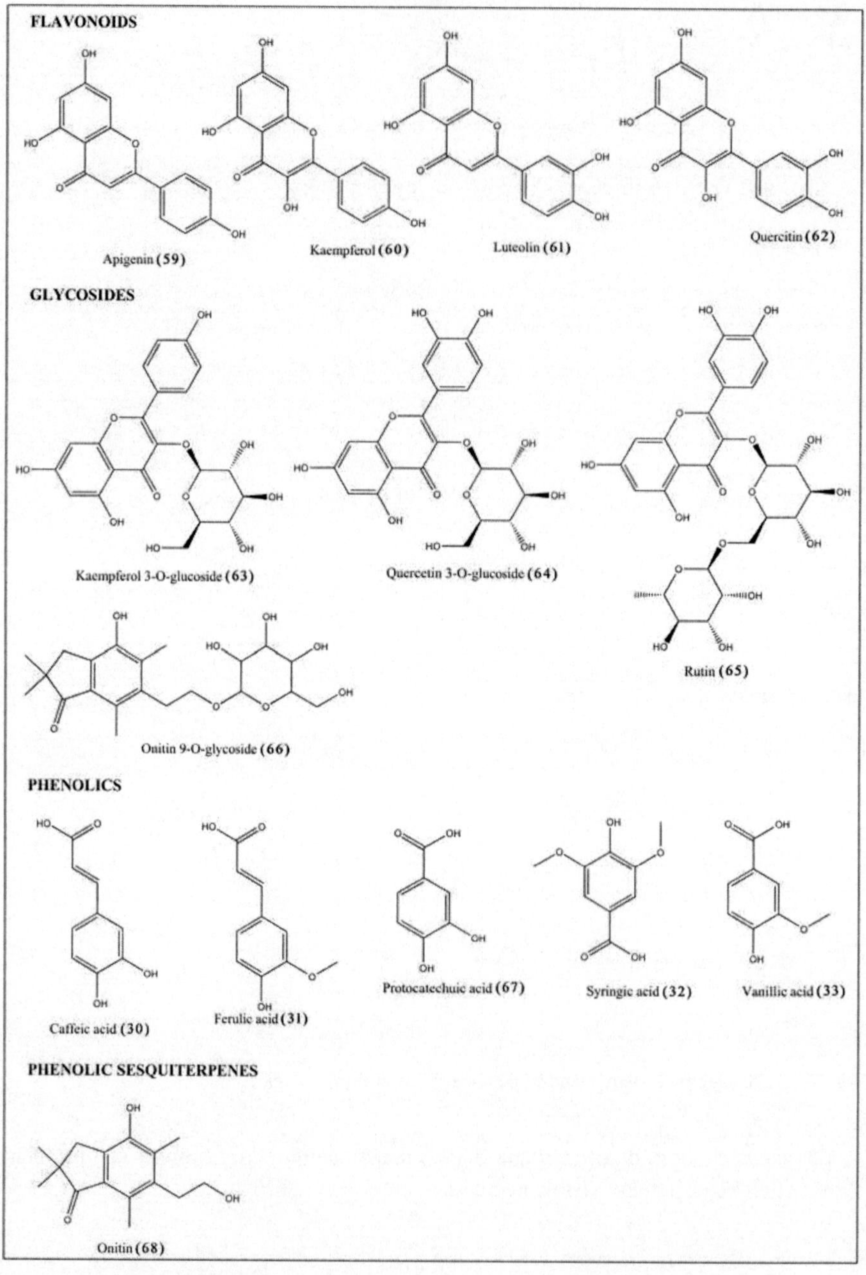

Fig. 8 Major flavonoids, glycosides, phenolics, and phenolic sesquiterpenes isolated from *Equisetum* spp.

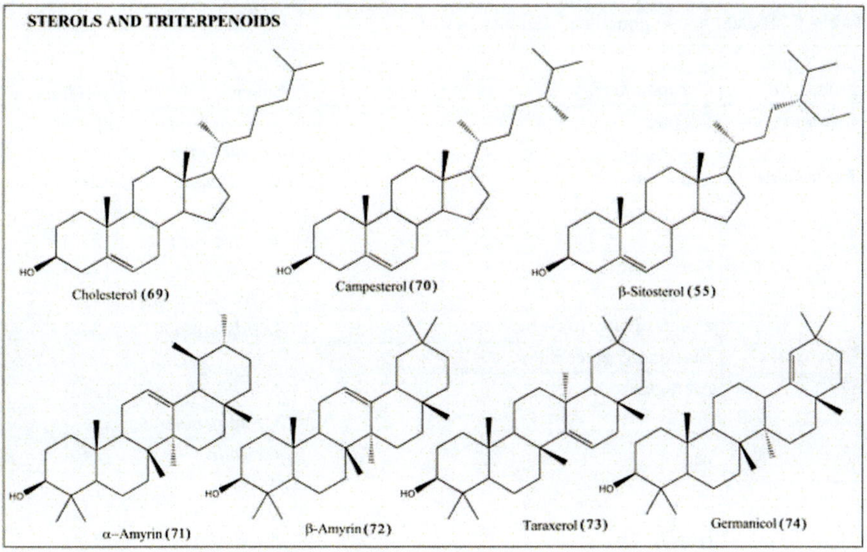

Fig. 9 Major sterols and terpenoids isolated from *Equisetum* spp.

different species of *Lycopodium* have demonstrated to possess antitumor activities. Mandal et al. [36] studied the effect of *L. clavatum* extract fraction containing lycopodine (**2**) and demonstrated that lycopodine (**2**) can inhibit the proliferation of HeLa cells through induction of apoptosis via caspase-3 activation. In another study, Bishayee et al. [75] displayed that lycopodine (**2**) could down regulate the expression of 5-lipoxygenase and 5-oxo-ETE receptor (OXE receptor 1) and epidermal growth factor (EGF), which eventually causing up-regulation of cytochrome-C with depolarization of mitochondrial membrane potential, finally leading to cell apoptosis. Ham et al. [48] conducted in vitro assay of *L. serratum* extract on several cancerous cell lines and showed that 100 μg/ml induced apoptosis of SK-Hep1 (75.7%), HT-29 (71.7%), A549 (53.8%), and HL-60 (89.2%) cells. Their subsequent investigations revealed that serratenediol (**9**) containing fraction was responsible for optimum inhibition of HL-60 cells with IC_{50} of 12.9 μM. A study by Ham et al. [48] further showed that caspase-9, − 3 activity was responsible for apoptotic events in HL-60 cells. Similarly, antitumor activities of lycojaponicumin A (**4**), lycojaponicumin B (**5**), lycojaponicumin C (**5**), and lycophlegmarin (**8**) have been reported [47].

4.4 Anti-metastasis Activity

Metastasis is the development of secondary malignant growths at a distance from the primary site of cancer. There has been evidence to suggest that some overexpressed proteolytic enzymes in cancer cells, such as matrix metalloproteinases (MMPs), play a

Table 6 Bioactive compounds of *Adiantum* spp.

Compound group	Compound	Species	References
Coumarins	Psoralen	*A. thalictroides* var. *hirsutum*	[243]
Flavonoids	Quercetin	*A. capillus-veneris*	[245]
		A. tetraphyllum	[232]
	Kaempferol	*A. aethiopicum* *A. monochlamys*	[246]
	2′,4′,6′-Trihydroxychalcone	*A. sulphureum*	[247]
Glycosides	**Flavonoid glycosides**		
	Astragalin	*A. aethiopicum*	[248]
		A. monochlamys	[248]
		A. capillus-veneris	[246]
		A. cuneatum	[246]
	Hyperin	*A. malesianum*	[249]
		A. monochlamys	[248]
	Isoquercitrin	*A. aethiopicum*	[248]
		A. capillus-veneris	[246]
		A. monochlamys	[248]
	Isovitexin	*A. malesianum*	[249]
	Kaempferol 3-glucuronide	*A. capillus-veneris* *A. cuneatum*	[246]
		A. capillus-veneris	[250]
	Kaempferol 3-O-α-D-galactopyranoside	*A. malesianum*	[249]
	Kaempferol 3-O-β-D-galactopyranoside	*A. malesianum*	[249]
	Naringin	*A. aethiopicum*	[248]
	Nicotiflorin	*A. capillus-veneris*	[246]
	Prunin	*A. aethiopicum* *A. monochlamys*	[248]
	Quercetin 3-galactoside	*A. monochlamys*	[246]
	Quercetin 3-glucoside	*A. aethiopicum*	[246]
		A. capillus-veneris	[245, 250]
		A. caudatum	[243]
		A. tetraphyllum	[232]
	Querciturone	*A. capillus-veneris* *A. cuneatum*	[246]
	Rutin	*A. capillus-veneris*	[245, 246]

(continued)

Table 6 (continued)

Compound group	Compound	Species	References
	Trifolin	*A. monochlamys*	[248]
	Vitexin	*A. malesianum*	[249]
	Lignan glycosides		
	Pinoresinol 4-O-β-D-glucopyranoside	*A. capillus-veneris*	[250]
	Phytosterol glycosides		
	Daucosterol	*A. caudatum*	[243]
Phenolics	Ferulic acid	*A. tetraphyllum*	[232]
	Caffeic acid	*A. tetraphyllum*	[232]
	Methyl-p-coumarate	*A. thalictroides* var. *hirsutum*	[243]
Phytosterols	Campesterol	*A. capillus-veneris*	[244]
	Stigmasterol	*A. capillus-veneris*	[244]
	β-Sitosterol	*A. capillus-veneris*	[244]
		A. caudatum	[243]
		A. tetraphyllum	[232]
Sulphate esters	1-*p*-Caffeylgalactose 6-sulphate	*A. capillus-veneris*	[251]
	1-*p*-Coumarylglucose 2-sulphate	*A. capillus-veneris*	[251]
Terpenoids	1-(5a,5b,8,8,11a,13b-Hexamethyleicosahydro-1H-cyclopenta[a]chrysen-3-yl)-1-ethanone	*A. incisum*	[252]
	13,14-Seco-8,22-cyclo-lanost-5(6), 24(25)-diene	*A. venustum*	[253]
	13-Epineohop-18-en-12α-ol	*A. cuneatum*	[254]
	17,29-Epoxyhopane	*A. capillus-veneris*	[239]
	17β,21β-Epoxyhopane	*A. caudatum*	[237]
	19α-Acetoxyadiantone	*A. edgeworthii*	[242]
	19α-Hydroxyadiantone	*A. edgeworthii*	[242]
		A. caudatum	[237]
	19α-Hydroxyfern-7-ene	*A. caudatum*	[237]
	19α-Hydroxyfern-9(11)-ene	*A. caudatum*	[237]
	19α-Hydroxyferna-7,9(11)-diene	*A. caudatum*	[237]
	19α-Hydroxyisoadiantone	*A. edgeworthii*	[242]
	19β-Hydroxyfern-9(11)-ene	*A. caudatum*	[237]
	21-Hydroxyadiantone	*A. venustum*	[238]

(continued)

Table 6 (continued)

Compound group	Compound	Species	References
	22,29ξ-epoxy-30-norhopane-13β-ol	*A. lunulatum*	[240]
	23-Hydroxyfernene	*A. incisum*	[255]
		A. pedatum	[235]
	25-Norfern-7-en-10β-yl formate	*A. cuneatum*	[236]
	28-Hydroxyfern-9(11)-ene	*A. capillus-veneris*	[241]
	30-Normethyl-lupan-20-one	*A. tetraphyllum*	[232]
	3,4-Dihydroxyfilicane	*A. capillus-veneris*	[245]
	3-Methoxy-4-hydroxyfilicane	*A. capillus-veneris*	[245]
	3α,4α-Epoxyfilicane	*A. capillus-veneris*	[256]
	3α-Hydroxy-4β-methoxyfilicane	*A. caudatum*	[237]
	3β,4-α-Dihydroxyfilicane	*A. capillus-veneris*	[257]
	4,23-Bisnor-3,3-dimethoxy-3,4-secofilic-5(24)-ene	*A. cuneatum*	[258, 259]
	4,23-Bisnor-3,4-secofilic-5(24)-en-3-al	*A. cuneatum*	[258, 259]
	4-Hydroxy-4,6a,6b,9,9,12a,14b-heptamethylperhydropicen-3-one hemihydrate	*A. incisum*	[260]
	4α -Hydroxyfilican-3-one	*A. capillus-veneris*	[241, 257]
	6-Oxofern-9(11)-ene	*A. lunulatum*	[240]
	6α-Acetoxy-16β,22-dihydroxy-3-ketoisohopane	*A. lunulatum*	[261]
	7-Fernene	*A. pedatum*	[262]
		A. monochlamys	[263]
	7α,8α-Epoxy-fernan-25-ol	*A. cuneatum*	[259]
	7β,25-Epoxyfern-8-ene	*A. cuneatum*	[236]
	7β,25-Epoxyfern-9(11)-en-8α-ol	*A. cuneatum*	[259]
	8α-Hydroxyfernan-25,7β-olide	*A. caudatum*	[237]
	Adian-5-en-25-ol	*A. cuneatum*	[254]
	Adian-5-ene ozonide	*A. monochlamys*	[264]
	Adian-5-ene	*A. capillus-veneris*	[239]
		A. monochlamys	[234]
	Adianene	*A. monochlamys*	[263]

(continued)

Table 6 (continued)

Compound group	Compound	Species	References
	Adianene ozonide	*A. monochlamys*	[234]
	Adiantol	*A. cuneatum*	[265]
	Adiantone	*A. capillus-veneris*	[239] [241]
		A. caudatum	[237]
		A. cuneatum	[236]
		A. edgeworthii	[242]
		A. incisum	[255]
		A. lunulatum	[240]
		A. monochlamys	[234]
		A. pedatum	[235, 262]
	Adiantoxide	*A. capillus-veneris*	[239, 241]
	Adiantulanosterol	*A. venustum*	[266]
	Adiantulupanone	*A. venustum*	[266]
	Adiantuoleanone	*A. venustum*	[266]
	Adian-5-en-3α-ol	*A. capillus-veneris*	[239]
	Adian-5(10)-en-3α-ol	*A. capillus-veneris*	[239]
	Adininaneone	*A. incisum*	[255]
	Adininaonol	*A. incisum*	[255]
	Adipedatol	*A. pedatum*	[235, 262]
	Capillirol B	*A. capillus-veneris*	[257]
	Capillirone	*A. capillus-veneris*	[257]
	Diploptene	*A. monochlamys*	[263]
	Epihakonanediol	*A. monochlamys*	[234]
	Fern-7-en-25-ol	*A. cuneatum*	[254]
	Fern-7-en-3α-ol	*A. capillus-veneris*	[239]
	Fern-7-ene	*A. capillus-veneris*	[239, 241]
		A. caudatum	[237]
		A. cuneatum	[236]
		A. edgeworthii	[242]
		A. monochlamys	[234]
		A. pedatum	[235]
	Fern-8-ene	*A. caudatum*	[237]
		A. monochlamys	[234]
		A. pedatum	[235]
	Fern-9(1 l)-en-25-oic acid	*A. venustum*	[267]
	Fern-9(11)-en-6-ol	*A. lunulatum*	[240]

(continued)

Table 6 (continued)

Compound group	Compound	Species	References
	Fern-9(11)-en-12-one	*A. capillus-veneris*	[239, 241]
	Fern-9(11)-en-12β-ol	*A. capillus-veneris*	[241]
	Fern-9(11)-en-25-oic acid	*A. edgeworthii*	[242]
		A. lunulatum	[240]
		A. venustum	[268]
	Fern-9(11)-en-25-ol	*A. cuneatum*	[254]
	Fern-9(11)-en-28-ol	*A. capillus-veneris*	[239]
		A. lunulatum	[240]
	Fern-9(11)-ene	*A. capillus-veneris*	[239]
		A. caudatum	[237]
		A. cuneatum	[236]
		A. edgeworthii	[242]
		A. lunulatum	[240]
		A. monochlamys	[234]
		A. pedatum	[235]
	Ferna-7,9(11)-diene	*A. capillus-veneris*	[239, 241]
		A. caudatum	[237]
		A. cuneatum	[236]
		A. monochlamys	[234]
		A. pedatum	[235]
	Fernene	*A. monochlamys*	[233]
		A. pedatum	[262]
	Ferrn-9(11)-ene	*A. capillus-veneris*	[241]
	Fern-9(11)-en-3α-ol	*A. capillus-veneris*	[239]
	Filic-3-ene	*A. capillus-veneris*	[239, 241]
		A. caudatum	[237]
		A. cuneatum	[236]
		A. edgeworthii	[242]
		A. monochlamys	[234]
		A. pedatum	[235]
	Filican-3-one	*A. monochlamys*	[263]
	Filican-3α-ol	*A. monochlamys*	[263]
	Filicenal	*A. cuneatum*	[236, 265]
		A. pedatum	[235, 262]

(continued)

Table 6 (continued)

Compound group	Compound	Species	References
	Filicene	*A. cuneatum*	[265, 269]
		A. monochlamys	[263]
		A. pedatum	[262]
	Filicenoic acid	*A. pedatum*	[235]
	Filicenol A	*A. monochlamys*	[234]
	Filicenol B	*A. lunulatum*	[240]
		A. monochlamys	[234]
	Glaucanol A	*A. pedatum*	[235]
	Glaucanol B acetate	*A. cuneatum*	[236]
	Hakonanediol	*A. monochlamys*	[234]
	Hopan-22-ol	*A. tetraphyllum*	[232]
		A. latifolium	[270]
	Hopan-28,22-olide	*A. capillus-veneris*	[239]
	Hop-22(29)-ene	*A. capillus-veneris*	[239, 241]
		A. edgeworthii	[242]
		A. monochlamys	[234]
	Hydoxyhopane	*A. capillus-veneris*	[241]
		A. cuneatum	[259]
	Hydroxyadiantone	*A. capillus-veneris*	[239, 241]
		A. cuneatum	[236]
		A. monochlamys	[234, 271]
		A. pedatum	[235]
	Hydroxyhopane	*A. capillus-veneris*	[239]
		A. edgeworthii	[242]
		A. pedatum	[235]
	Isoadiantane-19,22-dione	*A. edgeworthii*	[242]
	Isoadiantol	*A. capillus-veneris*	[241]
		A. cuneatum	[259]
	Isoadiantol B	*A. capillus-veneris*	[245]
		A. monochlamys	[234]
		A. pedatum	[235]
	Isoadiantone	*A. capillus-veneris*	[239, 241, 245]
		A. caudatum	[237]
		A. cuneatum	[236, 265]
		A. incisum	[255]

(continued)

Table 6 (continued)

Compound group	Compound	Species	References
		A. monochlamys	[234]
		A. pedatum	[235]
	Isofernene	*A. monochlamys*	[263]
		A. pedatum	[262]
	Isoglaucanone	*A. capillus-veneris*	[239, 241]
		A. cuneatum	[236]
		A. pedatum	[235]
	Ketohakonanol	*A. cuneatum*	[259]
		A. monochlamys	[234, 271]
		A. pedatum	[235]
	Llanost-20(22)-en-3,19-ether (=Adiantulanostene ether)	*A. venustum*	[272]
	Methyl fern-(11)-en-25-oate	*A. edgeworthii*	[242]
	Mollugogenol A	*A. lunulatum*	[261]
	Neohop-12-ene	*A. capillus-veneris*	[239, 241]
		A. cuneatum	[236]
		A. edgeworthii	[242]
		A. monochlamys	[234]
		A. pedatum	[235]
	Neohop-13(18)-ene	*A. caudatum*	[237]
		A. cuneatum	[236]
		A. monochlamys	[233, 234]
		A. pedatum	[235]
	Neohop-13(18)-en-19α-ol	*A. cuneatum*	[254]
	Neohop-18-en-12α-ol	*A. cuneatum*	[254]
	Neohopa-11,13(18)-diene	*A. cuneatum*	[236]
		A. monochlamys	[234]
		A. pedatum	[233, 235]
	Neohopene	*A. monochlamys* *A. pedatum*	[233]
	Olean-12-en-3-one	*A. capillus-veneris*	[241]
	Olean-18-en-3-one	*A. capillus-veneris*	[241]
	Phyten-3(20)-1,2-diol	*A. tetraphyllum*	[232]
	Phytol	*A. tetraphyllum*	[232]
	Pteron-14-en-7α-ol	*A. capillus-veneris*	[239]
	Pterosterone	*A. capillus-veneris*	[250]

(continued)

Table 6 (continued)

Compound group	Compound	Species	References
	Tetrahymanol	*A. monochlamys*	[234]
		A. pedatum	[235]
	Tirucall-8,22-cyclo-24(25)-ene (=Adiantutirucallene B)	*A. venustum*	[253]
	Trisnorhopane	*A. capillus-veneris*	[239]
		A. cuneatum	[236]
	Zeorin	*A. edgeworthii*	[242]

critical role in the migration, intravasation, and extravasation. Therefore, the discovery of effective agents to suppress cancer metastasis by inhibition of metastasis-associated proteins or signaling pathways is an efficient approach to developing a new cancer therapy [347]. Recent studies revealed that lignans obtained from various plant sources have inhibitory activity against MMP-3 and MMP-9, showing a significant effect on anti-drug resistance [348]. Zhu et al. [172] isolated several lignans from *S. moellendorffii* which have demonstrated potent matrix metalloproteinases-9 inhibition activities.(−)-(7*S*,8*S*,8*R*)-4,4-dihydroxy-3,3,5,5′-tetramethoxy-7,9-epoxylignan-9-ol-7-one (**34**), burseneolignan (**37**), 1-(4′-hydroxy-3′-methoxyphenyl)-2-[4″-(3-hydroxy-propyl)-2″,6″-dimethoxyphenoxy]propane-1,3-dio] (**36**), selaginellol (**40**), (8*R*)-3,5′-dimethoxy-8,3′-neoligna-4,4′,9,9′-tetraol (**35**), and dihydrobuddlenol B (**38**) were some of the lignans which exhibited antitumour activity against human cancer cell lines, viz., HepG2, T24, MGC-803, A549 [172]. In addition, in vitro enzyme inhibition (MMP-9), surface plasmon resonance, and molecular docking studies revealed that (8*R*)-3,5′-Dimethoxy-8,3′-neoligna-4,4′,9,9′-tetraol (**35**) as a potential drug for cancer therapy.

Selaginellins are a small group of pigments isolated and characterized by varied species of *Selaginella*. They have polyphenolic skeletons, commonly featuring tautomeric phenol-quinone methide, alkynylphenol, or fluorene moieties, and several investigations have shown that these compounds are having a broad range of bioactivities, such as cytotoxic, antimicrobial, and antiviral activities [349]. Cao et al. [175] isolated diselaginellin B (**45**) from *S. pulvinata*, which displayed apoptosis-inducing and antimetastatic activities against the human hepatocellular carcinoma cell line SMMC-7721. Microarray analysis [175] demonstrated that diselaginellin B (**45**) altered the expression of genes related to metabolism, angiogenesis, and metastasis.

4.5 Antifungal Activity

Dermatophytosis is an infection of the hair, skin, or nails caused by fungus *Trichophyton rubrum*. Yang et al. [327] demonstrated that the aspidin BB (**100**), a phloroglucinol derivative that was isolated from *Dryopteris fragrans*, showed significant antifungal properties against *Trichophyton rubrum*. They looked at how

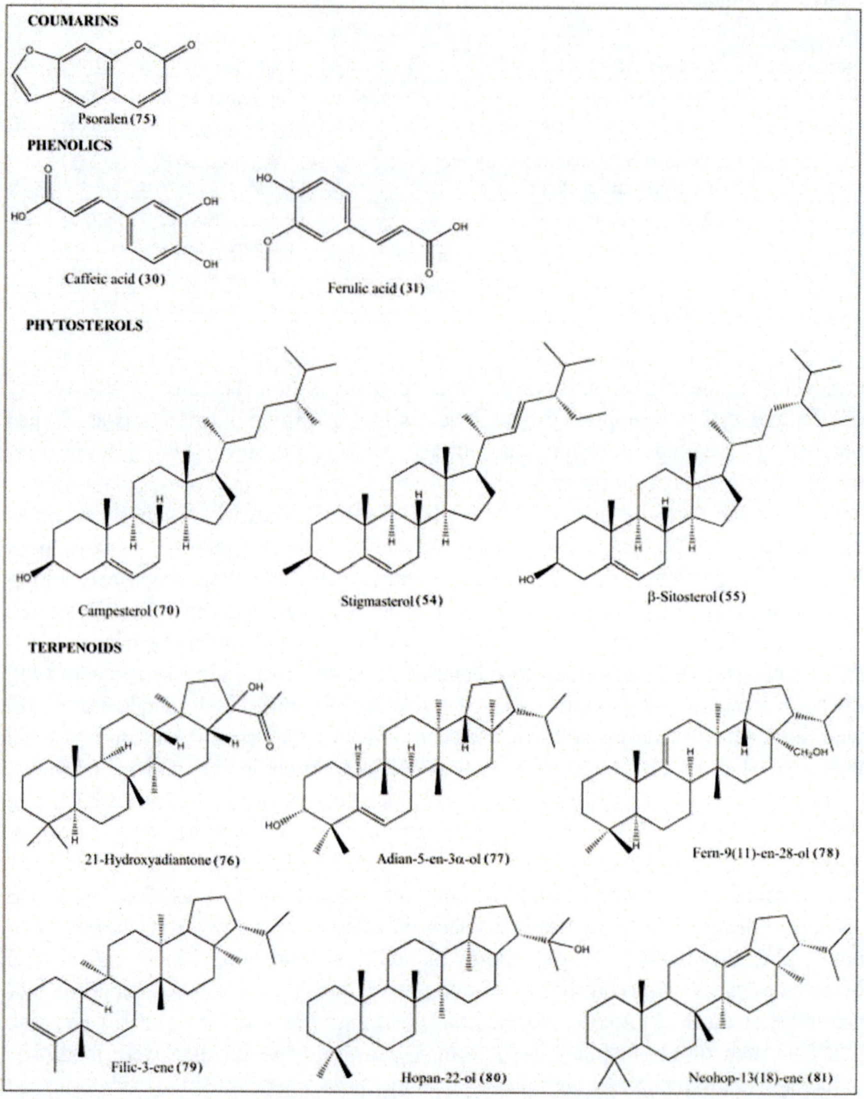

Fig. 10 Major coumarins, phenolics, phytosterols, and terpenoids isolated from *Adiantum* spp.

aspidin BB affected the synthesis of ergosterol, which is an important chemical for membrane integrity in *Trichophyton rubrum*. Yang et al. [327] results demonstrated the ergosterol inhibition with the treatment of aspidin BB which leads to the disintegration of *Trichophyton rubrum*.

Bioflavone compound isolated from *Selaginella tamariscina*, namely, amentoflavone (**22**), is reported to possess antifungal activity against many human pathogenic fungi. Amentoflavone was examined by Jung et al. [157] who also

Table 7 Bioactive compounds of *Dryopteris* spp.

Compound group	Compound	Species	References
Carbamic acids	(1,7a-Dihydro-1H-inden-2(7aH)-idene)methylcarbamic acid	*D. wallichiana*	[307]
Chromones	5,7-Dihydroxy-2-hydroxymethylchromone	*D. fragrans*	[278, 282]
Coumarins	Dryofracoumarin A	*D. fragrans*	[284, 285]
	Dryofracoumarin B	*D. fragrans*	[284]
	Esculetin	*D. fragrans*	[282, 285]
	Isoscopoletin	*D. fragrans*	[285]
Flavanoids	(+)-Catechin-8-acetic acid	*D. erythrosora*	[281]
	4β -Carboxymethyl-(−)-epicatechin	*D. crassirhizoma*	[308]
	4β-Carboxymethyl-(−)-epicatechin	*D. crassirhizoma*	[308]
	3, 5, 7-Trihydroxy-2-(p-tolyl) chorman-4-one	*D. cycadina*	[306]
	2,3,4,5,2′,4′,5′,6′-Octamethoxychalcone	*D. erythrosora*	[281]
	2′-Hydroxy-2,3,4,5,4′,5′,6′-heptamethoxychalcone	*D. erythrosora*	[281]
	(2S)-5-Hydroxy-7,8,6′-trimethoxyflavanone-2′-O-β-D-glucuronide	*D. erythrosora*	[281]
	2S-5,7,2′,5′-Tetrahydroxy-6-methoxyflavanone	*D. erythrosora*	[281]
	5, 7, 2′-Trihydroxy-6, 8-dimethylflavanon	*D. sublaeta*	[309]
	2′,4′-Dihydroxy-6′-methoxy-3′,5′-dimethylchalcone	*D. fragrans*	[284]
	2(S)-5, 7, 3′-Trihydroxy-6, 8-dimethyl-5′-methoxyflavanone	*D. sublaeta*	[309]
	Apigenin	*D. villarii*	[275]
	Biflorin	*D. crassirhizoma*	[308]
	Desmethoxymatteucinol	*D. sublaeta*	[309]
	Eriodictyol	*D. crassirhizoma*	[310]
		D. fragrans	[282]
	Isobiflorin	*D. crassirhizoma*	[308]
	Kaempferol	*D. villarii*	[275]
	Koreanoside B	*D. erythrosora*	[281]
	Matteucinol	*D. sublaeta*	[309]
	Quercetin	*D. erythrosora*	[273, 280]
		D. fragrans	[278]
		D. villarii	[275]
Glycosides	**Benzophenone glycoside**		
	Iriflophenone-3-C-β-d glucopyranoside	*D. ramosa*	[311]

(continued)

Table 7 (continued)

Compound group	Compound	Species	References
	Chromone glycosides		
	Frachromone A	*D. fragrans*	[312]
	Frachromone C	*D. fragrans*	[313]
	Undulatoside A	*D. fragrans*	[313]
	Coumarin glycosides		
	Dryofracoulin A	*D. fragrans*	[313]
	Flavanoid glycosides		
	(−)-5,7-O-Dimethyl-3′,4′,5′-O-trimethylepigallocatechin-3-O-(3″,4″,5″-O-trimethyl) gallate	*D. erythrosora*	[281]
	(+)-Catechin-6-C-β-D-glucopyranoside	*D. crassirhizoma*	[308]
	4‴-α-Rhamnopyranosyl-2″-O-β-D-galactopyranosylvitexin	*D. erythrosora*	[273]
	5, 7, 4′-Trihydroxyflavon-3-glucopyranoid	*D. cycadina*	[306]
	7-o-Glucoside 4′-*p*-coumarate	*D. villarii*	[314]
	Apigenin 4′-O-(caffeoylglucoside)	*D. villarii*	[315]
	Apigenin 4′-O-(feruloylglucoside)	*D. villarii*	[315]
	Apigenin 4′-O-glucoside	*D. villarii*	[276]
	Apigenin 7-O-(sulphatoglucoside)	*D. villarii*	[314]
	Apigenin 7-O-glucoside	*D. erythrosora*	[273, 280]
	Apigenin 7-O-rutinoside	*D. erythrosora*	[280, 281]
	Apigenin 6-C-arabinoside-8-C-glucoside	*D. erythrosora*	[281]
	Apigenin-C-pentoside	*D. erythrosora*	[281]
	Astragalin	*D. villarii*	[275]
	Biochanin A-7-O-glucoside-6″-O-malonate	*D. erythrosora*	[281]
	Crassirhizomoside A	*D. crassirhizoma*	[279]
	Crassirhizomoside B	*D. crassirhizoma*	[279]
	Crassirhizomoside C	*D. crassirhizoma*	[279]
	Formononetin 7-O-(6″-malonylglucoside)	*D. erythrosora*	[281]
	Gliricidin 7-O-hexoside	*D. erythrosora*	[273]
	Globularin	*D. erythrosora*	[273]
	Isocarthamidin 7-O-glucuronide	*D. erythrosora*	[281]
	Isoliquiritin apioside	*D. erythrosora*	[281]
	Isoorientin	*D. fragrans*	[282]
	Kaempferide 3-Rhamnoside-7-(6″-succinylglucose)	*D. erythrosora*	[281]

(continued)

Table 7 (continued)

Compound group	Compound	Species	References
	Kaempferitrin	*D. crassirhizoma*	[277]
	Kaempferol 3, 4-di-O-α-L-rhamnopyranoside	*D. cycadina*	[316, 317]
	Kaempferol 3,5-di-O-α-L-rhamnoside	*D. cycadina*	[317]
	Kaempferol 3,7-di-O-α-L-rhamnopyranoside	*D. cycadina*	[317]
	Kaempferol 3-O-(acetylrutinoside)	*D. villarii*	[275]
	Kaempferol 3-O-(caffeoylrhamnoside)	*D. villarii*	[315]
	Kaempferol 3-O-rutinoside	*D. erythrosora*	[280]
	Kaempferol 3-O-α-L-arabinopyranoside	*D. erythrosora*	[281]
	Keampferol 3-O-rhamnoside	*D. erythrosora*	[273, 280]
	Keampferol 3-O-rutinoside	*D. erythrosora*	[273]
	Keampferol 7-O-gentiobioside	*D. erythrosora*	[273]
	Keampferol 7-O-rutinoside	*D. erythrosora*	[273]
	Matteuorienate A	*D. sublaeta*	[318]
	Matteuorienate C	*D. sublaeta*	[318]
	Myricetin 3-O-glucoside	*D. erythrosora*	[280, 281]
	Myricetin 3-O-rhamnoside	*D. erythrosora*	[273, 280]
	Nicotiflorin	*D. villarii*	[275]
	Quercetin 3-O-(acetylglucoside)	*D. villarii*	[275]
	Quercetin 3-O-(acetylrutinoside)	*D. villarii*	[275]
	Quercetin 3-O-(X″-acetyl-X″-cinnamoyl-glucoside)	*D. villarii*	[276]
	Quercetin 3-O-glucosylrhamnoside	*D. villarii*	[276]
	Quercetin 3-O-rhamnoside-7-O-glucoside	*D. villarii*	[314]
	Quercetin 3-O-glucoside	*D. fragrans*	[278]
		D. fragrans	[312]
		D. villarii	[275]
	Quercetin 7-O-galactoside	*D. erythrosora*	[273]
	Quercetin 7-O-rutinoside	*D. erythrosora*	[273]
	Quercetin 3-O-galactoside	*D. erythrosora*	[280]
	Quercetin 3-O-β-D-glucopyranoside (3′ → O-3‴)-β-D- Quercetin −3-O-β-D-galactopyranoside	*D. cycadina*	[306]
	Quercetin 3-O-β-D-xylopyranoside	*D. erythrosora*	[281]
	Quercetin O-dihexoside	*D. erythrosora*	[281]
	Quercitrin	*D. erythrosora*	[273, 280]
		D. filix-mas	[274]
		D. villarii	[276]

(continued)

Table 7 (continued)

Compound group	Compound	Species	References
	Rutin	*D. erythrosora*	[281]
		D. fragrans	[278]
		D. villarii	[275]
	Scutellarein 7-O-glucobioside	*D. erythrosora*	[281]
	Sublaetentin A	*D. sublaeta*	[318]
	Sublaetentin B	*D. sublaeta*	[318]
	Sublaetentin C	*D. sublaeta*	[318]
	Sublaetentin D	*D. sublaeta*	[318]
	Sutchuenoside A	*D. crassirhizoma*	[277, 279]
	Vitexin	*D. fragrans*	[312]
	Phenolic glycosides		
	(E)-Caffeic acid-9-O-β-D-xylpyranosyl-(1 → 2)-β-D-glucopyranosyl ester	*D. fragrans*	[283]
	1–1,3-Dihydroxy-5-methoxyphenyl-4-O-β-D-glucopyranoside	*D. erythrosora*	[280]
	1-β-D-glucopyranosyloxy-3-methoxy-5-hydroxybenzene	*D. crassirhizoma*	[308]
	3,4-Dimethoxyphenyl-1-O-β-D-glucopyranoside	*D. sublaeta*	[318]
	3,5-Dimethyl-6-hydroxy-2-methoxy-4-O-D-glucopyranosyl-oxy-acetophenone	*D. fragrans*	[319]
	3-Methoxy-4-hydroxyphenyl-1-O-β-D-glucopyranoside	*D. sublaeta*	[318]
	4-O-β-D-Glucopyranosyl-(1‴ → 3″)-glucopyranosyl-2-hydroxy-6-methoxy-5-methylphenyl-1-butanone	*D. erythrosora*	[280]
	Arbutin	*D. sublaeta*	[318]
	Divarin-3-O-β-glucopyranoside (3-hydroxy-5-propylphenyl-O-β-D-glucopyranoside	*D. fragrans*	[312]
	Dryopteroside	*D. crassirhizoma*	[308]
	Monogalloyl glucose	*D. erythrosora*	[273]
	Sesquiterpene glucosides		
	Xianglinmaojueside A	*D. fragrans*	[320]
	Xianglinmaojueside B	*D. fragrans*	[320]
	Xianglinmaojueside C	*D. fragrans*	[320]
	Steroidal glycosides		
	β-Sitosterol 3-O-β-D-glucopyranoside	*D. cycadina*	[306]

(continued)

Table 7 (continued)

Compound group	Compound	Species	References
	Stilbene glucosides		
	3,5,4′-Trihydroxy-bibenzyl-3-O-β-D-glucoside	*D. sublaeta*	[321]
	3,5-Dihydroxy-stilbene-3-O-neohesperidoside	*D. sublaeta*	[321]
	3,5-Dihydroxy-stilbene-3-O-β-D-glucoside	*D. sublaeta*	[321]
	Polydotin peceid	*D. sublaeta*	[321]
	Xanthone glycosides		
	Mangiferin	*D. ramosa*	[322]
	Isomangiferin	*D. ramosa*	[322]
Phenolics	(*E*)-4-(3,4-Dimethoxyphenyl) but-3-en-1-ol	*D. fragrans*	[285]
	1-β-D-Glucopyranosyloxy-3-methoxy-5-hydroxybenzene	*D. crassirhizoma*	[308]
	1,3-Dihydroxyl-5-propylbenzene	*D. fragrans*	[283]
	2-Ethyl-6-hydroxybenzoic acid	*D. fragrans*	[283]
	3,4-Dihydroxyacetophenone	*D. fragrans*	[283]
	3,4-Dihydroxybenzaldehyde	*D. fragrans*	[283]
	4-Hydroxyacetophenone	*D. fragrans*	[283]
	Caffeic acid	*D. fragrans*	[283]
	cis-3-(3,4-Dimethoxyphenyl)-4-[(E)-3,4-dimethoxystyryl] cyclohex-1-ene	*D. fragrans*	[285]
	Dihydroconiferyl alcohol	*D. fragrans*	[283]
	Dihydroconiferylalcohol	*D. fragrans*	[282]
	Dryofragone	*D. fragrans*	[284]
	trans-3-(3,4-Dimethoxyphenyl)-4-[(E)-3,4-dimethoxystyryl] cyclohex-1-ene	*D. fragrans*	[285]
Phloroglucinols	1-(2, 4, 6-Trihydroxy-3-methylphenyl)butanone	*D. crassirhizoma*	[300]
	1-(2, 4, 6-Trihydroxy-3-methylphenyl)pentanone	*D. crassirhizoma*	[300]
	1-(2,4,6-Trihydroxy-phenyl)-propan-1-one	*D. crassirhizoma*	[300]
	1-(2,4,6-Trihydroxy-3-methylphenyl) propanone	*D. crassirhizoma*	[300]
	1,3 -(2,4,6-Trihydroxyphenyl) dibutanone	*D. crassirhizoma*	[300]
	1,3-(2,4,6-Trihydroxyphenyl) dipropanone	*D. crassirhizoma*	[300]
	2,4,6-Trihydroxy-acetophenone	*D. crassirhizoma*	[300]

(continued)

Table 7 (continued)

Compound group	Compound	Species	References
	2′,4′,6′-Trihydroxy-5′-methyl acetate-3′-methyl-1′-butyrophenone	*D. fragrans*	[323]
	3-Methyl-butyrylphloroglucinol	*D. crassirhizoma*	[302]
	3-Methyl-phlorbutyrophenon	*D. remota*	[303]
	5-Acetyl-2,4,6-trihydroxyacetophenone	*D. crassirhizoma*	[300]
	Abbreviatin BB	*D. abbreviata*	[286]
	Abbreviatin PB	*D. abbreviata*	[286]
	Aemulin BB	*D. aitoniana*	[287]
		D. championii	[293]
		D. crassirhizoma	[302]
	Albaspidin	*D. chrysocoma*	[324]
		D. fragrans	[297]
	Albaspidin AA	*D. aitoniana*	[287]
		D. crassirhizoma	[300, 302]
		D. hawaiiensis	[299]
		D. wallichiana	[325]
	Albaspidin AB	*D. aitoniana*	[287]
		D. crassirhizoma	[302]
		D. robertiana	[304]
		D. wallichiana	[325]
	Albaspidin AP	*D. crassirhizoma*	[298, 302]
		D. fragrans	[295]
		D. robertiana	[304]
	Albaspidin BA	*D. hawaiiensis*	[299]
	Albaspidin BB	*D. aitoniana*	[287]
		D. fragrans	[295]
		D. robertiana	[304]
		D. spinulosa	[296]
	Albaspidin iBiB	*D. subtriangularis*	[326]
	Albaspidin PB	*D. crassirhizoma*	[298, 302]
		D. fragrans	[295]
		D. robertiana	[304]
	Albaspidin PP	*D. crassirhizoma*	[298, 302]
		D. fragrans	[295]
		D. robertiana	[304]
	Albaspidin-1	*D. remota*	[303]
	Albaspidin-2	*D. remota*	[303]
	Araspidin BB	*D. crassirhizoma*	[300]
	Aspidin	*D. fragrans*	[297]
	Aspidin AA	*D. gymnosora*	[288]
	Aspidin AB	*D. aitoniana*	[287]

(continued)

Table 7 (continued)

Compound group	Compound	Species	References
		D. championii	[293]
		D. fragrans	[290, 291, 295]
		D. intermedia	[296]
		D. patula	[301]
		D. remota	[303]
	Aspidin BB	*D. aitoniana*	[287]
		D. championii	[293]
		D. crassirhizoma	[298, 302]
		D. fragrans	[290, 291, 294, 295, 327]
		D. gymnosora	[288]
		D. intermedia	[296]
		D. remota	[303]
	Aspidin PB	*D. fragrans*	[290, 291, 327]
	Aspidin PP	*D. subimpressa*	[328]
	Aspidinol	*D. dilatata*	[296]
		D. fragrans	[284, 285, 289, 291, 297]
		D. hawaiiensis	[299]
		D. remota	[303]
		D. robertiana	[304]
	Aspidinol B	*D. aitoniana*	[287]
		D. crassirhizoma	[298, 302, 310]
		D. fragrans	[323]
	Aspidinol P	*D. crassirhizoma*	[310]
	Atrata-phloroglucinol A	*D. atrata*	[329]
	Atrata-phloroglucinol B	*D. atrata*	[329]
	Bisphlorobutyrophenone	*D. crassirhizoma*	[300]
	Butyryl-3-methylphloroglucinol	*D. crassirhizoma*	[310]
	Butyrylphloroglucinol	*D. crassirhizoma*	[310]
	Desaspidin	*D. remota*	[303]
	Desaspidin AB	*D. aitoniana*	[287]
		D. robertiana	[304]
	Desaspidin AP	*D. robertiana*	[304]
		D. subimpressa	[328]
	Desaspidin BB	*D. aitoniana*	[287]
		D. assimilis	[296]
		D. patula	[301]
		D. robertiana	[304]

(continued)

Table 7 (continued)

Compound group	Compound	Species	References
	Desaspidin PP	*D. robertiana*	[304]
		D. subimpressa	[328]
	Desaspidinol	*D. austriaca*	[330]
		D. championii	[293]
	Dimethylphlorobutyrophenone	*D. abbreviata*	[286]
	Dryocrasin	*D. fuscoatra*	[299]
	Dryocrassin ABBA	*D. crassirhizoma*	[300, 302]
	Dryofragin	*D. crassirhizoma*	[302]
		D. fragrans	[290, 291, 294, 295, 331]
	Dryopteroside	*D. crassirhizoma*	[302]
	Filicinsaureacetylon	*D. remota*	[303]
	Filicinsaurebutanon	*D. remota*	[303]
	Filixic acid	*D. abbreviata*	[286]
		D. chrysocoma	[324]
	Filixic acid ABA	*D. aitoniana*	[287]
		D. parallelogramma	[301]
		D. commixta	[332]
		D. crassirhizoma	[300, 302]
		D. dickinsii	[332]
		D. fuscoatra	[299]
		D. tasiroi	[332]
		D. wallichiana	[325]
	Filixic acid ABB	*D. aitoniana*	[287]
		D. parallelogramma	[301]
		D. commixta	[332]
		D. crassirhizoma	[302]
		D. dickinsii	[332]
		D. fuscoatra	[299]
		D. tasiroi	[332]
		D. wallichiana	[325]
	Filixic acid ABP	*D. crassirhizoma*	[298, 302]
	Filixic acid BBB	*D. aitoniana*	[287]
		D. commixta	[332]
		D. dickinsii	[332]
		D. fuscoatra	[299]
		D. remota	[303]
		D. tasiroi	[332]
	Flavaspidic acid	*D. abbreviata*	[286]
		D. chrysocoma	[324]
	Flavaspidic acid AA	*D. crassirhizoma*	[300]

(continued)

Table 7 (continued)

Compound group	Compound	Species	References
	Flavaspidic acid AB	*D. abbreviata*	[286]
		D. aitoniana	[287]
		D. parallelogramma	[301]
		D. crassirhizoma	[292, 298, 300, 302, 305]
		D. fragrans	[295]
		D. fuscoatra	[299]
		D. patula	[301]
	Flavaspidic acid AP	*D. crassirhizoma*	[300, 305]
		D. goldieana	[296]
	Flavaspidic acid BB	*D. aitoniana*	[287]
		D. parallelogramma	[301]
		D. crassirhizoma	[305]
		D. fragrans	[295]
		D. fuscoatra	[299]
		D. patula	[301]
	Flavaspidic acid PB	*D. crassirhizoma*	[292, 298, 300, 302, 305]
		D. fragrans	[295]
	Flavaspidic acid PP	*D. subimpressa*	[328]
	Flavaspidsaure	*D. remota*	[303]
	Fraginol B	*D. aitoniana*	[287]
	Margaspidin	*D. marginalis*	[296]
	Margaspidin A	*D. championii*	[333]
	Margaspidin AB	*D. hawaiiensis*	[299]
	Margaspidin B	*D. championii*	[333]
	Margaspidin BB	*D. aitoniana*	[287]
		D. hawaiiensis	[299]
	Methylene bis aspidinol BB	*D. aitoniana*	[287]
		D. fragrans	[295]
	Methylene bis desaspidinol	*D. championii*	[293]
	Methylene bis desaspidinol BB	*D. aitoniana*	[287]
	Methylene bis methylphlorobutyrophenone	*D. crassirhizoma*	[302]
	Methylene bis methylphlorobutyrophenone	*D. crassirhizoma*	[298]
	Methylene bis phlorobutyrophenone	*D. crassirhizoma*	[300]
	Methylphlorbutyrophenone	*D. championii*	[293]
		D. fragrans	[285]
	Norflavaspidic acid AA	*D. aitoniana*	[287]

(continued)

Table 7 (continued)

Compound group	Compound	Species	References
	Norflavaspidic acid AB	*D. aitoniana*	[287]
		D. commixta	[332]
		D. crassirhizoma	[298, 300, 302]
		D. dickinsii	[332]
		D. fuscoatra	[299]
		D. tasiroi	[332]
	Norflavaspidic acid AP	*D. crassirhizoma*	[300]
	Norflavaspidic acid BB	*D. crassirhizoma*	[300]
	Norflavaspidic acid PB	*D. crassirhizoma*	[300]
		D. fuscoatra	[299]
	Nortrisflavaspidic acid ABB	*D. crassirhizoma*	[300]
	ortho-Desaspidin BB	*D. aitoniana*	[287]
	ortho-Desaspidin AB	*D. aitoniana*	[287]
	para-Aspidin	*D. remota*	[303]
	para-Aspidin AA	*D. robertiana*	[304]
	para-Aspidin AB	*D. aitoniana*	[287]
		D. hawaiiensis	[299]
	para-Aspidin BB	*D. aitoniana*	[287]
		D. campyloptera	[296]
		D. hawaiiensis	[299]
	Penta-albaspidin ABBBA	*D. crassirhizoma*	[302]
	Phlopyron	*D. campyloptera*	[296]
	Phloraspidinol	*D. austriaca*	[330]
		D. hawaiiensis	[299]
	Phloraspidinol BB	*D. aitoniana*	[287]
	Phloraspin	*D. marginalis*	[296]
	Phloraspin BB	*D. aitoniana*	[287]
	Phlorobutyrophenone	*D. crassirhizoma*	[300]
		D. remota	[303]
	Phloropyron A	*D. championii*	[333]
	Phloropyron B	*D. championii*	[333]
	Phloropyron BB	*D. aitoniana*	[287]
		D. crassirhizoma	[302]
	Phloropyron C	*D. championii*	[333]
	Phloraspyron	*D. austriaca*	[330]
	Picraquassioside D	*D. crassirhizoma*	[302]
	Propionyl 3-methylphloroglucinol	*D. crassirhizoma*	[310]
	Pseudoaspidinol A	*D. championii*	[333]
	Pseudoaspidinol B	*D. championii*	[293]
	Pulvinuliferin VV	*D. pulvinulifera*	[326]

(continued)

Table 7 (continued)

Compound group	Compound	Species	References
	Saroaspidin A	*D. fragrans*	[295]
	Subtriangularin iB	*D. subtriangularis*	[326]
	Tributyrylphloroglucinol	*D. crassirhizoma*	[300]
	Tripropionylphloroglucinol	*D. crassirhizoma*	[300]
	Trisabbreviatin BBB	*D. abbreviata*	[286]
	Trisaspidin BBB	*D. remota*	[303]
	Trisdesaspidin PBP	*D. subimpressa*	[328]
	Trisdesaspidin	*D. remota*	[303]
	Trisdesaspidin BBB	*D. aitoniana*	[287]
	Trisflavaspidic acid ABB	*D. crassirhizoma*	[300, 302]
	Trisflavaspidic acid BBB	*D. crassirhizoma*	[300]
	Trispara-aspidin	*D. remota*	[303]
	Trispara-aspidin BBB	*D. aitoniana*	[287]
	Wallichin A	*D. wallichiana*	[325]
	Wallichin B	*D. wallichiana*	[325]
	Wallichin C	*D. wallichiana*	[325, 334]
	Wallichin D	*D. wallichiana*	[325, 334]
	Wallichin E	*D. wallichiana*	[334]
	Wallichin F	*D. wallichiana*	[334]
	ψ-Aspidinol	*D. remota*	[303]
Phytosterols	β-Sitosterol	*D. championii*	[293]
		D. cycadina	[306]
		D. fragrans	[278, 291]
Terpenoids	**Monoterpenes**		
	Geniposide	*D. fragrans*	[335]
	Sesquiterpenes		
	3-O-β-D-Glucopyranosylalbicanol-11-O-β-D-glucopyranoside	*D. fragrans*	[282]
	Albicanol	*D. fragrans*	[284, 285, 290, 291]
	Albicanyl acetate	*D. fragrans*	[291]
	Conicumol	*D. fragrans*	[291]
	Dryofraterpene A	*D. fragrans*	[336]
	α-Cadinene	*D. fragrans*	[291]
	Triterpenes		
	17αH-Trisnorhopan-21-one	*D. crassirhizoma*	[337]
	Dryocrassol	*D. crassirhizoma*	[337]
	Dryocrassyl acetate	*D. crassirhizoma*	[337]
	Fern-9(11)-en-12-one	*D. crassirhizoma*	[337]
	Fern-9(11)-ene	*D. crassirhizoma*	[337]
	Hop-22(29)-ene	*D. championii*	[293]
		D. crassirhizoma	[337]

(continued)

Table 7 (continued)

Compound group	Compound	Species	References
	Hydroxyhopane	*D. crassirhizoma*	[337]
	Isoadiantone	*D. crassirhizoma*	[337]
Others	vitamin E Quinone	*D. fragrans*	[284]
	3S,5R,6R,7E,9S-Megastigman-7-ene-3,5,6,9- tetrol-3-O-β-D-glucopyranoside	*D. fragrans*	[335]
	(6S,9R)-3-oxo-α-ionol-9-O-β-D-glucopyranoside	*D. fragrans*	[335]
	(E)-3-(4-Hydroxyphenyl)acrylic acid	*D. fragrans*	[282]
	(E)-3-Nonacosene-2-ketone	*D. championii*	[293]
	12-Ursen-28-oic acid-3-O-β-D-glucopyranoside	*D. wallichiana*	[307]
	12-Ursen-3-O-β-D-glucopyranoside	*D. wallichiana*	[307]
	3,7,11,15-tetramethyl-2-hexadecen-1-ol	*D. wallichiana*	[307]
	Fragranoside A	*D. fragrans*	[335]
	Norflavesone	*D. fragrans*	[284]

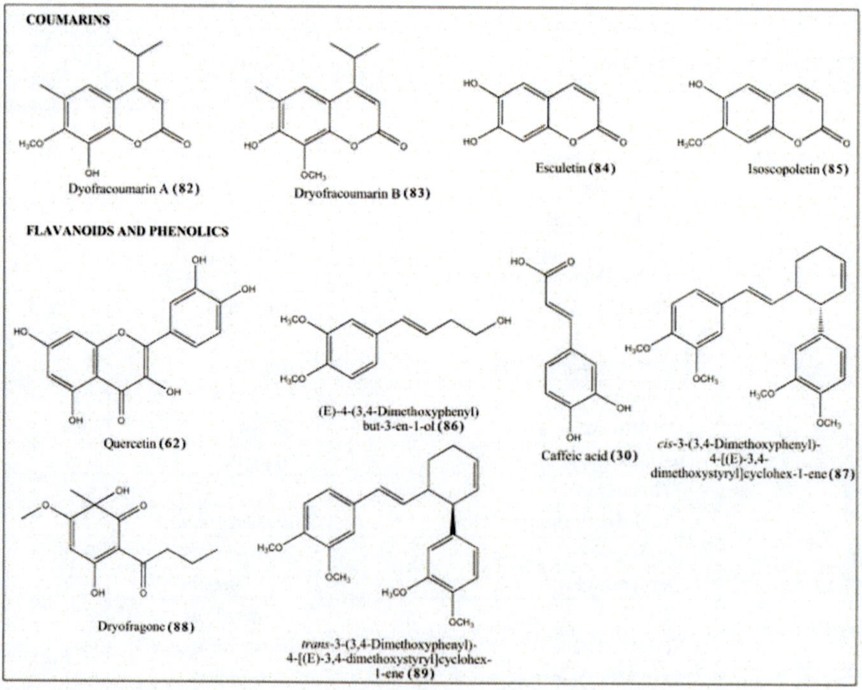

Fig. 11 Major coumarins, flavonoids, and phenolics isolated from *Dryopteris* spp.

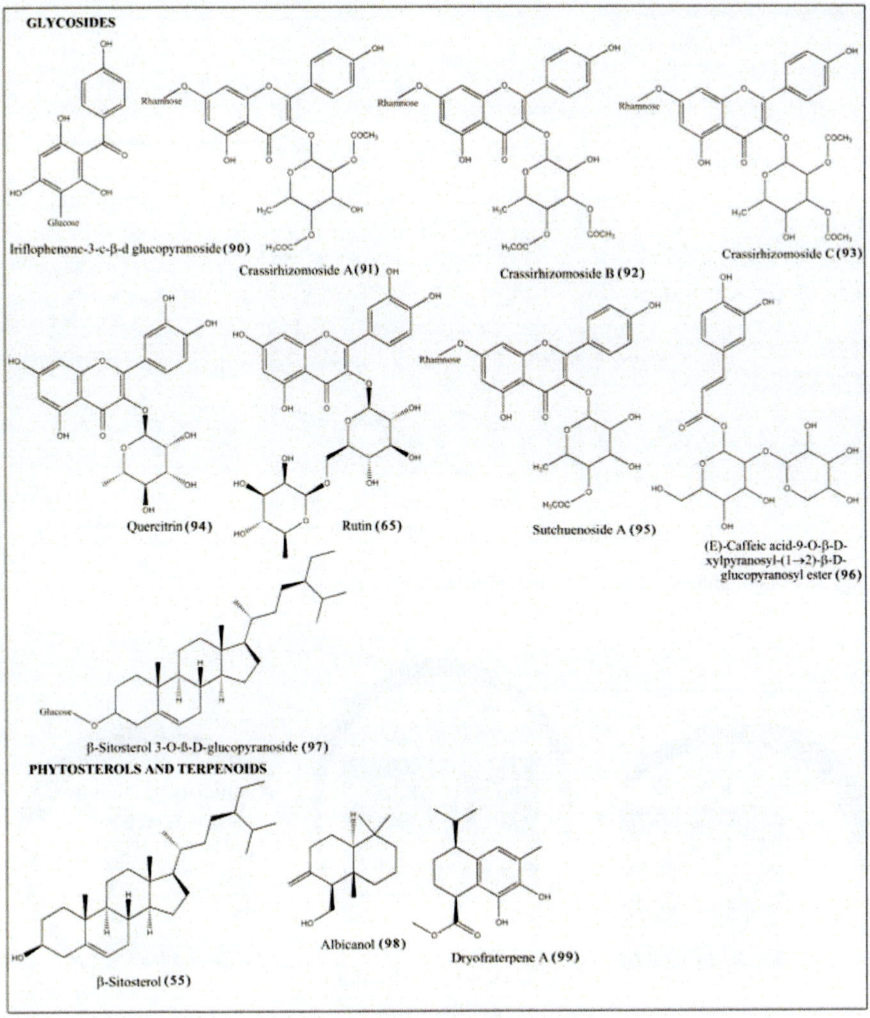

Fig. 12 Major glycosides, phytosterols, and terpenoids isolated from *Dryopteris* spp.

showed that it had an anticandidal action on *Candida albicans*. A cell cycle study was carried out by Jung et al. [157] to examine the effects of amentoflavone on the cellular physiology of *C. albicans*, and their findings revealed that amentoflavone considerably halted cell cycles during the S-phase. Isocryptomerin (**25**), another biflavonoid that was isolated from *Selaginella tamariscina*, was shown by Lee et al. [195] to have antifungal properties. Bis-(1,3-dibutylbarbituric acid) trimethine oxonol (DiBAC4), a translational membrane potential dye, was used in a flow cytometric analysis on the regeneration of the wall material of fungal protoplasts to better understand the method of action of isocryptomerin (**25**). They conducted fluorescence study with 1,6-diphenyl-1,3,5-hexatriene (DPH), a probe for membrane

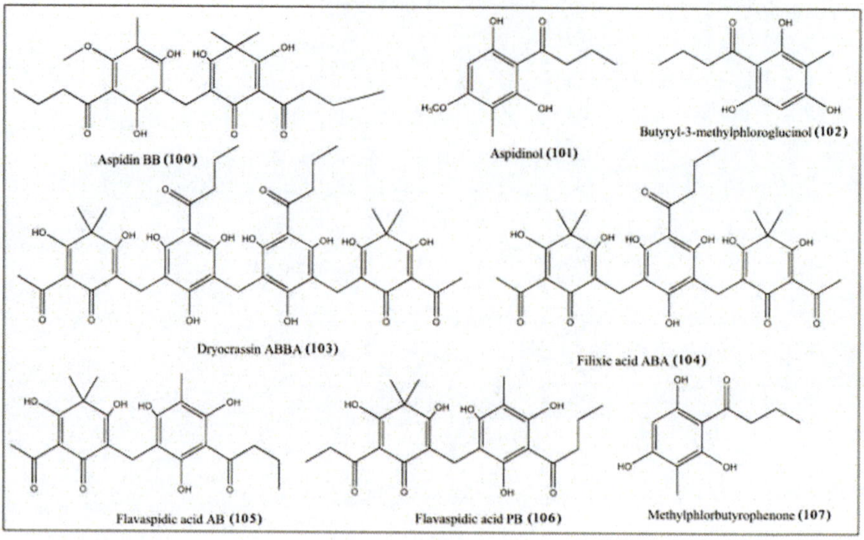

Fig. 13 Major phloroglucinols isolated from *Dryopteris* spp.

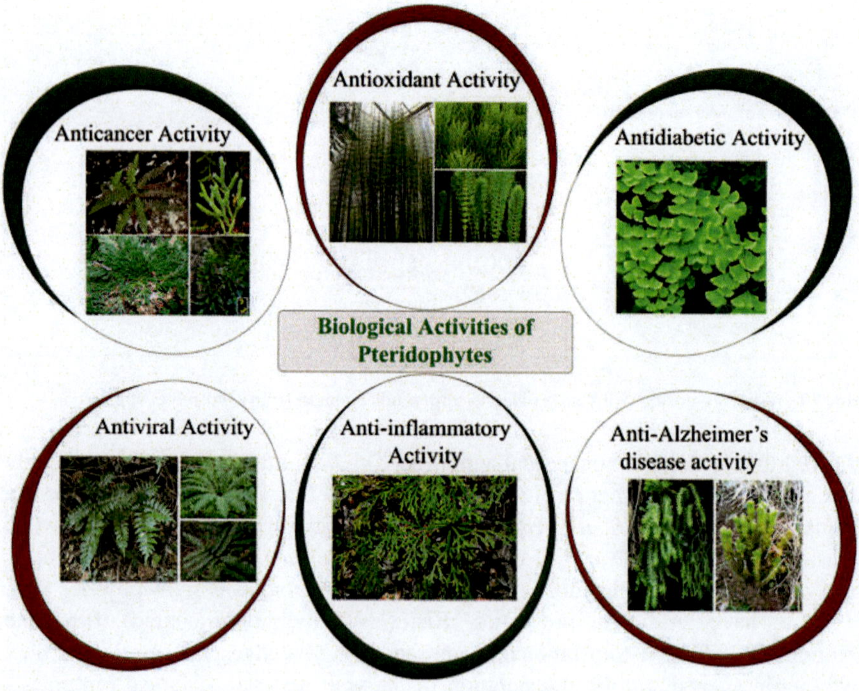

Fig. 14 Important biological activities of selected pteridophytes

studies by depolarization, revealing that isocryptomerin could depolarize fungal membrane supporting the antifungal actions of isocryptomerin (**25**).

4.6 Antibacterial Activity

Ishaque et al. [322] isolated the xanthone C-glycoside isomers mangiferin (MF) and isomangiferin (IsoMF) from *Dryopteris romosa* and tested their antibacterial activities against *Bacillus subtilis* (ATCC 6633), *Staphylococcus aureus* (ATCC 65380), and *Klebsiella pneumoniae* by using the agar disc diffusion method, which demonstrated a very good antibacterial activity of the isolated compounds. Similarly, Ishaque et al. [311] isolated and identified iriflophenone-3-C-D glucopyranoside (**90**) from aqueous fractions of *Dryopteris ramosa* (Hope) C. Chr. Iriflophenone-3-C-D glucopyranoside was tested for antibacterial activity using the agar well diffusion method against five bacterial strains, including *Bacillus subtilis*, *Escherichia coli*, *Klebsiella pneumoniae*, *Salmonella enterica*, and *Staphylococcus aureus*. It was then compared to the widely used antibiotic cefixime. With minimum inhibitory concentrations of 31.1 7.2, 62.5 7.2, and 62.5 7.2 μg/mL against *Klebsiella pneumoniae*, *Staphylococcus aureus*, and *Escherichia coli*, respectively, the iriflophenone-3-C-D glucopyranoside demonstrated significant potential.

Hwang et al. [193] established the antibacterial effectiveness of amentoflavone (**22**) against *Streptococcus mutans*. They looked into how the antibiotics ampicillin, cefotaxime, and chloramphenicol interacted with the amentoflavone. The generation of hydroxyl radicals was also detected using 3′-(p-hydroxyphenyl) fluorescein, and the NAD+ cycling assay was utilized to calculate the NAD+/NADH ratio. Their research suggested that hydroxyl radical generation would have a synergistic effect and this oxidative stress was the result of a transient NADH deficiency. All of the aforementioned research findings demonstrate the phytochemical potential of various pteridophytes.

4.7 Anti-Human Immunodeficiency Virus (HIV-1) Activity

Min et al. [279] isolated kaempferol glycosides, viz., crassirhizomoside A (**91**), crassirhizomoside B (**92**), crassirhizomoside C (**93**), and sutchuenoside A (**95**) (Fig. 12), from rhizomes of *D. crassirhizoma* and tested these chemicals on human immunodeficiency virus reverse transcriptase-associated DNA polymerase [RNA-dependent DNA polymerase (RDDP) and DNA-dependent DNA polymerase (DDDP)] and RNase H activities. Crassirhizomoside A (**91**), crassirhizomoside C (**93**), and sutchuenoside A (**95**) inhibited RDDP with an IC_{50} value of 215, 240, and 405 μM, respectively, and DDDP with an IC_{50} value of 25, 28, and 23 μM, respectively. These are promising results in terms of anti-HIV activities of plant-based phytochemicals.

4.8 Anti-influenza Virus (H5N1) Activity

Wang et al. [302] isolated several phloroglucinols from *D. crassirhizoma* and tested their effect on inhibitory effect on neuraminidase (NA) in vitro. The phloroglucinol compounds, namely, dryocrassin ABBA (**103**) and filixic acid ABA (**104**) exhibited inhibitory effects on NA with IC_{50} as 18.59 ± 4.53 and 29.57 ± 2.48 μM, respectively. These studies suggest the use of these selected compounds to control influenza virus (H5N1) infection.

4.9 Anti-inflammatory Activity

Inflammatory illnesses like inflammatory bowel disease and asthma may be treated using several secondary metabolites that have been identified from *Selaginella* species. From *S. tamariscina*, Shim et al. [162] extracted the bioflavonoids, namely, hinokiflavone (**23**) and 7′-O-methyl hinokiflavone (**21**) and tested their anti-inflammatory properties in colon epithelial cells and lipopolysaccharide (LPS)-mediated murine macrophages (RAW 264.7) (HT-29). They demonstrated that the inflammatory mediator's nitric oxide (NO), interleukin (IL)-6, IL-8, and tumor necrosis factor (TNF), which are most highly active in inflammatory bowel disease, were inhibited by both hinokiflavone (**23**) and 7′-O-methyl hinokiflavone, respectively. Additionally, they showed through Western blot analysis that hinokiflavone and 7′-O-methyl hinokiflavone inhibited the expression of inducible nitric oxide synthase (iNOS) and cyclooxygenase (COX)-2 as well as the activation of nuclear factor-B (NF-B) and extracellular regulated kinases (ERK) 1/2 when LPS was present. These results show that hinokiflaovne (**23**) and 7-O-methyl hinokiflavone (**21**) are substances possessing powerful anti-inflammatory actions that could be employed to treat different anti-inflammatory-related disorders.

4.10 Antioxidant Activity

Among varied *Equisetum* species, *E. arvense* is popular in culinary and medicinal preparations. The aerial portions of the plant are consumed as food and also used in the preparation of herbal tea in Japan [350]. *E. arvense* is reported to be rich in phenolics, alkaloids, and phytosterols, which are accountable for antioxidant, anti-inflammatory, and vasorelaxant activities [212, 351]. Čanadanović-Brunet et al. [352] evaluated the radical scavenging activity of *n*-butanol, ethyl acetate, and water extracts of *E. arvense* and reported the highest 2,2-diphenyl-1-picrylhydrazyl ($EC_{50} = 0.65$ mg/ml) and hydroxyl radical scavenging activities ($EC_{50} = 0.74$ mg/ml) with *n*-butanol extract. Through high-performance liquid chromatographic (HPLC) analysis, they showed the presence of caffeic acid (**30**), ferulic acid (**31**), syringic acid (**32**), vanillic acid (**33**), rutin (**65**), and procatechuic acid (**67**) in the *n*-butanol extract which are responsible for antioxidant activities. Another study [221] tested the antioxidant activity of *E. telmateia* and characterized the polyphenols present in extracts by HPLC analysis. Correia et al. [221] evaluated *E. telematia* aqueous and ethyl acetate extracts by 1,1-dipehnyl-2-picrylhydrazyl (DPPH), trolox equivalent antioxidant capacity (TEAC), and thiobarbituric acid reactive substances (TBARS) assays. They reported a high and significant antioxidant activity in the ethyl acetate fraction and through HPLC analysis they showed the presence of phenolic compounds, such as kaempferol (**60**) and its derivatives which were responsible for the antioxidant activities. Olazarán-Santibañez et al. [353] demonstrated antioxidant activity of ethanolic extract of *E. myriochaetum* using DPPH assay. Through ultra-performance liquid chromatography analysis, they showed that a particular ethanol fraction of *E. myriochaetum* contained the flavonoids apigenin (**59**), kaempferol (**60**), and quercetin (**62**) which were responsible for antioxidant activities.

4.11 Hepatoprotective Activities

The liver is one of the most important organs in the human body and performs a fundamental role in the regulation of vital functions, such as metabolism, secretion, and storage. Varied biological factors, such as bacteria, viruses, and parasites; autoimmune diseases; and toxic substances are responsible for hepatic diseases. Many of the plant-based phytochemicals are reported to have hepatoprotective activities. Oh et al. [212] demonstrated the hepatoprotective activity of methanolic extract of *Equisetum arvense*. They isolated and characterized two phenolic petrosins, onitin (**68**) and onitin-9-*O*-glucoside (**66**), and four flavonoids, apigenin (**59**), luteolin (**61**), kaempferol-3-O-glucoside (**63**), and quercetin-3-O-glucoside (**64**) in methanolic fractions of *E. arvense*. Among them, onitin (**68**) and luteolin (**61**) containing fractions exhibited hepatoprotective activities on tacrine-induced cytotoxicity in human liver-derived HepG2 cells, displaying EC_{50} values of 85.8 ± 9.3 μM and 20.2 ± 1.4 μM, respectively. These results support the use of *E. arvense* for the treatment of hepatitis in different systems of traditional medicine.

4.12 Antidiabetic Activity

In streptozotocin-induced diabetic rats, *Adiantum capillus-veneris* was discovered to have good anti-diabetic action. In streptozotocin-induced diabetic rats, Ranjan et al. [354] tested the effectiveness of various concentrations (100–400 mg/kg/day) of aqueous and methanol extracts of *A. capillus-veneris* and the common drug metformin (50 mg/kg/day), and they found that the aqueous extract (100 mg/kg/day) had the strongest anti-diabetic effects. An oral glucose tolerance test on rabbits loaded with glucose was used in another investigation to determine the impact of the anti-hyperglycemic activity of alcoholic extracts of *A. capillus-veneris* [245]. Glipizide (8 mg/kg body weight) was utilized as the standard reference medication. The alcoholic extract demonstrated a considerable hypoglycemic impact when administered (600 mg/kg body weight) 30 min before glucose loading. The inclusion of flavonoids, which are recognized for their hypoglycemic effects, may be the cause of the alcoholic extract's hypoglycaemic effects [245].

4.13 Larvicidal Activity

One of the most destructive pests of palms, especially the coconut palm, is the *Oryctes rhinoceros* L. (Coleoptera) rhinoceros beetle. The adult beetles bore holes in the palms' crowns to feed on the young fronds and inflorescences, which stunt the growth, yield, and productivity of the palms. Pradeep Kumar et al. [270] isolated the triterpenoid component 22-hydroxyhopane (hopan-22-ol) (**80**) from *Adiantum latifolium* L., which exhibited larvicidal efficacy against the pest (LC_{50} value 20.81 μg/g). Additionally, this substance showed antibacterial action against the symbiotic gut bacteria of the midgut tissues and prevented the secretion of digesting enzymes like protease, amylase, and trehalose, which caused weight loss in the larvae and hampered their ability to transform. As a result, 22-hydroxyhopane (hopan-22-ol) (**80**) is a promising phytochemical for the management of the rhinoceros beetle pest; however, an in-depth research is required for the effective application of this compound for pest management.

In a different investigation, Pradeep Kumar and Siddique [355] used the molecular docking technique to assess the binding affinity of 22-hydroxyhopane (hopan-22-ol) (**80**) for proteins required for SARS-CoV-2 growth in host cells. According to their research, 22-hydroxyhopane is unique to six enzymes, an RNA binding protein, a spike protein, a membrane protein, and the ACE2 receptor of SARS-CoV-2. These findings imply that 22-hydroxyhopane might be useful as a treatment medication against the SARS virus, but more research is necessary.

4.14 Regulation of Hyperthyroidism

A disorder known as hypothyroidism (underactive thyroid) occurs when the thyroid gland fails to generate enough of the number of important hormones. Early on, hypothyroidism may not show any obvious signs. Obesity, joint discomfort, infertility, and heart disease are just a few of the health issues that untreated hypothyroidism can lead to over time. To determine thyroid gland weight, thyroid peroxidase activity, and an estimate of the concentration of total thyroid hormones, such as thyroxine (T4), triiodothyronine (T3), and thyroid-stimulating hormone (TSH) in the serum of experimental mice, Vijayalakshmi and Kiran Kumar [356] studied the effect of an ethanol extract of *A. capillus-veneris* on thyroid dysfunction – hypothyroidism. They gave the mice a 500 mg/kg dose of an ethanol extract of *A. capillus-veneris*, and found that this treatment reduced the weight of the thyroid gland while increasing thyroid peroxidase activity, serum T4 and T3 levels, and decreasing serum TSH levels significantly ($p \geq 0.01$) when compared to hypothyroid control animals. Vijayalakshmi and Kiran Kumar [356] demonstrated using the high-performance thin-layer chromatography method that the ethanol extract of

A. capillus-veneris contains polyphenolics including quercetin and gallic acid. According to the investigations, the ethanolic extract of *A. capillus-veneris* may be used to treat hypothyroidism.

4.15 Antinociceptive Activity

The body's reaction to potentially hazardous stimuli, such as dangerous chemicals (such as formalin), mechanical injury (such as cutting or crushing), or extreme temperatures (such as heat and cold), is known as antinociception, sometimes known as nociception or nociperception. Researchers are looking at phytochemicals that are powerful analgesic agents without any side effects because many analgesic pharmaceuticals created through chemical synthesis have possible unwanted effects. Ali et al. [316] extracted kaempferol-3,4′-di-O-L-rhamnopyranoside from *Dryopteris cycadina* and assessed its *in vivo* antinociceptive efficacy in experimental mice. With a peak antinociceptive activity of 46.12% at 10 mg/kg i.p. against acetic acid-induced writhing, kaempferol-3,4′-di-O-L-rhamnopyranoside demonstrated dose-dependent antinociceptive effects. Additionally, it demonstrated dose-dependent blocking of noxious stimulation in both phases of the formalin test, with respective percentages of 40.78 and 43.44 in the first and second phases at 10 mg/kg i.p. With more research, it may be possible to turn this substance into a painkiller.

4.16 Anti-platelet Activity

Platelets are an important component of the initial response to vascular endothelial injury; however, platelet dysfunction induces the acute clinical symptoms of thrombotic disorders, which trigger severe cardiovascular diseases, such as myocardial infarction, ischemia, and stroke. Yim et al. [310] isolated a phloroglucinol derivative butyryl-3-methylphloroglucinol (**102**) from *Dryopteris crassirhizoma* and investigated its inhibitory activity in the collagen and arachidonic acid -AA) induced platelet aggregation. Butyryl-3-methylphloroglucinol (**102**) showed inhibition ratios of 92.36% and 89.51% in the collagen and AA-induced platelet aggregation, respectively, without any cytotoxicity. The above results support the use of butyryl-3-methylphloroglucinol for antiplatelet remedies.

5 Conclusions

Pteridophytes are popular as edible plants especially fiddlehead, which are proved to be rich in nutrients, such as carbohydrates, proteins, fats, minerals, and amino acids. Pteridophytes are abundant with phytochemicals including polyphenols, alkaloids, and terpenoids. Phytochemicals isolated from pteridophytes have proven to possess antioxidant, anti-cancer, anti-diabetic, anti-inflammatory, anti-microbial, and neuroprotective effects. Because of the above, pteridophytes are proved to be the future plants for human usage.

References

1. The Plant List (2022). http://www.theplantlist.org/browse/P/. Accessed 6 June 2022
2. Smith AR, Pryer KM, Schuettpelz E, Korall P, Schneider H, Wolf PG (2006) A classification for extant ferns. Taxon 55:705–731. https://doi.org/10.2307/25065646
3. Kawano T (2015) Pteridophytes as active components in gardening, agricultural and horticultural ecosystems in Japan. Adv Hortic Sci 29:41–47
4. Dvorakova M, Pumprova K, Antonínova Z, Rezek J, Haisel D, Ekrt L, Vanek T, Langhansova L (2021) Nutritional and antioxidant potential of fiddleheads from European ferns. Foods 10: 460. https://doi.org/10.3390/foods10020460
5. Maroyi A (2014) Not just minor wild edible forest products: consumption of pteridophytes in sub-Saharan Africa. J Ethnobiol Ethnomed 10:78. https://doi.org/10.1186/1746-4269-10-78
6. Yumkham SD, Chakpram L, Salam S, Bhattacharya MK, Singh PK (2017) Edible ferns and fern–allies of North East India: a study on potential wild vegetables. Genet Resour Crop Evol 64:467–477. https://doi.org/10.1007/s10722-016-0372-5
7. Baskaran X, Geo Vigila A, Zhang S, Feng S, Liao W (2018) A review of the use of pteridophytes for treating human ailments. J Zhejiang Univ-Sci B 19:85–119. https://doi.org/10.1631/jzus.B1600344
8. Goswami HK, Sen K, Mukhopadhyay R (2016) Pteridophytes: evolutionary boon as medicinal plants. Plant Genetic Resour 14:328–355. https://doi.org/10.1017/S1479262116000290
9. Cao H, Chai TT, Wang X, Morais-Braga MFB, Yang JH, Wong FC, Wang R, Yao H, Cao J, Cornara L, Burlando B, Wang Y, Xiao J, Coutinho HDM (2017) Phytochemicals from fern species: potential for medicine applications. Phytochem Rev 16:379–440. https://doi.org/10.1007/s11101-016-9488-7
10. Zhu QF, Zhao QS (2019) Chemical constituents and biological activities of lycophytes and ferns. Chin J Nat Med 17:887–891. https://doi.org/10.1016/S1875-5364(19)30108-6
11. DeLong JM, Hodges DM, Prange R, Forney C, Toivenon P, Bishop MC, Elliot M, Jordan M (2011) The unique fatty acid and antioxidant composition of ostrich fern (*Matteuccia struthiopteris*) fiddleheads. Can J Plant Sci 91:919–930. https://doi.org/10.4141/cjps2010-042
12. Liu Y, Wujisguleng W, Long C (2012) Food uses of ferns in China: a review. Acta Soc Bot Pol 81:263–270
13. Lee CH, Shin SL (2011) Functional activities of ferns for human health. In: Kumar A, Fernandez H, Revilla MA (eds) Working with ferns. Springer, New York, pp 347–359
14. Madeja J, Harmata K, Kolaczek P, Karpinska-Kolaczek M, Piatek K, Naks P (2009) Bracken [*Pteridium aquilinum* (L.) (Kuhn)], mistletoe [*Viscum album* (L.)] and bladder-nut [*Staphylea pinnata* (L.)] - mysterious plants with unusual applications. Cultural and ethnobotanical studies. In: Morel JP, Mercuri AM (eds) Plants, and culture; seeds of the cultural heritage of Europe. pp 207–215
15. Bassey ME, Etuk EUI, Ibe MM, Ndon BA (2001) *Diplazium sammatii*: Athyraceae ('Nyama idim'): age-related nutritional and antinutritional analysis. Plant Foods Hum Nutr 56:7–12. https://doi.org/10.1023/A:1008185513685
16. Sareen B, Bhattacharya A, Srivatsan V (2021) Nutritional characterization and chemical composition of *Diplazium maximum* (D. Don) C. Chr. J Food Sci Technol 58:844–854. https://doi.org/10.1007/s13197-020-04598-w
17. Anderson JW, Baird P, Davis RH Jr, Ferreri S, Knudtson M, Koraym A, Waters V, Williams CL (2009) Health benefits of dietary fiber. Nutr Rev 67:188–205. https://doi.org/10.1111/j.1753-4887.2009.00189.x
18. Chettri S, Manivannan S, Muddarsu VR (2018) Nutrient and elemental composition of wild edible ferns of the Himalaya. Am Fern J 108:95–106. https://doi.org/10.1640/0002-8444-108.3.95

19. Oloyede F, Makinde A, Ajayi O (2012) Proximate analysis, nutritional and anti-nutritional compositions of a tropical fern, *Nephrolepis furcans* in Nigeria. Acta Bot Hungar 54:345–354. https://doi.org/10.1556/ABot.54.2012.3-4.12
20. Oloyede FA, Alafe BB, Oloyede FM (2008) Nutrient evaluation of *Nephrolepis biserrata* (nephrolepidiaceae, Pteridophyta). Botanica Lithuanica 14:207–210
21. Lognay G, Haubruge E, Delcarte E, Wathelet B, Mathieu F, Marlier M, Malaisse F (2008) *Ophioglossum polyphyllum* A. Braun in Seub. (Ophioglossaceae, Pteridophyta), a rare potherb in south central Tibet (T.A.R., P.R. China). Geo-Eco-Trop 32:9–16
22. Longvah T, Ananthan R, Bhaskaracharya K, Venkaiah K (2017) Indian food composition tables. National Institute of Nutrition, Hyderabad
23. Ayer WA, Trifonov LS (1994) Lycopodium alkaloids. In: Cordell GA, Brossi A (eds) The alkaloids: chemistry and pharmacology. Academic Press, San Diego, pp 233–266
24. Hirasawa Y, Morita H, Kobayashi J (2004) Nankakurine A, a Novel C16N2 -type alkaloid from *Lycopodium hamiltonii*. Org Lett 6:3389–3391. https://doi.org/10.1021/ol048621a
25. Ayer WA, Altenkirk B, Valverde-Lopez S, Douglas B, Raffaus RF, Weisbach JA (1968) The alkaloids of *Lycopodium alopecuroides* L. Can J Chem 46:15–20
26. Burnell RH, Mootoo BS (1961) Lycopodium alkalloids. Part IV. Alkaloids of jamaican *Lycopodium clavatum* Linn. Can J Chem 39:1090–1093
27. Castillo M, Loyola LA, Morales G, Singh I, Calvo C, Holland HL, Maclean DB (1976) The alkaloids of *L. magellanicum* and the structure of magellanine. Can J Chem 54:2893–2899
28. Castillo M, Morales G, Loyola LA, Singh I, Calvo C, Holland HL, MacLean DB (1976) The alkaloids of *L. paniculatum* and the structure of paniculatine. Canadian J Chem 54:2900–2908
29. Chen Y, He H-W, Mei Z-N, Yang GZ (2014) Lycopodium alkaloids from *Lycopodium obscurum* L. Helv Chim Acta 97:519–523. https://doi.org/10.1002/hlca.201300243
30. He J, Chen XQ, Li MM, Zhao Y, Xu G, Cheng X, Peng LY, Xie MJ, Zheng YT, Wang YP, Zhao QS (2009) Lycojapodine A, a novel alkaloid from *Lycopodium japonicum*. Org Lett 11: 1397–1400. https://doi.org/10.1021/ol900079t
31. Inubushi Y, Tsuda Y, Sano T (1962) Studies on the constituents of domestic *Lycopodium* genus plants. I. Yakugaku Zasshi 82:1537–1541. https://doi.org/10.1248/yakushi1947.82.11_1537
32. Ishiuchi K, Jiang WP, Fujiwara Y, Wu JB, Kitanaka S (2016) Serralongamines B–D, three new Lycopodium alkaloids from *Lycopodium serratum* var. *longipetiolatum* , and their inhibitory effects on foam cell formation in macrophages. Bioorg Med Chem Lett 26:2636–2640. https://doi.org/10.1016/j.bmcl.2016.04.019
33. Jiang WW, Liu YC, Zhang ZJ, Liu YC, He J, Su J, Cheng X, Peng LY, Shao L-D, Wu XD, Yang J-H, Zhao QS (2016) Obscurumines HP, new Lycopodium alkaloids from the club moss *Lycopodium obscurum*. Fitoterapia 109:155–161. https://doi.org/10.1016/j.fitote.2015.12.017
34. Johns SR, Lamberton JA, Sioumis AA (1969) Alkaloids of *Lycopodium volubile* (Lycopodiaceae). Aust J Chem 22:1317–1317
35. Loyola LA, Morales G, Castillo M (1979) Alkaloids of *Lycopodium magellanicum*. Phytochemistry 18:1721–1723. https://doi.org/10.1016/0031-9422(79)80193-4
36. Mandal SK, Biswas R, Bhattacharyya SS, Paul S, Dutta S, Pathak S, Khuda-Bukhsh AR (2010) Lycopodine from *Lycopodium clavatum* extract inhibits proliferation of HeLa cells through induction of apoptosis via caspase-3 activation. Eur J Pharmacol 626:115–122. https://doi.org/10.1016/j.ejphar.2009.09.033
37. Manske RHF (1953) The alkaloids of *Lycopodium* species. XII. *Lycopodium densum*. Canadian J Chem 31:894–895
38. Miller N, Mees F, Braekman JC (1971) Alcaloïdes de *Lycopodium alpinum*. Phytochemistry 10:1931–1934. https://doi.org/10.1016/S0031-9422(00)86462-6
39. Takayama H, Katakawa K, Kitajima M, Yamaguchi K, Aimi N (2003) Ten new Lycopodium alkaloids having the lycopodane skeleton isolated from *Lycopodium serratum* THUNB. Chem Pharm Bull 51:1163–1169. https://doi.org/10.1248/cpb.51.1163
40. Wang XJ, Li L, Si YK, Yu SS, Ma SG, Bao XQ, Zhang D, Qu J, Liu YB, Li Y (2013) Nine new lycopodine-type alkaloids from *Lycopodium japonicum* Thunb. Tetrahedron 69:6234–6240. https://doi.org/10.1016/j.tet.2013.05.028

41. Wang LJ, Xiong J, Wang W, Zhang HY, Yang GX, Hu JF (2016) Lycopodium alkaloids from *Lycopodium obscurum* L. f. *strictum*. Phytochem Lett 15:260–264. https://doi.org/10.1016/j.phytol.2016.02.001
42. Yang Q, Zhu Y, Peng W, Zhan R, Chen Y (2016) A New Lycopodine-type alkaloid from *Lycopodium japonicum*. Nat Prod Res 30:2220–2224. https://doi.org/10.1080/14786419.2016.1146885
43. Zhang XY, Dong LB, Liu F, Wu XD, He J, Peng LY, Luo HR, Zhao QS (2013) New Lycopodium alkaloids from *Lycopodium obscurum*. Nat Products Bioprospect 3:52–55. https://doi.org/10.1007/s13659-013-0015-x
44. Zhu Y, Dong LB, Zhang ZJ, Fan M, Zhu QF, Qi YY, Liu YC, Peng LY, Wu XD, Zhao QS (2019) Three new Lycopodium alkaloids from *Lycopodium japonicum*. J Asian Nat Prod Res 21:17–24. https://doi.org/10.1080/10286020.2018.1427075
45. Hirasawa Y, Tanaka T, Kobayashi J, Kawahara N, Goda Y, Morita H (2008) Malycorins A-C, new Lycopodium alkaloids from *Lycopodium phlegmaria*. Chem Pharm Bull 56:1473–1476. https://doi.org/10.1248/cpb.56.1473
46. Ayer WA, Browne LM, Orszanska H, Valenta Z, Liu JS (1989) Alkaloids of *Lycopodium selago.* On the identity of selagine with huperzine A and the structure of a related alkaloid. Can J Chem 67:1538–1540
47. Wang B, Guan C, Fu Q (2022) The traditional uses, secondary metabolites, and pharmacology of *Lycopodium* species. Phytochem Rev 21:1–79. https://doi.org/10.1007/s11101-021-09746-4
48. Ham YM, Yoon WJ, Park SY, Jung YH, Kim D, Jeon YJ, Wijesinghe WAJP, Kang SM, Kim KN (2012) Investigation of the component of *Lycopodium serratum* extract that inhibits proliferation and mediates apoptosis of human HL-60 leukemia cells. Food Chem Toxicol 50:2629–2634. https://doi.org/10.1016/j.fct.2012.05.019
49. Zhang Z, ElSohly HN, Jacob MR, Pasco DS, Walker LA, Clark AM (2002) Natural products inhibiting *Candida albicans* secreted aspartic proteases from *Lycopodium cernuum*. J Nat Prod 65:979–985. https://doi.org/10.1021/np0200616
50. Yang Q, Zhu Y, Zhan R, Chen Y (2018) A new Fawcettimine-related alkaloid from *Lycopodium japonicum*. Chem Nat Compd 54:729–731. https://doi.org/10.1007/s10600-018-2456-2
51. Yin S, Fan CQ, Wang XN, Yue JM (2006) Lycodine type alkaloids from *Lycopodium casuarinoid*es. Helv Chim Acta 89:138–143. https://doi.org/10.1002/hlca.200690006
52. Pan K, Luo JG, Kong LY (2013) Two new Lycopodium alkaloids from *Lycopodium obscurum*. Helv Chim Acta 96:1197–1201. https://doi.org/10.1002/hlca.201200505
53. Pongpamorn P, Wan-erlor S, Ruchirawat S, Thasana N (2016) Lycoclavatumide and 8β,11-α-dihydroxylycopodine, a new fawcettimine and lycopodine-type alkaloid from *Lycopodium clavatum*. Tetrahedron 72:7065–7069. https://doi.org/10.1016/j.tet.2016.09.046
54. Ayer WA, Altenkirk B, Masaki N, Valverde-Lopez S (1969) Alkaloids of *Lycopodiurn alopecuroides*. Part 2. Alopecurine, a new type of lycopodium alkaloid. Can J Chem 47: 2449–2455
55. Ayer WA, Law DA (1962) Lycopodium alkaloids. IV. The constituents and stereochemistry of lycoclavine, an alkaloid of *Lycopodium clavatum* var. *megastachyon*. Can J Chem 40:2088–2100
56. Morales G, Loyola LA, Castillo M (1979) Alkaloids of *Lycopodium paniculatum*: the structure of paniculine. Phytochemistry 18:1719–1720. https://doi.org/10.1016/0031-9422(79)80192-2
57. Morita H, Hirasawa Y, Kobayashi J (2005) Lycopodatines A–C, C16N alkaloids from *Lycopodium inundatum*. J Nat Prod 68:1809–1812
58. Anet FAL, Khan NH (1959) Alkaloids of *Lycopodium annotinum*. Part II. Isolation of four new alkaloids. Can J Chem 37:1589–1596
59. Koyama K, Morita H, Hirasawa Y, Yoshinaga M, Hoshino T, Obara Y, Nakahata N, Kobayashi J (2005) Lannotinidines A–G, new alkaloids from two species of *Lycopodium*. Tetrahedron 61: 3681–3690. https://doi.org/10.1016/j.tet.2005.02.016
60. Tang Y, Xiong J, Zhang JJ, Wang W, Zhang HY, Hu JF (2016) Annotinolides A–C, Three lycopodane-derived 8,5-lactones with polycyclic skeletons from *Lycopodium annotinum*. Org Lett 18:4376–4379. https://doi.org/10.1021/acs.orglett.6b02132

61. Morita H, Ishiuchi K, Haganuma A, Hoshino T, Obara Y, Nakahata N, Kobayashi J (2005) Complanadine B, obscurumines A and B, new alkaloids from two species of *Lycopodium*. Tetrahedron 61:1955–1960. https://doi.org/10.1016/j.tet.2005.01.011
62. Pan K, Luo JG, Kong LY (2013) Three new Lycopodium alkaloids from *Lycopodium obscurum*. J Asian Nat Prod Res 15:441–445. https://doi.org/10.1080/10286020.2013.780045
63. Burnell RH, Mootoo BS, Taylor DR (1960) Alkaloids of *Lycopodium fawcettii*. Part II. Can J Chem 38:1927–1932
64. Katakawa K, Kogure N, Kitajima M, Takayama H (2009) A new Lycopodium alkaloid, lycoposerramine-R, with a novel skeleton and three new fawcettimine-related alkaloids from *Lycopodium serratum*. Helv Chim Acta 92:445–452. https://doi.org/10.1002/hlca.200800327
65. Ayer WA, Altenkirk B, Burnell RH, Moinas M (1969) Alkaloids of *Lycopodium lucidulum* Michx. structure and properties of alkaloid L.23. Can J Chem 47:449–455
66. Sun Y, Yan J, Meng H, He CL, Yi P, Qiao Y, Qiu MH (2008) A new alkaloid from *Lycopodium japonicum* Thunb. Helv Chim Acta 91:2107–2109. https://doi.org/10.1002/hlca.200890225
67. Ayer WA, Altenkirk B (1969) Structure of lucidioline. An alkaloid of *Lycopodium lucidulum*. Can J Chem 47:499–502
68. Ayer WA, Iverach GG (1962) The structure and stereochemistry of lycodoline (lycopodium alkaloid L.8). Tetrahedron Lett 3:87–92. https://doi.org/10.1016/S0040-4039(00)71105-1
69. Hirasawa Y, Morita H, Kobayashi J (2002) Lyconesidines A–C, new alkaloids from *Lycopodium chinense*. Tetrahedron 58:5483–5488. https://doi.org/10.1016/S0040-4020(02)00520-3
70. Morita H, Hirasawa Y, Shinzato T, Kobayashi J (2004) New phlegmarane-type, cernuane-type, and quinolizidine alkaloids from two species of *Lycopodium*. Tetrahedron 60:7015–7023. https://doi.org/10.1016/j.tet.2003.09.106
71. Anet FAL, Haq MZ, Khan NH, Ayer WA, Hayatsu R, Valverde-Lopez S, Deslongchamps P, Riess W, Ternbah M, Valenta Z, Wiesner K (1964) The structure of lyconnotine: a novel lycopodium alkaloid. Tetrahedron Lett 5:751–757. https://doi.org/10.1016/0040-4039(64)83031-8
72. Ayer WA, Masaki N (1971) Alkaloids of *Lycopodium alopecuroides*. Part 4. The structure of lycopecurine. Can J Chem 49:524–527
73. Zhang ZJ, Zhu QF, Su J, Wu XD, Zhao QS (2018) Lycoplanines B-D, three Lycopodium alkaloids from *Lycopodium complanatum*. Nat Products Bioprospect 8:177–182. https://doi.org/10.1007/s13659-018-0161-2
74. Kogure N, Maruyama M, Wongseripipatana S, Kitajima M, Takayama H (2016) New lycopodine-type alkaloids from *Lycopodium carinatum*. Chem Pharm Bull 64:793–799. https://doi.org/10.1248/cpb.c16-00171
75. Bishayee K, Chakraborty D, Ghosh S, Boujedaini N, Khuda-Bukhsh AR (2013) Lycopodine triggers apoptosis by modulating 5-lipoxygenase, and depolarizing mitochondrial membrane potential in androgen sensitive and refractory prostate cancer cells without modulating p53 activity: signaling cascade and drug–DNA interaction. Eur J Pharmacol 698:110–121. https://doi.org/10.1016/j.ejphar.2012.10.041
76. Morita H, Arisaka M, Yoshida N, Kobayashi J (2000) Serratezomines A–C, new alkaloids from *Lycopodium serratum* var. *serratum*. J Org Chem 65:6241–6245. https://doi.org/10.1021/jo000661e
77. Kobayashi J, Hirasawa Y, Yoshida N, Morita H (2001) Lyconadin A, a novel alkaloid from *Lycopodium complanatum*. J Org Chem 66:5901–5904. https://doi.org/10.1021/jo0103874
78. Choo CY, Hirasawa Y, Karimata C, Koyama K, Sekiguchi M, Kobayashi J, Morita H (2007) Carinatumins A–C, new alkaloids from *Lycopodium carinatum* inhibiting acetylcholinesterase. Bioorg Med Chem 15:1703–1707. https://doi.org/10.1016/j.bmc.2006.12.005
79. Liu F, Wu XD, He J, Deng X, Peng LY, Luo HR, Zhao QS (2013) Casuarines A and B, Lycopodium alkaloids from *Lycopodium casuarinoides*. Tetrahedron Lett 54:4555–4557. https://doi.org/10.1016/j.tetlet.2013.06.083
80. Kobayashi J, Hirasawa Y, Yoshida N, Morita H (2000) Complanadine A, a new dimeric alkaloid from *Lycopodium complanatum*. Tetrahedron Lett 41:9069–9073. https://doi.org/10.1016/S0040-4039(00)01630-0

81. Ishiuchi K, Kubota T, Mikami Y, Obara Y, Nakahata N, Kobayashi J (2007) Complanadines C and D, new dimeric alkaloids from *Lycopodium complanatum*. Bioorg Med Chem 15:413–417. https://doi.org/10.1016/j.bmc.2006.09.043
82. Ishiuchi K, Kubota T, Ishiyama H, Hayashi S, Shibata T, Mori K, Obara Y, Nakahata N, Kobayashi J (2011) Lyconadins D and E, and complanadine E, new Lycopodium alkaloids from *Lycopodium complanatum*. Bioorg Med Chem 19:749–753. https://doi.org/10.1016/j.bmc.2010.12.025
83. Gerard RV, MacLean DB, Fagiani R, Lock CJ (1986) Fastigiatine, a *Lycopodium* alkaloid with a new ring system. Can J Chem 64:943–949
84. Burnell RH, Chin CG, Mootoo BS, Taylor DR (1963) Lycopodium alkaloids. Part VIII. New alkaloids from Jamaican *Lycopodium* species. Can J Chem 41:3091–3094
85. Morita H, Hirasawa Y, Kobayashi J (2003) Himeradine A, a Novel C27N3-type alkaloid from *Lycopodium chinense*. J Org Chem 68:4563–4566. https://doi.org/10.1021/jo034294t
86. Yeap JSY, Lim KH, Yong KT, Lim SH, Kam TS, Low YY (2019) Lycopodium alkaloids: lycoplatyrine A, an unusual lycodine–piperidine adduct from *Lycopodium platyrhizoma* and the absolute configurations of lycoplanine D and lycogladine H. J Nat Prod 82:324–329. https://doi.org/10.1021/acs.jnatprod.8b00754
87. Anet FAL, Eves CR (1958) Lycodine, a new alkaloid of *Lycopodium annotinum*. Can J Chem 36:902–909
88. Hirasawa Y, Kato E, Kobayashi J, Kawahara N, Goda Y, Shiro M, Morita H (2008) Lycoparins A–C, new alkaloids from *Lycopodium casuarinoides* inhibiting acetylcholinesterase. Bioorg Med Chem 16:6167–6171. https://doi.org/10.1016/j.bmc.2008.04.044
89. Ishiuchi K, Kubota T, Hayashi S, Shibata T, Kobayashi J (2009) Lycopladines F and G, new C16N2-type alkaloids with an additional C4N unit from *Lycopodium complanatum*. Tetrahedron Lett 50:4221–4224. https://doi.org/10.1016/j.tetlet.2009.04.139
90. Shen YC, Chen CH (1994) Alkaloids from *Lycopodium casuarinoides*. J Nat Prod 57:824–826. https://doi.org/10.1021/np50108a021
91. Ayer WA, Habgood TE, Deulofeu V, Juliani HR (1965) Lycopodium alkaloids - Sauroxine. Tetrahedron 21:2169–2172. https://doi.org/10.1016/S0040-4020(01)98352-8
92. Ayer WA, Berezowsky JA, Iverach GG (1962) Lycopodium alkaloids–II. Tetrahedron 18:567–573. https://doi.org/10.1016/S0040-4020(01)92707-3
93. Wang XJ, Li L, Yu SS, Ma SG, Qu J, Liu YB, Li Y, Wang Y, Tang W (2013) Five new fawcettimine-related alkaloids from *Lycopodium japonicum* Thunb. Fitoterapia 91:74–81. https://doi.org/10.1016/j.fitote.2013.08.007
94. Katakawa K, Nozoe A, Kogure N, Kitajima M, Hosokawa M, Takayama H (2007) Fawcettimine-related alkaloids from *Lycopodium serratum*. J Nat Prod 70:1024–1028. https://doi.org/10.1021/np0700568
95. Katakawa K, Mito H, Kogure N, Kitajima M, Wongseripipatana S, Arisawa M, Takayama H (2011) Ten new fawcettimine-related alkaloids from three species of *Lycopodium*. Tetrahedron 67:6561–6567. https://doi.org/10.1016/j.tet.2011.05.107
96. Ayer WA, Altenkirk B (1969) Alkaloids of *Lycopodium alopecuroides*. Part 3. Structure of alolycopine. Can J Chem 47:2457–2459
97. Zhang ZJ, Qi YY, Wu XD, Su J, Zhao QS (2018) Lycogladines A-H, fawcettimine-type Lycopodium alkaloids from *Lycopodium complanatum* var. *glaucum* Ching. Tetrahedron 74: 1692–1697. https://doi.org/10.1016/j.tet.2018.02.034
98. Wang XJ, Zhang GJ, Zhuang PY, Zhang Y, Yu SS, Bao XQ, Zhang D, Yuan YH, Chen NH, Ma S, Qu J, Li Y (2012) Lycojaponicumins A–C, three alkaloids with an unprecedented skeleton from *Lycopodium japonicum*. Org Lett 14:2614–2617. https://doi.org/10.1021/ol3009478
99. Ishiuchi K, Kubota T, Morita H, Kobayashi J (2006) Lycopladine A, a new C16N alkaloid from *Lycopodium complanatum*. Tetrahedron Lett 47:3287–3289. https://doi.org/10.1016/j.tetlet.2006.03.027
100. Ishiuchi K, Kubota T, Hoshino T, Obara Y, Nakahata N, Kobayashi J (2006) Lycopladines B–D and lyconadin B, new alkaloids from *Lycopodium complanatum*. Bioorg Med Chem 14: 5995–6000. https://doi.org/10.1016/j.bmc.2006.05.028

101. Takayama H, Katakawa K, Kitajima M, Seki H, Yamaguchi K, Aimi N (2001) A new type of Lycopodium alkaloid, lycoposerramine-A, from *Lycopodium serratum* Thunb. Org Lett 3: 4165–4167. https://doi.org/10.1021/ol0167762
102. Katakawa K, Kitajima M, Aimi N, Seki H, Yamaguchi K, Furihata K, Harayama T, Takayama H (2005) Structure elucidation and synthesis of lycoposerramine-B, a novel oxime-containing Lycopodium alkaloid from *Lycopodium serratum* Thunb. J Org Chem 70:658–663. https://doi.org/10.1021/jo0483825
103. Takayama H, Katakawa K, Kitajima M, Yamaguchi K, Aimi N (2002) Seven new Lycopodium alkaloids, lycoposerramines-C, -D, -E, -P, -Q, -S, and -U, from *Lycopodium serratum* Thunb. Tetrahedron Lett 43:8307–8311. https://doi.org/10.1016/S0040-4039(02)02026-9
104. Ayer WA, Ma YT, Liu JS, Huang MF, Schultz LW, Clardy J (1994) Macleanine, a unique type of dinitrogenous Lycopodiurn alkaloid. Can J Chem 72:128–130
105. Breakman JC, Hootele C, Miller N, Declercq JP, Germain G, Meerssche MV (1979) Megastachine, a new alkaloid from *Lycopodium megastachyum*. Can J Chem 57:1691–1693
106. Hu T, Chandler RF, Hanson AW (1987) Obscurinine: a new lycopodium alkaloid. Tetrahedron Lett 28:5993–5996. https://doi.org/10.1016/S0040-4039(00)96845-X
107. Inubushi Y, Tsuda Y, Hasashi I, Sano T, Hosokawa M, Harayama T (1964) Studies on the constituents of domestic *Lycopodium* genus plants. II. on the constituents *of lycopodium serratum* Thunb. var. *Thunberghii* Makino (from Mt. Hira). Yakugaku Zasshi 84:1108–1113
108. Ayer WA, Jenkins JK, Valverde-Lopez S (1967) The alkaloids of *Lycopodium cernuum* L. I. The structures of cernuine and lycocernuine. Can J Chem 45:433–443
109. Ayer WA, Jenkins JK, Piers K, Valverde-Lopez S (1967) The alkaloids of *Lycopodium cernuum* L. II. The stereochemistry of cernuine and lycocernuine. Can J Chem 45:445–450
110. Marion L, Manske RHF (1948) The alkaloids of *Lycopodium* species. X. *Lycopodium cernuum* L. Canadian J Res Sec B26:1–2
111. Morita H, Hirasawa Y, Yoshida N, Kobayashi J (2001) Senepodine A, a novel $C_{22}N_2$ alkaloid from *Lycopodium chinense*. Tetrahedron Lett 42:4199–4201. https://doi.org/10.1016/S0040-4039(01)00688-8
112. Koyama K, Hirasawa Y, Kobayashi J, Morita H (2007) Cryptadines A and B, novel $C_{27}N_3$-type pentacyclic alkaloids from *Lycopodium cryptomerinum*. Bioorg Med Chem 15:7803–7808. https://doi.org/10.1016/j.bmc.2007.08.043
113. Ayer WA, Browne LM, Nakahara Y, Tori M, Delbaere LT (1979) A new type of Lycopodium alkaloid. $C_{30}N_3$The alkaloids from *Lycopodium lucidulum*. Can J Chem 57:1105–1107
114. Tori M, Shimoji T, Shimura E, Takaoka S, Nakashima K, Sono M, Ayer WA (2000) Four alkaloids, lucidine B, oxolucidine A, lucidine A, and lucidulinone from Lycopodium lucidulum. Phytochemistry 53:503–509. https://doi.org/10.1016/S0031-9422(99)00592-0
115. Ayer WA, Masaki N, Nkunika DS (1968) Luciduline: a unique type of Lycopodium alkaloid. Can J Chem 46:3631–3642
116. Hirasawa Y, Tanaka T, Koyama K, Morita H (2009) Lycochinines A–C, novel C27N3 alkaloids from *Lycopodium chinense*. Tetrahedron Lett 50:4816–4819. https://doi.org/10.1016/j.tetlet.2009.05.072
117. Hirasawa Y, Kobayashi J, Morita H (2006) Lycoperine A, a novel $C_{27}N_3$-type pentacyclic alkaloid from *Lycopodium hamiltonii*, inhibiting acetylcholinesterase. Org Lett 8:123–126. https://doi.org/10.1021/ol052760q
118. Ishiuchi K, Kubota T, Hayashi S, Shibata T, Kobayashi J (2009) Lycopladine H, a novel alkaloid with fused-tetracyclic skeleton from *Lycopodium complanatum*. Tetrahedron Lett 50: 6534–6536. https://doi.org/10.1016/j.tetlet.2009.09.035
119. Shigeyama T, Katakawa K, Kogure N, Kitajima M, Takayama H (2007) Asymmetric total syntheses of two phlegmarine-type alkaloids, lycoposerramines-V and -W, Newly isolated from *Lycopodium serratum*. Org Lett 9:4069–4072. https://doi.org/10.1021/ol701871v
120. Katakawa K, Kitajima M, Yamaguchi K, Takayama H (2006) Three new phlegmarine-type lycopodium alkaloids, lycoposerramines-X, -Y, and -Z, having a nitrone residue, from *Lycopodium serratum*. Heterocycles 69:223–229. https://doi.org/10.3987/COM-06-S(O)16

121. Cheng JT, Liu F, Li XN, Wu XD, Dong LB, Peng LY, Huang SX, He J, Zhao QS (2013) Lycospidine A, a new type of lycopodium alkaloid from *Lycopodium complanatum*. Org Lett 15:2438–2441. https://doi.org/10.1021/ol400907v
122. Jiang WP, Ishiuchi K, Wu JB, Kitanaka S (2014) Serralongamine A, a new lycopodium alkaloid from *Lycopodium serratum* var. *longipetiolatum*. Heterocycles 89:747–752. https://doi.org/10.3987/COM-13-12928
123. Kubota T, Yahata H, Yamamoto S, Hayashi S, Shibata T, Kobayashi J (2009) Serratezomines D, and E, new Lycopodium alkaloids from *Lycopodium serratum* var. *serratum*. Bioorg Med Chem Lett 19:3577–3580. https://doi.org/10.1016/j.bmcl.2009.04.146
124. Ayer WA, Ball LF, , Browne LM, Tori M, Delbaere LTJ, Silverberg A (1984) Spirolucidine, a new Lycopodium alkaloid. Can J Chem 62:298–302
125. Ansari FR, Ansari WH, Rahman W, Seligmann O, Chari VM, Wagner H, Osterdahl BG (1979) A new acylated apigenin 4′-O-β-D-glucoside from the leaves of *Lycopodium clavatum* L. J Med Plant Res 36:196–199
126. Zhang Y, Yi P, Chen Y, Mei Z, Hu X, Yang G (2014) Lycojaponicuminol A–F: cytotoxic serratene triterpenoids from *Lycopodium japonicum*. Fitoterapia 96:95–102. https://doi.org/10.1016/j.fitote.2014.04.012
127. Zhao YH, Deng TZ, Chen Y, Liu XM, Yang GZ (2010) Two new triterpenoids from *Lycopodium obscurum* L. J Asian Nat Prod Res 12:666–671. https://doi.org/10.1080/10286020.2010.493881
128. Yan J, Zhang XM, Li ZR, Zhou L, Chen JC, Sun LR, Qiu MH (2005) Three new triterpenoids from *Lycopodium japonicum* Thunb. Helv Chim Acta 88:240–244. https://doi.org/10.1002/hlca.200590004
129. Nguyen VT, Zhao BT, Seong SH, Kim JA, Woo MH, Choi JS, Min BS (2017) Inhibitory effects of serratene-type triterpenoids from *Lycopodium complanatum* on cholinesterases and β-secretase 1. Chem Biol Interact 274:150–157. https://doi.org/10.1016/j.cbi.2017.07.006
130. Wang X, Yu D, Yu S (2014) Two new triterpenoids from *Lycopodium japonicum* Thunb. Chin J Chem 32:1007–1010. https://doi.org/10.1002/cjoc.201400456
131. Inubushi Y, Tsuda Y, Sano T, Nakagawa R (1965) 21-Episerratenediol, isolation, and its structure. Chem Pharm Bull 13:104–105
132. Miller N, Hootele C, Braekman JC (1973) Triterpenoids of *Lycopodium megastachyum*. Phytochemistry 12:1759–1761. https://doi.org/10.1016/0031-9422(73)80398-X
133. Tsuda Y, Hatanaka M (1969) Triterpenoids of *Lycopodium clavatum*: the structure of 21-episerratriol. J Chem Soc D 1969:1040b–11042. https://doi.org/10.1039/c2969001040b
134. Shi H, zhen-yu L, Guo YW (2005) A new serratane-type triterpene from *Lycopodium phlegmaria*. Nat Prod Res 19:777–781. https://doi.org/10.1080/14786410500044906
135. Sun ZH, Li W, Tang GH, Yin S (2017) A new serratene triterpenoid from *Lycopodium japonicum*. J Asian Nat Prod Res 19:299–303. https://doi.org/10.1080/10286020.2016.1208182
136. Zhou W, Kang F, Huang L, Li J, Wang W, Xiao L, Wen Q, Yu X, Xu Y, Zou Z, Zhou H, Zang H, Chen S, Xu K (2020) Serratane triterpenoids from *Lycopodium complanatum* and their anti-cancer and anti-inflammatory activities. Bioorg Chem 101:103959. https://doi.org/10.1016/j.bioorg.2020.103959
137. Trofimova NN, Gromova AS, Semenov AA (1996) Serratene triterpenoids from *Lycopodium clavatum* L. (Lycopodiaceae). Russ Chem Bull 45:961–963. https://doi.org/10.1007/BF01431333
138. Sano T, Fujimoto T, Tsuda Y (1970) Clavatol: a novel triterpenoid of the bisnoronocerane type isolated from *Lycopodium clavatum*. J Chem Soc D 1970:1274–1275. https://doi.org/10.1039/c29700001274
139. Inubushi Y, Hibino T, Harayama T, Hasegawa T, Somanathan R (1971) Triterpenoid constituents of *Lycopodium phlegmaria* L. J Chem Soc C: Organic 1971:3109–3114. https://doi.org/10.1039/j39710003109
140. Li XL, Zhao Y, Cheng X, Tu L, Peng LY, Xu G, Zhao QS (2006) Japonicumins A–D: four new compounds from *Lycopodium japonicum*. Helv Chim Acta 89:1467–1473. https://doi.org/10.1002/hlca.200690148
141. Tsuda Y, Fujimoto T (1970) The structure of lycoclavanin: triterpenoid of *Lycopodium clavatum*. J Chem Soc D 260. https://doi.org/10.1039/c29700000260

142. Burnell RH, Mo L, Moinas M (1972) Le lycoxanthol, nouveau diterpenoide de *Lycopodium lucidulum*. Phytochemistry 11:2815–2820. https://doi.org/10.1016/S0031-9422(00)86518-8
143. Inubushi Y, Sano T, Price JR (1967) Triterpene constituents of *Lycopodium complanatum* L. from new guinea. Aust J Chem 20:387–388
144. Sano T, Tsuda Y, Inubushi Y (1970) Structures of tohogenol and tohogeninol: triterpenoids of *Lycopodium serratum*. Tetrahedron 26:2981–2986. https://doi.org/10.1016/S0040-4020(01)92878-9
145. Orhan I, Terzioglu S, Şener B (2003) α-Onocerin: An acetylcholinesterase inhibitor from *Lycopodium clavatum*. Planta Med 69:265–267. https://doi.org/10.1055/s-2003-38489
146. Yao CP, Zou ZX, Zhang Y, Li J, Cheng F, Xu PS, Zhou G, Li XM, Xu KP, Tan GS (2019) New adenine analogues and a pyrrole alkaloid from *Selaginella delicatula*. Nat Prod Res 33:1985–1991. https://doi.org/10.1080/14786419.2018.1482892
147. Zhuo JX, Wang YH, Su XL, Mei RQ, Yang J, Kong Y, Long CL (2016) Neolignans from *Selaginella moellendorffii*. Nat Prod Bioprospect 6:161–166. https://doi.org/10.1007/s13659-016-0095-5
148. Ma LY, Ma SC, Wei F, Lin RC, But PPH, Lee SHS, Lee SF (2003) Uncinoside A and B, two new antiviral chromone glycosides from *Selaginella uncinata*. Chem Pharm Bull 51:1264–1267. https://doi.org/10.1248/cpb.51.1264
149. Zou Z, Xu P, Wu C, Zhu W, Zhu G, He X, Zhang G, Hu J, Liu S, Zeng W, Xu K, Tan G (2016) Carboxymethyl flavonoids and a chromone with antimicrobial activity from *Selaginella moellendorffii* Hieron. Fitoterapia 111:124–129. https://doi.org/10.1016/j.fitote.2016.04.022
150. Bi Y, Zheng X, Feng W, Shi S (2004) Isolation and structural identification of chemical constituents from *Selaginella tamariscina* (Beauv.) Spring. Acta Pharm Sin 39:41–45
151. Liu JF, Xu KP, Li FS, Shen J, Hu CP, Zou H, Yang F, Liu GR, Xiang HL, Zhou YJ, Li YJ, Tan GS (2010) A new flavonoid from *Selaginella tamariscina* (Beauv.) Spring. Chem Pharm Bull 58:549–551. https://doi.org/10.1248/cpb.58.549
152. Weng JK, Noel JP (2013) Chemodiversity in *Selaginella*: a reference system for parallel and convergent metabolic evolution in terrestrial plants. Front Plant Sci 4. https://doi.org/10.3389/fpls.2013.00119
153. Aguilar MI, Romero MG, Chávez MI, King-Díaz B, Lotina-Hennsen B (2008) Biflavonoids isolated from *Selaginella lepidophylla* inhibit photosynthesis in spinach chloroplasts. J Agric Food Chem 56:6994–7000. https://doi.org/10.1021/jf8010432
154. Cao Y, Chen JJ, Tan NH, Wu YP, Yang J, Wang Q (2010) Structure determination of selaginellins G and H from *Selaginella pulvinata* by NMR spectroscopy. Magn Resonance Chem: MRC 48:656–659. https://doi.org/10.1002/mrc.2623
155. Cao Y, Tan NH, Chen JJ, Zeng GZ, Ma YB, Wu YP, Yan H, Yang J, Lu LF, Wang Q (2010) Bioactive flavones and biflavones from *Selaginella moellendorffii* Hieron. Fitoterapia 81:253–258. https://doi.org/10.1016/j.fitote.2009.09.007
156. Chen B, Wang X, Lin D, Xu D, Li S, Huang J, Weng S, Lin Z, Zheng Y, Yao H, Lin X (2019) Proliposomes for oral delivery of total biflavonoids extract from *Selaginella doederleinii*: formulation development, optimization, and *in vitro–in vivo* characterization. Int J Nanomedicine 14:6691–6706. https://doi.org/10.2147/IJN.S214686
157. Jung HJ, Park K, Lee IS, Kim HS, Yeo SH, Woo ER, Lee DG (2007) S-Phase accumulation of *Candida albicans* by anticandidal effect of amentoflavone isolated from *Selaginella tamariscina*. Biol Pharm Bull 30:1969–1971. https://doi.org/10.1248/bpb.30.1969
158. Lee CW, Choi HJ, Kim HS, Kim DH, Chang IS, Moon HT, Lee SY, Oh WK, Woo ER (2008) Biflavonoids isolated from *Selaginella tamariscina* regulate the expression of matrix metalloproteinase in human skin fibroblasts. Bioorg Med Chem 16:732–738. https://doi.org/10.1016/j.bmc.2007.10.036
159. Lin LC, Kuo YC, Chou CJ (2000) Cytotoxic biflavonoids from *Selaginella delicatula*. J Nat Prod 63:627–630. https://doi.org/10.1021/np990538m
160. Long HP, Zou H, Li FS, Li J, Luo P, Zou ZX, Hu CP, Xu KP, Tan GS (2015) Involvenflavones A-F, six new flavonoids with 3′-aryl substituent from *Selaginella involven*. Fitoterapia 105:254–259. https://doi.org/10.1016/j.fitote.2015.07.013

161. López-Sáez JA, Alonso MJP, Negueruela AV (1994) Biflavonoids of *Selaginella denticulata* growing in Spain. Zeitschrift für Naturforschung C 49:267–270. https://doi.org/10.1515/znc-1994-3-417

162. Shim SY, Lee S, Lee M (2018) Biflavonoids isolated from *Selaginella tamariscina* and their anti-inflammatory activities via ERK 1/2 signaling. Molecules 23:926. https://doi.org/10.3390/molecules23040926

163. Swamy RC, Kunert O, Schühly W, Bucar F, Ferreira D, Rani VS, Kumar BR, Appa Rao AVN (2006) Structurally unique biflavonoids from *Selaginella chrysocaulos* and *Selaginella bryopteris*. Chem Biodivers 3:405–413. https://doi.org/10.1002/cbdv.200690044

164. Zheng JX, Wang NL, Liu HW, Chen HF, Li MM, Wu LY, Fan M, Yao XS (2008) Four new biflavonoids from *Selaginella uncinata* and their anti-anoxic effect. J Asian Nat Prod Res 10: 945–952. https://doi.org/10.1080/10286020802181166

165. Xu J, Yang L, Wang R, Zeng K, Fan B, Zhao Z (2019) The biflavonoids as protein tyrosine phosphatase 1B inhibitors from *Selaginella uncinata* and their antihyperglycemic action. Fitoterapia 137:104255. https://doi.org/10.1016/j.fitote.2019.104255

166. Zheng JX, Zheng Y, Zhi H, Dai Y, Wang NL, Fang YX, Du ZY, Zhang K, Li MM, Wu LY, Fan M (2011) New 3′,8″-linked biflavonoids from *Selaginella uncinata* Displaying protective effect against anoxia. Molecules 16:6206–6214. https://doi.org/10.3390/molecules16086206

167. Zou H, Yi ML, Xu KP, Sheng XF, Tan GS (2016) Two new flavonoids from *Selaginella uncinata*. J Asian Nat Prod Res 18:248–252. https://doi.org/10.1080/10286020.2015.1063617

168. Zou Z, Xu K, Xu P, Li X, Cheng F, Li J, Yu X, Cao D, Li D, Zeng W, Zhang G, Tan G (2017) Seladoeflavones A–F, six novel flavonoids from *Selaginella doederleinii*. Fitoterapia 116:66–71. https://doi.org/10.1016/j.fitote.2016.11.014

169. Chen X, Xu PS, Zou ZX, Liu Y, Zhou WH, Ren Q, Li D, Li X-M, Xu KP, Tan GS (2019) Sinensiols B-G, six novel neolignans from *Selaginella sinensis*. Fitoterapia 134:256–263. https://doi.org/10.1016/j.fitote.2019.02.034

170. Cheng F, Xu K, Liu L, Yao C, Xu P, Zhou G, Li D, Li X, Chen K, Zou Z, Tan G (2018) New neolignans from *Selaginella picta* and their protective effect on HT-22 cells. Fitoterapia 127: 69–73. https://doi.org/10.1016/j.fitote.2018.02.008

171. He XR, Xu LY, Jin C, Yue PF, Zhou ZW, Liang XL (2019) Tamariscinols U–W, new dihydrobenzofuran-type norneolignans with tyrosinase inhibitory activity from *Selaginella tamariscina*. Phytochem Lett 34:79–83. https://doi.org/10.1016/j.phytol.2019.08.013

172. Zhu Y, Huang RZ, Wang CG, Ouyang XL, Jing XT, Liang D, Wang HS (2018) New inhibitors of matrix metalloproteinases 9 (MMP-9): Lignans from *Selaginella moellendorffii*. Fitoterapia 130:281–289. https://doi.org/10.1016/j.fitote.2018.09.008

173. Zhang LP, Liang YM, Wei XC, Cheng DL (2007) A new unusual natural pigment from *Selaginella sinensis* and its noticeable physicochemical properties. J Org Chem 72:3921–3924. https://doi.org/10.1021/jo0701177

174. Cao Y, Chen JJ, Tan NH, Oberer L, Wagner T, Wu YP, Zeng GZ, Yan H, Wang Q (2010) Antimicrobial selaginellin derivatives from *Selaginella pulvinata*. Bioorg Med Chem Lett 20: 2456–2460. https://doi.org/10.1016/j.bmcl.2010.03.016

175. Cao Y, Zhao M, Zhu Y, Zhu ZH, Oberer L, Duan JA (2017) Diselaginellin B, an unusual dimeric molecule from *Selaginella pulvinata*, inhibited metastasis and induced apoptosis of SMMC-7721 human hepatocellular carcinoma cells. J Nat Prod 80:3151–3158. https://doi.org/10.1021/acs.jnatprod.7b00404

176. Cheng XL, Ma SC, Yu JD, Yang SY, Xiao XY, Hu JY, Lu Y, Shaw PC, But PPH, Lin RC (2008) Selaginellin A and B, two novel natural pigments isolated from *Selaginella tamariscina*. Chem Pharm Bull 56:982–984. https://doi.org/10.1248/cpb.56.982

177. Le DD, Nguyen DH, Zhao BT, Seong SH, Choi JS, Kim SK, Kim JA, Min BS, Woo MH (2017) PTP1B inhibitors from *Selaginella tamariscina* (Beauv.) Spring and their kinetic properties and molecular docking simulation. Bioorg Chem 72:273–281. https://doi.org/10.1016/j.bioorg.2017.05.001

178. Liu X, Luo HB, Huang YY, Bao JM, Tang GH, Chen YY, Wang J, Yin S (2014) Selaginpulvilins A–D, New phosphodiesterase-4 inhibitors with an unprecedented skeleton from *Selaginella pulvinata*. Org Lett 16:282–285. https://doi.org/10.1021/ol403282f

179. Liu X, Tang GH, Weng HZ, Zhang JS, Xu YK, Yin S (2018) A new selaginellin derivative and a new triarylbenzophenone analog from the whole plant of *Selaginella pulvinata*. J Asian Nat Prod Res 20:1123–1128. https://doi.org/10.1080/10286020.2017.1378646
180. Liu R, Zou H, Zou ZX, Cheng F, Yu X, Xu PS, Li XM, Li D, Xu KP, Tan GS (2020) Two new anthraquinone derivatives and one new triarylbenzophenone analog from *Selaginella tamariscina*. Nat Prod Res 34:2709–2714. https://doi.org/10.1080/14786419.2018.1452008
181. Nguyen PH, Zhao BT, Ali MY, Choi JS, Rhyu DY, Min BS, Woo MH (2015) Insulin-mimetic selaginellins from *Selaginella tamariscina* with protein tyrosine phosphatase 1B (PTP1B) inhibitory activity. J Nat Prod 78:34–42. https://doi.org/10.1021/np5005856
182. Tan GS, Xu KP, Li FS, Wang CJ, Li TY, Hu CP, Shen J, Zhou YJ, Li YJ (2009) Selaginellin C, a new natural pigment from *Selaginella pulvinata* Maxim (Hook et Grev.). J Asian Nat Prod Res 11:1001–1004. https://doi.org/10.1080/10286020903207043
183. Yang C, Shao Y, Li K, Xia W (2012) Bioactive selaginellins from *Selaginella tamariscina* (Beauv.) Spring. Beilstein J Org Chem 8:1884–1889. https://doi.org/10.3762/bjoc.8.217
184. Yao WN, Huang RZ, Hua J, Zhang B, Wang CG, Liang D, Wang HS (2017) Selagintamarlin A: A selaginellin analogue possessing a 1H-2-benzopyran core from *Selaginella tamariscina*. ACS Omega 2:2178–2183. https://doi.org/10.1021/acsomega.7b00209
185. Zhang G, Jing Y, Zhang H, Ma E, Guan J, Xue F, Liu H, Sun X (2012) Isolation and cytotoxic activity of selaginellin derivatives and biflavonoids from *Selaginella tamariscina*. Planta Med 78:390–392. https://doi.org/10.1055/s-0031-1298175
186. Zheng J, Zheng Y, Zhi H, Dai Y, Wang N, Wu L, Fan M, Fang Y, Zhao S, Zhang K (2013) Two new steroidal saponins from *Selaginella uncinata* (Desv.) Spring and their protective effect against anoxia. Fitoterapia 88:25–30. https://doi.org/10.1016/j.fitote.2013.04.003
187. Chen JJ, Duh CY, Chen JF (2005) New cytotoxic biflavonoids from *Selaginella delicatula*. Planta Med 71:659–665. https://doi.org/10.1055/s-2005-871273
188. Ke LY, Zhang Y, Xia MY, Zhuo JX, Wang YH, Long CL (2018) Modified abietane diterpenoids from whole plants of *Selaginella moellendorffii*. J Nat Prod 81:418–422. https://doi.org/10.1021/acs.jnatprod.7b00909
189. Wang YH, Long CL, Yang FM, Wang X, Sun QY, Wang HS, Shi YN, Tang GH (2009) Pyrrolidinoindoline alkaloids from *Selaginella moellendorfii*. J Nat Prod 72:1151–1154. https://doi.org/10.1021/np9001515
190. Lee NY, Min HY, Lee J, Nam JW, Lee YJ, Han AR, Wiryawan A, Suprapto W, Lee SK, Seo EK (2008) Identification of a new cytotoxic biflavanone from *Selaginella doederleinii*. Chem Pharm Bull 56:1360–1361. https://doi.org/10.1248/cpb.56.1360
191. Xu JC, Liu XQ, Chen KL (2009) A new biflavonoid from *Selaginella labordei* Hieron. ex Christ. Chin Chem Lett 20:939–941. https://doi.org/10.1016/j.cclet.2009.03.046
192. Wang HS, Sun L, Wang YH, Shi YN, Tang GH, Zhao FW, Niu HM, Long CL, Li L (2011) Carboxymethyl flavonoids and a monoterpene glucoside from *Selaginella moellendorffii*. Arch Pharm Res 34:1283–1288. https://doi.org/10.1007/s12272-011-0807-7
193. Hwang JH (2013) Antibacterial effect of amentoflavone and its synergistic effect with antibiotics. J Microbiol Biotechnol 23:953–958. https://doi.org/10.4014/jmb.1302.02045
194. Yao W, Lin Z, Wang G, Li S, Chen B, Sui Y, Huang J, Liu Q, Shi P, Lin X, Liu Q, Yao H (2019) Delicaflavone induces apoptosis via mitochondrial pathway accompanying G2/M cycle arrest and inhibition of MAPK signaling cascades in cervical cancer HeLa cells. Phytomedicine 62: 152973. https://doi.org/10.1016/j.phymed.2019.152973
195. Lee J, Choi Y, Woo ER, Lee DG (2009) Isocryptomerin, a novel membrane-active antifungal compound from *Selaginella tamariscina*. Biochem Biophys Res Commun 379:676–680. https://doi.org/10.1016/j.bbrc.2008.12.030
196. Zou Z, Xu P, Zhang G, Cheng F, Chen K, Li J, Zhu W, Cao D, Xu K, Tan G (2017) Selagintriflavonoids with BACE1 inhibitory activity from the fern *Selaginella doederleinii*. Phytochemistry 134:114–121. https://doi.org/10.1016/j.phytochem.2016.11.011
197. Zeng W, Yao CP, Xu PS, Zhang GG, Liu ZQ, Xu KP, Zou ZX, Tan GS (2017) A new neolignan from *Selaginella moellendorffii* Hieron. Nat Prod Res 31:2223–2227. https://doi.org/10.1080/14786419.2017.1297935

198. Wang YZ, Chen H, Zheng XK, Feng WS (2007) A new sesquilignan from *Selaginella sinensis* (Desv.) Spring. Chin Chem Lett 18:1224–1226. https://doi.org/10.1016/j.cclet.2007.08.016
199. Zheng X, Shi S, Bi Y, Feng W, Wang J, Niu J (2004) The isolation and identification of a new lignanoside from *Selaginella tamariscina* (Beauv.) Spring. Acta Pharm Sin 39:719–721
200. Zhu QF, Shao LD, Wu XD, Liu JX, Zhao QS (2019) Isolation, structural assignment of isoselagintamarlin A from *Selaginella tamariscina* and its biomimetic synthesis. Nat Products Bioprospect 9:69–74. https://doi.org/10.1007/s13659-018-0195-5
201. Xu KP, Zou H, Tan Q, Li FS, Liu JF, Xiang HL, Zou ZX, Long HP, Li YJ, Tan GS (2011) Selaginellins I and J, two new alkynyl phenols, from *Selaginella tamariscina* (Beauv.) Spring. J Asian Nat Prod Res 13:93–96. https://doi.org/10.1080/10286020.2010.536535
202. Xu KP, Zou H, Li FS, Xiang HL, Zou ZX, Long HP, Li J, Luo YJ, Li YJ, Tan GS (2011) Two new selaginellin derivatives from *Selaginella tamariscina* (Beauv.) Spring. J Asian Nat Prod Res 13:356–360. https://doi.org/10.1080/10286020.2011.558840
203. Cao Y, Yao Y, Huang XJ, Oberer L, Wagner T, Guo JM, Gu W, Liu WD, Lv GX, Shen YN, Duan JA (2015) Four new selaginellin derivatives from *Selaginella pulvinata*: mechanism of racemization process in selaginellins with quinone methide. Tetrahedron 71:1581–1587. https://doi.org/10.1016/j.tet.2015.01.017
204. Xu KP, Li J, Zhu GZ, He XA, Li FS, Zou ZX, Tan LH, Cheng F, Tan GS (2015) New selaginellin derivatives from *Selaginella tamariscina*. J Asian Nat Prod Res 17:819–822. https://doi.org/10.1080/10286020.2015.1016001
205. Zhu B, Wang TB, Hou LJ, Lv HX, Liu AM, Zeng P, Li AH (2016) A New selaginellin from *Selaginella moellendorffii* inhibits hepatitis B virus gene expression and replication. Chem Nat Compd 52:624–627. https://doi.org/10.1007/s10600-016-1725-1
206. Woo S, Bin KK, Kim J, Sung SH (2019) Molecular networking reveals the chemical diversity of selaginellin derivatives, natural phosphodiesterase-4 inhibitors from *Selaginella tamariscina*. J Nat Prod 82:1820–1830. https://doi.org/10.1021/acs.jnatprod.9b00049
207. Zhang JS, Liu X, Weng J, Guo YQ, Li QJ, Ahmed A, Tang GH, Yin S (2017) Natural diarylfluorene derivatives: isolation, total synthesis, and phosphodiesterase-4 inhibition. Organic Chem Front 4:170–177. https://doi.org/10.1039/C6QO00623J
208. Nguyen PH, Ji DJ, Han YR, Choi JS, Rhyu DY, Min BS, Woo MH (2015) Selaginellin and biflavonoids as protein tyrosine phosphatase 1B inhibitors from *Selaginella tamariscina* and their glucose uptake stimulatory effects. Bioorg Med Chem 23:3730–3737. https://doi.org/10.1016/j.bmc.2015.04.007
209. Gao L, Yin S, Li Z, Sha Y, Pei Y, Shi G, Jing Y, Hua H (2007) Three novel sterols isolated from *Selaginella tamariscina* with antiproliferative activity in leukemia cells. Planta Med 73:1112–1115. https://doi.org/10.1055/s-2007-981562
210. Jin M, Zhang C, Zheng T, Yao D, Shen L, Luo J, Jiang Z, Ma J, Jin XJ, Cui J, Lee JJ, Li G (2014) A new phenyl glycoside from the aerial parts of *Equisetum hyemale*. Nat Prod Res 28: 1813–1818. https://doi.org/10.1080/14786419.2014.947491
211. Milovanović V, Radulović N, Todorović Z, Stanković M, Stojanović G (2007) Antioxidant, antimicrobial and genotoxicity screening of hydro-alcoholic extracts of five Serbian *Equisetum* Species. Plant Foods Hum Nutr 62:113–119. https://doi.org/10.1007/s11130-007-0050-z
212. Oh H, Kim DH, Cho JH, Kim YC (2004) Hepatoprotective and free radical scavenging activities of phenolic petrosins and flavonoids isolated from *Equisetum arvense*. J Ethnopharmacol 95:421–424. https://doi.org/10.1016/j.jep.2004.08.015
213. Wei Z, Pan Y, Li L, Huang Y, Qi X, Luo M, Zu Y, Fu Y (2014) Simultaneous determination of phenolic compounds in *Equisetum palustre* L. by ultra high-performance liquid chromatography with tandem mass spectrometry combined with matrix solid-phase dispersion extraction. J Sep Sci 37:3045–3051. https://doi.org/10.1002/jssc.201400575
214. Ganeva Y, Chanev C, Dentchev T (2001) Triterpenoids and sterols from *Equiseturn arvense*. Comptes Rendus de l'Academie Bulgare des Sciences 54:53–56
215. Camacho MR, Mata DCR, Palacios-Rios M (1992) Constituents of *Equisetum myriochaetum*. Fitoterapia 63:471

216. Cramer L, Ernst L, Lubienski M, Papke U, Schiebel HM, Jerz G, Beuerle T (2015) Structural and quantitative analysis of *Equisetum* alkaloids. Phytochemistry 116:269–282. https://doi.org/10.1016/j.phytochem.2015.03.003
217. Tan JM, Qiu YH, Tan XQ, Tan CH, Xiao K (2011) Chemical constituents of *Equisetum debile*. J Asian Nat Prod Res 13:811–816. https://doi.org/10.1080/10286020.2011.596829
218. Tipke I, Bücker L, Middelstaedt J, Winterhalter P, Lubienski M, Beuerle T (2019) HILIC HPLC-ESI-MS/MS identification and quantification of the alkaloids from the genus *Equisetum*. Phytochem Anal 30:669–678. https://doi.org/10.1002/pca.2840
219. Kolomiets NE, Yusubov MS, Kalinkina GI (2012) Flavonoid composition of *Equisetum arvense* and *E. x litorale* studied by high-performance liquid chromatography-mass spectrometry. Chem Nat Compd 48:135–136. https://doi.org/10.1007/s10600-012-0181-9
220. Yeganegi M, Tabatabaei Yazdi F, Mortazavi SA, Asili J, Alizadeh Behbahani B, Beigbabaei A (2018) *Equisetum telmateia* extracts: chemical compositions, antioxidant activity and antimicrobial effect on the growth of some pathogenic strain causing poisoning and infection. Microb Pathog 116:62–67. https://doi.org/10.1016/j.micpath.2018.01.014
221. Correia H, González-Paramás A, Amaral MT, Santos-Buelga C, Batista MT (2005) Characterisation of polyphenols by HPLC-PAD-ESI/MS and antioxidant activity in *Equisetum telmateia*. Phytochem Anal 16:380–387. https://doi.org/10.1002/pca.864
222. Kanchanapoom T, Otsuka H, Ruchirawat S (2007) Megastigmane glucosides from *Equisetum debile* and *E. diffusum*. Chem Pharm Bull 55:1277–1280. https://doi.org/10.1248/cpb.55.1277
223. Mimica-Dukic N, Simin N, Cvejic J, Jovin E, Orcic D, Bozin B (2008) Phenolic compounds in field Horsetail (*Equisetum arvense* L.) as natural antioxidants. Molecules 13:1455–1464. https://doi.org/10.3390/molecules13071455
224. Jun C, Li-Jiang X, Ya-Ming X (2001) Three new phenolic glycosides from the fertile sprouts of *Equisetum arvense*. Acta Bot Sin 43:193–197
225. Francescato LN, Debenedetti SL, Schwanz TG, Bassani VL, Henriques AT (2013) Identification of phenolic compounds in *Equisetum giganteum* by LC–ESI-MS/MS and a new approach to total flavonoid quantification. Talanta 105:192–203. https://doi.org/10.1016/j.talanta.2012.11.072
226. Wiedenfeld H, Andrade Cetto A, Perez Amador C (2000) Flavonol glycosides from *Equisetum myriochaetum*. Biochem Syst Ecol 28:395–397. https://doi.org/10.1016/S0305-1978(99)00074-5
227. Gurbuz I, Yesilada E, Ito S (2009) An anti-ulcerogenic flavonol diglucoside from *Equisetum palustre* L. J Ethnopharmacol 121:360–365. https://doi.org/10.1016/j.jep.2008.11.016
228. de Queiroz GM, Politi FAS, Rodrigues ER, Souza-Moreira TM, Moreira RRD, Cardoso CRP, Santos LC, Pietro RCLR (2015) Phytochemical characterization, antimicrobial activity, and antioxidant potential of *Equisetum hyemale* L. (Equisetaceae) extracts. J Med Food 18:830–834. https://doi.org/10.1089/jmf.2014.0089
229. Xu XH, Tan CH, Jiang SH, Zhu DY (2006) Debilosides A–C: three new megastigmane glucosides from *Equisetum debile*. Helv Chim Acta 89:1422–1426. https://doi.org/10.1002/hlca.200690142
230. Veit M, Geiger H, Wray V, Abou-Mandour A, Rozdzinski W, Witte L, Strack D, Czygan FC (1993) Equisetumpyrone, a styrylpyrone glucoside in gametophytes from *Equisetum arvense*. Phytochemistry 32:1029–1032. https://doi.org/10.1016/0031-9422(93)85249-Q
231. Veit M, Geiger H, Kast B, Beckert C, Horn C, Markham KR, Wong H, Czygan FC (1995) Styrylpyrone glucosides from *Equisetum*. Phytochemistry 39:915–917. https://doi.org/10.1016/0031-9422(95)00941-Y
232. Melos JLR, Silva LB, Peres MTLP, Mapeli AM, Faccenda O, Anjos HH, Torres TG, Tiviroli SC, Batista AL, Almeida FGN, Flauzino NS, Tibana LA, Hess SC, Honda NK (2007) Constituintes químicos e avaliação do potencial alelopático de *Adiantum tetraphyllum* Humb. & amp; Bonpl. Ex. Wild (Pteridaceae). Química Nova 30:292–297. https://doi.org/10.1590/S0100-40422007000200010
233. Ageta H, Shiojima K, Arai Y (1968) Fern constituents: neohopene, hopene-II, neohopadiene, and fernadiene isolated from *Adiantum* species. Chem Commun (Camb) 1968:1105–1107. https://doi.org/10.1039/c19680001105

234. Shiojima K, Arai Y, Kasama T, Ageta H (1993) Fern constituents: triterpenoids Isolated from the leaves of *Adiantum monochlamys*. filicenol A, filicenol B, isoadiantol B, hakonanediol and epihakonanediol. Chem Pharm Bull 41:262–267. https://doi.org/10.1248/cpb.41.262
235. Shiojima K, Sasaki Y, Ageta H (1993) Fern constituents: triterpenoids isolated from the leaves of *Adiantum pedatum*. 23-Hydroxyfernene, glaucanol A and filicenoic acid. Chem Pharm Bull 41:268–271. https://doi.org/10.1248/cpb.41.268
236. Shiojima K, Arai Y, Nakane T, Ageta H (1997) Fern constituents: *Adiantum cuneatum*. I. Three new triterpenoids, glaucanol B acetate, 7β,25-epoxyfern-8-ene and 25-norfern-7-en-10β-yl formate. Chem Pharm Bull 45:636–638. https://doi.org/10.1248/cpb.45.636
237. Tsuzuki K, Ôhashi A, Arai Y, Masuda K, Takano A, Shiojima K, Ageta H, Cai SQ (2001) Triterpenoids from *Adiantum caudatum*. Phytochemistry 58:363–367. https://doi.org/10.1016/S0031-9422(01)00198-4
238. Zaman A, Prakash A, Berti G, Bottari F, Macchia B, Marsili A, Morelli I (1966) A new nortriterpenoid ketol from two *adiantum* species. Tetrahedron Lett 7:3943–3947. https://doi.org/10.1016/S0040-4039(00)70007-4
239. Nakane T, Arai Y, Masuda K, Ishizaki Y, Ageta H, Shiojima K (1999) Fern constituents: Six new triterpenoid alcohols from *Adiantum capillus-veneris*. Chem Pharm Bull 47:543–547. https://doi.org/10.1248/cpb.47.543
240. Reddy NVL, Ravikanth V, Prabhakar Rao T, Diwan PV, Venkateswarlu Y (2001) A new triterpenoid from the fern *Adiantum lunulatum* and evaluation of antibacterial activity. Phytochemistry 56:173–175. https://doi.org/10.1016/S0031-9422(00)00334-4
241. Nakane T, Maeda Y, Ebihara H, Arai Y, Masuda K, Takano A, Ageta H, Shiojima K, Cai SQ, Abdel-Halim OB (2002) Fern constituents: triterpenoids from *Adiantum capillus-veneris*. Chem Pharm Bull 50:1273–1275. https://doi.org/10.1248/cpb.50.1273
242. Shiojima K, Ageta H (1994) Fern constituents: triterpenoids isolated from the leaves of *Adiantum edgeworthii*. Structures of 19α-hydroxyadiantone and Fern-9(11)-en-25-oic acid. Chem Pharm Bull 42:45–47. https://doi.org/10.1248/cpb.42.45
243. Gupta SK, Bagchi A, Roy SK, Ray AB (1990) Chemical constituents of a member of *Adiantum caudatum* complex. J Indian Chem Soc 67:86–88
244. Marino A, Elberti MG, Cataldo A (1989) Phytochemical investigation of *Adiantum capillus-veneris*. Bollettino della Societa Italiano di Biologia Sperimentale 65:461–463
245. Ibraheim ZZ, Ahmed AS, Gouda YG (2011) Phytochemical and biological studies of *Adiantum capillus-veneris* L. Saudi Pharmaceutical J 19:65–74. https://doi.org/10.1016/j.jsps.2011.01.007
246. Akabori Y, Hasegawa M (1969) Flavanoid pattern in the Pteridaceae II. Flavanoid constituents in the fronds of *Adiantum capillus-veneris* and *A. cuneatum*. Bot Mag Tokyo 82:294–297
247. Wollenweber E (1979) Some new external flavonoids from American ferns. Flora 168:138–145
248. Hasegawa M, Akabori Y (1968) Flavanoid pattern in Pteridaceae, I Flavonoid glycosides obtained from the fronds of *Adiantum aethiopicum* and *A. monochlamys*. Bot Mag Tokyo 81: 469–472
249. Murakami T, Wada H, Tanaka N, Kido T, Iida H, Saiki Y, Chen CM (1986) Chemical and chemotaxonomical studies of Filices. LXV. A few new flavonoid glycosides. Yakugaku Zasshi 106:982–988
250. Sallam MM, Mostafa AE, Genady EAM, Ismail SK (2019) Phytochemical and biological study of *Adiantum capillus-veneris* L. growing in Egypt. Al-Azhar J Pharmaceutical Sci 59:9–26. https://doi.org/10.21608/ajps.2019.64101
251. Imperato F (1982) Sulphate esters of hydroxycinnamic acid – sugar derivatives from *Adiantum capillus-veneris*. Phytochemistry 21:2717–2718. https://doi.org/10.1016/0031-9422(82)83105-1
252. Hussain A, Siddiqui HL, Rashid R, Khan KM, Parvez M (2008) 1-(5a,5b,8,8,11a,13b-Hexamethyleicosahydro-1 *H*-cyclopenta [*a*]chrysen-3-yl)-1-ethanone. Acta Crystallographica Sect E 64(4):o723. https://doi.org/10.1107/S1600536808006831
253. Chopra N, Alam MS, Ali M (2001) Isolation and characterization of two novel triterpenes from *Adiantum venustum*. Indian J Chem 40:350–353

254. Shiojima K, Arai Y, Nakane T, Ageta H (1997) Fern constituents: *Adiantum cuneatum*. II. six new triterpenolds, neohop-18-en-12α-ol, 13-epineohop-18-en-12α-ol, neohop-13(18)-en-19α-ol, fern-7-en-25-ol, Fern-9(11)-en-25-ol, and adian-5-en-25-ol. Chem Pharm Bull 45: 639–642. https://doi.org/10.1248/cpb.45.639
255. Hayat S, Rahman A, Choudhary MI, Khan KM, Latif H, Bayer E (2002) Two new triterpenes from fern *Adiantum incisum*. Zeitschrift für Naturforschung B 57:233–238. https://doi.org/10.1515/znb-2002-0215
256. Berti G, Bottari F, Marsili A (1969) Structure and stereochemistry of a triterpenoid epoxide from *Adiantum capillus-veneris*. Tetrahedron 25:2939–2947. https://doi.org/10.1016/S0040-4020(01)82826-X
257. Haider S, Kharbanda C, Alam MS, Hamid H, Ali M, Alam M, Nazreen S, Ali Y (2013) Anti-inflammatory and anti-nociceptive activities of two new triterpenoids from *Adiantum capillus-veneris* Linn. Nat Prod Res 27:2304–2310. https://doi.org/10.1080/14786419.2013.828292
258. Shiojima K, Nakane T, Ageta H, Cai SQ (1996) Fern constituents: two new secofilicane triterpenoids from *Adiantum cuneatum*. Chem Pharm Bull 44:630–632. https://doi.org/10.1248/cpb.44.630
259. Shiojima K, Arai Y, Nakane T, Ageta H, Cai SQ (1997) Fern constituents: *Adiantum cuneatum*. III. four new triterpenoids, 4,23-bisnor-3,4-secofilic-5(24)-en-3-al, 4,23-bisnor-3,3-dimethoxy-3,4-secofilic-5(24)-ene, 7β,25-epoxyfern-9(11)-en-8α-ol and 7α,8-α-epoxyfernan-25-ol. Chem Pharm Bull 45:1608–1610. https://doi.org/10.1248/cpb.45.1608
260. Hussain A, Siddiqui HL, Zia-ur-Rehman M, Elsegood MRJ, Khan MK (2008) 4-Hydroxy-4,6a,6b,9,9,12a,14b-heptamethylperhydropicen-3-one hemihydrate isolated from *Adiantum incisum*. Acta Crystallographica Sect E Struct Rep Online 64:o264–o264. https://doi.org/10.1107/S1600536807064021
261. Brahmachari G, Chatterjee D (2002) Triterpenes from *Adiantum lunulactum*. Fitoterapia 73: 363–368. https://doi.org/10.1016/S0367-326X(02)00119-3
262. Ageta H, Iwata K (1966) Fern constituents: adipedatol, filicenal and other triterpenoids isolated from *Adiantum pedatum*. Tetrahedron Lett 7:6069–6074. https://doi.org/10.1016/S0040-4039(01)84172-1
263. Ageta H, Iwata K, Natori S (1964) Fern constituents: adianene, filicene, 7-fernene, isofernene and diploptene. Triterpenoid hydrocarbons isolated from *Adiantum monochlamys*. Tetrahedron Lett 5:3413–3418. https://doi.org/10.1016/S0040-4039(01)89403-X
264. Ageta H, Shiojima K, Kamaya R, Masuda K (1978) Fern constituent: naturally occurring adian-5-ene ozonide in the leaves of *Adiantum monochlamys* and *Oleandra wallichii*. Tetrahedron Lett 19:899–900. https://doi.org/10.1016/S0040-4039(01)91430-3
265. Bresciani LFV, Priebe JP, Yunes RA, Dal Magro J, Delle Monache F, de Campos F, de Souza MM, Cechinel-Filho V (2003) Pharmacological and phytochemical evaluation of *Adiantum cuneatum* growing in Brazil. Zeitschrift für Naturforschung C 58:191–194. https://doi.org/10.1515/znc-2003-3-409
266. Alam MS, Chopra N, Ali M, Niwa M (2000) Normethyl pentacyclic and lanostane-type triterpenes from *Adiantum venustum*. Phytochemistry 54:215–220. https://doi.org/10.1016/S0031-9422(00)00063-7
267. Banerjee J, Datra G, Dutta CP, Eguchi T, Fujimoto Y, Kakinuma K (1991) Fern-9(11)-en-25-oic acid, a triterpene from *Adiantum venustum*. Phytochemistry 30:3478–3480. https://doi.org/10.1016/0031-9422(91)83238-G
268. Chopra N, Alam MS, Ali M, Niwa M (1997) A novel tirucallane triterpene from *Adiantum venustum*. Pharmazie 52:412–413
269. de Souza MM, Pereira MA, Ardenghi JV, Mora TC, Bresciani LF, Yunes RA, Delle Monache F, Cechinel-Filho V (2009) Filicene obtained from *Adiantum cuneatum* interacts with the cholinergic, dopaminergic, glutamatergic, GABAergic, and tachykinergic systems to exert antinociceptive effect in mice. Pharmacol Biochem Behav 93:40–46. https://doi.org/10.1016/j.pbb.2009.04.004
270. Pradeep Kumar R, Dinesh Babu KV, Evans DA (2019) Isolation, characterization and mode of action of a larvicidal compound, 22-hydroxyhopane from *Adiantum latifolium* Lam. against

Oryctes rhinoceros Linn. Pestic Biochem Physiol 153:161–170. https://doi.org/10.1016/j.pestbp.2018.11.018
271. Ageta H, Iwata K, Arai Y, Tsuda Y, Isobe K, Fukushima S (1966) Fern constituents: hydroxyadiantone and ketohakohanol isolated from *Adiantum monochlamys*. Tetrahedron Lett 7:5679–5684. https://doi.org/10.1016/S0040-4039(01)84177-0
272. Chopra N, Alam MS, Ali M, Niwa M (2000) A new lanostane triterpenic ether from *Adiantum venustum*. Pharmazie 55:538–539
273. Cao J, Xia X, Chen X, Xiao J, Wang Q (2013) Characterization of flavonoids from *Dryopteris erythrosora* and evaluation of their antioxidant, anticancer, and acetylcholinesterase inhibition activities. Food Chem Toxicol 51:242–250. https://doi.org/10.1016/j.fct.2012.09.039
274. Erhirhie EO, Emeghebo CN, Ilodigwe EE, Ajaghaku DL, Umeokoli BO, Eze PM, Ngwoke KG, Okoye FBC (2019) *Dryopteris filix-mas* (L.) Schott ethanolic leaf extract and fractions exhibited profound anti-inflammatory activity. Avecinna J Phytomed 9:396–409
275. Imperato F (2006) Kaempferol 3-O-(acetylrutinoside), a new flavonoid and two new fern constituents, quercetin 3-O-(acetylglucoside) and 3-O-(acetylrutinoside) from *Dryopteris villarii*. Am Fern J 96:93–99
276. Imperato F (2007) A new flavonoid, quercetin 3-O-(X″-acetyl-X″-cinnamoyl-glucoside) and a new fern constituent, quercetin 3-O- (glucosylrhamnoside) from *Dryopteris villarii*. Am Fern J 97:124–126. https://doi.org/10.1640/0002-8444(2007)97[124:SN]2.0.CO;2
277. Jiang B, Chi C, Fu Y, Zhang Q, Wang G (2013) *In vivo* anthelmintic effect of flavonol rhamnosides from *Dryopteris crassirhizoma* against *Dactylogyrus intermedius* in goldfish (*Carassius auratus*). Parasitol Res 112:4097–4104. https://doi.org/10.1007/s00436-013-3600-3
278. Li B, Zhu JF, Zou ZJ, Yin YQ, Shen ZB (2009) Studies on the chemical constituents of *Dryopteris fragrans*. Zhong Yao Cai 32:1232–1233
279. Min BS, Tomiyama M, Ma CM, Nakamura N, Hattori M (2001) Kaempferol acetylrhamnosides from the rhizome of *Dryopteris crassirhizoma* and their inhibitory effects on three different activities of human immunodeficiency virus-1 reverse transcriptase. Chem Pharm Bull 49:546–550. https://doi.org/10.1248/cpb.49.546
280. Yoo G, Park S, Yang H, Nguyen X, Kim N, Park J, Kim S (2017) Two new phenolic glycosides from the aerial part of *Dryopteris erythrosora*. Pharmacogn Mag 13:673–676. https://doi.org/10.4103/pm.pm_326_16
281. Zhang X, Wang X, Wang M, Cao J, Xiao J, Wang Q (2019) Effects of different pretreatments on flavonoids and antioxidant activity of *Dryopteris erythrosora* leave. PLoS One 14: e0200174. https://doi.org/10.1371/journal.pone.0200174
282. Huang YH, Zeng WM, Li GY, Liu GQ, Zhao DD, Wang J, Zhang YL (2014) Characterization of a new sesquiterpene and antifungal activities of chemical constituents from *Dryopteris fragrans* (L.) Schott. Molecules 19:507–513. https://doi.org/10.3390/molecules19010507
283. Liu ZD, Zhao DD, Jiang S, Xue B, Zhang YL, Yan XF (2018) Anticancer phenolics from *Dryopteris fragrans* (L.) Schott. Molecules 23:680. https://doi.org/10.3390/molecules23030680
284. Zhang T, Wang L, Duan DH, Zhang YH, Huang SX, Chang Y (2018) Cytotoxicity-guided isolation of two new phenolic derivatives from *Dryopteris fragrans* (L.). Schott Mol 23:1652. https://doi.org/10.3390/molecules23071652
285. Zhao DD, Zhao QS, Liu L, Chen ZQ, Zeng WM, Lei H, Zhang YL (2014) Compounds from *Dryopteris Fragrans* (L.) Schott with cytotoxic activity. Molecules 19:3345–3355. https://doi.org/10.3390/molecules19033345
286. Coskun N, Sakushima A, Nishibe S, Hisada S (1982) Phloroglucinol derivatives of *Dryopteris abbreviata*. Chem Pharm Bull 30:4102–4106
287. Euw J, Reichstein T, Widén CJ (1985) The Phloroglucinols of *Dryopteris aitoniana* PICHI SERM. (Dryopteridaceae, Pteridophyta). Helv Chim Acta 68:1251–1275. https://doi.org/10.1002/hlca.19850680522
288. Hisada S, Inoue O, Inagaki I (1974) A new acylphloroglucinol of *Dryopteris gymnosora*. Phytochemistry 13:655–655

289. Hua X, Yang Q, Zhang W, Dong Z, Yu S, Schwarz S, Liu S (2018) Antibacterial activity and mechanism of action of aspidinol against multi-drug-resistant methicillin-resistant *Staphylococcus aureus*. Front Pharmacol 9:619. https://doi.org/10.3389/fphar.2018.00619
290. Ito H, Muranaka T, Mori K, Jin ZX, Yoshida T (1997) Dryofragin and aspidin pb, piscicidal components from *Dryopteris fragrans*. Chem Pharm Bull 45:1720–1722. https://doi.org/10.1248/cpb.45.1720
291. Ito H, Muranaka T, Mori K, Jin Z-X, Tokuda H, Nishino H, Yoshida T (2000) Ichthyotoxic phloroglucinol derivatives from *Dryopteris fragrans* and their anti-tumor promoting activity. Chem Pharm Bull 48:1190–1195. https://doi.org/10.1248/cpb.48.1190
292. Lee HB, Kim JC, Lee SM (2009) Antibacterial activity of two phloroglucinols, flavaspidic acids AB and PB, from *Dryopteris crassirhizoma*. Arch Pharm Res 32:655–659. https://doi.org/10.1007/s12272-009-1502-9
293. Li Z, Hong-qing W, Ruo-yun C (2005) Chemical constituents in roots of *Dryopteris championii*. Chin Tradit Herb Drug 36:177
294. Li XJ, Fu YJ, Luo M, Wang W, Zhang L, Zhao CJ, Zu YG (2012) Preparative separation of dryofragin and aspidin BB from *Dryopteris fragrans* extracts by macroporous resin column chromatography. J Pharm Biomed Anal 61:199–206. https://doi.org/10.1016/j.jpba.2011.12.003
295. Liu X, Liu J, Jiang T, Zhang L, Huang Y, Wan J, Song G, Lin H, Shen Z, Tang C (2018) Analysis of chemical composition and *in vitro* antidermatophyte activity of ethanol extracts of *Dryopteris fragrans* (L.) Schott. J Ethnopharmacol 226:36–43. https://doi.org/10.1016/j.jep.2018.07.030
296. Lounasmaa M, Karjalainen A, Widen CJ, Huhtikangas A (1972) Mass spectral studies on some naturally occurring phloroglucinols derivatives. Part III. The mass spectra of some mono- and bicyclic phloroglucinol derivatives from rhizomes of different *Dryopteris* species. Acta Chem Scand 26:89–101
297. Molodozhnikov LM, Ban'kovskii AI, Sergeev NM, Shreter NM (1971) Derivatives of phloroglucinol from *Dryopteris fragrans* (L.) Schott. Khimiko-Farmatseveticheskii Zhurnal 5:32–36
298. Na M, Jang J, Min BS, Lee SJ, Lee MS, Kim BY, Oh WK, Ahn JS (2006) Fatty acid synthase inhibitory activity of acylphloroglucinols isolated from *Dryopteris crassirhizoma*. Bioorg Med Chem Lett 16:4738–4742. https://doi.org/10.1016/j.bmcl.2006.07.018
299. Patama TT, Widen CJ (1991) Phloroglucinol derivatives from *Dryopteris fusco-atra* and *D. hawaiiensis*. Phytochemistry 30:3305–3310. https://doi.org/10.1016/0031-9422(91)83198-T
300. Phong NV, Oanh VT, Yang SY, Choi JS, Min BS, Kim JA (2021) PTP1B inhibition studies of biological active phloroglucinols from the rhizomes of *Dryopteris crassirhizoma*: kinetic properties and molecular docking simulation. Int J Biol Macromol 188:719–728. https://doi.org/10.1016/j.ijbiomac.2021.08.091
301. Tryon R, Widén C-J, Huhtikangas A, Lounasmaa M (1973) Phloroglucinol derivatives in *Dryopteris parallelogramma* and *D. patula*. Phytochemistry 12:683–687. https://doi.org/10.1016/S0031-9422(00)84464-7
302. Wang J, Yan YT, Fu SZ, Peng B, Bao LL, Zhang YL, Hu JH, Zeng ZP, Geng DH, Gao ZP (2017) Anti-influenza virus (H5N1) activity screening on the phloroglucinols from rhizomes of *Dryopteris crassirhizoma*. Molecules 22:431. https://doi.org/10.3390/molecules22030431
303. Widén CJ, von Euw J, Reichstein T (1970) Trispara-aspidin, ein neues phloroglucid aus dem farn *Dryopteris remota* (A. Br.) Hayek. Helv Chim Acta 53:2176–2188. https://doi.org/10.1002/hlca.19700530830
304. Wollenweber E, Stevens JF, Ivanic M, Deinzer ML (1998) Acylphloroglucinols and flavonoid aglycones produced by external glands on the leaves of two *dryopteris* ferns and *Currania robertiana*. Phytochemistry 48:931–939. https://doi.org/10.1016/S0031-9422(97)01003-0
305. Yuk HJ, Kim JY, Sung YY, Kim DS (2020) Phloroglucinol derivatives from *Dryopteris crassirhizoma* as potent xanthine oxidase inhibitors. Molecules 26:122. https://doi.org/10.3390/molecules26010122

306. Amin S, Ullah B, Ali M, Rauf A, Khan H, Uriarte E, Sobarzo-Sánchez E (2019) Potent *in vitro* α-glucosidase inhibition of secondary metabolites derived from *Dryopteris cycadina*. Molecules 24:427. https://doi.org/10.3390/molecules24030427
307. Kang WY, Li CQ, Ji ZQ (2011) A new carbamic acid from *Dryopteris wallichiana*. Chem Nat Compd 47:91–93. https://doi.org/10.1007/s10600-011-9837-0
308. Chang X, Li W, Koike K, Wu L, Nikaido T (2006) Phenolic constituents from the rhizomes of *Dryopteris crassirhizoma*. Chem Pharm Bull 54:748–750. https://doi.org/10.1248/cpb.54.748
309. Feng W, Cao X, Zheng X, Kuang H (2005) A new flavanone from *Dryopteris sublaeta*. Yao Xue Xue Bao 40:443–446
310. Yim NH, Lee JJ, Lee B, Li W, Ma JY (2019) Antiplatelet activity of acylphloroglucinol derivatives isolated from *Dryopteris crassirhizoma*. Molecules 24:2212. https://doi.org/10.3390/molecules24122212
311. Ishaque M, Bibi Y, Ayoubi S, Masood S, Nisa S, Qayyum A (2021) Iriflophenone-3-C-β-d glucopyranoside from *Dryopteris ramosa* (Hope) C. Chr. with promising future as a natural antibiotic for gastrointestinal tract infections. Antibiotics 10:1128. https://doi.org/10.3390/antibiotics10091128
312. Bing P, Zu-ping Z, Ping L, Wei H, Hong W, Zeng-ping G (2013) A new chromone glycoside from *Dryopteris fragrans*. Chinese Herbal Med 44:2347–2349
313. Peng B, Bai RF, Li P, Han XY, Wang H, Zhu CC, Zeng ZP, Chai XY (2016) Two new glycosides from *Dryopteris fragrans* with anti-inflammatory activities. J Asian Nat Prod Res 18:59–64. https://doi.org/10.1080/10286020.2015.1121853
314. Imperato F (2008) A new flavone glucoside, apigenin 7-O-glucoside 4′-acetate and a New fern constituent, quercetin 3-O-rhamnoside-7-O-glucoside from *Dryopteris villarii*. Am Fern J 98: 251–253
315. Imperato F (2007) Three new flavonoid glycosides, kaempferol 3-O-(caffeoylrhamnoside), apigenin 4′-O- (caffeoylglucoside) and 4′-O-(feruloylglucoside) from *Dryopteris villarii*. Am Fern J 97:233–236
316. Ali M, Khan SA, Rauf A, Khan H, Shah MR, Ahmad M, Mubarak MS, Ben Hadda T (2015) Characterization and antinociceptive activity (*in vivo*) of kempferol-3,4′-di-O-α-L-rhamnopyranoside isolated from *Dryopteris cycadina*. Med Chem Res 24:3218–3229. https://doi.org/10.1007/s00044-015-1373-1
317. Amin S, Ullah B, Ali M, Khan H, Rauf A, Khan SA, Sobarzo-Sánchez E (2020) *In Vitro* α-glucosidase inhibition and computational studies of kaempferol derivatives from *Dryopteris cycanida*. Curr Top Med Chem 20:731–737. https://doi.org/10.2174/1568026620666200130161033
318. Feng W, Cao X, Kuang H, Zheng X (2007) Flavanone O-glycosides from the rhizomes of *Dryopteris sublaeta*. Yao Xue Xue Bao 42:867–871
319. Kuang H, Zhang Y, Li G, Zeng W, Wang H, Song Q (2008) A new phenolic glycoside from the aerial parts of *Dryopteris fragrans*. Fitoterapia 79:319–320. https://doi.org/10.1016/j.fitote.2008.01.003
320. Kuang H, Sun C, Zhang Y, Zhang Y, Chen D, Yang B, Xia Y (2009) Three drimane sesquiterpene glucoside from the aerial parts of *Dryopteris fragrans* (L.) schot. Fitoterapia 80:134–137. https://doi.org/10.1016/j.fitote.2008.12.004
321. Wei-sheng F, Xin-wei C, Hai-xue K, Xiao-ke Z (2005) A new stilbene glycoside from *Dryopteris sublaeta*. Acta Pharm Sin 40:1131–1134
322. Ishaque M, Bibi Y, Masood S, al Ayoubi S, Qayyum A, Nisa S, Ahmed W (2022) Xanthone C-glycosides isomers purified from *Dryopteris ramosa* (Hope) C. Chr. with bactericidal and cytotoxic prospects. Saudi J Biol Sci 29:1191–1196. https://doi.org/10.1016/j.sjbs.2021.09.047
323. Zhu CC, Peng B, Zeng ZP, Han XY, Wang H, Wang TY (2021) A new phloroglucinol compound from *Dryopteris fragrans*. Zhongguo Zhong Yao Za Zhi 46:388–390
324. Puri HS, Widén CJ, Lounasmaa M (1976) Phloroglucinol derivatives in *Dryopteris chrysocoma*. Phytochemistry 15:343–344. https://doi.org/10.1016/S0031-9422(00)89031-7

325. Socolsky C, Domínguez L, Asakawa Y, Bardón A (2012) Unusual terpenylated acylphloroglucinols from *Dryopteris wallichiana*. Phytochemistry 80:115–122. https://doi.org/10.1016/j.phytochem.2012.04.017
326. Widen CJ, Ayras P, Neuvonen K, Reichstein T (1993) New phloroglucinol derivatives in *Dryopteris pulvinulifera* and *D. subtriangularis* (Pteridophyta, Dryopteridaceae). Ann Bot Fenn 30:285–297
327. Yang S, Chen W, Shan F, Jia X, Deng R, Tang C, Shen Z (2017) Antifungal activity of aspidin BB from *Dryopteris fragrans* against *Trichophyton rubrum* involved inhibition of ergosterol biosynthesis. Chinese Herbal Med 9:63–68. https://doi.org/10.1016/S1674-6384(17)60077-7
328. Widén CJ, Fraser-Jenkins CR, Reichstein T (1997) New phloroglucinol derivatives in *Dryopteris subimpressa* (Pteridophyta, Dryopteridaceae). Ann Bot Fenn 34:21–26
329. Fuchino H, Nakamura H, Wada H, Hakamatsuka T, Tanaka N (1997) Chemical and chemotaxonomical studies of ferns. Part XCI. two new acyl-phloroglucinols from *Dryopteris atrata*. Chem Pharm Bull 45:1101–1102. https://doi.org/10.1248/cpb.45.1101
330. Penttila A, Sundman J (1963) Phloraspyron and phloraspidinol, new phloroglucinol derivatives from *Dryopteris Ferns*. Acta Chem Scand 17:1866–1890
331. Su Y, Wan D, Song W (2016) Dryofragin inhibits the migration and invasion of human osteosarcoma U2OS cells by suppressing MMP-2/9 and elevating TIMP-1/2 through PI3K/AKT and p38 MAPK signaling pathways. Anti-Cancer Drugs 27:660–668. https://doi.org/10.1097/CAD.0000000000000381
332. Hisada S, Shiraishi K, Inagaki I (1972) Phloroglucinol derivatives of *Dryopteris dickinsii* and some related ferns. Phytochemistry 11:2881–2882. https://doi.org/10.1016/S0031-9422(00)86531-0
333. Chen NH, Qian YR, Li W, Zhang YB, Zhou YD, Li GQ, Li YL, Wang GC (2017) Six new acylphloroglucinols from *Dryopteris championii*. Chem Biodivers 14:e1700001. https://doi.org/10.1002/cbdv.201700001
334. Hou B, Zhang YM, Liao HY, Fu F, Li DD, Zhao X, Qi JX, Yang W, Xiao GF, Yang L, Zuo ZY, Wang L, Zhang XL, Bai F, Yang L, Gao GF, Song H, Hu JM, Shang WJ, Zhou J (2022) Target-based virtual screening and LC/MS-guided isolation procedure for identifying phloroglucinol-terpenoid inhibitors of SARS-CoV-2. J Nat Prod 85:327–336. https://doi.org/10.1021/acs.jnatprod.1c00805
335. Yan-long Z, Hai-yan F, Ying-ying Z, Qing-yu S, Wen-hua X, Hai-xue K (2008) Chemical components from *Dryopteris fragrans* and their cytotoxicity. Chin Tradit Herb Drug 39:648–651
336. Zhong ZC, Zhao DD, Liu ZD, Jiang S, Zhang YL (2017) A new human cancer cell proliferation inhibition sesquiterpene, dryofraterpene A, from medicinal plant *Dryopteris fragrans* (L.). Schott Mol 22:180. https://doi.org/10.3390/molecules22010180
337. Shiojima K, Arai Y, Ageta H (1990) Seasonal fluctuation of triterpenoid constituents from dried leaflets of *Dryopteris crassirhizoma*. Phytochemistry 29:1079–1082. https://doi.org/10.1016/0031-9422(90)85406-6
338. Sureshkumar J, Silambarasan R, Bharati KA, Krupa J, Amalraj S, Ayyanar M (2018) A review on ethnomedicinally important pteridophytes of India. J Ethnopharmacol 219:269–287. https://doi.org/10.1016/j.jep.2018.03.024
339. Maroyi A (2017) Utilization of pteridophytes as herbal medicines in sub-Saharan Africa. In: Neffati M, Najjaa H, Máthé Á (eds) Medicinal and aromatic plants of the world - Africa. Springer, Dordrecht, pp 383–408
340. Kumar R, Viktorova J, Krizkovska B, Lipov J, Ruml T (2021) Structural diversity and biological activities of secondary metabolites isolated from the genus *Selaginella*. Phytochem Rev 20:1209–1243. https://doi.org/10.1007/s11101-021-09743-7
341. Ma X, Gang DR (2004) The *Lycopodium* alkaloids. Nat Prod Rep 21:752–772. https://doi.org/10.1039/b409720n
342. Cheng DH, Ren H, Tang XC (1996) Huperzine A, a novel promising acetylcholinesterase inhibitor. Neuroreport 8:97–101. https://doi.org/10.1097/00001756-199612200-00020

343. Liang YQ, Tang XC (2004) Comparative effects of huperzine A, donepezil, and rivastigmine on cortical acetylcholine level and acetylcholinesterase activity in rats. Neurosci Lett 361:56–59. https://doi.org/10.1016/j.neulet.2003.12.071
344. Raves ML, Harel M, Pang YP, Silman I, Kozikowski AP, Sussman JL (1997) Structure of acetylcholinesterase complexed with the nootropic alkaloid,(−)-huperzine A. Nat Struct Mol Biol 4:57–63. https://doi.org/10.1038/nsb0197-57
345. Ou LY, Tang XC, Cai JX (2001) Effect of huperzine A on working memory in reserpine- or yohimbine-treated monkeys. Eur J Pharmacol 433:151–156. https://doi.org/10.1016/S0014-2999(01)01500-X
346. Zhang Z, Wang X, Chen Q, Shu L, Wang J, Shan G (2002) Clinical efficacy and safety of huperzine Alpha in treatment of mild to moderate Alzheimer disease, a placebo-controlled, double-blind, randomized trial. Zhonghua Yi Xue Za Zhi 82:941–944
347. Friedl P, Wolf K (2003) Tumour-cell invasion and migration: diversity and escape mechanisms. Nat Rev Cancer 3:362–374. https://doi.org/10.1038/nrc1075
348. Hausott B, Greger H, Marian B (2003) Naturally occurring lignans efficiently induce apoptosis in colorectal tumor cells. J Cancer Res Clin Oncol 129:569–576. https://doi.org/10.1007/s00432-003-0461-7
349. Li W, Tang GH, Yin S (2021) Selaginellins from the genus *Selaginella*: isolation, structure, biological activity, and synthesis. Nat Prod Rep 38:822–842. https://doi.org/10.1039/D0NP00065E
350. Nagai T, Myoda T, Nagashima T (2005) Antioxidative activities of water extract and ethanol extract from field horsetail (tsukushi) *Equisetum arvense* L. Food Chem 91:389–394. https://doi.org/10.1016/j.foodchem.2004.04.016
351. do Monte FHM, dos Santos JG, Russi M, Bispo Lanziotti VMN, Leal LKAM, de Andrade Cunha GM (2004) Antinociceptive and anti-inflammatory properties of the hydroalcoholic extract of stems from *Equisetum arvense* L. in mice. Pharmacol Res 49:239–243. https://doi.org/10.1016/j.phrs.2003.10.002
352. Čanadanović-Brunet JM, Ćetković GS, Djilas SM, Tumbas VT, Savatović SS, Mandić AI, Markov SL, Cvetković DD (2009) Radical scavenging and antimicrobial activity of horsetail (*Equisetum arvense* L.) extracts. Int J Food Sci Technol 44:269–278. https://doi.org/10.1111/j.1365-2621.2007.01680.x
353. Olazarán-Santibañez F, Rivera G, Vanoye-Eligio V, Mora-Olivo A, Aguirre-Guzmán G, Ramírez-Cabrera M, Arredondo-Espinoza E (2021) Antioxidant and Antiproliferative activity of the ethanolic extract of *Equisetum myriochaetum* and molecular docking of its main metabolites (apigenin, kaempferol, and quercetin) on β-tubulin. Molecules 26:443. https://doi.org/10.3390/molecules26020443
354. Ranjan V, Vats M, Gupta N, Sardana S (2014) Antidiabetic potential of the whole plant of *Adiantum capillus veneris* Linn. in streptozotocin-induced diabetic rats. Int J Pharmaceutical Clin Res 6:341–347
355. Pradeep Kumar R, Siddique S (2021) 22-Hydroxyhopane, a novel multitargeted phytocompound against SARS-CoV-2 from *Adiantum latifolium* Lam. Nat Prod Res:1–6. https://doi.org/10.1080/14786419.2021.1976177
356. Vijayalakshmi A, Kiran Kumar Y (2013) Evaluation of goitrogenic and antithyroidal effect of the fern *Adiantum capillus-veneris*. Rev Bras 23:802–810. https://doi.org/10.1590/S0102-695X2013000500013

Anticancer Properties of Pteridophytes and Derived Compounds: Pharmacological Perspectives and Medicinal Use

9

Vartika Jain, Mimosa Ghorai, Protha Biswas, and Abhijit Dey

Contents

Abstract

Pteridophytes are primitive vascular plants with sporophytic generation as a dominant phase. Moist and shady habitats are preferred for their luxuriant growth. Lycopods and Ferns are the two major groups of pteridophytes. These plants are used by ethnic communities for a variety of uses, such as food, beverage, and therapeutics. Cancer is a dreadful disease and modern synthetic drugs have their side effects along with high cost. Many plant species have been screened for their anticancer potential and pteridophyte flora is among one of them. Pteridophytes have been found effective against breast, brain, cervical, colorectal, liver, lung, ovary, pancreas, prostrate, gastric, blood, squamosal cancer cells, etc., and are also rich in many phytoconstituents which have shown to possess significant anticancer activity as demonstrated in various in vivo and in vitro studies. The major cytotoxic phytochemicals derived from pteridophytes belong to categories of flavanoids, phenolics, terpenoids, steroids, benzenoids, glycosides, and their

V. Jain
Department of Botany, Government Meera Girls College, Udaipur, Rajasthan, India

M. Ghorai · P. Biswas · A. Dey (✉)
Department of Life Sciences, Presidency University, Kolkata, West Bengal, India
e-mail: abhijit.dbs@presiuniv.ac.in

H. N. Murthy (ed.), *Bioactive Compounds in Bryophytes and Pteridophytes*,
Reference Series in Phytochemistry, https://doi.org/10.1007/978-3-031-23243-5_12

derivatives. The present chapter summarizes the anticancer property of pteridophytes and isolated bioactive molecules effective against various cancer cell lines.

Keywords

Biflavanoids · Fern · Pimpinellin · Protoapigenone · Pterosin · Tetracosane

Abbreviations

5-LOX	5 lipoxygenase
Akt	Protein kinase B
ATCC	American Type Culture Collection
Bax	B-cell lymphoma 2 Associated X
BPH	Benign prostatic hyperplasia
Caco-2	Cancer Coli
CC_{50}	50% Cytotoxic concentration
Cdk2	Cyclin-dependent kinase
CNS	Central Nervous System
COX-2	Cyclooxygenase 2
DNA	Deoxyribonucleic acid
DPPH	2,2-diphenyl-1-picrylhydrazyl
EA cells	Ehrlich ascites tumor cells
EC_{50}	Half maximal effective concentration
ED_{50}	Median effective dose
FAS	Fatty acid synthase
FITC	Fluorescein isothiocyanate
FL	Follicular lymphoma
GAPDH	Glyceraldehyde-3-phosphate dehydrogenase
HaCaT	Human epidermal keratinocyte
HCC	Hepatocellular carcinoma
HeLa	Henrietta Lacks
HepG2	Hepatocellular Carcinoma
HIV	Human Immunodeficiency Virus
hnRNP-A2/B1	Heterogeneous nuclear ribonucleoprotein A2/B1
HSP90	Heat-Shock Protein 90
HT-29	Human colorectal adenocarcinoma cell line
IC_{50}	Half maximal inhibitory concentration
IGF	Immortalized gingival fibroblasts
IUCN	International Union for Conservation of Nature
LC_{50}	Lethal Concentration (50%)
LNCaP	Lymph Node Carcinoma of the Prostate
M	Mitosis
MAPK	Mitogen-activated protein kinase
mRNA	Messenger RNA
MTT	(3-(4, 5-dimethylthiazolyl-2)-2, 5-diphenyltetrazolium bromide)
n-BuOH	Butanol

NF-κB	Nuclear Factor Kappa B
NO	Nitric Oxide
NPM	Nucleophosmin
Nrf2	Nuclear factor-erythroid factor 2-related factor 2
NRU	Neutral Red Uptake
PARP	Poly adenosine diphosphate-ribose polymerase
PC-3	Prostrate cancer cell line
PI	Propidium Iodide
PI3K	Phosphatidylinositol 3-kinase
ROS	Reactive Oxygen Species
TRAIL	Tumor necrosis factor-related apoptosis inducing ligand
UV	Ultra Violet

1 Introduction

Pteridophytes as primitive vascular plants were the dominant flora from 280 to 230 million years ago during the Triassic period. This group is the second-largest group of vascular plants present on earth having about 13,600 species distributed all over the world. Earlier, these were classified into four groups, Psilopsida, Lycopsida, Equisetopsida, and Pteropsida, and recently, in two, Lycopods and ferns [1]. Pteridophytes belong to the category of spore-bearing, non-seed plants [2–5]. Pteridophytes have been primary colonizers for various reasons, such as numerous spores which are desiccation resistant and disperse to newer areas and colonize barren lands as well as due to scattering of rhizome fragments especially in Horse tails after various catastrophic events that happened on the earth [6]. However, a recent assessment of 5% of known pteridophytes by the IUCN (Red List version 2019–3) reveals that 32% of species come into a threatened category which needs serious attention from scientists and conservationists [7].

Bryophytes and pteridophytes are jointly called “Amphibians of Plant Kingdom” and both groups do not produce seeds. Though pteridophytes have a sporophytic dominant generation with a vascular plant body separated into the root, stem, and leaves, Bryophytes had gametophytic dominant generation and mainly thalloid, avascular plant body. Habitat of both groups is almost similar to moist, damp, and shady places. Pteridophytes have been used for food, beverage, medicine, art, ornamentation, cosmetics, building material, domestic utensils purpose, etc. since ancient times [1, 8, 9]. Few fern species have been employed as biofertilizers and bioremediation agents. Pteridophyte flora has been long used in traditional medicinal systems, such as Ayurveda, Unani, Chinese, Korean, and Homeopathy [4, 10]. Several pharmacological activities, for example, anti-inflammatory, antioxidant, antifungal, antibacterial and antiviral, anti-HIV, antimalarial, anti- Alzheimer’s, anti-diabetic, wound healing, cytotoxic, acetylcholinesterase inhibitory, neuroprotective, antitrypanosomal, anthelmintic, bronchodilator, immunomodulatory, CNS stimulant, mast cell stabilizing, anti-anaphylactic, etc. have been demonstrated from various pteridophyte plant species [11, 12]. Many secondary metabolites have also been isolated from its various plant extracts such as triterpenoids, diterpenoids, sesquiterpenoids, phenolic compounds, flavonoids, alkaloids, steroids, amino acids and fatty acids [12, 13].

Cancer is one of the scariest diseases implicated with a painful death. Data suggests that the number of cancer instances is going to increase to 22 million by 2032. The modern-day available cancer drugs are armed with cytotoxic/chemopreventive potential along with some serious side effects and increasing drug resistance. Plants are excellent sources of several phytochemicals which show bioactivity against cancer cells. It is estimated that more than 60% of anti-cancer medications are derived either directly or indirectly from plants [14, 15]. The wider safety profile of plant species makes them appropriate candidates for screening of various phytopharmaceutical compounds. Seed plants have long been used for this purpose and Pteridophytes do not lag. Many of the pteridophyte plant species have been evaluated for their cytotoxic/antiproliferative potential in both in vivo and in vitro studies which are described in this chapter with emphasis on the specific bioactive molecule.

2 Anticancer Activity of Pteridophytes

Many pteridophyte species have been screened for their anticancer potential [4, 12] as described briefly in the following paragraphs:

Ahn et al. [16] have demonstrated in vitro anti-cancer efficacy of *Selaginella tamariscina* using the MTT assay. Apoptosis was induced by plant extract dose-dependently in human leukemia HL-60 cells and induced nuclear condensation and DNA fragmentation, caspase activation, and the specific proteolytic cleavage of PARP (poly(ADP-ribose) polymerase) with the increase in the proapoptotic Bax levels, and decrease in Bcl-2 expression. Extracts in ethyl acetate of *Selaginella labordei*, *Selaginella tamariscina*, and *Selaginella uncinata* have shown to significantly reduce the viability of HeLa cells in a way that depends on the dose in the MTT assay. Cytotoxic efficacy was observed in the following manner for both Bel-7402 and HeLa cells: *S. pulvinata* < *S. remotifolia* < *S. delicatula* < *S. moellendorfii* < *S. uncinata* < *S. tamariscina* < *S. labordei*. Apoptosis of these extracts was less prominent on HT-29 cells than on HeLa cells and IC_{50} values of <10 μg/ml for ethyl acetate extracts of *S. labordei*, *S. tamariscina*, and *S. uncinata* in cell growth inhibition and apoptosis [17]. Chloroform, ethyl acetate, and hexane, fractions of *Selaginella convoluta* were evaluated for cytotoxicity potential by MTT assay against cancer cell line of human ovarian cancer cell (OVCAR-8), human glioblastoma (SF-295), human colorectal (HCT116) at 50 μg/ml dose. Minimal cytotoxic activity was detected with all the fractions excluding the chloroform fraction, which showed moderate cytotoxicity (62.23%) against the SF-295 cell line after 72 h of incubation [18].

Methanolic extract of *Actiniopteris semiflabellata* has been shown to scavenge DPPH radical 55.39% at 50 μg/ml concentration and also exhibited an IC_{50} value of 950 μg/ml using FL-Cells; the human amniotic epithelial cell line in neutral red uptake assay [19].

Leaves of *Blechnum orientale* L., collected from Kuala Lumpur were tested for cytotoxic potential toward four malignant cell lines, namely, HT-29 (human colonic adenocarcinoma), HCT-116 (human colonic carcinoma), MCF-7 (human breast adenocarcinoma), K562 (human leukemia), and one non-malignant Chang liver cells using Thiazolyl Blue Tetrazolium Bromide (MTT) assay. The methanolic extract was partitioned with various solvents, such as *n*-butanol, petroleum ether, ethyl acetate, chloroform, and water. However, cytotoxicity analysis was performed only with butanol, water, and ethyl acetate fractions due to solubility issues. Cytotoxic activity of butanol, water, and ethyl acetate fractions was obtained toward colonic HT-29 (adenocarcinoma cell line) with IC_{50} of 27.5, 33.4, and 42.8 μg/ml, respectively, which were higher than IC_{50} of 5.4 μg/ml obtained for the positive control curcumin. Butanol and water fractions have shown mild cytotoxic action toward human HCT-116 (colonic carcinoma cells) with IC_{50} values of 72.7 ± 4.1 and 93.9 ± 3.2 μg/ml, respectively. No significant anticancer efficacy was observed against K562, MCF-7, and the non-malignant Chang liver cells. This selective cytotoxicity is favorable for patients with colon cancer and emphasizes the need for the isolation of bioactive molecules from the butanol fraction of *B. orientale* [20].

Ethyl acetate extract of *Equisetum arvense* L. has demonstrated an antiproliferative efficacy on the human cancer MCF7, HeLa, and HT-29 without stimulating any cell development in human tumor cell lines in the sulforhodamine B colorimetric assay lines [21]. *Equisetum arvense* L. is rich in phenolic acids, flavonoids, and flavonoid-O-glycosides. Recently, Zia-Ur-Rehman [22] isolated 22 compounds from the polarity-based fractions and crude extracts of *E. arvense* and which are quercetin, kaempferol, luteolin, apigenin, tricin, sakuranetin, genkwanin, 7-O-methylchrysin, kaempferol 3-O-glucoside, daidzein-4′,7-diglucoside, isoquercetin, and apigenin-4′-O-glucoside, 6″-O-acetylgenistin, *trans*-ferulic acid, 6″-O-acetylglycitin, 2′,4′-dihydroxy-4-prenyloxy chalcone, 8-acetlyharpagide, pinolenic acid, monolinolenin, myrcene, dihydroactinidiolide, and 4-(*sec*-butoxy) benzoic acid. The absorbance assay was used to assess the crude extract's cytotoxic activity and that of its fractions (CellTiter-Blue® Cell Viability Assay) against MDA-MB-435 (human melanoma) and OVCAR3 (ovarian cancer cells). At 500 μg/ml concentration of every sample, ~ 95% inhibition was observed against MDA-MB-435 cells whereas the *n*-hexane fraction exhibited 95.5% inhibition against OVCAR3 cancer cells at 500 μg/ml concentration. Recently, silver nano-particles were developed from aqueous extract of *Equisteum arvense* and have shown cytotoxic efficacy toward HepG2 cancer cell lines in a manner dependent on concentration [23].

The ethanolic extract of *Adiantum venustum* Don. leaves and stem in two doses (150 and 250 mg/kg) were administered intraperitoneally to Ehrlich Ascites Carcinoma bearing male Swiss albino mice which significantly inhibited the packed cell volume, tumor volume, tumor cell count dose dependently and increasing proportion of dead cells in tumor-bearing mice stained with trypan blue. Moreover, the median survival time was increased to 24 ± 0.33 and 29 ± 0.49 with doses of 150 and 250 mg/kg, respectively, as compared with control mice (22 ± 0.25 days). Besides, hematological parameters, antioxidant enzymes, and hepatic lipid peroxidation were also restored indicating that the antitumorigenic effect of *Adiantum venustum* could be due to the antioxidant property of the plant extract [24].

Methanolic and chloroform extracts of leaves of *Diplazium esculentum* have shown brine shrimp nauplii mortality percentage increased in a concentration-dependent manner with LC_{50} values of 1.87 μg/ml and 1.62 μg/ml, respectively, in comparison to LC_{50} of 0.66 μg/ml of the standard anticancer drug vincristine sulphate exhibiting the occurrence of cytotoxic principle in the plant [25]. In another experiment, the methanol extract of *D. esculentum* has also been found to be toxic against brine shrimp where LC_{50} was 18.56 μg/ml [26]. No cytotoxic efficacy of ethanolic extract of *D. esculentum* shoots was observed against MCF-7 and MDA-MB-231 (breast cancer), Caco-2 (colon cancer), HepG2 (liver cancer), and Chang's liver (normal liver) cell lines [27].

Phymatopteris triloba (Houtt) Pichi Serm of the *Polypodiaceae* family is rich in phenolic chemicals, such as protocatechuic acid, p-hydroxybenzoic acid, sinapic acid, gallic acid, and myricetin. Rhizomes and leaf extracts of the plant were subjected to evaluate the antiproliferative action against the HeLa (human cervix cancer cell line) and K562 (human chronic myelogenous leukemia cell line) using an MTT assay. Both the plant part extracts dose-dependently showed antiproliferative activity against the tested cell lines and about 30% and 42% antiproliferative activity was observed against K562 cells and HeLa cells, respectively, at 500 μg dry matter/ml concentration of leaf extract [28].

Methanol and ethyl acetate extracts of leaves of *Lygodium venustum* have demonstrated 67% and 53% potential for killing mammalian fibroblast (NCTC-929) cells, respectively, in a concentration of 500 μg/ml. However, no cytotoxic effect of dichloromethane extract was observed in the same concentration [29].

Ethyl acetate, methanol, and aqueous leaf extracts of *Marsilea quadrifolia* were investigated for in vitro cytotoxic potential on MCF-7 (human breast cancer) cell line using MTT assay. Cell viability and growth were inhibited dose-dependently with IC_{50} values of 39.065 μg/ml, 47.825 μg/ml, and 187.5 μg/ml for methanol, ethylacetate, and aqueous extracts, respectively [30].

Cytotoxicity of leaves of *Nephrolepis exaltata* has been demonstrated against prostate cancer cell lines – LNCaP and PC-3. Extracts from the *Cibotium barometz, Cyrtomium falcatum,* and *Nephrolepis exaltata* have also been shown to elicit hormonal effects on PC-3 and LNCap prostate cancer cells [31].

Cytotoxic activity of a fern *Asplenium aethiopicum* (Burm. f.) Becherer was studied in the lethality test in brine shrimp. The acetone extract was the most effective extract exhibiting dose-dependent brine shrimp nauplii mortality with an LC_{50} of 192.8 ppm and LC_{90} of 434.3 ppm as compared to chloroform, methanol, and petroleum ether extracts [32].

Cytotoxic activity of six ferns, *Christella arida* (D. Don) Holtt (IC_{50} 478.62 ± 39.11 μg/ml), *Christella dentata* (Forsk.) Brownsey & Jermy (194.50 ± 14.74 μg/ml), *Cyclosorus interruptus* (Willd.) H. Ito (314.52 ± 6.34 μg/ml), *Microsorum punctatum* (L.) Copel. (399.68 ± 48.60 μg/ml), *Nephrolepis acutifolia* (Desv.) Christ. (190.82 ± 5.52 μg/ml), and *Pleocnemia irregularis* (C. Presl.) Holtt. (253.85 ± 22.82 μg/ml) has been demonstrated toward

K562 (human chronic myelogenous leukemia cell line) in MTT assay. The EC_{50} value of 212.86 ± 7.89 μg/ml was obtained for the positive control 5-fluorouracil which was lesser than the EC_{50} value obtained for *Christella arida*. These fern species, therefore, could be further analyzed for the isolation of bioactive anti-cancer molecules [33].

Methanolic extracts of leaves of *Pteris cretica* and *Athyrium filix-femina* have shown strong cytotoxic efficacy in the lethality assay in the brine shrimp where LC_{50} values were 15.5 and 6.1 μg/ml correspondingly as compared with podophyllotoxin as a positive control with LC_{50} value of 30 μg/ml. Besides these plants, methanolic extracts of rhizome and leaves of *Polystichum woronowii* Fomin, *Polypodium interjectum* Shivas, *Polystichum aculeatum* (L.) Schott, *Dryopteris affinis* (Lowe) Fraser-Jenk, *Asplenium scolopendrium* L., and *Asplenium adiantum-nigrum* L. were also tested for the lethality assay in brine shrimp and a dose-dependent larva lethality rate was observed. Among these ferns, leaf extracts of *Asplenium adiantum-nigrum* and *Asplenium scolopendrium* have shown LC_{50} values of 112 and 158.9 μg/ml, respectively, and *Polystichum woronowii* and *Dryopteris affinis* have shown LC_{50} values of 52.2 and 85.5 μg/ml, correspondingly [34].

Araiostegia yunnanensis (Christ) Copel possesses many flavanoids, such as eriodictyol 7-O-rutinoside, myricetin 3-O-rhamnosylglucoside, quercetin 3-O-rutinoside, luteolin 7-O-glucoside, luteolin-7-O-apiosylglucoside, and quercetin 3-O-rhamnosylgalactoside with total flavonoids content of 84.90 mg/g. Its methanolic extract has also shown cytotoxic activity against A549 cells without any effect on acetylcholinesterase [35].

A study by George and Josekumar [36] has shown very little cytotoxic activity of a fern *Acrostichum heterophyllum* L. IC_{50} values of 254.18, 266.99, and 191.93 μg/ml were obtained for hexane, methanol, and ethyl acetate extracts, respectively, of the fern in MTT assay using L929 normal mouse fibroblast cell line. Moreover, all these extracts were non-toxic toward the brine shrimp napulii at the concentration of 1 mg/ml.

Reshi et al. [37] have also shown cytotoxic action of crude extract in the water of *Adiantum capillus-veneris* against HELA cell line where IC_{50} value was 36.96 μg/ml. Cytotoxicity of various extracts prepared from *Adiantum capillus-veneris* L. aerial parts against HELA cells has been demonstrated by Roy et al. [38]. The IC_{50} value of 52.53 μg/ml, 17.15 μg/ml, 32.35 μg/ml, and 5.68 μg/ml were obtained for methanolic, n-hexane, ethyl acetate, and chloroform extracts, respectively as compared to IC_{50} of 0.87 μg/ml shown by 5-fluorouracil as a positive control. Lower IC_{50} values of ethyl acetate and n-hexane extracts indicate the existence of bioactive substances, such as triterpenoids which might be responsible for the cytotoxic action.

Methanol and petroleum ether extracts of *Dicranopteris linearis* (Burm.f.) Underw. of the family Gleicheniaceae, leaves were investigated for in vitro cytotoxic potential against MCF-7 and MDA-MB-231 (breast), HeLa (cervical), HT-29 (colon), HepG2 (hepatocellular), and A549 (lung) cancer cell lines using MTT assay. The most notable growth restriction was observed toward MDA-MB-231 cells with an IC_{50} of 22.4 μg/ml by methanolic extract whereas petroleum ether extract did not show any cytotoxic effect. Apoptosis was induced by methanolic extract in MDA-MB-231 cells in a temporal fashion as validated by PI staining and annexin V-FITC staining [39].

Recently, six differentially expressed proteins were isolated from HeLa (Human cervical cancer cells) and K562 (human myelogenous leukemia cancer cells), after being exposed to cytotoxic extracts of two ferns, namely, *Blechnum orientale* and *Phymatopteris triloba*. Four cancer cell proteins are differently expressed, namely, elongation factor 1-γ (EF1-γ), GAPDH (glyceraldehyde-3-phosphate dehydrogenase), HSP90-β, and hnRNP-A2/B1 (heterogeneous nuclear ribonucleoprotein A2/B1) were isolated after exposure with *Blechnum orientale* extract. This suggests that the lethal effects of *B. orientale* extract may result from altering the mechanism for protein synthesis in cancer cells and by lowering EF1- expression levels [40].

Two differentially expressed cancer proteins, namely, nucleophosmin (NPM) (truncated nucleolar phosphoprotein B23) and tubulin-β (tubulin beta chain) were isolated after exposure to the *Phymatopteris triloba* extract. Downregulation of NPM expression indicates toward detection and creation of chemotherapeutic drugs with particular target activity from other fern species [40].

Additionally, a brine shrimp lethality test was utilized to evaluate the possible cytotoxicity of the whole plant of *Cheilanthes tenuifolia* collected from Dhaka. Chloroform and ethyl acetate extracts have shown more cytotoxic potential with the LC_{50} values of 34.493, and 66.235 μg/ml, correspondingly as compared with LC_{50} values of 205.984, and 751.169 μg/ml obtained for methanol and N-hexane extracts, respectively in terms of brine shrimp nauplii mortality. The LC_{50} value was 2.254 μg/ml of the reference cytotoxic drug vincristine sulfate [41].

Methanol extract of fronds of *Polypodium vulgare* is rich in phenolic compounds and used for evaluation of cytotoxicity against 3 T3 and HaCaT (non-tumoral) and A549, MCF-7, HeLa, and HepG2 (tumoral cells lines) using NRU and MTT assays. The extract did not show cytotoxicity at physiological doses as determined in the six cell lines. Among the non-tumoral cell lines slightly higher cytotoxicity activity was observed against HaCaT cell lines (35.3%) than in 3 T3 (46.4%) at a 1 mg/ml concentration of methanolic extract. A decrease in cell viability was observed only in the concentrations of 1 and 2 mg/ml in HepG2, 3 T3, HeLa, and, HaCaT cell lines [42]. Another *Polypodium* species, *Polypodium leucotomos* has also shown its photoprotective role against skin cancer, photoaging, and photocarcinogenesis [43]. Phenolic chemicals are abundant in *P. leucotomos* and possess strong antioxidant properties. Its extract has been shown to inhibit AP1 and NF-B, stop UV and ROS from causing DNA damage and decreasing cellular apoptosis and inflammation. It triggers tumor suppressor p53 as well as suppresses the production of the COX-2 (cyclooxygenase-2) enzyme when UV is present and helps in decreasing UV radiation-induced cell proliferation [44].

Phenolic and flavanoid rich herb *Stenoloma chusanum* has demonstrated greater suppression of tyrosinase than of arbutin as well as the greatest apoptosis and proliferation of K562 cells [45].

Extract of *Athyrium multidentatum* (Doll.) Ching was discovered to have cytotoxic activity against hepatocellular carcinoma (HepG2) cells where IC_{50} values after 24 h and 48 h were 220.32 ± 10.32 and 113.51 ± 7.45 μg/ml, respectively. It also increased the protein production of Fas-L, Fas, Bax/Bcl-2, PARP, cleaved caspase-3, and cyto-c in a dose-dependent manner to cause apoptosis and cell cycle arrest and created oxidative stress by the generation of ROS, NO, and H_2O_2, and modifying the PI3K/Akt, Nrf2, MAPK, and NFκB pathways leading to apoptosis in HpeG2 cells. In vivo antitumor efficacy of *Athyrium multidentatum* was also observed in the xenograft tumor mice model [46].

Extract of *Dryopteris crassirhizoma* at a dose of 50 μg/ml and 100 μg/ml shown antiproliferative activity toward human metastatic prostate PC-3 and PC3-MM2 cells without any cytotoxic action against normal spleen cells from BALB/C mice. At 100 μg/ml concentration, it also triggered apoptosis through the extrinsic and intrinsic pathway by the initiation of caspase-3, caspase −8, caspase −9, PARP, and bid in PC3-MM2 cells and causing cell cycle arrest. The combined effect with TRAIL (tumor necrosis factor-related apoptosis-inducing ligand) and effective inhibition of cell proliferation was observed [47]. Recently, ethanolic extract of *Dryopteris crassirhizoma* rhizome was evaluated for cytotoxicity in a mouse xenograft model of mandibular invasion by gingival squamous cell carcinoma. In the concentration of 5 and 30 μg/ml, the plant extract had no impact on the viability of immortalized gingival fibroblasts (IGF) and HSC-3 OSCC cells as determined through the MTT assay. IC_{50} values at 24 h for HSC-3 OSCC and IGF were 90 and 82.1 μg/ml whereas the viability of normal gingival fibroblasts was not affected even at >130 μg/ml concentration at 24 h. However, migration and invasion of gingival cancer cells were inhibited at 5 or 10 μg/ml concentrations and a decrease in tumor growth and osteolytic mandibular bone lesions was noted in the group receiving treatment after 5 weeks [48]. Hence, *Dryopteris crassirhizoma* could be more investigation done for the isolation of anti-cancer bioactive molecules.

Tectaria cicutaria rhizome ethanolic extract has shown anticancer efficacy toward K562 (Human Leukemia Cell Line) with a GI_{50} value of 11.9 μg/ml [49]. Cytotoxic activity of different extracts of whole plant material of the *Tectaria paradoxa* (Fee.) Sledge. was also assessed through brine shrimp lethality bioassay. Dose-dependent mortality in *Artemia salina* was observed along with induction of morphological changes affecting the loss and deformation of antennae, ability of swimming, feed, and enlargement of the intestine. The LC_{50} values obtained for chloroform, petroleum ether, methanol, and acetone extracts were 25.52, 36.99, 44.26, and 55.9 μg/ml, respectively [50].

Aerial portions of the three species of fern *Asplenium*, namely, *Asplenium trichomanes* L., *Asplenium ceterach* L., and *Asplenium scolopendrium* L., were screened for cytotoxicity against three human lung carcinoma (A549); human cervical adenocarcinoma (HeLa); human amnion origin (FL), and three murine cell lines, namely, RAW 264.7, designated as TIB-71, mouse monocyte/macrophage cell line generated from a tumor caused by the Abelson murine leukemia virus; mouse embryonic fibroblasts (NIH/3 T3) and mouse fibroblast-like permanent cell line (LS48) using MTT and Trypan blue assays. The strongest cytotoxic activity against human cervical cancer cells was observed for an extract of *A. ceterach* with a mechanism of high proapoptotic potential and the ability to induce oxidative damage. The IC_{50} values obtained for HeLa cells after 24 h of treatment with extracts of

A. ceterach were 40.48 ± 6.47 μg/ml and for *A. trichomanes* 120.68 ± 4.7 μg/ml and for *A. scolopendrium* was 204.83 ± 3.6 μg/ml. However, the A54 9 (cell line of lung adenocarcinoma) was not much impacted by *Asplenium* extracts. The three mouse cell lines exposed to *A. scolopendrium* extract exhibited the greatest cytotoxicity with the most sensitive effect on tumor-derived TIB-71 cell lines. *A. scolopendrium* and *A. trichomanes* extracts were also found to be most sensitive against NIH/3 T3 cell lines and both these extracts were observed as potent inducers of necrotic cell death. Marked reduction in cell viability was also observed for all three extracts in Trypan blue vitality assay [51].

3 Anticancer Compounds from Pteridophytes

3.1 Lycopods

3.1.1 *Lycopodium*

Pentacyclic and serratene triterpenoids isolated from *Lycopodium phlegmaria*, namely, Lycophlegmariol B (21β,24,29-trihydroxyserrat-14-en-3β-yl dihydrocaffeate), 14β,21α,29-trihydroxyserratan-3β-yl dihydrocaffeate (Lycophlegmariol D), and 21-β-hydroxy-serrat-14-en-3α-ol has shown that the T-lymphoblast (MOLT-3 acute lymphoblastic leukemia) has been susceptible to inhibition, where IC_{50} values were 14.6 μM, 3.0 μM, and 2.9 μM, respectively [52]. Shi et al. [53] have reported in vitro growth-inhibitory activity of Lycophlegmarin; a serratane-type triterpene, toward BEL 7402 (human hepatoma) cells.

3.1.2 *Selaginella*

Ginkgetin; isolated from *Selaginella moellendorffii* ethanolic extract has shown to inhibit OVCAR-3 (human ovarian adenocarcinoma) cells dose-dependently where IC_{50} value was 1.8 μg/ml [54] and IC_{50} values were 3.0, 5.2, and 8.3 μg/ml against OVCAR-3, HeLa (cervical carcinoma), and FS-5 (foreskin fibroblast), respectively [55].

Selaginella doederleinii Hieron extract in ethyl acetate has shown dose-dependent cytotoxicity against hepatocellular carcinoma (HepG2), cervical carcinoma (Hela), lung cancer (A549), prostatic carcinoma (DU145), pheochromocytoma (PC12), and African green monkey kidney (Vero) cells with IC_{50} values as 65.8 ± 4.4, 76.1 ± 1.9, 51.9 ± 1.5, 70.5 ± 2.6, >150 and >150 μg/ml, correspondingly in the MTT assay. The higher IC_{50} value against pheochromocytoma indicates that plant extract can do little damage to the central nervous system as well as low cytotoxicity for Vero cells. In vivo antitumor potential of *Selaginella doederleinii* extracts was confirmed by a significant decrease in the weights of tumors of H-22 hepatoma carrying Kunming mice. The plant extract having many bioflavonoids induced cell apoptosis which might decrease the bax and bcl-2 ratio, and mRNA levels, activate caspase-3, suppress survivin, and reduce the COX-2, 12-LOX, 5-LOX, and FLAP mRNAs expression to promote cell apoptosis [56]. Cytotoxic activity of *S. doederleinii* has also been demonstrated toward NCI-H358, CNE,

HCT, and K562, and cells [57]. Biflavonoids, like robustaflavone, amentoflavone, 2″,3″-dihydro-3′,3‴-biapigenin, 3′,3‴-binaringenin, 7,4′,7″,4‴-tetra-O-methyl-amentoflavone and have aflavone have shown strong antitumor activities [17, 58]. *S. doederleinii* ethanolic extract has also been demonstrated to cause human nasopharyngeal cancer CNE cells to undergo mitochondria-related apoptosis [57].

Cytotoxic bioflavonoids, such as 4′,7″-di-O-methylamentoflavone, isocryptomerin, and 7″-O-methylrobustaflavone, have also been isolated from leaves of *Selaginella willdenowii* [59]. *Selaginella delicatula* is rich in various biflavonoids, such as robustaflavone 7,4′-dimethyl ether, 2″,3″ -dihydrorobustaflavone 7,4′, –dimethyl ether, robustaflavone 4′-methyl ether, 2″,3‴-dihydrorobustaflavone 7,4′,7″-trimethyl ether, robustaflavone 7,4′,4‴-trimethyl ether, robustaflavone 4′,4‴-dimethyl ether, 2,3-dihydroamentoflavone 7,4′-dimethyl ether, 2,3-dihydroamentoflavone 7,4′,7″ -trimethyl ether, 2″,3″-dihydroisocryptomerin 7-methyl ether, robustaflavone, amentoflavone, and three caffeoylquinic acids, 3,5-di-O-caffeoylquinic acid, 4,5-di-O-caffeoylquinic acid, and 3, 4-di-O-caffeoylquinic acid. Complexes, such as robustaflavone 4′-methyl ether and 2″,3″ -dihydrorobustaflavone 7,4′,-dimethyl ether have shown to inhibit the Calu-1 and Raji tumor cell lines growth as well as robustaflavone 4′,4‴-dimethyl ether, alpha-tocopheryl quinone, and 2,3-dihydroamentoflavone 7,4′-dimethyl ether also demonstrated cytotoxic efficacy toward HT-29 and/or P-388 cell lines in vitro with ED50 values <4 μg/ml [60, 61].

Zhang et al. [62] have also found the cytotoxicity of the *Selaginella tamariscina* (Beauv.) Spring against human cancer cell lines, human glioma cells (U251), human cervical carcinoma cells (HeLa), and human breast cancer cells (MCF-7). Two compounds, namely, Selaginellin M and Selaginellin N showed moderate efficacy in the tested cancer cell lines.

3.2 Ferns

3.2.1 *Abacopteris*

The cytotoxic activity of triphyllin A and abacopterins A-D, isolated from methanolic extract of *Abacopteris penangiana* rhizome was observed against HepG2 human hepatoma cells [63]. A significant in vitro anticancer activity of two compounds, namely, (2R,4S)-5,7-O-beta-D-di-glucopyranosyloxy-4′ -methoxy-6,8-dimethyl-4,2″-oxidoflavane and (2R,4S)-6,8-dimethyl-7-hydroxy-4′ -methoxy-4,2″-oxidoflavan-5-O-beta-D-6″-O-acetyl-glucopyranoside derived from *A. penangiana* rhizomes has been demonstrated toward L929 and HeLa cell lines [64]. Protective effect of total flavanol glycosides, isolated from *Abacopteris penangiana* (Hook.) Ching and its acid hydrolysate (100 or 200 mg/kg/d) were evaluated on testosterone-induced BPH (benign prostatic hyperplasia) in rats after 4 weeks of administration. Both the chemicals decreased serum 5-alpha-reductase and dihydrotestosterone levels as well as inflammatory cytokines in prostatic tissue and inhibited the expressions of the proteins like NF-κB, Bcl-2, and p-Akt indicating their anti-BPH action through anti-proliferative, anti-inflammatory, and antioxidant effects [65].

3.2.2 *Acrostichum*

Uddin et al. [66] have derived two new sesquiterpenes and their two derivatives from methanol extract of the *Acrostichum aureum* aerial part which were evaluated against human cancer cell lines, such as AGS, MDA-MB-231, HT-29, and MCF-7 and the normal fibroblast cell line of the mouse (NIH3T3) using the MTT assay. The compound (2*S*,3*S*)-sulfated pterosin C exhibited cytotoxicity against the tested cells with IC_{50} values ranging between 23.9 μM and 68.8 with the lowest IC_{50} value of 23.9 μM toward AGS (gastric adenocarcinoma) cells by exerting an apoptotic activity after treatment within 24 h on AGS cells.

Methanol extract of *Acrostichum aureum* L. (Pteridaceae) whole plant was utilized to isolate tetracosane, patriscabratine, and five flavonoids, namely, quercetin-3-*O*-α-l-rhamnoside, quercetin-3-*O*-β-d-glucoside, kaempferol, quercetin-3-*O*-β-d-glucosyl-(6 → 1)-α-l-rhamnoside, and quercetin-3-*O*-α-l-rhamnosyl-7-O-β-d-glucoside. These compounds were investigated for their in vitro cytotoxic action on normal mouse fibroblast cells and four human cancer cell lines (ATCC: CRL-1658, NIH 3 T3; ATCC: CRL-1739, AGS; ATCC: HTB-38, HT-29; ATCC: HTB-22, MCF-7; and, ATCC: HTB-26, MDA-MB-231) using the MTT assay. Significant cytotoxic activity of Patriscabratine against gastric (AGS) with IC_{50} of 133.6 μM and MCF-7 and MDA-MB-23 (breast cancer cells) with IC_{50} of 69.8 μM and 197.3 μM, respectively, and non-cytotoxic toward NIH3T3 (normal mouse fibroblasts) and HT-29 (colon cancer cells) was observed. Tetracosane exhibited significant cytotoxicity against HT-29 (colon cancer) cells where the IC_{50} value was 128.7 μM and moderate cytotoxic activity against gastric cancer cells (AGS), estrogen-dependent breast cancer cells (MDA-MB-231), NIH 3 T3, and HT-29 cells with IC_{50} values varying from 128.7 μM – >250 μM but there was no cytotoxicity toward MCF-7 (breast cancer) cells independent of estrogen. Both the active compounds tetracosane and patriscabratine exhibited apoptosis at a rate of about 10% after 24 h and 20% apoptosis after 48 h of treatment against human gastric adenocarcinoma (AGS) cells using the FITC Annexin V apoptosis assay which was greater than the apoptosis effects obtained for positive control cycloheximide. However, the other five flavanoids did not show noticeable cytotoxicity with IC_{50} of >500 μM toward the tested cell lines [67].

3.2.3 *Asplenium*

Twelve flavanoids have been derived from the various fractions of methanolic extract of whole plant material of *Asplenium nidus*, such as gliricidin 7-O-hexoside, globularin, apigenin 7-O-glucoside, keampferol 7-O- gentiobioside, quercetin 7-O-rutinoside, quercetin 7-O-galactoside, keampferol-3-O-rutinoside, myricetin 3-O-rhamnoside, linoleic acid dimer, keampferol 3-O-rhamnoside, quercetin, and keampferol-7-O-rutinoside. Two of these flavonoids, quercetin-7-O-rutinoside, and gliricidin-7-O-hexoside have shown a down-regulating effect on the HeLa and HepG2 cells growth in MTT assay at all the concentrations from 240 to 600 μg /ml with the IC_{50} value 507 lg/ml. The maximum viability of gliricidin-7-O-hexoside was observed for HepG2 cells at 76.15 ± 0.04 and HeLa cells at 66.91 ± 1.93 at 600 μg/ml concentration [68].

3.2.4 *Cheilanthes*

Two flavanoids, namely, rutin and quercetin were extracted from the *Cheilanthes tenuifolia* plant's methanolic extract by the use of column Sephdex LH and screened for possible anticancer effects against human cancer HeLa and human hepatoma HepG2 cells using MTT assay. Both the bioactive molecules prevented HepG2 and HeLa cells from growing at all the selected concentrations ranging from 240 μg/ml to 600 μg/ml. Quercetin showed maximum cytotoxicity, 78.16 ± 0.04% against HepG2 and 80.91 ± 1.93% against HeLa cells at an 600 μg/ml concentration. Comparatively, rutin showed weak anti-cancer efficacy against HepG2(−5.67 ± 2.59%) and HeLa cells (11.10 ± 2.10%) at 600 μg/ml [69].

3.2.5 *Cyclosorus*

Two coumarin compounds, 5,7-dihydroxy-6-methyl-4-phenyl-8-(3-phenyl-trans-acryloyl)-1- benzopyran-2-one and 5,7-dihydroxy-6-methyl-4-phenyl-8-(3-phenylpropionyl)-1-benzopyran-2-one isolated from *Cyclosorus interruptus* (Willd.) H. Itô has also shown cytotoxic potential toward KB (human nasopharyngeal carcinoma) cell line [70].

Seven chalcone components are isolated from the leaves of *Cyclosorus parasiticus* out of which parasiticin C and 2′,4′-dihydroxy-6′-methoxy-3′,5′ -dimethylchalcone have shown in vitro cytotoxicity toward the six human cancer cell lines. Both were specifically cytotoxic against HepG2 cells where the IC_{50} values were 1.60 μM and 2.82 μM, correspondingly, and also incite the HepG2 cell line to undergo apoptosis [71].

3.2.6 *Cyrtomium*

Twenty phytoconstituents, namely, woodwardinsauremethylester, physcion, pimpinellin, trans-2-coumaric acid, protocate chaldehyde, ursolic acid, betulin, sitost-4-en-3-one, 3′,4′,5-trihydroxy-3,7-dimethoxyflavone, sitosterol-3-O-β-D-glucopyranoside, woodwardinic acid, sutchuenoside A, kaempferol-3,7-O-α-L-dirhamnoside, β-sitosterol,(−)-epicatchin, kaempferol, (+)-catechin hydrate, asiatic acid, crassirhizomoside A, 2β,3β,23-tihydroxy-12-oleanen-28-oic acid, kaempferol-3-O-(3-O-acetyl-α-L-rhamnopyranoside), kaempferol-3-O-α-L-rhamnopyranoside-7-O-α-L-rhamopyranoside, and 2α,3α,24-trihydroxyurs-12-en-28-oic acid have been isolated from the *n*-BuOH (Butanol) and ethyl acetate extracts of *Cyrtomium fortumei* (J.) Smith rhizomes. In vitro cytotoxic activity of these compounds and extracts (20 μM or 50 μg/ml) was evaluated on Stomach cancer (MGC-803); Prostate cancer (PC3); Malignant melanoma (A375) and Mouse fibroblasts (NIH3T3) by MTT (thiazolyl blue tetrazolium bromide) assay. *n*-BuOH and ethyl acetate and extracts had shown potent antitumor efficacy and the compound pimpinellin had also shown concentration-dependent cytotoxicity against the three tumor cell lines, where the IC_{50} values were 14.4 ± 0.3, 29.2 ± 0.6, and 20.4 ± 0.5 μM, against MGC-803,

A375, and PC3 cells, respectively; lower compared to NIH3T3 cells where the IC_{50} value is >100 μM. The inhibition percentage of pimpinellin after treatment of 72 h was 57.2%, 67.1%, 24.8%, and 45.8% on PC3, MGC-803, NIH3T3, and A375 cells, respectively. Moreover, it also promoted the death of MGC-803 cells, with the maximum apoptosis ratio occurring 72 h after treatment, at 27.44% at a concentration of 20 μM [72].

3.2.7 *Davallia*

Davallia cylindrica Ching is also rich in flavanoid content with total flavanoid content of 164.41 mg/g with some major identified flavonoids as quercetin-3-O-rutinoside, quercitrin, quercetin 7-O-glucoside, and kaempferol 3-O-rutinoside. The flavonoid-rich extract has shown cytotoxic action against A549 cells along with dose-dependent inhibition of acetylcholinesterase [73].

3.2.8 *Dryopteris*

Dryopteris erythrosora is rich in flavonoids, such as apigenin 7-O-glucoside, gliricidin 7-O-hexoside, quercetin 7-O-rutinoside, keampferol 7-O-gentiobioside, quercetin 7-O-galactoside, keampferol-3-O-rutinoside, quercitrin, and myricetin 3-O-rhamnoside. Flavonoids extracted from the plant have shown cytotoxic activity against A549 cells along with dose-dependent inhibition of acetylcholinesterase [74].

FAS (Fatty acid synthase) inhibition is considered a potential cancer treatment target. FAS inhibitory activity of 10 phloroglucinol derivatives was reported from methanolic extract of *Dryopteris crassirhizoma* rhizome. Two acylphloroglucinol derivatives, namely, methylene-bis-methylphlorobutyrophenone and flavaspidic acid PB have shown the highest FAS inhibitory activity [75]. This indicates the possible role of *D. crassirhizoma* in the development of anti-cancer molecules.

Nine compounds were derived from the *Dryopteris fragrans* (L.) Schott. whole plants' ethanol extract and subjected for cytotoxicity evaluation against three Human HepG2, A549, and MCF7 cell lines through MTT assay. Compounds, Dryofracoumarin A, and Aspidinol were cytotoxic toward A549 and MCF7 cell lines where IC_{50} values were 6.56 and 10.14 μM for dryofracoumarin A and 12.59 and 10.58 μM for aspidinol, respectively. Compound, Albicanol was cytotoxic against MCF7 cells with an IC50 of 24.14 μM. Compounds, Esculetin, Iso-scopoletin, Cis-3-(3, 4-dimethoxyphenyl)-4-[(E)-3,4-dimethoxystyryl] cyclohex-1-ene and trans-3-(3, 4-dimethoxyphenyl)-4-[(E)-3,4-dimethoxystyryl] cyclohex-1-ene were cytotoxic against all the three tested cell lines with IC_{50} values varies from 2.73 to 23.75 μM [76].

3.2.9 *Isoetes*

A flavone, 5,7,2′,4′,5′-pentahydroxy flavone (Isoetin) has been discovered from *Isoetes durieui* and *Isoetes sinensis* [77]. Isoetin 5′-methyl ether has shown in vitro cytotoxic effect on A549 (human lung cancer) cell line, Sk-Mel-2 (human melanoma), and B16F1 (mouse melanoma) cell lines with IC_{50} values of 0.92 μg/ml, 8.0 μg/ml, and 7.23 μg/ml [78]. *Isoetes sinensis* was also found to be rich in flavonoids. Four flavones, namely, apigenin-7-glucuronide, apigenin, homoplantageninisoetin, and acacetin-7-O-glcopyranoside; four flavonols, such as limocitrin-Neo, kaempferol-3-O-glucoside, isoetin, and quercetin-3-O-[6″-O-(3-hydroxy-3-methylglutaryl)-β-D-glucopyranoside]; a prodelphinidin: procyanidins; and a nothofagin: dihydrochalcone have been isolated from *I. sinensis* and some of these compounds also possess anticancer potential [77, 79].

3.2.10 *Macrothelypteris*

A significant in vitro antitumor efficacy of Protoapigenone isolated from *Macrothelypteris torresiana* has been demonstrated against Tca-8113, HepG2, MCF-7, K562, and M5 cell lines with IC_{50} values of 0.6, 2.3, 0.8, 0.9, and 0.3 μg/ml, respectively. Total flavonoid fraction isolated from *M. torresiana* roots and utilizing sodium carboxymethyl cellulose to dissolve and initiated by hydroxypropyl-β-cyclodextrin have demonstrated a high tumor growth inhibition ratio in BALB/c mice employing the mouse sarcoma S-180 with low acute oral toxicity (LD_{50} of 2.76 g/kg and 0.87 g/kg body wt) respectively [80]. Protoapigenone has also shown potential anti-tumor efficacy against Hep 3B, HepG2, MCF-7, MDA-MB-231, and A549, and cell lines with IC_{50} values of 0.23, 1.60, 0.78, 0.27, and 3.88 μg/ml, correspondingly [81]. Cytotoxicity of protoapigenone has also been reported against SKOV3 and MDAH-2774 (human ovarian cancer cells) and cells whose growth was arrested at G2/M and S phases via lowering the Cdk2, p-Cdk2, Cyclin B1, and p-Cyclin B1 expression and enhancing the inactive p-Cdc25C expression. At the S and G(2)/M stages, protoapigenone has also slowed the development of human prostate cancer cells and also increased the quantities of caspase-3 and cleaved poly(ADP-ribose) polymerase to cause apoptosis [82, 83]. In vivo and in vitro antitumor efficacy of protoapigenone isolated from *Macrothelypteris oligophlebia* has also been reported [84]. Weak cytotoxicity of flavones derivative derived from aerial parts of *M. torresiana* has been demonstrated toward human tumor cell lines K562, HepG2, and MCF-7 [85].

Protoapigenone, 5,7-dihydroxy-2-(1-hydroxy-2,6-dimethoxy-cyclohex-4-oxo)-chromen-4-one, and DICO [5,7-dihydroxy-2-(1,2-isopropyldioxy-4-oxo-cyclohex-5-enyl) -chromen-4-one] derived from *Macrothelypteris viridifrons* (Tagawa) Ching have shown antiproliferative activities in concentration-dependent manner against HepG2, MOLT4, A-549, MCF-7, PC-3, and HT-29 tumor cell lines [86]. DICO; a flavonoid having a nonaromatic B-ring has been shown to impede the development of human hepatoma HepG2 cells dose and time-dependently. Its significant anti-tumor activity is explained by causing apoptosis through a ROS-dependant mito-chondrial cascade, causing Bax translocation, lowering Bcl-2 levels, halting the G2/M phase of the cell cycle, releasing cytochrome c, activating caspase-9 and caspase-3, and changing the levels of cyclin A and cyclin B1, p-cdc25c, and p-CDK1 [87].

3.2.11 *Osmunda*

The aromatic oil derived from aerial components of *Osmunda regalis* L. has 11.82% of hexahydrofarnesyl acetone as the main compounds, 6.80% of 2,4-di-tert-butylphenol and 6.46% of phytol has shown no cytotoxic effect on Human epithelial (HEp-2) cell line with CC_{50} of 1772.41 ± 0.95 μg/ml [88].

3.2.12 *Palhinhaea*

Six serratene triterpenoids have been obtained from the whole plants of *Palhinhaea cernua* and evaluated for cytotoxicity toward SMMC-7721, SGC7901, and K562, three human cancer cell lines in vitro. With an IC_{50} value of 56.1 ug/ml, the compound 3′,21′,24′-trihydroxyserrat-14-en-24-(4′-hydroxybenzoate) was cytotoxic to K562 cell lines [89].

3.2.13 *Pityrogramma*

Cytotoxicity of *Pityrogramma calomelanos* and its isolated DHCs (dihydrochalcones) were studied on DLA cells (Dalton's lymphoma ascites tumor cells) and EA cells (Ehrlich ascites tumor cells) utilizing the trypan blue exclusion assay which exhibited an IC_{50} of 16 μg/ml and 18 μg/ml, correspondingly whereas IC_{50} values of 6.1 μg/ml and 11.5 μg/ml, correspondingly were obtained for DHCs. Additionally, they had cytotoxic effects on KB (human nasopharyngeal) cells and K562 (human myelogenous leukemia) cells, with IC_{50} values of 1.1 g/ml and 8 g/ml, correspondingly. Antitumor activity of DHC was also observed where IC_{50} value was found as 8 μg/ml [90].

3.2.14 *Pteridium*

In vitro antitumor activity of a bihomoflavanonol pteridium III, isolated from *Pteridium aquilinum* has been demonstrated against melanoma cells (A375), lung cancer (NCI-H46) cells, and glioma cells (U-7MG) with IC_{50} values of 106.7, 22.9 and 1540.5 μmol/L, respectively [91, 92].

3.2.15 *Pteris*

From an extract of ethyl acetate, 12 novel chemicals were discovered from the dried whole plant of *Pteris ensiformis* Burm. and evaluated for cytotoxic potential against human liver cancer (HepG2), human lung carcinoma (A549), breast carcinoma (MDA-MB-231), breast carcinoma (MCF-7), human oral squamous carcinoma (Ca9–22), and human leukemia (HL 60) cell lines utilizing MTT assay. Out of which two compounds, 2*R*,3*R*-pterosin L 3-*O*-*β*-D-glucopyranoside, and pterosin B have shown cytotoxicity toward human leukemia HL 60 cells (IC_{50} of 3.7 μg/ml and 8.7 μg/ml, respectively) [93].

Pteris semipinnata L. is rich in ent-kaurane diterpenoids, for example, ent-11-α-hydroxy-15-oxo-kaur-16-en-19-oic-acid has revealed noteworthy cytotoxic and anticancer potential in various in vivo and in vitro studies. It has also been shown to induce apoptosis in gastric cancer cell line MKN-45 and also impedes the proliferation of the human lung cancer cell lines CRL-2066, A549, and NCI-H23 as well as releases cytochrome c into the cytosol and activates caspase-3 by causing the overexpression of Bax and its translocation into the mitochondria [94–96]. Ent-11α-hydroxy-15-oxo-kaur-16-en-19-oic-acid has been found to inhibit hepatocellular carcinoma (HCC) as observed by minimizing adverse effects while decreasing the number of tumor foci and tumor size in a mouse model of diethylnitrosamine-induced HCC and stabilizing IkB to impede NF-B [97]. Ent-11α-hydroxy-15-oxo-kaur-16-en-19-oic-acid also causes the G2 cell cycle, which prevents the growth of CNE-2Z (nasopharyngeal cancer) and triggers apoptosis by raising the Bax/Bcl-2 ratio and the concentration of cytochrome C in the cytosol while lowering NF-κB-p65 levels and raising IκB levels [98].

Another diterpenoid, 7,11-dihydroxy-15-oxo-ent-kaur-16-en-19,6-olide, and 7,9-dihydroxy-15-oxo-ent-kaur-16-en-19,6-olide, isolated from *P. semipinnata,* act as DNA topoisomerase II inhibitors and inhibitors for tyrosine-protein kinase, at 0.01 mg/L concentration, inhibit adenocarcinoma cells of the lung [99] as well as reducing the oncogene c-myc expression [100].

Thirteen compounds were isolated from *Pteris multifida* among which significant cytotoxic activity of 4,5-dicaffeoylquinic acid and pterosin C 3-O-beta-D-glucopyrannoside was observed with IC_{50} values of 2.35 μg/ml and 5.38 μg/ml correspondingly and medium efficacy was exhibited in 4-caffeoyl quinic acid 5-O-methyl ether (IC_{50} 12.3) against human KB cell line [101]. The entire plants of *Pteris multifida* were also used to isolate three novel C 14 pterosin-sesquiterpenoids, known as multifidoside A-C among which, multifidoside A and B exhibited cytotoxic activity (IC_{50} < 10 μM) toward the HepG2 and K562 tumor cell line with IC_{50} values of 10.63 μM and 9.57 μM, correspondingly [102]. Dehydropterosin B; derived from *Pteris multifida* Poir. aerial parts have shown significant cytotoxic activity toward human pancreatic cancer (PANC-1) and human small-cell lung cancer (NCI-H446) cell lines [103]. A pair of isomers as C_{14} pterosin dimers, namely, A and bimutipterosins B, have been obtained from the whole plant of *Pteris multifida* and exhibited cytotoxic activity against human leukemia (HL 60) cell line with IC_{50} values of 12.8 μM and 26.6 μM, correspondingly [104]. Shu et al. [105] have reported cytotoxic activity of pterosin sesquiterpenes, namely, 2R,3R-13-hydroxypterosin L 3-O–D-glucopyranoside, 2S,3S-acetylpterosin C, and 2R,3S-acetylpterosin C isolated from whole plants of *Pteris multifida* against human leukemia (HL60) cell line with IC_{50} values of 14.6 mM, 35.7 mM, and 48.3 mM, correspondingly.

Nine pterosin components were derived from *Pteris cretica* aerial parts among which four new compounds 13-hydroxy-2(R),3(R)-pterosin L, creticolactone A, spelosin 3-O-β-d-glucopyranoside, and creticoside A were studied for cytotoxic efficacy toward neuroblastoma cell line (SH-SY5Y), gastric cancer cell line (SGC-7901), colon cancer cell line (HCT-116), and colorectal cancer cell line (Lovo) using the MTT assay. All four compounds showed no activity (IC_{50} value of >100 μM) against SH-SY5Y, Lovo, and SGC-7901, and cell lines. However, creticolacton A and spelosin 3-O-β-d-glucopyranoside displayed cytotoxic action toward HCT-116 cells where IC_{50} values were found at 22.4 and 15.8 μM, correspondingly [106].

Gou et al. [107] have isolated a new compound, 5,11,12-trihydroxy-15-oxo-ent-kuar-16-en-19-oic acid (diterpene), and 3-dihydroxylnorpterosin C (sesquiterpene 1) from *Pteris dispar* both of which have exhibited in vitro cytotoxic activity against KB cells, with IC_{50} values of 59.8 mol/l and 36.5 mol/l.

3.2.16 *Salvinia*

From an ethanol extract, 50 different chemicals have been found from *Salvinia molesta* out of which salviniol, a rare abietane diterpene along with 15 other abietane diterpenes exhibited anti-tumour activity in vitro [108].

4 Pharmacological Perspectives

Pteridophytes are rich in flavonoids; terpenoids including diterpenoids, triterpenoids, and sesquiterpenoids; steroids; alkaloids; benzenoids; glycosides; and their various derivatives [11]. Several of the compounds isolated from pteridophytes have shown cytotoxic activity (Table 1). Besides, other isolated compounds, such as isoquercetin, kaempferol-3-*O*-glucoside, apigenin-4′-glucoside, kaempferol, quercetin, luteolin, tricin, apigenin, sakuranetin, dihydroactinidiolide, myrcene, genkwanin, pinolenic acid, *trans*-ferulic acid, and β-sitosterol have also shown anticancer potential against cancer cell lines of melanoma, colorectal, liver, ovarian, and breast [109–114].

Not only anticancer effect, but the phytoconstituents of various pteridophyte species also possess antioxidant, antidiabetic, acetylcholinesterase inhibitory, thrombolytic, neuroprotective, antimicrobial, antiprotozoal, antidiarrheal, antiulcerogenic, anti-inflammatory, and antitubercular activities [11, 12]. In this regard of having multifarious bioactivities and a number of phytopharmaceutical molecules, pteridophytes are not lesser than higher seed plants and could be further explored for the isolation of novel pharmaceutical molecules useful for humanity.

5 Conclusion

Pteridophytes are non-seed-bearing plants with a limited distribution range to angiosperms. However, pteridophytes are shown to possess various phytochemical compounds having a wide range of pharmacological activities. These species have been divided into two phylogenetically distinct groups lycophytes and ferns and both groups possess important phytopharmaceutical compounds. Lycophytes, such as *Lycopodium* and *Selaginella* species possess lycophlegmariol, ginkgetin, selaginellin, amentoflavone, robustaflavone, heveaflavone, alpha-tocopheryl quinine, and many derivatives as cytotoxic molecules. Ferns; globally having approximately 10,535 species possess several important anticancer compounds, for example, pterosin, gliricidin-7-O-hexoside, tetracosane, patriscabratine, pimpinellin, dryofracoumarin A, aspidinol, albicanol, isoscopoletin, protoapigenone, pteridium III, multifidoside, creticolacton, salviniol, etc. Mechanism of anticancer activity has also been revealed for some of these molecules. However, detailed research is warranted, especially, in large-scale clinical studies so that these molecules could be effectively used as an anticancer drug with less cost and side effects. The present chapter briefly describes anticancer efficacy and anticancer compounds

Table 1 Some anticancer compounds isolated from pteridophytes

Names of anticancer compound
Lycophlegmariol B
Lycophlegmariol D
21β-Hydroxy-serrat-14-en-3α-ol
Ginkgetin
Amentoflavone
Robustaflavone
Robustaflavone 4′-methyl ether
Robustaflavone 4′,4‴-dimethyl ether
Heveaflavone
7″-O-Methylrobustaflavone
2″,3″-Dihydro-3′,3‴-biapigenin
3′,3‴-Dinaringenin
2″,3″ -Dihydrorobustaflavone 7,4′, −dimethyl ether
Alpha-tocopheryl quinone
4′,7″-di-O-methylamentoflavone
Isocryptomerin
Selaginellin M
Selaginellin N
(2R,4S)-6,8-Dimethyl-7-hydroxy-4′-methoxy-4,2″-oxidoflavan-5-O-beta-D-6″-O-acetyl-glucopyranoside
(2R,4S)-5,7-O-beta-D-di-Glucopyranosyloxy-4′-methoxy-6,8-dimethyl-4,2″-oxidoflavane
(2*S*,3*S*)-Sulfated pterosin C
Patriscabratine
Tetracosane
Gliricidin-7-O-hexoside
Quercetin-7-O-rutinoside
Quercetin
Rutin
5,7-Dihydroxy-6-methyl-4-phenyl-8-(3-phenylpropionyl)-1-benzopyran-2-one
Parasiticin C
2′,4′-Dihydroxy-6′-methoxy-3′,5′-dimethylchalcone
Pimpinellin
Dryofracoumarin A
Aspidinol
Albicanol
Esculetin
Isoscopoletin
Trans-3-(3, 4-dimethoxyphenyl)-4-[(E)-3,4-dimethoxystyryl] cyclohex-1-ene
Cis-3-(3, 4-dimethoxyphenyl)-4-[(E)-3,4-dimethoxystyryl] cyclohex-1-ene
Protoapigenone
DICO [5,7-dihydroxy-2-(1,2-isopropyldioxy-4-oxo-cyclohex-5-enyl) -chromen-4-one]
5,7-dihydroxy-2-(1-hydroxy-2,6-dimethoxy-cyclohex-4-oxo)-chromen-4-one
3β,21β,24-Trihydroxyserrat-14-en-24-(4′-hydroxybenzoate)

(continued)

Table 1 (continued)

Names of anticancer compound
Pteridium III
2*R*,3*R*-Pterosin L 3-*O*-*β*-D-glucopyranoside
Pterosin B
Ent-11α-Hydroxy-15-oxo-kaur-16-en-19-oic-acid
7,9-Dihydroxy-15-oxo-ent-kaur-16-en-19,6-olide
7,11-Dihydroxy-15-oxo-ent-kaur-16-en-19,6-olide
Pterosin C 3-O-beta-D-glucopyrannoside
4,5-Dicaffeoylquinic acid
Multifidoside A
Multifidoside B
Dehydropterosin B
Bimutipterosins A
Bimutipterosins B
2R,3R-13-Hydroxypterosin L 3-O–D-glucopyranoside
2R,3S-Acetylpterosin C
2S,3S-Acetylpterosin C
Creticolacton A
Spelosin 3-O-β-d-glucopyranoside
5,11,12-Trihydroxy-15-oxo-ent-kuar-16-en-19-oic acid
1, 3-Dihydroxylnorpterosin C
Salviniol
Sakuranetin
Genkwanin
Pinolenic acid

isolated from various lycophyte and fern species the world over and would be useful for scholars interested in cancer drug development from phytochemicals.

References

1. Christenhusz MJM, Chase MW (2014) Trends and concepts in fern classification. Ann Bot 1–24. https://doi.org/10.1093/aob/mct299
2. Kato M, Imaichi R (1997) Morphological diversity and evolution of vegetative organs in pteridophytes. In: Iwatsuki K, Raven PH (eds) Evolution and diversification of land plants. Springer, Tokyo. https://doi.org/10.1007/978-4-431-65918-1_2
3. Sureshkumar J, Ayyanar M, Silambarasan R (2020) Pteridophyte species richness along elevation gradients in Kolli Hills of the Eastern Ghats. India J Asia-Pac Biodivers 13(1):92–106. https://doi.org/10.1016/j.japb.2019.11.008
4. Goswami H, Sen K, Mukhopadhyay R (2016) Pteridophytes: evolutionary boon as medicinal plants. Plant Gene Res 14(4):328–355. https://doi.org/10.1017/S1479262116000290
5. Nitta JH, Ebihara A (2019) Virtual issue: ecology and evolution of pteridophytes in the era of molecular genetics. J Plant Res 132:719–721. https://doi.org/10.1007/s10265-019-01139-1

6. Thomas BA, Cleal CJ (2022) Pteridophytes as primary colonisers after catastrophic events through geological time and in recent history. Palaeobio Palaeoenv 102:59–71. https://doi.org/10.1007/s12549-021-00492-1
7. Khine PK, Schneider H (2020) First assessment of pteridophytes' composition and conservation status in Myanmar. Global Ecol Conserv 22:e00995. https://doi.org/10.1016/j.gecco.2020.e00995
8. Keller H, Prance GT (2015) The ethnobotany of ferns and lycophytes. Fern Gazette 20:1–14
9. Jain V, Jain SK (2016) Compendium of Indian folk medicine and ethnobotany (1991–2015). Deep Publications, New Delhi
10. Sureshkumar J, Silambarasan R, Bharati KA, Krupa J, Amalraj S, Ayyanar M (2018) A review on ethnomedicinally important pteridophytes of India. J Ethnopharmacol 219:269–287. https://doi.org/10.1016/j.jep.2018.03.024
11. Cao H, Chai TT, Wang X et al (2017) Phytochemicals from fern species: potential for medicine applications. Phytochem Rev 16:379–440. https://doi.org/10.1007/s11101-016-9488-7
12. Baskaran XR, Geo Vigila AV, Zhang SZ, Feng SX, Liao WB (2018) A review of the use of pteridophytes for treating human ailments. J Zhejiang Univ Sci B 19(2):85–119. https://doi.org/10.1631/jzus.B1600344
13. Ho R, Teai T, Bianchini J-P, Lafont R, Raharivelomanana P (2010) Ferns: from traditional uses to pharmaceutical development, chemical identification of active principles. In: Bahillo MAR et al (eds) Working with ferns: issues and applications. https://doi.org/10.1007/978-1-4419-7162-3_23
14. Wang H, Khor TO, Shu L, Su ZY, Fuentes F, Lee JH, Kong AN (2012) Plants vs. cancer: a review on natural phytochemicals in preventing and treating cancers and their druggability. Anti-Cancer Agents Med Chem 12(10):1281–1305. https://doi.org/10.2174/187152012803833026
15. Fridlender M, Kapulnik Y, Koltai H (2015) Plant derived substances with anti-cancer activity: from folklore to practice. Front Plant Sci 6. https://doi.org/10.3389/fpls.2015.00799
16. Ahn SH, Mun YJ, Lee SW, Kwak S, Choi M-K, Baik S-K et al (2006) Selaginella tamariscina induces apoptosis via a caspase-3-mediated mechanism in human promyelocytic leukemia cells. J Med Food 9:138–144
17. Li J, Lei X, Chen KL (2014) Comparison of cytotoxic activities of extracts from *Selaginella* species. Pharmacogn Mag 10(40):529–535. https://doi.org/10.4103/0973-1296.141794
18. Macêdo LARO, Oliveira RG Jr, Souza GR, Oliveira AP, Lavor EM, Silva MG et al (2018) Chemical composition, antioxidant and antibacterial activities and evaluation of cytotoxicity of the fractions obtained from *Selaginella convoluta* (Arn.) Spring (Selaginellaceae). Biotech Biotechnol Equip 32(2):506–512. https://doi.org/10.1080/13102818.2018.1431055
19. Mothana RAA, Abdo SAA, Hasson S, Althawab FMN, Alaghbari SAZ, Lindequist U (2010) Antimicrobial, antioxidant and cytotoxic activities and phytochemical screening of some Yemeni medicinal plants. eCAM 7(3):323–330. https://doi.org/10.1093/ecam/nen004
20. Lai HY, Lim YY, Kim KH (2010) *Blechnum orientale* Linn.- a fern with potential as antioxidant, anticancer and antibacterial agent. BMC Complement Altern Med 10:15. http://www.biomedcentral.com/1472-6882/10/15
21. Cetojević-Simin DD, Canadanović-Brunet JM, Bogdanović GM, Djilas SM, Cetković GS, Tumbas VT, Stojiljković BT (2010) Antioxidative and antiproliferative activities of different horsetail (*Equisetum arvense* L.) extracts. J Med Food 13(2):452–459. https://doi.org/10.1089/jmf.2008.0159
22. Zia-Ur-Rehman Gurgul A, Youn I, Maldonado A, Wahid F, Che CT, Khan T (2022) UHPLC-MS/MS-GNPS based phytochemical investigation of *Equisetum arvense* L. and evaluation of cytotoxicity against human melanoma and ovarian cancer cells. Saudi J Biol Sci 29(6):103271. https://doi.org/10.1016/j.sjbs.2022.03.021
23. Das G, Patra JK, Shinb H-S (2020) Biosynthesis, and potential effect of fern mediated biocompatible silver nanoparticles by cytotoxicity, antidiabetic, antioxidant and antibacterial, studies. Mater Sci Eng C 114:111011. https://doi.org/10.1016/j.msec.2020.111011

24. Viral D, Shivanand P, Jivani N (2011) Anticancer evaluation of *Adiantum venustum* Don. J Young Pharm 3(1):48–54. https://doi.org/10.4103/0975-1483.76419
25. Akter S, Hossain MM, Ara I, Akhtar P (2014) Investigation of *in vitro* antioxidant, antimicrobial and cytotoxic activity of *Diplazium esculentum* (RETZ). SW. Int J Adv Pharm Biol Chem 3(3):723–733
26. Ullah MO, Haque M, Urmi KF, Md Zulfiker AH, Anita ES, Begum M, Hamid K (2013) Anti–bacterial activity and brine shrimp lethality bioassay of methanolic extracts of fourteen different edible vegetables from Bangladesh. Asian Pac J Trop Biomed 3:1–7. https://doi.org/10.1016/S2221-1691(13)60015-5
27. Rahmat A, Kumar V, Fong LM, Endrini S, Sani HA (2003) Determination of total antioxidant activity in three types of local vegetables shoots and the cytotoxic effect of their ethanolic extracts against different cancer cell lines. Asia Pac J Clin Nutr 12(3):292–295
28. Chai TT, Quah Y, Keng FO, Ismail M, Ismaliza N, Ang Y-V, Sanmugapriya E, Yew Y, Ong H-C, Wong F-C (2013) Anti-proliferative, antioxidant and iron-chelating properties of the tropical Highland Fern, *Phymatopteris triloba* (Houtt) Pichi Serm (Family Polypodiaceae). Trop J Pharm Res 12:747–753. https://doi.org/10.4314/tjpr.v12i5.13
29. Morais-Braga MF, Souza TM, Santos KK, Guedes GM, Andrade JC, Vega C, Rolón M, Costa JG, Saraiva AA, Coutinho HD (2013) Phenol composition, cytotoxic and anti-kinetoplastidae activities of *Lygodium venustum* SW. (Lygodiaceae). Exp Parasitol 134(2):178–182. https://doi.org/10.1016/j.exppara.2013.03.014
30. Uma R, Pravin B (2013) *In vitro* cytotoxic activity of *Marsilea quadrifolia* Linn. of MCF-7 cells of human breast cancer. Int Res J Med Sci 1(1):10–13
31. Bobach C, Schurwanz J, Franke K, Denkert A, Sung TV, Kuster R et al (2014) Multiple readout assay for hormonal (androgenic and antiandrogenic) and cytotoxic activity of plant and fungal extracts based on differential prostate cancer cell line behavior. J Ethnopharmacol 155(1):721–730
32. Antonysamy JMA, Janarthanan G, Arumugam S, Narayanan J, Mani N (2014) Antioxidant, larvicidal, and cytotoxic studies on *Asplenium aethiopicum* (Burm. f.) Becherer. Int Sch Res Notices:876170, 6. https://doi.org/10.1155/2014/876170
33. Chai T-T, Yeoh L-Y, Ismail NIM, Ong HC, Wong F-C (2015) Cytotoxicity and antiglucosidase potential of six selected edible and medicinal ferns. Acta Poloniae Pharm- Drug Res 72(2): 297–401
34. Valizadeh H, Sonboli A, Kordi FM, Dehghan H, Bahadori MB (2015) Cytotoxicity, antioxidant activity and phenolic content of eight fern species from North of Iran. Pharm Sci 21:18–24. https://doi.org/10.15171/PS.2015.12
35. Chen L, Xu W, Shao R, Du X (2015) Bioactivities of phytochemicals in *Araiostegia yunnanensis* (Christ). Food Chem 186:37–45. https://doi.org/10.1016/j.foodchem.2014.07.113
36. George M, Josekumar VS (2016) *In vitro* cytotoxicity screening, phytochemical profile and heavy metal analysis of different extracts of *Acrostichum heterophyllum* L. Indian J Nat Prod Resour 7(1):19–24
37. Reshi IA, Sarkar TK, Malik HU, Muhee A, Bhat AH (2017) Evaluation of the cytotoxic activity of crude aqueous extracts of some medicinal plants of Kashmir. J Cell Tissue Res 17: 6047–6050
38. Roy SC, Sajeeb BK, Muhit MA, Bachar SC (2019) Evaluation of antioxidant and cytotoxic activities of aerial parts of *Adiantum capillus-veneris* L. growing in Bangladesh. Dhaka Univ J Pharma Sci 18(2):217–222. https://doi.org/10.3329/dujps.v18i2.43264
39. Baharuddin AA, Roosli R, Zakaria ZA, Md Tohid SF (2018) *Dicranopteris linearis* extract inhibits the proliferation of human breast cancer cell line (MDA-MB-231) via induction of S-phase arrest and apoptosis. Pharm Biol 56(1):422–432. https://doi.org/10.1080/13880209.2018.1495748
40. Tan ST, Ong HC, Chai TT, Wong FC (2018) Identification of potential anticancer protein targets in cytotoxicity mediated by tropical medicinal fern extracts. Phcog Mag 14:227–230

41. Mahfuz A, Salam FBA, Deepa KN, Hasan AHMN (2019) Characterization of in-vitro antioxidant, cytotoxic, thrombolytic and membrane stabilizing potential of different extracts of *Cheilanthes tenuifolia* and Stigmasterol isolation from n-hexane extract. Clin Phytosci 5:39. https://doi.org/10.1186/s40816-019-0135-x
42. Farràs A, Mitjans M, Maggi F, Caprioli G, Vinardell MP, López V (2021) *Polypodium vulgare* L. (*Polypodiaceae*) as a source of bioactive compounds: polyphenolic profile, cytotoxicity and cytoprotective properties in different cell lines. Front Pharmacol 12:727528. https://doi.org/10.3389/fphar.2021.727528
43. Gonzalez S, Gilaberte Y, Philips N (2010) Mechanistic insights in the use of a *Polypodium leucotomos* extract as an oral and topical photoprotective agent. Photochem Photobiol Sci 9(4):559–563. https://doi.org/10.1039/b9pp00156e
44. Parrado C, Mascaraque M, Gilaberte Y, Juarranz A, Gonzalez S (2016) Fernblock (*Polypodium leucotomos* extract): molecular mechanisms and pleiotropic effects in light-related skin conditions, photoaging and skin cancers, a review. Int J Mol Sci 17:1026. https://doi.org/10.3390/ijms17071026
45. Wu S, Li J, Wang Q, Cao H, Cao J, Xiao J (2017) Seasonal dynamics of the phytochemical constituents and bioactivities of extracts from *Stenoloma chusanum* (L.) Ching. Food Chem Toxicol 108(Pt B):458–466. https://doi.org/10.1016/j.fct.2016.10.003
46. Qi G, Liu Z, Fan R et al (2017) *Athyrium multidentatum* (Doll.) Ching extract induce apoptosis via mitochondrial dysfunction and oxidative stress in HepG2 cells. Sci Rep 7:2275. https://doi.org/10.1038/s41598-017-02573-8
47. Chang SH, Bae JH, Hong DP, Choi KD, Kim SC, Her E, Kim SH, Kang CD (2010) *Dryopteris crassirhizoma* has anti-cancer effects through both extrinsic and intrinsic apoptotic pathways and G0/G1 phase arrest in human prostate cancer cells. J Ethnopharmacol 130(2):248–254. https://doi.org/10.1016/j.jep.2010.04.038
48. Lee J, Nho YH, Yun SK, Hwang YS (2019) Anti-invasive and anti-tumor effects of *Dryopteris crassirhizoma* extract by disturbing actin polymerization. Integr Cancer Ther 18: 1534735419851197. https://doi.org/10.1177/1534735419851197
49. Preeti K, Namdeo J (2018) Phytochemical screening and *in-vitro* anticancer activity of extracts of *Tectaria cicutaria*. IJPSR 9:3463–3468
50. Manivannan V, Johnson M (2020) Total accepted phenolic, tannin, triterpenoid, flavonoid and sterol contents, anti-diabetic, anti-inflammatory and cytotoxic activities of *Tectaria paradoxa* (Fee.) Sledge. Toxicol Rep 7:1465–1468. https://doi.org/10.1016/j.toxrep.2020.10.013
51. Petkov V, Batsalova T, Stoyanov P, Mladenova T, Kolchakova D, Argirova M, Raycheva T, Dzhambazov B (2021) Selective anticancer properties, proapoptotic and antibacterial potential of three *Asplenium* species. Plan Theory 10:1053. https://doi.org/10.3390/plants10061053
52. Wittayalai S, Sathalalai S, Thorroad S, Worawittayanon P, Ruchirawat S, Thasana N (2012) Lycophlegmariols A-D: cytotoxic serratene triterpenoids from the club moss *Lycopodium phlegmaria* L. Phytochemistry 76:117–123. https://doi.org/10.1016/j.phytochem.2012.01.006
53. Shi H, Li ZY, Guo YW (2005) A new serratane-type triterpene from *Lycopodium phlegmaria*. Nat Prod Res 19(8):777–781. https://doi.org/10.1080/14786410500044906
54. Sun CM, Syu WJ, Huang YT, Chen CC, Ou JC (1997) Selective cytotoxicity of ginkgetin from *Selaginella moellendorffii*. J Nat Prod 60:382–384
55. Su Y, Sun CM, Chuang HH, Chang PT (2000) Studies on the cytotoxic mechanisms of ginkgetin in a human ovarian adenocarcinoma cell line. Naunyn-Schmiedeberg's Arch Pharmacol 362(1):82–90. https://doi.org/10.1007/s002100000240
56. Wang JZ, Li J, Zhao P et al (2015) Antitumor activities of ethyl acetate extracts from *Selaginella doederleinii* Hieron *in vitro* and *in vivo* and its possible mechanism. ECAM 2015:1–12
57. Liu H, Peng H, Ji Z et al (2011) Reactive oxygen species-mediated mitochondrial dysfunction is involved in apoptosis in human nasopharyngeal carcinoma CNE cells induced by *Selaginella doederleinii* extract. J Ethnopharmacol 138(1):184–191

58. Li S, Zhao M, Li Y et al (2014) Preparative isolation of six anti-tumour biflavonoids from *Selaginella doederleinii* Hieron by high-speed counter-current chromatography. Phytochem Anal 25(2):127–133
59. Silva GL, Chai H, Gupta MP, Farnsworth NR, Cordell GA, Pezzuto JM, Beecher CW, Kinghorn AD (1995) Cytotoxic biflavonoids from *Selaginella willdenowii*. Phytochemistry 40(1):129–134. https://doi.org/10.1016/0031-9422(95)00212-p
60. Lin LC, Kuo YC, Chou CJ (2000) Cytotoxic biflavonoids from *Selaginella delicatula*. J Nat Prod 63(5):627–630. https://doi.org/10.1021/np990538m
61. Chen JJ, Duh CY, Chen JF (2005) New cytotoxic biflavonoids from *Selaginella delicatula*. Planta Med 71(7):659–665. https://doi.org/10.1055/s-2005-871273
62. Zhang GG, Jing Y, Zhang HM et al (2012) Isolation and cytotoxic activity of selaginellin derivatives and biflavonoids from *Selaginella tamariscina*. Planta Med 78:390–392
63. Zhao Z, Ruan J, Jin J, Zou J, Zhou D, Fang W, Zeng F (2006) Flavan-4-ol glycosides from the rhizomes of *Abacopteris penangiana*. J Nat Prod 69(2):265–268. https://doi.org/10.1021/np050191p
64. Fang JB, Chen JC, Duan HQ (2010) Two new flavan-4-ol glycosides from *Abacopteris penangiana*. J Asian Nat Prod Res 12:355–359
65. Yang X, Yuan L, Xiong C, Yin C, Ruan J (2014) *Abacopteris penangiana* exerts testosterone-induced benign prostatic hyperplasia protective effect through regulating inflammatory responses, reducing oxidative stress and anti-proliferative. J Ethnopharmacol 157:105–113. https://doi.org/10.1016/j.jep.2014.09.025
66. Uddin SJ, Grice ID, Tiralongo E (2011) Cytotoxic effects of bangladeshi medicinal plant extracts. Evid Based Complement Alternat Med 2011:578092. https://doi.org/10.1093/ecam/nep111
67. Uddin SJ, Grice D, Tiralongo E (2012) Evaluation of cytotoxic activity of patriscabratine, tetracosane and various flavonoids isolated from the Bangladeshi medicinal plant *Acrostichum aureum*. Pharm Biol 50(10):1276–1280. https://doi.org/10.3109/13880209.2012.673628
68. Jarial R, Thakur S, Sakinah M, Zularisam AW, Sharad A, Kanwar SS, Singh L (2018) Potent anticancer, antioxidant and antibacterial activities of isolated flavonoids from *Asplenium nidus*. J King Saud Univ Sci 30:185–192. https://doi.org/10.1016/j.jksus.2016.11.006
69. Jarial R, Shard A, Thakur S, Sakinah M, Zularisam AW, Rezania S, Kanwar SS, Singh L (2018) Characterization of flavonoids from fern *Cheilanthes tenuifolia* and evaluation of antioxidant, antimicrobial and anticancer activities. J King Saud Univ Sci 30:425–432. https://doi.org/10.1016/j.jksus.2017.04.007
70. Quadri-Spinelli T, Heilmann J, Rali T, Sticher O (2000) Bioactive coumarin derivatives from the fern *Cyclosorus interruptus*. Planta Med 66(8):728–733. https://doi.org/10.1055/s-2000-9908
71. Wei H, Zhang X, Wu G, Yang X, Pan S, Wang Y, Ruan J (2013) Chalcone derivatives from the fern *Cyclosorus parasiticus* and their anti-proliferative activity. Food Chem Toxicol 60:147–152. https://doi.org/10.1016/j.fct.2013.07.045
72. Yang S, Liu M, Liang N, Zhao Q, Zhang Y, Xue W, Yang S (2013) Discovery and antitumor activities of constituents from *Cyrtomium fortumei* (J.) Smith rhizomes. Chem Cent J 7(1):24. https://doi.org/10.1186/1752-153X-7-24
73. Cao J, Xia X, Dai X, Wang Q, Xiao J (2014) Chemical composition and bioactivities of flavonoids-rich extract from *Davallia cylindrica* Ching. Environ Toxicol Pharm 37(2):571–579. https://doi.org/10.1016/j.etap.2014.01.011
74. Cao J, Xia X, Chen X, Xiao J, Wang Q (2013) Characterization of flavonoids from *Dryopteris erythrosora* and evaluation of their antioxidant, anticancer and acetylcholinesterase inhibition activities. Food Chem Toxicol 51:242–250. https://doi.org/10.1016/j.fct.2012.09.039
75. Na M, Jang J, Min BS, Lee SJ, Lee MS, Kim BY, Oh WK, Ahn JS (2006) Fatty acid synthase inhibitory activity of acylphloroglucinols isolated from *Dryopteris crassirhizoma*. Bioorg Med Chem Lett 16(18):4738–4742. https://doi.org/10.1016/j.bmcl.2006.07.018

76. Zhao DD, Zhao QS, Liu L, Chen ZQ, Zeng WM, Lei H, Zhang YL (2014) Compounds from *Dryopteris fragrans* (L.) Schott with cytotoxic activity. Molecules 19(3):3345–3355. https://doi.org/10.3390/molecules19033345
77. Wang X, Ding G, Liu B, Wang Q (2020) Flavonoids and antioxidant activity of rare and endangered fern: *Isoetes sinensis*. PLoS One 15(5):e0232185. https://doi.org/10.1371/journal.pone.0232185
78. Rahman MAA, Moon S-S (2007) Isoetin 5′-methyl ether, a cytotoxic flavone from *Trichosanthes kirilowii*. Korean Chem Soc 28(8):1261–1264. https://doi.org/10.5012/bkcs.2007.28.8.1261
79. Madunić J, Madunić IV, Gajski G, Popićd J, Vrhovac VG (2018) Apigenin: a dietary flavonoid with diverse anticancer properties. Cancer Lett. https://doi.org/10.1016/j.canlet.2017.10.041
80. Huang XH, Xiong PC, Xiong CM, Cai YL, Wei AH, Wang JP, Liang XF, Ruan JL (2010) *In vitro* and *in vivo* antitumor activity of *Macrothelypteris torresiana* and its acute/subacute oral toxicity. Phytomedicine 17(12):930–934. https://doi.org/10.1016/j.phymed.2010.03.006
81. Lin AS, Chang FR, Wu CC, Liaw CC, Wu YC (2005) New cytotoxic flavonoids from *Thelypteris torresiana*. Planta Med 71(9):867–870. https://doi.org/10.1055/s-2005-871292
82. Chang HL, Su JH, Yeh YT, Lee YC, Chen HM, Wu YC, Yuan SS (2008) Protoapigenone, a novel flavonoid, inhibits ovarian cancer cell growth *in vitro* and *in vivo*. Cancer Lett 267(1):85–95. https://doi.org/10.1016/j.canlet.2008.03.007
83. Chang HL, Wu YC, Su JH, Yeh YT, Yuan SS (2008) Protoapigenone, a novel flavonoid, induces apoptosis in human prostate cancer cells through activation of p38 mitogen-activated protein kinase and c-Jun NH2-terminal kinase 1/2. J Pharmacol Exp Ther 325(3):841–849. https://doi.org/10.1124/jpet.107.135442
84. Wu GH, Wei AH, Cai YL et al (2011) Chemical constituents of *Macrothelypteris oligophlebia* and their antitumor activity *in vitro* and *in vivo*. Chin Pharm J 5:004
85. Fang W, Ruan J, Cai Y, Wei A, Zhou D, Zhang W (2011) Flavonoids from the aerial parts of *Macrothelypteris torresiana*. Nat Prod Res 25(1):36–39. https://doi.org/10.1080/14786411003766888
86. Wei A, Wu G, Xiong C, Zhou D, Cai Y, Ruan J (2011) Flavonoids with special B-ring from *Macrothelypteris viridifrons* and their anti-proliferative effects on tumor cell. Zhongguo Zhong Yao Za Zhi 36(5):582–584
87. Zhou D, Wei A, Cao C, Ruan J (2013) DICO, a novel nonaromatic B-ring flavonoid, induces G2/M cell cycle arrest and apoptosis in human hepatoma cells. Food Chem Toxicol 57:322–329. https://doi.org/10.1016/j.fct.2013.03.032
88. Bouazzi S, Jmii H, ElMokni R, Faidi K, Falconieri D, Piras A, Jaïdane H, Porcedda S, Hammami S (2018) Cytotoxic and antiviral activities of the essential oils from Tunisian Fern, *Osmunda regalis*. S Afr J Bot 118:52–57
89. Yan J, Zhou ZY, Zhang M, Wang J, Dai HF, Tan JW (2012) New serratene triterpenoids from *Palhinhaea cernua* and their cytotoxic activity. Planta Med 78(12):1387–1391. https://doi.org/10.1055/s-0032-1314999
90. Sukumaran K, Kuttan R (1991) Screening of 11 ferns for cytotoxic and antitumor potential with special reference to *Pityrogramma calomelanos*. J Ethnopharmacol 34:93–96
91. Chen ND, Chen ND, Chen CW et al (2013) A novel bihomoflavanonol with an unprecedented skeleton from *Pteridium aquilinum*. Chin Herb Med 5:96–100
92. Chen JH, Wang TC, Yang CK et al (2013) New pterosin sesquiterpenes and antitubercular constituents from *Pteris ensiformis*. Chem Biodivers 10:1903–1908
93. Chen YH, Chang FR, Lu MC et al (2008) New benzoyl glucosides and cytotoxic pterosin sesquiterpenes from *Pteris ensiformis* Burm. Molecules 13:255–266
94. Liu Z, Chen G, Vlantis A et al (2005) Cell death induced by ent-11α-hydroxy-15-oxo-kaur-16-en-19-oic-acid in anaplastic thyroid carcinoma cells is via a mitochondrial mediated pathway. Apoptosis 10:1345–1356
95. Li MY, Leung J, Kong AW et al (2010) Anticancer efficacy of 5F in NNK-induced lung cancer development of A/J mice and human lung cancer cells. J Mol Med 88:1265–1276

96. Li L, Chen GG, Lu YN et al (2012) Ent-11α-hydroxy-15-oxokaur-16-en-19-oic-acid inhibits growth of human lung cancer A549 cells by arresting cell cycle and triggering apoptosis. Chin J Cancer Res 24:109–115
97. Chen GG, Leung J, Liang NC et al (2012) Ent-11α-hydroxy-15-oxo-kaur-16-en-19-oic-acid inhibits hepatocellular carcinoma *in vitro* and *in vivo* via stabilizing IkBα. Invest New Drugs 30:2210–2218
98. Wu K, Liu Y, Lv Y et al (2013) Ent-11α-hydroxy-15-oxo-kaur-16-en-19-oic-acid induces apoptosis and cell cycle arrest in CNE-2Z nasopharyngeal carcinoma cells. Oncol Rep 29: 2101–2108
99. Tomšík P (2014) Ferns and lycopods – a potential treasury of anticancer agents but also a carcinogenic hazard. Phytother Res 28:798–810
100. Li JH, Liang NC, Mo LE et al (2001) Effect of active compounds isolated from *Pteris semipinnata* L. on DNA topoisomerases and tyrosine protein kinase and expression of C-MYC in lung adenocarcinoma cells. Chin J Cancer Res 13:105–109
101. Harinantenaina L, Matsunami K, Otsuka H (2008) Chemical and biologically active constituents of *Pteris multifida*. J Nat Med 62(4):452–455. https://doi.org/10.1007/s11418-008-0265-9
102. Ge X, Ye G, Li P, Tang WJ, Gao JL, Zhao WM (2008) Cytotoxic diterpenoids and sesquiterpenoids from *Pteris multifida*. J Nat Prod 71(2):227–231. https://doi.org/10.1021/np0706421
103. Ouyang DW, Ni X, Xu HY, Chen J, Yang PM, Kong DY (2010) Pterosins from *Pteris multifida*. Planta Med 76(16):1896–1900. https://doi.org/10.1055/s-0030-1249934
104. Liu J, Shu J, Zhang R et al (2011) Two new pterosin dimmers from *Pteris mutifida* Poir. Fitoterapia 82:1181–1184
105. Shu J, Liu J, Zhong Y et al (2012) Two new pterosin sesquiterpenes from *Pteris multifida* Poir. Phytochem Lett 5:276–279
106. Lu J, Peng C, Cheng S, Liu J, Ma Q, Shu J (2019) Four new pterosins from *Pteris cretica* and their cytotoxic activities. Molecules 24(15):2767. https://doi.org/10.3390/molecules24152767
107. Gou ZP, Liang NC, Hou J et al (2011) Two new diterpene and sesquiterpene from *Pteris dispar*. Chin Chem Lett 22:1451–1453
108. Li S, Wang P, Deng G, Yuan W, Su Z (2013) Cytotoxic compounds from invasive giant salvinia (*Salvinia molesta*) against human tumor cells. Bioorg Med Chem Lett 23(24):6682–6687. https://doi.org/10.1016/j.bmcl.2013.10.040
109. Chen S-J, Hsu C-P, Li C-W, Lu J-H, Chuang L-T (2011) Pinolenic acid inhibits human breast cancer mda-mb-231 cell metastasis *in vitro*. Food Chem 126(4):1708–1715. https://doi.org/10.1016/j.foodchem.2010.12.064
110. Stompor M (2020) A review on sources and pharmacological aspects of sakuranetin. Nutrients 12:513. https://doi.org/10.3390/nu12020513
111. Tavsan Z, Kayali HA (2019) Flavonoids showed anticancer effects on the ovarian cancer cells: involvement of reactive oxygen species, apoptosis, cell cycle and invasion. Biomed Pharmacother 116:109004. https://doi.org/10.1016/j.biopha.2019.109004
112. Tomko AM, Whynot EG, Ellis LD, Dupré DJ (2020) Anti-cancer potential of cannabinoids, terpenes, and flavonoids present in *Cannabis*. Cancers 12(7):1985. https://doi.org/10.3390/cancers12071985
113. Wang X, Song ZJ, He X, Zhang RQ, Zhang CF, Li F, Wang CZ, Yuan CS (2015) Antitumor and immunomodulatory activity of genkwanin on colorectal cancer in the pc(min/+) mice. Int Immunopharmacol 29:701–707. https://doi.org/10.1016/j.intimp.2015.09.006
114. Vo TK, Ta QTH, Chu QT, Nguyen TT, Vo VG (2020) Anti-hepatocellular-cancer activity exerted by β-sitosterol and β-sitosterol-glucoside from *Indigofera zollingeriana* Miq. Molecules 25(13):3021. https://doi.org/10.3390/molecules25133021

On the Bioactive Potential of Ferns: An Overview

10

Kandikere Ramaiah Sridhar

Contents

Abstract

Ferns are a prominent entity in our ecosystem as one of the valuable resources of phytochemicals, medicinal, nutritional, and industrial attributes. They have multiple applications from the cottage industry to health-promoting metabolites. Ferns are relatively ignored flora compared to the angiosperms, especially for their impact on nutrition, health, and ecosystem services. They have attracted the attention recently towards harnessing their nutritional, biochemical, and industrial values based on the ethnic knowledge and analysis of bioactive potential using sophisticated methods. Ferns are known for their novelties in human and livestock nutrition and protective or curative potential against several diseases (e.g., cancer, malaria, gastrointestinal, neurological, gynecological, dermal, diabetes, rheumatism, rickets, and respiratory). Many ferns are known to produce ecdysteroids, which have a high potential to serve as bioinsecticides. Based on various studies, the species of the genus *Diplazium* have potent edible as well as

K. R. Sridhar (✉)
Department of Biosciences, Mangalore University, Mangalagangotri, Mangalore, India

H. N. Murthy (ed.), *Bioactive Compounds in Bryophytes and Pteridophytes*,
Reference Series in Phytochemistry, https://doi.org/10.1007/978-3-031-23243-5_11

medicinal attributes. This chapter consolidates the bioactive potential of ferns with an emphasis on their nutraceutical and pharmacological attributes of the species of *Diplazium*.

Keywords

Diplazium · Ethnic knowledge · Insecticides · Nutraceuticals · Phytochemicals · Therapeutics

Abbreviations

AD	Alzheimer's disease
APP	Amyloid precursor protein
Ca/P	Calcium/phosphorus
CNS	Central nervous system
EAA	Essential amino acids
FAO-WHO	Food and Agriculture Organization-World Health Organization
L-DOPA	3,4-dihydroxyphenylalanine
Na/K	Sodium/potassium
NRC-NAS	National Research Council-National Academy of Sciences
PER	Protein efficiency ratio
ROS	Reactive oxygen species
SFA	Saturated fatty acids
TAA	Total amino acids
TEAA	Total essential amino acids
TSFA	Total saturated fatty acids
TUFA	Total unsaturated fatty acids
UFA	Unsaturated fatty acids

1 Introduction

After angiosperms, ferns constitute the second largest group of plants. Currently, over 12,000 species of pteridophytes are known globally, while the Indian subcontinent is endowed up to 10% (1200 species with 235 endemic species) and the Himalayas support up to 800 species [1–4]. They are the most ancient assemblage of nonflowering vascular plants (~400 mya) cosmopolitan in the distribution in a wide range of humid tropical, subtropical, temperate, alpine, and arid ecosystems (terrestrial, aquatic, epiphytic, lithophytic, and polluted) [5, 6]. Similar to their widespread distribution, they also become broadly extremophilic as they are capable to live in extreme conditions like excessive carbon dioxide, harsh drought, high salinity, and contaminated soils with heavy metals [7–9]. They are versatile in their morphology as well as bioactive potential. Besides nutritional and medicinal attributes, they have ample socioeconomic usefulness in developing gardens, landscapes, ornamentals, lawns, avenue trees, handicrafts, utensils, fodder, biofertilizers, and bioremediation [4, 6, 10–16]). Ferns are the ideal plant resource to follow their uniqueness in their diversity, human nutrition, bioactive potential, insect deterrence, and the impact of climate change [17]. Based on the ethnic knowledge since time immemorial, ferns are used for various purposes, especially for nutrition, medicine, and bioremediation purposes. They occupied a prominent domain in Ayurveda and homeopathic systems of medicine in India as well as in Chinese medicine. Besides human needs, ferns are also used to feed and maintain livestock. The task of the present chapter is to provide a brief outline of the usefulness of ferns based on ethnic knowledge, bioactive potential (nutraceutical and pharmacological), and eco-friendly applications.

2 Nutraceutical Attributes

Being the earliest group of vascular plants, ferns occupied an important place in human nutrition as well as medicine for several centuries. The historical perspectives of edible ferns in China go back to about 3000 years [18]. There are several reports on their potential nutritional and nutraceutical attributes to humans as well as livestock (Table 1). Edible ferns possess considerable quantities of proximal components, mineral constituents, vitamins, amino acids, and fatty acids of nutraceutical significance.

Yumkham et al. [11] recorded ferns as a potential source of carbohydrates, minerals, vitamins, proteins, amino acids, and fatty acids along with the medicinal value of their active principles. Petkov et al. [19] assessed many ferns for their nutritional components like carbohydrates, minerals, and fatty acids. The proximal composition of nine ferns from different geographic locations has been compiled by Greeshma and Sridhar [20]. Himalayan edible ferns are a potential source of proximal components, minerals, and vitamin C [21]. The Himalayas are known for their highly rich fern population, and croziers of six species serve as highly popular vegetables (*Diplazium dilatatum, D. esculentum, D. maximum, D. spectabile, Dryopteris cochleata,* and *Tectaria coadunata*) [21]. Dvorakova et al. [22] assessed the nutritional potential of fiddleheads of 24 edible ferns from Europe for fatty acid methyl esters. These fiddleheads possess considerable antioxidant potential along with essential fatty acids and desired ratio of ω-6/ω-3 fatty acids. Giri et al. [23] have evaluated ω-3 as well as ω-6 fatty acids in ferns and their significance as nutraceuticals, pharmaceuticals, and cosmeceuticals. Tender fronds of the shuttlecock fern (*Matteuccia struthiopteris*) have been recommended as a feasible vegetable in a human diet based on the fatty acid composition and antioxidant activity.

Owing to nutritional and medicinal value, many ferns serve as nutraceuticals in the human diet [20]. Kholia and Balkrishna [4] assessed the nutraceutical potential of many ferns and their usefulness as beverages, flavors, and fodders. Besides, some ferns serve as nutritional products like soup, cake, and noodles [18]. Mehltreter et al. [24] found that 101 species of ferns (11 genera in 6 families) produce nectars in leaves and petioles and nourish nectariferous fauna (e.g., ants, aphids, and snails). Such nectars possess sugars (3.8–15.3%), and some are sucrose-rich or hexose-rich.

Table 1 Selected recent literature on bioactive properties of ferns

Property	Fern	Component/application	Reference	Remarks
Phytochemistry and ethnomedicinal	Various ferns	Traditional uses and pharmaceuticals	[25]	Compiled
Ethnomedicinal and nutritional	Various ferns	Diet (fresh and dried) (soup, cake, salad, and noodles)	[18]	Compiled
Medicinal	Various ferns	Flavonoids, hydroxycinnamic acid, proanthocyanidin; radical-scavenging and ferric-reducing power; antioxidants, antidiabetic, and antibacterial	[48]	Studied
Phytochemistry and medicinal	*Achrostichum* spp.	Wound-healing	[75]	Studied
Phytochemistry and medicinal	Various ferns	Pterosins and antidiabetic	[92]	Studied
Phytochemistry and medicinal	Various ferns	Alkaloids, terpenoids, flavonoids, flavones, sesquiterpenes, terpene glycosides, sitosterols, salaginellins, kaempferol glycosides, procyanidins, and phenolic acids; antioxidants, anticancer, antiprotozoal, antidiabetic, antituberculosis, antimicrobial, insecticidal, anti-inflammatory, anticancer, antitumor, neuroprotective, antidiarrheal, cytotoxic, hepatoprotective, and molluscicidal	[43]	Compiled
Medicinal, phytochemistry, and nutritional	Various ferns	Phenolics, flavonoids, triterpenoids, triterpenes, phytosterols, saponins, alkaloids, arbutin, kaempferol, coumaric acid, hexadecanoic acid, daucosteril, beta-sitosterol, hydroxycinnamic acid, proanthocyanidins, cardiac glycosides, alcohols, b-sitosterol, and adenine; carbohydrates, proteins, xanthoproteins, glutamic acid, vitamins, phosphorus, potassium, and fatty acids; used to treat diseases like dermal, microbial, wounds, inflammation, Alzheimer's, respiratory, snakebites, rheumatism, arthritis, headache, fever, and ulcer	[11]	Complied

(continued)

Table 1 (continued)

Property	Fern	Component/application	Reference	Remarks
Phytochemistry and medicinal	Various ferns	Antioxidants; anticancer, antidiabetic, anti-inflammation, wound-healing, antiviral, antimicrobial, and anti-Alzheimer's	[93]	Compiled
Nutritional	Various ferns	Proximal components, minerals, and vitamin C	[21]	Studied
Nutritional	Various ferns	Nutraceutical	[20]	Compiled
Phytochemistry and medicinal	*Cyathea contaminans*	Total phenolics; antioxidants; and antibacterial	[69]	Studied
Phytochemistry and nutraceutical	*Isoetes sinensis*	Flavonoids; antioxidants; and nutraceutical	[56]	Studied
Phytochemistry	*Isoetes sinensis*	Various bioactive compounds	[56]	Compiled
Medicinal and nutritional	Various ferns	Total phenolics, vitamin C, carotenoids, xanthophylls, fatty acid methyl esters; antioxidants; and nutraceutical	[22]	Studied
Medicinal	Various ferns	Used to treat wounds, burns, dysentery, gastrointestinal, gynecological, bone fracture, rheumatism, dermal ailments, herpes, respiratory, headache, inflammation, jaundice, snake bite, and helminthic; and nutraceutical	[33]	Compiled
Medicinal and nutritional	Various ferns	Minerals, carbohydrates, and fatty acids; phenolics, flavonoids, tannins' antioxidant; antimicrobial; and nutraceutical	[19]	Studied
Ethnomedicinal	Various ferns	Used to treat diseases like urinary, helminthic, cancer, dermal, respiratory, jaundice, malaria, gastrointestinal, diabetes, gynecological, gonorrhea, rheumatism, inflammation, and antidote for zoo toxins	[44]	Compiled
Ethnomedicinal and nutritional	Various ferns	Used to treat dysentery, fever, inflammation, stomach ache, jaundice, hepatic ailments, burns, diabetes, helminthic, ulcer, snake bite, leucorrhoea, and urinary infection; and nutraceutical	[34]	Surveyed

(continued)

Table 1 (continued)

Property	Fern	Component/application	Reference	Remarks
Essential fatty acids	Various ferns	Fatty acids; medicinal; cosmeceutical; and nutraceutical	[23]	Compiled
Medicinal	Various edible ferns	Used to treat constipation, dermal, gastrointestinal, inflammation, jaundice, malaria, typhoid, rheumatism, snakebite, respiratory, epilepsy, Alzheimer's, Parkinson's, dementia, wound, rickets, gynecological, bone fracture, and ulcers; and nutraceutical	[32]	Compiled
Ethnomedicinal, nutritional	Various ferns	Used to treat dysentery, helminthic, pneumonia, dermal, gonorrhea, piles, wounds, respiratory, rheumatism, inflammation, diabetes, cancer, and Alzheimer's; beverage, flavor, and fodder; and nutraceutical	[4]	Compiled
Phytochemistry	Various ferns	Carbohydrates, reducing sugars, amino acids, proteins, steroids, saponins, terpenoids, triterpenoids, alkaloids, phenolics, tannins, flavonoids, catechins, glycosides, cardiac glycosides, anthraquinone, coumarin, betacyanin, and quinone; and nutraceutical	[29]	Compiled
Medicinal	Various ferns	Antimicrobial; cytotoxic, anticancer, anti-inflammatory, antidiabetic, hepatoprotective, and wound-healing	[41]	Compiled
Ethnomedicine	*Adiantum* spp.	Used to treat respiratory, pox, dermal, diabetes, fever, gastrointestinal, headache, helminthic, snake bite, kidney stones, influenza, pneumonia, wounds, and gynecological diseases	[94]	Compiled
Antiproliferation	*Angiopteris evecta*	Used as remedy for colon cancer	[95]	Studied
Ecdysteroids	Different ferns	Serve as bioinsecticides	[83]	Compiled
Phytochemistry	*Nephrolepis auriculata*	Antioxidants and antidiabetic	[28]	Studied

2.1 Ethnonutritional Knowledge

The recent prospects in ferns stem from the ethnic knowledge of different geographic regions [25]. Tribals of different parts of the world have enormous folk knowledge of the uses of ferns as nutritional sources [26–29]. Edible ferns are valuable in the human diet in China and an inventory revealed 144 species of ferns as a traditional food source [18]. *Pteris multifida* is the most widely used vegetable in China and its beverages in Taiwan are based on ethnic knowledge [30]. Indigenous people in the Peruvian Amazon use *Lygodium venustum* to make hallucinogenic beverages [31]. Giri and Uniyal [32] surveyed the medicinal uses of 50 edible ferns in northern India based on the ethnic knowledge of different tribes. Ojha and Devkota [33] reviewed the literature on ethnically edible medicinal pteridophytes of Nepal. Among the 55 ferns, 14 were used as food as well as medicine, while the rest were used for nutritional or medicinal purposes. Antony and Suresh [34] reported ethnonutritional uses of ferns by the 14 tribals in Kerala in India to fulfill human nutritional needs. They listed 19 different edible ferns used by the tribals and local people of Kerala.

3 The Wonder Fern *Diplazium*

The genus *Diplazium* has a special significance in nutraceutical potential. It has over 400 species with pantropical distribution (Malesia, Afro-Madagascar, Neotropics, and Eurasia) possessing value-added nutraceutical and bioactive attributes [20, 27]. It has a high potential for application in the human diet, livestock feed, and production of industrially valued pharmaceutical products.

3.1 Nutritional Values

In the entire Himalayas, about 35 species of *Diplazium* are known to be edible [4]. *Diplazium esculentum* and *D. maximum* were the most commonly used vegetables in local markets. The *D. esculentum* is a well-known vegetable in Sikkim, Darjeeling, and Nepal local markets. Similarly, the other 12 *Diplazium* spp. are also popular vegetables in the Himalayas (*Diplazium doederleinii, D. dilatatum, D. forrestii, D. heterophlebium, D. javanicum, D. kawakamii, D. laxifrons, D. maximum, D. sikkimense, D. spectabile, D. stoliczkae,* and *D. succulentum*). In the Kumaun region of Uttarakhand, India croziers of *Botrychium lanuginosum, B. multifidum, B. ternatum, Dryopteris cochleata,* and *Helminthostachys zeylanica* are the popular vegetables. In Nepal, five *Diplazium* spp. are edible and three are nutraceutical (edible and medicinal) [33].

Diplazium esculentum grows in the swampy riparian regions of the Western Ghats are traditionally used for several nutritional products (soup, curry, sides, and others). Five different species of *Diplazium* possess crude protein matching with the legume seeds with low lipid, high crude fiber, a moderate amount of carbohydrates, and high vitamin C content (Table 2). Besides vitamin C, fiddleheds of

Table 2 The proximal components of *Diplazium* species (%) (–, not determined)

Species	Habitat	Crude protein	Total lipids	Crude fiber	Ash	Carbohydrates	Vitamin C	Reference
Diplazium dilatatum (fiddleheads)	India (Himalayas)	6.1	0.2	4.9	1.5	–	21.3	[21]
Diplazium esculentum (fiddleheads)	Sikkim (Himalayas)	2.6	2.0	–	1.3	1.0	–	[96]
D. esculentum (leaves)	Indonesia	2.2	0.04	4.8	1.4	–	–	[97]
D. esculentum (leaves)	Nepal (Himalayas)	1.0	0.2	1.0	1.1	–	6.2	[98]
D. esculentum (leaves)	India (Himalayas)	14.4	0.1	3.9	12.2	8.4	–	[99]
D. esculentum (leaves)	India (Himalayas	17.4	5.6	12.7	17.6	37.7	–	[100]
D. esculentum (leaves)	Philippines	0.9–10.7	0.3–3.4	0.7–9.1	1.4–17.4	–	–	[101]
D. esculentum (fiddleheads)	India (Himalayas)	31.2	8.3	4.6	16.2	44.3	21.0	[102]
D. esculentum (leaves)	India (Himalayas)	18.3	0.3	4.5	14.4	5.5	–	[104]
D. esculentum (leaves)	India (Himalayas)	0.2	–	–	–	0.02	–	[104]
D. esculentum (fiddleheads)	India (Himalayas)	3.8	0.3	5.1	1.3	–	21.4	[21]
D. esculentum (fiddleheads)	India (Western Ghats)	16.1	7.5	2.3	12.1	19.3	–	[105]

D. esculentum (leaves)	Indonesia	6.2–8.3	0.5–0.7	–	1.9–2.1	–	–	[106]
D. esculentum (fiddleheads)	Bangladesh	8.7	2.2	15.6	5.1	59.6		[107]
D. esculentum (fiddleheads, young twigs, premature twigs, and mature twigs)	India (Himalayas)	1.9–12.8	2.9–4.0	–	11.7–13.1	45.9–52.0	–	[42]
D. esculentum (young fronds)	India (Uttarakhand)	8.9	2.5	3.1	–	18.8	–	[108]
Diplazium maximum (leaves)	India (Himalayas)	0.2	–	–	–	0.02	–	[103]
D. maximum (fiddleheads)	India (Himalayas)	5.6	0.2	3.9	1.0	–		[21]
Diplazium sammatii (young pinna and crozier)	Nigeria	10.2	14.1	0.4	10	62.3	–	[109]
D. sammatii (mature pinna)	Nigeria	10.3	9.5	0.4	11.2	68.6	–	[109]
Diplazium spectabile (fiddleheads)	India (Himalayas)	3.6	0.2	3.4	1.6	–	21.6	[21]

D. esculentum also possess B complex vitamins like thiamine, riboflavin, and niacin. They have a sweet taste owing to the rich content of vitamin C. Fiddleheads, leaves, young twigs, premature twigs and mature twigs, young fronds, young pinna, and crozier were assessed for proximal and mineral components. They are also a potent sources of flavonoids especially α-carotene and β-carotene, which are known as human health promotions. Shoots of *Diplazium* spp. possess good quantities of micro- and macro-minerals with adequate Na/K ratio (<1) and Ca/P ratio (>1) (Table 3). Many minerals are comparable to the NRC-NAS stipulation (required for adults, children, and infants).

Leaves and fiddleheads of *Diplazium* spp. are also endowed with indispensable amino acids and many of them are comparable with FAO-WHO stipulated standards (Table 4). Interestingly, between raw and cooked fiddleheads of *D. esculentum,* no drastic differences were seen. The ratio of total essential amino acids vs. total amino acids increased in cooked *D. esculentum* fiddleheads indicates favorable changes owing to cooking. Similarly, the protein efficiency ratios (PER) of fiddleheads were also favorable indicating the superior quality of proteins [20].

3.2 Nutraceutical and Medicinal Values

Diplazium spp. being widely used as vegetables, they are endowed with a variety of nutritional components as well as medicinal attributes [29]. Nutritional, phytochemical, and metabolites of a variety of *Diplazium* spp. have been projected in the review by Marimuthu et al. [29]. *Diplazium esculentum* is nutraceutical versatile owing to the presence of minerals, vitamins (A, B, and C), essential amino acids, and essential fatty acids [20]. Besides, flavonoids (α- and β-carotenes) serve as potential scavengers of free radicals to protect from cold, cough, inflammation, and cancers. Many *Diplazium* spp. possess desired Na/K ratio (<1) (to combat blood pressure) as well as Ca/P ratio (>1) (to prevent the drain of calcium in urine and restoration of calcium in bones) [35, 36] (see Table 3). Being an edible and potent inhibitor of α-glucosidase, *D. esculentum* serves as a powerful antidiabetic agent that could be accomplished through diet management [37, 38]. Different parts of *Diplazium esculentum* are also known for antidiabetic, CNS stimulation, immunomodulatory, anti-inflammatory, and anti-anaphylactic potential [27].

No drastic differences between the raw and cooked fiddleheads of *D. esculentum* could be seen in the quantities of essential amino acids [20]. Interestingly, Semwal et al. [27] reported that pressurized hot water extraction of *D. esculentum* flour at 175 °C up to 21 min treatment also has optimum antioxidant potential. The total phenolics and flavonoid contents in fiddleheads of *D. esculentum* were not affected by cooking [39]. Similarly, the antioxidant activities (total antioxidant activity and ferrous ion-chelation capacity) were not decreased by cooking. Palmitic acid was highest among saturated fatty acids in raw and cooked fiddleheads of *D. esculentum* (Table 5), which has industrial applications especially cosmeceuticals (cleanser, lubricant, toner, moisturizer, conditioner, and surface-active agents) [40]. Table 5 projects the nutraceutical and medicinal properties of *D. esculentum* from recent

Table 3 Mineral constituents of *Diplazium* species (mg/100 g) in comparison with NRC-NAS standards (*, Range for adults, children, and infants; –, Not determined)

Mineral and ratio	*Diplazium* species and parts used								
	Diplazium dialatatum (fiddleheads) [21]	*Diplazium esculentum* (fronds) [110]	*D. esculentum* (fiddleheads) [95]	*D. esculentum* (leaves) [96]	*D. esculentum* (leaves) [97]	*D. esculentum* (young fronds) [111]	*D. esculentum* (leaves) [98]	*D. esculentum* (fiddleheads) [99]	*Diplazium esculentum* (fiddleheads) [104]
Sodium	1.1	79.0	8.1	–	–	–	118.0	360.0	145
Potassium	–	2370.0	927.4	0.04	–	–	4373.0	1120.0	3351
Calcium	1.1	1020.0	200.5	0.4	–	–	873.0	1290.0	436
Magnesium	13.4	505.0	–	0.1	–	10.0–12.1	–	–	481
Phosphorus	–	500.0	–	0.1	117.0	–	–	80.0	1050
Manganese	6.4	–	–	–	–	0.04–0.4	5.1	–	123
Iron	10.6	560.0	–	44.6	1.03	20.2–23.4	25.7	–	52
Zinc	0.1	58.0	–	–	–	–	16.7	–	194
Copper	19.3	4.0	–	4.2	–	1.0–1.3	2.6	–	509
Na/K ratio	–	0.03	0.009	–	–	–	0.03	0.32	0.04
Ca/P ratio	–	2.04	–	4.0	–	–	–	16.10	0.41
Mineral and ratio	*Diplazium* species and parts used								
	D. esculentum (leaves) [101]	*D. esculentum* (fiddleheads) [102]	*D. esculentum* (fiddleheads) [21]	*D. esculentum* (fiddleheads) [106]	*Diplazium maximum* (fiddleheads) [21]	*Diplazium sammatti* (young pinna) [108]	*D. sammatti* (mature pinna) [108]	*Diplazium spectabile* (fiddleheads) [21]	NRC-NAS standards* [112]
Sodium	9.5	29.0	0.5	20.21	3.6	520.0	560.0	0.5	120–500
Potassium	914.4	74.5	–	7.9		1600.0	1600.0		500–2000
Calcium	192.7	52.7	0.7	–	1.0	190.0	190.0	1.0	600–800
Magnesium	0.4	15.3	9.6	–	6.4	–	–	12.0	60–350
Phosphorus	–	–	–	1.6		6.8	7.2		500–800
Manganese	–	21.1	11.9	–	11.1	–	–	11.1	

(continued)

Table 3 (continued)

Mineral and ratio	*Diplazium* species and parts used								
	Diplazium dialatatum (fiddleheads) [21]	*Diplazium esculentum* (fronds) [110]	*D. esculentum* (fiddleheads) [95]	*D. esculentum* (leaves) [96]	*D. esculentum* (leaves) [97]	*D. esculentum* (young fronds) [111]	*D. esculentum* (leaves) [98]	*D. esculentum* (fiddleheads) [99]	*Diplazium esculentum* (fiddleheads) [104]
Iron	11.2	38.2	14.4	–	16.4	4.3	6.7	10.4	10–15
Zinc	2.7	4.3	–	–	2.8	3.6	4.5	0.2	12–15
Copper	0.3	1.7	13.4		18.8	3.5	2.5	14.2	0.6–3
Na/K ratio	0.01	0.39	–	2.6	–	0.33	0.35	–	0.24–0.25
Ca/P ratio	–	–	–	–	–	27.94	26.39	–	1.2–1.00

Table 4 Amino acids in comparison with FAO-WHO [116] standard and fatty acids composition of *Diplazium esculentum* (*, Methionine + Cystine; **, Phenylalanine + Tyrosine; –, Not detectable)

	Amino acids (g/100 g protein)				Fatty acids (g/100 g lipid)		
		D. esculentum (fiddleheads) [113]			*D. esculentum* (fiddleheads) [113]		
Amino acid	*D. esculentum* (leaves) [102]	Uncooked	Cooked	FAO-WHO standard [114]		Uncooked	Cooked
Essential amino acids (EAA)					Saturated fatty acids (SFA)		
Histidine	0.2	2.3	2.1	1.9	Pentadecanoic acid	4.6	6.6
Isoleucine	0.6	5.3	4.9	2.8	Palmitic acid	29.6	19.4
Leucine	0.7	8.1	8.3	6.6	Stearic acid	0.7	10.9
Lysine	0.3	8.4	11.8	5.8	Total	34.9	35.9
Methionine	2.1	1.4	0.6	2.5*	Unsaturated fatty acids (UFA)		
Cystine	4.2	0.5	0.3		Oleic acid	4.7	–
Phenylalanine	0.8	6.2	6.2	6.3**	Eicosanoic acid	–	1.0
Tyrosine	0.6	3.3	3.3		Total	4.7	1.0
Threonine	0.6	4.3	4.3	3.4	Ratio: TUFA/TSFA	0.13	0.02
Tryptophan	–	–	–	1.1			
Valine	0.2	6.3	6.4	3.5			
Other amino acids							
Alanine	0.4	7.7	8.2				
Arginine	0.2	5.2	4.6				
Aspartic acid	0.3	6.3	7.1				
Glutamic acid	4.6	8.0	8.4				
Glycine	0.1	10.5	10.1				
Proline	–	6.8	6.1				
Serine	–	5.5	5.3				
Ratio: TEAA/TAA	0.64	0.47	0.49				

Table 5 Nutraceutical and medicinal properties of *Diplazium esculentum*

Property	Fern (geographic region)	Components	Activity	Reference	Remarks
Nutritional	*D. esculentum* (Western Ghats, India)	Proteins, lipids, fiber, ash, and carbohydrates	Nutraceutical	[104]	Studied
Nutritional	*D. esculentum* (Western Ghats, India)	Amino acids and proteins	Nutraceutical	[104]	Studied
Nutritional	*Diplazium* spp. (India, Philippines, and Nigeria)	Proteins, lipids, fiber, ash, carbohydrates, and calorific value; minerals; amino acids, proteins, and fats	Nutraceutical	[39]	Compiled
Nutritional	*D. esculentum* (Western Ghats, India)	Lipids	Nutraceutical	[104]	Studied
Cytotoxic	*D. esculentum* (Bidor, Malaysia)	α-Glucosidase inhibition	Antidiabetic; cytotoxic against K562 cells	[47]	Studied
Medicinal	*D. esculentum* (Western Ghats, India)	Minerals	Nutraceutical	[104]	Studied
Medicinal	*D. esculentum* (Assam, India)	Total phenolics, flavonoids, and antioxidant enzymes	Antioxidants, antidiabetic, and hepatoprotection	[115]	Studied
Medicinal	*D. esculentum* (Western Ghats, India)	Total phenolics, tannins, flavonoids, vitamin C, phytic acid, L-DOPA, trypsin inhibition, and hemagglutination	Antioxidant	[39]	Studied

Medicinal	*D. esculentum* (different regions)	Bioactive principles	Antioxidants; cytotoxic; antibacterial, antidiabetic, antihelminthic, hepatoprorective, immunomodulatory, neuromodulatory, and larvicidal	[116]	Compiled
Medicinal	*D. esculentum* (Kuantan, Malaysia)	Aqueous extract of fiddleheads	Antimalarial	[117]	Studied
Medicinal and Nutritional	*D. esculentum* (Assam, India)	Proteins, lipids, fiber, ash, and carbohydrates	Nutraceutical	[42]	Studied
Medicinal	*D. esculentum* (Chiang Mai, Thailand)	Total phenolics and AD-related enzymes of ethanolic extract	Antioxidants; anti-Alzheimer's disease (inhibition of AD-related enzymes)	[118]	Studied
Medicinal and nutritional	*D. esculentum* (Himalayas, India)	Proteins, lipids, fiber, ash, carbohydrates, and calorific value; minerals; and vitamin C	Antioxidants; CNS stimulation; antimicrobial, antidiabetic, anti-inflammatory, and anti-anaphylactic	[27]	Compiled

investigations (antioxidant, antidiabetic, hyperlipidimic, cytotoxic, antibacterial, antihelminthic, hepatoprotective, immunomodulatory, neuromodulatory, larvicidal, antimalarial, CNS stimulation, and anti-anaphylactic potential). Marimuthu et al. [41] mainly discussed the bioactive principles as well as the bioactive potential of *D. esculentum*.

Nutritional and biochemical composition of *Diplazium* spp. (different parts) of having been evaluated by some workers [11, 42]. It is evident that the primary as well as secondary metabolites in *D. esculentum* differ depending on the part and growth stage [42]. Yamkham et al. [11] discussed the nutraceutical potential of young fronds of cultivated *D. esculentum.* Extracts of different parts of *D. esculentum* are known to be effective against pathogenic microbes (bacteria and fungi) [27]. They possess antibacterial activity against pathogenic bacteria which is comparable to tetracycline. It is interesting to note that *D. esculentum* also inhibits acetylcholinesterace in vitro, which prevents neurotoxic peptide production to ameliorate AD (Alzheimer's disease) prevention based on the *Drosophila* model.

4 Pharmacological Attributes

Ferns are endowed with a wide array of phytochemicals with significant medicinal and pharmacological applications [43] (see Table 1). Some of their bioactive potential includes cytotoxic, anticancer, neuroprotective, antiproliferative, wound-healing, antimicrobial, antiviral, hepatoprotective, leishmanicidal, trypanocidal, antinociceptive, anti-inflammatory, immunomodulatory, and chemopreventive functions. Such knowledge has been gained mainly by the ethnopharmacological and pharmacological inventions to formulate new value-added medicinal products. Cao et al. [43] discussed extensively the phytochemical constituents of ferns belonging to different families. Abraham and Thomas [44] reviewed the pharmacological uses of ferns disease-wise and also different parts of the fern used in formulations. Some of the important bioactive compounds isolated from 18 ferns include: *Abacopteris penangiana* (abacopterin K and abacopterin L); *Adiantum lunulatum* (filicenol-B and adiantone); *Drynaria fortunei* (flavan-3-ols); *Equisetum palustre* (palustrine and palustridiene); *Helminthostachys zeylanica* (ugonin E, ugonin M, ugonin L, ugonin N, and ugonin T); *Lygodium japonicum* (1,4-naphthoquinone, ecdysteroside, and lygodiumsteroside A); *Lygodium microphyllum* (quercetin, isoquercetin, β-sitosterol, and stigmast-4-en-3-one); *Ophioglossum pedunculosum* (pedunculosumoside A, pedunculosumoside B, pedunculosumoside C, pedunculosumoside D, pedunculosumoside E, and pedunculosumoside G); *Ophioglossum petiolatum* (ophioglonin, ophioglonin 7-obeta-Dglucopyranoside, ophioglonol, ophioglonol prenyl ether, and ophioglonol 40-obeta-Dglucopyranoside); *Pityrogramma calomelanos* (calomelanol A, calomelanol B, and calomelanol C); *Pteridium aquilinum* var. *caudatum* (ptaquiloside, isoptaquiloside, caudatoside, pteridanoside, and pteridanone), *Pteridium aquilinum* var. *latiusculum* (palmitylpterosin A and palmitylpterosin B); *Pteridium esculentum* (ptesculentoside); *Pteris cretica* (creticoside A); *Pteris ensiformis* (henrin A,

20-hydroxy-40-methoxychalcone, and cyclolaudenol); *Pteris multifida* (multifidoside A, multifidoside B, multifidoside C, (2R)-pterosin P, and dehydropterosin B); *Pteris semipinnata* (pterisolic acid A, pterisolic acid B, pterisolic acid C, pterisolic acid D, pterisolic acid E, pterisolic acid F, semipterosin A, and (2R)-norpterosin B); *Pteris semipinnata* ((2R)-pterosin B, (2S, 3S)-pterosin C, and norpterosin C); and *Pteris vittata* (kaempferol and quercetin).

Pteris multifida is useful to treat many ailments (bacterial dysentery, hepatitis, hematuria, and eczema) [45]. *Pteris vittata* is useful in the treatment of diarrhea, abdominal pains, and flu [46], which is also cytotoxic against K562 leukemic cells [47]. Pterosins of *P. multifida* showed cytotoxicity against HL60 human leukemic cell lines, and *Pteris semipinnata* is known for its significant cytotoxicity as well as anticancer potential [43]. Pterosins are low-molecular-weight products that occur in several ferns and are known for antidiabetic activity [43]. Pterosins are plentiful in ferns and pterosin A has the capacity to regulate blood glucose, prevent cell death, and reduce reactive oxygen species (ROS). Five ferns with edible and medicinal attributes (*Blechnum orientale, Davallia denticulata, Diplazium esculentum, Nephrolepis biserrata,* and *Pteris vittata*) showed inhibition of α-glucosidase as well as cytotoxic potential [47]. The edible fern *Diplazium esculentum* possesses higher α-glucosidase inhibitory activity compared to the standard drug myricetin. Aqueous extracts of *Dicranopteris curranii* and *Gleichinia truncate* also possess antidiabetic properties [48].

Ethanol, hexane, ethyl acetate, and methanol extracts of *Lygodium venustum* exhibited cytotoxicity against mammalian fibroblasts [49]. Strong cytotoxicity against K562 cells was shown by *Pteris vittata* [43]. Onitin, a phenolic petrosin, showed hepatoprotective function by tacrine-induced cytotoxicity of the human liver, hence justifying its use in treating hepatitis in oriental medicine [50]. Different parts of *Pityrogramma calomelanos* are used to treat renal, digestive, respiratory, hypertension, and bleeding problems [45]. In Pakistan, *Adiantum capillus-veneris* is employed to treat measles, inflammatory, and skin diseases [51, 52]. Extracts of *Equisetum arvense* have the capacity to prevent stroke [53], while *Lygodium venustum* is used to combat emotional instability as well as nervousness [54]. Extracts of *E. arvense* possess cosmeceutical potential against antiaging (e.g., moisturizers, anti-acne, anti-wrinkle, and hair conditioning) [55]. The endangered fern *Isoetes sinensis* has been evaluated for its nutraceutical values by Wang et al. [56] and found a strong antioxidant activity owing to the presence of flavonoids. Flavonoids in *Stenoloma chusanum* showed seasonal periodicity by peaking during February in China [57] indicating its use in a specific time frame.

Methanolic extract of *Lygodium venustum* effectively inhibits *Entamoeba histolytica* as well as *Giardia lamblia* [58]. This extract has also inhibitory activity against *Trypanosoma cruzi* [59]. *Pteris calomelanos* possess antiplasmodial potential against *Plasmodium falciparum* [60]. A wide array of ferns possess antibacterial and antifungal activities. *Pteris multifida* has versatile antibacterial activity against several pathogenic bacteria [61]. One of the flavonoids isolated from *Pteris calomelanos* using ethyl acetate showed inhibition of bacteria [43]. *Pteris ensiformis* possesses antituberculosis potential against *Mycobacterium tuberculosis* [62]. *Stenoloma chusanum* has potent antifungal properties against many pathogenic fungi (e.g., *Aspergillus, Cryptococcus, Epidermophyton, Microsporum,* and *Trichophyton*) [63]. Ethanol, methanol, and hexane extracts of *P. calomelanos* showed antifungal activity against many *Candida* species (*C. albicans, C. krusei*, and *C. tropicalis*) [43].

4.1 Ethnomedicinal Knowledge

Ethnic knowledge of the medicinal uses of ferns is the basis to follow the value of ferns in treating several human ailments. Ethnic approaches and herbal products occupied the prime place in primary health care owing to many economic reasons. Ethnic uses of ferns in uplifting the primary human health will have two important approaches: (1) use of a specific fern to treat several diseases; and (2) use of several ferns to treat a specific disease. For instance, for the production of the Chinese medicine the Gusuibu, up to six ferns are used (*Davallia divaricata, D. mariesii, D. solida, Drynaria fortunei, Humata griffithiana*, and *Pseudodrynaria coronans*) [45]. Similarly, differences in the medicinal applications of ferns are also dependent on geographical variations [25]. For example, in the Himalayas, *Lygodium flexuosum* is used to treat expectorants, rheumatism, sprains, scabies, and others, while the same fern in China is used to treat rheumatoid lumbago and gallstone [64, 65]. Likely, such differences are also dependent on the availability of a fern species and its parts in a specific geographic area.

In Chinese herbal medicine, an aqueous extract of *Pteris ensiformis* is used for immunomodulatory effect [66]. *Dravallia* species are widely used in Chinese medicine to treat bone injuries, osteoporosis, inflammation, and cancers [45]. Ethnic knowledge pertains to the part of fern used in medicine and is also important in following the distribution and extraction of active principles. Rhizome extracts of *Helminthostachys zeylanica* are extensively used in oriental medicine in India, Sri Lanka, and China (as an analgesic, antipyretic, hepatoprotective, and anti-inflammatory agent) and used in treating diseases like malaria, jaundice, and syphilis [67, 68]. Fronds of the fern *Cyathea contaminants* are used in herbal medicine showed good antioxidant activity as well as antibacterial activity (*Escherichia coli* as well as *Staphylococcus aureus*) [69].

Sureshkumar et al. [70] discussed the ethnomedicinal applications of ferns by the Malayali tribe in southern India. Antony and Suresh [34] reported ethnomedicinal ferns used by the 14 tribals in Kerala, India to combat various human ailments. The members of the genus *Adiantum* are used in ethnic Chinese medicine to promote urination, relieve swelling, and combat fevers [71]. In Chinese medicine, *Pteris semipinnata* is used in the treatment of venomous snake bites [43]. *Botrychium ternanum* serves as folk medicine in Japan as well as China for many diseases (headache, dizziness, cough, asthma, and fever) [72]. *Phymatopteris hastata* is traditionally used in China to treat diseases like diarrhea, bronchitis, and influenza [73]. Members of the family Gleicheniaceae (e.g., *Dicranopteris dichotoma*) in

Malaysia serve as folk medicine to treat urinary ailments, trauma, and fever in children [74]. Herman et al. [75] demonstrated the wound-healing capability of ethanolic extracts of two *Acrostichum* spp. obtained from the Matang mangroves of Malaysia in the rat model. *Ophioglossum vulgatum* is a commonly used oriental medicine to resolve dermatological, hemostatic, and antiparasitic ailments in Bangladesh [76]. Fierascu et al. [77] have demonstrated the enhancement of antimicrobial, cytotoxic, and phytotoxic properties of the hydroalcoholic extract of *Asplenium scolopendrium* collected from Valsan Valley (Romania) by the gamma irradiation at the dose of 0.6 kGy.

4.2 Pharmaceutical Products

Ethnic medicinal uses of ferns maintain the primary human health or treat or cure specific ailments owing to folk knowledge of the tribal populations around the world. Recent developments in phytochemical investigations yielded some appreciable formulations of drugs with potent medicinal value. Some of these drugs derived from ferns are available based on clinical trials. The drug Anapsos, a Spanish pharmaceutical product obtained from the rhizomes of *Polypodium leucotomos*, has dermatological applications especially to treat dermatitis and psoriasis [78]. It has the potential immunomodulating effect, stimulation of proliferation, and activation of lymphocytes (T and natural killer cells). It is also found that this drug serves as an antioxidant, immunomodulatory, and photoprotectant useful as an effective cosmeceutical [79]. The product of aqueous extract from the aerial parts of *P. leucotomos* with the trade name Fernblock® is a strong antioxidant used in topical gels, creams, sprays, and oral dietary supplements [80]. It has vital in the treatment of dermatitis, melisma, psoriasis, vitiligo, and has the capacity to minimize infection in athletes. The Fernblock® is a polyphenol-enriched product that possesses photoprotective (harmful effects of UV light, sunburn, skin cancer, and carcinogenesis) as well as nutraceutical value devoid of toxic or mutagenic effects [80, 81].

5 Environmental Attributes

Besides nutritional and medicinal potential, the ecosystem services of ferns include phytoremediation and biosorption of heavy metals [6, 82]. Ferns are known to produce steroids called ecdysone, which is similar to the steroids produced by plants (phytoecdysones) and animals (zooecdysones) [83, 84]. Production of phytoecdysones has been reported in 27 families in pteridophytes. Various field applications of ecdysteroids against the pests have been discussed by Sahayaraj [83]. Ferns belong to the genera *Adiantum, Asplenium, Cheilanthes, Cyclosorus, Dicranopteris, Diplazium, Dryopteris, Microsorum, Polypodium, Pteridium,* and *Schizaea*, are known to synthesize ecdysteroids [83]. Five ecdysteroids were found in ethanol extract of *Dicranopteris rufopilosum* from the Yunnan province of China [85]. Fronds and rhizomes of *Microsorum membranifolium* and *M. scolopendria* are used in ethnic medicine in Polynesia and are rich in ecdysteroids [25]. Ecdysteroids possess wide applications to control insects as feed deterrents as well as endocrine disrupters [84]. Besides, they have a potential pharmacological impact on mammals by influencing the membrane-bound receptors [84]. Ecdysones are also known for inducing cell regeneration and skin texture refinement. Among the ecdysones, 20-hydroxyecdysone is a main active principle that acts against insects, whereas its analogs serve as either storage forms or pro-hormones. Table 6 projects the list of widely distributed fern species producing different kinds of phytoecdysones. Members of the Polypodiaceae stand first (15 species) followed by Pteridaceae (10 species) and Athyriaceae (7 species).

Ecdysteroids are triterpenoids like triterpenes, saponins, and phytosterols. They play an important role as biopesticides as deterrents, repellents, and toxicants, and they interfere with the insect oviposition similar to insect-molting hormones. Phytoecdysones of ferns at a very low concentration are known to control a wide range of insects [86]. The family Polypodiaceae is known for phytoecdysones in high concentrations and is ideal for insect pest control.

Insecticidal ecdysteroids of ferns are also miticidal and useful to protect products from storage pests and other insects. Caution needs to be exercised to prevent the death of beneficial insects while using fern-based ecdysteroids. Several species of *Elaphoglossum* are known for molluscicidal activities against *Biomphalaria peregrina* [87]. Methanolic extracts of some ferns (e.g., whole plants of *Onychium japonicum* and leaves of *Pteris vittata*) possess insecticidal activities against houseflies as well as mosquitoes [88]. Ecdysteroids are highly valuable metabolites owing to their wide range of biological functions (e.g., anabolic, hypocholesterolemic, hepatoprotective, hypoglycemic, antidepressant, and purgative functions) [89]. The drug Filixsäure obtained from the rhizomes of the male fern *Aspidium filix-mas* acts against the phytophagous insects, while it has no toxicity against the pollinator *Aphis rumicis* [83]. The acetone solution of the Filixsäure has the capacity to control the *Culex fatigans* mosquitoes. It has also been shown to control housefly *Musca domestica* [90].

Ferns are also effective flora in the phytoremediation of pollutants in soils by various mechanisms [6, 82, 91]. *Pteris vittata* is well-known for arsenic-hyperaccumulation, while other arsenic hyperaccumulators include *Pteris multifida* found in Asian countries [43]. The phytochemicals (e.g., enzymes and other compounds) involved in various phytoremediation and hyperaccumulation potential warrants further study.

6 Conclusions

Ferns are highly valued biota in the biosphere involved in several ecosystem services (nutritional, medicinal, and environmental restoration). They are the natural repository of the plethora of valuable metabolites of human, livestock, industrial, and environmental significance. The modern pharmacological advances of ferns have progressed owing to enormous ethnic knowledge gained in different parts of the

Table 6 Some of the widely distributed potential fern species producing various types of phytoecdysones

Species	Family	Reference
Acrophorus stipellatus	Dryopteridaceae	[119]
Acrostichum aureum	Pteridaceae	[120]
Adiantum aethiopicum	Pteridaceae	[121]
Adiantum capillus-veneris	Pteridaceae	[121]
Adiantum cunninghamii	Pteridaceae	[121]
Adiantum flabellulatum	Pteridaceae	[121]
Adiantum hispidulum	Pteridaceae	[121]
Adiantum philippense	Pteridaceae	[121]
Anemia phyllitidis	Schizaeaceae	[119]
Athyrium aphanoneuron	Athyriaceae	[121]
Athyrium arisanense	Athyriaceae	[122]
Athyrium atkinsonii	Athyriaceae	[120]
Athyrium crenulata serratum	Athyriaceae	[119]
Athyrium decurrentialatum	Athyriaceae	[119]
Athyrium yokoscense	Athyriaceae	[123]
Azolla imbricata	Salviniaceae	[121]
Blechnum minus	Blechnaceae	[124]
Blechnum vulcanicum	Blechnaceae	[125]
Brainea insignis	Blechnaceae	[126]
Cheilanthes farinosa	Pteridaceae	[127]
Cheilanthes tenuifolia	Pteridaceae	[128]
Chingia sakayensis	Thelypteridaceae	[129]
Cyathea cooperi	Cyatheaceae	[119]
Diplazium esculentum	Athyriaceae	[121]
Lepidogrammitis drymoglossoides	Polypodiaceae	[130]
Lepisorus contortus	Polypodiaceae	[131]
Microsorum commutatum	Polypodiaceae	[132]
Microsorum grossum	Polypodiaceae	[133]
Microsorum insigne	Polypodiaceae	[134]
Microsorum maximum	Polypodiaceae	[132]
Microsorum membranifolium	Polypodiaceae	[132]
Microsorum punctatum	Polypodiaceae	[132]
Microsorum scolopendria	Polypodiaceae	[133]
Neocheiropteris multiflorins	Polypodiaceae	[135]
Phymatosorus membranifolum	Polypodiaceae	[135]
Phymatosorus scolopendria	Polypodiaceae	[133]
Podocarpus nakaii	Podocarpaceae	[136]
Polypodium leucotomos	Polypodiaceae	[137]
Polypodium vulgare	Polypodiaceae	[119, 138]
Pteridium aquilinum	Dennstaedtiaceae	[139, 140]
Pteris inaequalis	Pteridiaceae	[141]

world. The livelihood and socioeconomic status of tribals are partly dependent on the diversity, distribution, and availability of ferns in different geographic regions. Traditional medicine generally uses concoction instead of purified compound, if such admixture results in curing an ailment or multiple ailments probably owing to a blend of active principles than a single compound. Future pharmacological investigations have to focus on such curative properties by admixture of active principles. Ethnic or traditional knowledge is the basis to understand the nutritional and pharmacological significance of ferns to shape the herbal industries. The ethnic knowledge of a specific fern, its parts, specific season, mode of preparation, and prophylactic attributes are relevant in modern pharmacology. Special emphasis needs to be exercised to propagate and conserve the ferns that possess active principles in their underground parts (roots and rhizomes). Owing to multifarious socioeconomic applications (e.g., aesthetic, landscape, handicrafts, nutrition, medicine, and industrial), the conservation and rehabilitation of ferns should not be overlooked.

Acknowledgments I am indebted to Mangalore University and the Department of Biosciences for academic encouragement. I am benefitted from Dr. Mahadevakumar, Kerala Forest Research Institute, Peechi, Kerala (India), for constructive suggestions and discussion to draft this chapter. Improvement of the presentation was possible by the constructive comments of referees.

References

1. Dixit RD (2000) Conspectus of pteridophytic diversity in India. Indian Fern J 17:77–91
2. Chapman AD (2009) Numbers of living species in Australia and the world. Australian Biological Resources Study, Canberra, Australia
3. Fraser-Jenkins CR (2012) Rare and threatened Pteridophytes of Asia 2. Endangered species of India- the higher IUCN categories. Bull Nat Mus Nat Sci B 38:153–181
4. Kholia RS, Balkrishna A (2022) Pteridophytes used by peoples of Indian Himalayan region and northern India. In: Johnson M, Fernández H, Kumar A, Shibila A (eds) Ferns - biotechnology, propagation, medicinal uses and environmental regulation, Springer, Singapore, pp 379–412
5. Fernádez H, Kumar A (2022) Introduction. In: Marimuthu J, Fernández H, Kumar A, Thangaiah S (eds) Ferns - biotechnology, propagation, medicinal uses and environmental regulation. Springer Nature, Singapore, pp 1–10
6. Sajeev S, Roshni PT, Mathias RC, Morajkar S, Prabhu S, Hegde S (2022) Pteridophytes: effective agents of phytoremediation. In: Marimuthu J, Fernández H, Kumar A, Thangaiah S (eds) Ferns - biotechnology, propagation, medicinal uses and environmental regulation. Springer Nature, Singapore, pp 627–650
7. Markham K, Chalk T, Neal Stewart C Jr (2006) Evaluation of fern and moss protein-based defenses against phytophagous insects. Int J Plant Sci 167:111–117
8. Rathinasabapathi B (2006) Ferns represent an untapped biodiversity for improving crops for environmental stress tolerance. New Phytol 172:385–390. https://doi.org/10.1111/j.1469-8137.2006.01889.x
9. Tosens T, Nishida K, Gago J, Coopman RE, Cabrera HM et al (2016) The photosynthetic capacity in 35 ferns and fern allies: mesophyll CO_2 diffusion as a key trait. New Phytol 209: 1576–1590. https://doi.org/10.1111/nph.13719

10. Morais-Braga MF, Souza TM, Santos KK, Guedes GM, Andrade JC, Tintino SR (2012) Phenolic compounds and interaction between aminoglycosides and natural products of *Lygodium venustum* SW against multiresistant bacteria. Chemotherapy 58:337–340
11. Yumkham SD, Chakpram L, Salam S, Bhattacharya MK, Singh PK (2017) Edible ferns and fern-allies of North East India: a study on potential wild vegetables. Genet Resour Crop Evol 64:467–477
12. Balkrishna A, Joshi B, Srivastva A, Shankar R, Tiwari S et al (2019) Some economic aspects of ferns and fern-allies of Seijosa forest area of Pakke-Kessang District, Arunachal Pradesh. Int J Adv Res Bot 5:14–20
13. Ranil RHG, Bussmann RW (2021) Potential uses of lycophytes and ferns in Sri Lanka: an ethnopteridological perspective. Enthnobot Res Appl 21:36. https://doi.org/10.32859/era.21.36.1-11
14. Bujak A, Bujak J (2022) Azolla's use as a biofertilizer and livestock feed. In: Marimuthu J, Fernández H, Kumar A, Thangaiah S (eds) Ferns - biotechnology, propagation, medicinal uses, and environmental regulation. Springer Nature, Singapore, pp 671–695
15. Singh LS, Singh PK, Singh EJ (2001) Ethnobotanical uses of some pteridophytic species in Manipur. Indian Fern J 18:14–17
16. Sambantham V (2022) In Vitro propagation of *Histiopteris incisa* (Thunb.) J. Sm.: an ornamental fern. In: Marimuthu J, Fernández H, Kumar A, Thangaiah S (eds) Ferns - biotechnology, propagation, medicinal uses, and environmental regulation. Springer Nature, Singapore, pp 243–248
17. Marimuthu J, Fernádez H, Kumar A, Thangaiah S (2022) Ferns: biotechnology, propagation, medicinal uses, and environmental regulation. Springer Nature, Singapore
18. Liu Y, Wujisguleng W, Long C (2012) Food uses of ferns in China: a review. Acta Soc Bot Pol 81:263–270
19. Petkov V, Slavova I, Teneva D, Mladenova T, Stoyanov P, Argirova M (2021) Phytochemical study and biological activity of three fern species of the *Asplenium* genus growing in Bulgaria. Nat Prod J 11:e120521193308. https://doi.org/10.2174/2210315511666210512024716
20. Greeshma AA, Sridhar KR (2019) Nutraceutical and bioactive significance of ferns with emphasis on the medicinal fern *Diplazium*. In: Egamberdieva D, Tiezzi A (eds) Medicinally important plant biomes: source of secondary metabolites. Springer Nature, Singapore, pp 115–131
21. Chettri S, Manivannan S, Muddarsu VR (2018) Nutrient and elemental composition of wild edible ferns of the Himalaya. Am Fern J 108:95–106
22. Dvorakova M, Pumprova K, Antonínová Ž, Rezek J, Haisel D et al (2021) Nutritional and antioxidant potential of fiddleheads from European ferns. Foods 10:460. https://doi.org/10.3390/foods10020460
23. Giri P, Kumar A, Uniyal PL (2022) Omega-3 and omega-6 fatty acids distribution in pteridophytes and its significance in nutraceutical, pharmacology, and cosmetic industry. In: Marimuthu J, Fernández H, Kumar A, Thangaiah S (eds) Ferns - biotechnology, propagation, medicinal uses, and environmental regulation. Springer Nature, Singapore, pp 521–535
24. Mehltreter K, Tenhaken R, Jansen S (2021) Nectaries in ferns: their taxonomic distribution, structure, function, and sugar composition. Am J Bot 109:1–12. https://doi.org/10.1002/ajb2.1781
25. Ho R, Teai T, Binchini J-P, Lafont R, Raharivelomanana P (2011) Ferns: from traditional uses to pharmaceutical development, chemical identification of active principles. In: Kumar A, Fernández H, Revilla M (eds) Working with ferns. Springer, New York. https://doi.org/10.1007/978-1-4419-7162-3_23
26. Singh A, Johari D (2016) Sustainable utilization of biodiversity: role of pteridophytes in livelihoods of Tharu Tribes of Uttar Pradesh, pp 19–30. https://doi.org/10.13140/RG.2.2.32813.79841

27. Semwal P, Painuli S, Painuli KM, Antika G, Tumer TB et al (2021) *Diplazium esculentum* (Retz.) Sw.: ethnomedicinal, phytochemical, and pharmacological overview of the Himalayan ferns. Oxidative Med Cell Longev 1917890. https://doi.org/10.1155/2021/1917890
28. Sureshkumar J, Ayyanar M (2022) Phytochemical composition and in vitro antioxidant and antidiabetic activities of *Nephrolepis auriculata* (L.) Trimen: an unexplored ethnomedicinal fern. In: Marimuthu J, Fernández H, Kumar A, Thangaiah S (eds) Ferns - biotechnology, propagation, medicinal uses, and environmental regulation. Springer Nature, Singapore, pp 571–584
29. Marimuthu J, Janakiraman N, Saleride JC, Sivaraman A, Shivananthini B, Paulraj K (2022) Phytochemistry of Indian pteridophytes: a review. In: Marimuthu J, Fernández H, Kumar A, Thangaiah S (eds) Ferns - biotechnology, propagation, medicinal uses, and environmental regulation. Springer Nature, Singapore, pp 433–480
30. Lu H, Hu J, Zhang LX, Tan RX (1999) Bioactive constituents from *Pteris multifida*. Planta Med 65:586–587
31. Rivier L, Lindgren J-E (1972) Ayahuasca, the South American hallucinogenic drink: an ethnobotanical and chemical investigation. Econ Bot 26:101–129
32. Giri P, Uniyal PL (2022) Edible ferns in India and their medicinal uses: a review. Proc Natl Acad Sci India Sect B Biol Sci. https://doi.org/10.1007/s40011-021-01293-4
33. Ojha R, Devkota HP (2021) Edible and medicinal pteridophytes of Nepal: a review. Ethnobot Res Appl 22:16. https://doi.org/10.32859/era.22.16.1-16
34. Antony R, Suresh S (2022) Ethnobotanical uses of ferns and lycophytes of Kerala. In: Marimuthu J, Fernández H, Kumar A, Thangaiah S (eds) Ferns - biotechnology, propagation, medicinal uses, and environmental regulation. Springer Nature, Singapore, pp 413–431
35. Shills MEG, Young VR (1988) Modern nutrition in health and disease. In: Neiman DC, Buthepodorth DE, Nieman CN (eds) Nutrition. WmC Brown, Dubugue, pp 276–282
36. Yusuf AA, Mofio BM, Ahmed AB (2007) Proximate and mineral composition of *Tamarindus indica* Linn 1753 seeds. The Sci World J 2:1–4
37. Bahadoran Z, Mirmiran P, Azizi F (2013) Dietary polyphenols as potential nutraceuticals in management of diabetes: a review. J Diabetes Metab Disord 12:43. https://doi.org/10.1186/2251-6581-12-43
38. Chai TT, Yeoh LY, Ismail NIM, Ong H-C, Manan FA, Wong F-C (2015) Evaluation of glucosidase inhibitory and cytotoxic potential of five selected edible and medicinal ferns. Trop J Pharma Res 14:449–454
39. Greeshma AA, Sridhar KR, Pavithra M (2019) Biologically active components of the Western Ghats medicinal fern *Diplazium esculentum*. In: Egamberdieva D, Tiezzi A (eds) Medicinally important plant biomes: source of secondary metabolites. Springer Nature, Singapore, pp 67–83
40. Adhoni SA, Thimmappa SC, Kaliwal BB (2016) Phytochemical analysis and antimicrobial activity of *Chorella vulgaris* isolated from Unkal Lake. J Coast Life Med 4:368–373
41. Marimuthu J, Janakiraman N, Saleride JC, Sivaraman A, Shivananthini B, Paulraj K (2022) Biopotency of pteridophytes: a review. In: Marimuthu J, Fernández H, Kumar A, Thangaiah S (eds) Ferns – biotechnology, propagation, medicinal uses, and environmental regulation. Springer Nature, Singapore, pp 481–519
42. Gogoi LJ, Upadhyay S, Baruah M, Konwar A, Nath MP et al (2021) A comparative and biochemical evaluations of edible parts of *Diplazium esculentum* (Retz.) SW. a locally available fern of Assam at different growing stages. Eco Environ Cons 27:S173–S177
43. Cao H, Chai T-T, Wang X, Morais-Braga MFB, Yang J-H et al (2017) Phytochemicals from fern species: potential for medicine applications. Phytochem Rev 16:79–440
44. Abraham S, Thomas T (2022) Ferns: a potential source of medicine and future prospects. In: Marimuthu J, Fernández H, Kumar A, Thangaiah S (eds) Ferns – biotechnology, propagation, medicinal uses and environmental regulation. Springer Nature, Singapore, pp 445–378

45. Chang H-C, Huang G-J, Agrawal DC, Kuo C-L, Wu C-R, Tsay H-S (2007) Antioxidant activities and polyphenol contents of six folk medicinal ferns used as Gusuibu. Bot Stud 48: 397–406
46. Li TS (2006) Taiwanese native medicinal plants: phytopharmacology and therapeutic values. CRC Press, London
47. Chai T-T, Yeoh L-Y, Ismail NIM, Ong H-C, Manan FA, Wong F-C (2015) Evaluation of glucosidase inhibitory and cytotoxic potential of five selected edible and medicinal ferns. Trop J Pharmaceut Res 14:449–454
48. Chai T-T, Elamparuthi S, Yong A-L, Quah Y, Ong H-C, Wong F-C (2013) Antibacterial, anti-glucosidase, and antioxidant activities of selected highland ferns of Malaysia. Bot Stud 54:55. http://www.as-botanicalstudies.com/content/54/1/55
49. Morais-Braga MFB, Souza TM, Santos KKA, Guedes GMM, Andrade JC et al (2013) Phenol composition, cytotoxic and anti-kinetoplastidae activities of *Lygodium venustum* SW. (Lygodiaceae). Exp Parasitol 134:178–182
50. Oh H, Kim DH, Cho J-H, Kim Y-C (2004) Hepatoprotective and free radical scavenging activities of phenolic petrosins and flavonoids isolated from *Equisetum arvense*. J Ethnopharmacol 95:421–424
51. Abbasi AM, Khan MA, Ahmad M, Zafar M, Jahan S, Sultana S (2010) Ethnopharmacological application of medicinal plants to cure skin diseases and in folk cosmetics among the tribal communities of North-West Frontier Province. Pak J Ethnopharmacol 128:322–335
52. Haider S, Nazreen S, Alam M, Gupta A, Hamid H, Alam MS (2011) Anti-inflammatory and anti-nociceptive activities of ethanolic extract and its various fractions from *Adiantum capillus veneris* Linn. J Ethnopharmacol 138:741–747
53. Qi Z-M, Wang Q (2004) Effects of extracts of *Equisetum hiemale* on platelet aggregation and thrombosis in rats. Chin J Clin Rehab 8:7738–7739
54. Rahman AU (2008) Studies in natural products chemistry. Elsevier Science, London
55. Sandhu NS, Kaur S, Chopra D (2010) *Equisetum arvense*: pharmacology and phytochemistry – a review. Asian J Pharm Clin Res 3:146–150
56. Wang X, Ding G, Liu B, Wang Q (2020) Flavonoids and antioxidant activity of rare and endangered fern: *Isoetes sinensis*. Flavonoids and antioxidant activity of rare and endangered fern: *Isoetes sinensis*. PLoS One 15:e0232185. https://doi.org/10.1371/journal.pone.0232185
57. Xia X, Cao J, Zheng Y, Wang Q, Xiao J (2014) Flavonoid concentrations and bioactivity of flavonoid extracts from 19 species of ferns from China. Ind Crop Prod 58:91–98
58. Calzada F, Cervantes-Martínez A, Yépez-Mulia L (2005) In vitro antiprotozoal activity from the roots of *Geranium mexicanum* and its constituents on *Entamoeba histolytica* and *Giardia lamblia*. J Ethnopharmacol 98:191–193
59. Morais-Braga MFB, Souza TM, Santos KKA, Andrade JC, Guedes GMM et al (2013) Citotocixidade e atividade antiparasitária de *Lygodium venustum* SW. Acta Toxicol Arg 21: 1–12
60. Valadeau C, Pabon A, Deharo E, Alba'n-Castillo J, Estevez Y et al (2009) Medicinal plants from the Yanesha (Peru): evaluation of the leishmanicidal and antimalarial activity of selected extracts. J Ethnopharmacol 123:413–422
61. Zhou R, Li S (1998) Study on antibacterial effects of pteridophyta. Nat Prod Res Dev 11: 53–56
62. Chen CK, Chen IS, Peng CF, Chen JF (2013) Bioactive constituents from *Pteris ensiformis*. Plant Medica 79:P120. https://doi.org/10.1055/s-0033-1352110
63. Ren B, Xia B, Li W, Wu J, Zhang H (2009) Two novel phenolic compounds from Stenoloma chusanum and their antifungal activity. Chem Nat Comp 45:182–186
64. Lee S, Xiao C, Pei S (2008) Ethnobotanical survey of medicinal plants at periodic markets of Honghe prefecture in Yunnan Province, SW China. J Ethnopharmacol 117:362–377
65. Upreti K, Jalal JS, Tewari LM, Joshi GC, Pangtey YPS, Tewari G (2009) Medicinal uses of pteridophytes of Kumaun Himalaya, Uttarakhand, India. J Am Sci 5:167–170

66. Wu MJ, Weng CY, Wang L, Lian TW (2005) Immunomodulatory mechanism of the aqueous extract of sword brake fern (*Pteris ensiformis* Burm.). J Ethnopharmacol 98:73–81
67. Huang Y-L, Yeh P-Y, Shen C-C, Chen C-C (2003) Antioxidant flavonoids from the rhizomes of *Helminthostachys zeylanica*. Phytochemistry 64:1277–1283
68. Suja SR, Latha PG, Pushpangadan P, Rajasekharan S (2003) Aphrodisiac property of *Helminthostachys zeylanica* in male mice. J Trop Med Plants 3:191–195
69. Faizal A, Taufik I, Rachmani AF, Azar AWP (2020) Antioxidant and antibacterial properties of tree fern *Cyathea contaminans*. Biodiversitas 21:2201–2205
70. Sureshkumar J, Ayyanar M, Silambarasan R (2021) Ethnomedicinal uses, phytoconstituents and pharmacological importance of pteridophytes used by Malayalis in Kolli hills, India: a quantitative survey. J Herb Med 25:100418
71. Pan C, Chen YG, Ma XY, Jiang JH, He F, Zhang Y (2011) Phytochemical constituents and pharmacological activities of plants from the genus *Adiantum*: a review. Trop J Pharmaceut Res 10:681–692
72. Warashina T, Umehara K, Miyase T (2012) Flavonoid glycosides from *Botrychium ternatum*. Chem Pharm Bull 60:1561–1573
73. Su W, Li P, Huo L, Wu C, Guo N, Liu L (2011) Phenolic content and antioxidant activity of *Phymatopteris hastata*. J Serb Chem Soc 76:1485–1496
74. Kamisan FH, Yahya F, Mamat SS, Kamarolzaman MFF, Mohtarrudin N et al (2014) Effect of methanol extract of *Dicranopteris linearis* against carbon tetrachloride-induced acute liver injury in rats. BMC Compl Alt Med 14:123. https://doi.org/10.1186/1472-6882-14-123
75. Herman HP, Susanti D, Saad S, Taher M, Ramli N (2013) Wound healing properties of ethanolic extract of *Acrostichum aureum* and *Acrostichum speciosum* rhizome in rats. J Trop Resour Sustain Sci 1:42–48
76. Sarker SK, Hossain AE (2009) Pteridophytes of greater Mymensingh district of Bangladesh used as vegetables and medicines. Bangladesh J Plant Taxon 16:47–56
77. Fierascu I, Ditu L-M, Sutan AN, Drăghiceanu OA, Fierascu RC et al (2021) Influence of gamma irradiation on the biological properties of *Asplenium scolopendrium* L. hydroalcoholic extracts. Rad Phys Chem 181:109175. https://doi.org/10.1016/j.radphyschem.2020.109175
78. Sempere-Ortells JM, Campos A, Velasco I, Marco F, Ramirez-Bosca A et al (2002) Anapsos (*Polypodium leucotomos*) modulates lymphoid cells and the expression of adhesion molecules. Pharmacol Res 46:185–190
79. Gombau L, Garcia F, Lahoz A, Fabre M, Roda-Navarro P et al (2006) *Polypodium leucotomos* extract: antioxidant activity and disposition. Toxicol In Vitro 20:464–471
80. Choudhry SZ, Bhatia N, Ceilley R, Hougeir F, Lieberman R et al (2014) Role of oral *Polypodium leucotomos* extract in dermatologic diseases: a review of the literature. J Drugs Dermatol 13:148–153
81. Parrado C, Juarranz A, Gilaberte Y, Phillips N, Gonzalez S (2014) Fern extract, oxidative stress, and skin cancer. In: Preedy VR (ed) Cancer: oxidative stress and dietary antioxidants. Academic Press, San Diego, pp 255–264
82. Prabhu S, Hegde S (2022) Pteridophytes as effective biosorption agents of heavy metals. In: Marimuthu J, Fernández H, Kumar A, Thangaiah S (eds) Ferns – biotechnology, propagation, medicinal uses and environmental regulation. Springer Nature, Singapore, pp 651–670
83. Sahayaraj K (2022) Ferns, a source of phytoecdysones, and their applications in pestiferous insect management. In: Marimuthu J, Fernández H, Kumar A, Thangaiah S (eds) Ferns – biotechnology, propagation, medicinal uses and environmental regulation. Springer Nature, Singapore, pp 181–198
84. Lafont R, Balducci C, Dinan L (2021) Ecdysteroids. Encyclopedia 1:1267–1302. https://doi.org/10.3390/encyclopedia1040096
85. Hu J, Shi X, Mao X, Li H, Chen J, Shi J (2014) Ecdysteroids from the ethanol extract of *Diplopterygium rufopilosum*. Phytochem Lett 8:73–76

86. Das N, Siddhartha KM, Anusha B, Eunüs SA, Anupam B (2020) The phytochemical, biological, and medicinal attributes of phytoecdysteroids: an updated review. Acta Pharm Sin B 11:1740–1766. https://doi.org/10.1016/j.apsb.2020.10.012
87. Socolsky C, Borkosky S, Bardon A (2011) Structure-molluscicidal activity relationships of acylphloroglucinols from ferns. Nat Prod Comm 6:387–391
88. Huang Y-C, Hwang T-L, Yang Y-L, Wu S-H, Hua HM et al (2010) Acetogenin and prenylated flavonoids from *Helminthostachys zeylanica* with inhibitory activity on superoxide generation and elastase release by neutrophils. Planta Med 76:447–453
89. Lafont R, Dinan L (2003) Practical uses for ecdysteroids in mammals and humans: an update. J Insect Sci 3:7. http://www.insectscience.org/3.7
90. Gupta MK, Marangmei L, Sarma AK (2018) Efficacy of extracts of some ferns against *Plutella xylostella*. Indian J Entomol 80:1667–1671
91. Prabhu SG, Srinikethan G, Hegde S (2021) Pelletization of pristine *Pteris vittata* L. pinnae powder and its application as a biosorbent of Cd(II) and Cr(VI). SN Appl Sci 2:132. https://doi.org/10.1007/s42452-019-1906-1
92. Chen C-Y, Chiu F-Y, Lin Y, Huang W-J, Hsieh P-S, Hsu F-L (2015) Chemical constituents analysis and antidiabetic activity validation of four fern species from Taiwan. Int J Mol Sci 16: 2497–2516. https://doi.org/10.3390/ijms16022497
93. Barkaran X-R, Vigila A-VG, Zhang S-Z, Feng S-X, Liao W-B (2018) A review of the use of pteridophytes for treating human ailments. J Zhejiang Univ Sci B 19:85–119
94. Patil S, Lavate R, Raole V, Rajput K (2022) *Adiantum* L.: overview on taxonomy, distribution, conservation status, ethnomedicinal uses, phytochemistry, and pharmacognosy from India. In: Marimuthu J, Fernández H, Kumar A, Thangaiah S (eds) Ferns – biotechnology, propagation, medicinal uses and environmental regulation. Springer Nature, Singapore, pp 553–570
95. Sara SC, Ruby RGD (2022) In vitro antiproliferative effect of *Angiopteris evecta* (G. Forst.) Hoffm. extracts against cultured HT-29 colon cancer cells. In: Marimuthu J, Fernández H, Kumar A, Thangaiah S (eds) Ferns – biotechnology, propagation, medicinal uses and environmental regulation. Springer Nature, Singapore, pp 537–551
96. Rai AK, Sharma RM, Tamang JP (2005) Food value of common edible wild plants of Sikkim. J Hill Res 18:99–103
97. Irawan D, Wijaya CH, Limin SH, Hashidoko Y, Osaki M, Kulu IP (2006) Ethnobotanical study and nutrient potency of local traditional vegetables in Central Kalimantan. Tropics 15: 441–448
98. Shrestha S (2011) Use and chemical analysis of wild food in protected areas of Nepal. Nep J Agric Sci 9:103–106
99. Seal T (2012) Evaluation of nutritional potential of wild edible plants, traditionally used by the tribal people of Meghalaya state in India. Am J Pl Nutr Fert Technol 2:19–26
100. Tag H, Tsering J, Hui PK, Gogoi BJ, Veer V (2014) Nutritional potential and traditional uses of high altitude wild edible plants in Eastern Himalayas, India. Int J Nutr Food Eng 8:238–243
101. Tongco JV, Villaber RA, Aguda R, Razal R (2014) Nutritional and phytochemical screening, and total phenolic and flavonoid content of *Diplazium esculentum* (Retz.) Sw. from Philippines. J Chem Pharmaceut Res 6:238–242
102. Pradhan S, Manivannan S, Tamang JP (2015) Proximate, mineral composition and antioxidant properties of some wild leafy vegetables. J Sci Ind Res 74:155–159
103. Saha S, Sharma MC, Murari MK, Owen LA, Caffee MW (2015) Geomorphology, sedimentology, and minimum exposure ages of streamlined subglacial landforms in the NW Himalaya, India. Boreas 45:284–303. https://doi.org/10.1111/bor.12153
104. Wali A, Sharma S, Walia M, Kumar P, Thakur S et al (2016) Two edible ferns of Western Himalaya: a comparative *in vitro* nutritional assessment, antioxidant capacity and quantification of lutein by UPLC-DAD. Int J Food Nutr Sci 5:9–18
105. Greeshma AA, Sridhar KR, Pavithra M (2018) Nutritional prospects of edible fern of the Western Ghats of India. In: Öztürk M, Hakeem KR, Ashraf M, Ahmed MSA (eds), Global perspectives on underutilized crops. Springer International, Switzerland p 151–164

106. Koniyo Y, Lumenta C, Olii AH, Mantiri RO (2019) The characteristic and nutrients concentrated leaves of vegetable fern (*Diplazium esculentum* (Retz.) Swartz) live in different locations. J Phy (Conf Ser) 1387:012003. https://doi.org/10.1088/1742-6596/1387/1/012003
107. Zihad SMNK, Gupt Y, Uddin SJ, Islam MT, Alam MR et al (2019) Nutritional value, micronutrient and antioxidant capacity of some green leafy vegetables commonly used by southern coastal people of Bangladesh. Heliyon 5:e02768. https://doi.org/10.1016/j.heliyon.2019.e02768
108. Naik B, Maurya VK, Kumar V, Kumar V, Upadhyay S, Gupta S (2021) Phytochemical analysis of *Diplazium esculentum* reveals the presence of medically important components. Curr Nutr Food Sci 17:210–215
109. Bassey ME, Etuk EUI, Ibe MM, Ndon BA (2001) *Diplazium sammatii*, Athyraceae (Nyama Idim): age-related nutritional and antinutritional analysis. Pl Foods Hum Nutr 56:7–12
110. Sundriyal M, Sundriyal RC (2001) Wild edible plants of the Sikkim Himalaya: nutritive values of selected species. Econ Bot 55:377–390
111. Archana GN, Pradeesh S, Chinmayee MD, Mini I, Swapna TS (2012) *Diplazium esculentum*: a wild nutrient-rich leafy vegetable from Western Ghats. In: Proceedings of prospects in bioscience: addressing the issues, Springer, pp 293–301. https://doi.org/10.1007/978-81-322-0810-5_35
112. NRC-NAS (1989) Recommended dietary allowances. National Academy Press, Washington, DC
113. Greeshma AA (2017) Studies on non-conventional nutritional sources of the Western Ghats and west coast of India. PhD Dissertation, Department of Biosciences, Mangalore University, Mangalore
114. FAO-WHO (1991) Protein quality evaluation. Reports of a joint FAO-WHO expert consultation. Food and nutrition paper # 51. Food and Agriculture Organization of the United Nations, Rome
115. Junejo JA, Gogoi G, Islam J, Rudrapal M, Mondal P et al (2018) Exploration of antioxidant, antidiabetic and hepatoprotective activity of *Diplazium esculentum* – a wild edible plant from North Eastern India. Future J Pharmaceut Sci 4:93–101
116. Roy S, Chaudhuri TK (2020) A comprehensive review on the pharmacological properties of *Diplazium esculentum*, an edible fern. J Pharmaceut Phamacol Res 3:1–9. https://doi.org/10.31579/2693-7247/014
117. Zulkifli NZM, Ramli N, Othman MZ, Zawawi NS, Baba MS (2020) Antimalarial activities of *Diplazium esculentum* (Retz.) Sw. aqueous extract in *Plasmodium berghei*-infected mice. Int J Allied Health Sci 4:1657–1664
118. Kunkeaw T, Suttisansanee U, Trachootham D, Karinchai J, Chantong B et al (2021) *iplazium esculentum* (Retz.) Sw. reduces BACE-1 activities and amyloid peptides accumulation in *Drosophila* models of Alzheimer's disease. Sci Rep 11:23796. https://doi.org/10.1038/s41598-021-03142-w
119. Lafont R, Raimana H, Phila R, Laurie D (2011) Ecdysteroids in ferns: distribution, diversity, biosynthesis, and functions. In: Working with ferns. Springer, New York, pp 305–319
120. Vetter J (2018) Secondary metabolites of ferns. In: Current advances in fern research. Springer, Switzerland pp 305–327
121. Masato W, Tomoka M, Hari Prasad D (2021) Phenolic compounds and ecdysteroids of *Diplazium esculentum* (Retz) Sw (Athyriaceae) from Japan and their chemotaxonomic significance. Biochem Syst Ecol 94:104211. https://doi.org/10.1016/j.bse.2020.104211
122. Yen KY, Yang LL, Tom WM (1971) Studies on physiological active phytohormones of ferns in Taiwan I. J Taiwan Pharm Assoc 23:83–92
123. Ohta S, Guo J-R, Hiraga Y, Suga T (1996) 24-epi-pterosterone: a novel phytoecdysone from the roots of *Athyrium yokoscense*. Phytochemistry 41:745–747
124. Chong YK, Galbraith MN, Horn DHS (1970) Isolation of deoxyecrustecdysone, deoxyecdysone, and α-ecdysone from the fern *Blechnum minus*. J Chem Soc D Chem Comm 18:1217–1218

125. Russell GB, Greenwood DR, Lane GA, Blunt JW, Munro MH (1981) 2-Deoxy-3-epiecdysone from the fern *Blechnum vulcanicum*. Phytochemistry 20:2407–2410
126. Wu P, Xie H, Tao W, Miao S, Wei X (2010) Phytoecdysteroids from the rhizomes of *Brainea insignis*. Phytochemistry 71:975–981
127. Josephrajkumar A, Subrahmanyam B, Devakumar C (2000) Growth-regulatory activity of silver fern extract on the cotton bollworm, *Helicoverpa armigera* (Hübner). Int J Trop Insect Sci 20:295–302
128. Faux A, Galbraith MN, Horn DHS, Middleton EJ, Thomson JA (1970) The structures of two ecdysone analogues, cheilanthones A and B, from the fern *Cheilanthes tenuifolia*. J Chem Soc D 4:243–244
129. Sutoyo S, Indrayanto G, Zaini NC (2007) Chemical constituents of the fern *Chingia sakayensis* (Zeiller) Holtt. Nat Prod Comm 2:1934578X0700200513
130. Yao JN, Li ZF, Lou HY, Huang L, Liang GY, Cao PX (2014) A new ecdysteroidal glycoside from *Lepidogrammitis drymoglossoides* (Bak.). Ching J Carbohydr Chem 33:206e11
131. Yang JH, Kondratyuk TP, Jermihov KC, Marler LE, Qiu X et al (2011) Bioactive compounds from the fern *Lepisorus contortus*. J Nat Prod 74:129–136
132. Ho R, Teai T, Loquet D, Bianchini J-P, Girault J-P et al (2007) Phytoecdysteroids in the genus *Microsorum* (Polypodiaceae) of French Polynesia. Nat Prod Comm 2:1934578X0700200803
133. Snogan E, Vahirua-Lechat I, Ho R, Bertho G, Girault JP et al (2007) Ecdysteroids from the medicinal fern *Microsorum scolopendria* (Burm. f.). Phytochem Anal 18:441–450
134. Varangkana J, Somnuk P, Namfon T, Sahanat P (2016) Screening of some species of Thai microsoroid ferns for phytoecdysteroid. SDU Res J 9:81–97
135. Aulakh MK, Kaur N, Saggoo MIS (2019) Bioactive phytoconstituents of pteridophytes – a review. Ind Fern J 36:37–79
136. Nakanishi K, Koreeda M, Sasaki S, Chang ML, Hsu HY (1996) Insect hormones. The structure of ponasterone A, insect-moulting hormone from the leaves of *Podocarpus nakaii* Hay. Chem Commun (Camb) 24:915–917
137. Garcia F, Pivel JP, Guerrero A, Brieva A, Martinez-Alcazar MP et al (2006) Phenolic components and antioxidant activity of Fernblock (R), an aqueous extract of the aerial parts of the fern *Polypodium leucotomos*. Methods Find Exp Clin Pharmacol 28(3):157–160
138. Simon A, Vanyolos A, Beni Z, Dekany M, Toth G, Bathori M (2011) Ecdysteroids from *Polypodium vulgare* L. Steroids 76:1419e24
139. Macek T, Vaněk T (1994) *Pteridium aquilinum* (L.) Kuhn (bracken fern): in vitro culture and the production of ecdysteroids. In: Medicinal and aromatic plants VI. Springer, Berlin, pp 299–315
140. Sahayaraj K, Selvaraj P, Balasubramanian R (2007) Cell mediated immune response of *Helicoverpa armigera* Hubner and *Spodoptera litura* Fab. to fern phytoecdysone. J Entomol 4:289–298
141. Murakami T, Minoru K, Satoshi T, Nobutoshi T, Yasuhisa S, Chiuming C (1978) Weitere Inhaltsstoffe aus *Pteris inaequalis* Baker var. *aequata* (MIQ.) TAGAWA. Chem Pharm Bull 26:643–645

Fern Fatty Acids: From Diversity to Dietary Value

11

Eduard V. Nekrasov

Contents

Abstract

Fatty acids are essential components of all living cells and play vital functions in plant and animal organisms. Literature data demonstrate a diverse set of fatty acids in ferns: about 90 structures of fatty acids have been reported for ferns so far. From their ancestors, ferns inherited long-chain polyunsaturated fatty acids, like arachidonic and eicosapentaenoic acids, which are rare or completely absent in the evolutionarily later lineages of gymnosperms and angiosperms. These fatty acids are valuable nutrients for humans making ferns their potential sources. The review summarizes different aspects of fatty acids in ferns (class Polypodiopsida): structural diversity, distribution in complex lipids, developmental and organ specificity, and ecological and environmental factors affecting fatty

E. V. Nekrasov (✉)
Amur Branch, Botanical Garden-Institute of the Far Eastern Branch of the Russian Academy of Sciences, Blagoveshchensk, Russia
e-mail: garden@ascnet.ru

H. N. Murthy (ed.), *Bioactive Compounds in Bryophytes and Pteridophytes*,
Reference Series in Phytochemistry, https://doi.org/10.1007/978-3-031-23243-5_27

acid composition and content. The biosynthesis of fatty acids in ferns is poorly investigated, however, and can be deduced from the data established for other plant groups. Finally, the nutritional value of ferns as dietary sources of the long-chain polyunsaturated fatty acids and some other practical applications of fern fatty acids are considered.

Keywords

Arachidonic acid · Biosynthesis · Eicosapentaenoic acid · Fatty acids · Ferns · Lipids · Nutritional value · Polypodiopsida

Abbreviations

ACP	Acyl-carrier protein
ARA	Arachidonic acid (20:4n-6)
DGDG	Digalactosyldiacylglycerol
DGTS	Diacylglyceryltrimethylhomoserine
DHA	Docosahexaenoic acid (22:6n-3)
DW	Dry weight
EPA	Eicosapentaenoic acid (20:5n-3)
FAD	Fatty acid desaturase
LC-PUFA	Long-chain polyunsaturated fatty acid
MGDG	Monogalactosyldiacylglycerol
PC	Phosphatidylcholine
PG	Phosphatidylglycerol
PE	Phosphatidylethanolamine
PI	Phosphatidylinositol
PS	Phosphatidylserine
PUFA	Polyunsaturated fatty acid
SQDG	Sulfoquinovosyldiacylglycerol
TG	Triacylglycerol
WW	Wet weight

1 Introduction

Fatty acids are monocarboxylic acids with hydrocarbon chains of different lengths. Fatty acids are key constituents of simple and complex lipids thus providing hydrophobic and other fundamental properties of cell membranes. They are the most effective form of energy deposition in cells (as a part of triacylglycerols). Fatty acids are components of waxes, cutin, and suberin having protective and barrier functions against environmental stresses. Finally, they are precursors of oxylipins and related mediators playing signaling roles. Some vital fatty acids cannot be synthesized in animal and human organisms and must be consumed with food. Such fatty acids are called essential fatty acids.

The structural diversity of fatty acids is determined by the number of carbon atoms in a chain, the number, position, and configuration of double bonds, and different substitutions along a chain. Besides straight-chain fatty acids, branched-chain and cyclic ones occur also in nature. Depending on the number of double bonds, fatty acids are divided into saturated (no double bonds), monoenoic or monounsaturated with one double bond, and polyunsaturated (di-, tri-, tetra-, penta-, etc. unsaturated) fatty acids with two and more double bonds. In natural fatty acids, the double bonds are mostly of the *cis*- (or *Z*-) configuration although there are also some fatty acids with *trans*- (or *E*-) double bonds. Due to the regular and specific insertion of double bonds into an acyl chain, polyunsaturated fatty acids (PUFAs) have the double bonds separated by a single methylene group or, in other words, they are with methylene-interrupted double bonds. Following the steps of elongation and desaturation in the process of biosynthesis, PUFAs are grouped into families (or series) which are designated by the position of a terminal double bond relative to the methyl end (ω) of the acyl chain. Principal families of PUFA are omega-6 (ω6 or (n-6)) and omega-3 (ω3 or (n-3)), though (n-9), (n-7), and other families of PUFA are also known. Double bonds can form conjugated systems (often in the *trans*-configuration) or they can be separated by two or more methylene groups in fatty acids which are referred to as conjugated fatty acids or polymethylene-interrupted fatty acids, respectively. For further reading on the subject, readers are referred to the online resources [1, 2]. The structures of some fatty acids discussed in this chapter are shown in Fig. 1.

Interest in fern fatty acids is related to their long-chain polyunsaturated fatty acids (LC-PUFAs) with an acyl chain of 20 carbon atoms or longer, first of all, arachidonic (20:4n-6, ARA) and eicosapentaenoic (20:5n-3, EPA) acids which distinguish them from the higher plants and, in the opposite, link to the lower lineages. These and other LC-PUFAs are quite common in the lower plants including marine algae [3–6] and bryophytes [7, 8]. It should be noted that LC-PUFAs are located in the lower plants in vegetative tissues, i.e., thalli for macroalgae and bryophytes, and fronds for ferns. On the contrary, the seed plants (gymnosperms and angiosperms) do not contain the LC-PUFAs in the vegetative tissues; however, these and some polymethylene-interrupted LC-PUFAs have been found in seeds of several species of gymnosperms [9, 10] and even angiosperms (the family Ranunculaceae) [11]. Thus, ferns are the most advanced vascular lineage which retains the ability to synthesize ARA, EPA, and some other LC-PUFAs in vegetative tissues. Taking into account their terrestrial habitats and relatively high biomass productivity, the ferns deserve to be considered as a source of these LC-PUFAs especially since many fern species have a history of food consumption in different parts of the world [12–18]. LC-PUFAs play important roles in human physiology and have the bioactive potential [19–21]. This chapter summarizes literature data on fern fatty acids and discusses their possible applications.

Structure	Shorthand designation	Systematic / trivial name
I. Saturated acids		
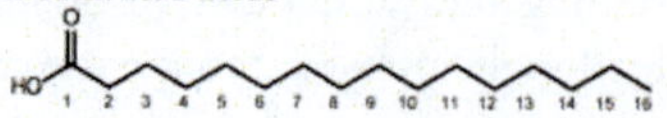	16:0	hexadecanoic / palmitic acid
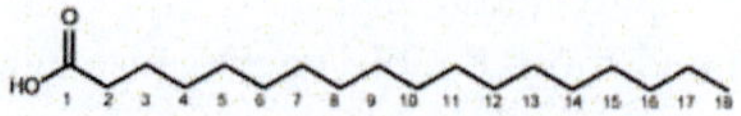	18:0	octadecanoic / stearic acid
II. Monounsaturated (monoenoic) acids		
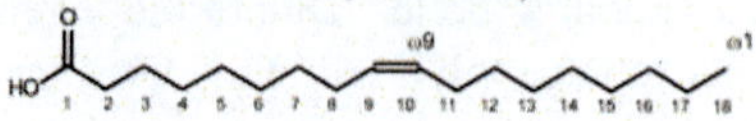	9*c*-18:1 / 18:1ω9 / 18:1n-9	(9*Z*)-octadec-9-enoic / oleic acid
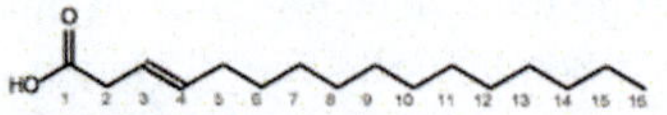	3*t*-16:1	(2*E*)-hexadec-2-enoic / 3-*trans*-hexadecenoic acid
III. Polyunsaturated acids		
dienoic acid		
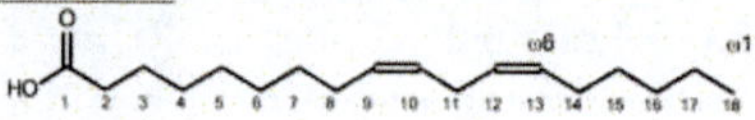	9*c*12*c*-18:2 / 18:2ω6 / 18:2n-6	(9*Z*,12*Z*)-octadeca-9,12-dienoic / linoleic acid
trienoic acids		
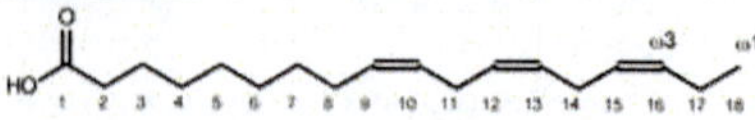	9*c*12*c*15*c*-18:3 / 18:3ω3 / 18:3n-3	(9*Z*,12*Z*,15*Z*)-octadeca-9,12,15-trienoic / α-linolenic acid
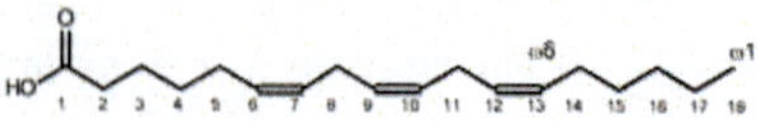	6*c*9*c*12*c*-18:3 / 18:3ω3 / 18:3n-6	(6*Z*,9*Z*,12*Z*)-octadeca-6,9,12-trienoic / γ-linolenic acid
tetraenoic acid		
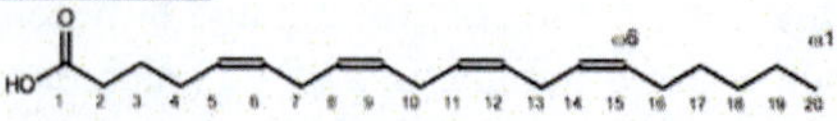	5*c*8*c*11*c*14*c*-20:4 / 20:4ω6 / 20:4n-6	(5*Z*,8*Z*,11*Z*,14*Z*)-icosa-5,8,11,14-tetraenoic / arachidonic acid
pentaenoic acid		
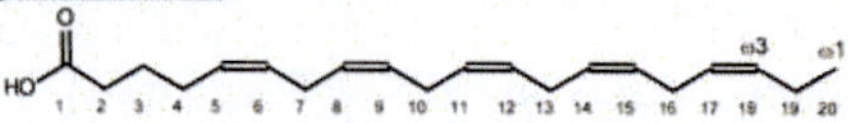	5*c*8*c*11*c*14*c*17*c*-20:5 / 20:5ω3 / 20:5n-3	(5*Z*,8*Z*,11*Z*,14*Z*,17*Z*)-icosa-5,8,11,14,17-pentaenoic / eicosapentaenoic acid
Δ5-unsaturated polymethylene-interrupted fatty acids		
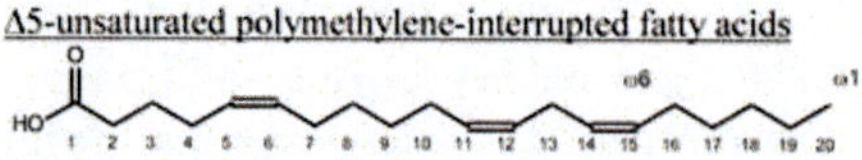	5*c*11*c*14*c*-20:3 / 5,11,14-20:3	(5*Z*,11*Z*,14*Z*)-icosa-5,11,14-trienoic / sciadonic acid
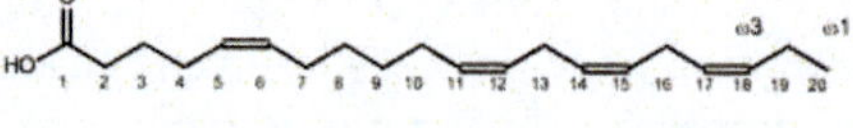	5*c*11*c*14*c*17*c*-20:4 / 5,11,14,17-20:4	(5*Z*,11*Z*,14*Z*,17*Z*)-icosa-5,11,14,17-tetraenoic / juniperonic acid

Fig. 1 Structures of some fatty acids found in ferns. Fatty acid carbon atoms are numbered from the carboxyl end. The position of a terminal double bond is also denoted by counting from the methyl end (ω1)

2 Historical Retrospective

Studies on the fatty acid composition of ferns were started as early as the nineteenth century and were related to the development of an anthelmintic drug from ferns. A German pharmacologist J. Katz investigated fatty material extracted from rhizomes of *Aspidium filix-mas* (current name *Dryopteris filix-mas* (L.) Schott) H.P. Fuchs) and identified oleic (18:1), palmitic (16:0), cerotic (26:0), and butyric (4:0) acids [22].

A few years later, P. Farup [23] reported two octadecatrienoic acids, linolenic acid and probably its unidentified isomer, in rhizomes of *Aspidium spinulosum* (current name *Dryopteris carthusiana* (Vill.) H.P. Fuchs). In the research published in 1942 [24], J. Maizīte from Latvia reported fatty acids for rhizomes of four local ferns (*Nephrodium filix mas* (L.) Rich., *Nephrodium spinulosum* (Miiller) Strempel subsp. *eu-spinulosum* (Ascherson) Hayek, *N. spinolosum* (Muller) Strempel subsp. *austriacum* (Jacq.) Woynar, and *Nephrodium cristatum* (L.) Mich.). The fat from the ferns consisted mainly of glycerides of linoleic (18:2) and oleic acids in a ratio of 1:4–5. He also found butyric, palmitic, and cerotic acids. The author concluded that the fern fat does not have any particular anthelmintic action, but its role seems to be like a solvent for active compounds [24].

The next period of fatty acids studies in ferns began with the development of gas-liquid chromatography (GLC) and its application for fatty acid analysis. The key research belongs to J.L. Gellerman and H. Schlenk from the University of Minnesota (USA) who first identified ARA and EPA in some plants including ferns, thus opposing the conventional view of that time on arachidonic acid as characteristic for animals only [25, 26]. The presence of ARA was first established by GLC retention times and then verified by chemical methods. Moreover, they were the first who reported the presence of polymethylene-interrupted LC-PUFAs (5,11,14–20:3 and 5,11,14,17–20:4) in the *Equisetum* species [26].

The 60–70s were a time of high interest to fern fatty acids. The ferns were investigated in different aspects: fatty acid composition of different fern species [27–29], their distribution in lipid classes [30, 31], fatty acid biosynthesis and metabolism [31, 32], fatty acids of fern spores [33].

Different ecological aspects of fern lipids including fatty acids were studied only this century mainly by Russian researchers. Effects of heavy metals, ecological, coenotic, and environmental conditions were considered in a number of studies [34–40].

The first report on the nutritional composition of an edible fern, *Matteucia struthiopteris* var. *pensylvanica* (Willd.) Morton or the ostrich fern from the North-Eastern United States and Eastern Canada is dated 1982 [41]. While the authors also gave information on its fatty acid composition, they did not indicate any presence of LC-PUFA in the plant material. Although it had been already demonstrated [26] the ostrich fern does contain ARA in significant quantities. The interest in ferns as sources of the valuable LC-PUFAs, and the ostrich fern, in particular, is still under way [42–44].

Another practical interest in fern fatty acids lies in the area of biodiesel production so popular in recent years. The aquatic fern *Azolla* was suggested as a producer of fatty acids for this purpose [45–47]. However, 30 years before, the fern constituents including fatty acids were analyzed with the aim of using *Azolla* biomass as an animal feedstuff [48]. Recently, we described an application of near-critical fluids for the extraction of fatty acids from wet fern fronds [49] which can be scaled up and used for the production of fern fatty acids both for food and biofuel technologies.

3 Diversity of Fern Fatty Acids

About 92 species of the class Polypodiopsida (or ferns [50]) have been analyzed for fatty acid composition to date. The species represent 20 families of four subclasses: Equisetidae (horsetails), Ophioglossidae, Marattiidae, and Polypodiidae (leptosporangiates). Green parts of sporophytes (aerial parts, fronds, leaves, pinna) are predominating material used for fatty acid analysis. Young fronds (fiddleheads or crosiers) mostly became objects of nutritional studies. A few papers dealt with whole sporophytic plants, while gametophytes have been analyzed rarely. Despite the fact that rhizomes happened to be the first fern organ where fatty acids were discovered, they, with few exceptions, have not been objects of systematic research. Finally, spores of several fern species have been analyzed for fatty acid composition. Available literature data on fern species and plant parts used for fatty acid analysis are shown in Table 1.

The found diversity of fatty acids in ferns looks very impressive: in total, up to 90 structures of fatty acids have been reported for ferns (Table 2). Many of them are minor or found in trace quantities, and some of them are reported just in one or two papers.

The data on the length of fern fatty acids available in the literature are ranged from C4-C6 [43, 54] to C28 [32, 39] and even longer: C30 fatty acid was found in some fern species [29]. As can be expected for plant fatty acids, the fern fatty acids are a mostly straight chain with an even number of carbons. The presence of a branched-chain fatty acid (branched-16:0) in ferns was reported only by one group of researchers [32] and it was not identified accurately.

Fatty acids with an odd number of carbons are usually minor, but quite common constituents of ferns. The reported length of the fatty acids is in the range of C11–C29 [29], among which C15 and C17 are the most widespread [53]. They are mostly saturated though monoenoic ones have been also reported [53, 60]. A strikingly high content (up to 10% of total fatty acids, here and further the reported values are rounded to whole numbers if not indicated otherwise) of the fatty acid 13:0 was reported for the whole plants of *Osmunda cinnamomea* [29]. Also, very high content of 15:1 (10–13%) was reported for the epiphytic filmy ferns *Hymenophyllum caudiculatum* and *H. plicatum* [54]. Nevertheless, the majority of papers on fern fatty acids deal with even C12–C22 fatty acids (Table 2).

The major saturated fatty acid in ferns is palmitic acid (16:0). When whole plants of some ferns were analyzed, its percentages exceeded 50% of total fatty acids [29, 48]. Surprisingly low content of palmitic acid (0.5–1.2% of total fatty acids) was found in the fronds of the filmy ferns, *H. caudiculatum,* and *H. plicatum* [54]. All other saturated fatty acids were minor constituents among which only 14:0, 18:0, 20:0, 22:0, and 24:0 reached a few percent [26–28, 31, 32, 37, 52, 53]. High contents of saturated and, particularly, very long-chain saturated fatty acids ($C \geq 22$) were found in the whole plants of ferns [29] and the leaves of epiphytic *Asplenium nidus* [39].

Monoenoic acids of ferns include mostly hexadecenoic and octadecenoic acids, particularly oleic acid (18:1n-9), which might reach a quite substantial level as found for whole plants of *Lygodium japonicum* (33%) [29], leaves of *Botrychium lunaria* (39%) [37], and fronds with sori of *Phymatosorus pustulatus* (60%) [53]. Isomers of hexadecenoic acids were found in significant quantities in the aquatic ferns *Salvinia*

Table 1 Classification of fern species used for fatty acid analysis

Species taxonomy[a]	Plant parts analyzed	Reference
Class Polypodiopsida		
Subclass Equisetidae		
Family Equisetaceae		
Equisetum arvense L.	Green parts, leaves, aerial parts	[26, 27, 38]
	Strobili[DA]	[51]
Equisetum hyemale L.	Green parts	[26]
	Strobili[DA]	[51]
Equisetum fluviatile L.	Leaves	[27]
	Strobili[DA]	[51]
Equisetum litorale Kühlew. ex Rupr. (accepted name *Equisetum* × *litorale* Kühlew. ex Rupr.)	Strobili[DA]	[51]
Equisetum palustre L.	Strobili[DA]	[51]
Equisetum pratense Ehrh.	Strobili[DA]	[51]
Equisetum ramosissimum Desf.	Strobili[DA]	[51]
Equisetum scirpoides Michx.	Aerial parts	[38]
Equisetum sylvaticum L.	Strobili[DA]	[51]
Equisetum telmateia Ehrh.	Strobili[DA]	[51]
Equisetum variegatum Schleich. ex F. Weber & D. Mohr	Aerial parts	[38]
	Strobili[DA]	[51]
Subclass Ophioglossidae		
Family Psilotaceae		
Psilotum nudum (L.) P. Beauv.	Aerial parts, rhizomes	[52]
Psilotum triquetrum Sw. (synonym of *P. nudum*)	Leaves	[27]
Family Ophioglossaceae		
Botrychium lunaria (L.) Sw.	Leaves	[37]
Subclass Marattiidae		
Family Marattiaceae		
Ptisana salicina (Sm.) Murdock	Fronds with sori	[53]
Subclass Polypodiidae		
Family Osmundaceae		
Osmunda regalis L.	Leaves	[27]
	Fiddleheads	[43]
	Spores	[33]
O. regalis var. *spectabilis* (Wield.)A.Gray	Whole plants	[29]
Osmunda claytoniana L.	Green parts, fronds	[26, 53]
Osmunda cinnamomea L. (synonym of *Osmundastrum cinnamomeum* (L.) C.Presl)	Whole plants	[29]
Osmundastrum asiaticum Tagawa	Fiddleheads	[44]
Family Hymenophyllaceae		
Hymenophyllum caudiculatum Mart.	Fronds	[54]
Hymenophyllum plicatum Kaulf.	Fronds	[54]

(continued)

Table 1 (continued)

Species taxonomy[a]	Plant parts analyzed	Reference
Family Lygodiaceae		
Lygodium japonicum (Thunb.) Sw.	Whole plants	[29]
	Spores	[33]
Lygodium scandens (L.) Sw.	Spores	[33]
Lygodium volubile Sw.	Spores	[33]
Family Anemiaceae		
Anemia phyllitidis (L.) Sw.	Spores	[33]
	Germinating spores	[55, 56]
Family Salviniaceae		
Azolla caroliniana Willd.	Whole plants	[48]
	Whole plants[HA]	[57]
Azolla filiculoides Lam.	Whole plants	[45, 46, 58]
	Leaves, roots[HA]	[59]
	Whole plants, leaves, roots, microsporocarps	[47]
	Whole plants[HA]	[57]
Azolla rubra R. Br. (synonym of *A. filiculoides* var. *rubra* (R. Br.) Strasb.)	Whole plants[HA]	[57]
Azolla mexicana C. Presl	Whole plants[HA]	[57]
Azolla microphylla Kaulf.	Whole plants[HA]	[57]
Azolla nilotica Mett. (Status: Ambiguous)	Whole plants[HA]	[57]
Azolla pinnata R. Br.	Whole plants	[47]
	Whole plants[HA]	[57]
Salvinia cucullata Roxb. (Status: Ambiguous)	Whole plants[HA]	[57]
Salvinia natans (L.) All.	Leaves, sori	[60]
Salvinia molesta D.S. Mitch. (synonym of *Salvinia adnata* Desv.)	Whole plants[HA]	[57]
Salvinia minima Baker	Whole plants[HA]	[57]
Salvinia oblongifolia Martius	Leaves, roots[HA]	[57]
Family Cyatheaceae		
Cyathea dealbata Sw.	Fronds with sori	[49, 53]
	Wax of fronds	[61]
Family Pteridaceae		
Ceratopteris thalictroides (L.) Brongn.	Spores	[33]
Adiantum capillus-veneris L.	Gametophytes, parts of fronds	[62]
Adiantum pedatum L.	Green parts	[26]
	Whole plants	[29]
	Fronds	[53]
Pityrogramma argentea (Willd.) Domin	Spores	[33]
Pteris longifolia L.	Spores	[33]

(continued)

Table 1 (continued)

Species taxonomy[a]	Plant parts analyzed	Reference
Family Dennstaedtiaceae		
Pteridium aquilinum (L.) Kuhn	Pinnae	[28]
	Young fronds	[43, 44, 63]
Pteridium esculentum (G. Forst.) Cockayne	Fronds	[53]
Family Cystopteridaceae		
*Cystopteris dickieana*R. Sim (synonym of *Cystopteris fragilis* (L.) Bernh.)	Leaves	[37]
Cystopteris fragilis (L.) Bernh.	Fronds	[53]
Gymnocarpium dryopteris (L.) Newman	Leaves	[37, 64]
Family Aspleniaceae		
Asplenium nidus L.	Leaves	[39]
Asplenium oblongifolium Colenso (Status: Ambiguous)	Fronds	[53]
Asplenium scolopendrium L.	Leaves	[40]
	Fiddleheads	[43]
Phyllitis scolopendrium (L.) Newman (synonym of *Asplenium scolopendrium* L.)	Leaves	[27]
Asplenium trichomanes L.	Young leaves	[39]
Scolopendrium vulgare Sm. (Status: Ambiguous)	Fronds	[32]
	Sporiferous fronds	[30]
Family Woodsiaceae		
Woodsia glabella R. Br. ex Richardson	Leaves	[37, 64]
Family Onocleaceae		
Matteuccia struthiopteris (L.) Tod.	Green parts, fronds	[26, 36, 53]
	Fronds, buds, rhizomes, roots	[65]
	Fiddleheads	[36, 42–44]
	Gametophytes, young fronds	[66]
Onoclea sensibilis L.	Green parts, fronds	[26, 53]
	Fiddleheads	[43]
Family Athyriaceae		
Athyrium distentifolium Tausch ex Opiz (synonym of *Athyrium alpestre* (Hoppe) Clairv.)	Leaves	[37]
	Fiddleheads	[43]
Athyrium filix-femina (L.) Roth	Fronds	[32, 53]
	Fiddleheads	[43]
	Spores	[33]
Athyrium sinense Rupr.	Fronds	[53]
Athyrium yokoscense (Franch. & Sav.) Christ	Fronds	[53]
Athyrium spinulosum (Maxim.) Milde	Fronds	[53]

(continued)

Table 1 (continued)

Species taxonomy[a]	Plant parts analyzed	Reference
Cornopteris crenulatoserrulata (Makino) Nakai	Fronds	[53]
Deparia pycnosora (Christ) M. Kato	Fronds	[53]
Family Thelypteridaceae		
Lastrea limbosperma (All.) Ching (synonym of *Oreopteris limbosperma* Holub)	Fiddleheads	[43]
Dryopteris oreopteris (Ehrh.) Maxon (synonym of *Oreopteris limbosperma* Holub)	Spores	[33]
Parathelypteris noveboracensis (L.) Ching	Fronds	[53]
Phegopteris connectilis (Michx.) Watt	Fronds	[53]
	Fiddleheads	[43]
Thelypteris palustris (A. Gray) Schott (synonym of *Thelypteris confluens* (Thunb.) C.V. Morton)	Fiddleheads	[43]
Family Dryopteridaceae		
Dryopteris aemula Kuntze (Status: Ambiguous)	Fiddleheads	[43]
Dryopteris affinis Fraser-Jenk. (Status: Ambiguous)	Fiddleheads	[43]
Dryopteris austriaca (Jacq.) Woyn. ex Schinz & Thell.	Pinnae	[28]
Dryopteris borreri V.I. Krecz. (synonym of *Dryopteris pseudomas* (Woll.) Holub bis & Pouzar)	Fiddleheads	[43]
Dryopteris cambrensis Beitel & W.R. Buck (Status: Ambiguous)	Fiddleheads	[43]
Dryopteris carthusiana (Vill.) H.P. Fuchs	Fiddleheads	[43]
Dryopteris spinulosa (O.F. Müll.) Watt (synonym of *Dryopteris carthusiana* (Vill.) H.P. Fuchs)	Spores	[33]
Dryopteris caucasica (A. Braun) Fraser-Jenk. & M.F.V. Corley	Fiddleheads	[43]
Dryopteris crassirhizoma Nakai	Fronds	[53]
Dryopteris dilatata (Hoffm.) A. Gray	Fiddleheads	[43]
Dryopteris expansa (C. Presl) Fraser-Jenk. & Jermy	Leaves, fronds	[37, 53]
	Fiddleheads	[43]
Dryopteris filix-mas (L.) Schott	Fronds, pinnae	[27, 28, 32]
	Fiddleheads	[43]
	Spores	[33]
Dryopteris goeringiana (Kunze) Koidz.	Fronds	[53]
Dryopteris hirtipes (Blume) Kuntze	Spores	[33]
Dryopteris oreades Fomin (Status: Ambiguous)	Fiddleheads	[43]
Dryopteris remota (A. Braun) Hayek	Fiddleheads	[43]
Polystichum aculeatum (L.) Roth ex Mert.	Fiddleheads	[43]
Polystichum lonchitis (L.) Roth	Fiddleheads	[43]
	Spores	[33]
Polystichum tripteron (Kunze) C. Presl	Fronds	[53]
Family Polypodiaceae		
Platycerium bifurcatum (Cav.) C. Chr.	Generative leaves without sori	[39]

(continued)

Table 1 (continued)

Species taxonomy[a]	Plant parts analyzed	Reference
Pleopeltis polypodioides (L.) E.G. Andrews & Windham	Fronds	[67, 68]
Pyrrosia eleagnifolia (Bory) Hovenkamp	Fronds	[53]
Polypodium meyenianum (Schott) Hook. (synonym of *Aglaomorpha meyeniana* Schott)	Spores	[33]
Polypodium crassifolium L. (synonym of *Niphidium crassifolium* (L.) Lellinger)	Spores	[33]
Polypodium vulgare L.	Fronds, pinnae	[28, 31]
	Fiddleheads	[43]
Polypodium polypodioides (L.) Watt	Whole plants	[29]
Phymatosorus pustulatus (G. Forst.) Large, Braggins & P.S. Green	Fronds	[53]

[a]Family name and their order follow [50]. Species names and their current status are according to [69]
[DA]Dicarboxylic acids only were analyzed
[HA]Hydroxy fatty acids and structurally related hydroxyl compounds were analyzed

natans (7% 16:1n-9 and 4% 16:1n-7) [60] and *Azolla caroliniana* (2% 16:1n-7) [48], and also rhizomes of *Psilotum nudum* (about 5% 16:1n-5) [52]. A monoenoic acid characteristic for green tissue of different plants, 3*t*-16:1, made up consistent amounts in ferns (0.2–2.7%) [27, 28, 31, 32, 53, 60], and its lower homolog, 3*t*-14:1, was also reported [32]. The *trans*-isomer of oleic acid (*t*-18:1n-9) was reported for some pteridophytes from the Polar Urals, and it constituted a significant portion of total fatty acids, up to 6% [37]. Other monoenoic acids were found in relatively minor quantities [28, 32, 53, 60]. Monoenoic fatty acids with C > 20 (as well as other unsaturated ones) were also reported for ferns in some studies [32, 40, 43, 54, 60].

Chain length of fern polyunsaturated fatty acids (PUFAs) is limited to C16-C20 with a few exceptions: 14:3 in the fronds of *Polypodium vulgare* [31], 22:3 in the fronds of *Onoclea sensibilis* [26], 22:2n-6 as a very minor constituent of young fronds of many fern species [43] and *Salvinia natans* [60]. Also, the aquatic fern contained 14:2n-5, 22:3n-6, 22:3n-3, 22:4n-3, 22:5n-3, 22:6n-3, 24:3n-3, and 26:3n-3 [60]. The finding of docosahexaenoic acid (DHA, 22:6n-3) in ferns is of particular interest: besides *S. natans*, it was found in the filmy ferns (up to 7% of total fatty acids in *H. plicatum*) [54] and all the tested European ferns in small quantities [43]. DHA was also reported for *Azolla filiculoides* without details of its content [45]. However, if the possible metabolic precursor of DHA, docosapentaenoic acid (22:5n-3), has been reported for *S. natans* along with DHA [60], it was not reported for other ferns with discovered DHA.

PUFAs of ferns mostly belong to the omega-6 and omega-3 families and the major of them are hexadecatrienoic (16:3n-3), linoleic (18:2n-6), α-linolenic (18:3n-3), and arachidonic (20:4n-6) acids. The presence of 16:3n-3 in ferns confirmed their belonging to "16:3 plants" which are characterized by the "prokaryotic"

Table 2 List of fatty acids reported for ferns (class Polypodiopsida)[a]

Shorthand designation	Trivial name	Reference	Shorthand designation	Trivial name	Reference
1. Saturated			**2. Monoenoic (*cont.*)**		
4:0	Butyric acid	[54]	21:1n-12		[60]
6:0		[43, 54]	22:1n-15		[60]
8:0		[43, 54]	22:1n-13		[60]
10:0		[27, 29, 40, 43, 48]	22:1n-11		[60]
11:0		[27, 29, 54]	22:1n-9	Erucic acid	[40, 43, 54]
12:0	Lauric acid	[27–29, 32, 38–40, 43, 48, 52, 53, 60]	23:1[b]		[60]
13:0		[29, 39, 40, 60]	24:1[b]		[32]
14:0	Myristic acid	[26–29, 32, 38–40, 43, 48, 52–54, 60, 65]	24:1n-9	Nervonic acid	[43, 54]
15:0		[27–29, 32, 38–40, 43, 52, 53, 58, 60, 64]	24:1n-13		[60]
16:0	Palmitic acid	all references in this table	25:1n-14		[60]
branched-16:0		[32]	26:1n-15		[60]
17:0	Margaric acid	[27–29, 38, 43, 48, 52, 53, 60]	27:1[b]		[60]
18:0	Stearic acid	all references in this table	**3. Dienoic**		
19:0		[29, 48, 39, 40, 60]	14:2n-5		[60]
20:0	Arachidic acid	[26–29, 32, 37–40, 43, 48, 52, 53, 58, 60, 64]	16:2[b]		[26, 27, 32, 48]
21:0		[32, 40, 54, 60]	16:2n-4	Palmitolinoleic acid	[48, 60]
22:0	Behenic acid	[28, 29, 31, 32, 37–40, 43, 53, 60, 64]	16:2n-6		[28, 38–40, 53]

23:0		[29, 32, 39, 40, 43, 60, 64]	18:2[b]		[26, 27, 29, 31, 32, 58]
24:0	Lignoceric acid	[29, 31, 37, 39, 40, 43, 48, 53, 54, 58, 60, 64]	18:2n-5		[39, 40]
25:0		[29, 32, 39, 40]	18:2n-6	Linoleic acid	[28, 37–40, 43, 48, 52–54, 60, 64, 65]
26:0	Cerotic acid	[29, 32, 39, 40, 60]	5,9–18:2	Taxoleic acid	[38]
27:0		[29]	20:2[b]		[26, 32, 64]
28:0		[29, 32, 39, 40]	20:2n-6		[28, 39, 40, 43, 52, 53, 60]
29:0		[29]	22:2n-6		[43, 60]
30:0		[29]	**4. Trienoic**		
2. Monoenoic			14:3[b]		[31]
12:1[b]		[27, 60]	16:3[b]		[26, 27, 31, 32, 48, 58]
13:1[b]		[60]	16:3n-6		[60]
14:1[b]		[28]	16:3n-3		[28, 38, 40, 53, 60]
14:1n-5	Myristoleic acid	[43, 52, 60]	18:3[b]		[27, 29, 58]
3t-14:1		[32]	18:3n-6	γ-Linolenic acid, GLA	[28, 32, 37, 40, 43, 48, 52–54, 60, 64]
15:1[b]		[27, 54]	18:3n-3	Linolenic acid, α-linolenic acid	[26, 28, 32, 37–40, 43, 48, 52, 53, 60, 64, 65]
15:1n-8		[60]	20:3[b]		[27, 31, 48, 52, 58]
15:1n–6		[60]	20:3n-9		[60]
16:1[b]		[26–28, 32, 37, 58, 64, 65]	20:3n-6	Dihomo-γ-linolenic acid	[26, 28, 32, 39, 40, 43, 53, 54, 60, 64]

(continued)

Table 2 (continued)

Shorthand designation	Trivial name	Reference	Shorthand designation	Trivial name	Reference
16:1n-12		[40]	20:3n-3		[38, 40, 43, 52–54, 60]
16:1n-9		[38–40, 52, 53, 60]	5,11,14–20:3	Sciadonic acid	[26, 44, 53]
16:1n-7	Palmitoleic acid	[31, 38–40, 43, 48, 52, 53, 60]	$22:3^{b}$		[26]
16:1n-5		[52, 53]	22:3n-6		[60]
16:1n-2		[40]	22:3n-3		[60]
3*t*-:16:1		[27, 28, 31, 32, 53, 60]	24:3n-3		[60]
$17:1^{b}$		[32, 53, 54]	26:3n–3		[60]
17:1n-10		[60]	**5. Tetraenoic**		
17:1n-7		[39, 43]	16:4n-3		[60]
17:1n-8		[52, 60]	$18:4^{b}$		[26, 32]
$18:1^{b}$		[26, 27, 29, 31, 32, 58, 65]	18:4n-3	Stearidonic acid	[28, 53, 60]
18:1n-11		[38, 39, 60]	$20:4^{b}$		[27, 31, 32, 37, 58, 64]
18:1n-9	Oleic acid	[28, 37–40, 43, 48, 52–54, 60, 64]	20:4n-6	Arachidonic acid	[26, 28, 32, 39, 40, 43, 48, 52, 53, 60, 65]
18:1n-7	*Cis*-vaccenic acid	[37, 39, 40, 52, 53, 60]	20:4n-3		[28, 40, 53, 60]

18:1n-5		[52, 53]	5,11,14,17–20:4	Juniperonic acid	[26, 38, 44, 53]
t-18:1n-9	Elaidic acid	[37, 64]	22:4n-3		[60]
19:1[b]		[53]	**6. Pentaenoic**		
19:1n-10		[60]	20:5[b]		[27, 31, 32, 58, 64]
20:1[b]		[27, 32]	20:5n-3	Eicosapentaenoic acid, timnodonic acid, EPA	[26, 28, 40, 43, 52, 53, 60, 65]
20:1n-13		[60]	22:5n-3		[60]
20:1n-11		[38, 60]	**7. Hexaenoic**		
20:1n-9	Gondoic acid	[39, 40, 43, 52, 53, 60]	22:6n-3	Docosahexaenoic acid, cervonic acid, DHA	[43, 45, 54, 60]
20:1n-7		[52]			

[a]Monocarboxylic acid are only shown. Hydroxy and dicarboxylic fatty acids are not shown
[b]Positions of double bonds were not specified

pathway of galactolipid biosynthesis [70]. However, neither 16:3 and 18:3n-3 nor 20:4n-6 were reported for the fronds of the filmy ferns *H. caudiculatum* and *H. plicatum* [54]. Also, surprisingly low levels of 18:3 (0–6% of total fatty acids) without any 16:3 and 20:4 were reported for five common terrestrial species of ferns in the paper [29]. This may reflect the fact that whole plants were taken for extraction [29], thus other fatty acids dominated in the samples.

Except for the abovementioned DHA, EPA (20:5n-3) is the most unsaturated fatty acid found in significant amounts in the majority of tested fern species (up to 5.5% of total fatty acids) [26, 27, 49, 53, 60, 65]. The potential precursors of ARA and EPA are common constituents of fern fronds: γ-linolenic (18:3n-6), dihomo-γ-linolenic (20:3n-6), 18:4n-3, 20:3n-3, and 20:4n-3 [28, 32, 40, 43, 44, 53, 60, 66]. Although these acids are usually minor (less than 1% of total fatty acids), there are a few remarkable exceptions related to the fatty acids: *S. natans* was found to contain significant percentages of 18:3n-6, 18:4n-3, and 20:4n-3 (in the range of 1–4% of total fatty acids) [60]; extremely high contents of 18:3n-6 (up to 28%) and 20:3n-3 (up to 35%) were reported for the filmy ferns with also relatively high levels of 20:3n-6 (3–5%) [54].

PUFAs of other fatty acid families include omega-4 (16:2n-4 [48, 60]), omega-5 (14:2n-5 [60] and 18:2n-5 [39, 40]), and omega-9 (20:3n-9 [60]). The unusual PUFAs were always minor (<1%) and were reported for the aquatic ferns *S. natans* (14:2n-5, 16:2n-4, 20:3n-9) [60] and *A. caroliniana* (16:2n-4) [48], and two *Asplenium* species (18:2n-5) [39, 40].

Among PUFAs found in ferns, there are two fatty acids that form a specific group: sciadonic (5,11,14–20:3) and juniperonic (5,11,14,17–20:4) acids. Strictly speaking, these fatty acids belong to the omega-6 and omega-3 families, respectively. However, they differ from other PUFAs in the reciprocal locations of double bonds. The majority of natural PUFAs have methylen-interrupted double bonds i.e., double bonds are separated by a methylene group(–CH=CH–**CH2**–CH=CH–). Sciadonic and juniperonic acids have two double bonds (at the 5- and 11-carbon atoms) separated by a butylene spacer(–CH=CH–**CH2–CH2–CH2–CH2**–CH=CH–). Such fatty acids are called Δ5-unsaturated polymethylene-interrupted fatty acids (Δ5-UPIFA) [10]. They were first found in some conifers and horsetails but not in ferns [26]. Recently, we found both fatty acids as minor constituents in a number of fern species as well [53]. Since the content of Δ5-UPIFA in fern fronds is low (0.1–1.9% for sciadonic acid, and 0.0–0.4% for juniperonic acid), it is not surprising they were overlooked in other studies. In horsetails (*Equisetum*), the content of the fatty acids was found to be higher: 0.6–3.6% sciadonic acid and 1.6–2.5% juniperonic acid [26, 38].

Fatty acids modified with additional functional groups are mostly uninvestigated in ferns. However, hydroxy fatty acids were analyzed in the aquatic ferns: six species of the genus *Azolla* and four *Salvinia* species (Table 1). The two genera were found to differ in some hydroxy compounds from each other and both were different from the closely related ferns from the family Marsileaceae [57]. The hydroxy lipids were suggested as biomarkers of the species in sedimentary records and the past occurrence of the plants [57, 59]. According to the definition given in the introduction to this paper, fatty acids are monocarboxylic acids. However, dicarboxylic acids with 10–35 carbon atoms have been found in spores of different horsetail species [51].

4 Distribution of Fatty Acids in Ontogenetic Stages and Different Organs

Alike other land plants, the fern life cycle is an alternation of two generations (or phases): the sporophyte phase produces spores that germinate to give haploid gametophytes and begin the gametophyte phase, which in turn aimed at gamete production with the subsequent initiation of new sporophytes via sexual reproduction. In ferns, gametophytes and sporophytes are independent multicellular organisms with the sporophytes being the dominant (larger, longer-lived) phase [71]. Fern gametophytes are much less differentiated and simple in morphology as compared to sporophytes. On the contrary, sporophytic plants are differentiated into true vegetative organs: roots, stem (rhizome), and leaves (megaphylls or generally known as fronds). The fern frond is often complex in morphology and divided into a number of parts (stipe, lamina, rachis, pinna) and may bear spore-producing organs: capsule-like sporangia which are usually aggregated into sori [72].

Fern spores contain high levels of lipids (24–79% of spore fresh weight) indicating their reserve role in the spores [31, 33, 55, 62]. However, some fern species (*Osmunda regalis*, *Ceratopteris thalictroides*) contain low levels of lipids in the spores (14 and 4%, respectively) [33]. Analysis of fatty acid composition in the spores of 17 fern species [31, 33] revealed three major fatty acids: 16:0 (mean = 15%, range = 5–27% of total fatty acids), 18:2 (mean = 29%, range = 5–46%), and 18:1 (mean = 52%, range = 19–75%). Thus, unsaturated fatty acids are predominant in fern spores resembling fatty acids of many seed oils [73]. Other fatty acids (14:0, 16:1, 18:0, 18:3, 20:0, 20:3, and 20:4) were also detected in minor quantities [31, 33].

In contrast, sorus lipids of *Salvinia natans* contained more saturated (53%) than monoenic (37%) acids [60]. As expected, the sori were depleted in PUFAs (4%), thus many of those found in leaves (including all $C \geq 20$ PUFAs) were absent in the sori.

Spore germination and early stages of gametophyte growth were accompanied by a drop in total lipids; however, the fatty acid composition was not much different from dormant spores [31, 55]. With gametophyte growth, the percentage of 18:3 increased and those of 18:1 and 18:2 decreased [55]. The fatty acid composition of mature gametophytes of *Matteuccia struthiopteris* was similar to young growing fronds of the fern sporophytes [66, 44]: the major fatty acids were 16:0 (26% of total fatty acids in the gametophytes vs. 22–27% in the young fronds), 18:3n-3 (24% vs. 14–27%), 18:2n-6 (19% vs. 15–21%), 18:1n-9 (8% vs. 6–10%), and 20:4n-6 (10% vs. 5–13%). In contrast to the sporophyte fronds, the gametophytes contained less 20:5n-3 (0.6% vs. 2.3–3.3%) [66].

Most papers on fern fatty acids have been dealing with the green tissues of sporophyte fronds (Table 1). Figure 2 summarizes the literature data on the content of fatty acids in fronds of fern species from the subclass Polypodiidae (with

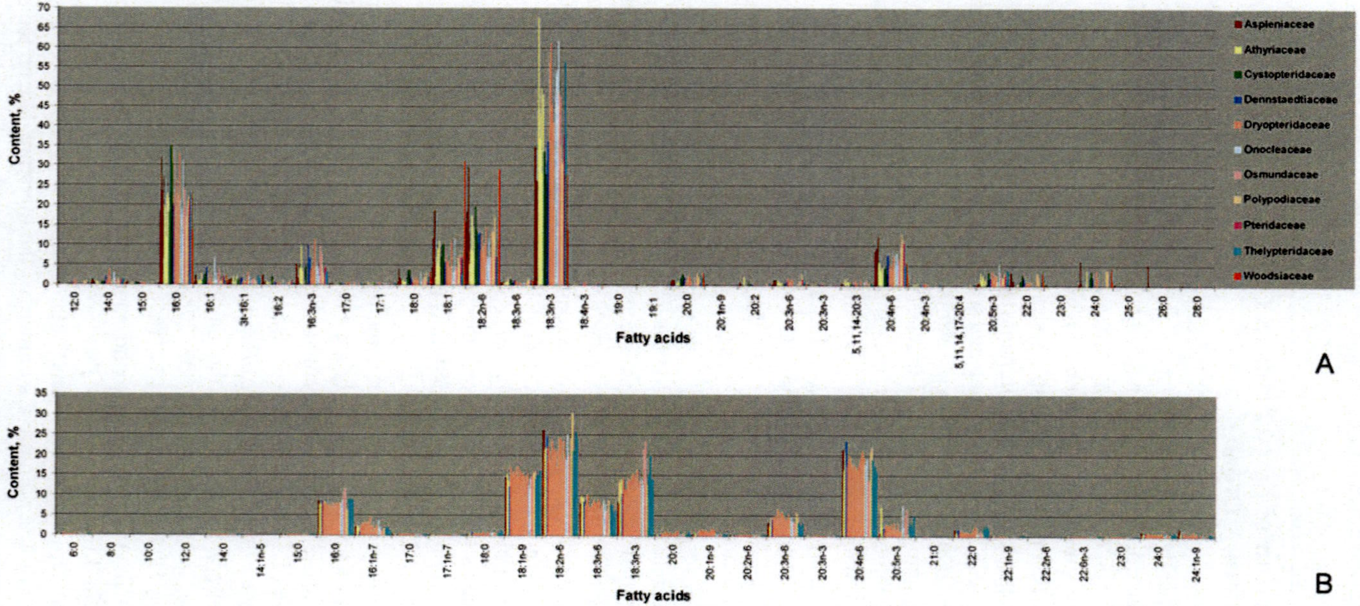

Fig. 2 Relative abundance (%) of individual fatty acids in the mature (**a**) and young (**b**) fronds of the ferns belonging to the subclass Polypodiidae. The charts are built on the data reported in [26–28, 31, 32, 37, 39, 40, 53, 65] for the mature fronds (**a**) and calculated from the values given in [43] for the young fronds (**b**). The percentages of the isomers of 16:1 (16:1n-9, 16:1n-7, 16:1n-5) and 18:1 (18:1n-11, 18:1n-9, 18:1n-7, 18:1n-5) are combined

exceptions of *Salvinia natans* [60], *Hymenophyllum caudiculatum,* and *H. plicatum* [54], which are discussed in Sect. 6.1), presuming the used fronds were mature, containing no sori/spores, or at least it is not indicated in a paper. It can be seen that linolenic acid (18:3n-3) is the predominant one (mean/median = 38/37%, range = 9–68% of total fatty acids) followed by palmitic (16:0) (20/20, 6–35) and linoleic (18:2n-6) (11.5/11%, 5–30%) acids. ARA (20:4n-6) and oleic acid (18:1n-9) have close percentages: (6/5%, 2–13% and 6/5%, 0.8–31%, respectively). Being the "16:3 plants" [70], ferns contain significant portions of 16:3n-3 in the leaf lipids: 4.5/4%, 0.6–12%. Alike all photosynthetic plants, fern fatty acids include 3*t*-16:1:1.2/1.2%, 0.3–2.7%. EPA (20:5n-3) is a minor, but a prominent constituent of fern fronds: 2.0/2.0%, 0.3–5.5% of total fatty acids. A few percent (1.1–2.8% on average) are made up of saturated fatty acids: 18:0, 20:0, 22:0, 24:0, 26:0. All other fatty acids, except some unusual fatty acids (14:3, branched-16:0, 22:3) which were reported only for 1–3 species, are on average below 1% (Fig. 2a).

Young growing fronds are different from mature ones in fatty acid percentage. Curled fronds, also known as fiddleheads or crosiers, usually contain more linoleic (18:2n-6) than α-linolenic (18:3n-3) acid [36, 42–44, 63]. Other fatty acids associated with chloroplasts (16:3n-3; 3*t*-16:1) are also low in content [44]. On the contrary, ARA is high reaching up to 18% of total fatty acids and even higher in some samples [42, 43]. At this stage of frond development, their precursors of ARA, 18:3n-6 and 20:3n-6, often exceed 1% of total fatty acids [42, 44, 63]; their very high percentages (respectively, 8 and 4% on average) were reported for 28 European fern species [43]. EPA is within the range found for the mature fronds (Fig. 2b).

Data on fatty acids in organs of fern sporophytes other than fronds are scarce. In the early studies it was found that rhizomes of three fern species (*Nephrodium filix-mas*, *N. spinulosum*, and *N. cristatum*) contain mainly 18:2 and 18:1, and also 16:0 and 26:0 [24]. Rhizomes of *Psilotum nudum* contained less 16:0 (18% vs. 22–30%), 18:1n-9 (5% vs. 12%), 18:3n-3 (10% vs. 15–22%) and some minor fatty acids, but more 18:2n-6 (35% vs. 19–22%) and much more 16:1n-5 (5% vs. 0.1%) and 18:1n-7 (4% vs. 0.6–0.9%) as compared to the aerial parts of the plant [52]. Moreover, ARA was only found in minor quantities in rhizomes and not in the aerial parts [52]. The most comprehensive comparative study of the fatty acid composition of different organs (fronds, buds, rhizomes, roots) was performed on *Matteuccia struthiopteris* [65]. As expected, the omega-3 fatty acids were more abundant in the fronds than in other parts, whereas the reverse was found for the omega-6 fatty acids. The predominant fatty acids were 16:0 (31% of total fatty acids in the fronds) or 18:2n-6 (34–40% in all other organs). Buds and rhizomes were close in fatty acid percentages with a higher content of 18:3n-3 and less content of 18:2n-6 in the buds as compared to the rhizomes. Roots were different from rhizomes in higher percentages of 16:0 and 16:1 and lower ones of 18:2n-6 and 18:1. ARA was present in all organs with the lowest percentage in the fronds (6%) as compared to other organs (about 9%). EPA was found in the fronds, buds, and rhizomes but not in the roots. There was a surprisingly low content of 18:3n-3 in the fronds which was less (25%) when compared with the results reported by the authors in their previous research (54% [26]) and our study (62% [53]) for this fern species.

However, it was comparable with the plants grown under laboratory conditions (16–23%) in the study [36]. An explanation may be that, according to the authors' description [65], the fern was collected from a very shady area, and the lower portion of the fronds was brownish. A contribution of different parts of fronds (lamina and stalks) into a sample used for analysis should be also taken into account. Fatty acid ratios can also vary in different parts of fern fronds as was demonstrated for upper, middle, and lower segments of fronds of *M. struthiopteris* [36].

In the aquatic fern *A. filiculoides*, roots were higher in the percentages of saturated 16:0 (up to 51% of total fatty acids), 18:0 (up to 20%), and lower in 24:0 and the unsaturated 18:1, 18:3, 20:4 as compared with leaves. Male microsporocarps were enriched in 18:2 and 18:1 and contained less 16:0 and 18:3 [47].

5 Distribution of Fatty Acids in Lipid Classes

In fern spores, fatty acids are mostly components of triacylglycerols while free fatty acids can also accumulate [33].

The distribution of fatty acids among lipid classes in fern fronds is similar to angiosperm plants in common fatty acids. The major polar lipids of plant green tissue, monogalactosyldiacylglycerol (MGDG) and digalactosyldiacylglycerol (DGDG), are enriched in 18:3n-3 [28, 31, 62, 74]. However, MGDG and DGDG of ferns were found to have a somewhat lower content of this fatty acid (49–78% of total fatty acids) as compared to angiosperm plants (62–84% [74] and up to 90–97% [73, 75]). There is a prominent difference between fern MGDG and DGDG: high levels of 16:3 (20–30% and up to 47%) in MGDG as compared to DGDG (0.5–5%) [28, 62, 74] which is similar to "16:3 plants" of angiosperms [73, 74] and mosses [75]. On the other hand, MGDG from frond tissue of *P. vulgare* contained only 6% 16:3, which was anyway higher than the content of the fatty acid in DGDG (3%) [31]. It was found that the content of 16:3 in MGDG increased upon expansion of fronds of *A. capillus-veneris* reaching the maximum level in the mature pinnae and correlated with chloroplast development [62]. The presence of saturated acids in these glycolipids is rather controversial. While MGDG of all tested species was depleted in 16:0 (2–5%), DGDG contained a high level of 16:0 in the fern *Adiantum capillis-veneris* (33%) [62], but, in other species, the percentage of this fatty acid in DGDG was below its average content (Fig. 2a) in fern green tissues: 16% for *Dryopteris filix-mas* [28], 11% for *Polypodium vulgare* [31]. For DGDG of *Pteridium aquilinum*, both high (33% [74]) and low (5% [28]) percentages have been reported. The C20 PUFAs (20:3n-6, 20:4n-6, 20:4n-3, 20:5n-3), which are absent in angiosperm MGDG and DGDG, were usually minor components in the fern glycolipids reaching maximum 4–8% in MGDG (20:4n-6, 20:5n-3) and 5% in DGDG (20:4n-6) of some fern species [28, 31, 62]. In the moss *Hypnum cupressiforme*, MGDG and DGDG were higher in both 20:4 and 20:5 [75].

Predominant fatty acids of another glycolipid, sulfoquinovosyldiacylglycerol (SQDG), are 16:0 and 18:1n-9, while 18:3n-3 is in the third place as found for three species: *P. aquilinum* [76], *P. vulgare* [31], and *A. capillis-veneris* [62]. In the angiosperm plants and the moss, the predominant were 16:0 and 18:3 [73, 75, 76].

Fatty acid distribution in phospholipids was varying among phospholipid classes in two tested species: *A. capillus-veneris* [62] and *P. vulgare* [31]. For phosphatidylcholine (PC), 16:0, 20:4, 18:2, and 18:3n-3 were major fatty acids. Phosphatidylglycerol (PG) was the only phospholipid containing 3*t*-16:1 being one of its major fatty acids which included also 18:1, 16:0, and 18:3n-3. Phosphatidylinositol (PI) had the identical order of the major fatty acid abundance in the two tested fern species: 16:0 > 20:4n-6 > 18:3n-3 > 18:2n-6. Fatty acid composition of phosphatidylethanolamine (PE) and phosphatidylserine (PS) were investigated only for *A. capillus-veneris*. Both phospholipids were particularly enriched in ARA (37–40%) making them the most "arachidonic" classes among lipids. Their other major fatty acids were 16:0 (PE and PS), 22:0, and 20:5n-3 (PS).

The fern betaine lipid, diacylglyceryltrimethylhomoserine (DGTS), showed similarity with PC in fatty acid content with the major fatty acids being 16:0, 18:2n-6, and 18:3n-3, but was different from PC in the higher percentage of 16:0 (48% vs. 38% in PC) and the lower one of 20:4n-6 (5% vs. 22%) [62].

Triacylglycerol (TG), while a minor component of fern pinna, was enriched in the PUFAs 20:4n-6 (17%) and 18:3n-3 (15%) with the predominant 16:0 (20%) [62]. It is different from TG of fern spores where 18:3n-3 usually is a very minor component and 20:4n-6 absents [33].

The ratio of fatty acids in different lipid classes depends on the stage of fern development. In 10-day-old gametophytes of *A. capillus-veneris*, the percentages of fatty acids were similar to sporophytes in some lipid classes (DGDG, SQDG, to some extent DGTS) but different in others (TG, MGDG, PG, PE, PS, PI, PC) [62]. However, the distribution of the major fatty acids in lipid fractionations was similar for sporophyte young fronds and mature gametophytes of *Matteuccia struthiopteris* [66]. In both tissues, the fractions enriched in phospholipids contained the major portions of 16:0, 18:2n-6, ARA, EPA, *t*16:1n13, and also the potential metabolic precursors of ARA (18:3n-6, 20:3n-6) and EPA (20:4n-3). The fractions enriched in glycolipids were high in 18:3n-3, 18:1n-9, 16:3n-3, and also included significant portions of 16:0, 18:2n-6, EPA, and ARA. Nonpolar lipids in gametophytes and sporophytes contained all the major fatty acids in very close percentages [66].

The gametophytes in the early stages of development (6–10 days) of *A. capillus-veneris* were not much different in fatty acid ratios in four lipid classes (MGDG, PG, DGTS, PC) [62]. In sporophyte fronds of the fern, the fatty acid ratios in the lipid classes depended on a stage of development and a part of the fronds, which were partly associated with chloroplast development: a low level of 16:3n-3 in MGDG and absence of 3*t*-16:1 in PG were observed for the young leaves and mature frond petioles. Linoleic acid was often higher in the lipids (except PG) of young leaves as compared to the mature and old pinna [62].

6 Factors Affecting Fatty Acid Content in Fern Fronds

6.1 Fern Taxonomy

In our previous study [53], analysis of fatty acid content in fronds of 23 fern species representing 12 families revealed no strict connection between fatty acid percentages and fern taxonomy. Species belonging to one family often formed clusters with species from different families but not the same one. The species from the subclass Marattiidae, *Ptisana salicina*, was within a cluster with the species from different families (Polypodiaceae, Aspleniaceae, Dennstaedtiaceae, Cystopteridaceae) belonging to the subclass Polypodiidae and that cluster was distant from some other cluster of the subclass Polypodiidae. However, the literature data indicate that there are a few exceptions when ferns of particular systematic groups have a significant difference in fatty acid composition.

First of all, the subclass Equisetidae is represented by only one genus *Equisetum* where green tissues contain neither ARA nor EPA. Instead, Δ5-UPIFAs were found in appreciable quantities [26, 38]. Unfortunately, due to the limited number of researches on horsetail fatty acids, there are still some uncertainties and controversial data in their regard. H. Schlenk and J.L. Gellerman [26] identified both sciadonic (5,11,14–20:3) and juniperonic (5,11,14,17–20:4) acids in *Equisetum* green parts with no other 20:3 or 20:4 isomers found. Radunz [27] reported only 20:3 and 20:4 fatty acids with no position of their double bonds established. The author noted that these fatty acids had retention times different from the authentic 5,8,11–20:3 and ARA, respectively, when analyzed by gas chromatography. More recently, Dudareva and coworkers [38] reported only juniperonic acid of the two Δ5-UPIFAs in three species of *Equisetum* while eicosatrienoic acid was identified as the common for ferns methylene-interrupted 20:3n-3. According to our investigation of *E. arvense*, it contains both sciadonic and juniperonic acids along with 20:3n-3 as minor components [77]. Other fatty acids of the *Equisetum* were within the ranges found for fern species of the subclass Polypodiidae (Fig. 2) with two exceptions: relatively high content of 20:3n-3 (6% of total fatty acids) in *E. variegatum*, and 5,9–18:2 (6%) in *E. scirpoides* [38]. The last one, also known as taxoleic acid, belongs to Δ5-UPIFA. It was found in lipids of some gymnosperm seeds [10] but not anywhere else in other horsetails or ferns.

A possible loss of ARA and EPA during the evolution of Equisetidae will be discussed in Sect. 7; however, the presence of Δ5-UPIFAs in horsetails does not make them different from other fern subclasses. Both 5,11,14–20:3 and 5,11,14,17–20:4 were found as minor components in all 23 fern species tested in our research [53].

Species of two genera of the subclass Ophioglossidae, *Psilotum* and *Botrychium*, have been analyzed for fatty acid composition (Table 1). No ARA was found in the aerial parts of *Psilotum nudum* but it occurred in small quantities in rhizomes (0.7% [52]). According to [27], eicosatetraenoic acid (20:4) of the green tissue of *Psilotum triquetrum* is different from ARA in gas chromatographic properties presuming a different position of double bonds. Besides ARA, there are two other isomers of 20:4

found in ferns: 20:4n-3 and 5,11,14,17–20:4 (Table 2) which both have retention times longer on a polar stationary phase. Since *Psilotum* does contain EPA [27, 52], its 20:4 is likely to be 20:4n-3 which is a metabolic precursor of 20:5n-3. Other prominent differences in PUFAs of *Psilotum* in comparison with other fern species are very low content of 16:3 (0.7% [27]) or even its absence [52], and the presence of at least two isomers of 20:3. One of the 20:3 isomers was identified as 20:3n-3, while another remained unidentified despite its appreciable quantities (5% in the aerial parts and up to 8% in the rhizomes) [52]. A. Radunz [27] reported about another 20:3 isomer as having the same retention time as 5,8,11–20:3. The last fatty acid was found also only in the leaves of *S. natans* in small quantities [60].

No ARA was detected in leaves of *Botrychium lunaria* from the family Ophioglossaceae, subclass Ophioglossidae [37]. Also, no other PUFAs intrinsic to many fern species (16:3, 20:3, EPA) were reported for the species, although the latter fatty acids were not reported for other fern species of the subclass Polypodiidae in this research (while ARA was reported). It can be concluded from these scarce data that ARA is absent or a minor component in the green tissue of plants belonging to the subclass Ophioglossidae. *B. lunaria* was found to contain monoenic 18:1n-9 as the predominant fatty acid (39% of total fatty acids) in the leaves [37] which also distinguishes the species from other ferns (Fig. 2a).

Within the subclass Polypodiidae, there are two prominent exceptions that may be attributed to fern taxonomy. The aquatic fern *Salvinia natans* (family Salviniaceae) was found to have the highest diversity of fatty acids among fern species studied so far [60]. Along with all common fern fatty acids, its leaves contained unusual unsaturated fatty acids, which were absent or rarely found in other fern species, as already mentioned in Sect. 3: 14:2n-5, 15:1n-8, 15:1n-6, 16:2n-4 (also found in *A. caroliniana* [48]), 16:3n-6, 16:4n-3, 17:1n-10, 17:1n-8 (also found in *Psilotum nudum* [52]), 18:1n-11 (also found in *Asplenium nidus* [39] and some horsetails [38]), 19:1n-10, 20:1n-13, 20:1n-11 (also found in some horsetails [38]), 20:3n-9 (also probably in *Psilotum* [27]), 21:1n-12, 22:1n-15, 22:1n-13, 22:1n-11, 22:2n-6 (also found in young fronds of the European fern species [43]), 22:3n-6, 22:3n-3, 22:4n-3, 22:5n-3, 22:6n-3 (also found in *Hymenophyllum caudiculatum*, *H. plicatum* [54], and in the young fronds of the European fern species [43]), 23:1, 24:1n-13, 24:3n-3, 25:1n-14, 26:1n-15, 26:3n-3. While most of them are minor components ($\leq$1% of total fatty acids), others were found in significant quantities: 16:4n-3 (3%), 22:5n-3 (5%), 22:6n-3 (5%). Some other peculiarities of *S. natans* fatty acids included the highest percentages of 16:1 isomers (7% 16:1n-9 and 4% 16:1n-7) and the low level of 18:3n-3 (6%) among the ferns [60]. Such fatty acid diversity was not reported for *Azolla* species, *A. filiculoides* [46, 47, 58], *A. caroliniana* [48], and *A. pinnata* [47], which currently belong to the same family Salviniaceae [50]. With few exceptions (16:2n-4 and an unsaturated C22 fatty acid in *A. caroliniana* [48]), the *Azolla* species are similar to many other fern species in fatty acid composition and content. Besides the less diverse composition of fatty acids, the *Azolla* species differed from *S. natans* in the higher percentages of 16:0 (39% in *A. filiculoides* and 36% in *A. pinnata* [47], and up to 52% in *A. caroliniana* [48] vs.

7% in *S. natans*) and 18:3n-3 (13–25% in *A. caroliniana* [48] and *A. pinnata* [47], and up to 34% in *A. filiculoides* [58] vs. 6% in *S. natans* [60]).

The most striking set of fatty acids in ferns was reported for the fronds of *Hymenophyllum caudiculatum* and *H. plicatum* from the family Hymenophyllaceae [54]. Of the major fatty acids common for ferns (16:0, 18:1n-9, 18:2n-6, 18:3n-3, and ARA), these species contained only 18:1n-9 in significant quantities (8–10% of total fatty acids); 16:0 and 18:2n-6 became minor fatty acids (0.5–1.2% and 0–2%, respectively), but 18:3n-3 and ARA were not reported at all. Instead, 20:3n-3, 18:3n-6, and 15:1 were found to be major fatty acids (24–35%, 21–28%, and 10–13%, respectively). Homo-γ-linolenic acid (20:3n-6) and DHA (22:6n-3) were also among the PUFAs with an unusually high content (3–5% and 1.4–7%, respectively). Since there are no other reports on the fatty acid composition of plants from the family Hymenophyllaceae, and because such a fatty acid profile has not been reported for other fern groups, it seems to be exclusive among ferns and even wider in vascular plants, particularly taking into account the fatty acid composition of glycerolipids involved in photosynthesis, which often determine fatty acid profile in green tissue [78, 79].

Very low levels or even absence of EPA (0–0.8% of total fatty acids) were found for several species of the Aspleniaceae family [27, 32, 39, 43, 53] which may be a characteristic feature of the family.

Except for the abovementioned particularities of the specific fern groups, fern taxonomy has a limited predictive value for a fatty acid profile of a fern species. Significant variations in fatty acid percentages within a family make little difference among families. It was found as for the mature fronds as for the young fronds (Fig. 2).

6.2 Developmental and Seasonal Changes

As it was discussed in Sect. 4, young fronds differ in fatty acid profiles from the mature ones (Fig. 2). Frond growth and maturation are accompanied by the changes in percentages of fatty acids with an increase of those fatty acids which are associated with lipids of chloroplast membranes. The content of omega-3 acids (16:3 and, especially, 18:3) rapidly increased in the early weeks of frond development of *Pteridium aquilinum* and *Dryopteris filix-mas* and even further to the middle of the vegetative season for 18:3n-3, while 18:2n-6 decreased up to the middle of the vegetative season [28]. A similar tendency was found for the aquatic fern *Azolla filiculoides* cultured for 3 weeks [58]. During this period content of palmitic, oleic, and linoleic acids gradually decreased with an increase in the proportions of 16:3 and 18:3. The percentage of ARA decreased at the beginning of the frond development of the terrestrial ferns [28] and slightly increased in *A. filiculoides* [58].

By the end of the vegetative season, the percentage of 18:2n-6 increased in the fronds of *P. aquilinum* and *D. filix-mas,* and those of 18:3n-3, 16:3n-3, and 20:4n-6 decreased [28]. In another aquatic fern, *Azolla caroliniana*, biomass collected from the midsummer to the late autumn was characterized by a decrease in 16:0 (from

52% of total fatty acids in the summer to 39% at the beginning of autumn and 37% in November) and an increase in the percentages of unsaturated fatty acids: 16:3 (1→3→4%, respectively), 18:1n-9 (6→10→8%), 18:3n-6 (0.3→0.6→0.7%), 18:3n-3 (13→21→22%), and 20:4n-6 (1→3→4%) [48]. The reasons behind these changes remained unclear: plant development, a lowering of the water temperature, or a contribution from lipids of the endosymbiotic cyanobacterium *Anabaena azollae*.

A more accurate picture of the developmental/seasonal changes may be drawn on the basis of the absolute (or weight) content of fatty acids in frond tissues. In the fronds of five species of New Zealand ferns [53], the chloroplast-associated fatty acids, 18:3n-3, 16:3n-3, and 3*t*-16:1, usually increased in quantities from the spring samples (young fronds) to the summer ones (mature fronds) and then remained constant or slightly elevated in the winter time (aged fronds). *Pyrrosia eleagnifolia* was an exemption with very minor changes of 18:3n-3 and 16:3n-3 over the vegetative season (Fig. 3). The changes in ARA and EPA content have an opposite trend: while ARA usually decreased in the absolute content from the spring to the summer, and then remained unchanged or slightly increased in the winter, the content of EPA was increasing (Fig. 3). It should be noted that the content of ARA was found to be the least variable among the major fatty acids of the ferns during the vegetation period [53]. Highly variable content was found for 16:0, 18:1n-9, and 18:2n-6. The levels of one, two, or all three of these fatty acids increased many folds in the summer or winter fronds with sporangia as compared to their corresponding young fronds or mature fronds without sporangia [53]. Similar, fertile plants of *A. filiculoides* with lipid-rich male microsporocarps contained higher percentages of 18:1 and 18:2 than the plants at the vegetative stage [47]. Such variations are explained by the contribution from the spores containing lipids enriched in these

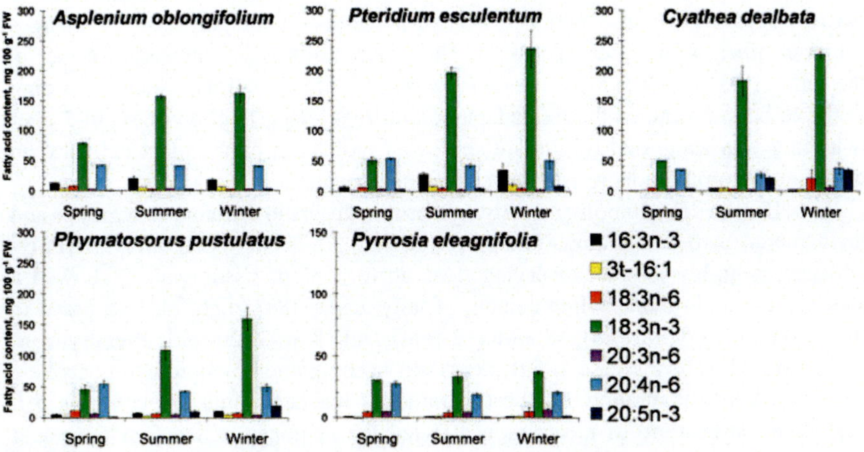

Fig. 3 Content of selected fatty acids (mg 100 g^{-1} fresh weight) in the fronds of New Zealand ferns in different seasons. The graphs are built based on the data published in [53]. Bars indicate an average of two samples ± initial values

fatty acids (see Sect. 4). Thus, developmental changes of fern vegetative and reproductive organs (frond growth and maturation, formation and maturation of sporangia/spores) may affect fatty acid profiles of fronds.

Seasonal differences in fatty acid distribution in lipid classes were found for the horsetail *Equisetum arvense* [77]. The Δ5-UPIFAs, 5,11,14–20:3 and 5,11,14,17–20:4, were minor components of its shoots during the vegetative season (0.8–1.3% and 2.6–3.5% of total fatty acids, respectively). The fatty acids were mostly associated with phospholipid fraction (50–68%) in the spring and autumn shoots but in the midsummer, their major portion was found in the fraction of neutral lipids (67–73%). The seasonal differences may reflect an effect of temperature or insolation.

6.3 Ecological and Environmental Factors

Effects of temperature or solar irradiation on fatty acid composition have not been investigated for ferns in detail. Few studies have just some information on their possible impacts. A possible effect of shading [65] was mentioned in Sect. 4. Effect of temperature was questioned when fatty acid content was compared in the summer and winter fronds of the New Zealand ferns: PUFA content was alike for the fronds collected in the different seasons when similar samples were compared, i.e., with or without sporangia, irrespective of temperature difference in the two seasons [53]. Little changes in fatty acid composition were reported for fronds of *Polypodium vulgare* during the winter months [28].

On the other hand, fronds of *Asplenium scolopendrium*, an evergreen fern that tolerated sub-zero temperatures, had higher percentages of the saturated fatty acids (16:0, 18:0, 22:0) in the spring after wintering [40]. In the summer fronds, the levels of unsaturated fatty acids (16:3n-3, 18:2n-6, 18:3n-3, 20:4n-6) were higher that is partly (regarding 16:3n-3 and 18:3n-3) in agreement with the seasonal changes found in other studies (see Sect. 6.2). The unsaturation index increased further in the autumn fronds, when the temperature was stable at below 0 °C, which, according to the authors, promoted the fern adaptation to frost [40]. The drop of 18:3n-3 after wintering was suggested by the authors as an indication of the plant dehydration which might promote its resistance to freezing as well.

The effect of dehydration on a fatty acid profile differs in different fern groups and their tolerance to desiccation. The epiphytic fern *Pleopeltis polypodioides* (Polypodiaceae), highly tolerant to drying (loss up to 95% of tissue water), showed a steady increase in the absolute content of fatty acids: 18:0 (eightfold compared to fresh fronds), 16:0 (18-fold), 18:2n-6 (12-fold), and 18:3n-3 (20-fold) after drying at 25 °C for 72 h. The increase in fatty acids was accompanied by lipid hydroperoxide content. Under rehydration, all these parameters reached initial levels within 5 h [67]. The temperature of desiccation affected the response of the fern fatty acid metabolism: absolute content of all major fatty acids (16:0, 18:0, 18:1, 18:2, 18:3) increased up to 35 °C of desiccation temperature, afterward it decreased but some fatty acids were still higher in the content at 40 °C and 45 °C (16:0, 18:0, 18:3) than after drying at 25 °C [68]. Desiccation at 50 °C led to the drop of all fern fatty acids.

The percentage of total unsaturated fatty acids in the dried fronds was lower at 30 °C (68%) as compared with 25 °C (80%) and further decreased more slowly up to 50 °C (64%). These researchers [68] also investigated the effects of heat stress and oxygen deprivation (full or partial submersion in water) on fatty acid profiles in hydrated fronds of the fern. The increased temperatures (30–35 °C and up to 40 °C in the case of fully submersed fronds) caused also an accumulation of all or some fatty acids but afterward (40–50 °C) the content of fatty acids dropped as compared with 25 °C. At the same time, the level of unsaturation decreased significantly only at 50 °C in the case of fully submersed fronds and at 35 °C in the case of partly submersed fronds suggesting oxygen deprivation does not affect fatty acid unsaturation and may act in a protective manner. It should be noted that overall estimation of the fern under heat stress indicated dehydration as a protective mechanism from heat damage in *Pleopeltis* [68], although the response of fatty acid metabolism seems to be similar in dehydrated and hydrated fronds.

Effect of desiccation–rehydration on fatty acid profiles were studied in the fronds of two species of the filmy ferns differing in their desiccation tolerance: *Hymenophyllum caudiculatum* (less tolerant) and *H. plicatum* (more tolerant) [54]. The peculiarities of the fatty acid composition of this group of ferns were discussed above (see Sect. 6.1). However, the lack of the common and usually major fatty acid in plant green tissues, 18:3n-3, is the most bizarre feature of the ferns. In the initial hydrated state, the ferns were different in fatty acid profile: more saturated (16:0, 18:0, 21:0) and monoenic (15:1, 17:1, 18:1n-9) acids were found in the fronds of *H. plicatum* as compared to *H. caudiculatum*. For PUFAs, some of them were higher in *H. caudiculatum* (18:3n-6, 20:3n-3, and the total PUFA level) while others (20:3n-6, 22:6n-3) in *H. plicatum*. No linoleic acid (18:2n-6) was detected in *H. caudiculatum*, which was also a minor component of *H. plicatum* (2% of total fatty acids). Under desiccation, the levels of their major fatty acids changed similarly: 18:3n-6 decreased while 20:3n-3 increased in both species. After plant rehydration, these fatty acids also changed similarly in the two species: they both declined to reach levels below the initial ones. Other less abundant fatty acids changed often in different manners in *H. caudiculatum* and *H. plicatum*. The desiccation caused the appearance of unusual 22:1n-9 and 24:1n-9 in *H. plicatum* but not in *H. caudiculatum*. These fatty acids remained in the fronds of *H. plicatum* after rehydration as well. The authors concluded that membrane stability during the desiccation–rehydration cycle does not rely on membrane fluidity due to the unsaturation of fatty acids but may be dependent on the presence of some minor fatty acids, particularly, 20:3n-6, 22:6n-3, 22:1n-9, and 24:1n-9. Such differences in fatty acid profiles may determine the differences in desiccation tolerance which, in turn, can lead to different microhabitat preferences of the Hymenophyllaceae species [54].

A link between a fern habitat and fatty acid composition was studied on two epiphytic ferns, *Platycerium bifurcatum* and *Asplenium nidus*, and a terrestrial species, *Asplenium trichomanes* [39]. Both the epiphytic species contained much less total fatty acids in their fronds than the terrestrial one: 9.9 μmol g^{-1} WW in *P. bifurcatum* and 7.9 in *A. nidus* versus 67.1 μmol g^{-1} WW in *A. trichomanes*. The epiphytes were characterized by a much higher content of saturated fatty acids (about 50% of total fatty acids) and less all unsaturated ones (mono-, di-, tri-, and to a lesser extent of tetraenoic acids) than the terrestrial *A. trichomanes* (only about 20% of saturated fatty acids). The differences between the two epiphytic species were in a higher diversity of fatty acids in *A. nidus* (27) than in *P. bifurcatum* (14), the presence of very long-chain saturated fatty acids in significant amounts, particularly 24:0 (6% of total fatty acids) and 26:0 (5%) in *A. nidus*. In *P. bifurcatum*, the high level of saturated fatty acids was achieved due to 18:0 and 20:0. The researchers did not detect 20:5n-3 in all three species including the terrestrial one, although EPA is common for ferns as mentioned above. The low level of EPA can be a feature of the genus *Asplenium* or the whole Aspleniaceae family (see Sect. 6.1).

An example of the New Zealand fern *Pyrrosia eleagnifolia* is of interest as a plant of extremely diverse habitats which is ranged from sheltered forests to exposed coastal situations. The species is described as "a very tough and adaptable fern which can survive the driest conditions" [72]. Similar to the abovementioned epiphytic ferns, mature fronds of *P. eleagnifolia* without sori contained less total lipids (about 0.6% of WW) and fatty acids (1.7 mg g^{-1} WW) than other ferns [53]. The fronds were found to have relatively low levels of 18:3n-3 (22% of total fatty acids) and EPA (0.3–0.7%), whereas the percentage of 18:2n-6 and ARA was high (17% and 12%, respectively) [53].

The effect of heavy metals, as a negative environmental factor, has also been studied in relation to fern fatty acids. *Matteuccia struthiopteris* grown in the presence of lead ions (1–1000 μM) was found to have increased unsaturation indices of fatty acids in the fern fiddleheads. However, there was no correlation between the lead concentration and the index value; and such an effect was not always observed in the mature leaves. By the way, no lead accumulation was detected in the fern organs [34]. Low accumulation of cadmium occurred in the fern fiddleheads and fronds when *M. struthiopteris* was grown in the presence of 100 μM cadmium nitrate, but the presence of the salt in the plant medium even stimulated the fern growth [36]. As a result, the effect of cadmium on the fatty acid profile was small and often contradictory in fiddleheads and different parts of fronds. For example, cadmium ions caused accumulation of saturated 14:0, 20:0, and 23:0 in the lower parts of fronds, but not in its upper and middle parts as well as the fiddleheads. Oleic acid increased in the fiddleheads, the upper and middle parts of the fronds, while linolenic decreased in the middle and lower parts of the fronds in the presence of cadmium. Also, the percentages of ARA were usually slightly higher, and those of EPA were sometimes lower in plants grown in the presence of cadmium nitrate. The authors concluded that the heavy metal effect depends on a stage of frond formation [36].

Some variations in the content of the major fatty acids (16:0, 18:1n-9, 18:2n-6, 18:3n-3, 20:4n-6) were found in fiddleheads of *Mattueccia strutiopteris* collected from different sites in Canada [42]. There were regional differences in the fatty acid percentages in leaves of *Gymnocarpium dryopteris* and *Woodsia glabella* from two northern regions in Russia [37] and in fiddleheads of three fern species (*M. strutiopteris*, *Pteridium aquilinum*, and *Osmundastrum asiaticum*) from two regions of Russian Far East [44]. However, the fatty acid percentages were very close in young fronds of *P. aquilinum* grown under different climatic conditions [63].

7 Fatty Acid Biosynthesis in Ferns

Fatty acid biosynthesis in plants has been intensively studied and general metabolic pathways are well established [80–82], including some particular groups of fatty acids [83–86]. The general pathway of de novo fatty acid biosynthesis in ferns is expected to follow the same reactions and enzyme systems as in other higher plants. Briefly, the biosynthesis is started by the formation of malonyl-CoA from acetyl-CoA catalyzed by acetyl-CoA carboxylase with subsequent cycles of condensation and acyl chain elongation catalyzed by fatty acid synthase (FAS). FAS is a multi-subunit complex consisting of six catalytic polypeptides and one low molecular weight acyl-carrier protein (ACP). Both of these systems are located in the chloroplast. Once a chain length of acyl-ACP reaches C16–C18, the acyl chain can undergo desaturation to give mono-unsaturated acyl-ACP, or the acyl group can be incorporated into plastid glycerolipids or released as a free fatty acid [80–82]. The incorporation of fatty acids into plastid glycerolipids follows the "prokaryotic" pathway when fatty acids esterified to the *sn*-1 position of glycerol-3-phosphate by a soluble glycerol-3-phosphate-1-acyltransferase and then to the *sn*-2 position by membrane-bound 1-acyl-*sn*-glycerol-3-phosphate acyltransferase. The resulting phosphatidic acid gives origin to PG and, in the "16:3 plants," via diacylglycerol to glycolipids (MGDG, DGDG, SQDG). As constituents of these lipids, fatty acids are subjected to desaturation by a number of fatty acid desaturases (FADs). Such desaturation provides a specific distribution of fatty acids which depends on the lipid substrate [83]. The export of synthesized fatty acids from the plastids into cytosol occurs in the form of acyl-CoA which in turn participates in the formation of different complex lipids or can be converted into long-chain fatty acids by a membrane-bound elongase complex. The elongase complex produces (also in combination with acyl-CoA desaturase-like enzymes) saturated, monoenic, and, to a much lesser extent, dienic fatty acids often with long and very long chains [85]. Another portion of fatty acids exported to the cytoplasm is involved in the "eukaryotic" pathway associated with the endoplasmic reticulum: the fatty acids esterify PC via de novo synthesis or by the interconversion between PC and lysoPC. The second route may also occur in chloroplast membranes by means of the plastid-associated lysoPC acyltransferase. The recycled PC is then transported to the endoplasmic reticulum thus making the PC a carrier for fatty acids of the plastid origin [83]. Oleoyl (18:1) of PC is desaturated by omega-6 fatty acid desaturase (FAD2). The resulting product containing now linoleoyl (18:2) can be desaturated further by omega-3 FAD3 to linolenoyl (18:3) or transferred back to plastids for glycolipid production [83].

The distribution of fatty acids in lipid classes, the presence of 16:3 in ferns, and its association with MGDG and DGDG (Sect. 5) indicate the existence of the same general pathways for fatty acid biosynthesis in ferns as well. Moreover, transcriptome analysis of *Azolla filiculoides* revealed that the fern expresses genes encoding enzymes of fatty acid biosynthesis (including very long-chain fatty acids) which are often homologous to those of seed plants [87], see Supporting Information Notes S2 therein]. The point of interest is the biosynthesis of the fatty acids specific for ferns, particularly, the LC-PUFAs with C20 and more. Such fatty acids require involvement of an elongase and additional FADs for their synthesis. In general, biosynthesis of ARA and EPA in eukaryotic organisms occurs by the subsequent action of Δ^6-desaturase, elongase, and Δ^5-desaturase starting from 18:2n-6 or 18:3n-3, respectively. If elongation of 18:2n-6 or 18:3n-3 precedes the desaturation with the production of 20:2n-6 or 20:3n-3, respectively, Δ^8-desaturase can be involved leading to the formation of the direct precursors of ARA and EPA, 20:3n-6 and 20:4n-3 [92].

For ferns, there is not much information about LC-PUFA biosynthesis and metabolism, and the available information came from the early studies [31, 32]. Developing gametophytes readily incorporate exogenous acetate into lipids as shown with spores of *P. vulgare* germinating in the presence of Na-[2-^{14}C]acetate [31]. At the early stage of fern gametophyte development (10 days after the beginning of spore imbibition), the major portion of the total ^{14}C activity among fatty acids was found in the 20:3 and 20:4 (21 and 28% of total incorporation in fatty acids, respectively) while the percentages of these fatty acids did not exceed 1% in the gametophytes. It seems that exogenous acetate is much spent on the elongation with the formation of the unsaturated C20 fatty acids of the omega-6 family. However, the high biosynthetic rate of ARA does not lead to its accumulation as the low level of the fatty acid shows, thus indicating a high rate of ARA turnover during the early stages of fern development.

The pathway leading to the production of ARA was investigated in the work [32]. Young opening fronds of *Dryopteris filix-mas* were able to take up and elongate radiolabeled 16:0 and 18:0 but no desaturation was observed. Supplementation of fronds with [1-^{14}C]18:1n-9 or [1-^{14}C]18:2n-6 resulted in radioactivity incorporation into all other polyunsaturated fatty acids (18:2n-6, 18:3n-3, 18:3n-6, 20:3n-6, and 20:4n-6), but not into 16:0 and 18:0. It is in agreement with the general pathway of fatty acid biosynthesis in higher plants which suggests that the initial introduction of a double bond takes place in plastids. According to the results, further elongation and desaturation of fatty acids, particularly, ARA and its metabolic precursors occur in the extraplastid environment. The labeling of 18:3n-6 and no detectable labeling of 20:2n-6 led the authors to the conclusion that synthesis of ARA in *D. filix-mas* proceeds according to Scheme 1 [32]:

$$18:2\text{n-}6 \rightarrow 18:3\text{n-}6 \rightarrow 20:3\text{n-}6 \rightarrow 20:4\text{n-}6. \quad (1)$$

Although EPA biosynthesis has not been investigated in ferns, the major pathway in bryophytes was found as follows in Scheme 2 [88, 89]:

$$18:3\text{n-}3 \rightarrow 18:4\text{n-}3 \rightarrow 20:4\text{n-}3 \rightarrow 20:5\text{n-}3. \quad (2)$$

Δ^8-desaturase was also discovered in some algae [90] suggesting an alternative pathway (Scheme 3):

$$18:3\text{n-}3 \rightarrow 20:3\text{n-}3 \rightarrow 20:4\text{n-}3 \rightarrow 20:5\text{n-}3. \quad (3)$$

Finally, an ω3 desaturation of ARA with the direct production of EPA is discussed [91]. For ferns, all the potential intermediates of EPA biosynthesis (18:4n-3, 20:4n-3, 20:3n-3) have been found (Table 2). The simultaneous presence of 18:4n-3 and 20: 4n-3 have been found in a number of species: *Dryopteris austriaca*, *Dryopteris filix-mas*, *Pteridium aquilinum* (in traces), and *Polypodium vulgare* from Scotland [28], *Salvinia natans* [60], *Adiantum pedatum*, *Athyrium filix-femina*, *Athyrium sinense*, *A. spinulosum*, *A. yokoscense*, *Cornopteris crenulatoserrulata*, *Deparia pycnosora*, *Dryopteris crassirhizoma*, *D. expansa*, *D. goeringiana*, *Matteuccia struthiopteris*, *Osmunda claytoniana*, *Osmundastrum asiaticum*, *Parathelypteris noveboracensis*, *Phegopteris connectilis*, *Pteridium aquilinum* from the Russian Far East [44, 53, 66], *Cyathea dealbata*, *Pteridium esculentum*, *Phymatosorus pustulatus* from New Zealand [49, 53]. Some of the papers also reported 20:3n-3 for the same samples [49, 53, 60]. Some pteridophytes (*Psilotum nudum* [52], *Cystopteris fragilis*, *Ptisana salicina*, *Asplenium oblongifolium* [53], the young fronds of the 24 European fern species of 8 families [43]) contained 20:3n-3, whereas 18:4n-3 and 20:4n-3 were in traces or not reported. Finally, for a few fern species (*Polystichum tripteron*, *Pyrrosia eleagnifolia* [53], *Asplenium scolopendrium* [40]), only 20:4n-3 and 20:3n-3 were reported but not 18:4n-3. Significant correlations, from moderate to strong, were found between the absolute contents of 20:5n-3 and the fatty acids of the omega-3 family (18:3n-3, 18:4n-3, 20:4n-3, 20:3n-3) in the fronds of five ferns measured in three vegetative seasons [53]. All these data demonstrate that both pathways of EPA biosynthesis can exist in ferns.

Three papers reported the presence of DHA (22:6n-3) in ferns [43, 54, 60]. When considering the formation of 22:6n-3 from 20:5n-3, at least one elongation step and an introduction of one additional double bond must take place. However, only one of the papers reports the presence of a potential intermediate along with 20:5n-3 and 22: 6n-3 in lipids, and this intermediate was 22:5n-3 [60]. For the filmy ferns, there is a gap of any intermediate between 20:3n-3 and 22:6n-3 [54], and only 20:5n-3 and 22: 6n-3 are reported for the fern fiddleheads [43]. The biosynthetic pathway of DHA from 20:5n-3 known for mammals via C24 intermediates followed by β-oxidation in peroxisomes [92] seems unlikely since no C24 with five or six double bonds have been detected in ferns. Thus, if DHA is present, the involvement of elongase and subsequent action of Δ^4-desaturase is more realistic [91].

Metabolic pathways for the biosynthesis of Δ5-UPIFAs with polymethylene-interrupted double bonds were proposed for gymnosperm seeds which are characterized by a high structural diversity [10]. According to the suggested scheme, the two Δ5-UPIFAs found in Polypodiopsida, sciadonic (5,11,14–20:3) and juniperonic (5,11,14,17–20:4) acids, are originated from 11,14–20:2 (or 20:2n-6) and 11,14,17–20:3 (or 20:3n-3), respectively, by the action of Δ^5-desaturase. Δ^5-desaturases, which are capable to produce Δ5-UPIFAs, have been characterized in seeds of the angiosperm *Anemone leveillei* [91] and the gymnosperm *Torreya grandis* [93, 94]. In the case of ferns, both potential precursors of sciadonic and juniperonic acids are mostly minor fatty acids, but they are quite common (Table 2). They were usually found in fern fronds along with 5,11,14–20:3 and 5,11,14,17–20:4 [53]. Interestingly, aerial parts and rhizomes of *Psilotum nudum* were found to contain relatively high levels of 20:2n-6 (1–1.7% of total fatty acids) and 20:3n-3 (0.2–1.5%), and there were also significant quantities of 20:3 with undetermined positions of double bonds (5–8%) as well as unidentified fatty acids [52].

Controversial data are available for horsetails (*Equisetum*) which are known to contain Δ5-UPIFAs for a long time. The only C20 PUFAs reported for *E. hyemale* and *E. arvense* were 5,11,14–20:3 and 5,11,14,17–20:4 in the study of H. Schlenk and J.L. Gellerman [26] but they did not report any potential precursor (20:2n-6 or 20:3n-3) of these fatty acids. In the study of L.V. Dudareva and coworkers [38], 5,11,14,17–20:4 and its intermediate precursor (20:3n-3) were found but neither 5,11,14–20:3 nor 20:2n-6 were reported for three species of horsetails including one (*E. arvense*) analyzed in the previous study [26]. Meanwhile, horsetails look like interesting models for LC-PUFA biosynthesis in ferns. If the biosynthesis of 5,11,14–20:3 and 5,11,14,17–20:4 proceeds via 20:2n-6 and 20:3n-3, respectively, then the last fatty acids, in turn, are produced from 18:2n-6 and 18:3n-3 by elongation. Both 18:2n-6 and 18:3n-3 are present in horsetails [26, 38]. On the other hand, 18:2n-6 and 18:3n-3 are precursors for 18:3n-6 and 18:4n-3, respectively, which are formed by the action of Δ^6-desaturase and serve as intermediates in the pathways leading to the formation of ARA and EPA [92]. The absence of 18:3n-6 and 18:4n-3 along with ARA, and EPA means the absence of Δ^6-desaturase in the horsetails which might be lost during evolution. Similarly, the horsetails do not contain 20:3n-6 and 20:4n-3, which can be formed from 20:2n-6 and 20:3n-3, respectively, by the action of Δ^8-desaturase, and which are also immediate precursors of ARA and EPA. Thus, the last enzyme (Δ^8-desaturase) is likely absent in the horsetails as well. Since Δ^5-desaturase should be preserved in the horsetails to produce 5,11,14–20:3 and 5,11,14,17–20:4 from 20:2n-6 and 20:3n-3, it is unlikely that both desaturases (Δ^6-and Δ^8-) are lost together in the horsetails. It may be speculated that Δ^8-desaturase is absent in the whole lineage of Polypodiopsida. If it is true, the biosynthesis of EPA does not occur via 20:3n-3 as it was discussed above (Scheme 3).

Possible pathways for the biosynthesis of LC-PUFAs in ferns are summarized in Fig. 4. The scheme can be only confirmed after the characterization of the enzymes involved in the fatty acid biosynthesis. The biosynthesis of hydroxy and dihydroxy very long-chain fatty acid derivatives in the Salviniaceae family is discussed in the paper [57].

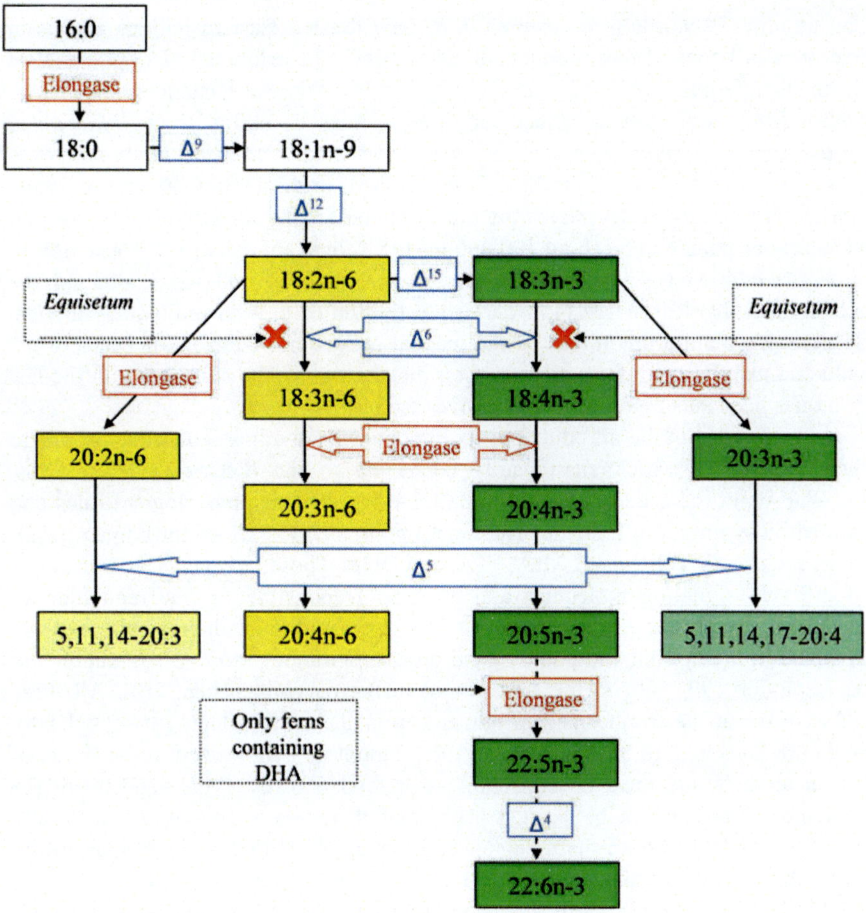

Fig. 4 Possible pathways for the biosynthesis of LC-PUFAs in ferns. Fatty acids of the omega-3 and omega-6 families are shown in green and yellow, respectively, and Δ5-UPIFAs are lighter in tone. Desaturases are indicated as Δ followed by the position of an inserted double bond. Elongases are not specified

8 Dietary Value of Fern Fatty Acids

There are numerous records on edible ferns and their consumption for food in different regions around the globe [12–17, 95]. However, for the application of ferns in the food industry, it is even more important to estimate the real economic value and resource potential of the edible ferns. In general, they are considered a valuable nontimber forest product contributing to local economies [95–97]. Young fronds of the bracken fern (*Pteridium aquilinum*) are harvested across the regions of Siberia and the Far East of Russia for export to China and Japan and also for internal consumption. According to customs data, only the Far Eastern regions of Russia exported 94.1 tons of fern products in 2016–2019. The estimated yield potential of the bracken fern is about 20.2 tons annually for one of the Far Eastern regions, Amur Oblast [98]. High market prices for three species of edible ferns (*Matteuccia struthiopteris*, *P. aquilinum* subsp. *japonicum*, and *Osmunda japonica*) were reported in Japan (1.2–22.1 million ¥ ton^{-1}) [18]. In the United States, where the ostrich fern (*M. struthiopteris*) is the most popular among the edible ferns, its total yields are estimated to be about 100,000 lbs (45.4 tons) annually [97]. These figures show that edible ferns have economic and market value as food plants, although the scale of their harvest remains low and is mostly limited to Asia and North America. The reasons behind this might be the manual collection of fern fiddleheads in the wild and insufficient information on their dietary value. The situation with the last point has been successfully improved over the past 10 years.

As mentioned in the introduction, fern fatty acids are of particular interest due to their LC-PUFAs which significantly contribute to the dietary value of ferns. LC-PUFAs are considered valuable nutrients due to their physiological roles and limited biosynthesis in humans. The most important LC-PUFAs for human physiology are usually considered ARA, EPA, and DHA. These fatty acids are precursors of different signaling molecules and are also responsible for the regulation of membrane properties [21]. As a result, the fatty acids and their metabolites are involved in the normal and pathological processes ranging from inflammation and its resolution [99, 100] to the functioning of the nervous system [19]. All these LC-PUFAs can be synthesized in humans from the corresponding essential fatty acids (18:2n-6 and 18:3n-3) which, in turn, cannot be synthesized in the tissue of higher animals and must be consumed with food. However, the capacity of the biosynthesis appears to be limited [101] and there are examples of potentially beneficial effects of both the omega-3 and omega-6 LC-PUFAs on human health under different conditions [19, 102, 103]. Fish is the major source of the omega-3 LC-PUFAs (EPA, DHA), while the omega-6 LC-PUFAs (ARA and its precursor, 20:3n-6) are consumed mainly with meat [102]. Intake of ARA with food is high enough for normal healthy adults and estimated to be 100–250 mg day^{-1} in advanced countries; however, vegetarians may consume as little as 3–44 mg day^{-1} [102]. EPA intake is about 36 mg day^{-1} for adults (20–55 years) in the USA based on a 2000 kcal diet [104].

The first report on the fatty acid composition of an edible fern, as part of its nutritional assessment, is dated as early as 1982 for fiddleheads of the ostrich fern [41]. Surprisingly, the authors reported neither ARA nor EPA in the samples but found common 16:0, 18:2n-6, 18:3n-3, 18:0, 18:1n-9, and the unusual and minor for ferns (Table 2) erucic acid (22:1n-9). Fat content was reported to decrease in processed (frozen and canned) products as compared to raw fiddleheads [105]. Later, Canadian researchers analyzing fatty acids in fiddleheads of the ostrich ferns collected from different sites in eastern Canada found the tissue contains ARA, EPA, γ-linolenic, and dihomo-γ-linolenic acids along with the common fatty acids and concluded that the ostrich fern has the most complete fatty acid spectrum of any edible green plant [42]. Postharvest storage of the ostrich fern fiddleheads in cold water resulted in a significant accumulation of almost all fatty acids (including 20:3n-6 and 20:4n-4) over 42 days. Only EPA insignificantly declined during the storage [106].

Significant percentages of ARA (14–15% of total fatty acids) were found in the young shoots of the bracken fern from different zones of Amur Oblast while the content of EPA was low (0.6–1.5%) [63]. The fatty acids are well preserved in the frozen fronds: ARA (15% of total fatty acids) and EPA (0.9%) [107]. When three species of edible ferns (*M. struthiopteris*, *P. aquilinum,* and *O. asiaticum*) from the Russian Far East were analyzed for fatty acids [44], the major ones of the fern fiddleheads were found to be 16:0 (25–30% of total fatty acids), 18:2n-6 (17–27%), 18:3n-3 (11–24%), 20:4n-6 (6–14%), and 18:1n-9 (5–10%). EPA was in the range of 0.8–3.2%. There were found significant variations among the ferns in the ratios of omega-6/omega-3 fatty acids and ARA/EPA: from 2.6–3.2 of omega-6/omega-3 and 11–17 of ARA/EPA in *P. aquilinum*, 1.8 of omega-6/omega-3 and 3.9–5.6 of ARA/EPA in *M. struthiopteris*, to 1 of omega-6/omega-3 and 2–3 of ARA/EPA in *O. asiaticum*. Commercially available products of fern fronds, which were processed in different ways for long-term storage, retained the valuable LC-PUFAs in varying degrees. The lowest content was found in salted products of the bracken fern which contained much less ARA and EPA (51–70% and 34–54%, respectively) as compared to the freshly harvested fronds. A dried product of the bracken fern contained about 84% ARA and 64% EPA of their levels found in the freshly harvested fronds. In the case of the ostrich fern, the content of EPA made up 85% and that of ARA was even slightly higher in a dried product as compared to the unprocessed freshly collected plant material. The LC-PUFAs were also well preserved in the cooked fern fronds [44].

There is no available information on current fern consumption in Europe; however, the nutritional value of the young fronds of ferns growing in Europe was recently investigated by the researchers from Czech Republic [43]. Fatty acid content was assessed in 24 fern species representing 8 families. The predominant fatty acid was 18:2n-6 (range = 2.8–7.3, median = 4.5 mg g^{-1} DW) followed by 20:4n-6 (2.3–6.3, 3.7 mg g^{-1} DW), 18:1n-9 (2.1–4.8, 3.0 mg g^{-1} DW), 18:3n-3 (1.2–4.8, 2.9 mg g^{-1} DW), 16:0 (1.0–2.3, 1.7 mg g^{-1} DW), and 18:3n-6 (1.0–2.6, 1.6 mg g^{-1} DW). EPA was in the range of 0.1–1.8 with a median = 0.5 mg g^{-1} DW. The highest content of ARA was found in *Polypodium vulgare* (5.3–6.3 mg g^{-1} DW) and *P. aquilinum* (5.7 mg g^{-1} DW). Fiddleheads of *M. struthiopteris* were the most enriched in EPA (1.8 mg g^{-1} DW). The ratios of omega-6/omega-3 and ARA/EPA varied in a very wide range: from 1.3 to 6.4 (median = 3.3) for omega-6/omega-3 and from 1.9 to 51 (median = 8.0) for ARA/EPA. Similar to the Far Eastern ferns, the fiddleheads of the European counterparts had the same order of omega-6/omega-3 and ARA/EPA: 4.8 and 11.3, respectively, for *P. aquilinum*, 2.3 and 2.4 for *M. struthiopteris*, and 1.3–1.6 and 1.9–2.7 for *Osmunda regalis* which also belongs to the family Osmundaceae like *O. asiaticum* in our study [44]. The most interesting ferns as sources of LC-PUFAs were found to be: *M. struthiopteris* (the highest level of EPA with a relatively high content of ARA and the low omega-6/omega-3 and ARA/EPA ratios); *Athyrium distentifolium* which is very similar in the parameters to *M. struthiopteris*; *P. aquilinum* as a rich source of the omega-6 fatty acids (18:2n-6, 20:4n-6, 18:3n-6, 20:3n-6); *Polypodium vulgare* also as a rich source of the omega-6 fatty acids; *Dryopteris expansa* which contains high levels of both omega-6 and omega-3 fatty acids including 20:4n-6, 18:3n-6, 20:3n-6, and 20:5n-3 [43]. Unfortunately, the data lack information on the potential metabolic precursors of EPA, 18:4n-3 and 20:4n-3.

The data on the content of the physiologically important PUFAs in different edible ferns are summarized in Table 3 along with the other sources of PUFAs which include green vegetables, macroalgae, fish, and meat. The content is shown on a dry and wet weight basis to compare the data from different sources. Table 3 includes only those fern species from the study [43] which are listed as edible [14].

Alike green vegetables, fern fronds contain PUFAs essential for human metabolism, linoleic (18:2n-6), and α-linolenic (18:3n-3) acids. While the ferns contain similar or higher levels of linoleic acid than green vegetables, the last are superior in the content of α-linolenic acid, which provides the predominance of the omega-3 fatty acids over omega-6 ones in the vegetables (Table 3). However, it is not so evident for some popular salad vegetables like spinach and arugula (Table 3). Moreover, the green vegetables do not contain the LC-PUFAs present in ferns (the traces of ARA found in *E. sativa* and *S. oleracea* [43] should be considered with precautions). Similar to the green vegetables, the edible algae-macrophytes have omega-3 fatty acids mostly prevailing over omega-6 fatty acids (n-6/n-3 $\leq$ 1), and they contain EPA and its metabolic precursors. In comparison with the ferns, only a few of the algae species have EPA content significantly higher than in ferns but many of them contain EPA in amounts comparable with the fern fiddleheads (Table 3). For ARA, it is usually lower in the algae than in the ferns (Table 3).

Fish is a valuable and established source of the omega-3 LC-PUFAs including EPA and, particularly, DHA. Their content usually substantially surpasses the content of the omega-6 fatty acids (n-6/n-3 $<$ 1, Table 3). The content of EPA may reach 0.4–1 g 100 g^{-1} raw tissue in oily fish like salmon [120]. Ferns are significantly inferior to fish in the content of EPA and other omega-3 LC-PUFAs. However, the omega-6 ARA and its precursors turned out to be much higher in the fern fiddleheads than in the commercial fish species (Table 3). Surprisingly, the fern fiddleheads resemble meat in the content of the omega-3 and omega-6 LC-PUFAs: n-6/n-3 $\geq$ 1, ARA prevails over EPA (ARA/EPA $>$ 1), the absolute content of ARA in the ferns is comparable with that in different meats (Table 3). The meat of ruminants has the amounts of EPA exceeding those in ferns, while others (pork, rabbit, chicken) are of the same level (Table 3). Animal meat contains other LC-PUFAs of the omega-3 and omega-6 families besides EPA and ARA [116–119]. However, an advance of ferns is that they are plants and they are vegetarian sources of "meat" fatty acids which are absent in common vegetables. In addition, ferns contain the essential fatty acids inherent to the vegetables. It makes fern fiddleheads a valuable starting material for the development of vegetarian products.

A way of fatty acid extraction from ferns for industrial applications has been demonstrated recently in our paper [49]. Dimethyl ether and its mixture with a water-ethanol co-solvent were tested under near-critical conditions for the lipid extraction

Table 3 Content of selected PUFAs in the young fronds of edible ferns in comparison with other food sources

Source	18:2n-6	18:3n-6	20:3n-6	20:4n-6	18:3n-3	18:4n-3	20:4n-3	20:5n-3	n-6/n-3 (ARA/EPA)	Reference
Fatty acid content on a dry weight basis, mg g^{-1} DW										
Fern fiddleheads (raw if not indicated otherwise)										
Athyrium filix-femina	4.93	2.12	0.96	4.67	3.15	–	–	0.55	3.4 (8.5)	[43]
Dryopteris expansa	5.01	2.62	1.77	4.89	4.46	–	–	0.96	2.6 (5.1[a])	[43]
Matteuccia struthiopteris										
Raw	7.72	1.12	0.75	4.06	6.08	0.08	0.06	1.09	1.8 (3.7)	[44]
Dried	7.45	0.87	0.93	4.36	7.32	0.07	0.07	0.94	1.6 (4.7)	[44]
Raw	11.16	1.19	1.43	8.07	7.55	–	–	0.58	2.7 (13.9)	[42]
Dried	10.71	0.79	1.33	7.18	7.01	–	–	1.38	2.4 (5.2)	[42]
Raw	4.60	2.15	1.15	4.30	3.35	–	–	1.84	2.3[a] (2.4[a])	[43]
Onoclea sensibilis	4.87	1.50	0.72	3.71	2.43	–	–	0.40	3.8 (9.3[a])	[43]
Osmunda regalis (different growth conditions)	3.83/ 2.79	1.10/ 1.06	0.63/ 0.57	2.68/2.30	4.15/4.12	–	–	1.00/ 1.19	1.6/1.3[a] (2.7/ 1.9[a])	[43]
Osmundastrum asiaticum										
Raw Dried at 110 °C Air-dried	4.1 3.9 3.6	0.58 0.53 0.52	0.31 0.27 0.27	2.15 1.76 1.86	5.8 5.5 5.2	0.14 0.12 0.12	0.08 0.07 0.06	0.78 0.66 0.69	1.0 (2.8) 1.0 (2.7) 1.0 (2.7)	[44]
Pteridium aquilinum										
Raw	11.59	1.16	0.74	5.51	5.99	0.09	0.05	0.49	2.8 (11.2)	[44]
Dried	8.30	0.64	0.60	4.63	3.91	0.05	0.04	0.32	3.2 (14.8)	[44]
Salted (two producers)	8.02/ 7.33	0.66/ 0.53	1.00/ 0.59	3.85/2.83	3.04/2.30	0.04/ 0.04	0.07/ 0.03	0.26/ 0.17	3.8/4.2 (14.7/ 16.7)	[44]
Raw	5.90	2.01	0.97	5.65	2.51	–	–	0.50	4.8[a] (11.3[a])	[43]

(continued)

Table 3 (continued)

Source	18:2n-6	18:3n-6	20:3n-6	20:4n-6	18:3n-3	18:4n-3	20:4n-3	20:5n-3	n-6/n-3 (ARA/EPA)	Reference
Green vegetables										
Purslane (*Portulaca oleracea* cv. *sativa*)	3.81	0	0	0	18.0	–	–	0	0.21	[42]
Arugula (*Eruca sativa*)	2.66	0	0	0.005	7.01	–	–	0	0.4[a]	[43]
Spinach (*Spinacia oleracea*)	1.97	0	0	0.007	5.77	–	–	0	0.3[a]	[43]
Cabbage (13 cultivars)	9.5–30.6	1.0–2.4	2.1–8.2	–	14.7–26.9	–	–	–	0.6–1.5[a]	[108]
Marine macroalgae										
Alaria esculenta	0.38	–	–	0.74	0.25	–	–	0.48	1.1[a]	[6]
Palmaria palmate										
	0.13	0	–	0.33	0.31	0.35	–	8.34	0.05	[4]
	0.14	–	–	0.13	–	–	–	1.50	0.16[a]	[6]
Porphyra dioica	0.23	–	–	1.06	0.10	–	–	2.79	0.53[a]	[6]
Porphyra purpurea	0.06	–	–	0.13	0.01	–	–	0.86	0.32[a]	[6]
Porphyra umbilicalis	0.06	–	–	0.13	0.08	–	–	0.70	0.28[a]	[6]
Saccharina latissima	0.33	–	–	0.48	0.24	–	–	0.39	0.77[a]	[6]
Ulva lactuca	5.55	0.48	–	0.36	4.46	1.85	–	0.23	1.0	[4]
Undaria pinnatifida										
Blade	1.75	0.35	0.20	3.70	1.84	3.31	–	2.16	0.82	[109]
Fronds	0.80	0.24	–	2.94	1.20	2.26	–	2.86	0.60	[4]
Fish										
Trout (*Salmo trutta*)										
Norway Siberia	3.47 2.18	0.16 0.06	0.09 0.07	0.37 0.32	1.36 0.64	0.73 0.63	0.66 0.53	3.76 2.79	0.24 (0.1[a]) 0.14 (0.1[a])	[110]

Herring (*Clupea harengus pallasi*)	0. 47	0.04	0.04	0.29	0.25	0.73	0.32	5.62	0.06 (0.05[a])	[110]
Sole (*Lepidopsetta bilineata*)	0.21	0.06	0.05	0.85	0.03	0.46	0.16	7.19	0.11 (0.1[a])	[110]
Cod (*Gadus morhua maris-albi*)	0.08	0	0.01	0.29	0.03	0.07	0.05	3.13	0.025 (0.1[a])	[110]
Fatty acid content on a fresh weight basis, mg 100 g^{-1} WW										
Fern fiddleheads and cooked dishes										
Matteuccia struthiopteris (raw)	99.8	14.5	9.8	52.5	78.5	1.0	0.8	14.1	2.0 (3.9)	Nekrasov, unpublished results
Osmundastrum asiaticum										Nekrasov, unpublished results
Raw Cooked	77.2 167.5[b]	10.9 2.3	5.9 2.7	40.2 7.1	109.1 45.7	2.5 0.5	1.4 0.8	14.5 3.2	1.0 (2.8) 3.4 (2.2)	
Pteridium aquilinum										Nekrasov, unpublished results
Raw Cooked	168.3 798.8[b]	16.9 12.6	10.8 11.9	80.1 72.0	86.7 52.1	1.3 0.8	0.8 0.9	7.1 5.0	2.9 (11.4) 14.8 (14.4)	
Green vegetables (leaves)										
Purslane (*P. oleracea*) cultivated/wild	96.8/ 70.4	–	–	–	341.2/ 322.1	–	–	–	–	[111]
Spinach (*S. oleracea*)	10.4	–	–	–	48.0	–	–	–	–	[111]
14 species from India	10–150	–	–	–	6–480	–	–	–	–	[112]
11 species from Australia	3–97	–	–	–	22–195	–	–	–	–	[113]
Fish										
Salmon (*Oncorhynchus tshawytscha*)										
Raw Baked	66 139	–	–	–	60 96	–	–	55 60	0.34 0.46	[114]
Carp (*Cyprinus carpio carpio*)										

(continued)

Table 3 (continued)

Source	18:2n-6	18:3n-6	20:3n-6	20:4n-6	18:3n-3	18:4n-3	20:4n-3	20:5n-3	n-6/n-3 (ARA/EPA)	Reference
Raw Baked	335 501	–	–	–	85 130	–	–	58 50	2.24 2.57	[114]
White sucker (*Catostomus commersonii*)										
Raw Baked	33 242	–	–	–	13 116	–	–	92 104	0.26 0.70	[114]
Lake trout (*Salvelinus namaycush*)										
Raw Baked	268 472	–	–	–	263 385	–	–	322 374	0.39 0.45	[114]
Walleye (*Sander vitreus*)										
Raw Baked	24 82	–	–	–	48 76	–	–	47 52	0.32 0.46	[114]
Meat										
Pork										
Different diets	130–160	–	–	23.0–33.0	4.3–51.1	–	–	1.6–15.8	2.20–11.2 (1.5–20.6[a])	[115]
Different muscle tissues	189/422	8.08/13.8	6.92/10.1	46.0/59.7	22.7/41.3	–	–	6.4/5.3	4.8/7.0 (7.2/11.3[a])	[116]
Beef (muscle tissue)										
Raw Cooked	–	–	–	70 106	41 51	–	–	29 50	– (2.4[a]) – (2.1[a])	[117]
Veal (muscle tissue)										
Raw Cooked	–	–	–	55 101	19 39	–	–	28 52	– (2.0[a]) – (1.9[a])	[117]

Lamb (muscle tissue)										
Raw Cooked	–	–	–	93 111	64 165	–	–	27 50	– (3.4[a]) – (2.2[a])	[117]
Mutton (muscle tissue)										
Raw Cooked	–	–	–	97 131	100 234	–	–	46 88	– (2.1[a]) – (1.5[a])	[117]
Rabbit										
Raw Cooked	944.4 1254.9	–	–	28.8 39.4	134.3 183.5	–	–	2.9 3.9	6.37 (9.9[a]) 6.21 (10.0[a])	[118]
Broiler chicken	565	4.1	10.6	59.2	19.1	0.73	–	1.38	17.6 (42.9[a])	[119]

"–" not reported
[a]The value is calculated from the data given in the paper
[b]The fatty acid content is mostly contributed by cooking oil

from fronds of the tree fern *Cyathea dealbata*. The process recovered 88–93% of the major fatty acids (18:3n-3, 16:0, 18:2n-6, 18:1n-9) and was more effective for EPA (95% recovery) and less for ARA (81% recovery). The obtained extracts after dehydration contained ARA 6.1–6.9 mg g^{-1} and EPA 13.4–14.2 mg g^{-1} corresponding to tenfold enrichment for ARA and more than 15-fold enrichment for EPA. Although the aged fronds of the fern were used in the study, the process can be equally applied to fern fiddleheads setting up a scalable and green technology for the food industry.

9 Other Possible Applications of Fern Fatty Acids

Other applications for fern fatty acids are related to the aqueous ferns of the genus *Azolla*. In the earlier studies, *A. caroliniana* [48] and *A. filiculoides* [58] were analyzed for fatty acids as a potential animal feedstuff. Supplementation of a diet for the fish *Tilapia* with increasing content of the fern biomass led to a decrease in the fish weight. Thus, the author came to the conclusion that the fish diet can be replaced with *Azolla* biomass by no more than 20% [58].

In the last 10 years, the interest in *Azolla* fatty acids aroused as a starting material for biofuel production [45–47]. Biomass of *A. filiculoides* was found to contain 7.9% lipids on a dry weight basis with 41% of the lipid fraction being fatty acids [46]. Plants collected at the reproductive stage had a higher yield of total lipids than at the vegetative stage (11.2% vs. 8.2% DW) which was explained by the production of lipid-rich male microsporocarps [47]. In the study [46], the major fatty acids of *A. filiculoides* were 16:0, 18:2, and 18:3 (5–14 mg g^{-1} DW), and all other fatty acids were below 2 mg g^{-1}. Some different results were found for *A. filiculoides* and *A. pinnata* in the study [47] where the major fatty acid was 16:0 (about 7 mg g^{-1} DW) followed by 18:3 (3–4 mg g^{-1}) and 18:0 (1–2 mg g^{-1}) while all others were about 1 mg g^{-1} or lower. Based on the fatty acid composition, it was concluded that *Azolla* lipid compositions meet the most of the important requirements for biodiesel standards [46, 47] but an additional fractionation step is to be applied to remove 24:0 and mid-chain (di)hydroxyl compounds and, thus, to decrease the cold filter plugging point [46].

It should be noted that *Azolla* is considered a candidate for domestication which can be used as a crop to yield food, fuels, and chemicals on an industrial scale [87].

10 Conclusions

It is evident from this review that there is a vast massive of data has been accumulated on fern fatty acids to date. The data mostly include fatty acid composition and to lesser extent factors able to affect the fatty acid composition and content. These data provide good insights for researchers on the diversity of fatty acids in ferns, which fatty acids can be expected in different taxonomic groups, species enriched in a particular component and, thus, representing potential models for further studies or as sources of a particular component. However, despite enough big data and the fact that numbers of species have been studied for some fern families or genera, there are many other groups that are investigated insufficiently or not analyzed at all but are still important from the evolutionary point of view. To my opinion, the species of the subclass Ophioglossidae (the families Psilotaceae and Ophioglossaceae) remain insufficiently studied, although may represent a link between the subclass Equisetidae, so different in fatty acid composition, and other fern subclasses. Especially considering the fact that not all unsaturated fatty acids were identified even in the studied *Psilotum*. Only one species of the subclass Marattiidae have been analyzed so far. Even within the subclass Polypodiidae, some surprises are possible as can be seen from the available literature on *Salvinia natans* (family Salviniaceae) and the *Hymenophyllum* species (family Hymenophyllaceae), which are characterized by very specific sets of fatty acids. Such interesting differences may reflect or even determine the ecological specificity of the ferns and should be investigated in terms of the physiological role of fatty acids.

The physiological roles of fatty acids in ferns are out of the scope of this review. Being components of storage and membrane structural lipids, they undoubtedly have the same functions as they do in other plants or, wider, in all living cells. The interest is in the functions of the fatty acids specific for ferns, particularly, LC-PUFAs. Unfortunately, there are no experimental data on this subject, and some suggestions may be drawn only from indirect evidence obtained for ferns and the knowledge gained from studies of other plant objects which produce the same LC-PUFAs.

The elucidation of fern fatty acid functions is strongly hindered by insufficient information on their biosynthesis. Enzymes of fatty acid biosynthesis in ferns and genes encoding the enzymes are almost unexplored. Once again, the comparative studies of fatty acids in different taxonomic groups of ferns will be helpful for our understanding of fatty acid metabolic routes and promote the discovery of a potentially diverse set of enzymes involved in their biosynthesis. Taking into account the domination of the omega-6 fatty acids in young fronds, the enzymes responsible for their biosynthesis might be a subject of biotechnological research aiming at the production of arachidonic acid and its valuable metabolic precursors.

Food biotechnology may be interested in fern young fronds as a raw material containing LC-PUFAs. With the accumulated data on the fatty acid content in the fern species, differing in the ratios of the omega-6 and omega-3 PUFAs and levels of ARA and EPA, the industry gains a plant resource for vegetarian products with "meat" fatty acids.

Acknowledgments This work was financed by the Ministry of Science and Higher Education of the Russian Federation (Project No. 122040800086-1).

References

1. Christie WW (2022) Fatty acids and oxylipins. The LipidWeb. https://www.lipidmaps.org/resources/lipidweb/index.php?page=lipids/fa-eic.html. Accessed 16 June 2022

2. Leray C (2022) Fatty acids. Cyberlipid. http://cyberlipid.gerli.com/description/simple-lipids/fatty-acids/#3. Accessed 16 June 2022
3. Khotimchenko SV, Vaskovsky VE, Titlyanova TV (2002) Fatty acids of marine algae from the Pacific coast of North California. Bot Mar 45:17–22
4. van Ginneken VJT, Helsper JPFG, de Visser W, van Keulen H, Brandenburg WA (2011) Polyunsaturated fatty acids in various macroalgal species from north Atlantic and tropical seas. Lipids Health Dis 10:104. https://doi.org/10.1186/1476-511X-10-104
5. Martins DA, Custodio L, Barreira L, Pereira H, Ben-Hamadou R, Varela J, Abu-Salah KM (2013) Alternative sources of n-3 long-chain polyunsaturated fatty acids in marine microalgae. Mar Drugs 11:2259–2281. https://doi.org/10.3390/md11072259
6. Biancarosa I, Belghit I, Bruckner CG, Liland NS, Waagbo R, Amlund H, Heesch S, Lock EJ (2018) Chemical characterization of 21 species of marine macroalgae common in Norwegian waters: benefits of and limitations to their potential use in food and feed. J Sci Food Agric 98: 2035–2042. https://doi.org/10.1002/jsfa.8798
7. Dembitsky VM (1993) Lipids of bryophytes. Prog Lipid Res 32:281–356
8. Lu Y, Eiriksson FF, Thorsteinsdottir M, Simonsen HT (2019) Valuable fatty acids in bryophytes–production, biosynthesis, analysis and applications. Plan Theory 8:524. https://doi.org/10.3390/plants8110524
9. Wolff RL, Christie WW, Pedrono F, Marpeau AM (1999) Arachidonic, eicosapentaenoic, and biosynthetically related fatty acids in the seed lipids from a primitive gymnosperm, *Agathis robusta*. Lipids 34:1083–1097. https://doi.org/10.1007/s11745-999-0460-y
10. Wolff RL, Christie WW (2002) Structures, practical sources (gymnosperm seeds), gas-liquid chromatographic data (equivalent chain lengths), and mass spectrometric characteristics of all-*cis* Δ5-olefinic acids. Eur J Lipid Sci Technol 104:234–244
11. Tsevegsuren N, Aitzetmuller K (1997) Unusual Δ5*cis*-fatty acids in seed oils of *Cimicifuga* species. J High Res Chromat 20:237–241. https://doi.org/10.1002/jhrc.1240200411
12. von Aderkas P (1984) Economic history of ostrich fern, *Matteuccia struthiopteris*, the edible fiddlehead. Econ Bot 38:14–23
13. Crowe A (1997) A field guide to the native edible plants of New Zealand. Godwit Publishing, Auckland
14. Khrapko OV (2007) Far Eastern ferns: opportunities for application. Bulletin of the Botanical Garden Institute 1:81–87. http://botsad.ru/media/oldfiles/journal/number1/08.pdf. Accessed 16 May 2022 (In Russian)
15. Liu Y, Wujisguleng W, Long C (2012) Food uses of ferns in China: a review. Acta Soc Bot Pol 81:263–270. https://doi.org/10.5586/asbp.2012.046
16. Maroyi A (2014) Not just minor wild edible forest products: consumption of pteridophytes in sub-Saharan Africa. J Ethnobiol Ethnomed 10:78. https://doi.org/10.1186/1746-4269-10-78
17. Sujarwo W, Lugrayasa IN, Caneva G (2014) Ethnobotanical study of edible ferns used in Bali Indonesia. Asia Pac J Sustain Agric Food Energy 2:1–4
18. Koide D, Kadoya T (2019) Resource amount and cultural legacy affect spatially unbalanced human use of Japan's non-timber forest products. Ecol Indic 97:204–210. https://doi.org/10.1016/j.ecolind.2018.10.017
19. Dyall SC (2015) Long-chain omega-3 fatty acids and the brain: a review of the independent and shared effects of EPA, DPA and DHA. Front Aging Neurosci 7:52. https://doi.org/10.3389/fnagi.2015.00052
20. Bazinet RP, Metherel AH, Chen CT, Shaikh SR, Nadjar A, Joffre C, Laye S (2020) Brain eicosapentaenoic acid metabolism as a lead for novel therapeutics in major depression. Brain Behav Immun 85:21–28. https://doi.org/10.1016/j.bbi.2019.07.001
21. Harayama T, Shimizu T (2020) Roles of polyunsaturated fatty acids, from mediators to membranes. J Lipid Res 61:1150–1160. https://doi.org/10.1194/jlr.R120000800
22. Katz J (1898) Das fette Oel des Rhizoms von Aspidium filix mas. Arch Pharm 236:655–662. https://doi.org/10.1002/ardp.18982360903

23. Farup P (1904) Ueber die Zusammensetzung des fetten Oeles von Aspidium spinulosum. Arch Pharm 242:17–24. https://doi.org/10.1002/ardp.19042420103
24. Maizite J (1942) Über das Farnfett. Arch Pharm 280:132–156. https://doi.org/10.1002/ardp.19422800403
25. Gellerman JL, Schlenk H (1964) The occurrence of arachidonic and related acids in plants. Experientia 20:426–427. https://doi.org/10.1007/BF02152126
26. Schlenk H, Gellerman JL (1965) Arachidonic, 5,11,14,17-eicosatetraenoic and related acids in plants - identification of unsaturated fatty acids. J Am Oil Chem Soc 42:504–511. https://doi.org/10.1007/BF02540092
27. Radunz A (1967) Über die Lipide der Pteridophyten – I. Die Isolierung und Identifizierung der Polyensäuren. Phytochemistry 6:399–406
28. Jamieson GR, Reid EH (1975) The fatty acid composition of fern lipids. Phytochemistry 14: 2229–2232. https://doi.org/10.1016/S0031-9422(00)91104-X
29. Lytle TF, Lytle JS, Caruso A (1976) Hydrocarbons and fatty acids of ferns. Phytochemistry 15: 965–970. https://doi.org/10.1016/S0031-9422(00)84381-2
30. Guervin CM, Laur M-H (1969) Lipochimie vegetale. – Les acides gras des frondes sporiferes du *Scolopendrium vulgare* Sm. C R Acad Sci Paris 269:583–585
31. Robinson PM, Smith DL, Safford R, Nichols BW (1973) Lipid metabolism in the fern *Polypodium vulgare*. Phytochemistry 12:1377–1381. https://doi.org/10.1016/0031-9422(73)80569-2
32. Haigh WG, Safford R, James AT (1969) Fatty acid composition and biosynthesis in ferns. Biochim Biophys Acta 176:647–650
33. Gemmrich AR (1977) Fatty acid composition of fern spore lipids. Phytochemistry 16:1044–1046. https://doi.org/10.1016/S0031-9422(00)86721-7
34. Rozentsvet OA, Bogdanova ES (2009) Influence of ions of lead on growth, the contents and structure of lipids of the membranes of the *Matteuccia sthruthiopteris* (L.) Todari. Izvestia of Samara Scientific Center of the Russian Academy of Sciences 11:205–211. (In Russian)
35. Rozentsvet OA, Bogdanova ES, Makurina ON (2010) Influence of cadmium ions on qualitative and quantitative composition of lipids and pigments in the forming leaves of *Matteuccia sthruthiopteris*. Vestnik Samarskogo Universiteta Estestvenno-Nauchnaya Seriya 6:232–239. (In Russian)
36. Rozentsvet OA, Bogdanova ES, Murzaeva SV (2011) Composition of lipids and fatty acids in the forming leaves of the fern *Matteuccia struthiopteris* in the presence of cadmium. Transactions of KarRC RAS 3(97–104) http://transactions.krc.karelia.ru/publ.php?plang=e&id=8769. Accessed 15 May 2022 (In Russian)
37. Rozentsvet OA, Bogdanova ES, Golovko TK, Zakhozhiy IG (2013) Composition of lipids of the leaves of some representatives of Polypodiophyta in the Polar Urals. Khimija Rastitel'nogo Syr'ja [Chemistry of Plant Raw Material] 4:127–134. http://journal.asu.ru/cw/article/view/jcprm.1304127/147. Accessed 21 June 2022 (In Russian)
38. Dudareva LV, Nokhsorov VV, Rudikovskaya EG, Petrov KA (2015) Fatty-acid profiles of aerial parts of three horsetail species growing in central and northern Yakutia. Chem Nat Comp 51:220–223. https://doi.org/10.1007/s10600-015-1247-2
39. Voronkov AS, Ivanova TV (2021) Fatty acids composition of the epiphytic ferns, *Platycerium bifurcatum* and *Asplenium nidus*, and the terrestrial fern, *Asplenium trichomanes*. Am Fern J 111:117–128. https://doi.org/10.1640/0002-8444-111.2.117
40. Voronkov A, Ivanova T (2022) Significance of lipid fatty acid composition for resistance to winter conditions in *Asplenium scolopendrium*. Biology 11:507. https://doi.org/10.3390/biology11040507
41. Bushway AA, Wilson AM, McGann DF, Bushway RJ (1982) The nutrient composition of fresh fiddlehead greens. J Food Sci 47:666–667. https://doi.org/10.1111/j.1365-2621.1982.tb10147.x
42. DeLong JM, Hodges DM, Prange RK, Forney CF, Toivenon PMA, Bishop MC, Elliot ML, Jordan MA (2011) The unique fatty acid and antioxidant composition of ostrich fern

(*Matteuccia struthiopteris*) fiddleheads. Can J Plant Sci 91:919–930. https://doi.org/10.4141/cjps2010-042
43. Dvorakova M, Pumprova K, Antoninova Z, Rezek J, Haisel D, Ekrt L, Vanek T, Langhansova L (2021) Nutritional and antioxidant potential of fiddleheads from European ferns. Foods 10: 460. https://doi.org/10.3390/foods10020460
44. Nekrasov EV, Svetashev VI (2021) Edible Far Eastern ferns as a dietary source of long-chain polyunsaturated fatty acids. Foods 10:1220. https://doi.org/10.3390/foods10061220
45. Salehzadeh A, Naeemi AS, Arasteh A (2014) Biodiesel production from *Azolla filiculoides* (water fern). Trop J Pharm Res 13:957–960. https://doi.org/10.4314/tjpr.v13i6.19
46. Brouwer P, van der Werf A, Schluepmann H, Reichart GJ, Nierop KGJ (2016) Lipid yield and composition of *Azolla filiculoides* and the implications for biodiesel production. Bioenergy Res 9:369–377. https://doi.org/10.1007/s12155-015-9665-3
47. Miranda AF, Liu Z, Rochfort S, Mouradov A (2018) Lipid production in aquatic plant *Azolla* at vegetative and reproductive stages and in response to abiotic stress. Plant Physiol Biochem 124:117–125. https://doi.org/10.1016/j.plaphy.2018.01.012
48. Paoletti C, Bocci F, Lercker G, Capella P, Materassi R (1987) Lipid composition of *Azolla caroliniana* biomass and its seasonal variation. Phytochemistry 26:1045–1047. https://doi.org/10.1016/S0031-9422(00)82347-X
49. Nekrasov EV, Tallon SJ, Vyssotski MV, Catchpole OJ (2021) Extraction of lipids from New Zealand fern fronds using near-critical dimethyl ether and dimethyl ether–water–ethanol mixtures. J Supercrit Fluids 170:105137. https://doi.org/10.1016/j.supfu.2020.105137
50. Pteridophyte Phylogeny Group I (2016) A community-derived classification for extant lycophytes and ferns. J Syst Evol 54:563–603. https://doi.org/10.1111/jse.12229
51. Rezanka T (1998) Branched and very long-chain dicarboxylic acids from *Equisetum* species. Phytochemistry 47:1539–1543. https://doi.org/10.1016/S0031-9422(97)00774-7
52. Akihisa T, Kawashima T, Takahashi S, Sahashi N, Okamoto T, Niiya I, Tamura T (1992) Sterols and fatty acids of a whisk fern *Psilotum nudum*. J Am Oil Chem Soc 69:1232–1235. https://doi.org/10.1007/BF02637687
53. Nekrasov EV, Svetashev VI, Khrapko OV, Vyssotski MV (2019) Variability of fatty acid profiles in ferns: relation to fern taxonomy and seasonal development. Phytochemistry 162: 47–55. https://doi.org/10.1016/j.phytochem.2019.02.015
54. Rabert C, Inostroza K, Bravo S, Sepulveda N, Bravo LA (2020) Exploratory study of fatty acid profile in two filmy ferns with contrasting desiccation tolerance reveal the production of very long chain polyunsaturated omega-3 fatty acids. Plants 9:1431. https://doi.org/10.3390/plants9111431
55. Gemmrich AR (1977) Mobilization of reserve lipids in germinating spores of the fern *Anemia phyllitidis* L. Plant Sci Let 9:301–307. https://doi.org/10.1016/0304-4211(77)90100-6
56. Gemmrich AR (1982) Effect of red light and gibberellic acid on lipid metabolism in germinating spores of the fern *Anemia phyllitidis*. Physiol Plant 54:58–62. https://doi.org/10.1111/j.1399-3054.1982.tb00577.x
57. Nierop KGJ, Brouwer P, Dekker R, Schluepmann H, Reichart GJ (2018) ω20-Hydroxy and ω9,ω10-dihydroxy biomarker lipids in ferns from the Salviniaceae family. Org Geochem 125: 229–242. https://doi.org/10.1016/j.orggeochem.2018.09.014
58. Shiomi N, Kitoh S (2001) Culture of *Azolla* in a pond, nutrient composition, and use as fish feed. Soil Sci Plant Nutr 47:27–34. https://doi.org/10.1080/00380768.2001.10408365
59. Speelman EN, Reichart G-J, de Leeuw JW, Rijpstra WIC, Damste JSS (2009) Biomarker lipids of the freshwater fern *Azolla* and its fossil counterpart from the Eocene Arctic Ocean. Org Geochem 40:628–637. https://doi.org/10.1016/j.orggeochem.2009.02.001
60. Rozentsvet OA, Rezanka T, Bosenko ES, Uzhametskaya EA, Dembitskii VM (2005) Fatty acids, phospholipids, and the betaine lipid DGTS from the aquatic fern *Salvinia natans*. Chem Nat Compd 41:487–490. https://doi.org/10.1007/s10600-005-0189-5
61. Franich RA, Goodin SJ, Hansen E (1985) Wax esters of the New Zealand silver fern, *Cyathea dealbata*. Phytochemistry 24:1093–1095. https://doi.org/10.1016/S0031-9422(00)83191-X

62. Sato N, Furuya M (1984) The composition of lipids and fatty adds determined at various stages of haploid and diploid generations in the fern *Adiantum capillus-veneris*. Physiol Plant 62: 139–147. https://doi.org/10.1111/j.1399-3054.1984.tb00361.x
63. Etenko OA, Ivankina NF, Anikina EA (2005) Fatty acid composition of bracken fern fronds from the southern and northern zones of Amur Oblast. In: Nekrasov EV (ed) Botanical investigation in Priamurie and borderline territory, proceedings of the region conference, AB BGI FEB RAS, Blagoveshchensk. (In Russian)
64. Rozentsvet OA, Bogdanova ES, Tabalenkova GN, Golovko TK, Zakhozhiy IG (2011) Ecological and biochemical properties of some typical representatives of cryptogramic plants of Southern Timan. Povolzhskiy J Ecol 10:489–498. (In Russian)
65. Gellerman JL, Anderson WH, Schlenk H (1972) Highly unsaturated lipids of *Mnium*, *Polytrichum*, *Marchantia*, and *Matteuccia*. Bryologist 75:550–557
66. Nekrasov EV, Shelikhan LA, Svetashev VI (2019) Fatty acid composition of gametophytes of *Matteuccia struthiopteris* (L.) Tod. (Onocleaceae, Polypodiophyta). Botanica Pacifica 8:63–66. https://doi.org/10.17581/bp.2019.08104
67. John SP, Hasenstein KH (2018) Biochemical responses of the desiccation-tolerant resurrection fern *Pleopeltis polypodioides* to dehydration and rehydration. J Plant Physiol 228:12–18. https://doi.org/10.1016/j.jplph.2018.05.006
68. John SP, Hasenstein KH (2020) Desiccation mitigates heat stress in the resurrection fern *Pleopeltis polypodioides*. Front Plant Sci 11:597731. https://doi.org/10.3389/fpls.2020.597731
69. WFO (2022) World Flora Online. http://www.worldfloraonline.org. Accessed 16 June 2022
70. Mongrand S, Bessoule J-J, Cabantous F, Cassagne C (1998) The $C_{16:3}/C_{18:3}$ fatty acid balance in photosynthetic tissues from 468 plant species. Phytochemistry 49:1049–1064. https://doi.org/10.1016/S0031-9422(98)00243-X
71. Haufler CH, Pryer KM, Schuettpelz E, Sessa EB, Farrar DR, Moran R, Schneller JJ, Watkins JE, Windham MD (2016) Sex and the single gametophyte: revising the homosporous vascular plant life cycle in light of contemporary research. Bioscience 66:928–937. https://doi.org/10.1093/biosci/biw108
72. Brownsey PJ, Smith-Dodsworth JC (2000) New Zealand ferns and allied plants. David Bateman, Auckland
73. Gunstone FD, Harwood JL (2007) Occurrence and characterisation of oils and fats. In: Gunstone FD, Harwood JL, Dijkstra AJ (eds) The lipid handbook with CD-ROM, 3rd edn. CRC Press Taylor & Francis Group, Boca Raton
74. Auling G, Heinz E, Tulloch AP (1971) Combination and positional distribution of fatty acids in plant galactolipids. Hoppe Seylers Z Physiol Chem 352:905–912
75. Nichols BW (1965) The lipids of a moss (*Hypnum cupressiforme*) and of the leaves of green holly (*Ilex aquifolium*). Phytochemistry 4:769–779. https://doi.org/10.1016/S0031-9422(00)86248-2
76. Tulloch AP, Heinz E, Fischer W (1973) Combination and positional distribution of fatty acids in plant sulfolipids. Hoppe Seylers Z Physiol Chem 354:879–889
77. Nekrasov EV, Svetashev VI (2019) Distribution of sciadonic and juniperonic acids in lipid classes in shoots of the horsetail *Equisetum arvense* during the vegetative season. IX Congress of society physiologists of plants of Russia, Kazan, 18–24 Sept 2019. https://congresskazan2019.ofr.su/viewthesis?id=605. Accessed 18 June 2022 (In Russian)
78. Boudière L, Michaud M, Petroutsos D, Rébeillé F, Falconet D, Bastien O, Roy S, Finazzi G, Rolland N, Jouhet J, Block MA, Maréchal E (2014) Glycerolipids in photosynthesis: composition, synthesis and trafficking. Biochim Biophys Acta 1837:470–480. https://doi.org/10.1016/j.bbabio.2013.09.007
79. Hernández ML, Cejudo FJ (2021) Chloroplast lipids metabolism and function. A redox perspective. Front Plant Sci 12. https://doi.org/10.3389/fpls.2021.712022
80. Ohlrogge JB, Jaworski JG (1997) Regulation of fatty acid synthesis. Ann Rev Plant Physiol Plant Mol Biol 48:109–136. https://doi.org/10.1146/annurev.arplant.48.1.109

81. Brown AP, Slabas AR, Rafferty JB (2009) Fatty acid biosynthesis in plants – metabolic pathways, structure and organization. In: Wada H, Murata N (eds) Lipids in photosynthesis: essential and regulatory functions. Springer, Dordrecht. https://doi.org/10.1007/978-90-481-2863-1_2
82. Troncoso-Ponce MA, Nikovics K, Marchive C, Lepiniec L, Baud S (2016) New insights on the organization and regulation of the fatty acid biosynthetic network in the model higher plant *Arabidopsis thaliana*. Biochimie 120:3–8. https://doi.org/10.1016/j.biochi.2015.05.013
83. He M, Qin C-X, Wang X, Ding N-Z (2020) Plant unsaturated fatty acids: biosynthesis and regulation. Front Plant Sci 11:390. https://doi.org/10.3389/fpls.2020.00390
84. Batsale M, Bahammou D, Fouillen L, Mongrand S, Joubès J, Domergue F (2021) Biosynthesis and functions of very-long-chain fatty acids in the responses of plants to abiotic and biotic stresses. Cell 10:1284. https://doi.org/10.3390/cells10061284
85. Kazaz S, Miray R, Lepiniec L, Baud S (2022) Plant monounsaturated fatty acids: diversity, biosynthesis, functions and uses. Prog Lipid Res 85:101138. https://doi.org/10.1016/j.plipres.2021.101138
86. Zhukov A, Popov V (2022) Synthesis of C_{20-38} fatty acids in plant tissues. Int J Mol Sci 23: 4731. https://doi.org/10.3390/ijms23094731
87. Brouwer P, Brautigam A, Kulahoglu C, Tazelaar AOE, Kurz S, Nierop KGJ, van der Werf A, Weber APM, Schluepmann H (2014) *Azolla* domestication towards a biobased economy? New Phytol 202:1069–1082. https://doi.org/10.1111/nph.12708
88. Kaewsuwan S, Cahoon EB, Perroud P-F, Wiwat C, Panvisavas N, Quatrano RS, Cove DJ, Bunyapraphatsara N (2006) Identification and functional characterization of the moss *Physcomitrella patens* Δ^5-desaturase gene involved in arachidonic and eicosapentaenoic acids biosynthesis. J Biol Chem 281:21988–21997. https://doi.org/10.1074/jbc.M603022200
89. Kajikawa M, Matsui K, Ochiai M, Tanaka Y, Kita Y, Ishimoto M, Kohzu Y, Shoji S-i, Yamato KT, Ohyama K, Fukuzawa H, Kohchi T (2008) Production of arachidonic and eicosapentaenoic acids in plants using bryophyte fatty acid Δ6-desaturase, Δ6-elongase, and Δ5-desaturase genes. Biosci Biotechnol Biochem 72:435–444. https://doi.org/10.1271/bbb.70549
90. Meesapyodsuk D, Qiu X (2012) The front-end desaturase: structure, function, evolution and biotechnological use. Lipids 47:227–237. https://doi.org/10.1007/s11745-011-3617-2
91. Khozin-Goldberg I, Leu S, Boussiba S (2016) Microalgae as a source for VLC-PUFA production. In: Nakamura Y, Li-Beisson Y (eds) Lipids in plant and algae development, subcellular biochemistry 86. Springer, Cham. https://doi.org/10.1007/978-3-319-25979-6_19
92. Lee JM, Lee H, Kang SB, Park WJ (2016) Fatty acid desaturases, polyunsaturated fatty acid regulation, and biotechnological advances. Nutrients 8:23. https://doi.org/10.3390/nu8010023
93. Sayanova O, Haslam R, Caleron MV, Napier JA (2007) Cloning and characterization of unusual fatty acid desaturases from *Anemone leveillei*: identification of an acyl-coenzyme A C_{20} Δ^5-desaturase responsible for the synthesis of sciadonic acid. Plant Physiol 144:455–467. https://doi.org/10.1104/pp.107.098202
94. Wu J, Huang J, Hong Y, Zhang H, Ding M, Lou H, Hu Y, Yu W, Song L (2018) *De novo* transcriptome sequencing of *Torreya grandis* reveals gene regulation in sciadonic acid biosynthesis pathway. Ind Crop Prod 120:47–60. https://doi.org/10.1016/j.indcrop.2018.04.041
95. DeLong JM, Prange RK (2008) Fiddlehead fronds: nutrient rich delicacy. Chron Hortic 48:12–15
96. Gerasimenko N, Korol A, Pikhanova S, Gochachko S (2003) Marketing evaluation of non-wood products of Far Eastern south forests. Practical Marketing 3:17–25. https://www.cfin.ru/press/practical/2003-04/04.shtml Accessed 20 June 2022 (In Russian)
97. Chamberlain JL, Emery MR, Patel-Weynand T (2018) Assessment of nontimber forest products in the United States under changing conditions. USDA Forest Service, Southern Research Station. https://doi.org/10.2737/SRS-GTR-232
98. Government of Amur Oblast (2021). https://economy.amurobl.ru/posts/print.php?id=100152. Accessed 08 June 2022 (In Russian)

99. Haeggstrom JZ, Funk CD (2011) Lipoxygenase and leukotriene pathways: biochemistry, biology, and roles in disease. Chem Rev 111:5866–5898. https://doi.org/10.1021/cr200246d
100. Serhan CN, Petasis NA (2011) Resolvins and protectins in inflammation resolution. Chem Rev 111:5922–5943. https://doi.org/10.1021/cr100396c
101. Burdge GC (2018) Polyunsaturated fatty acid biosynthesis and metabolism in adult mammals. In: Burdge GC (ed) Polyunsaturated fatty acid metabolism. AOCS Press. https://doi.org/10.1016/B978-0-12-811230-4.00002-8
102. Kawashima H (2019) Intake of arachidonic acid-containing lipids in adult humans: dietary surveys and clinical trials. Lipids Health Dis 18:101. https://doi.org/10.1186/s12944-019-1039-y
103. Deckelbaum RJ, Calder PC (2020) Editorial: is it time to separate EPA from DHA when using omega-3 fatty acids to protect heart and brain? Curr Opin Clin Nutr Met Care 23:65–67. https://doi.org/10.1097/MCO.0000000000000632
104. Thompson M, Hein N, Hanson C, Smith LM, Anderson-Berry A, Richter CK, Stessy Bisselou K, Kusi Appiah A, Kris-Etherton P, Skulas-Ray AC, Nordgren TM (2019) Omega-3 fatty acid intake by age, gender, and pregnancy status in the United States: national health and nutrition examination survey 2003–2014. Nutrients 11:177. https://doi.org/10.3390/nu11010177
105. Bushway AA, Serreze DV, McGann DF, True RH, Work TM, Bushway RJ (1985) Effect of processing method and storage time on the nutrient composition of fiddlehead greens. J Food Sci 50:1491–1492. https://doi.org/10.1111/j.1365-2621.1985.tb10508.x
106. DeLong JM, Hodges DM, Prange RK, Forney CF, Fan L, Bishop MC, Elliot ML, Jordan MA, Doucette C (2013) The influence of cold water storage on fatty acids, antioxidant content and activity, and microbial load in ostrich fern (*Matteuccia struthiopteris*) fiddleheads. Can J Plant Sci 93:683–697. https://doi.org/10.4141/CJPS2012-165
107. Shalisko IV, Dmitrichenko MI, Pelenko VV, Shakhov SV (2016) Changes in consumer properties of bracken using different storage methods. Proceed Voronezh State Univ Eng Technol 3:151–158. https://doi.org/10.20914/2310-1202-2016-3-151-158. (In Russian)
108. Znidarcic D, Vidrih R (2009) Content of essential fatty acids in cabbage (*Brassica oleracea* L.). Acta Agric Slov 93:225–230. https://www.dlib.si/details/URN:NBN:SI:doc-RT6HT80K. Accessed 19 June 2022 (In Slovenian)
109. Boulom S, Robertson J, Hamid N, Ma Q, Lu J (2014) Seasonal changes in lipid, fatty acid, α-tocopherol and phytosterol contents of seaweed, *Undaria pinnatifida*, in the Marlborough Sounds, New Zealand. Food Chem 161:261–269. https://doi.org/10.1016/j.foodchem.2014.04.007
110. Gladyshev MI, Sushchik NN, Gubanenko GA, Demirchieva SM, Kalachova GS (2007) Effect of boiling and frying on the content of essential polyunsaturated fatty acids in muscle tissue of four fish species. Food Chem 101:1694–1700. https://doi.org/10.1016/j.foodchem.2006.04.029
111. Simopoulos AP, Norman HA, Gillaspy JE, Duke JA (1992) Common purslane: a source of omega-3 fatty acids and antioxidants. J Am Coll Nutr 11:374–382. https://doi.org/10.1080/07315724.1992.10718240
112. Ghafoorunissa PJ (1993) Vegetables as sources of α-linolenic acid in Indian diets. Food Chem 47:121–124. https://doi.org/10.1016/0308-8146(93)90231-4
113. Pereira C, Li D, Sinclair AJ (2001) The α-linolenic acid content of green vegetables commonly available in Australia. Int J Vitam Nutr Res 71:223–228. https://doi.org/10.1024/0300-9831.71.4.223
114. Neff MR, Bhavsar SP, Braekevelt E, Arts MT (2014) Effects of different cooking methods on fatty acid profiles in four fresh water fishes from the Laurentian Great Lakes region. Food Chem 164:544–550. https://doi.org/10.1016/j.foodchem.2014.04.104
115. Haak L, Sioen I, Raes K, Van Camp J, De Smet S (2007) Effect of pan-frying in different culinary fats on the fatty acid profile of pork. Food Chem 102:857–864. https://doi.org/10.1016/j.foodchem.2006.06.054

116. Dugan ME, Vahmani P, Turner TD, Mapiye C, Juarez M, Prieto N, Beaulieu AD, Zijlstra RT, Patience JF, Aalhus JL (2015) Pork as a source of omega-3 (n-3) fatty acids. J Clin Med 4: 1999–2011. https://doi.org/10.3390/jcm4121956
117. Droulez V, William PG, Levy G, Stobaus T, Sinclair A (2006) Composition of Australian red meat 2002. 2. Fatty acid profile. Food Australia 58:335–341. https://ro.uow.edu.au/hbspapers/1/. Accessed 19 June 2022
118. Kouba M, Benatmane F, Blochet JE, Mourot J (2008) Effect of a linseed diet on lipid oxidation, fatty acid composition of muscle, perirenal fat, and raw and cooked rabbit meat. Meat Sci 80:829–834. https://doi.org/10.1016/j.meatsci.2008.03.029
119. de Souza VJ, Alvarenga TIRC, Andrew NR, McPhee M, Kolakshyapati M, Hopkins DL, Ruhnke I (2021) Technological quality, amino acid and fatty acid profile of broiler meat enhanced by dietary inclusion of black soldier fly larvae. Foods 10:297. https://doi.org/10.3390/foods10020297
120. Larsen D, Quek SY, Eyres L (2010) Effect of cooking method on the fatty acid profile of New Zealand King Salmon (*Oncorhynchus tshawytscha*). Food Chem 119:785–790. https://doi.org/10.1016/j.foodchem.2009.07.037

Ferns and Lycophytes with Insecticidal Activity: An Overview

12

Gabriela Pereira Lima, Jamilly Bignon de Souza,
Selma Ribeiro Paiva, and Marcelo Guerra Santos

Contents

Abstract

Despite their economic and ecological importance, some insect species are human disease vectors. In addition, agricultural pests cause significant losses in a large number of important crops worldwide. Insect control has traditionally used synthetic insecticides, but their indiscriminate use has damaged the environment and compromised human health. Thus, the search for plant-based insecticides has prompted an increase in studies. Seedless vascular plants are included in two different lineages: ferns and lycophytes. Several biological activities, including insecticidal, are attributed to the extracts/compounds of these plant groups. The

G. P. Lima · J. B. de Souza
Programa de Pós-Graduação em Ciências Aplicadas a Produtos para a Saúde, Universidade Federal Fluminense, Niterói, Brazil
e-mail: gabrielapl@id.uff.br; jamillybignon@id.uff.br

S. R. Paiva
Programa de Pós-Graduação em Ciências Aplicadas a Produtos para a Saúde, Universidade Federal Fluminense, Niterói, Brazil

Laboratório de Botânica Estrutural e Funcional, Departamento de Biologia Geral, Instituto de Biologia, Universidade Federal Fluminense, Niterói, Brazil
e-mail: selmapaiva@id.uff.br

M. G. Santos (✉)
Laboratório de Biodiversidade, Departamento de Ciências, Faculdade de Formação de Professores, Universidade do Estado do Rio de Janeiro, São Gonçalo, Brazil
e-mail: marceloguerrasantos@gmail.com

H. N. Murthy (ed.), *Bioactive Compounds in Bryophytes and Pteridophytes*,
Reference Series in Phytochemistry, https://doi.org/10.1007/978-3-031-23243-5_13

aim of the present study was to analyze the literature on the insecticidal potential of fern and lycophyte extracts. The review was conducted on the PubMed, ScienceDirect, Scielo, and Web of Science databases. No time or language restrictions were established, and the search was concluded in December 2021. A total of 43 studies were found between 1994 and 2020. Of the 80 species assessed, 47 fern and 2 lycophyte species exhibited important medical or agricultural insecticidal, repellent, or insect growth-regulating activity. The insects most widely used as experimental models were *Spodoptera litura* and *Helicoverpa armigera* (Lepidoptera) and *Aedes aegypti*, *Aedes albopictus,* and *Anopheles stephensi* (Diptera). Most of the insect species (71.8%) assessed exhibited holometabolous development, with the highest percentage of studies performed with insect pests (69.8%). Tests normally involved adult insects or in the larval stage, contact tests being the most frequently applied. The most widely used extracts were aqueous and ethanolic, demonstrating better insecticidal activity for the species tested. Despite the scarcity of studies, it is believed that these plants show potential as a source of substances with insecticidal and repellent activity.

Keywords

Bioinsecticides · Botanical insecticides · Insects · Plant extracts · Pteridophytes

Abbreviations

AE	Acetonic Extract
EMBRAPA	Brazilian Agricultural Research Corporation
EtOAcE	Ethyl Acetate Extract
EtOHE	Ethanolic Extract
IGR	Insect Growth Regulator
LC_{50}	Average lethal concentration
PPG I	Pteridophyte Phylogeny Group

1 Introduction

The most numerous and diversified animals in nature, insects, exhibit wide geographic distribution and adaptations related to different habitats and feeding habits [1]. Although insects are beneficial to humans and the ecosystem in general, some species play an important role in pathogen transmission [2, 3]. In addition, many insects are agricultural pests that attack a variety of important crops and stored grains, causing significant losses [4].

The most common insect control strategy is the use of synthetic insecticides [5]. However, the indiscriminate use of these substances has caused concern due to the development of insect populations resistant to insecticides and the potential adverse effects on the environment and human health [6]. Thus, plant-based products are natural alternatives to synthetic insecticides for reducing their negative effects [7]. Plant-derived substances can act in different ways and are effective as insecticides [8], ovicides and larvicides [9], food inhibitors [10], insect growth regulators (IGRs) [11], and repellents [12].

Seedless vascular plants (pteridophytes) are included in two monophyletic lineages: ferns and lycophytes [13]. They are found worldwide [14], and it is estimated that there are approximately 11,196 species in 337 genera and 51 families, including 10,578 ferns and 1338 lycophyte species [15]. Several biological activities are attributed to ferns and lycophytes, such as antibiotics, antioxidants, analgesics, anti-inflammatory, sedatives, and anticonvulsants [16–19]. In addition, substances with insecticidal and/or repellent potential have been described in the chemical profile assessment of these plant groups [20, 21]. As such, the primary aim of the present study was to analyze the scientific literature on the insecticidal potential of ferns and lycophytes.

2 Methods

The review was conducted on the PubMed, ScienceDirect, Scielo, and Web of Science databases. The following keywords were applied, in Portuguese and English, with different combinations: pteridophytes, ferns, lycophytes, biological activities, and insecticidal activity. No specific time or language restrictions were established, and the database search was concluded in December 2021. Inclusion criteria were studies on the insecticide and/or repellent activity of ferns and lycophytes. After duplicates were excluded, the eligible articles were read in their entirety and the information extracted was stored in Microsoft Excel® spreadsheets (version 2019) for subsequent analysis. The nomenclature of fern and lycophyte species was updated using the Pteridophyte Collections Consortium (https://www.pteridoportal.org/portal/index.php), Word Ferns. Synonymic Checklist and Distribution of Ferns and Lycophytes of the World (https://www.worldplants.de/ferns/), Flora e Funga do Brasil (http://floradobrasil.jbrj.gov.br), and the PPG I classification system [15]. Insect species nomenclature was updated using the Illinois Natural History Survey: Insect Collection and Natural History Museum at South Kensington databases (https://insect.inhs.illinois.edu/).

3 Results and Discussion

A total of 41 original articles were found, published between 1994 and 2020, in addition to a master's dissertation and doctoral thesis published in 2011 and 2012, respectively, on the insecticidal action of ferns and lycophytes. A total of 75 fern species belonging to 26 families and 5 lycophyte species belonging to two families (Lycopodiaceae and Selaginellaceae) were studied for their insecticidal and/or repellent properties. Of the 80 fern and lycophyte species studied, 31 did not obtain significant results for the insects tested, corresponding to 39% (Tables 1 and 2).

The families with the highest proportion of assessed species were Osmundaceae (22.2%), Equisetaceae, and Onocleaceae (both with 20% each). The families with the lowest proportion of assessed species were Dryopteridaceae (0.1%),

Table 1 Families and number of fern and lycophyte species assessed for insecticidal and/or repellent activity and the number of ferns and lycophytes according to PPG I [15]

Family	N° of species in the world	N° of species assessed	The proportion of species assessed (%)	N° of species with activity	The proportion of species assessed with activity (%)
Ferns					
Anemiaceae	115	1	0.9	1	100
Aspleniaceae	730	5	0.7	4	80
Athyriaceae	650	3	0.5	2	67
Blechnaceae	265	5	1.9	2	40
Cibotiaceae	9	1	11	0	0
Cyatheaceae	643	5	0.8	4	80
Cystopteridaceae	37	2	5.4	1	50
Davalliaceae	65	1	1.5	1	100
Dennstaedtiaceae	265	1	0.4	1	100
Dicksoniaceae	35	1	2.9	1	100
Diplaziopsidaceae	4	1	25	1	100
Dryopteridaceae	2115	2	0.1	2	100
Equisetaceae	15	3	20	1	33
Gleicheniaceae	157	3	1.9	2	67
Hymenophyllaceae	434	1	0.2	1	100
Lygodiaceae	40	3	7.5	0	0
Metaxyaceae	6	1	16.7	0	0
Nephrolepidaceae	19	3	15.8	2	67
Onocleaceae	5	1	20	1	100
Ophioglossaceae	112	1	0.9	0	0
Osmundaceae	18	4	22.2	2	50
Polypodiaceae	1652	5	0.3	3	60
Pteridaceae	1211	11	0.9	7	64
Salviniaceae	21	2	9.5	1	50
Tectariaceae	250	1	0.4	1	100
Thelypteridaceae	1034	8	0.8	6	75
Lycophytes					
Lycopodiaceae	388	3	0.8	1	33
Selaginellaceae	700	2	0.3	1	50

Selaginellaceae (0.3%), Hymenophyllaceae (0.2%), and Polypodiaceae (0.3%). Dryopteridaceae, Polypodiaceae, Pteridaceae, and Thelypteridaceae are the fern families with the greatest species richness in the world [15] (Table 1). However, they belong to the group of 17 families with less than 5% of the species assessed, considering ferns and lycophytes, which represent 63% of the families studied.

Among the families with more than 2 species, Aspleniaceae, Cyatheaceae (both with 80%), and Thelypteridaceae (75%), contained the highest proportion of species with insecticidal activity. The Lygodiaceae, Metaxyaceae, and Ophioglossaceae

Table 2 Fern and lycophyte species described in the literature, tested for insecticidal and repellent activity

Family and species	Extract/plant part used	Insect tested/order	Insect stage	Type of test	Result/lethal dose	References
Anemiaceae						
Anemia tomentosa var. *anthriscifolia* (Schrad.) Mickel	Essential oils /not informed	*Aedes aegypti* Linnaeus, 1762 /Diptera	Adult	Contact (Repellency)	Repellent for 60 min (90%)	[38]
Aspleniaceae						
Asplenium aethiopicum (Burm. f.) Bech.	Acetonic, chloroformic, ethereal, and methanolic/ entire plant	*Culex quinquefasciatus* Say, 1823/Diptera	Larva/ fourth instar	Contact	LC_{50}(AE) = 166 ppm	[49]
	Aqueous/leaves	*Helopeltis theivora* Waterhouse, 1886/ Hemiptera	Adult	Contact	No significant results	[50]
Asplenium griffithianum Hook.	Methanolic/root	*Musca domestica* Linnaeus, 1758, and *Aedes albopictus* Skuse, 1894/Diptera	Adult and larva/ fourth instar	Ingestion	50% mortality (10,000 mg L^{-1}) for *M. domestica*	[22]
Asplenium nidus L.	Methanolic/entire plant	*Musca domestica* Linnaeus, 1758, and *Aedes albopictus* Skuse, 1894/Diptera	Adult and larva/ fourth instar	Ingestion	No significant results	[22]
Asplenium normale D. Don	Methanolic/entire plant	*Musca domestica* Linnaeus, 1758, and *Aedes albopictus* Skuse, 1894/Diptera	Adult and larva/ fourth instar	Ingestion	96.7% mortality (10,000 mg L^{-1}) for *M. domestica*	[22]
Asplenium platyneuron (L.)	Proteic (HEPES buffer)/ entire plant	*Helicoverpa zea* Boddie, 1850, and	Larva/ second instar	Ingestion	100% mortality for *H. zea* (0.6–2.1 μg μL^{-1}) and *S. frugiperda* (2 μg μL^{-1})	[51]

(continued)

Table 2 (continued)

Family and species	Extract/plant part used	Insect tested/order	Insect stage	Type of test	Result/lethal dose	References
Britton, Sterns, & Pogg.		*Spodoptera frugiperda* Smith, 1808/Lepidoptera				
Athyriaceae						
Athyrium filix-femina (L.) Roth	Proteic (HEPES buffer)/ entire plant	*Helicoverpa zea* Boddie, 1850 and *Spodoptera frugiperda* Smith, 1808, / Lepidoptera	Larva/ second instar	Ingestion	No significant results	[51]
	Aqueous/entire plant	*Leptinotarsa decemlineata* Say, 1824/Coleoptera	Larva/first and second instar	Contact	96.1% mortality ($25\ g\ L^{-1}$)	[52]
	Aqueous/entire plant	*Hyalopterus pruni* Geoffrey, 1762/Hemiptera	Adult	Contact	No significant results	[53]
Diplazium esculentum (Retz.) Sw.	Methanolic /leaves	*Culex* Linnaeus, 1758/ Diptera	Larva/third and fourth instar	Contact	LC_{50} of 149.279 ppm	[54]
Diplazium polypodioides Blume	Aqueous/leaves	*Helopeltis theivora* Waterhouse, 1886 / Hemiptera	Adult	Contact	No significant results	[50]
Homalosorus pycnocarpus (Spreng.) Pic.Sem.	Proteic (HEPES buffer)/ entire plant	*Helicoverpa zea* Boddie, 1850, and *Spodoptera frugiperda* Smith, 1808/ Lepidoptera	Larva/ second instar	Ingestion	100% mortality ($2.1\ \mu g\ \mu L^{-1}$) for *H. zea* and 70% ($2\ \mu g\ \mu L^{-1}$) for *S. frugiperda*	[51]
Blechnaceae						
	Methanolic/stem			Ingestion		[22]

Blechnopis orientalis (L.) C. Presl		*Musca domestica* Linnaeus, 1758, and *Aedes albopictus* Skuse, 1894/Diptera	Adult and larva/ fourth instar		100% mortality (10,000 mg L^{-1}) for *M. domestica*	
Lorinseria areolata (L.) C.Presl	Proteic (HEPES buffer)/ entire plant	*Helicoverpa zea* Boddie, 1850 and *Spodoptera frugiperda* Smith, 1808, / Lepidoptera	Larva/ second instar	Ingestion	No significant results	[51]
Parablechnum cordatum (Desv.) Gasper & Salino	Hexanic and in ethyl acetate/ leaves	*Drosophila melanogaster* Meigen, 1830/Diptera	Larva/first instar	Contact	66.7% mortality (EtOAcE = 800 ppm) and premature pupation (500 and 200 ppm) dead adults exhibited malformations	[42]
Woodwardia japonica (L. f.) Sm.	Methanolic/roots and leaves	*Musca domestica* Linnaeus, 1758, and *Aedes albopictus* Skuse, 1894/Diptera	Adult and larva/ fourth instar	Ingestion	No significant results	[22]
Woodwardia martinezii Maxon ex Weath	Methanolic/entire plant	*Musca domestica* Linnaeus, 1758, and *Aedes albopictus* Skuse, 1894/Diptera	Adult and larva/ fourth instar	Ingestion	No significant results	[22]
Cibotiaceae						
Cibotium barometz (L.) J. Sm.	Methanolic/stem and leaves	*Musca domestica* Linnaeus, 1758, and *Aedes albopictus* Skuse, 1894/Diptera	Adult and larva/ fourth instar	Ingestion	No significant results	[22]
Cyatheaceae						
Alsophila crinita Hook.	Ethereal, chloroformic, acetonic, and ethanolic/not informed	*Culex quinquefasciatus* Say, 1823/Diptera	Larva/ fourth instar	Contact	LC_{50} (EtOHE) = 320.7 μg ml^{-1}	[55]

(continued)

Table 2 (continued)

Family and species	Extract/plant part used	Insect tested/order	Insect stage	Type of test	Result/lethal dose	References
Alsophila gigantea Wall. ex Hook	Ethereal, chloroformic, acetonic, and ethanolic/not informed	*Culex quinquefasciatus* Say, 1823/Diptera	Larva/ fourth instar	Contact	LC_{50} (EtOHE) = 361.07 μg ml^{-1}	[55]
Alsophila nilgirensis (Holttum) R.M.Tryon	Ethereal, chloroformic, acetonic, and ethanolic/not informed	*Culex quinquefasciatus* Say, 1823/Diptera	Larva/ fourth instar	Contact	LC_{50} (EtOHE) = 373.99 μg ml^{-1}	[55]
Alsophila podophylla (Hook.)	Methanolic/entire plant	*Musca domestica* Linnaeus, 1758, and *Aedes albopictus* Skuse, 1894/Diptera	Adult and larva/ fourth instar	Ingestion	No significant results	[22]
Alsophila spinulosa (Wall. ex Hook.) R.M.Tryon	Methanolic/leaves	*Musca domestica* Linnaeus, 1758, and *Aedes albopictus* Skuse, 1894/Diptera	Adult and larva/ fourth instar	Ingestion	100% mortality (10,000 mg L^{-1}) for *M. domestica*	[22]
Cystopteridaceae						
Cystopteris fragilis (L.) Bernh.	Proteic (HEPES buffer)/ entire plant	*Helicoverpa zea* Boddie, 1850, and *Spodoptera frugiperda* Smith, 1808/ Lepidoptera	Larva/ second instar	Ingestion	No significant results	[51]
Gymnocarpium dryopteris (L.) Newman	Proteic (HEPES buffer)/ entire plant	*Helicoverpa zea* Boddie, 1850, and *Spodoptera frugiperda* Smith, 1808/ Lepidoptera	Larva/ second instar	Ingestion	70% mortality (2.1 μg μL^{-1}) for *H. zea* and 90% (2 μg μL^{-1}) for *S. frugiperda*	[51]
Davalliaceae						
	Methanolic/entire plant			Ingestion		[22]

Davallia tyermannii (T. Moore) H.J.Veitch		*Musca domestica* Linnaeus, 1758, and *Aedes albopictus* Skuse, 1894/Diptera	Adult and larva/ fourth instar		50% mortality (10,000 mg L^{-1}) for *M. domestica*	
Dennstaedtiaceae						
Pteridium aquilinum (L.) Kuhn	Aqueous/leaves	*Cerotoma tingomarianus* Bechyné, 1951/Coleoptera	Adult	Contact	No significant results	[56]
	Aqueous, ethanolic, chloroformic, hexanic, and phytoecdysteroids fractions/ leaves	*Helicoverpa armigera* Hübner, 1809, and *Spodoptera litura* Fabricius, 1775/Lepidoptera	Larva/ fourth instar	Ingestion	LC_{50} (EtOHE) of 0.198% and 0.141% for *H. armigera* and *S. litura*, respectively	[37]
	Proteic (HEPES buffer)/ entire plant	*Helicoverpa zea* Boddie, 1850, and *Spodoptera frugiperda* Smith, 1808/ Lepidoptera	Larva/ second instar	Ingestion	No significant results	[51]
	Acetonic and aqueous/ leaves	*Sitophilus zeamais* Motschulsky, 1855/ Coleoptera	Adult	Contact	No significant results	[57]
	Methanolic /not informed	*Tribolium castaneum* Herbst, 1797/Coleoptera	Larva/fifth instar	Ingestion	No significant results	[58]
	Methanol/entire plant	*Musca domestica* Linnaeus, 1758, and *Aedes albopictus* Skuse, 1894/Diptera	Adult and larva/ fourth instar	Ingestion	No significant results	[22]
	Aqueous and ethanolic/ leaves	*Planococcus citri* Risso, 1813/Hemiptera	Nymph/ third instar	Contact	No significant results	[59]
	Aqueous/not informed	*Helicoverpa armigera* Hubner, 1808, and *Spodoptera litura* Fabricius, 1775/Lepidoptera	Larva/ fourth instar	Ingestion	56.73% population reduction in the field (0.5–16.0%) for *S. litura*	[60]

(continued)

Table 2 (continued)

Family and species	Extract/plant part used	Insect tested/order	Insect stage	Type of test	Result/lethal dose	References
	Ethanolic/entire plant	*Helicoverpa armigera* Hubner, 1808, and *Spodoptera litura* Fabricius, 1775/Lepidoptera	Larva	Contact	0.3% population reduction in the field for *S. litura* (0.1%)	[20]
	Aqueous/leaves	*Myzus persicae* Sulzer, 1776/Hemiptera and *Ascia monuste orseis* Godart, 1818/Lepidoptera	Adult and larva/second instar	Topical	63% mortality for *M. persicae* and 60% for *A. monuste orseis* (30% $p\ v^{-1}$)	[61]
	Aqueous/leaves	*Brevicoryne brassicae* Linnaeus, 1758/Hemiptera	Adult	Contact (repellency)	78% mortality, repellent action, reduced survival, and offspring (10% $p\ v^{-1}$)	[39]
	Aqueous/leaves	*Brevicoryne brassicae* Linnaeus, 1758, and *Myzus persicae* Sulzer, 1776/Hemiptera	Adult	Contact (repellency)	Repellent and insecticidal action, nonpreferential food effects, and reduced survival and offspring (10% $p\ v^{-1}$)	[21]
	Silver nanoparticles from aqueous extract/leaves	*Anopheles stephensi* Liston, 1901/Diptera	Larva/first instar	Contact	$LC_{50} = 7.48$ ppm. 100% larva reduction in the field (10×7.48 ppm). Reduced fertility and longevity	[62]
	Aqueous/leaves	*Helopeltis theivora* Waterhouse, 1886/Hemiptera	Adult	Contact	No significant results	[50]
Dicksoniaceae						
Dicksonia sellowiana Hook.	Ethanolic/leaves	*Oncopeltus fasciatus* Dallas, 1852/Hemiptera	Ninfa/fifth instar	Topical	50% mortality ($p < 0.0001$)	[63]

Dryopteridaceae						
Dryopteris filix-mas (L.) Schott	Aqueous/entire plant	*Leptinotarsa decemlineata* Say, 1824/Coleoptera	Larva/first and second instar	Contact	No significant results	[51]
	Aqueous/entire plant	*Hyalopterus pruni* Geoffrey, 1762/Hemiptera	Adult	Contact	No significant results	[53]
	Aqueous/roots and rhizome	*Corcyra cephalonica* Stainton, 1866/Lepidoptera	Larva/third instar	Ingestion	100% mortality (0.16% p v^{-1})	[64]
	Ethanolic/roots and rhizome	*Corcyra cephalonica* Stainton, 1866/Lepidoptera	Larvicide/third instar	Contact	100% mortality (0.20% p v^{-1})	[65]
	Aqueous/leaves	*Aphis fabae* Scopoli, 1763/*Hemiptera*	Adult	Contact	75.1% mortality (5 ml 5 ml^{-1})	[66]
Polystichum acrostichoides (Michx.) Schott	Proteic (HEPES buffer)/entire plant	*Helicoverpa zea* Boddie, 1850, and *Spodoptera frugiperda* Smith, 1808/Lepidoptera	Larva/second instar	Ingestion	70% mortality (2.1 μg μL^{-1}) for *H. zea*	[51]
Equisetaceae						
Equisetum arvense L.	Ethanolic/shoots	*Schizaphis graminum* Rondani, 1847/Hemiptera and *Frankliniella occidentalis* Pergande, 1895/Thysanoptera	Adult and larva/second instar	Contact	No significant results	[67]
Equisetum arvense subsp. *diffusum* (D. Don) Fraser-Jenkins	Methanolic/entire plant	*Musca domestica* Linnaeus, 1758, and *Aedes albopictus* Skuse, 1894/Diptera	Adult and larva/fourth instar	Ingestion	No significant results	[22]
Equisetum hyemale L.	Methanolic/stem	*Plutella xylostella* Linnaeus, 1767, and *Spodoptera litura* Fabricius, 1775/Lepidoptera	Adult and larva/second instar	Contact	Food inhibition for *S. litura* (5000 ppm)	[68]

(continued)

Table 2 (continued)

Family and species	Extract/plant part used	Insect tested/order	Insect stage	Type of test	Result/lethal dose	References
	Methanolic/entire plant	*Lymantria dispar* Linnaeus, 1758, *Hyphantria cunea* Drury, 1770 *Dendrolimus spectabilis* Butler, 1877/ Lepidoptera, and *Acantholyda posticalis* Matsumura, 1912/ Hymenoptera	Larva*	Contact	> 80% mortality for *A. posticalis* (5000 ppm)	[69]
	Proteic (HEPES buffer)/ entire plant	*Helicoverpa zea* Boddie, 1850, and *Spodoptera frugiperda* Smith, 1808/ Lepidoptera	Larva/ second instar	Ingestion	No significant results	[51]
	Ethanolic/entire plant	*Lycoriella ingenua* Dufour, 1839, and *Coboldia fuscipes* Meigen, 1830/Diptera	Larva*	Contact	93% mortality for *L. ingenua* (0.07 mg cm^{2-1})	[70]
Gleicheniaceae						
Dicranopteris splendida (Hand.-Mazz.) Tag.	Methanolic/entire plant	*Musca domestica* Linnaeus, 1758, and *Aedes albopictus* Skuse, 1894/Diptera	Adult and larva/ fourth instar	Ingestion	No significant results	[22]
Dicranopteris linearis (Burm. f.) Underw.	Methanolic/entire plant Aqueous/entire plant	*Musca domestica* Linnaeus, 1758, and *Aedes albopictus* Skuse, 1894/Diptera *Helopeltis theivora* Waterhouse, 1886/ Hemiptera	Adult and larva/ fourth instar Adult	Ingestion Contact	96.7% mortality (10,000 mg L^{-1}) for *M. domestica* No significant results	[22] [50]

Diplopterygium chinensis (Rosenst.) De Vol	Methanolic/entire plant	*Musca domestica* Linnaeus, 1758, and *Aedes albopictus* Skuse, 1894/Diptera	Adult and larva/ fourth instar	Ingestion	53% mortality (10,000 mg L^{-1}) for *M. domestica*	[22]
Hymenophyllaceae						
Hymenophyllum barbatum (Bosch) Baker	Methanolic/entire plant	*Musca domestica* Linnaeus, 1758, and *Aedes albopictus* Skuse, 1894/Diptera	Adult and larva/ fourth instar	Ingestion	53% mortality (10,000 mg L^{-1}) for *M. domestica*	[22]
Lycopodiaceae						
Huperzia lucidula (Michx.) Trevis.	Proteic (HEPES buffer)/ entire plant	*Helicoverpa zea* Boddie, 1850, and *Spodoptera frugiperda* Smith, 1808/ Lepidoptera	Larva/ second instar	Ingestion	No significant results	[51]
Lycopodiastrum casuarinoides (Spring) Holub ex Dixit	Methanolic/entire plant	*Musca domestica* Linnaeus, 1758, and *Aedes albopictus* Skuse, 1894/Diptera	Adult and larva/ fourth instar	Ingestion	No significant results	[22]
Lycopodium clavatum L.	Methanolic/leaves	*Aedes aegypti* Linnaeus, 1762, and *Anopheles stephensi* Liston, 1901/ Diptera	Larva/third and fourth instar	Contact	*A. aegypti* (LC_{50} III-IV instar 388.7, 960.3 ppm), *A. stephensi* (LC_{50} III-IV instar 177.8, 283.5 ppm)	[71]
Lygodiaceae						
Lygodium flexuosum (L.) Sw. *Lygodium japonicum* (Thunb.) Sw.	Methanolic/entire plant	*Musca domestica* Linnaeus, 1758, and *Aedes albopictus* Skuse, 1894/Diptera	Adult and larva/ fourth instar	Ingestion	No significant results	[22]

(continued)

Table 2 (continued)

Family and species	Extract/plant part used	Insect tested/order	Insect stage	Type of test	Result/lethal dose	References
Lygodium palmatum (Bernh.) Sw.	Proteic (HEPES buffer)/ entire plant	*Helicoverpa zea* Boddie, 1850, and *Spodoptera frugiperda* Smith, 1808/ Lepidoptera	Larva/ second instar	Ingestion	No significant results	[51]
Metaxyaceae						
Metaxya rostrata (Kunth) C. Presl	Methanolic/roots	*Spodoptera littoralis* Boisduval, 1833/ Lepidoptera	Larva/first instar	Ingestion	No significant results	[72]
Nephrolepidaceae						
Nephrolepis cordifolia (L.) C. Presl	Methanolic/entire plant	*Musca domestica* Linnaeus, 1758, and *Aedes albopictus* Skuse, 1894/Diptera	Adult and larva/ fourth instar	Ingestion	83.3% mortality (10,000 mg L^{-1}) for *M. domestica*	[22]
	Phytoecdysone fractions/ entire plant	*Spodoptera litura* Fabricius, 1775/Lepidoptera	Larva/third instar	Ingestion	$LC_{50} = 0.9177\%$, reduced pupation and adult emergences, larval, pupal, and adult deformities	[43]
	Methanolic/entire plant	*Spodoptera litura* Fabricius, 1775/Lepidoptera	Larva/third instar	Ingestion	$LC_{50} = 0.9177\%$, decreased pupation and adult emergence, larval, pupal, and adult deformities	[73]
	Ethanolic/leaves	*Oncopeltus fasciatus* Dallas, 1852/Hemiptera	Ninfa/fifth instar	Topical	63% mortality ($p < 0.0001$)	[63]
Nephrolepis exaltata (L.) Schott	Aqueous/leaves	*Nauphoeta cinerea* Olivier, 1789/Blattodea	Adult	Topical application on the semi-isolated heart	1:1 and 1: 2 aqueous extracts completely paralyzed the heartbeats of *N. cinerea*	[74]

Nephrolepis pectinata (Willd.) Schott	Hexanic, ethanolic, and aqueous/not informed	*Blattella germanica* Linnaeus, 1767/Blattodea	Ninfa/ second and third instar	Ingestion	No significant results	[75]
Onocleaceae						
Onoclea sensibilis L.	Proteic (HEPES buffer)/ entire plant	*Helicoverpa zea* Boddie, 1850, and *Spodoptera frugiperda* Smith, 1808 / Lepidoptera	Larva/ second instar	Ingestion	100% mortality (0.9–2.1 µg µL^{-1}) for *H. zea* and 90% (2 µg µL^{-1}) for *S. frugiperda*	[51]
Ophioglossaceae						
Botrypus virginianum (L.) Michx.	Proteic (HEPES buffer)/ entire plant	*Helicoverpa zea* Boddie, 1850, and *Spodoptera frugiperda* Smith, 11,808/ Lepidoptera	Larva/ second instar	Ingestion	No significant results	[51]
Osmundaceae						
Osmunda japonica Thunb.	Methanolic/entire plant	*Musca domestica* Linnaeus, 1758, and *Aedes albopictus* Skuse, 1894/Diptera	Adult and larva/ fourth instar	Ingestion	93% (10,000 mg L^{-1}) for *M. domestica*	[22]
Osmunda regalis L.	Proteic (HEPES buffer)/ entire plant	*Helicoverpa zea* Boddie, 1850, and *Spodoptera frugiperda* Smith, 1808/ Lepidoptera	Larva/ second instar	Ingestion	No significant results	[51]
Osmundastrum cinnamomeum (L.) C. Presl	Proteic (HEPES buffer)/ entire plant	*Helicoverpa zea* Boddie, 1850, and *Spodoptera frugiperda* Smith, 1808/ Lepidoptera	Larva/ second instar	Ingestion	No significant results	[51]
Plenasium vachellii (Hook.) C. Presl	Methanolic/entire plant	*Musca domestica* Linnaeus, 1758, and *Aedes albopictus* Skuse, 1894/Diptera	Adult and larva/ fourth instar	Ingestion	50% mortality (10,000 mg L^{-1}) for *M. domestica*	[22]

(continued)

Table 2 (continued)

Family and species	Extract/plant part used	Insect tested/order	Insect stage	Type of test	Result/lethal dose	References
Polypodiaceae						
Lepisorus buergerianus (Miq.) C.F.Zhao, R. Wei & X.C.Zhang	Methanolic/entire plant	*Musca domestica* Linnaeus, 1758, and *Aedes albopictus* Skuse, 1894/Diptera	Adult and larva/ fourth instar	Ingestion	No significant results	[22]
Lepisorus fortunei (T. Moore) C.M. Kuo	Methanolic/entire plant	*Musca domestica* Linnaeus, 1758, and *Aedes albopictus* Skuse, 1894/Diptera	Adult and larva/ fourth instar	Ingestion	50% mortality (10,000 mg L^{-1}) for *M. domestica*	[22]
Microsorum punctatum (L.) Copel.	Methanolic/entire plant	*Musca domestica* Linnaeus, 1758, and *Aedes albopictus* Skuse, 1894/Diptera	Adult and larva/ fourth instar	Ingestion	83.% mortality (10,000 mg L^{-1}) for *M. domestica*	[22]
Pleopeltis polypodioides (L.) E.G.Andrews Windham	Proteic (HEPES buffer)/ entire plant	*Helicoverpa zea* Boddie, 1850, and *Spodoptera frugiperda* Smith, 1808/ Lepidoptera	Larva/ second instar	Ingestion	No significant results	[51]
Selliguea hastata (Thunb.) Fraser-Jenk.	Methanolic/entire plant	*Musca domestica* Linnaeus, 1758, and *Aedes albopictus* Skuse, 1894/Diptera	Adult and larva/ fourth instar	Ingestion	50% mortality (10,000 mg L^{-1}) for *M. domestica*	[22]
Pteridaceae						
	Methanolic/leaves	*Aedes aegypti* Linnaeus, 1762, and *Anopheles*		Contact	*A. aegypti* (LC_{50} III-IV instar 37.5, 67.6 ppm),	[71]

Actiniopteris radiata (J. Koening ex Sw.) Link		*stephensi* Liston, 1901/ Diptera	Larva/third and fourth instar		*A. stephensi* (LC_{50} III-IV instar 70.4, 113.8 ppm)	
Adiantum capillus-veneris L.	Silver nanoparticles from aqueous extract/leaves	*Anopheles stephensi* Liston, 1901/Diptera	Adult, pupa and larva/first to fifth instar	Contact (repellency) and ingestion	100% mortality (10%) of first instar larvae 95.1% repellency (8%)	[40]
Adiantum caudatum L.	Methanolic/leaves	*Aedes aegypti* Linnaeus, 1762, and *Anopheles stephensi* Liston, 1901/ Diptera	Larva/third and fourth instar	Contact	*A. aegypti* (LC_{50} III-IV instar 74.5, 95.9 ppm), *A. stephensi* (LC_{50} III-IV instar 112.1, 175.3 ppm)	[71]
	Methanolic/entire plant	*Musca domestica* Linnaeus, 1758, and *Aedes albopictus* Skuse, 1894/Diptera	Adult and larva/ fourth instar	Ingestion	73.3% mortality (10,000 mg L^{-1}) for *M. domestica*	[22]
Adiantum flabellulatum L.	Methanolic/entire plant	*Musca domestica* Linnaeus, 1758, and *Aedes albopictus* Skuse, 1894/Diptera	Adult and larva/ fourth instar	Ingestion	No significant results	[22]
Adiantum latifolium Lam.	Methanolic/leaves	*Oryctes rhinoceros* Linnaeus, 1758/Coleoptera	Larva/third instar	Ingestion	$LC_{50} = 5018$ mg kg^{-1} and reduced pupation	[45]
Adiantum raddianum C. Presl	Aqueous/leaves	*Helopeltis theivora* Waterhouse, 1886/ Hemiptera	Adult	Contact	No significant results	[50]
Hemionitis arifolia (Burm. f.) T. Moore	Aqueous/not informed	*Helicoverpa armigera* Hubner, 1809, and *Spodoptera litura* Fabricius, 1775/Lepidoptera	Larva/ fourth instar	Ingestion	53% population reduction in the field for *S. litura* (0.5–16.0%)	[60]

(continued)

Table 2 (continued)

Family and species	Extract/plant part used	Insect tested/order	Insect stage	Type of test	Result/lethal dose	References
	Ethanolic/entire plant	*Helicoverpa armigera* Hubner, 1809, and *Spodoptera litura* Fabricius, 1775/Lepidoptera	Larva*	Contact	0.3% population reduction in the field (0.4%)	[20]
	Methanolic/leaves	*Aedes aegypti* Linnaeus, 1762, and *Anopheles stephensi* Liston, 1901/ Diptera	Larva/third and fourth instar	Contact	*A. aegypti* (LC_{50} III-IV instar 494.6, 630.5 ppm), *A. stephensi* (LC_{50} III-IV instar 127.5, 148.8 ppm)	[71]
Oeosporangium elegans (Poir.) Fraser-Jenk. & Pariyar	Methanolic/leaves	*Aedes aegypti* Linnaeus, 1762, and *Anopheles stephensi* Liston, 1901/ Diptera	Larva/third and fourth instar	Contact	*A. aegypti* (LC_{50} III-IV instar 152.4, 271.5 ppm), *A. stephensi* (LC_{50} III-IV instar 301.2, 315.2 ppm)	[71]
Pteris multifida Poir. *Pteris semipinnata* L.	Methanolic/entire plant	*Musca domestica* Linnaeus, 1758, and *Aedes albopictus* Skuse, 1894/Diptera	Adult and larva/ fourth instar	Ingestion	No significant results	[22]
Pteris vittata L.	Methanolic/roots	*Musca domestica* Linnaeus, 1758, and *Aedes albopictus* Skuse, 1894/Diptera	Adult and larva/ fourth instar	Ingestion	73.3% mortality (10,000 mg L^{-1}) for *M. domestica*	[22]
Salviniaceae						
Azolla caroliniana Willd.	Proteic (HEPES buffer)/ entire plant	*Helicoverpa zea* Boddie, 1850, and *Spodoptera frugiperda* Smith, 1808/ Lepidoptera	Larva/ second instar	Ingestion	No significant results	[51]

Azolla pinnata R. Br.	Methanolic/entire plant	*Aedes aegypti* Linnaeus, 1762, and *Aedes albopictus* Skuse, 1894/Diptera	Larva/ fourth instar	Contact	LC_{50} of 1093 and 1035 mg L^{-1} for *A. aegypti* and *A. albopictus*, respectively	[47]
	Methanolic/entire plant	*Aedes aegypti* Linnaeus, 1762, and *Aedes albopictus* Skuse, 1894/Diptera	Egg and adult	Contact	LC_{50} of 2572.45 and 2329.34 ppm for *A. aegypti* and *A. albopictus*, respectively. 100% egg mortality (1500–125 ppm). Egg-laying inhibition	[48]
Selaginellaceae						
Selaginella apoda (L.) Spring	Proteic (HEPES buffer)/ entire plant	*Helicoverpa zea* Boddie, 1850, and *Spodoptera frugiperda* Smith, 1808/ Lepidoptera	Larva/ second instar	Ingestion	No significant results	[51]
Selaginella uncinata (Desv. ex Poir.) Spring	Methanolic/entire plant	*Musca domestica* Linnaeus, 1758, and *Aedes albopictus* Skuse, 1894/Diptera	Adult and larva/ fourth instar	Ingestion	50% mortality (10,000 mg L^{-1}) for *M. domestica*	[22]
Tectariaceae						
Tectaria coadunata (Wall. ex Hook. & Grev.) C.Chr.	Proteic (Tma12) /leaves e roots	*Bemisia tabaci* Gennadius, 1889/Lepidoptera	Adult	Contact	> 90% mortality (5.0 μg ml^{-1})	[76]
Thelypteridaceae						
Christella dentata (Forssk.) Brownsey & Jermy	Aqueous and ethanolic/ leaves	*Aedes aegypti* Linnaeus, 1762, and *Aedes albopictus* Skuse, 1894/Diptera	Larva/third instar	Contact	No significant results	[77]
	Phytoecdysteroids fractions (from ethanolic extract/ entire plant	*Spodoptera litura* Fabricius, 1775/Lepidoptera	Larva/third instar	Ingestion	$LC_{50} = 0.765\%$, decreased pupation and adult emergence, larval, pupal, and adult deformities	[43]

(continued)

Table 2 (continued)

Family and species	Extract/plant part used	Insect tested/order	Insect stage	Type of test	Result/lethal dose	References
	Methanolic/entire plant	*Spodoptera litura* Fabricius, 1775/Lepidoptera	Larva/third instar	Ingestion	LC_{50} = 0.765%, decreased pupation and adult emergence, larval, pupal, and adult deformities	[73]
Christella parasitica (L.) Lév.	Chloroformic, ethanolic, and phytoecdysteroids fractions (from ethanolic extract) /not informed	*Helicoverpa armigera* Hubner, 1809, and *Spodoptera litura* Fabricius, 1775/Lepidoptera	Larva/third instar	Ingestion	LC_{50} (EtOHE) of 0.140% and 0.198% for *H. armigera* and *S. litura*, respectively	[78]
	Aqueous/not informed	*Helicoverpa armigera* Hubner, 1809, and *Spodoptera litura* Fabricius, 1775/Lepidoptera	Larva/fourth instar	Ingestion	52.1% population reduction in the field for *S. litura* (0.5–16.0%)	[60]
	Ethanolic/entire plant	*Helicoverpa armigera* Hubner, 1809, and *Spodoptera litura* Fabricius, 1775/Lepidoptera	Larva	Ingestion	No significant results	[20]
Cyclosorous interruptus (Willd.) H. Itô	Phytoecdysteroids fractions/ entire plant	*Spodoptera litura* Fabricius, 1775/Lepidoptera	Larva/third instar	Ingestion	LC_{50} = 0.890%, decreased pupation and adult emergence, larval, pupal, and adult deformities	[43]
	Methanolic/entire plant	*Spodoptera litura* Fabricius, 1775/Lepidoptera	Larva/third instar	Ingestion	LC_{50} = 0.890%, decreased pupation and adult emergence, larval, pupal, and adult deformities	[73]
	Aqueous/leaves	*Helopeltis theivora* Waterhouse, 1886/ Hemiptera	Adult	Contact	No significant results	[50]

Grypothrix megacuspis (Baker) S.E.Fawc. & A.R. Sm.	Methanolic and in ethyl acetate/entire plant	*Solenopsis invicta* Buren, 1972/Hymenoptera	Adult	Contact	No significant results	[79]
Menisciopsis lakhimpurensis (Rosenst.) S.E. Fawc. & A.R.Sm.	Methanolic/entire plant	*Musca domestica* Linnaeus, 1758, and *Aedes albopictus* Skuse, 1894/Diptera	Adult	Ingestion	76.7% mortality (10,000 mg $L-^{1}$) for *M. domestica*	[22]
Sphaerostephanos heterocarpus (Blume) Holttum	Methanolic/entire plant	*Musca domestica* Linnaeus, 1758, and *Aedes albopictus* Skuse, 1894/Diptera	Adult and larva/ fourth instar	Ingestion	63.3% mortality (10,000 mg L^{-1}) for *M. domestica*	[22]
Stegnogramma griffithii (T. Moore) K. Iwats.	Methanolic/entire plant	*Musca domestica* Linnaeus, 1758, and *Aedes albopictus* Skuse, 1894/Diptera	Adult and larva/ fourth instar	Ingestion	70% mortality (10,000 mg L^{-1}) for *M. domestica*	[22]
Phegopteris hexagonoptera (Michx.) Fée	Proteic (HEPES buffer)/ entire plant	*Helicoverpa zea* Boddie, 1850, and *Spodoptera frugiperda* Smith, 1808/ Lepidoptera	Larva/ second instar	Ingestion	No significant results	[51]

Legend: *larval stage not identified by the author of the study, *AE* acetonic extract, *EtOAcE* ethyl acetate extract, *EtOHE* ethanolic extract, LC_{50} average lethal concentration, *ppm* parts per million, $mg\ L^{-1}$ milligram per liter, $\mu g\ \mu L^{-1}$ microgram per microliter, $g\ L^{-1}$ gram per liter, $\mu g\ ml^{-1}$ microgram per milliliter, $p\ v^{-1}$ weight per volume, *ml* milliliter, $mg\ cm^{2-1}$ milligram per square centimeter, > greater than species showed no insecticidal activity. In this case, the low number of species assessed should be considered (Table 1).

The species *Pteridium aquilinum* (Dennstaedtiaceae), *Dryopteris filix-mas* (Dryopteridaceae), and *Equisetum hyemale* (Equisetaceae) were the most studied. *P. aquilinum* was discussed in 14 articles, *D. filix-mas* in 5, and *E. hyemale* in 4 (Table 2).

The orders of the studied insects were Blattodea, Coleoptera, Diptera, Hemiptera, Hymenoptera, Lepidoptera, and Thysanoptera, order Diptera being the most studied (Fig. 1). Huang et al. [22] evaluated the insecticidal potential of 33 ferns and three lycophyte species as well as two insect species belonging to this order. The second most studied order was Lepidoptera, followed by Hemiptera, while Thysanoptera was assessed in only one study. The main orders that cause economic damage are Diptera, Lepidoptera, Hemiptera, and Coleoptera [23], justifying the interest in studying them.

A total of 30 studies with insect pests were conducted, corresponding to 69.8%, and 13 studies with vector insects, accounting for 30.2%. Of the 39 insect species assessed, 28 (71.8%) exhibited holometabolous development, that is, complete metamorphosis, passing through the pupal phase, and 11 (28.2%) hemimetabolous development, passing through incomplete metamorphous. The most widely studied insects were *Spodoptera litura* (Lepidoptera) (7 studies), *Aedes aegypti* (Diptera) (5 studies), *Helicoverpa armigera* (Lepidoptera) (4 studies), *Aedes albopictus,* and *Anopheles stephensi* (Diptera) (3 studies each) (Table 2). The highest percentage of studies investigated insect pests (69.8%). The caterpillar of *S. litura* is a globally important agricultural pest and may cause significant damage to different crops [24]. The larvae of the genus *Helicoverpa* (Lepidoptera) are pests in Asia, Australia, and the Americas. They cause yield losses in many important crops, such as cotton, chickpeas, maize, and tomato [25]. Fewer studies were found on vector insects (30.2%). Despite being discussed in only one study, the species *Musca domestica* (Diptera) was assessed in 36 fern and lycophyte species. Domestic flies (*M. domestica*) are well known as bird and cattle pests and are generalized vectors of human pathogens [26]. The mosquito *A. aegypti* is a vector of important arboviruses, such as dengue, Zika, and chikungunya [27], while *A. albopictus* is currently the most invasive in the world, exhibiting medical importance due to its ability to

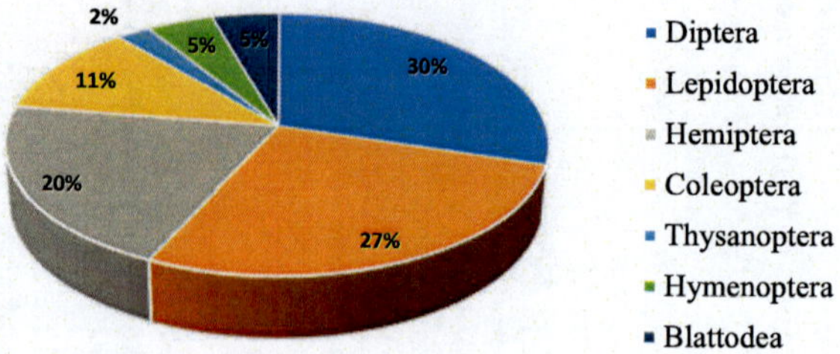

Fig. 1 Percentage of studies with insects of different orders used to assess the insecticidal/repellent potential of ferns and lycophytes

transmit viruses to humans, including dengue [28]. The mosquito *A. stephensi* is an important malaria vector in urban areas of Africa and Asia [29].

The fern and lycophyte extracts were tested in insects in the following development stages: egg, larval (first, second, third, fourth, and fifth instar), pupal, nymph (second and third instar), and adult. The larval stage was the most assessed, primarily third and fourth instar larvae, followed by the adult stage. The egg and pupal stages were assessed in only one study each (Fig. 2).

The types of extracts assessed were ethyl acetate, acetonic, aqueous, chloroformic, ethanolic, ethereal, hexanic, methanolic, and HEPES buffer (proteic), in addition to essential oil. The aqueous extract was the most widely used for holometabolous and hemimetabolous insects, followed by ethanolic and methanolic extracts. Three studies isolated phytoecdysteroid fractions by successive partitions of crude ethanolic extract. Phytoecdysteroids are plants' secondary metabolites analogous to molting insect hormones. They are usually involved in insect signaling pathways, causing interruptions or blockages that can damage the metabolism of insects, which can also lead to their death. Once they mimic the insects molting hormones and have no toxic effects on vertebrates, these compounds can be used as pest control agents. The function of these substances for plants is not yet evident; however, it is believed that they provide some degree of protection against phytophagous insects or soil nematodes [30]. Ethereal extract and essential oil were assessed in only one study each (Fig. 3). Despite the scarcity of studies that evaluated the insecticidal activity of fern and lycophyte essential oils, many angiosperm oils have been tested with promising results, such as oils of *Cymbopogon martini* (Roxb.), W.Watson (Poaceae), *Piper aduncum* L. (Piperaceae), and *Lippia gracilis* Schaue (Verbenaceae) [31]. In addition, a number of studies on the chemical characterization of fern oils identified insecticidal activity in some of the main

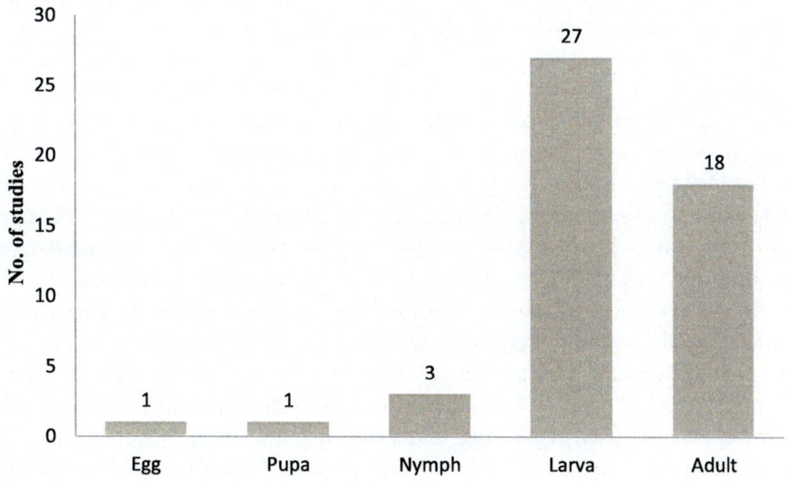

Fig. 2 Developmental stages of insects during the studies in order to assess insecticidal and/or repellent activity of fern and lycophyte extracts

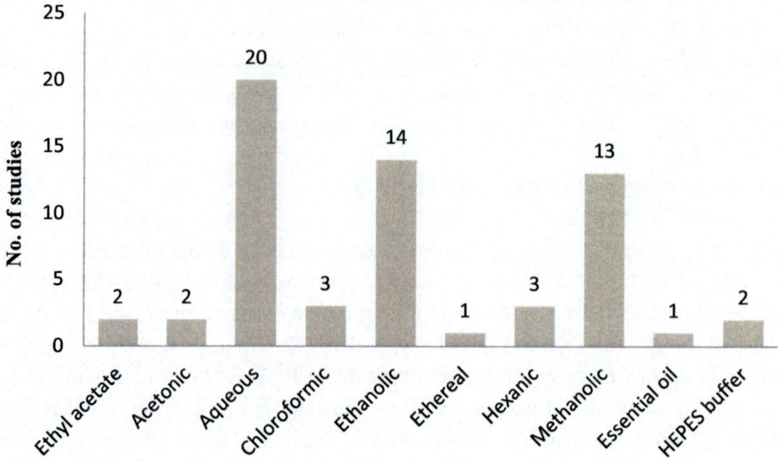

Fig. 3 Types of extracts obtained from the ferns and lycophytes used in the studies to assess insecticidal and/or repellent activity

components of these plants, such as the monoterpene α-pinene, present in the species *Diplazium esculetum* (Athyriaceae) and *Polystichum montevidense* (Spreng.) Rosenst. (Dryopteridaceae) [32, 33]. As a consequence of their ability to mimic the actions of arthropod molting hormones while being nontoxic to vertebrates, phytoecdysteroids are interesting lead compounds for the development of environmental friendly invertebrate pest control agents.

Some studies compared the efficacy of the extracts of different polarities obtained from fern species, demonstrating that the most polar extracts displayed the best insecticidal activity, likely due to the polar nature of the substances present in the species tested. Ethanolic extracts of *Alsophila crinita*, *A. gigantea*, and *A. nilgirensis* (Cyatheaceae) were more effective with the mosquito *Culex quinquefasciatus* when compared to ethereal, chloroformic, and acetonic extracts. A test conducted with *Pteridium aquilinum* indicated that the ethanolic extract of the species was more efficient in controlling the insects *H. armigera* and *S. litura*, with LC_{50} values of 0.198% and 0.141%, respectively, when compared to aqueous, chloroformic, and hexanic extracts. The ethanolic extract of *Christella parasitica* was more effective at killing third instar larvae of *H. armigera* and *S. litura* when compared to the chloroformic extract. The acetonic extract of *Asplenium aethiopicum* caused more death in the fourth instar larvae of *Culex quinquefasciatus* when compared to chloroformic, ethereal, and methanolic extracts.

Phenolic substances, normally present in polar extracts, may exhibit different mechanisms of insecticidal action, such as food deterrence, reduced nutrient absorption efficiency, decreased egg-laying, and a change in nutritional parameters [34]. In addition, these substances may also act on the central nervous system of insects, through acetylcholinesterase inhibition, which promotes an accumulation of acetylcholine in the synapse, causing continuous transmission of nerve impulses, and may provoke insect paralysis and death [35, 36].

The extracts were evaluated using three types of tests: contact, ingestion, and topic. The contact test was the most widely used for holometabolous and hemimetabolous insects (63%), conducted by spraying the extract directly on the insect, the surface where it will be allocated or solubilized and dispersed in the environment where the insects are located. Among the contact tests performed, four assessed repellent activity, recording the time that the insects remained off the surface sprayed with the extract. The ingestion test involved offering the insect food embedded in the extract to be analyzed. The topical test was used in only three studies, where the extract was applied directly on the dorsum of the insect, and in one study it was applied directly to the semi-isolated heart of the insect (Fig. 4).

The extracts of the following species provoked 100% death in the first, second, third, and fourth instar larval stages: *Adiantum capillus-veneris* (Pteridaceae), *Asplenium platyneuron* (Aspleniaceae), *Homalosorus pycnocarpus* (Diplaziopsidaceae), *Blechnopis orientalis* (Blechnaceae), *Alsophila spinulosa* (Cyatheaceae), *Dryopteris filix-mas* (Dryopteridaceae), *Pteridium aquilinum* (Dennstaedtiaceae), and *Onoclea sensibilis* (Onocleaceae). An assessment of the chemical profile of *P. aquilinum* demonstrated the presence of alpha ecdysteroids and beta ecdysones with a structure analogous to the hormones responsible for insect molting [37]. Thus, when administered to insects, they may cause malformation, sterility, and death. In addition, Lovatto, Schiedeck, and Mauch [21] associate *P. aquilinum* bioactivity with the presence of other chemical classes, such as flavonoids (quercetin and kaempferol), terpenoids (ptaquiloside), cyanogenic glycosides (prunasin), and carboxylic acids (shikimic acid).

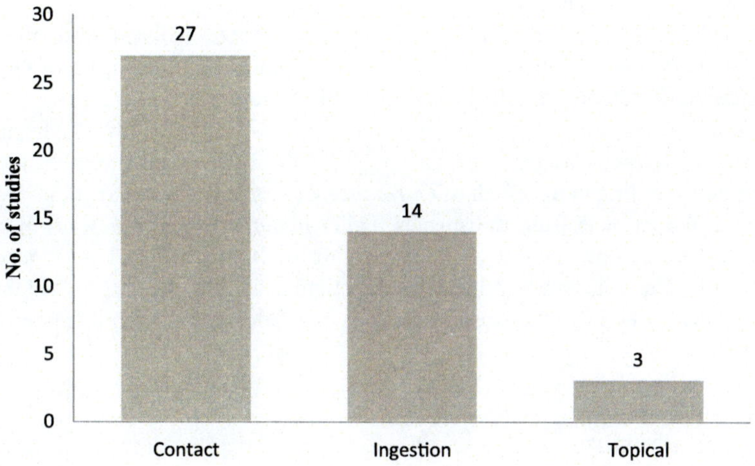

Fig. 4 Types of tests used in the studies to assess the insecticidal and/or repellent activity of fern and lycophyte species

Extracts of the species *Asplenium normale* (Aspleniaceae), *Athyrium filix-femina* (Athyriaceae), *Dicranopteris linearis* (Gleicheniaceae), *Nephrolepis cordifolia* (Nephrolepidaceae), *Osmunda japonica* (Osmundaceae), *Microsorum punctatum* (Polypodiaceae), and *Tectaria coadunata* (Tectariaceae) (Table 2) caused more than 80% insect death in the adult and larval stages (second and third instar).

The species *Anemia tomentosa* var. *anthriscifolia* (Anemiaceae), *Pteridium aquilinum* (Dennstaedtiaceae), and *Adiantum capillus-veneris* (Pteridaceae) displayed repellent activity [38–40]. Assessment of the volatile constituents of *A. tomentosa* revealed the significant presence of sesquiterpenes, with alpha-bisabolol as the main component [38]. Different sesquiterpenes present in plants have demonstrated insecticidal activity through acetylcholinesterase inhibition and as insect growth regulators (IGRs) [41]. *Parablechnum cordatum* (Blechnaceae) caused premature pupation of first instar larvae of *Drosophila melanogaster,* and the dead adults showed malformations. Phytochemical analysis of the hexanic extract of the species demonstrated the presence of four phytoecdysteroids: ecdysone, ponasterone, shidasterone, and 2-desoxycrustecdysone [42]. The species *Nephrolepis cordifolia* (Nephrolepidaceae), *Christella dentata* (Thelypteridaceae), and *Cyclosorous interruptus* (Thelypteridaceae) provoked a decrease in pupation, and adults in *Spodoptera litura* and larva emerged for the adult phase at a 1.5% concentration of methanolic extract. In addition, they caused deformities in larvae, pupae, and adults [43]. The presence of secondary metabolites, such as alkaloids, steroids, tannins, flavonoids, terpenoids, glycosides, and phenolic compounds was detected in the crude extract of these plants [43]. Adebiyi et al. [44] investigated the bioactive substances present in the hexanic extract of *N. cordifolia*. The main components related to the biological activities of the species were *n*-hexadecenoic acid (24.42%), *cis*-13-octadecanoic acid (10.31%), octadecanoic acid (11.05%), bis (2-ethylhexyl) phthalate (8.72%), 1,4-benzenedicarboxylic acid, bis(2-ethylhexyl) ester (18.35%), and squalene (5.82%).

The substance adiantobischrysene, a triterpene dimer, isolated from the fern *Adiantum latifolium*, exhibited larvicidal activity against the third instar of *Oryctes rhinoceros* and is more effective than the methanolic extract of the leaves of the species, with LC_{50} values of 8.4 and 5018 mg kg^{-1}, respectively [45]. The hopanoid triterpene 22-hydroxypropan was also isolated from the methanolic extract of *A. latifolium*, leading to the death of *O. rhinoceros* larvae (LC_{50} 20.81 μg g^{-1}) [46].

Methanolic extracts from the species *Azolla pinnata* caused the death of eggs, larvae, and adults and inhibited the egg-laying of *A. aegypti* and *A. Albopictus* mosquitoes. The substances related to insecticidal activity were 1- (O-alpha-D-glucopyranosyl) (1,3*R*, 25*R*)-hexacosanetriol, pyridate, and *N*-nicotinamide oxide [47, 48].

4 Conclusion

The fern and lycophyte families that stood out for their insecticidal potential were Aspleniaceae, Cyatheaceae, and Thelypteridaceae. Of the 80 species studied, 47 ferns and two lycophytes exhibited medically and agriculturally important insecticidal, repellent, or insect growth-regulatory activity. Although few studies performed the chemical characterization of fern and lycophyte extracts, terpenes and phenolic substances were present and can be involved in the insecticide/repellent action observed. The most widely used insects in the tests were *Spodoptera litura*, *Helicoverpa armigera*, *Aedes aegypti*, *Aedes albopictus,* and *Anopheles stephensi*, belonging to the orders Lepidoptera and Diptera. Most of the insects used in the tests were considered pests (69.8%) and displayed holometabolous development (71.8%). The contact tests involving adult insects were used preferentially to assess the insecticidal potential of pteridophyte species. Polar extracts, such as aqueous and ethanolic, were used, demonstrating better insecticidal activity for the species tested. Despite the scarcity of studies, only 0.7 and 0.4% of ferns and lycophyte species, respectively, were assessed, and it is believed that these groups of plants show potential as a source of substances with insecticidal and repellent activity.

References

1. Ruppert EE, Fox RS, Barnes RD (2005) Zoologia de Invertebrados: uma abordagem funcional – evolutiva. Roca, São Paulo
2. Núncio MS, Alves MJ (2014) Doenças associadas a artrópodes vetores e roedores. Instituto Nacional de Saúde Doutor Ricardo Jorge, Lisboa
3. Lourenço AF, Rodrigues FM (2017) Diseases transmitted by the *Aedes aegypti* (Linnaeus, 1762) in Brazil in the last ten years. Revista EVS 44:72–77. https://doi.org/10.18224/evs.v0i0.5788
4. Gallo D, Nakano O, Silveira Neto S, Carvalho RPL, Baptista GC, Berti Filho E, Parra JRP, Zucchi RA, Alves SB, Vendramim JD, Marchini LC, Lopes JRS, Omoto C (2002) Entomologia agrícola. FEALQ, Piracicaba
5. Paul A, Harrington LC, Scott JG (2006) Evaluation of novel insecticides for control of dengue vector *Aedes aegypti* (Diptera: Culicidae). J Med Entomol 43:55–60. https://doi.org/10.1093/jmedent/43.1.55
6. Ranson H, Burhani J, Lumjuan N, Black WC IV (2010) Insecticide resistance in dengue vectors. TropIKAnet 1:1–12. https://archive.lstmed.ac.uk/999/1/Ranson_et_al_a03v1n1.pdf
7. Koul O, Walia SE, Dhaliwal GS (2008) Essential oils as green pesticides: potential and constraints. Biopestic Int 4:63–84. https://doi.org/0973-483X/08/63-84
8. Plata-Rueda A, Martínez LC, Dos-Santos MH, Fernandes FL, Wilcken CF, Soares MA, Zanuncio JC (2017) Insecticidal activity of garlic essential oil and their constituents against the mealworm beetle, *Tenebrio molitor* Linnaeus (Coleoptera: Tenebrionidae). Sci Rep 7:1–11. https://doi.org/10.1038/srep46406
9. Muturi EJ, Ramirez JL, Zilkowski B, Flor-Weiler LB, Rooney AP (2018) Ovicidal and larvicidal effects of garlic and asafoetida essential oils against West Nile virus vectors. J Insect Sci 18:43–49. https://doi.org/10.1093/jisesa/iey036
10. Kanda D, Kaur S, Koul O (2017) A comparative study of monoterpenoids and phenylpropanoids from essential oils against stored grain insects: acute toxins or feeding deterrents. J Pest Sci 90:531–545. https://doi.org/10.1007/s10340-016-0800-5
11. Lee SH, Ha KB, Park DH, Fang Y, Kim JH, Park MG, Je YH (2018) Plant-derived compounds regulate formation of the insect juvenile hormone receptor complex. Pestic Biochem Physiol 150:27–32. https://doi.org/10.1016/j.pestbp.2018.06.013
12. Plata-Rueda A, Martínez LC, Rolim GDS, Coelho RP, Dos-Santos MH, De Tavares WS, Zanuncio JC, Serrão JE (2020) Insecticidal and repellent activities of *Cymbopogon citratus*

(Poaceae) essential oil and its terpenoids (citral and geranyl acetate) against *Ulomoides dermestoides*. Crop Prot 137:105299. https://doi.org/10.1016/j.cropro.2020.105299
13. Pryer KM, Schneider H, Magallón S (2004) The radiation of vascular plants. In: Cracraft J, Donoghue MJ (eds) Assembling the tree of life. Oxford University Press, London
14. Moran RC (2008) Diversity, biogeography, and floristics. In: Ranker TA, Haufler CH (eds) Biology and evolution of ferns and lycophytes. Cambridge University Press, New York
15. PPG I (2016) A community-derived classification for extant lycophytes and ferns. J Syst Evol 54:563–603. https://doi.org/10.1111/jse.12229
16. May LW (1978) The economic uses and associated folklore of ferns and fern allies. Bot Rev 44: 491–528. https://doi.org/10.1007/BF02860848
17. Santos MG, Kelecom A, Paiva SR, Moraes MG, Rocha L, Garrett R (2010) Phytochemical studies in pteridophytes growing in Brazil: a review. Am J Plant Sci Biotechnol 4:113–125. http://www.globalsciencebooks.info/Online/GSBOnline/images/2010/AmJPSB_4(SI1)/AmJPSB_4(SI1)113-125o.pdf
18. Keller HA, Prance GT (2015) The ethnobotany of ferns and lycophytes. Fern Gaz 20:1–13. https://ebps.org.uk/bigmedia/Fern%20Gazette%20V20.pdf
19. Cao H, Chai TT, Wang X, Morais-Braga MFB, Yang JH, Wong FC, Wang R, Yao H, Cao J, Cornara L, Burlando B, Wang Y, Xiao J, Coutinho HDM (2017) Phytochemicals from fern species: potential for medicine applications. Phytochem Rev 16:379–440. https://doi.org/10.1007/s11101-016-9488-7
20. Sahayaraj K, Selvaraj P (2013) Field evaluation of three fern extracts on *Helicoverpa armigera* (Hubner) and *Spodoptera litura* (Fab.) incidence and infestation and groundnut production. Legume Res 36:84–86. https://arccjournals.com/journal/legume-research-an-international-journal/ARCC211
21. Lovatto PB, Schiedeck G, Mauch CR (2013) Efeito do extrato de *Pteridium aquilinum* (Dennstaedtiaceae) sobre a preferência alimentar e biologia de *Brevicoryne brassicae* e *Myzus persicae* (Hemiptera: Aphididae). Cadernos de Agroecologia 8:1–5. https://revistas.aba-agroecologia.org.br/cad/article/view/13917
22. Huang SQ, Zhang ZX, Li YZ, Li YX, Xu HH (2010) Anti-insect activity of the methanol extracts of fern and gymnosperm. Agric Sci China 9:249–256. https://doi.org/10.1016/S1671-2927(09)60090-0
23. Gillott C (2005) Entomology, 3rd edn. Springer, p 831
24. Tharamak S, Yooboon T, Pengsook A, Ratwatthananon A, Kumrungsee N, Bullangpoti V, Pluempanupat W (2020) Synthesis of thymyl esters and their insecticidal activity against *Spodoptera litura* (Lepidoptera: Noctuidae). Pest Manag Sci 76:928–935. https://doi.org/10.1002/ps.5598
25. Volpicella M, Ceci LR, Cordewener J, America T, Gallerani R, Bode W, Beekwilder J (2003) Properties of purified gut trypsin from *Helicoverpa zea*, adapted to proteinase inhibitors. Eur J Biochem 270:10–19. https://doi.org/10.1046/j.1432-1033.2003.03368.x
26. Lecuona RE, Turica M, Tarocco F, Crespo DC (2005) Microbial control of *Musca domestica* (Diptera: Muscidae) with selected strains of *Beauveria bassiana*. J Med Entomol 42:332–336. https://doi.org/10.1093/jmedent/42.3.332
27. Maniero VC, Santos MO, Ribeiro RL, De Oliveira PA, Da Silva TB, Moleri AB, Cardozo SV (2016) Dengue, chikungunya e zika vírus no brasil: situação epidemiológica, aspectos clínicos e medidas preventivas. Almanaque multidisciplinar de pesquisa 3:118–145. http://publicacoes.unigranrio.edu.br/index.php/amp/article/view/3409
28. Benedict MQ, Levine RS, Hawley WA, Lounibos LP (2007) Spread of the tiger: global risk of invasion by the mosquito *Aedes albopictus*. Vector Borne Zoonotic Dis 7:76–85. https://doi.org/10.1089/vbz.2006.0562
29. Miazgowicz KL, Shocket MS, Ryan SJ, Villena OC, Hall RJ, Owen J, Murdock CC (2020) Age influences the thermal suitability of *Plasmodium falciparum* transmission in the Asian malaria vector *Anopheles stephensi*. Proc R Soc B 287:20201093. https://doi.org/10.1098/rspb.2020.1093

30. Dinan L (2010) Phytoecdysteroids: biological aspects. Phytochemistry 57:325–339. https://doi.org/10.1016/S0031-9422(01)00078-4
31. Pereira ACRL, Oliveira JVD, Gondim Junior MGC, Câmara CAGD (2008) Atividade inseticida de óleos essenciais e fixos sobre *Callosobruchus maculatus* (FABR., 1775) (Coleoptera: Bruchidae) em grãos de caupi [*Vigna unguiculata* (L.) WALP.]. Ciênc Agrotec 32:717–724. https://doi.org/10.1590/S1413-70542008000300003
32. Labucka OD, Lopez LM, Gurvich ED, Martini C, Zygadlo AJ, Rotman A, Ahumada O (2005) Aromatic plants of Yungas, Part V. essential oils composition of *Polystichum montevidense* (Spreng.) Rosenst.(Dryopteridaceae) and *Podocarpus parlatorei* Pilg.(Podocarpaceae). J Essent Oil Bear Pl 8:183–186. https://doi.org/10.1080/0972060X.2005.10643441
33. Essien EE, Ascrizzi R, Flamini G (2019) Characterization of volatile compounds of *Diplazium esculentum*. Chem Nat Compd 55:958–959. https://doi.org/10.1007/s10600-019-02860-y
34. Simmonds MS (2003) Flavonoid–insect interactions: recent advances in our knowledge. Phytochemistry 64:21–30. https://doi.org/10.1016/S0031-9422(03)00293-0
35. Freitas TCD (2012) Caracterização da atividade entomotóxica induzida pelo extrato metanólico de *Araucaria angustifolia* em baratas da espécie *Phoetalia pallida*. Dissertation, Universidade Federal do Pampa
36. Morais SM, Lima KSB, Siqueira SMC, Cavalcanti ESB, Souza MST, Menezes JESA, Trevisan MTS (2013) Correlação entre as atividades antiradical, antiacetilcolinesterase e teor de fenóis totais de extratos de plantas medicinais de farmácias vivas. Rev Bras Pl Med 15:575–582. https://doi.org/10.1590/S1516-05722013000400014
37. Selvaraj P, De Britto AJ, Sahayaraj K (2005) Phytoecdysone of *Pteridium aquilinum* (L.) Kuhn (Dennstaedtiaceae) and its pesticidal property on two major pests. Arch Phytopathol Plant Prot 38:99–105. https://doi.org/10.1080/0323540040007517
38. Gillij YG, Gleiser RM, Zygadlo JA (2008) Mosquito repellent activity of essential oils of aromatic plants growing in Argentina. Bioresour Technol 99:2507–2515. https://doi.org/10.1016/j.biortech.2007.04.066
39. Lovatto PB (2012) As plantas bioativas como estratégia à transição agroecológica na agricultura familiar: análise sobre a utilização empírica e experimental de extratos botânicos no manejo de afídeos em hortaliças. Thesis, Universidade Federal de Pelotas
40. Mary SC, Murugan K, Roni M, Sivapriyajothi S, Suganya NA (2013) Larvicidal potential of silver nanoparticles synthesized using *Adiantum capillus verenis* against *Anopheles stephensi* (Diptera, Culicidae). Int J Curr Trop Med Health Res 1:09–18
41. Viegas-Júnior C (2003) Terpenos com atividade inseticida: uma alternativa para o controle químico de insetos. Quím Nova 26:390–400. https://doi.org/10.1590/S0100-40422003000300017
42. Hincapie CA, Monsalve Z, Parada K, Lamilla C, Alarcón J, Cespedes CL, Seigler D (2011) Insect growth regulatory activity of *Blechnum chilense*. Nat Prod Commun 6:1085–1088. https://doi.org/10.1177/1934578X1100600808
43. Xavier GSA, Selvaraj P, John N (2016) Impact of phytoecdysone fractions of the ferns *Cyclosorous interruptus*, *Christella dentata* and *Nephrolepis cordifolia* on the biology of *Spodoptera litura* (Fab.). J Biopest 9:125. http://www.jbiopest.com/users/LW8/efiles/vol_9_2_125-134.pdf
44. Adebiyi AO, Oyeyemi SD, Tedela PO, Ojo VI (2019) GC-MS analysis of bioactive compounds from n-hexane leaf extract of a tropical fern, *Nephrolepis cordifolia* (L.) C. Presl. East Afr Scholars J Biotechnol Genet 1:118–123. https://doi.org/10.36349/easjbg.2019.v01i05.004
45. Kumar RP, John A, Kumar P, Babu KVD, Evans DA (2018) Larvicidal efficacy of Adiantobischrysene from *Adiantum latifolium* against *Oryctes rhinoceros* through disrupting metamorphosis and impeding microbial mediated digestion. Pest Manag Sci 74:1821–1828. https://doi.org/10.1002/ps.4880
46. Kumar RP, Babu KD, Evans DA (2019) Isolation, characterization and mode of action of a larvicidal compound, 22-hydroxyhopane from *Adiantum latifolium* Lam. against *Oryctes rhinoceros* Linn. Pestic Biochem Phys 153:161–170. https://doi.org/10.1016/j.pestbp.2018.11.018

47. Ravi R, Zulkrnin NSH, Rozhan NN, Nik Yusoff NR, Mat Rasat MS, Ahmad MI, Amin MFM (2018) Chemical composition and larvicidal activities of *Azolla pinnata* extracts against *Aedes* (Diptera: Culicidae). PLoS One 13:e0206982. https://doi.org/10.1371/journal.pone.0206982
48. Ravi R, Rajendran D, Oh WD, Rasat MSM, Hamzah Z, Ishak IH, Amin MFM (2020) The potential use of *Azolla pinnata* as an alternative bio-insecticide. Sci Rep 10:1–8. https://doi.org/10.1038/s41598-020-75054-0
49. Antonysamy MAJ, Janarthanan G, Arumugam S, Narayanan J, Mani N (2014) Antioxidant, larvicidal, and cytotoxic studies on *Asplenium aethiopicum* (Burm. f.) Becherer. Int Sch Res Notices 2014:1–6. https://doi.org/10.1155/2014/876170
50. Prabhakaran P, Radhakrishnan B, Srikumar KK, Kumar BS (2017) Efficacy of certain common ferns against red spider mite *Oligonychus coffeae* and tea mosquito bug *Helopeltis theivora* infesting tea. Plant Prot Sci 53:232–242. https://doi.org/10.17221/23/2015-PPS
51. Markham K, Chalk T, Stewart CN Jr (2006) Evaluation of fern and moss protein-based defenses against phytophagous insects. Int J Plant Sci 167:111–117. https://doi.org/10.1086/497651
52. Brudea V, Rišca I, Enea C, Tomescu C (2007) Researches concerning the efficacy of some vegetal metabolites in the control of Colorado potato beetle (coleoptera-*Leptinotarsa decemlineata* say). Lucrări Ştiinţifice 51:258–263
53. Tomescu CV, Brudea V, Rišca M (2009) Preliminary research on the efficiency of some vegetal metabolites in fighting the mealy plum aphid (*Hyalopterus pruni* Geoffroi – Ord. Homoptera). Analele Universitatii Stefan Cel Mare Suceava Sectiunea Silvicultura 11
54. Halimatussakdiah H, Amna U, Wahyuningsih P (2018) Preliminary phytochemical analysis and larvicidal activity of edible fern (*Diplazium esculentum* (retz.) sw.) extract against culex. J Nat 18:141–147. https://doi.org/10.24815/jn.v18i3.11335
55. Narayanan J, Antonysamy MAJ (2017) Larvicidal potential of *Cyathea* species against *Culex quinquefasciatus*. Pharm Biomed Res 3:48–51. https://doi.org/10.18869/acadpub.pbr.3.1.48
56. Fazolin M, Estrela JLV, Lima AP, Argolo VM (2002) Avaliação de plantas com potencial inseticida no controle da vaquinha-do-feijoeiro (*Cerotoma tingomarianus* Bechyné). Embrapa Acre, Rio Branco
57. Potenza MR, Justi Junior J, Alves JN (2006) Evaluation of contact activities of plant extracts against *Sitophilus zeamais* Motschulsky (Coleoptera: Curculionidae). Paper presented at the Proceedings of 9th International Working Conference on Stored Product Protection, Campinas, São Paulo, 15–18 October 2006
58. Jbilou R, Amri H, Bouayad N, Ghailani N, Ennabili A, Sayah F (2008) Insecticidal effects of extracts of seven plant species on larval development, α-amylase activity and offspring production of *Tribolium castaneum* (Herbst) (Insecta: Coleoptera: Tenebrionidae). Bioresour Technol 99:959–964. https://doi.org/10.1016/j.biortech.2007.03.017
59. Santa-Cecília LVC, Santa-Cecília FV, Pedroso EDC, Sousa MVD, Abreu FA, Oliveira DF, Carvalho GA (2010) Extratos de plantas no controle de *Planococcus citri* (Risso, 1813) (Hemiptera: Pseudococcidae) em cafeeiro. Coffee Sci 5:283–293. http://www.coffeescience.ufla.br/index.php/Coffeescience/article/view/171
60. Sahayaraj K (2011) Aqueous and water extracts of chosen botanicals on *Helicoverpa armigera* Hubner and *Spodoptera litura* Fab. J Agric Technol 7:339–347. http://ijat-aatsea.com/pdf/April_v7_n2_11/13%20IJAT2010_30FT.pdf
61. Gerhardt A, Putzke MTL, Lovatto PB (2012) Atividade inseticida de extratos botânicos de três espécies silvestres do Rio Grande do Sul, Brasil, sobre *Myzus persicae* (Hemiptera: Aphididae) e *Ascia monuste orseis* (Lepidoptera: Pieridae). Caderno de Pesquisa 24:55–64. https://doi.org/10.17058/cp.v24i2.3590
62. Panneerselvam C, Murugan K, Roni M, Suresh U, Rajaganesh R, Madhiyazhagan P, Alarfaj AA (2016) Fern-synthesized nanoparticles in the fight against malaria: LC/MS analysis of *Pteridium aquilinum* leaf extract and biosynthesis of silver nanoparticles with high mosquitocidal and antiplasmodial activity. Parasitol Res 115:997–1013. https://doi.org/10.1007/s00436-015-4828-x

63. De Souza JB, Lima GP, Feder MD, Lancellotti IR, Santos MG (2020) *Dicksonia sellowiana* Hook. and *Nepholepis cordifolia* (L.) C. Presl. extracts as potential green pesticides: insecticidal activity. Res. Soc Dev 9:e120985182. https://doi.org/10.33448/rsd-v9i8.5182
64. Shukla S, Tiwari SK (2011) Insecticidal activity of *Dryopteris filix-mas* (Linn.) schott ethanolic extract against *Corcyra cephalonica* Staint. (Lepidoptera: Pyralidae). J Biopest 4:138–143. http://www.jbiopest.com/users/lw8/efiles/vol_4_2_253c.pdf
65. Shukla S, Tiwari SK (2011) Toxicological effects of *Dryopteris filix-mas* against the ontogeny of rice-moth, *Corcyra cephalonica* (Staint). World Appl Sci J 12:16–20. http://citeseerx.ist.psu.edu/viewdoc/download?doi=10.1.1.390.3751&rep=rep1&type=pdf
66. Benoufella-Kitous K, Doumandji S, Fellag H, Hance T (2014) Bio-insecticide effect of the extracts of *Urtica dioïca* and *Dryopteris filix-mas* on *Aphis fabae* (Homoptera: Aphididae). Int J Biol Technol 5:14–20
67. Chermenskaya TD, Stepanycheva EA, Shchenikova AV, Chakaeva AS (2010) Insectoacaricidal and deterrent activities of extracts of Kyrgyzstan plants against three agricultural pests. Ind Crop Prod 32:157–163. https://doi.org/10.1016/j.indcrop.2010.04.009
68. Kweon JH, Ahn YJ, Known HW, Jang KS, Cho KY (1994) Larvicidal and antifeeding activities of oriental medicinal plant extracts against *Plutella xyslotella* (Lepidoptera: Yponomeutoidae) and *Spodoptera litura* (Lepidoptera: Noctuidae). Korean J Appl Entomol 33:225–229. https://www.koreascience.or.kr/article/JAKO199403042974516.pdf
69. Park SJ, Lee SG, Shin SC, Lee BY, Ahn YJ (1997) Larvicidal and antifeeding activities of oriental medicinal plant extracts against four species of forest insect pests. Appl Entomol Zool 32:601–608. https://doi.org/10.1303/aez.32.601
70. Yi JH, Park IK, Choi KS, Shin SC, Ahn YJ (2008) Toxicity of medicinal plant extracts to *Lycoriella ingenua* (Diptera: Sciaridae) and *Coboldia fuscipes* (Diptera: Scatopsidae). J Asia Pac Entomol 11:221–223. https://doi.org/10.1016/j.aspen.2008.09.002
71. Kamaraj C, Deepak P, Balasubramani G, Karthi S, Arul D, Aiswarya D, Perumal P (2018) Target and non-target toxicity of fern extracts against mosquito vectors and beneficial aquatic organisms. Ecotoxicol Environ Saf 161:221–230. https://doi.org/10.1016/j.ecoenv.2018.05.062
72. Kainz K (2011) Phytochemical investigation and bioactivity-guided isolation of cytotoxic compounds from *Metaxya rostrata*. Dissertation, Universität Wien
73. Selvaraj P, Jayaseeli PJR, Ansilin M (2018) Pest control potential of *Cyclosorous interruptus*, *Christella dentata* and *Nephrolepis cordifolia* on the biology of *Spodoptera litura* (fab.). J Biopest 11:76–81. http://www.jbiopest.com/users/LW8/efiles/vol_11_1_76-81.pdf
74. Sanchez DME, De Souza J, Oshima-Franco Y, Rostealto-Ferreira S (2018) Pharmacological effects of *Nephrolepis exaltata* L.(fern) aqueous extract on an insect-based model (*Nauphoeta cinerea*). J Plant Sci Phytopathol 2:031–036. https://doi.org/10.29328/journal.jpsp.1001017
75. Potenza MR, Silva RD, Arthur V, Felicio JD, Rossi MH, Sakita NM (2004) Avaliação de produtos naturais irradiados para o controle de *Blattella germanica* (L.) (Dictyoptera: Blattellidae). Arq Inst Biol 71:485–492
76. Shukla AK, Upadhyay SK, Mishra M, Saurabh S, Singh R, Singh H, Srivastava S (2016) Expression of an insecticidal fern protein in cotton protects against whitefly. Nat Biotechnol 34: 1046–1051. https://doi.org/10.1038/nbt.3665
77. Alvarez MRS, Quiming NS, Heralde FM (2015) Screening for larvicidal activity of aqueous and ethanolic extracts of fourteen selected plants and formulation of a larvicide against *Aedes aegypti* (Linn.) and *Aedes albopictus* (Skuse) larvae. Int J Pharm Pharm Sci 9:1183–1189. https://doi.org/10.5281/zenodo.1129836
78. Balasubramanian R, Selvaraj P, Sahayaraj K (2008) Partial purification and characterization of phytoecdysone from *Chrystella parasitica* (L.) and screening its pesticidal properties on lepidopteran pests. J Biopest 1:201–205. http://www.jbiopest.com/users/LW8/efiles/R.Balasubramanian_5_2.pdf
79. Huang SQ, Fu JT, Wang K, Xu HH, Zhang ZX (2016) Insecticidal activity of the methanol extract of *Pronephrium megacuspe* (Thelypteridaceae) and its active component on *Solenopsis invicta* (hymenoptera: Formicidae). Fla Entomol 99:634–638. https://doi.org/10.1653/024.099.0408

Phytochemicals from the Pteridaceae Family and Their Prospects as Future Drugs 13

Shreeta Singha, Rajat Nath, Subrata Das, Sibashish Kityania, Deepa Nath, and Anupam Das Talukdar

Contents

Abstract

Pteridaceae is the most ethnomedicinally significant fern family among pteridophytes. This family possesses numerous phytochemically sound and potent species. There are 58 species altogether that bear the potential to treat a great

S. Singha · R. Nath · S. Das · S. Kityania · A. Das Talukdar (✉)
Department of Life Science and Bioinformatics, Assam University, Silchar, Assam, India
e-mail: anupam@bioinfoaus.ac.in

D. Nath
Department of Botany, Gurucharan College, Silchar, Assam, India

H. N. Murthy (ed.), *Bioactive Compounds in Bryophytes and Pteridophytes*,
Reference Series in Phytochemistry, https://doi.org/10.1007/978-3-031-23243-5_16

number of human ailments. Traditional ethnic minorities staying in a remote locality hold indigenous wisdom helping us to get enlightened about the significance of traditional knowledge of plants. Since the ancient day's *Adiantum capillus-veneris*, a species from the family Pteridaceae was used as an Indian herb-based medicine mentioned in Samhitas. An insight phytochemical analysis unwrapped the presence of iso-adiantol, rutin, quercetin, and iso-adiantone in this species, which act as an anticoagulant, antipyretic, and antivenom against wounds, fever, and snake bites, respectively. Further, a number of genera – *Acrostichum, Actiniopteris, Aleuritopteris, Antrophyum, Ceratopteris, Cheilanthes, Cheilosoria, Coniogramme, Doryopteris, Hemionitis, Notholaena, Onychium, Parahemionitis, Pityrogramma, Pteris, and Vittaria* – elongate the list revealing the essence of family Pteridaceae being pharmacologically active possessing rich secondary metabolites, such as tannins, saponins, cardiac glycosides, sterols, oils, flavonoids, quinones, glycosides, terpenoids, phenol, coumarins, steroids, alkaloids, anthocyanin, and betacyanin addressing a variety of severe human ailments like diarrhea, fever, bronchitis, tuberculosis, uterine problem, wounds, sprains, typhoid, bone fractures, blood pressure, helminthiasis, and rheumatism and a plethora of ailments and health disorders. This chapter emphasizes the various phytochemicals present in all medicinally important known species of the Pteridaceae family which are pharmacologically active against numerous human disorders.

Keywords

Ethnomedicine · Fern · Phytochemicals · Pteridaceae · Pteridophytes

Abbreviations

ACE2	Angiotensin converting enzyme 2
Bax	BCL2-Associated X Protein
Bcl-2	B-cell lymphoma 2
CAT	Catalase
CD40	Cluster of Differentiation 40
COX	Cyclooxygenase
ENT	Ear, nose, and throat problems
FITC	Fluorescein isothiocyanate
GLUT4	Glucose transporter type 4
GSH	Glutathione
IL6	Interleukin 6
NADPH	Nicotinamide adenine dinucleotide phosphate
NF-kB	Nuclear factor kappa
Nrf2	Nuclear factor erythroid 2–related factor 2
ROS	Reactive oxygen species
SOD	Superoxide dismutase
STAT1	Signal transducer and activator of transcription 1
STAT-3	Signal transducer and activator of transcription 3

STZ	Streptozotocin
WHO	World Health Organization

1 Introduction

Plants have been nourishing us selflessly for millions of years. One of the most important divisions of plants emerged in the Silurian period ~438 Mya called pteridophyte. Pteridophytes are the first group of the vascular plant on Earth from where land flora originated. Rhyniophytes, horneophytes, and psilophytes, the most primitive dichotomously branched pteridophytes, emerged undergoing morphophysiological changes from non-vascular bryophyte *Anthoceros* from the class Anthocerotopsida. Due to adaptation and evolutionary mechanisms, it initiated the blossoming of thousands of species listed and grouped under the Pteridaceae family. The extensive studies of literature, significant information gathered by proper surveys, and deep study, analysis, and exploration of papers and articles conveyed a sound group of ethnomedicinally important genera of Pteridaceae. The plant parts herbal formulation of Pteridaceae genera started anciently from the era of Sushruta and Charaka (~100 CE). The therapeutic potential of Pteridophytes like *Adiantum capillus-veneris* was mentioned in the ancient book "Samhitas." This gives us convincing clue that India has sound biodiversity and medicinally important plants belonging to the family Pteridaceae. There are 58 species from the family which are found to be active therapists curing vast human ailments. A thorough study of literature on ethnomedicinally important pteridophytes revealed that the most significant family of Pteridaceae embraces 18 genera, namely *Acrostichum*, *Actiniopteris*, *Adiantopsis*, *Adiantum*, *Aleuritopteris*, *Antrophyum*, *Ceratopteris*, *Cheilanthes*, *Cheilosoria*, *Coniogramme*, *Doryopteris*, *Hemionitis*, *Notholaena*, *Onychium*, *Parahemionitis*, *Pityrogramma*, *Pteris*, *and Vittaria* [1].

2 Pteridaceae Family

The Pteridaceae family is distributed among 53 genera and 1000 species [2]. Normally, they are spread among the aquatic and terrestrial habitats. Mostly, the plant body is small in size. There are some species where leaves grow 6 meters in length. The stem position is erect and sometimes creeps on the substratum where it grows. Stele types generally found in species of Pteridaceae are protostele, solenostele, or sometime dictyostele. Trichomes are present, scales also present, or sometimes both trichomes and scales coexist. The lamina is generally pinnate, pedate, or entire, palmate, or radiate. Veins are mostly free. Sporangia bear long stalk, 2 or 3 rows of cells form the stalk. Soral lines can be observed along the veins and sometimes at the marginal line. Spores are chlorophyll-less, mostly trilete shaped, and rarely alete shaped [3, 4]. Silica bodies of Pteridaceae are elongated, linear to elliptical, with apices mostly blunt or acute [3, 5]. The important species from the 18 different genera of the Pteridaceae family are ethnomedicinally important and have a high therapeutic effect [6–10]. All together from these 18 genera, 58 species represent the medicinal significance of the very Pteridaceae family for human health [1, 11]. The secondary metabolites, viz., tannins, saponins, flavonoids, quinones, glycosides, phenols, coumarins, betacyanins and other active constituents present in those species, increase the bioactivity and therapeutic potential to a high fold. Globally the indigenous people and local people have been practicing herbal formulation of pharmacologically active genera of the family Pteridaceae [2, 3, 5, 11–17].

3 Ethnobotanical Importance of Pteridaceae

Tribal communities residing vicinity of a biodiverse forest is solely dependent on natural remedies for healthcare. The medicinal plants from the Pteridaceae family are grown and used by various ethnic groups scattered in many states of India. *Adiantum capillus-veneris*, *Adiantum lunulatum*, *Actiniopteris radiata*, and *Coniogramme japonica* are predominantly used in the Northeastern and Northern zone of India by Naga, Gara, Apatani, Abor, Mishmi, Mizo, Khasi, Dafla, and Galong. *Aleuritopteris farisona*, *Adiantum venustum*, *Acrostichum radiata*, *Pteris biaurita*, and *Vittaria microlepis* are commonly used by the tribal community residing in Southern India, namely, Badaga, Irula, Toda, Kanikar, Koya, Gadaba, Karar, Muthuvam, Yenadi, and Uradi [18–20].

The questionnaire, interviews, and survey done by researchers across the nook and corner of the world with experienced elderly traditional knowledgeable practitioners and local knowledgeable people revealed the significant bioactivity and therapeutic role held by plenty of important species belonging to the family Pteridaceae. Nigerian local people, Mexican indigenous people, Amazonian indigenous people, and Northeast Brazilians use *Adiantum* and *Pteris* to cure respiratory diseases, sensory organ diseases, intestinal diseases, reproductive-related diseases, and other human ailments [21]. In India, also states like Uttarakhand, Tamil Nadu, Rajasthan, Andhra Pradesh, and Madhya Pradesh use *Actiniopteris radiata* to cure tuberculosis, bronchitis, uterine problems, dysentery, fever, sore throat, bronchitis, and piles [22]. Local people of Assam, Mizoram, Arunachal Pradesh, Rajasthan, Goa, Maharashtra, Madhya Pradesh, and Uttarakhand use *Adiantum lunulatum* to treat chest pain, bronchitis, indigestion, menstrual disorders, dysentery, fever, epilepsy, ulcer, hair fall, leprosy, diabetes, wounds, swelling, and gastritis [5, 23, 24]. *Antrophyum plantagineum* is used against fever. *Aleuritopteris* sp. is used to cure urinary problems, skin irritation, and stomach acid menstrual disorders. *Adiantopsis chlorophylla* is used against cough. *Ceratopteris thalictroides* is used to heal cuts, wounds, bleeding, piles and diuretic. *Cheilanthes farinosa* is used to treat fever, nose-ear ache, jaundice, epilepsy, nervous problem, pus, ulcer, throat infection, and hair tonic. *Cheilosoria tenuifolia* is used as an antitumor, to treat the bone fracture, throat pain, and burns. *Coniogramme caudata* is used to treat headache. *Doryopteris concolor* is used to cure chronic disorders. *Hemionitis arifolia* is used to treat hypertension, diabetes, cancer, rheumatism, migraine, wounds, skin infection, smallpox, and chicken pox. *Notholaena standleyi* is used to treat post-natal disorders. *Onychium japonicum* is used as a hair tonic. *Parahemionitis cordata* is used to heal menstrual problems and earache. *Pityrogramma calomelanos* is used against renal problem, hypertension, cough, measles, boils, chest congestion, flu, and fever and as antihelminth. *Pteris biaurita* is used against body pain, cuts, malaria, and chronic disorders. *Pteris confusa* is used against scabies, leprosy, burns, and boils, and *Vittaria elongata* is used to cure knee pains, cuts, wound, and rheumatism [1, 6]. The knowledge held by ethnic traditional people thriving closely associated with nature provides depth and significant information about nature's love. Indigenous and local people inhabiting closely to the natural biodiversity residing in developing country like India depend upon natural plant resources for medicinal use. The database analysis by the World Health Organization (WHO) revealed that 80% of people rely upon nature's therapy in form of crude products which have therapeutic potential to cure a huge number of human disorders [1, 25].

4 Therapeutic Potential of the Species Belonging to Pteridaceae

More than 80% of traditional medicine is a formulation of plant extracts. The Pteridaceae family is the richest family among the pteridophytes division possessing rich nutraceuticals and therapeutic potent phytochemicals. These phytochemicals are the source of medicine having bioactivity and therapeutic value which is capable of curing hundreds of human ailments and disorders. Several severe disorders of human beings are categorized as hemorrhoids; central nervous system/brain and memory-related disorder; kidney problems; circulatory system/cardiovascular disorder; dermatological infection and diseases; liver problems; ear, nose, and throat problems (ENT); neoplasms; endocrinal disorders; ophthalmological disorder; fever; poisonous bites; gastrointestinal ailments; respiratory system diseases; musculoskeletal system disorders; gynecological/reproductive health diseases; hair care; and general health that can be cured by the use of active chemical ingredients present in significant medicinal plants of the Pteridaceae family. *Actiniopteris dichotoma* whole plant methanolic extract possesses phytochemicals like sugar, sterols, glycosides, fixed oil, saponins, and flavonoids which have pharmacological activity against diabetes, heart disease, cancer, analgesic, antitubercular, and skin disease [26–29]. *Adiantopsis chlorophylla* whole frond extracts show bioactivity against expectorant, heart problems, cough, and stomach refresher [30]. *Aleuritopteris farinosa* shows pharmaceutical values against urinary problems, stomach aches, skin irritation, and menstrual disorders [31]. *Ceratopteris thalictroides* are used against skin diseases, wounds, and piles. *Cheilanthes rufa* is utilized to treat pus. *Coniogramme caudata* is used against headaches and burns. *Hemionitis arifolia* is utilized to treat wounds, female fertility, hypertension, diabetes, cancer, ulcer, smallpox, migraine, and skin infection. *Notholaena standleyi* is used to cure post-natal disorders. *Pityrogramma calomelanos* can be used against kidney trouble, antiparasitic, hypertension, cough, measles, flu, fever, asthma, mouth cancer, and gastrointestinal problem [1, 21, 23].

5 Phytochemicals from the Pteridaceae Family

Globally spread, 58 species from different 18 genera residing in various climatic zone across the world of the family Pteridaceae are found to have extensive use. The highly potential ethnomedicinally important 58 species and their phytochemicals constituent via immense data search, literature survey, and analysis have been revealed in Table 1.

6 Structure of Some Essential Therapeutic Potential Compounds Belonging to the Pteridaceae

Some important common flavonoids present in *Acrostichum aureum and Acrostichum heterophyllum* are tetracosane, patriscabratine, and kaempferol.

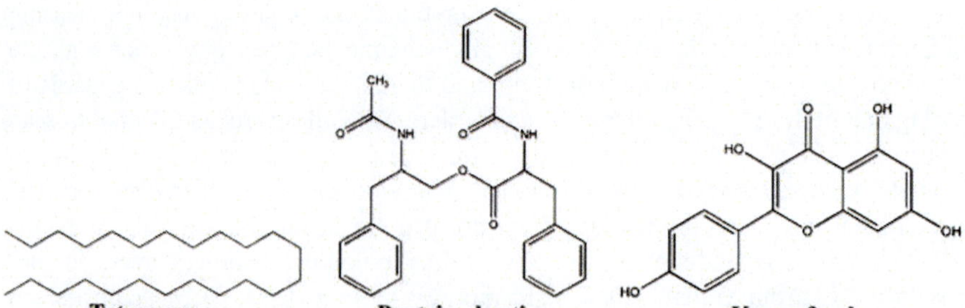

Tetracosane **Pantriscabratine** **Kaempferol**

Some important common triterpenoids present in *Adiantum lunulatum*, *Adiantum capillus-veneris*, *Adiantum caudatum*, *Adiantum venustum*, and *Adiantum edgeworthii* are neohopene, hopene, and filicane.

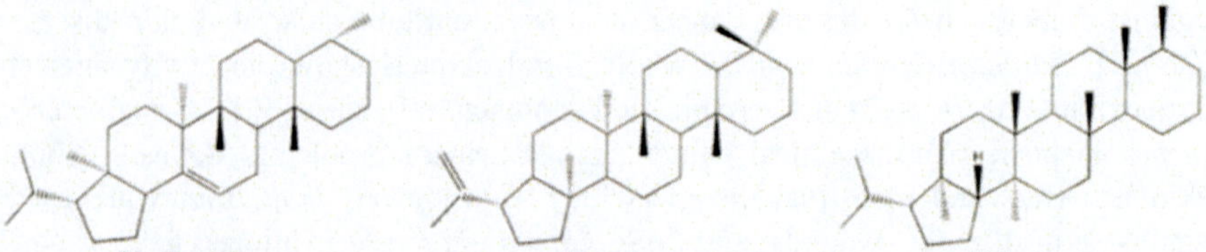

Neohopene **Hopene** **Filicane**

Table 1 Active phytochemicals and their respective therapeutic activity exhibited by 68 species of Pteridaceae

Scientific name	Type of solvent used	Active phytochemicals	Therapeutic activity	References
Acrostichum aureum	Petroleum ether extract	Saponins, phenols, glycosides, proteins, flavonoids (patriscabratine, tetracosane), and terpenoids	Wounds, boils, ulcers, and headaches	[25, 32]
Acrostichum heterophyllum	Methanol extract	Sugars, phenolics, cardiac glycosides, tannins, saponins, alkaloids, flavonoids, and proteins	Bone fracture, eczema, itching, cough, gonorrhea, constipation, chicken pox, and smallpox	[33]
Actiniopteris australis	Ethanol extract	Glycosides and phenols	Body cooling, edema, and wounds	[33]
Actiniopteris dichotoma	Ethyl acetate extract	Tannins, glycosides, and phenols	Antidiabetic, heart disease, cancer, diarrhea, anti-helminthic, leprosy, and skin disease	[27, 34]
Actiniopteris radiata	Hydro alcoholic extract	Glycosides, tannins, carbohydrates, flavonoids, sterols, terpenoids, alkaloids, phenolics, and saponins	Tuberculosis, bronchitis, epilepsy, typhoid, rheumatism, aphrodisiac, joint pain, uterine problems, diarrhea, blood pressure, cardiac tonic, sprains, and fever	[35–37]
Adiantopsis chlorophylla	Methanol extract	Alkaloids, sugar present, sterols, glycosides, fixed oil, saponins, and flavonoids	Heart problems, stomatal refresher, and cough	[27, 28]
Adiantopsis radiata	Methanol extract	Phenolics, tannins, saponins, and glycosides	Nose bleeding, headache, and nausea	[27]
Adiantum capillus-veneris	Methanol extract	Carotenoids, carbohydrates, steroids (beta-sitosterol, stigmasterol, and campesterol), flavonoids (rutin, quercetin, naringin, astragalin, populin),	Rheumatism, diarrhea, kidney problem, optical disorder, analgesic, snake bite fever, cough, hair loss, piles, bronchitis, and leucorrhea	[38, 39]

(continued)

Table 1 (continued)

Scientific name	Type of solvent used	Active phytochemicals	Therapeutic activity	References
		triterpenoids (iso-glauconome, isoadiantone, isoadiantol)		
Adiantum caudatum	Methanol extract	Phenols, terpenoids (fernane and filicane), flavonoids (isoquercitin, rutin, keampferol), and tannins	Diabetes, skin disease, cough, fever, and wounds	[40]
Adiantum edgeworthii	Methanol extract	Terpenoids (isohopane, neohopane, fernane, isofernane, filicane, pteronane, and adiane), steroids, lipids, phenolic propanoids, flavonoids, alicyclic acid, and long-chain compounds	Headache and mouth blisters	[41]
Adiantum flabellulatum	Methanol extract	Flavonoids, steroids, triterpenoids, and lipids	Anti-helminthic and cough	[41]
Adiantum incisum	Methanol extract	Steroids, triterpenoids, and phenolics	Urinary problems, asthma, skin diseases, hair loss, malaria, diabetes, jaundice, menstrual diseases, and intestinal pain	[41]
Adiantum lunulatum	Methanol extract	Triterpenoids, steroids, lipids, phenolic propanoids, flavonoids, alicyclic acid, and long-chain compounds	Dysentery, epilepsy, leprosy, chest pain, ulcer, irregular menstruation, snake bite, bronchitis, indigestion, rheumatism, gastric disorders, diabetes, and bone fracture	[41]
Adiantum pedatum	Methanol extract	Phenolic propanoids, flavonoids, and alicyclic acid	Chronic disorders	[41]
Adiantum peruvianum	Methanol extract	Steroids, Terpenoids (filicane and fernane), and phenolics	Demulcent, expectorant, and emmenagogue	[42]

(continued)

Table 1 (continued)

Scientific name	Type of solvent used	Active phytochemicals	Therapeutic activity	References
Adiantum poiretti	Methanol extract	Triterpenoids, steroids, lipids, phenolic propanoids, flavonoids, alicyclic acid, and long-chain compounds	Skin disease, diabetes, fever, and cough	[41]
Adiantum raddianum	Methanol extract	Steroids, triterpenoids, and phenol (daphnoretin and psoralen)	Asthma, indigestion, leprosy, hair growth, fertility control, cough, blood clotting, and sprain	[41]
Adiantum venustum	Methanol extract	Lipids, triterpenoids (adiantulanostene), phenolics, and alicyclic acid	Ophthalmic, headache, fever, expectorant, tumors, chest pain, hydrophobia, cold headache, and diuretic	[42]
Adiantum zollingeri	Methanol extract	Triterpenoids, alkaloids, and phenolics	Fever, skin irritation, and cold.	[41]
Aleuritopteris farinosa	Methanol extract	Lipids, phenolics, and terpenes	Skin diseases, menstrual disorders, urinary problems, and stomachache	[1]
Antrophyum plantagineum	Methanol extract	Phenols, triterpenoids, and lipids	Fever	[42]
Ceratopteris thalictroides	Methanol extract	Alkaloids, terpenoids, phytosterols, glycosides, coumarins, proteins, tannins, flavonoids, saponins, quinones, and phenols	Skin diseases, wounds, and piles	[43]
Cheilanthes albomarginata	Methanol extract	Glycosides, tannins, steroids, and alkaloids	Cuts, tuberculosis, wounds, chest pain, and burns	[43]
Cheilanthes argentea	Methanol extract	Glycosides, steroids, alkaloids	Lowering body temperature	[1]
Cheilanthes bicolor	Methanol extract	Glycosides, phenols, tannins, and alkaloids	Fever and ear most throat problems	[43]
Cheilanthes farisona	Methanol extract	Glycosides, alkaloids, and tannins	Eczema, jaundice, wounds, ulcer, skin irritation, and central nervous disorders	[1, 44]

(continued)

Table 1 (continued)

Scientific name	Type of solvent used	Active phytochemicals	Therapeutic activity	References
Cheilanthes grisea	Methanol extract	Tannins, steroids, and alkaloids	Tonic	[1]
Cheilanthes rufa	Methanol extract	Glycosides, tannins, alkaloids, and steroids.	Pus	[43]
Cheilanthes swartzii	Methanol extract	Glycosides, flavonoids, steroids, alkaloids	Throat infection and antioxidant	[43, 45]
Cheilosoria belangeri	Methanol extract	Flavonoids, glycerides, and phenols	Antitumor	[1]
Cheilosoria mysurensis	Methanol extract	Glycerides, lipids, and phenolics	Burns, fracture, and throat pain	[1]
Cheilosoria tenuifolia	Methanol extract	Flavonoids, phenols, and alkaloids	Tonic, giddiness, wounds, cuts, and swellings	[44]
Coniogramme caudata	Ethanol extract	Glycerol, apigenin and beta-sitosterol	Headache, anticancer, and anti-inflammatory	[46–49]
Coniogramme fraxinea	Ethanol extract	Sterols and glycerol	Burns and hepatic protective	[1, 48, 50, 51]
Doryopteris concolor	Methanol extract	Flavonoids, phenolics, and lipids	Chronic disorders	[1, 32, 52]
Hemionitis arifolia	Methanol extract	Flavonoids and phenolics	Hypertension, cuts, jaundice, dysentery, aphrodisiac, malaria, smallpox, and migraine	[35, 53]
Notholaena standleyi	Methanol extract	Flavonoids, chalcones, and phenols	Post-natal problems, antioxidant activity, and antimicrobial activity	[1, 54, 55]
Onychium contiguum	Methanol extract	Alkaloids, sterols, and phenolics	Tonic	[1, 56]
Onychium japonicum	Methanol extract	Sterols, phenols (caffeic acid), terpenoids, flavonoids (luteolin), and pteroside	Hair fall, antioxidant, anti-inflammatory, anticarcinogenic, and antidiabetic	[1, 57, 58]
Onychium siliculosum	Methanol extract	Alkaloids and sterols	Skin diseases and dysentery	[1]
Parahemionitis cordata	Methanol extract	Phenols and saponins	Menstruation problem and ear pain	[1, 59]
Pityrogramma calomelanos	Methanol extract	Flavonoids	Kidney problems, hypertension, boils,	[1, 59]

(continued)

Table 1 (continued)

Scientific name	Type of solvent used	Active phytochemicals	Therapeutic activity	References
			gastric problem, mouth cancer, and asthma	
Pteris biaurita	Methanol extract	Anthraquinones, triterpenoids, alkaloids, tannins, amino acids, saponins, and sugars	Malaria, pain, dysentery, and wounds	[60]
Pteris confuse	Methanol extract	Steroids, flavonoids, reducing sugar, saponins, phenolics, and tannins	Leprosy, scabies, and wounds	[60]
Pteris cretica	Methanol extract	Saponins, phenolics, and tannins	Cuts and wounds	[1]
Pteris ensiformis	Methanol extract	Steroids, flavonoids, reducing sugar	Fracture, pains, and diarrhea	[1]
Pteris geminate	Methanol extract	Flavonoids, reducing sugar, saponins, phenolics, and tannins	Antiviral and tonic	[1]
Pteris longifolia	Methanol extract	Saponins, phenolics, and tannins	Blood dysentery and joint pain	[1]
Pteris longipes	Methanol extract	Flavonoids, phenolics, and tannins	Throat problem	[1]
Pteris camerooniana	Methanol extract	Sugar, phenolics, and tannins	Boils and burns	[1]
Pteris quadriaurita	Methanol extract	Glucosides, phenolics, flavonoids, and tannins	Pus, wounds, boil, and irregular period	[1]
Pteris scabripes	Methanol extract		Poisonous bites.	[1]
Pteris semipinnata	Methanol extract	Saponins, phenolics, and tannins	Tooth pain and body pain	[1]
Pteris vittata	Methanol extract	Triterpenoids, alkaloids, flavonoids, reducing sugar, saponins, tannins, and amino acids	Hypertension, diarrhea, fever, cold, cough, and tonic	[60]
Pteris wallichiana	Methanol extract	Flavonoids, reducing sugar, saponins, phenolics, and tannins	Wounds, skin diseases, and glandular swells	[1]
Vittaria elongata	Methanol extract	Alkaloids, saponins, phenolics, and tannins	Rheumatism, knee pain, and wounds	[61]
Vittaria microlepis	Methanol extract	Alkaloids, alkyne, and phenolics	Inflammation	[61]
Vittaria zosterifolia	Methanol extract	Alkaloids, allyl halides, alkynes, and aromatic compounds	Skin problem	[61]

Some sterols present in *Coniogramme caudata* and *Coniogramme fraxinea* are beta-sitosterol, stigmasterol and apigenin.

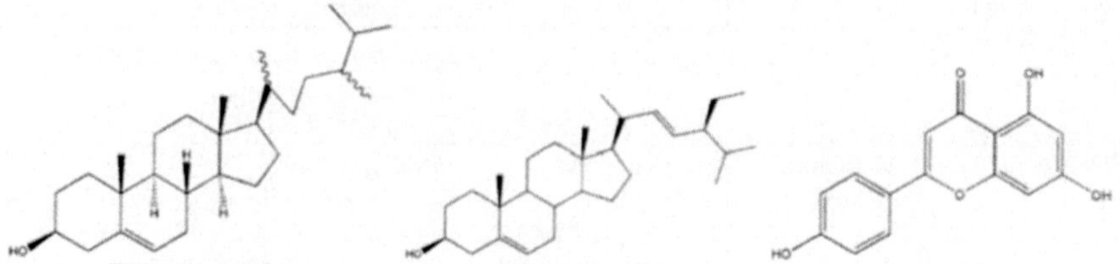

Beta-sitosterol **Stigmasterol** **Apigenin**

Some flavonoids found in *Notholaena standleyi* are chalcones, flavone, and flavonol.

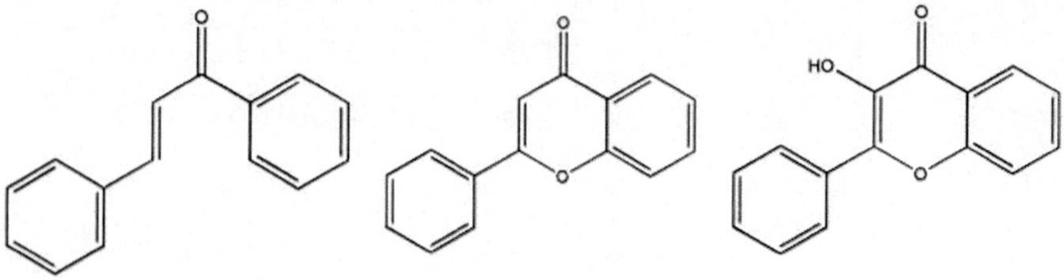

Chalcones **Flavone** **Flavonol**

Some phenolics present in *Onychium contiguum* and *Onychium japonicum* are luteolin, butin, and caffeic acid (Fig. 1).

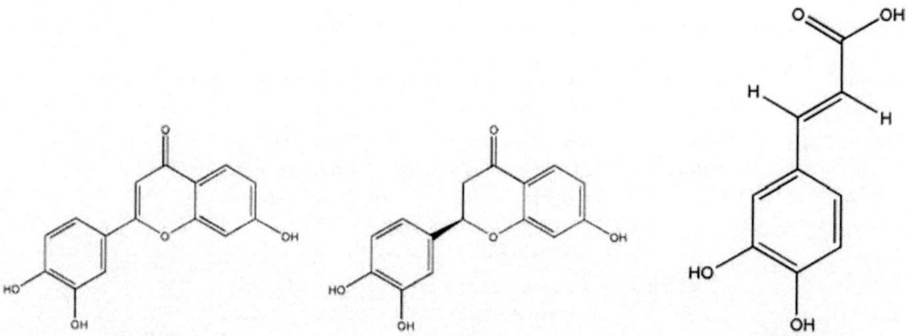

Luteolin **Butin** **Caffeic acid**

Fig. 1 Structures of some important phytochemicals isolated from therapeutically important genera of the Pteridaceae family – *Acrostichum, Adiantum, Coniogramme, Notholaena,* and *Onychium*

7 Phytochemicals of Some Important Species from the Pteridaceae Family Have Significant Pharmacological Activity

7.1 Analgesic Activity

The plant extracted from *Adiantum capillus-veneris* with hexane solvent and petroleum ether extract revealed the presence of filicane and filicenal [41]. Filicane has also been isolated from *Adiantum venustum* ethanolic extract. Filicane and filicenal isolated from hexane extract of *Adiantum capillus-veneris* showed analgesic activity which inhibits pain in mice model induced with formalin. Analgesic activity had been observed when tested in a model organism.

7.2 Anti-inflammatory Activity

Iso adiantone retrieved from petroleum ether extract of *Adiantum capillus-veneris* showed anti-inflammatory activity when induced in rats [41]. *Adiantum venustum* ethanolic extract exhibits anti-inflammatory action. Luteolin, a compound isolated from the methanolic extract, has anti-inflammatory action against human ailments [62]. Luteolin had evidence to act as an anti-inflammatory in the upregulation of Nrf2 in STZ-induced diabetic rats and inhibition of NF-kB [63]. Apigenin induces sedation and relaxation of muscle; it inhibits the cyclooxygenase (COX) pathway [64].

7.3 Cytotoxic Activity

Apigenin found from methanolic extract of *Coniogramme japonica* and *Coniogramme fraxinea* is believed to have therapeutic bioactivity against human ailments [50]. The polyphenol flavonoids present, namely, apigenin isolated from methanolic extract, have the cytotoxic ability when studied in mice. Apigenin activates apoptosis modulating Bcl-2, Bax, STAT-3, and Akt protein expression [47, 49, 51, 65]. A study in human cell cultures showed that the apigenin and luteolin induce angiogenesis and metastasis inhibition by the process of interaction with the signal molecules in three important mitogen-activated protein kinase pathways: (1) c-Jun N-terminal kinases, (2) extracellular-signal-regulated kinase, and (3) p38 [66]. A study by Rezai-Zadeh et al. [67] revealed that the apigenin suppresses cluster of differentiation 40 (CD 40), IL-6, and tumor necrosis factor by inhibiting signal transducer phosphorylation and by inhibiting the expression of activators of STAT1 in murine microglia.

Kaempferol exhibits downregulation of the protein expressions which are involved in the progression of cancer and induces apoptosis. Quercetin is a flavonoid isolated from *Pityrogramma calomelanos* that shows cytotoxic activity by an exhibition of proapoptotic effect against tumors and numerous cancer cells [68, 69].

Patriscabratine and tetracosane flavonoids extracted from *Acrostichum aureum* methanolic extract have been found to exhibit cytotoxic activity when assayed with FITC Annexin V apoptosis assay by inducing necrosis and apoptosis against gastric cancer cells [32].

Luteolin, a flavonoid isolated from the ethanolic extract of the *Onychium japonica* plant, has a therapeutic effect against numerous types of cancer [70].

7.4 Antioxidant Activity

Apigenin present in *Coniogramme japonica* and *Coniogramme fraxinea* shows activity that reduces the expression of adhesion molecules helping to manage oxidative stress. Apigenin enhances ROS scavenging enzyme production of GSH-synthase, catalase, and SOD to combat cellular electrophilic and oxidative stress (Fig. 2). Apigenin blocks the expression of NADPH oxidase complex and the downstream targeting of inflammatory genes resulting in the enhancement of gene expression encoding phase 2 enzymes –glutathione S-transferase and N-acetyltransferase [22, 71–73]. Caffeic acid, a phenolic compound present in a methanolic extract from the plant *Onychium japonica*, has antioxidant activity against ROS.

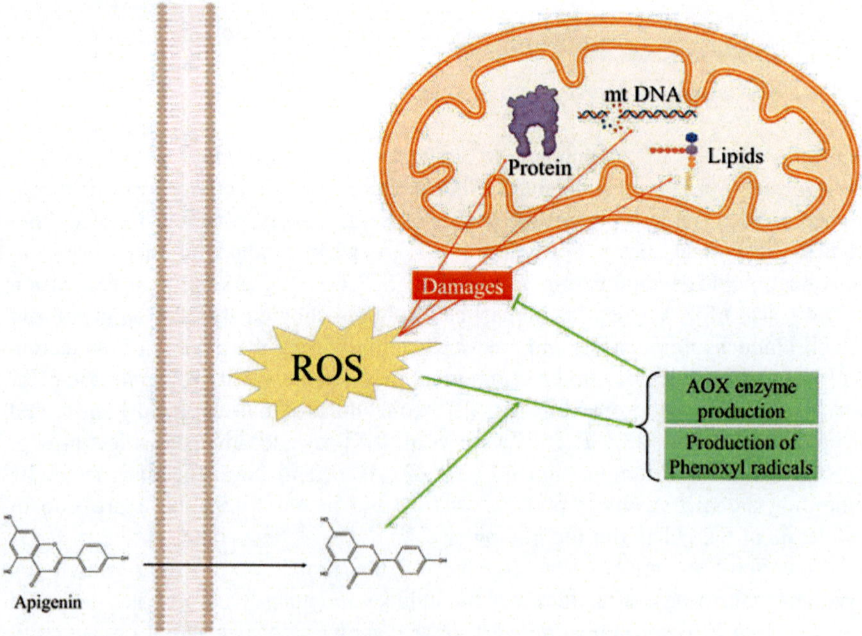

Fig. 2 Antioxidant activity of the compound apigenin

The flavones and chalcones isolated from the methanolic extract of the plant *Notholaena standleyi* have shown activity regulating oxidation stress. Flavonoids can scavenge free radicals, and they can reduce free radical formation by donating an atom of H from the OH group thus resulting in the scavenging of the free radicals. This reaction results in the production of the stable molecule and flavonoid phenoxyl radical. Subsequently, the unpaired electron of the aromatic core undergoes redistribution resulting in resonance structure. Flavonoids have an inhibitory effect on enzymes like xanthine oxidase and protein kinase which are responsible for giving rise to superoxide anion radicals [55].

7.5 Hepatoprotective Activity

Apigenin isolated from *Coniogramme japonica* exhibits hepatoprotective activity. The compound showed effects on angiotensin-converting enzyme 2(ACE2) which regulates and protects kidney functions in the rat. Rats having low blood pressure when induced with apigenin resulted in ACE2 upregulation [74]. Apigenin can increase antioxidant enzyme production like CAT, GSH, and SOD activity during hepatic stress.

7.6 Antidiabetic Activity

Pterosin and pteroside isolated from the plant *Onychium japonicum* with methanol solvent have pharmacological activity in diabetes mellitus [58].

Pterosin regulates the functioning of GLUT 4 (an insulin-responsive glucose transporter protein) and plays a significant role in maintaining the glucose appearance and glucose disappearance homeostasis in normal blood sugar circulation. Pterosin has been identified in diabetic mice in proper regulations of GLUT 4 transporter which results in an improvement in diabetic mice by inhibiting gluconeogenesis and enhancing glucose disposal. Thus, glucose homeostasis is maintained [58].

Apigenin isolated from *Coniogramme japonica* shows an antidiabetic effect by increasing insulin secretion, inhibiting the activity of alpha-glycosides, and neutralizing ROS activity in the cell. Apigenin reduces the risk of injury and dysfunction of the endothelial cells by enhancing the supply of nitric oxide to the endothelial cells in the case of hyperglycemia [75] (Fig. 3). Another study in diabetic mice confirmed that apigenin can regulate thyroid dysfunction, hyperglycemia, and lipid peroxidation [76].

7.7 Antibacterial Activity

Chalcones isolated from the plant *Notholaena standleyi* with little derivation when studied in vitro microbes showed antibacterial activity. Chalcones were found to exhibit antimicrobial activity with minimum inhibitory concentration ranging from

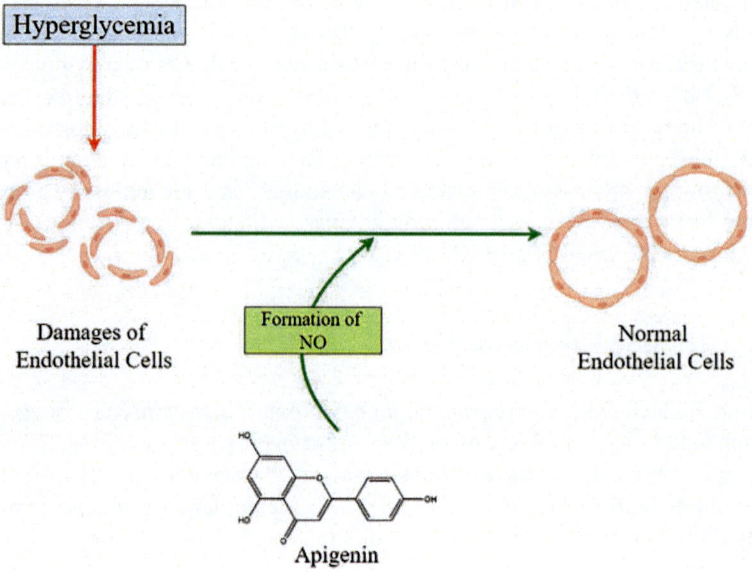

Fig. 3 Antidiabetic activity of the compound apigenin

0.4 to 0.6 mg/mL when determined by the micro broth dilution technique on *Bacillus subtilis* and *Staphylococcus aureus* [77, 78]. Methanolic extract of *Adiantum caudatum*, *A. peruvianum*, and *A. capillus-veneris* has exhibited antimicrobial activity when tested against *Staphylococcus aureus*, *Klebsiella pneumoniae*, *and Escherichia coli* grown in agar well containing the plant extract in the media.

7.8 Antifungal Activity

Saponin isolated from methanolic extract of the plant *Pteris biaurita* shows antifungal activity. Saponin has a membranolytic effect on the mycelium of fungal species – *Botrytis cinerea* and *Rhizoctonia solani*. Further saponin gets converted into aglycones near the cell membrane by enzymatic activity resulting in leakage of free amino acids. Thus, the compound may be used to cure dermal diseases [79] (Fig. 4).

7.9 Antiviral Activity

Cardiac glycosides like digoxin and digitoxin isolated from methanolic extract of the plant *Pteris biaurita* show activity against the virus. Digoxin inhibits viral protein translation against the herpes simplex virus and also inhibition of Src pathway against chikungunya virus. Digitoxin alters viral pre-RNA splicing against human immunodeficiency virus [80]. The plant extracts of *Adiantum capillus-veneris*

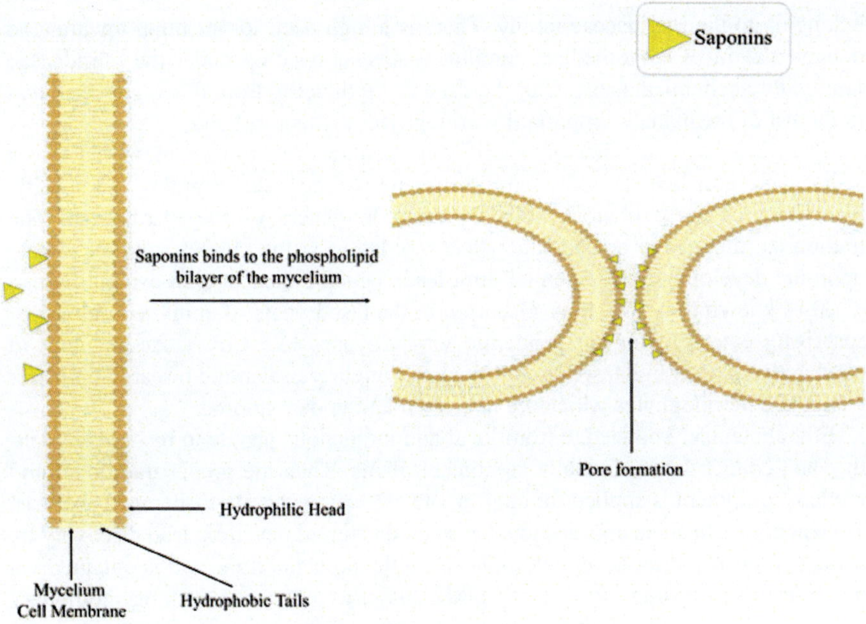

Fig. 4 Antifungal activity of the compound saponin

showed antiviral activity against vesicular stomatitis virus induced in cell lines of monkeys [41].

7.10 Antitubercular Activity

Flavone-5-O glycosides and quercetin isolated from the methanolic extract of the plant *Cheilanthes albomarginata* and *Cheilanthes swartzii* had shown activity against *Mycobacterium tuberculosis*. The flavone-5-O glycosides inhibit the production of mycolic-acid-producing fatty acid synthase II. Thus, inhibiting the growth of *Mycobacterium tuberculosis* exhibiting antitubercular activity [81].

8 Conclusions

The Pteridaceae family is the most ethnomedicinally significant bearing the richest active phytochemicals among the pteridophytes division. The family has a total of 58 species according to the literature survey to date which has shown bioactivity against a high number of human diseases categorized in 17 different categories. There are several unexplored and under-explored regions globally and also in India. In India, Northeastern states have not been much explored. As there is rich biodiversity in Northeastern states, there is a light of hope to obtain several useful species belonging to the Pteridaceae family. There is a high need for attention towards the documentation of ethnomedical significant species residing under the Pteridaceae family. Phytochemical analysis of the flora of Pteridaceae from this region will add to the list of medicinally important pteridophytes present in India.

Though the ethnomedicinally significant Pteridaceae family is widespread across India, there is a low rate of scientific research in the field regarding the specific bioactivity of those phytochemicals present in genera of the Pteridaceae. The traditional knowledge consideration and validation is the correct solution to rely upon the development of potential drug leads which are capable to combat human diseases known to us. We have also seen in the last 3 years of uninvited-unknown-surprising emergent human pandemic viral diseases like Covid. As an input to combat diseases in India, we should focus on intensive scientific research on lower plants like pteridophytes which are neglected and underexplored.

Ethnobotanical knowledge from local and indigenous people to recognize, identify, and collect the medicinally significant plants. Then the proper traditional and modern equipment is applied for appropriate extraction and phytochemical analysis. Then applying in silico approach will ease up the aim of new drug lead discovery by using appropriate drug designing tools. Then the potential drug lead is obtained via depth analysis of drug likeliness, ADME toxicity, molecular docking, homology modeling, and molecular dynamics scans. Lastly, testing the in vitro and in vivo analysis gives the best pharmacologically potential lead which may be used as a potential drug lead for human disease combating and management. The ultimate goal is to use the naturally derived organic compounds from plants to use as a medicine for human welfare limiting the ill or side effects.

References

1. Sureshkumar J, Silambarasan R, Bharati KA, Krupa J, Amalraj S, Ayyanar M (2018) A review on ethnomedicinally important pteridophytes of India. J Ethnopharmacol 219:269–287
2. Schuettpelz E, Schneider H, Huiet L, Windham MD, Pryer KM (2007) A molecular phylogeny of the fern family Pteridaceae: assessing overall relationships and the affinities of previously unsampled genera. Mol Phylogent Evol 44(3):1172–1185
3. Tryon RM (1990) Pteridaceae. In: Pteridophytes and gymnosperms, vol 1. Springer, Berlin/Heidelberg, pp 230–256
4. Vaganov AV, Gureyeva II, Kuznetsov AA, Shmakov AI, Romanets RS, König VA (2017) Spore morphology of the representatives of the subfamily Ceratopteridoideae (J. Sm.) RM Tryon from the family Pteridaceae EDM Kirchn.(Pteridophyta). Ukr J Ecol 7(2):124–129
5. Shah SN, Celik A, Ahmad M, Ullah F, Zaman W, Zafar M, Bahadur S (2019) Leaf epidermal micromorphology and its implications in systematics of certain taxa of the fern family Pteridaceae from northern Pakistan. Microsc Res Tech 82(3):317–332
6. Khan SA, Hossain MA, Panthi S, Asadujjaman M, Hossin A (2013) Assessment of antioxidant and analgesic activity of *Acrostichum aureum* Linn. (Family-Pteridaceae). Pharmacologyonline 1:166–171
7. Sundue M (2009) Silica bodies and their systematic implications in Pteridaceae (Pteridophyta). Bot J Linn Soc 161(4):422–435
8. Hou M, Hu W, Hao K, Xiu Z, Sun J, Cao S (2020) Chemical constituents from *Pteris ensiformis* Burm. (Pteridaceae). Biochem Syst Ecol 92:104–107

9. Cao H, Chai TT, Wang X, Morais-Braga MFB, Yang JH, Wong FC, Wang R, Yao H, Cao J, Cornara L, Burlando B (2017) Phytochemicals from fern species: potential for medicine applications. Phytochem Rev 16(3):379–440
10. Ghorpade PN, Thakar SB, Dongare MM, Kale M (2015) Phytochemical analysis of four *Cheilanthes* species from Northern Western Ghats of India. RJLBPCS 1(2):92
11. Abotsi KE, Radji AR, Rouhan G, Dubuisson JY, Kokou K (2015) The Pteridaceae family diversity in Togo. Biodiver J 3:e5078
12. Hossan S, Agarwala B, Sarwar S, Karim M, Jahan R, Rahmatullah M (2010) Traditional use of medicinal plants in Bangladesh to treat urinary tract infections and sexually transmitted diseases. Ethnobot Res Appl 8:061–074
13. Pattanaik C, Reddy CS, Dhal NK, Das R (2008) Utilisation of mangrove forests in Bhitarkanika wildlife sanctuary, Orissa. CSIR 7(4):598–603
14. DeFilipps RA, Maina SL, Pray LA (1988) The Palauan and Yap medicinal plant studies of Masayoshi Okabe. Atoll Res Bull 317:1941–1943
15. Benjamin A, Manickam VS (2007) Medicinal pteridophytes from the Western Ghats. CSIR 6(4):611–618
16. Ahmed ZU (ed) (2008) Encyclopedia of Flora and Fauna of Bangladesh: angiosperms: dicotyledons: Balsaminaceae-Euphorbiaceae, vol 7. Asiatic Society of Bangladesh, Dhaka, p 546
17. Creek W (2010) United States Department of Agriculture Natural Resources Conservation Service. City 29(26):25
18. Jain SK (2010) Manual of ethnobotany, vol 2. Scientific Publishers, p 241
19. Ignacimuthu S, Ayyanar M, Sivaraman KS (2006) Ethnobotanical investigations among tribes in Madurai district of Tamil Nadu (India). J Ethnobiol Ethnomed 2(1):1–7
20. Rao RR (1981) Ethnobotany of Meghalaya: medicinal plants used by Khasi and Garo tribes. Econ Bot 35(1):4–9
21. Rout SD, Panda T, Mishra N (2009) Ethno-medicinal plants used to cure different diseases by tribals of Mayurbhanj district of North Orissa. Stud Ethno-Med 3(1):27–32
22. Manickam VS, Irudayaraj V (1992) Pteridophyte flora of the Western Ghats, South India, vol 1. BI Publications, pp 357–369
23. Sen A, Ghosh PD (2011) A note on the ethnobotanical studies of some pteridophytes in Assam. NISCAIR-CSIR 10(2):292–295
24. Gaur RD, Bhatt BP (1994) Folk utilization of some pteridophytes of Deoprayag area in Garhwal Himalaya: India. Econ Bot 48(2):146–151
25. Badhsheeba MA, Vadivel V (2020) Physicochemical and phytochemical contents of the leaves of *Acrostichum aureum* L. J Global Biosci 9(4):7003–7018
26. Rao PS, Farheen M, Diwan PV (2018) Alloxan induced diabetes and impairment of oxidative defense system in rat pancreas: protective effect of *Actinopteris dichotoma*. Int J Ayur Herbal Med 8(1):3104–3111
27. Singh H, Krishna MC, Kumar S, Mangal AK (2016) Pharmacognostic and preliminary phytochemical analysis of *Actiniopteris dichotoma* Bedd. Int J Ayur Med 7(3):160–164
28. Jain J (2015) Evaluation of anti-inflammatory and analgesic activity on *Actiniopteris dichotoma* Bedd. Asian J Pharmacol Toxicol 3(4):39–44
29. Jain J (2015) Phytochemical analysis and toxicity study of *Actiniopteris dichotoma* Bedd. Asian J Pharmacol Toxicol 3(6):43
30. Keller HA, Torres EIM, Prance GT (2011) Ethnopteridology of the Guaranís of Misiones Province, Argentina. Am Fern J 101(3):193–204
31. Joshi B, Panda SK, Jouneghani RS, Liu M, Parajuli N, Leyssen P, Neyts J, Luyten W (2020) Antibacterial, antifungal, antiviral, and anthelmintic activities of medicinal plants of Nepal selected based on ethnobotanical evidence. Evid Based Complementary Altern Med 2020: 1043471
32. Uddin SJ, Grice D, Tiralongo E (2012) Evaluation of cytotoxic activity of patriscabratine, tetracosane and various flavonoids isolated from the Bangladeshi medicinal plant *Acrostichum aureum*. Pharm Biol 50(10):1276–1280

33. George M, Josekumar VS (2016) In vitro cytotoxicity screening, phytochemical profile and heavy metal analysis of different extracts of *Acrostichum heterophyllum* L. NISCAIR-CSIR 7(1):19–24
34. Krishnasamy R, Jeyapal G, Chandrasekar MJN, Dhanbal SP (2017) Pharmacognostical studies on whole plant of *Actiniopteris dichotoma*. Int J Res Pharm Sci 8(3):364–372
35. Kalpana Devi R, Rajesh NV, Vasantha S, Sathia Geetha V (2015) Anti-parasitic action of Actinopteris radiata, *Acrostichum aureum* and *Hemionitis arifolia*. Pteridol Res 4(1):1–9
36. Singh BPSR, Upadhyay R (2013) Ethnobotanical uses of *Isoetes coromandelina* LF and actiniopteris radiata (Sw.) At Aravali Hills Rampura Region. Life Sci Leafl 41:30–34
37. Manjunath M, Sharma PVGK, Reddy OVS (2008) In vitro evaluation of antibacterial activity of *Actiniopteris radiata* (Sw.) Link. J Pharm Chem 2(2):112–117
38. Al-Snafi AE (2015) The chemical constituents and pharmacological effects of Adiantum capillus-veneris-a review. Asian J Pharma Sci Technol 5(2):106–111
39. Sallam MM (2019) Phytochemical and biological study of *Adiantum capillus-veneris* l. Growing in Egypt. Al-Azhar J Pharma Sci 59(1):9–26
40. Ahmed D, Khan MM, Saeed R (2015) Comparative analysis of phenolics, flavonoids, and antioxidant and antibacterial potential of methanolic, hexanic and aqueous extracts from *Adiantum caudatum* leaves. Antioxidants 4(2):394–409
41. Pan C, Chen YG, Ma XY, Jiang JH, He F, Zhang Y (2011) Phytochemical constituents and pharmacological activities of plants from the genus *Adiantum*: a review. Trop J Pharm Res 10(5):681–692
42. Singh M, Singh N, Khare PB, Rawat AKS (2008) Antimicrobial activity of some important *Adiantum* species used traditionally in indigenous systems of medicine. J Ethnopharmacol 115(2):327–329
43. Karthik V, Raju K, Ayyanar M, Gowrishankar K, Sekar T (2011) Ethnomedicinal uses of pteridophytes in kolli hills, Eastern Ghats of Tamil Nadu, India. J Nat Prod Plant Resour 1(2):50–55
44. Mahfuz A, Salam FBA, Deepa KN, Hasan AHM (2019) Characterization of in-vitro antioxidant, cytotoxic, thrombolytic and membrane stabilizing potential of different extracts of *Cheilanthes tenuifolia* and stigmasterol isolation from n-hexane extract. Clinical Phytosci 5(1):1–10
45. Chowdhary S, Verma DL, Pande R, Kumar H (2010) Antioxidative properties of flavonoids from *Cheilanthes anceps* Swartz. J Am Sci 6(5):203–207
46. Fang CW, Chen JJ, Liu SJ (2010) Studies on the chemical constituents of rhizoma of Coniogramme japonica. J Chin Med Mater 33(4):557–559
47. Seo HS, Choi HS, Kim SR, Choi YK, Woo SM, Shin I, Woo JK, Park SY, Shin YC, Ko SK (2012) Apigenin induces apoptosis via extrinsic pathway, inducing p53 and inhibiting STAT3 and NFκB signaling in HER2-overexpressing breast cancer cells. Mol Cell Biochem 366(1): 319–334
48. Lees AM, Mok HY, Lees RS, McCluskey MA, Grundy SM (1977) Plant sterols as cholesterol-lowering agents: clinical trials in patients with hypercholesterolemia and studies of sterol balance. Atherosclerosis 28(3):325–338
49. Karmakar S, Davis KA, Choudhury SR, Deeconda A, Banik NL, Ray SK (2009) Bcl-2 inhibitor and apigenin worked synergistically in human malignant neuroblastoma cell lines and increased apoptosis with activation of extrinsic and intrinsic pathways. Biochem Biophys Res Commun 388(4):705–710
50. Fang CW, Chen JJ, Liu SJ (2010) Studies on the chemical constituents of rhizoma of *Coniogramme japonica*. J Chin Med Mat 33(4):557–559
51. Sui H, Yu Q, Zhi Y, Geng G, Liu H, Xu H (2010) Effects of apigenin on the expression of angiotensin-converting enzyme 2 in kidney in spontaneously hypertensive rats. J Hygi Res 39(6):693–696

52. Neira DA, Andrada AR, Páez VDLÁ, Rodriguez AMDV, Ríos NF, Martínez OG, Hernández MA (2017) Anatomical, histochemical and cytogenetic features of *Doryopteris triphylla* (Pteridaceae). Am J Plant Sci 8:907–920
53. Nair SA, Shylesh BS, Gopakumar B, Subramoniam A (2006) Anti-diabetes and hypoglycaemic properties of *Hemionitis arifolia* (Burm.) Moore in rats. J Ethnopharmacol 106(2):192–197
54. Wollenweber E, Schneider H (2000) Lipophilic exudates of Pteridaceae–chemistry and chemotaxonomy. Biochem Syst Ecol 28(8):751–777
55. Pietta PG (2000) Flavonoids as antioxidants. J Nat Prod 63(7):1035–1042
56. Li M, Tang S, Duan H (1994) Chemical constituents from *Onychium japonicum*. Chin Trad Herbal Drugs 24(1994):580897
57. Akabori S, Akabori Y, Hasegawa M (1980) Further structural studies of 1-indanone derivatives obtained from *Onychium japonicum*. Chem Pharm Bull 28(4):1311–1313
58. Hsu FL, Huang CF, Chen YW, Yen YP, Wu CT, Uang BJ, Yang RS, Liu SH (2013) Antidiabetic effects of pterosin A, a small-molecular-weight natural product, on diabetic mouse models. Diabetes 62(2):628–638
59. Iinuma M, Tanaka T, Takenaka M, Mizuno M, Asai F (1992) Five complex flavonoids in the farinose exudate of *Pityrogramma calomelanos*. Phytochemistry 31(7):2487–2490
60. Gracelin DHS, Britto A, Kumar BJRP (2013) Qualitative and quantitative analysis of phytochemicals in five Pteris species. Int J Pharm Pharm Sci 5(1):105–107
61. Sahaya Sathish S, Vijayakanth P, Palani R, Thamizharasi T, Vimala A (2014) UV-vis spectroscopis studies on the medicinal fern *Vittaria elongata* Sw. In: National level conference on recent innovation and future trends in biology, vol 1, pp 127–132
62. Aziz N, Kim MY, Cho JY (2018) Anti-inflammatory effects of luteolin: a review of in vitro, in vivo, and in silico studies. J Ethnopharmacol 225:342–358
63. Zhang M, He L, Liu J, Zhou L (2021) Luteolin attenuates diabetic nephropathy through suppressing inflammatory response and oxidative stress by inhibiting STAT3 pathway. Exp Clin Endocrinol Diabetes 129(10):729–739
64. Liang YC, Huang YT, Tsai SH, Lin-Shiau SY, Chen CF, Lin JK (1999) Suppression of inducible cyclooxygenase and inducible nitric oxide synthase by apigenin and related flavonoids in mouse macrophages. Carcinogenesis 20(10):1945–1952
65. Seo HS, Ku JM, Choi HS, Woo JK, Jang BH, Shin YC, Ko SG (2014) Induction of caspase-dependent apoptosis by apigenin by inhibiting STAT3 signaling in HER2-overexpressing MDA-MB-453 breast cancer cells. Anticancer Res 34(6):2869–2882
66. Peng Q, Deng Z, Pan H, Gu L, Liu O, Tang Z (2018) Mitogen-activated protein kinase signaling pathway in oral cancer. Oncol Lett 15(2):1379–1388
67. Rezai-Zadeh K, Ehrhart J, Bai Y, Sanberg PR, Bickford P, Tan J, Shytle RD (2008) Apigenin and luteolin modulate microglial activation via inhibition of STAT1-induced CD40 expression. J Neuroinflammation 5(1):1–10
68. Suyatno NH, Umami K, Sari IP (2014) Chemical constituents of Indodnesian silver fern (*Pityrogramma calomelanos*) and their cytotoxicity against murin leukemia P-388 cells. In: Proceeding of international conference on research, implementation and education of mathematics and sciences, vol 1, pp 229–236
69. Chudzik M, Korzonek-Szlacheta I, Król W (2015) Triterpenes as potentially cytotoxic compounds. Molecules 20(1):1610–1625
70. Lin Y, Shi R, Wang X, Shen HM (2008) Luteolin, a flavonoid with potential for cancer prevention and therapy. Curr Cancer Drug Targets 8(7):634–646
71. Nicholas C, Batra S, Vargo MA, Voss OH, Gavrilin MA, Wewers MD, Guttidge DC, Grotewold E, Doseff AI (2007) Apigenin blocks lipopolysaccharide-induced lethality in vivo and proinflammatory cytokines expression by inactivating NF-κB through the suppression of p65 phosphorylation. J Immunol 179(10):7121–7127
72. Lai HY, Lim YY, Tan SP (2009) Antioxidative, tyrosinase inhibiting and antibacterial activities of leaf extracts from medicinal ferns. Biosci Biotechnol Biochem 73(6):1362–1366

73. Kashyap D, Sharma A, Tuli HS, Sak K, Garg VK, Buttar HS, Setzer WN, Sethi G (2018) Apigenin: a natural bioactive flavone-type molecule with promising therapeutic function. J Funct Foods 48:457–471
74. Tamayose CI, Romoff P, Toyama DO, Gaeta HH, Costa CR, Belchor MN, Ortolan BD, Velozo LSM, Kaplan MAC, Ferreire MJP, Toyama MH (2017) Non-clinical studies for evaluation of 8-C-rhamnosyl apigenin purified from *Peperomia obtusifolia* against acute edema. Int J Mol Sci 18(9):1972
75. Wang QQ, Cheng N, Yi WB, Peng SM, Zou XQ (2014) Synthesis, nitric oxide release, and α-glucosidase inhibition of nitric oxide donating apigenin and chrysin derivatives. Bioorg Med Chem 22(5):1515–1521
76. Panda S, Kar A (2007) Apigenin (4', 5, 7-trihydroxyflavone) regulates hyperglycaemia, thyroid dysfunction and lipid peroxidation in alloxan-induced diabetic mice. J Pharm Pharmacol 59(11):1543–1548
77. Das BC, Mariappan G, Sudip S, Debjit B (2010) Anthelmintic and anti-microbial activity of some novel chalcone derivatives. J Chem Pharm Res 2(1):113–120
78. Heera Vinod A, Joseph E (2021) A study on the phytochemical and antibacterial properties of three selected ferns. Int J Res Eng Sci 9(7):68–73
79. Takechi M, Matsunami S, Nishizawa J, Uno C, Tanaka Y (1999) Haemolytic and antifungal activities of saponins or anti-ATPase and antiviral activities of cardiac glycosides1. Planta Med 65(06):585–586
80. Reddy D, Kumavath R, Barh D, Azevedo V, Ghosh P (2020) Anticancer and antiviral properties of cardiac glycosides: a review to explore the mechanism of actions. Molecules 25(16):3596
81. Brown AK, Papaemmanouil A, Bhowruth V, Bhatt A, Dover LG, Besra GS (2007) Flavonoid inhibitors as novel antimycobacterial agents targeting Rv0636, a putative dehydratase enzyme involved in *Mycobacterium tuberculosis* fatty acid synthase II. Microbiology 153(10):3314–3322

Phytochemicals and Their Bioactivity from Plants of Dryopteridaceae Family 14

Shreeta Singha, Rajat Nath, Subrata Das, Sibashish Kityania, Anupam Das Talukdar, and Deepa Nath

Contents

Abstract

The division of the plant kingdom pteridophytes is the first vascular plant that appeared 440 million years ago during the Silurian Period. Pteridophytes are distributed across the world, and there are around 14,000 species in total. The Ancient Traditional Indian Medicine System has adopted lower medicinal plants as an effective curing strategy to treat a number of pathogenic attacks causing disease in humans. Because there were no advanced medical facilities at the time, indigenous traditional gurus used to cure disease by using bark, fruit, flowers,

S. Singha · R. Nath · S. Das · S. Kityania · A. D. Talukdar
Department of Life Science and Bioinformatics, Assam University, Silchar, Assam, India
e-mail: anupam@bioinfoaus.ac.in

D. Nath (✉)
Department of Botany, Gurucharan College, Silchar, Assam, India

H. N. Murthy (ed.), *Bioactive Compounds in Bryophytes and Pteridophytes*, Reference Series in Phytochemistry, https://doi.org/10.1007/978-3-031-23243-5_23

roots, or the entire plant to heal unhealthiness. The group pteridophytes possess an extremely significant family called "Dryopteridaceae." This family, despite holding bioactive potential, is ignored and less documented as a resource for therapy. Less exploration of this family kept the therapeutic wealth wrapped for a long time. Thorough and extensive attention to this group by modern scientists has unveiled the essence and medicinal value of these plants due to the presence of phytochemicals. The phytochemicals like saponin, glycosides, steroids, essential oils, and others have increased the potential of the plants belonging to the group as medicinally significant. This chapter focuses on bringing out the bioactive phytochemicals of the family Dryopteridaceae and gives perspective for considering a potential family of the division pteridophytes as a future drug lead in the process of drug development.

Keywords

Drug · Dryopteridaceae · Ethnomedicine · Ferns · Phytochemicals

Abbreviations

AOX	Alternative oxidase
Bcl-2	B-cell lymphoma 2
Caspase 3	Cysteine-aspartic proteases 3
Caspase 8	Cysteine-aspartic proteases 8
CAT	Catalase
CD44	Cluster of differentiation 44
COX	Cyclooxygenase
COX1	Cyclooxygenase 1
COX2	Cyclooxygenase 2
GSH	Glutathione
HIV	Human immunodeficiency virus
HT-29	human colorectal adenocarcinoma cell line
NO	Nitric oxide
ROS	Reactive oxygen species
SOD	Superoxide dismutase
SOX2	SRY-Box Transcription Factor 2
T cells	T lymphocyte
TNF-alpha	Tumor necrosis factor-alpha

1 Introduction

India is a country known to have been practicing a conventional medicinal system using medicinal plants for long days back [1]. Samhitas and Vedas have documentation of using pteridophytes as a source of medicine to cure diseases. The ethnic, indigenous, and traditional tribal groups who live near woods have been documented to completely rely on plant sources for disease treatment [2]. This traditional knowledge as folklore has been inherited generation after generation. In India as well as globally, traditional knowledge plays a major role in ethnotherapy [3]. This system is a powerful and significant technique to be inferred for the development of new potential leads which may result in antibiotics and various drugs to tackle innumerable diseases [4]. People mostly do not count lower plants like ferns as being economically essential and do not give any attention to the group. An extensive survey and literature review have revealed the therapeutic significance of those ferns [5]. Despite possession of highly bioactive chemical ingredients in the species of the Dryopteridaceae family, they are less explored by people. The family Dryopteridaceae possesses highly bioactive plant chemical constituents [6]. The phytochemicals flavonoids, alkaloids, saponins, phenols, steroids, coumarins, glycosides, terpenoids, and essential oils present in Dryopteridaceae species make it the best choice for ethnic, indigenous, and elderly people as a food and medicine [7]. Some of the important genera belonging to the family Dryopteridaceae are *Arachniodes*, *Aspidium*, *Bolbitis*, *Cyrtomium*, *Dryopteris*, *Hypodematium*, *Leucostegia*, *Poecilopteris*, *Polystichum*, and *Tectaria* and have been used as traditional medicinal plants to treat pus, rheumatism, cuts, wounds, snakebite, leprosy, dysentery, ulcers, epilepsy, skin irritation, cholera, gonorrhea, swells, backache, headache, stomach disorders, intestinal worms, and indigestion and also as an anti-helminthic, anti-venom, antimicrobial, and anticoagulant against numerous diseases [1]. The fronds of *Bolbitis costata* were used to treat cuts and wounds, and the frond extract of *Bolbitis contaminans* was used to treat helminthiasis [5, 8]. The extracts of *Cyrtomium falcatum* and *Cyrtomium hookerianum* are used to treat intestinal worms and helminthiasis [9]. *Dryopteris atrata* can be used against epilepsy [10]; *Dryopteris cochleata* against leprosy, cuts, and wounds [6]; and *Dryopteris sparsa* against helminth intestinal disorder [11]. *Hypodematium crenatum* can be used as an antibacterial. The rhizome possesses phytochemicals that resist bacterial growth [12]. The fronds of the plant *Leucostegia immersa* can be used to treat swellings, and the frond extract can be used to cure constipation; *Poecilopteris virens* can be used to treat wounds. *Polystichum aculeatum* can be used to cure diarrhea. The extract of *Polystichum nepalense* is used to heal indigestion [13]. The plant *Tectaria coadunata* is used to treat respiratory diseases. The rhizome extract of the plant cures diarrhea in children. *Tectaria wightii* is another medicinally important species of the *Tectaria* genus. The rhizome of the plant *Tectaria wightii* is used to treat helminthiasis in humans [12]. Thus, the significance of the species of Dryopteridaceae in the field of ethnobotany and the ethnomedicinal uses of ferns can be drawn from the above instances. Recent workers are inclined towards the phytochemical study of the Dryopteridaceae species, believing that there is a highly interesting element within the plant to be explored [13].

2 Dryopteridaceae Family

The family Dryopteridaceae belongs to the order Polypodiales. It consists of 20 genera and 1700 species distributed in the subtropical and temperate regions of Asia. Plants of the family are terrestrial, epiphytic, hemi-epiphytic, and epilithic in their habitat. The plant size is normally medium-sized. The plant leaves may persist in the plant body as evergreen or may sometimes fall off, as in the case of deciduous. The fronds of the plant are pinnate, proximate, 1–4 in number, papery, herbaceous, green in color when viewed adaxially, and a little pale green when viewed abaxially. The rhizome of the plant body is short, robust, erect, and sometimes creepy and possesses dense scales. Scales are brownish, reddish, or black in appearance, and the shapes are versatile. The spore-producing structures called sori of the plant are hemispherical, circular, or kidney-shaped. In most of the species, sori are arranged along the marginal position of the leaf divisions or leaflets. Sori are sometimes exposed or naked or possess indusium. Indusium is a protective membranous tissue that protects sori. The fertile leaves bear sporangia under the surface, and sometimes the entire leaves get covered by sporangia ventrally. The shape of the spores varies from bean-shaped to globose [6, 14]. Many species of the family Dryopteridaceae possess highly bioactive phytochemicals. Some of the active phytochemicals present in some medicinally important species of the family Dryopteridaceae are mentioned here. The rhizome of the plant *Aspidium cicutarium* bears alkaloids, phenols, glycosides, and steroids, which are used to treat swelling, wounds, cough, and fever [8]. The rhizome of the plant *Dryopteris ramosa* bears mangiferin and iso-mangiferin, which can be used to treat cytotoxic cells [15]. The rhizome of the plant *Dryopteris crassirhizoma* bears triterpenes, phenols, and phloroglucinol, which can act as antioxidants and can be used in HIV-affected individual treatment [16]. The fronds of *Dryopteris corleyi* bear albaspidin, which can be used to treat cancerous growth in humans [17], and *Dryopteris filix-mas* fronds bear filixic acid, which can be used as an anti-helminthic. The frond extract of *Cyrtomium caryotideum* bears cardiac glycosides and phlobatannins, which can be used to treat cardiac attacks and as antioxidants, respectively [18]. The frond extract of *Dryopteris cycadina* bears kaempferol, steroids, and terpenoids which show anticancerous activity. The frond extract of *Dryopteris thelypteris* squamigerum bears linalool, palmitic acid, and terpenes which show antioxidant activity, cytotoxic activity, and antimicrobial activity, respectively. The fronds of the plant *Dryopteris fragrans* bear glycosides showing antioxidant activity, and coumarin shows anti-inflammatory activity. The rhizome of the plant *Polystichum braunii* bears quercetin, which acts as an antioxidant; gallic acid acts as an anti-inflammatory, and kaempferol acts as an anticancer in humans. The frond of the plant *Dryopteris ardechensis* bears flavaspidic acid, which shows antibacterial activity. The fronds of the plant *Dryopteris dilatata* bear flavonoids, phenol, and tannins, which show antimicrobial, antioxidant, and anticancer properties, respectively [19]. The high significance of the Dryopteridaceae family is now well understood, and yet more focus should be driven to other closely related ferns of the division pteridophyte.

3 Ethnomedicinal Importance of the Family Dryopteridaceae

The people of the earlier days, when there was no advancement in the field of medicine like nowadays, were majorly dependent and relied upon the Ayurvedic System. Plants were the traditional people's primary source of food and medicine.

Till today, there are many areas globally as well as particularly in India, where there is no hospital or pharmacy available to them when there is an extreme emergency or health issue. The inhabitants of those remote, less developed regions have adapted to survive without those facilities in their world by acquiring traditional knowledge from elderly people. The traditional knowledge of using medicinal plants is found in a forest wild to treat various human ailments and disorders. They are extremely knowledgeable about the use of plants and their parts, such as fronds and rhizomes against which ailment. The medicinal plant and the methodology for preparation as a medicine can be surveyed by querying those tribes, elderly people, and medicine man. The surveying of those local people by researchers in their published literature study has given the idea that the people of India, especially the tribes, have been using the plant species of the Dryopteridaceae family as a source for medicine. The family Dryopteridaceae bears a number of ethnomedicinally important species belonging to various genera. The ferns and fern allied plants belonging to the family Dryopteridaceae have been used as a source of instant natural medicinal resources by the inhabitants of the north-eastern area, the Himalayan tribes, the tribes of Madhya Pradesh, and many local tribes of various parts of India. The fern *Dryopteris cochleata* is used as an anticoagulant and to treat gonorrhea by various tribes of Assam, namely, Chakma, Cachari, Jaintia, Rheang, Vaiphei, Bodo, Khasi, Mishing, Rankhol, Murah, H'mar, and Garo. Many other species of fern are used as a strategy to heal diseases in the form of extracts, pastes, or decoctions by the normal non-tribal people residing in Assam. The local inhabitants of Himachal Pradesh, particularly in Mandi District, also use *Dryopteris* sp. as a source of medicine to cure disorders like fever, cough, stomach disorders, and leprosy [20]. The female tribes of Bhil, Korku, Bharia, Mabasi, and Gond of Madhya Pradesh's Pachmarhi Biosphere reserve have revealed the use of various species of Dryopteridaceae as a source of medicine and food. The tribes of Similipal Biosphere Reserve in Orissa, namely, Kol, Santhal, Bhuyan, Bhumija, Saharas, Mahalis, and Sounti, use *Dryopteris filix-mas* as an aphrodisiac and *Dryopteris wallichiana* to treat rheumatism. The Western Ghats tribes use *Dryopteris cochleata* extract of the whole plant as a remedy to treat snakebite. The rhizome extract is used as an antibiotic, and the dried rhizome powder mixed with water is used to treat epilepsy, leprosy, and rheumatism. The tribes of the Western Ghats also use *Hypodematium crenatum* as a source of medicine. The rhizome of the plant possesses antibacterial activity. The fronds of *Hypodematium crenatum* help in conceiving a woman. The plant species *Polystichum moluccense* is also used by the tribes of the Western Ghats as a source of medicine. The extract of the sporophyll of the plant *Polystichum moluccense* shows antibacterial activity. The Western Ghats tribe uses sporophyll extracted from the plant *Polystichum squarrosum*, an important medicinal fern belonging to the family Dryopteridaceae that is utilized by the tribes of the Western Ghats. The plant is used to treat asthma, honeybee bites, bacterial attacks, and bronchitis. The tender fronds are cooked and utilized by those tribes to treat stomach disorders. The freshly plucked rhizome extract of the plant *Tectaria coadunata* is also used by the inhabitants of Darjeeling to cure problems of diarrhea in children. The rhizome extract of *Tectaria wightii*, another species of the genus *Tectaria*, is utilized by the local inhabitants of the Western Ghats to cure the problem of helminthiasis [12]. The tribes residing in the Eastern Ghats of Tamil Nadu in Kolli Hills use the rhizome of *Leucostegia immersa* to treat wounds and swellings [21]. There is an indication that the use of such a large number of species of the Dryopteridaceae family must be due to the possession of large numbers of diverse phytochemicals within the plant's frond, sporophyll, and rhizome [11].

4 Phytochemicals Reported in the Dryopteridaceae Family

The plants belonging to the family Dryopteridaceae are distributed across the globe and in many parts of India [2]. Highly rich amounts of phytochemicals are distributed in more than 50 species of the family Dryopteridaceae, which show bioactivity immensely [4]. The diverse phytochemicals of the plant group after extensive database collection, interpretation, and analysis have shown that those chemicals in the plant bear the potential to act as antioxidants, anticancer, antidiabetic, antimicrobial, anti-inflammatory, anticoagulant, and many more in curing human diseases [5]. The different genera of the family bear medicinal plant species that contain highly bioactive molecules that can be manifested as a potential lead against an innumerable number of diseases that affect human beings. The medicinal significance of the family Dryopteridaceae is tabulated in Table 1 and Fig. 1.

5 Bioactivity of Phytochemicals from Dryopteridaceae Family

5.1 Antioxidant Activity

Phloroglucinol compounds isolated from the methanolic extract of the plant *Aspidium cicutarium* rhizome and the methanolic frond extract of the plant *Dryopteris ardechensis* show activity against oxidative stress [39, 54]. The compound plays a significant role in evoking an antistress defense mechanism by inducing AOX production. The compound also inhibits lipid peroxidation [23, 39]. A compound, kaempferol, isolated from the *Dryopteris erythrosora* frond methanolic extract induces the generation of ROS, NO, and T cells. Thus, showing free radical scavenging activity lowers the risk of antioxidant stress [6]. *Dryopteris erythrosora* bears another significant compound called quercetin, which was isolated from the methanolic extract of fronds. This compound showed antioxidant activity by generating superoxide dismutase, peroxidase, glutathione peroxidase, and catalase. A compound called apigenin, isolated from the *Dryopteris juxtaposita* frond methanolic extract in an experiment where H_2O_2 is generated, showed antioxidant activity (Fig. 2) [4].

Table 1 The bioactive compounds and their respective therapeutic potential in the medicinal plant species belonging to the family Dryopteridaceae

Scientific name	Parts of the plant used (solvent used)	Bioactive phytochemicals	Therapeutic potential	References
Arachniodes exilis	Fronds and rhizome (methanol)	Phloroglucinol, desaspidin	Antibacterial, antifungal, anti-inflammatory, sedative, and anticancerous	[22–24]
Aspidium cicutarium	Rhizome (water)	Steroids, saponin, flavonoid	Antioxidant, anticancer	[8]
Bolbitis appendiculata	Fronds (acetone)	Tannin, gallic acid, terpenoids	Antidiabetic, anti-inflammatory	[25–27]
Bolbitis contaminans	Fronds (water)	Phenol, flavonoid	Rheumatism	[13, 28]
Bolbitis costata	Sporophyte (methanol)	Terpene, tannin, sterol	Wounds and cuts	[26, 29, 30]
Cyrtomium caryotideum	Fronds (ethanol)	Phlobatannin, glycosides, terpenes, saponin	Anti-pyretic, anti-helminthic, pus, and antifungal	[18, 31, 32]
Cyrtomium falcatum	Fronds (aqueous)	Phytosterol, organosulfur, carotenoid	Intestinal worms, anticancer, antioxidant, anti-inflammatory	[31, 33]
Cyrtomium fortunei	Rhizome (methanol)	Flavonoids, phenol	Melanogenesis	[34, 35]
Cyrtomium hookerianum	Fronds (methanol)	Kaempferol, daucosterol	Anti-helminthic, antioxidant	[36]
Cyrtomium macrophyllum	Fronds (ethanol)	Flavonol, tannin, polyphenol	Immunomodulator	[37, 38]
Dryopteris ardechensis	Fronds (methanol)	Flavaspidic acid, trisflavaspidic, albaspidin	Antibacterial, anticancer	[39]
Dryopteris affinis	Fronds (methanol)	Flavonoid, terpene, aristolene	Antibacterial, antioxidant	[10, 40–43]
Dryopteris atrata	Fronds (methanol)	Phloroglucinol	Epilepsy, rheumatism	[44, 45]
Dryopteris barbigera	Rhizome, fronds (methanol)	Flavonoids, tannin, oleoresin, filicene	Anticancer, antimicrobial, dysentery, intestinal worms	[46–50]
Dryopteris blanfordii	Fronds (aqueous)	Phlobatannin, saponin, alkaloid, terpene	Analgesic, body cooler, anti-inflammatory, helminthiasis	[51–53]

(continued)

Table 1 (continued)

Scientific name	Parts of the plant used (solvent used)	Bioactive phytochemicals	Therapeutic potential	References
Dryopteris borreri	Fronds (methanol)	Phloroglucinol, aspidinol, filicinic acid.	Antioxidant, antimicrobial	[54–57]
Dryopteris cambrensis	Fronds (methanol)	Terpene, nerolidol	Anti-inflammatory, antioxidant	[54, 58, 59]
Dryopteris chrysocoma	Fronds (methanol)	Phloroglucinol	Sciatica, helminthiasis, antimicrobial	[60, 61]
Dryopteris cochleata	Rhizome (methanol)	Aspidinol, hexadecane and germacrene D	Anti-venom, cholera, ulcer, leprosy, rheumatism, epilepsy	[1]
Dryopteris corleyi	Fronds (methanol)	Aspidin, trisflavaspidic acid	Antifungal, antibacterial	[13]
Dryopteris crassirhizoma	Rhizome (methanol)	Phloroglucinol, triterpene, and phenol	Antioxidant	[13]
Dryopteris cycadina	Fronds (methanol)	Kaempferol, gallic acid, and quercetin	Antinociceptive, anticancer	[13]
Dryopteris dilatata	Fronds (ethyl acetate)	Saponins, terpenes, phenolic compounds	Antidiabetic, antioxidant	[13]
Dryopteris erythrosora	Fronds (methanol)	Apigenin, beta-sterol, quercetin, kaempferol	Antioxidant, anticancer	[13]
Dryopteris filix-mas	Fronds (ethanol)	Tannin, saponin, flavonoids	Ulcer, anti-helminthic, antimicrobial	[13]
Dryopteris fragrans	Rhizome (petroleum ether)	Glycosides, stigmasterol, phloroglucinol, aspidinol	Cytotoxic, anti-inflammatory	[13]
Dryopteris hirtipes	Fronds (methanol)	Flavonoid, alkaloid, phloroglucinol.	Anti-inflammatory, anti-helminthic	[13]
Dryopteris juxtaposita	Fronds (methanol)	Quercetin, filicene	Dysentery	[13]
Dryopteris marginata	Fronds (methanol)	Phloroglucinol	Anti-helminthic	[13]

(continued)

Table 1 (continued)

Scientific name	Parts of the plant used (solvent used)	Bioactive phytochemicals	Therapeutic potential	References
Dryopteris ramosa	Fronds and rhizome (aqueous)	Quinone	Anticancer, anti-helminthic, and pus	[13]
Dryopteris schimperiana	Rhizome (n-hexane)	Saponin, flavonoids, anthraquinones	Antioxidant, anti-helminthic	[13]
Dryopteris stewartii	Fronds (methanol)	Phenol	Body coolant, wounds	[13]
Dryopteris wallichiana	Rhizome (diethyl ether)	Phloroglucinols, carbamic acid	Dysentery, antifungal	[13]
Hypodematium crenatum	Fronds (methanol)	Phenol, alkaloid	Antibacterial	[13]
Leucostegia immersa	Fronds (methanol)	Terpene, phenol	Constipation and swellings.	[13]
Poecilopteris virens	Fronds (ethanol)	Phenol, alkaloid	Cuts, wounds, and backache.	[13]
Polystichum aculeatum	Rhizome (methanol)	Polyphenol, flavonoid	Antioxidant and antibacterial	[13]
Polystichum auriculatum	Fronds (aqueous)	Phenol, tannin	Swelling, wounds	[13]
Polystichum braunii	Rhizome (methanol)	Kaempferol, quercetin, gallic acid	Anti-inflammatory, antioxidant	[13]
Polystichum fibrillosopaleaceum	Rhizome (methanol)	Sterols, phenolic compounds	Antimicrobial, wounds	[13]
Polystichum lepidocaulon	Fronds (n-hexane)	Flavonoids, tannins	Antioxidant	[13]
Polystichum munitum	Fronds (methanol)	Phenol, flavonoid	Digestion, ulcer	[13]
Polystichum moluccense	Sporophyll (ethanol)	Tannin, phenol	Anti-helminthic, wounds	[62]
Polystichum nepalense	Rhizome (methanol)	Steroids, phenol	Skin diseases, antioxidants, and indigestion.	[63]
Polystichum polyblepharum	Rhizome (methanol)	Sterols, terpenes	Antioxidant, microbial, wounds	[64, 65]
Polystichum pungens	Rhizome (methanol)	Flavonoids, steroids	Anti-helminthic, antibacterial	[66, 67]
Polystichum squarrosum	Rhizome (methanol)	Polyphenol, flavonoid	Pus, stomach disorder, and rheumatism.	[68, 69]

(continued)

Table 1 (continued)

Scientific name	Parts of the plant used (solvent used)	Bioactive phytochemicals	Therapeutic potential	References
Polystichum woronowii	Rhizome (methanol)	Polyphenol, triterpenes	Antibacterial	[41]
Tectaria coadunata	Rhizome and fronds (methanol)	Beta-sterols, flavonoid, terpenes	Bronchitis, antibacterial, asthma, and stomach problem	[70]
Tectaria wightii	Rhizome and fronds (methanol)	Quinone, glycoside	Anti-helminthic	[70]

5.2 Antidiabetic Activity

The acetonic extract of the plant *Bolbitis appendiculata* bears gallic acid that shows properties of cardiovascular protection and reduces the risk of diabetes by being antidiabetic. In an experiment with rabbit-induced diabetes by alloxan injection, it was shown that the compound kaempferol induces a reduction in blood glucose levels. It also inhibits alpha-glucosidase enzyme and alpha-amylase enzyme activity [25].

5.3 Antibacterial Activity

The methanolic extract of the fronds of the plant *Dryopteris fragrans* bears an aspidinol compound that showed antibacterial activity when tested against *Staphylococcus aureus*. The compound targets amino acid and ribosome synthesis inhibition. Another compound, quercetin, isolated from the plant *Dryopteris erythrosora* exhibits antibacterial activity when tested against Gram +ve and Gram -ve bacteria. It showed growth inhibition by targeting cell membrane and cell wall damage [54].

5.4 Antifungal Activity

When experimented with *Cladosporium herbarum*, methanolic extract of rhizome bears a compound called phloroglucinol and when experimented with *Arachniodes exilis* showed prevention of growth [24, 39]. Carbamic acid isolated from *Dryopteris wallichiana* diethyl ether extract of the plant rhizome showed activity against *Fusarium* and *Botrytis subtilis* by inhibiting the growth, resulting in a decrease in fungal load. *Cyrtomium caryotideum* ethanolic frond extract bears a compound called saponin that shows activity in killing fungi by disrupting the integrity of the cell membrane. The bilayer cell membrane nature is disintegrated by the activity of saponin, resulting in the death of the microorganism [38].

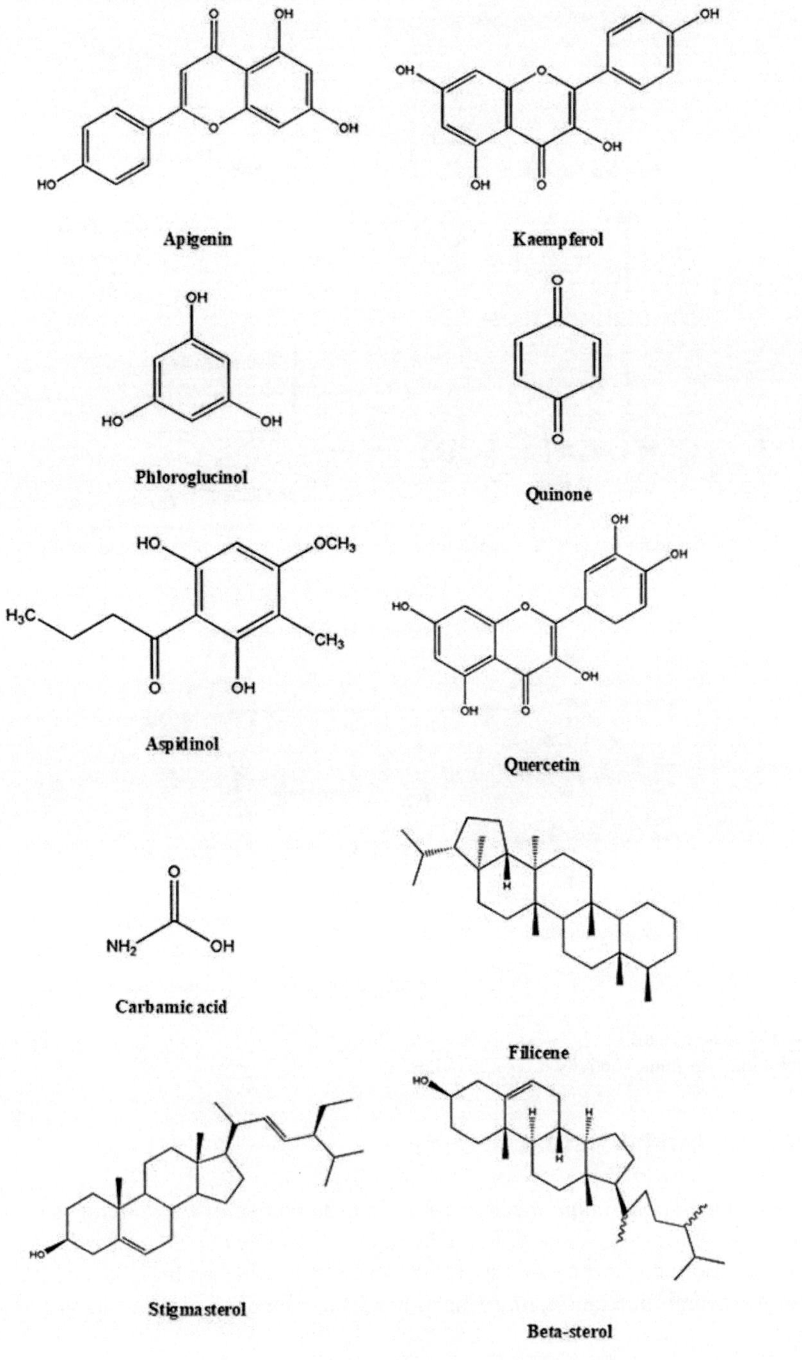

Fig. 1 Some important phytochemicals and their structure reported in the Dryopteridaceae family

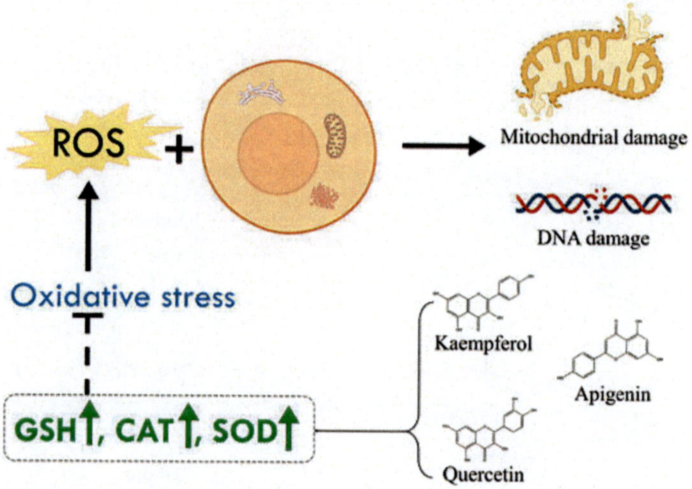

Fig. 2 Antioxidant activity of some compounds present in the Dryopteridaceae family

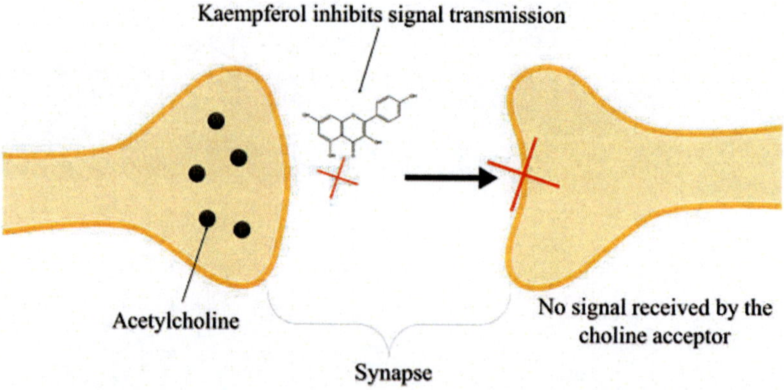

Fig. 3 Kaempferol acting as anti-helminthic when administered to *Raillietina echinobothrida* by inhibiting the neural activity of acetylcholine

5.5 Anti-helminthic Activity

Cyrtomium hookerianum bears a compound derivative of kaempferol that induces acetyl cholinesterase alteration in a helminth, *Raillietina echinobothrida*. The compound also induces a change in the enzyme nitric oxide synthase. Thus, it disrupts proper neural functioning in the helminth. The compound also results in ultrastructural alteration of *Raillietina echinobothrida* (Fig. 3) [36].

5.6 Anticancer Activity

The methanolic extract of rhizome of the plant *Arachniodes exilis* bears a compound called phloroglucinol. This compound in an experiment with stem cells of breast cancer induced reduction of CD44+ cells and expresses regulators of cancer stem cells like Oct4, beta-catenin, SOX2, CD44, and Notch2. The compound phloroglucinol induces cancerous cell death via apoptosis by altering Bcl-2 proteins, caspase 3, caspase 8, and cytochrome C activity in cancer cells of HT-29 colon cancer [23].

The plant rhizome aqueous extract of *Dryopteris ramosa* possesses a compound quinone that plays a significant role in the carcinogenesis initiation phase. In an experiment with mice induced with hepatoma, the quinone reductase mechanism activates cytotoxic activity by apoptosis induction (Fig. 4).

5.7 Anti-Inflammatory Activity

The methanolic extract of the plant *Arachniodes exilis* fronds bears a compound called phloroglucinol that shows anti-inflammatory activity when experimented in vitro. The compound phloroglucinol inhibits cyclooxygenase (COX) enzymes. COX-1 regulates physiological function and COX-2 induces inflammation; the compound works by inhibiting the COX activity. This compound also induces the production of TNF-alpha, prostaglandin, interleukin-6, and interleukin -beta [22, 23].

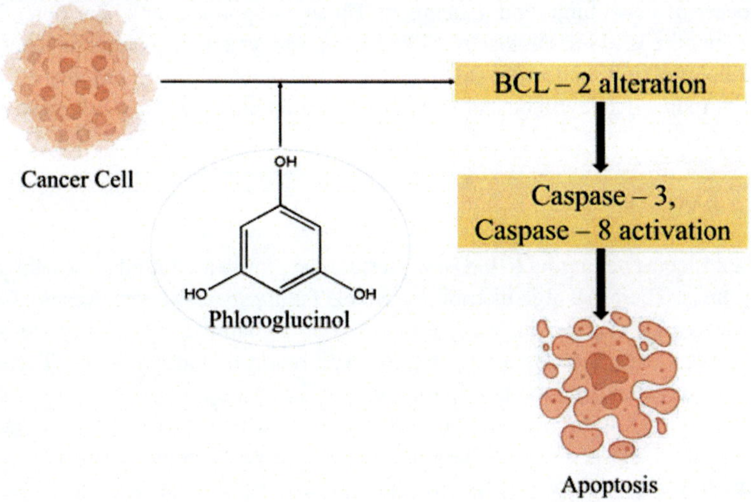

Fig. 4 Anticancer activity of phloroglucinol

5.8 Antinociceptive Activity

Dryopteris cycadina bears the compound kaempferol that is isolated from the methanolic extract of the fronds. The compound shows inhibiting pain sensation in a mice experiment where the pain was induced by acetic acid. Sensory channel activity got altered with the addressing of the compound. This activity hints at kaempferol being antinociceptive [57].

6 Future Prospects

The work on phytochemicals data collection and analysis for the family Dryopteridaceae needs more attention. More focus must be served on extensive active chemical ingredient recognition. The species of the Dryopteridaceae family despite its wider spread in India still lack ample resources like research papers, articles, and documentation of the plants ethnomedicinally important. Many tribes across India and the ethnic groups residing in Himachal Pradesh, Rajasthan, Assam, Tripura, Sikkim, Meghalaya, and the Western Ghats use the species of the family as a source of medicine. The knowledge is being passed from ancestors to a descendant as folklore; thus, efficient ethnomedicinal knowledge can be gained by proper questionnaire and surveying those ethnically knowledgeable people. The gap between ethnic groups and recent researchers working in the field of Phytochemistry, Taxonomy, and Ethnobotany must be examined and filled up. Parallelly, the medicinal plant identification via ethnobotany, phytochemical characterization, computational drug designing, and pharmacology study should merge and work together in the process of new compound generation. Those compounds may bear potential in tackling hundreds of human diseases like cancer, opportunistic pathogens (bacterial, fungal, helminthic), diabetes, oxidative stress, and many more in a highly effective way with reduced side effects compared to chemotherapy and physiotherapy.

7 Conclusions

The plant species belonging to the Dryopteridaceae family are found to be habituated in hilly areas. There is a high distribution of this family species in the Western Ghats, North-eastern states, and the Eastern Ghats. Due to habituating in hilly areas and being lower plants, there is a very scanty exploration of those species. There has been an observation that modern human society has set up a trend of using readily available medicines in pharmacies and hospitals which had led to the gap in communication between ethnic groups and people residing in urban areas. Low exposure to traditional knowledge has thus resulted. Proper data survey, collection, and analysis by the recent workers of that ethnomedicinal knowledge may result in better recognition of bioactive compounds from medicinal plants. This will lead to exposure to lower plants' medicinal values. More than 50 species from the family Dryopteridaceae have been showing medicinal aspects against human diseases.

Dryopteris and *Polystichum* genera bear a maximum number of medicinal species from the family Dryopteridaceae. From many medicinal plant species, phloroglucinol, kaempferol, apigenin, quercetin, carbamic acid, filixic acid, germacrene, aspinidol, quinone, and many more compounds have been isolated from the plant extract. These active compounds exhibit antifungal, antibacterial, anti-helminthic, antioxidant, antidiabetic, cytotoxic, and anti-inflammation activities. Thus, resolving health issues in human society with high efficacy and reduced side effects. Thus, the right approach in the process of drug discovery should be a combination of ethnobotany, phytochemical analysis, computational drug designing, and pharmacology study that will result in providing potential drug lead. Lastly, more concern must be given to the underexplored group of plants as they may possess highly bioactive phytochemicals that may result in effective therapy.

References

1. Dubal KN, Patil SM, Dongare MM, Kale MV (2015) Investigation of chemical composition from *Dryopteris cochleata* (D. Don) C. Chr. (Dryopteridaceae). Asian J Pharma Clin Res 8(4): 1–4
2. Deshpande M, Khilari S, Mahaja A (2014) Pharmacognostical and phytochemical studies of Kukutnakhi-*Aspidium Cicutarium*-A Folklore Plant. IOSR J Pharmacy 4(11):49–52
3. Thomas T (2009) Preliminary antibacterial and phytochemical evaluation of *Dryopteris cochleata* (D. don) C. CHR. J. Global Pharma Technol 3(6):4–8
4. Kathirvel A, Rai AK, Maurya GS, Sujatha V (2014) *Dryopteris cochleata* rhizome: a nutritional source of essential elements, phytochemicals, antioxidants and antimicrobials. Int J Pharm Pharm Sci 6:179–188
5. Cao H, Chai TT, Wang X, Morais-Braga MFB, Yang JH, Wong FC, Coutinho HD (2017) Phytochemicals from fern species: potential for medicine applications. Phytochem Rev 16(3): 379–440
6. Kramer KU, Holttum RE, Moran RC, Smith AR (1990) Dryopteridaceae. In: Pteridophytes and gymnosperms. Springer, Berlin/Heidelberg, pp 101–144
7. Gao Z, Ali Z, Zhao J, Qiao L, Lei H, Lu Y, Khan IA (2008) Phytochemical investigation of the rhizomes of *Dryopteris crassirhizoma*. Phytochem Lett 1(4):188–190
8. Kulkarni M, Tambe R, Bhise K (2013) Preliminary phytochemical screening and HPTLC studies of extracts of dried rhizomes of *Aspidium cicutarium*. J Pharmacog Phytochem 2(3): 50–54
9. Ishaque M, Bibi Y, Ayoubi SA, Masood S, Nisa S, Qayyum A, Valant-Vetschera K (2021) Iriflophenone-3-C-β-d Glucopyranoside from *Dryopteris ramosa* (Hope) C. Chr. with promising future as natural antibiotic for gastrointestinal tract infections. Antibiotics 10(9):1128
10. Valizadeh H, Sonboli A, Kordi FM, Dehghan H, Bahadori MB (2016) Cytotoxicity, antioxidant activity and phenolic content of eight fern species, from North of Iran. Pharm Sci 21(1):18–24
11. Kale M (2019) Study of bioactive compounds in rhizome extract of *Dryopteris cochleata* (D. DON) C. Chr. by using GC-MS analysis. Indian Fern J 30:18–29
12. Benjamin A, Manickam VS (2007) Medicinal pteridophytes from the Western Ghats. Indian J Tradit Knowl 6(4):611–618
13. Sureshkumar J, Silambarasan R, Bharati KA, Krupa J, Amalraj S, Ayyanar M (2018) A review on ethnomedicinally important pteridophytes of India. J Ethnopharmacol 219:269–287
14. Xu Z, Deng M (2017) Dryopteridaceae. In: Identification and control of common weeds, vol 2. Springer, Dordrecht, pp 75–82

15. Khurana RK, Kaur R, Lohan S, Singh KK, Singh B (2016) Mangiferin: a promising anticancer bioactive. Pharma Pat Anal 5(3):169–181
16. Gendaram O, Choi YH, Kim YS, Ryu SY (2011) Anti-oxidative and antibacterial constituents from *Sedum hybridum*. Nat Prod Sci 17(4):279–284
17. Na M, Jang J, Min BS, Lee SJ, Lee MS, Kim BY, . . . Ahn JS (2006) Fatty acid synthase inhibitory activity of acylphloroglucinols isolated from *Dryopteris crassirhizoma*. Bioorg Med Chem Lett 16(18): 4738–4742
18. Ullah S, Jan G, Jan FG, Khan S, Khattak M, Bibi H (2018) Phytochemical analysis, antipyretic and antifungal activities of *Cyrtomium caryotideum*. Biosci Biotechnol Res Asia 15(3):577–589
19. Van Vuuren SF (2008) Antimicrobial activity of South African medicinal plants. J Ethnopharmacol 119(3):462–472
20. Patil SM, Dubal K (2015) Investigation of chemical composition from *Dryopteris chochleata* (D. Don) C. CHR.(Dryopteridaceae). Asian J Pharma Res 8(4):311–314
21. Karthik V, Raju K, Ayyanar M, Gowrishankar K, Sekar T (2011) Ethnomedicinal uses of pteridophytes in kolli hills, Eastern Ghats of Tamil Nadu, India. J Nat Prod Plant Resour 1 (2):50–55
22. Schröder S, Lee S, Efferth T, Motoo Y (2013) Acupuncture and herbal medicine for cancer patients. Evid Based Complementary Altern Med 2013:313751
23. Widén CJ, Huure A, Sarvela J, Iwatsuki K (1978) Chemotaxonomic studies on Arachniodes (Dryopteridaceae) II. Phloroglucinol derivatives and taxonomic evaluation. Bot Mag 91(4): 247–254
24. Schröder S (2014) Acupuncture and herbal medicine for cancer patients 2014. Hindawi Publishing Corporation. London, United Kingdom
25. Manivannan V, Almeida RS, Coutinho HD (2021) Chemical profiling of Tectaria paradoxa (Fee.) Sledge and *Bolbitis appendiculata* (Willd.) K. Iwats using UHPLC. Biocatl Agric Biotechnol 34:102043
26. Manivannan V, Johnson M, Almeida RS, Coutinho HDM (2021) Phytochemical profile and bio-activity of Bolbitis appendiculata (Willd.) K. Iwats. Extracts. S Afr J Bot 137:236–241
27. Rathod VN, Pardeshi VN (2010) Occurrence of Bolbitis appendiculata (Willd.) Iwats. in Sahyadri hills of Western Ghats, Maharashtra. BIOINFOLET-J Life Sci 7(1):54–56
28. Mandal KK, Kar T, Pattanaik C, Reddy CS (2020) A census of pteridophytes in Eastern Ghats, India. Trop Plant Res 7(1):117–125
29. Xiang JY, Su-Gong W, Loc PK, Souliya O (2011) Studies on the genus *Bolbitis* (Dryopteridaceae) from Vietnam and Laos. Am Fern J 101(4):282–294
30. Kandel DR (2020) Pteridophytes of Nepal. Plant diversity of Nepal. Bot Soc Nepal:10–22
31. Kim H, Kim H, Kong CS, Lee BH, Sim HB, Seo Y (2021) Antioxidant and anti-inflammatory activities of the halophyte *Cyrtomium falcatum*. Ocean Polar Res 43(3):113–126
32. Lee CH, Shin SL (2011) Functional activities of ferns for human health. In: Working with ferns. Springer, New York, pp 347–359
33. Mironowicz A, Kromer K, Pawłowicz P, Siewiński A (1994) Abilities of some higher plants to hydrolyze the acetates of phenols and aromatic-aliphatic alcohols. Acta Soc Bot Pol 63(1):43–48
34. Choi SY (2013) Inhibitory effects of *Cyrtomium fortunei* J. Smith root extract on melanogenesis. Pharmacogn Mag 9(35):227
35. Horrocks J (1995) Cyrtomium fortunei. Hardy Fern Found News Lett 5(1):2–8
36. Suksathan P, Lindsay S, Middleton DJ (2010) *Polystichum hookerianum* (C. Presl) C. Chr. (Dryopteridaceae), a new record for Thailand. Thai Forest Bull Bot 38:120–123
37. Ren Z, He C, Fan Y, Si H, Wang Y, Shi Z, . . . Zhang H (2014) Immune-enhancing activity of polysaccharides from *Cyrtomium macrophyllum*. Int J Biol Macromol 70:590–595
38. Libing Z, Barrington D (2013) Taxonomic treatment of Cyrtomium for Flora of China. Cyrtomium C Presl Tent Pterid 86:1836
39. Frasser-Jenkins CR, Widen CJ (1993) Phloroglucinol derivatives in *Dryopteris ardechensis* and *D. corleyi* (Pteridophyta, Dryopteridaceae) and their putative ancestors. In: Annales Botanici Fennici. The Finnish Botanical Publishing Board, pp 43–51. Finland

40. Fraser-Jenkins CR (2007) The species and subspecies in the *Dryopteris affinis* group. Fern Gaz 18(1):1
41. Bahadori MB, Mahmoodi Kordi F, Ali Ahmadi A, Bahadori S, Valizadeh H (2015) Antibacterial evaluation and preliminary phytochemical screening of selected ferns from Iran. Res J Pharmacog 2(2):53–59
42. Rakkimuthu R, Naveenraj B, Benitta L, Dhanya K (2018) Phytochemical screeining and antibactiral activity of the fern Hemionitis arifolia (Burm.) Moore. World J Pharma Res 7(13):603–609
43. Rajesh KD, Subramani V, Annamalai P, Rajesh NV, Narayanaperumal J (2016) Phytochemical analysis, in vitro antioxidant potential and gas chromatography-mass spectrometry studies of *Dicranopteris linearis*. Asian J Pharma Clin Res 9(8):220–225
44. Fuchino H, Nakamura H, Wada H, Hakamatsuka T, Tanaka N (1997) Two new acylphloroglucinols from *Dryopteris atrata*. Chem Pharm Bull 45(6):1101–1102
45. Kim MM, Kim SK (2010) Effect of phloroglucinol on oxidative stress and inflammation. Food Chem Toxicol 48(10):2925–2933
46. Mehra PN, Mittal TC (1962) Pharmacognosy of Indian substitutes of male fern II: *Dryopteris barbigera* and *Dryopteris cochleata*. J Pharm Sci 51(4):367–371
47. Jana F, Pareeka S, Srivastavab RP, Zahoora I, Sharmaa A, Shrivastavaa D (2022) Anticancerous and anti-bacterial potential of silver nanoparticles synthesized using leaf extract of fern-*Dryopteris barbigera*. Dig J Nanomater Biostruct 17(1):285–299
48. Singh C, Kumar B, Rani A, Dhyani K, Singh R (2017) Biodiversity and conservation ferns diversity in different forests of Dehradun district. Int J Pharma Res Technol 7(2):01–07
49. Samant SS, Joshi HC, Arya SC, Pant S (2005) Diversity, distribution and conservation of pteridophytes in Nanda Devi Biosphere Reserve, West Himalaya. Indian Fern J 22:100–111
50. De Souza MM, Pereira MA, Ardenghi JV, Mora TC, Bresciani LF, Yunes RA, Cechinel-Filho V (2009) Filicene obtained from *Adiantum cuneatum* interacts with the cholinergic, dopaminergic, glutamatergic, GABAergic, and tachykinergic systems to exert antinociceptive effect in mice. Pharmacol Biochem Behav 93(1):40–46
51. Khan MS, Ullah S (2018) Analgesic, anti-inflammatory, antioxidant activity and phytochemical screening of *Dryopteris blanfordii* plant. J Pharmacog Phytochem 7(3):536–541
52. Mir SA, Mishra AK, Reshi ZA, Sharma MP (2013) Preliminary phytochemical screening of some pteridophytes from district Shopian (J & K). Int J Pharm Pharm Sci 5(4):632–637
53. Fraser-Jenkins CR (2016) A revised checklist of Indian Pteridophytes-1. Indian Fern J 33:193–205
54. Froissard D, Rapior S, Bessière JM, Fruchier A, Buatois B, Fons F (2014) Volatile organic compounds of six French *Dryopteris* species: natural odorous and bioactive resources. Nat Prod Commun 9(1):201401
55. Hegnauer R (1967) Chemical characters in plant taxonomy: some possibilities and limitations. Pure Appl Chem 14(1):173–188
56. Bosman AA, Combrinck S, Roux-Van der Merwe R, Botha BM, McCrindle RI, Houghton PJ (2004) Isolation of an anthelmintic compound from *Leucosidea sericea*. S Afr J Bot 70(4):509–511
57. Kim HW, Cho N, Huh J, Woo S, Sung SH (2015) Anti-neuroinflammatory compounds from *Athyrium yokoscense*. Planta Med 81(16):PW_129
58. De Carvalho RBF, De Almeida AAC, Campelo NB, Lellis DROD, Nunes LCC (2018) Nerolidol and its pharmacological application in treating neurodegenerative diseases: a review. Recent Pat Biotechnol 12(3):158–168
59. Widén CJ, Fraser-Jenkins C, Reichstein T, Gibby M, Sarvela J (1996) Phloroglucinol derivatives in *Dryopteris* sect. Fibrillosae and related taxa (Pteridophyta, Dryopteridaceae). In: Annales Botanici Fennici, vol 33. Finnish Zoological and Botanical Publishing Board Helsinki, pp 69–100
60. Ahmad M, Jahan N, Mehjabeen ABR, Ahmad M, Ullah O, Mohammad N (2011) Differential inhibitory potencies of alcoholic extract of different parts of *Dryopteris chrysocoma* on inflammation in mice and rats. Pak J Pharm Sci 24(4):559

61. Alam SM, Qureshi M, Jahan N (2010) Antimicrobial screening of some medicinal plants of Pakistan. Pak J Bot 42(6):4281–4284
62. Şuţan NA, Fierăscu I, Fierăscu R, Ionica D, Soare LC (2019) Phytochemical analysis and in vitro assessment of *Polystichum setiferum* extracts for their cytotoxic and antimicrobial activities. Caryologia 72(2):53–61
63. Goswami HK, Sen K, Mukhopadhyay R (2016) Pteridophytes: evolutionary boon as medicinal plants. Plant Genet Resour 14(4):328–355
64. Shinozaki J, Nakene T, Takano A (2018) Squalene cyclases and cycloartenol synthases from *Polystichum polyblepharum* and six allied ferns. Molecules 23(8):1843
65. Shin SL, Lee CH (2010) Antioxidant effects of the methanol extracts obtained from aerial part and rhizomes of ferns native to Korea. Kor J Plant Resour 23(1):38–46
66. Grierson DS, Afolayan AJ (1999) Antibacterial activity of some indigenous plants used for the treatment of wounds in the Eastern Cape, South Africa. J Ethnopharmacol 66(1):103–106
67. Afolayan AJ, Grierson DS, Kambizi L, Madamombe I, Masika PJ, Jäger AK (2002) In vitro antifungal activity of some South African medicinal plants. S Afr J Bot 68(1):72–76
68. Sivasankar S, Somvanshi R (2001) Pathological evaluation of *Polystichum squarrosum* (D. Don) fern in laboratory rats. NISCAIR-CSIR 39(8):772–776
69. Tapwal A, Garg S, Gautam N, Kumar R (2011) In vitro antifungal potency of plant extracts against five phytopathogens. Braz Arch Bio Tchnol 54(6):1093–1098
70. Reddy MN, Adnan M, Alreshidi MM, Saeed M, Patel M (2020) Evaluation of anticancer, antibacterial and antioxidant properties of a medicinally treasured fern *Tectaria coadunata* with its phytoconstituents analysis by HR-LCMS. Anti Cancer Agents Med Chem 20(15): 1845–1856

Bioactive Compounds and Biological Activities of *Dipteris wallichii*

15

Pankaj Chetia, Damini Dey, Minakshi Puzari, and Manabendra Dutta Choudhury

Contents

Abstract

Bryophytes and pteridophytes have been very communal among the tribal people in various parts of the world since time immemorial and have been evidenced to be useful in various grounds like food, tea, and medicines. Improvement in technologies in the human health sector has supported the inspections of the significant functional activities of these groups of plants, which have in turn dilated the scope of using these groups of neglected plants as a great boon for the pharmaceutical companies and their akin divisions. Different systems of medicines like Ayurvedic, homeopathic, and Unani have fruitfully utilized the therapeutic properties of fern and their allies in their system of medicines, after the in vitro screening of the latter. *Dipteris wallichii* (R.Br.) T. Moore, an endemic

P. Chetia (✉)
Department of Life Sciences, Dibrugarh University, Dibrugarh, Assam, India

Department of Life Science & Bioinformatics, Assam University, Silchar, Assam, India
e-mail: chetiapankaj@dibru.ac.in

D. Dey · M. Puzari
Department of Life Sciences, Dibrugarh University, Dibrugarh, Assam, India
e-mail: rs_daminidey@dibru.ac.in; minakshipuzari@dibru.ac.in

M. Dutta Choudhury
Department of Life Science & Bioinformatics, Assam University, Silchar, Assam, India
e-mail: drmdc@bioinfoaus.ac.in

H. N. Murthy (ed.), *Bioactive Compounds in Bryophytes and Pteridophytes*,
Reference Series in Phytochemistry, https://doi.org/10.1007/978-3-031-23243-5_15

fern of the northeastern region of India, is considered one of the rare genera with very few records available, concerning its phytochemical and ethnomedicinal importance. The rhizome of this species is recorded to be useful as a hepatoprotective agent. Very delimited distribution and lack of easy approachability to this plant permits it to be called an endangered species, which has in turn called upon to study the chemical and molecular characterization including the pharmacological evaluation of this particular plant that is anticipated to hold high probabilities to be used as drugs in the medical world. In this chapter, bioactive compounds isolated from *D. wallichii* have been discussed, which are shown to have high therapeutic efficacy. Among these compounds, an alkaloid has been reported to be one of the most promising inhibitors against the human β secretase enzyme, a potent and highly anticipated drug target for Alzheimer's disease. In addition to pondering upon the necessary conservatory measures to save this hidden treasure, there is a lot more to explore about the biological activity of the chemical compounds found in this interesting plant before it gets completely extinct.

Keywords

Alzheimer's disease · *Dipteris wallichii* · Ethnomedicinal · Hepatoprotective · Human β secretase · Pteridophytes

Abbreviations

ADMET	Absorption, distribution, metabolism, excretion, and toxicity
cpDNA	Chloroplast DNA
HIV	Human immunodeficiency virus
KEGG	Kyoto Encyclopedia of Genes and Genomes
NCBI	National Centre for Biotechnology Information
NMR	Nuclear magnetic resonance

1 Introduction

Since time immemorial, biodiversity has played a noteworthy role in assuring to fulfillment of several daily fundamental needs of both humans and livestock. Even though human civilization has gone through several decades of advancement, its inclination toward nature and natural products never got averted. Millions of people mostly in developing countries are still dependent and derive a substantial part of their subsistence and income from nature. Forests harbor most of Earth's terrestrial biodiversity, which is now threatened with extinction at an unprecedented rate due to the need for fulfillment of the demand of increasing population and also climatic changes that in turn bring about drastic changes in the ecological balance of the entire world. Among some of the self-achieved skills of humans, the proficiency and practice of extracting drugs from nature are one of the most important activities that humans have mastered [1].

Members of the lower cryptogams Bryophyta (nonvascular cryptogams) and the higher cryptogams Pteridophyta (vascular cryptogams) have been known to provide many health benefits to the ancient civilizations who have used them for various purposes like food, tea, and drugs. Plants yielding a good amount of biologically active compounds, such as polysaccharides, lipids, steroids, terpenoids, organic acids, and polyphenols are selected as the best-suited plants for the extraction of drugs to cure various ailments, including cancer. A combination of multidisciplinary technologies in the modern world has resulted in the extraction and identification of these various chemical compounds for producing medicines in a very particulate amount from different plant parts. Present-day inspections on the functional activities of both pteridophyte and bryophyte in the human health sector, followed by the discoveries of certain biologically active compounds and their usage in medicines, have dilated the scopes of using these groups of neglected plants as a great privilege for the pharmaceutical companies and many related industries. Bryophytes are very popular among the tribal people in various parts of the world to cure hepatic disorders, skin ailments, and cardiovascular disorders and are also found to possess antipyretic antimicrobial and wound-healing properties. Apart from these ethnomedicinal utilities, a number of naturally derived components from them have been found to be useful as anticancer and antitumor agents against different cancer cell lines. In vivo screening of different fern and fern allies revealed their drug properties, which have been successfully used in various systems of medicines like Ayurvedic, Unani, and homeopathic. Pteridophytes have been reported to contain medicinal properties against bacteria, fungi, cancer, inflammation, diabetes, rheumatism, and infertility.

2 Pteridophytes and Its Therapeutic Potential

Pteridophytes are known to be the most primitive vascular plant on Earth that appeared in the mid-Paleozoic era of the Silurian period (430 million years ago), dominated the Earth in the Triassic period (230–280 million years), and gradually began to get replaced by the flowering plants in the late Cretaceous period [2]. Spanning nearly one-third of the world's diversity, about 4500 species of fern and fern allies are found in Asia, out of which India is known to harbor around 1000 species from 70 families and 192 genera [3]. In India, pteridophytes are mainly found to occur along the Western Ghats, the Eastern Ghats, Eastern and Western Himalayas, Central India, and Andaman and Nicobar Islands. The Greek botanist Theophrastus (387–287 BC) in one of his books mentioned the ethnomedicinal uses of pteridophytes that have been known to the human race for more than 2000 years [4]. *Marsilea minuta* L. and *Adiantum capillus-veneris* L. were recommended for their exclusive medicinal uses by Charaka and Sushruta, the two medical giants of ancient India in their Samhitas [5]. Also, in "de Materia Medica" by Discorides (ca. 50 AD) a number of ferns including the bracken fern *Pteridium aquilinum* (L.) Kuhn and the male fern *Dryopteris filix-mas* (L.) Schott was reported to have medicinal values [6]. In the Ayurvedic and homeopathic system of medicine, the lycophytes *Selaginella ciliaris, Lycopodium clavatum,* and *Selaginella bryopteris* are well known, wherein *Selaginella* furthermore in the prior studies is shown to possess multiple pharmacological activities, such as antioxidant, anti-inflammatory, anticancer, antiviral, and to be able to cure certain neurological disorders like Alzheimer's disease, and also having alleviating properties to treat heat stroke effects [7]. Surprisingly, the "fern weeds" like *Azolla, Ceratopteris, Marsilea,* and *Salvinia,* which are generally known to reduce the freshwater wealth of a lake by invading them, are very good reservoirs of many organic compounds that can be employed to produce lifesaving drugs. The dried fronds and rhizomes of *Polystichum lepidocaulon* and *Polystichum polyblepharum* on screening were found to have more than 13% of total polyphenols out of a total of 37 fern and fern allies, which were useful phytochemicals bearing antioxidant activity. Decoction of the entire plant of *Adiantum aethiopicum* L. is useful as a tonic against bruises, gangrenous, wounds, diuretic disorders, cough expectorant, and parturition. Phytochemicals from medicinally important pteridophytes have been studied internationally to accentuate their roles in medicine. These trials have substantiated the presence of many bio-activities like anti-inflammatory, antimicrobial, antibacterial, antiviral, antitumor, and anti-HIV. Fern members of the genus *Adiantum* were found to show antibacterial activities against Gram-positive bacteria. In several cases, the active substances from pteridophytes were reported to show resistivity against penicillin-resistant *Staphylococcus aureus, Mycobacterium phlei, Pseudomonas aeruginosa, Vibrio cholerae,* and *Salmonella typhi.* Methanol extract of rhizome of *Drynaria quercifolia* has been reported to show concentration-dependent and broad-spectrum antibacterial activity against most of the tested bacterial strains. Similar to all the ferns mentioned above, the genus *Dipteris* is considered one of the rare genera with very few records available, concerning its phytochemical and ethnomedicinal importance.

2.1 *Dipteris*: A Neglected Genus with a High Curative Value

The fern family Dipteridaceae falls under the order Gleicheniales of the class Polypodiopsida, and are commonly known as umbrella ferns as they resemble the former in their appearance. Fossil records reveal that this family constituted at least four genera in the Jurassic period, viz., *Clathopteris, Camptopteris, Dictyophyllum,* and *Dipteris* [1]. Currently, *Cheiropleuria* and *Dipteris* are the two genera under this family with their species spread all over Asia, northern Australia, and New Guinea [8]. Primarily, the genera *Dipteris* differ from *Cheiropleuria* in having bilateral, monolete spores in contrast to tetrahedral and trilete spores in the latter. The leaves are monomorphic in both the genera with distinct sori for *Dipteris* and acrostichoid sporangia in *Cheiropleuria* [9].

Considering an early-diverging leptosporangiate fern lineage, the genus *Dipteris* comprises about eight species whose distribution is confined to the areas of the northeastern part of India, Indo-Malaysian islands, southern China, and from the southern Ryukyus to selected parts of Australia and Fiji [10]. Phenotypically, the genus *Dipteris* is characterized by having long creeping stems enfolded with

Fig. 1 *Dipteris wallichii* in natural habitat

Fig. 2 The ventral surface of *Dipteris wallichii* frond showing the arrangement of sori

articulate hairs, fronds resembling an umbrella or fan, elaborate reticulate veins, and included veinlets in the aereoles. The ventral surface of the fronds is covered with sporangia. The sori are exin-dusiate, discrete, and compital (borne at the junction of many veins) (Fig. 2). The vascular bundles found in the petioles are of both single and polystelic in nature, arranged proximally and distally, respectively. A comparative study of the vascular organization in the three species of *Dipteris,* viz., *D. conjugata* Reinw., *D. lobbiana* (Hook.) T. Moore, and *D. wallichii* (R.Br.) T. Moore (Fig. 1), revealed some morphological anomalies in the vasculature of their rhizome. On the basis of this study, gradual evolutionary changes in the leaf traces of these species, from the simplest condition in *D. chinensis*, *D. conjugata, D. nieuwenhuisii,* and *D. novoguineensis* to the most advanced and complex state in *D. wallichii*, through *D. lobbiana,* were likely reported to exhibit a derived condition. Moreover, the presence of triploidy condition in *D. wallichii* has additionally supported the cause of being the most evolutionary advanced species of the genus *Dipteris*. In Singapore, the only extant species of *Dipteris* currently found is *D. conjugata*, which is also now a critically endangered species in this country [11]. Another species record of *Dipteris,* namely, *D. papilioniformis* Kjellberg has been reported from Sulawesi, Indonesia [12]. According to Ouyang et al. [13], *D. chinensis*, also commonly known as octopus fern in Chinese, is a relict species that is endemic to China and is on the verge of extinction due to human intervention and climate changes [13]. Lately, a very new and unique species of *Dipteris* has been discovered, identified, and described in Shenzhen, Guangdong, southern China, by Wei et al. [14]. This new species, *D. shenzhenensis*, is often confused with *D. chinensis* due to its very close morphological resemblance with the latter, besides the molecular evidence, which has shown its similarity to *D. conjugata* [10]. Thus, an amalgamation of both morphological evidence and the result of cpDNA helped it to get authenticated as a distinct new species with several recognizable characteristics, out of which the presence of immensely long rhizome scales is the most striking one.

3 Bioactive Compounds in *Dipteris wallichii*

Dipteris wallichii (R. Br.) T. Moore (Fig. 1) is an endemic fern of the northeastern region of India, whose distribution is confined to Arunachal Pradesh, Tripura, Meghalaya, and southern Assam. Due to various causes like infrastructural developments, intensification of agriculture, and an increase in human intervention, many important species of plants are under threat or already extinct in the western, eastern, and central parts of Meghalaya, among which the species *Dipteris wallichii* (R.Br.) T. Moore needs a special mention. Similar to most of the ferns mentioned previously, the rhizomes of *D. wallichii* are bestowed with ethnomedicinal properties that are used as hepatoprotective agents to prevent damage to the liver [15]. The highly endemic distribution of this plant in the northeast region has helped it to gain the tag of "endangered species" that has essentially called upon to study the chemical and molecular characterization, including the pharmacological evaluation of this particular plant, which is anticipated to hold high probabilities to be used as drugs in the medical world [1].

According to Smith et al. [9], the systematic position of this plant is as follows:

Subphylum: Euphyllophytina
Infraphyllum: Moniliformopses
Class: Polypodiopsida
Order: Gleicheniales
Family: Dipteridaceae
Genus: *Dipteris*
Species: *Dipteris wallichii*

The Herbarium at the Royal Botanic Gardens located in Kew houses 44 records of *D. wallichii,* out of which all the records are from the eastern region of India, including the northeast. Due to very restricted distribution and lack of easy accessibility to this plant, very little is known about its ethnobotanical utilities, and not many reports are available so far, as its medicinal uses are concerned. As reported by MD Choudhury [14], it is used as a hepatoprotective agent by the *Reangs* of Tripura. The plant is an extremely endemic and endangered species, is thought to be a virgin so far, and its phytochemical screening is considered. As far as investigation is concerned, plant material could not be collected in larger amounts due to its rare availability, and to avoid its exploitation, only a handful of compounds have been reported in Ref. [1].

Crude extracts of *D. wallichii* leaves using different solvents for phytochemical screening majorly reported the presence of alkaloid, saponin, and tannin [1] as important phytochemicals (Table 1). Acetone and methanol extracts revealed the presence of alkaloids. The presence of saponin and tannin was confirmed by screening the crude leaf extracts with solvents like petroleum ether, ethyl acetate, acetone, and methanol. The existence of alkaloids in the plant probably indicated antimicrobial resistance against Gram-negative bacteria *Escherichia coli*. Alkaloids are also known to provide analgesic effects and work as effective painkillers. Apart from antimalarial, antihypertensive, and antitumor activities, to some extent, anticancer activity is also known to be shown by alkaloids [1]. Saponins are reported to show activities like expectorative, hemolytic, anti-inflammatory, and immune-stimulatory. These phytochemicals are also reported to show antimicrobial activity obligately against fungi and facultatively against bacteria and protozoa. Similarly, the presence of tannin might be responsible for showing antibacterial properties

Table 1 List of phytochemicals reported in *Dipteris wallichii*

Part used	Compound name	Medicinal uses	Reference
	Alkaloid	Antimicrobial, analgesic, anticancer, antihypertensive, antimalarial, anesthetic, antitumor, and antiarrhythmic activities	[1]
	Saponin	Hemolytic, antimicrobial, expectorative, anti-inflammatory, and immune-stimulating activity	
Extract from leaf	Tannin	Antibacterial, anti-inflammatory, antiviral, and antiparasitic activities	
	Alkaloid: (E)-4-amino-1-(5-((1E,4E)-hexa-1,4-dienyl)-1-methylpyrrolidin-2-yl) pent-2-en-1-one	Anti-Alzheimer drug and antibacterial activity against pathogenic bacterial strains like *E. coli* and *Klebsiella* spp.	
	Glycerol ester: (E)-3-((E)-undeca-5,10-dienoyloxy)-2-(undecanoyloxy) propyl dodec-8-enoate	Antibacterial activity against pathogenic bacterial strains like *E. coli* and *Klebsiella* spp.	

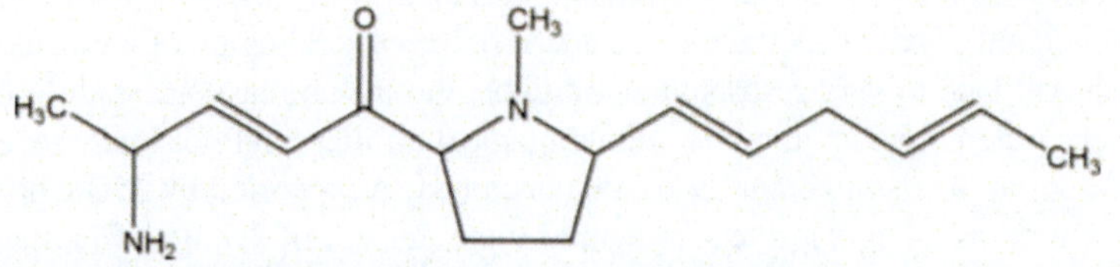

Fig. 3 (E)-4-amino-1-(5-((1E,4E)-hexa-1,4-dienyl)-1-methylpyrrolidin-2-yl)pent-2-en-1-one

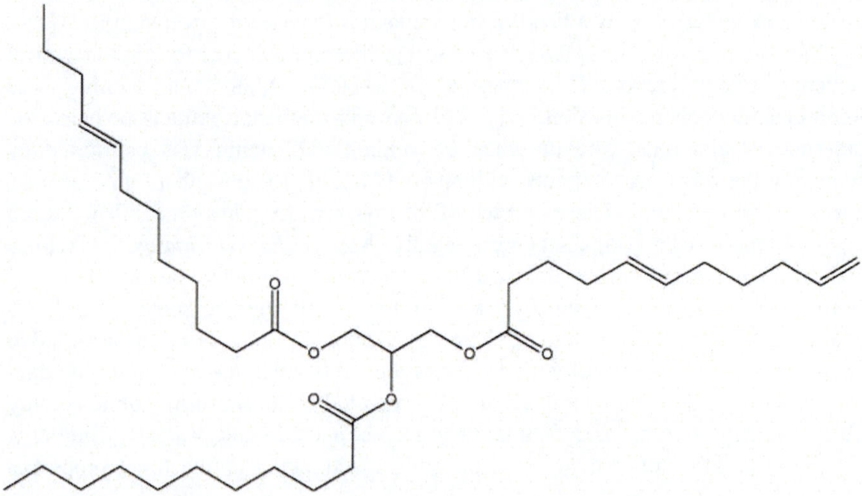

Fig. 4 (E)-3-((E)-undeca-5,10-dienoyloxy)-2-(undecanoyloxy) propyl dodec-8-enoate

against pathogenic strains of bacteria, although there is a need for a more detailed study of secondary metabolites using various other extracts to evaluate specific biological activities of *D. wallichii* [1].

Nuclear magnetic resonance (NMR) spectroscopy for structural elucidation of natural compounds derived from this plant disclosed the presence of two such novel compounds, which had no previous mention in the records available in major chemical databases like NCBI PubChem Compound, KEGG Ligand. Out of those two compounds (Figs. 3 and 4), one is an alkaloid, viz., (E)-4-amino-1-(5-((1E,4E)-hexa-1,4-dienyl)-1-methylpyrrolidin-2-yl) pent-2-en-1-one, and another is a glycerol ester, viz., (E)-3-((E)-undeca-5,10-dienoyloxy)-2-(undecanoyloxy) propyl dodec-8-enoate, although presence of glycerol ester is somewhat unusual in plants [1].

In pharmacokinetics, the disposition of any pharmaceutical compound within a living organism requires to be tested for its ADMET property (absorption, distribution, metabolism, and excretion), to estimate the drug-likeness of those compounds, in terms of these properties. A superior drug nominee, besides showing satisfactory efficiency against the drug target, should also abide by the ADMET rules at a remedial dose [15]. In *D. wallichii*, out of the two compounds isolated, the glycerol ester was reported to be toxic in ADME/Tox screening, although it showed antibacterial activity against *E. coli* and *Klebsiella* sp. [1]. On the other hand, the alkaloid compound derived from the plant successfully abided by the ADME/Tox rules at a therapeutic dose. Additionally, this compound was found to show a very strong rapport for the active site of human β secretase-1, as the highly anticipated drug target for Alzheimer's disease, when the Pharmmapper server was used to analyze its probable drug target [16]. Alzheimer's disease is biologically characterized by the progressive formation of insoluble β-amyloid plaques and abnormal accumulations of tau proteins inside the neurons. This genetic and sporadic neurodegenerative disease in its classical variant causes amnestic cognitive mutilation, whereas causes non-amnestic or mild cognitive disability in the less common alternates [17]. Therefore, the development of inhibitors for β secretase-1 protein could be an effective therapeutic target for dealing with Alzheimer's disease.

4 Conclusion

Usually, a plant irrespective of its origin is referred to be favorable to living organisms primarily due to the secondary metabolites present in it, and through generations, humans have developed their cognizance on how to avoid the detrimental and lethal compounds, and instead utilize the fruitful ones that are valuable to mankind. The most primitive sources from which the natural and bioactive products have been extracted to date are the terrestrial plants, dictating their proper preservation and collection of all kinds of chemical and biological information on such bases. *D. wallichii*, on which this chapter is based, also falls among those groups of terrestrial plants that are bestowed with major active constituents with their probable therapeutic uses in the medical world. Being one of the lesser-known and less studied plants due to its endemic distribution, there are lots more to explore about the biological activity of the chemical compounds found in this interesting plant before it gets completely extinct. The discovery of inhibitors for human β secretase 1, from *D. wallichii,* directs that there might be more such significant bioactive compounds that would be useful for treating not only Alzheimer's but other neurodegenerative ailments too. Additional in silico and in vitro studies of the compounds already isolated or to be isolated from this plant demand to be studied for exploring new dimensions regarding the therapeutic potentialities of the plant. Regardless of its high utility in the curative world, the current status of the plant is also quite alarming and an area of equal concern. Henceforth, proper conservation of this hidden treasure in its natural habitat should be proximately pondered upon by taking all the obligatory measures in doing the same.

References

1. Chetia P (2011) Therapeutic potential of *Dipteris wallichii* (R.Br.) Moore: an endemic plant of northeast India. PhD thesis, Assam University, India
2. Dudani SN, Mahesh MK, Chandran MDS, Ramachandra TV (2014) Pteridophyte diversity in wet evergreen forests of Sakleshpur in Central Western Ghats. Ind J Plant Sci 3:28–39
3. Sureshkumar J, Silambarasan R, Bharati KA, Krupa J, Amalraj S, Ayyanar M (2018) A review on ethnomedicinally important pteridophytes of India. J Ethnopharmacol 219:269–287
4. Pathak A, Gaurav N, Singh A, Kumar A, Singh A (2017) A review on ethnomedicinal important species of pteridophytes. https://doi.org/10.13140/rg.2.2.26033.43366
5. Chandra S (2000) The ferns of India: enumeration, synonyms, and distribution. International Book Distributros, Dehra Dun
6. Goswami HK, Sen K, Mukhopadhyay R (2016) Pteridophytes: evolutionary boon as medicinal plants. Plant Genet Resour 14:328–355
7. Baskaran XR, Vigila AVG, Zhang SZ, Feng SX, Liao WB (2018) A review of the use of pteridophytes for treating human ailments. J Zhejiang Univ Sci B 19:85–119
8. Christenhusz MJM, Byng JW (2016) The number of known plant species in the world and its annual increase. Phyotaxa 261:201–217
9. Smith AR, Pryer KM, Schuettpelz E, Korall P, Schneider H, Wold PG (2006) A classification for extant ferns. Taxon 55:705–731
10. Wei ZY, Gu YF, Xia ZQ, Chen LJ, Wang T, Zhang SZ, Zhao GH, Chen JB, Cao JG, Yan YH (2021) *Dipteris shenzhenensis*, a new endangered species of Dipteridaceae from Shenzhen, southern China. PhytoKeys 186:111–120
11. Lok AFSL, Ang WF, Tan HTW (2009) The status and distribution in Singapore of *Dipteris conjugata* Reinw. (Dipteridaceae). Nat Singapore 2:339–345
12. Catalogue of Life: *Dipteris papilioniformis* Kjellberg. http://www.catalogueoflife.org/annualchecklist/2018/details/species/id/a90e8e1cd1b156d0ff5fccd7e4742784
13. Ouyang K et al (2021) Habitat characteristics and population structure of *Dipteris chinensis*, a relict plant in China. Appl Ecol Environ Res 19:1–13
14. Choudhury MD (1999) Ethnomedico botanical aspects of reang tribe of Assam a comprehensive study. PhD dissertation, University of Gauhati, India
15. Guan L et al (2019) ADMET-score – a comprehensive scoring function for evaluation of chemical drug-likeness. Med Chem Commun 10:148–157
16. Chetia P, Mazumder MK, Mahanta S, De B, Choudhury MD (2020) A novel phytochemical from *Dipteris wallichii* inhibits human β-secretase 1: implications for the treatment of Alzheimer's disease. Med Hypotheses 143:109839
17. Knopman DS, Amieva H, Petersen RC, Chetelat G, Holzman DM, Hyman BT, Nixon RA, Jones DT (2021) Alzheimer disease. Nat Rev Dis Primer 7:1–21

Bioactive Compounds and Biological Activities of *Cyathea* Species

16

Johnson Marimuthu alias Antonysamy, Vidyarani George,
Silvia Juliet Iruthayamani, and Shivananthini Balasundaram

Contents

Abstract

Cyatheaceae family consists of most of the tree fern diversity with 600 species of tree ferns approximately. It includes three genera, namely, *Alsophila*, *Cyathea*, and *Sphaeropteris*. Many works focused on taxonomy and systematic treatments and molecular phylogenetic works of Cyatheaceae due to its puzzling morphological diversities in identifying species. *Cyathea* species are being used in traditional medicine worldwide to treat various ailments, such as dysentery, body pain, antidiabetic, hepatoprotective, inflammation, respiration, anti-helmintic, headache, and respiration. The phytochemical studies confirmed the presence of carbohydrates, starch, glycosides, saponins, alkaloids, tannins, flavonoids, sterols, terpenes, polyuronides, steroids, terpenoids, triterpenoids, anthraquinone, polyphenol, glycoside, coumarin, and cardiac glycosides. Hence, the present review aims to summarize the phytochemical, biological activities, and medicinal importance of the *Cyathea* species and provides an update for further research. Although some *Cyathea* species have demonstrated antimicrobial, antioxidant, antidiabetic, anti-inflammatory, cytotoxic, and anticancer potential, phytochemical studies and in vivo and in vitro biological activities remain unexplored for most of the *Cyathea* species.

J. Marimuthu alias Antonysamy (✉) · V. George · S. J. Iruthayamani · S. Balasundaram
Department of Botany, Centre for Plant Biotechnology, St. Xavier's College (Autonomous), Palayamkottai, India

H. N. Murthy (ed.), *Bioactive Compounds in Bryophytes and Pteridophytes*,
Reference Series in Phytochemistry, https://doi.org/10.1007/978-3-031-23243-5_17

Keywords

Cyathea · Biopotency · Antioxidant · Antimicrobial · Antidiabetic · Wound healing

Abbreviations

1D	One direction
2D	Two direction
ALP	Alkaline phosphatase
AQF	Aqueous fraction
BCB	β-carotene scavenging activity
BUF	Butanolic fraction
CCl_4	Carbon tetrachloride
DNA	Deoxyribonucleic acid
DPPH	2,2-Diphenyl-1-picryl-hydrazyl-hydrate
EAF	Ethyl acetate fraction
EDX	Energy dispersive X-ray analysis
FIC	Fractional inhibitory concentration
FRAP	Ferric reducing ability of plasma
FT-IR	Fourier transform infrared spectroscopy
GA	Gibberellic acid
GC-MS	Gas chromatography-mass spectrometry
HIV	Human immunodeficiency viruses
HPLC/GC	High performance liquid chromatography/gas chromatography
HPTLC	High performance thin liquid chromatography
LC_{50}	Lethal concentration 50%
MAO	Monoamine oxidase
MBC	Maximum breathing capacity
MCF	Material Collection Facility
MIC	Minimum inhibitory concentration
MS	Mass spectroscopy
MTT	3-(4,5-Dimethylthiazol-2-yl)-2,5-diphenyltetrazolium bromide
NMR	Nuclear magnetic resonance
PASS	Prediction of activity spectra for substances
PPM	Parts per million
PSA	Prostate-specific antigen
SEM-EDX	Scanning electron microscopy with energy dispersive X-ray spectroscopy
SGOT	Serum glutamic-oxaloacetic transaminase
SGPT	Serum glutamic pyruvic transaminase
TB	Tuberculosis
TLC	Thin layer chromatography
TP	Total protein test
UV-Vis	Ultraviolet-visible spectrophotometry
XRD	X-ray diffraction

1 Introduction

Pteridophytes are the most primitive vascular plants that appeared on earth in the mid-paleozoic era during the Silurian period. The pteridophytes have specialized water and food-conducting tissues which greatly colonized the terrestrial system so that they could attain the flowering trees' great heights. With the successful establishment of themselves as land plants, the ferns were stimulated at a rapid rate for evolution and so they dominated all the land plants in the carboniferous era. Later with the domination of the flowering plants in the late cretaceous, the pteridophytes started declining and started occupying various niches on land, marsh, and swamps [1].

Cyatheaceae are tree ferns with their tree-like habit characteristic. They have large compound leaves. *Cyatheaceae* has approximately 600 species of tree ferns at the global level. In India, there are 12 species. *Cyathea* species are arborescent with a height of up to 20 m and leaves several meters in length. Two different types of scales occur, namely, conforms scales and marginate scales. Indusium is either absent or small in size. They may be disc-shaped or completely cover the sori.

A number of biological activities are reported in *Cyathea* species, viz. antioxidant [2–8], anticancer and cytotoxic activity [9], antimicrobial activity [6, 10–15], antidiabetic activity [16, 17], hepatoprotectivity [18, 19], larvicidal activity [20], phytotoxicity [21], antichemotactic potential [3], and anti-edematogenic [10]. In addition, the green synthesis of different nanoparticles and their biological activities were also assessed [12, 22, 23].

Man has been using plants as a source of food, medicines, and many other necessities of life from ancient times. Even to this day, many primitive tribal societies that exist depend on the plant life in their surroundings. The pteridophytes were used in ayurvedic, homoeopathic, tribal, and unani medicines and provide food, ornamentations, and insecticides [24]. Also, they studied a lot of medicinal value pteridophytes which treat diseases like anthelmintic, cough, fever, throat infection, diabetes, and many other treatments. Sunil and Shalu [25] studied 25 medicinal pteridophytes from Himachal Pradesh, which were used to treat 46 types of human disorder. Medicinal uses of nearly 61 species of pteridophytes belonging to 31 families were used by tribal people to treat stomach disorders, snake bites, cough, asthma, and fever [26].

Phytochemicals are bioactive non-nutrient components of plants, commonly found in the human diet that may have beneficial or harmful to health. It is well established that chemicals extracted from plants have a wide range of pharmacological applications [27]. Compared to angiosperms, pteridophytes were less distributed, but they are widely used for medicinal purposes in places where they do occur. Phytochemical studies on different studies have been done, and the results revealed that ferns contain a wide range of alkaloids [28], polyphenols, terpenoids [29, 30], and steroids [18].

2 Phytochemistry of *Cyathea*

Hiraoka and Maeda [31] isolated a new aclated flavonol glucoside from the tree fern species, *Cyathea contaminan*. They chemically characterized the compound and found it to be kaempferol-7-(6-succinyl)-glucoside. Yamane et al. [32] identified ten gibberellins from the fern *Cyathea australis* sporophytes. The gibberellins are GA_1, GA_4, GA_9, GA_{15}, GA_{24}, GA_{35}, GA_{58}, 12β-hydroxyGA_9, 12α-hydroxyGA_9, and 12-β-hydroxyGA_4. Arai et al. [33] isolated four new triterpenoids, hopan-29, 17α-olide, hopan-17α, 29-epoxide, 3x-hydroxyfilic-4(23)-ene, and 2-oxofilic-3-ene from the dried leaflets of *Cyathea spinulosa*. Additionally, they isolated hop-22(29)-ene, fern-7-ene, fern-9(11)-ene, hydroxyhopane, filic-3-ene, dryocrassol, tetrahymanol, and cyclolaudenyl palmitate (Fig. 1). Newman [34] characterized fibrous material cut from the midrib of a fern frond *Cyathea dealbata* using Solid-state ^{13}C NMR spectroscopy. Signals associated with cellulose crystallites were separated by exploiting differences in proton rotating-frame relaxation time constants from those associated with the lignin-hemicellulosic matrix. The heights of signals assigned to C-4 in cellulose Iα (90.2) and Iβ (88.5) indicated similar proportions of the two crystalline forms.

Pizzolatti et al. [35] isolated Cyathenosin A, a spiropyranosyl derivative of protocatechuic acid from the *Cyathea phalerata* stem pith. It is found to be the naturally occurring compound that contained a spirocyclic orthoester pyranosidic structure. MS, 1D, and 2D-NMR were used to determine the compound structure. Talukdar et al. [36] executed the phytochemical screening and TLC profiling of *Cyathea gigantea* and *Cyathea brunoniana*. The plants were sequentially extracted with ethyl acetate, petroleum ether, methanol, and acetone solvents. Qualitative phytochemical analysis of caudex and leaves of *Cyathea gigantea* and *Cyathea brunoniana* revealed the presence of steroid, flavonoid, and saponin in both species. TLC profiling of both the species indicated the presence of various phytochemicals and different Rf values of the compound reflected ideas about their polarity. This information can be used for further separation of the compound from these species using a selection of the appropriate solvent system.

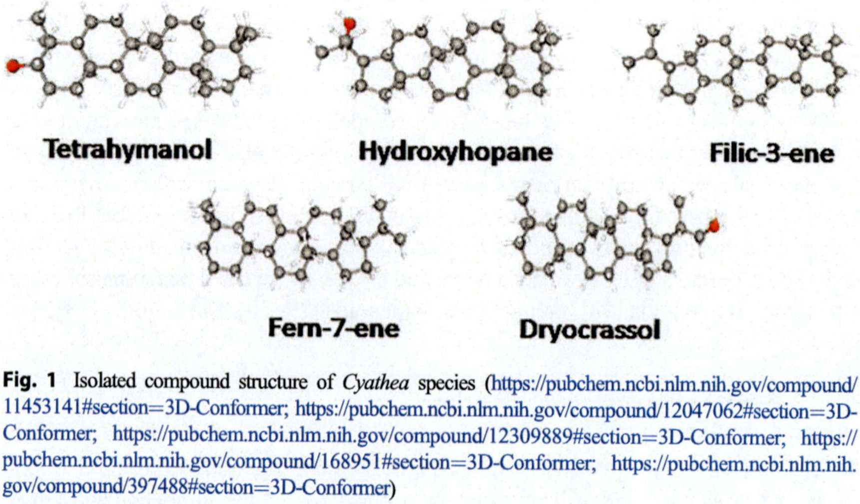

Fig. 1 Isolated compound structure of *Cyathea* species (https://pubchem.ncbi.nlm.nih.gov/compound/11453141#section=3D-Conformer; https://pubchem.ncbi.nlm.nih.gov/compound/12047062#section=3D-Conformer; https://pubchem.ncbi.nlm.nih.gov/compound/12309889#section=3D-Conformer; https://pubchem.ncbi.nlm.nih.gov/compound/168951#section=3D-Conformer; https://pubchem.ncbi.nlm.nih.gov/compound/397488#section=3D-Conformer)

May [37] studied the structural characterization to determine monosaccharide composition and glycosyl linkages in tree black fern, *Cyathea medullaris* using methylation, HPLC/GC, GC-MS, and NMR techniques. The characterization were glucoronomannan backbone, i.e., methylesterified 4-GlcpA (27.9 mol.%) with 2,3- (9.2 mol.%) and 2,3,4-linked Manp (10.9 mol.%) with branched sugar side-chains of galactose, arabinose, xylose, non methylesterified glucuronic acid (8.2 mol.%), and other simple sugars at the O-3 and O-4 of the mannose residues. Preliminary phytochemical analysis and fluorescence analysis were carried out by Janakiraman and Johnson [38] to study the interspecific variation studies on *Cyathea nilgirensis*, *Cyathea gigantea*, and *Cyathea crinita*. The plants were extracted with aqueous, chloroform, petroleum ether, and ethanol solvents. Steroids, alkaloids, phenolic groups, cardiac glycosides, flavonoids, saponins, tannins, and terpenoids were found in all three species. Catechin has existed only in *C. nilgirensis*, and coumarin was present in *C. gigantea* only. On the basis of the results of preliminary phytochemical analysis, cladogram was constructed. The *Cyathea* species showed two clades C_1 and C_2. The clade C_1 was shared by *C. nilgirensis* and *C. gigantea*, whereas in clade C_2, there is a unique presence of *C. crinita*. The fluorescence analysis of *Cyathea* species revealed that similar and distinguished color characteristics depending on various chemical reagents and solvents. Janakiraman and Johnson [39] performed the HPTLC studies in *Cyathea gigantea*, *Cyathea nilgirensis*, and *Cyathea crinita* to profile the phenol, tannin, and flavonoid in the ethanol plant extract. The phenolic profile of HPTLC revealed the presence of 27 different types of phenolic bands and R_f values ranged from 0.05 to 0.92. Ethanolic extract of *C. crinita* showed a maximum number of phenolic compounds. The phenolic bands with R_f values 0.05, 0.32, 0.43, 0.75, and 0.83 were observed in all three *Cyathea* species. The existence of quercetin and catechin was confirmed by the phenolic band with R_f values of 0.75, and 0.28 respectively. The presence of flavonoids represented with R_f values of 0.05, 0.23, 030, 0.43, and 0.73 in all three species of *Cyathea*. The presence of tannins was confirmed with R_f values 0.05, 0.09, and 0.82. The presence of gallic acid was confirmed with an R_f value of 0.49. The HPTLC profile of *Cyathea* species can be used for quality control purposes and chemotypic assessment.

GC-MS analysis of *Cyathea nilgirensis*, *Cyathea gigantea*, and *Cyathea crinita* ethanolic extracts was carried out by Janakiraman and Johnson [40]. The identified major components are methyloctadecyldichlorosilane (29.19%) in *C. nilgirensis* at the retention time of 38.75 min, 2-methylbutane-1,4-diol and 3-(1-ethoxyethoxy (42.37%) in *C. gigantea*, and 2-hydroxy-5-methylbenzaldehyde (55.45%) in *C. crinita* at the RT of 16.68 min. PASS software was used to predict the biological properties of the *Cyathea* species. The result revealed that all the predicted compounds exhibited various biological activities. Pradheesh et al. [12] investigated petroleum ether extract of *Cyathea nilgirensis* using GC-MS analysis. GC-MS analysis showed the presence of pentacosane, squalene, 5,10-dimethy-l6, 8-bisdehydropentapentadecafulvalene, triacontane, and 4-[10-(3,4-Dimethoxy) phenanthryl] morpholine compounds in the petroleum ether extract. Johnson et al. [22] synthesized silver nanoparticles using *Cyathea nilgirensis* and also evaluated their phytotoxic potentials. Silver nanoparticles were synthesized by mixing the aqueous extract with 1 mM $AgNO_3$ solution, and the color change was observed from pale yellow to yellowish brown color. The synthesized silver nanoparticles were characterized by using UV-Vis, FT-IR, and SEM-EDX. The absorption spectra of silver nanoparticles showed a maximum plasmon absorption band of 3.806 and 1.208, with absorbance of 311 and 440 nm, respectively. FTIR analysis confirmed the presence of functional groups, such as alcohols, phenols, amines, carbonyl groups, alkyl halide, and the carbonyl group of flavonoids. The size of silver nanoparticles ranged from 45.0 to 74.0 nm and was crowded together with good uniformity. The EDX spectra showed the existence of Si and a strong Ag peak.

Suresh et al. [23] carried out phytochemical screening and characterization of biosynthesized zinc oxide nanoparticles using *Cyathea nilgirensis* plant extract. Ethanolic extracts of plants were subjected to GC-MS analysis to assess various phytochemical constituents. Zinc oxide nanoparticles were synthesized using 0.1 M of $Zn (NO_3)_2{\cdot}6H_2O$ and is confirmed by XRD, FT-IR, SEM, TEM, EDX, SAED, and PSA analysis. Preliminary phytochemical screening revealed the presence of saponin, tannin, flavonoids, terpenoids, steroids, alkaloids, carbohydrates, triterpenoids, protein, anthraquinone, polyphenol, glycoside, and coumarin in the ethanolic extracts of *C. nilgirensis*. Hexadecanoic acid ethyl ester ($C_{18}H_{36}O_2$), [bis(trimethylsilyl)amino]-3,3-di-*t*-butyl-1-fluoro-3 [(fluorodiisopropyl)amino]-1-methyldisiloxane($C_{21}H_{54}F_2N_2OSi_5$), 1{[(bis[tri methyl silyl]amino) fluoro methyl silyl]amino}-1,1-di-*t*-butyl-3-fluoro-3, 3-diisopropyldisiloxane ($C_{21}H_{54}F_2N_2OSi_5$), and *t*-butyl 5-acetamido-4,6,8-tri-O-acetyl-2,3,5-trideoxy-7,9-O-isopropylidene-2-methylidene-D-glycero-D-talonononate ($C_{25}H_{39}NO_{11}$) showed peak area range of 10–15%. FT-IR analysis showed O–H, N–H, (NH_2), C–H, C=O, C–F, and C–O bonds. XRD result showed Bragg reflections with 2θ values of 31.70°, 34.38°, 36.20°, 47.52°, 56.54°, 62.85°, 67.95°, and 69.05° and these locations were indexed to (1 0 0), (0 0 2), (1 0 1),(1 0 2), (1 1 0), (1 0 3), (1 1 2), and (2 0 1) planes of ZnO hexagonal phase structure, respectively. Further occurrence of zinc nanoparticles in their oxide form was confirmed by EDX analysis. Most of the ZnO particles are hexagonal shaped, and their size were ranged between 20 and 40 nm.

Cyathea contaminans are used in many traditional practices and have lots of medicinal uses [6]. The fern was collected from Tangkuban Perahu Nature Park, West Java, and Indonesia. The Gas Chromatography-Mass Spectrometry (GC-MS) analysis was carried out [6]. The active compounds in the extracts observed were 2-thiophene acetic acid, 2H-tetrazole, 5-(thiophen-2yl) methyl (14.29%), phenol, 2,6-bis(1,1-dimethyl ethyl)-4-methyl, and 2-methyl phenyl ester (14.54%) (10.54%).

Ikpa et al. [7] evaluated the phytochemical composition of chloroform extract of *Cyathea latebrosa* leaves. Preliminary phytochemical studies revealed the presence of starch, glycosides, carbohydrates, saponins, tannins, flavonoids, alkaloids, sterols, polyuronides, and terpenes. Quantitative phytochemical analysis showed the estimation of the alkaloid ($0.08 \times 10^2 \pm 0.02$), saponins ($0.16 \times 10^2 \pm 0.02$), tannin ($0.11 \times 10^2 \pm 0.01$), flavonoid ($0.16 \times 10^2 \pm 0.01$), and phenol ($0.18 \times 10^2 \pm 0.01$). FT-IR (KBr) cm-1 analysis revealed –NH– (3433.41 weak), –OH– (3851.01 sharp), –C–F (1408.8 strong) –SO– (1101.39), and Ar (1646.3 weak) stretches;

GC/MS characterization showed the [Benzeneethamine 2-Aziridinylethyl]amine (70.7%), N-Methyltaurine (10.7%), 3-fluoro-β,5-dihydroxy-N-methyl (14.0%), and (R)-(−)-2-Amino-1-propanol (3.15%) in the extract. These compounds have the potential to cure alleviated cold, malaria, and cancer and regulate immune functions.

Primary and secondary metabolites are widely found at different levels in many medicinal plants. Of them, secondary metabolites serve as defense compounds against biotic and abiotic components [41]. The phytochemical studies confirmed the presence of starch, carbohydrates, glycosides, alkaloids, saponins, tannins, sterols, terpenes, flavonoids, polyuronides, terpenoids, steroids, triterpenoids, anthraquinone, polyphenol, glycoside, coumarin, and cardiac glycosides in the studied *Cyathea* species (Table 1) [7, 23, 36, 38, 39, 42]. Triterpenoids possess antioxidant, antimicrobial, antiviral, anti-inflammatory, hepatoprotective, cytotoxicity, and antitumor activity [43–45]. The plant-based tannin contains antimicrobial, antioxidant activity, cytotoxic, antineoplastic, anti-HIV, diuretic, antidiabetic, anti-inflammatory, and antitumor properties [43–45]. Alkaloid compounds may possess analgesic, antispasmodic, and bacterial activities. Flavonoids contain anti-allergic, antioxidant, antimicrobial, anti-inflammatory, and anticancer properties [43–45]. Steroids present in the plant have properties for nutrition, herbal medicine, cosmetics, antimicrobial, anti-inflammatory, and cardiotonic activities [43–45]. Plant-based saponin compounds contain hemolytic factor, anti-inflammatory, antimicrobial, insecticidal, anticancer, cytotoxic, and molluscicidal action [43–45]. Phenol and phenolic metabolites hold antioxidant, antibacterial, anticancer, cardioprotective, wound healing, immune system promoting and anti-inflammatory, anticarcinogenic, antimutagenic, inhibiting DNA binding and cell adhesion, migration, proliferation or differentiation, and blocking signaling pathways [43–45]. The available literature on the phytochemical studies validated the existence of varied metabolites with varied frequency and confirmed the metabolites responsible for the biological properties.

3 Biological Activities

Elavarasi et al. [17] studied the antidiabetic effect of a concoction of *Cyathea nilgirensis* (Holttum) and *Pterocarpus marsupium* Roxb. in diabetic rat model of streptozotocin-induced. The mixture was prepared by mixing the pith of *C. nilgirensis* powder and *P. marsupium* milk in a 1:1 ratio was used for concoction preparation. The toxicity of the mixture was determined by a modified method of Lorke. Antidiabetic activities of herbal test drugs were evaluated by analyzing changes in the increasing and decreasing level of abnormalities in serum glucose, serum insulin, and glycosylated hemoglobin. The result exhibited no toxic-oriented effects in the rat and no mortality, or animal behavior, and they noted physiological changes in the herbal concoction-treated rats. For the antidiabetic activity, 200 mg/kg dose of the herbal concoction was applied. Herbal concoction-treated rats revealed a remarkable depletion in blood glucose level, and a significant rise in serum insulin level. The author also identified the antidiabetic efficiency of the herbal

Table 1 Phytochemical studies on *Cyathea* species

Name of the species	Type of phytochemical studies undertaken	Extracts	References
Cyathea nilgirensis, Cyathea gigantea, and *Cyathea crinita*	HPTLC	Ethanolic extract	[39]
Cyathea nilgirensis, Cyathea gigantea, and *Cyathea crinita*	GC-MS	Ethanolic extract	[40]
Cyathea nilgirensis, Cyathea gigantea, and *Cyathea crinita*	Preliminary phytochemistry, fluorescence	Aqueous, petroleum ether, chloroform, and ethanol solvents	[38]
Cyathea nilgirensis	UV-Vis, FT-IR, and SEM-EDX	Silver nanoparticles	[22]
Cyathea nilgirensis	GC-MS	Petroleum extract	[12]
Cyathea gigantea and *Cyathea brunoniana*	Phytochemical screening and TLC profiling	Petroleum ether, ethyl acetate, acetone, and methanol solvents	[36]
Cyathea nilgirensis	Phytochemical screening, SEM, FT-IR, XRD, TEM, EDX, SAED, and PSA analysis	Zinc oxide nanoparticles	[23]
Cyathea latebrosa	Phytochemical composition	Chloroform extract	[7]
Cyathea contaminans	GC-MS		[6]
Cyathea medullaris	Methylation, HPLC/GC, GC-MS and NMR		[37]
Cyathea contaminans	Characterization		[31]
Cyathea spinulosa	Isolation of triterpenoids		[33]
Cyathea dealbata	Solid-state ^{13}C NMR spectroscopy		[34]
Cyathea australis	Isolation of compounds		[32]
Cyathea phalerata	MS, 1D, and 2D-NMR		[35]

mixture by its restoring capacity of the tissue damage in the liver, pancreas, and kidney of diabetic rats.

Ikpa et al. [7] determined in vitro antioxidant potential of chloroform extract of *Cyathea latebrosa* leaves. At 400 μg/ml concentration, ascorbic acid and the extract produced 96.02% and 54.34% antioxidant activity seperatively. The Ferric Reducing Antioxidant Power (FRAP) values of plant extracts were significantly ($p < 0.05$) lower than the value of ascorbic acid (2 μM). Even at higher concentrations, the plant extracts showed low antioxidant potential when compared to reference standard ascorbic acid.

Antimicrobial and anticancer activity of silver oxide nanoparticles synthesized from *Cyathea nilgirensis* was investigated by Pradheesh et al. [46]. The antibacterial activity of the silver oxide nanoparticle was studied against *M. luteus*, *S. aureus*, *E. coli*, *B. subtilis*, *S. paratyphi*, *K. pnueomonia*, *A. niger*, and *C. albicans* using disc diffusion method. *K. pneumonia* has exhibited a larger zone of inhibition (21 mm) comparatively. Anticancer activity was determined by Trypan blue dye assay. The result revealed that increased concentrations of Ag_2O nanoparticles from 10 to 200 μg have increased the anticancer efficiency of biosynthesized Ag_2O nanoparticles.

Faizal et al. [6] determined the antioxidant and antibacterial activities of polar and non-polar extracts of *Cyathea contaminans*. The extract was prepared using 80% methanol or n-hexane, and DPPH assay was employed for antioxidant activity. The result showed strong antioxidant activity in mature leaves extract with IC_{50} 37.13–225.19 μg/ml, IC_{50} value of whilst hairs were in the range of 179.50–255.49 μg/ml, and IC_{50} value of young fronds in the range of 544.27–> 2000 μg/ml and both exhibited weak activity.

Faizal et al. [6] measured the antibacterial activity of *C. contaminans* by using inhibition concentration of *Escherichia coli* and *Staphylococcus aureus*. The hexane fraction of frond showed activity against *E. coli* and *S. aureus* (43.92% and 46.8%), and from hairs against *E. coli* (48.1%) in the concentration of 250 μg/ml.

Ida et al. [21] investigated the phytotoxicity and phytotoxic substances of *Cyathea lepifera*. Plant fronds were extracted with 2.5 L of 70% aqueous methanol. p-coumaric acid and (−)-3-hydroxy-β-ionone were isolated by purifying extract through several chromatographic steps. Phytotoxicity was evaluated using garden cress bioassay. Ten seeds each of alfalfa, lettuce, and garden cress for monocotyledonous and ten seedlings each of barnyard grass, timothy, ryegrass, and timothy after germination in the dark at 25 °C for 36–48 h were kept in the petridishes. After 48 h of incubation in darkness at 25 °C, the length of the seedlings percentage was determined in reference to the control seedlings length. p-Coumaric acid and (−)-3-hydroxy-β-ionone showed inhibition in the growth of the garden cress roots and shoots at concentrations >1 mM and 0.3 μM with LC_{50} values of 1240, 11.2 for root and 1120, 10.7 for shoot, respectively. The results suggested that *C. lepifera* may possess phytotoxicity and might be the reason for fern's survival partly.

Nath et al. [14] determined the diameters of growth inhibition zones, MIC, and MBC using antibacterial activity of *Cyathea gigantea* against Multidrug-Resistant (MDR) bacteria. The crude extract form of *C. gigantea* was concluded to be active against all tested organisms. MIC was recorded against Gram-positive organisms at 200 mg/ml and Gram-negative at 400 mg/ml. Disc diffusion method was applied for synergistic activity with antibiotics, and it revealed that the plant extract with antibiotics increased the antibacterial property and its FIC index showed significant synergistic activity showed by ciprofloxacin followed by ampicillin, tetracycline, and oxacillin. The fractionated extract revealed that the FR II, FR III, and FR IV were active against both Gram-negative and Gram-positive. FR I, FR V, and FR VI didn't show any results.

Johnson et al. [22] synthesized silver nanoparticles using *Cyathea nilgirensis* and also evaluated their cytotoxic and phytotoxic potentials. Brine shrimp lethality bioassay was utilized for cytotoxic activity. The LC_{50} values of aqueous extract and AgNP's were 1533.28 μl/10 ml and 869.4 μl/10 ml, respectively. Phytotoxicity results revealed the seed germination inhibition of *V. radiata* and *S. vulgare* occurred only at the higher concentrations of 500 µl/10 ml of *Cyathea nilgirensis* AgNPs and germination reductions ranged between 38.65% and 100%.

Suresh et al. [23] carried out antimicrobial and the anticancer activity of zinc oxide nanoparticles biologically synthesized from *Cyathea nilgirensis* plant extract. Antimicrobial activity of ZnO nanoparticles was assayed by disc diffusion method. *S. paratyphi*, *Bacillus subtilis*, *Micrococcus luteus*, *Escherichia coli*, *Klebsiella pneumoniae*, and *Staphylococcus aureus* and, in fungi, *Candida albicans* and *Aspergillus niger* were employed for the test. Anticancer activity was determined using the trypan blue dye inclusion method. For antibacterial activity higher zone of inhibition was noted in *B. subtilis* at 16 mm (Gram-positive bacteria) and *K. pneumonia* at 20 mm (Gram-negative bacteria) compared to other strains. *C. albicans* (15 mm) showed a higher zone of inhibition compared to *A. niger* species. Anticancer activity was evaluated against the DLA cell line. At 200 µg, ZnO nanoparticles showed maximum inhibition rate.

The larvicidal potential of *Cyathea nilgirensis*, *C. gigantea*, and *C. crinita* against *Culex quinquefasciatus* was evaluated by Janakiraman and Johnson [20]. The plants were sequentially extracted with chloroform, acetone, petroleum ether, and ethanol. Larvicidal activities of different extracts were determined in accordance with the standard method described by WHO. The mortality rate of larval was recorded after 24 h. The result showed more mortality in ethanolic extracts of *Cyathea* species than in other extracts. The LC_{50} values varied from 320.72 to 657.03 µg/ml.

Pradheesh et al. [12] determined the antimicrobial activity of *Cyathea nilgirensis* petroleum ether extract. It was carried out by disc diffusion assay. The plant extract was tested against *Bacillus subtilis*, *Staphylococcus aureus*, *Salmonella paratyphi*, *Micrococcus luteus*, *Escherichia coli*, *Klebsiella pneumoniae*, *Candida albicans*, and *Aspergillus niger*. The petroleum ether fraction exhibited antimicrobial activity.

Antibacterial activity of *Angiopteris helferiana*, *Cyathea brunoniana*, and *Pronephrium nudatum* was determined by Nath et al. [14]. Water, methanol, and ethyl acetate were used for extraction. The antibacterial activity was assayed by a good diffusion method against multi-drug resistance organisms, i.e., *Staphylococcus aureus*, *Pseudomonas aeruginosa*, and *Escherichia coli*. The ethyl acetate extract of *A. helferiana* and *P. nudatum* exhibited antibacterial activity against both methicillin-resistant *S. aureus* and *S. aureus* (ATCC 25923) whereas *A. helferiana* showed inhibition only against *S. aureus* (ATCC 25923). *Cyathea brunoniana* and *Pronephrium nudatum* exhibited inhibition with a MIC concentration of 400 mg/ml.

Longtine and Tejedor [47] screened the effect of antimicrobial activity in the ethanolic and aqueous extracts of *Alsophila cuspidata* and *Cyathea microdonta*. The stipe and caudex were extracted with 96% ethanol and boiled water. Antibacterial activity was assayed by Kirby bauer disk diffusion assay. Ethanolic extract showed the highest antimicrobial activity and plant extract showed more active against Gram-positive than Gram-negative bacteria. Plant extracts made from the caudex, stipe, or leaves showed no variation in antimicrobial activity. The result validated the ethnomedicinal value of tree fern mucilage for leishmaniasis lesions and cuts and was a promising source for futuristic antimicrobial compounds.

Brine shrimp lethality bioassays and MCF 7 cell cultures were performed to study the cytotoxic and anticancer potential of *Cyathea nilgirensis*, *Cyathea gigantea*, and *Cyathea crinita* [42]. The plants were extracted with aqueous, petroleum ether, chloroform, and ethanol solvents. Compared to chloroform, petroleum ether, acetone extracts, ethanolic extracts of *C. nilgirensis*, *C. crinita*, and *C. gigantea* showed potential activity against brine shrimps with LC_{50} values of 304.73 mg/ml, 287.44 mg/ml, and 277.45 mg/ml, respectively. MTT assay revealed decreased cell viability and increased growth inhibition in MCF 7 cells. *C. crinita* showed maximum percentage cell inhibition (IC_{50} <400 μg/ml) followed by *C. nilgirensis* (IC_{50} 714.28 μg/ml) and *C. gigantea* (IC_{50} 806.45 μg/ml). The result revealed that *Cyathea* species were a valuable source for cytotoxic compounds.

Banerjee and Sen [11] studied the antibiotic activity in 114 species of pteridophytes. The extracts used were methanol, water, acetone, ether, and 70% ethanol, and they were assayed against three Gram-positive, five Gram-negative bacteria, one acid-fast, and three fungal plant pathogens. *Cyathea crinita* was tested against *Staphylococcus aureus*, *Mycobacterium phlei*, and *Salmonella typhi*, and the antibody was localized in sporophyll and rhizome in the extracts of water, methanol, and ethanol.

Mary and Mahesh [48] studied the anti-inflammatory activity of *Cyathea nilgirensis* by using the carrageenan-induced paw edema method. The plants were extracted with petroleum ether, chloroform, and ethanol. Among that, ethanol and chloroform extracts of *C. nilgirensis* showed a notable anti-edematogenic result on carrageenan-induced paw edema. The percentage of inhibition for chloroform extract was varying from value 13.1 ± 1.25 to 7.25 ± 1.52 at 100 mg/kg and for ethanol extract was varying from 13.86 ± 2.12 to 6.7 ± 2.64 at 100 mg/kg. The plant extract exhibited better results compared to standard diclofenac sodium. They found that *C. nilgirensis* can be used for acute anti-inflammatory disorders as an effective remedy.

Antibacterial and Antifungal activities of *Cyathea nilgirensis* were determined using the disc diffusion method by Mary and Mahesh [10]. The plants were sequentially extracted with chloroform, hexane, petroleum ether, methanol, and ethanol. Preliminary phytochemical analysis of *C. nilgirensis* showed the existence of carbohydrate, alkaloid, and flavonoid in the hexane, chloroform, and methanol extract. Antibacterial activity of *C. nilgirensis* was determined against *Klebsiella pneumonia*, *Staphylococcus aureus*, *Escherichia coli*, and *Bacillus subtilis*. The results exhibited inhibition zone ranging from 7.00 ± 1.00 to 10.00 ± 1.00 in the ethanol, methanol, chloroform, and petroleum ether extract. Antifungal activity was inspected against *Fusarium solani*, *F. verticillioides*, *Candida albicans*, and *Aspergillus ochraceus*. The plant exhibited a zone of inhibition ranging from 7.00 ± 0.50 to 12.50 ± 0.50. The identified secondary metabolites revealed the antimicrobial potency of the plant and can be used in ethnomedicine and as a source of antibiotics for the treatment of various diseases.

Ethanolic extracts of *Cyathea nilgirensis*, *Cyathea gigantea*, and *Cyathea crinita* were subjected to antioxidant activity to find natural antioxidants. The antioxidant activity was carried out using various assays, viz., DPPH, metal chelating, superoxide, hydroxyl radical scavenging, and nitric oxide activity. *Cyathea nilgirensis* showed the finest free radical scavenging activity in all the tested assays. The *Cyathea* species showed a scavenging effect in the following order: *C. nilgiresnsis* > *C. gigantea* > *C. crinita*. The result revealed prominent antioxidant activity in all three *Cyathea* species [4].

Turnip et al. [15] determined the effect of inhibition extract of *Cyathea contaminans* to *Microsporum gypseum* fungi that cause horn or keratin-eating substances, skin disease, and it damages the nails and hairs. Different concentrations like 500, 400, 300, 200, 100, and 10 mg/ml extracts were used for *M. gypseum* tests, respectively. The results showed a significant effect at 400 mg/ml and the smallest inhibition at 300 mg/ml extract. The smallest concentration is the value of Minimum Inhibitory Concentration (MIC).

Andrade et al. [3] investigated the antioxidant, monoamine oxidase, and anti-chemotactic inhibition effects of some pteridophytes from Brazil. Ferns are extracted with ethanol in a turbo extraction system. The total reactive antioxidant potential method was performed for evaluating in vitro antioxidant activity. For anti-chemotactic potential, the Boyden chamber method was used and lactate dehydrogenase levels, monoamine oxidase (MAO); a fluorescence-based method was used to assay cytotoxicity. The highest antioxidant activity was identified in *Asplenium serra*, *Lastreopsis amplissima*, and *Cyathea dichromatolepsis* at 10 μg/ml. *A. serra* (94.06%) and *Didymochlaena truncatula* (93.41%) recorded high antichemotactic activity. No cytotoxic activity was observed at the highest concentration. *D. truncatula* (82.61%), *Alosphila setosa* (82.21%), *Cyathea phalerata* (74.07%), and *C. delgadii* (70.32%) were the most active extracts (100 μg/ml) against MAO-A.

Chai et al. [16] evaluated the antibacterial, anti-glucosidase, and antioxidant activities of selected highland ferns of Malaysia. Aqueous extract of leaf and rhizome samples were prepared. For antibacterial activity, Minimum inhibitory concentration was calculated, for antiglucosidase activity the glucose inhibitory activity, and for antioxidant activity DPPH assay, FRAP assay was carried out. Hydroxycinnamic acid, total flavonoid, and proanthocyanidin contents in the leaf and rhizome extracts of *C. latebrosa* were 56.32 ± 0.79 and 4.84 ± 0.14 (mg CAE/g), 101.67 ± 6.52 and 5.85 ± 0.58(mg CE/g), 1.43 ± 0.01 and 2.53 ± 0.09 (mg LE/g) respectively. The EC_{50} value of leaf and rhizome extract of *C. latebrosa* was noted at 1413 and 755 μg DM/ml, respectively for anti-glucosidase activity. In DPPH activity, the EC_{50} value was noted at 383 μg DM/ml, MIC value of *C. latebrosa* was >50 mg/ml, 39 μmol Fe^{2+} equivalents/g DM were noted for rhizome extract whereas 571 μmol Fe^{2+} equivalents/g DM were noted in leaf extract.

The hepatoprotective activity of *Cyathea gigantea* was tested against paracetamol-induced liver damage in rats, and Silymarin was used as a reference standard by Kiran et al. [19]. Wistar albino of both sex was divided into five groups of six animals each. Two doses of 100 and 200 mg/kg of *Cyathea gigantea* were tested for hepatoprotective activity. After the treatment period, the blood was collected from the retro-orbital plexus for SGOT, SGPT, ALP, TB, and TP analysis in different groups. The levels of SGOT, SGPT, ALP, and TB were reduced due to the treatment with its methanolic leaf extract. It also retrieved the hepatic damage towards normal.

Lai and Lim [2] aimed to study the natural sources of antioxidants among the ferns in Malaysia. Fifteen fern species were selected. The Folin-Ciocalteu method was used for total phenolic content while the antioxidant activity was analyzed using DPPH, FRP, and BCB assays. Five ferns showed high total phenol content (2000 mg GAE/100 g). *Cyathea laterosa*, *Cibotium barometez*, *Drynaria quercifolia*, *Blechnum orientale*, and *Dicranopteris linearis* exhibited strong antioxidant activity.

The antioxidant and hepatoprotective effects of *Cyathea phalerata* using in vivo assays were investigated [49]. The plants were extracted with 8:2 ethanol: water and the further crude extract was separated into four fractions precipitate (PPT), ethyl acetate fraction (EAF), aqueous fraction (AQF), and butanolic fraction (BUF). The in vitro antioxidant potential was evaluated through DPPH, hydroxyl radicals (OH), lipid peroxidation, and superoxide anion (O_2^-) in rat liver homogenate. In vivo assay was evaluated through oxidative stress (Membrane lipoperoxidation, DNA fragmentation and carbonyl protein formation) and antioxidant defenses (Catalase, concentration of reduced glutathione, and glutathione S-transferase activities) were measured in pre-treated mice with EAF (10, 30, or 100 mg/kg orally) and further it was treated with carbon tetrachloride (CCl_4). The result showed that EAF reduced DNA damage, thiobarbituric acid reactive substances levels, and carbonyl protein contents and raised glutathione S-transferase activities and catalase. They concluded that *C. phalerata* prevents liver oxidative stress induced by CCl_4 in mice.

Antioxidant activity and total phenolic content of some Brazilian species were determined [5]. *Cyathea phalerata* were extracted with 96% EtOH after the aqueous phase was partitioned and fractioned as ethyl acetate, *n*-hexane, and *n*-butanol. Aqueous and alcoholic fractions were subjected to the estimated total phenolics and flavonoid contents, and antioxidant activity was determined using the DPPH assay. The result ranged between 37.97 ± 1.27 and 305.57 mg GAE/g dry extract for phenol, 9.03 ± 0.72 and 58.64 ± 0.32 for flavonoids. The IC_{50} value for antioxidant activity ranged between 9.0 and 20.0 μg/mL.

The reported biological potential of studied *Cyathea* species (Fig. 2) are, namely, antioxidant [2–7], anticancer and cytotoxic activity [42], antimicrobial activity [6, 10–15], antidiabetic activity [16, 17], hepatoprotectivity [19, 49], larvicidal activity [20], phytotoxicity [21], antichemotactic potential [3], and anti-edematogenic [48]. The frequency of the biological activities of studied *Cyathea* species is directly corresponded with the concentration of secondary metabolites existing in the tree ferns.

4 Medicinal Properties

Ethnobotanical study on pteridophytes which were used by the indigenous groups of Tacana and Huaorani from Amazonian peoples of Bolivia and Ecuador [50]. Eleven species out of 24 pteridophytes species were useful for Bolivia and 14 species to Ecuador were studied. *Cyathea pungens* is the only species used by both places. The

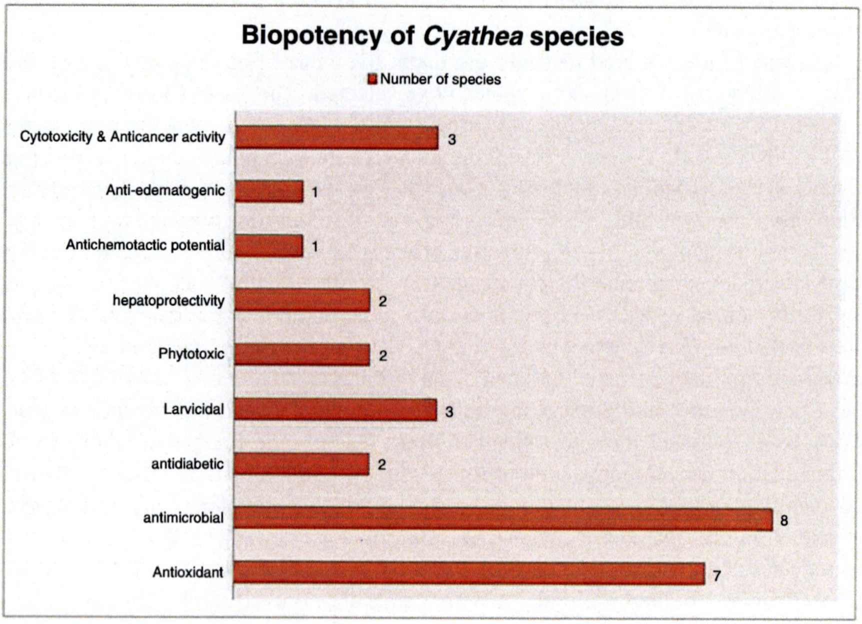

Fig. 2 Biopotency of *Cyathea*

apical part of *Alsophila cuspidata* mucilage and *Cyathea delgadii* is applied on boils. A poultice prepared from the plant is applied externally to reduce swelling in human body. The crushed apical part of *Cyathea amazonica* stipe was used to cure scabby dogs. The stipe of *Cyathea pungens* and *Cyathea lasiosora* were used to alleviate toothache by directly placing it on the gum.

Sixty-one pteridophytes used as an traditional method of treatment of various diseases by the tribal peoples of Western Ghats of India were reported [26]. Frond of *Cyathea gigantea* was used as an anti-inflammatory, and the rhizome part of the plant was used against snake bites.

Local traditional healers and villagers documented indigenous knowledge of medicinal plants and their usage to treat various diseases collected by Focho [51]. *Cyathea maniana* is one of the important medicinal ferns they have used. The burnt powdered dry leaves are applied topically to treat filariasis, respiration difficulty, and cardiac palpitation with the addition of *Aframomum pruinosum* seeds.

The ethnomedicinally important pteridophytes used by the Reang tribes, Tripura, India, were studied [52]. Sixteen pteridophyte species were studied, including two *Cyathea* species, *C. contaminans* and *C. henryi*. The pastes of apical portion of the fern were applied on cuts and wounds for immediate blood clotting, and this method controls microbial infections in the wound areas.

Aremu et al. [53] screened different medicinal fern extracts to determine the anthelmintic efficacy. *Cyathea dregei* is one of the rich medicinal plants they were screened and the dried root of this plant treats anthelmintic.

Cyathea spinulosa is one of the commonly used ferns by the native people, tourists, and tribal peoples of Satpura hills in Madhya Pradesh for the treatment of numerous diseases [54]. The fern had an important role in different medicines like Homeopathic, Unani, Ayurvedic and Siddha. The stipe powder of *Angiopteris helferiana* mixture were given to cattle to treat dyspepsia and baldness. To treat human hair loss and graying hair treatment, the tonic or oil form of plant was taken. The powder form of fronds were used for sudorific and aphrodisiac.

Tripura reported some medicinal plants investigated by gathering information from around 29 different villages [55]. *Cyathea contaminans* are one of the absolute choices of most traditional plant specialists for treating different types of illness as an antiseptic agent. Also, the paste made of *Cyathea* was applied for immediate blood clotting on cuts and wounds.

The rural and tribal peoples of Madhya Pradesh use a lot of medicinal pteridophytes, and those ferns were studied by Balendra and Ravi [56]. Of 61 studied medicinal pteridophytes, three tree ferns were used as a hair tonic. *Cyathea balakrishnanii* and *Cyathea spinulosa* are used to treat sudorific and aphrodisiac, *Cyathea gigantea* decreases white discharges and increases indigestion.

Traditional medicinal plants utilized by the Matigsalug tribes living in Marilog District, Davao City, were studied by Charisse and Melanie [57]. The stem part of *Cyathea contaminans* was used as medicine in the form of poultice to treat inflammation by the tribes.

In India, there were 12 tree ferns of the genus *Cyathea*. Shweta and Sahu [58] observed two species *C. gigantea* and *C. spinulosa* were naturally grown in Pachmarhi Biosphere Reserve (MP). The species were used as traditional medicine and as food. Dried powder of *C. spinulosa* fronds and stems were orally taken for Rheumatic Arthritis. Soft pith and roots were edible and also used to treat wounds. *C. gigantea* rhizome was given to women against white discharge. Rhizome extract treats snake bites also, it was used locally on cuts and boils. Rhizome has antitumor, antiviral, and hepatoprotective activity by nature.

Ethnomedicinal values of 50 species in Palani hills in South India were surveyed by Sathiyaraj et al. [59]. The medicinal pteridophytes were traditionally followed by the tribal people of Palani hills. The rhizome of *Cyathea gigantea* was used to cure dysentery and antidiabetic. The endemic tree fern *Cyathea nilgirensis* was used as an antidiabetic agent for a long period by the tribal people.

Ethnopharmacological uses of medicinal plants that cure skin illness in various districts of Manipur covered 35 different ethnic communities [60]. Totally 241 plants were reported for therapeutic use against skin disease. *Cyathea gigantea* is one of the medicinal ferns which were studied. The fern species has high hair promoter values and it treats urticaria.

Adorico [61] documented different species of pteridophytes which were used by the tribal peoples of Davao city, Philippines. Seventy-one species of pteridophyte species were collected, and out of 71 species, 37 ferns have different uses, such as source for various income, utility, medicine, food, and ritualistic materials in the tribal community. The tree fern *Cyathea contaminans* was one of the medicinal plants used to cure headaches. The leaves/fronds of the fern were pounded and rubbed on the forehead to cure headaches.

Fifty herbal pteridophyte species were found in New Zealand and Taiwan. In the *Cyatheaceae* family, *Cyathea dealbata* in New Zealand is called *Cyathea spinulosa* in Taiwan [62]. In New Zealand, the whole plant was used to treat oversensitive skin. In Taiwan, the plant is used to treat common cold, dampness removal, strengthening joints and muscles, cleaning pulmonary and gastric fire, used to stop bleeding, and detoxifying. It also treats dyspnea, asthma, lower back pain, injuries, swelling in abdomen, toothache, rheumatism, osteodynial, stomach pain, and arthritis.

Some pteridophytes are used as food and medicine by the Local tribes of Lakhimpur and Assam. Ayam [63] encountered 25 different pteridophytes. *Cyathea spinulosa* is one of the medicinal tree ferns. It was used as a general hair tonic and helps in preventing the loss of hair and hair graying. The powdered form of fronds treats sudorific and aphrodisiac.

Suraj et al. [64] encountered to study the traditional medicine importance in the Western Ghats in Virajpet taluq, Kodagu District. Totally 33 medicinal pteridophytes affiliated to 19 families were documented by them. From that *Cyathea gigantea* was used to treat dysentery, body pain, cough, and stomach ache. The whole plant was used orally for various treatments. The reported medicinal properties (Table 2) may be due to the presence of various metabolites.

Table 2 Medicinal values of *Cyathea*

Species	Diseases treats
C. gigantea	Dysentery, body pain, cough, stomach ache [64], white discharge [58], increase digestion [56], dysentery, antidiabetic [59], anti-inflammatory, snake bite [26], cuts and boils, antitumor, antiviral, hepatoprotective [58], hair promoter, urticaria [60]
C. spinulosa	Hair tonic, hair loss, hair graying, sudorific, aphrodisiac [63], indigestion [54], rheumatic arthritis, wound healing [58], clean pulmonary and gastric fire, cold, remove dampness, muscle strengthening, stop bleeding, dyspnea, asthma, toothache, internal injuries, abdomen pain, small intestine swelling, osteodynial, back pain, falling injuries [62]
C. contaminans	Inflammation [57], antiseptic, blood clotting [55], headache [61]
C. nilgirensis	Antidiabetic [59]
C. balakrishnanii	Sudorific, aphrodisiac [56]
C. dregei	Anthelmintic [53]
C. maniana	Filariasis, respiration difficulty, cardiac palpitation [51]
C. henryi	Blood clotting, antiseptic [52]
C. pungens	Toothache, swellings [50]
A. cuspidata	Treat boils, and swelling [50]
C. delgadii	Treat boils [50]
C. lasiosora	Toothache [50]
C. amazonica	Cure scabby dogs [50]

5 Conclusion

This review summarized the phytochemistry, biological activities, and medicinal importance of *Cyathea*. The authors have compiled 17 *Cyathea* species phytoconstituents. The NMR, GC-MS, HPLC/GC, SEM, FTIR, XRD, TEM, SAED, PSA, TLC profiling, SEM-EDX, UV-Vis, HPTLC, and Fluorescence studies are adopted to reveal the phytoprofile of *Cyathea*. Totally 11 *Cyathea* species have been investigated for their biological activities, such as cytotoxicity, anticancer, anti-edematogenic, anti-chemotactic potential, hepatoprotectivity, phytotoxic, larvicidal activity, and antidiabetic. The available literature confirmed the ethnopharmacological and medicinal uses of 13 *Cyathea* species. However, further in-depth studies on the tree ferns are required to reveal the biopotential and clinical values.

References

1. Dudani SN, Mahesh MK, Chandran SMD, Ramachandra TV (2014) Pteridophytie diversity in wet evergreen forests of Sakleshpur in Central Western Ghats. Indian J Plant Sci 3(1):28–39
2. Lai H, Lim Y (2011) Evaluation of antioxidant activities of the methanolic extracts of selected ferns in Malaysia. Int J Environ Sci Dev 2(6):442–447. https://doi.org/10.7763/IJESD.2011.V2.166
3. Andrade JMM, Passos CS, Dresch RR, Kieling-Rubio MA, Moreno PRH, Henriques AT (2014) Chemical analysis, antioxidant, antichemotactic and monoamine oxidase inhibition effects of some pteridophtes from Brazil. Pharmacogn Mag 10(37):S100–S109
4. Janakiraman N, Johnson M (2015) In vitro antioxidant properties of natural products isolated from selected species of *Cyathea*. J Clin Nephrol Res 2(2):1027
5. Brighente IMC, Dias M, Verdi LG, Pizzolatti MG (2007) Antioxidant activity and total phenolic content of some Brazilian species. Pharm Biol 45(2):156–161. https://doi.org/10.1080/13880200601113131
6. Faizal A, Taufik I, Rchman AF, Azar AWP (2020) Antioxidant and antibacterial properties of tree fern *Cyathea contaminans*. Biodiversitas 21:2201–2205
7. Ikpa CBC, Maduka TDO, Ikpa CU (2021) Evaluation of phytochemical composition and in vitro antioxidant potential of *Cyathea latebrosa* leaves. World Sci News 157:25–37
8. Mustacisa LMM, Dejarme AB, Albina MB (2021) Preparation, sensory evaluation and effectiveness of Philippine tree fern (*Cyathea contaminans*) as anti-spasm oil. Int J Agric Technol 17(4):1461–1470
9. Jose LGG, Pablo C (2009) Cytotoxicity screening of endemic plants from Guayana highlands. Trop Biomed 26(2):149–154
10. Mary S, Mahesh SK (2015) Determination of antibacterial and antifungal activities of *Cyathea nilgirensis* plant extract. Int J Curr Res 7(7):18276–18279
11. Banerjee RD, Sen SP (1980) Antibiotic activity of pteridophytes. Econ Bot 34:284–298. https://doi.org/10.1007/BF02858649
12. Pradheesh G, Suresh G, Suresh S, Alexramani V, Hong SI (2017) Antimicrobial activity, phytochemical screening and GC-MS analysis of petroleum ether extract of the medicinal plant *Cyathea nilgirensis* Holttum. Eur J Pharm Med Res 4(9):409–418
13. Nath K, Anupam DT, Mrinal B, Deepshikha B, Shiela C, Debarati C, Abhijit M, Nargis AC (2019) *Cyathea gigantea* (Cyatheaceae) as an antimicrobial agent against multidrug resistant organisms. BMC Complement Altern Med 12:279

14. Nath K, Talukdar AD, Bhattacharya MK, Dheeshikha B, Shiela C, Debarati C, Abhijit M, Amitabha B (2017) Antibacterial activity of certain ferns against multi drug resistant organisms. J Nat Remedies 17(1):144–153
15. Turnip CD, Batubara R, Ginting H (2014) The inhibition test of extract Paku Pohan (*Cyathea contaminans* (Hook.) Copel.) to fungi *Microsporum gypseum* in vitro. Academia 3(1):43–47
16. Chai T-T, Elamparuthi S, Yong AL, Quah Y, Ong H-C, Wong F-C (2013) Antibacterial, anti-glucosidase, and antioxidant activities of selected highland ferns of Malaysia. Bot Stud 54:55
17. Elavarasi S, Revathi G, Saravanan K, Averal HI (2021) Antidiabetic effect of mixture of *Cyathea nilgiriensis* (Holttum) and *Pterocarpus marsupium* Roxb. in streptozotocin induced diabetic rat model. Int J Pharm Sci Res 12(4):2147–2157
18. Ho R, Girault JP, Raharivelomanana P, Lafont R (2012) E-and Z-isomers of new phytoecdysteroid conjugates from French Polynesian *Microsorum membranifolium* (Polypodiaceae) fronds. Molecules 17:11598–11606
19. Kiran MA, Raju VB, Rao G (2012) Investigation of hepatoprotective activity of *Cyathea gigantea* (Wall. Ex. Hook.) leaves against paracetamol-induced hepatotoxicity in rats. Asian Pac J Trop Biomed 2(5):352–356
20. Janakiraman N, Johnson M (2017) Larvicidal potential of *Cyathea* species against *Culex quinquefasciatus*. Pharm Biomed Res 3(1):48–51
21. Ida N, Iwasaki A, Teruya T, Suenaga K, Kato-Noguchi H (2020) Tree fern *Cyathea lepifera* may survive by its phytotoxic property. Plants 9:46. https://doi.org/10.3390/plants9010046
22. Johnson M, Amutha S, Shibila T, Janakiraman N (2018) Green synthesis of silver nanoparticles using *Cyathea nilgirensis* Holttum and their cytotoxic and phytotoxic potentials. Part Sci Technol 36(5):578–582
23. Suresh J, Pradheesh G, Alexramani V, Hong SI (2018) Phytochemical screening, characterization and antimicrobial, anticancer activity of biosynthesized zinc oxide nanoparticles using *Cyathea nilgiriensis* Holttum plant extract. J Bionanosci 12(1):37–48
24. Mannar MM, Maridass M, Victor B (2008) A review on the potential uses of ferns. Ethnobot Leafl 12:281–285
25. Sunil KV, Shalu K (2020) Medicinal pteridophytes used in the treatment of various diseases by the inhabitatnts of Sarkaghat Tehsil, Mandi District, Himachal Pradesh. J Pharm Sci Res 12(3): 360–364
26. Benjamin A, Manickam VS (2007) Medicinal pteridophytes from the Western Ghats. Indian J Tradit Knowl 6(4):611–618
27. Zheng YF, Jassbi AR, Xiao JB (2016) Introduction to the 1st International Symposium on Phytochemicals in Medicine and Food (ISPMF 2015). J Agric Food Chem 64:2439–2441
28. Dong L, Yang J, He J, Luo HR, Wu XD, Deng X, Peng L, Cheng X, Zhao QS (2012) Lycopalhine A, a novel sterically congested *Lycopodium* alkaloid with an unprecedented skeleton from *Palhinhaea cernua*. Chem Commun 48:9038–9040
29. Socolsky C, Asakawa Y, Bardo'n A (2007) Diterpenoid glycosides from the bitter fern *Gleichenia quadripartita*. J Nat Prod 70:1837–1845
30. Socolsky C, Domı'nguez L, Asakawa Y, Bardon A (2012) Unusual terpenylated acylphloroglucinols from *Dryopteris wallichiana*. Phytochemistry 80:115–122
31. Hiraoka A, Maeda M (1979) A new acylated flavonol glycoside from *Cyathea contaminans* Copel. and its distribution in the Pterophyta. Chem Pharm Bull 27(12):3130–3136
32. Yamane H, Yamaguchi I, Kobayashi M, Takahashi M, Sato Y, Takahashi N, Iwatsuki K, Phinney BO, Spray CR, Gaskin P, Macmillan J (1985) Identification of ten gibberellins from sporophytes of the tree fern, *Cyathea australis*. Plant Physiol 78:899–903
33. Arai Y, Koide N, Ohki F, Ageta H, Yang L-L, Yen K-Y (1994) Fern constituents: triterpenoids isolated from leaflets of *Cyathea spinulosa*. Chem Pharm Bull 42(2):228–232
34. Newman RH (1997) Crystalline forms of cellulose in the silver tree fern *Cyathea dealbata*. Cellulose 4:269–279

35. Pizzolatti MG, Brighente IMC, Bortoluzzi AJ, Schripsema J, Verdi LG (2007) Cyathenosin A, a spiropyranosyl derivative of protocatechuic acid from *Cyathea phalerata*. Phytochemistry 68: 1327–1330
36. Talukdar AD, Choudhary DM, Chakraborty M, Dutta BK (2010) Phytochemical screening and TLC profiling of plant extracts of *Cyathea gigantea* (Wall. Ex. Hook.) Haltt. and *Cyathea brunoniana* Wall. Ex. Hook. (Cl. & Bak.). Assam Univ J Sci Technol Biol Environ Sci 5(I):70–74
37. May WSM (2015) Thesis. Physico-chemical characterization and functionality of the polysaccharide extracted from the New Zealand Black tree fern, *Cyathea medullaris* (Mamaku). Massey University, Palmerston North. http://hdl.handle.net/10179/6941
38. Janakiraman N, Johnson M (2015) Inter specific variation studies on *Cyathea* species using phytochemical and fluorescence analysis. J Pharm Toxicol Stud 3(4):11–17
39. Janakiraman N, Johnson M (2016) HPTLC fingerprint profile (phenolics) of selected *Cyathea* species from Western Ghats, South India. Chin J Biol 2016:6420371, 7 pages
40. Janakiraman N, Johnson M (2016) GC-MS analysis of ethanolic extracts of *Cyathea nilgiresis*, *Cyathea gigantea* and *Cyathea crinita*. Egypt Pharm J 15(1):43–47
41. Bennett RN, Wallsgrove RM (1994) Secondary metabolites in plant defense mechanisms. New Phytol 127:617–633
42. Janakiraman N, Johnson M (2016) Ethanol extracts of selected *Cyathea* species decreased cell viability and inhibited growth in MCF 7 cell line cultures. J Acupunct Meridian Stud 9(3): 151–155
43. Sakat S, Juvekar AR, Gambhire MN (2010) In-vitro antioxidant and anti-inflammatory activity of methanol extract *Ofoxalis corniculate* Linn. Int J Pharm Pharm Sci 2:146–155
44. Van-Wyk BE, Wink C, Wink M (2015) Handbuch der Arzneipflanzen, 3rd edn. Die Wissenschaftliche Verlagsgesellschaft, Stuttgart
45. Van-Wyk BE, Wink M (2015) Phytomedicines, Herbal drugs and Poisons. Briza/Kew Publishing, Cambridge, UK
46. Pradheesh G, Suresh S, Suresh J, Alexramani V (2020) Antimicrobial and anticancer activity studies on green synthesized silver oxide nanoparticles from the medicinal plant *Cyathea nilgiriensis* Holttum. Int J Pharm Invest 10(2):146–150
47. Longtine C, Tejedor A (2017) Antimicrobial activity of ethanolic and aqueous extracts of medicinally used tree ferns *Alsophila cuspidata* and *Cyathea microdonta*. Acta Bot Malacit 42(1):119–123
48. Mary S, Mahesh SK (2015) Anti-inflammatory activity of *Cyathea nilgirensis* Holttum against carrageenan induced paw edema. Int J Recent Sci Res 6(8):5807–5809
49. Hort MA, DalBo S, Brighente MC, Pizzolatti MG, Pedrosa RC, Ribeiro-do-Valle RM (2008) Antioxidant and Hepatoprotective effects of *Cyathea phalerata* Mart. (Cyatheaceae). Basic Clin Pharmacol Toxicol 103:17–24
50. Macia MJ (2004) A comparison of useful Pteridophytes between two Amerindian groups from Amazonian Bolivia and Ecuador. Am Fern J 94(1):39–46
51. Focho DA, Ndam WT, Fonge BA (2009) Medicinal plants of Aguambu–Bamumbu in the Lebialem highlands, Southwest province of Cameroon. Afr J Pharm Pharmacol 3(1):1–13
52. Shil S, Dutta CM (2009) Ethnomedicinal importance of pteridophytes used by Reang tribe of Tripura, North East India. Ethnobot Leafl 13:634–643
53. Aremu AO, Ndhlala AR, Fawole OA, Light ME, Finnie JF, Staden JV (2010) In vitro pharmacological evaluation and phenolic content of ten South African medicinal plants used as anthelmintics. S Afr J Bot 76:558–566
54. Ravi U, Balendra PS, Sharad TU (2011) Ethno-medicinal observations on a threatened tree fern *Cyathea spinulosa* Wall. Ex. Hook. in Satpura Hills. Indian Fern J 28:129–136
55. Sanjib S, Manabendra DC, Soumita D (2014) Indigenous knowledge of medicinal plants used by the Reang tribe of Tripura State of India. J Ethnopharmacol 152:135–141
56. Balendra PS, Ravi U (2014) Medicinal pteridophytes of Madhya Pradesh. J Med Plants Stud 2(4):65–68

57. Charisse PBG, Melanie MG (2018) Ethnobotanical practices of Matigsalug tribe on medicinal plants at Barangay Banganihan, Marilog District, Davao City. J Complement Altern Med Res 6(3):1–14
58. Shweta S, Sahu TR (2015) Tree ferns of Pachmarhi Biosphere Reserve, Madhya Pradesh, India: taxonomy, ethnobotany and conservation. Int J Adv Res 3(8):566–577
59. Sathiyaraj G, Muthukumar T, Ravindran KC (2015) Ethnomedicinal importance of fern and fern allies traditionally used by tribal people of Palani Hills (Kodaikanal), Western Ghats, South India. J Med Herbs Ethnomed 1:4–9
60. Radhapiyari DW (2016) Traditional medicinal plants used for various skin diseases and cosmoceuticals in Manipur, North-East India. In: Medicinal plants and its therapeutic uses. OMICS Group eBooks, Hyderabad
61. Adorico MA (2016) Ethnobotany of ferns and fern allies in Mount Macabol, Marilog District, Davao City, Philippines. Int Appl Bus Econ Res 14(2):1127–1137
62. Lin HH, Wen HC, Yu-Hwei R, Chung HH (2018) A comparative review on 50 herbal plants found in New Zealand and Taiwan. J Bot Sci 7(3):17–58
63. Ayam VS (2018) Ethnobotany of ferns of Lakhimpur, Assam, India. Int Res J Biol Sci 7(7): 27–34
64. Suraj RH, Jadeyegowda M, Kushalappa CG, Maheshwarappa V, Chandrashekar SY (2020) An ethnobotanical study of medicinal pteridophytes in kodagu region of Western Ghats, India. Int J Curr Microbiol Appl Sci 9(10):397–380

Phytochemicals of *Adiantum capillus-veneris* 17

Alam Zeb

Contents

Abstract

Adiantum capillus-veneris is an important plant of the family *Adiantum* commonly called Venus hair fern. The plant is widely distributed worldwide ranging from cool to hot tropical regions. Among the *Adiantum* species, the phytochemistry of *Adiantum capillus-veneris* is broadly reported. The plant is rich in important phytochemicals like triacylglycerols, phospholipids, carotenoids, chlorophylls, phenolic acids, flavonoids, phytosterols, terpenoids, and volatile organic compounds. *A. capillus-veneris* possess antimicrobial, wound healing, hearing loss, antidiabetic, nephroprotective, antidiarrheal, antispasmodic, antigoitrogenic, antiurolithiatic, antioxidant, anti-inflammatory, antinociceptive, and analgesic properties.

A. Zeb (✉)
Department of Biochemistry, University of Malakand, Chakdara, Pakistan

H. N. Murthy (ed.), *Bioactive Compounds in Bryophytes and Pteridophytes*,
Reference Series in Phytochemistry, https://doi.org/10.1007/978-3-031-23243-5_21

Keywords

Adiantum capillus-veneris · *Carotenoids* · *Flavonoids* · Lipids · Phenolic acids · Phytosterols · Terpenoids · Volatile compounds

Abbreviations

BPA	Bisphenol A
GC-MS	Gas chromatography-mass spectrometry
LC-MS	Liquid chromatography-mass spectrometry
MDA	Malondialdehyde

1 Introduction

The word "*Adiantum*" has been derived from the Greek word "*Adiantos*" which has meaning of "unwetted." This name was given to this genus due to their presence, and the fronds repel water; nevertheless, the herb also grows in an environment where shade and mist are common. The plant species in the genus *Adiantum* have diverse occurrences and spread worldwide, i.e., ranging from cold zones to hot regions. The covering range starts from the Pacific Islands to northeast Australia, Asia, South Europe, Madagascar, Tropical Africa, and America [1]. The tropical region of South America has shown the highest diversity of these species. Similarly, Eastern Asia has also rich diversity with forty species reported in China [2].

The plant species in the genus *Adiantum* are known to have been used for the treatment of various diseases. The genus has been broadly classified into twelve subgroups. There are more than sixty studies reported the traditional uses of this genus as medicine. The leaves of these plant species are widely known for these medicinal properties. Traditionally the plant leaves have been used for the treatment of lung disorders including simple cough with or without cold, bronchial diseases, lung irritation, and as expectorant. The species in the genus *Adiantum* had been used to treat fever, liver and kidney disorders, skin infections, and hair diseases [1].

The common name of the *Adiantum capillus-veneris* is Venus hair fern. It is commonly grown ranging in our gardens to the forest rich in coniferous trees with a humid environment. It is also commonly known as *Hansraj* or *Sumbal* or *Khati booty* in the Indian subcontinent. It is a herbaceous plant that grows up to a height of 30 cm (Fig. 1) and has extensive propagation ability [3]. *Adiantum capillus-veneris* belongs to the family Adiantaceae [4]. The genus has been used as ethnomedicine. It is used as a tonic (general and hair tonic); as a diuretic; and for the treatment of cough, colds, fever, and bronchial disorders. In addition to skin diseases, it is used in the treatment of tumors of the spleen, liver, and other viscera, in the treatment of jaundice and hepatitis, and many other ailments [5].

Fig. 1 The leaves of *Adiantum capillus-veneris*

2 Phytochemistry of *Adiantum capillus-veneris*

The chemical investigations of *A. capillus-veneris* revealed the presence of several secondary metabolites, such as carotenoids, chlorophylls, phenolic acids, flavonoids, triterpenoids, and phytosterols. More than 100 compounds have been reported from 17 species of *Adiantum*, which include 87 triterpenoids and 20 flavonoids as major classes of compounds [1]. The following reported compounds had been reported in *A. capillus-veneris* belonging to different origins.

2.1 Lipids

Lipids are a large group of primary metabolites present in the cell. Plants are rich in a variety of lipids in their various parts. Triacylglycerols, fatty acids, and phospholipids are major lipids present in plants like *Adiantum*, from the fronds of the fern *A. capillus-veneris*, phospholipids, namely, diacylglyceryl-*O*-4′-(*N, N, N*-trimethyl)-homoserine isolated [6]. Palmitic acid was present at the *sn*-1 position of glycerol, whereas *sn*-2 was occupied by either linoleic, linolenic, or arachidonic acids. These positions were confirmed by the treatment of lipase enzymes and subsequent fatty acids analysis.

Among the lipids, triacylglycerols and their derivatives especially glycolipids were the major components in *A. capillus-veneris*. The amounts of monogalactosyldiacylglycerol, (*2E,4E,6E*)-hexadeca-2,4,6-trienoic acid in monogalactosyldiacylglycerol, and *3Z*-hexadecenoic acid in phosphatidylglycerol were higher in the pinnae of sporophytes than in young leaves of sporophytes or gametophytes [7].

2.2 Carotenoids and Chlorophylls

Carotenoids are important natural antioxidants present in *A. capillus-veneris.* Zeb and Ullah [3] identified and quantified eight carotenoids in *A. capillus-veneris* together with chlorophylls (Fig. 2). Among the carotenoids, higher amount of lutein (806.0 μg/g), 9′-*Z*-neoxanthin (142.8 μg/g), and all-*E*-violaxanthin (82.2 μg/g) were present. Other carotenoids were β-carotene-5,6-epoxide, neo chrome and its isomer, and two isomers of lutein. Lutein is a type of carotenoid containing oxygen moiety. One of the key indicators of beneficial properties of the plant, such as antioxidant and human vision health and protection properties can be attributed to the presence of the high amount of lutein.

Chlorophyll *b* (410.0 μg/g) and chlorophyll *a* (162.4 μg/g) were present in higher amounts. Four pheophytins were also present [3]. Chlorophyll plays a significant role in the light-capturing for photosynthesis and maintaining proper soil moisture and growth performance in *A. capillus-veneris*.

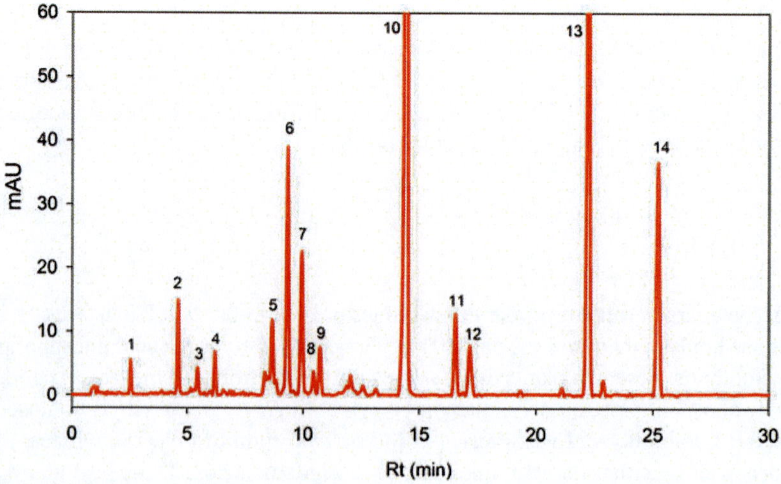

Fig. 2 Reversed-phase HPLC-DAD profile of carotenoids and chlorophylls in the *Adiantum capillus-veneris* leaves at 450 nm. Each peak represents individually identified compounds. (Adopted from Zeb and Ullah [3])

2.3 Phenolic Acids and Phenylpropanoids

Phenolic acids are an important class of phenolic compounds having a carboxylic acid-functional group attached to the phenol ring. The leaves of *A. capillus-veneris* had been found to contain different phenolic acids and phenylpropanoids. For example, 1-coumarylglucose-6-sulfate, l-coumaroylglucose-2-sulfate, l-caffeoylglucose-3-sulfate, 1-caffeylgalactose–6-sulphate, 1-caffeylglucose, shikimic acid, and quinic acid had been reported [8]. In another study, 5-*O*-caffeoylquinic acid, 3-*p*-coumaroylquinic acid, 4-*p*-coumaroylquinic acid, and hydroxycinnamic derivatives had been reported in Chinese samples [9]. The concentration of gallic in the Indian samples was 0.398 mg/g [10] proving helpful for the regulation of hypothyroidism.

Al-Hallaq et al. [11] determined ferulic acid at a concentration of 3.88 mg/g, gallic acid of 3.44 mg/g, and caffeic acid of 1.55 mg/g in samples from Jordon using LC-MS. Benzoic acid and syringic acid were among the main phenolic acids in Turkish samples [12]. In *A. capillus-veneris,* samples from Pakistan, Zeb, and Ullah [3] reported 4-hydroxybenzoic acid (6.51 mg/g), 3-*O*-caffeoylquinic acid (28.5 mg/g), caftaric acid (0.69 mg/g), *p*-coumaric acid (11.2 mg/g), rosmarinic acid (24.4 mg/g), and 5-*O*-caffeoylquinic acid (18.7 mg/g) as shown in Table 1.

2.4 Flavonoids

Flavonoids identified in Japanese samples of *A. capillus-veneris* fronds were kaempferol-3-*O*-glucoside, quercetin-3-*O*-glucoside, 3-*O*-glucuronides of quercetin and kaempferol, and 3-*O*-rhamnoglucosides of kaempferol and quercetin [13]. The

Table 1 Reversed phase HPLC-DAD profiling and quantification of bioactive phenolic compounds in *Adiantum capillus-veneris* leaves [3]

Peak	Rt (min)	Identity	Composition (mg/g)[a]
1	1.3	4-Hydroxybenzoic acid	6.51 ± 0.12
2	5.8	3-*O*-Caffeoylquinic acid	28.5 ± 0.23
3	7.2	2-Caffeoyl tartaric acid	0.69 ± 0.2
4	7.6	Kaemferol-3-feruloylsophoroside-7-glucoside	2.45 ± 0.05
5	8.1	*p*-Coumaric acid	11.2 ± 0.15
6	8.9	Rosmarinic acid	24.4 ± 0.07
7	10	Coumaric acid derivative	33.1 ± 0.06
8	11.1	5-*O*-Caffeoylquinic acid	18.7 ± 0.01
9	14.3	Quercetin hexoside derivatives	2.54 ± 0.02
10	16.5	Caffeic acid hexoside	2.57 ± 0.02
11	16.8	Kaempferol-3-*O*-sophorotrioside	58.7 ± 0.21
12	17.1	Quercetin rhamnoside-hexoside	1.65 ± 0.01
13	18.5	Quercetin-3-galactoside	0.89 ± 0.01

[a]Values are expressed as mean ± standard deviation (SD) of replicate readings based on fresh weight

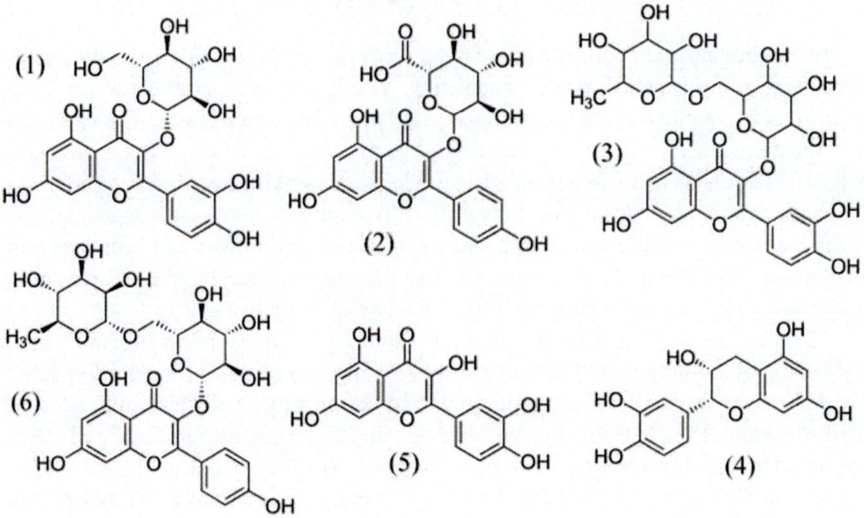

Fig. 3 Chemical structures of some flavonoids present in *Adiantum capillus-veneris*

concentration of quercetin in the Indian samples was less than 180 mg/g [10] proving to be used for the regulation of hypothyroidism.

Ibraheim et al. [5] isolated three flavonoids from the ethyl acetate fraction, namely, quercetin, quercetin-3-*O*-glucoside, and quercetin-3-*O*-rutinoside (rutin). These compounds were found to possess strong anti-inflammatory and hypoglycemic properties. In Chinese *A. capillus-veneris* samples, Yuan et al. [9] showed the presence of quercetin-3-*O*-glucoside, epicatechin-7-*O*-rutinoside, and kaempferol-3-*O*-glucoside. The amounts of quercetin-3-*O*-rutinoside (4.77 mg/g), quercetin-3-*O*-glucoside (3.96 mg/g), epicatechin (1.34 mg/g), and quercetin (0.43 mg/g) were reported to be present in the Jordanian *A. capillus-veneris* samples [11]. Similarly, kaempferol-3-*O*-D-glucuronide and quercetin-3-*O*-D-glucoside were isolated and identified in *A. capillus-veneris* samples from Egypt [14]. Structures of some of these flavonoids are shown in Fig. 3. These studies showed that *A. capillus-veneris* samples of different origins contain important flavonoids, most of which are known for their biological properties.

2.5 Phytosterols

In plant cells, phytosterols are important components of membranes. β-Sitosterol is an important phytosterol present in *A. capillus-veneris* [15]. In *A. capillus-veneris,* samples from Japan, γ-sitosterol, ergosta-5,22-dien-3-ol acetate, and 9,10-secocholesta-5,7,10(19)-triene-1,3-diol were reported [16, 17]. β-Sitosterol had been reported to possess several biological actions both at in vitro- and in vivo-level studies. These include antioxidant, antidiabetic, analgesic, antimicrobial, anti-inflammatory, anxiolytic, and sedative effects, immunomodulatory, anticancer, lipid-lowering effect, and hepato-protective properties. Other medicinal properties, such as in the treatment of respiratory diseases and wound healing had been attributed to the phytosterols [18].

2.6 Terpenoids

Several terpenoids and triterpenoids, such as hopan-22-ol, hopan-28,22-olide, neohop-12-ene, 22,29,30-trisnorhopane, 17αH-trisnorhopan-21-one, 21-hydroxy-30-norhopan-22-one, 30-norhopan-22-one, 30-norhopan-22-one, fern-9(11)-en-28-ol, fern-9(11)-en-3α-ol, fern-9(11)-ene, ferna-7,9(11)-diene, fern-9(11)-en-28-ol, adian-5-en-3α-ol, adian-5(10)-en-3α-ol, 4α-hydroxyfilican-3-one, 3α,4α-epoxyfilicane, 3β,4α-dihydroxyfilicane, 3-methoxy-4-hydroxyfilicane, olean-18-en-3-one, olean-12-en-3-one, and pteron-14-en-7α-ol had been reported in *A. capillus-veneris* samples from Japan [16, 17].

Four compounds of terpenoids moiety were isolated belonging to groups of adiantane and filicane using chromatographic fractionation of the alcoholic extract of the dried fronds of *A. capillus-veneris*. Compounds like isoadiantone, isoadiantol-B, 3-methoxy-4-hydroxyfilicane, and 3,4-dihydroxyfilicane were identified from the hexane fraction [5]. Similarly, olean-12-ene-3,15,16,21,22,28-hexol, (3β,15α,16α,21β,22α), (22S)-21-acetoxy-6α,11β-dihydroxy-16α,17αpropylmethylenedioxy, ethyl iso-allocholate, olean-12-ene 3,15,16,21,22,28-hexol, (3β,15α,16α,21β,22α), and olean-13(18)-ene were reported in samples from Iraq [19]. Three new hopane-type triterpenoids were isolated from the aerial parts of Chinese samples of *A. capillus-veneris*; these were fern-7(8)-en-19α, 28-diol, pteron-14-ene-7α,19α,28-triol, and 3β,4α,25-trihydroxyfilican [20]. The structures of these compounds are shown in Fig. 4. These compounds showed remarkable antifungal activity.

2.7 Volatile Organic Compounds

A. capillus-veneris of French origin revealed the presence of 2-phenylethanal, benzoic acid, (*E*)-2-heptenal, 1-octen-3-ol, (*E*)-2-nonenal, (*Z*)-2-decenal, (*E*)-2-decenal, (2*E*,4*Z*)-decadienal, (2*E*,4*E*)-decadienal, octanoic amide, nonanoic amide, and lauric amide [21]. Similarly, the percentage of carvacrol (13.7%), hexadecanoic acid (5.8%), thymol (4.0%), hexahydrofarnesyl acetone (3.16%), and n-nonanal (2.99%) were more than other identified constituents in Iranian samples [22]. In another study, the presence of 5,7-dodecadiyn-1,12-diol, 3-trifluoroacetoxypentadecane, 3-trifluoroacetoxypentadecane, pterin-6-carboxylic acid, imidazole-4-carboxylic acid, 2-fluoro-1-methoxymethyl- ethyl ester, D-carvone, pyrrolizin-1,7-dione-6-carboxylic acid, methyl (ester), estragole, 2-methyl-5-(1-methylethyl) phenol, 3-allyl-6-methoxyphenol, propiolic acid, 7-epi-trans-sesquisabinene hydrate, tetra-acetyl-D-xylonic nitrile, curan-17-oic acid, 2,16-didehydro-20-hydroxy-19-oxo methyl ester, *E*-vaccenic acid, tributyl acetyl-citrate, 10,13-dioxatricyclo[7.3.1.0(4,9)]tridecan-5-ol-

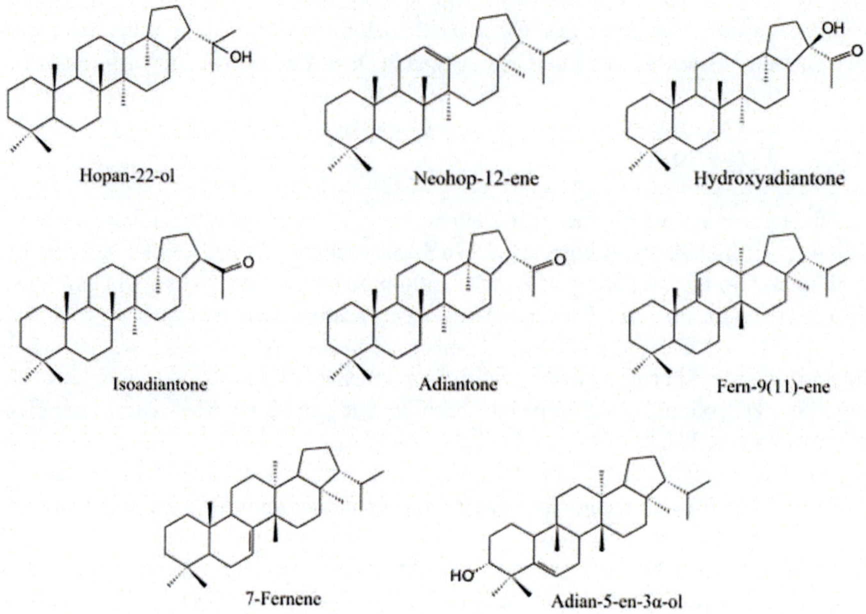

Fig. 4 Chemical structures of some important terpenoids present in *Adiantum capillus-veneris.* (1) Suercetin-3-glucoside, (2) kaempferol-3-glucuronide, (3) quercetin-3-rutinoside, (4) epicatechin, (5) quercetin, and (6) kaempferol-3-rhamnoglucoside

2-carboxylic acid, 18,19-secoyohimban-19-oic acid, and 16,17,20,21-tetradehydro-16,9-octadecenamide were identified using GC-MS in leaves of *A. capillus-veneris* [19]. These volatile compounds possess several biological properties.

2.8 Medicinal Properties of *Adiantum capillus-veneris*

The plant species of *Adiantum* have been reported to possess several pharmacological properties. Studies had demonstrated that these species have antimicrobial, wound healing, antidiabetic, kidney-protective, antifertility, and liver-protective properties. Rastogi et al. [1] reviewed the medicinal properties of *Adiantum* species including *A. capillus-veneris.* Among these, *A. capillus-veneris* possess antimicrobial, wound healing, hearing loss, antidiabetic, nephroprotective, antidiarrheal, antispasmodic, antigoitrogenic, antiurolithiatic, antioxidant activity, anti-inflammatory, anti-nociceptive, and analgesic properties. Recent studies revealed these new medicinal properties.

The extract of *A. capillus-veneris* has been shown to enhance wound healing in a rat model of diabetic type I [23]. The estrogenic effects of bisphenol A (BPA) on the reproductive system of rats had been reduced by the extract of *A. capillus-veneris*. The protective function against BPA-induced injury occurred through the antioxidative mechanism of the extract [24]. The supplementation of the optimal ratio of phytochemicals present in *A. capillus-veneris* has inhibited important enzymes involved in carbohydrate and lipid digestion and absorption in the gastrointestinal system [25], thus supporting glycemic control of obesity and diabetes.

Verma [26] showed the hypolipidemic and antiatherogenic effects of *A. capillus-veneris* in rats. It was found that *A. capillus-veneris* reduces the serum lipid profile. Similarly, Mutar et al. [27] reported that supplementation of the aqueous extract significantly reduces cholesterol and genotoxic effects in white rats. The extract of *A. capillus-veneris* showed a reduction in the levels of MDA and an improvement of levels of antioxidants in serum and brain of rats [28]. The extract was thus proving to have significant antidepressant and antianxiety effects in rats, since *A. capillus-veneris* contains large amounts of phenolic compounds, which are well-known for their antioxidant activity [29]. Thus, the above properties probably are due to its antioxidant and anti-inflammatory activities.

The methanolic extract of *A. capillus-veneris* was evaluated for its anticancer activities in MCF7 and BT47 cell lines by Rautray et al. [30]. The findings of these authors demonstrated that the extract possesses anticancer activities by modulating proteins involved in the cell cycle and apoptosis and thus acting as an encouraging treatment for cancer.

The findings of Yadegari et al. [31] suggested that the antioxidant properties of the extract of *A. capillus-veneris* reduce the destructive structural as well as molecular events associated with hypoxia. The research of Khoramian et al. [32] showed that aqueous extract and hydroalcoholic extract of *A. capillus-veneris* had beneficial effects in dose-dependent manneron acetic acid-induced colitis. The authors attributed these effects to the anti-inflammatory, ulcer healing, and antioxidant activities of extracts. The carbendazim toxicity in rats has been ameliorated by *A. capillus-veneris* in the female reproductive system [33]. The anti-inflammatory effect was found to be due to a reduction in the NF-κB-P65 markers.

3 Conclusion

Adiantum capillus-veneris is an important species of the family *Adiantum*. It is extensively distributed ranging from cold temperate zones to hot regions of the world. This chapter presented available literature on the phytochemistry of *Adiantum capillus-veneris*. Studies showed that this plant contains important phytochemicals like triacylglycerols, phospholipids, carotenoids, chlorophylls, phenolic acids, flavonoids, phytosterols, terpenoids, and volatile organic compounds. The reported compounds had significant medicinal properties. The whole plant is known for several medicinal properties like antimicrobial, wound healing, hearing loss, antidiabetic, nephroprotective, antidiarrheal, antispasmodic, antigoitrogenic, antiurolithiatic, antioxidant activity, anti-inflammatory, antinociceptive, and analgesic properties.

References

1. Rastogi S, Pandey MM, Rawat AKS (2018) Ethnopharmacological uses, phytochemistry and pharmacology of genus *Adiantum*: A comprehensive review. J Ethnopharmacol 215:101–119
2. Boonkerd T, Pollawatn R (2018) *Adiantum capillus-junonis* Rupr.: An additional species of *Adiantum* L. (Pteridaceae) for Thailand. Sci Asia 37:370–372
3. Zeb A, Ullah F (2017) Reversed phase HPLC-DAD profiling of carotenoids, chlorophylls and phenolic compounds in *Adiantum capillus-veneris* leaves. Front Chem 5:29
4. Singh M, Singh N, Khare P, Rawat A (2008) Antimicrobial activity of some important *Adiantum* species used traditionally in indigenous systems of medicine. J Ethnopharmacol 115:327–329
5. Ibraheim ZZ, Ahmed AS, Gouda YG (2011) Phytochemical and biological studies of *Adiantum capillus-veneris* L. Saudi Pharm J 19:65–74
6. Sato N, Furuya M (1983) Isolation and identification of diacylglyceryl-O-4′-(N, N, N-trimethyl)-homoserine from the fern *Adiantum capillus-veneris* L. Plant Cell Physiol 24:1113–1120
7. Sato N, Furuya M (1984) The composition of lipids and fatty acids determined at various stages of haploid and diploid generations in the fern *Adiantum capillus-veneris*. Physiol Plant 62:139–147
8. Imperato F (1982) New phenolic glycosides in the fern *Adiantum capillus-veneris* L. Chem Indust 23:957–958
9. Yuan Q, Wang J, Ruan J (2012) Screening for bioactive compounds from *Adiantum capillus–veneris* L. J Chem Soc Pak 34:207–216
10. Vijayalakshmi A, Kiran Kumar Y (2013) Evaluation of goitrogenic and antithyroidal effect of the fern *Adiantum capillus-veneris*. Rev Bras Farm 23:802–810
11. Al-Hallaq EK, Litescu SC, Kasabri V, Abdul-Razzak KK, Afifi F (2015) Hypocholesterolemic effects of *Adiantum capillus veneris* L. aqueous extract in high cholesterol diet-fed rats and HPLC-MS determination of its polyphenolics. Rom J Chem 60:357–365
12. Abdulqadir A, Cakmak YS, Zengin G (2018) Phenolic compounds, antioxidant properties and enzyme inhibition ability of *Adiantum capillus-veneris* L. linked to Alzheimer's disease, diabetes mellitus and skin disorders. Curr Org Chem 22:1697–1703
13. Akabori Y, Hasegawa M (1969) Flavonoid pattern in the Pteridaceae II. Flavonoid constituents in the fronds of *Adiantum capillus-veneris and* A. cuneatum. Bot Mag Tokyo 82:294–297
14. Sallam MM (2019) Phytochemical and biological study of *Adiantum capillus-veneris* L. growing in Egypt. Al-Azhar J Pharm Sci 59:9–26
15. Marino A, Elberti M, Cataldo A (1989) Phytochemical investigation of *Adiantum capillus veneris*. Boll Soc Ital Biol Sper 65:461–463
16. Nakane T, Arai Y, Masuda K, Ishizaki Y, Ageta H, Shiojima K (1999) Fern constituents : six new triterpenoid alcohols from *Adiantum capillus-veneris*. Chem Pharm Bull 47:543–547
17. Nakane T, Maeda Y, Ebihara H, Arai Y, Masuda K, Takano A, Ageta H, Shiojima K, Cai S-Q, Abdel-Halim OB (2002) Fern constituents: triterpenoids from *Adiantum capillus-veneris*. Chem Pharm Bull 50:1273–1275
18. Babu S, Jayaraman S (2020) An update on β-sitosterol: A potential herbal nutraceutical for diabetic management. Biomed Pharmacother 131:110702
19. Hussein HM, Hameed IH, Ibraheem OA (2016) Antimicrobial activity and spectral chemical analysis of methanolic leaves extract of *Adiantum capillus-veneris* using GC-MS and FT-IR spectroscopy. Int J Pharmacogn Phytochem Res 8:369–385
20. Zhang X, Chen H-L, Hong L, Xu L-L, Gong X-W, Zhu D-L, Xu X-H, Zhao W, Wang F, Yang X-L (2019) Three new hopane-type triterpenoids from the aerial part of *Adiantum capillus-veneris* and their antimicrobial activities. Fitoterapia 133:146–149
21. Fons F, Froissard D, Bessière J-M, Buatois B, Rapior S (2010) Biodiversity of volatile organic compounds from five french ferns. Nat Prod Commun 5:1655–1658
22. Khodaie L, Esnaashari S, Moghaddam SB (2015) Essential oil of arial parts of *Adiantum capillus-veneris*: chemical composition and antioxidant activity. Jund J Nat Pharm Prod 10:3–7

23. Galehdari H, Negahdari S, Kesmati M, Rezaie A, Shariati G (2016) Effect of the herbal mixture composed of *Aloe Vera*, Henna, *Adiantum capillus-veneris*, and Myrrha on wound healing in streptozotocin-induced diabetic rats. BMC Complement Altern Med 16:386
24. Yousaf B, Amina LG, Wang R, Qadir A, Ali MU, Kanwal Q, Munir B, Asmatullah AZ (2016) Bisphenol A exposure and healing effects of *Adiantum capillus-veneris* L. plant extract (APE) in bisphenol A-induced reproductive toxicity in albino rats. Environ Sci Pollut Res 23:11645–11657
25. Kasabri V, Al-Hallaq EK, Bustanji YK, Abdul-Razzak KK, Abaza IF, Afifi FU (2017) Anti-obesity and antihyperglycaemic effects of *Adiantum capillus-veneris* extracts: in vitro and in vivo evaluations. Pharm Biol 55:164–172
26. Verma N (2020) Hypolipidemic and antiatherogenic effects of *Adiantum capillus-veneris* extract in cholesterol-fed rats. Atherosclerosis 315:e167
27. Mutar WM, Al-Hayawi AY, Asaad MM (2021) Reduction of cholesterol and estimation of genotoxic effects of an aqueous infusion of *Adiantum capillus-veneris* leaves in male white rats' testicular tissue. System Rev Pharm 12:266–272
28. Rabiei Z, Setorki M (2019) Effect of ethanol *Adiantum capillus-veneris* extract in experimental models of anxiety and depression. Braz J Pharm Sci 55:e18099
29. Zeb A (2020) Concept, mechanism, and applications of phenolic antioxidants in foods. J Food Biochem 44:e13394
30. Rautray S, Panikar S, Amutha T, Rajananthini AU (2018) Anticancer activity of *Adiantum capillus veneris* and *Pteris quadriureta* L. in human breast cancer cell lines. Mol Biol Rep 45: 1897–1911
31. Yadegari M, Riahy S, Mirdar S, Hamidian G, Afkhami SM, Saeidi A, Rhibi F, Abderrahman AB, Hackney AC, Zouhal H (2019) The TNF-α, P53 protein response and lung respiratory changes to exercise, chronic hypoxia and *Adiantum capillus-veneris* supplementation. Adv Respir Med 87:226–234
32. Khoramian L, Sajjadi S-E, Minaiyan M (2020) Anti-inflammatory effect of *Adiantum capillus-veneris* hydroalcoholic and aqueous extracts on acetic acid-induced colitis in rats. Avic J Phytomed 10:492–503
33. Madboli AE-NA, Seif MM (2021) *Adiantum capillus-veneris* Linn protects female reproductive system against carbendazim toxicity in rats: immunohistochemical, histopathological, and pathophysiological studies. Environ Sci Pollut Res 28(16):19768–19782

Phytochemicals and Biological Activities of *Stenochlaena palustris* 18

Yixian Quah, Shi-Ruo Tong, Sheri-Ann Tan, Yit-Lai Chow, and Tsun-Thai Chai

Contents

Y. Quah
Developmental and Reproductive Toxicology Research Group, Korea Institute of Toxicology, Daejeon, Republic of Korea
e-mail: quah.yixian@kitox.re.kr

S.-R. Tong
Department of Physical Science, Faculty of Applied Sciences, Tunku Abdul Rahman University of Management and Technology, Kuala Lumpur, Malaysia
e-mail: tongsr@tarc.edu.my

S.-A. Tan
Department of Bioscience, Faculty of Applied Sciences, Tunku Abdul Rahman University of Management and Technology, Kuala Lumpur, Malaysia
e-mail: tansw@tarc.edu.my

Y.-L. Chow
Center for Agriculture and Food Research, Universiti Tunku Abdul Rahman, Kampar, Malaysia

Department of Biological Science, Faculty of Science, Universiti Tunku Abdul Rahman, Jalan Universiti, Kampar, Malaysia
e-mail: chowyl@utar.edu.my

T.-T. Chai (✉)
Center for Agriculture and Food Research, Universiti Tunku Abdul Rahman, Kampar, Malaysia

Department of Chemical Science, Faculty of Science, Universiti Tunku Abdul Rahman, Jalan Universiti, Kampar, Malaysia
e-mail: chaitt@utar.edu.my

H. N. Murthy (ed.), *Bioactive Compounds in Bryophytes and Pteridophytes*,
Reference Series in Phytochemistry, https://doi.org/10.1007/978-3-031-23243-5_26

Abstract

Stenochlaena palustris (family Blechnaceae) is a fern species that occurs in the tropics and subtropics. They have been used as edible wild vegetables and folk medicine by the indigenous people in the Asian region. Phytochemical analyses revealed the presence of flavonoids, phenolics, tannins, saponins, gums, steroids, glycosides, terpenoids, and alkaloids in *S. palustris*. Nonetheless, the main secondary metabolites identified from the fern are kaempferol glycosides, fatty acids and phytosterols. Phytocompounds and solvent extracts derived from the fern were demonstrated to have antioxidant, antiglucosidase, cytotoxic, antimicrobial, anti-butyrylcholinesterase, anti-metalotoxic, antipyretic, and termiticidal activities. In nursing mothers, the juice of the fern stimulated breast milk production. The antioxidant and antimicrobial potentials of the fern also contributed to their applications in food, cosmetics, and food packaging material. Thus, current literature indicates that *S. palustris* is a promising source of phytochemicals with potential applications in health promotion, food, and cosmetics, which deserves future research attention. This review presents an overview of the current knowledge pertaining to the biological activities of phytocompounds and extracts of *S. palustris*.

Keywords

Antioxidant activity · Antiglucosidase activity · Kaempferol glycoside · *Stenochlaena palustris*

Abbreviations

ABTS	2,2′-Azino-bis-(3-ethylbenzothiazoline-6-sulfonic acid)
AGI	alpha-glucosidase inhibitor
BChE	butyrylcholinesterase
DPPH	2,2-diphenyl-1-picrylhydrazyl
DU145	Duke University 145
EC_{50}	half-maximal effective concentration
EGFR-TK	epidermal growth factor receptor protein tyrosine
FRAP	ferric reducing antioxidant power
GAE	gallic acid equivalent
GC-MS	gas chromatography-mass spectrometry
H_2O_2	hydrogen peroxide
HD	hydrolysis degrees

HeLa	Henrietta Lacks
LC_{50}	lethal concentration 50
LC-MS	liquid chromatography-mass spectrometry
MCF-7	Michigan Cancer Foundation-7
MDA-MB-231	MD Anderson-Metastatic Breast-231
MIC	minimum inhibitory concentration
NMR	nuclear magnetic resonance
PVA	poly (vinyl alcohol)
ROS	reactive oxygen species
TF	total flavonoid
TH	total hydroxycinnamic acid
TP	total phenolic

1 Introduction

Pteridophytes also referred to as ferns and fern allies are seedless and flowerless plants [1]. There are approximately 12,000 species of pteridophytes throughout the world [1, 2]. Even so, this group of flora remains underexplored in terms of its medicinal values as compared to other plant species [3].

The genus *Stenochlaena* belongs to the family of Blechnaceae, comprising six to seven different dimorphic fern species [4]. Among these species, *Stenochlaena palustris* is the most abundant. It is widely distributed in tropical and subtropical countries, including Malaysia, Indonesia, India, and even northern Australia [5–7]. *S. palustris* thrives in both peat and sandy soils [8] in lowlands and even in mountainous regions [9, 10]. This fern possesses very long rhizomes (round in cross-section, with a diameter of 1 to 2.5 cm) that creep or climb over other vegetations [1, 11]. While it normally reaches the crowns of the trees in the rainforest, it always maintains a strong connection to the ground [4]. Scales surround the apical region of the rhizome; they are often shed off when the rhizome starts to grow [12]. The young fronds express purplish-red hues [6, 12], while their monolete spores are in the shades of yellow-brown to pale yellow [11]. *S. palustris* does not possess a woody trunk or caudex [4]. Figure 1 shows the contrasting appearances of the young and mature sterile fronds of the fern.

This endemic fern is consumed by indigenous people in the Borneo region as part of their diet [13]. Furthermore, its leaves and roots are also ingredients in their traditional medicinal system [6, 14]. However, this fern has a limited shelf life, thus preventing them to be sold outside the local markets [15]. Since *S. palustris* could be a promising export commodity, improved postharvest handling and packaging procedures were researched to ensure prolonged storage duration for this fern [16].

S. palustris has been shown to exhibit antioxidant [17], antibacterial [18, 19], antifungal [20], and antiglucosidase activities [17, 21]. Phytochemically, *S. palustris* is rich in flavonoids [2, 22] as well as terpenoids and tannins [6]. Since *S. palustris* is consumed as food and even utilized as traditional/folk medicine, the fern is a treasure

Fig. 1 The young (red arrow) and mature (white arrow) sterile fronds of *Stenochlaena palustris*. (Photo credit: Tsun-Thai Chai)

trove of bioactive compounds that may contribute to the search for novel functional ingredients in the food industry or as a new therapy for various ailments. The following sections delve deeper into the nutritional and phytochemical profiles of the fern, along with its bioactivities and prospective applications.

2 Ethnobotany

This edible creeping fern is known by various local names in different parts of Borneo. It is called "midin," "lambiding," "lemiding," and "kalakai" in Sarawak, Sabah, Brunei, and Kalimantan, respectively [13]. Interestingly, only the young sterile fronds (Fig. 1) of *S. palustris* are edible, while the spore-producing fertile fronds (Fig. 2) are not [14, 23]. The light, green-colored young fronds are generally selected for traditional treatment of skin problems, fever, diarrhea, and even gastric ulcers [6, 14]. Furthermore, *S. palustris* is also consumed by anemic pregnant mothers to enhance their hemoglobin levels and by nursing mothers to boost breast milk production [10, 24].

In Bangladesh, its rhizomes and roots are used as treatments for gastric tumors and rheumatism [25], whereas in Papua New Guinea, the leaves are utilized as a contraceptive [7]. This fern is called "Lumpeng" in Thailand where the whole plant is prepared as a decoction and orally consumed to reduce fever [26]. The local ethnic communities in Assam, India, on the other hand, consume the young fronds as vegetables [27]. Similarly, the many aboriginal groups in the Borneo region and peninsular Malaysia have also included this fern in their diet usually as a vegetable salad or "ulam." Table 1 summarizes information about how the fern is consumed by the aboriginal communities of Malaysia and Indonesia.

Fig. 2 The narrow leaf blades of the fertile fronds of *Stenochlaena palustris*, covered in sporangia. (Photo credit: Tsun-Thai Chai)

Table 1 Local names, parts consumed by aboriginal communities, and methods of preparation

Country	Aboriginal community	Local name	Plant part used	Method of preparation	References
Malaysia (Sabah)	Kadazandusun, Murut	Lambiding/ lembiding	Young shoots and leaves buds	Boiled or fried	[13, 28]
Malaysia (Sarawak)	Iban	Kemiding, Midin	Young fronds	Boiled or fried	[13, 29]
	Kelabit	Pauh ira	Young fronds	Fried	[30]
Malaysia (peninsular Malaysia)	Orang Asli (proto-Malays, Senoi, Negritos)	Paku nyai	Tender shoots	Braised or fried	[31]
Indonesia (Kalimantan)	Dayak	Kalakai	Young leaves	Boiled or fried	[13]

3 Nutritional Profile

The nutritional benefits of *S. palustris* in human health are mainly due to the presence of high phytochemical contents, such as phenolic compounds, saponins, terpenoids, and alkaloids [2, 5, 7]. Phenolic constituents identified in *S. palustris* include flavonoids, hydroxycinnamic acids as well as anthocyanins, and pro-anthocyanidins (condensed tannins) which are known as strong antioxidants [23]. A detailed explanation of the extraction, purification, and identification of bioactive phytochemicals from *S. palustris* is provided in Sect. 4. Moreover, this plant is also rich in other minerals like potassium, phosphorus [5, 31], and iron [31]. When compared to other local Indonesian vegetables, *S. palustris* was reported to have higher contents of vitamin C, folic acid, and minerals [24]. This may be the reason it is prescribed traditionally for the well-being of pregnant and nursing mothers to stimulate hemoglobin and milk production. A significantly higher level of anthocyanins was detected in the young sterile fronds of this fern as compared to the matured or fertile fronds [13, 23], thus supporting the usage of the former as herbal medicine. Besides, it is revealed that young fronds possessed more diverse aliphatic constituents, such as fatty acids when compared to matured ferns [14]. Furthermore, this fern is a good source of fiber and β-carotene, with their levels comparable to common vegetables in Malaysia [31].

4 Extraction, Purification, and Identification of Phytocompounds

Phytochemicals belonging to the classes of flavonoids, phenolics, tannins, saponins, gums, steroids, glycosides, terpenoids, and alkaloids were reported to be present in various *S. palustris* extracts [2, 5–7, 17, 23, 32]. However, the identification of specific secondary metabolites in *S. palustris* is still scarce. Phytocompounds identified from *S. palustris* are mainly flavonoids (Table 2).

Organic solvents are common media used to extract the phytochemicals of natural products, a process depending on the solubility of secondary metabolites in solvents. Referring to Table 2, several studies found that extraction of *S. palustris* with methanol and ethanol produced more yield, in addition to a higher number of extracted phytochemicals and higher bioactivities relative to other solvents [6, 14, 21, 22, 33, 34]. To a lesser extent, hexane was also used for extracting lipophilic compounds, such as sterols and fatty acids from *S. palustris* [14, 34]. Although the choice of extraction solvent is essential, post-harvest processing may contribute to maximizing phytochemical extraction. Keeping the harvested *S. palustris* in moist and cold conditions, especially during the drought season, can prevent the senescence of plant cells and reduce the degradation of the phytochemicals in the fern samples [21]. Besides, the use of a fine powder of *S. palustris* in <250 um particle size could also enhance extraction efficiency [21]. On the other hand, Lodevico et al. [32] reported that oven-drying pretreatment at 70 °C increased TP and TF contents in the methanolic extracts of *S. palustris*. The authors proposed that the observation could be attributed to the enhanced release of bound phenolics from the *S. palustris* samples following thermal treatment. Thus, thermal preprocessing of *S. palustris* samples before extraction may enhance phytochemical yield in the resulting extracts.

Fractionation is usually carried out to separate a crude plant extract into various fractions each comprising a less complex pool of compounds having similar properties. Fractionation steps may be performed multiple times using different methodologies till a pure compound is isolated [35]. Hence, fractionation is an important process that contributes to the identification of phytochemicals that exhibit the desired bioactivities. For *S. palustris*, bio- and chemical assay-guided fractionations have been reported as an approach to isolating a pure compound having the desired activity [21]. Such an assay-guided fractionation approach could facilitate the

Table 2 Phytochemical compounds identified from *S. palustris*

Compounds	Bioactivity	Source	Extraction solvent	Fractionation/ purification	Methods of identification	References
1. Kaempferol 3-*O*-(3″-*O*-*E*-*p*-coumaroyl)-(6″-*O*-*E*-feruloyl)-β-D-glucopyranoside	Antioxidant (**1, 2, 6, 7**) Anti-BChE (**1, 2, 3, 4, 5, 7**)	Mature fronds	Methanol	• Vacuum silica gel liquid chromatography	• 1D-NMR • 2D-NMR	[14, 22]
2. Kaempferol 3-*O*-(3″, 6″-di-*O*-*E*-*p*-coumaroyl)-β-D-glucopyranoside	Cytotoxicity against human breast and prostate cancers (**1, 2**)			• Flash-column chromatography • Sephadex LH-20		
3. Kaempferol 3-*O*-(6″-*O*-*E*-*p*-coumaroyl)-β-D-glucopyranoside				• Preparative-TLC		
4. Kaempferol 3-*O*-(3″-*O*-*E*-*p*-coumaroyl)-β-D-glucopyranoside						
5. Kaempferol 3-*O*-α-L-rhamnopyranoside						
6. Kaempferol 3-*O*-(6″-*O*-α-rhamnopyranoside)-β-D-glucopyranoside						
7. Kaempferol 3-*O*-β-D-glucopyranoside						
1. Kaempferol-3-*O*-β-D-glucopyranoside	Antiglucosidase (**1**)	Young and mature fronds	Methanol	• Liquid-liquid sequential separation • Silica N60 gel column chromatography • Normal phase silica plate TLC	• LCMS • Proton NMR	[21]
1. 20-hydroxyecdysone	Termiticidal (**1, 2, 3**)	Stem	Methanol	Silica gel column chromatography	• ^{1}H NMR	[33]
2. Ajugasterone C					• ^{13}C NMR	
3. Stenopalustroside A [kaempferol 3-*O*-(3″,6″-di-*O*-*E*-*p*-coumaroyl)-β-D-glucopyranoside]					• UPLC-MS	

(continued)

Table 2 (continued)

Compounds	Bioactivity	Source	Extraction solvent	Fractionation/ purification	Methods of identification	References
1. Campesterol	Anti-BChE	Young and mature fronds	Hexane	N/A	GC-MS	[14]
2. Ethyl arachidonate						
3. Fucosterol						
4. Linoleic acid						
5. Methyl linoleate						
6. Methyl palmitate						
7. Neophytadiene						
8. Palmitic acid						
9. *Trans*-phytol						
10. α-tocopherol,						
11. β-sitosterol						
12. Stigmasterol						
1. Stigmasterol,	N/A	Plant part used not specified	Methanol	N/A	GC-MS	[34]
2. Stigmast-5-en-3-ol, oleic						
3. Stigmasta-5,22-dien-3-ol, acetate						
4. Ethyl linoleate						

N/A: Information not available; anti-BChE: anti-butyrylcholinesterase; NMR: nuclear magnetic resonance; UPLC-MS: ultra-performance liquid chromatography-mass spectrometry; GC-MS: gas chromatography-mass spectrometry; TLC: thin layer chromatography recognition of selected fractions putatively containing the phytochemicals of interest after each round of fractionation. The fractionation methods often used for the purification of bioactive phytochemicals from solvent extracts of *S. palustris* are column chromatography [14, 21, 22, 33] and thin-layer chromatography [14, 21, 22, 33].

The identification of bioactive compounds purified from the fern has relied on technologies, such as nuclear magnetic resonance (NMR), liquid chromatography-mass spectrometry (LC-MS), and gas chromatography-mass spectrometry (GC-MS). As shown in Table 2, the bioactive phytocompounds identified from *S. palustris* include classes, such as ecdysteroids (e.g., 20-hydroxyecdysone) [33], kaempferol glycosides (e.g., kaempferol 3-*O*-β-D-glucopyranoside) [14], fatty acids (e.g., linoleic acid) [14], and phytosterol (e.g., fucosterol) [14]. To date, the number of bioactive compounds identified from *S. palustris* remains scarce. A few studies have reported *S. palustris* solvent extracts and/or fractions exhibiting antiglucosidase and antioxidant [17, 23], as well as cancer-cytotoxic [7] activities. Nevertheless, the actual active compounds responsible for the reported activities remain to be identified. On the other hand, although certain metabolites were identified in *S. palustris*, their actual bioactivities remain to be ascertained. For instance, Erwin et al. [34] reported that the ethanol extract of *S. palustris* possessed antibacterial activity against *S. aureus* and *E. coli*. Although the authors identified stigmasterol and two derivatives as the major constituents in the extract, they did not test the isolated compounds for antibacterial activity [34].

5 Biological Activities

5.1 Antioxidant Activity

Numerous antioxidant assays have been employed to measure the ability of *S. palustris* extracts to prevent the oxidation process. These assays include radical (2,2-diphenyl-1-picrylhydrazyl (DPPH), 2,2′-azino-bis-(3-ethylbenzothiazoline-6-sulfonic acid (ABTS), superoxide anion radical, and hydrogen peroxide (H_2O_2)) scavenging assays, ferric reducing antioxidant power (FRAP) assay, and metal chelating assay. Based on reports in the literature, *S. palustris* is an excellent source of antioxidative natural products (Table 3).

Chai et al. [23] compared the radical scavenging, ferric reducing, and metal chelating activities of the fern's edible and non-edible leaves. Using water as the extraction solvent, the authors sought to distinguish these bioactivities and phytochemical compositions between the fern's young sterile, mature sterile, young fertile, and mature fertile fronds. Total polyphenol (TP), total flavonoid (TF), and total hydroxycinnamic acid (TH) contents were the highest in the mature sterile frond extract. The extract of the edible young sterile frond contained 20 times more anthocyanins (51.32 mg/100 g dry matter) than the other extracts. The therapeutic properties pertaining to the intake of anthocyanin-rich foods have been exhaustively described [37–41]. The antioxidant mechanisms of anthocyanin have also been

Table 3 Summary of antioxidant activities of *S. palustris* extracts prepared from different parts of the fern and by using different solvents

Part of fern	Extraction solvent	Antioxidant activities[a]	References
Young sterile fronds	Water	DPPH: 1.801 mg TE/mg GAE	[23]
		FRAP: 9.58 μmol Fe^{2+}/mg GAE	
		Metal chelating activity: 101.22 μg EDTA/mg GAE	
Mature sterile fronds		DPPH: 2.071 mg TE/mg GAE	
		FRAP: 13.87 μmol Fe^{2+}/mg GAE	
		Metal chelating activity: 16.54 μg EDTA/mg GAE	
Young fertile fronds		DPPH: 1.977 mg TE/mg GAE	
		FRAP: 10.56 μmol Fe^{2+}/mg GAE	
		Metal chelating activity: 82.14 μg EDTA/mg GAE	
Mature fertile fronds		DPPH: 2.299 mg TE/mg GAE	
		FRAP: 10.82 μmol Fe^{2+}/mg GAE	
		Metal chelating activity: 176.37 μg EDTA/mg GAE	
Unspecified edible part	Methanol	DPPH: 600 μg/mL (EC_{50})	[31]
Mature fronds	Hexane	DPPH: 464.70 μg/mL (EC_{50})	[17]
		FRAP: 155 μmol Fe^{2+}/g	
	Chloroform	DPPH: 218.65 μg/mL (EC_{50})	
		FRAP: 318 μmol Fe^{2+}/g	
	Ethyl acetate	DPPH: 49.68 μg/mL (EC_{50})	
		FRAP: 1248 μmol Fe^{2+}/g	
	Methanol	DPPH: 11.65 μg/mL (EC_{50})	
		FRAP: 8366 μmol Fe^{2+}/g	
	Water	DPPH: 19.30 μg/mL (EC_{50})	
		FRAP: 4583 μmol Fe^{2+}/g	
		H_2O_2 scavenging activity: 364 μg/mL (EC_{50})	
		Superoxide scavenging: 220 μg/mL (EC_{50})	
		Copper chelating: 436 μg/mL (EC_{50})	
Mature fronds	Ethanol (fraction 11)	DPPH: 120 μg/mL (EC_{50})	[7]
	Ethyl acetate (fraction 4)	DPPH: 650 μg/mL (EC_{50})	
Combination of unspecified stem and fronds	Water	DPPH: 919.95 μg/mL (EC_{50})	[10]

(continued)

Table 3 (continued)

Part of fern	Extraction solvent	Antioxidant activities[a]	References
Unspecified edible part	Ethanol	DPPH: 140.75 ppm (EC_{50})	[2]
Stem	Ethanol	DPPH: 89.96 μg/mL (EC_{50})	[6]
		FRAP: 3.82 mM Fe^{2+}/g	
		ABTS: 0.62 mM TE/g	
	Ethyl acetate	DPPH: 178.27 μg/mL (EC_{50})	
		FRAP: 6.79 mM Fe^{2+}/g	
		ABTS: 0.56 mM TE/g	
	Hexane	DPPH: 323.5 μg/mL (EC_{50})	
		FRAP: 2.93 mM Fe^{2+}/g	
		ABTS: 0.56 mM TE/g	
Unspecified frond type	Ethanol	DPPH: 24.24 μg/mL (EC_{50})	
		FRAP: 17.95 mM Fe^{2+}/g	
		ABTS: 0.61 mM TE/g	
	Ethyl acetate	DPPH: 159.1 μg/mL (EC_{50})	
		FRAP: 7.54 mM Fe^{2+}/g	
		ABTS: 0.46 mM TE/g	
	Hexane	DPPH: 323.43 μg/mL (EC_{50})	
		FRAP: 2.94 mM Fe^{2+}/g	
		ABTS: 0.57 mM TE/g	
Unspecified leaf type	Water	DPPH: 838.26 μg/mL (EC_{50})	[36]

[a]mg TE/mg GAE, mg Trolox equivalents/mg gallic acid equivalents; μmol Fe^{2+}/mg GAE, μmol Fe^{2+} equivalents/mg gallic acid equivalents; μg EDTA/mg GAE, μg EDTA equivalents/mg gallic acid equivalents; mM TE/g, mM Trolox equivalents/g dry sample

thoroughly described elsewhere [37]. The discovery of high anthocyanin content in the young sterile frond of *S. palustris* may potentiate its application as a functional natural food colorant with concurrent antioxidant activity. The DPPH free radical scavenging activity of *S. palustris* aqueous extract was also reported by Samad et al. [36] in their attempt in developing edible coating material. They reported that *S. palustris* aqueous extract possessed scavenging activity with an EC_{50} of 838.26 μg/mL.

Abdul Wahab et al. [31] prepared a methanolic extract of *S. palustris* extract and assessed its contents of TP and TF, as well as DPPH radical scavenging activity. The authors determined that *S. palustris* methanolic extract possessed scavenging activity with an EC_{50} of 0.60 mg/mL. Nurhasnawati et al. [2] conducted a screening study on the antioxidant activity, TP, and TF contents of several indigenous ferns. This study determined the antioxidant activity of the fern extract using only the DPPH assay, and the EC_{50} value obtained for *S. palustris* ethanolic leaf extract was 140.75 ppm. *S. palustris* was considered to exhibit relatively moderate antioxidant activity compared to the other fern species tested.

The antioxidant activity of *S. palustris* extracts may be affected by the types of solvents used for extraction and the sample preparation methods. Chai et al. [17]

compared the DPPH radical scavenging activity and FRAP values of mature frond extracts prepared with sequential solvent extraction using hexane, chloroform, ethyl acetate, methanol, and deionized water. They found that the water fraction was the strongest antioxidant. Besides, their findings implied that the presence of hydroxycinnamic acids may have contributed to the antioxidant activities detected in *S. palustris*. Wijaya et al. [10] applied a multiple-stage extraction process aiming to obtain *S. palustris* extract with the highest antioxidant activity. The use of a multiple-stage process has proven to increase the antioxidant activity of *S. palustris* by up to 72.43% compared to single-stage extraction. The multistage-extracted *S. palustris* sample showed antioxidant activity with EC_{50} of 404.0 mg/mL. Their findings indicated that the method of extract preparation affected the recovery of antioxidant activity. Arullappan et al. [7] determined the radical scavenging activities of fractions isolated from ethanol and ethyl acetate *S. palustris* leaf extracts. The authors discovered that fraction 11 of ethanol extract and fraction 4 of ethyl acetate extract had the highest DPPH radical scavenging activity. Between the two fractions, the ethanolic extract exhibited a stronger scavenging activity. Ndanusa et al. [6] reported that the ethanolic *S. palustris* leaf extract displayed the highest antioxidant activity compared to the other extracts prepared from various plant parts using various solvents. Recently, pickled *S. palustris* produced by lactic fermentation was shown to have greater radical scavenging activities relative to the unfermented fern [42]. This finding suggests that pickled *S. palustris* may have more health benefits than fresh *S. palustris*.

The relationship between antioxidant activity and phytochemical composition in *S. palustris* extracts was also investigated. According to Chai et al. [23], the TP content had a strong correlation with radical scavenging activity and ferric reducing activity, but only moderately with metal chelating activity. A weak correlation was found between polyphenol content and metal chelating activity among the four extracts prepared from the young and mature fronds. Two additional studies measured the TP and TF contents, as well as antioxidant activities of the extracts [2, 6]. The studies, however, did not investigate the correlation between these data. Consequently, we have performed a correlation analysis on their data. In the study conducted by Ndanusa et al. [6], ethanolic *S. palustris* leaf extract had the highest TP content (3.80 ± 0.22 mg gallic acid equivalents/g of extract), while ethyl-acetate leaf extract had the highest concentration of TF content (3.96 ± 0.072 mg quercetin equivalent). Interestingly, the DPPH and FRAP scavenging capacities of *S. palustris* extracts were strongly correlated with their TP content, whereas the ABTS scavenging capacities were strongly and negatively correlated with their TF content (Fig. 3A). In the study by Nurhasnawati et al. [2], the DPPH scavenging capacity of *S. palustris* extracts was more strongly correlated with the TP content than with TF (Fig. 3B). Together, both studies pointed to a strong, positive correlation between the TP contents of *S. palustris* extracts and their DPPH scavenging activities.

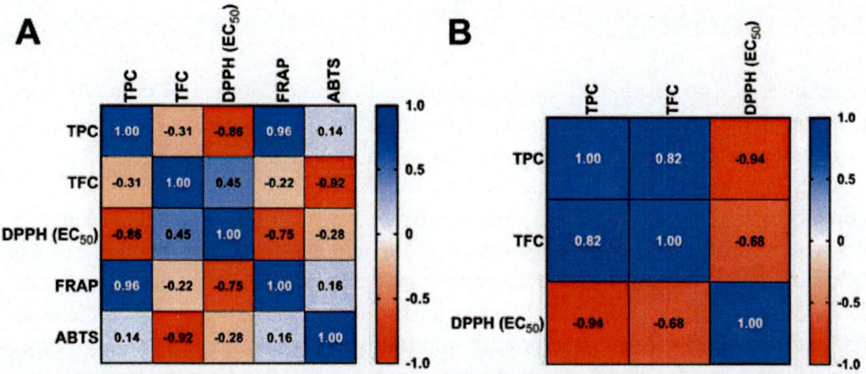

Fig. 3 Pearson correlation coefficients for total phenolic content (TPC), total flavonoid content (TFC), 1,1-diphenyl 1–2-picryl-hydrazyl (DPPH) EC_{50}, ferric reducing antioxidant power (FRAP), and 2,2′-azino-bis-(3-ethylbenzothiazoline-6-sulfonic acid (ABTS) of *S. palustris* extracts reported by (**a**) Ndanusa et al. [6] and (**b**) Nurhasnawati et al. [2]

5.2 Antiglucosidase Activity

Alpha-glucosidase inhibitors (AGIs) are one class of drugs used in the treatment of diabetes. AGIs function by slowing down carbohydrate digestion, thereby reducing glucose absorption in the gastrointestinal tract [43]. Among the variously reported bioactivities of *S. palustris* extracts, antiglucosidase is one that gains increasing interest in natural product research due to its potential as a hyperglycemia remedy. Alpha-glucosidase inhibitory activity can be evaluated through in vitro enzymatic assays [44]. In a study by Chai et al. [17], the water fraction of methanolic extract of *S. palustris* mature fronds was found to have a notably stronger alpha-glucosidase inhibitory effect with an EC_{50} value of 2.9 μg/mL compared to the reference AGI compound, quercetin ($EC_{50} = 115.05$ μg/mL).

Several classes of phytochemicals including anthocyanidins, cinnamic acids, flavanols, and phytosterols were reported to have an antidiabetic effect [45]. Phytochemical analysis of the *S. palustris* extracts by Chai et al. [17] showed that the water fraction contained polar and water-soluble components, including high amounts of TP and TH. The TH content of the extract correlated with its antiglucosidase activity. In fact, the bioactive compounds in the water fraction were found to have an alpha-glucosidase inhibitory effect by binding reversibly to the enzyme's catalytic site. Moreover, the water fraction showed higher specificity towards alpha-glucosidase than alpha-amylase.

Gunawan-Puteri et al. [21] reported that alkaloids and phenolic compounds in the methanolic extract of *S. palustris* contributed to its AGI activity. Through activity-guided fractionation and characterization of the isolated compounds by LC-MS and H^+ NMR, they identified a flavonoid glycoside, kaempferol 3-*O*-β-D-glucopyranoside, in the extract which accounted for its AGI activity.

5.3 Cytotoxicity

Kormin et al. [46] evaluated the phytochemical contents of several epiphytic fern species, including *S. palustris*, found on the trunks of oil palm trees. They found that *S. palustris* fronds had the highest TP content in both the water extract (639.4 mg GAE/g dry weight) and the ethanol extract (271.60 mg GAE/g dry weight) in comparison with six other fern species. Brine shrimp (*Artemia salina*) lethality assay was carried out on the water and ethanolic fern extracts. The ethanolic extracts of all epiphytic ferns tested showed moderate brine shrimp larvicidal activity. The LC_{50} value of *S. palustris* ethanolic extract was found to be 430 μg/mL. On the other hand, the water extract showed poor or non-toxic activity with LC_{50} values of more than 1000 μg/mL for all other fern species except for *S. palustris* frond with an LC_{50} value of 825 μg/mL. The toxic metabolites and flavonoids were postulated to contribute to the cytotoxicity of the fern extract as *S. palustris* frond had the highest total flavonoid content in both water and ethanolic extracts among the seven fern species evaluated [46].

The cytotoxic effects of *S. palustris* leaf extracts have also been tested on human epithelial carcinoma cells. According to the study by Arullappan et al. [7], fractions from ethyl acetate extract showed IC_{50} values ranging from 8.6 μg/mL to 19.4 μg/mL, whereas fractions from ethanolic extract showed more potent activity with IC_{50} values ranging from 5.83 μg/mL to 10.16 μg/mL. The IC_{50} value of doxorubicin hydrochloride, which was used as a reference cytotoxic compound in their study, was found to be 2.21 μg/mL in HeLa cells.

Acylated derivatives of flavonoid glycosides have been reported to have enhanced bioactivities [47, 48]. Chear et al. [22] isolated a series of flavonoids from the mature, sterile fronds of *S. palustris*. They tested the cytotoxicity of seven acylated and non-acylated kaempferol glycosides on human breast cancer cell lines (MCF-7 and MDA-MB-231) as well as prostate cancer cell line (DU-145) (Duke University 145). Their results showed only two glycosides, kaempferol-*O*-(3″-*O*-*E*-*p*-coumaroyl)-(6″-*O*-*E*-feruloyl)-β-D-glucopyranoside and kaempferol 3-*O*-(3″, 6″di-*O*-*E*-*p*-coumaroyl)-β-D-glucopyranoside, were cytotoxic to the cancer cells. By comparison of their structures, it was found that di-acylated kaempferol 3-*O*-β-D-glucopyranoside has stronger cytotoxicity. Meanwhile, the other five mono-acylated or non-acylated kaempferol 3-*O*-β-D-glucopyranoside were not able to kill the cancer cells. A molecular docking study was carried out by Chear et al. [22] to evaluate the in silico binding interactions of the glycosides with epidermal growth factor receptors. Their results suggested that di-acylation at 3″-*O* and 6″-*O* positions of kaempferol 3-*O*-β-D-glucopyranoside increased its cytotoxicity [22].

Apart from flavonoid glycosides, plant sterols have also been associated with anticancer properties. Phytosterols, such as β-sitosterol and campesterol are reported to have an effect on the growth and metastasis of breast cancer cells [49, 50]. The anticancer property of campesterol on ovarian cancer cells was associated with its effect on disrupting mitochondrial function, thereby elevating reactive oxygen species (ROS) production as well as increasing calcium concentrations [51]. Another phytosterol, fucosterol, was reported to have multiple biological activities, including anticancer activity. It was found to induce apoptosis in human promyelocytic leukemia HL-60 cells and human colon cancer HCT-116 cells [52]. Marisa et al. [53] identified several phytosterols from *S. palustris* leaf extract including α-tocopherol, campesterol, stigmasterol, β-sitosterol, and fucosterol to be potent antibreast cancer compounds. They used the molecular docking method to elucidate the anticancer action mechanism. Results showed those phytosterols could potentially inhibit estrogen receptors, thereby inhibiting estrogen-dependent malignancy.

5.4 Antimicrobial Activity

Another bioactivity closely associated with cytotoxicity is the antimicrobial effect. Liu et al. [18] identified five new acylated flavonol glycosides, and four of those novel glycosides exhibited inhibitory effects against *Bacillus cereus*, *Micrococcus luteus*, *Staphylococcus aureus*, and *Staphylococcus epidermidis*. Structure-activity relationship of the four antibacterial compounds showed that they all contain one or two *cis-p*-coumaroyl groups as opposed to other structurally similar compounds that have only *trans-p*-coumaroyl or *trans*-feruloyl groups. The latter group of compounds did not exhibit antibacterial activity against the bacteria tested. From those results, Liu et al. [18] postulated that the antibacterial effect may be due to the doubly-bonded coumaroyl substitution on the flavonol glycosides.

Antibacterial activity was also reported for phytosterols found in *S. palustris*. Through GC-MS analysis, Erwin et al. [34] found that the major components in *S. palustris* ethanolic extract were 97.47% stigmasterol and its derivatives whereby 46.51% was stigmasterol. They tested the antibacterial activity of the extract against *Escherichia coli* and *S. aureus* and reported a minimum inhibitory concentration (MIC) value of 10% (w/v) of the ethanolic extract on both bacteria strains.

Sumathy et al. [54] evaluated the methanolic extract of *S. palustris* leaf against a foodborne pathogen, *Aspergillus niger*. They found the MIC against the fungus was 50 mg/mL, whereas the IC_{50} value for fungal inhibition was 17.41 mg/mL. Although the mechanism of action was not fully elucidated, a scanning electron micrograph of the *A. niger* treated with *S. palustris* extract showed that the antifungal activity was due to disruption of the hyphae cell wall.

A more extensive antimicrobial activity investigation using *S. palustris* extracts was done by Zuraini et al. [19]. They tested the *S. palustris* root, stem, and leaf methanolic extracts against various types of bacteria, fungi, and yeast using the disk diffusion assay. The microbes tested included *S. aureus*, *B. subtilis*, *Micrococcus* sp., *Enterobacter aerogenes*, *E. coli*, *Proteus mirabilis*, *Klebsiella pneumoniae*, *Salmonella typhi*, *Azospirillum lipoferum*, Azobacter, *Penicillium chrysogenum*, *Rhizopus stolonifer*, *A. niger*, *Fusarium* sp., and *Saccharomyces cerevisiae*. Their results showed all the extracts had an antimicrobial effect on all gram-positive and gram-negative bacteria, yeast, and fungi except for *R. stolonifer*. The results from various studies suggested that *S. palustris* extracts are a broad-spectrum antimicrobial agent.

5.5 Other Bioactivities

Biworo et al. [55] reported that *S. palustris* leaves had anti-metalotoxic properties. The cadmium pollution in the river water in South Kalimantan contributed to the discovery of the anti-cadmium property in *S. palustris*, as it is a plant that frequently grows in South Kalimantan's wetland areas. The ethanolic extract of *S. palustris* demonstrated anti-metalotoxic activity against cadmium-induced oxidative stress in the liver of rats [55]. It is well known that cadmium induces lipid peroxidation via an indirect mechanism, i.e., ROS formation [56]. The anti-cadmium activity of *S. palustris* extract can be attributed to its antioxidant properties, which inhibit lipid peroxidation in the liver and reduce the formation of ROS.

The anti-cholinesterase properties of young and mature fronds of *S. palustris* were evaluated [14]. In both the hexane extract of young fronds and the methanolic extract of mature fronds, butyrylcholinesterase (BChE)-inhibitory effect was observed. By inhibiting both acetylcholinesterase and BChE, the breakdown of acetylcholine into choline and acetic acid can be delayed [57]. This finding suggests that this plant may be useful in the treatment of Alzheimer's disease.

An ethyl acetate fraction of *S. palustris* stem was studied for its anti-termite activity [33]. To evaluate anti-termite activity, samples at the concentration of 1% (sample weight/filter paper weight x 100%) were applied onto paper discs and were placed into a Petri dish containing 30 workers and three soldiers of *Coptotermes curvignathus* for 7 days. The anti-termite activity of the ethyl acetate fraction was attributed to the major constituents, namely, 20-hydroxyecdysone, ajugasterone C, and stenopalustroside. This study showed that *S. palustris* stem could be used as a natural alternative to termite control agents.

S. palustris plant was commonly consumed by the Dayak people in Kalimantan and was known as a plant that can stimulate breast milk production in nursing mothers. In a study by Mahdiyah et al. [24], *S. palustris* juice was administered to the women daily in increasing amounts for 7 days. There was a significant increase in breast milk production in women who were given *S. palustris* juice. The average volume of breast milk collected each day for the first 6 days after ingesting the fern juice increased linearly and then decreased by 2.5 mL on the final day of treatment. This plant is easily accessible to local women and is nourishing for anemic pregnant and postpartum mothers.

S. palustris aqueous extract showed antipyretic effects by reducing hepatic peroxidative stress in *Marmota caligata* with induced fever [58]. Briefly, the animals were induced with diphtheria, tetanus toxoids, and pertussis vaccine, followed by oral administration of *S. palustris* aqueous extract at different doses. Peroxidative stress arises when inflammatory reactions occur upon infection. Oxidative damage will cause the disruption of the oxidative phosphorylation chain and finally result in mitochondria damage and apoptosis. *S. palustris* aqueous extract significantly reduced hepatic peroxidative stress in animals with induced fever, according to the authors. This discovery may explain the mechanism by which *S. palustris* treats fever.

Suhartono et al. [59] recently reported the protective role of *S. palustris* ethanolic extract against DNA damage caused by malathion, a carcinogenic organophosphate pesticide. In the *S. palustris* ethanolic extract, the authors identified 6.1% hexadecanoic acid and 6.5% neophytadiene. Neophytadiene was able to bind to DNA and interact with hexadecanoic acid at the C-17 position. This interaction could prevent malathion from binding to DNA. This result indicated that *S. palustris* ethanolic extract could be used to prevent malathion toxicity. With all the evidence of the bioactivities of *S. palustris* discussed in this and the preceding sections, it is clear that this fern has many health benefits. Thus, it is unsurprising that scientists have been working to find new ways to incorporate the fern as a natural ingredient in food, food packaging material, and cosmetics.

6 Applications in Food, Cosmetics, and Packaging Material

Researchers incorporated leaf extract of *S. palustris* into ice cream and discovered that the quality and the antioxidant activity of the confectionery could be preserved for up to 4 weeks [60]. The extract's stability was improved by keeping it at a low temperature. Eventually, the bioactivity of the extract was maximized for the consumer. The ethanolic extract of *S. palustris* roots was also incorporated into lotion [61] and emulgel, as sunscreen [8]. Both studies reported the physical properties of the products. These products used the root extract taking into consideration its known antioxidant activity and traditional medicinal applications.

Three groups of researchers have attempted to develop biodegradable and functional packaging materials using *S. palustris* extract as a key ingredient. The development of antioxidant films using blends of *S. palustris* flour and poly (vinyl alcohol) (PVA) was reported by Ma'Radzi et al. [62]. The use of 88% HD PVA (hydrolysis degrees (HD) of PVA) blends with *S. palustris* flour produced the highest antioxidant activity among the other PVAs tested in their study. In addition to developing biodegradable packaging material, the objective of this study was to incorporate antioxidant films into cosmetics. The edible film was developed from *S. palustris* leaves extract blended with cassava starch/eggshell powder [63]. The antioxidant activity of edible films containing 5% and 10% *S. palustris* extract was tested, yielding DPPH radical scavenging activity of 5.19% and 13.93%, respectively. This research has successfully developed biodegradable packaging material that is edible. A recent attempt to develop an edible coating to preserve figs using *S. palustris* extract revealed that it helped to maintain pH and total soluble solids level, preserve the color of the fruit, and reduce microbial growth [36]. This innovation could be used to extend the shelf life of not only figs but also other foods.

7 Structure-Activity Relationship

Current knowledge about the structure-activity relationship of bioactive compounds identified from *S. palustris* is still limited. A series of kaempferol glycosides exhibiting different levels of anti-BChE activities were identified from *S. palustris* [14]. The studies found that the parental structure of these flavonoids is kaempferol 3-*O*-β-D-glucopyranoside (Fig. 4a), which underwent acylation to produce more complex derivatives, such as di-acylated kaempferol glycosides and mono-acylated kaempferol glycosides [14, 22]. The di-acylation at 3″-*O* and 6″-*O* position of glucose unit with hydroxycinnamoyl groups (*p*-coumaroyl and *p*-feruloyl group moieties) as shown in Fig. 4b and c with blue highlight led to the highest anti-BChE activities among the kaempferol glycosides [14]. When compared with kaempferol 3-*O*-(3″-*O*-*E*-*p*-coumaroyl)-(6″-*O*-*E*-feruloyl)-β-D-glucopyranoside (Fig. 4c), kaempferol 3-*O*-(3″, 6″-di-*O*-*E*-*p*-coumaroyl)-β-D-glucopyranoside (Fig. 4b) showed a greater anti-BChE activity and cytotoxicity to cancer cells [14, 22]. This showed that the substitution of meta-hydrogen with methoxy groups on the feruloyl moiety in kaempferol 3-*O*-(3″-*O*-*E*-*p*-coumaroyl)-(6″-*O*-*E*-feruloyl)-β-D-glucopyranoside (as highlighted in red in Fig. 4c) reduced its efficiency inhibiting BChE and in causing cancer cell toxicity [14, 22]. Nevertheless, kaempferol 3-O-(3″-*O*-*E*-*p*-coumaroyl)-(6″-*O*-*E*-feruloyl)-β-D-glucopyranoside (Fig. 4c) still exhibited a stronger free radical

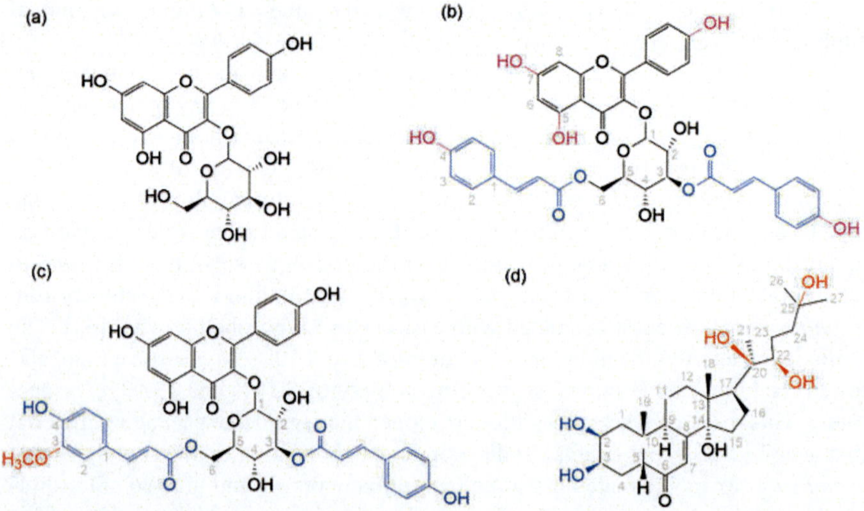

Fig. 4 Chemical structures of (**a**) kaempferol 3-*O*-β-D-glucopyranoside, (**b**) kaempferol 3-*O*-(3″, 6″-di-*O*-*E*-*p*-coumaroyl)-β-D-glucopyranoside, (**c**) kaempferol 3-*O*-(3″-*O*-*E*-*p*-coumaroyl)-(6″-*O*-*E*-feruloyl)-*β*-D-glucopyranoside, and (**d**) 20-hydroxyecdysone. The colors in the structures indicate the active groups contributing to bioactivities

scavenging activity relative to kaempferol 3-*O*-(3″,6″-di-*O*-*E*-*p*-coumaroyl)-β-D-glucopyranoside [22].

Through molecular docking simulations, it was found that kaempferol 3-*O*-(3″, 6″-di-*O*-*E*-*p*-coumaroyl)-β-D-glucopyranoside (Fig. 4b) was cytotoxic to cancer cells probably through binding to the epidermal growth factor receptor protein tyrosine (EGFR-TK) [22]. The di-acylated kaempferol glycoside was predicted to form five hydrogen bonds between the phenol groups of its kaempferol aglycone and *p*-coumaroyl moiety (highlighted in purple in Fig. 4b), to the five kinase domains of EGFR-TK. The domains include the Met-769 in adenine/hinge region of EGFR, Leu-764 in the hydrophobic region I, and Phe-771 in the hydrophobic region II. The other two hydrogen bonds were predicted to form through the interaction of 5-OH of kaempferol aglycone and 4-OH of 6″-*O*-*p*-coumaroyl moieties to the Glu-738 in the helix sugar pocket of EGFR-TK. Meanwhile, another cation-π hydrophobic interaction between the aromatic rings of the *p*-coumaroyl group at 6″-O sites of kaempferol 3-*O*-(3″,6″-di-*O*-*E*-*p*-coumaroyl)-β-D-glucopyranoside and the basic residue of Lys-721 in the phosphate-binding region of EFGR plays important role in cytotoxicity [22].

20-hydroxyecdysone (Fig. 4d) is an ecdysteroid identified from an ethyl acetate fraction/methanolic extract of *S. palustris* [33]. The termiticidal effect of 20-hydroxyecdysone is mostly attributed to the presence of 2,3-diol as highlighted in blue in Fig. 4d. The 20,22-diol and the 25-OH moiety of the phytosterol, highlighted in red in Fig. 4d, contributed only weakly to the termiticidal activity of 20-hydroxyecdysone [33].

8 Conclusions

A review of the literature revealed that more attention was given to characterizing the bioactivities of *S. palustris* than to identifying the bioactive phytocompounds in the fern. Experimental evidence, mostly obtained from in vitro studies, revealed that the fern possessed antioxidant, antiglucosidase, anti-butyrylcholinesterase, cytotoxic, antibacterial, antifungal, and other bioactivities. Applications of *S. palustris* in the production of food, cosmetics, and food packaging material also produced promising outcomes, benefitting from the antioxidant and antimicrobial effects of the fern. To date, a number of *S. palustris* phytocompounds, mostly kaempferol glycosides, were identified. Such phytochemicals exhibited antioxidant, anti-butyrylcholinesterase, cytotoxic, antiglucosidase, and termiticidal activities. However, the specific phytocompounds responsible for other reported bioactivities of the fern remain to be ascertained. Identification of bioactive phytochemicals from *S. palustris* is still a current gap in knowledge. More intensive future research is warranted to not only fill this knowledge gap but also to better understand the in vitro and in vivo modes of action underlying the bioactivities of previously identified phytocompounds of *S. palustris*.

References

1. Pranjal B, Barukial J (2021) On the pteridophytes of Bherjan-Borajan-Padumoni wildlife sanctuary, Assam. India J Threat Taxa 13(12):19781–19790. https://doi.org/10.11609/jott.7488.13.12.19781-19790
2. Nurhasnawati H, Sundu R, Sapri SR, Kuspradini H, Arung ET (2019) Antioxidant activity, total phenolic and flavonoid content of several indigenous species of ferns in East Kalimantan. Indonesia Biodiversitas 20(2):576–580. https://doi.org/10.13057/BIODIV/D200238
3. Cao H, Chai T-T, Wang X, Morais-Braga MFB, Yang J-H, Wong F-C et al Phytochemicals from fern species: potential for medicine applications. Phytochem Rev 16(3):379–440. https://doi.org/10.1007/s11101-016-9488-7
4. Chambers TC (2013) A review of the genus *Stenochlaena* (Blechnaceae, subfamily Stenochlaenoideae). Telopea 15
5. Debnath S, Kundu P, Ahad M, Saha L, Biswas NN, Sadhu S (2021) Investigation of phytochemical and pharmacological assessment of ethanol extract of *Stenochlaena palustris* – an edible fern of Sundarbans. J Med Plants Stud 9:226–232. https://doi.org/10.22271/plants.2021.v9.i3c.1294
6. Ndanusa AH, Cicuzza D, Siddique MM (2020) Analysis of the phytochemical contents and anti-oxidative properties of *Stenochlaena palustris*. Int Food Res J 27(5):798–804
7. Arullappan S, Sawai S, Chee L, Mahandan M, Sanmugavelan R (2017) Phytochemical screening and evaluation of cytotoxic effect and antioxidant activity of fractions isolated from *Stenochlaena palustris* (Burm.F.) Bedd. Leaves. Indian J Pharm Educ Res 51:s735–ss40. https://doi.org/10.5530/ijper.51.4s.106
8. Paramawidhita RY, Adawiyah R, Umaternate A (2021) The formulation and physical evaluation of emulgel the kalakai (*Stenochlaena palustris* Bedd) roots ethanol extract as a sunscreen. J Phys Conf Ser 1764(1):012021. https://doi.org/10.1088/1742-6596/1764/1/012021
9. Sulasmi ES, Suhadi SMS, Mawaddah K (2019) Alkaloids identification from the fronds of 5 species pterydophyta in Baluran Nasional Park. AIP Conf Proc 2120(1):030015. https://doi.org/10.1063/1.5115619
10. Wijaya E, Widiputri DI, Rahmawati D (2017, 1904) Optimizing the antioxidant activity of kelakai (*Stenochlaena palustris*) through multiplestage extraction process. AIP Conf Proc (1): 020034. https://doi.org/10.1063/1.5011891
11. Sofiyanti N, Iriani D, Wati F, Marpaung A (2019) Morphology, palynology, and stipe anatomy of four common ferns from Pekanbaru, Riau Province, Indonesia. Biodivers J 20:327–336. https://doi.org/10.13057/biodiv/d200138
12. Sofiyanti N, Iriani D, Fitmawati D, Roza AA (2015) *Stenochlaena riauensis* (Blechnaceae), a new fern species from Riau. Indonesia Bangladesh J Plant Taxon 22(2):137–141. https://doi.org/10.3329/bjpt.v22i2.26075
13. Awang-Kanak F, Abu Bakar MF (2020) Traditional vegetable salad (ulam) of Borneo as source of functional food. Food Res 4(1):1–12. https://doi.org/10.26656/fr.2017.4(1).138
14. Chear NJ-Y, Khaw K-Y, Murugaiyah V, Lai C-S (2016) Cholinesterase inhibitory activity and chemical constituents of *Stenochlaena palustris* fronds at two different stages of maturity. J Food Drug Anal 24(2):358–366. https://doi.org/10.1016/j.jfda.2015.12.005
15. Uda SK, Hein L, Adventa A (2020) Towards better use of Indonesian peatlands with paludiculture and low-drainage food crops. Wetl Ecol Manag 28(3):509–526. https://doi.org/10.1007/s11273-020-09728-x
16. Nicholas D, Chua HP, Saniah K (2013) Postharvest treatment and packaging system for Sarawak's indigenous midin fern (*Stenochlaena palustris*). Acta Hortic 979:131–137. https://doi.org/10.17660/ActaHortic.2013.979.11
17. Chai T-T, Kwek M-T, Ong H-C, Wong F-C (2015) Water fraction of edible medicinal fern *Stenochlaena palustris* is a potent α-glucosidase inhibitor with concurrent antioxidant activity. Food Chem 186:26–31. https://doi.org/10.1016/j.foodchem.2014.12.09

18. Liu H, Orjala J, Sticher O, Rali T (1999) Acylated flavonol glycosides from leaves of *Stenochlaena palustris*. J Nat Prod 62(1):70–75. https://doi.org/10.1021/np980179f
19. Zuraini Z, Sasidharan S, Roopin Kaur S, Nithiyayini M (2010) Antimicrobial and antifungal activities of local edible fern *Stenochlaena palustris* (Burm. F.) Bedd. Pharmacologyonline 1: 233–237
20. Zakaria Z, Sanduran S, Sreenivasan S (2010) Antifungal activity of the edible ferns: application for public health. Int J Humanit 8(8):113–118. https://doi.org/10.18848/1447-9508/CGP/v08i08/43008
21. Gunawan-Puteri MDPT, Kato E, Rahmawati D, Teji S, Santoso JA, Pandiangan FI et al (2021) Post-harvest and extraction conditions for the optimum alpha glucosidase inhibitory activity of *Stenochlaena palustris*. Int J Technol 12(3):291–319. https://doi.org/10.14716/ijtech.v12i3.4409
22. Chear NJ-Y, Fauzi AN, Khaw K-Y, Choi S-B, Yaacob NS, Lai C-S (2019) Free radical scavenging and cytotoxic properties of acylated and non-acylated kaempferol glycosides from *Stenochlaena palustris*: a perspective on their structure – activity relationships. Pharm Chem J53(3):188–193. https://doi.org/10.1007/s11094-019-01977-2
23. Chai T-T, Panirchellvum E, Ong H-C, Wong F-C (2012) Phenolic contents and antioxidant properties of *Stenochlaena palustris*, an edible medicinal fern. Bot Stud 53:439–446
24. Mahdiyah D, Rahmawati D, Waty DS (2021) The effect of giving Kelakai (*Stenochlaena palustris*) juice on the volume of breastfeeding for postpartum mothers in the working area of the Pekauman health Center, Banjarmasin. IOP Conf Ser Earth Environ Sci 755(1):012040. https://doi.org/10.1088/1755-1315/755/1/012040
25. Rudra S, Islam KN, Rahman MM, Uddin SB (2021) Medicinal plant diversity and their therapeutic uses in selected village common forests in Chittagong Hill tracts, Bangladesh. Int J Geogr Inf Syst 27(1):83–107. https://doi.org/10.1080/10496475.2020.1786874
26. Neamsuvan O, Sengnon N, Seemaphrik N, Chouychoo M, Rungrat R, Bunrasri S (2015) A survey of medicinal plants around upper Songkhla lake, Thailand. Afr J Tradit Complement Altern Med 12:133. https://doi.org/10.4314/ajtcam.v12i2.20
27. Sen A, Ghosh P (2011) A note on the ethnobotanical studies of some pteridophytes in Assam. Indian J Tradit Knowl 10:292–295
28. Kulip J (2014) The ethnobotany of Dusun people in Tikolod village, Tambunan district, Sabah, Malaysia. Reinwardtia 14:101–121
29. Noorasmah S, Ainul Asyira S, Muta Harah ZY, Shahrul Razid S, Nurul Aisyah Y (2020) An ethnobotanical study of indigenous leafy vegetables among local communities in Bintulu, Sarawak, Malaysia. Borneo J Resour Sci Technol 10(2). https://doi.org/10.33736/bjrst.2623.2020
30. Chai PPK (2006) Medicinal plants of Sarawak. Paul Chai P.K
31. Abdul Wahab N, Ahdan R, Ahmad Aufa Z, Kong KW, Johar MH, Shariff Mohd Z et al (2015) Nutritional values and bioactive components of under-utilised vegetables consumed by indigenous people in Malaysia. J Sci Food Agric 95(13):2704–2711. https://doi.org/10.1002/jsfa.7006
32. Lodevico DP, Enot MM, Mendez RA, Abarquez VR, Monisit GFB, Coritico FP et al (2018) Phytochemical and antioxidant activity variation of processed edible ferns. CMU J Sci 22(2): 57–67
33. Pardede A, Adfa M, Kusnanda AJ, Ninomiya M, Koketsu M (2018) Isolation of secondary metabolites from *Stenochlaena palustris* stems and structure-activity relationships of 20-hydroxyecdysone derivatives on antitermite activity. Holzforschung 72(10):899–904. https://doi.org/10.1515/hf-2017-0212
34. Erwin E, Anggeraini D, Suryani S (2016) Chemical analysis and antibacterial activity of the ethanolic extract of *Stenochlaena palustris*. Pharm Lett 8(1):233–236
35. Abubakar AR, Haque M (2020) Preparation of medicinal plants: basic extraction and fractionation procedures for experimental purposes. J Pharm Bioallied Sci 12(1):1–10. https://doi.org/10.4103/jpbs.JPBS_175_19

36. Samad A, Makhbar S, Sharifulazar H, Basri AM, Lim SA (2022) The effects of *Diplazium esculentum* Retz. And *Stenochlaena palustris* incorporated with sodium alginate as edible coating on packaged figs (*Ficus carica* L.): a preliminary study. J Food Process Preserv: e16611. https://doi.org/10.1111/jfpp.16611
37. Tena N, Martín J, Asuero AG (2020) State of the art of anthocyanins: antioxidant activity, sources, bioavailability, and therapeutic effect in human health. Antioxidants 9(5):451. https://doi.org/10.3390/antiox9050451
38. Ahmad A, Kaleem M, Ahmed Z, Shafiq H (2015) Therapeutic potential of flavonoids and their mechanism of action against microbial and viral infections – a review. Int Food Res J 77:221–235. https://doi.org/10.1016/j.foodres.2015.06.021
39. Tsuda T, Ueno Y, Aoki H, Koda T, Horio F, Takahashi N et al (2004) Anthocyanin enhances adipocytokine secretion and adipocyte-specific gene expression in isolated rat adipocytes. Biochem Biophys Res Commun 316(1):149–157. https://doi.org/10.1016/j.bbrc.2004.02.031
40. Liu Y, Li D, Zhang Y, Sun R, Xia M (2014) Anthocyanin increases adiponectin secretion and protects against diabetes-related endothelial dysfunction. Am J Physiol Endocrinol Metab 306(8):E975–EE88. https://doi.org/10.1152/ajpendo.00699.2013
41. Pojer E, Mattivi F, Johnson D, Stockley CS (2013) The case for anthocyanin consumption to promote human health: a review. Compr Rev Food Sci 12(5):483–508. https://doi.org/10.1111/1541-4337.12024
42. Minh NP (2021) Effectiveness of pickling on the phytochemical composition and antioxidant ability in kelakai (*Stenochlaena palustris*) creeping fern. Res on Crops 22(4):948–952. https://doi.org/10.31830/2348-7542.2021.155
43. Kalra S (2014) Alpha glucosidase inhibitors. J Pak Med Assoc 64(4):474–476
44. Chai T-T, Mohan M, Ong HC, Wong F-C (2014) Antioxidant, iron-chelating and anti-glucosidase activities of *Typha domingensis* Pers (Typhaceae). Trop J Pharm Res 13(1):67–72. https://doi.org/10.4314/tjpr.v13i1.10
45. Graf BL, Raskin I, Cefalu WT, Ribnicky DM (2010) Plant-derived therapeutics for the treatment of metabolic syndrome. Curr Opin Investig Drugs (London, England: 2000) 11(10):1107–1115
46. Kormin F, Khan M, Iwansyah AC (2016) Microwave assisted extraction; phytochemical evaluation of Malaysian palm oil trunk epiphytes ferns. Int J Pharm Pharm Sci 8(4):174–180
47. Parejo I, Bastida J, Viladomat F, Codina C (2005) Acylated quercetagetin glycosides with antioxidant activity from *Tagetes maxima*. Phytochemistry 66(19):2356–2362. https://doi.org/10.1016/j.phytochem.2005.07.004
48. Mellou F, Lazari D, Skaltsa H, Tselepis AD, Kolisis FN, Stamatis H (2005) Biocatalytic preparation of acylated derivatives of flavonoid glycosides enhances their antioxidant and antimicrobial activity. J Biotechnol 116(3):295–304. https://doi.org/10.1016/j.jbiotec.2004.12.002
49. Awad AB, Williams H, Fink CS (2003) Effect of phytosterols on cholesterol metabolism and MAP kinase in MDA-MB-231 human breast cancer cells. J Nutr Biochem 14(2):111–119. https://doi.org/10.1016/s0955-2863(02)00274-7
50. Shahzad N, Khan W, Md S, Ali A, Saluja SS, Sharma S et al (2017) Phytosterols as a natural anticancer agent: current status and future perspective. Biomed Pharmacother 88:786–794. https://doi.org/10.1016/j.biopha.2017.01.068
51. Bae H, Park S, Yang C, Song G, Lim W (2021) Disruption of endoplasmic reticulum and ROS production in human ovarian cancer by campesterol. Antioxidants 10(3):379
52. Abdul QA, Choi RJ, Jung HA, Choi JS (2016) Health benefit of fucosterol from marine algae: a review. J Sci Food Agric 96(6):1856–1866. https://doi.org/10.1002/jsfa.7489
53. Marisa D, Hayatie L, Juliati S, Suhartono E, Komari N (2021) Molecular docking of phytosterol compounds from kelakai (*Stenochlaena palustris*) as anti-breast cancer. Acta Bio Ina 4(2):59. https://doi.org/10.32889/actabioina.59

54. Sumathy V Jr, Jothy Lachumy S, Zuraini Z, Sasidharan S (2010) Effects of *Stenochlaena palustris* leaf extract on growth and morphogenesis of food borne pathogen, *aspergillus Niger*. Malays J Nutr 16(3):439–446
55. Biworo A, Azizi NA, Padelia R, Raharja MA, Azima O, Suhartono E (2018) Anti-metalotoxic properties of kelakai (*Stenochlaena palustris*) leaves extract against cadmium-induced liver tissue damage. Asian J Pharm Clin Res 11(3):43–46. https://doi.org/10.22159/ajpcr.2018.v11s3.30028
56. Branca JJV, Fiorillo C, Carrino D, Paternostro F, Taddei N, Gulisano M et al (2020) Cadmium-induced oxidative stress: focus on the central nervous system. Antioxidants 9(6):492. https://doi.org/10.3390/antiox9060492
57. Colović MB, Krstić DZ, Lazarević-Pašti TD, Bondžić AM, Vasić VM (2013) Acetylcholinesterase inhibitors: pharmacology and toxicology. Curr Neuropharmacol 11(3):315–335. https://doi.org/10.2174/1570159X11311030006
58. Adenan A, Suhartono E (2016) *Stenochlaena palustris* aqueous extract reduces hepatic peroxidative stress in *Marmota caligata* with induced fever. Univ Med 29(3):123–128. https://doi.org/10.18051/UnivMed.2010.v29.123-128
59. Suhartono E, Muthmainah N, Marisa D, Siahaan SCPT, Komari N (2022) Protective role of kelakai (*Stenochlaena Palustris*) extract on malathion-induced genotoxic: FTIR spectroscopy study. Int J Drug Deliv Technol 12(1):15–18. https://doi.org/10.25258/ijddt.12.1.3
60. Hadhiwaluyo K, Rahmawati D, Gunawan Puteri MDPT (2017) Development of antioxidative effect in ice cream with Kalakai (*Stenochlaena palustris*) water extract. AIP Conf Proc 1904(1): 020007. https://doi.org/10.1063/1.5011864
61. Adawiyah R, Apriliyanti R, Umaternate A (2021) Formulation and physical properties of lotion Kalakai root ethanol extract (*Stenochlaena palustris* Bedd). Pharm Educ 21(2):235–240. https://doi.org/10.46542/pe.2021.212.235240
62. Ma'Radzi AH, Kasim KF, Mahmud N, Rahman N, Jullok N (2018) Development of antioxidant film based on blends of *Stenochlaena palustris* flour and poly (vinyl alcohol). IOP Conf Ser Mater Sci Eng 429:012066. https://doi.org/10.1088/1757-899x/429/1/012066
63. Nata IF, Irawan C, Adawiyah M, Ariwibowo S (2020) Edible film cassava starch/eggshell powder composite containing antioxidant: preparation and characterization. IOP Conf Ser Earth Environ Sci 524(1):012008. https://doi.org/10.1088/1755-1315/524/1/012008

Allelochemicals from *Pteridium arachnoideum* 19

Luciana de Jesus Jatoba

Contents

Abstract

Pteridium arachnoideum, also known as the Neotropical bracken fern, is a plant species native to South America and an aggressive invader that, as well as other bracken species around the globe, represent a challenge for the management and conservation of biodiversity as they become dominant in degraded or anthropic areas, causing significant environmental changes. Much interest in the phytochemistry of *Pteridium* species exists especially because of the toxic and carcinogenic action of these plants when consumed by mammals, effects caused by the norsesquiterpene glycoside ptaquiloside. Polyphenolic compounds are the main chemical class of secondary metabolites in bracken species, although they are not considered the main toxic substances produced by them. Among these compounds, condensed tannins, such as proanthocyanidins and prodelphins derivatives, are the main ones, both being associated with anti-herbivory and UV radiation protection. Efforts to understand bracken super-dominance underlying

L. d. J. Jatoba (✉)
Federal Institute of Education, Science, and Technology of São Paulo – IFSP, Hortolândia, Brazil
e-mail: lujatoba@ifsp.edu.br

H. N. Murthy (ed.), *Bioactive Compounds in Bryophytes and Pteridophytes*,
Reference Series in Phytochemistry, https://doi.org/10.1007/978-3-031-23243-5_24

mechanisms date back to the beginning of the twentieth century and are still not well understood. Early evidence pointed to the role of bracken phenolics as allelochemicals, secondary metabolites with interfering activity on neighboring plant development, soil microbiota, or soil properties. The production of such allelochemicals could lead to better competitive fitness, explaining bracken super-dominant behavior. The proanthocyanidin selliguein A is the only allelochemical isolated and identified to this day for all bracken species, found as the major secondary metabolite of *P. arachnoideum* green fronds and litter, present in the soil under Neotropical bracken patches in bioactive phytotoxic concentrations. The relation between abiotic stress and condensed tannin content in this plant and its possible relation to plant dominance mediated by allelopathy is discussed, as well as other open questions important to understanding and controlling one of the most common and aggressive plant species complexes in the world, from the point of view of its allelochemicals.

Keywords

Allelopathy · Bracken fern · Condensed tannins · Native invader · Phenolics · Phytochemistry · Proanthocyanidins · Plant ecology · Super-dominant species

Abbreviations

IAS	International Allelopathy Society
IC50	Minimal concentration needed of a given bioactive substance to promote 50% of the desired effect
UV	Ultraviolet radiation

1 Introduction

The influence of a plant on the development of another neighboring plant is a phenomenon observed since Ancient Greece [1], with major impacts and applications on agriculture [2] and, more recently, environmental conservation management [3]. Despite being an ecological phenomenon known since antiquity, occurring both in natural and cultivated environments, the first attempt at its definition occurred only in 1937, with the proposal of the term allelopathy by Molisch. It originates from the Greek words "allelon" = mutual and "pathós" = prejudice, defining it as an interaction of mutual prejudice between plant individuals [4]. Since then, the definition of the term has been improved. For instance, Rice defined allelopathy in 1984 as any direct or indirect malefic effect of one plant on another, through the production of chemical compounds and their release into the environment [1]. In 1999, Rizvi and collaborators expanded these initial definitions, encompassing malefic and beneficial effects of a plant on other organisms, which could be another plant, soil fungus or bacteria, herbivorous insects, etc. [5]. Currently, the International Allelopathy Society (IAS) defines allelopathy as a phenomenon occurring between neighboring plants and their associated microflora and macroflora, mediated directly or indirectly by the production of chemical compounds, the allelochemicals, which result mainly in interferences on plant development, including germination and early growth inhibition or stimulus. This definition includes interactions in terrestrial or aquatic environments and involves the production of allelochemicals directly by plants or modified by microorganisms [6]. Thus, in allelopathy as a biological phenomenon, there is a donor plant, the one that produces allelochemicals, and a target organism, which suffers its effects [7]. Even though this phenomenon may be triggered by other organisms besides plants, i.e., with microorganisms as the donors in their interaction with plants, most of the relevant literature in the area focuses on plants as allelochemicals-producing organisms, mediating their relation to other plants in natural or agricultural environments or the soil microbiota [3].

Most allelochemicals derive from the activity of the plant's secondary metabolism and may be present in all plant organs. However, their distribution is not uniform and may also vary in quality and quantity in the different organs of the plant and over time [7, 8]. These compounds are, in general, products of secondary metabolism, that is, compounds that are not directly linked to plant physiology and metabolism like those of primary metabolism, being important mediators of plant interactions with their environment. The secondary metabolites that act as allelochemicals belong to different chemical classes, with diverse biosynthetic origins (Fig. 1). In general, they are classified into three main categories: terpenes, phenolic, and nitrogen-containing compounds [9]. Specifically, according to Rice's (1984) initial proposal, the main categories of allelochemicals are water-soluble organic acids, straight-chain alcohols, aliphatic aldehydes, and ketones; simple unsaturated lactones; long-chain

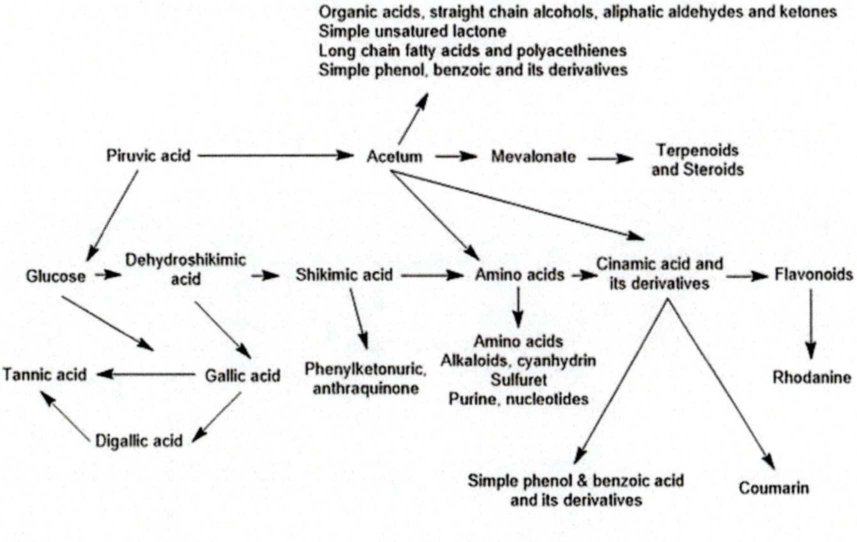

Fig. 1 Major biosynthetic pathways for the production of secondary metabolites with allelopathic activity

fatty acids and polyacetylenes; quinines, as benzoquinone, anthraquinone, and complex quinines; phenolics; cinnamic acid and its derivatives; coumarins; flavonoids; tannins; and steroids and terpenoids, such as sesquiterpene lactones, diterpenes, and triterpenoids [10, 11]. Of these, phenolic compounds are the most abundant secondary metabolites in vascular plants, performing different functions associated with the defense of the producing plant, and are reported as the most abundant allelochemicals in ecosystems [9, 10].

To act as mediators of ecological interaction, allelochemicals must be released into the environment, becoming available to act on target plants. Thus, these substances must be released directly into the environment by living plants through root exudation or leaching, through the decomposition of plant residues, or by volatilization [7], as illustrated in Fig. 2. In addition, the compounds produced by donor plants can be modified by the soil microbiota, and their activity may be increased or reduced by it, including activation of otherwise inert metabolites due structural modifications by microbial activity [12].

Given the great structural diversity among allelochemicals, it is expected that their modes of action on the target plant are also very diverse, and in many cases, these modes of action are not known [11, 13]. In general, we have that these modes of action occur in two main ways: environmental disturbances or phytotoxic activity. Environmental disturbances include means of indirect action, such as interference with the soil microbiota [12, 14], on the dynamics of nutrient cycling and its availability [15], or soil pH or organic matter content [16], among others. The phytotoxic activity consists of the direct inhibitory effect of an allelochemical on

Fig. 2 Main mechanisms of allelochemical release

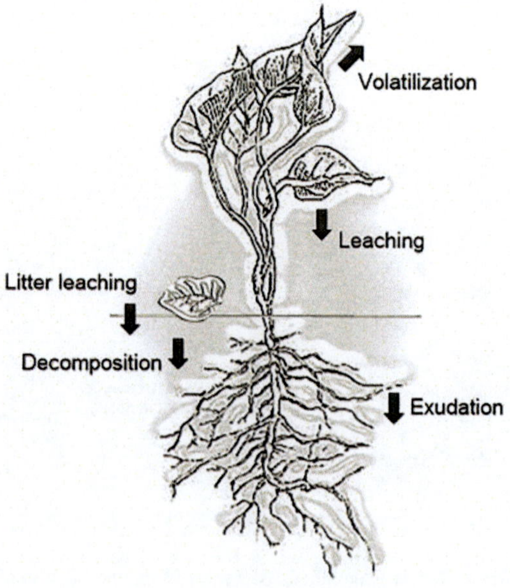

the germination and/or early development of the target plant, which can occur through interference in the stability and functioning of biological membranes, on the cell cycle or cell structure, synthesis of pigments or enzymes, action on photosynthesis rate or chlorophyll content, among other possible effects [11, 13, 17].

Back to the definition of allelopathy as an ecological phenomenon, positive interferences are also expected to be found, such as germination promotion or early growth stimulation of the target plant, or the improvement of critical environmental conditions [5, 6]. However, most of the allelopathy research to date has focused on its negative effects, especially due to interest in applying these principles in cultivated environments or the production of bioherbicides [18–20]. In addition to generally presenting allelopathy as a phenomenon of ecological relevance, this chapter aims to discuss how studies on this subject can contribute to the understanding of dominance mechanisms and the environmental effects of invasive species on the sites in which they become dominant through the production and release of allelochemicals. The focus will be given to what is known and what remains to be discovered about these mechanisms in the Neotropical bracken fern, *Pteridium arachnoideum*.

2 Allelopathy: A How to Guide

Due to the great diversity in the chemical structure of allelochemicals, their large diversity of action modes on target plants, and the synergy with other forms of ecological interaction, proving allelopathy has been a challenge with theoretical and methodological implications.

As for some of the theoretical implications, a question that has already been raised is about the use of phytotoxicity as an important criterion in the allelopathy phenomenon analysis from the point of view of the consequences on the germination or development of the target plant. This was, for a long time, the most evaluated criterion to determine a species as allelopathic in studies published in the area, with the use of the terms "allelopathic" and "phytotoxic" as synonyms and attracting much criticism since this approach ignores aspects of allelopathy as an ecological phenomenon [18, 21, 22].

Several criteria are needed to prove allelopathy, in which the isolation and structural elucidation of an allelochemical is crucial. Considering the most widely accepted concept, it is necessary, at least, to gather evidence of the production of a bioactive secondary metabolite by a donor plant, and that this metabolite is present in the environment in relevant concentrations. Ideally, it is necessary to isolate and identify the chemical compounds responsible for the observed effects on target plant germination and development first. Then, the evaluation of the allelopathic potential of these compounds should be established considering the ecological context, using cohabitating target species. These experiments cannot be restricted to laboratory conditions and should be reinforced by data gathered in the field and/or greenhouse experiments. Also, it is crucial to determine production and environment release modes, as well as evidence of the presence of these compounds in the environment in relevant, active concentrations gathered in situ [23–25]. Thus, the use of extracts or leachates from donor plants under laboratory conditions on the development of target species that do not coexist with the donor in nature, as in most phenomenological works produced in the area, is not enough to prove the phenomenon of allelopathy [3, 18].

In the cases in which meeting all the ideal criteria for proving allelopathy is not possible, experimental designs in this field must meet enough criteria for allelopathy to be the best explanation for the phenomena observed under field conditions, especially regarding spatial distribution and the dominance of certain species [18, 24]. In this context, classical germination and early seedling growth experiments still have their value, as long as the target plants chosen are significant, either because they are good bioindicator species or because they are species that effectively coexist with the donor plant in natural or cultivated environments [11, 21]. From these experiments, if the phytotoxic activity – or allelopathic potential, as it has been called more recently – is observed, the isolation and identification of the compound(s) responsible for it are the next steps. If these compound(s) can be found in the environment in sufficient quantities to affect the development of neighboring plants, then we may be facing an allelochemical, produced by an allelopathic plant that alters the environment in which it lives through allelopathy [24].

To make It possible to go beyond phenomenological studies, with the isolation and structural identification of allelochemicals, the use of chemical analysis techniques and methodologies is needed, which bring the studies in this area closer to those of phytochemistry, drug development, and chemistry of natural products [2, 26]. Recent advances in the field have much to thank for chromatographic and spectrometric techniques, among other analytical chemistry techniques and methodologies, making allelopathy an interdisciplinary field [11, 23]. Therefore, we can use well-established chemical methods to perform the extraction, isolation, and structural determination of allelochemicals also understood as bioactive compounds.

Regarding extraction, the use of organic solvents was criticized for a long time, especially concerning the direct use of crude solvent extracts in germination and early seedling growth bioassays, since the use of these solvents would cause the release of compounds that could hardly be found in the environment by natural mechanisms of leaching or decomposition [27]. However, the use of these organic solvents allows for the preparation of samples that can be subjected to chromatographic separation and spectrometric analysis in a more simplified way than those based on water. For that, the preparation of these extracts allows for the application of methodologies, such as bio-guided isolation with the use of activity bioindicators, as developed by Macías [28, 29] with the use of etiolated wheat coleoptiles. These quick and low resource-consuming experiments can be easily repeated and indicate which fractions are the most active to be fractionated next, until allowing the isolation of a metabolite with allelopathic potential. Attention must be given, however, to the polarity of the obtained extracts, since the highly apolar ones or their fractions may present activity due to mechanical rather than physiological modes, and is rich in compounds that would hardly be available in the environment in the field conditions [2].

From the isolation of a bioactive compound with allelopathic potential, we can proceed with the structural determination of the isolated substance and its use in germination and early seedling growth experiments, to observe its action on the development of significant target species. The structural determination and identification of the metabolite with allelochemical potential allow for the development of analytical assays to evaluate the presence of this substance in the environment, for example, in soil solution [30]. It also allows for advances in studies on the structure-activity relationships, which can unfold in the synthetic production of compounds with structural modifications that present similar and/or enhanced desirable activity for agricultural use [2, 26]. As for the use of these metabolites in experiments on target plants, as long as the principles of the ecological significance of the chosen target species and the use of positive and negative controls are respected [3, 18], the development of reproducible and accurate experiments could lead to the establishment of concentration-effect relationships and the calculation of indices, such as IC_{50}, the minimal concentration needed of a given bioactive substance to promote 50% of the desired effect. It would allow for the analysis of the putative allelochemical presence in the environment, as well as for its presence and availability in active concentrations, explaining the environmental effects of the donor plant that could not be explained by other ecological mechanisms [26, 30].

3 Allelopathy and Plant Community Structure

Allelopathy has long been recognized as an important factor in defining the structure of plant communities, with the spatial distribution of plant species being used as an indicator of the allelopathic potential of dominant species since the beginning of the studies in this area [3]. By disrupting the germination, growth, or survival of neighboring plants, an allelopathic species can result in changes in species richness and their distribution in the ecosystem, acting as one of the mechanisms for its dominance [31]. In addition to this direct action, allelochemicals can affect the structure of plant communities indirectly, by disrupting soil microbiota composition and/or soil physicochemical properties and nutrient cycling [32]. A plant's dominance in its environment occurs due to life history-related factors, such as high growth and reproduction rates [33], as well as due to ecological factors, such as resource-mediated interactions and competition in the strict sense [34], and also interactions, such as allelopathy [31, 35].

Plants with aggressive behavior tend to dominate areas that have been through anthropic impacts, such as fire or deforestation, forming monospecific patches, and affecting ecological succession, which results in structure and composition changes of plant communities with lasting effects [36, 37]. Taking allelopathy as an important underlying factor organizing ecological mechanisms and plant local dominance, many studies have been developed to understand the possible contribution of allelopathy to the success of exotic invasive plants. These studies led to the "new weapons hypothesis," proposed by Callaway and Ridenour [38] after a biogeographic approach study of invasion by *Centaurea maculosa*, a species native to the Mediterranean region, western North American plant communities, through the mediation of the phytotoxic compound catechin. According to this proposal, invasive alien species could gain advantages in the invaded environments by producing secondary metabolites with allelopathic activity, for which the native plants and soil microbiota of the new environment would not have defenses. In this way, invasive plant species would gain a competitive advantage over native ones, placing allelopathy as an important dominance mechanism [38–40].

Since then, several studies have been published indicating allelopathy as a mechanism of dominance in plants with invasive behavior. For example, Nielsen and colleagues [41] determined the importance of allelopathy in the spatial pattern of *Thymus vulgaris* dominance in New Zealand, through the production of terpene-rich essential oils modulated by abiotic factors. Fabbro and Prati [42] determined the importance of the "allelopathy legacy" by analyzing the continuity of the allelopathic effect over time of 11 invasive donor species for compounds still present in the soil after their manual removal. In a recent literature review, Kalisz et al. [31] concluded that allelopathy is a ubiquitous dominance mechanism of invasive species, concluding that most of the 524 analyzed species, and more than 72% of the 113 plant families of which they are part of, produce substances with defined allelopathic potential. Only in the case of monilophyte (ferns) species, do we have the involvement of allelopathy in the formation of monospecific banks of *Gleichenia japonica* (Gleicheniaceae) in Southeast Asia, mediated by the production and release of glycosylated diterpenoids [43]; as well as the formation of monospecific banks of *Dicranopteris linearis* (Gleicheniaceae) in mountainous terrain in Takamatsu, Japan, mediated by condensed tannins produced by these plants [44].

4 *Pteridium* as Problem Species: Behavior and Impacts

Pteridium arachnoideum (Kaulf.) Maxon., also known as the neotropical bracken fern, is a species belonging to the monilophytes clade, which comprises ferns (formerly classified as Pteridophytes) and other relatives [45, 46]. Bracken fern is a common name in use to refer to different species of the *Pteridium* genus, Dennstaedtiaceae family, cosmopolitan heliophyte ferns present in all continents except for Antarctica, being one of the most abundant plant taxa on the planet [45, 47–49].

Plants of the *Pteridium* genus show high phenotypic plasticity, with many morphological intermediates between definable morphotypes that can hybridize with each other, even over long distances. Until recently, this genus was considered monotypic, that is, made up of a single species: *Pteridium aquilinum*. This single species was divided into two subspecies (*Pteridium aquilinum* ssp. *aquilinum* and *Pteridium aquilinum* ssp. *caudatum*), each one with different varieties. For the subspecies ssp. *aquilinum*, eight European varieties have been recognized, while for the subspecies ssp. *caudatum* four varieties, found in South and Central America,

Asia, and Oceania are known. Among these subspecies and varieties, there was *Pteridium aquilinum* ssp. *caudatum* var. *arachnoidum* [48, 49]. The growing need to review the *Pteridium* genus species complex culminated in the proposal of an infrageneric taxonomic organization currently in use, based on analyzes of chromosomic ploidy and chloroplast phylogeny, in which *Pteridium arachnoideum* is considered an independent species [48] distributed in eastern South America, extending throughout the southern and southeastern regions of Brazil, native to two of the main Brazilian biomes, Atlantic Forest and Brazilian Cerrado (Tropical Savanna) [46].

Up till now, there is a lot of confusion in the literature about the correct use of these plants' classification, with the popular term bracken fern being used to refer to all the different *Pteridium* species. In addition, there is also much discussion about the consequences of these plants' dominance on ecosystems, and they are often referred to as invasive plants. Thus, it is important to clarify that for the present work, the most recent nomenclature was adopted, although many references in the literature on the biology, ecology, and phytochemistry of the species have been published according to the classification into subspecies and varieties.

In addition to the confusion regarding their biological classification, bracken ferns also stand out for being problematic for the conservation and management of natural areas. *Pteridium* taxa represent a challenge for the management and conservation of biodiversity around the world, as they become dominant in degraded or anthropic-impacted areas [50–53], causing significant environmental changes in the areas where they are established [54–56]. Among the environmental impacts where they become dominant are changes in soil physical-chemical parameters, and cycling and nutrient availability [54, 57, 58]. A large amount of litter they produce acts as fuel biomass, contributing to the occurrence of large, high-temperature fires [56], which they resist through their robust and resistant underground rhizome system [59, 60]. These plants also represent a challenge in ecological restoration areas, as they interfere in the composition of the seed bank [51], in the regrowth of interest species [56], in the course and result of secondary ecological succession [58], and forest restoration dynamics [52]. In Brazil, *P. arachnoideum* is described as an aggressive pioneer species that is difficult to manage, invading and dominating abandoned or recently burnt areas [49, 61].

Ferns are known for being the first plants to colonize an environment in primary or secondary ecological succession, for the formation of monospecific banks, and for containing allelochemicals that allow them to dominate and expand in the environments in which they became established, especially after anthropic impacts, such as fire [33, 37, 62]. Although *P. arachnoideum* has many characteristics associated with the success of invasive plants, such as rapid growth, high production of upper soil biomass, high production of buds and seedlings, soil surface litter coverage, providing space and light competition [45], and the formation of extensive monospecific banks after fires or soil-use changes [36], these plants are native to the environments in South America were they became problematic and, therefore, do not fit into the new weapons hypothesis to explain allelopathy as one of the mechanisms of their dominance. Although recent, studies on biological invasions have given rise to a well-established research field, which perceives invasive species as those that, once established, reproduce and spread rapidly through the environment, persisting as dominant species and causing important ecological, environmental, and socioeconomic impacts. To meet the invasive species criterion, it must also be non-native and introduced by human activity into the new environment. This limitation of origin for the definition of an invasive species ignores the potential that some native species have to cause deep ecological and socioeconomic impacts when they become super-dominant due to the same anthropogenic mechanisms that allow biological invasions, such as habitat fragmentation and wildfires. These create niche opportunities or vacancies through the loss or decline of other native species, facilitating the reproduction, survival, and biomass production by species with morpho-physiological characteristics that allow them to become dominant and cause ecosystem structural and functional changes, even without breaking any biogeographical barriers [63–65]. In addition to the functional aspect, defining whether a species is native or exotic can prove to be a challenge, especially for species that have been introduced long ago, or have a cosmopolitan distribution, as is the case of *Pteridium* species. Even from the new weapons hypothesis point of view, it can be argued that under these conditions an exotic species cannot remain new in the environment for long. Although the co-evolutionary process can lead to the naturalization of an exotic species, which then loses its novelty status, in general, when we analyze the characteristics necessary for a species to succeed and become dominant in a given environment, there are no significant differences between native and exotic species [64, 66]. However, the differentiation of biogeographic origin leads to less attention being given to the problem, of super-dominant or native invasive species. This happens due to poor academic discussion on the subject, leading to a poor design of management strategies and specific control policies [65]. These issues, besides the closed position of the invasive alien species field regarding the criticisms discussed here [67], may be a possible explanation for the invasive alien species title attributed to bracken fern species in the literature, despite the absence of human introduction records and available literature of phylogenetic studies attesting these plants' native status in the studied environments, adding to the confusion of systematics and nomenclature. Despite being a native species to the Brazilian Cerrado and Atlantic Forest, *P. arachnoideum* causes important environmental changes in the areas where it becomes dominant, characterizing an invasive behavior for which it is classified as an important problem species in these biomes [49, 68], which are considered biodiversity conservation hotspots [69]. Although the Brazilian Cerrado and the Atlantic Forest consist of two quite distinct biomes under structural and climatic terms, *P. arachnoideum* occurs in both, forming monospecific patches and becoming super-dominant after fires [70, 71]. Its effects on these biomes include seed bank depletion [45], interference with tree species establishment [61], and reduction of secondary ecological succession after a wildfire in the Atlantic Forest [37]. In the Brazilian Cerrado, although there have been observations of interference in the vegetation structure [59], soil cover by *P. arachnoideum* litter did not cause changes in the seed bank [72].

Although the new weapons hypothesis is not adequate to explain the success of *P. arachnoideum* in the areas where it becomes invasive, allelopathy may contribute to the dominance of this species in addition to the effects caused by its large biomass production, which serves as a physical barrier that prevents seed rain from reaching the soil and impairs seedling development by shading [45, 59, 73]. According to the stress gradient hypothesis, the strength of the allelopathic interaction between plants and its impact on the competitive fitness of the donor species varies according to environmental biotic and abiotic stress gradients. This is because of the production of many secondary metabolites of plants, including those with allelopathic potential, increases with the increment of environmental stress [3]. For example, Knuesting and colleagues [74] observed a change in the competitive fitness of *P. arachnoideum* along an environmental stress gradient in the tropical Andes. Along an altitudinal gradient, these authors observed that the competitive power of *P. arachnoideum* over *Setaria sphacelata*, a planted forage grass and main competitor of bracken fern in tropical pasture areas in the Ecuadorian Andes, increased with altitude, correlated with the increase of ultraviolet radiation (UV) protective flavonoids present in the fronds of these plants, which led *P. arachnoideum* to outgrow the grass in altitudes over 1800 m. Condensed tannins present in the fronds of *P. arachnoideum*, which are flavonoid oligopolymers also related to UV radiation protection, were identified as allelochemicals of this species [30]. Thus, under the conditions of biotic and abiotic stress imposed in areas under ecological succession process after anthropic impacts, such as fires, the ability to produce and release substances with phytotoxic activity can be an important trait to the dominance of a native plant with invasive behavior, as is the case to *P. arachnoideum*.

5 Phytochemistry and Allelopathy of Bracken Ferns

Many studies on the phytochemistry of *Pteridium* species were carried out due to the pharmaceutical and toxicological interests in the several secondary metabolites with different biological activities it produces. Bracken has the greatest diversity of secondary metabolism compounds among all ferns [60, 75]. This interest was brought about, especially by the toxic and carcinogenic action of these plants' consumption by mammals, especially cattle and horses. Animals that feed on these plants, especially their green fronds, have different syndromes, such as thiamine deficiency, acute poisoning, progressive retinal degeneration, and digestive tract cancer [49, 76, 77].

Among the bioactive compounds isolated and identified from these plants, there is a wide variety of phenolic compounds, such as tannins with carcinogenic activity in rats, causing the appearance of bladder cancer under laboratory conditions [78, 79]; flavonoids, such as quercetin, with carcinogenic activity, also observed in rats, causing intestinal and bladder cancer [80]; pterosins and pterosides with cytotoxic and carcinogenic activity under in vivo and in vitro conditions [81–83]. Among all, sesquiterpene ptaquiloside is considered the main secondary

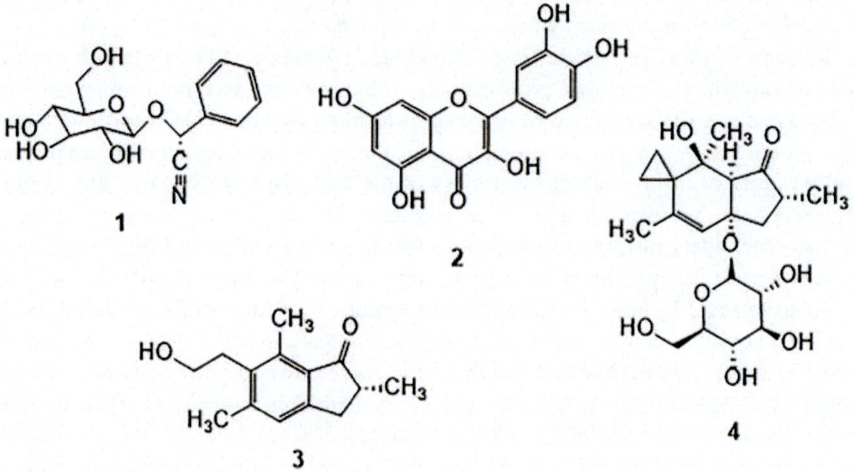

Fig. 3 Main bioactive secondary metabolites from *Pteridium* spp: prunasin, a cyanogenic glycoside (1), quercetin (2), pterosin B (3), and ptaquiloside (4)

metabolite of bracken responsible for the carcinogenic activity of alimentary *P. aquilinum* consumption by mammals. Its action mechanism is well-established, and it occurs through structural changes in the DNA chain by the alkylation of the nitrogenous base adenine, causing H-*ras* mutations [49, 84] (Fig. 3).

Specifically for *Pteridium arachnoideum*, similar bioactivities were observed in studies carried out in recent decades in Brazil. However, there was no isolation nor identification of the bioactive compounds responsible for the specific effects observed, so they have been generally attributed to the ptaquiloside present in these plants. The observation of cattle poisoning, chromosomal aberrations, and carcinogenicity promoted by *P. arachnoideum* in mammals, under field and laboratory conditions, raises concerns about the risk to human health, as bracken croziers are used in the traditional cuisine in Southeast Brazil [75].

Despite the effort to understand biological, ecological, and toxicological aspects of the bracken fern, further studies are still needed to better understand the different biological activities of the large number of secondary compounds produced by these plants, especially regarding the ecological role they play. The literature points to the action of some of these compounds in bracken defense against insect herbivory and pathogenic microorganisms [60, 85]. For instance, insect food deterrent activity has been observed for bracken cyanogenic glycosides, tannins, and condensed tannins [86–88]. Polyphenolic compounds are the main chemical class of secondary metabolites in plants of the *Pteridium* genus [89], although they are not considered the main toxic substances produced by the Neotropical bracken. Among these compounds, condensed tannins, such as proanthocyanidins and prodelphins derivatives, are the main ones, associated with both anti-herbivory protection and protection against UV radiation [90].

5.1 Bracken Fern Allelopathy

Efforts to understand the underlying mechanisms of bracken fern super-dominance date back to the beginning of the twentieth century, and they have not yet been well understood. At first, it was thought that the main mechanism would be through direct competition, but a large number of reports of these plants' toxicity raised the possibility of allelopathy's role as an underlying dominance mechanism in bracken. Gleissman and Muller dedicated their work to demonstrating allelopathy as a mechanism of dominance in *Pteridium aquilinum*, having observed the phytotoxic activity of aqueous extracts from bracken litter over *Bromus rigidus* grass germination [91] and demonstrated through a series of field experiments and in greenhouse conditions in southern California that the dominance of this plant cannot be explained solely by competition, but also by the inhibitory action of the litter leachate on co-occurring herbs [92]. The authors concluded that this activity would be due to the presence of phenolic compounds that were not isolated and identified, most likely due to the state of the art of techniques necessary for this purpose at that time. Thus, this activity was attributed to the possible presence of caffeic acid, ferulic acid, and *p*-coumaric acid in these leachates, phenolic compounds that had been previously identified in the bracken fern [92]. Allelopathy has also been suggested as a mechanism of bracken dominance in other environments, from the Pacific Northwest to Costa Rica [93].

Other authors have also worked to evaluate the allelopathic potential of the bracken fern. Phytotoxic activity on the root development of barley (*Hordeum vulgare*) was observed for solutions containing phenolic acids associated with the rhizosphere of *P. aquilinum* in New Zealand, *p*-hydroxybenzoic acid, vanillic acid, *p*-hydroxycinnamic acid, and ferulic acid [94]. Taylor and Thomson [95] analyzed the phytotoxicity of leachate from young, mature, and senescent *P. esculentum* fronds in Sydney, Australia. The selected target species were the bioindicator alfalfa (*Medigago sativa*), the native white clover (eudicot, *Trifolium repens*), and ryegrass (monocot, *Lolium perene*). Leachates from young and mature fronds caused root development inhibition on alfalfa and white clover, while leachates from senescent fronds caused root development inhibition of both white clover and ryegrass. More recently, the allelopathic potential of aqueous solutions of green fronds of *P. aquilinum* was observed on different bioindicator plant species, showing a reduction in root size [16]. Inhibition of germination and root length of economically important weeds was also observed under laboratory and greenhouse conditions, using extracts and dry residues of *P. aquilinum* (aerial parts and rhizome). The most abundant secondary metabolites in the extracts used were flavonoids, tannins, and polyphenols [96]. Specifically for *P. arachnoideum*, Matos and Belinato [61] observed inhibition of germination and initial development of pioneer and secondary Brazilian Cerrado tree species exerted by leachate from green fronds, also with inhibition of root length. All these results suggest that allelopathy is an important feature as an underlying mechanism for the super-dominance and invasive behavior of bracken fern worldwide, due to root development inhibition of neighboring plants mediated by phenolic compounds. However, no compounds with phytotoxic activity or any other allelopathy-related activity have been adequately and individually isolated, identified, and tested for *P. arachnoideum* or any of the other bracken species.

5.2 Selliguein A as *Pteridium arachnoideum* Allelochemical

Almost a century after the first experiments attesting to the phytotoxicity of secondary metabolites of bracken fern [91], Jatoba and collaborators [30] managed to isolate, identify, and establish the mode of action of bracken allelochemical from extracts of green fronds and leaf litter of *P. arachnoideum*, using the bio-guided isolation method with etiolated wheat coleoptiles. This allelochemical, the proanthocyanidin selliguein A (Fig. 4), was identified as the major secondary metabolite in the green fronds and leaf litter of this plant. The same substance was also observed as the major compound in the soil solution under *P. arachnoideum* patches.

Proanthocyanidins are condensed tannins, three-ring flavonoid oligopolymers with C-C bonds, characterized by the monomers catechin and epi-catechin and two hydroxyl radicals attached to the B ring [97]. Tannins are considered the most important group of biochemical substances produced by vascular plants, representing a significant part of terrestrial carbon biomass. As they are energy-intensive substances, they are believed to play an important role in the functioning and evolution of producing plants, with important activity related to protection against UV radiation [89, 90]. Several biological activities have been described for proanthocyanidins or materials rich in these compounds, such as antioxidant, anti-inflammatory, and antibiotic activity [98–100]. From an ecological point of view, defensive activity against herbivory [101], influence on litter decomposition

Fig. 4 Selliguein A

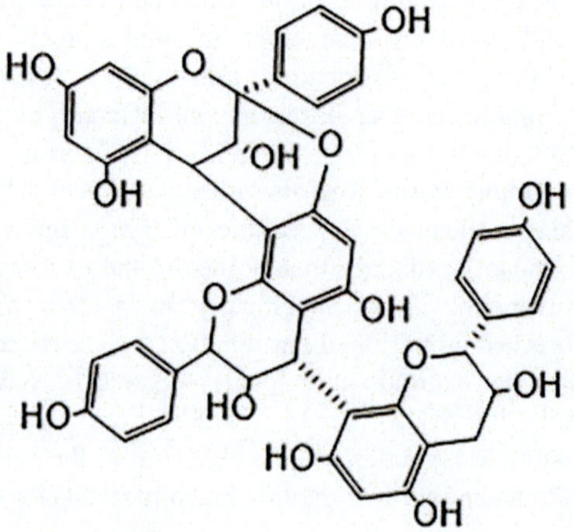

[102, 103], nutrient cycling, and soil microbial activity [104], among other activities, were observed for condensed tannins.

As an allelochemical, the phytotoxic activity of selliguein A isolated from the Neotropical bracken fern was observed in a dose-response relationship over sesame (*Sesamum indicum*) early growth, chosen as a bioindicator species, under laboratory conditions. Sesame seedlings growing in the presence of selliguein A presented root and stem growth inhibition at concentrations between 1 and 0.03 mM. The suggested mode of action for the activity observed, especially on root development, would be an interference in metaxyllem cell elongation, since metaxyllem cell length inhibition was observed between concentrations of 1 and 0.3 mM, with $IC50 = 0.98$ mM [30]. Analyses of selliguein A presence in soil solution under *P. arachnoideum* patches showed that this substance was present in an unaltered state at concentrations between 0.106 mM and 0.648 mM, an amount sufficient to potentially act on co-occurring species, favoring bracken dominance. Furthermore, selliguein A presence in soil solution would be positively correlated with litter biomass [105]. Therefore, selliguein A is first produced by *P. arachnoideum* in its green fronds. Seliguein A is a proanthocyanidin, a condensed tannin. Tannins are derived from cinnamic acid via the shikimic acid pathway, the biochemical pathway that represents the major link between primary and secondary metabolism in plants [8]. In this route, from the precursor phenylalanine, cinnamic acid is formed and then chalcone, which serves as a base for flavonoids and proanthocyanidins biosynthesis [8, 106] (Fig. 5).

The status of the major secondary metabolite in green fronds of *P. arachnoideum* remains unaltered for selliguein A during the senescence process, and it is still the major secondary compound present in the litter. As in the case of other polyphenolic compounds, biosynthesis of proanthocyanidins is induced mainly by biotic and abiotic stresses, such as herbivory, pathogens, soil salinity and pH, and UV radiation [8]. However, considering the metabolic cost of these compounds, the presence of selliguein A in leaf litter, maintained in high amounts in organs that no longer require protection from herbivory or UV radiation, suggests other important roles for this substance, as an allelochemical [30]. From bracken litter, selliguein A is released and kept unchanged in the soil, again as the major compound. Considering the different mechanisms of allelochemicals release to the environment, leaching of green fronds and litter, and litter decomposition are the most likely means involved in the release of selliguein A by *P. arachnoideum* [7]. Thus, by producing, maintaining, and releasing a phytotoxic secondary metabolite, selliguein A, mainly from its leaf litter, *Pteridium arachnoideum* uses allelopathy as one of its dominance mechanisms.

5.3 Bracken Fern Allelopathy and Its Consequences on the Restauration of Invaded Areas

Efforts made to control *Pteridium* invasion need to take into account the importance of litter removal as a source of allelochemicals, which is supported by evidence provided in control and ecological restoration experiments in areas invaded by the

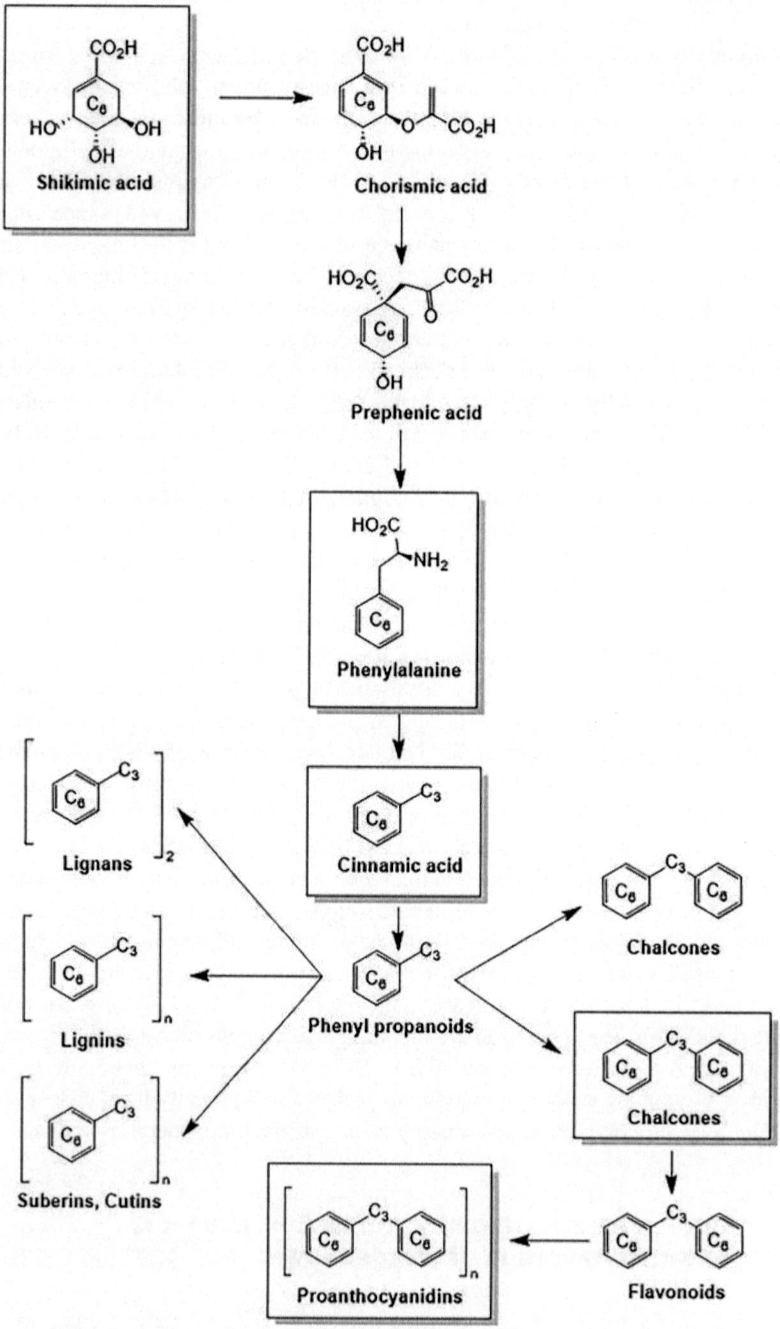

Fig. 5 Phenolic compounds synthesized by the shikimic acid pathway

Neotropical bracken fern. For instance, in the Atlantic Forest, the effects of *P. arachnoideum* on seed bank composition [45] may be associated with the mechanical effect of litter cover. However, in addition to it, the release of the allelochemical selliguein A from bracken litter suggests that its removal may also be associated with the higher species richness obtained when this method is used for invasiveness control and recovery of invaded areas [53]. A similar effect was also observed in the Brazilian Cerrado, where changes in the vegetation structure were observed upon *P. arachnoideum* invasion, even without any mechanical litter effect on seed bank composition [59]. In this biome, the best restoration strategy proved to be the removal of green fronds and litter associated with planting drought-tolerant tree species [107]. Similar results were also achieved for other bracken species in different geographic contexts. The cover and depth of the bracken litter layer influenced the seed bank composition in areas characterized by herbaceous and shrubby vegetation at acid grasslands and heathlands in the United Kingdom, acting as a barrier that interferes with germination or prevents seeds from reaching the soil [51], while experiments carried out in the United Kingdom and Norway have shown that, in the long-term, mechanical cutting performed twice a year and leaf litter removal are the best approaches to control *P. aquilinum* and promote greater species richness [52, 53, 108]. With this, comes to light the need to establish whether the phenotypic plasticity of the *Pteridium* genus could influence its chemical composition, with special attention to the condensed tannins present in the green fronds and leaf litter [48, 49]. This knowledge could lead to the development of more efficient bracken control protocols for ecological restoration and biodiversity conservation.

6 Conclusions

There is still a long way ahead to fully understand the role of bracken fern allelochemicals on this plant's dominance and invasiveness. In addition to establishing the activity of selliguein A on the development of natural co-occurring target species under laboratory and field conditions, further evidence on the environmental release modes, activity mechanisms on target species, and the dynamics of this allelochemical in the soil and over its microbiota, many other important questions are still open.

In the first studies trying to identify bracken fern allelochemicals, Gliessman and Miller [91, 92] pointed to derivatives of cinnamic acid, present in the soil under *Pteridium aquilinum* patches, as likely responsible for this plant allelopathic, although they were not able to isolate and identify the molecular structures of those compounds. Selliguein A, a cinnamic acid derivative with phytotoxic activity, was identified in soil solution under *Pteridium arachnoideum* [30]. Could seliguein A be a polyphenolic compound ubiquitous to this species complex in its wide geographic distribution or would it be restricted only to the Neotropical bracken in Southwest Brazil? Would the allelopathic potential observed in other bracken species in different geographical locations be produced by other condensed tannins or polyphenols from the same biosynthetic pathway as proanthocyanidins or by metabolites of completely different chemical classes?

Considering that proanthocyanidins biosynthesis is induced mainly by biotic and abiotic stress [8] and that the concentration of condensed tannins of *P. arachnoideum* varies according to abiotic stress gradients, such as UV radiation [74], the evaluation of fluctuations on selliguein A concentration in green fronds, litter, and soil in the face of different sources and gradients of environmental stress, such as salinity and soil pH or water availability, may corroborate the role of allelopathy as a mechanism of the dominance of native species with invasive behavior according to the stress gradient hypothesis [3]. With the significant accumulation of selliguein A in the soil solution [30] and the observation of changes in soil physical-chemical parameters, cycling, and nutrient availability in areas dominated by bracken fern [54, 57, 58], could this or other allelochemicals produced by these plants be responsible for the observed effects? Would this action take place directly or indirectly through interactions with soil microbiota?

The intention here was not to propose an exhaustive list of future research possibilities on the ecology and allelopathy of the bracken fern. The raised points are just some examples of how much we can still advance and how much remains to be learned about one of the most common and aggressive plants in the world, from the point of view of its allelochemicals.

References

1. Rice EL (1984) Allelopathy. Academic, New York
2. Macías FA, Mejías FJ, Molinillo JM (2019) Recent advances in allelopathy for weed control: from knowledge to applications. Pest Manag Sci 75:2413–2436. https://doi.org/10.1002/ps.5355
3. Hierro JL, Callaway RM (2021) The ecological importance of allelopathy. Annu Rev Ecol Evol Syst 52:25–45. https://doi.org/10.1146/annurev-ecolsys-051120-030619
4. Molisch H (2001) The influence of one plant on another: allelopathy. Scientific Publishers, Jodhpur
5. Rizvi SJH, Tahir M, Rizvi V, Kohli RK, Ansari A (2010) Allelopathic interactions in agroforestry systems. 18:773–796. https://doi.org/10.1080/07352689991309487
6. IAS (2021) What is allelopathy? In: International allelopathy society. Available at https://allelopathy-society.osupytheas.fr/about/. Accessed 23 Oct 2022
7. Macías FA, Durán AG, Molinillo JMG (2020) Allelopathy: the chemical language of plants. In: Kinghorn AD, Falk H, Gibbons S, Kobayashi J, Asakawa Y, Liu JK (eds) Progress in the chemistry of organic natural products. Springer, Cham
8. Santos-Sánchez NF, Salas-Coronado R, Hernández-Carlos B, Villanueva-Cañongo C, Santos-Sánchez NF, Salas-Coronado R, Hernández-Carlos B, Villanueva-Cañongo C (2019) Shikimic acid pathway in biosynthesis of phenolic compounds. In: Soto-Hernández M, García-Mateos R, Palma-Tenango M (eds) Plant physiological aspects of phenolic compounds. IntechOpen, London
9. Anulika NP, Ignatius EO, Raymond ES, Osasere OI, Abiola AH (2016) The chemistry of natural product: plant secondary metabolites. Int J Technol Enhancements Emerg Eng Res 4:1–9
10. Li ZH, Wang Q, Ruan X, de Pan C, Jiang DA (2010) Phenolics and plant allelopathy. Molecules 15:8933–8952. https://doi.org/10.3390/MOLECULES15128933

11. Lotina-Hennsen B, King-Diaz B, Aguilar MI, Hernandez Terrones MG (2006) Plant secondary metabolites. Targets and mechanisms of allelopathy. In: Reigosa M, Pedrol N, González L (eds) Allelopathy. Springer, Dordrecht
12. Bonanomi G, Zotti M, Idbella M, Mazzoleni S, Abd-ElGawad AM (2021) Microbiota modulation of allelopathy depends on litter chemistry: mitigation or exacerbation? Sci Total Environ 776:145942. https://doi.org/10.1016/J.SCITOTENV.2021.145942
13. Lambers H, Oliveira RS (2019) Biotic influences: ecological biochemistry: allelopathy and defense against herbivores. Plant Physiol Ecol 541–581. https://doi.org/10.1007/978-3-030-29639-1_13
14. Qu T, Du X, Peng Y, Guo W, Zhao C, Losapio G (2021) Invasive species allelopathy decreases plant growth and soil microbial activity. PLoS One 16:e0246685. https://doi.org/10.1371/JOURNAL.PONE.0246685
15. Kraus TEC, Dahlgren RA, Zasoski RJ (2003) Tannins in nutrient dynamics of forest ecosystems – a review. Plant Soil 256:41–66. https://doi.org/10.1023/A:1026206511084
16. Wang HH, Chen BJ, Hsu LM, Cheng YM, Liou YJ, Wang CY (2012) Allelopathic effects of bracken fern (*Pteridium aquilinum* L. Kuhn) in Taiwan. Allelopath J 27:97–110
17. Inderjit, Duke SO (2003) Ecophysiological aspects of allelopathy. Planta 217:529–539. https://doi.org/10.1007/s00425-003-1054-z
18. Silva ER, Overbeck GE, Soares GLG (2017) Something old, something new in allelopathy review: what grassland ecosystems tell us. Chemoecology 27:217–231. https://doi.org/10.1007/S00049-017-0249-X/FIGURES/5
19. Scavo A, Mauromicale G (2021) Crop Allelopathy for sustainable weed management in agroecosystems: knowing the present with a view to the future. Agronomy 11:2104. https://doi.org/10.3390/AGRONOMY11112104
20. Hanada K, Manghwar H, Zaman W, Ullah R, Aslam Z, Attia H, Sultan K, Alamer KH, Mansha MZ, Althobaiti AT, Amin N, al Kashgry T, Algethami B, Uz Zaman Q (2022) Sorghum allelopathy: alternative weed management strategy and its impact on mung bean productivity and soil rhizosphere properties. Life 12:1359. https://doi.org/10.3390/LIFE12091359
21. Başaran F (2020) Ecological aspects of allelopathy. Int J Agric For Life Sci 5:80–86
22. Sinkkonen A (2006) Ecological relationships and allelopathy. In: Reigosa MJ, González L, Pedrol N (eds) Allelopathy. Springer, Dordrecht
23. Blair AC, Weston LA, Nissen SJ, Brunk GR, Hufbauer RA (2008) The importance of analytical techniques in allelopathy studies with the reported allelochemical catechin as an example. Biol Invasions 11:325–332. https://doi.org/10.1007/s10530-008-9250-1
24. Duke SO (2015) Proving allelopathy in crop–weed interactions. Weed Sci 63:121–132. https://doi.org/10.1614/WS-D-13-00130.1
25. Romeo JT (2011) Raising the beam: moving beyond phytotoxicity. J Chem Ecol 26:2011–2014
26. Colegate SM, Molyneux RJ (2008) Bioactive natural products: detection, isolation, and structural determination, 2nd edn. CRC Press, Boca Raton
27. Inderjit, Callaway RM (2003) Experimental designs for the study of allelopathy. Plant Soil 256:1–11
28. Matsumoto S, Varela RM, Palma M, Molinillo JMG, Lima IS, Barroso CG, Macías FA (2014) Bio-guided optimization of the ultrasound-assisted extraction of compounds from *Annona glabra* L. leaves using the etiolated wheat coleoptile bioassay. Ultrason Sonochem 21:1578–1584. https://doi.org/10.1016/j.ultsonch.2014.01.024
29. Cala A, Salcedo JR, Torres A, Varela RM, Molinillo JMG, Macías FA (2021) A study on the phytotoxic potential of the seasoning herb marjoram (*Origanum majorana* L.) leaves. Molecules 26:3356. https://doi.org/10.3390/molecules26113356
30. Jatoba LJ, Varela RM, Molinillo JMG, Din ZU, Gualtieri SCJ, Rodrigues-Filho E, Macías FA (2016) Allelopathy of bracken fern (*Pteridium arachnoideum*): new evidence from green fronds, litter, and soil. PLoS One 11:e0161670. https://doi.org/10.1371/JOURNAL.PONE.0161670

31. Kalisz S, Kivlin SN, Bialic-Murphy L (2021) Allelopathy is pervasive in invasive plants. Biol Invasions 23:367–371. https://doi.org/10.1007/S10530-020-02383-6/FIGURES/1
32. Adedeji AA, Babalola OO (2020) Secondary metabolites as plant defensive strategy: a large role for small molecules in the near root region. Planta 252:1–12. https://doi.org/10.1007/S00425-020-03468-1/TABLES/2
33. Kato-Noguchi H (2015) Involvement of allelopathy in the formation of monospecific colonies of ferns. Nat Prod Commun 10:811–814. https://doi.org/10.1177/1934578X1501000526
34. Orrock JL, Dutra HP, Marquis RJ, Barber N (2015) Apparent competition and native consumers exacerbate the strong competitive effect of an exotic plant species. Ecology 96:1052–1061. https://doi.org/10.1890/14-0732.1
35. Greer MJ, Wilson GWT, Hickman KR, Wilson SM (2014) Experimental evidence that invasive grasses use allelopathic biochemicals as a potential mechanism for invasion: chemical warfare in nature. Plant Soil 385:165–179. https://doi.org/10.1007/s11104-014-2209-3
36. Suazo-Ortuño I, Lopez-Toledo L, Alvarado-Díaz J, Martínez-Ramos M (2015) Land-use change dynamics, soil type and species forming mono-dominant patches: the case of *Pteridium aquilinum* in a neotropical rain forest region. Biotropica 47:18–26. https://doi.org/10.1111/btp.12181
37. Carvalho TF, Carvalho AC, Zanuncio JC, de Oliveira MLR, Machado ELM, José AC, Santos JB, Pereira IM (2022) Does invasion by *Pteridium aquilinum* (Dennstaedtiaceae) affect the ecological succession in Atlantic Forest areas after a fire? Environ Sci Pollut Res 29:14195–14205. https://doi.org/10.1007/S11356-021-16761-7/FIGURES/4
38. Callaway RM, Ridenour WM (2004) Novel weapons: invasive success and the evolution of increased competitive ability. Front Ecol Environ 2:436. https://doi.org/10.2307/3868432
39. Bais HP, Vepachedu R, Gilroy S, Callaway RM, Vivanco JM (2003) Allelopathy and exotic plant invasion: from molecules and genes to species interactions. Science 301:1377–1380. https://doi.org/10.1126/science.1083245
40. Inderjit, Wardle DA, Karban R, Callaway RM (2011) The ecosystem and evolutionary contexts of allelopathy. Trends Ecol Evol 26:655–662. https://doi.org/10.1016/j.tree.2011.08.003
41. Nielsen JA, Whigham PA, Frew RD, Callaway RM, Dickinson KJM (2013) Invasion essentials: does secondary chemistry plasticity contribute to the invasiveness of *Thymus vulgaris* L.? Chemoecology 24:15–27. https://doi.org/10.1007/s00049-013-0142-1
42. Fabbro C, Prati D (2015) The relative importance of immediate allelopathy and allelopathic legacy in invasive plant species. Basic Appl Ecol 16:28–35. https://doi.org/10.1016/j.baae.2014.10.007
43. Kato-Noguchi H, Saito Y, Ohno O, Suenaga K (2013) Allelopathy is involved in the formation of pure colonies of the fern *Gleichenia japonica*. J Plant Physiol 170:577–582. https://doi.org/10.1016/j.jplph.2012.11.015
44. Kato-Noguchi H, Saito Y, Suenaga K (2012) Involvement of allelopathy in the establishment of pure colony of *Dicranopteris linearis*. Plant Ecol 213:1937–1944. https://doi.org/10.1007/s11258-012-0096-3
45. Silva USR, Matos DMS (2006) The invasion of *Pteridium aquilinum* and the impoverishment of the seed bank in fire prone areas of Brazilian Atlantic Forest. Biodivers Conserv 15:3035–3043. https://doi.org/10.1007/s10531-005-4877-z
46. Schwartsburd PB, de Moraes PLR, Lopes-Mattos KLB (2014) Recognition of two morphotypes in eastern South American brackens (*Pteridium*-Dennstaedtiaceae-Polypodiopsida). Phytotaxa 170:103–117. https://doi.org/10.11646/phytotaxa.170.2.3
47. Alonso-Amelot ME (2002) The chemistry and toxicology of bioactive compounds in bracken fern (*Pteridium* spp), with special reference to chemical ecology and carcinogenesis. In: Atta-ur Rahman (ed) Studies in natural products chemistry, vol 26. Elsevier, Berlin
48. Der JP, Thomson JA, Stratford JK, Wolf PG (2009) Global chloroplast phylogeny and biogeography of bracken (*Pteridium*; Dennstaedtiaceae). Am J Bot 96:1041–1049. https://doi.org/10.3732/ajb.D0800333

49. Gil da Costa RM, Bastos MMSM, Oliveira PA, Lopes C (2012) Bracken-associated human and animal health hazards: chemical, biological and pathological evidence. J Hazard Mater 203–204:1–12. https://doi.org/10.1016/j.jhazmat.2011.12.046
50. Royo AA, Carson WP (2006) On the formation of dense understory layers in forests worldwide: consequences and implications for forest dynamics, biodiversity, and succession. Can J For Res 36:1345–1362. https://doi.org/10.1139/X06-025
51. Ghorbani J, le Duc MG, McAllister HA, Pakeman RJ, Marrs RH (2006) Effects of the litter layer of *Pteridium aquilinum* on seed banks under experimental restoration. Appl Veg Sci 9: 127–136
52. Maren I, Vandvik V, Ekelund K (2008) Restoration of bracken-invaded *Calluna vulgaris* heathlands: effects on vegetation dynamics and non-target species. Biol Conserv 141:1032–1042. https://doi.org/10.1016/j.biocon.2008.01.012
53. Stewart G, Cox E, le Duc M, Pakeman R, Pullin A, Marrs R (2008) Control of *Pteridium aquilinum*: meta-analysis of a multi-site study in the UK. Ann Bot 101:957–970. https://doi.org/10.1093/aob/mcn020
54. Griffiths RP, Filan T (2007) Effects of bracken fern invasions on harvested site soils in Pacific Northwest (USA) Coniferous Forests. Northwest Sci 81:191–198
55. Olivares E, Peña E, Benítez M (2007) *Pteridium caudatum* (L.) Maxon behaves as a potassium plant and accumulates aluminum in the subterranean organs. Am Fern J 97:81–94
56. Adie H, Richert S, Kirkman KP, Lawes MJ (2011) The heat is on: frequent high intensity fire in bracken (*Pteridium aquilinum*) drives mortality of the sprouting tree *Protea caffra* in temperate grasslands. Plant Ecol 212:2013–2022. https://doi.org/10.1007/s11258-011-9945-8
57. DeLuca TH, Zewdie SA, Zackrisson O, Healey JR, Jones DL (2012) Bracken fern (*Pteridium aquilinum* L. kuhn) promotes an open nitrogen cycle in heathland soils. Plant Soil 367:521–534. https://doi.org/10.1007/s11104-012-1484-0
58. Baptiste AJ, Macario PA, Islebe GA, Vargas-Larreta B, Pool L, Valdez-Hernández M, López-Martínez JO (2019) Secondary succession under invasive species (*Pteridium aquilinum*) conditions in a seasonal dry tropical forest in southeastern Mexico. Peer J 2019:e6974. https://doi.org/10.7717/PEERJ.6974/SUPP-1
59. Miatto RC, Silva IA, Silva-Matos DM, Marrs RH (2011) Woody vegetation structure of Brazilian Cerrado invaded by *Pteridium arachnoideum* (Kaulf.) Maxon (Dennstaedtiaceae). Flora – Morphol Distrib Funct Ecol Plants 206:757–762. https://doi.org/10.1016/j.flora.2010.12.001
60. Agarwal K, Haldar S, Boland W, Venkatesan R (2018) Chemical ecology of bracken ferns. In: Nowicki L, Kowalska A (eds) Ferns: ecology, importance to humans and threats. Nova Science Publishers, Hauppauge
61. Matos DMDS, Belinato TA (2010) Interference of *Pteridium arachnoideum* (Kaulf.) Maxon. (Dennstaedtiaceae) on the establishment of rainforest trees. Braz J Biol 70:311–316
62. Ooka JK, Owens DK (2018) Allelopathy in tropical and subtropical species. Phytochem Rev 17:1225–1237. https://doi.org/10.1007/S11101-018-9596-7/TABLES/1
63. Carey MP, Sanderson BL, Barnas KA, Olden JD (2012) Native invaders – challenges for science, management, policy, and society. Front Ecol Environ 10:373–381. https://doi.org/10.1890/110060
64. Cassini MH (2020) A review of the critics of invasion biology. Biol Rev 95:1467–1478. https://doi.org/10.1111/BRV.12624
65. Pivello VR, Vieira MV, Grombone-Guaratini MT, Matos DMS (2018) Thinking about super-dominant populations of native species – examples from Brazil. Perspect Ecol Conserv 16:74–82. https://doi.org/10.1016/J.PECON.2018.04.001
66. Funk JL, Standish RJ, Stock WD, Valladares F (2016) Plant functional traits of dominant native and invasive species in mediterranean-climate ecosystems. Ecology 97:75–83. https://doi.org/10.1890/15-0974.1
67. Guiaşu RC, Tindale CW (2018) Logical fallacies and invasion biology. Biol Philos 33(5):1–24. https://doi.org/10.1007/S10539-018-9644-0

68. Pejchar L, Mooney HA (2009) Invasive species, ecosystem services and human well-being. Trends Ecol Evol 24:497–504. https://doi.org/10.1016/j.tree.2009.03.016
69. Myers N, Mittermeier RA, Mittermeier CG, da Fonseca GA, Kent J (2000) Biodiversity hotspots for conservation priorities. Nature 403:853–858. https://doi.org/10.1038/35002501
70. Menezes GSC, Cazetta E, Dodonov P (2019) Vegetation structure across fire edges in a Neotropical rain forest. For Ecol Manag 453:117587. https://doi.org/10.1016/J.FORECO.2019.117587
71. Tiberio FCS, Xavier RO, Dodonov P, Silva Matos DM (2022) Fire has short-term negative effects on a super-dominant native fern, *Pteridium arachnoideum* (Dennstaedtiaceae), in a Brazilian savanna. Nat Conserv Res 7:15–25. https://doi.org/10.24189/ncr.2022.027
72. Xavier RO, Alday JG, Marrs RH, Matos DMS (2016) The role of *Pteridium arachnoideum* (Kaulf) on the seed bank of the endangered Brazilian Cerrado. Braz J Biol 76:256–267. https://doi.org/10.1590/1519-6984.21814
73. Silva Matos DM, Xavier RO, Tiberio FCS, Marrs RH (2014) A comparative study of resource allocation in *Pteridium* in different Brazilian ecosystems and its relationship with European studies. Braz J Biol 74:156–165. https://doi.org/10.1590/1519-6984.22012
74. Knuesting J, Brinkmann MC, Silva B, Schorsch M, Bendix J, Beck E, Scheibe R (2018) Who will win where and why? An ecophysiological dissection of the competition between a tropical pasture grass and the invasive weed Bracken over an elevation range of 1000 m in the tropical Andes. PLoS One 13:e0202255. https://doi.org/10.1371/journal.pone.0202255
75. Santos MG, Kelecom A, Paiva SR, Moraes MG, Rocha L, Garrett R (2010) Phytochemical studies in pteridophytes growing in Brazil: a review. Am J Plant Sci Biotechnol 4:113–125
76. Finnie J, Windsor P, Kessell A (2011) Neurological diseases of ruminant livestock in Australia. II: toxic disorders and nutritional deficiencies. Aust Vet J 89:247–253. https://doi.org/10.1111/j.1751-0813.2011.00793.x
77. Furlan FH, Mendes ERS, Ducatti KR, Marcon GC, Dombrosky T, Amorim TM, Riet-Correa F (2014) Acute poisoning by *Pteridium arachnoideum* and *Pteridium caudatum* in cattle and distribution of the plants in Mato Grosso. Pesqui Vet Bras 34:343–348. https://doi.org/10.1590/S0100-736X2014000400008
78. Wang CY, Chiu CW, Pamukcu AM, Bryan GT (1976) Identification of carcinogenic tannin isolated from bracken fern (*Pteridium aquilinum*). J Natl Cancer Inst 56:33–36. https://doi.org/10.1093/jnci/56.1.33
79. Pamukcu AM, Wang CY, Hatcher J, Bryan GT (1980) Carcinogenicity of tannin and tannin-free extracts of bracken fern (*Pteridium aquilinum*) in rats. J Natl Cancer Inst 65:131–136. https://doi.org/10.1093/jnci/65.1.131
80. Pamukcu AM, Yalciner S, Hatcher JF, Bryan GT (1980) Quercetin, a rat intestinal and bladder carcinogen present in bracken fern (*Pteridium aquilinum*). Cancer Res 40:3468–3472
81. Saito M, Umeda M, Enomoto M, Hatanaka Y, Natori S, Yoshihira K, Fukuoka M, Kuroyanagi M (1975) Cytotoxicity and carcinogenicity of pterosins and pterosides, 1-indanone derivatives from bracken (*Pteridium aquilinum*). Experientia 31:829–831. https://doi.org/10.1007/BF01938490
82. Van der Hoeven JCM, Lagerweij WJ, Posthumus MA, Veldhuizen A, Holterman HAJ (1983) Aquilide A, a new mutagenic compound isolated from bracken fern (*Pteridium aquilinum* (L.) Kuhn). Carcinogenesis 4:1587–1590. https://doi.org/10.1093/carcin/4.12.1587
83. Fletcher MT, Brock IJ, Reichmann KG, McKenzie RA, Blaney BJ (2011) Norsesquiterpene glycosides in bracken ferns (*Pteridium esculentum* and *Pteridium aquilinum* subsp. *wightianum*) from Eastern Australia: reassessed poisoning risk to animals. J Agric Food Chem 59:5133–5138. https://doi.org/10.1021/jf104267c
84. Prakash AS, Pereira TN, Smith BL, Shaw G, Seawright AA (2006) Mechanism of bracken fern carcinogenesis: evidence for *H*-ras activation via initial adenine alkylation by ptaquiloside. Nat Toxins 4:221–227. https://doi.org/10.1002/(SICI)(1996)4:5<221::AID-NT4>3.0.CO;2-Q
85. Hojo-Souza NS, Carneiro CM, dos Santos RC (2010) *Pteridium aquilinum*: what we know and what is yet to be learnt. Biosci J 26:798–808

86. Alonso-Amelot ME, Oliveros A, Calcagno MP, Arellano E (2001) Bracken adaptation mechanisms and xenobiotic chemistry. Pure Appl Chem 73:549–553. https://doi.org/10.1351/pac200173030549
87. Alonso-Amelot ME, Oliveros-Bastidas A (2005) Kinetics of the natural evolution of hydrogen cyanide in plants in neotropical *Pteridium arachnoideum* and its ecological significance. J Chem Ecol 31:315–331. https://doi.org/10.1007/s10886-005-1343-z
88. Tempel AS (1981) Field studies of the relationship between herbivore damage and tannin concentration in bracken (*Pteridium aquilinum* Kuhn). Oecologia 51:97–106. https://doi.org/10.1007/BF00344659
89. Alonso-Amelot ME, Oliveros-Bastidas A, Calcagno-Pisarelli MP (2007) Phenolics and condensed tannins of high altitude *Pteridium arachnoideum* in relation to sunlight exposure, elevation, and rain regime. Biochem Syst Ecol 35:1–10. https://doi.org/10.1016/j.bse.2006.04.013
90. Alonso-Amelot ME, Oliveros A, Calcagno-Pisarelli MP (2004) Phenolics and condensed tannins in relation to altitude in neotropical *Pteridium* spp. Biochem Syst Ecol 32:969–981. https://doi.org/10.1016/j.bse.2004.03.005
91. Gliessman SR, Muller CH (1972) The phytotoxic potential of bracken, *Pteridium aquilinum* (L.) Kuhn. Madrono 21:299–304
92. Gliessman SR, Muller CH (1978) The allelopathic mechanisms of dominance in bracken (*Pteridium aquilinum*) in Southern California. J Chem Ecol 4:337–362
93. Gliessman SR (1976) Allelopathy in a broad spectrum of environments as illustrated by bracken. Bot J Linn Soc 73:95–104. https://doi.org/10.1111/J.1095-8339.1976.TB02015.X
94. Glass ADM (1976) The allelopathic potential of phenolic acids associated with the rhizosphere of *Pteridium aquilinum*. Can J Bot 54:2440–2444. https://doi.org/10.1139/b76-259
95. Taylor JE, Thomson JA (1990) Allelopathic activity of frond run-off from *Pteridium esculentum*. In: Taylor JA, Smith RT (eds) Bracken biology and management. Australian Institute of Agricultural Science – AIAS, Sydney
96. Mira YD, Ramírez LA, Londoño LF, Castañeda DA, Mira YD, Ramírez LA, Londoño LF, Castañeda DA (2021) Allelopathic effects of common bracken (*Pteridium aquilinum* (L.) Kuhn) on germination and growth of economically important weeds. Chil J Agric Anim Sci 37:290–300. https://doi.org/10.29393/CHJAAS37-30AEYD40030
97. de la Rosa LA, Moreno-Escamilla JO, Rodrigo-García J, Alvarez-Parrilla E (2019) Phenolic compounds. In: Yahia E, Carrilo-Lopez A (eds) Postharvest physiology and biochemistry of fruits and vegetables. Woodhead Publishing, Elsevier, Amsterdam
98. Sarg TM, Abbas FA, El-Sayed ZI, Mustafa AM (2011) Two new polyphenolic compounds from *Ficus retusa* L. "variegata" and the biological activity of the different plant extracts. J Pharmacogn Phytother 3:89–100
99. Wang L, Lou G, Ma Z, Liu X (2011) Chemical constituents with antioxidant activities from litchi (*Litchi chinensis* Sonn.) seeds. Food Chem 126:1081–1087. https://doi.org/10.1016/j.foodchem.2010.11.133
100. Nakashima S, Oda C, Masuda S, Tagashira M, Kanda T (2012) Isolation and structure elucidation of tetrameric procyanidins from unripe apples (*Malus pumila* cv. Fuji) by NMR spectroscopy. Phytochemistry 83:144–152. https://doi.org/10.1016/j.phytochem.2012.07.011
101. Moctezuma C, Hammerbacher A, Heil M, Gershenzon J, Méndez-Alonzo R, Oyama K (2014) Specific polyphenols and tannins are associated with defense against insect herbivores in the tropical oak *Quercus oleoides*. J Chem Ecol 40:458–467. https://doi.org/10.1007/s10886-014-0431-3
102. Hättenschwiler S, Jørgensen HB (2010) Carbon quality rather than stoichiometry controls litter decomposition in a tropical rain forest. J Ecol 98:754–763. https://doi.org/10.1111/j.1365-2745.2010.01671.x
103. Coq S, Souquet J-M, Meudec E, Cheynier V, Hättenschwiler S (2010) Interspecific variation in leaf litter tannins drives decomposition in a tropical rain forest of French Guiana. Ecology 91:2080–2091. https://doi.org/10.1890/09-1076.1

104. Smolander A, Kanerva S, Adamczyk B, Kitunen V (2011) Nitrogen transformations in boreal forest soils – does composition of plant secondary compounds give any explanations? Plant Soil 350:1–26. https://doi.org/10.1007/s11104-011-0895-7
105. Jatoba LJ (2016) Alelopatia em *Pteridium arachnoideum* (Kaulf.) Maxon. (Dennstaedtiaceae). São Carlos Federal University, São Carlos
106. Naczk M, Shahidi F (2004) Extraction and analysis of phenolics in food. J Chromatogr A 1054:95–111. https://doi.org/10.1016/J.CHROMA.2004.08.059
107. Xavier R, Marques U, Pivello V, Marrs R, Abilleira P, Nascimento JL, Silva Matos D (2022) Combining mechanical removal and tree planting to restore montane atlantic forests dominated by the Neotropical Bracken (*Pteridium arachnoideum*). SSRN Electronic Journal. https://doi.org/10.2139/SSRN.4237683
108. Ghorbani J, le Duc MG, Mcallister HA, Pakeman RJ, Marrs RH (2007) Temporal responses of propagule banks during ecological restoration in the United Kingdom. Restor Ecol 15: 103–117

Bioactive Compounds in *Polypodium vulgare* L. (Polypodiaceae)

20

Adrià Farràs, Montserrat Mitjans, and Víctor López

Contents

A. Farràs
Department of Pharmacy, Faculty of Health Sciences, Universidad San Jorge, Villanueva de Gállego (Zaragoza), Spain

Department of Biochemistry and Physiology, Faculty of Pharmacy and Food Sciences, Universitat de Barcelona, Barcelona, Spain
e-mail: afarrama7@alumnes.ub.edu

M. Mitjans
Department of Biochemistry and Physiology, Faculty of Pharmacy and Food Sciences, Universitat de Barcelona, Barcelona, Spain
e-mail: montsemitjans@ub.edu

V. López (✉)
Department of Pharmacy, Faculty of Health Sciences, Universidad San Jorge, Villanueva de Gállego (Zaragoza), Spain
e-mail: ilopez@usj.es

H. N. Murthy (ed.), *Bioactive Compounds in Bryophytes and Pteridophytes*, Reference Series in Phytochemistry, https://doi.org/10.1007/978-3-031-23243-5_22

Abstract

Polypodium vulgare L. (*Polypodiaceae*), a fern with a wide distribution worldwide and in different continents, is also known as a polypody due to the shape of its frond as feet; since ferns are poorly studied from a phytochemical and ethnopharmacological perspective, these species may also represent a source of bioactive compounds with interesting properties for potential pharmacological, cosmetic, or agrochemical applications. *Polypodium vulgare* has been used traditionally for medicinal and food purposes. In certain regions, the young fronds are used as edible foods and traditional uses include this species as a medicinal plant with hepatoprotective, wound-healing, expectorant, or laxative properties. According to chemical studies, two main classes of phytochemicals have been detected: polyphenols and steroids. Among polyphenols, caffeoylquinic acids, such as chlorogenic acid and flavonoids, such as rutin have been quantified. Among steroids, some studies reveal the presence of osladin, a compound with sweetener capacity.

Keywords

Antioxidants · Ferns · Polyphenols · Polypod · Steroids · Traditional medicine

Abbreviations

3T3	NIH 3T3 mouse fibroblast cell line
DPPH	2,2-diphenyl-1-picrylhydrazyl
EMA	European Medicines Agency
EU	European Union
HaCaT	Spontaneously immortalized human keratinocyte cell line
HPLC-DAD	High-performance liquid chromatography-diode array detector
HPLC-MS/MS	High-performance liquid chromatography-tandem mass spectrometry
ORAC	Oxygen Radical Absorbance Capacity
PLE	Aqueous extract of *Polypodium leucotomos* fronds
PVM	Methanolic extract of *Polypodium vulgare* L. fronds
ROS	Reactive Oxygen Species
UVA	Ultraviolet A
WHO	World Health Organization

1 Introduction

Since ancient times humans have used substances from nature for the treatment and prevention of different types of disorders. Many of the current medicines used worldwide have originated from medicinal plants, and in many countries, traditional medicine is still the first line of treatment against many diseases. Within plant sciences, experimental studies based on ferns are not abundant, but some works reveal that ferns contain phytochemicals with antioxidant, antidiabetic, antiviral, antibacterial, or anti-inflammatory properties. *Polypodium vulgare* L. (*Polypodiaceae*), despite its wide distribution, has not been extensively studied; however, this chapter summarizes the main findings about the ethnobotanical, phytochemical, and pharmacological aspects of this species.

2 Botany

Three classes of *Pteridophyta* have been described: *Ophioglossopsida*, *Marattiopsida*, and *Filicopsida* or *Polypodiopsida*. They are differentiated by the type of the sporangium wall because of the development and structure of this reproductive organ by spores through meiosis, obtaining eusporangia (eusporangiate pteridophytes) and leptosporangia (leptosporangiate pteridophytes) that present a pluristrata wall (thick wall) or with a single cell stratum (thin wall), respectively. Figure 1 represents the classes of the division *Pteridophyta*, also known as true ferns [1, 2].

The *Filicopsida* class is the taxon that brings together most of the living species of pteridophytes and is characterized by being leptosporangiate ferns, mostly terrestrial, despite the existence of a small portion of aquatic ones with exceptional characteristics represented by the order *Salviniales*. The *Filicopsida* class presents trophosporophiles and isospores, except for the species of the order *Salviniales*, aquatic leptosporangiate ferns, which are heterospores.

The sporophytic generation of the terrestrial *Filicopsida* species presents herbaceous ferns distributed mainly in temperate zones, such as *Polypodium vulgare* L., up to arborescent ferns distributed mainly in the tropical zone. For example, in the

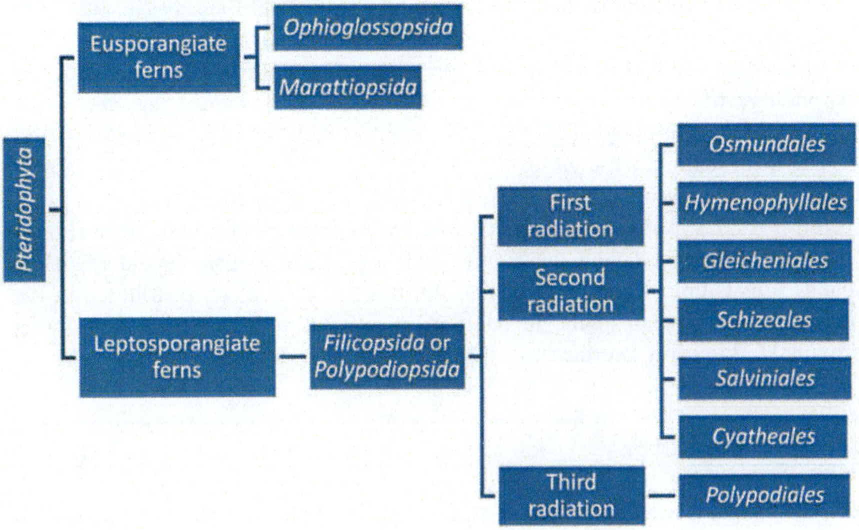

Fig. 1 Classes and orders of the *Pteridophyta* division

Canary Islands and the Northeast of the Iberian Peninsula there is one of the few European tree ferns, *Dryopteris guanchica* Gibby & Jermy (*Dryopteridaceae*), which is an allotetraploid derived from *Dryopteris aemula* (Aiton) Kuntze (*Dryopteridaceae*) and *Dryopteris intermedia* (Muhl. ex Willd.) A. Gray (*Dryopteridaceae*) [3]. Phylogenetic studies have certified the existence of three taxa of leptosporangiate ferns that differ due to the successive radiation that this class has suffered. Currently, of these three taxa, only species of the second and third generation of leptosporangiate ferns survive.

The ferns of the third radiation of the leptosporangiates, also called polypods (from the Greek multiarmed), are made up of a single order, the order *Polypodiales*. They arose at the beginning of the Cretaceous and present an incomplete ring of dehiscence of the spores that receives the name of stomium. They are classified into six suborders: *Polypodiineae*, *Aspleniineae*, *Dennstaedtiineae*, *Pteridineae*, *Lindsaeineae*, and *Saccolomatineae*. Due to the few similarities between these taxa, there is no clear classification accepted by the taxonomic organisms from the families of these suborders. Currently, thanks to molecular studies, Eupolypods are distinguished from Non-Eupolypods [4].

The suborder *Polypodiineae* (Eupolypod I) makes up a total of nine families (*Dryopteridaceae*, *Polypodiaceae*, *Didymochlaenaceae*, *Hypodematiaceae*, *Lomariopsidaceae*, *Nephrolepidaceae*, *Tectariaceae*, *Oleandraceae*, and *Davalliaceae*), while the suborder *Aspleniineae* (Eupolypod II) groups a total of 11 families (*Cystopteridaceae*, *Rhachidosoraceae*, *Diplaziopsidaceae*, *Desmophlebiaceae*, *Hemidictyaceae*, *Aspleniaceae*, *Theylpteridaceae*, *Woodsiaceae*, *Athyriaceae*, *Onocleaceae*, and *Blechnaceae*). Currently, the in chemico and in vitro activity of different ferns have been determined; however, the *Polypodiaceae* family presents few studies of the activity of its species in comparison with other families of ferns, such as *Aspleniaceae* [5].

Polypodium vulgare is the most widely distributed fern species in the world, occurring mainly in warm and humid forests [6]. The taxonomic key for the characterization-identification of the true ferns is mainly due to the structure-morphology of the frond (veining, arrangement, and divisions), as well as the structure of the sporophyte [7].

The main taxonomic keys for the identification of the frond *P. vulgare* are the frond lobe, the presence or absence of sorus indusium, and sorus shape [8]. As can be seen in Fig. 2, *P. vulgare* has a green frond with entire pinnae with an oval blade attached at their base with the rachis. Spores grouped in sori devoid of yellow rounded indusium on the underside of the frond. The sori are distributed in two parallel rows for each pinna. In addition, it presents a creeping rhizome buried proximal to the earth's surface.

3 Ethnobotany

The diversity of traditional uses of these ferns in our environment requires more studies to be able to consider these ferns as medicinal plants [10]. For this reason, ethnobotany can be an important tool in the revaluation of depopulated territories, as

Fig. 2 Fresh fronds (**a**) and desiccated frond (**b**) of *Polypodium vulgare* [9]. (© Adrià Farràs Martínez)

well as being a claim to promote environmental interest, since ferns are considered ecological environmental indicators because they are sensitive to environmental changes [11]. The transmission of its traditional uses is usually a source of discussion for various reasons, such as the use of different common names, in different regions for the same species of fern. It is the case that in the mountains of Prades (Tarragona, Spain) the *herba pigotera* (popular catalan name) corresponds to the scientific name of *Polypodium vulgare* [12], or else in translations of ancient Latin texts, such as the case of *Asplenium trichomanes* L. [13]. The selection of the plant drug is decisive in the isolation of phytochemicals, derived from the fact that each plant drug is specialized in certain plant functions. In the case of ferns, the use of all its parts as plant drugs have been described: fronds, stem/rachis, rhizome, and even the entire plant. In certain studies, only the spores are used, as in the case of *P. vulgare*. But in most ethnobotanical and bioprospecting studies of ferns from the *Polypodiaceae* and *Aspleniaceae* families, fronds are the major plant drug used for medicinal purposes [14].

Currently, the traditional use of the fronds of *P. vulgare* in traditional Spanish medicine has been described by the use of infusions in humans due to its hepatoprotective power [12, 15] and for the treatment of jaundice in veterinary [16]. In the case of traditional Polish medicine, the use of the infusion of the rhizome of *P. vulgare* has been described for its expectorant and diuretic action, and it is used exclusively to treat chronic nephritis and pyelonephritis [17]. From the traditional uses of *P. vulgare* in traditional Spanish and Polish medicine, it can be deduced that the phytochemicals responsible for the corresponding activities must be of polar origin since the herbal drug is administered as an infusion. The polarity of the solvent is key for the extraction of phenolic compounds as well as the contact time of the solvent with the plant drug [18, 19]. Another conclusion is how different cultures influence the use of the plant drug of the same species in their respective traditional medicines. Thus, in two countries of the European continent (Spain and Poland), the plant drug used for *P. vulgare* is different, being the frond and the rhizome, respectively.

In addition, *P. vulgare* has been used as human food and also as a sweetener [17]. However, ferns, due to their value as edible foods and their content of long-chain polyunsaturated fatty acids, represent an essential part of the local economy of the Russian Far East formed by the regions of Khabarovsky Krai, Primorsky Krai, and Sakhalin Oblast [20]. Eating young fern fronds has been linked to improving inflammatory and nervous processes. However, the degree of maturation in which the fronds are ingested is important since the presence of sporangia/spores on the fatty acid content in the fronds is presumably associated with the lipid contribution of the spores. So generally, a greater number of sporangia increases the number of lipids [21].

One observation in ferns is that they are generally not affected by bacteria, a fact which may be due to the presence of phytoalexins [22]. For example, in certain species of ferns of the genus *Asplenium*, such as the case of *Asplenium adiantum-nigrum* L., *Asplenium ruta-muraria* L., *Asplenium trichomanes* L., and *Ceterach officinarum* Willd., the positive antibacterial activity of its fronds against gram positive/negative bacteria has been determined [23, 24]. Therefore, it is not improbable that the fronds of *P. vulgare* also present antibacterial activity. Currently, the aqueous extract of *P. vulgare* rhizome has been validated for its antibacterial action against the gram-negative bacterium *Escherichia coli* [25]; in this study, this extract turned out to be an agent that enhances the action of chlorin-e6 in photodynamic therapy and also as an agent that enhances the antibiotic ciprofloxacin (quinolone family). This fact strengthens the use of this fern in traditional medicine in Sobrarbe (Huesca, Spain) as an antiseptic and wound-healing remedy [26].

Cancer is a multicausal disease with an increasing trend due to the aging of the population in developed countries. Redox medicine has established that ROS plays a key role in the early stages of carcinogenesis establishment. On the other hand, the World Health Organization (WHO) has defined biomarkers as measurable and key agents to predict the incidence of diseases [27]. In the case of the rhizome and fronds of *P. vulgare*, the presence of ecdysteroids, such as 20-hydroxyecdysone, has been described [6], also isolated from fronds of *Microsorum scolopendria* (Burm. f.) Copel. (*Polypodiaceae*) [28]. Ecdysteroids are phytochemicals that may be a trigger for antiproliferative activity [29]. Although these ecdysteroids are terpenes, they can be isolated from the methanolic extract of *P. vulgare* fronds [30].

The skin, the largest organ in the body, has multiple functions, among which the barrier between the external and internal environment stands out. Thus, the myriad of factors that converge in this tissue means that skin cancer is currently the main cancer worldwide [31]. Predictions for projected skin cancer incidence in the coming years suggest an upward trend globally [32]. One way to reduce exposure is the use of photoprotection strategies, either physical or topical or systemic, or oral [33, 34]. The first is based on the use of physical elements and topical agents to protect us from ultraviolet radiation, while the second, which should be complementary to the first, is based on the intake of potentially protective components against ultraviolet radiation. Among these agents, some vitamins or phytochemical compounds predominate, in isolated or complex forms, which through their antioxidant capacity contribute to redox homeostasis. In this sense, ferns, being a reservoir for obtaining phytochemicals (especially polyphenols), are potential candidates for obtaining compounds or products with photoprotective capacity at a systemic level [35]. Up to now, the aqueous extract of *Polypodium leucotomos* (*Polypodiaceae*) fronds is the fern extract with the greatest scientific evidence in the treatment of human skin disorders derived from exposure to ultraviolet rays [36–38]. Currently, it has been described that certain ferns are potential reservoirs of polyphenols justifying ethnopharmacological and bioprospecting studies in research in the field of diseases related to oxidative stress.

4 Phytochemistry

In the case of the phytochemicals of *P. vulgare*, there is no complete phytocharacterization of its frond and rhizome, which are the main plant drugs used in traditional medicine. Currently, the fern with the greatest phytocharacterization of the *Polypodiaceae* family is the aqueous extract of *Polypodium leucotomos* fronds (PLE) [39]. In this case, the main phytochemicals of PLE determined by HPLC-MS/MS are phenolic acids, specifically *p*-coumaric acid, ferulic acid, caffeic acid, and chlorogenic acid (also called 3-*O*-caffeoylquinic acid) which are cinnamic acids and vanillic acid which is benzoic acid. In 2019, Farràs et al. [40] published the first phytocharacterization of the methanolic extract of *P. vulgare* fronds (PVM) using the HPLC-DAD methodology; using HPLC-DAD, 3-*O*-caffeoylquinic acid was one of the main phytochemicals. Thus, cinnamic acid derivatives are some of the main phytochemicals in PVM and PLE extracts, both extracts from fern fronds of the *Polypodiaceae* family. HPLC-DAD is a technique widely used in the initial phytocharacterization of phytochemicals. However, identification by overlapping peaks may lead to incorrect identification since the peak is not specific to each phytochemical but may be of a structurally related one. Therefore, the use of another methodology is required, such as the HPLC-MS/MS technique. In addition, this type of double detection has already been carried out with the PLE [41] and allows the expansion of both the quantity and the groups of families of phytochemicals to be determined. In the doctoral thesis entitled "Fitocaracterización y actividad biológica de extractos de helechos de las familias *Polypodiaceae* y *Aspleniaceae* procedentes de las montañas de Prades (Tarragona)" recently presented [9], the phytocharacterization by HPLC-MS/MS of PVM was also published. In this characterization by HPLC-MS/MS, chlorogenic acid represents the major phytochemical representing 94.5% of the total determined

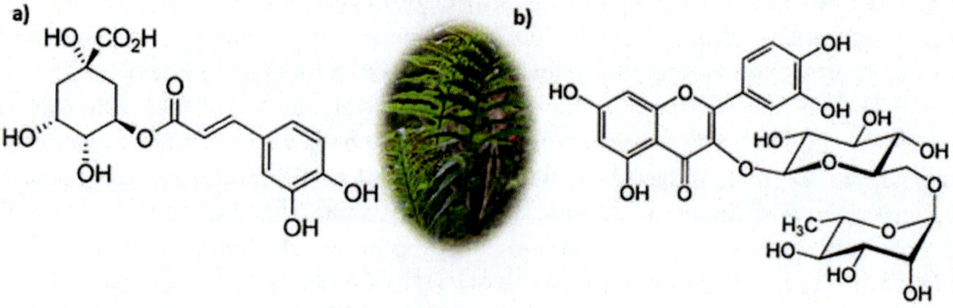

Fig. 3 Chlorogenic acid (**a**) and rutin (**b**) were isolated in the methanolic extract of *Polypodium vulgare* fronds detected by HPLC-MS/MS [9]. (© Adrià Farràs Martínez)

phytochemicals. With both HPLC phytocharacterization techniques (HPLC-DAD and HPLC-MS/MS) used by Farràs in PVM, has been evaluated the presence of different phenolic acids: gallic and vanillic acids, 5-*O*-caffeoylquinic acid (neochlorogenic acid), 3-*O*-caffeoylquinic acid (chlorogenic acid), 3,5-di-*O*- caffeoylquinic acid, and ferulic, caffeic, and cumaric acids.

Among flavonoids, different classes have also been detected by HPLC-MS/MS: rutin, quercitrin, isoquercitrin, kaempferol 3-*O*-glucoside, hyperoside, catechin, epicatechin, and hesperidin. It is also interesting that certain anthocyanins, such as delphinidin-3,5-diglucoside and cyanidin-3-*O*-glucoside are also detected. However, it must be stated that different analytical techniques may lead to different phytochemicals. In addition, the majority of phytochemicals in both HPLC assays with a high amount in proportion to the rest of the phytochemicals are chlorogenic acid. Figure 3 shows the most representative polyphenolic compounds of the methanolic extract of *P. vulgare* fronds in the thesis of Farràs [9] according to the family of phenolic acids and flavonoids by HPLC-MS/MS.

In relation to the study of antioxidant properties, the use of the fronds turns out to be more interesting than the rhizome. For example, in the ferns *A. adiantum-nigrum*, *A. trichomanes*, and *C. officinarum*, species of the *Aspleniaceae* family, it has been determined that the content of phytochemicals in the aglycone form is higher than in the glycoside form in the fronds than in the rhizome [42]. When studying the antioxidant property between the polar and apolar extract, the polar extract turns out to be more interesting. As concluded from the study of the antioxidant activity of the methanolic and hexanic extracts from the fronds of *P. vulgare*, *A. adiantum-nigrum*, *A. trichomanes*, and *C. officinarum*, the methanolic extracts present notable antioxidant activity compared to the corresponding hexanic ones [40]. The similarity of the behavior on cytotoxicity and cytoprotection in the 3T3 and HaCaT cell lines of the methanolic extracts of the fronds of *P. vulgare* and *C. officinarum* may be the result of their similar phytochemical profile in which the major phytochemicals are phenolic acids, especially highlighting the presence of chlorogenic acid [30, 43]. It currently remains to be determined whether the fronds of *P. vulgare* contain xanthones since, as in the case with the methanolic extract of the *A. adiantum-nigrum* fronds, the greatest amount of phytochemicals correspond to xanthones

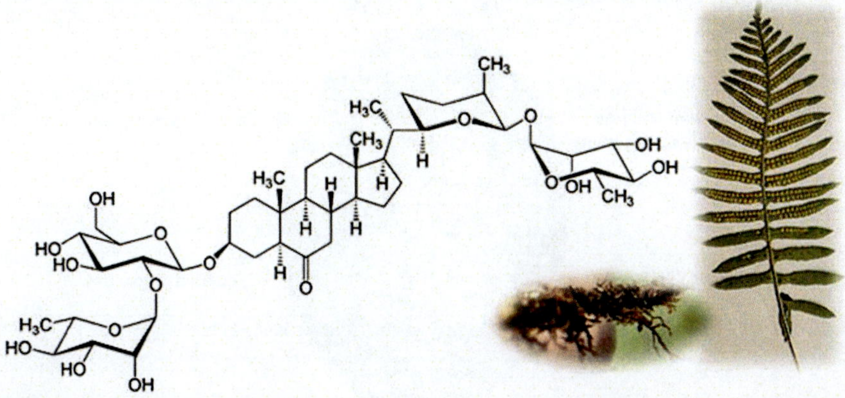

Fig. 4 Chemical structure of osladin, a glycoside steroid found in the rhizome of *Polypodium vulgare*

[23], a situation that does not report Farràs et al. [44] since xanthones were not analyzed. Xanthones have been proposed as chemical markers and elements of significant importance in the taxonomy of ferns [45].

In the case of the phytochemistry of polar extracts from the rhizome of *P. vulgare*, it has been found that different ecdysteroids have been determined [46]. Previously, other authors determined the presence of different ecdysteroids, such as 20*E*-hydroxyecdysone [6]. The main characteristic of the ecdysteroids determined from this plant drug is the hemiacetal group in its side chain that is characteristic of the species of the genus *Polypodium*. The glycosidic forms of these phytochemicals exhibit much higher sweetness intensity than sucrose [46]. These properties put forward the idea of using *Polypodium* ferns as powerful sweeteners. An example of these phytochemicals is osladin, responsible for the traditional use of the rhizome of *P. vulgare* as a sweetener [17].

Figure 4 represents one of the most representative phytochemicals of the *P. vulgare* rhizome.

In the case of apolar fractions, we find the studies of the hexanic extract of the rhizome of *P. vulgare* by Arai et al. [47, 48]. These authors determined that triterpenoid hydrocarbons and fatty acid esters were isolated in the *P. vulgare* rhizome. Of the different triterpenoid hydrocarbons, hop-22(29)-ene stands out, which has also been isolated in other fern species, such as *Polypodium polypodioides* (*Polypodiaceae*) [49] and *Asplenium trichomanes* (*Aspleniaceae*) [50].

5 Biological and Pharmacological Properties of Polypody

Although there is a reduced number of studies regarding the biological properties of polypody extracts, we can find a wide range of bioactivities reported (Fig. 5).

Fig. 5 Biological and pharmacological properties of *Polypodium vulgare*

5.1 Antioxidant Activity and Protective Effects Against Oxidative Stress

Free radicals and oxidative stress are known to be implicated in a lot of biological and physio-pathological processes, being responsible for certain molecular aspects of the disease. *P. vulgare* has shown antioxidant activity and cytoprotective properties in different in vitro models. For example, methanolic and hexanic extracts obtained from *P. vulgare* fronds have exerted antioxidant and radical scavenging capacity against free radicals using the DPPH, ORAC, and xanthine oxidase methodologies. The methanolic extract of *P. vulgare* demonstrated better antioxidant activity than the hexanic extract due to the presence of polyphenols; in general, the methanolic extract had a better extraction yield and scavenging capacity against DPPH, peroxyl, and superoxide radicals than the hexanic extract. Considering that superoxide radicals generated by xanthine oxidase may exist in biology, it is interesting that the activity was superior to Trolox in this assay [40].

Additionally, polypody has been tested against oxidative stress in different cell lines, such as fibroblasts (3T3 cells) and keratinocytes (HaCaT); since polypody fronds methanolic extract contains polyphenols, such as flavonoids and caffeoylquinic acid derivatives, it could protect cell lines from oxidative stress induced by hydrogen peroxide, showing cytoprotective and cellular repair activity in 3T3 cells. In the cytoprotective experiments, cells were pretreated with 0.01, 0.1, 1, and 2 mg/mL of polypody fronds methanolic extract while in the cell repair assays cells were treated after the toxic insult with hydrogen peroxide (2 mM during 2.5 h) [30]. All these results may be explained by the presence of phenolic compounds, which have extensively been positioned as natural antioxidants [51].

It is interesting and remarkable that other *Polypodium* species have also exerted antioxidant and protective effects; this is the case of aqueous extract of *P. leucotomos* fronds, which has demonstrated cytoprotective properties in human fibroblast from cytoskeletal disarrangements induced by UVA light [52].

5.2 Enzyme-Inhibitory Activity

Tyrosinase is currently found to act as a key enzyme in skin pigmentation. For this reason, certain natural products have been also proposed to act as tyrosinase inhibitors and are used in cosmetic products. In a recent study published in 2019, the authors found that methanolic and hexanic polypody fronds extracts acted as tyrosinase inhibitors, showing better capacity than other fern species, such as *A. adiantum-nigrum*, *A. trichomanes*, or *C. officinarum* [40]. Although the best tyrosinase inhibitor was the methanolic *P. vulgare* fronds extract, its IC_{50} value (0.107 mg/mL) was not superior to the reference substance kojic acid (0.063 mg/mL). The interesting results of this study in relation to tyrosinase and free radicals put forward the idea of ferns as natural substances with potential cosmetic applications.

Regarding the activity of polypody against other enzymes of relevant interest in human biology and disease, we found that both aqueous and hydroalcoholic extracts from common polypody rhizomes showed very low activity against acetylcholinesterase [53]. The authors of this study selected plants traditionally used in Iran for Alzheimer's disease symptoms; however, the best extract in that study was betel (*Areca catechu* L.).

5.3 Central Nervous System and Neuropharmacological Activity

There is only one study regarding the sedative properties of polypody root aqueous extract. Mannan et al. [54] stated that in rats, polypody aqueous extract administered intraperitoneally reduced motor activity, prolonged sleeping time of pentobarbitone-induced hypnosis, reduced body temperature, and increased the reaction time to pain stimuli; additionally, 10–60 mg/kg of the extract produced a fall in blood pressure in anesthetized dogs revealing a potential activity toward ß-adrenergic receptors.

5.4 Antimicrobial Activity

Antibiotic resistance is a big issue in human and veterinary health today. Owing to the increasing number of resistant bacteria and ineffective treatments, many research groups and pharmaceutical companies are being involved in the development of new antimicrobial substances of natural origin.

The antibacterial activity of polypody has been tested against different uropathogenic *Escherichia coli* strains. The former study by Glensk et al. [17] showed that the rhizome aqueous extract as well as purified osladin from the extract showed inhibitory effects on the survival and biofilm formation of *E. coli* strains. In a more recent study from the same authors, it is demonstrated that polypody rhizome aqueous extract enhances the photodynamic therapy and ciprofloxacin activity against uropathogenic *Escherichia coli* decreasing bacterial survival, biofilm mass synthesis, and morphological changes of the bacteria under the combined treatments [25]. Antiviral activity has also been suggested [55].

5.5 Laxative and Expectorant Properties

As stated in the previous section, herbal preparations in solid dosage forms for oral use are approved by the European Medicines Agency (EMA) as an expectorant in cough and cold and for short-term use in cases of occasional constipation [56]. These uses are based on their "traditional use," which means that, although there is insufficient evidence from clinical trials, the effectiveness of these herbal medicines is plausible and there is evidence that they have been used safely in this way for at least 30 years (including at least 15 years within the EU).

6 Conclusion

Polypodium vulgare has traditionally been used and approved by EMA as a laxative and expectorant traditional medicine, but in addition, some in vitro studies reveal antioxidant and antibacterial properties. The wide range of bioactivities together with the presence of different classes of phytochemicals, such as caffeoylquinic acids, flavonoids, and steroids make us conclude that *Polypodium vulgare* is a source of bioactive compounds with potential pharmaceutical applications. The absence of toxicity is also a good starting point for industrial applications of this fern.

Acknowledgments Universidad San Jorge and Gobierno de Aragón are thanked for financial and technical support (B44_20D-Phyto-Pharm group).

References

1. Díaz González TE, Fernández-Carvajal Álvarez MC, Fernández Prieto JA (2004) Curso de Botánica. Ciencias, 1st edn. Gráficas Summa, S.A, Gijón
2. Sessa EB (2018) Evolution and classification of ferns and lycophytes. In: Fernández H (ed) Current advances in fern research. Springer, Cham, pp 179–200
3. Juslen A, Vare H, Wikstrom N (2011) Relationships and evolutionary origins of polyploid *Dryopteris* (*Dryopteridaceae*) from Europe inferred using nuclear *pgiC* and plastid *trnL-F* sequence data. Taxon 60(5):1284–1294. https://doi.org/10.1002/tax.605005
4. Jin D-M, Zhou X-L, Schneider H, Wei H-J, Wei H-Y, Yan Y-H (2020) Functional traits: adaption of ferns in forest. J Syst Evol 59:1040. https://doi.org/10.1111/jse.12669
5. Langhansova L, Pumprova K, Haisel D, Ekrt L, Pavicic A, Zajickova M et al (2021) European ferns as rich sources of antioxidants in the human diet. Food Chem 356:12. https://doi.org/10.1016/j.foodchem.2021.129637
6. Messeguer J, Melé E, Reixach N, Irurre-Santilari J, Casas J (1998) *Polypodium vulgare* L. (Wood Fern): In vitro cultures and the production of phytoecdysteroids. In: Bajaj YPS

(ed) Medicinal and aromatic plants X. Biotechnology in agriculture and forestry. Springer, Berlin/Heidelberg, pp 333–348

7. Vasco A, Moran RC, Ambrose BA (2013) The evolution, morphology, and development of fern leaves. Front Plant Sci 4:16. https://doi.org/10.3389/fpls.2013.00345
8. Bonnier G, De Layens G (2002) Claves para la determinación de plantas vasculares, Primera edn. Ediciones Omega, S.L., Barcelona
9. Farràs Martínez A (2022) Fitocaracterización y actividad biológica de extractos de helechos de las familias *Polypodiaceae* y *Aspleniaceae* procedentes de las montañas de Prades (Tarragona). Universitat de Barcelona, Barcelona
10. Heinrich M (2015) Quality and safety of herbal medical products: regulation and the need for quality assurance along the value chains. Br J Clin Pharmacol 80(1):62–66. https://doi.org/10.1111/bcp.12586
11. Silva VL, Mehltreter K, Schmitt JL (2018) Ferns as potential ecological indicators of edge effects in two types of Mexican forests. Ecol Indic 93:669–676. https://doi.org/10.1016/j.ecolind.2018.05.029
12. Batet D, Cartanyà J, Castells R, Piñas I, Salat X (2011) Entrevista n°10. Etnobotànica a les muntanyes de Prades, 1st edn. Centre d'Història Natural de la Conca de Barberà, Montblanc, pp 127–129
13. Fiorin E, Saez L, Malgosa A (2019) Ferns as healing plants in medieval Mallorca, Spain? Evidence from human dental calculus. Int J Osteoarchaeol 29(1):82–90. https://doi.org/10.1002/oa.2718
14. Sureshkumar J, Silambarasan R, Bharati KA, Krupa J, Amalraj S, Ayyanar M (2018) A review on ethnomedicinally important pteridophytes of India. J Ethnopharmacol 219:269–287. https://doi.org/10.1016/j.jep.2018.03.024
15. Bonet MÀ, Agelet A, Vallès J, Villar Pérez L (2001) Contribution à la connaissance ethnobotanique des ptéridophytes dans les Pyrénées. Bocconea 13:605–612
16. Bonet MÀ, Vallès J (2007) Ethnobotany of Montseny biosphere reserve (Catalonia, Iberian Peninsula): plants used in veterinary medicine. J Ethnopharmacol 110(1):130–147. https://doi.org/10.1016/j.jep.2006.09.016
17. Glensk M, Tichaczek-Goska D, Sroda-Pomianek K, Wlodarczyk M, Wesolowski CA, Wojnicz D (2019) Differing antibacterial and antibiofilm properties of *Polypodium vulgare* L. Rhizome aqueous extract and one of its purified active ingredients-osladin. J Herb Med 17–18. https://doi.org/10.1016/j.hermed.2019.100261
18. Zhou MM, Chen P, Lin Y, Fang SZ, Shang XL (2019) A comprehensive assessment of bioactive metabolites, antioxidant and antiproliferative activities of *Cyclocarya paliurus* (Batal.) Iljinskaja leaves. Forests 10(8):16. https://doi.org/10.3390/f10080625
19. Dimcheva V, Karsheva M (2018) *Cistus incanus* from Strandja Mountain as a source of bioactive antioxidants. Plants-Basel 7(1):14. https://doi.org/10.3390/plants7010008
20. Nekrasov EV, Svetashev VI (2021) Edible far eastern ferns as a dietary source of long-chain polyunsaturated fatty acids. Foods 10(6):16. https://doi.org/10.3390/foods10061220
21. Nekrasov EV, Svetashev VI, Khrapko OV, Vyssotski MV (2019) Variability of fatty acid profiles in ferns: relation to fern taxonomy and seasonal development. Phytochemistry 162: 47–55. https://doi.org/10.1016/j.phytochem.2019.02.015
22. Beckman CH (2000) Phenolic-storing cells: keys to programmed cell death and periderm formation in wilt disease resistance and in general defence responses in plants? Physiol Mol Plant Pathol 57(3):101–110. https://doi.org/10.1006/pmpp.2000.0287
23. Zivkovic S, Milutinovic M, Maksimovi V, Ciric A, Ivanov M, Bozunovic J et al (2020) Antioxidant and antimicrobial activity of two *Asplenium* species. S Afr J Bot 132:180–187. https://doi.org/10.1016/j.sajb.2020.03.034
24. Petkov V, Batsalova T, Stoyanov P, Mladenova T, Kolchakova D, Argirova M et al (2021) Selective anticancer properties, proapoptotic and antibacterial potential of three *Asplenium* species. Plants-Basel 10(6):14. https://doi.org/10.3390/plants10061053
25. Tichaczek-Goska D, Glensk M, Wojnicz D (2021) The enhancement of the photodynamic therapy and ciprofloxacin activity against uropathogenic *Escherichia coli* strains by

Polypodium vulgare rhizome aqueous extract. Pathogens 10(12):15. https://doi.org/10.3390/pathogens10121544
26. Villar L, Bonet MÀ (2018) *Polypodium* L. In: Pardo de Santayana M, Morales R, Tardío J, Molina M (eds) Inventario Español de los Conocimientos Tradicionales relativos a la Biodiversidad, 2nd edn. Ministerio de Agricultura y Pesca, Alimentación y Medio Ambiente de España, Madrid, pp 51–55
27. Frijhoff J, Winyard PG, Zarkovic N, Davies SS, Stocker R, Cheng D et al (2015) Clinical relevance of biomarkers of oxidative stress. Antioxid Redox Signal 23(14):1144–1170. https://doi.org/10.1089/ars.2015.6317
28. Snogan E, Vahirua-Lechat I, Ho R, Bertho G, Girault JP, Ortiga S et al (2007) Ecdysteroids from the medicinal fern *Microsorum scolopendria* (Burm. f.). Phytochem Anal 18(5):441–450. https://doi.org/10.1002/pca.1000
29. Mamadalieva NZ (2012) Phytoecdysteroids from *Silene* plants: distribution, diversity and biological (antitumour, antibacterial and antioxidant) activities. Boletin Latinoamericano Del Caribe Plantas Medicinales Aromaticas 11(6):474–497
30. Farràs A, Mitjans M, Maggi F, Caprioli G, Vinardell MP, López V (2021) *Polypodium vulgare* L. (*Polypodiaceae*) as a source of bioactive compounds: polyphenolic profile, cytotoxicity and cytoprotective properties in different cell lines. Front Pharmacol 12:15. https://doi.org/10.3389/fphar.2021.727528
31. Rubió-Casadevall J, Hernandez-Pujol AM, Ferreira-Santos MC, Morey-Esteve G, Vilardell L, Osca-Gelis G et al (2016) Trends in incidence and survival analysis in non-melanoma skin cancer from 1994 to 2012 in Girona, Spain: a population-based study. Cancer Epidemiol 45:6–10. https://doi.org/10.1016/j.canep.2016.09.001
32. Rundle CW, Militello M, Barber C, Presley CL, Rietcheck HR, Dellavalle RP (2020) Epidemiologic burden of skin cancer in the US and worldwide. Curr Dermatol Rep 9(4):309–322. https://doi.org/10.1007/s13671-020-00311-4
33. Taylor JH, Rosen CF (2020) Systemic photoprotection. Curr Dermatol Rep 9(3):181–188. https://doi.org/10.1007/s13671-020-00306-1
34. Krutmann J, Passeron T, Gilaberte Y, Granger C, Leone G, Narda M et al (2020) Photoprotection of the future: challenges and opportunities. J Eur Acad Dermatol Venereol 34(3): 447–454. https://doi.org/10.1111/jdv.16030
35. Evans JA, Johnson EJ (2010) The role of phytonutrients in skin health. Nutrients 2(8):903–928. https://doi.org/10.3390/nu2080903
36. Parrado C, Mascaraque M, Gilaberte Y, Juarranz A, Gonzalez S (2016) Fernblock (*Polypodium leucotomos* extract): molecular mechanisms and pleiotropic effects in light-related skin conditions, photoaging and skin cancers, a review. Int J Mol Sci 17(7). https://doi.org/10.3390/ijms17071026
37. Parrado C, Nicolas J, Juarranz A, Gonzalez S (2020) The role of the aqueous extract *Polypodium leucotomos* in photoprotection. Photochem Photobiol Sci 19(6):831–843. https://doi.org/10.1039/d0pp00124d
38. Aguilera J, Vicente-Manzanares M, de Gálvez MV, Herrera-Ceballos E, Rodríguez-Luna A, González S (2021) Booster effect of a natural extract of *Polypodium leucotomos* (Fernblock®) that improves the UV barrier function and immune protection capability of sunscreen formulations. Front Med 8:10. https://doi.org/10.3389/fmed.2021.684665
39. Gombau L, García F, Lahoz A, Fabre M, Roda-Navarro P, Majano P et al (2006) *Polypodium leucotomos* extract: antioxidant activity and disposition. Toxicol In Vitro 20(4):464–471. https://doi.org/10.1016/j.tiv.2005.09.008
40. Farràs A, Cásedas G, Les F, Terrado EM, Mitjans M, López V (2019) Evaluation of anti-tyrosinase and antioxidant properties of four fern species for potential cosmetic applications. Forests 10(2):14. https://doi.org/10.3390/f10020179
41. García F, Pivel JP, Guerrero A, Brieva A, Martínez-Alcázar M, Caamaño-Somoza M et al (2006) Phenolic components and antioxidant activity of Fernblock®, an aqueous extract of the

aerial parts of the fern *Polypodium leucotomos*. Methods Find Exp Clin Pharmacol 28(3):157–160. https://doi.org/10.1358/mf.2006.28.3.985227

42. Durdevic L, Mitrovic M, Pavlovic P, Bojovic S, Jaric S, Oberan L et al (2007) Total phenolics and phenolic acids content in leaves, rhizomes and rhizosphere soil under *Ceterach officinarum* DC, *Asplenium trichomanes* L. and *A. adiantum nigrum* L. in the gorge of sicevo (Serbia). Ekologia (Bratislava) 26(2):164–173
43. Farràs A, Mitjans M, Maggi F, Caprioli G, Vinardell MP, López V (2022) Exploring wild *Aspleniaceae* ferns as safety sources of polyphenols: the case of *Asplenium trichomanes* L. and *Ceterach officinarum* Willd. Front Nutr 9:18. https://doi.org/10.3389/fnut.2022.994215
44. Farràs A, López V, Maggi F, Caprioli G, Vinardell MP, Mitjans M (2022) Chemical composition and cytoprotective activities of methanolic extract of *Asplenium adiantum-nigrum* L. (*Aspleniaceae*). Horticulturae 8(9):19. https://doi.org/10.3390/horticulturae8090815
45. Richardson PM (1984) The taxonomic significance of Xanthones in ferns. Biochem Syst Ecol 12(1):1–6. https://doi.org/10.1016/0305-1978(84)90002-4
46. Simon A, Vanyolos A, Beni Z, Dekany M, Toth G, Bathori M (2011) Ecdysteroids from *Polypodium vulgare* L. Steroids 76(13):1419–1424. https://doi.org/10.1016/j.steroids.2011.07.007
47. Arai Y, Shiojima K, Ageta H (1989) Fern constituents: cyclopodmenyl acetate, a cycloartanoid having a new 33-carbon skeleton, isolated from *Polypodium vulgare*. Chem Pharm Bull 37(2):560–562. https://doi.org/10.1248/cpb.37.560
48. Arai Y, Yamaide M, Yamazaki S, Ageta H (1991) Fern constituents – triterpenoids isolated from *Polypodium-vulgare*, *Polypodium-fauriei* and *Polypodium-virginianum*. Phytochemistry 30 (10):3369–3377. https://doi.org/10.1016/0031-9422(91)83212-4
49. Ageta H, Arai Y (1990) Chemotaxonomy of ferns, 3. Triterpenoids from *Polypodium – Polypodioides*. J Nat Prod 53(2):325–332. https://doi.org/10.1021/np50068a009
50. Bottari F, Morelli I, Marsili A, Pacchiani M (1972) Aliphatic and triterpenoid hydrocarbons from ferns. Phytochemistry 11(8):2519–2523. https://doi.org/10.1016/s0031-9422(00)88528-3
51. Virgili F, Marino M (2008) Regulation of cellular signals from nutritional molecules: a specific role for phytochemicals, beyond antioxidant activity. Free Radic Biol Med 45(9):1205–1216. https://doi.org/10.1016/j.freeradbiomed.2008.08.001
52. Alonso-Lebrero JL, Dominguez-Jimenez C, Tejedor R, Brieva A, Pivel JP (2003) Photoprotective properties of a hydrophilic extract of the fern *Polypodium leucotomos* on human skin cells. J Photochem Photobiol B-Biol 70(1):31–37. https://doi.org/10.1016/s1011-1344(03)00051-4
53. Saeedi M, Babaie K, Karimpour-Razkenari E, Vazirian M, Akbarzadeh T, Khanavi M et al (2017) *In vitro* cholinesterase inhibitory activity of some plants used in Iranian traditional medicine. Nat Prod Res 31(22):2690–2694. https://doi.org/10.1080/14786419.2017.1290620
54. Mannan A, Khan RA, Asif M (1989) Pharmacodynamic studies on *Polypodium-vulgare* Linn. Indian J Exp Biol 27(6):556–560
55. Husson GP, Vilagines R, Delaveau P (1986) Recherche de Propriétés Antivirales de quelques Extraits D'orgine Naturelle. Ann Pharmaceutiques Françaises 44:41–48
56. European Medicines Agency (2008). Community herbal monograph on *Polypodium vulgare* L., rhizoma. London. EMEA/HMPC/600668/2007

Phytochemicals and Biological Activities of *Asplenium ceterach* 21

Suzana Živković, Milica Milutinović, and Marijana Skorić

Contents

Abstract

Asplenium ceterach L. (family Aspleniaceae) is a perennial herbaceous fern. The lower surface of its fronds is covered with scales, which over time mature to a rusty color, giving the plant a characteristic appearance, hence the common name "rustyback" fern. It inhabits almost all of Europe, including the Mediterranean region, and Central Asia to the Himalayas, usually occurring in cracks in rocks. Rustyback fern is belonging to resurrection plants due to its ability to withstand desiccation and consequently completely recover after rehydration. It is thought that high concentrations of phenolic acids (mainly chlorogenic acid and caffeic acid) enable *A. ceterach* to reduce the damage of reactive oxygen species formed during dehydration. In addition to phenolic acids, the presence of other phenolic

S. Živković (✉) · M. Milutinović · M. Skorić
Institute for Biological Research "Siniša Stanković"-National Institute of Republic of Serbia, University of Belgrade, Belgrade, Serbia
e-mail: suzy@ibiss.bg.ac.rs; milica.milutinovic@ibiss.bg.ac.rs; mdevic@ibiss.bg.ac.rs

H. N. Murthy (ed.), *Bioactive Compounds in Bryophytes and Pteridophytes*, Reference Series in Phytochemistry, https://doi.org/10.1007/978-3-031-23243-5_19

compounds is detected in *A. ceterach* including flavonoids, tannins, coumarins, and quinones. Phytochemical studies have shown that these phenolic compounds are mainly found in sporophytes of rustyback fern, while *A. ceterach* gametophyte is characterized predominantly by xanthones, primarily mangiferin and its glycoside. Additionally, rustyback fern represents a rich source of different volatile organic compounds, mainly lipid derivatives. Owing to the presence of various bioactive metabolites, it is not surprising that *A. ceterach* has long been known for its healing properties, and numerous studies have proven its various biological activities, including antimicrobial, antioxidant, cytotoxic, anticancer, diuretic, as well as anti-inflammatory properties.

Keywords

Asplenium ceterach · Bioactivity · Gametophyte · Lipid derivatives · Phenolics · Sporophyte · Volatile compounds

Abbreviations

ABTS	2,2′-azino-bis(3-ethylbenzothiazoline-6-sulfonic acid) assay
AE	Aqueous extract
ALP	Alkaline phosphatase
ALT	Alanine aminotransferase
AST	Aspartate aminotransferase
BHA	Butylated hydroxyanisole
BHT	Butylated hydroxytoluene
CAT	Catalase
CC	Column chromatography
COD	Calcium oxalate dyhydrate
COM	Calcium oxalate monohydrate
DdH_2O	Deionized water
DPPH	1,1-diphenyl-2-picrylhydrazyl
DW	Dry weight
EPR	Electron paramagnetic resonance
Et_2O	Diethyl ether
EtOAc	Ethyl acetate
EtOH	Ethanol
FAs	Fatty acids
FRAP	Ferric reducing antioxidant power assay
FW	Fresh weight
G	Gametophyte
GAE	Gallic acid equivalent
GC-MS	Gas chromatography-mass spectrometry
HDL	High-density lipoprotein
HPLC-DAD	High-performance liquid chromatography with diode array detection

HS GC-MS	Headspace gas chromatography-mass spectrometry
LDL	Low-density lipoprotein
MeOH	Methanol
MPLC	Medium pressure liquid chromatography
N-BuOH	Butanol
ORAC	Oxygen radical absorbance capacity assay
PC	Paper chromatography
PTR-MS	Proton transfer reaction-mass spectrometry
PUFAs	Polyunsaturated fatty acids
QE	Quercetin equivalent
RBC	Red blood cell count
ROS	Reactive oxygen species
S	Sporophyte
SEM	Scanning electron microscopy
SOD	Superoxide dismutase
STZ	Streptozotocin
TE	Trolox equivalent
TEMPONE	2,2,6,6-tetramethyl-4-oxo-piperidin-1-oxyl
TF	Total flavonoid content
TLC	Thin layer chromatography
TPC	Total phenolics content
UHPLC/HESI-MS^2	Ultra-high performance liquid chromatography/heated electrospray ionization mass spectrometry
UHPLC-DAD/HESI-MS^2	Ultra-high performance liquid chromatography-diode array detection/heated electrospray ionization mass spectrometry
UV	Ultraviolet
UVS	Ultraviolet spectroscopy
VLC	Vacuum liquid chromatography
VOCs	Volatile organic compounds
WBC	White blood cell count

1 Introduction

Ferns first appeared more than 300 million years ago and are among the oldest higher plants. During the Carboniferous period, ferns were the dominant part of the vegetation. Although their presence has been significantly reduced in the Cretaceous, ferns continue to flourish and show their enormous diversity and worldwide distribution. Modern ferns represent the second largest group of vascular plants. They are classified into 300 genera with over 10,000 species [1].

Ferns have a characteristic life cycle with two, morphologically distinct, alternating generations: a diploid sporophyte and a haploid free-living gametophyte (prothallus or prothallium). A sporophyte (Fig. 1a) represents the dominant phase of the

Fig. 1 *Asplenium ceterach* L. Mature sporophyte (**a**) and gametophyte (**b**); Young frond, pinnately divided almost to the central midrib, with a leathery, deep green upper surface and dense scales, entirely covering the lower surface. In time, scales mature to the rusty color, hence the common name "rustyback fern" (**c**); detailed view of young frond covered with silver scales (**d**); a cluster of mature sporangia (sori), linear in shape, with vestigial indusia replaced by scales (**e-f**); scanning electron microscopy (SEM) micrographs of the abaxial side of young fronds (**g**) and sori (**h-i**). Scale bars and magnification are indicated on micrographs

fern life cycle that produces spores in sporangia, which typically occurs on the lower surface of fern leaves-fronds (Fig. 1e-f, h-i) [2, 3]. Gametophyte (Fig. 1b) develops from the haploid spore and it is tinny, one-cell thick, usually without surface protective covering. In most fern species, prothallium is heart-shaped, but it may be ribbon-like or even filamentous, varying depending on the species' habitat. In addition to the above morphological differences, generations of sporophyte and gametophyte are nutritionally and physically independent during most of their life cycle. In almost all other terrestrial plants (bryophytes and seed plants) due to nutritional dependence, these two halves of the life cycle overlap and do not establish different ecological entities [4]. The family Aspleniaceae belongs to the order of Polypodiales or Aspleniales. It comprises approximately 700 species grouped in 2 genera, *Asplenium* Linn. and *Hymenasplenium* Hayata [5]. Genus *Asplenium* is a cosmopolitan genus that consists of both epiphytic and terrestrial species. Many species belonging to the *Asplenium* genus can be found in tropical and northern temperate areas, inhabiting forest soils, river banks, and mountain ravines [6].

Asplenium ceterach L., formerly known as *Ceterach officinarum,* is a perennial fern [7]. The sporophyte of this fern consists of pinnate leaves (fronds), about 2 cm wide and up to 20 cm long, which have a photosynthetic and reproductive function. They are solid, simply divided, and have 6 to 12 pairs of pins, arranged alternately (Fig. 1a). The adaxial leaf surface is dark green and without scales. The abaxial surface of mature fronds is covered with densely distributed, orange-brown scales, which is why it is popularly known as "scale" or "rustyback" fern (Fig. 1c). Clusters of sporangia, called sori, are formed on the underside of the fronds and concealed by the abundant cloak of scales (Fig. 1e-f). Gametophytes of *A. ceterach,* as of all other isosporic ferns, are bisexual, but their antheridia often mature before archegonia, thus achieving intergametophyte fertilization.

Asplenium ceterach is a poikilohydric plant that grows on a limestone base, in places with good drainage. It is widespread in Europe, North Africa, and the Middle East, and the range of this species extends to western China [8]. The ability to enter a state of anabiosis allows it to survive severe droughts. During drying, the leaves twist, which gives them mechanical strength [9]. About 60 species are known as desiccation-resistant ferns [9], which are mainly distributed in dry regions around the world, being particularly numerous and diverse in Mexico and the southwestern United States [10]. These ferns have developed a diversity of morphological adaptations that allow them to withstand drying more easily. Compared to mosses and angiosperms, ferns have an intermediate mechanism of tolerance to desiccation. There are only several studies regarding the resistance of *A. ceterach* to water stress. Schwab and Heber [11] examined the stability of *A. ceterach* thylakoid membranes in desiccation caused by osmotic stress and freezing and found that membrane stability depends on a number of stabilizing compounds, such as sugars and amino acids. Schwab et al. [12] found that the respiratory process in *A. ceterach* is less sensitive to dehydration than in spinach. According to Živković et al. [13], the content of chlorogenic acid in rustyback fern fronds decreased during short-term dehydration, while a significant increase in the activity of peroxidase and polyphenol oxidase enzymes was detected [13]. These changes in enzyme activity and phenolic content may play an important role in the adaptation of the rustyback fern to water deficit and upsurge the overall plant resistance to stress conditions. Furthermore,

Fig. 2 In vitro sporophytes and gametophytes of *Asplenium ceterach*

Bogdanović et al. [14] found that changes in phenolic content in rustyback fern gametophytes subjected to salt stress may suggest that these plants use the antioxidative properties of phenolics as a part of salt tolerance mechanisms. One should bear in mind that *A. ceterach* completely recovers from dryness after re-watering and restores physiological activity within 72 h [15]. The rehydration process involves various metabolic changes, including increased emissions of volatile substances, mainly fatty acid derivatives. Interestingly, lipoxygenase activity was not stimulated during rehydration [15].

Methods of in vitro culture have been extensively used in research of *Pteridopsida* for years. Although ferns, as well as bryophytes, have distinct gametophyte and sporophyte stages, both stadiums can be in vitro cultivated (Fig. 2) [14, 16, 17]. Propagation methods are predominantly used for mass production of ferns with ornamental, medicinal, and ethnobotanical importance [16, 18], but also for the endangered fern species, for their ex situ preservation.

Although ferns are quite underestimated as medicinal plants and their use in traditional medicine is limited despite their diversity, in recent years there has been an increasing number of studies on their phytochemical composition and various biological activities.

2 Phytochemical Profile *of Asplenium ceterach*

Studies on the chemical composition of *Asplenium ceterach* revealed the presence of various bioactive metabolites, including phenolic compounds (such as phenolic acids, flavonoids, tannins, coumarins, quinones), xanthones, terpene/terpenoids, and fatty acids. Furthermore, rustyback fern represents a rich source of different volatile organic compounds (VOCs) with a wide array of biosynthetic origins. Most studies on the phytochemical constituents of *A. ceterach* have been performed on the aerial part of mature sporophytes (fresh or dry) and, to a much lesser extent, on gametophytes [19], using a plethora of analytical techniques, as summarized in Tables 1, 2 and 3.

Sidhoum et al. [20] have confirmed that *A. ceterach* contains different classes of secondary metabolites, with leucoanthocyanins and tannins as the most abundant, followed by the noticeable level of free quinones, coumarins, and flavonoids. Nevertheless, the presence of anthocyanins, combined quinones, saponosides, and alkaloids were not confirmed by the authors.

The results of the phytochemical screening have revealed a significantly high proportion of phenolics in sporophytes and gametophytes of rustyback fern. Phenolic compounds are a group of molecules characterized by a specific structure, which includes at least one phenol unit. Based on their chemical structures, phenolics can be divided into different subgroups [31]. Depending on the form in which they are present in plants, phenolics are categorized as free (soluble, mainly stored in vacuoles) or bound phenolic compounds (insoluble, participate in the cell wall formation) [31].

Bound phenolics have not been as thoroughly investigated as free phenolics in rustyback fern. Đurđević et al. [32] have reported that the amounts of the total free phenolics exceeded their bound forms in *A. ceterach,* with the leaves containing 11.7 mg g^{-1} DW and 9.13 mg g^{-1} DW, respectively [32]. A similar pattern was also observed in rustyback fern rhizomes, which contained 2.81 mg g^{-1} DW of free phenolics and 1.96 mg g^{-1} DW of phenolics in bound form. Additionally, the same authors have conducted a comprehensive analysis of the rhizosphere soil under the *A. ceterach* and concluded that the number of bound forms of phenolics exceeded that of their free forms up to 280 times, with bound forms of cinnamic acid derivatives (*p*-coumaric and ferulic acid) being the most abundant phenolics determined in rhizosphere soil [15].

According to the literature data, the Folin-Ciocalteu reagent is frequently used for the determination of the total phenolics content (TPC), free and bound, in rustyback fern sporophytes, with the calculated concentrations given in gallic acid equivalent units (GAE). Depending on the different methodological approaches (plant material source, extraction details, etc.), TPC values varied significantly between 7.5 mg g^{-1} DW [20] and 275 mg g^{-1} DW [33]. Conversely, there is only one record of the TPC in *A. ceterach* gametophytes grown in vitro [19], based on which the TPC value in gametophytic tissue reached 51 mg GAE g^{-1} FW.

Rustyback fern belongs to resurrection plants and may survive desiccation, passing quickly from anabiosis to full biological activity once water becomes available again [34]. Considering that *A. ceterach* can tolerate severe water stress, its phytochemical profile is closely related to both environmental conditions in the natural habitat and specific desiccation tolerance metabolism. In order to cope with desiccation resurrection, plants have to overcome a number of obstacles, among them the most critical being oxidative stress [35]. The role of phenolic compounds in

Table 1 Phytochemical composition of *Asplenium ceterach*

	Compound	Generation	Analytical approach	Solvent	Reference
	Phenolic acids and derivatives				
	Hydroxybenzoic acids and derivatives				
1	Gallic acid	S	HPLC-DAD	MeOH	[20]
2	Vanillic acid	S	HPLC-DAD	MeOH	[20]
3	Benzoic acid	S	HPLC-DAD, GC-MS	MeOH, Et_2O	[20, 21]
4	Tannic acid	S	HPLC-DAD	MeOH	[20]
5	Protocatechuic acid	S, G	UHPLC-DAD/ HESI-MS^2	MeOH	[19]
6	Gentisic acid	S, G	UHPLC-DAD/ HESI-MS^2	MeOH	[19]
7	*p*-Hydroxybenzoic acid	S, G	UHPLC-DAD/ HESI-MS^2, GC-MS	MeOH, Et_2O	[19, 21]
8	3-Methoxy-4-hydroxybenzoic acid	S	GC-MS	Et_2O	[21]
9	4-β-*O*-D-glucosyl-3,4-dihydroxybenzyl alcohol	S	MPLC	MeOH/water/ *n*-BuOH	[22]
10	4-vinyl-phenol-1-*O*-β-D-glucopyranoside	S	MPLC, VLC, CC, TLC	MeOH/water/ *n*-BuOH	[22]
	Hydroxycinnamic acids and derivatives				
11	Chlorogenic acid	S, G	UHPLC-DAD/ HESI-MS^2, HPLC-DAD, CC, HPLC, MPLC	MeOH, MeOH/ water/EtOAc, MeOH/water/ *n*-BuOH	[13, 19, 20, 22–24]
12	*p*-Coumaric acid	S	HPLC-DAD, UHPLC-DAD/ HESI-MS^2	MeOH	[19, 20]
13	p-Coumaroyl-4-*O*-β-D-glucoside	S	MPLC, VLC, CC	MeOH/water/ *n*-BuOH	[22]
14	3-(1-*O*-β-D-glucopyranosyl-4-hydroxy-phenyl)-propionic acid	S	MPLC, VLC, CC	MeOH/water/ *n*-BuOH	[22]
15	Caffeic acid	S, G	UHPLC-DAD/ HESI-MS^2, CC, HPLC	MeOH, MeOH/ water/EtOAc	[13, 19, 22, 24]
16	1-Caffeoylglucose 2-sulfate	S	PC, TLC, UVS, GC-MS	EtOH	[25]
17	1-Caffeoylglucose 3-sulfate	S	PC, TLC, UVS, GC-MS	EtOH	[25]
18	1-Caffeoylglucose 6-sulfate	S	PC, TLC, UVS, GC-MS	EtOH	[25]

(continued)

Table 1 (continued)

	Compound	Generation	Analytical approach	Solvent	Reference
19	Rosmarinic acid	S, G	UHPLC-DAD/ HESI-MS^2	MeOH	[19]
20	Aesculin	S, G	UHPLC-DAD/ HESI-MS^2	MeOH	[19]
21	Sinapic acid	S	HPLC-UV-VIS	MeOH	[26]
	Flavonoids				
	Flavones and derivatives				
22	Luteolin	S	HPLC-DAD	MeOH	[27]
	Flavanols and derivatives				
23	Gallocatechin	S, G	UHPLC-DAD/ HESI-MS^2	MeOH	[19]
24	Epicatechin	S	UHPLC-DAD/ HESI-MS^2	MeOH	[19]
25	Epigallocatechin gallate	S, G	UHPLC-DAD/ HESI-MS^2	MeOH	[19]
26	Epigallocatechin	G	UHPLC-DAD/ HESI-MS^2	MeOH	[19]
27	Catechin	S, G	UHPLC-DAD/ HESI-MS^2, HPLC-DAD	MeOH	[19, 24]
	Flavanones and derivatives				
28	Naringenin 7-*O*-(2-*O*-α-L-rhamnosyl)-β-D-glucoside (Naringin)	S	PC, TLC, UVS	EtOH	[25]
29	Naringenin 7-*O*-(6-*O*-L-arabinosyl)-D-glucoside	S	PC, TLC, UVS	EtOH	[25]
	Flavonols and derivatives				
30	Rutin	G	UHPLC-DAD/ HESI-MS^2	MeOH	[19]
31	Quercetin	S	HPLC-DAD, UHPLC/HESI-MS^2	MeOH	[20, 24, 26]
32	Kaempferol-3-*O*-glucoside (astragalin)	S	HPLC-DAD	MeOH	[27]
33	Kaempferol	S	HPLC-DAD, UHPLC/HESI-MS^2	MeOH	[26, 27]
34	Kaempferol-3-*O*-β-D-gentiobioside	S	MPLC, VLC, HPLC	MeOH/water/ *n*-BuOH	[22]
35	Kaempferol 3-(6-malonyl)-D-glucoside	S	PC, TLC, UVS	EtOH	[28]
36	Kaempferol 3-(6-malonyl)-D-galactoside	S	PC, TLC, UVS	EtOH	[28]

(continued)

Table 1 (continued)

	Compound	Generation	Analytical approach	Solvent	Reference
37	Quercetin 3-*O*-β-D-glucoside (Isoquercitrin)	S	PC, TLC, UVS	EtOH	[28]
38	Quercetin-3-*O*-β-D-gentiobioside	S	PC, TLC, UVS, MPLC, VLC, CC	EtOH, MeOH/water/*n*-BuOH	[22, 28]
39	Kaempferol 3-glucuronide	u.d.	u.d.	u.d.	[29]
40	Kaempferol 3-digalactoside	u.d.	u.d.	u.d.	[29]
41	Quercetin-3-*O*-β-D-glucuronide	S	MPLC, VLC, CC	MeOH/water/*n*-BuOH	[22, 29]
42	Quercetin 3-digalactoside	u.d.	u.d.	u.d.	[29]
43	Isorhamnetin	S	UHPLC/HESI-MS^2	MeOH	[26]
	Xanthones				
44	Mangiferin glucoside	S, G	UHPLC-DAD/HESI-MS^2	MeOH	[19]
45	Mangiferin	S, G	UHPLC-DAD/HESI-MS^2	MeOH	[19]
	Terpenes and terpenoids				
	Sesquiterpenes				
46	Pterosin B	S	HPLC-DAD	MeOH	[24]
	Triterpenes				
47	Diploptene	S	u.d.	u.d.	[30]
48	(24R)-Cyclolaudenol	S	u.d.	u.d.	[30]
49	Squalen	S	HS GC-MS	ddH_2O	[22]

S-sporophyte, G-gametophyte, CC-Column Chromatography, PC-Paper Chromatography, VLC-Vacuum Liquid Chromatography, TLC-Thin Layer Chromatography, HPLC-DAD-High-Performance Liquid Chromatography with Diode Array Detection, MPLC-Medium Pressure Liquid Chromatography, UVS-Ultraviolet Spectroscopy, GC-MS-Gas Chromatography-Mass Spectrometry, HS GC-MS- Headspace Gas Chromatography-Mass Spectrometry, UHPLC-DAD/HESI-MS^{2}Ultra-high Performance Liquid Chromatography-Diode Array Detection combined with Heated Electrospray Ionization Mass Spectrometry, UHPLC/HESI-MS^{2}Ultra-high Performance Liquid Chromatography combined with Heated Electrospray Ionization Mass Spectrometry, EtOH-ethanol, MeOH-methanol, Et_2O-diethyl ether, EtOAc-ethyl acetate, *n*-BuOH-butanol, ddH_2O-deionized water, u.d.-unavailable data.

the ability of rustyback fern to withstand the oxidative stress imposed by free radicals has been investigated by EPR spectroscopy [36]. The authors have noticed an increased abundance of phenolics in *A. ceterach* fronds in response to higher rates of hydroxyl radical ($^{\bullet}OH$) formation during the rehydration process, as revealed by the reduction of TEMPONE free radical. Moreover, Živković et al. [13] observed

Table 2 Volatile organic compounds (VOCs) detected in *Asplenium ceterach*

	Compound	Generation	Analytical approach	Solvent	Reference
50	*n*-Heptanal	S	HS GC-MS	ddH_2O	[22]
51	Hex-1-ene-3-ol	S	HS GC-MS	ddH_2O	[22]
52	(*E*)-2-Heptenal	S	HS GC-MS, GC-MS, PTR–MS	Et_2O, ddH_2O	[21, 22]
53	1-Octen-3-ol	S	HS GC-MS, GC-MS, PTR–MS	Et_2O, ddH_2O	[21, 22]
54	3-Octanone	S	HS GC-MS	ddH_2O	[22]
55	2,3-Octanedione	S	GC-MS	Et_2O	[21]
56	2-Pentyl furan	S	HS GC-MS, GC-MS, PTR–MS	Et_2O, ddH_2O	[21, 22]
57	(2*E*,4*Z*)-Heptadienal	S	HS GC-MS, PTR-MS	ddH_2O	[22]
58	Octanal	S	GC-MS	Et_2O	[21]
59	(2*E*,4*E*)-Heptadienal	S	HS GC-MS, PTR-MS	ddH_2O	[22]
60	3-Octen-2-one	S	GC-MS	Et_2O	[21]
61	(2*E*)-Octen-1-al	S	HS GC-MS	ddH_2O	[22]
62	Oct-3-ene-1-ol	S	HS GC-MS, PTR-MS	ddH_2O	[22]
63	*n*-Octanol	S	HS GC-MS	ddH_2O	[22]
64	(*E*)-2-Octenal	S	GC-MS	Et_2O	[21]
65	(*E*)-2-Octenol (2*E*)-Octen-1-ol	S	HS GC-MS, GC-MS, PTR–MS	Et_2O, ddH_2O	[21, 22]
66	*n*-Undecane	S	HS GC-MS	ddH_2O	[22]
67	*n*-Nonanal	S	HS GC-MS, GC-MS, PTR–MS	Et_2O, ddH_2O	[21, 22]
68	*N*-Acetylpyrrolidone	S	GC-MS	Et_2O	[21]
69	(2*E*)-Nonen-1-al	S	HS GC-MS, GC-MS	Et_2O, ddH_2O	[21, 22]
70	*n*-Dodecane	S	HS GC-MS	ddH_2O	[22]
71	*n*-Decanal	S	HS GC-MS	ddH_2O	[22]
72	2,3-Dihydrobenzofuran	S	GC-MS	Et_2O	[21]
73	(*E*,*E*)-2,4-Nonadienal	S	GC-MS	Et_2O	[21]
74	g-Octalactone	S	HS GC-MS	ddH_2O	[22]
75	2-Amino-4-methoxyphenol	S	GC-MS	Et_2O	[21]
76	(*E*)-2-Decenal	S	HS GC-MS, GC-MS	Et_2O, ddH_2O	[21, 22]
77	Phenylacetic acid	S	GC-MS	Et_2O	[21]
78	2-Decenol	S	GC-MS	Et_2O	[21]
79	Nonanoic acid	S	GC-MS	Et_2O	[21]
80	3,4-Dihydroxystyrene	S	GC-MS	Et_2O	[21]
81	(2*E*,4*Z*)-Decadienal	S	HS GC-MS, PTR-MS	ddH_2O	[22]
82	*n*-Undecanal	S	HS GC-MS	ddH_2O	[22]

(continued)

Table 2 (continued)

	Compound	Generation	Analytical approach	Solvent	Reference
83	(2*E*,4*E*)-Decadienal	S	HS GC-MS, GC-MS, PTR–MS	Et_2O, ddH_2O	[21, 22]
84	(*E*)-2-Undecenal	S	HS GC-MS, PTR-MS	ddH_2O	[22]
85	*n*-Dodecanal	S	HS GC-MS	ddH_2O	[22]
86	*n*-Tridecanal	S	HS GC-MS	ddH_2O	[22]
87	Propionylfilicinic acid	S	GC-MS	Et_2O	[21]
88	(6*Z*)-Pentadecen-2-one	S	HS GC-MS	ddH_2O	[22]
89	3,4-Dihydroxybenzaldehyde	S	GC-MS	Et_2O	[21]
90	(2*E*)-Tridecenol acetate	S	HS GC-MS	ddH_2O	[22]
91	Hexahydrofarnesylacetone	S	GC-MS	Et_2O	[21]
92	4-Hydroxy-ß-ionone	S	GC-MS	Et_2O	[21]
93	3-Hydroxy-5,6-epoxy-ß-ionone	S	GC-MS	Et_2O	[21]

S-sporophyte, GC-MS-Gas Chromatography-Mass Spectrometry, HS GC-MS-Headspace Gas Chromatography-Mass Spectrometry, PTR-MS-Proton Transfer Reaction-Mass Spectrometry, Et_2O-diethyl ether, ddH_2O-deionized water.

that short-term dehydration may induce changes in the phenolic metabolism in *A. ceterach* by increasing the activity of phenol oxidizing enzymes (such as peroxidase and polyphenol oxidase). These results implicated that phenolic compounds could take an essential part in overall *A. ceterach* tolerance to water stress.

Phytochemical studies have shown that phenolic acids and flavonoids, with their corresponding derivatives, comprise the major components of the *A. ceterach* phenolic profile (Table 1).

Phenolic acids represent a group of powerful natural compounds with substantial lipid and water solubility [37] that include two distinctive carbon frameworks, namely, the hydroxycinnamic acid and hydroxybenzoic acid structures. They all maintain the same basic structure, while the number and position of hydroxyl groups are different among the two. Cinnamic acids, especially hydroxycinnamic acids, have the role of basic precursors in the biosynthesis of various plant phenols [38]. Generally, ten different hydroxycinnamic acids and corresponding derivatives, along with eleven compounds belonging to the group of hydroxybenzoic acids and derived compounds, were found in *A. ceterach* (**1–21**). Within the identified phenolic acids, there are only seven, protocatechuic acid (**5**), gentisic acid (**6**), *p*-Hydroxybenzoic acid (**7**), chlorogenic acid (**11**), caffeic acid (**15**), rosmarinic acid (**19**), and aesculin (**20**) present in rustyback fern gametophyte [19] (Fig. 3). Interestingly, several studies suggest that chlorogenic acid represents the most abundant component of the *A. ceterach* sporophyte phytochemical profile [13, 20, 22, 24].

Hydroxycinnamic acids (such as *p*-coumaric, caffeic, and ferulic acids) and hydroxybenzoic acids (*p*-Hydroxybenzoic, protocatechuic, and vanillic acids) often occur as sugar esters, acylated glycosides, and/or various alcohol esters.

Table 3 Fatty acids composition of *Asplenium ceterach*

	Common name	Structural formula	C:D[a]	Reference
		Saturated fatty acids		
94	Caprylic	$CH_3(CH_2)_6COOH$	8:0	[21]
95	Pelargonic	$CH_3(CH_2)_7COOH$	9:0	[21]
96	Capric	$CH_3(CH_2)_8COOH$	10:0	[21]
97	Lauric	$CH_3(CH_2)_{10}COOH$	12:0	[26]
98	Myristic	$CH_3(CH_2)_{12}COOH$	14:0	[21, 26]
99	Pentadecylic	$CH_3(CH_2)_{13}COOH$	15:0	[26]
100	Palmitic	$CH_3(CH_2)_{14}COOH$	16:0	[22, 26]
101	Margaric	$CH_3(CH_2)_{15}COOH$	17:0	[26]
102	Stearic	$CH_3(CH_2)_{16}COOH$	18:0	[26]
103	Behenic	$CH_3(CH_2)_{20}COOH$	22:0	[26]
		Unsaturated fatty acids		
	Monounsaturated			
104	(*E*)-2-Tridecenoic	$CH_3(CH_2)_9\mathbf{CH{=}CH}COOH$	13:1	[21]
105	Myristoleic	$CH_3(CH_2)_3\mathbf{CH{=}CH}(CH_2)_7COOH$	14:1	[26]
106	Palmitoleic	$CH_3(CH_2)_5\mathbf{CH{=}CH}(CH_2)_7COOH$	16:1	[26]
107	Oleic	$CH_3(CH_2)_7\mathbf{CH{=}CH}(CH_2)_7COOH$	18:1	[26]
	Polyunsaturated			
108	Linoleic	$CH_3(CH_2)_4\mathbf{CH{=}CH}CH_2\mathbf{CH{=}CH}(CH_2)_7COOH$	18:2 (ω-6)	[22, 26]
109	Linolenic	$CH_3CH_2\mathbf{CH{=}CH}CH_2\mathbf{CH{=}CH}CH_2\mathbf{CH{=}CH}(CH_2)_7COOH$	18:3 (ω-3)	[26]
110	Arachidonic	$CH_3(CH_2)_4\mathbf{CH{=}CH}CH_2\mathbf{CH{=}CH}CH_2\mathbf{CH{=}CH}CH_2\mathbf{CH{=}CH}(CH_2)_3COOH$	20:4 (ω-6)	[26]

a Total amount of carbon atoms and double bonds

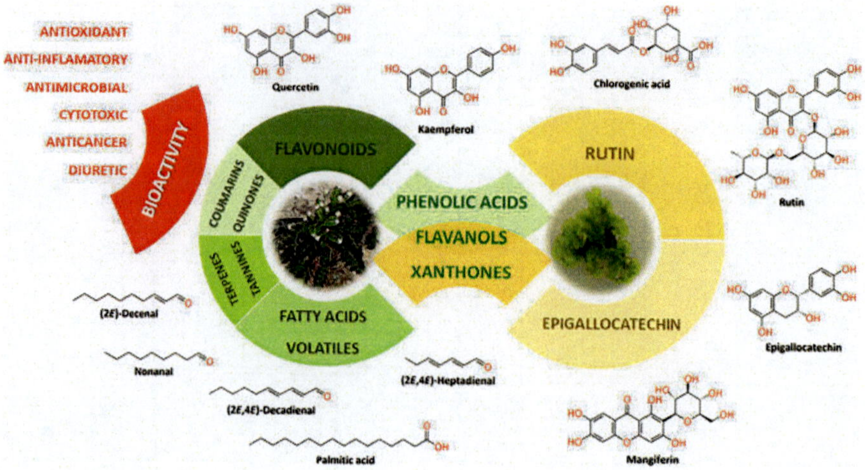

Fig. 3 Phytochemicals detected in *Asplenium ceterach*. Different colors indicate compounds/groups of compounds identified in sporophyte (left, green) and/or gametophyte (right, yellow), while biological activities of sporophyte are indicated in red. Chemical structures of most representative bioactive components of *A. ceterach* are also shown

Sugar esters (acyl sugars) are very common plant constituents, but only 15 are identified in ferns, among which eight of them belong to a group of sulfate esters of hydroxycinnamic acid-sugar derivatives [39]. A mixture of 1-caffeoylglucose 2-sulfate (**16**), 1-caffeoylglucose 3-sulfate (**17**), and caffeoylglucose 6-sulfate (**18**) has been identified in *A. ceterach* [28] that seems to have a protective role in all fern species [40].

Total flavonoid content (TF) in rustyback fern varies within individual experimental procedures, ranging from 8 mg QE (quercetin equivalents) g^{-1} DW [41] to 46 mg QE g^{-1} DW [23]. Most of the flavonoids identified in *A. ceterach* belong to a group of flavonols, predominantly quercetin and kaempferol and their glycosides (a total of 13 compounds, **30–43**), followed by the flavanol catechin (**27**) and derived compounds, that is, gallocatechin (**23**), epicatechin (**24**), epigallocatechin gallate (**25**), and epigallocatechin (**26**). At the same time, a diminished number of flavones (luteolin, **22**) and flavanones derivatives (naringin, **28**, and naringenin 7–*O*-(6–*O*-L-arabinosyl)-D-glucoside, **29**) were detected in *A. ceterach*. It should be noted that flavanol epigallocatechin (**26**), as well as flavonol glycoside rutin (**30**), was identified only in the gametophytes of *A. ceterach* [19] (Fig. 3). Currently, phenolic profiling provides a valuable database of phenolic acids and flavonoids in the plant group studied and could suggest the applicability of targeted compounds as markers for chemotaxonomic studies [42].

A broad range of xanthones, which are biosynthetically related to flavonoids, have also been found in some *Asplenium* species [43, 44]. Nevertheless, mangiferin (**44**) and mangiferin glucoside (**45**) were the unique xanthones determined in *A. ceterach* [19].

Among terpenoids identified in rustyback fern, there are three compounds belonging to the group of triterpenes: diploptene (**47**), (24R)-cyclolaudenol (**48**) [30] and squalene (**49**) [15], and one sesquiterpenoid denoted as pterosin B (**46**). The pterosins represent a large group of naturally occurring sesquiterpenes with an indanone skeleton, firstly isolated as characteristic constituents of bracken fern [45].

Živković et al. [19] focused attention on the apparent phytochemical difference between distinct life stages of the *A. ceterach* ontogenetic cycle. As revealed by the authors, phenolic acids (in particular chlorogenic acid) were the most prominent in the rustyback fern sporophytes, whereas gametophytes were mainly characterized by the prevalence of xanthones.

Up to date, detailed phytochemical analysis of volatiles was performed only on *A. ceterach* sporophytes, including both the detection of VOCs emission from fresh (intact) plants and/or during the rehydration process, namely, the transition from dormant (desiccated) to a physiologically active state (Table 2). This was achieved by using different techniques, such as GC-MS, coupled Headspace GC–MS, and online PTR-MS analysis [15, 21]. Lipid derivatives were designated as the most prevalent components of the *A. ceterach* VOCs profile, reaching values that ranged around 80% (varying from 77.4% to 82.1%) of the total volatiles detected.

According to Froissard et al. [21], the VOCs pattern of *A. ceterach* is dominated mainly by (*E*)-2-Decenal (**76**), *n*-Nonanal (**67**), and (*E*)-2-Heptenal (**52**), while significantly lower content of shikimic acid derivatives (21.3%) was revealed. The same authors also confirmed the presence of several aromatic compounds (24.9%) in fresh aerial parts of rustyback fern, such as 3,4-Dihydroxybenzaldehyde (**89**) and *p*-Hydroxybenzoic acid (**7**), as well as a small amount of ionone-type carotenoid derivatives (**92** and **93**). Živković et al. [15] demonstrated that aerial parts of *A. ceterach* contain a broad range of isomeric heptadienals and decadienals (**57**, **59**, **81**, and **83**), representing more than 45% of the total VOCs content. In addition, considerable amounts of other lipid derivatives were also found (**50**, **51**, **52**, **53**, **62**, **67**, **76**, and **84**) comprising 35% of all identified volatiles, whereas green odor C_6-compounds, terpene-type hydrocarbons, and corresponding terpenoids were not detected. In the same study, authors screened VOCs emission from the rustyback fern sporophytes during the process of rehydration. In addition, online (real-time) monitoring of volatiles by using the PTR-MS method revealed a substantial enhancement of VOCs emission directly after re-watering that was followed by the gradual reduction of volatiles rates to the control levels within the next 12 h of rehydration [15].

Plants synthesize more than 450 different fatty acids (FAs) [46] that constitute a reserve of carbon and energy, ingredients and modulators of cell membranes, and precursors of various bioactive molecules and also take an important part in plant defense against various biotic and abiotic stresses [47]. Nevertheless, only a few of them represent common constituents of membrane lipids. Two saturated FAs, palmitic acid (16:0) and stearic acid (18:0), together with unsaturated oleic acid (18:1), linoleic acid (18:2, ω-6), and linolenic acid (18:3, ω-3), represent between 30 and 70% of total FAs in higher plants [48]. All previously mentioned FAs that belong to a group of so-called common fatty acids were identified in *A. ceterach* (Table 3) [15, 21, 26].

Based on the lipidomic analysis of three ferns from the family Aspleniaceae, Petkov et al. [26] reported that saturated FAs prevailed in *A. ceterach* sporophyte (75.4% of the total FAs content), with palmitic acid (**100**) as the most abundant FA species. As stated by the same authors, the de novo synthesis of FAs, as the mode of energy storage, reflects the adaptation of rustyback fern to the environment by giving preference to saturated FAs. Moreover, *A. ceterach* was highlighted as a rich source of polyunsaturated omega-3 and omega-6 FAs (**108**, **109**, and **110**) in comparison to other analyzed species from the same family [26]. It should be noted that a higher degree of polyunsaturated FAs in membrane phospholipids results in better membrane fluidity, which could be very important for cell and tissue flexibility during the dehydration/rehydration cycle in *A. ceterach* [49].

From the various studies, it has been reported that *A. ceterach* also contains a certain amount of "unusual fatty acids" (**94, 95, 96, 97, 98, 99, 101, 103, 104, 105**, and **110**), including FAs with less than 16 and/or more than 18 C-atoms [21, 26], the variant position of the double bond and cis/trans orientation [21], and odd number of C-atoms [21, 26]. Although unusual FAs are excluded from the membrane, they commonly occur as major components of seed oils, or in the polymerized form [50].

3 Biological Activities of *Asplenium ceterach*

From ancient times, people used almost all parts of the ferns as foods and beverages, ornamental plants, fertilizers, and handicrafts, for cosmetic and medicinal purposes. A wide variety of natural bioactive compounds with different pharmacological activities, including phenolics, flavonoids, alkaloids, terpenoids, and steroids, have been isolated from various fern species [45, 51, 52].

In recent years, there is a growing interest in phytochemicals with antioxidant, antimicrobial, and/or anti-inflammatory properties based on their potential use for the treatment of various chronic and infectious diseases [53]. Among the secondary metabolites detected in ferns, phenolic acids, flavonoids, and xanthones are shown to have considerable antioxidant activity, radical scavenging capacity, as well as anti-inflammatory, antimicrobial, and anticancer properties [54–57], while anti-inflammatory and antimicrobial effects of fatty acids are well known [58, 59].

Herein are summarized all biological activities of *A. ceterach* confirmed so far, including antioxidant potential, antimicrobial activity, cytotoxic and anticancer activity, anti-inflammatory properties, and diuretic activity (Fig. 3).

3.1 Antioxidant Potential

Plants represent a rich source of antioxidants that might protect human cells against oxidative stress, which is connected with aging, development, and progression of chronic diseases, including neurodegenerative, cardiovascular, cancer cardiovascular, etc. Exogenous antioxidants or enhancing endogenous antioxidant defenses of the human body can prevent or slow undesirable effects on cells of reactive oxygen species (ROS), which are induced by oxidative stress. Among the main classes of plant secondary metabolites, phenylpropanoid and associated phenolic compounds seem to be the most important, as they have displayed significant antioxidant activity in both in vivo and in vitro investigations. Seedless vascular plants, in particular ferns, are also known to synthesize these useful phytochemicals [60–62] and have therefore attracted the attention of researchers as a promising source of valuable antioxidants.

The antioxidant potential of *A. ceterach* has so far been the most studied of all biological activities. This is not surprising considering the results of phytochemical analyses of rustyback fern, which indicated the high content of phenolic compounds. Berk et al. [63] performed an extensive analysis of the antioxidant activity of *A. ceterach* aqueous extract using five complementary assays: β-carotene/linoleic acid, 1,1-diphenyl-2-picrylhydrazyl (DPPH) free radical scavenging, reducing power, chelating effect, and total antioxidant capacity (evaluated by the phosphomolybdenum method). According to the results obtained by the β-carotene/linoleic acid method, AE of *A. ceterach* revealed notable antioxidant activity (~95%) very similar to the synthetic antioxidants BHA and BHT used as a positive control. DPPH scavenging capacity of the extract reached approximately 85% for the concentrations $\geq$0.4 mg ml^{-1}, while the reducing power of the aqueous rustyback fern extract increased in a dose-dependent manner, with maximal values detected at 1.0 mg ml^{-1}. On the other hand, it was observed that aqueous extract of *A. ceterach* displayed rather weak chelating effects toward ferrous ions, reaching barely a 10% level of standard compound activity (EDTA), as compared to the positive control. Karadeniz et al. [64] have reported that water extracts of *A. ceterach* possess high scavenging activities for superoxide and nitric oxide radicals. The capacity to scavenge specified radicals, expressed as a percentage of inhibition, increased with the rising concentrations of the extract, which suggests that the radical scavenging ability of the fern extracts is concentration-dependent, with an IC_{50} value of 145.17 mg ml^{-1} for superoxide radicals and 372.03 mg ml^{-1} for nitric oxide radicals. In the extensive study of antioxidant activity of *A. ceterach*, Pekgöz and Çinbilgel [24] analyzed fern samples collected at 22 sites with different altitudes, within two provinces in Turkey. DPPH radical scavenging activity of tested methanol extracts varied considerably, ranging from 47.91 μg ml^{-1} to 294.66 μg ml^{-1}. However, the authors confirmed that the variations in antioxidant activity values were not significantly correlated to the altitude, but are more likely related to other environmental factors.

De Bellis et al. [23] determined the antioxidant activity of *Asplenium ceterach* (*C. officinarum*) aqueous extract (AE) in terms of radical scavenging ability by ORAC and DPPH radical scavenging assays. The ORAC value was 2798 $\pm$ 14.1 μmol TE (Trolox Equivalents) g AE^{-1} DW, while DPPH radical scavenging activity was dose-dependent, displaying a percentage of DPPH radical inhibition ranging from 11.26 $\pm$ 1.54% at 5.95 μg DW ml^{-1} assay mix, to 85.0 $\pm$ 1.6% at 96.5 μg DW ml^{-1} assay mix. Kazazic et al. [33] applied three different methods to evaluate the antioxidant activity of AE of rustyback fern (DPPH, ABTS, and FRAP method) and found a positive correlation between the total phenol content (TPC) and determined antioxidant activity. EC_{50} for DPPH radical scavenging activity was 5.49 ± 0.72 mg 100 g^{-1} DW of water extract of *A. ceterach*, while its ability to bind ABTS radicals was 5.79 ± 0.74 mM TE 100 g^{-1} DW. Reductive capacity of AE determined by the FRAP method was 12.26 ± 0.15 mM Fe^{2+} 100 g^{-1} DW [33], which was nearly three-fold lower than values reported by Fatima et al. [65] (29.36 ± 0.32 mM Fe^{2+} 100 g^{-1} DW). However, the dissimilarities in obtained data should be taken with caution, because, when performing the DPPH test, there are noticeable differences in terms of reagent concentration, solvent, and/or way of presenting the results [66]. Thus, Farràs Adrià et al. [67] determined the capacity of *A. ceterach* extracts to scavenge peroxyl radicals and proved that this effect is strongly correlated to the type of extract. Based on the results obtained by ORAC and DPPH assays, methanol extract of dry rustyback fronds showed 3.5- to ten-fold enhanced radical absorbance capacity, respectively, when compared to extracts prepared with nonpolar (lipophilic) solvent hexane. In addition, the same authors [67] stated that the antioxidant activity of fern methanol extracts against superoxide radicals (generated by xanthine/xanthine oxidase) was twice as high as the reference compound Trolox. Notable variation in evaluated antioxidant capacity might be caused by the way of testing, so Petkov and coworkers [26], who used DPPH in the endpoint mode, actually detected antioxidant capacity of *A. ceterach* rather than antioxidant activity.

In contrast to the extensive study and diverse reports related to the antioxidant activity of mature rustyback fern plants (sporophytes), similar data related to the *A. ceterach* gametophytes can be rarely found in the literature. Živković et al. [19] evaluated the scavenging activity of rustyback fern gametophytes using DPPH and ABTS assays and recorded IC_{50} values of 30.47 mg ml^{-1} and 2.22 mg ml^{-1}, respectively. Concurrently, the scavenging activity of the fronds of *A. ceterach* sporophyte was more efficient, reaching the IC_{50} value of 0.46 mg ml^{-1} (DPPH assay) and 0.18 mg ml^{-1} (ABTS assay). Authors concluded that a significant difference in antioxidant activity between these two ontogenetic stages of the rustyback fern could be attributed to the higher content of phenolic acids in sporophytes, particularly chlorogenic acid [19].

3.2 Antimicrobial Activity

The increased incidence of antibiotic-resistant pathogens has drawn the attention of the pharmaceutical and scientific communities to the examination of the antimicrobial activity of bioactive substances of plant origin. Plant extracts, or compounds from plant sources, often demonstrate high activity against pathogens, and they rarely have severe side effects. Plant secondary metabolites, due to their diverse chemical structure, exhibit various pharmacological properties and represent a useful and rich source of antimicrobial agents [68]. Based on the chemical structures, they can be classified into several major groups, including alkaloids, sulfur-containing compounds, terpenoids, and polyphenols [69]. Numerous studies of the antimicrobial activities of fern extracts indicate that they have significant antimicrobial potential and can be a noteworthy source of pharmacologically active compounds.

However, reports of the effectiveness of *Asplenium* species, especially *A. ceterach*, are limited. Berk et al. [37] examined the antimicrobial activity of the AE of *A. ceterach* against *Candida albicans* (ATCC 10231) and 11 bacterial strains, using two complementary tests. No activity of AE was observed against *Pseudomonas aeruginosa* (ATCC 27853), *Klebsiella pneumoniae* (NCTC 5046), *Escherichia coli* (ATCC 35218), *Proteus vulgaris* (RSHM 96022), *Corynebacterium diphtheriae* (RSHM 633), and *C. albicans*. Among tested microorganisms, the most sensitive was *Staphylococcus aureus* (ATCC 25923) (MIC value 9.00 μg ml^{-1}), followed by *Shigella dysenteriae* (NCTC 9762) (MIC value 18.00 μg ml^{-1}). Comparing obtained results with the positive control (gentamicin and nystatin), Berk et al. [63] described the antimicrobial activity of rustyback AE as moderate and nonselective. Contrary to these results, according to Petkov et al. [70], *A. ceterach* extract showed prominent effects against *P. aeruginosa* and inhibited the growth of *Bacillus cereus* (ATCC 11778). The same authors, in parallel with phytochemical analysis, performed the most comprehensive examination of in vitro antimicrobial activity of *A. ceterach* [26]. They tested an in vitro antimicrobial activity of rustyback extract against even fourteen microorganisms: four Gram-positive bacteria (*Bacillus subtilis* ATCC 19659, *S. aureus* ATCC 25923, *Listeria monocytogenes* ATCC 19111, *Enterococcus faecalis* ATCC 19433), four Gram-negative bacteria (*P. vulgaris* ATCC 6380, *E. coli* ATCC 25922, *Salmonella abony* NCTC 6017, *P. aeruginosa* NBIMCC 1370), four fungi (*Penicillium chrysogenum* ATCC 28089, *Rhizopus arrhizus* ATCC 11145, *Aspergillus niger* ATCC 1015, *Fusarium moniliforme* ATCC 38932), and two yeasts (*C. albicans* NBIMCC 74, *Candida utilis* ATCC 4240). Rustback fern AE displayed antimicrobial activity against: *E. coli*, *P. vulgaris*, *S. aureus*, *B. subtilis*, *C. utilis*, and *F. moniliforme*, and no selectivity was detected [26]. It is known that the antimicrobial activity of plant secondary metabolites could be dissimilar in different microbial organisms, that is, between Gram-positive/Gram-negative bacteria and/or bacterial/fungal cells [71]. Antimicrobial potential and mechanism of action of different secondary metabolites are influenced by several factors; for instance, their solubility in water, concentration, temperature, and pH [68, 71, 72]. Secondary metabolites, such as polyphenols and flavonoids, which have phenolic hydroxyl functional groups in their structure, may be responsible for the antimicrobial activity of ferns (including *A. ceterach*), although the contribution of other isolated compounds from fern species (alkaloids, steroids, triterpenes, and polysaccharides) should not be neglected.

3.3 Cytotoxic and Anticancer Activity

Plants have been known for a long time as a valuable source of compounds with anticancer activity and therefore used in pharmaceutics [73]. Although the focus of research on natural compounds as anticancer agents has been extended to many seedless vascular plants, data available on anticancer or cytotoxic activity of rustyback fern is rather scarce. Fatima et al. [65] determined the antiproliferative activity of AE of *A. ceretach* by clonogenic cell survival assay. Namely, the AE of

A. ceterach led to the declined clonogenic survival of PC3 and LNCaP, two highly metastatic prostate cancer cell lines. Additionally, treatment with different concentrations of AE of *A. ceterach* inhibited proliferative/colony formation capacity of both PC3 and LNCaP cancer cell lines, as well as their migration in a dose- and time-dependent manner, as compared to the untreated control cells. Petkov et al. [70] examined in vitro cytotoxicity and anticancer potential of three *Asplenium* species, including *A. ceterach*, against six different mammalian cell lines: three humans (HeLa, ATCC CCL-2; A549, ATCC CCL-185; FL, NBIMCC 94) and three mouse (LS48, DSM ACC 2455; NIH/3 T3, ATCC CRL-1658; RAW 264.7, ATCC TIB-71) cell lines. The extract from *A. ceterach* displayed the strongest inhibitory properties against human cervical cancer cells, where approximately 40 μg ml^{-1} of the extract induced 50% inhibition of HeLa cell cultures. Two other cancerous cell lines (A549-human lung carcinoma and TIB-71-mouse tumor-derived cell lines) were less affected by treatment with extracts of *A. ceterach*, as well as all noncancerous cells (FL-human and NIH/3 T3-mouse cells), which were used as control. Additionally, authors have noticed that treatment of HeLa cells with *A. ceterach* extract was associated with a significant proapoptotic effect, thus indicating the possible mechanism of anticancer action [70]. Recent studies have shown that *A. ceterach* contains significantly high levels of tannins, ω-3 and ω-6 polyunsaturated fatty acids (PUFAs) [26], which are well known for their various positive bioactivities, including anticancer effects [74, 75].

3.4 Anti-Inflammatory Properties

Results of the evaluation of the anti-inflammatory activity of *A. ceterach* have been reported by Sidhoum et al. [20]. Acute plantar and topical inflammations were induced in mice by using carrageenan and xylene, respectively. Authors indicated that treatment with inflammatory agents induced edema formations, but the oral application of the *A. ceterach* extract, in the concentration of 200 mg ml^{-1}, resulted in a significant decrease in the inflammatory symptoms. Namely, plant infusion administered orally reduced the formation of plantar edema (inflammations) by 32.75%, while topical anti-inflammation properties of the *A. ceterach* extract were close to the range of reference compound Ibuprofen (200 mg), used as a positive control (44.18%).

3.5 Diuretic Activity

Recent studies on the diuretic activity of *A. ceterach* have been conducted by Sidhoum et al. [20]. The plant extract was administered orally to the Wistar albino rats, at a dose of 50 mg kg^{-1}. It was found that the accumulated volumetric urinary excretion increased continuously after the treatment with *A. ceterach* infusion, showing fine differences with respect to both the control (0.9% saline solution) and the reference samples (diuretic drug furosemide). Accordingly to the experimental results obtained on the total urinary volumes, the authors concluded that the extract of *A. ceterach* induced significantly high diuretic activity, compared to the control treatments, with volumetric urinary excretions values of 83.53% [20].

Asplenium ceterach is traditionally used in southern Italy for its diuretic properties and to treat kidney stones (mineral salts, mucilages, and tannins); hence, the potential mechanism of the antilithiatic action of rustyback fern extracts has been proposed recently by De Bellis et al. [23]. The authors explained the major impact of *A. ceterach* AE on kidney stone formation and/or elimination throughout several effects. Firstly, treatment with AE of rustyback fern led to inhibition of calcium oxalate monohydrate (COM) growth and aggregation, and subsequent reduction of the size and shape of COM crystals, which became less adherent to the renal tubular cells and more easily excreted through the urinary tract, thus preventing the kidney stone formation. Furthermore, *A. ceterach* AE promoted the formation of calcium oxalate dihydrate (COD) rather than monohydrate, which are known to be less dangerous than the COM in urolithiasis pathogenesis. Based on the obtained results, De Bellis et al. [23] suggested that *A. ceterach* AE represents promising natural therapy for the urolithiasis treatment.

3.6 Other Activities

It is well known that UV irradiation and the secondary reaction of free radicals can cause serious damage to the structure of DNA, leading to gene coding and protein expression errors and subsequent metabolic disorders that cause pathological issues. Therefore, the identification of new natural products that can provide protection against oxidative stress may have important implications for human health [76]. Berk et al. [63] reported that the AE of *A. ceterach* sporophytes showed noteworthy DNA damage protection potential. Namely, the addition of 20, 40, and 50 mg ml^{-1} of rustyback fern AE to the reaction mixture provided significant protection against the damage of native supercoiled DNA (derived from pBR322 plasmid) after treatment with UV and H_2O_2. Based on these data, the authors highlighted the possible use of *A. ceterach* extract in the food and cosmetic industries.

Zangeneh et al. [77] demonstrated that *A. ceterach* extract has hepatoprotective and hematoprotective properties, thus reducing the causality of diabetes in experimental mice. Streptozotocin (STZ), a diabetogenic agent, produces liver damage in mice, which is manifested by the alteration levels of several serum parameters (ALP, AST, ALT, cholesterol). Different doses of rustyback fern AE significantly decreased the raised levels of ALP, AST, ALT, cholesterol, LDL, WBC, and platelet and at the same time induced increment of HDL, SOD, CAT, and RBC values. Furthermore, the application of *A. ceterach* extract protected the liver against STZ-induced hepatocellular injuries. The observed hepatoprotective effect of *A. ceterach* extract, according to the authors, could be correlated to its phytochemical profile, particularly to bioactive compounds that possess significant anti-inflammatory and antioxidant potential [77].

Petkov et al. [41] have recently reported **receptor-mediated activity** of *A. ceterach* extract on the contractility and bioelectrogenesis of rat gastric smooth muscle tissues. It was shown that rustyback fern AE acts as agonist of 5-HT_2 receptors and causes contraction in muscle tissues by triggering serotonergic signaling systems. The noted effect was concentration-dependent in the range 0.015–0.150 mg ml^{-1} of *A. ceterach* AE, whereas the contractions were reduced by ketanserin (5×10^{-7} and 5×10^{-6} mol l^{-1}), an antagonist of serotonin 5-HT_2 receptor.

4 Ethnomedicinal Uses of *Asplenium ceterach*

The distribution of the ferns from the spleenwort family (Aspleniaceae) is subcosmopolitan, being the most numerous in the tropics. Their economic significance includes many important cultivars and some traditional medicinal uses [78, 79].

Both the scientific name of the family (and its major genus *Asplenium* as well) and the common name "spleenwort" derive from the Greek word "asplênon" meaning the remedy for the spleen. Indeed, various species assigned to the genus *Asplenium* have been used for centuries in different regions of the world for the treatment of spleen diseases [21, 80]. *Asplenium ceterach* used to be called "Miltwaste," most likely due to the lobular milt-like shape appearance of its fronds resembling the form of the spleen, for which it was thought to have healing powers. This species was used as a remedy for enlarged spleen [81], but also to cure liver and other visceral obstructions [82].

Whole rustyback fern plants have diuretic and astringent properties, while rhizomes are used for the treatment of calculus, jaundice, and incontinence of urine [18, 83]. An infusion prepared of aerial fern parts is widely used in folk medicine against kidney ailments [84–86], hemorrhoids, as well as a natural remedy for respiratory and intestinal disorders [87, 88]. Diverse ethnobotanical studies indicated a great frequency of use of these plant species for the treatment of kidney stones [89–93], bladder pains, and infections [94]. The whole plant extract has been reported as a demulcent, expectorant, and laxative, with anticatarrhal and anthelmintic properties, and most of the listed effects were directly attributed to the high phenolics content [21, 63, 92, 95, 96]. Aerial parts of *A. ceterach* sauteed in olive oil or eaten revealed hemostatic effects and have been recommended as a remedy for ulcers and wounds healing [97]. Ethnopharmacological studies have suggested the use of rustyback fern leaves decoctions against constipation [98], hypertension, joint aches, malaria, and renal colic [90, 99]. Wild medicinal plants, like *A. ceterach*, have been shown to play a key role in treatments related to women's healthcare in rural communities [99, 100]. Since *A. ceterach* contains active sterols analogous to those identified in *A. trichomanes*, its use for regulation of menstruation could be ascribed to a similar phytochemical profile [101]. Young leaves extract of *A. ceterach* L. subsp. *ceterach* was commonly utilized against enema in children, while women used rhizome decoction as a traditional herbal remedy to treat miscarriage and infertility [102]. Decoction and leaf paste (applied as a poultice) prepared from *A. ceterach* x *punjabense* Chahala Nakaike roots and leaves showed antiviral activity and were used for the treatment of skin diseases and leucorrhoea [103]. Rustyback fern has been considered an important part of folk veterinary phytotherapy and was utilized as a remedy for gastrointestinal complaints [104]. It must be pointed out that despite the species belonging to the group of ferns may exhibit noxious effects, no toxicity has been reported for *A. ceterach* L. so far [105]. Ethnobotanical study regarding medicinal plants and health is one of the sources for future scientific studies and represents a valuable guide to their exploration for use as modern medicines [85].

5 Conclusions

The qualitative phytochemical analyses of the *Asplenium ceterach* gametophyte and sporophyte have revealed the presence of a wide variety of bioactive constituents, including phenolic acids, flavonoids, xanthones, terpenes, coumarins, and tannins, followed by the free quinones and leucoanthocyanins. Further, *A. ceterach* represents a rich source of volatile organic compounds, especially lipid derivatives. Among the 110 phytochemicals identified so far, phenolic acids and flavonoids are the most abundant and have been recognized as the representative ingredients in terms of the antioxidant and biochemical potential of rustyback fern, although a role of fatty acids and derived compounds in the overall biological activity of *A. ceterach* should not be neglected. Based on diverse ethnobotanical uses and numerous biological effects (antioxidant, antimicrobial, cytotoxic, anticancer, anti-inflammatory, diuretic, etc.), the rustyback fern may be considered a promising source of bioactive compounds. In addition, listed bioactivities highlight the great potential of *A. ceterach* in the development of novel therapeutics for medicinal applications.

Acknowledgments This work was supported by the Ministry of Education, Science and Technological Development of the Republic of Serbia, grant numbers OI173024, 451-03-9/2021-14/200007, and 451-03-68/2022-14/200007. The authors would like to thank Dr. Uroš Gašić (University of Belgrade, Serbia) for preparing chemical structures.

References

1. Schneider H, Schuettpelz E, Pryer KM, Cranfill R, Magallón S, Lupia R (2004) Ferns diversified in the shadow of angiosperms. Nature 428(6982):553–557. https://doi.org/10.1038/nature02361
2. The Pteridophyte Phylogeny Group I (2016) A community-derived classification for extant lycophytes and ferns. J Syst Evol 54(6):563–603. https://doi.org/10.1111/jse.12229
3. Anderson OR (2021) Physiological ecology of ferns: biodiversity and conservation perspectives. Int J Biodivers Conserv 13(2):49–63. https://doi.org/10.5897/IJBC2021.1482
4. Nitta JH, Meyer JY, Taputuarai R, Davis CC (2017) Life cycle matters: DNA barcoding reveals contrasting community structure between fern sporophytes and gametophytes. Ecol Monogr 87(2):278–296. https://www.jstor.org/stable/26358571

5. Xu Z, Deng M (2017) Aspleniaceae. In: Identification and control of common weeds, vol 2. Springer, Dordrecht. https://doi.org/10.1007/978-94-024-1157-7_11
6. Xu KW, Zhang L, Rothfels CJ, Smith AR, Viane R, Lorence D, Wood KR, Chen CW, Knapp R, Zhou L, Lu NT, Zhou XM, Wei HJ, Fan Q, Chen SF, Cicuzza D, Gao XF, Liao WB, Zhang LB (2020) A global plastid phylogeny of the fern genus *Asplenium* (Aspleniaceae). Cladistics 36(1):22–71. https://doi.org/10.1111/cla.12384
7. Van den Heede CJ, Viane RL, Chase MW (2003) Phylogenetic analysis of *Asplenium* subgenus *Ceterach* (Pteridophyta: Aspleniaceae) based on plastid and nuclear ribosomal ITS DNA sequences. Am J Bot 90(3):481–495. https://www.jstor.org/stable/4124166
8. Trewick SA, Morgan-Richards M, Russell SJ, Henderson S, Rumsey FJ, Pinter I, Barrett JA, Gibby M, Vogel JC (2002) Polyploidy, phylogeography and Pleistocene refugia of the rockfern *Asplenium ceterach*: evidence from chloroplast DNA. Mol Ecol 11(10):2003–2012. https://doi.org/10.1046/j.1365-294X.2002.01583.x
9. Oliver MJ, Tuba Z, Mishler BD (2000) The evolution of vegetative desiccation tolerance in land plants. Plant Ecol 151(1):85–100. https://doi.org/10.1023/A:1026550808557
10. Grantham M (1997) Desert ferns. Spring Newsletter 22(2):1–2
11. Schwab KB, Heber U (1984) Thylakoid membrane stability in drought-tolerant and drought-sensitive plants. Planta 161(1):37–45. https://doi.org/10.1007/BF00951458
12. Schwab KB, Schreiber U, Heber U (1989) Response of photosynthesis and respiration of resurrection plants to desiccation and rehydration. Planta 177(2):217–227. https://doi.org/10.1007/BF00392810
13. Živković S, Popović M, Dragišić-Maksimović J, Momčilović ID, Grubišić D (2010) Dehydration-related changes of peroxidase and polyphenol oxidase activity in fronds of the resurrection fern *Asplenium ceterach* L. Arch Biol Sci 62(4):1071–1081. https://doi.org/10.2298/ABS1004071Z
14. Bogdanović M, Ilić M, Živković S, Sabovljević A, Grubišić D, Sabovljević M (2012) Comparative study on the effects of NaCl on selected moss and fern representatives. Aust J Bot 59(8):734–740. https://doi.org/10.1071/BT11059
15. Živković S, Skorić M, Ristić M, Filipović B, Milutinović M, Perišić M, Puač N (2021) Rehydration process in rustyback fern (*Asplenium ceterach* L.): profiling of volatile organic compounds. Biology 10(7):574. https://doi.org/10.3390/biology10070574
16. Soare LC (2008) *In vitro* development of gametophyte and sporophyte in several fern species. Not Bot Hort Agrobot Cluj 36(1):13–19. https://doi.org/10.15835/nbha36183
17. Sabovljević M, Vujičić M, Sekulić JŠ, Segarra-Moragues JG, Papp B, Skorić M, Dragačević L, Sabovljević A (2012) Reviving, *in vitro* differentiation, development, and micropropagation of the rare and endangered moss *Bruchia vogesiaca* (Bruchiaceae). HortScience 47(9):1347–1350
18. Mannan MM, Maridass M, Victor B (2008) A review on the potential uses of ferns. Ethnobot leafl 2008(1):33. https://opensiuc.lib.siu.edu/ebl/vol2008/iss1/33
19. Živković S, Skorić M, Šiler B, Dmitrović S, Filipović B, Nikolić T, Mišić D (2017) Phytochemical characterization and antioxidant potential of rustyback fern (*Asplenium ceterach* L.). Lekovite sirovine 37:15–20. https://doi.org/10.5937/leksir1737015Z
20. Sidhoum R, Bey MB, Halli L, Yalaoui O, Belkadi A (2020) Phenolic profile, anti-inflammatory and diuretic properties of *Asplenium ceterach* tested on albino mice and Wistar albino rats. Asian J Research Chem 13(6):449–454. https://doi.org/10.5958/0974-4150.2020.00080.2
21. Froissard D, Rapior S, Bessière JM, Buatois B, Fruchier A, Sol V, Fons F (2015) Asplenioideae species as a reservoir of volatile organic compounds with potential therapeutic properties. Nat Prod Commun 10(6):1079–1083. https://doi.org/10.1177/1934578X1501000671
22. Tomou EM, Skaltsa H (2018) Phytochemical investigation of the fern *Asplenium ceterach* (Aspleniaceae). Nat Prod Commun 13(7):849–850. https://doi.org/10.1177/1934578X1801300715

23. De Bellis R, Piacentini MP, Meli MA, Mattioli M, Menotta M, Mari M, Valentini L, Palomba L, Desideri D, Chiarantini L (2019) In vitro effects on calcium oxalate crystallization kinetics and crystal morphology of an aqueous extract from *Ceterach officinarum*: analysis of a potential antilithiatic mechanism. PLoS One 14(6):e0218734. https://doi.org/10.1371/journal.pone.0218734
24. Pekgöz AK, Çinbilgel I (2019) Phytochemical contents and antioxidant activities of fern, *Asplenium ceterach* L. in different altitudes. Bangladesh J Bot 48(2):315–320. https://doi.org/10.3329/bjb.v48i2.47674
25. Imperato F (1983) A flavanone glycoside from the fronds of *Ceterach officinarum*. Phytochemistry 22(1):312–313
26. Petkov V, Slavova I, Teneva D, Mladenova T, Stoyanov P, Argirova M (2021) Phytochemical study and biological activity of three fern species of the *Asplenium* genus growing in Bulgaria. Nat Prod J 11:e120521193308. https://doi.org/10.2174/2210315511666210512024716
27. Ismail AM, Hamdi BA, Maulood BK, Al-Khasrejy TO, Maythem AA (2021) Phytochemical study of *Asplenium ceterach* L. (Aspleniaceae) from Erbil province, Kurdistan of Iraq. ZJPAS 33(s1):187–191. https://doi.org/10.21271/ZJPAS.33.s1.20
28. Imperato F (1981) New sulphate esters of hydroxycinnamic acid-sugar derivatives in ferns. Chem Ind 19:691–692
29. Iwashina T, Matsumoto S, Nishida M, Nakaike T (1993) The flavonoids from three Pakistani *Asplenium* subgenera *Ceterach* and *Ceterachopsis* species and their chemotaxonomic consideration. In: Nakaike T, Malik S (eds) Cryptogamic Flora of Pakistan, vol 2. National Science Museum Tokyo, Japan
30. Berti G, Bottari F (1968) Constituents of ferns. In: Rheinhold L, Liwschitz Y (eds) Progress in phytochemistry. Interscience, London
31. Gan RY, Chan CL, Yang QQ, Li HB, Zhang D, Ge YY, Gunaratne A, Ge J, Corke H (2019) 9-bioactive compounds and beneficial functions of sprouted grains. In: Feng H, Nemzer B, DeVries JW (eds) Sprouted grains, nutritional value, production and applications. AACC International Press & Elsevier Inc. https://doi.org/10.1016/B978-0-12-811525-1.00009-9
32. Đurđević L, Mitrović M, Pavlović P, Bojović S, Jarić S, Oberan L, Gajić G, Kostić O (2007) Total phenolics and phenolic acids content in leaves, rhizomes and rhizosphere soil under *Ceterach officinarum* DC, *Asplenium trichomanes* L. and *A. adiantum nigrum* L. in the gorge of Sićevo (Serbia). Ekológia (Bratislava) 26(2):164–173
33. Kazazic M, Djapo M, Ademovic E (2016) Antioxidant activity of water extracts of some medicinal plants from Herzegovina region. Int J Pure App Biosci 4(2):85–90. https://doi.org/10.18782/2320-7051.2251
34. Bewley JD, Krochko JE (1982) Desiccation-tolerance. In: Lange OL, Nobel PS, Osmond CB, Ziegler H (eds) Physiological plant ecology II. Encyclopedia of plant physiology, vol 12/B. Springer, Berlin, Heidelberg. https://doi.org/10.1007/978-3-642-68150-9_11
35. Oliver MJ, Farrant JM, Hilhorst HWM, Mundree S, Williams B, Bewley JD (2020) Desiccation tolerance: avoiding cellular damage during drying and rehydration. Annu Rev Plant Biol 71(1):435–460. https://doi.org/10.1146/annurev-arplant-071219-105542
36. Živković S, Popović Bijelić A, Mojović M (2012) EPR investigation of free radical formation in rustyback fern (*Asplenium ceterach* L.). Proceedings of the Regional Biophysics Conference, September 3–7, 2012 held at Kladovo, Serbia
37. Lisete-Torres P, Losada-Barreiro S, Albuquerque H, Sánchez-Paz V, Paiva-Martins F, Bravo-Díaz C (2012) Distribution of hydroxytyrosol and hydroxytyrosol acetate in olive oil emulsions and their antioxidant efficiency. J Agric Food Chem 60:7318–7325. https://doi.org/10.1021/jf301998s
38. Pratyusha S (2022) Phenolic compounds in the plant development and defense: an overview. In: Hasanuzzaman M, Nahar K (eds) Plant stress physiology-perspectives in agriculture. IntechOpen, London. https://doi.org/10.5772/intechopen.102873
39. Murakami T, Tanaka N (1988) Occurrence, structure and taxonomic implications of fern constituents. In: Herz W, Grisebach H, Kirby GW, Tamm C (eds) Fortschritte der Chemie organischer Naturstoffe/progress in the chemistry of organic natural products, vol 54. Springer, Vienna. https://doi.org/10.1007/978-3-7091-8999-3_1

40. San Francisco M, Cooper-Driver G (1984) Anti-microbial activity of phenolic acids in *Pteridium aquilinum*. Am Fern J 74(3):87–96. https://doi.org/10.2307/1546543
41. Petkov VH, Ardasheva RG, Prissadova NA, Kristev AD, Stoyanov PS, Argirova MD (2021) Receptor-mediated biological effects of extracts obtained from three *Asplenium* species. Z Naturforsch C J Biosci 76(9–10):367–373. https://doi.org/10.1515/znc-2020-0223
42. Mišić D, Šiler B, Gašić U, Avramov S, Živković S, Nestorović Živković J, Milutinović M, Tešić Ž (2015) Simultaneous UHPLC/DAD/(+/−)HESI-MS/MS analysis of phenolic acids and nepetalactones in methanol extract of *Nepeta* species: a possible application in chemotaxonomic studies. Phytochem Anal 26(1):72–85. https://doi.org/10.1002/pca.2538
43. Iwashina T, Matsumoto S (2011) Flavonoid properties of six *Asplenium* species in Vanuatu and New Caledonia, and distribution of flavonoid and related compounds in *Asplenium*. Bull Natl Mus Nat Sci Ser B 37(3):133–145
44. Živković S, Milutinović M, Maksimović V, Ćirić A, Ivanov M, Božunović J, Banjanac T, Mišić D (2020) Antioxidant and antimicrobial activity of two *Asplenium* species. S Afr J Bot 132:180–187. https://doi.org/10.1016/j.sajb.2020.03.034
45. Cao H, Chai TT, Wang X, Morais-Braga MF, Yang JH, Wong FC, Wang R, Yao H, Cao J, Cornara L, Burlando B (2017) Phytochemicals from fern species: potential for medicine applications. Phytochem Rev 16(3):379–440. https://doi.org/10.1007/s11101-016-9488-7
46. Ohlrogge J, Thrower N, Mhaske V, Stymne S, Baxter M, Yang W, Liu J, Shaw K, Shorrosh B, Zhang M, Wilkerson C (2018) PlantFAdb: a resource for exploring hundreds of plant fatty acid structures synthesized by thousands of plants and their phylogenetic relationships. Plant J 96 (6):1299–1308. https://doi.org/10.1111/tpj.14102
47. He M, Ding NZ (2020) Plant unsaturated fatty acids: multiple roles in stress response. Front Plant Sci 11:562785. https://doi.org/10.3389/fpls.2020.5627851378
48. Tjellström H, Yang Z, Allen DK, Ohlrogge JB (2012) Rapid kinetic labeling of Arabidopsis cell suspension cultures: implications for models of lipid export from plastids. Plant Physiol 158(2):601–611. https://doi.org/10.1104/pp.111.186122
49. Bartels D, Hussain SS (2011) Resurrection plants: physiology and molecular biology. In: Lüttge U, Beck E, Bartels D (eds) Plant desiccation tolerance. ecological studies, vol 215. Springer, Berlin, Heidelberg
50. Pollard M, Beisson F, Li Y, Ohlrogge JB (2008) Building lipid barriers: biosynthesis of cutin and suberin. Trends Plant Sci 13(5):236–246. https://doi.org/10.1016/j.tplants.2008.03.003
51. Ho R, Teai T, Bianchini JP, Lafont R, Raharivelomanana P (2010) Ferns: from traditional uses to pharmaceutical development, chemical identification of active principles. In: Fernández H, Revilla MA, Kumar A (eds) Working with ferns: issues and applications. Springer, New York
52. Winter A (2003) Plant resources of South-East Asia no 15 (2). In: de Winter WP, Amoroso VB (eds) Cryptogams: ferns and fern allies. Backhuys Publishers, Leiden
53. Halliwell B (1996) Antioxidant in human health and diseases. Annu Rev Nutr 16:33–50. https://doi.org/10.1146/annurev.nu.16.070196.000341
54. Kumar N, Goel N (2019) Phenolic acids: natural versatile molecules with promising therapeutic applications. Biotechnol Rep (Amst) 24:e00370. https://doi.org/10.1016/j.btre.2019.e00370
55. Kumar S, Pandey AK (2013) Chemistry and biological activities of flavonoids: an overview. Sci World J 2013:162750. https://doi.org/10.1155/2013/162750
56. Ahmad I (2016) Recent insight into the biological activities of synthetic xanthone derivatives. Eur J Med Chem 116:267–280. https://doi.org/10.1016/j.ejmech.2016.03.058
57. Baskaran XR, Geo Vigila AV, Zhang SZ, Feng SX, Liao WB (2018) A review of the use of pteridophytes for treating human ailments. J Zhejiang Univ Sci B 19(2):85–119. https://doi.org/10.1631/jzus.B1600344
58. Bremner P, Heinrich M (2002) Natural products as targeted modulators of the nuclear factor-KB pathway. J Pharm Pharmacol 54(4):453–472. https://doi.org/10.1211/0022357021778637
59. Casillas-Vargas G, Ocasio-Malavé C, Medina S, Morales-Guzmán C, Del Valle RG, Carballeira NM, Sanabria-Ríos DJ (2021) Antibacterial fatty acids: an update of possible

mechanisms of action and implications in the development of the next-generation of antibacterial agents. Prog Lipid Res 82:101093. https://doi.org/10.1016/j.plipres.2021.101093
60. Kasote DM, Katyare SS, Hegde MV, Bae H (2015) Significance of antioxidant potential of plants and its relevance to therapeutic applications. Int J Biol Sci 11(8):982–991. https://doi.org/10.7150/ijbs.12096
61. Šamec D, Karalija E, Šola I, Vujčić Bok V, Salopek-Sondi B (2021) The role of polyphenols in abiotic stress response: the influence of molecular structure. Plan Theory 10(1):118. https://doi.org/10.3390/plants10010118
62. Dvorakova M, Pumprova K, Antonínová Ž, Rezek J, Haisel D, Ekrt L, Vanek T, Langhansova L (2021) Nutritional and antioxidant potential of fiddleheads from European ferns. Foods 10 (2):460. https://doi.org/10.3390/foods10020460
63. Berk S, Tepe B, Arslan S, Sarikurkcu C (2011) Screening of the antioxidant, antimicrobial and DNA damage protection potentials of the aqueous extract of *Asplenium ceterach* DC. Afr J Biotechnol 10(44):8902–8908. https://doi.org/10.5897/AJB11.1011
64. Karadeniz A, Çinbilgel I, Gün SŞ, Çetin A (2015) Antioxidant activity of some Turkish medicinal plants. Nat Prod Res 29(24):2308–2312. https://doi.org/10.1080/14786419.2015.1005618
65. Fatima A, Laila O, Murtaza I, Masoodi K (2020) Nutraceutical composition and anticancerous potential of an unexplored herb *Asplenium ceterach* from Kashmir region. Indian J Pure Appl Biosci 8(2):289–297. https://doi.org/10.18782/2582-2845.8036
66. Xie J, Schaich KM (2014) Re-evaluation of the 2, 2-diphenyl-1-picrylhydrazyl free radical (DPPH) assay for antioxidant activity. J Agric Food Chem 62(19):4251–4260. https://doi.org/10.1021/jf500180u
67. Farràs A, Cásedas G, Les F, Terrado EM, Mitjans M, López V (2019) Evaluation of anti-tyrosinase and antioxidant properties of four fern species for potential cosmetic applications. Forests 10(2):179. https://doi.org/10.3390/f10020179
68. Savoia D (2012) Plant-derived antimicrobial compounds: alternatives to antibiotics. Future Microbiol 7(8):979–990. https://doi.org/10.2217/fmb.12.68
69. Khameneh B, Iranshahy M, Soheili V, Bazzaz BS (2019) Review on plant antimicrobials: a mechanistic viewpoint. Antimicrob Resist Infect Control 8:118. https://doi.org/10.1186/s13756-019-0559-6
70. Petkov V, Batsalova T, Stoyanov P, Mladenova T, Kolchakova D, Argirova M, Raycheva T, Dzhambazov B (2021) Selective anticancer properties, proapoptotic and antibacterial potential of three asplenium species. Plan Theory 10(6):1053. https://doi.org/10.3390/plants10061053
71. Radulovic NS, Blagojevic PD, Stojanovic-Radic ZZ, Stojanovic NM (2013) Antimicrobial plant metabolites: structural diversity and mechanism of action. Curr Med Chem 20(7): 932–952
72. Gorlenko CL, Kiselev HY, Budanova EV, Zamyatnin AA, Ikryannikova LN (2020) Plant secondary metabolites in the battle of drugs and drug-resistant bacteria: new heroes or worse clones of antibiotics? Antibiotics 9(4):170. https://doi.org/10.3390/antibiotics9040170
73. Skorić M, Gligorijević N, Čavić M, Todorović S, Janković R, Ristić M, Mišić D, Radulović S (2017) Cytotoxic activity of *Nepeta rtanjensis* Diklić & Milojević essential oil and its mode of action. Ind Crop Prod 100:163–170. https://doi.org/10.1016/j.indcrop.2017.02.027
74. Dierge E, Debock E, Guilbaud C, Corbet C, Mignolet E, Mignard L, Bastien E, Dessy C, Larondelle Y, Feron O (2021) Peroxidation of n-3 and n-6 polyunsaturated fatty acids in the acidic tumor environment leads to ferroptosis-mediated anticancer effects. Cell Metab 33(8): 1701–1715.e5. https://doi.org/10.1016/j.cmet.2021.05.016
75. Teodor DE, Ungureanu O, Gatea F, Radu LG (2020) The potential of flavonoids and tannins from medicinal plants as anticancer agents. Anti Cancer Agents Med Chem 20(18):2216–2227. https://doi.org/10.2174/1871520620666200516150829
76. Lu HP, Jia YN, Yu Y, Xu L (2017) DNA protective activity of a hydroethanol extract and six polyphenol monomers from *Morus alba* L. (mulberry) twig. Int J Food Prop 20(sup2): 2207–2219. https://doi.org/10.1080/10942912.2017.1368554

77. Zangeneh MM, Zangeneh A, Bahrami E, Almasi M, Amiri-Paryan A, Tahvilian R, Moradi R (2018) Evaluation of hematoprotective and hepatoprotective properties of aqueous extract of *Ceterach officinarum* DC against streptozotocin-induced hepatic injury in male mice. Comp Clin Path 27(6):1427–1436. https://doi.org/10.1007/s00580-018-2754-x
78. Simpson MG (2010) Evolution and diversity of vascular plants. In: Simpson MG (ed) Plant systematics, 2nd edn. Academic Press, Elsevier Inc. https://doi.org/10.1016/C2009-0-02260-0
79. Muhammad M, Ismail ZS, Schneider H, Hawkins JA (2020) Medicinal use of ferns: an ethnobotanical review. Sains Malays 49(5):1003–1014. https://doi.org/10.17576/jsm-2020-4905-05
80. Malamas M, Marselos M (1992) The tradition of medicinal plants in Zagori, Epirus (northwestern Greece). J Ethnopharmacol 37(3):197–203. https://doi.org/10.1016/0378-8741(92)90034-o
81. May LW (1978) The economic uses and associated folklore of ferns and fern allies. Bot Rev 44 (4):491–528. https://www.jstor.org/stable/4353942
82. Grieve M (1971) A modern herbal: the medicinal, culinary, cosmetic and economic properties, cultivation and folk-lore of herbs, grasses, fungi, shrubs, & trees with all their modern scientific uses, vol I. Courier Corporation, Chelmsford
83. Guarrera PM, Lucia LM (2007) Ethnobotanical remarks on central and southern Italy. J Ethnobiol Ethnomedicine 3:23. https://doi.org/10.1186/1746-4269-3-23
84. Brussell DE (2004) Medicinal plants of Mt. Pelion, Greece. Econ Bot 58(1):S174–S120. https://www.jstor.org/stable/4256917
85. Bulut G, Tuzlacı E (2015) An ethnobotanical study of medicinal plants in Bayramiç. Marmara Pharm J 19(3):268–282. https://doi.org/10.12991/mpj.201519392830
86. Kültür Ş, Gürdal B, Sari A, Melikoğlu G (2021) Traditional herbal remedies used in kidney diseases in Turkey: an overview. Turk J Bot 45(4):269–287. https://doi.org/10.3906/bot-2011-32
87. Hanlidou E, Karousou R, Kleftoyanni V, Kokkini S (2004) The herbal market of Thessaloniki (N Greece) and its relation to the ethnobotanical tradition. J Ethnopharmacol 91(2–3):281–299. https://doi.org/10.1016/j.jep.2004.01.007
88. González-Tejero MR, Casares-Porcel M, Sánchez-Rojas CP, Ramiro-Gutiérrez JM, Molero-Mesa J, Pieroni A, Giusti ME, Censorii E, de Pasquale C, Della A, Paraskeva-Hadijchambi D, Hadjichambis A, Houmani Z, El-Demerdash M, El-Zayat M, Hmamouchi M, Eljohrig S (2008) Medicinal plants in the Mediterranean area: synthesis of the results of the project Rubia. J Ethnopharmacol 116(2):341–357. https://doi.org/10.1016/j.jep.2007.11.045
89. Tuzlacı E, Erol MK (1999) Turkish folk medicinal plants. Part II: Eğirdir (Isparta). Fitoterapia 70(6):593–610. https://doi.org/10.1016/S0367-326X(99)00074-X
90. Guarrera PM, Salerno G, Caneva G (2005) Folk phytotherapeutical plants from Maratea area (Basilicata, Italy). J Ethnopharmacol 99(3):367–378. https://doi.org/10.1016/j.jep.2005.01.039
91. Polat R, Satıl F (2012) An ethnobotanical survey of medicinal plants in Edremit Gulf (Balıkesir-Turkey). J Ethnopharmacol 139(2):626–641. https://doi.org/10.1016/j.jep.2011.12.004
92. Pieroni A, Sõukand R (2017) The disappearing wild food and medicinal plant knowledge in a few mountain villages of North-Eastern Albania. J Appl Bot Food Qual 90:58–67. https://doi.org/10.5073/JABFQ.2017.090.009
93. Meddour R, Sahar O, Abdoune N, Dermouche M (2022) Quantitative ethnobotanical investigation of medicinal plants used by the local population in the rural municipalities of Haizer and El Asnam, province of Bouira. Northern Algeria Mediterr Bot 43:e72688. https://doi.org/10.5209/mbot.71190
94. Abu-Rabia A (2005) Palestinian plant medicines for treating renal disorders: an inventory and brief history. Altern Complement Ther 11(6):295–300. https://doi.org/10.1089/act.2005.11.295

95. Guarrera P, Lucchese F, Medori S (2008) Ethnophytotherapeutical research in the high Molise region (Central-Southern Italy). J Ethnobiol Ethnomed 4:7. https://doi.org/10.1186/1746-4269-4-7
96. Pieroni A, Quave CL, Santoro RF (2004) Folk pharmaceutical knowledge in the territory of the Dolomiti Lucane, in land southern Italy. J Ethnopharmacol 95(2–3):373–384. https://doi.org/10.1016/j.jep.2004.08.012
97. Said O, Khalil K, Fulder S, Azaizeh H (2002) Ethnopharmacological survey of medicinal herbs in Israel, the Golan Heights and the West Bank region. J Ethnopharmacol 83(3):251–265. https://doi.org/10.1016/s0378-8741(02)00253-2
98. Bulut G, Tuzlaci E (2013) An ethnobotanical study of medicinal plants in Turgutlu (Manisa—Turkey). J Ethnopharmacol 149(3):633–647. https://doi.org/10.1016/j.jep.2013.07.016
99. Passalacqua NG, Guarrera PM, De Fine G (2007) Contribution to the knowledge of the folk plant medicine in Calabria region (Southern Italy). Fitoterapia 78(1):52–68. https://doi.org/10.1016/j.fitote.2006.07.005
100. De Natale A, Pollio A (2007) Plants species in the folk medicine of Monte corvino Rovella (inland Campania, Italy). J Ethnopharmacol 109(2):295–303. https://doi.org/10.1016/j.jep.2006.07.038
101. Dall'Acqua S, Tomè F, Vitalini S, Agradi E, Innocenti G (2009) In vitro estrogenic activity of *Asplenium trichomanes* L. extracts and isolated compounds. J Ethnopharmacol 122(3):424–429. https://doi.org/10.1016/j.jep.2009.02.012
102. Irfan M, Jan G, Murad W, Jan FG, Rauf A, Alsayari A, Almarhoon ZM, Mabkhot YN (2021) Ethnomedicinal and traditional uses of the ferns of Khyber Pakhtunkhwa. Pakistan Braz J Biol 84:e250256. https://doi.org/10.1590/1519-6984.250256
103. Gul A, Alam J, Ahmad H, Shah GM, Hussain M, Dogan Y, Rahman KU (2016) Traditional, medicinal and food uses of Pteridophytes of district Mansehra (Pakistan) and their some adjacent areas. Int J Biosci 9:116–133. https://doi.org/10.12692/ijb/9.5.116-133
104. Viegi L, Pieroni A, Guarrera PM, Vangelisti R (2003) A review of plants used in folk veterinary medicine in Italy as basis for a databank. J Ethnopharmacol 89(2–3):221–244. https://doi.org/10.1016/j.jep.2003.08.003
105. Petropoulos SA, Karkanis A, Martins N, Ferreira IC (2018) Halophytic herbs of the Mediterranean basin: an alternative approach to health. Food Chem Toxicol 114:155–169. https://doi.org/10.1016/j.fct.2018.02.031

Part III

Applications of Bryophytes and Pteridophytes in the Fields of Biotechnology, Nanotechnology, and Allied Fields

Bryophytes and the Nanotechnology: Recent Developments and Perspectives

22

Irina Fierascu, Anda Maria Baroi, Toma Fistos, Roxana Ioana Brazdis, Ionela Daniela Sardarescu (Toma), and Radu Claudiu Fierascu

Contents

Abstract

Nanotechnology represents the science of nanomaterials (materials with at least one dimension under 100 nm). The technological development in the recent

I. Fierascu · A. M. Baroi
National Institute for Research & Development in Chemistry and Petrochemistry – ICECHIM Bucharest, Bucharest, Romania

University of Agronomic Sciences and Veterinary Medicine of Bucharest, Bucharest, Romania
e-mail: irina.fierascu@icechim.ro; anda.baroi@icechim.ro

T. Fistos · R. I. Brazdis · R. C. Fierascu (✉)
National Institute for Research & Development in Chemistry and Petrochemistry – ICECHIM Bucharest, Bucharest, Romania

University "Politehnica" of Bucharest, Bucharest, Romania
e-mail: toma.fistos@icechim.ro; roxana.brazdis@icechim.ro; fierascu.radu@icechim.ro

I. D. Sardarescu (Toma)
University "Politehnica" of Bucharest, Bucharest, Romania

National Research and Development Institute for Biotechnology in Horticulture, Ștefănești, Argeș, Romania
e-mail: ionela.toma93@yahoo.com

H. N. Murthy (ed.), *Bioactive Compounds in Bryophytes and Pteridophytes*, Reference Series in Phytochemistry, https://doi.org/10.1007/978-3-031-23243-5_29

decades led to an increased implication of nanotechnology and nanomaterials in our everyday life. The application of plant extracts, acting both as a reduction and as a capping agent for the phytosynthesis of nanomaterials, led to the development of alternative, green materials, with controlled morphologies, as well as with enhanced properties. The present chapter aims to bring to the attention of the reader the application of the plants from the *Bryophytes* taxonomic division in this area, as well as in other related areas.

Keywords

Bryophyte application · General composition · Metal uptake · Model plant · Nanotechnology · Nanoparticle phytosynthesis · Toxicity

Abbreviations

Cit	Citrate
DLS	Dynamic light scattering
DNA	Deoxyribonucleic acid
EDX	Energy-dispersive X-ray analysis
FTIR	Fourier transformed infrared spectroscopy
GC-MS	Gass chromatography-mass spectrometry
HPLC	High-performance liquid chromatography
MDA	Malondialdehyde
NP(s)	Nanoparticle(s)
PVP	Poly (N-vinyl-2-pyrrolidone)
ROS	Reactive oxygen species
SEM	Scanning electron microscopy
TEM	Transmission electron microscopy
UV-Vis	Ultraviolet-visible spectroscopy
XRD	X-ray Diffraction
XRF	X-ray Fluorescence

1 Introduction

Nanotechnology represents the science of nanomaterials (materials with at least one dimension under 100 nm). The technological development in the recent decades led to an increased implication of nanotechnology and nanomaterials in our everyday life. Nanomaterials are encountered all over, from food packaging to medicine, and from cosmetic products to the automotive industry [1]. Being such a common aspect of our daily life, it is no wonder that the nanotechnology and nanomaterials area was aligned with the principles of "green chemistry" [2]. In order to answer the principles developed by Anastas and Warner [2], a new area of nanomaterials was born, respectively, nanomaterials "phytosynthesis" (synthesis of nanomaterials assisted by plants) [3]. The application of plant extracts, acting both as a reduction and as a capping agent for the phytosynthesis of nanomaterials, led to the development of alternative, "green materials," with controlled morphologies, as well as with enhanced properties, even when compared with nanoparticles obtained by other routes, also considered green chemistry approaches (i.e., radiation-assisted nanoparticles synthesis) [4, 5]. The present chapter aims to bring to the attention of the reader the application of the plants from the *Bryophytes* division in this area, as well as in other related areas.

2 Characteristics of the *Bryophyta* Division

2.1 General Aspects

The *Bryophytes* division, with the name originating from the Greek *βρύον (brúon)* (which translated to "tree moss, liverwort"), and *φυτόν (phutón)* meaning "plant," consists of around 20,000 plant species (Fig. 1). Without having as a goal to exhaustively present the division (this not being the goal of this chapter), the Bryophyta plants share several characteristics [6]:

- Life cycles with a multicellular gametophyte stage.
- Unbranched sporophytes.
- The absence of a lignin-containing vascular tissue.

Bryophytes can be found all over the world, from the arctics to deserts, from sea-level to alpine environments, and from deserts to rain forests (although they do prefer moist environments) [7]. As bryophytes do not depend on the root system for

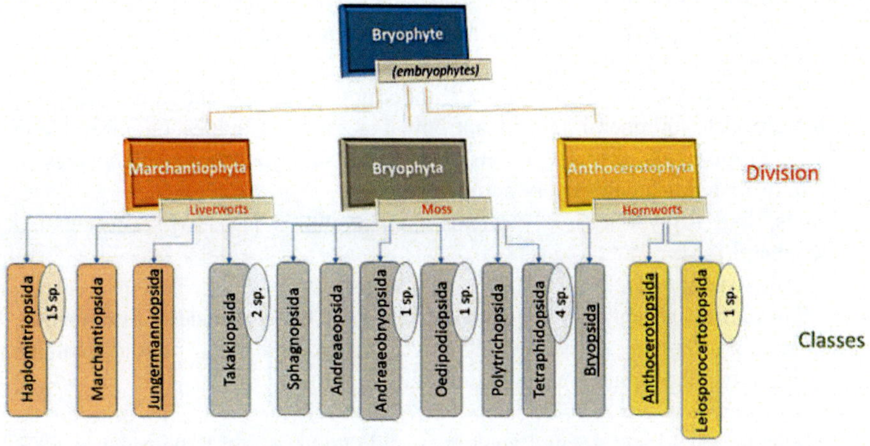

Fig. 1 An overview of the bryophytes. The dominant class in each division is underlined. The number of species in each class is presented only for classes with a reduced number of species. The dominant classes usually contain hundreds of species, some new species being currently discovered (i.e., for the *Anthocerotopsida* class). The highest number of species is recorded for moss

the nutrients uptake from the soil, they can be encountered in which the vascularized plants cannot survive, even on rocks [8]. Recent studies [9] revealed that bryophyte diversity is related to vascular plant diversity, being proposed as indicators of the overall plant diversity and site conservation status, or even for monitoring the quality of aqueous ecosystems [10].

2.2 Chemical Composition

As already presented, the bryophytes contain a huge number of species, each with its particular chemical profile, thousands of chemical constituents of bryophytes being described in the literature. Also, as is the case for all the living plants, their composition strongly depends on the environmental conditions and harvesting time [11]. However, a general composition can be briefly defined as:

- Fatty acids: acetylenic, arachidonic, and eicosapentaenoic acids, compounds rarely found on other plants [12].
- Terpenoids (over 1000 compounds being recorded), including the following:
  - *sesquiterpenes* (seco-africanes, noraristolanes, 1,10-seco- and 2,3-seco-aromadendranes, seco-cuparanes, tridensanes, pinguisanes, ricciocarpanes, neotrifaranes, chenopodanes, riccardiphanes, maaliol, diplophyllolide, and veticadinoxide),
  - *mono-* (β-citrocitral, α- and β-pinene, limonene, camphene, bornyl acetate, β-sabinene, bornylferulate, and limonen),
  - *di-* (momilactones, ent-16-hydroxykaurene, isophytol, and geranylgeraniol),
  - *triterpenoids* (ursane, fernane, hopan, lupan, cycloartane, friedelane, taraxane, dammarane, polypodane, obtusifolane, serratane, and A-amyrane),
  - *acetogenins* [13–15].

  The bryophyte terpenoids present important biological activities, including phytotoxicity, antimicrobial and antifungal activity, activities on insects and mollusks, piscicidal activity, antileishmanial and antitrypanosomal activity, cytotoxicity, and anti-inflammatory activity [14].
- Lipids, like triglycerides, glycolipids, phospholipids [11, 15], ecosane, and tocopherol [16].
- Phenolic compounds:
  - *flavonoids* (present as C-glycosides, O- and O,C-glycosides), bryoflavone, heterobryoflavone, philonotisflavone, dicranolomin, anthocyanidins (3-deoxyanthocyans), apigenine, and luteolin,
  - *structural variants of bibenzyl and bis(bibenzyl)* (lunularic acid, brittonin A, prenylbibenzyl and geranyl analogues, radulanins A and F, perrottetins E-G, marchantins A, C and I, riccardins B and C, plagiochin D, 6,7-dihydroxy-4-(3,4-dihydroxyphenyl) naphthalene-2-carboxylic acid, scapaniapyrone, and megacerosic and anthocerosic acids) [13–16],
- Plant hormones (auxins, cytokinin, gibberellins, and ethylene) [15].

- Phytosterols (campesterol, 22-dihydrobrassicasterol, sitosterol, clionasterol, stigmasterol, cholesterol, 24-methyl-5,22-cholestadienol, and 24-methyl-5,7,22-cholestatrienol) [17].
- Nitrogen-containing compounds (coriandrins and the methyl tridentatols) [18].

Further details on the bryophyte chemical composition can be found in several works, including the excellent work of Asakawa and Ludwiczuk [19], which we recommend for further thoroughgoing study on this topic.

3 Application of Bryophytes in Nanotechnology

3.1 Nanomaterials Phytosynthesis

As already mentioned, the goal of the "green chemistry" approach for the synthesis of nanomaterials is to replace the use of hazardous compounds with safer, natural, alternatives. When speaking of nanomaterials phytosynthesis, the most important aspect is the presence of secondary metabolites which could act as reducing and capping agents. As already known, the most important classes of compounds for this application are the phenolic compounds and the terpenoids [20], compounds found in abundance in the bryophytes, as previously shown. The general mechanism for the phytosynthesis of metallic nanoparticles is presented in Fig. 2.

The application of the phytosynthesis approach for the development of nanoparticles leads to the obtaining of materials with enhanced properties not only due to the capacity to develop materials with controlled morphologies and dimensions but also as a consequence of the contribution of the secondary metabolites, which,

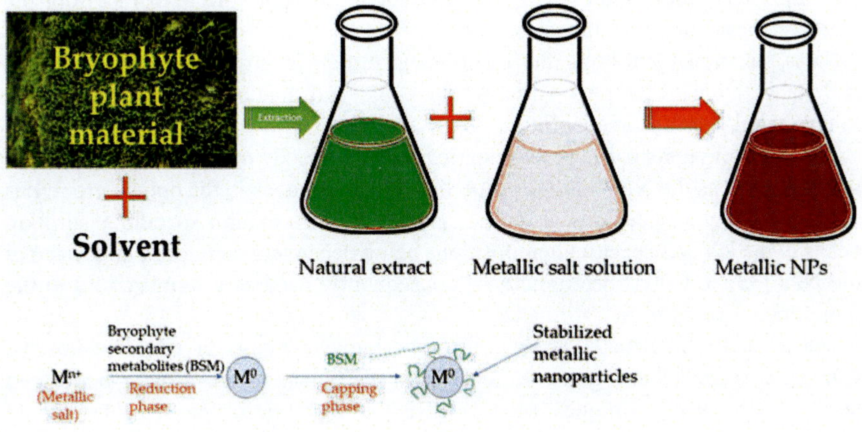

Fig. 2 A general mechanism for the phytosynthesis of metallic nanoparticles using bryophytes extracts

besides their use as capping agents, also often possess significant biological activities.

Table 1 presents a short overview regarding the application of bryophytes in the nanotechnology area, as emerging from the literature data.

As it can be seen from Table 1, the data regarding the nanoparticle phytosynthesis using bryophytes are scarce, and the topic is not present in the journals from the main scientific stream. However, from the examples provided, it can be noticed that the phytosynthesis of metallic nanoparticles (especially not only silver nanoparticles, the most encountered nanoparticles in the literature, but also copper NPs) is possible using all classes of bryophytes (mosses, liverworts, and hornworts). The provided data presents nanoparticles with different shapes (spherical, triangular, cuboidal, etc.) and with a wide range of dimensions, suggesting the possibility to select for phytosynthesis the plant species most appropriate for the envisaged application. The antimicrobial potential of the phytosynthesized nanoparticles using bryophytes was evaluated against several gram-positive and gram-negative bacterial strains and yeasts, with satisfactory results.

Also, the NPs seem not to present any mutagenic effect, while showing pBR322 DNA cleavage activity at different concentrations.

3.2 Characterization of Phytosynthesized Nanoparticles

When dealing with phytosynthesized nanoparticles, an important aspect is represented by the characterization of the plant extracts used for the process. Several analytical techniques can be used for this purpose, including, but not limited to, HPLC (high-performance liquid chromatography), GC-MS (gass chromatography – mass spectrometry), or phytochemical screening assays (such as total phenolic content, total flavonoids, total terpenoids, etc.), as several authors presented [4, 5, 31, 32]. This approach allows the characterization not only of the extract prior the phytosynthesis process, but also of the changes in their composition after the process, as such identifying the compounds primarily involved in the reduction and stabilization of the nanoparticles, due to the information provided, as well as to the technique-wide availability.

Once the phytosynthesis process started, it can be easily monitored using UV-Vis spectrometry, as the NPs' unique properties induce changes in the optical properties of the colloidal suspensions; as such, most of the NPs exhibit specific adsorption peaks in the UV-Vis spectra, their positions being dependent on the size and shape of the NPs [33]. UV-Vis spectrometry is consequently used as a primary tool in the monitorization of the NPs formation.

As a next step usually applied for the characterization of phytosynthesized nanoparticles, the metal presence and phase identification are usually performed using X-ray-based techniques, such as XRF (X-ray fluorescence, with the variant EDX – energy dispersive X-ray analysis, when used as an electron microscopy accessory) and XRD (X-ray diffraction), respectively. If XRF is usually applied for the confirmation of metal presence, XRD can provide information not only on the

Table 1 Application of bryophytes for the phytosynthesis of metallic nanoparticles

Bryophytes class	Plant/part or pre-extraction treatment	Type of NP	Characteristics of NP	Application	Ref.
Moss	*Taxithelium nepalense* Brotherus, 1899/ gametophyte	AuNPs	NPs size: 42–145 nm; average diameter 70 ± 5 nm	–	[21]
Liverworts	*Conocephalum conicum* (L.) Dum. / isolate	AgNPs	NPs size: 25–50 nm; subglobose to a broadly ellipsoidal shape	–	[22]
Liverworts	*Riccia huebeneriana* Lindenb./fresh, green mature thalli	AgNPs	NPs size: 20–50 nm; shape: Cuboidal and triangular	Antibacterial activity against *Pseudomonas aeruginosa*	[23]
Liverworts	*Targionia indica* Udar and A. Gupta, 1983; *Cyathodium tuberosum* Kashyap, 1914/ dried plant	AgNPs	Formation confirmed by UV-vis spectrometry	Antimicrobial activity against bacterial strains (*Escherichia coli* – Inhibition zone 16/11 mm, *P. aeroginosa* – Inhibition zone 23/26 mm, and *Staphylococcus aureus* – Inhibition zone 11/13 mm), and fungal strains (*Rhizopus* spp. – Inhibition zone 8/6 mm, *Alternaria* spp. – Inhibition zone 11/13 mm)	[24]
Hornworts	*Anthoceros erectus* Kashyap, 1915/dried plant	AgNPs	Formation confirmed by UV-vis spectrometry	Antimicrobial activity against bacterial strains (*Escherichia coli* – Inhibition zone 8 mm, *P. aeroginosa* – Inhibition zone 14 mm, and *Staphylococcus aureus* – Inhibition zone 14 mm), and fungal strains	[24]

(continued)

Table 1 (continued)

Bryophytes class	Plant/part or pre-extraction treatment	Type of NP	Characteristics of NP	Application	Ref.
				(*Rhizopus* spp. – Inhibition zone 12 mm, *Alternaria* spp. – Inhibition zone 8 mm)	
Moss	*Fissidens minutus* Thwaites & Mitt./ fresh thallus	AgNPs	Formation confirmed by UV-vis spectrometry	Antimicrobial activity against bacterial strains (*E. coli* – Inhibition zone 6 mm, *Bacillus subtilis* – Inhibition zone 4.1 mm, *P. aeruginosa* – Inhibition zone 8 mm, *Klebsiella pneumonia* – Inhibition zone 6.3 mm)	[25]
Hornwort	*Anthoceros agrestis* (Paton) Damsholt / fresh mature thalli	AgNPs	NPs size: 20–50 nm; shape: Cuboidal and triangular	Antibacterial activity against pathogenic strains (*E. coli* – Inhibition zone 11 mm, *K. pneumoniae* – Inhibition zone 8 mm, *P. aeruginosa* – Inhibition zone 12 mm, and *B. subtilis* – Inhibition zone 9 mm)	[26]
Moss	*Campylopus flexuosus* (Hedw.) bird/ dried powder	AgNP	NPs size: 50 – 70 nm; particle size (by XRD) 51 nm	–	[27]
Liverwort	*Riccia fluitans* L./ refrigerated sample	CuNPs; AgNPs	–	Antimicrobial activity against gram-positive bacteria: *E. coli, K. pneumoniae, S. aureus,* and *B. subtilis.* Best results (CuNP	[28]

(continued)

Table 1 (continued)

Bryophytes class	Plant/part or pre-extraction treatment	Type of NP	Characteristics of NP	Application	Ref.
				50 mM, inhibition zones): *E. coli* – 22 mm, *K. pneumoniae* – 18 mm, *S. aureus* – 24 mm, and *B. subtilis* – 27 mm	
Moss	*Dicranum majus* Turner, 1804/ shade dried, powder	AgNPs	Spherical nanoparticles, hydrodynamic diameter 268.6 ± 8.86 nm	Antibacterial activity (minimum inhibitory concentration): *Enterococcus faecalis* (0.004 mg/L), *listeria monocytogenes* (0.078 mg/L), *Proteus vulgaris* (0.009 mg/L), *P. aeruginosa* (2.5 mg/L), *B. subtilis* (1.25 mg/L), *E. coli* (5 mg/L); pBR322 DNA cleavage activity at 1.25–20 ppm; no mutagenic effect; and decreased antioxidant effect (DPPH assay), compared with the plant extract	[29]
Moss	*Bryum salakense*; Cardot, 1912 / shade dried, powdered	AgNPs	35–40 nm	Antibacterial activity (inhibition zones): *E. coli* (6 mm), *K. pneumoniae* (4 mm), *Serratia marcescens* (4 mm), *S. lentus* (4 mm), and *S. aureus* (5 mm)	[30]

Table 2 Applied analytical techniques for the characterization of Bryophita-phytosynthesized NPs

Type of NP	Analytical techniques applied in the study	Ref.
AuNPs	UV-vis (maximum peak observed at 540 nm), EDX, DLS, XRD, FTIR, and TEM	[21]
AgNPs	UV-vis (maximum peak observed at 415/480 nm), SEM, and EDX	[22]
AgNPs	UV-vis (maximum peak observed at approx. 412 nm), SEM, and EDX	[23]
AgNPs	UV-vis (monitoring of the 200 nm - 600 nm range)	[24]
AgNPs	UV-vis (maximum peak observed at approx. 413 nm), SEM, and EDX	[25]
AgNPs	UV-vis (maximum peak observed at 425 nm), SEM, and EDX	[26]
AgNP	UV-vis (maximum peak observed at 436 nm), SEM, EDX, DLS, and XRD	[27]
CuNPs	UV-vis (maximum peak observed at 565 nm)	[28]
AgNPs	UV-vis (maximum peak observed at 435 nm)	[28]
AgNPs	UV-vis (maximum peak observed at 422 nm), FTIR, TEM, ZP, and DLS	[29]
AgNPs	UV-vis (maximum peak observed at 443 nm), FTIR, SEM, ZP, and XRD	[30]

phases present in the sample (thus confirming the obtaining of metallic or metal oxide nanoparticles, for example) but also on the crystallite size (using the Scherrer equation), in order to place them in the "nanomaterials" category [4, 5, 32]. XRD is currently the primary method for the confirmation of NPs formation.

The morphology of the phytosynthesized nanoparticles is evaluated using electron microscopy, usually TEM (transmission electron microscopy), and in some cases SEM (scanning electron microscopy) [5, 32, 34]. The wide application of those techniques for the characterization of NPs is due to the importance that their morphology has for the envisaged application.

As supplemental methods that can be applied for the characterization of phytosynthesized NPs, FTIR spectroscopy (for the identification of the phytoconstituents involved in the phytosynthesis) [21], DLS (dynamic light scattering, for evaluation of the hydrodynamic diameter of the nanoparticles) [21], determination of the Zeta potential (ZP – applied for the evaluation of nanoparticle stability) [29, 30], or the evaluation of thermal behavior of the NPs can be mentioned [35].

An overlook of the main analytical techniques applied for the characterization of Bryophita-phytosynthesized NPs is presented in Table 2.

3.3 Other Applications of Bryophytes in Nanotechnology and Related Areas

As the closest living relatives to the first land plants, bryophytes were traditionally used in a series of different applications, including environmental pollution control (by up-taking both inorganic and organic pollutants) [36, 37]. The bryophytes were proven to be able to remediate different types of environments, such as the filtering of heavy metals (Pb, Cd, Zn, Cu, As, and Cr) from contaminated water by *Taxiphyllum barbieri* (Cardot & Copp.) Z. Iwats. (aquatic moss) [31], up-take of

Cd in Milieu Bristol growth medium by *Physcomitrella patens* (Hedw.) Bruch & Schimp. (moss) [38], removal of Fe and Cr from water by *Bryum muehlembeckii* Bruch & Schimp., *Ricciocarpos natans* (L.) Corda, and *Sphagnum perichaetiale* Hampe [39]. This uptake capacity leads to another important application of bryophytes, their potential application as model plants for the evaluation of nanoparticle effects. Upon the development of different nanoparticles, their potentially harmful effects should be explored, including their environmental effect. Although divergent from their early evolution stages, mosses and vascular plants share fundamental genetic and physiological processes [40].

Some examples regarding this application are provided in Table 3.

Unlike nanoparticle phytosynthesis, the nanoparticle uptake represents the subject of several works published in mainstream scientific journals, an indicator of the importance of the subject. As emerging from the literature data presented in Table 3, the bryophytes (especially mosses) can be used for the biomonitoring/biosensing of a wide range of organic (polystyrene) and inorganic nanoparticles (TiO_2, ZnO, Ni, Co, or Cu). This could be of particular importance for the assessment of the environment quality and for signaling the presence of the targeted pollutants in different ecosystems (water, soil, and air).

Serving as model organisms, bryophytes were subjected to a series of tests to reveal the changes induced by the exposure to metallic nanoparticles. Canivet et al. [41] evaluated the effects of iron nanoparticles, a representative of the metallurgical industrial emissions, on *Physcomitrella patens* (Hedw.) Bruch & Schimp upon foliar exposure at different concentrations (5 ng to 50 μg per plant). No significant changes were observed for adenosine-5′-triphosphate concentration reactive oxygen species (ROS) production, malondialdehyde (MDA) production, or glutathione regulation. The conclusion of the study was that, at the studied doses and in the application conditions, the FeNP did not induce any toxic effect on the bryophyte.

The influence of AgNPs, a type of nanoparticles widely applied at the industrial level, could present harmful effects on the environment, due to their massive presence in products. The influence of spherical silver nanoparticles, with or without different coatings (poly (N-vinyl-2-pyrrolidone, or citrate) was studied on *Physcomitrella patens* (Hedw.) Bruch & Schimp at the gametophyte stages: protonema and leafy gametophyte [42]. The study found that all the nanoparticles presented toxicity toward the model plant, in terms of growth and development of the gametophyte, in a concentration-dependent and surface functionalization manner. The conclusion of the study was that the protonema was more sensitive to Ag NPs than the gametophyte. The protonema growth was inhibited in the trend AgNPs > AgNPs citrate coated > AgNPs- PVP coated, the nanoparticles affecting the thylakoid and chlorophyll content. At the gametophyte stage, the growth was inhibited in the trend AgNPs-Citrate coated > AgNPs ≈ AgNPs-PVP coated. It was found that the nanoparticles decreased the chlorophyll b content. The study brought to attention the possible harmful effects of the silver nanoparticles [42].

Another type of widely used nanoparticles (ZnO) was applied on *Pleurozium schreberi* (Brid.) Mitt. model plant, in order to evaluate the L-ascorbic acid content [43]. The bryophytes were exposed for 5 weeks to a concentration of 0.1 g/L nanoparticles. The nanoparticles-induced oxidative stress was evaluated by the

Table 3 Application of mosses for the evaluation of the nanoparticles' properties

Plant	Type of NP	Characteristics of NP	Study	Ref.
Physcomitrella patens (Hedw.) Bruch & Schimp	FeNPs	NPs: 20–80 nm spherical shape	Study of cytotoxicity, oxidative stress, lipid peroxidation of the membrane of iron nanoparticles through the axenic moss culturing at five concentrations (5 ng, 50 ng, 500 ng, 5 μg, and 50 μg per plant)	[41]
Physcomitrella patens (Hedw.) Bruch & Schimp	AgNPs	Spherical particles with the size of 37.4 ± 13.4 nm for AgNPs-B (ag NPs without surface coating), 29.0 ± 6.0 nm for AgNPs-PVP (ag NPs coated with poly (N-vinyl-2-pyrrolidone), and 21.5 ± 4.2 nm for AgNPs-Cit (ag NPs coated with citrate)	Physiological and biochemical responses of bryophytes to ag NPs with different surface coatings at the gametophyte stages: Protonema and leafy gametophyte	[42]
Pleurozium schreberi (Brid.) mitt.	ZnO NPs	–	The stress-induced by the exposure of plants to nanoparticles via the determination of the content of L-ascorbic acid	[43]
Sphagnum palustre L.	Polystyrene NPs	Red fluorescent unmodified polystyrene NPs, with diameters of about 50 nm	Biomonitoring – Detection of microplastics in fresh water systems	[44]
Hylocomium splendens (Hedw.) Schimp., *Pleurozium schreberi* (Brid.) mitt.	TiO_2 NPs, ZnO NPs	–	Biomonitoring of the nanoparticle pollution	[45]
Physcomitrella patens (Hedw.) Bruch & Schimp	Mn_3O_4 NPs	Commercial nanoparticles (30 nm), specific surface area 65 m^2/g, bulk density 0.78 g/cm^3, and 20 μg/mL concentration	Genotoxicity and DNA hypomethylation studies	[46]

(continued)

Table 3 (continued)

Plant	Type of NP	Characteristics of NP	Study	Ref.
Fontinalis antipyretica Hedw.	ZnNPs, NiNPs, CoNPs, and CuNPs	10–20 nm – ZnNPs/ 20–80 nm – NiNPs/ ≤20 nm – Co and cu NPs, obtained by laser ablation	Biosensing of metal NPs by spectrally and time-resolved imaging of endogenous fluorescence for short-time exposure (3–5 days)	[47]

determination of L-ascorbic acid. The study showed that the nanoparticles significantly reduced the L-ascorbic acid content. Also, as a preventive pretreatment, washing was found preferable to irrigation, as increased concentrations of L-ascorbic acid were found in this case. The decrease in ascorbic acid can be correlated to the nanoparticles-induced stress, as the acid is involved in the plant stress reaction pathways [43].

The genotoxicity of Mn_3O_4 30 nm spherical to cubic-shaped nanoparticles was evaluated on *Physcomitrella patens* (Hedw.) Bruch & Schimp model plant [46]. The study revealed the induction of DNA strand breaks (as confirmed by comet assay) at several nanoparticle concentrations (5, 10, and 20 μg/mL), all results being corroborated with the enhanced generation of reactive oxygen species. The phenomenon was influenced by the NPs internalization, adsorption, and dissolution. The epigenetic alterations induced by the nanoparticles were evaluated by comet assay. The nanoparticles induced DNA hypomethylation, due to its clastogenic potential. The recorded DNA hypomethylation registered was comparable with that of the DNA methylation blocker 5-azacytidine. Although not as common as other nanoparticles, the Mn_3O_4 NPs were thus shown to induce epi-genotoxicity [46].

A combination of the two potential applications in nanotechnology of bryophytes was reported by Lu and Fong [48]. Thus, frozen and lyophilized *Leucobryum glaucum* (Hedw.) Ångstr. plants were used to uptake iron and cobalt from solutions. The annealing of the metal-containing moss led to the obtaining of corresponding metal oxide nanoparticles embedded within the carbon frames (10 nm nanoparticles decorating the 300 nm iron oxide octahedrons, respectively, 20–50 nm Co_3O_4 nanospheres), further used for the development of ultra-high-performance metal-ion batteries.

4 Conclusion and Future Perspectives

Bryophytes represent not only a "beautiful miniature forest" (as some authors describe those species) but also the basis of superior plant development, colonizing sterile soils [49]. Important from an evolutionary point of view, they are also very important to various ecosystems, not only through their water and nutrients adsorption and release but also considering their pollutant uptake capacity. As the bryophyte species can be encountered in multiple environments, they can remove hazardous pollutants from both aqueous and land ecosystems, including nanoparticles, as described in the present chapter. On this topic, future perspectives should focus on the application of other bryophyte species in environmental studies involving nanoparticles, similar to those presented in the previous paragraphs, as well as continuing using as model plants the species currently under study.

When considering their direct application in nanotechnology, the lack of data regarding the phytosynthesis of nanoparticles using bryophytes is surprising, as this area represents not only an emerging field of research but also a very attractive one, considering the very high number of papers published in the last decade on this subject. At first sight, this would seem rather hard to explain, considering their very wide habitats, bryophyte species being encountered all over the world, in practically any climate. However, when considering the interdisciplinarity of the research groups working on this area, a possible explanation can be glimpsed: In order to perform a solid phytosynthesis study, bryophyte specialists should be definitively involved. As new species are currently discovered all over the world, the few specialists in this area are probably more interested in taxonomic studies than in transdisciplinary ones.

This, in turn, reveals a huge potential development regarding the application of bryophytes in nanoparticle phytosynthesis. This area is a subject worth investigating, and, in our opinion, the results would be spectacular. The small dimensions of the plants, as well as the lack of beautiful colored flowers, should not be a reason to overlook the potential of these species. Another interesting aspect that should clarify is the fate of metals once the uptake is completed. A thorough study could provide interesting results, as was the case with alfalfa, which started a new area of science [50].

In conclusion, the present chapter should be considered as an invitation for the researchers working in the area of nanotechnology to discover and find practical applications for the amazing bryophyte plants.

Acknowledgments This chapter was supported by a grant from the Ministry of Research, Innovation and Digitization, CNCS – UEFISCDI, project number PN-III-P4-PCE-2021-0292, within PNCDI III. The authors also gratefully acknowledge the financial support of a grant of the Ministry of Research, Innovation and Digitization, CNCS/CCCDI – UEFISCDI, project number PN-III-P2-2.1-PED-2021-0042, and project number PN-III-P2-2.1-PED-2021-0423, within PNCDI III. I.F., A.M.B., T.F., R.I.B. and R.C.F. also acknowledge the support provided by the Ministry of Research, Innovation and Digitization through Program 1 - Development of the national research and development system, Subprogram 1.2 -Institutional performance - Projects to finance excellence in RDI, Contract no. 15PFE / 2021.

References

1. Chaurasia N (2017) Nanotechnology and nanomaterials in everyday life. Int J Sci Res 6:1560–1562
2. Anastas PT, Warner JC (1998) Green chemistry: theory and practice. Oxford University Press, New York
3. Fierascu RC, Ion RM, Dumitriu I (2010) Noble metals nanoparticles synthesis in plant extracts. Optoelectron Adv Mat RC 4(9):1297–1300

4. Şuţan NA, Fierăscu I, Fierăscu RC, Manolescu DT, Soare LC (2016) Comparative analytical characterization and in vitro cytogenotoxic activity evaluation of Asplenium scolopendrium L. leaves and rhizome extracts prior to and after Ag nanoparticles phytosynthesis. Ind Crop Prod 83:379–338
5. Fierascu RC, Fierascu I, Lungulescu EM, Nicula N, Somoghi R, Diţu LM, Ungureanu C, Sutan AN, Drăghiceanu OA, Paunescu A, Soare LC (2020) Phytosynthesis and radiation-assisted methods for obtaining metal nanoparticles. J Mater Sci 55:1915–1932
6. Lucas WJ, Groover A, Lichtenberger R, Furuta K, Yadav SR, Helariutta Y, He XQ, Fukuda H, Kang J, Brady SM, Patrick JW (2013) The plant vascular system: evolution, development and functions F. J Integrat Plant Biol 55(4):294–388
7. Levetin MMK (2012) Plants and society. McGraw-Hill, New York
8. https://www.anbg.gov.au/bryophyte/ecology-habitats.html. Accessed 22 May 2022
9. Ren H, Wang F, Ye W, Zhang Q, Han T, Huang Y, Chu G, Hui D, Guo Q (2021) Bryophyte diversity is related to vascular plant diversity and microhabitat under disturbance in karst caves. Ecol Indic 120:106947
10. Debén S, Aboal JR, Carballeira A, Cesa M, Fernández JA (2017) Monitoring river water quality with transplanted bryophytes: a methodological review. Ecol Indic 81:461–470
11. Fierascu I, Dinu-Pirvu CE, Fierascu RC, Velescu BS, Anuta V, Ortan A, Jinga V (2018) Phytochemical profile and biological activities of *Satureja hortensis* L.: a review of the last decade. Molecules 23(10):2458
12. Dembitsky VM, Rezanka T, Bychek IA, Afonina OM (1993) Polar lipid and fatty acid composition of some bryophytes. Phytochemistry 33:1009–1014
13. Ćosić MV, Janošević DA, Oaldje MM, Vujičić MM, Lang I, Sabovljević MS, Sabovljević AD (2021) Terpenoid evidences within three selected bryophyte species under salt stress as inferred by histochemical analyses. Flora 285:151956
14. Ludwiczuka A, Asakawa Y (2021) Bryophytes as a source of bioactive volatile terpenoids – a review. Food Chem Toxicol 132:110649
15. Aditya Rao SJ (2021) Structural information of natural product metabolites in bryophytes. In: Srivastava AK, Kannaujiya VK, Singh RK, Singh D (eds) Evolutionary diversity as a source for anticancer molecules. Academic Press, London
16. Horn A, Pascal A, Lončarević I, Volpatto Marques R, Lu Y, Miguel S, Bourgaud F, Thorsteinsdóttir M, Cronberg N, Becker JD, Reski R, Simonsen HT (2021) Natural products from bryophytes: from basic biology to biotechnological applications. Crit Rev Plant Sci 40(3): 191–217
17. Chiu PL, Patterson GW, Fenner GP (1985) Sterols of bryophytes. Phytochemistry 24(2):263–266
18. Sabovljević MS, Sabovljević AD, Ikram NKK, Peramuna A, Bae H, Simonsen HT (2016) Plant Gen Res 14(SI 4):314–327
19. Asakawa Y, Ludwiczuk A (2018) Chemical constituents of bryophytes: structures and biological activity. J Nat Prod 81(3):641–660
20. Fierascu I, Fierascu IC, Brazdis RI, Baroi AM, Fistos T, Fierascu RC (2020) Phytosynthesized metallic nanoparticles – between nanomedicine and toxicology. A brief review of 2019's findings. Materials 13(3):574
21. Acharya K, Sarkar J (2014) Bryo-synthesis of gold nanoparticles. Int J Pharm Sci Rev Res 29 (1):82–86
22. Raudabaugh DB, Tzolov MB, Calabrese JP, Overton BE (2013) Synthesis of silver nanoparticles by a Bryophilous Rhizoctonia species. Nanomat nanotechnol 3:2
23. Kulkarni AP, Srivastava AA, Nagalgaon RK, Zunjarrao RS (2012) Phytofabrication of silver nanoparticles from a novel plant source and its application. Int J Biol Pharm Res 3(3):417–421
24. Khamgal A, Wangikar H, Chavan S, Bankar P (2021) Green synthesis and antimicrobial studies of silver nanoparticles from some ex-situ conserved Bryophytes in coastal sites of Raigad district of Maharashtra, India. Int J Botany Stud 6:896–900

25. Srivastava AA, Kulkarni AP, Harpale PM, Zunjarrao RS (2011) Plant mediated synthesis of silver nanoparticles using a bryophyte: *Fissidens minutus* and its anti-microbial activity. Int J Eng Sci Technol 3(12):8342–8347
26. Kulkarni AP, Srivastava AA, Harpale PM, Zunjarrao RS (2011) Plant mediated synthesis of silver nanoparticles – tapping the unexploited sources. J Nat Prod Plant Resour 1(4):100–107
27. Vimala A, Sathish SS, Thamizharasi T, Palani R, Vijayakanth P, Kavitha R (2017) Moss (Bryophyte) mediated synthesis and characterization of silver nanoparticles from *Campylopus flexuosus* (Hedw.) Bird. J Pharm Sci Res 9(3):292–297
28. Ganga G (2021) Antimicrobial activity of copper and silver metal nanoparticles synthesized from *Riccia fluitans*. GSC Biol Pharmaceut Sci 18(1):77–82
29. Yayintas OT, Demir N, Yilmaz S, Cicekliyurt MM (2021) Green synthesis, characterization and biological activity of silver nanoparticles from *Dicranum majus* Turner. Digest J Nanomat Biostruct 16(4):1487–1499
30. Vimala A, Sahaya Sathish S (2017) Green synthesis characterization and antibacterial activity of silver nanoparticles from *Bryum Salakense* card. A Moss Plant (Bryophyte). World J Pharm Pharm Sci 6(7):1880–1892
31. Siakavella IK, Lamari F, Papoulis D, Orkoula M, Gkolfi P, Lykouras M, Avgoustakis K, Hatziantoniou S (2020) Effect of plant extracts on the characteristics of silver nanoparticles for topical application. Pharmaceutics 12(12):1244
32. Ungureanu C, Fierascu I, Fierascu RC, Costea T, Avramescu SM, Călinescu MF, Somoghi R, Pirvu C (2021) In vitro and in vivo evaluation of silver nanoparticles Phytosynthesized using Raphanus sativus L. Waste Extracts. Materials 14(8):1845
33. George G, Wilson R, Joy J (2017) Ultraviolet spectroscopy: a facile approach for the characterization of nanomaterials. In: Thomas S, Thomas R, Zachariah AK, Mishra RK (eds) Spectroscopic methods for nanomaterials characterization, Micro and nano technologies. Elsevier, Amsterdam
34. Fierascu I, Georgiev MI, Ortan A, Fierascu RC, Avramescu SM, Ionescu D, Sutan A, Brinzan A, Ditu LM (2017) Phyto-mediated metallic nano-architectures via Melissa officinalis L.: synthesis, characterization and biological properties. Sci Rep 7:12428
35. Lagashetty A, Patil MK, Ganiger SK (2019) Green synthesis, characterization, and thermal study of silver nanoparticles by Achras sapota, Psidium guajava, and Azadirachta indica plant extracts. Plasmonics 14:1219–1226
36. Papadia P, Barozzi F, Migoni D, Rojas M, Fanizzi FP, Di Sansebastiano GP (2020) Aquatic mosses as adaptable bio-filters for heavy metal removal from contaminated water. Int J Mol Sci 21(13):4769
37. Zynter R, Biswas N, Bewtra JK (1989) Adsorption and desorption of perchloroethylene in soils, peat moss, and granular activated carbon. Can J Civ Eng 16:786–806
38. Kováčik J, Dresler S, Babula P (2020) Long-term impact of cadmium in protonema cultures of *Physcomitrella patens*. Ecotoxicol Environ Safety 193:110333
39. Tesser TT, Bordin J, Da Rocha CM, Da Silva A (2021) Application of the dry and wet biomass of bryophytes for phytoremediation of metals: batch experiments. Environ Challeng 5:100382
40. Rensing SA, Lang D, Zimmer AD, Terry A, Salamov A, Shapiro H, Nishiyama T, Perroud PF, Lindquist EA, Kamisugi Y, Tanahashi T, Sakakibara K, Fujita T, Oishi K, Shin-I T, Kuroki Y, Toyoda A, Suzuki Y, Hashimoto S, Yamaguchi K, Sugano S, Kohara Y, Fujiyama A, Anterola A, Aoki S, Ashton N, Barbazuk WB, Barker E, Bennetzen JL, Blankenship R, Cho SH, Dutcher SK, Estelle M, Fawcett JA, Gundlach H, Hanada K, Heyl A, Hicks KA, Hughes J, Lohr M, Mayer K, Melkozernov A, Murata T, Nelson DR, Pils B, Prigge M, Reiss B, Renner T, Rombauts S, Rushton PJ, Sanderfoot A, Schween G, Shiu SH, Stueber K, Theodoulou FL, Tu H, Van de Peer Y, Verrier PJ, Waters E, Wood A, Yang L, Cove D, Cuming AC, Hasebe M, Lucas S, Mishler BD, Reski R, Grigoriev IV, Quatrano RS, Boore JL (2008) The Physcomitrella genome reveals evolutionary insights into the conquest of land by plants. Science 319(5859): 64–69

41. Canivet L, Dubot P, Garçon G, Denayer FO (2015) Effects of engineered iron nanoparticles on the bryophyte, *Physcomitrella patens* (Hedw.) Bruch & Schimp, after foliar exposure. Ecotoxicol Environ Safety 113:499–505
42. Liang L, Tang H, Deng Z, Liu Y, Chen X, Wang H (2018) Ag nanoparticles inhibit the growth of the bryophyte, Physcomitrella patens. Ecotoxicol Environ Safety 164:739–748
43. Motyka O, Štrbová K, Olšovská E, Seidlerová J (2019) Influence of Nano-ZnO exposure to plants on L-ascorbic acid levels: indication of nanoparticle-induced oxidative stress. J Nanosci Nanotechnol 19:3019–3023
44. Capozzi F, Carotenuto R, Giordano S, Spagnuolo V (2018) Evidence on the effectiveness of mosses for biomonitoring of microplastics in fresh water environment. Chemosphere 205:1–7
45. Motyka O, Chlebíková L, Kutláková KM, Seidlerová J (2019) Ti and Zn content in Moss shoots after exposure to TiO_2 and ZnO nanoparticles: biomonitoring possibilities. Bull Environ Contam Toxicol 102:218–223
46. Ghosh I, Sadhu A, Moriyasu Y, Bandyopadhyay M, Mukherjee A (2019) Manganese oxide nanoparticles induce genotoxicity and DNA hypomethylation in the moss *Physcomitrella patens*. Mutat Res Gen Toxicol Environ Mutagen 842:146–157
47. Chorvatova AM, Uherek M, Chorvat D (2021) Biosensing the presence of metal nanoparticles by spectrally- and time-resolved endogenous fluorescence in water Moss Fontinalis antipyretica. Front Phys 9:634324
48. Lu Y, Fong E (2017) Botanic chemistry enabled synthesis of 3D hollow metal oxides/carbon hybrids for ultra-high performance metal-ion batteries. Mat Today Energ 4:89–96
49. Crooks V (2021) Bryophytes. Tiny plants in a big changing world. https://stri.si.edu/story/bryophytes. Accessed 1 June 2022
50. Gardea-Torresdey JL, Parsons JG, Gomez E, Peralta-Videa J, Troiani HE, Santiago P, Jose Yacaman M (2002) Formation and growth of au nanoparticles inside live alfalfa plants. Nano Lett 2(4):397–401

Biotechnology Investigations in Bryophytes and Pteridophytes 23

Wojciech J. Szypuła

Contents

Abstract

The phyla bryophytes, pteridophytes, and lycophytes are embryophytes distinguished among all land plants by the life cycle that consists of an alteration of generations with the dominant gametophyte (bryophytes) or sporophyte (pteridophytes). As axenic cultures of gametophytes and/or sporophytes are obtained, these plants may be used as model systems in basic research, e.g., on phylogenesis and evolution of the primary aquatic ancestors, developmental studies, genetics, or biotechnology. Some bryophytes and pteridophytes are used commercially, and biotechnology techniques are applied to maximize their biomass yield for the horticulture industry or to biosynthesize secondary metabolites for the pharmaceutical industry (e.g., the alkaloid huperzine A isolated from some club mosses). Also, in vitro cultures are tools for the ex situ conservation of threatened species. As there is an alteration of generations, cultures of bryophytes, pteridophytes, and lycophytes, unlike those of seed plants, can be initiated from spores, shoot fragments, or other plant organs acquired from natural sources. Initiating and maintaining such axenic cultures is difficult as both gametophytes and sporophytes are colonized by bacteria and fungi, endophytic and/or mycorrhizal. Representatives of the Zygomycota and Ascomycota, e.g., the

W. J. Szypuła (✉)
Department of Pharmaceutical Biology, Faculty of Pharmacy, Medical University of Warsaw, Warsaw, Poland
e-mail: wszypula@wum.edu.pl

H. N. Murthy (ed.), *Bioactive Compounds in Bryophytes and Pteridophytes*, Reference Series in Phytochemistry, https://doi.org/10.1007/978-3-031-23243-5_28

Dothideomycetes class, are common club moss endophytes. Endophyte contamination of the vegetative organs is responsible for culture losses, especially affecting commercial micropropagation. It is then necessary to identify the endophytic organisms to use specific fungicides and antibiotics, especially if methods routinely used in cultures of seed plants have failed. This chapter summarizes studies on endophytes in bryophyte and pteridophyte cultures, and on endophytes and endosymbionts forming mycorrhizae on roots and shoots of various pteridophyte species, in vivo and in vitro. Procedures to obtain axenic cultures of sporophytes and gametophytes of *Huperzia selago* (L.) Bernh. ex Schrank & Mart. (*Lycopodium selago* L., *Lycopodiaceae*) are presented based on the results of the studies and other available reports.

Keywords

Axenic cultures · Biotechnology · Bryophytes · Endophyte · Lycophytes · Micropropagation · Mycorrhiza · Pteridophytes

1 Introduction

Bryophytes, pteridophytes, and lycophytes (Fig. 1) are characterized by several features that indicate that they may be used as model plant systems in basic research. Bryophytes which are the oldest nonvascular land plants occupy a key position in the studies of plant evolution, especially those aimed at reconstructing the evolution of their primary aquatic ancestors or the process of leaving their aquatic habitats for terrestrial life, considering that the bryophytes were dominant land plants for the initial 100 million years of plant evolution. Nowadays, the bryophytes are represented by three phyla of embryophytes, i.e., hepatics or liverworts

Fig. 1 Exemplary species of bryophytes (**a**), pteridophytes (**b**), and lycophytes (**c**). (**a**) *Polytrichum* Hedw.; (**b**) *Dryopteris cristata* (L.) A. Gray; and (**c**) *Diphasiastrum alpinum* (L.) Holub

(Hepatophyta), mosses (Bryophyta), and hornworts (Anthocerophyta) which include some 20,000 species.

The clade of vascular plants comprises over 93% of all contemporary plant species, and in addition to seed, plants include two divergent evolutionary groups, lycophytes and ferns. Modern phylogenetics using molecular markers has changed the systematics and taxonomy of these plant groups as considerable differences in their evolution have been revealed between club mosses and the remaining pteridophytes and vascular plants. According to systematic botanists collaborating within the Pteridophyte Phylogeny Group (PPG) [67], the monophyletic class of pteridophytes includes some 11,916 species in existence in 337 genera, 51 families, 14 orders, and two classes of Lycodiopsida Bartl. (approximately 1388 species) and Polypodiopsida Cronquist, Takht. & Zimm. (approximately 10, 597 species, i.e., most of the pteridophytes). The phylogenetic tree of vascular plants, showing that club mosses are relatives of ferns, and used in the PPG I classification [41, 43, 44, 46, 59, 60, 62, 63, 68, 94], demonstrates that Lycopodiopsida is a monophyletic group [38, 39, 59, 60, 88]. They represent a group of vascular plants which early diverged from an ancient, diverse group of fossil plants noted over 420 million years ago [7, 8, 64]. The group was dominant in the Devonian and Carboniferous periods but was subsequently replaced by ferns and coniferous plants, and later by seed plants (Angiospermae). Now, lycophytes are evolutionary relicts, account for less than 1% of the world's flora [40], and include three extant orders: Lycopodiales DC. ex Bercht. & J. Presl (16 genera and approximately 388 species), Isoëtales Prantl (approximately 250 species), and Selaginellales Prantl (approximately 700 species) [67].

Biotechnology, including in vitro plant culture systems, applies to bryophytes and lycophytes as procedures have been developed for their micropropagation, clonal propagation of ferns in vitro, biosynthesis of secondary metabolites in tissue culture, or the use of bioreactors to increase culture yield. Micropropagation and the application of tissue culture techniques to preserve biodiversity play a major role in the ex situ conservation of many threatened and endangered pteridophyte species, such as the ferns *Asplenium adulterinum* Milde and *Asplenium trichomanes* [51, 91]. Tissue culture techniques can be also used for the biosynthesis of secondary metabolites and generally biotechnology allows use on an industrial scale of many plant species which are either rare or threatened with extinction.

Biotechnologies, including in vitro techniques in use for the propagation of ferns, club mosses, and psylophytes have been reviewed by Szypuła and Pietrosiuk [72, 73]. They report that many pteridophyte species have been thoroughly investigated and procedures have been developed which allow the production of some of them, such as *Asplenium* or *Platycerium* ferns, on an industrial scale. Without tissue, and culture techniques, preservation of many plant species would not be possible although micropropagation is difficult or has not been adequately investigated for most club moss, spikemosses, quillwort, or psylophyte species [73–75].

The use of bryophytes as a model plant system to study plant genetics and development is by no means new, and it has been amply presented by Duckett et al. [19] and Cove et al. [13, 14]. Research into bryophyte biotechnologies has been conducted since at least the 1920s using a variety of techniques, from traditional agar cultures to biomass growth in bioreactors. A paper by Duckett et al. [19] is a comprehensive review of in vitro cultivation of bryophytes. So far, in vitro, culture methods have been developed for over 300 mosses and nearly 50 hepatics. Duckett et al. [19] emphasize that in vitro cultivation of bryophytes not only serves the purposes of developmental, molecular, and genetic research but also is of key importance for the elucidation of the roles of juvenile stages in reproductive biology. The life cycle of mosses consists of the free-living haploid gametophyte and the diploid sporophyte which is nutritionally dependent on the gametophyte and partially or completely heterotrophic. Unlike in the case of vascular plants, the haploid gametophyte is the dominant generation in all three bryophyte phyla. Most studies in plant development, especially those concerned with bryophytes, focus on the gametophyte just as studies of club mosses and other pteridophytes utilize a prothallus. For research purposes, the haploid gametophyte appears to be superior to the sporophyte as it is possible to assess the direct effects of recessive mutations. In bryophytes, the germinating spore produces the protonema, a branched thread-like structure with filaments that are one cell thick. Under natural conditions, this is a short-lived stage, but in in vitro cultures, it may survive much longer [13, 14, 19]. Other important uses of in vitro cultivation of bryophytes include molecular research to establish systematic and taxonomic relationships within the division of bryophytes and between bryophytes and other embryophytes, ex situ preservation of rare and endangered taxa, or elucidation of symbiotic relationships, including mycorrhizae.

2 Microorganisms in In Vitro Plant Tissue Cultures: A Brief Review

So far, cultures of pteridophytes have been initiated from spores or shoot fragments collected in natural habitats (Fig. 2). Shoot fragments and propagules require complicated disinfection procedures to eliminate microbiological contamination, which considerably limits the use of explants to initiate cultures. In their natural habitats, bryophytes and pteridophytes are colonized by numerous microorganisms which should be eliminated before cultures are initiated as microorganisms can potentially interfere with the hormone balance of the plant or produce some unpredictable, varied effects on the morphogenesis and organogenesis of explants [28, 56, 91, 92]. As a result, stabilization of explants and controlling the processes of organogenesis and further micropropagation is difficult. Bacteria are present on the surface, in conducting tissues, intercellular spaces, and inside cells. Some are plant pathogens that multiply abundantly, destroy cells, and produce toxic substances. Most bacteria, however, are endo- or exophytic and beneficial to their host and do not induce any hypersensitive response or signs of disease [56]. Hallmann et al. [28] defined endophytes as those nonpathogenic bacteria which do not visibly harm the plant, and can be isolated from surface-disinfected plant tissue or extracted from within the plant. They may reside latently or actively colonize plant tissues.

Fig. 2 Fir club moss *Huperzia selago* in its natural habitat (**a**) and in vitro culture of sporophytes and gametophytes; (**b**) Bulbil of *H. selago* after surface disinfection and (**d**) young sporophyte after 3 months of culture; and (**c**) mature gametophyte of *H. selago* after 6 months of culture

Endophytes can be transmitted by spores and vegetatively propagated plant material and gain entrance to tissues through stomata, glandular cells, wounds, hairs, or natural fissures related to plant growth and development. Endogenous bacteria which have remained in a latent or dormant state in the disinfected explants of club mosses and ferns are very likely to multiply and will reveal their presence after several passages in the stage of intense propagation and rooting. Some of these bacteria may be classified as saprobionts present in cells. They feed on organic substances from dead cells, propagate vigorously, and can be seen in aging cultures. As a result of their excessive proliferation, saprobiotic bacteria may decrease the efficacy of plant tissue propagation, rooting, or adventitious regeneration.

Apart from bacteria, saprophytic, parasitic, and symbiotic fungi colonize the epiphyte cover and frequently are present as endophytes in parenchymal tissues (Fig. 3). Mycorrhiza as an evolutionary developed association between the roots or gametophytes of ferns with the mycelium is common in angiosperms and gymnosperms as well as club mosses [21, 26, 31, 32, 34, 66]. The mycorrhizal fungus hyphae obtain nutrients from the soil and deliver them to the plant roots [31, 32]. There are three types of mycorrhiza: ect-, endo-, and ectendomycorrhiza. Endomycorrhiza includes arbuscular (AM), vesicular-arbuscular (VAM), ericoid, monotropoid, and orchid mycorrhiza, and ectendomycorrhiza includes arbutoid mycorrhiza (op. cit.). Ectomycorrhizae are nearly ubiquitous in trees growing in the boreal and Alpine zones and fairly common in temperate climates. Endomycorrhizae, on the other hand, are found in most plants in all climate zones. Zenkteler [91] investigating the presence of endophytes in fern rhizome tissue observed that they reach up to the parenchyma of the major axis of the rhizome

Fig. 3 (**a**) *Phialocephala fortinii* hyphae growing from *Huperzia selago* shoots after 6 months of culture and (**b**) isolates in plate culture; (**c**) hyphae of *P. fortinii* under magnification; and (**d**) hyphae of an endophytic fungus of the genus *Mucor* or *Rhizomucor* under the leaf epidermis of *H. selago* from an in vitro culture. The photos were taken after 4 weeks of incubation of shoots on medium with the addition of Benzol 50 WP. (Photo W. Szypuła (**a**, **c**), J. Pawłowska (**b**) and M. Wrzosek (**d**))

and the filopodia and considered classifying this location of the hyphae as mycorrhiza as highly controversial.

The relationship between endophytic and mycorrhizal fungi in ferns and club mosses has been poorly investigated, and further research is necessary [32, 34, 66]. For culturing purposes, colonization of plant tissues by numerous endophytes, including mycorrhizal fungi, makes appropriate disinfection procedures mandatory. Usually, surface disinfection of the explants is insufficient, and to obtain sterility of cultures, two or three antibiotics are used in combination with systemic and contact fungicides [74, 75, 91], which is effective against endophytes in the intercellular spaces and inside the cells. Obtaining aseptic explants from the plant material with endophytes present within the conducting system is difficult. Long-term disinfection is used in cases of persistent contamination. Antibiotics or fungicides previously tested for their antimicrobial activity are added to the medium on which the explants are to be cultured. The procedure often reduces the regenerative properties of plant tissues although carbendazim, a widely used fungicide, is an alternative plant growth regulator with cytokinin-like properties which powerfully stimulate regeneration in vitro [18].

To make antibiotic treatment, and hence the decontamination, effective, the bacteria must be isolated and identified, their antibiotic susceptibility determined, and the phytotoxicity of the most effective product assessed [92]. Falkiner [22] pointed out that an antibiotic that was bactericidal at a dose of 50 $\mu g \cdot mL^{-1}$ and not noxious to explants at a dose of 100 $\mu g \cdot mL^{-1}$ could be safely used in in vitro cultures. Some antibiotics are phytotoxic. It has been observed that some β-lactam antibiotics inhibit chloroplast division when used at higher doses [92].

On the other hand, contrary to the opinions voiced by some researchers on the noxious effects produced by the presence of endophytic bacteria and fungi in the tissues of plants cultivated and propagated in vitro, there has been a growing number of studies demonstrating beneficial associations between endophytic fungi and vascular plants [32, 70, 78, 89, 91, 92], possibly due to the production of growth regulators by the microorganisms and their involvement in the biosynthesis of secondary metabolites. Also, some papers suggest both beneficial and noxious effects produced by the same endophytic bacteria and fungi [1].

Since the interactions may vary depending on the species of the fungus and the host plant, each case should be considered individually [15]. Environmental conditions and the presence of other species of fungi and endophytic bacteria may also influence these interactions [36]. The beneficial effect of endophytes on the host plants is supported by the observation of their frequent presence on the roots of plants growing in harsh conditions, such as the Arctic and Alpine regions. Their presence in the tissues of the host plant is likely to facilitate its survival. Studies have been conducted on the absorption of nutrient elements by plants depending on the presence or absence of endophytes. Barrow and Osuna [5] found that dark septate endophytes (DSE) which are the most common fungal colonizers of roots increased the absorption of phosphorus by plants while Upson et al. [80] observed that plants which were not colonized by DSEs were not able to absorb organic nitrogen. According to Zhang et al. [93] and Li et al. [47], DSEs increase plant resistance to heavy metal stress. Porras-Alfaro et al. [58] found that DSEs contribute to soil nutrient cycling and plant survival during droughts, and Lv et al. [90] reported similar findings.

Numerous DSEs have been isolated from many plant species (Fig. 3), and experiments have been conducted on inoculating plant roots with DSEs [50], and there have been trials to utilize the effects of DSEs in agriculture to increase crop yields or for plant protection. Andrade-Linares et al. [3] conducted studies to find out whether plants inoculated with DSEs would be more resistant to fungal infections, a frequent cause of reduced crop yields. Some findings were promising, and the authors emphasized that the younger the plant, the more potent the protective effect. Narisawa et al. [54, 55] investigated the potential of DSEs to protect plants against virulence.

The beneficial effects of DSEs on plants include facilitating the host acquisition of mineral nutrients (P, N, Zn, Cu, Ni, S, Mn, B, Fe, Ca, and K) and the use of organic sources (C, N, and P), effect on the host acquisition of water and environmental tolerance, improved tolerance to drought stress and extreme temperatures, herbivore deterrence, and resistance to pathogens [49]. Potential negative effects of DSEs on the host plant are described in the review papers by Alberton et al. [2] and Tellenbach et al. [77] and are thought to depend on the host species, the species of the fungi colonizing the roots, and the environmental conditions.

3 Endophytic and Mycorrhizal Microorganisms Colonizing Club Mosses and Procedures to Obtain Axenic Cultures

The studies conducted to develop culture methods for the fir club moss *Huperzia selago* included an assessment of the diversity of fungal species colonizing the explant tissue (sporophyte generation, shoots) (Fig. 3; [74, 75]) and were undertaken when club moss cultures, especially axenic cultures, proved difficult to initiate and maintain. From 420 tissue fragments of *H. selago* shoots that underwent standard disinfection, 120 isolates representing 13 species of endophytic fungi were recovered. The isolates belonged to the Zygomycota (1 species) and Ascomycota (12 species) (Fig. 3). The representatives of the Ascomycota phylum belonged to three classes: Dothideomycetes, Soradriomycetes, and Eurotiomycetes [11]. Additionally, from *H. selago* vegetative propagules a representative of the order Pleosporales, the *Paraconiothyrium* genus was isolated, which proved to be a new species described as *Paraconiothyrium babiogorense*, which is the first reported *Paraconiothyrium* species from fir club moss [12]. This recently reported species closely resembles the earlier described *Coniothyrium lycopodinum* Sacc. & Paol., isolated from *Lycopodium annotinum* L. Some other reported fungi include *Phialocephala fortinii* C.J.K. Wang & H.E. Wilcox. (Fig. 3), a representative of the Ascomycota, the order *Helotiales*, which is a dark septate endophytic fungus [37]. The above species were observed at different rates in agar cultures usually within a few weeks of initiation. The knowledge of fungal endophytes in *H. selago* and other club mosses remains fragmentary and derived from the studies on mycorrhiza. These endophytes are less widely known, saprophytic, parasitic, and symbiotic fungi colonizing different plant tissues.

A review of the literature on *Lycopodium* spores and prothalli by Pacyna [57] shows that studies to determine the systematic affiliation of symbiotic fungi found on the prothalli of various species of *Lycopodium* and their role in the prothallus development were conducted as early as the 1880s. The author cites the studies by Treub [79], Bruchmann [9, 10], Lang [45], Goebel [27], Spessard [69], Barrows [6], and Freeberg [24] and describes *Pythium* Pringsh (*Lycopodium cernuum*, *L. annotinum*, *L. lucidulum*, and *L. obscurum var. dendroideum* endophyte), *Phycomycetes*, *Ascomycetes* (*L. obscurum* endophyte), and *Perenosporaceae* (*L. clavatum* endophyte). According to Pacyna [57], the presented findings are not consistent, but some indicate that different fungal species may facultatively coexist with the prothalli of various club mosses. This is confirmed by Freeberg [24] who infected *L. complanatum*, *L. selago*, and *L. cernuum* gametophytes with a fungus isolated from a prothallus of *L. obscurum* and observed a more rapid weight gain of prothalli grown with the fungus. The role of the endophytic fungi has not been fully elucidated. It is known that a gametophyte acquires nutrients from the fungus [24, 57]. A gametophyte developing from a spore does not have an independent existence after the stage of a few cell divisions [10]. Freeberg [24] suggests that fungi are involved in converting nutrients into forms that are available to gametophytes.

So far, the mechanism of communication between fungi and club mosses remains unknown, and very little is known about a potential association between the mycotrophism of *Lycopodiaceae* prothalli and sporophytes as only a few studies on the mycorrhizal associations of club mosses have been conducted. Higgis et al. [34] from tissue fragments of the North American fir club moss *H. selago* sporophytes isolated endophytes that they recognized as Ascomycota. Forty-three (43) isolates were recovered which subsequently, based upon the analysis of nuclear internal transcribed spacer (ITS) sequences, were divided into genotypes belonging to the genus *Dothideomycetes*. Wang and Qiu [82] studied the phylogeny and evolution of mycorrhizae in land plants, including numerous species of the *Lycopodiaceae s.l.* and the types of their mycorrhizal associations. Arbuscular mycorrhiza (AM) was found in *Diphasiastrum alpinum*, *D. issleri*, *Lycopodium annotinum* (*Spinulum annotinum*), *L. cernuum*, *L. clavatum*, *L. japonicum*, *Huperzia australiana*, and *H. phyllantha.* AM or fungal hyphae which were an instance of myco-heterotrophy were observed in *Diphasiastrum complanatum* and *Huperzia selago*. Wang and Qiu [82] consider myco-heterotrophy to be the earliest stage in the evolution of the symbiotic association of plants and fungi, preceding full-form mycorrhizae. These early evolutionary forms of plant-fungal symbiosis probably evolved many times and in many plant groups (e.g., in the *Orchidaceae*, *Gentianaceae*, and *Polygalaceae* families). In the early stage of their development club, moss gametophytes may be also described as myco-heterotrophic as there is a shift from a mutualistic symbiotic relationship toward controlled exploitation of the fungi by a not yet fully autotrophic plant. In the next stage, gametophytes become fully photosynthetic and develop normal mycorrhizae.

Schmid and Oberwinkler [66] found in underground *Lycopodium* gametophytes an infrequent host-fungus association which had some features of vesicular-arbuscular mycorrhiza (VAM), but without arbuscules and with very well-developed, septate hyphae located inside cells. They believe that gametophytes of other club moss species are similar and contain structures that resemble intracellular arbuscules formed by VAM-like fungi of the order Glomales. Harley and Harley [29] and Gemma et al. [26] described the mycorrhizal associations of *Lycopodium* sporophytes as typical vesicular-arbuscular. Treu et al. [78] examined some species of North American vascular plants from Denali National Park and preserve (Alasca) for mycorrhizal colonization and observed a variety of intracellular colonization which they described as dark septate endophytes, without the features of VAM, but with intracellular sclerotes. The structures closely resembled those found by Read and Haselwandter [61] in different species in Austrian Alpine plant communities, the *Cyperaceae* family in particular. Haselwandter [31] described these fungi as representatives of the genera *Rhizoctonia* DC and *Phialopora* Medlar. Stoyke and Currah [71] identified the species *Phialocephala fortinii* C.J.K. Wang & H.E. Wilcom. as dark septate fungi. Väre et al. [81], in their extensive studies of mycotrophism in Arctic plant species from Spitzbergen, described similar fungi which they classified as the genus *Microdochium* H. Sydon. According to Treu et al. [78], identifying *Phialophora finlandia* J. K. Wang & H. E. Wilcom and *Phialocephara fortinii* in the *mycelium radicis atrovirens* complex, including many other species, confirms the mutualistic nature of some of these relationships [83]. *Phialophora finlandia* is known to form ectendomycorrhizae in many tree species [89]. *Phialocephara fortinii* has been recovered from the roots of some orchid species [16].

Dark septate endophytic (DSE) fungi have been found in approximately 600 species representing nearly 320 genera and 100 families of vascular plants [37]. The abundance of these fungi in the roots of plants growing in harsh climates of the Arctic or at high altitudes proves their mutualistic associations [61, 65, 78]. In plant communities at altitudes of more than 3000 m above sea level, mycorrhizal fungi are replaced by a specific group of fungi initially named *mycelium radicis atrovirens* and now known as DSE fungi. Presumably, DSE fungi have taken over the role of VAM fungi in harsh climates [32]. This is confirmed by the examples of a close, mutually beneficial association between the plant host and the fungus observed in taxonomically distant plants. Thus, Haselwandter and Read [33] observed a significant increase in the content of phosphorus in the shoots of two sedge species and increased growth of the shoots in one species following their inoculation with DSE fungi. A protective effect of the genus *Phialaphora* colonization of grassroots was observed by Deacon [17] who found that the hyphae protected the grasses from infection by the parasite *Gaeumannomyces graminis* (Sacc) Arx & D. L. Olivier. According to Treu et al. [78], the presence of DSE fungi in the roots of plants is a new example of mycorrhiza which, however, is not included in the classification of mycorrhizae by Harley and Smith [30].

Methods of plant tissue culture using pteridophytes are discussed in review papers by Szypuła and Pietrosiuk [72–75] who find that in vitro cultures have been investigated for only a few club moss species. The very first papers by Freeberg and Wetmore [25] and Freeberg [23] described gametophyte cultures and apogamous development of sporophytes of *Lycopodium complanatum*, *L. cernuum*, and *L. selago* (=*Huperzia selago*) grown from sporelings in vitro. Later published studies by Atmane et al. [4] presented the procedure of culturing the callus and sporophytes grown from the somatic embryos of *Lycopodiella inundata*. There are also reports of gametophyte cultures of *L. obscurum* [84], *L. digitatum* [85], *L. lucidulum* [87], and *H. selago* [76, 86]. In the literature, there are only two protocols specifying the procedures for the propagation of club mosses, i.e., *Huperzia selago* and *Phlegmariurus squarrosus* sporophyte cultures [48, 74, 75]. The methods to initiate and maintain axenic cultures of these species depended on the type of explants used [35, 74–76]. Some were not adequately described [4, 35, 48], and it was difficult to employ them successfully in practice.

Own research into initiating and maintaining axenic sporophyte cultures included testing the methods of disinfection commonly employed for in vitro cultures of angiosperms which, however, proved somewhat ineffective. Atmane et al. [4] in the only published study on in vitro cultures of the sporophytes of *Lycopodiella inundata* (*Lycopodiaceae s.s.* family) did not give any details concerning the preparation of explants. That is why the present authors had to develop their methods [74, 75]. Based on reports on in vitro culturing of ferns, various methods of surface and internal sterilization and a variety of chemical agents were tested including changing the sequence at which they were applied and the length of explant exposure [91, 92]. We found that the treatments to eliminate microbial contamination from explants had to differ depending on the explant type, i.e., shoots or vegetative propagules. In each case, the major steps included surface disinfection, short-term antibiotic treatment in vitro and antifungal treatment, and long-term internal disinfection in vitro. The sequence of steps required to obtain aseptic explants of shoot fragments and propagules was established.

The main substances we recommended for surface disinfection of explants are ethanol and inorganic compounds releasing active chlorine, such as calcium or sodium hypochlorite or hydrogen peroxide. Two-step disinfection was proved to be the most effective in *H. selago* shoots [74]. In each step, the surface disinfection proper was preceded by rinsing the explants for 5–60 min with running tap water followed by soaking in water with Tween® 20 detergent added (0.1% v/w) for 5–15 min. The first step of disinfection proper consisted of soaking the explants in solutions of contrasting pH (0.01 M HCl: pH 2.36; ACE with H_2O: pH 12.38; and 2.4 mM citric acid: pH ca. 3.0). In the second step, the following were used sequentially: 70% ethanol (v/v), 5% sodium hypochlorite suspension or ACE bleach (Procter & Gamble) solution in water (1:5 v/v), and hydrogen peroxide solution. The use of solutions of contrasting pH values improved the effectiveness of disinfection. Zenkteler [91, 92] did not recommend using the method for very delicate and young leaves, but at the stage of initiating in vitro cultures of *H. selago* no damage to the explants was observed, and after 2–4 weeks of incubation on the initial culture medium, 90% of shoots were free from contamination. When the first step of soaking the explants in solutions with contrasting pH values was omitted, the disinfection was less effective and only 60% of explants showed no signs of contamination.

When the length of shoot exposure to 7% hydrogen peroxide solution was extended from 10 to 15 min and Step one omitted, after 2–4 weeks in axenic cultures the plants were dying without any signs of contamination which was considered to result from excessive exposure to 7% hydrogen peroxide solution. Such surface disinfection, however, was not successful in the case of shoots as in the subsequent weeks in culture some sporophytes which had earlier appeared aseptic were found to become infected. Then, surface disinfection was combined with long-term internal disinfection and/or short-term antibiotic treatment in vitro altering the order in which they were used. Various antibiotics or bactericidal agents (e.g., streptomycin, penicillin, malachite green, amphotericin B, and nystatin) were combined with benzimidazole fungicides including thiophanate-methyl or its derivatives (e.g., Topsin M 500 S.C. and Benazol 50 WP withdrawn from sale in the EU in 2021 or earlier) and dithiocarbamate fungicides (Dithane M-45) [74, 75].

The use of these disinfection procedures considerably decreased the contamination rates to 5% of all explant cultures considered to be axenic [74]. The contaminants were fungi and bacteria which were not revealed in the first few passages following comprehensive disinfection of the explants. According to the definition of endophytes, they can live within plant tissue for at least part of their life cycle without causing apparent disease as nonpathogenic, virtually harmless microorganisms which can be removed from plants by surface disinfection [28]. Endophytic bacteria in *H. selago* in vitro culture were seen as cloudiness, often under the explants, discoloration, or infiltrates on the medium surface. They varied in color, but most were white, pink, or red. The pink bacterial colonies, isolated from shoot cultures and transferred onto a fresh medium, e.g., half-strength Murashige and Skoog [53], grew well. After culturing on differential media, they were classified as the genus *Acetobacter*.

Long-term disinfection to eliminate particularly persistent contamination, using bactericidal agents, antibiotics, and/or fungicides added to the medium, may decrease explants' ability to regenerate or their death [91]. In in vitro culture of *H. selago*, phytotoxicity was observed after shoots were incubated on a medium with 500 mg $\cdot$ L^{-1} of Antibiotic/Antimycotic (AAS) solution (Sigma), which contained penicillin, streptomycin, and amphotericin B, and 0.5 mg $\cdot$ L^{-1} malachite green (Table 1). That could be due to the synergistic effect of the solution ingredients on the explant cells. Streptomycin alone was not phytotoxic. It is an aminoglycoside antibiotic that works by interfering with the functioning of the bacterial 30S ribosomal subunit, whose composition is similar to that of chloroplast ribosome. The use of aminoglycosides in in vitro cultures is likely to interfere with protein biosynthesis in chloroplasts with the resulting reduction in the regenerative potential of the explants. Table 1 presents the efficacy of the reagents used and the duration of their effect from trials to determine the optimal methods of disinfection for *H. selago* shoots [74].

Vegetative propagules, which are young sporophytes borne singly in the axils of microphylls, characteristic for some species of the genus *Huperzia* Rothm. are a better material to initiate in vitro cultures than shoots. They are diaspores compensating for the long cycle of generative growth and allow plants to fairly quickly

Table 1 The efficacy of the reagents used and the duration of their effect from trials to determine the optimal methods of disinfection for *H. selago* shoots

	Substance added to ½ MS medium with 20 mg · L^{-1} saccharose				
Internal disinfection variant[a]	Malachite Green (mg · L^{-1})	Antibiotic/ Antimycotic (AAS) (Sigma) (mg · L^{-1})	Benazol 50 WP (g · L^{-1})	Topsin M 500 S.C. (g · L^{-1})	% viable free from microbiological contamination after 4 weeks of culture
1.	0.1	–	–	–	0%
2	0.5	–	–	–	0%
3.	1	–	–	–	All shoots dead after 4 weeks of culture without any signs of contamination
4.	–	50	–	–	0%
5.	–	100	–	–	0%
6.	–	500	–	–	0%
7.	**0.5**	**50**	–	–	**20%**
8.	**0.5**	**100**	–	–	**20%**
9.	0.5	500			All shoots dead after 4 weeks of culture without any signs of contamination
10.	–	–	**2**	–	**95%**
11.	–	–	–	2	All shoots dead after 4 weeks of culture without any signs of contamination

[a]In each variant, 100 shoots were tested

propagate and disperse in the wild. Propagules cut from parent shoots are effectively disinfected using the simplified methods described above. Short-term antifungal treatment is successful, and it consists of incubating propagules with contact fungicides that contain mancozeb, a dithiocarbamate agricultural fungicide, thiophanate-methyl derivatives, or preferably 2-methyl-3(2H)-isothiazolone or 5-chloro-2methyl-3(H) isothiazolone. The last two compounds are the active ingredients of the product Plant Preservative Mixture (PPM®) (Plant Cell Technology, USA). The length of incubation is selected experimentally, and in the precultures of *H. selago* it was 3–4 weeks at 18 ± 1 °C (day) and 16 ± 1 °C (night), in light at 50 μM m^{-2} · s^{-1} and 14 h photoperiod. After incubation, young sporophytes growing from propagules were transferred onto appropriate culture media with the addition of PPM 2 ml/L. PPM effectively protected against microorganisms in the cultures which were maintained for several months, and it did not interfere with the regenerative potential of sporophytes grown from the propagules. No phytotoxic effect of contact dithiocarbamate fungicides was observed on the cultures of

H. selago shoots and propagules. Phytotoxicity manifested as explant death was observed on the media with the addition of 2 g · L^{-1} of the systemic fungicide Topsin M 500 S.C, which contains thiophanate-methyl. Benomyl [1-(Butylcarbamoyl)-1*H*-1,3-benzimidazole-2-yl methylcarbamate] at a dose of 2 g · L^{-1} was well tolerated by the explants. Maxwell and Brody [52] described the conversion of the active substance in Benomyl, i.e., benzimidazole to carbendazim, which was fungi toxic and found to possess a cytokinin-like activity of a growth regulator. Beneficial effects of the compound have been observed in tissue culture systems, such as accelerated growth and regeneration of plant tissues [18]. In the *H. selago* cultures, no differences were observed in the growth rate between shoots cultured on media with and without Benopmyl.

Using spores to start cultures was much simpler and cheaper than the use of sporophytes and their fragments. Acquisition of spores from populations growing in the wild does not reduce their number or produce damage to the shoots of parent plants and is safe for critically endangered club moss species or small local populations [76]. Surface disinfection of spores is simple and consists of their incubation in solutions sequenced as follows: 70% ethanol (v/v) for 1 min., ACE with H_2O (1:5 v/v) for 10 min (ACE, Procter & Gamble, Cincinnati, Ohio, USA) with H_2O (1:5 v/v, i.e., 0.8% sodium hypochlorite solution), in 7% v/v H_2O_2, or 5 min. No bacteria and fungi were observed over the following months in the incubated outgrowths. Also, spores incubated on culture media over a few months prior to germination did not show any signs of microbial contamination. The prothallus cultures thus generated are axenic, free from endosymbionts, and may be achieved within a few months of spore seeding. Under natural conditions, club moss gametophytes may require 12–15 years for maturation, and this period is considerably shortened when tissue culture systems are used [20, 76, 84–87]. As indicated by the consistent findings reported by Freeberg and Wetmore [25], Freeberg [23], Whittier [84, 85], Whittier and Webster [87], Whittier and Strochova [86], and Szypuła et al. [76], the best results are obtained when media poor in mineral salts are used, such as Moore [25] or Knudson [42]. As demonstrated by recent studies, gametophyte in vitro cultures are an excellent model of secondary metabolite biosynthesis [76] and potentially of use for genetic studies, for example, to investigate recessive mutations or to analyze microtranscriptome, especially microRNA, to get new insights into phylogenetic relationships or gene expression regulation in secondary metabolite biosynthesis.

4 Conclusion

In vitro cultures of spermatophytes (seed plants) are easily initiated from seeds, and usually, there are no special problems encountered in the initial stage of disinfection and obtaining axenic cultures. In contrast, cultures of pteridophyte sporophytes are difficult to obtain, especially when shoot fragments are used as starting material. This results from high levels of endophyte accumulation as bacteria and fungi may appear even after many months in cultures previously considered axenic and cause serious damage. Of over 11,000 pteridophyte species, only a few have been grown in culture. Each of the species usually requires individually tailored methods of plant material disinfection. Research has demonstrated that successful disinfection of explants (shoots and propagules) for culture initiation is a multistage process consisting of surface disinfection, short-term antibiotic treatment, or short-term antifungal treatment in vitro or long-term internal disinfection in vitro [74, 75]. Treating explants with agents of contrasting pH values as the first step helps to remove exophytic contaminants and improves the effectiveness of further long-term internal disinfection of the explants in vitro, preferably with antibiotics and fungicides, such as thiophanate-methyl, 2-methyl-3(2H)-isothiazolone, 5-chloro-2methyl-3(H) isothiazolone, or Benomyl. Spores are effectively disinfected using standard procedures and may be used to initiate axenic gametophyte cultures.

References

1. Addy H, Piercey M, Currah R (2005) Microfungal endophytes in roots. Can J Bot 83:1–13
2. Alberton O, Kuyper T, Summerbell R (2010) Dark septate root endophytic fungi increase growth of *Scots pine* seedlings under elevated CO2 through enhanced nitrogen use efficiency. Plant Soil 328:459–470
3. Andrade-Linares DR, Grosch R, Restrepo S et al (2011) Effects of dark septate endophytes on tomato plant performance. Mycorrhiza 21:413–422. https://doi.org/10.1007/s00572-010-0351-1
4. Atmane N, Blervacq AS, Michaux-Ferriere N, Vasseur J (2000) Histological analysis of indirect somatic embryogenesis in the Marsh clubmoss *Lycopodiella inundata* (L.) Holub (Pteridophytes). Plant Sci 156:159–167
5. Barrow JR, Osuna P (2002) Phosphorus solubilization and uptake by dark septate fungi in fourwing saltbush, *Atriplex canescens* (Pursh). Nutt J Arid Environ 51(3):449–459
6. Barrows FL (1935) Propagation of *Lycopodium*. II. Endophytic fungus in gametophyte and sporophyte. Contrib Boyce Thompson Inst 7:295–309
7. Bateman RM (1996) An overview of lycophyte phylogeny. In: Camus JM, Gibby M, Johns RJ (eds) Pteridology in perspective. Royal Bot. Gardens, Kew, pp 405–417
8. Bateman RM, Di Michelle WA, Willard DA (1992) Experimental cladistican analyses of anatomically preserved arborescent Lycopsids from the Carboniferous of Euramerica: an essey in paleobotanical phylogenetics. Ann Mo Bot Gard 79:500–550
9. Bruchmann H (1885) Das Prothallium von *Lycopodium*. Bot Tidsskr 21(1):23–28
10. Bruchmann H (1910) Die Keimung der Sporen und die Entwiklung der Prothallien von *Lycopodium clavatum*, *L. annotinum*, und *L. selago*. Flora 101:220–267
11. Budziszewska J, Szypula W (2010) Influence of site conditions on the diversity of endophytic fungi of fir clubmoss *Huperzia selago* (L.) Bernh. ex Schrank et Mart. Pol J Ecol 58(4):619–626
12. Budziszewska J, Szypuła W, Wilk M, Wrzosek M (2011) *Paraconiothyrium babiogorense* sp. nov., a new endophyte from fir club moss *Huperzia selago* (Huperziaceae). Mycotaxon 115: 457–468. https://doi.org/10.5248/115.457
13. Cove DJ, Knight CD, Lamparter T (1997) Mosses as model systems. Trends Plant Sci 2 (3):99–105
14. Cove D, Bezanilla M, Harries P, Quatrano R (2006) Mosses as model systems for the study of metabolism and development. Annu Rev Plant Biol 57(1):497–520
15. Curlevski NJA, Chambers SM, Anderson IC, Cairney JWG (2009) Identical genotypes of an ericoid mycorrhiza-forming fungus occur in roots of *Epacris pulchella* (*Ericaceae*) and

Leptospermum polygalifolium (*Myrtaceae*) in an Australian sclerophyll forest. FEMS Microbiol Ecol 67(3):411–420. https://doi.org/10.1111/j.1574-6941.2008.00637.x
16. Currah RS, Sigler L, Hambleton S (1987) New records and new taxa of fungi from the mycorrhizae of terrestrial orchids of Alberta. Can J Bot 65:2473–2482
17. Deacon JW (1973) *Phialophora radicicola* and *Gaeumannomyces graminis* on root of grasses and cereals. Trans Br Mycol 61:471–485
18. Debergh PC, Coster DE, Steurbaut W (1993) Carbendazim as an alternative plant growth regulator in tissue culture systems. In Vitro Cell Dev Biol 29:89–91
19. Duckett JG, Burch J, Fletcher PW, Matcham HW, Read DJ, Russell AJ, Pressel S (2004) In vitro cultivation of bryophytes: a review of practicalities, problems, progress and promise. J Bryol 26 (1):3–20
20. Eames AJ (1942) Illustrations of some Lycopodium gametophytes. Am Fern J 32:1–12
21. Ellis MB, Ellis JP (1997) Microfungi on land plants: an identification handbook. The Richmond Publishing, Slough
22. Falkiner FR (1997) In: Cassels AC (ed) Pathogen and microbial contamination management in micropropagation. red. Kluwer Academic, Dordrecht, pp 155–161
23. Freeberg JA (1957) The apogamous development of sporelings *of Lycopodium cernuum* L., L. complanatum var. Flabelliforme Fernald and L. selago L. in vitro. Phytomorphology 7:217–229
24. Freeberg JA (1962) *Lycopodium* prothalli and their endophytic fungi as studied in vitro. Am J Bot 49:530–535
25. Freeberg JA, Wetmore RH (1957) Gametophytes of *Lycopodium* as grown in vitro. Phytomorphology 7:204–217
26. Gemma JN, Flyn T, Koske RE (1992) Mycorrhizae in Hawaiian Pteridophytes: occurrence and evolutionary significance. Am J Bot 79:843–852
27. Goebel K (1887) Ueber Prothallien und Keimpflanzen von *Lycopodium inundatum*. Bot Zeit 45 (11):161–167
28. Hallmann J, Quandt-Hallmann A, Mahaffee WF, Kloepper JW (1997) Bacterial endophytes in agricultural crops. Can J Microbiol 43:895–914
29. Harley JL, Harley EL (1987) A check-list of mycorrhiza in the British flora. New Phytol 105(2): 1–102
30. Harley JL, Smith SE (1983) Mycorrhizal symbiosis. Academic, London/New York
31. Haselwandter K (1987) Mycorrhizal infection and its possible ecological significance in climatically and nutritionally stressed alpine plant communities. Angew Bot 61:107–114
32. Haselwandter K, Read DJ (1980) Fungal associations of roots of dominant and sub-dominant plants in high-alpine vegetation systems with special reference to mycorrhiza. Oecologia 45:57–62
33. Haselwandter K, Read DJ (1982) The significance of a root-fungus associations in two *Catex* species of high-alpine plant communities. Oecologia 53:352–354
34. Higgins KL, Arnold AE, Miadlikowska J, Servate SD, Lutzoni F (2006) Phylogenetic relationships, host affinity, and geographic structure of boreal and arctic endophytes from three major plant lineages. Mol. Phylogenet Evol 42:543–555
35. Ishiuchi K, Park JJ, Long RM, Gang DR (2013) Production of huperzine A and other *Lycopodium* alkaloids in *Huperzia* species grown under controlled conditions and in vitro. Phytochemistry 91:208–219
36. Jumpponen A (2001) Dark septate endophytes – are they mycorrhizal? Mycorrhiza 11:207–211
37. Jumpponen A, Trappe J (1998) Dark septate endophytes: a review of facultative biotrophic root-colonizing fungi. New Phytol 140:295–310
38. Kenrick P, Crane PR (1997a) The origin and early diversification of land plants: Achladistic study, vol 389. Smithsonian, Washington, DC, pp 33–39
39. Kenrick P, Crane PR (1997b) The origin and early evolution plants on land. Nature 389:33–39
40. Kenrick P, Davis P (2004) Fossil plants. Natural History Museum, London. 216 pp

41. Knie N, Fischer S, Grewe F, Polsakiewicz M, Knoop V (2015) Horsetails are the sister group to all other monilophytes and Marattiales are sister to leptosporangiate ferns. Mol Phylogenet Evol 90:140–149
42. Knudson L (1922) Nonsymbiotic germination of orchid seeds. Bot Gaz 73:1–25
43. Korall P, Pryer KM, Metzgar JS, Schneider H, Conant DS (2006) Tree ferns: monophyletic groups and their relationships as revealed by four protein-coding plastid loci. Mol Phylogenet Evol 39:830–845
44. Korall P, Conant DS, Metzgar JS, Schneider H, Pryer KM (2007) A molecular phylogeny of scaly tree ferns (Cyatheaceae). Am J Bot 94:873–886
45. Lang WH (1899) The prothallus of *Lycopodium clavatum*. Ann Bot 13:279–318
46. Lehtonen S (2011) Towards resolving the complete fern tree of life. PLoS One 6:e24851
47. Li T, Liu M, Zhang H, Zhang T, Sha T, Zhao Z (2011) Improved tolerance of maize (*Zea mays* L.) to heavy metals by colonization of a dark septate endophyte (DSE) *Exophiala pisciphila*. Sci Total Environ 409:1069–1074
48. Ma X, Gang DR (2008) In vitro production of huperzine A, a promising drug candidate for Alzheimer's disease. Phytochemistry 69:2022–2028
49. Mandyam K, Jumpponen A (2005) Seeking the elusive function of the root-colonising dark septate endophytic fungi. Stud Mycol 53:173–189
50. Mandyam K, Fox C, Jumpponen A (2011) Septate endophyte colonization and host responses of grasses and forbs native to a tallgrass prairie. Mycorrhiza 22:109–119
51. Marszał-Jagacka J, Kromer K (2010) In vitro propagation of rare and endangered serpentine fern species. In: Kumar A, Fernández H, Revilla MA (eds) Working with ferns: issues and applications. Springer, New York, pp 149–164
52. Maxwell WA, Brody G (1971) Antifungal activity of selected benzimidazole compounds. Appl Microbiol 21:944–945
53. Murashige T, Skoog F (1962) A revised medium for rapid growth and bioassays with tobacco tissue culture. Plant Physiol 15:473–497
54. Narisawa K, Kawamata H, Currah R, Hashiba T (2002) Suppression of Verticillium wilt in eggplant by some fungal root endophytes. Eur J Plant Pathol 108:103–109
55. Narisawa K, Usuki F, Hashiba T (2004) Control of Verticillium Yellows in Chinese cabbage by the Dark Septate endophytic Fungus LtVB3. Phytopathology 94:412–418
56. Orlikowska T, Zawadzka M (2006) Bacteria in plant tissue cultures. Biotechnologia 4:64–77
57. Pacyna A (1972) Zarodniki i przedrośla widłaków. Wiad Bot 16(1):57–63
58. Porras-Alfaro A, Herrera J, Sinsabaugh R, Odenbach K, Lowrey T, Natvig D (2008) Novel root fungal consortium associated with a Dominant Desert grass. Appl Environ Microbiol 74(9): 2805–2813
59. Pryer KM, Schneider H, Smith AR, Cranfill R, Wolf PG, Hunt JS (2001) Horseitals and ferns are a monophyletic group and the closest living relatives to seed plants. Nature 409:618–621
60. Rai HS, Graham SW (2010) Utility of a large, multigene plastid data set in inferring higher-order relationships in ferns and relatives (monilophytes). Am J Bot 96:1444–1456
61. Read DJ, Haselwandter K (1981) Observations on the mycorrhiza status of some alpine plant communities. New Phytol 88:341–352
62. Rothfels CJ, Larsson A, Kuo L-Y, Korall P, Chiou W-L, Pryer KM (2012) Overcoming deep roots, fast rates, and short internodes to resolve the ancient rapid radiation of eupolypod II ferns. Syst Biol 61:490–509
63. Rothfels CJ, Li F-W, Sigel EM, Huiet L, Larsson A, Burge DO, Ruhsam M, Deyholos M, Soltis DE, Stewart CN Jr, Shaw SW, Pokorny L, Chen T, de Pamphilis C, DeGironimo L, Chen L, Wei X, Sun X, Korall P, Stevenson DW, Graham SW, Wong GK-S, Pryer KM (2015) The evolutionary history of ferns inferred from 25 low-copy nuclear genes. Am J Bot 102:1–19
64. Ruggiero MA, Gordon DP, Ornell TM, Bailly N, Bourgoin T, Brusca RC, Cavalier-Smith T, Guiry MD, Kirk PM (2015) A higher level classification of all living organisms. PLoS One 10: eo119248

65. Schadt CW, Mullen RB, Schmidt SK (2001) Isolation and phylogenetic identification of a dark-septate fungs associated with the alpine plants *Ranunculus adoneus*. New Phytol 150(3):747–755
66. Schmid E, Oberwinkler F (1993) Mycorrhiza – like interaction between the achlorophyllus gametophyte of *Lycopodium clavatum* L. and its fungal endophyte studied by light and electron microscopy. New Phytol 124:69–81
67. Schuettpelz E (2016) A community – derived classification for extant lycophytes and ferns. J Syst Evol 54(6):563–603
68. Schuettpelz E, Pryer KM (2007) Fern phylogeny inferred from 400 leptosporangiate species and three plastid genes. Taxon 56:1037–1050
69. Spessard EA (1922) Prothallia of *Lycopodium* in America. Bot Gaz 74:392–412
70. Steiner U, Mahalia A, Ahimsa-Müller MA, Markert A, Kucht S, Groß J, Kauf N, Kuzma M, Zych M, Lamshöft M, Furmanowa M, Knoop V, Drewke C, Leistner E (2006) Molecural characterization of seed transmitted clavicipitaceeous fungs occurring on dicotyledoneous plants (Convolvulaceae). Planta 224:533–544
71. Stoyke G, Currah RS (1991) Endophytic fungi from the mycorrhizae of alpine ericoid plants. Can J Bot 69:347–352
72. Szypuła WJ, Pietrosiuk A (2020) Production of cholinesterase-inhibiting compounds in in vitro cultures of Club mosses. In: Ramawat KG, Ekiert HM, Goyal S (eds) Plant cell and tissue differentiation and secondary metabolites: fundamentals and applications. Springer International Publishing, Cham, pp 1–40. https://doi.org/10.1007/978-3-030-30185-9_30
73. Szypuła WJ, Pietrosiuk A (2021) Biology, phytochemistry, pharmacology, and biotechnology of European ferns, Club mosses, and horsetails: a review. In: Ekiert HM, Ramawat KG, Arora J (eds) Medicinal plants. Sustainable development and biodiversity, vol 28. Springer, Cham. https://doi.org/10.1007/978-3-030-74779-4_19
74. Szypuła W, Pietrosiuk A, Suchocki P, Olszowska O, Furmanowa M, Kazimierska O (2005) Somatic embryogenesis and in vitro culture of *Huperzia selago* shoots as a potential source of Huperzine A. Plant Sci 168:1443–1452. https://doi.org/10.1016/j.plantsci.2004.12.021
75. Szypuła WJ, Mistrzak P, Olszowska O (2013) A new and fast method to obtain in vitro cultures of *Huperzia selago* (Huperziaceae) sporophytes, a club moss which is a source of Huperzine A. Acta Soc Bot Pol 82:313–320. https://doi.org/10.5586/asbp.2013.034
76. Szypuła WJ, Wileńska B, Misicka A, Pietrosiuk A (2020) Huperzine A and Huperzine B production by Prothallus cultures of *Huperzia selago* (L.) Bernh. ex Schrank et Mart. Molecules 25(14):3262. https://doi.org/10.3390/molecules25143262u
77. Tellenbach C, Grünig C, Sieber T (2011) Negative effects on survival and performance of Norway spruce seedlings colonized by dark septate root endophytes are primarily isolate-dependent. Environ Microbiol 13:2508–2517
78. Treu R, Laursen GA, Stephenson SL, Landolt JC, Densmore R (1996) Mycorrhiza from Denali National Park and Preserve Alaska. Mycorrhiza 6:21–29
79. Treub M (1884) Études sur les Lycopodiacées. II. Le prothalie du *Lycopodium phlegmaria* L. Ann Jard Bot Buitenz 5:87–114
80. Upson R, Newsham K, Bridge P, Pearce D, Read D (2009) Taxonomic affinities of dark septate root endophytes of *Colobanthus quitensis* and *Deschampsia Antarctica*, the two native Antarctic vascular plant species. Fungal Ecol 2:184–196
81. Väre H, Vestberg M, Eurda S (1992) Mycorrhiza and root-associated fungi in Spitsbergen. Mycorrhiza 1:93–104
82. Wang B, Qiu YL (2006) Phylogenetic distribution and evolution of mycorrhizas in land plants. Mycorrhiza 16:299–363
83. Wang CJK, Wilcox HE (1985) New species of ectoendomycorrhizal and pseudomycorrhizal fungi: *Phialophora finlandia, Chloridium paacisporum,* and *Phialocephala fortinii*. Mycologia 77:951–958
84. Whittier DP (1977) Gametophytes of *Lycopodium obscurum* as grown in axenic culture. Can J Bot 55:563–567

85. Whittier DP (1981) Gametophytes of *Lycopodium digitatum* (formerly *L. complanatum* var. *flalellifare*) as grown in Axenix Culture. Bot Gaz 142:519–524
86. Whittier DP, Strochova H (2007) The gametophyte of Huperzia selago in culture. Am Fern J 97 (3):149–154
87. Whittier DP, Webster TR (1986) Gametophytes of *Lycopodium lucidulum* from Axenic Culture. Am Fern J 76(2):48–55
88. Wickett NJ, Mirarab S, Nguyen N, Warnow T, Carpenter E, Matasci N, Ayyampalayam S, Barker MS, Burleigh JG, Gitzendanner MA, Ruhfel BR (2014) Phylotranscriptomic analysis of the origin and early diversification of land plants. Proc Natl Acad Sci U S A 111:E4859–E4868
89. Wilcox HE, Wang CJK (1987) Mycorrhizal and pathological associations of dematiaceous fungi in roots of 7-month-old tree seedlings. Can J For Res 17:884–899
90. Ya-Li L, Zhang F, Chrn J, Cui J, Xing Y, Li X (2010) Diversity and antimicrobial activity of endophytic fungi associated with the alpine plant Saussurea involucrata. Biol Pharm Bull 33(8): 1300–1306
91. Zenkteler E (2000) Systemy wegetatywnego rozmnażania paproci *in vivo* oraz *in vitro*. Wydawnictwo Naukowe Uniwersytetu im. A. Mickiewicza w Poznaniu, Poznań, pp 88–109
92. Zenkteler E (2001) Znaczenie prekultur w mikrorozmnażaniu roślin zasiedlonych przez endofity. Biotechnologia 4(55):87–99
93. Zhang Y, Zhang Y, Liu M, Shi X, Zhao Z (2008) Dark septate endophyte (DSE) fungi isolated from metal polluted soils their taxonomic position, tolerance, and accumulation of heavy metals *in vitro*. J Microbiol 46:624–632
94. Zhou XM, Zhang LB (2015) A classification of *Selaginella* (Selaginellaceae) based on molecular (chloroplast and nuclear), macromorphological, and spore features. Taxon 64:1117–1140

Ecometabolomics Studies of Bryophytes

24

Kristian Peters, Yvonne Poeschl, Kaitlyn L. Blatt-Janmaat, and Henriette Uthe

Contents

K. Peters (✉)
German Centre for Integrative Biodiversity Research (iDiv) Halle-Jena-Leipzig, Leipzig, Germany

Institute of Biology/Geobotany and Botanical Garden, Martin Luther University Halle-Wittenberg, Halle (Saale), Germany

Bioinformatics and Scientific Data, Leibniz Institute of Plant Biochemistry, Halle (Saale), Germany
e-mail: kpeters@ipb-halle.de

Y. Poeschl · H. Uthe
German Centre for Integrative Biodiversity Research (iDiv) Halle-Jena-Leipzig, Leipzig, Germany

Institute of Biodiversity, Friedrich Schiller University, Jena, Germany
e-mail: yvonne.poeschl@idiv.de; henriette.uthe@idiv.de

K. L. Blatt-Janmaat
Bioinformatics and Scientific Data, Leibniz Institute of Plant Biochemistry, Halle (Saale), Germany

Department of Chemistry, University of New Brunswick, Fredericton, NB, Canada

H. N. Murthy (ed.), *Bioactive Compounds in Bryophytes and Pteridophytes*,
Reference Series in Phytochemistry, https://doi.org/10.1007/978-3-031-23243-5_30

Abstract

Bryophytes are the largest group of non-vascular plants that occur in almost any land ecosystem and have remarkable impact on ecosystem functioning at a global level. Despite that they have evolved an extraordinary chemical diversity, only a few bryophytic species have been studied using metabolomic techniques. Ecometabolomics systematically investigates the composition of metabolic compounds in bryophytes and relates these to organismal and environmental interactions. The application of ecometabolomics to bryophytic organisms can lead to new insights into their molecular biology, can identify novel bioactive natural products, can shed light on the phylogenetic and evolutionary mechanisms bryophytes realize in order to sustain ecological change, or can greatly improve the mechanistic understanding of ecological processes that are mediated by metabolic compounds at various levels. In this chapter, we first describe ecometabolomics and provide an introduction to how it can be performed. We then focus on case studies covering the various research fields of natural product chemistry, chemodiversity, chemotaxonomy/chemophenetics, functional ecology and plant traits, bioindication and biomonitoring, bioactivities, and the molecular biology of bryophytes. Finally, we present the latest advancements in analytic and computational methods to show the tremendous potential of the emerging technology of ecometabolomics for research with bryophytic organisms.

Keywords

Bioactivity · Bioindicators · Biomonitors · Biotechnology · Chemical diversity · Ecological metabolomics · Molecular biology · Mass spectrometry · Molecular traits

Abbreviations

(C)CA	(Canonical) Correspondence Analysis
(N)MDS	(Non-metric) Multi-Dimensional Scaling
4CL	4-coumarate CoA ligase
ACP	Acyl carrier protein
ANOVA	Analysis of Variances
C	Net plant carbon
C4H	Cinnamate 4-hydrolase
CAWG	Chemical Analysis Working Group
CBGA	Cannabigerolic acid
CHS	Chalcone synthase
CoA	Coenzyme A

DBR	Double bond reductase
DDA	Data-dependent acquisition
DIA	Data-independent acquisition
FAIR	Findable, Accessible, Interoperable, Reusable
FT-ICR-MS	Fourier Transform Ion Cyclotron Resonance Mass-Spectrometry
GC	Gas chromatography
GC/MS	Gas chromatography coupled to mass-spectrometry
H′	Shannon diversity index
IPP	Isopentyl diphosphate delta isomerase
J	Pielou's evenness
LC	Liquid chromatography
LC/MS-MS	liquid chromatography coupled with tandem mass-spectrometry
m/z	Mass-to-charge ratio
MEP	Non-mevalonate
MS	Mass-spectrometry
MSI	Metabolomics Standards Initiative
MTPSL	Microbial terpene synthase-like
MVA	Mevalonate
N	Nutrients
NMR	Nuclear magnetic resonance
PAL	Phenylalanine ammonia-lyase
PCA	Principal Component analyses
PDH	Pyruvate dehydrogenase
PLS	Partial Least Squares regression
PLSDA	PLS coupled with Discriminant Analysis
QC	Quality control
RDA	ReDundancy Analysis
S	Compound richness
SOM	Soil organic matter
STCS	Stilbene carboxylate synthase
THC	Tetrahydrocannabinol
TIMS	Ion-Mobility Mass-Spectrometry
ToF	Time of flight
U	Number of unique compounds
VOCs	Volatile organic compounds

1 Introduction

Bryophytes occur in almost any land ecosystem and have a remarkable ecological impact despite their small size. For instance, bryophytes are major contributors to global biogeochemical cycles and ecosystem functioning. They further influence vegetation dynamics and are distinctive bioindicators [1–5]. Bryophytes constitute a major part of the biodiversity in rainforests, marshes, high mountain areas, peatlands, and tundra ecosystems. For example, in Northern peatland ecosystems, bryophytes cover approx. 90% of the land area. In these areas bryophyte species store about 30% of the global carbon and approx. 10% of the global soil nitrogen [5]. In boreal forests, bryophytes can account for up to 50% of the annual gross primary productivity [6]. Moreover, due to their remarkable water retention capabilities, bryophytes control hydrology and temperature regimes in many ecosystems [7]. Bryophytes are also among the first plants that (re)colonize disturbed or newly exposed environments. As they stabilize the soil and accumulate humus in the early successional stages, they prepare habitats for secondary plant establishment [1]. In this context, bryophytes are keystone species in biocrust communities in dryland ecosystems, and *Sphagnum* species structure entire northern peatland ecosystems [5, 6].

Despite the ecological impact of bryophytes, their biochemistry is relatively unstudied – especially with regard to the ecological function of compounds. Terpenoids and phenolic compounds constitute the largest groups of compounds in bryophytes. They also produce a large variety of glycosides, flavonoids, and other aromatic compounds [8, 9]. Restricted to liverworts are specialized organelles called "oil bodies" that appear under the microscope as oil droplets [10]. They contain not only fatty acids and terpenoids but many other specialized compounds, such as cyclic neolignans ((bis)-bibenzyls), diverse sesquiterpene lactone derivatives, and flavonoids. Many of these compounds are considered to be bioactive and are produced in response to particular ecological stimuli [8, 11]. The research field of ecometabolomics investigates these and other ecological relationships of metabolic compounds [12].

Metabolomics is defined as the systematic study of all chemical processes concerning metabolic compounds. This provides characteristic chemical fingerprints of cellular processes and molecular fingerprints, i.e., responses to environmental changes or biotic interactions of species [12, 13]. The use of the chemical-analytical tool metabolomics to answer ecological questions is therefore called ecometabolomics [26].

Untargeted metabolomics is a biochemical approach used to measure many thousands of metabolites in biological organisms without focusing on particular metabolites or groups of compounds. In the past 10 years, tremendous advances in the technology of (untargeted) metabolomics, both analytical and computational, have been made. Using mass-spectrometry, it is now possible to measure and annotate nearly all low molecular weight (typically <1000 Da) polar and semi-polar compounds in bryophytes, plants, microbials, or even mammals without targeting specific compounds [14, 15]. Despite the fact that there are several analytical metabolomics acquisition techniques available, liquid chromatography coupled with tandem mass-spectrometry (LC/MS-MS) has been most often used in the context of ecology to link different spatiotemporal scales [12].

The virtue of ecometabolomics is that this technology can be applied to a wide range of research questions in the field of ecology [16, 17]. These can include analyzing the chemical diversity and natural product chemistry of bryophytes [8, 18], assisting phenotyping approaches [19, 20], performing "full-metabolome" annotations (as was recently shown for the moss *Ceratodon purpureus*) [21], exploring the mechanistic basis of biotic and environmental interactions, [4] such as the molecular effect of UV-B radiation [22], investigating intra- and interspecific seasonal variations [23], assessing the differences in compound classes between fresh samples and herbarium specimens [24], performing chemotaxonomy (see Sect. 3.3), and many more.

Compared to other chemical-analytical methods, the analysis of the metabolome is the closest we can come to analyzing the chemical phenotype and its response to internal and external stressors [25, 26]. From a single sample, it is theoretically possible to reveal large numbers of different metabolites and their abundances at specific time points in one single analytical process. This huge amount of chemical information can be used for specific questions about single metabolites but also the discovery of unknown natural products and evolutionary mechanisms. As shown for the peat moss *Sphagnum fallax*, characterizing metabolite profiles in bryophytes provides a baseline to understand the mechanisms in which bryophytes interact with their environment and can further reveal key biomarkers that predict ecological function [1, 27]. As metabolomics has not often been applied to bryophytic organisms, there is a huge research opportunity to discover unknown bioactive compounds, to further elucidate the role of chemical diversity in bryophytes, or to characterize unique compounds in bryophytes with potential for practical use in, i.e., medicine [28].

In the following, we describe the most prominent research in the past years and show how ecometabolomics can improve our mechanistic understanding of ecological and biological processes at various scales.

2 Integrating Metabolomics and Ecology

The integration of metabolomics in ecology opens up completely new possibilities and depths to understanding the chemical interactions of organisms and the environment. This technology enables the generation of large and complex datasets, which often turn out to be a double-edged sword. In order to create valuable data sets with metabolomics and to evaluate them in a target-oriented manner, teamwork from several parties (ecologists, chemists, bioinformaticians) is required to plan the optimal metabolomic approach. Usually, the ecologist needs support here and good project management is the key [29]. Furthermore, scientists should be aware of the different steps of the metabolomics workflow (see Fig. 1) and their challenges, i.e., when it comes to compound annotation (Sect. 2.9). Nowadays, there are many high-quality computational resources available for metabolomics, such as the many statistical analysis tools [30], open spectral libraries such as MassBank [31], MetFrag, and SIRIUS software suites enabling computational compound identification [32, 33], or tools that relate compounds to molecular pathways (GNPS, BioCyc) [34, 35]. The following sections give a short introduction to the individual steps of the analytical and bioinformatics workflows.

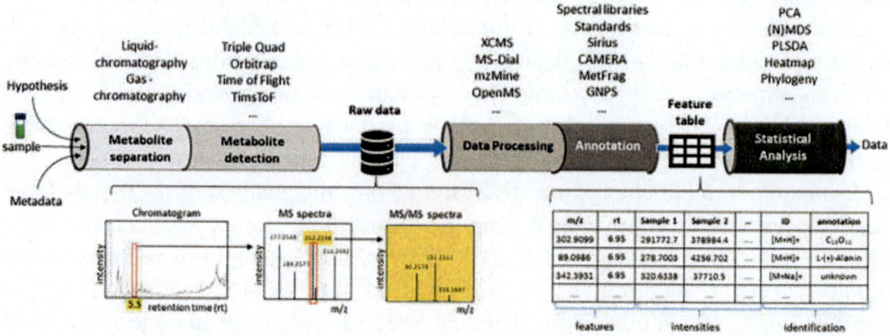

Fig. 1 Exemplary Mass-spectrometric metabolomic workflow: The compounds in the extracted sample are separated by chromatography and detected via mass-spectrometry (MS). The raw data includes the chromatogram (intensity vs. retention time), MS (intensity vs. mass-to-charge ratio (m/z)), and (if MS/MS is applied) the spectra of the fragmented metabolites (intensity vs. m/z). Retention time, m/z value and MS/MS spectra (in yellow) are the three properties needed to assign an identity to a feature. Raw data is transformed into a feature table applying feature detection, retention time correction, feature grouping, and annotation. The feature table consists of the samples (columns), detected features (rows), and the respective intensities of every feature in every sample. Depending on the tools used (i.e., CAMERA), the feature table includes information about the identity, i.e., adducts, isotope peaks, and water losses or spectral library matches. According to the hypothesis and metadata, statistical approaches are applied

2.1 Metadata and Its Important Role in the Whole Workflow

Meaningful metadata is essential to producing high-quality metabolomic data. Ecological and environmental metadata documents every step of data collection including the design, sampling, extraction, mass-spectrometric analysis, MS data pre-processing, and even statistical analyses. The metadata should be complete enough to be used by other scientists to independently replicate the experiment and reproduce its results. Throughout the study, metadata is collected in the form of various tables, lists, and protocols. Metadata is also generated by software tools during computational analyses. This information is highly needed for statistical analysis, modeling, and even for the publication of the study. Many journals require submitting data and associated metadata to a specialized or general data platform that stores data permanently and provides a persistent identifier for datasets. For metabolomic data, MetaboLights [36] and MassIVE and ReDU (GNPS) [37, 38] are currently the most commonly used repositories. Good metadata also enables other scientists to use mass-spectrometry data for meta-analysis, which is quite common in the field of ecology. To ensure interoperability it is recommended that the uploaded metadata is structured to be human and machine-readable, to comply with the Minimal Information Standard for metabolomic data [39], to be "FAIR" (Findable, Accessible, Interoperable, Reusable) [40], and to be suitable for the use of appropriate metabolomic or ecological-specific controlled vocabularies (ontologies) ([29], https://youtu.be/rL_nC-56oVY).

In the following, we show how expressive metadata is generated throughout the entire metabolomics workflow and how it can be used.

2.2 Experimental Design

Ecometabolomics approaches often produce big and complex data sets of several gigabytes. Therefore, it is necessary to design the whole experiment with the expected outcome in mind. The whole experimental design from sampling to the annotation strategy needs to fit the hypothesis. In general, there are two strategies to analyze data: (1) Metabolite fingerprinting (where patterns of metabolites are related to ecological function) and (2) Metabolite profiling (where specific metabolites are related to ecological function). Depending on which strategy is pursued, the experimental design must be adapted and also clarified in advance with those responsible for metabolomics and biostatistics analyses.

The most challenging part of ecometabolomics is the high variability of the metabolite profiles that may produce results where the effect is not recognizable anymore. A preliminary experiment helps to identify the factors causing biological and technical variability (https://youtu.be/2MUMcLh8xO8). By removing other influences (biotic or abiotic), genetically determined metabolomic variation can be identified much better.

Natural variation can also be handled with a sufficient number of biological replicates. Generally, biological replicates are defined as measurements of biologically distinct samples that show biological variation but with very similar growing conditions (age, herbivory, host tree, light conditions, etc.). Most statistical models require a minimum of 3 biological replicates to fit a model to the data, but for reliable results, 6–10 replicates are needed. This relatively high number of replicates covers natural variation, mixing up species, and sample loss. This is even more important when the samples do not come from a controlled environment (climate chamber or greenhouse). When sampling in the field, it might not be possible to take enough true replicates. Reasons can be export limitations (i.e., the Nagoya protocol, https://www.cbd.int/abs/) or the absence of enough species in the experimental field. In that case, the expectations about which questions can be addressed or if the study remains rather descriptive must be communicated from the beginning.

Another aspect of the experimental design in ecometabolomics experiments is the use of blanks and quality controls (QC) to determine background noise, contaminations, and technical variance. Metabolomics is highly sensitive but not selective, therefore every metabolite from the processing might appear in the data later on. To distinguish between contaminants (i.e., from the soil, plastic tubes, gloves, or solvents) and the metabolites of the actual sample, blanks of the materials and chemicals used should be included in the analytical processing pipeline [41]. Samples for QC are prepared by pooling samples, i.e., all samples or sets of samples of predefined conditions. QC samples do not only help to estimate technical variability but also play a role in the MS data processing (see Sect. 2.7).

2.3 Sampling and Storage

As metabolomics can be challenging, we recommend standardizing sampling and storage to avoid high variability due to uncontrolled experimental factors. Thus, sampling should be carried out using the same methods and sterile tweezers to avoid any contamination. Furthermore, a long sampling duration should be avoided as the metabolome of vascular plants is known to be changing over the day due to differences in light availability and temperature [42]. We recommend collecting moss material (usually sampling of several moss plants is needed) in sterile non-chlorinated tea bags, plastic petri dishes, or Eppendorf tubes. Samples should be clearly labeled, including the species name, date, GPS coordinates, and edaphic and climatic conditions (if applicable in field sampling). A hand lens or a microscope is often necessary to identify the species. As it is not clear how gametophytes and sporophytes differ with regard to their metabolism, we recommend sampling only gametophytes, as these are the usual development stage of bryophytes, unless the sporophyte is part of the research focus (i.e., for chemotaxonomic analysis).

The golden standard for metabolomic sample storage is the immediate transfer of samples directly in liquid nitrogen or dry ice and transfer in the −80 °C freezer to quench metabolomic processes. This will stop all enzyme activities by freezing. Other methods include destroying (denaturing) the enzymes by boiling or putting the material into an organic solvent [43]. When collecting bryophytes in the field, we recommend putting the samples in sterile transport vessels (see above) and transporting the samples as fast as possible to the wet lab. Putting samples on dry ice also slows down enzyme activity. Processing frozen samples always involve the risk of such materials thawing during processing and this too may lead to the loss of labile compounds. For safe sample handling, freeze-drying is recommended since it removes all water. A freeze-dried sample can be stored and processed at room temperature. Nevertheless, some volatile compounds might get lost during lyophilization [44].

If liquid nitrogen or dry ice is not available at the sampling site, a second-best option is cooling the samples on dry ice packs until freezing or by immersing them in solvents (methanol, ethanol) for short-term storage. It should be noted that long-term storage in methanol will lead to methylated artifacts [45]. Finally, the sample collection must be done in the same way for all samples.

2.4 Metabolomics Analysis Itself, Targeted Versus Untargeted

The ecometabolomic workflow (Fig. 1) can be divided into five parts: (1) metabolite separation, (2) metabolite detection, (3) data processing, (4) annotation, and (5) statistical analysis. The different parts, i.e., the devices or software used, can be put together according to the scientific question. There is no "one workflow" and the approach needs to be taken to fit the research question.

Usually, the first question to be clarified is whether there is a targeted or untargeted approach necessary. Untargeted means that the aim of the experiment is the detection of as many metabolites as possible to get a holistic impression of the chemical status of a sample [8]. The extraction methods, the instruments, and the analytical method are chosen for minimal losses of metabolites. Nevertheless, no matter which extraction method or instrument is chosen, the results will only show a part of the metabolome. Metabolites are often present in complex mixtures at extremely different concentrations and sometimes dissolve poorly or not at all in the extraction agent. By contrast, a targeted approach is the investigation of one or more already known metabolites or a group of compounds in a sample, also referred to as metabolic profiling [46]. In targeted studies, specific compounds are quantified and compared to established reference ranges [47]. For this approach, the extraction and analytical workflow is chosen to focus on the metabolite of interest. Optimally, the metabolite is available as a purified standard to be used for the development of specific analytical methods. An example would be the analysis of phytohormones as a response to environmental factors [48, 49].

The extraction methods are generally very similar for bryophytes and for vascular plants [8, 50]. Protocols developed for vascular plants like *Arabidopsis* also work for bryophytes [23, 50]. However, they differ depending on the choice of approach. As the chemical properties of all metabolites of a sample vary heterogeneously, the scientific question decides the extraction method. For example, specialized metabolites like phenols or sesquiterpenes (which might be involved in plant-herbivore interaction) are often extracted from the plant material by using a mixture of methanol and water [51]. For the analysis of signal lipids, dichlormethane is the solvent of choice [52].

2.5 Platforms (GC, LC, NMR)

There are thousands of chemically different (i.e., molecular weight, structure, polarity) molecules in a sample or plant extract. A comprehensive analysis that aims to distinguish and detect as many molecules as possible requires two prerequisites: separation and detection (Fig. 1). The separation, or chromatography, is based on the principle that molecules in a mixture are applied onto the surface or into the solid stationary phase (stable phase) and then separated from each other while moving with the aid of a mobile phase. Depending on whether the mobile phase is a gas or a liquid, the separation is achieved by either gas (GC) or liquid chromatography (LC) (for more information see [53]). Due to the interaction with the stationary phase (column), the metabolites are separated and are given a specific property: the retention time. While LC can theoretically be used for all metabolites soluble in water and organic solvent, GC can only be applied for volatile organic compounds (i.e., monoterpenes, sesquiterpenes) [54] or primary metabolites (i.e., sugars, amino acids, fatty acids) after derivatization.

After chromatographic separation, the compounds are analyzed further by MS or nuclear magnetic resonance (NMR) [55], whereby the focus will be on MS in this chapter. MS facilities like Quadrupoles, ToFs (Time of flight), Orbitrap [56], and TIMs-ToF [57] determine the m/z value which corresponds directly to the exact molecular mass, the second specific property. The m/z value is defined as the mass of a metabolite (in Dalton) in relation to its charge. Depending on the facility, high- and low-resolution MS can be distinguished. The mass resolution refers to the ability to separate two narrow mass-spectral peaks, i.e., between m/z = 243.0456 and m/z = 243.0012 which might be two completely different metabolites (see [29]). In an untargeted approach, thousands of metabolites are analyzed simultaneously, including those with very similar molecular masses. In order to distinguish between them with certainty, an instrument with high mass resolution must be used for an untargeted analysis.

2.6 Tandem MS (MS/MS)

With the two properties of retention time and molecular mass (m/z), it is only possible to annotate adducts, isotope peaks, and in-source fragments of certain features (i.e., with CAMERA [58]). They do not provide enough information to describe an unknown metabolite further or to certainly identify it. For a more comprehensive annotation, information at the MS/MS level (tandem MS) is necessary. "In brief, molecules are ionized and the resulting mass/charge ratio is measured (m/z). During this ionization step, the molecules are brought to higher energy states and become unstable. This leads to chemical bond breaking and the fragmentation of the original molecule (precursor ion). The m/z values of the original molecule and the newly formed fragments (product ions) are visualized in the mass spectrum of the metabolite" [29]. A second mass spectrum of such fragments is characteristic of a metabolite and therefore the third property for the metabolite identity. Not all metabolites in a sample are fundamentally fragmented and generate MS/MS spectra. This depends on the instrumental settings (voltage) and the applied acquisition mode: data-dependent or data-independent. In Data-dependent acquisition (DDA) mode only the most (i.e., 10, 15, 20) intense metabolites within a certain range are selected, isolated, fragmented, and analyzed further. In contrast, in data-independent acquisition (DIA) mode all metabolites within a certain mass range are fragmented in tandem mass spectrometry [59]. At this point of the metabolomic workflow, every feature is assigned to a retention time, m/z value, and - depending on the acquisition mode - an MS/MS spectrum.

2.7 Data Processing

Raw MS data is not easily readable by humans. It needs some treatments, referred to as processing, to make it easily accessible, interpretable, and analyzable by visualization or statistics software. Raw MS data are acquired and stored in vendor-specific file formats and are processed using the corresponding licensed vendor software. Hence, vendor-specific file formats can be converted with freely available tools (i.e., Proteowizard [60]) into open file formats like mzML. The mzML file format is widely used in the community and is also suitable for sharing and storing raw data in databases (see Sect. 2.10). mzML-formatted raw MS data can be processed with various free software, i.e., desktop-based mzMine [61], MS-Dial [62], OpenMS [63], browser-based Galaxy [64] or XCMS online [65], or script-based XCMS [66, 67], an R package to be used within the statistical framework R [68]. A comprehensive overview of metabolomics data formats and how they are used is given by Hoffmann et al. (2022) [69].

MS data processing comprises all steps from raw data to the final feature table. Algorithms addressing the single steps in the processing workflow are provided by the aforementioned software. As these algorithms rely on different assumptions and have different parameters, they therefore will yield different results. Providing metadata for the corresponding samples to the processing tools would make it easier to set parameter values and therefore would improve the processing results. First, signals differing from background noise are detected in the raw data. Signals are referred to as features that are characterized by their corresponding mass-to-charge value (m/z value), retention time, and intensity. Signals have a peak-like shape that begins and ends at certain retention time points. Often the mean retention time is reported for the feature. The corresponding intensity of the feature could either be the area covered by the peak-like shape or its maximum intensity value.

In a second step, features are grouped or aligned across all measured samples or groups of samples defined in the metadata by allowing for some tolerance in m/z and/or retention time depending on the algorithm used. The grouping results in the feature table where the number of columns corresponds to the number of measured samples and the number of rows corresponds to the number of detected features. Very often, thousands of features are detected in tens to hundreds of samples. So, both dimensions differ considerably. The interpretability and completeness of the feature table can be increased by applying the following optional steps.

Optionally, in a third step retention times of features can be corrected. When many samples are measured it is very likely that shifts in retention time are observed. These shifts can have a disturbing impact on the grouping step. Shifts can be corrected by available algorithms that incorporate signals of external standards or QC samples. After retention time correction the grouping results in a more accurate feature table.

Also optionally, gaps can be filled. A gap corresponds to a feature for which no signal could be detected in a measured sample in the first run. This could be for many reasons, such as the detection of the signal being disturbed by the background noise. By applying appropriate algorithms, the gap signal might be rescuable despite not being properly measured. There are several algorithms available making use of different strategies to do the gap filling or if needed, data imputation.

Filtering the feature table could also be a necessary step. Features belonging to external standards, putative contaminations, or having intensity values too low for statistical analyses can be filtered out. On the other hand, redundant (feature) information can be filtered out, as a feature can be either the mono isotopic ion of a compound or an isotope, adduct, or fragment ion of a compound. Therefore, very often a set of features corresponding to one compound leads to redundant information in the feature table and could disturb statistical analyses. Tools (i.e., the R package CAMERA [58]) are available to annotate these types of ions to perform the filtering for redundancies.

The final feature table containing the m/z values, retention times, and intensities of thousands of features (therefore we refer to it as the *intensity matrix*, see Fig. 1) together with the corresponding metadata provides a good starting point for data exploration and statistical analyses.

Depending on the parameter settings of the MS instrument, tandem mass (MS/MS) spectra are acquired for a certain number of features. MS/MS spectra need to be linked to the corresponding MS1 precursors and depend on the analytical MS/MS acquisition mode (usually either DDA or DIA, see Sect. 2.6). Linking is already provided by processing software like mzMine, MS-Dial, or XCMS, but also by Global Natural Products Social Molecular Networking (GNPS, [35]), an open-source and open-access platform for sharing MS/MS spectrometry data.

2.8 Metabolite Annotation

The annotation of metabolites based on MS data can be done on different levels using the three properties m/z, retention time, and MS/MS fragmentation pattern.

- The m/z value is directly related to the molecular weight. Using a high-resolution mass spectrometer, can also be used for the calculation of the molecular formula [70].
- The retention time reflects the behavior of the metabolite on the used column and therefore relates to the chemical properties like polarity and molecular size.
- The MS/MS fragmentation pattern can be used as a molecular fingerprint and allows structural insights.

Comparing these properties with the values of standard compounds which were analyzed by the same instrument is the most reliable way to annotate metabolites. However, with an untargeted approach, this way of annotation is only possible for a fraction of the metabolites. On the one hand, few labs own many standards, and on the other hand, many metabolites are not commercially available or simply not yet described or isolated.

Another way of feature annotation is the comparison of the molecular mass and the MS/MS spectra only. Those spectral libraries (commercial like NIST, or public ones like MassBank [31], or Weizmass [71]) contain MS/MS spectra from labs all over the world and they grow every day. As bryophytes are non-model species, reference data is rare and can only be found for a fraction of the spectra. However, the comparison of spectra from one's data set with spectra from other working groups should be treated with caution, as different mass spectrometry and fragmentation mode settings can lead to different MS/MS spectra even using the same metabolite. Another strategy is the comparison of the own MS/MS data with in silico MS/MS data, such as the MetFrag [33], SIRIUS [32], or GNPS [35] software.

Compounds can also be classified and sorted into different groups of compounds, i.e., phenolics, and sesquiterpenoids. Computational approaches can use the MS/MS information in several ways to classify compounds into compound classes and this can provide a much quicker way to assess differences in these classes. There have been described numerous approaches with MS2LDA, which decomposes fragment data to annotate molecular substructures [72], and CANOPUS, which can classify unknown fragment spectra using a deep neural network [14], to be the currently most commonly used *in silico* classification tools. Software, such as SIRIUS and MS-Dial [62] also provide user-friendly classification tools at different levels. Classification is usually based on the Chemical Ontology CHEMONT [73]. Lately, there has also been the NP-Classifier, which is less chemistry-focused and more related to biological function [74]. Classification is useful in many aspects. It can give a striking overview of the chemical diversity and the complexity of natural products in bryophytes via sunburst plots (Fig. 2). As ecological studies commonly aim to identify a broader functioning than in molecular biology, compound classification can provide an abstract and more generalized description that is often sufficient, and, in the context of ecometabolomics, can be even more powerful [18, 29].

The different approaches for compound or compound class annotation come along with different levels of confidence. For example, the direct comparison of MS/MS spectra of an unknown compound with the m/z, retention time, and MS/MS spectra of a standard compound analyzed under the same conditions is more reliable than the comparison with MS/MS spectra from public libraries only. The golden standard here is the NMR analysis of isolated compounds. This approach can even give information about the chirality and isoform. Therefore, scientists should always be aware of what kind of information they actually can provide about a compound they "identified." According to the Metabolomics Standards Initiative of the Chemical Analysis Working Group (CAWG), four levels of annotation can be defined [75]: identified compound, putatively annotated, putatively characterized, and unknown compound. The necessary level of annotation highly depends on the scientific question and should be discussed at the very beginning of a project.

2.9 Data Analysis from Exploration to Statistical Analyses

A multitude of statistical methods from other fields were made available and have already been successfully applied to MS data. The review written by Stanstrup et al. [30] provides a good overview of the methods that are available in statistical framework R. In addition, browser-based platforms like MetaboAnalyst [76] are available that provide a large set of tools to perform many different analyses on the MS intensity data.

Depending on the biological question and the data analysis method, data contained in the intensity matrix may need to be normalized, scaled, or transformed [77]. To achieve a better comparability of the samples, they can be normalized to, i.e., their carbon content or a total number of ions (sum of all feature intensities). Scaling needs to be applied to, i.e., make all features equally important or reduce the

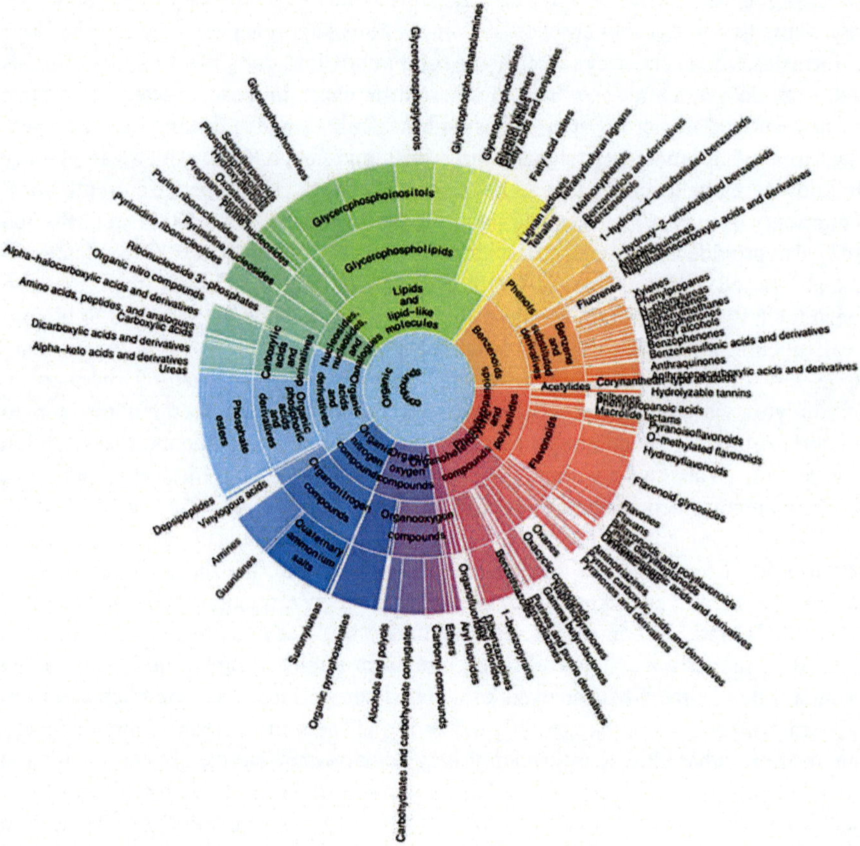

Fig. 2 Exemplary sunburst plot of the chemical profile of *Marchantia polymorpha*. Data and the description of how to produce sunburst plots are described in [18]. Sunburst plots show an overview of the chemical diversity. They are interpreted from the inside out starting with organic compounds in the center. The classes further to the outside represent the more specialized classes. For example, superclass Lipids and lipid-like molecules can be found in the center, whereas more specialized subclasses like Fatty acid esters can be found in the outside region of the plot. The classes are arranged at different levels usually based on the CHEMONT ontology [73]

importance of large values, whereas transformation can reduce the heteroscedasticity of the data. Applying normalization, scaling, or transformation to the data will affect the results of analyses, therefore appropriate choices must be made for the specific analyses [77].

Related to the hypotheses, data in the intensity matrix can be analyzed by samples or by compounds. Usually, the data is analyzed by samples first to perform quality checks like sample identity checks or relation to the QC samples by applying Principal Component analyses (PCA) or (non-metric) Multi-Dimensional Scaling ((N)MDS). It is straightforward to visualize their outcomes in two-dimensional plots for exploratory analyses.

Untargeted metabolomics favors two strategies to analyze the intensity matrix, metabolite profiling and metabolite fingerprinting. In metabolite profiling PCA, MDS, NMDS, or hierarchical clustering are applied to relate specific compounds or pre-defined sets of compounds to biological or ecological functions. By contrast, metabolite fingerprinting is used in ecometabolomics to identify patterns across metabolite profiles and link them to ecological functions. Hence, ordination methods like ReDundancy Analysis (RDA), Partial Least Squares regression (PLS) also coupled with Discriminant Analysis (PLSDA) or (Canonical) Correspondence Analysis ((C)CA) are used to link the compound's intensity matrix to explanatory variables like traits or environmental properties [12, 78].

Complementary feature or variable selection can be performed to identify candidate compounds (metabolites) that could play a key role in answering the research question. Usually, univariate statistical methods like Analysis of Variances (ANOVA) or Kruskal-Wallis' Test, depending on the normalization and transformation of the data, are applied to perform this task, but also results from ordination methods like PLSDA can be analyzed to select variables that drive the separation of groups. A more sophisticated way to extract a restricted and hopefully meaningful set of compounds that still has a high prediction performance would be to use methods like Genetic Algorithms, Lasso, or Elastic Nets [79].

Often, i.e., in the case of different (plant) species or tissues, the compound's intensity matrix suffers from a heavy zero content. This issue has a strong impact on the result of the applied data analysis methods, i.e., PCA, PLSDA, (N)MDS. When applying NMDS analyses for data exploration, appropriate distance measures like Bray-Curtis [80], which is widely used for ecological data analysis, could be selected to circumvent this problem. The resulting outcome is more biological or ecological meaningful.

Missing values can occur not only for biological reasons but also for technical reasons that cause the intensity of features to drop under the detection level, leaving a gap in the feature matrix. For statistical analyses, it could be inevitable to fill these gaps [81]. Tools like XCMS provide gap-filling methods that attempt to recover the signal intensity from the corresponding raw data. Missing values can also be imputed by making use of observed intensity values from other samples, i.e., by computing mean values or more sophisticated methods like random forests or k-nearest neighbor's approaches. An overview and comparison of the widely applied methods are given by Do et al. 2018 [82].

2.10 Data Deposition and Sharing

Making data available via public data repositories will allow scientists from different communities, i.e., ecology, biochemistry, and bioinformatics, to easily access them and to connect each other for interdisciplinary data science. Data can be re-used for methods validation or reproduction of results in scientific publication, or in a more comprehensive context of meta-analyses or data synthesis. Therefore, meaningful metadata along with the MS data is indispensable.

There is no dedicated data repository for ecometabolomics data yet. Widely used in the ecometabolomics community are MetaboLights ([83], https://www.ebi.ac.uk/metabolights/) and Mass Spectrometry Interactive Virtual Environment (MassIVE [37], https://massive.ucsd.edu/ProteoSAFe/static/massive.jsp), which are dedicated to storing the memory-hogging raw mass spectrometry data in open file formats and corresponding meaningful metadata with controlled vocabulary. Everything is uploaded in a structural and controlled manner and is therefore easily searchable and reusable for other scientists. MetaboLights also provides to store related compound annotations. MassIVE is tightly linked to Global Natural Products Social Molecular Networking (GNPS [35], http://gnps.ucsd.edu) and ReDu [38]. GNPS provides open-access sharing of raw, processed, or identified MS/MS data. Also mentioned is MassBank [31], one of the first public data repositories to store and share (MS/MS) spectra of unknown or annotated compounds. The (annotated) primary metabolomics data is then aggregated into secondary libraries, such as LOTUS ([80], https://lotus.naturalproducts.net) which stores occurrences of natural products, and their spectra, and additional information about different organisms, including bryophytes.

An overview of repositories that store and share primary metabolomics or ecological data is provided by the portals http://www.re3data.org and FAIRsharing.org. Another way is to upload files related to ecometabolomics, i.e., feature intensity matrix and annotation in multiple file formats to more general data repositories like Zenodo (https://zenodo.org). A DOI will be assigned after upload so data can be referenced in publications. To allow for long-term reusability of the data, open file formats like *csv* or *tsv* for feature tables, *maf* for feature tables containing annotation information in form of CHEBI- or Inchi-keys, and *mgf* or *msp* for spectra are appreciated.

3 Approaches of Ecometabolomics Using Bryophytes

The application of metabolomics can unravel the function of bryophyte metabolic compounds as responses to environmental changes, mechanical damage, or stress, and in interactions of bryophytes with other organisms, i.e., to defend themselves against pathogens or to attract mutualistic microorganisms [1, 9, 84, 85]. For example, it has been shown that some bryophytes can inhibit the germination of vascular plant seeds [86, 87]. Some species such *Dicranum scoparium* also produce specialized acetylic oxylipins as a defense against herbivorous slugs [88, 89]. By contrast, it has been shown for the cosmopolitan moss *Ceratodon purpureus* emits sex-specific volatile organic compounds (VOCs) to attract mutualistic micro-arthropods that significantly increase moss fertilization rates [85].

Many bryophytes are considered to be desiccation-tolerant and, thus, to sustain long periods without water and light [90]. They have developed anatomical and morphological structures, such as hyaline hairs or convoluting leaves to reduce evaporation and exposure to radiation, but specialized metabolites are required to support cells in their mechanical stability and to sustain osmotic stress involved with desiccation [91]. Ecometabolomics can help to identify the chemical constituents that are associated with desiccation and associated changes in the intracellular water regime and explore the ecological or environmental circumstances under which bryophytes produce these natural products [8, 92, 93]. For example, some remarkable desiccation tolerances were discovered in *Tortula ruralis* and *Riccia* species that can survive periods without water for over 20 years and can still recover within a few hours [94]. While desiccation tolerance is a relatively widespread trait in bryophytes, only a few vascular species (approx. 70 fern and 60 angiosperms out of approx. 300,000 species) are known that can withstand comparable conditions [90]. Differences in water availability can have the most profound effects on plant metabolism as cells are aiming at maintaining homeostasis [95]. Desiccation impacts osmotic adjustments, cavitation, the production of inorganic solutes, and the general nutrient availability to cells, such as the K metabolism affecting the C:K ratio, hydrogen and oxygen bonds of molecules, their charge and acidic or basic behavior, or the chemical benzene-motifs relating to changes in the polyphenol production [96–101]. These changes can be measured using metabolomics and linked to physiology and ecology using traits, such as the leaf water content [102] and, thus, providing mechanistic linkages [103–105]. However, the molecular traits that enable bryophytes to carry out desiccation tolerance and evolutionary relationships are still largely unexplored. Thus, ecometabolomics has great potential to get new insights into the molecular mechanisms used by bryophytes to achieve this remarkable quality [93].

Terrestrial bryophytes are similar to vascular plants as they both produce flavonoids for protection against UV-B radiation [106]. Although flavonoid biosynthesis is less characterized in bryophytes than in other plants, studies that have investigated flavonoids have found significant differences between vascular plants and bryophytes [106]. Metabolomics has been used on the model species *Physcomitrella patens* to uncover its flavonoid profile and flavonoid compounds have been linked to biochemical pathways and the strength of UV-B radiation [107, 108]. Furthermore, transcriptomics has been integrated with metabolomics to investigate the impact of UV-B radiation on the biosynthesis of flavonoids in the moss *Leucobryum pyriforme* under high levels of UV-B, which are typical for arctic environments during summer [22]. As shown by the authors, bryophytes need to invest a large fraction of their metabolism to produce flavonoid structures indicating the large ecological importance of molecular flavonoid pathways [109].

Ecometabolomics is also a useful tool to elucidate the importance of bryophytes in carbon and nitrogen cycling [110, 111] and the biochemical role of bryophytes within biological soil crusts [112–114]. It is estimated that biological soil crusts cover approx. 40–50% of the total land surface area and here, bryophytes are major components interacting closely with algae, fungi, lichens, bacteria, and archaea, which are embedded together within a matrix of exopolysaccharides and soil particles to form a surface crust [111, 112]. The role of bryophytes has been intensively studied with regard to Carbon fixation and Nitrogen deposition rates [110, 113]. While there has been research on identifying the mono- and polysaccharides that species produce to form the soil crusts [115, 116], the biochemical interactions between the different organisms, including bryophytes, are still not well understood and approaches are needed that combine several omics technologies, including metabolomics [117].

As with other organisms, bryophytes are important biogeochemical components that modify soil organic matter (SOM) and alter the net plant carbon (C) and nutrient (N) inputs and outputs. These relationships were first described by Löhnis [118] and are known as the priming effect. Bryophytes contain high concentrations of specialized carbon compounds that delay decomposition in the soil and usually promote soil carbon storage [119]. However, in Northern peatlands, where peat mosses store the majority of C, rapid warming due to climate change can accelerate ecosystem respiration which could lead to a release of millennia-old carbon from the peatland organic matter stock. Walker et al. [120] have found that this effect is less pronounced in bryophytes when compared to vascular plants probably due to higher atmospheric Nitrogen deposition rates and leaching of labile substrates [121]. Most prominently, hydrolysable tannins usually linked to glycosylated compounds and (poly-)terpenes are characterizing bryophytes just like vascular plants in habitats with low soil fertility and low pH, such as peatlands [122]. More research is needed to identify the chemical constituents of bryophytes in these processes and the role of these compounds in their interactions with microorganisms. Metabolomics techniques can help to untangle the complex biogeochemical processes in which bryophytes play a substantial part.

In a striking experiment utilizing an air-flow system, Vicherova et al. [123] showed that the pleurocarpous moss *Hamatocaulis vernicosus* increases its growth when exposed to volatile organic compounds (VOCs) of its natural competitor the peat moss *Sphagnum flexuosum*. VOCs are commonly produced in vascular plants to, i.e., repel pathogens or herbivores, attract pollinators, or communicate with other organisms at various tropic levels [124–126]. VOCs are usually determined using gas chromatography coupled to mass-spectrometry (GC/MS). The study by Vicherova et al. [123] is a great example of the ecometabolomics of bryophytes as it determined the biotic interactions of two mosses that are mediated by volatile compounds and clearly shows the enormous potential of metabolomics techniques. In the following, we want to give a more detailed overview of the role of metabolomics techniques in different areas of ecology and related research fields.

3.1 Natural Product Chemistry

Bryophytes produce a high diversity of specialized metabolites, some of which are rare or absent in vascular plants [127]. Liverworts are by far the most chemically studied bryophyte phylum, with thousands of specialized metabolites identified to date [128]. The high chemodiversity of liverworts can be attributed to specialized organelles called oil bodies (Fig. 3). They are single membrane-bound structures that synthesize and store these compounds, allowing them to be enriched in the plant and easier to isolate [10, 129–131]. Mosses and hornworts do not appear to possess

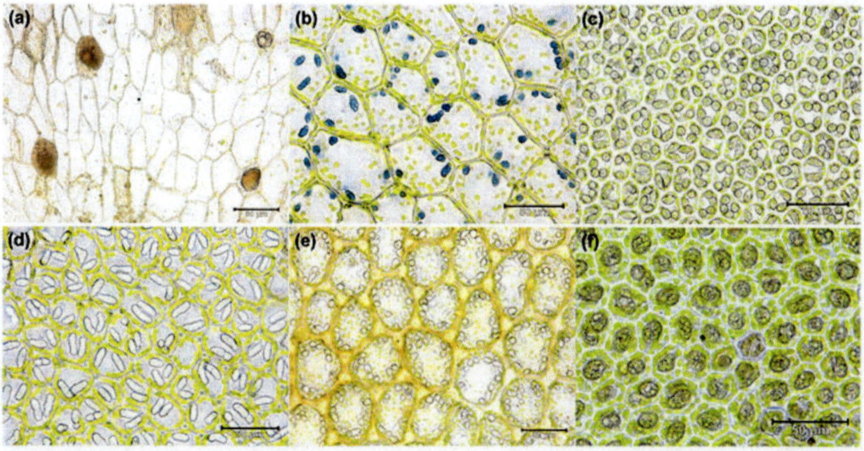

Fig. 3 Different types of oil bodies. (**a**) Oil cells with one large oil body fill the entire cell in *Lunularia cruciata.* Typical for some Marchantiales. (**b**) Segmented blue oil bodies in *Calypogeia azurea*. The blue color of the oil bodies is largely due to the aromatic sesquiterpenoid 1,4-dimethylazulene [132, 133]. (**c**) Structured oil bodies in *Plagiochila killarniensis*. Typical for Jungermanniales. (**d**) Clear cucumber-shaped oil bodies in *Nardia scalaris*. (**e**) Segmented oil bodies in *Mylia anomala*. (**f**) A singular large oil body in the cells of *Radula complanata*. Images are reproduced with permission from Kristian Peters. The images are licensed under Creative Commons CC-BY-SA. Permissions to use them in the book are granted

structures analogous to the oil bodies of liverworts and are much less chemically studied overall. Hornworts in particular are poorly studied.

The most prominent classes of specialized metabolites found in bryophytes are terpenoids, flavonoids, (bis)bibenzyls, and lipids [134, 135]. Terpenoids, natural products composed of isoprene units, are a diverse class of metabolites found in almost all living species [134]. Bryophytes have a contrasting terpenoid profile compared to vascular plants; potentially due to unique biosynthetic pathways that will be discussed later in the chapter [136]. Sesquiterpenoids are the most abundant and diverse terpenoid class in bryophytes, followed by the diterpenoids and monoterpenoids. Mosses also have a high diversity of triterpenoids that are not seen in liverworts or hornworts [136]. Liverworts have the highest terpenoid diversity of all the bryophyte phyla, likely due to the oil bodies described above [9]. Flavonoids are pigmented compounds derived from the phenylpropanoid pathway and enable land plants to interact with their environment [106]. Interestingly, flavonoids are completely absent in hornworts, potentially due to an acquired mutation that led to the loss or modification of the biosynthetic pathway. Acrocarpous mosses were reported to generally have more flavonoids than pleurocarpous mosses; however, liverworts had a higher flavonoid content overall [137]. Flavonoids are diverse enough in liverworts that they can be utilized to assign chemotypes, as was done with *Metzgeria furcata* [127]. Bibenzyls, and their dimerized form the (bis)-bibenzyls, are steroidal ethane derivatives that branch off the phenylpropanoid biosynthetic pathway [138]. While bibenzyls are found in many plants, bisbibenzyls

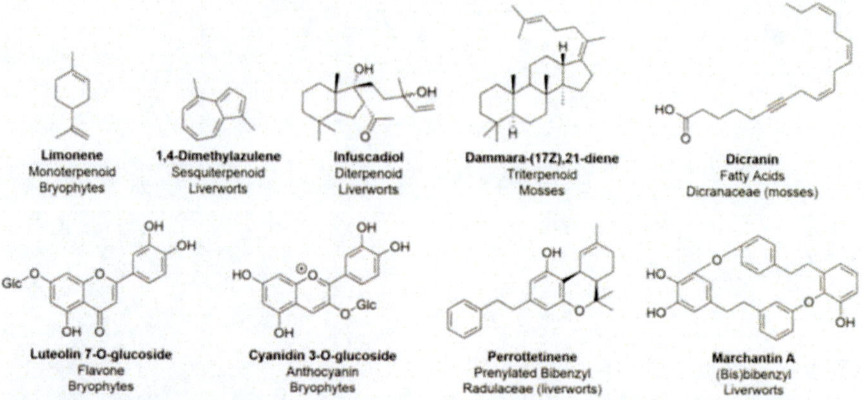

Fig. 4 Some exemplary compounds selected from bryophytes

are largely found only in liverworts [139]. The *Marchantia* and *Radula* genera have been extensively studied due to their prolific production of macrocyclic (bis)-bibenzyls (marchantins) [139], and prenylated bibenzyls [140], respectively. Prenylated bibenzyls are of particular interest due to their structural similarity to known bioactive cannabinoids, such as tetrahydrocannabinol (THC) and cannabigerolic acid (CBGA) [141]. Lipids and fatty acids are aliphatic carboxylic acids that are necessary for forming lipid membranes. Living bryophytes do not accumulate free fatty acids, instead, they are commonly localized in membranes or stored as triglycerides [142]. Bryophytes readily produce very long chain fatty acids (C20 and above) which are absent from vascular land plants. Some bryophytes produce species-specific fatty acids, such as dicranin (octadeca-6-yn-9, 12, 15-trienoic acid) which is commonly found in the Dicranaceae family but is uncommon in other taxonomic groups [142]. Example structures of all these compounds can be found in Fig. 4. In general, one of the most complete overview on natural products of bryophytes is provided by [8].

3.2 Chemodiversity

Ecometabolomics is also a great tool to discover novel natural products and assessing the diversity of chemical compounds in bryophytes [8, 143]. Targeted metabolomics has most often been applied by focusing on particular compound classes [144]. But it can also be applied untargeted, determining the (almost) entire breadth of compounds produced in different species (Fig. 2).

In ecology, diversity is quantified by Hill numbers [145], which are also known as effective numbers of types. Types are, in ecology, usually species, which correspond to features or compounds in our case for ecometabolomicsdata, and diversity is computed per site, which corresponds to samples here. Hill numbers can be computed for different orders of diversity, where increasing the order increases the

$$H' = -\sum_{i=1}^{S} p_i ln(p_i) \quad and \quad J = \frac{H'}{\log S}$$

Fig. 5 Definitions of the Shannon diversity index (H′) and Pielou's evenness (J), where the relative compound abundance $\boldsymbol{p_i}$ is calculated as the ratio of the abundance of the ith compound and the sum of the abundances of all compounds in a species ($\boldsymbol{p_i} = N_i/N.$); and $\boldsymbol{S}$ is the species richness

weight given to the most abundant compounds. Based on the Hill numbers, various diversity measures related to information theory can be derived.

The chemodiversity of bryophytic species is usually assessed by calculating four diversity measures. First, the compound richness $\boldsymbol{S}$, which is the Hill number of order 0, is the number of compounds that can be found in a species. It is calculated by simply counting the chemical compounds per species. Second, the number of unique compounds $\boldsymbol{U}$ defines the number of compounds that are present in one species but not the other tested species. The Hill number of order 1 is related to the Shannon diversity index $\boldsymbol{H'}$ (the third measure) and is defined as shown in Fig. 5 [145, 146].

Finally, the fourth measure of diversity, Pielou's evenness $\boldsymbol{J}$ describes the homogeneity of the distribution of the intensity or abundance of compounds present in a species (Fig. 5). Species in which all compounds have the same intensity are perfectly even. By contrast, species having only one very abundant and many low-abundant compounds have a low evenness.

In ecology, alpha, gamma, and beta diversity are computed to examine the hierarchy of diversity at different scales. Alpha diversity is the average diversity at the scale of a single sampling unit, i.e., the average diversity of a species within a community. Whereas gamma diversity is the diversity at the scale of the all-sampling units, i.e., the whole community. Hence, beta diversity is the variation in diversity across the scales. The definition of the scales depends on the biological question [147]. Beta and gamma diversities allow an assessment of the status of species within the community and relate mechanistic molecular processes to biogeochemistry [1]. So far, to our knowledge, beta and gamma diversities have not been determined for bryophytes at the molecular level. These diversity measures can be calculated at different levels providing different resolutions, i.e., for compounds, or at the level of compound classes [18].

For example, it has been found that liverworts (Marchantiophyta) are chemically more diverse than mosses *s.str.* (Bryophyta) [8]. As a result, much emphasis is given to liverworts and generally, liverworts are the best-studied group of bryophytes with regard to the diversity of chemical compounds. The most diverse chemical classes in bryophytes are terpenoids, (bis)bibenzyls (aromatic compounds), and flavonoids [8]. Although the chemical richness of bryophytes usually is not larger than in other plants, their biochemistry is often different due to a unique specialized metabolism. This fact makes bryophytes fascinating study subjects as the many unique compounds can exert very specialized molecular functions [148].

Furthermore, species of the family Marchantiaceae produce a large variety of macrocyclic or acyclic bisbibenzyls like Marchantins. These substances have been intensively studied as they possess biological activities, including cytotoxicity against cancer cells. Other species groups have larger diversity in other chemical classes, such as members of the Ricciaceae which have extraordinarily high diversity in acetylenic fatty acids when compared to other Marchantiales [149, 150]. Plants of the *Radula* genera are prolific producers of prenylated bibenzyls which are currently being studied as potential bioactive compounds due to their structural similarities to cannabinoids [140, 151]. In short, the different cannabinoids are present in different abundances in the different species.

As demonstrated for cannabinoids in bryophytes and vascular plants, metabolomics can be used to unravel the plasticity in the evolution of specialized compounds and compound classes in different taxa and can reveal phylogenetic relationships based on the distribution of certain specialized compounds throughout the plant kingdom [52]. Metabolomics can also be used to screen for certain compounds in different taxa, revealing how diverse compounds and chemical classes are. For example, targeted phytohormone profiling to relate chemical diversity in phytohormones with the plant kingdom [152]. The study found that phytohormones in 30 bryophyte species belonging to nine different orders correlated rather with prevailing environmental conditions and plant survival strategy rather than with plant evolution [152].

3.3 Chemotaxonomy and Chemophenetics

As the ancestors of bryophytes were among the first plants that colonized the land approx. 500 Million years ago, they share both the phylogenetic similarities between algae and vascular plants, which makes them very appealing study subjects to analyze phylogenetic relationships [6, 153, 154]. Traditionally, taxonomy focuses on describing species based on their anatomical and morphological properties. Phylogenomic barcoding and transcriptomic sequencing have been applied to a wide range of species spanning a large phylogenetic diversity in a broad sense, including bryophytes [155]. However, some bryophyte groups, such as liverworts have been found to be very challenging to sequence mainly due to their small size and resulting difficulties in extracting sufficient amounts of DNA or RNA. Bryophytes are also difficult to sequence due to a lack of detailed genetic maps [156], a high abundance of repetitive sequences that impede computational sequence alignment and assembly of next-generation sequencing [157], a high abundance of glycosides, polyphenols, flavonoids, tannins, fatty acids, and other specialized metabolites that co-precipitate with some common DNA and RNA extraction procedures [158, 159], or the presence of DNAses and RNAses, which are proteins that break down continuous DNA or RNA sequences [159]. Metabolomics techniques can also be used to assist phylogenetic and taxonomic studies when samples are difficult to sequence, when DNA/RNA sequencing methods are not available, or when integration of morphology and sequencing is still inconclusive [160].

Despite their haploid-dominant life cycle (gametophyte), bryophytes have evolved a unique diversity of bioactive compounds that can also be investigated with regard to phylogenetic relationships between closely or distantly related species, or species groups. Chemotaxonomy involves the classification of biological species and reconstruction of their phylogenetic relationships using chemical compounds as marker molecules in addition to genetic loci and morphological and anatomical properties [161, 162]. Lately, different terms, such as chemophenetics have been proposed that better capture the extent to which the natural product chemistry represents phylogenetic relatedness of species [163]. Chemotaxonomy and chemophenetics are based on the assumption that most morphologically defined species have a constant core metabolome, irrespective of their geographic origin, or ecology [164, 165], and that robust morphological properties correspond to chemical differences as first shown for pigments of the arthropod genus *Daphnia* [161]. While chemotaxonomy has been applied for a wide range of organisms [162, 166], including vascular plants, algae, and other cryptogams, such as lichens for nearly 50 years [161, 164, 167–169], for bryophytes chemotaxonomy has rarely been investigated [127, 170–174]. However, due to the enormous progress in analytical and computational metabolomics (see above), chemotaxonomy can now be considered a viable alternative in bryophytes.

Here, two strategies for phylogenetic analyses based on chemical information can be distinguished. Compounds can either be identified and then passively mapped onto existing taxonomies or compounds can be used as biomarkers themselves either creating new phylogenetic trees (when, i.e., genetic methods are not available) or assisting existing phylogeny to improve the taxonomic classification. The latter strategy is also known as chemotaxonomy but has not often been pursued bryophytes so far. By contrast, for vascular plants, targeted metabolomics has been frequently used to assist taxonomic classifications [175]. Here, N-glycan structures and the N-glycosylation have been found to uniquely characterize bryophytes and confirm their intermediate position between algae, which are prolific producers of a large diversity of glycan structures, and the extremely conservative vascular plants [176].

Similarly, as with vascular plants, chemotaxonomy is often performed targeted by focusing on a particular compound class that is representative or chemically diverse between the investigated species groups [163]. For example, in bryophytes terpenoids constitute the largest compound class and are considered to be extraordinarily suited for phylogenetic analyses, especially when incorporating chemical biosynthesis pathways [136, 174]. Moreover, acetylenic fatty acids have been used to reconstruct phylogenetic relationships in mosses of the order Dicranales [170]. However, metabolomics techniques are not restricted to compound classes and when used untargeted (without focusing on a particular chemical class), metabolomics can provide a deeper and more complete look into the composition of various metabolic compounds and the chemical diversity in different taxa. The virtue of untargeted metabolomics is, that, when investigating profiles of different bryophyte species, the profiles also inherently include phylogenetic information which can be used even without explicitly focusing on taxonomy [177]. Thus, integrating untargeted metabolomics with molecular or morphological taxonomy will allow for a taxonomic re-evaluation of taxa and more realistic species circumscriptions [163]. It will also allow us to better elucidate the ecology and evolution of the species and can enlighten their origin and biogeographic history [178].

3.4 Molecular Traits

In general, plant traits are the morphological, physiological, and molecular features that determine the phenotype and represent the ecological strategies of plants. The most commonly used phenotypic traits in bryophytes involve autecological attributes, such as indicator values defined by Ellenberg [179–181], which assign categorical numbers from 1 to 10 with regard to the response of bryophyte species to light, temperature, continentality, moisture, soil reaction, and nutrient availability. Vegetative and sexual regeneration traits (such as presence/absence, or attributes of gemmae, or spores), morphology-based traits, such as growth stature, estimated shoot length, and life-history traits, such as growth form, life-form, life strategy have also been devised [180, 182–184]. Large trait databases have been shaped to provide ecologists with a backbone for functional analyses. The largest plant trait database is the plant trait database TRY [185], including more than 42 traits of more than 720 species of bryophytes [7].

In ecological research, functional trait analyses are among the most important ecological studies aiming at understanding how interspecific variations in trait attributes predict ecosystem processes and changes mechanistically [6, 186] as traits impact plant fitness and individual performance via their effects on growth, reproduction, and survival [103]. However, functional trait studies using bryophytes are still rare as the few phenotypic traits that have been identified often explain little ecological variance or represent imprecise estimates of highly variable species [6, 7]. As bryophytes respond to ecological change largely at the metabolic scale [23] molecular traits have been discussed to greatly improve the predictions and our mechanistic understanding of bryophytes [177].

Molecular traits in bryophytes do not only include primary and specialized metabolites. Metabolic compounds and natural products are also integral components of the so-called proxy traits that generalize over a wide range of physiological functions. Here, the Leaf Economic Spectrum describes the need of organisms to balance their investment in particular functions due to limited resources or environmental filtering [187]. These trade-offs constrain the range of possible trait combinations, allowing for a generalization of closely related proxy traits across species groups or ecosystems, such as forests. Leaf mass per area is such a proxy trait generalizing as a predictor of leaf attributes, such as leaf lifespan and photosynthesis rates, but also the metabolites that are produced by the cells to provide leaf functioning [6]. Other traits have been found that describe relationships for growth and reproduction in individual bryophytes, such as shoot water holding capacity, chlorophyll fluorescence, or above-ground net primary production [6, 185]. There are also mixed response-effect traits, such as tissue C:N, or dry matter content influencing both the individual performance of bryophytes and the ecosystem productivity [5]. In general, identifying traits in bryophytes can be a great gain for functional ecology, as they provide a direct link between individual performance and ecosystem processes.

As bryophytes are small and inconspicuous, they have developed anatomical and morphological characteristics at a microscopic scale, such as toothed leaves, or papillosity that are often different from vascular plants. To this end, analogous traits from vascular plants may not be 1:1 transferred and it is still unresolved whether an adaptation of trait concepts is needed for bryophytes. Moreover, thallose liverworts, such as *Marchantia polymorpha* do not have leaves. In order to identify relationships to water-regime, it was shown for lichens (which have similar thallus growth statures) that the Specific Thallus Mass is an accurate predictor of water relations [188, 189]. Furthermore, Stanton & Coe [6] discuss Dynamic Water Content and Dehydration Tolerance to be major proxy traits in bryophytes regarding desiccation tolerances and maintenance of the water regime. As shown by Fernández-Martínez et al. [190] for (semi-)aquatic bryophytes, molecular elements, such as C, P, and Zn show a significant correlation with cell length and cell form indicating that there is a molecular relationship to form and function that can be determined with metabolomics techniques.

Bryophytes can display large plasticity and the resulting traits can appear highly variable [191]. This is in part due to trait multi-functionality [192]. Here, traits (and in particular metabolic compounds) can exert several (molecular) functions at the same time. On the other hand, traits can also be impacted by multiple ecosystems and environmental processes and multiple intra-specific developmental, genetic, or phylogenetic constraints [6, 193]. While these processes can aggravate the identification of ecological function in bryophytes, these challenges can either be addressed with proper experiment designs that capture a wide array of ecological variance or with the use of sophisticated (statistical) analysis methods. In Sect. 2 we provide recommendations on how experiment design and analyses can improve the identification of molecular traits.

In summary, molecular traits aggregate and abstract patterns of metabolic compounds and chemical classes to describe mechanistic molecular functioning at broader levels. Identifying and linking them can be a great gain for bryophyte ecology. Research just started to relate compounds to the form and function of bryophytes [190]. Here, bryophytes are ideal study organisms as their metabolism is largely controlled at the metabolic scale.

3.5 Bioindication and Biomonitoring

As bryophytes usually are highly sensitive and respond suddenly to environmental changes, they have been considered for bioindication and biomonitoring of, i.e., pollution, climate change, or ecosystem function [5, 110, 194]. While research has traditionally focused on the anatomical and physiological properties of bryophytes, current approaches also consider biochemical and genetic adaptations of bryophytes to anthropogenic changes to explore the mechanistic basis of bioindication [4, 135, 190].

As bryophytes only have rhizomes and, apart from a few exceptions like *Polytrichum* species, they also do not have specialized transport systems like xylem or phloem cells, nutrient uptake predominantly takes place through the entire plant and above-ground by atmospheric depositions. As a result, bryophytes have developed many specialized phenotypic structures like papillae (epidermal surface structures that contain many waxes and fatty acids compounds). As epicuticular waxes are strongly water-repellent and absorb UV-B radiance they are also important components in desiccation tolerance and are produced by many dryland bryophytes [195, 196]. Lately, waxes have also been found in cell walls of the vascular conduits of *Polytrichum commune*, effectively isolating gas exchange and enabling the transport of solubles through these specialized cells [197]. Furthermore, the transport is facilitated by the incorporation of lignin-like structures ('proto-lignin') in the conducting cell wall tissues [24]. Proto-lignin compounds are functionally similar but structurally simpler than in vascular plants. In bryophytes they are restricted to the family Polytrichaceae [197–199]. Galacturonic acid and mannuronic acid were reported to be present in the cell walls of many bryophytes that facilitate the uptake of inorganic nutrients like K or Na [200, 201]. Internal translocation of elements is mainly steered by specialized metabolic compounds that allow compounds to be transported between cells, once again indicating the importance of metabolism and homeostasis in bryophytes [202].

Due to the large subjection to atmospheric nutrients, bryophytes often respond very sensitively to changes in environmental conditions [203]. As a result, they are considered good bioindicators. Epiphytic bryophytes are especially sensitive to sulfuric compounds resulting from the pollution that changes the acidity of the tree host substrate and interferes with cellular function [204–206]. Although the uptake of certain elements, such as ^{15}N, ^{35}S, or ^{14}C through labeling has been intensively studied, there is still a knowledge gap in which compounds these labeled elements are incorporated [205, 207]. Here, ecometabolomics techniques, such as Ion-Mobility or Fourier Transform Ion Cyclotron Resonance Mass-Spectrometry (TIMS, FT-ICR-MS) can be used to assess the metabolic compounds in vivo which are involved in the responses of bryophytes as bioindicators [208–211]. These powerful analytical techniques possess a great potential to unravel the mechanistic molecular processes behind bioindication.

As bryophytes take up pollutants and heavy metals like Cd, Cu, Zn, or Pb similarly quickly as other inorganic elements, they have been used as bioindicators and biomonitors in terrestrial and semi-aquatic habitats [212, 213]. The accumulation of pollutants often quickly leads to a visible damaging effect. This sensitivity can be assessed by determining the coloration of leaves (which is itself also a phenotypic trait, see section above), or by assessing the elemental status within the species. For example, one of the most commonly used biomonitoring species in semi-aquatic systems is the moss *Fontinalis antipyretica* [214]. The species accumulate a variety of inorganic heavy metals and changes in its status have been frequently linked to water pollution [213]. Fernandez-Martinez et al. [190] have also related elemental concentration to changes in phenotypic properties, proving that there are pronounced and complex physiological and metabolic adaptations underlying bioindication. *Fontinalis* has also been used to monitor and detect organic pharmacological pollutants, such as oxolinic acid, flumequine, or oxytetracycline in aquatic environments [215]. The large uptake of these organic substances by *Fontinalis* has been considered for the decontamination of water ecosystems. Ecometabolomics can help to assess the metabolic changes at a large scale, i.e., identifying enzymes and compounds that metabolize pollutants, to further investigate the molecular fate of pollutants and their ecological roles.

3.6 Bioactivities

The many unique bioactive natural products and other molecular compounds make bryophytes particularly interesting for biotechnology, pharmaceutics, and medicine [92, 134]. While applications in biotechnology and pharmaceutical potential have been explored somewhat, the ecological impact and the questions of why and under which circumstances bryophytes produce these extraordinary natural products are still largely unexplored.

One of the main *planta* roles of specialized metabolites is to allow the plant to interact with its environment. These interactions can be with the physical conditions of the environment (abiotic factors) or with other organisms in the environment (biotic factors). Bryophytes adapt well to differing environments, which could in part be attributed to the high chemodiversity of these plants [134]. Seasonal variation in the metabolite profile of different bryophyte species has also been observed [23]. Environmental factors were found to be largely responsible for the stress phytohormone variation found in some bryophyte species, which could in turn directly alter the specialized metabolite profile of these plants [152]. Changes in light intensity, temperature, and nutrition were found to alter the production of the polyunsaturated fatty acids arachidonic acid and eicosapentaenoic acid in some mosses and liverworts [142]. In vascular plants, lower temperatures are correlated with an increase in fatty acid content to increase freezing tolerance [216]. Terpenoids are involved in a wide variety of ecological interactions, such as UV-B irradiation protection, desiccation resistance, and allelopathic activity [136]. Unfortunately, there is little information regarding the specific functions of these compounds in bryophytes, with most terpenoid functions being inferred from their roles in vascular plants. Flavonoids, similar to terpenoids, are also involved in mediating a variety of abiotic stresses. Light exposure was found to have a negative effect on flavonoid production, with dimmer light producing more flavonoids than harsh light [137]. Curiously, this directly contrasts one of the major roles of flavonoids in vascular plants: photoprotection against UV-B irradiation [106]. Interestingly, UV-C irradiation was found to increase the (bis)bibenzyl content in *Marchantia polymorpha*, suggesting that phenolic compounds may have taken on the photoprotective role usually played by flavonoids [217].

Two major types of biotic stress experienced by all plants are herbivore attack and microbial infection. Bryophytes have different methods of defending themselves against herbivory that can be broadly classified as either chemical or mechanical defenses [218]. Due to the simple morphology of bryophytes, mechanical defense options are limited which has potentially led to an increase in the diversity of antifeeding specialized metabolites. In liverworts, oil bodies have been identified as crucial for deterring insect herbivory [219, 220]. While a single compound or compound class has not been identified as the reason for this activity, many of the metabolite classes stored in these structures (specifically flavonoids, terpenoids, and phenolics) have been reported to serve as anti-herbivory agents [221, 222]. In the moss *Dicranum scoparium*, slug feeding is deterred by acetylenic oxylipins; extremely potent bioactive fatty acids that are produced through the oxidation of polyunsaturated fatty acids [89]. In *Sphagnum spp*., herbivore damage resulted in a ~ 300% increase in phenolic compounds, demonstrating that these mosses have both innate and induced herbivore defenses [223]. Analysis of the hornwort genome identified the same microbial terpene synthase-like (MTPSL) genes found in liverworts which could be used for the synthesis of anti-herbivory terpenes [224]. In addition to serving as antifeeding agents, many of the terpenoids and phenolic compounds present in bryophytes also possess antifungal activity [134, 135]. A recent study on *Physcomitrium patens* demonstrated that infection with *Colletotrichum gloeosporioides* upregulated genes involved in specialized metabolite production and downregulated genes involved in primary metabolism [225]. While this observation has been made repeatedly in vascular plants, the identification of induced stress responses in bryophytes is still relatively new. The phenylpropanoid pathway, discussed in more detail below, has been identified as one of the upregulated pathways in response to mechanical damage in *Marchantia polymorpha* [226]. Flavonoids and (bis)bibenzyls were accumulated due to increased phenylalanine ammonia-lyase (PAL) activity, suggesting that these phenylpropanoids may serve as antifeeding agents.

Overall, more work is needed to identify the stress response of metabolites produced by bryophytes and to determine their modes of action. More studies on the induced responses of bryophytes could also lead to the identification of novel natural products not innately produced by these plants.

3.7 Molecular Biology

In general, there is little information available about the molecular biology of bryophytes. The literature that is available largely focuses on the biosynthesis of specialized metabolites due to their potential bioactivity. This is slowly shifting, in part due to the use of the moss *Physcomitriella patens* and the liverwort *Marchantia polymorpha* as model species, allowing for a more thorough analysis of their biochemistry [134].

While information regarding the primary metabolic pathways in bryophytes is currently lacking, some interesting observations have been made. Unique sugars in the cell walls of bryophytes have been identified, suggesting that modifications in the biosynthetic pathways for carbohydrates may be present [227]. Analysis of the phytohormones found in bryophytes has demonstrated that mosses and liverworts differ in the conjugation and catabolism of their growth hormones [152]. Unfortunately, hornworts were not included in the study, so it is unknown how their phytohormone profile compares to other bryophyte phyla. Stress hormones were also assessed; however, their expression appears to be more dependent on environmental stressors than phylogeny. Deviations in phytohormones between bryophytes and other land plants have also been detected, specifically regarding the hormones utilized. In the jasmonate signaling pathway for *M. polymorpha*, dn-OPDA, an intermediate in jasmonate-isoleucine biosynthesis, is utilized as the primary ligand for the COI1 receptor [228]. Lunularic acid, a bibenzyl metabolite commonly found in liverworts, has been demonstrated to serve as an endogenous growth inhibitor with similar activity to abscisic acid [229].

In contrast to the primary metabolic pathways, the biosynthetic pathways for the major specialized metabolite classes have been thoroughly investigated either through traditional biochemistry, transcriptomic analysis, or a combination of these approaches. A brief overview of these pathways can be found in Fig. 6. Overall, the precursor pathways for these compounds are conserved between bryophytes and vascular plants. For terpenoid biosynthesis, the two pathways responsible for isoprenoid production, the mevalonate (MVA) and non-mevalonate (MEP) pathways are completely conserved [134]. Terpenoid biosynthesis in bryophytes deviates from vascular plants in the later stages due to the presence of microbial terpene synthase-like genes (MTPSL) [136]. These genes have been found in all lineages of seedless plants and are suspected to be responsible for the diversity of terpenoids found in bryophytes. The phenylpropanoid pathway, the core of flavonoid biosynthesis, is conserved in all land plants, including hornworts even though they lack flavonoids [106]. Hornworts possess other products of phenylpropanoid biosynthesis, such as lignans, suggesting that the modification of the flavonoid pathway occurs in the later steps. Mosses and liverwort possess flavonoids that are uncommon in other land plants; however, deviations to the core pathway to identify the origin of these deviations have not yet been identified. The bibenzyl and (bis)bibenzyl biosynthetic pathway is a deviation of the phenylpropanoid pathway discussed above [138]. Based on transcriptome analysis, it does not appear that unique enzyme classes are involved in the production of these compounds; however, their functions differ slightly to produce (bis)bibenzyls [138]. In vascular plants, a stilbene synthase is utilized for the cyclization step while a stilbene carboxylate synthase is utilized in bryophytes, resulting in the preservation of the COOH moiety on the bibenzyl skeleton [230]. In the case of prenylated bibenzyls, such as those found in *Radula spp.*, the terpenoid biosynthetic pathway is also involved to provide the terpenoid moiety [141]. Fatty acid biosynthesis is similarly conserved, with the core pathway in bryophytes reflecting the core pathway found in other land plants [142]. Bryophyte fatty acid biosynthesis deviates due to the expression of Δ-6 desaturase, Δ-6 elongase, and Δ-5 desaturase which are responsible for the production of fatty acids not found in other land plants.

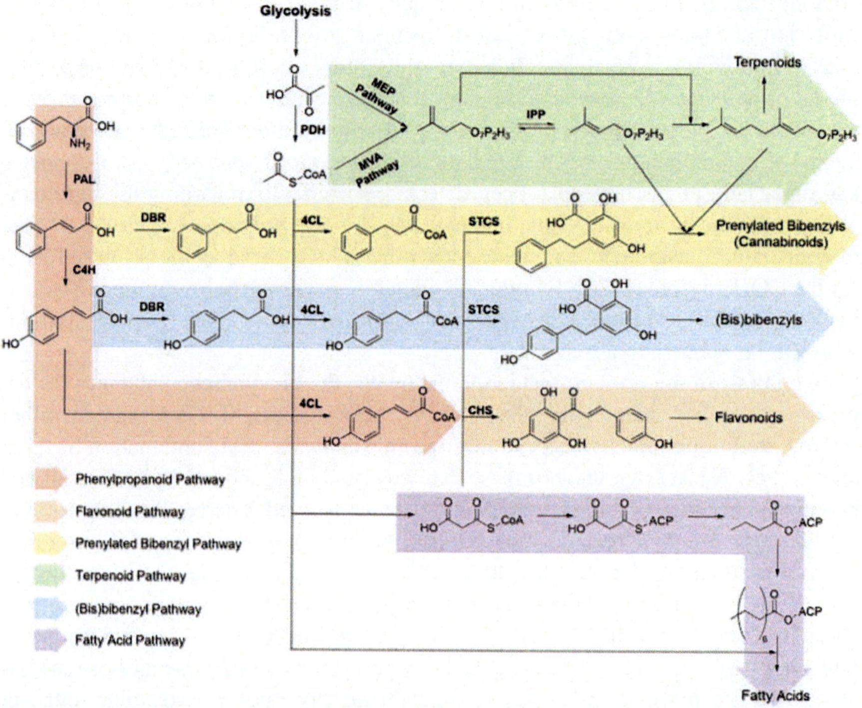

Fig. 6 Overview of the biosynthetic pathways used for specialized metabolite production in bryophytes. The phenylpropanoid and flavonoid pathways were outlined in [106], the terpenoid pathways were outlined in [136], the fatty acid pathway as outlined in [142] and the (bis)bibenzyl pathway was described in [138]. The abbreviations are as follows: **PAL**: Phenylalanine ammonia-lyase, **C4H**: cinnamate 4-hydrolase, **DBR**: double bond reductase, **4CL**: 4-coumarate CoA ligase, **STCS**: stilbene carboxylate synthase, **CHS**: chalcone synthase, **IPP**: isopentyl diphosphate delta isomerase, **PDH**: pyruvate dehydrogenase, **MVA**: mevalonate, **MEP**: non-mevalonate, **CoA**: Coenzyme A, **ACP**: Acyl carrier protein

4 Conclusion and Perspectives

Ecometabolomics is a transdisciplinary research discipline linking biochemistry and ecology using bioinformatics methods and connects several distinct spatiotemporal scales. Metabolomics is a technology that when applied in the context of ecology can reveal potentially groundbreaking insight into the molecular functioning of bryophytes. When scaled down at spatiotemporally fine scales, such as the physiological or cellular levels, ecometabolomics can reveal molecular effects, such as metabolic fluxes or changes of cellular pathways underlying ecological processes, such as environmental changes, desiccation, or the molecular responses of bryophytes to pathogens [12]. When scaled up, ecometabolomics can reveal the molecular mechanisms underlying fundamental ecological processes ranging from individuals, populations and communities to whole ecosystems, such as biological soil crusts and peatlands.

Ecometabolomics has seen substantial analytical and computational advancements in the past years that allow it not just to capture the majority of non-polar and polar (<1000 Da) compounds using DDA and DIA but also to annotate many fragment spectra with ever-increasing confidence. By improving computational methods and reusing an ecometabolomics bryophyte dataset, it was shown that compound structures can be generated *de novo* from fragment spectra increasing both the annotation and annotation accuracy [231]. For developing and improving bioinformatics, bryophytes are perfectly suited as a reference as they are non-model species and as such usually contain many "unknown unknown" molecular structures [232]. Moreover, the annotation at the compound class level using computational classification tools like CANOPUS [14] may be even more powerful in ecology as compound classes provide a generalized and abstracted view of the broader functioning of compounds.

While there has been tremendous research on bryophytes using some analytical methods, there is still a considerable lack of use of metabolomics technologies despite the fact that metabolomics is commonly used in plant ecology nowadays [144, 233, 234]. Although bryophytes are not necessarily chemically richer than vascular plants in terms of compound numbers, their biochemical pathways are often very much distinct from vascular plants - especially in liverworts. This makes bryophytes and liverworts interesting for various applications in biotechnology, such as the production of compounds to be used as pharmaceutical products [134]. Furthermore, the discovery of novel bioactive natural products as part of their specialized metabolome with diverse functions, such as phytotoxic, antibacterial, antifungal, antifeedant, or molluscicidal activities have great potential for practical use in medicine [28, 128]. Here, it was shown for (bis)bibenzyls in thallose liverworts that can be used as therapeutic agents against cancer [139]. Acetylenic oxylipins extracted from the moss *Dicranum scoparium* have agricultural relevance as they showed antifeeding activity against herbivorous slugs [89].

In summary, many of the ecological circumstances bryophytes experience to produce these unique compounds with extraordinary functions are still unexplored and there is huge potential for ecometabolomics to investigate the "why" and "when" compounds are produced by bryophytes. Here, the biodiversity and the role of bryophytes in ecosystems may have even been underestimated. For example, Davey et al. [235] show how bryophyte-associated fungal communities structure different successional stages of forests. As metabolic compounds play a fundamental role in the communication between bryophytes and their fungal partners, and as bryophytes are mainly responding metabolically to ecological change, ecometabolomics has great potential to further estimate the ecological impact of bryophytes and to unravel the molecular mechanisms behind the extraordinary capabilities of bryophytes not just in forest ecosystems but also in many others such as biological soil crusts.

Acknowledgments KP, YP, and HU acknowledge the support of iDiv (funded by the German Research Foundation, DFG-FZT 118, 202548816). KBJ was funded by NSERC via the CGS-MSFSS (Application No. 566822-2021). Further, we like to thank the Leibniz Foundation for supporting this study. Lastly, we like to thank Harald Zechmeister for providing valuable feedback and for improving the manuscript.

References

1. Cornelissen JHC, Lang SI, Soudzilovskaia NA, During HJ (2007) Comparative cryptogam ecology: a review of bryophyte and lichen traits that drive biogeochemistry. Ann Bot 99:987–1001. https://doi.org/10.1093/aob/mcm030
2. Streitberger M, Schmidt C, Fartmann T (2017) Contrasting response of vascular plant and bryophyte species assemblages to a soil-disturbing ecosystem engineer in calcareous grasslands. Ecol Eng 99:391–399. https://doi.org/10.1016/j.ecoleng.2016.11.037
3. Markert BA, Breure AM, Zechmeister HG (2003) Bioindicators & biomonitors: principles, concepts and applications, 1st edn. Elsevier, Amsterdam
4. Roos RE, Zuijlen K, Birkemoe T et al (2019) Contrasting drivers of community-level trait variation for vascular plants, lichens and bryophytes across an elevational gradient. Funct Ecol 33:2430–2446. https://doi.org/10.1111/1365-2435.13454
5. Vitt DH, House M (2021) Bryophytes as key indicators of ecosystem function and structure of northern peatlands. Bryophyt Divers Evol 43:10.11646/bde.43.1.18
6. Stanton DE, Coe KK (2021) 500 million years of charted territory: functional ecological traits in bryophytes. Bryophyt Divers Evol 43:10.11646/bde.43.1.17
7. Bernhardt-Römermann M, Poschlod P, Hentschel J (2018) BryForTrait – a life-history trait database of forest bryophytes. J Veg Sci 29:798–800. https://doi.org/10.1111/jvs.12646
8. Asakawa Y, Ludwig A, Nagashima F (2013) Chemical constituents of bryophytes. Prog Chem Org Nat Prod 95. https://doi.org/10.1007/978-3-7091-1084-3_1
9. Ludwiczuk A, Asakawa Y (2019) Bryophytes as a source of bioactive volatile terpenoids – a review. Food Chem Toxicol 132:110649. https://doi.org/10.1016/j.fct.2019.110649
10. He X, Sun Y, Zhu RL (2013) The oil bodies of liverworts: unique and important organelles in land plants. CRC Crit Rev Plant Sci 32:293–302
11. Xie Y, Sun X, Feng Q et al (2019) Comparative physiological and metabolomic analyses reveal mechanisms of aspergillus aculeatus-mediated abiotic stress tolerance in tall fescue. Plant Physiol Biochem 142:342–350. https://doi.org/10.1016/j.plaphy.2019.07.022
12. Peters K, Worrich A, Weinhold A et al (2018) Current challenges in plant eco-metabolomics. Int J Mol Sci 19:1385–1423. https://doi.org/10.3390/ijms19051385
13. Rivera R, Garrido N (2019) Metabolomics. In: Oxidants, antioxidants and impact of the oxidative status in male reproduction. Elsevier, pp 277–285, eBook ISBN: 9780128125670
14. Dührkop K, Nothias LF, Fleischauer M et al (2021) Systematic classification of unknown metabolites using high-resolution fragmentation mass spectra. Nat Biotechnol 39:462–471. https://doi.org/10.1038/s41587-020-0740-8
15. Jarmusch SA (2021) Advancements in capturing and mining mass spectrometry data are transforming natural products research. Nat Prod Rep 38(11):2066–2082. https://doi.org/10.1039/d1np00040c
16. Roessner U, Dias DA (2013) Metabolomics tools for natural product discovery. Humana Press, Totowa
17. Wolfender J-L, Rudaz S, Choi YH, Kim HK (2013) Plant metabolomics: from holistic data to relevant biomarkers. Plant Metabolomics 20:1056–1090
18. Peters K, Balcke G, Kleinenkuhnen N et al (2021) Untargeted in silico compound classification – a novel metabolomics method to assess the chemodiversity in bryophytes. Int J Mol Sci 22:3251. https://doi.org/10.3390/ijms22063251

19. Hall RD, D'Auria JC, Silva Ferreira AC et al (2022) High-throughput plant phenotyping: a role for metabolomics? Trends Plant Sci:S1360138522000309. https://doi.org/10.1016/j.tplants.2022.02.001
20. Nagler M, Nägele T, Gilli C et al (2018) Eco-metabolomics and metabolic modeling: making the leap from model systems in the lab to native populations in the field. Front Plant Sci 871:1–13. https://doi.org/10.3389/fpls.2018.01556
21. Sala-Carvalho WR, Montessi-Amaral FP, Esposito MP et al (2022) Metabolome of Ceratodon purpureus (Hedw.) Brid., a cosmopolitan moss: the influence of seasonality. Planta 255(77). https://doi.org/10.1007/s00425-022-03857-8
22. Liu S, Fang S, Liu C et al (2021) Transcriptomics integrated with metabolomics reveal the effects of ultraviolet-B radiation on flavonoid biosynthesis in Antarctic Moss. Front Plant Sci 12:788377
23. Peters K, Gorzolka K, Bruelheide H, Neumann S (2018) Seasonal variation of secondary metabolites in nine different bryophytes. Ecol Evol 8:9105–9117. https://doi.org/10.1002/ece3.4361
24. Peters K, Treutler H, Döll S et al (2019) Chemical diversity and classification of secondary metabolites in nine bryophyte species. Meta 9:222. https://doi.org/10.3390/metabo9100222
25. Fiehn O, Kopka J, Dörmann P et al (2000) Metabolite profiling for plant functional genomics. Nat Biotechnol 18:1157–1161. https://doi.org/10.1038/81137
26. Sardans J, Peñuelas J, Rivas-Ubach A (2011) Ecological metabolomics: overview of current developments and future challenges. Chemoecology 21:191–225. https://doi.org/10.1007/s00049-011-0083-5
27. Fudyma JD, Lyon J, AminiTabrizi R et al (2019) Untargeted metabolomic profiling of Sphagnum fallax reveals novel antimicrobial metabolites. Plant Direct 3:e00179. https://doi.org/10.1002/pld3.179
28. Dziwak M, Wroblewska K, Szumny A, Galek R (2022) Modern use of bryophytes as a source of secondary metabolites. Agronomy 12:1456. https://doi.org/10.3390/agronomy12061456
29. Uthe H, van Dam NM, Hervé MR, Sorokina M, Peters K, Weinhold A (2021) Chapter Six – A practical guide to implementing metabolomics in plant ecology and biodiversity research, Editor(s): Pierre Pétriacq, Alain Bouchereau, Advances in Botanical Research, Academic Press, Volume 98, Pages 163–203. https://doi.org/10.1016/bs.abr.2020.09.019
30. Stanstrup J, Broeckling CD, Helmus R et al (2019) The metaRbolomics Toolbox in Bioconductor and beyond. Metabolites 9(10):200–255. https://doi.org/10.3390/metabo9100200
31. Horai H, Arita M, Kanaya S et al (2010) MassBank: a public repository for sharing mass spectral data for life sciences. J Mass Spectrom 45:703–714. https://doi.org/10.1002/jms.1777
32. Dührkop K, Fleischauer M, Ludwig M et al (2019) SIRIUS 4: a rapid tool for turning tandem mass spectra into metabolite structure information. Nat Methods 16:299–302
33. Ruttkies C, Schymanski EL, Wolf S et al (2016) MetFrag relaunched: incorporating strategies beyond in silico fragmentation. J Cheminform 8:3. https://doi.org/10.1186/s13321-016-0115-9
34. Allard P-M, Genta-Jouve G, Wolfender J-L (2017) Deep metabolome annotation in natural products research: towards a virtuous cycle in metabolite identification. Curr Opin Chem Biol 36:40–49. https://doi.org/10.1016/j.cbpa.2016.12.022
35. Wang M, Carver JJ, Phelan VV et al (2016) Sharing and community curation of mass spectrometry data with global natural products social molecular networking. Nat Biotechnol 34:828–837. https://doi.org/10.1038/nbt.3597
36. Haug K, Cochrane K, Nainala VC et al (2020) MetaboLights: a resource evolving in response to the needs of its scientific community. Nucleic Acids Res 48:D440–D444. https://doi.org/10.1093/nar/gkz1019
37. Leao TF, Clark CM, Bauermeister A et al (2021) Quick-start infrastructure for untargeted metabolomics analysis in GNPS. Nat Metab 3:880–882. https://doi.org/10.1038/s42255-021-00429-0

38. Jarmusch AK, Wang M, Aceves CM et al (2020) ReDU: a framework to find and reanalyze public mass spectrometry data. Nat Methods 17:901–904. https://doi.org/10.1038/s41592-020-0916-7
39. Spicer RA, Salek R, Steinbeck C (2017) Compliance with minimum information guidelines in public metabolomics repositories. Sci Data 4:170137. https://doi.org/10.1038/sdata.2017.137
40. Wilkinson MD, Dumontier M, IjJ A et al (2016) The FAIR guiding principles for scientific data management and stewardship. Sci Data 3:160018. https://doi.org/10.1038/sdata.2016.18
41. Broadhurst D, Goodacre R, Reinke SN et al (2018) Guidelines and considerations for the use of system suitability and quality control samples in mass spectrometry assays applied in untargeted clinical metabolomic studies. Metabolomics 14:72. https://doi.org/10.1007/s11306-018-1367-3
42. Kim S-G, Yon F, Gaquerel E et al (2011) Tissue specific diurnal rhythms of metabolites and their regulation during herbivore attack in a native tobacco, Nicotiana attenuata. PLoS One 6: e26214. https://doi.org/10.1371/journal.pone.0026214
43. Villas-Bas SG, Roessner U, Hansen MAE et al (2007) Metabolome analysis. Wiley, Hoboken
44. Díaz-Maroto MC, Pérez-Coello MS, González Viñas MA, Cabezudo MD (2003) Influence of drying on the flavor quality of spearmint (Mentha spicata L.). J Agric Food Chem 51:1265–1269. https://doi.org/10.1021/jf0208051
45. Sauerschnig C, Doppler M, Bueschl C, Schuhmacher R (2017) Methanol generates numerous artifacts during sample extraction and storage of extracts in metabolomics research. Meta 8:1. https://doi.org/10.3390/metabo8010001
46. Fiehn O (2002) Metabolomics – the link between genotypes and phenotypes. Plant Mol Biol 48:155–171. https://doi.org/10.1023/A:1013713905833
47. Gertsman I, Barshop BA (2018) Promises and pitfalls of untargeted metabolomics. J Inherit Metab Dis 41:355–366. https://doi.org/10.1007/s10545-017-0130-7
48. Davies A (2010) Close-up and macro photography. Focal Press, Burlington
49. Chu J, Fang S, Xin P, Guo Z, Chen Y (2017) 14 – Quantitative analysis of plant hormones based on LC-MS/MS, Editor(s): Jiayang Li, Chuanyou Li, Steven M. Smith, Hormone Metabolism and Signaling in Plants, Academic Press, Pages 471–537. https://doi.org/10.1016/B978-0-12-811562-6.00014-1.53
50. Lu Y, Eiriksson FF, Thorsteinsdóttir M, Simonsen HT (2021) Effects of extraction parameters on lipid profiling of mosses using UPLC-ESI-QTOF-MS and multivariate data analysis. Metabolomics 17. https://doi.org/10.1007/s11306-021-01847-7
51. Papantoniou D, Vergara F, Weinhold A et al (2021) Cascading effects of root microbial symbiosis on the development and metabolome of the insect herbivore Manduca sexta L. Meta 11:731. https://doi.org/10.3390/metabo11110731
52. Gachet MS, Schubert A, Calarco S et al (2017) Targeted metabolomics shows plasticity in the evolution of signaling lipids and uncovers old and new endocannabinoids in the plant kingdom. Sci Rep 7:41177. https://doi.org/10.1038/srep41177
53. Harris DC, Lucy CA (2020) Quantitative chemical analysis, 10th edn. Macmillan learning
54. Negi K, Asthana AK, Chaturvedi P (2020) GC–MS analysis and antifungal activity of acetone extract of Conocephalum conicum (L) Underw (liverwort) against aflatoxins producing fungi. South Afr J Bot 131:384–390. https://doi.org/10.1016/j.sajb.2020.02.035
55. Rivas-Ubach A, Pérez-Trujillo M, Sardans J et al (2013) Ecometabolomics: optimized NMR-based method. Methods Ecol Evol 4:464–473. https://doi.org/10.1111/2041-210X.12028
56. Haag AM (2016) Mass analyzers and mass spectrometers. In: Mirzaei H, Carrasco M (eds) Modern proteomics – sample preparation, analysis and practical applications. Springer International Publishing, Cham, pp 157–169
57. Willems S, Voytik E, Skowronek P et al (2021) AlphaTims: indexing trapped ion mobility spectrometry–TOF data for fast and easy accession and visualization. Mol Cell Proteomics 20: 100149. https://doi.org/10.1016/j.mcpro.2021.100149

58. Kuhl C, Tautenhahn R, Böttcher C et al (2012) CAMERA: an integrated strategy for compound spectra extraction and annotation of liquid chromatography/mass spectrometry data sets. Anal Chem 84:283–289. https://doi.org/10.1021/ac202450g
59. Guo J, Huan T (2020) Comparison of full-scan, data-dependent, and data-independent acquisition modes in liquid chromatography–mass spectrometry based untargeted metabolomics. Anal Chem 92:8072–8080. https://doi.org/10.1021/acs.analchem.9b05135
60. Kessner D, Chambers M, Burke R et al (2008) ProteoWizard: open source software for rapid proteomics tools development. Bioinformatics 24:2534–2536. https://doi.org/10.1093/bioinformatics/btn323
61. Katajamaa M, Miettinen J, Orešič M (2006) MZmine: toolbox for processing and visualization of mass spectrometry based molecular profile data. Bioinformatics 22:634–636. https://doi.org/10.1093/bioinformatics/btk039
62. Tsugawa H, Cajka T, Kind T et al (2015) MS-DIAL: data-independent MS/MS deconvolution for comprehensive metabolome analysis. Nat Methods 12:523–526. https://doi.org/10.1038/nmeth.3393
63. Röst HL, Sachsenberg T, Aiche S et al (2016) OpenMS: a flexible open-source software platform for mass spectrometry data analysis. Nat Methods 13:741–748. https://doi.org/10.1038/nmeth.3959
64. Giacomoni F, Le Corguillé G, Monsoor M et al (2014) Workflow4Metabolomics: a collaborative research infrastructure for computational metabolomics. Bioinformatics 31:1493–1495. https://doi.org/10.1093/bioinformatics/btu813
65. Tautenhahn R, Patti GJ, Rinehart D, Siuzdak G (2012) XCMS online: a web-based platform to process untargeted metabolomic data. Anal Chem 84:5035–5039. https://doi.org/10.1021/ac300698c
66. Smith CA, Want EJ, O'Maille G et al (2006) XCMS: processing mass spectrometry data for metabolite profiling using nonlinear peak alignment, matching, and identification. Anal Chem 78:779–787. https://doi.org/10.1021/ac051437y
67. Tautenhahn R, Bottcher C, Neumann S (2008) Highly sensitive feature detection for high resolution LC/MS. BMC Bioinformatics 9:1–16. https://doi.org/10.1186/1471-2105-9-504
68. R Core Team (2021) R: A Language and Environment for Statistical Computing. R Foundation for Statistical Computing, Vienna. https://www.R-project.org/
69. Hoffmann N, Mayer G, Has C, Kopczynski D, Al Machot F, Schwudke D, Ahrends R, Marcus K, Eisenacher M, Turewicz MA (2022) Current Encyclopedia of Bioinformatics Tools, Data Formats and Resources for Mass Spectrometry Lipidomics. Metabolites 12(7):584–611. https://doi.org/10.3390/metabo12070584
70. Pluskal T, Uehara T, Yanagida M (2012) Highly accurate chemical formula prediction tool utilizing high-resolution mass spectra, MS/MS fragmentation, heuristic rules, and isotope pattern matching. Anal Chem 84:4396–4403. https://doi.org/10.1021/ac3000418
71. Shahaf N, Rogachev I, Heinig U et al (2016) The WEIZMASS spectral library for high-confidence metabolite identification. Nat Commun 7:12423. https://doi.org/10.1038/ncomms12423
72. van der Hooft JJJ, Wandy J, Young F et al (2017) Unsupervised discovery and comparison of structural families across multiple samples in untargeted metabolomics. Anal Chem 89:7569–7577. https://doi.org/10.1021/acs.analchem.7b01391
73. Djoumbou Feunang Y, Eisner R, Knox C et al (2016) ClassyFire: automated chemical classification with a comprehensive, computable taxonomy. J Cheminform 8:1–20. https://doi.org/10.1186/s13321-016-0174-y
74. Kim HW, Wang M, Leber CA et al (2021) NPClassifier: a deep neural network-based structural classification tool for natural products. J Nat Prod 84:2795–2807. https://doi.org/10.1021/acs.jnatprod.1c00399
75. Sumner LW, Amberg A, Barrett D et al (2007) Proposed minimum reporting standards for chemical analysis: Chemical Analysis Working Group (CAWG) Metabolomics Standards Initiative (MSI). Metabolomics 3:211–221. https://doi.org/10.1007/s11306-007-0082-2

76. Pang Z, Zhou G, Ewald J et al (2022) Using MetaboAnalyst 5.0 for LC–HRMS spectra processing, multi-omics integration and covariate adjustment of global metabolomics data. Nat Protoc. https://doi.org/10.1038/s41596-022-00710-w
77. van den Berg RA, Hoefsloot HCJ, Westerhuis JA et al (2006) Centering, scaling, and transformations: improving the biological information content of metabolomics data. BMC Genomics 7:142. https://doi.org/10.1186/1471-2164-7-142
78. Rohart F, Gautier B, Singh A, Lê Cao K-A (2017) mixOmics: an R package for 'omics feature selection and multiple data integration. PLoS Comput Biol 13:e1005752
79. Ristok C, Poeschl Y, Dudenhöffer JH et al (2019) Plant species richness elicits changes in the metabolome of grassland species via soil biotic legacy. J Ecol 107:2240–2254. https://doi.org/10.1111/1365-2745.13185
80. Bray JR, Curtis JT (1957) An ordination of the upland forest communities of southern Wisconsin. Ecol Monogr 27:325–349. https://doi.org/10.2307/1942268
81. Tan W, Zhou H, Rong S et al (2018) Fusion of multi-focus images via a Gaussian curvature filter and synthetic focusing degree criterion. Appl Opt 57:10092. https://doi.org/10.1364/AO.57.010092
82. Do KT, Wahl S, Raffler J et al (2018) Characterization of missing values in untargeted MS-based metabolomics data and evaluation of missing data handling strategies. Metabolomics 14:128. https://doi.org/10.1007/s11306-018-1420-2
83. Haug K, Cochrane K, Nainala VC et al (2019) MetaboLights: a resource evolving in response to the needs of its scientific community. Nucleic Acids Res:gkz1019. https://doi.org/10.1093/nar/gkz1019
84. Klavina L (2015) A study on bryophyte chemical composition–search for new applications. Agron Res 13:969–978
85. Rosenstiel TN, Shortlidge EE, Melnychenko AN et al (2012) Sex-specific volatile compounds influence microarthropod-mediated fertilization of moss. Nature 489:431–433. https://doi.org/10.1038/nature11330
86. Donath TW, Eckstein RL (2010) Effects of bryophytes and grass litter on seedling emergence vary by vertical seed position and seed size. Plant Ecol 207:257–268. https://doi.org/10.1007/s11258-009-9670-8
87. Michel P, Burritt DJ, Lee WG (2011) Bryophytes display allelopathic interactions with tree species in native forest ecosystems. Oikos 120:1272–1280. https://doi.org/10.1111/j.1600-0706.2010.19148.x
88. Boch S, Prati D, Fischer M (2016) Gastropods slow down succession and maintain diversity in cryptogam communities. Ecology 97:2184–2191. https://doi.org/10.1002/ecy.1498
89. Rempt M, Pohnert G (2010) Novel Acetylenic Oxylipins from the Moss Dicranum scoparium with antifeeding activity against herbivorous slugs. Angew Chemie Int Ed 49:4755–4758. https://doi.org/10.1002/anie.201000825
90. Proctor MCF, Oliver MJ, Wood AJ et al (2007) Desiccation-tolerance in bryophytes: a review. Bryologist 110:595–621. https://doi.org/10.1639/0007-2745(2007)110[595:DIBAR]2.0.CO;2
91. Charron AJ, Quatrano RS (2009) Between a rock and a dry place: the water-stressed Moss. Mol Plant 2:478–486. https://doi.org/10.1093/mp/ssp018
92. Sabovljević MS, Sabovljević AD, Ikram NKK et al (2016) Bryophytes – an emerging source for herbal remedies and chemical production. Plant Genet Resour 14:314–327. https://doi.org/10.1017/S1479262116000320
93. Dinakar C, Bartels D (2013) Desiccation tolerance in resurrection plants: new insights from transcriptome, proteome and metabolome analysis. Front Plant Sci 4:482. https://doi.org/10.3389/fpls.2013.00482
94. Oliver MJ (2005) Desiccation tolerance in bryophytes: a reflection of the primitive strategy for plant survival in dehydrating habitats? Integr Comp Biol 45:788–799. https://doi.org/10.1093/icb/45.5.788

95. Weckwerth W (2011) Green systems biology — from single genomes, proteomes and metabolomes to ecosystems research and biotechnology. J Proteome 75:284–305. https://doi.org/10.1016/j.jprot.2011.07.010
96. Asner GP, Anderson CB, Martin RE et al (2015) Landscape biogeochemistry reflected in shifting distributions of chemical traits in the Amazon forest canopy. Nat Geosci 8:567–573. https://doi.org/10.1038/ngeo2443
97. Gargallo-Garriga A, Sardans J, Pérez-Trujillo M et al (2015) Warming differentially influences the effects of drought on stoichiometry and metabolomics in shoots and roots. New Phytol 207:591–603. https://doi.org/10.1111/nph.13377
98. Hättenschwiler S, Vitousek PM (2000) The role of polyphenols in terrestrial ecosystem nutrient cycling. Trends Ecol Evol 15:238–243. https://doi.org/10.1016/S0169-5347(00)01861-9
99. Jones HG (2014) Plants and Microclimate – A Quantitative Approach to Environmental Plant Physiology. Third Edition. Cambridge University Press. ISBN 978-0-521-27959-8
100. Markesteijn L, Poorter L, Paz H et al (2011) Ecological differentiation in xylem cavitation resistance is associated with stem and leaf structural traits: vulnerability to cavitation of tropical dry forest tree species. Plant, Cell Environ 34:137–148. https://doi.org/10.1111/j.1365-3040.2010.02231.x
101. Rivas-Ubach A, Gargallo-Garriga A, Sardans J et al (2014) Drought enhances folivory by shifting foliar metabolomes in textitQuercus ilex trees. New Phytol 202:874–885. https://doi.org/10.1111/nph.12687
102. Kergunteuil A, Descombes P, Glauser G et al (2018) Plant physical and chemical defence variation along elevation gradients: a functional trait-based approach. Oecologia 187:561–571. https://doi.org/10.1007/s00442-018-4162-y
103. Violle C, Navas M-L, Vile D et al (2007) Let the concept of trait be functional! Oikos 116:882–892. https://doi.org/10.1111/j.0030-1299.2007.15559.x
104. Seel WEF, Hendry GA, Lee JA (1992) Effects of desiccation on some activated oxygen processing enzymes and anti-oxidants in mosses. J Exp Bot 43:1031–1037. https://doi.org/10.1093/jxb/43.8.1031
105. Proctor MCF, Tuba Z (2002) Poikilohydry and homoihydry: antithesis or spectrum of possibilities? New Phytol 156:327–349. https://doi.org/10.1046/j.1469-8137.2002.00526.x
106. Davies KM, Jibran R, Zhou Y et al (2020) The evolution of flavonoid biosynthesis: a bryophyte perspective. Front Plant Sci 11:1–21. https://doi.org/10.3389/fpls.2020.00007
107. Hiraga Y, Ara T, Nagashima Y et al (2020) Metabolome analysis using multiple data mining approaches suggests luteolin biosynthesis in Physcomitrella patens. Plant Biotechnol 37:377–381. https://doi.org/10.5511/plantbiotechnology.20.0525b
108. Erxleben A, Gessler A, Vervliet-Scheebaum M, Reski R (2012) Metabolite profiling of the moss Physcomitrella patens reveals evolutionary conservation of osmoprotective substances. Plant Cell Rep 31:427–436. https://doi.org/10.1007/s00299-011-1177-9
109. Boelen P, de Boer MK, de Bakker NVJ, Rozema J (2006) Outdoor studies on the effects of solar UV-B on bryophytes: overview and methodology. Plant Ecol 182:137–152. https://doi.org/10.1007/s11258-005-9023-1
110. Turetsky MR (2003) The role of bryophytes in carbon and nitrogen cycling. Bryologist 106: 395–409. 10.1639/05
111. Mogul R, Vaishampayan P, Bashir M et al (2017) Microbial community and biochemical dynamics of biological soil crusts across a gradient of surface coverage in the Central Mojave Desert. Front Microbiol 8. https://doi.org/10.3389/fmicb.2017.01974
112. Weber B, Belnap J, Büdel B et al (2022) What is a biocrust? A refined, contemporary definition for a broadening research community. Biol Rev. https://doi.org/10.1111/brv.12862
113. Seppelt RD, Downing AJ, Deane-Coe KK et al (2016) Bryophytes within biological soil crusts. In: Weber B, Büdel B, Belnap J (eds) Biological soil crusts: an organizing principle in drylands. Springer International Publishing, Cham, pp 101–120

114. Rousk K, Pedersen PA, Dyrnum K, Michelsen A (2017) The interactive effects of temperature and moisture on nitrogen fixation in two temperate-arctic mosses. Theor Exp Plant Physiol 29: 25–36. https://doi.org/10.1007/s40626-016-0079-1
115. Ye C, Tao Y, Zhang Y et al (2020) Monosaccharide composition of primary cell wall polysaccharides as a developmental level indicator of biological soil crusts. Catena 195:104782. https://doi.org/10.1016/j.catena.2020.104782
116. Li J, Li X, Zhang P (2014) Micro-morphology, ultrastructure and chemical composition changes of *Bryum argenteum* from a desert biological soil crust following one-year desiccation. Bryologist 117:232–240. https://doi.org/10.1639/0007-2745-117.3.232
117. Baran R, Ivanova NN, Jose N, Garcia-Pichel F, Kyrpides NC, Gugger M, Northen TR (2013) Functional genomics of novel secondary metabolites from diverse cyanobacteria using untargeted metabolomics. Mar Drugs 11(10):3617–31. https://doi.org/10.3390/md11103617
118. Löhnis F (1926) Nitrogen availability of green manures. Soil Sci 22:253–290
119. De Deyn GB, Cornelissen JHC, Bardgett RD (2008) Plant functional traits and soil carbon sequestration in contrasting biomes. Ecol Lett 11:516–531. https://doi.org/10.1111/j.1461-0248.2008.01164.x
120. Walker TN, Garnett MH, Ward SE et al (2016) Vascular plants promote ancient peatland carbon loss with climate warming. Glob Chang Biol 22:1880–1889. https://doi.org/10.1111/gcb.13213
121. Koranda M, Michelsen A (2021) Mosses reduce soil nitrogen availability in a subarctic birch forest via effects on soil thermal regime and sequestration of deposited nitrogen. J Ecol 109: 1424–1438. https://doi.org/10.1111/1365-2745.13567
122. Adamczyk B, Adamczyk S, Smolander A, Kitunen V, Simon J (2018) Plant Secondary Metabolites–Missing Pieces in the Soil Organic Matter Puzzle of Boreal Forests. Soil Systems 2(1):2–12. https://doi.org/10.3390/soils2010002
123. Vicherová E, Glinwood R, Hájek T et al (2020) Bryophytes can recognize their neighbours through volatile organic compounds. Sci Rep 10:7405
124. Koeduka T (2018) Functional evolution of biosynthetic enzymes that produce plant volatiles. Biosci Biotechnol Biochem 82:192–199. https://doi.org/10.1080/09168451.2017.1422968
125. Holopainen JK (2004) Multiple functions of inducible plant volatiles. Trends Plant Sci 9:529–533. https://doi.org/10.1016/j.tplants.2004.09.006
126. Rosenkranz M, Chen Y, Zhu P, Vlot AC (2021) Volatile terpenes – mediators of plant-to-plant communication. Plant J 108:617–631. https://doi.org/10.1111/tpj.15453
127. Asakawa Y (2004) Chemosystematics of the hepaticae. Phytochemistry 65:623–669. https://doi.org/10.1016/j.phytochem.2004.01.003
128. Asakawa Y, Ludwiczuk A (2018) Chemical constituents of bryophytes: structures and biological activity. J Nat Prod 81:641–660. https://doi.org/10.1021/acs.jnatprod.6b01046
129. Suire C, Bouvier F, Backhaus RA et al (2000) Cellular localization of isoprenoid biosynthetic enzymes in Marchantia polymorpha. Uncovering a new role of oil bodies. Plant Physiol 124: 971–978. https://doi.org/10.1104/pp.124.3.971
130. Tanaka M, Esaki T, Kenmoku H et al (2016) Direct evidence of specific localization of sesquiterpenes and marchantin a in oil body cells of Marchantia polymorpha L. Phytochemistry 130:77–84. https://doi.org/10.1016/j.phytochem.2016.06.008
131. Flegel M, Becker H (2000) Characterization of the contents of oil bodies from the liverwort Radula complanata. Plant Biol 2:208–210. https://doi.org/10.1055/s-2000-9156
132. Meuche D, Huneck S (1966) Inhaltsstoffe der Moose, II. Azulene aus Calypogeia Trichomanis (L.) Corda. Chem Ber 99:2669–2674. https://doi.org/10.1002/cber.19660990836
133. Tazaki H, Okihara T, Koshino H et al (1998) An aromadendrane-type sesquiterpenoid from the liverwort Calypogeia azurea. Phytochemistry 48:147–149. https://doi.org/10.1016/S0031-9422(97)00751-6
134. Horn A, Pascal A, Lončarević I et al (2021) Natural products from bryophytes: from basic biology to biotechnological applications. CRC Crit Rev Plant Sci 40:191–217. https://doi.org/10.1080/07352689.2021.1911034

135. Commisso M, Guarino F, Marchi L et al (2021) Bryo-activities: a review on how bryophytes are contributing to the arsenal of natural bioactive compounds against fungi. Plan Theory 10: 1–28. https://doi.org/10.3390/plants10020203
136. Chen F, Ludwiczuk A, Wei G et al (2018) Terpenoid secondary metabolites in bryophytes: chemical diversity, biosynthesis and biological functions. CRC Crit Rev Plant Sci 37:210–231. https://doi.org/10.1080/07352689.2018.1482397
137. Wang X, Cao J, Dai X et al (2017) Total flavonoid concentrations of bryophytes from Tianmu Mountain, Zhejiang Province (China): phylogeny and ecological factors. PLoS One 12:2–11. https://doi.org/10.1371/journal.pone.0173003
138. Takahashi H, Asakawa Y (2017) Transcriptome analysis of Marchantin biosynthesis from the liverwort Marchantia polymorpha. Nat Prod Commun 12:1269–1272. https://doi.org/10.1177/1934578x1701200831
139. Nandy S, Dey A (2020) Bibenzyls and bisbybenzyls of bryophytic origin as promising source of novel therapeutics: pharmacology, synthesis and structure-activity. DARU, J Pharm Sci 28: 701–734. https://doi.org/10.1007/s40199-020-00341-0
140. Asakawa Y, Nagashima F, Ludwiczuk A (2020) Distribution of Bibenzyls, Prenyl Bibenzyls, Bis-bibenzyls, and Terpenoids in the liverwort genus radula. J Nat Prod 83:756–769. https://doi.org/10.1021/acs.jnatprod.9b01132
141. Gülck T, Møller BL (2020) Phytocannabinoids: origins and biosynthesis. Trends Plant Sci. https://doi.org/10.1016/j.tplants.2020.05.005
142. Lu Y, Eiriksson FF, Thorsteinsdóttir M, Simonsen HT (2019) Valuable fatty acids in bryophytes – production, biosynthesis, analysis and applications. Plan Theory 8:1–18. https://doi.org/10.3390/plants8110524
143. Asakawa Y, Ludwiczuk A (2013) Bryophytes: liverworts, mosses, and hornworts: extraction and isolation procedures. In: Roessner U, Dias DA (eds) BT – metabolomics tools for natural product discovery: methods and protocols. Humana Press, Totowa, pp 1–20
144. van Dam NM, van der Meijden E (2011) A role for metabolomics in plant ecology. In: Hall RD (ed) Annual plant reviews, vol 43. Wiley-Blackwell, Oxford, UK, pp 87–107
145. Hill MO (1973) Diversity and evenness: a unifying notation and its consequences. Ecology 54: 427–432. https://doi.org/10.2307/1934352
146. Washington HG (1984) Diversity, biotic and similarity indices. Water Res 18:653–694. https://doi.org/10.1016/0043-1354(84)90164-7
147. Wetzel WC, Whitehead SR (2020) The many dimensions of phytochemical diversity: linking theory to practice. Ecol Lett 23:16–32. https://doi.org/10.1111/ele.13422
148. Azar AWP, Rosleine D, Faizal A (2019) Secondary metabolite profiles in the methanolic extract of Leucobryum javense isolated from tropical montane forest in West Java, Indonesia. AIP Conf Proc 2120:30027. https://doi.org/10.1063/1.5115631
149. Kohn G, Vandekerkhove O, Hartmann E, Beutelmann P (1988) Acetylenic fatty acids in the ricciaceae (hepaticae). Phytochemistry 27:1049–1051. https://doi.org/10.1016/0031-9422(88)80271-1
150. Markham KRJ, Porter L (1975) Evidence of biosynthetic simplicity in the flavonoid chemistry of the ricciaceae. Phytochemistry 14:199–201. https://doi.org/10.1016/0031-9422(75)85037-0
151. Blatt-Janmaat K, Neumann S, Schmidt F et al (2022) Impact of in vitro hormone treatments on the bibenzyl production of Radula complanata. Botany. https://doi.org/10.1139/cjb-2022-0048
152. Drábková LZ, Dobrev PI, Motyka V (2015) Phytohormone profiling across the bryophytes. PLoS One 10:1–19. https://doi.org/10.1371/journal.pone.0125411
153. Bowman JL, Kohchi T, Yamato KT et al (2017) Insights into land plant evolution garnered from the Marchantia polymorpha genome. Cell 171:287–304.e15. https://doi.org/10.1016/j.cell.2017.09.030
154. Bowles AMC, Paps J, Bechtold U (2022) Water-related innovations in land plants evolved by different patterns of gene cooption and novelty. New Phytol:nph.17981. https://doi.org/10.1111/nph.17981

155. Initiative OTPT (2019) One thousand plant transcriptomes and the phylogenomics of green plants. Nature 574:679–685. https://doi.org/10.1038/s41586-019-1693-2
156. Saługa M (2020) At the crossroads of botanical collections and molecular genetics laboratory: a preliminary study of obtaining amplifiable DNA from moss herbarium material. PeerJ 8: e9109. https://doi.org/10.7717/peerj.9109
157. Treangen TJ, Salzberg SL (2012) Repetitive DNA and next-generation sequencing: computational challenges and solutions. Nat Rev Genet 13:36–46. https://doi.org/10.1038/nrg3117
158. Križman M, Jakše J, Baričevič D et al. (2006) Robust CTAB-activated charcoal protocol for plant DNA extraction. Acta agriculturae Slovenica 87(2):427–433
159. Jordon-Thaden IE, Chanderbali AS, Gitzendanner MA, Soltis DE (2015) Modified CTAB and TRIzol protocols improve RNA extraction from chemically complex Embryophyta. Appl Plant Sci 3:1400105. https://doi.org/10.3732/apps.1400105
160. Schlick-Steiner BC, Steiner FM, Seifert B et al (2010) Integrative taxonomy: a multisource approach to exploring biodiversity. Annu Rev Entomol 55:421–438. https://doi.org/10.1146/annurev-ento-112408-085432
161. Fox HM (1946) Chemical taxonomy. Nature 157:511. https://doi.org/10.1038/157511a0
162. Reynolds T (2007) The evolution of chemosystematics. Phytochemistry 68:2887–2895. https://doi.org/10.1016/j.phytochem.2007.06.027
163. Zidorn C (2019) Plant chemophenetics – a new term for plant chemosystematics/plant chemotaxonomy in the macro-molecular era. Phytochemistry 163:147–148. https://doi.org/10.1016/j.phytochem.2019.02.013
164. Brodo IM (1986) Interpreting chemical variation in lichens for systematic purposes. Bryologist 89:132. https://doi.org/10.2307/3242753
165. Rogers RW (1989) Chemical variation and the species concept in lichenized ascomycetes. Bot J Linn Soc 101:229–239. https://doi.org/10.1111/j.1095-8339.1989.tb00156.x
166. Ramawat KG (2019) Biodiversity and chemotaxonomy. Springer International Publishing, Cham
167. Culberson WL (1969) The use of chemistry in the systematics of the lichens. Taxon 18:152–166. https://doi.org/10.2307/1218673
168. Lumbsch HT (2002) Analysis of phenolic products in lichens for identification and taxonomy. In: Kranner IC, Beckett RP, Varma AK (eds) Protocols in lichenology. Springer, Berlin/Heidelberg, pp 281–295
169. Li N, Xu C, Li-Beisson Y, Philippar K (2016) Fatty acid and lipid transport in plant cells. Trends Plant Sci 21:145–158. https://doi.org/10.1016/j.tplants.2015.10.011
170. Kohn G, Demmerle S, Vandekerkhove O et al (1987) Distribution and chemotaxonomic significance of acetylenic fatty acids in mosses of the dicranales. Phytochemistry 26:2271–2275. https://doi.org/10.1016/S0031-9422(00)84699-3
171. Ludwiczuk A, Odrzykoski IJ, Asakawa Y (2013) Identification of cryptic species within liverwort Conocephalum conicum based on the volatile components. Phytochemistry 95: 234–241. https://doi.org/10.1016/j.phytochem.2013.06.011
172. Mues R, Zinsmeiter H (1988) The chemotaxonomy of phenolic compounds in bryophytes. J Hattori Bot Lab 141:109–141
173. Figueiredo AC, Sim-Sim M, Barroso JG et al (2009) Liverwort radula species from Portugal: chemotaxonomical evaluation of volatile composition. Flavour Fragr J 24:316–325. https://doi.org/10.1002/ffj
174. Gradstein SR, Matsuda R, Asakawa Y (1985) A chemotaxonomic survey of terpenoids and aromatic compounds in the Lejeuneaceae. Nov Hedwigia Beihefte 80:63–86
175. Wink M, Waterman PG (1999) Chemotaxonomy in relation to molecular phylogeny of plants. Annual Plant Reviews 2:295–335. https://doi.org/10.1002/9781119312994.apr0017
176. Stenitzer D, Mócsai R, Zechmeister H, Reski R, Decker EL, Altmann F (2022) O-methylated N-glycans Distinguish Mosses from Vascular Plants. Biomolecules 12(1):136–148. https://doi.org/10.3390/biom12010136

177. Walker TWN, Alexander JM, Allard P et al (2022) Functional traits 2.0: the power of the metabolome for ecology. J Ecol 110:4–20. https://doi.org/10.1111/1365-2745.13826
178. Lee GE, Bechteler J, Pócs T, Schäfer-Verwimp A, Tang HY, Chia PW (2022) Integrative Taxonomy Reveals a New Species of the Genus Lejeunea (Marchantiophya: Lejeuneaceae) from Peninsular Malaysia. Plants 11(13):1642–1654. https://doi.org/10.3390/plants11131642
179. Landolt E (2010) Flora indicativa: ökologische Zeigerwerte und biologische Kennzeichen zur Flora der Schweiz und der Alpen, 2., völlig. Haupt, Bern
180. Ellenberg H, Leuschner C (2010) Vegetation Mitteleuropas mit den Alpen – in ökologischer, dynamischer und historischer Sicht. 6th Edition. UTB. https://doi.org/10.36198/9783825281045
181. Simmel J, Ahrens M, Poschlod P (2021) Ellenberg N values of bryophytes in Central Europe*. J Veg Sci 32:e12957. https://doi.org/10.1111/jvs.12957
182. Hill MO, Preston CD, Bosanquet SDS, Roy DB (2007) BRYOATT: attributes of british and irish mosses, liverworts and hornworts with information on native status, size, life form, life history, geography and habitat. NERC, Huntingdon/Cambridgeshire
183. During HJ (1992) Ecological classification of bryophytes and lichens. In: Bryophytes and lichens in a changing environment. Clarendon Press, Oxford, pp 1–31
184. Frisvoll AA (1997) Bryophytes of spruce Forest stands in Central Norway. Lindbergia 22:83–97
185. Kattge J, DíAz S, Lavorel S et al (2011) TRY – a global database of plant traits. Glob Chang Biol 17:2905–2935. https://doi.org/10.1111/j.1365-2486.2011.02451.x
186. Valares Masa C, Alías Gallego J, Chaves Lobón N, Sosa Díaz T (2016) Intra-population variation of secondary metabolites in Cistus ladanifer L. Molecules 21:945. https://doi.org/10.3390/molecules21070945
187. Wright IJ, Reich PB, Westoby M et al (2004) The worldwide leaf economics spectrum. Nature 428:821–827. https://doi.org/10.1038/nature02403
188. Gauslaa Y, Coxson D (2011) Interspecific and intraspecific variations in water storage in epiphytic old forest foliose lichens. Botany 89:787–798. https://doi.org/10.1139/b11-070
189. Ellis CJ, Asplund J, Benesperi R, Branquinho C, Di Nuzzo L, Hurtado P, Martínez I, Matos P, Nascimbene J, Pinho P, Prieto M, Rocha B, Rodríguez-Arribas C, Thüs H, Giordani P (2021) Functional Traits in Lichen Ecology: A Review of Challenge and Opportunity. Microorganisms 9(4):766–793. https://doi.org/10.3390/microorganisms9040766
190. Fernández-Martínez M, Corbera J, Cano-Rocabayera O et al (2021) Do bryophyte elemental concentrations explain their morphological traits? Plan Theory 10:1581. https://doi.org/10.3390/plants10081581
191. Oke TA, Turetsky MR, Weston DJ, Shaw JA (2020) Tradeoffs between phenotypic plasticity and local adaptation influence the ecophysiology of the moss, Sphagnum magellanicum. Oecologia 193:867–877. https://doi.org/10.1007/s00442-020-04735-4
192. Sack L, Buckley TN (2020) Trait multi-functionality in plant stress response. Integr Comp Biol 60:98–112. https://doi.org/10.1093/icb/icz152
193. Hedenäs L (2012) Morphological and anatomical features associated with epiphytism among the pleurocarpous mosses – one basis for further research on adaptations and their evolution. J Bryol 34:79–100. https://doi.org/10.1179/1743282011Y.0000000049
194. Zechmeister HG, Grodzińska K, Szarek-Łukaszewska G (2003) Chapter 10 Bryophytes, Editor(s): B.A. Markert, A.M. Breure, H.G. Zechmeister, Trace Metals and other Contaminants in the Environment, Elsevier, Volume 6, Pages 329–375. https://doi.org/10.1016/S0927-5215(03)80140-6
195. Proctor MCF (1979) Surface wax on the leaves of some mosses. J Bryol 10:531–538. https://doi.org/10.1179/jbr.1979.10.4.531
196. Proctor MCF (1990) The physiological basis of bryophyte production. Bot J Linn Soc 104:61–77. https://doi.org/10.1111/j.1095-8339.1990.tb02211.x
197. Brodribb TJ, Carriquí M, Delzon S et al (2020) Advanced vascular function discovered in a widespread moss. Nat Plants 6:273–279. https://doi.org/10.1038/s41477-020-0602-x

198. Niklas KJ, Cobb ED, Matas AJ (2017) The evolution of hydrophobic cell wall biopolymers: from algae to angiosperms. J Exp Bot 68:5261–5269. https://doi.org/10.1093/jxb/erx215
199. Mishler BD, Churchill SP (1984) A Cladistic approach to the phylogeny of the "Bryophytes.". Brittonia 36:406. https://doi.org/10.2307/2806602
200. Verhoeven JTA, Liefveld WM (1997) The ecological significance of organochemical compounds in sphagnum. Acta Bot Neerl 46:117–130. https://doi.org/10.1111/plb.1997.46.2.117
201. Rühling Å, Tyler G (1970) Sorption and retention of heavy metals in the woodland Moss Hylocomium splendens (Hedw.) Br. Et Sch. Oikos 21:92–97. https://doi.org/10.2307/3543844
202. Herrmann A, Schauer N (2013) Metabolomics-Assisted Plant Breeding, In: The Handbook of Plant Metabolomics (Eds. Weckwerth W, Kahl G). Wiley. https://doi.org/10.1002/9783527669882.ch13
203. Gilbert OL (1968) Bryophytes as indicators of air pollution in the Tyne Valley. New Phytol 67: 15–30. https://doi.org/10.1111/j.1469-8137.1968.tb05450.x
204. Shi X-M, Song L, Liu W-Y et al (2017) Epiphytic bryophytes as bio-indicators of atmospheric nitrogen deposition in a subtropical montane cloud forest: response patterns, mechanism, and critical load. Environ Pollut 229:932–941. https://doi.org/10.1016/j.envpol.2017.07.077
205. Song L, Lu H-Z, Xu X-L et al (2016) Organic nitrogen uptake is a significant contributor to nitrogen economy of subtropical epiphytic bryophytes. Sci Rep 6:30408. https://doi.org/10.1038/srep30408
206. Czerepko J, Gawryś R, Szymczyk R et al (2021) How sensitive are epiphytic and epixylic cryptogams as indicators of forest naturalness? Testing bryophyte and lichen predictive power in stands under different management regimes in the Białowieża forest. Ecol Indic 125: 107532. https://doi.org/10.1016/j.ecolind.2021.107532
207. Koranda M, Kerschbaum S, Wanek W et al (2007) Physiological responses of bryophytes Thuidium tamariscinum and Hylocomium splendens to increased nitrogen deposition. Ann Bot 99:161–169. https://doi.org/10.1093/aob/mcl239
208. Freund DM, Hegeman AD (2017) Recent advances in stable isotope-enabled mass spectrometry-based plant metabolomics. Curr Opin Biotechnol 43:41–48. https://doi.org/10.1016/j.copbio.2016.08.002
209. Domenick TM, Gill EL, Vedam-Mai V, Yost RA (2021) Mass spectrometry-based cellular metabolomics: current approaches, applications, and future directions. Anal Chem 93:546–566. https://doi.org/10.1021/acs.analchem.0c04363
210. González-Riano C, Dudzik D, Garcia A et al (2020) Recent developments along the analytical process for metabolomics workflows. Anal Chem 92:203–226. https://doi.org/10.1021/acs.analchem.9b04553
211. Maia M, Figueiredo A, Cordeiro C, Sousa Silva M (2021) FT-ICR-MS-based metabolomics: a deep dive into plant metabolism. Mass Spectrom Rev. https://doi.org/10.1002/mas.21731
212. Shakya K, Chettri MK, Sawidis T (2008) Experimental investigations of five different mosses on accumulation capacities of Cu, Pb and Zn. Toxicol Environ Chem 90:585–601. https://doi.org/10.1080/02772240701609780
213. Stanković JD, Sabovljević AD, Sabovljević MS (2018) Bryophytes and heavy metals: a review. Acta Bot Croat 77:109–118. https://doi.org/10.2478/botcro-2018-0014
214. Gecheva G, Yurukova L (2014) Water pollutant monitoring with aquatic bryophytes: a review. Environ Chem Lett 12:49–61. https://doi.org/10.1007/s10311-013-0429-z
215. Delépée R, Pouliquen H, Le Bris H (2004) The bryophyte Fontinalis antipyretica Hedw. bioaccumulates oxytetracycline, flumequine and oxolinic acid in the freshwater environment. Sci Total Environ 322:243–253. https://doi.org/10.1016/j.scitotenv.2003.09.018
216. Singer SD, Zou J, Weselake RJ (2016) Abiotic factors influence plant storage lipid accumulation and composition. Plant Sci 243:1–9. https://doi.org/10.1016/j.plantsci.2015.11.003
217. Kageyama A, Ishizaki K, Kohchi T et al (2015) Abscisic acid induces biosynthesis of bisbibenzyls and tolerance to UV-C in the liverwort Marchantia polymorpha. Phytochemistry 117:547–553. https://doi.org/10.1016/j.phytochem.2015.05.009
218. Glime JM (2006) Bryophytes and herbivory. Cryptogam Bryol 27:191–203

219. Kanazawa T, Morinaka H, Ebine K et al (2020) The liverwort oil body is formed by redirection of the secretory pathway. Nat Commun 11. https://doi.org/10.1038/s41467-020-19978-1
220. Romani F, Banić E, Florent SN et al (2020) Oil body formation in Marchantia polymorpha is controlled by MpC1HDZ and serves as a defense against arthropod herbivores. Curr Biol 30: 2815–2828.e8. https://doi.org/10.1016/j.cub.2020.05.081
221. Mithöfer A, Boland W (2012) Plant defense against herbivores: chemical aspects. Annu Rev Plant Biol 63:431–450. https://doi.org/10.1146/annurev-arplant-042110-103854
222. War AR, Paulraj MG, Ahmad T et al (2012) Mechanisms of plant defense against insect herbivores. Plant Signal Behav 7:1306–1320. https://doi.org/10.4161/psb.21663
223. Da CY, Moles A, Bu ZJ et al (2021) Induced defense and its cost in two bryophyte species. Am J Bot 108:777–787. https://doi.org/10.1002/ajb2.1654
224. Zhang J, Fu XX, Li RQ et al (2020) The hornwort genome and early land plant evolution. Nat Plants 6:107–118. https://doi.org/10.1038/s41477-019-0588-4
225. Otero-Blanca A, Pérez-Llano Y, Reboledo-Blanco G et al (2021) Physcomitrium patens infection by colletotrichum gloeosporioides: understanding the fungal–bryophyte interaction by microscopy, phenomics and RNA sequencing. J Fungi 7:1–24. https://doi.org/10.3390/jof7080677
226. Yoshikawa M, Luo W, Tanaka G et al (2018) Wounding stress induces phenylalanine ammonia lyases, leading to the accumulation of phenylpropanoids in the model liverwort Marchantia polymorpha. Phytochemistry 155:30–36. https://doi.org/10.1016/j.phytochem.2018.07.014
227. Popper ZA, Fry SC (2003) Primary cell wall composition of bryophytes and charophytes. Ann Bot 91:1–12. https://doi.org/10.1093/aob/mcg013
228. Monte I, Franco-Zorrilla JM, García-Casado G et al (2019) A single JAZ repressor controls the Jasmonate pathway in Marchantia polymorpha. Mol Plant 12:185–198. https://doi.org/10.1016/j.molp.2018.12.017
229. Pryce RJ (1971) Lunularic acid, a common endogenous growth inhibitor of liverworts. Planta 97:354–357. https://doi.org/10.1007/BF00390214
230. Hussain T, Plunkett B, Ejaz M et al (2018) Identification of putative precursor genes for the biosynthesis of cannabinoid-like compound in Radula marginata. Front Plant Sci 9:537. https://doi.org/10.3389/fpls.2018.00537
231. Stravs MA, Dührkop K, Böcker S, Zamboni N (2022) MSNovelist: de novo structure generation from mass spectra. Nat Methods. https://doi.org/10.1038/s41592-022-01486-3
232. da Silva RR, Dorrestein PC, Quinn RA (2015) Illuminating the dark matter in metabolomics. Proc Natl Acad Sci 112:12549–12550. https://doi.org/10.1073/pnas.1516878112
233. van Dam NM, Bouwmeester HJ (2016) Metabolomics in the rhizosphere: tapping into belowground chemical communication. Trends Plant Sci 21:256–265. https://doi.org/10.1016/j.tplants.2016.01.008
234. Sardans J, Gargallo-Garriga A, Urban O et al (2021) Ecometabolomics of plant–herbivore and plant–fungi interactions: a synthesis study. Ecosphere 12:e03736. https://doi.org/10.1002/ecs2.3736
235. Davey ML, Kauserud H, Ohlson M (2014) Forestry impacts on the hidden fungal biodiversity associated with bryophytes. FEMS Microbiol Ecol 90:313–325. https://doi.org/10.1111/1574-6941.12386

Physiological Ecology of Ferns

25

O. Roger Anderson

Contents

Abstract

This is a review of research on the physiological ecology of ferns encompassing aquatic, terrestrial, and epiphytic species. Each of these is addressed within a focus of three themes: (i) development, growth, and life cycles, (ii) environmental adaptations, such as desiccation resilience and stress tolerance, and (iii) physiological aspects of light, photosynthesis, and respiration. To complement other contributions in this volume on bioactive compounds, fern research articles that included the role of biologically active compounds, such as phytohormones and regulatory molecules were particularly chosen for review. Fern species have a long geological history. They evolved in the Middle Devonian, 390 million years ago (mya), and expanded in diversity during the Cenozoic (65 mya), when

O. R. Anderson (✉)
Biology and Paleo Environment, Lamont-Doherty Earth Observatory, Columbia University, Palisades, NY, USA
e-mail: ora@ldeo.columbia.edu

H. N. Murthy (ed.), *Bioactive Compounds in Bryophytes and Pteridophytes*,
Reference Series in Phytochemistry, https://doi.org/10.1007/978-3-031-23243-5_33

angiosperms were becoming dominant. The increasing presence of forests provided suitable, shady habitats where ferns adapted and flourished beneath tree canopies and some diversified to become epiphytes on the trunks and branches of the overlying trees. During evolutionary history, and radiation into diverse environments spanning deserts and drylands to tropical rain forests, ferns have developed a wide range of adaptations to survive and succeed in these sometimes challenging habitats. This chapter explores some of these remarkable adaptations and concludes with recommendations for future research strategies that capitalize on emerging molecular genetic methods that promise to bring a greater depth of insight into the role of bioactive compounds in the physiological ecology of ferns.

Keywords

Adaptation · Antheridiogen · Biochemistry · Biomass · Climate · Crassulacean acid metabolism · Desiccation response · Environment · Gametophyte · Gas exchange · Hydrology · Illumination · Life cycles · Nutrients · Photosynthesis · Phytohormones · Precipitation · Respiration · Spore germination · Sporophyte · Transcriptomics · Transpiration

Abbreviations

ABA	Abscisic acid
A_{max}	Maximum rate of carbon dioxide assimilation
AN	Maximum net photosynthesis
A_{sat}	Light-saturated rate of photosynthesis
CAM	Crassulacean acid metabolism
Chl	Chlorophyll
CRY	Cryptochrome
DS	Desiccation sensitive
DT	Desiccation tolerant
F_v/F_m	Ratio of variable photosynthesis to maximum photosynthesis
GA	Gibberellic acid
GID1	GA-insensitive dwarf 1
g_m	Mesophyll conductance to CO_2
g_s	Stomatal conductance for gas exchange
IAA	Indole-acetic acid
JA	Jasmonic acid
LHC	Light-harvesting complex
mRNA	Messenger RNA
mya	Million years ago
E_{opt}	Optimum irradiance
PAR	Photosynthetically active radiation
PHY	Phytochrome
pI	Isoelectric point
P_{max}	Maximum rate of photosynthesis
POD	Peroxidase
PPO	Polyphenol oxidase

R_d	Dark respiration rate
RWC	Relative water content
$S_{C/O}$	Rubisco specificity factor
T_{90}	Time to reach 90% maximum net photosynthetic rate
TCA	Tricarboxylic acid
TDLAS	Tunable diode laser absorption spectroscopy
$V_{c,max}$	Apparent in vivo carboxylation activity
VPD	Vapor pressure deficit
WUE	Water use efficiency
μS	Microsiemens

1 Introduction

This is a review of the physiological ecology of aquatic, terrestrial, and epiphytic ferns that are largely homosporous, bearing only one size of spores, but also includes aquatic species of the Marsileaceae and Salviniaceae (two families in the Order Marsileales) that are heterosporous. The heterosporous ferns produce two sizes of spores: microspores (giving rise to male gametophytes) and megaspores (giving rise to female gametophytes). Most emphasis in this review is placed on leptosporangiate ferns, but also includes information on eusporangiate species, when it is particularly relevant within a particular section of the review. Ferns, bearing spores in clusters called sori, are taxonomically separated from the bryophytes (also known as lycophytes) that encompass mosses and liverworts. The bryophytes produce spores in capsules or cone-like strobili. Only ferns will be addressed in this review, although authors of other chapters in this publication consider bryophytes as well as ferns within the scope of their contributions.

Ferns, categorized generally as pteridophytes, evolved later than the bryophytes. Current scientific evidence indicates that ferns evolved in the Middle Devonian, 390 million years ago (mya). Modern descendants (ca. 10,500 species) evolved more recently in the last 70 million years, particularly during a geological time when angiosperms (seed plants) were proliferating. The ferns expanded in diversity during the Cenozoic (65 mya) when the angiosperms were increasing in dominance [1, 2]. Ferns were particularly dominant in the Carboniferous (ca. 359 to 299 mya) contributing to the formation of coal and other sources of carbon-based fuel.

Subsequently, during ensuing geological periods, they became more restricted and lost space to gymnosperms and angiosperms, especially when forests became increasingly established. However, forests provided suitable, shady habitats where ferns adapted and flourished beneath tree canopies, and some diversified to become epiphytes on the trunks and branches of the overlying trees [3].

Through geographic dispersal and adaptation to varied habitats, ferns have become established in almost every environmental niche, worldwide, spanning rain forests to deserts and dry lands, except in the harshest polar environments [4–6]. Consequently, they have evolved remarkably diverse and, in some cases, peculiarly specialized physiological adaptations that have contributed to their survival and exploitation of these diverse habitats. With expansive evolution and major radiation into diverse habitats, ferns have developed a capacity to synthesize a broad range of biologically active compounds. Some are basic components of metabolism, others are regulatory organic compounds (plant hormones), and yet others are defensive toxic or repellent compounds that deter predators or other invading organisms. Some of these bioactive compounds are also of value biotechnologically, or have medicinal properties, as more fully documented in other chapters of this volume. The purpose of this chapter is to provide a broad perspective on the physiological ecology of ferns, including metabolic and biochemical aspects, pertinent to the theme of this book "Bioactive Compounds in Bryophytes and Pteridophytes."

In comparison to other vascular plants, such as angiosperms that reproduce by seeds, ferns have a life cycle with two alternating generations. The leaf-bearing vascular stage that we typically observe growing in the natural environment is diploid (a sporophyte) and produces spores that are haploid (Fig. 1).

The haploid spores germinate, under sufficiently favorable conditions of light and moisture, and give rise to a haploid photosynthetic gametophyte known as a prothallus. Prothalli vary in morphology among species. The prothalli of some species are filamentous or web-like; others more commonly are relatively small (cm-size), strap-shaped, or cordate (heart-shaped) individuals (Fig. 1). The prothallus matures to produce sexual organs, female archegonia, and male antheridia. The archegonium contains a haploid egg, and the male antheridium produces motile haploid sperm. When there is a sufficient water film, sperm released from the antheridium swim to an archegonium of the same prothallus, or to the archegonium of a nearby prothallus, where the sperm enters the neck canal of the archegonium, migrates to the venter, and fuses with the basal egg producing a diploid zygote. The

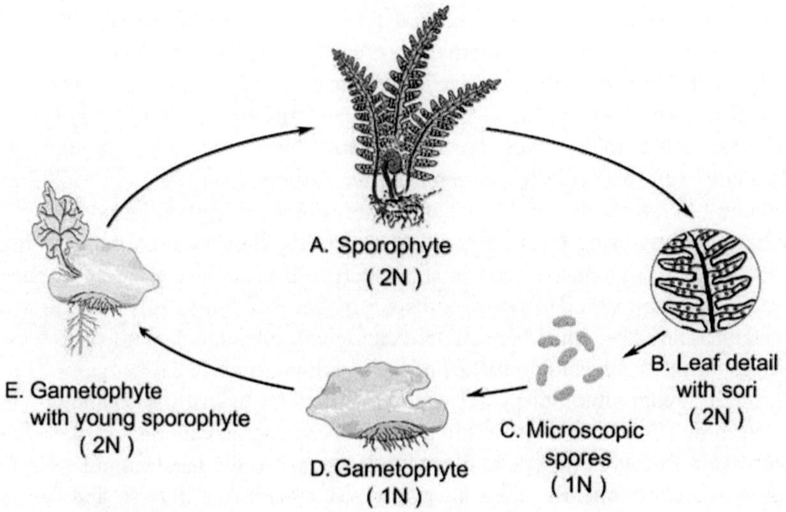

Fig. 1 Fern life cycle diagram

zygote matures and produces a young sporophyte containing roots and leaves while still attached to the prothallus that provides it some nourishment. The young sporophyte eventually becomes independent of the degenerating prothallus and becomes an autonomous member of a new sporophyte generation.

This reproductive strategy of two alternating generations provides some positive adaptive features to enhance dispersal, establishment, and growth of the fern species, but it also introduces some limitations. Among the positive adaptive features, the lightweight spores are easily dispersed by wind and water currents, thus potentially spreading a species beyond its current established habitat. The prothallus stage provides an initial hold on a new favorable habitat that may promote the establishment of the species, especially if the new habitat particularly supports the survival and growth of the prothallus. Moreover, in circumstances where there are sufficient densities of spores from different fern parents of the same species, there can be favorable circumstances for cross-fertilization among prothalli from different parents. This can promote greater genetic diversity that may confer enhanced survival value for the sporophyte that develops from the fertilized prothallus stage. However, the prothallus stage may also be less tolerant to some environmental stressors, and it is a critical link between the parent generation and that of the offspring. If the prothallus fails to adapt and survive, the next generation also fails. Thus, the alternation of generations in ferns introduces another layer of complexity in reproduction and natural selection that makes the study of ferns more challenging scientifically and also provides a greater opportunity to more fully appreciate the complexity of life and its adaptations to different environments, as encompassed in the study of fern physiological ecology.

The subsequent sections of this chapter address three broad topics based at the first level on the habitat of the fern: (i) aquatic and semi-aquatic ferns, (ii) terrestrial ferns, and (iii) epiphytic ferns. Some species of ferns are sufficiently adaptable that they can be found growing either terrestrially or epiphytically. Thus, where these species are addressed in the review (either terrestrial or epiphytic) depends on the locale where they were growing when studied. In addition, each of these three major topics is further subdivided into three subtopics focusing on appropriate ecophysiological variables: (i) development, growth, and life cycles, (ii) environmental adaptations, such as desiccation resilience and stress tolerance, and (iii) physiological aspects of light, photosynthesis, and respiration. These were chosen as potentially relevant to the foci of other chapters in this book on bioactive compounds produced by ferns. Broader areas of fern physiological ecology have been considered in other publications, including climate effects, threats to survival, and conservation of the species, among other aspects [e.g., 4, 7–10].

2 Aquatic and Semi-Aquatic Ferns

Two major groups of aquatic and semi-aquatic ferns are addressed: floating ferns (e.g., *Azolla* and *Salvinia*) and rooted ferns, growing in submerged water or in bogs or living at the margins of standing water (i.e., *Marsilea*, *Regnellidium*, *Pilularia*, and *Ceratopteris*).

2.1 Development, Growth, and Life Cycles, Including Environmental Aspects

The germination characteristics of spores and the sporeling success of *Azolla filiculoides*, an invasive floating fern in some wetlands, have been studied experimentally [11]. Batches of spores were subjected to different storage conditions, salinity concentrations, phosphorus availability, and drought effects. Cold pre-treatment at 4 °C for 3 months prior to germination did not stimulate spore germination compared to controls maintained at 20 °C. However, germination decreased under high salinities and reached values around 10% at salinities of 5000 and 8000 μS cm^{-1}. Phosphorus availability (0–185.8 mmol l^{-1}) had no significant correlation with major developmental events during germination, although the highest germination percentage was obtained with a concentration of 92.9 mmol l^{-1}. Spores that were subjected to almost a year of drought showed a significantly lower germination percentage when compared to spores germinated without drought effects.

Sadeghi et al. [12] published a review article addressing major ecological variables affecting the growth of *Azolla*, complementing an earlier review by Lumpkin and Plucknett [13] who emphasized some economic aspects. Among the findings reviewed by Sadeghi et al. were the following. Water availability is the key factor for growth. Optimal light intensity (15–18 Klux) and temperature (18°–28 °C) with relative humidity (55–83 %) are major influential variables. However, excessive wind and turbulent water can fragment and kill *Azolla*. The importance of both macro- (e.g., phosphorus, nitrogen, potassium, calcium, and magnesium) and micronutrients (e.g., molybdenum, cobalt, etc.) was also documented based on the literature. Various types of insects (e.g., caterpillars), bacteria, fungi, and viruses can adversely affect *Azolla* growth.

Azolla spp. are diazotrophs, harboring N_2-fixing cyanobacteria, and are strong phosphorus accumulators. Temmink et al. [14] examined the growth and N and P sequestration rates of *Azolla filiculoides* in N-free water at different P concentrations. The growth response appeared to be biphasic and highest at levels ≥10 μmol l^{-1} of phosphate. Diazotrophic nitrogen sequestration increased upon phosphorus addition, and rates were three times higher at high phosphorus than at low phosphorus concentrations. At concentrations of 10 μmol l^{-1} of phosphorus, nitrogen sequestration rates reached their maximum, and *A. filiculoides* growth became saturated. Due to luxury consumption, phosphorus sequestration rates increased until a concentration of 50 μmol l^{-1} was reached in the culture medium. At higher phosphorus concentrations (≥ 50 μmol l^{-1}), however, chlorosis occurred that appeared to be caused by limited iron and not by nitrogen deficiency.

A laboratory study of the effects of varying concentrations of NH_4^+ (0.25–15 mM) as an exclusive nitrogen source on the growth, morphology, and NH_4^+ uptake by *Salvinia natans* [15] showed that it grew well with relative growth rates of c. 0.25 g g^{-1} d^{-1} at external NH_4^+ concentrations up to 5 mM; but at higher levels, growth was suppressed and the plants had small leaves and short roots with stunted growth. High NH_4^+ levels also barely affected the concentrations of mineral cations and anions in the plant tissue. It was concluded that *S. natans* can be characterized as NH_4^+-tolerant, consistent with a number of other species of wetland plants because growth was unaffected at NH_4^+ concentrations as high as 5 mM. Moreover, symptoms of toxicity at higher concentrations were relatively mild.

Further studies on the effect of various forms of nitrate on the growth of *S. natans*, studied in laboratory aquatic cultures [16], found that among three forms of added nitrogen (N_2, NO_3^-, and NH_4^+), NO_3^- was the most productive producing a dry weight gain of 0.800 g, compared to an initial 0.220 g at the beginning of the experiment. The gain in weight for N_2 was 0.380 g and for NH_4^+ was 0.600 g. Additional similar evidence of the nutrient treatments on two other species (*Salvinia auriculata* and *Salvinia minima*) was also reported. Interpretations of the experiments using acid-base analyses and $\Delta^{13}C$ measurements were also addressed, emphasizing the importance of these methodologies as potential tools in the study of differences between habitats and nutrient acquisition, particularly N.

With increasing evidence of higher levels of atmospheric CO_2, some experimental evidence is emerging on the combined effects of atmospheric CO_2 and solar radiation on the growth of *Azolla* spp. In a greenhouse laboratory experiment [17], *Azolla ficuloides* grown at 1600 ppm CO_2 produced twice as much biomass as when grown at 400 ppm CO_2 in spring and autumn. However, high summer temperature and solar radiation seemed to significantly lower its maximum growth potential, presumably as a result of a metabolic shift as indicated by water loss. Nutrient availability became more important at higher atmospheric CO_2 concentrations. In addition, high temperature and solar radiation affected the nitrogen concentrations in *A. filiculoides*, most likely by inducing photorespiration and photoinhibition.

Further growth studies examined the effects of varying light intensity on the growth of the floating fern *Salvinia auriculata* in greenhouse cultures with natural light, under three treatments: control (0% shade), 30% shade, and 70% shade [18]. The number of new ramets (clonal growth) increased exponentially in the three treatments; however, this number was higher in shade than in full sun plots. The size of floating leaves, submerged leaves, and rhizomes was higher under shade treatments compared to individuals under sun treatment. These results showed that *S. auriculata* has a high growth performance under shade environments, including both clonal growth and the size of individual ramets, similar to most land ferns. This species appears highly plastic as it also thrives under the bright sun, which appears to be a consistent pattern among fern species in both terrestrial and aquatic environments.

A study of ecological factors affecting the growth of *Salvinia minima* in southern Florida over 39 months at four sites [19] reported that the most important factors included the abundance of other plant species, changes in water depth, water quality, and herbivory by insects. Abiotic factors like temperature, pH, dissolved oxygen, and conductivity varied among sites and, more importantly, over time. Principal component analysis identified four components which together explained 64% of the variability in *S. minima* biomass. The first component correlated strongly with herbivory as well as the abundance of the duckweed *Spirodela polyrrhiza*. Temperature effects were strongly represented in the second principal component. A stepwise regression model that best predicted *S. minima* biomass incorporated conductivity, insect herbivory, and interspecific plant abundance. Broader dry versus wet season influences were apparent and linked to temperature, water depth, and conductivity that covaried with *S. minima* biomass.

The ecology of rooted aquatic or semi-aquatic fern species encompasses additional aspects of complexity because the rhizomes and roots are anchored in the substratum where the chemical and physical properties of the rooting medium become an important factor. The effects of soil depth (5, 10, 15, 20, and 25 cm) on the growth of *Marsilea crenata* (commonly known as the clover fern because this genus has four leaves) were studied in experimental flooded planting pots [20]. Ferns grown at a soil depth of 15 cm possessed the longest internode length of 7.43 cm, while ferns at a soil depth of 20 cm possessed the longest stalk length of 16.16 cm, and those planted at a soil depth of 25 cm showed the widest leaf diameter at 4.00 cm. The ferns at a soil depth of 25 cm demonstrated the highest fresh and dry weights at 971.11 g and 146.11 g, respectively ($p < 0.05$).

The morphology, and concomitant physiological ecology of *Marsilea* spp., varies between plants that are submerged versus those that become exposed to the atmosphere, either due to the emergence of the leaves from above the water or when growing in open, wet locations at the periphery of water. Submerged leaves are fork-like with four leaves expanding parallel to the long axis of the petiole. When they are exposed to the atmosphere, they expand on a plane perpendicular to the petiole, resembling a four-leaf clover. Liu [21] studied the role of the plant hormone, abscisic acid (ABA), in the transformation of the leaf morphology from submerged to exposed when ABA was added. ABA at final concentrations ranging from 10^{-4} to 10^{-8} M, was added in a liquid medium containing submerged *M. quadrifolia* plantlets with two nodes. ABA is a plant hormone that mediates stress responses and dormancy, among other functions. Addition of the ABA to the liquid medium induced development of morphological characteristics distinctive for leaves that are emergent from water, or those of land forms of the plant, with four broad leaflets. Additional changes included an increase in surface area of the leaflets, differentiation of stomata and trichomes, elongation of petioles and roots, development of lateral roots, shortening of the rhizome, and reduction in the number of leaves and roots formed on each plant.

The effects of hydrology on spore germination and development of the aquatic fern *Pilularia minuta* (an aquatic fern with slender, grass-like leaves) were studied during a 5-year field study in Tunisia, including an experiment under controlled conditions [22]. The spatial distribution of *P. minuta* was recorded in the field annually, while the experiment tested the effects of water levels and flooding dates on germination (the emergence of new individuals) and development. The water level was found to be the major factor affecting the germination and the development of *P. minuta* with flooding date as a second limiting factor. For germination to occur, the sediment where the plant is growing must be completely waterlogged. Water depths of 5–10 cm were optimal for plant development. In addition, the plant required a minimal flooding period of 7 weeks to develop. Late-spring precipitation appeared to be more critical than earlier rains in promoting plant development.

Additional evidence of the importance of water depth in determining the floristic composition of plant communities containing *P. minuta* was found in field studies of locales in varied geographical regions of Sicily [23].

Laboratory studies of the environmental requirements for spore germination and growth of *Pilularia americana* in Texas (USA) [24] demonstrated that the spores germinated under a range of photo-period-temperature conditions, but spore germination was inhibited by calcareous soils and by certain other edaphic conditions, such as physical composition and pH. Germination was inhibited in soils ranging from sandy soils (pH 4.5) to calcareous soils (pH 7.35). The niche of *P. americana* in this geographic locale restricts it to semipermanent pools on acidic granitic soils and may explain a major role in the restricted distribution throughout its range.

Ceratopteris, a leafy aquatic to semi-aquatic fern, has been widely studied in laboratory ecophysiological studies. A laboratory study investigated the role of antheridiogen, a plant hormone that induces gender (male or female) in fern gametophytes of *Ceratopteris richardii* [25].

The gametophytes of *Ceratopteris richardii* (a homosporous fern) may develop as males with antheridia or hermaphrodites bearing both antheridia and archegonia. Hermaphrodites secrete antheridiogen that induces nearby, undifferentiated spores to develop as males. Given prior evidence that induction is not 100% in the presence of antheridiogen, it was hypothesized that resources may alter *C. richardii* gender status. An experiment was undertaken to examine possible factors determining gender. Among other aspects, spore size was not predictive of gender but was positively related to both male and hermaphrodite gametophyte size. Antheridiogen was found to slow the growth of male and hermaphrodite gametophytes of the wild type and male gametophytes of a *him1* mutant. These results supported the idea that gender may be determined indirectly through antheridiogen's effect on gametophyte growth.

The effects of nutrient enrichment on the growth of *Ceratopteris pteridoides* in laboratory culture [26] demonstrated that the addition of nutrients in natural sources of freshwater with adjusted electrical conductivity, pH (5.5–6.5), and major and minor inorganic ions produced enhanced growth compared to control medium without added nutrients (interpreted from graphical data published in the paper). The treatment plants had the greatest increase in fresh biomass at 35 days of growth (85 g) compared to controls in a non-enriched growth medium (10–24 g). Likewise, dry leaf biomass in the treatment group (65 g) was greater than in controls (8–13 g). The mean root length in the treatment group (17 cm) was longer than in the controls (7–11 cm).

Additional studies of *Marsilea* spp. include the following: (i) effects of flooding depth and canopy cover on the growth of *Marsilea villosa* in Hawaii [27], (ii) experimental evidence of elevated temperature tolerance of five species in India [28], and (iii) ecophysiological morphology and functional traits of three *Marsilea* species, including water use efficiency and leaf dissection, distributed in different geographical climate regimes (temperate to tropical) [29].

For further information on *Ceratopteris* spp., several sources focus on environmental factors. The role of spore germination and gametophyte development in habitat selection was found to be related to temperature responses in some temperate and tropical ferns, including comparisons among three fern species [30]. Effects of the nutrient stoichiometry of carbon, nitrogen, and phosphorus were studied on sex determination of gametophytes and C:N in dry mass at ambient CO_2 [31]. Furthermore, distribution of *Ceratopteris* spp. in China was reported across multiple habitats as related to water chemistry, particularly pH and conductivity [32].

2.2 Light, Photosynthesis, and Respiration

The most relevant published studies related to aspects of environmental influences on photosynthesis addressed physiological variables in species of *Marsilea* and *Ceratopteris*. The photosynthetic response of *Marsilea crenata* leaves to increasing light intensity and their respiration rate were studied and compared to terrestrial ferns in different habitats in laboratory studies of illuminated leaves [33]. Apparent photosynthesis of *M. crenata* increased with increasing light intensity, and light saturation was reached at 4×10^4 erg cm^{-2} s^{-1}. P_{max} was 5.0 mg of CO_2 $hour^{-1}$ g^{-1} fresh weight, comparable to some terrestrial ferns. Dark respiration was in the range of 0.5–0.7 mg of CO_2 $hour^{-1}$ g^{-1} fresh weight. Light compensation points were in the range of 1 to 4×10^3 erg cm^{-2} s^{-1}.

Evidence was presented for an active stomatal control of CO_2 exchange in *Marsilea crenata* leaves as related to CO_2 concentration and the role of the plant hormone ABA [34]. Stomatal conductance (g_s) of *M. crenata* significantly decreased, while photo-saturated photosynthetic rate (A_{max}) increased as [CO_2] increased from 0 to 600 ppm, resulting in increments in water use efficiency (WUE). The reduction in g_s when [CO_2] was elevated from 400 to 800 ppm was more in leaves of full light-grown than those of shade-grown plants; however, the increment in A_{max} was similar. The leaf stomata of *M. crenata* gradually closed after 30 minutes of application of exogenous ABA, resulting in a 52.1% reduction in g_s and a 40% in A_{max}, hence a 25% increase of WUE.

Marsilea species exhibit some of the highest photosynthetic rates and stomatal conductances recorded in non-angiosperms [35, 36]. An interesting study examined the stomatal and photosynthetic dynamics following changes in irradiance of the leaves in *Marsilea drummondii* as compared to some other terrestrial ferns, gymnosperms, and angiosperms [36]. In general, species with the fast stomatal opening (such as ferns) forgo less photosynthesis during photosynthetic induction. By contrast, there was no relationship between stomatal closure speed and the water lost by transiently more-open stomata, because species with higher rates of gas exchange also showed faster stomatal closure. Shade-adapted species of the aquatic and terrestrial ferns, among the vascular plants studied, possessed fast-opening but slow-closing stomata, consistent with ecological adaptation to maximize lightfleck use.

3 Terrestrial Ferns

Terrestrial ferns vary remarkably in size. For example, one of the smallest is the rare, eusporangiate fern *Ophioglossum malviae* (1–1.2 cm in height) discovered in the forests of Western Ghats, India. The tallest fern in the world is the tree fern *Cyathea australis*, which grows to a height of 20 m and has fronds up to 3 m long. The height of most terrestrial ferns, however, is typically in the range of 1 m or less. However, the large golden leather fern (*Acrostichum aureum*), an understory fern found in wetlands or mangrove forests, reaches a height of 1.2 to 1.8 m, and the crown of spreading leaves can be as broad as the plant is tall.

As may be expected, climate and particularly the geographic locale can be a major factor in plant size. Ferns adapted to desert or dryland locales are typically much smaller than species found in more mesic environments, such as temperate forests or tropical rain forests. As explained above, most species have a life cycle alternating between gametophyte and sporophyte. However, one species (*Vittaria appalachiana*) exists only as a gametophyte stage that propagates asexually by dispersal of gemmae. As the species name implies, it is found in the Appalachian mountain system of the USA.

3.1 Development, Growth, and Life Cycles

Spores of ferns vary in size, morphology, surface texture, and the presence or absence of chlorophyll [37]. Green spores containing chlorophyll typically germinate more rapidly but are less viable over time (mean, 48 days) compared to non-green spores (mean, 1045 days). Non-green spores typically require longer to germinate, and in some cases, the gametophyte develops less rapidly [38]. Based on current evidence, most fern species have non-green spores; but among those that have chlorophyll-bearing spores, many are members of the families Osmundaceae, Grammitidacease, and Hymenophyllaceae. All species of *Christiopteris*, *Marginariopsis*, *Matteuccia*, *Onoclea*, and *Onocleopsis* have chlorophyll-bearing spores. Generally, all fern spores produce some chlorophyll relatively early during spore germination. Germinating spores undergo cell division and produce a filamentous protonema containing chloroplasts and a colorless rhizoid for anchoring to the substratum. During maturation, the green protonema filament continues to divide and elongate, eventually producing the gametophyte stage. In those species with a flattened, thallus-shaped gametophyte, the maturing protonema undergoes repeated cell division largely in two dimensions (one plane), ultimately producing a leaf-like morphology characteristic of the fern species' gametophyte (Fig. 1).

Particular attention has been given to spore germination in ferns because it is the first stage of establishment of the next generation in the fern life cycle; and success of spore germination is necessary, but not sufficient, for the establishment of the next generation.

A useful general review of basic aspects was published by Raghavan [39] and more recently by Suo et al. [40]. Briefly, among other factors, light, gravity, phytohormones, metal ions, and temperature are important environmental variables. Most fern spores germinate at 25 °C, and an optimal CO_2 concentration is necessary for spore germination of certain fern plants. Light is required to initiate spore germination in many species, but the quality of the light is also important. The characteristics of the illumination required for germination among different species vary markedly; for example, (i) red light-stimulated and far-red light-inhibited spore germination; (ii) far-red light, uninhibited spore germination; (iii) blue light-inhibited spore germination; and (iv) some spores germinate in the dark. The temporal pattern of illumination, whether constant or fluctuating, can also affect spore germination, as exemplified in some species of tropical ferns [41]. The optimal light intensity and illumination time for spore germination are different among various fern species. Light response, which promotes spore germination, is initiated beginning with spore-cell mitosis. This initial stage is regulated by phytochromes (PHYs) and cryptochromes (CRYs). The following forms of phytochromes (AcPHY2, AcCRY3, and/or AcCRY4) are hypothesized to be involved in spore germination. In some cases, the presence of bioactive substances, such as antheridiogens in the environment can also induce spore germination in the dark, suggesting that the presence of antheridiogen-secretion by growth stages of other ferns in soils may induce spore germination, even if the spore is beneath the soil surface [42]. Additional evidence indicates that growth regulators, such as gibberellins can induce spore germination in darkness in some species [43].

Further evidence indicates possible interactive effects between abscisic acid (ABA) and gibberellic acid on the germination of spores and growth of protonemata in the fern *Mohria caffrorum* [44]. When grown on agar, ABA did not affect the initial divisions of the spore protoplast leading to the formation of rhizoid and protonema but inhibited the subsequent elongation of the protonemata. Applied GA promotes the overall growth of the protonemata, probably by direct stimulation of its growth, rather than due to its interaction with ABA. During its role as a growth promoter of protonemata, GA is accepted by a receptor called GID1 (GA-insensitive dwarf 1); when GID1 binds to a GA molecule, DELLA protein, an inhibitor of gibberellin response, is degraded, promoting GA biological activity [45].

Substantial attention has been given to fern gametophyte development, growth, and sexual maturation [46–48]. A homosporous fern spore can give rise potentially to (i) a male gametophyte (producing antheridia), (ii) a female gametophyte (producing archegonia), or (iii) a hermaphrodite gametophyte (producing both antheridia and archegonia). Typically, gametophytes of homosporous ferns differentiate antheridia earlier than archegonia. One of the signal research achievements was the elucidation of the role of an extractable biologically active compound (now known to be antheridiogens) in the induction of gametophyte antheridia [49, 50]. The addition of antheridiogen to immature prothalli can induce early differentiation of male gametophytes with antheridia. Antheridiogen, capable of inducing male development, is secreted by hermaphrodites once they become insensitive to its male-inducing effect. Thus, the hermaphrodite can promote the differentiation of other nearby prothalli to produce antheridia. When prothalli from different parents are in sufficient proximity, the release of antheridiogens from a nearby mature hermaphrodite can induce antheridium production in the neighboring prothalli, thus potentially producing sperm that could fertilize the egg of the hermaphrodite and contribute to cross-fertilization. This is a likely adaptation supporting gene mixing during fertilization, including possible hybridization, thus contributing to greater genetic diversity and adaptability. Some antheridiogens show considerable interspecific activity. However, others are more species-specific. One of the antheridiogens with a rather broad species activity is produced by *Pteridium aquilinum*; interestingly, this is the species studied by Döpp [49] in his seminal observation of antheridium induction in male gametophytes.

The role of antheridiogens in the fern life cycle, particularly its role in sex determination pathways, has been reviewed by Atallah and Banks [51]. Antheridiogen serves multiple functions in male gametophyte development. It represses divisions of cells in the prothallus that establish the lateral meristem; it promotes the rapid differentiation of antheridia, and it represses its biosynthesis. There is increasing evidence that some gibberellins, known particularly as growth regulatory compounds, have a biological activity to induce responses similar to antheridiogens; and there is increasing evidence that antheridiogens may be synthesized biologically from gibberellins in some fern species [52, 53]. Further detailed information on in vitro phytohormonal regulation of fern gametophytes, including the roles of cytokinins, auxins, gibberellins, ethylene, jasmonic acid, and brassinosteroids, is presented by Romanenko et al. [54].

Auxin, indole-3- acetic acid is a phytohormone promoting growth, among other responses, and interacts with other phytohormones (e.g., GA and ABA) in regulating the growth and sexual development of fern protonema.

Ohishi et al. [55] reported a detailed morphological study investigating the roles of multiple phytohormones in antheridium formation, protonema elongation, and prothallus formation in the climbing fern *Lygodium japonicum*. GA_4 methyl ester stimulated both protonema elongation and antheridium formation. However, these effects were inhibited by simultaneous application of abscisic acid (ABA). On the other hand, IAA (indole-3-acetic acid) promoted protonema elongation but reduced antheridium formation. These effects were partially recovered by transferring to an IAA-free medium. Their evidence suggests that the crosstalk between these plant hormones might regulate protonema elongation and antheridium formation in *L. japonicum*. Furthermore, their results indicate a possible involvement of auxin in the prothalli development of *L. japonicum*.

In addition to antheridium induction, antheridiogens have been implicated in gametophytic spatial competition. However, some evidence calls this into question and suggests that other bioactive chemical compounds are involved in intra- and inter-gametophytic competition as elucidated in laboratory studies [56]. The effects of intraspecific and interspecific chemical competition were studied for spore germination, sexual development, and gametophyte size in two closely related and co-occurring species of *Polypodium* from eastern North America (*Polypodium appalachianum* and *Polypodium virginianum*). Evidence was collected in a series of treatments representing different levels of competition within and between the two species. Mature gametophytes of both species reciprocally inhibited spore germination, sexual development, and growth of neighboring plants. These effects were attributed to chemical interactions and not competition for resources. Competitive effects on spore germination and sexual development were similar for within- and between-species interactions, but repression of growth and gametophyte size was greater when competing gametophytes were heterospecific. An additional mechanism of intergametophytic competition via proliferous vegetative growth was also reported. Overall, the evidence indicated that gametophytes of these species interact through a novel context-dependent chemical system that is independent from sex expression-regulating pheromones (e.g., antheridiogens).

Haufler et al. [57] published an interesting discussion of the role of gametophyte sexuality in fern life cycles and the status of heterozygosity among fern species. They incorporated a review of research on a variety of mechanisms that actively promote heterozygosity in ferns and lycophytes, including genetic load, asynchronous production of eggs and sperm, and pheromonal control of gamete production. Overall, the authors provide a contemporary understanding of fern reproduction, revealing ferns as evolutionarily dynamic and exploiting a wide range of mating systems.

In addition to community-level analyses, laboratory studies on spore germination, gametophyte development, and sporophyte growth have provided some insights into the temporal phases of each stage of growth within different fern species. In a study of *Adiantum capillus-veneris* [58], spores cultured on agar plates under cool white fluorescent illumination began germination in 6 days, and uniseriate, elongated filamentous rhizoids also appeared. Green, prothallial protonema emerged perpendicular to the rhizoid. The heart-shaped gametophyte developed between 27 and 60 days, with 12–16 rhizoids, and it contained mature archegonia and antheridia. Young sporophytes appeared 90 days after sowing. After 4 weeks of growth, the sporophyte height measured 0.70 cm; with further maturation, it increased in height and measured 1.90, 2.30, 2.50, and 2.60 cm after 6, 8, 10, and 12 weeks of growth, respectively.

The zygote of gametophytes and the subsequent developing sporophyte are genetically and structurally separated from the gametophyte: firstly as genetic products of fertilization and secondly due to their enclosure within the venter of the archegonium. However, there is evidence that the gametophyte provides photosynthetic nourishment to the young sporophyte through the transfer of organic compounds during its early stages of development before it reaches the one-leaf stage. The parent gametophyte provides sustenance until the sporophyte develops sufficient root system and size to grow autonomously [59].

Fern species vary substantially in life cycle duration (from spore germination to mature spore-bearing sporophyte), and there is a need for more detailed documentation of life cycle duration in ferns, both in the natural environment and laboratory studies. Klekowski [60] documented the duration of life cycle development in 10 leptosporangiate species in greenhouse culture, with detailed data on the number of days to the appearance of the sporophyte on the gametophyte and the total days from spore sowing to a viable, mature spore-producing sporophyte. The shortest duration for sporophyte appearance on the gametophyte was *Thelypteris normalis*

(136 days, and 206 days to full maturity), and the longest was for *Osmunda regalis* (498 days, and 563 days to full maturity). Species with intermediate duration for the appearance of the sporophyte on the gametophyte (in days) were *Adiantum capillus-veneris* (184), *Thelypteris puberula* (249), *Adiantum seemannii* (274), and *Asplenium adiantum-nigrum* (280).

3.2 Environmental Adaptations

The diverse habitats of terrestrial ferns, spanning deserts, and drylands to tropical rain forests and montane cloud forests have led to some remarkably diversified morphological and physiological traits to maximize survival. One of the more significant environmental factors is available moisture. Consequently, there has been substantial interest in studies on stress resilience to desiccation [61]. Some examples of physiological research studies are reviewed encompassing various taxonomic groups and different habitats varying in available moisture. Particular attention is given to studies that involve phytohormones or other regulatory, bioactive, molecules.

Fern gametophytes are typically small (mm range), relatively fragile, and one-cell thick, and they vary in desiccation tolerance, i.e., varying among taxa, environmental conditions, and the rate at which they are dehydrated [62, 63]. In a comparative study of desiccation tolerance of gametophytes and sporophytes among species of filmy ferns (Hymenophyllaceae), gametophytes had similar or less desiccation tolerance compared to sporophytes, depending on their physiology and niche requirements [64]. Terrestrial species were less tolerant of desiccation than epiphytic species in both life stages. Despite their lack of greater physiological tolerances, gametophytes of several species occurred over a wider elevational range than conspecific sporophytes.

Research attention has been given particularly to terrestrial sporophytes. Among desiccation resistant taxa, the sporophytes of "resurrection ferns" are of interest, because they are capable of complete desiccation, becoming dormant, but with the potential to rehydrate and resume active growth, within a reasonable duration of dormancy. The rock-dwelling, desiccation-tolerant fern, *Polypodium virginianum*, has been studied to document the biochemical and physiological response mechanisms to desiccation and rehydration. In laboratory studies [65], detached fronds of *P. virginianum* were examined for desiccation tolerance. They lost ca. 65–70% of water content during slow drying under laboratory conditions. Rehydration of the slowly dried fronds to 100% water content occurred between 12 and 24 h when placed on water-saturated filter paper. After 6 h of rehydration, when the water content was 78% relative to undesiccated controls, protein synthesis resumed rapidly and approached the level of synthesis exhibited by non-desiccated fronds. These proteins were expressed also when the fully hydrated fronds were treated with exogenous abscisic acid (ABA), suggesting that ABA may mediate some of the desiccation-tolerant biochemical responses. During rehydration from the air-dried state, unique rehydration-specific polypeptides were produced. They were transiently produced, perhaps indicating they were only necessary during the early stages of rehydration.

Anemia caffrorum produces fronds that are either desiccation tolerant (DT) in dry seasons or desiccation sensitive (DS) in wet seasons. To examine the physiological responses of DT and DS states, plants with either DT or DS fronds were acclimated in growth chambers [66]. DT and DS fronds were compared under well-watered conditions, and the results showed that the former (DT) presented higher leaf mass per area, minimum stomatal conductance, tissue elasticity, and lower CO_2 assimilation. Water deficit resulted in a similar induction of abscisic acid concentration in both frond types, but DT fronds maintained higher stomatal conductance and upregulated lipophilic antioxidants more prominently. Given the episodic water availability in this species' environment, the seasonal alternation in the production of DT and DS fronds in *A. caffrorum* represents a mechanism by which carbon gain can be maximized during the rainy season, but greater investment in protective mechanisms occurs during the hot dry season, enabling the exploitation of the varying water availability.

A study of the mechanism of desiccation tolerance in the DT "resurrection fern" as it is commonly known, *Adiantum raddianum*, was based on laboratory experiments with detached fronds that were dehydrated and rehydrated under in vitro conditions [67]. During the desiccation process, the relative water content (RWC) of detached fronds decreased to as low as 16% after 4 h with intense inward curving. Upon rehydration, within 4 h, the RWC of the desiccated fronds regained 85% of the initial water content, and the rehydrated fronds almost showed the original morphology. Furthermore, physiological activities of antioxidant enzymes (e.g., superoxide dismutase, peroxidase, catalase, glutathione reductase, lipid peroxidation) and the amino acid proline increased during desiccation, but sucrose and starch content showed differential responses. In this study, starch content decreased and sucrose content increased in desiccated fronds of *A. raddianum*. Therefore, it is evident that this principal carbohydrate (starch) is broken down during desiccation in this fern.

Additional evidence of dehydration-related biochemical changes has been reported for the resurrection fern *Asplenium ceterach* [68]. *A. ceterach* (a high desiccation-tolerant fern) was analyzed during 24 h desiccation at four different water potentials. The treatments resulted in moderate water loss (partial desiccation) in the fern tissue. Biochemical analyses indicated, for the first time, that there are peroxidase (POD) and polyphenol oxidase (PPO) isoforms in this species. Two PPO isoforms were identified (one anionic with a pI of 6.3 and one cationic with a pI of about 9.0). During short-term dehydration, a remarkable increase in POD and PPO activity was detected. The increase in these enzymes and substrates during dehydration may enhance the overall plant resistance to the stress conditions.

One of the adaptive responses of ferns, and other leaf-bearing plants, to increasing desiccation stress, is control of stomatal opening and hence regulation of water loss by reducing transpiration through the stomatal pores. There has been considerable debate whether ferns exhibit active control of stomatal closure through hormonal control, versus depending on closure due to decreasing osmotic pressure in the guard cells during desiccation stress. Hõrak et al. [69] examined the response of three temperate fern species (*A. filix-femina*, *Dryopteris carthusiana*, and *Dryopteris filix-mas*) to explore this issue. Two of these (*A. filix-femina* and *D. filix-mas*) responded to ABA and CO_2. As a result, the authors concluded that an active mechanism of stomatal regulation in response to reduced air humidity is present in some ferns.

Importantly, their evidence further indicated fern stomatal responses depend on growth conditions. The data suggest that the stomatal behavior of ferns is more complex than anticipated previously, and active stomatal regulation is present in some ferns and has possibly been lost in others. Thus, overall, the guard cell ABA response appears to be present in some fern species but may have been lost in some other fern lineages. The results of this study offer new incentives to examine the generality of these findings, and this topic remains an active area of discussion and research [70–73].

Recent research further confirms the absence of ABA-induced stomatal response in some fern species and implicates the role of secondary metabolites to explain differences in stomatal activity between angiosperms and ferns [74]. Metabolomics analysis highlighted that ferns have a higher accumulation of secondary rather than primary metabolites throughout the diel course of the experiment, with the opposite being observed in angiosperms.

In addition to controlling water loss by transpiration, stomata also regulate gas exchange (carbon dioxide and oxygen) between the leaf and the environment. Consequently, stomatal action also affects photosynthesis. This topic is addressed more fully in the next section.

3.3 Light, Photosynthesis, and Respiration

Current evidence indicates that in addition to differences in stomatal action, fern photosynthesis rates are generally lower than those of angiosperms. To further document the status of fern photosynthesis, a gas exchange/chlorophyll fluorescence analysis was performed in three fern species of high phylogenetic range within Polypodiopsida (*Osmunda regalis*, *Blechnum gibbum*, and *Nephrolepis exaltata*) [75]. Measurements were made on species' maximum net photosynthesis (AN), stomatal (g_s) and mesophyll (g_m) conductances to CO_2, and maximum velocity of carboxylation ($V_{c,max}$). Values of AN, g_s, g_m, and $V_{c,max}$ were within the lowest range of those observed in seed plants. However, in vitro Rubisco specificity factor ($S_{C/O}$) was also determined, and three species had values for $S_{C/O}$ similar to those typical of seed plants. In addition, g_s was unresponsive to light and CO_2, as already described in other fern species. On the contrary, g_m varied with changes in CO_2.

In contrast to some published reports that ferns have very low maximal rates of net CO_2 uptake (<2 μmol m^{-2} s^{-1}), further evidence (based on an analysis of 14 species) suggested a higher rate of 7 μmol m^{-2} s^{-1} at 25 °C [76]. Net CO_2 uptake reached 90% of saturation at an average PAR (400 to 700 nm) of only 240 μmol m^{-2} s^{-1}. This is consistent with the typically shaded habitats of most ferns. Maximal CO_2 uptake rates were positively correlated with several variables, i.e., PAR that produced 90% saturation ($r^2 = 0.59$), chlorophyll per unit leaf area ($r^2 = 0.30$), and water vapor conductance ($r^2 = 0.69$).

Additional evidence has been reported suggesting that the regulation of mesophyll conductance (g_m) and photosynthesis may differ between ferns and angiosperms [77]. Using concurrent measurements of leaf gas exchange and carbon isotope discrimination, stomatal conductance (g_s) and mesophyll conductance (g_m) were examined in response to atmospheric CO_2 concentration [CO_2] using two fern species: *Pteridium aquilinum* and *Thelypteris dentata*. In these experiments, the CO_2 concentration was experimentally varied widely. Concurrently, analyses by tunable diode laser absorption spectroscopy (TDLAS) revealed a rapid and continuous response in g_m within a few minutes. In most cases, both ferns showed rapid and significant responses of g_m to changes in [CO_2]. For example, when [CO_2] was increased from 200 to 700 μmol mol^{-1}, g_s did not change in either species, but g_m of *P. aquilinum* decreased by 1.1 mmol m^{-2} s^{-1} and increased in *T. dentata* to 4.9 mmol m^{-2} s^{-1}. There were marked contrasts to responses as reported in angiosperms. Photosynthetic parameters were similar in both fern species. Photosynthesis assimilation (A_{sat}) (μmol CO_2 m^{-2} s^{-1}) was 7.54 ± 0.93 in *P. aquilinum* and 6.24 ± 0.90 in *T. dentata*. Dark respiration (μmol m^{-2} s^{-1}) was 0.35 ± 0.06 and 0.46 ± 0.10 in *P. aquilinum* and *T. dentata*, respectively.

In general, ferns and fern allies have low photosynthetic rates compared to seed plants. A comprehensive assessment of the foliar gas exchange and mesophyll structural traits involved in photosynthetic function was made for 35 species of ferns and fern allies to elucidate possible ecophysiological explanations for these differences [78]. Low mesophyll conductance to CO_2 was the main cause for low photosynthesis in ferns and fern allies, which, in turn, was associated with thick cell walls and reduced chloroplast distribution towards intercellular mesophyll air spaces.

To further examine the role of light intensity effects on fern photosynthesis, ecophysiological traits of *Blechnum chilense* growing in a wide range of habitats, ranging from shaded forest understories to abandoned clearings and large, sunlit gaps, were studied in Chilean temperate rain forests [79]. The purpose of the study was to address if contrasting light environments can exert differential selection on ecophysiological traits of *B. chilense*. The authors measured phenotypic selection on functional traits related to carbon gain, namely, photosynthetic capacity (A_{max}), dark respiration rate (Rd), water use efficiency (WUE), leaf size, and leaf thickness in populations growing in gaps compared to those in understory environments. In gaps, survival of *B. chilense* was positively correlated with WUE and negatively correlated with leaf size. In contrast, survival in shaded understories was positively correlated with leaf size. By comparison, A_{max} (μmol CO_2 m^{-2} s^{-1}) of ferns in the gaps was 6.75 ± 0.33 and in the understory was 3.18 ± 0.16. Dark respiration (μmol CO_2 m^{-2} s^{-1}) in gaps was 0.94 ± 0.04 and in the understory was 0.34 ± 0.03. In understories, ferns of lower Rd. and greater leaf size showed greater fecundity. WUE was greater and leaves were smaller in gaps. Thus, whereas control of water loss was optimized in sunlit gaps, light capture and net carbon balance were optimized in shaded understories.

Variations in solar radiation within microenvironments (varying in dimensions of meters or less) can influence fern growth and photosynthesis. *Asplenium platyneuron*, a species frequently inhabiting rocky locations, was studied in an urban environment where its location within a brick and masonry wall produced variations in sunlight from direct incidence on the front face to moderate shade near the back, due to the geometry of the wall [80]. The intensity of illumination at the location near the back of the wall was ca. 40% of the intensity at the outer location, where the photosynthetic active radiation (PAR) at mid-day was ca. 1484 μmol m^{-2} s^{-1}. The diel temperature of the substratum was comparable between the two locations on the wall. Mean leaf length (cm) was longer for leaves of plants at the shaded inner location (30 ± 1.1) compared to those of plants at the more illuminated outer location (9 ± 1.2), but stomatal densities were similar. The maximum photosynthetic rate (Amax) for the outside sample of ferns (6.3 ± 0.25) was larger, and statistically significantly different, compared to the A_{max} for the inside sample (4.3 ± 0.72). The mean respiration in the light was significantly larger for the ferns growing outside (0.52 ± 0.03) compared to those at the inside (0.36 ± 0.04), indicating greater respiratory CO_2 loss outside. These results are similar to those reported above [79] for ferns growing in sunlit gaps compared to shaded understories and are comparable to other studies of fern photosynthesis in high- and low-light, natural environments [81–83].

The photosynthetic response of four fern species varying in habitats from relatively mesic, low light to drought-prone more high light was studied experimentally by withholding irrigation (to induce desiccation stress) in order to investigate the fern photosynthetic responses when compared to control plants without desiccation [84]. The four species and their habitats are as follows: *Adiantum pedatum* (terrestrial deciduous, shady moist environments), *Pteridium aquilinum* (terrestrial, sunny habitat), *Dryopteris erythrosora* (terrestrial evergreen species in forest borders and understories), and *Lepisorus thunbergianus* (epiphytic, evergreen species subject to periods of desiccation). In the last week of the drought treatment, gs for the drought-treated plants declined to 21–50% of the values of the control plants, with stomatal conductance (g_s) values (mmol H_2O m^{-2} s^{-1}) for the treated species as follows: *A. pedatum* (13.0 ± 1.5), *P. aquilinum* (48.6 ± 5.9), *D. erythrosora* (13.4 ± 1.3), and *L. thunbergianus* (12.1 ± 3.6). When photosynthetic parameters were assessed between species in the gs range < 30%, where drought stress was the highest, photosynthesis maximum (P_{max}) was higher in *P. aquilinum* (3.48 ± 0.37) than that in *D. erythrosora* (1.70 ± 0.2) and *A. pedatum* (1.18 ± 0.22). P_{max} for the epiphyte (*L. thunbergianus*) was intermediate (2.34 ± 0.61). *Lepisorus thunbergianus* had low stomatal density and showed high water-use efficiency (WUE) retaining photosynthetic activity with low relative frond water content under drought stress, which suggested their high adaptation to drought. On the other hand, low WUE with the low light-saturated photosynthetic rate in *A. pedatum* (terrestrial, shady environment) was associated with much lower photosynthesis than in the other species under drought stress, suggesting lower adaptation to drought-prone habitats.

4 Epiphytic Ferns

Current evidence indicates that epiphytic ferns evolved, largely in the tertiary, from terrestrial ferns through adaptation to the particular habitat requirements occasioned by growth on host tree trunks and limbs in forest canopies [85]. Epiphytic ferns are especially notable in tropical and semi-tropical environments, but they also occur in moderate temperate environments. Due to the varying distribution of rainfall in some climatic locations (or even in locales with moderate rainfall where the epiphyte species may be exposed to high solar insolation), epiphytic ferns are subjected to desiccation stress. Consequently, many species have adapted to periodic drying by exhibiting a variety of ways to tolerate or avoid desiccation, including survival of the gametophytic and sporophytic life stages.

4.1 Development, Growth, and Life Cycles

Homosporous epiphytic ferns (with the exception of some species that are gametophyte only) have typical alternating generations of sporophyte and gametophyte life stages (Fig. 1), although some have filamentous gametophytes. A general review of fern species that are gametophyte only has been published by Pinson et al. [86]. Some illustrative studies of the ecophysiology of epiphytic fern gametophytes are presented prior to a more detailed review of the physiological ecology of the sporophyte phase. Appreciably more research has been done on the development and ecophysiology of the gametophyte stage of epiphytic ferns compared to the early stages of spore germination and emergence of the protonema.

Spore germination and early gametophyte development have been documented in *Platycerium bifurcatum* [87]. In that investigation, the process of spore germination and early gametophyte development of *P. bifurcatum* was documented in detail when cultured on agar or in liquid media in laboratory preparations. The percentage of germination at 3 days and 6 days after sowing was reported. After 3 days of culture, among eight different media used, the percentage of germination varied from 47.3% to 68.2%. At 6 days, the percentage of germination varied from 59.8% to 87.2%, corresponding to the same two media as for the 3-day measurements. In general, emergent rhizoids were longer in the liquid medium than on agar; and for the latter, decreasing the amount of agar in the preparation increased rhizoid length. Likewise, on liquid media, gametophytes had a significantly higher number of cells than on solid media. The presence of sucrose had no effect on cell numbers growing on the surface of solid media, while in liquid media with sucrose, significantly higher cell numbers were obtained after 8 days.

Subsequent to the emergence of the rhizoid from the exine of the spore, the development of the protonema was documented at regular intervals as follows. Day 6: The larger cell (arising from the first asymmetric division of the spore) is divided asymmetrically by a wall parallel to the polar axis of the spore, giving rise to a protonemal cell. Day 8: After further cell divisions, the protonemal filament was three- to four-celled in length. Day 10: Longitudinal cell divisions were commonly observed in the protonemal filament and were initiated more often in the subterminal than in the terminal cell. After 10 days in culture, as many as 6.2% of germinated spores produced two protonemata. Further details of spore germination in other genera including *Drymoglossum*, *Platycerium*, and *Pyrrosia* (Polypodiaceae) have also been published [88].

In additional studies, the effects of the plant hormone jasmonic acid (JA) on gametophyte development in *Platycerium bifurcatum* were investigated [89, 90]. JA had no influence on spore germination and primary rhizoid initiation, but promoted early gametophyte development, including initial protoplast cell divisions, the number of rhizoids produced, and transition from filamentous to a planar growth. Except for the experiment in which the effect of JA on germination was tested, JA concentrations exceeding 1 μM did not promote cell elongation or cell division but were instead inhibitory. Moreover, the effects of JA on gametophyte growth were age-dependent. After 20 to 30 days, JA had no pronounced effect relative to a control, while after 40 days at concentrations exceeding 0.1 μM, growth was inhibited. Analysis of soluble proteins from JA-treated gametophytes at 2 months gave evidence of some alterations in polypeptide patterns compared to the control. Most marked was an increase in a 92–93 kDa polypeptide band when 10 and 100 μM JA was included in the medium.

The induction of apogamous production of sporophytes (without sexual reproduction) in *Platycerium coronarium* gametophytes by application of IAA was studied in laboratory culture [91]. The highest percentage of apogamy, as well as the total number of apogamous sporophytes per clump of gametophytes, was induced by the presence of 40 μM IAA. When ethylene (acting as a phytohormone) was allowed to accumulate in the culture vessel in the presence of an optimum level of IAA, the percentage and the total number of apogamous sporophytes that were produced decreased significantly.

4.2 Environmental Adaptations

The gametophyte stage of the epiphytic fern life cycle is particularly prone to desiccation and temperature stress, especially in relation to its position within tree branches in the canopy. Individuals situated in the outer margins of the canopy, or within positions where the canopy has opened, are likely to experience increased insolation and desiccation stresses. In the succeeding text, aspects of environmental adaptations of epiphytic fern gametophytes are presented first followed by environmental adaptations of the sporophyte phase of the life cycle.

In a broadly designed research study, the effects of desiccation stress on gametophytes and sporophytes of 19 species (22 taxa including varieties) of filmy ferns (Hymenophyllaceae) on the island of Moorea (French Polynesia) were studied [64]. Chlorophyll fluorescence was used to measure desiccation tolerance and light responses. Gametophytes had similar or less desiccation tolerance (ability to recover from 2 days of desiccation at −86 MPa) and lower photosynthetic optima (maximum electron transport rate of photosystem II and light level at 95% of that rate) than sporophytes. Epiphytes were more tolerant of desiccation than terrestrial species in both life stages. Despite their lack of greater physiological tolerances, gametophytes of several species occurred over a wider elevational range than conspecific sporophytes. Overall, these results demonstrate that filmy fern gametophytes and sporophytes differ in their physiology and niche requirements and point to the importance of microhabitats in shaping the evolution of water-use strategies in vascular plants.

Given increasing evidence for the role of ABA in desiccation tolerance and recovery during rehydration in vascular plants, some attention has been given to the role of ABA in epiphytic fern sporophyte responses to desiccation. Physiological and ultrastructure characteristics of the roles of ABA and translational regulation in the desiccation-tolerant fern *Hymenophyllum caudiculatum* were analyzed during a desiccation and rehydration cycle [92]. Cell structure was conserved during slow desiccation and rehydration, largely attributed to the constitutive properties of the cell wall, allowing invagination and folding of the membranes and an important change in chloroplast size. The use of a translational inhibitor of mRNA synthesis (cycloheximide) impeded recovery of photochemical efficiency during the first 80 min. of rehydration, but the transcriptional inhibitor (actinomycin D) had no effect. Exogenous ABA delayed photochemical inactivation, and endogenous ABA levels decreased during desiccation and rehydration.

A similar study examined the morphological and physiological responses of the epiphytic fern *Elaphoglossum luridum* to water stress in laboratory studies with three treatments (control, water deficit, and ABA) [93]. No changes in relative water content were observed among treatments. However, significant decreases in chlorophyll a content and photosynthetic parameters, including optimal irradiance and maximum electron transport rate, were determined by evidence from rapid light curves. No evidence of a crassulacean acid metabolism (CAM) pathway was observed in *E. luridum* in response to either water deficit or exogenous application of ABA. On the other hand, malate content decreased in the *E. luridum* frond after ABA treatment, seeming to downregulate malate metabolism at night, possibly through tricarboxylic acid (TCA) cycle regulation. Additional research on desiccation tolerance and physiology of epiphytic ferns has been investigated in a broad range of geographical locations including temperate climates of North America [94, 95], western Europe [96], subtropical and tropical regions [97, 98], and Mexican cloud forest [99].

Genes associated with desiccation-rehydration cycles in the epiphytic fern *Hymenoglossum cruentum* were studied using transcriptomic analysis [100]. *H. cruentum* plants were subjected to 7 days long desiccation-rehydration process and then analyzed to identify key expressed genes associated with its capacity to dehydrate and rehydrate. The relative water content (RWC) and maximum quantum efficiency (Fv/Fm) of *H. cruentum* fronds decayed to 6% and 0.04, respectively, at the end of the desiccation stage. After re-watering, the fern showed rapid recovery of RWC and Fv/Fm (ca. 73% and 0.8, respectively). Based on clustering and network analysis, the results revealed key genes, such as (i) UBA/TS-N, DYNLL, and LHC, that orchestrated intracellular motility and photosynthetic metabolism; (ii) strong balance between avoiding cell death and defense (CAT3,

AP2/ERF) when dehydrated, and (iii) detoxifying pathways and stabilization of photosystems (GST, CAB2, and ELIP9) during rehydration.

The effects of desiccation stress in some ferns interact with physiological processes supporting photosynthesis [101]. Drought stress was induced for 10 weeks in *Asplenium nidus* and *Davallia denticulata*, two tropical epiphytic ferns, with the aim to test the effects of drought and the recovery rate of the fern species and its relationship with stomatal conductance and chlorophyll content supporting photosynthesis. In the prolonged drought stress period, *D. denticulata* showed a decline in both stomatal conductance and chlorophyll content values, although *A. nidus* only showed a decline in stomatal conductance. In the rehydration interval, both species showed a steady recovery. The results indicated that the two species were negatively impacted by the 10 weeks of drought stress. However, *A. nidus* and *D. denticulata* recovered quickly in terms of stomatal conductance, whereas, for chlorophyll content, *D. denticulata* recovered at a slower rate compared to *A. nidus*. Therefore, this study highlighted that *A. nidus* is better adapted to low precipitation and performed better in relation to future changes in precipitation. *D. denticulata* ranked second due to having more difficulty recovering from drought.

4.3 Light, Photosynthesis, and Respiration

A comparative study of the distribution of vascular epiphytes (ferns and orchids) across gradients of light and humidity in Australian tropical rainforests, ranging from 800 to 1,180 m elevation, examined the roles of transmitted light at locations of individual epiphytes and vapor pressure deficit from the canopy and base of host trees. The aim was to determine their correlation with the resultant responses of ferns and orchids to variation in the physical environments [102]. The light was correlated with height within the host tree, and VPD differed significantly with respect to the position in the host tree and elevation. Orchids occurred in environments that had higher mean light levels and mean daily maximum VPD (27% and 0.43 kPa, respectively) than ferns (21% and 0.28 kPa). There was also strong microclimatic partitioning of species within taxonomic groups, suggesting that microclimatic factors play an important role in the realized niche spaces of epiphytes within the tropical Australian rainforest.

Variations in light can have strong effects on epiphytic ferns growing within the canopy of forest trees, particularly with respect to changing intensities of light flecks. The photosynthetic response of six fern species of different life forms was compared to seedlings of six fig species to test whether epiphytes and hemiepiphytes display a slower response to light flecks and fast induction loss after light fleck compared with their terrestrial counterparts [103]. Additionally, the objective was to determine if ferns display a slower response to light flecks and slower induction loss compared to figs. Compared with figs, ferns had thicker fronds, larger stomata with low density, and lower stomatal conductance and photosynthetic capacity. Ferns had a lower light compensation point and dark respiration rate, conferring a positive carbon gain under low diffuse light beneath the canopy. Epiphytic ferns had a slower induction time to reach 90% maximum net photosynthetic rate (T_{90}) than terrestrial ferns (19.9–26.3 vs. 5.9–16.3 min, respectively). Compared with figs, ferns showed a slower response to light flecks; and ferns had a significantly slower induction loss compared with fig plants.

Variable moisture and potential stress due to transpiration have promoted adaptations to conserve water while also maximizing photosynthetic gain. Crassulacean acid metabolism (CAM), involving CO_2 uptake and storage in organic acid molecules during the night to support leaf photosynthesis during the day (when stomata can be closed to conserve water), has been documented in a wide variety of tropical epiphytic fern species. These include *Pyrrosia* sp. [104, 105], *Microsorium*, and *Polypodium* spp. [106] and *Drymoglossum* and *Pyrrosia* spp. [107].

Comparative analysis of two epiphytic populations of *Pyrrosia piloselloides* (one in sun and the other in shade) was investigated to determine the regulation of CAM in the two fern populations [105]. Shade fronds were less succulent and had higher chlorophyll content, although the chlorophyll a:b ratio was lower, and light compensation points and dark respiration rates were reduced. Dawn-dusk variations in titratable acidity and carbohydrate pools were two to three times greater in fronds acclimated to high photosynthetically active radiation (PAR), indicating greater CAM-related productivity, although water deficits were also higher than in shade fronds. When external and internal CO_2 supply to attached fronds of the fern was varied to regulate CAM activity, a significant proportion of titratable acidity was derived from the refixation of respiratory CO_2 (27% and 35% recycling for sun and shade populations, respectively). Recycled respiratory CO_2 in leaves can alleviate the lack of CO_2 uptake from the atmosphere during the daylight hours when the stomata are closed.

Water stress and abscisic acid may induce the CAM pathway as demonstrated in the epiphytic fern *Vittaria lineata* [108]. In experimental studies, plants subjected to desiccation stress (DS) and treated with exogenous ABA showed a significant increase in malate concentration (a nocturnal carbon storage compound) demonstrating nocturnal acidification to produce the carbon storage compound, malate, when compared to controls. This evidence indicates that *V. lineata* may change its mode of carbon fixation from C_3 to CAM in response to drought. Furthermore, plants subjected to stress treatments showed a significant decline in the contents of Chlorophyll a and b, but the concentrations of carotenoids were stable. Photosynthetic parameters obtained from rapid light curves showed a significant decrease after DS and ABA treatments.

Other species of epiphytic ferns that have morphological adaptations to reduce stresses associated with an aerial habitat on host plants do not exhibit CAM. For example, the epiphytic resurrection fern, *Davallia angustata*, which grows commonly on tree trunks and palm trees in south-east Asia is capable of rapid recovery from desiccation but is not a CAM plant like some other epiphytic ferns [109]. Under well-watered conditions, *D. angustata* shows a diurnal cycle of photosynthesis with maxima in mid-morning ~0900 h (solar time). Under optimum conditions, the optimum irradiance (E_{opt}) = 879.3 ± 65.31 μmol photons m^{-2} s^{-1} or ~45% of full sunlight – qualifying it as a sun plant. The maximum photosynthetic electron transport rate (ETRmax) was 77.77 ± 3.423 μmol e^{-} m^{-2} s^{-1} or, on a Chl *a* basis, 350 ± 36.0 μmol g^{-1} (Chl *a*) s^{-1}. There was some accumulation of titratable acid in the morning, but no accumulation of organic acids at night. Hence, *D. angustata* is not a CAM plant.

Given the considerable biomass and unique life form, of epiphytic ferns, especially in subtropical and tropical geographic locales, additional research is warranted on their physiological ecology, particularly related to the role of phytohormones and primary productivity and carbon cycle parameters in forest ecosystems where they are abundant.

5 Conclusion

During millions of years of evolution, ferns have acquired remarkable adaptive features to survive in diverse environments spanning aquatic, terrestrial, and epiphytic habitats, on a global scale. Concomitantly, ferns have evolved a broad range of biologically active compounds (phytohormones and signaling molecules) that enhance their survival capacity within environments as broadly different as deserts and tropical cloud forests. Some of these molecular constituents have been highlighted above in aquatic, terrestrial, and epiphytic fern species to elucidate their physiological ecology, consistent with the theme of this book on bioactive compounds. Some of the phytohormones, such as abscisic acid, gibberellic acid, indole acetic acid, and jasmonic acid (among others), appear to have evolved early in the evolutionary origin of unicellular photosynthetic organisms, such as algae [110] and have increasingly become more highly integrated into the survival strategies of vascular plants, including ferns, as they invaded and exploited terrestrial environments.

In this review, three major themes were addressed that complemented the overall scope of topics on bioactive compounds included in this book: (i) development, growth, and life cycles, (ii) environmental adaptations, such as desiccation resilience and stress tolerance, and (iii) physiological aspects of light, photosynthesis, and respiration. More comprehensive treatments of the physiological ecology of ferns have been published elsewhere [e.g., 4, 8]. While there are some generalizable conclusions that can be made about the role of phytohormones, and other regulatory molecules, across fern taxa (and their different life forms in varied geographic and climatic environments), it is clear that adaptive radiation has led to varied ways in which these biologically active compounds have been integrated into biochemical and physiological pathways specific to different fern morphotypes and taxonomic groups. Hence, many more comparative physiological ecology studies are needed to clarify the commonality, as well as the particularity, of the various ways; plant bioactive compounds have been incorporated into survival strategies of ferns with different phylogenetic histories and varying geographic habitats.

Recent advances in molecular genetic techniques have markedly expanded our understanding of fern phylogeny and systematics; and modern high-throughput gene sequencing techniques have opened remarkable new avenues to elucidate the genetic basis of plant ecophysiology through transcriptomics, proteomics, and related "omic" approaches [e.g., 100, 111]. By combining these molecular techniques with modern field-based and laboratory research methods, we should be able to integrate fundamental genetic, physiological, and ecological insights into a more comprehensive understanding of fern form and function in varied environments and across different taxonomic groups.

Ultimately, a more profound understanding of the role of bioactive compounds in fern biology can benefit from an additional, detailed understanding of the genetic basis of molecular activity in relation to species differences and life forms of ferns in varied habitats. Much of prior research as reviewed above (based on established ecophysiological research methods) can serve as a context and guide for this exploration into the molecular genetic basis of fern physiological ecology – an enterprise that increasingly shows promise of substantially enhancing our understanding and appreciation of fern flora as unique members of biological diversity, worldwide.

References

1. Schneider H, Schuettpelz E, Pryer KM, Cranfill R, Magallón S, Lupia R (2004) Ferns diversified in the shadow of angiosperms. Nature 428:553–557
2. Schuettpelz E, Pryer KM (2009) Evidence for a Cenozoic radiation of ferns in angiosperm-dominated canopy. Proc Natl Acad Sci U S A 106:11200–11205
3. Watkins JE Jr, Cardelús CL (2012) Ferns in an angiosperm world: cretaceous radiation into the epiphytic niche and diversification on the forest floor. Int J Plant Sci 173:695–710
4. Anderson OR (2021) Physiological ecology of ferns: biodiversity and conservation perspectives. Int J Biodivers Conserv 13:49–63
5. Moran RC (2013) Diversity, biogeography, and floristics. In: Ranker TA, Haufler CH (eds) Biology and evolution of ferns and lycophytes. Cambridge University Press, New York
6. Kessler M (2010) Biogeography of ferns. In: Mehltretter K, Walker LR, Sharpe JM (eds) Fern ecology. Cambridge University Press, New York
7. Sharpe JM (2019) Fern ecology and climate change. Indian Fern J 36:179–199
8. Sharpe JM, Mehltreter K, Walker LR (2010) Ecological importance of ferns. In: Mehltreter K, Walker LR, Sharpe JM (eds) Fern ecology. Cambridge University Press, Cambridge
9. Mehltreter K (2010) Fern conservation. In: Mehltreter K, Walker LR, Sharpe JM (eds) Fern ecology. Cambridge University Press, Cambridge
10. Arcand NN, Ranker TA (2013) Conservation biology. In: Ranker TA, Haufler CH (eds) Biology and evolution of ferns and lycophytes. Cambridge University Press, Cambridge
11. Fernández-Zamudio R, García-Murillo P, Cirujano S (2010) Germination characteristics and sporeling success of *Azolla filiculoides* Lamarck, an aquatic invasive fern, in a Mediterranean temporary wetland. Aquat Bot 93:89–92
12. Sadeghi R, Zarkami R, Sabetraftar K, Van Damme P (2013) A review of some ecological factors affecting the growth of *Azolla* spp. Caspian J Env Sci 11:65–76
13. Lumpkin TA, Plucknett DL (1980) *Azolla*: botany, physiology, and use as a green manure. Econ Bot 34:111–153
14. Temmink RJM, Harpenslager SF, Smolders AJP, van Dijk G, Roy CJH, Peters RCJH, Lamers LPM, van Kempen MML (2018) *Azolla* along a phosphorus gradient: biphasic growth response linked to diazotroph traits and phosphorus-induced iron chlorosis. Sci Rep 8:4451
15. Jampeetong A, Brix H (2009) Effects of NH_4^+ concentration on growth, morphology and NH_4^+ uptake kinetics of *Salvinia natans*. Ecol Eng 35:695–702

16. Martins-Loucao MA, Wollenweber B, Raven JA (1993) Response of *Salvinia* spp. to different nitrogen sources: the acid-base regulation approach. Oecologia 93:524–530
17. van Kempen MML, Smolders AJP, Bögemann GM, Lamers LPM, Roelofs JGM (2016) Interacting effects of atmospheric CO_2 enrichment and solar radiation on growth of the aquatic fern *Azolla filiculoides*. Freshw Biol 61:596–606
18. Medeiros JCC, Silva JCF, Teodoro GS, Coelho FF (2017) Effects of shade on individual ramet growth and on clonal growth of the aquatic fern *Salvinia auriculata* (Salviniaceae). Am Fern J 107:21–29
19. Tipping PW, Martin MR, Bauer L, Pierce RM, Center TD (2012) Ecology of common *Salvinia*, *Salvinia minima* Baker, in southern Florida. Aquat Bot 102:23–27
20. Pimsuwan S, Wongsrisakulkaew Y (2019) The effects of soil depth on the growth of the clover fern and the uses of the clover fern on the germination of fern spores. Int J Geomate 17:50–55
21. Liu B-LL (1984) Abscisic acid induces land form characteristics in *Marsilea quadrifolia*. Am J Bot 71:638–644
22. Daoud-Bouattour A, Bottollier-Curtet M, Jamaa HF-B, Ghrabi-Gammar Z, Saad-Limam SB, Rhazi L, Muller SD (2014) Effects of hydrology on recruitment of *Pilularia minuta* Durieu (Marsileaceae), an endangered plant of Mediterranean temporary pools. Aquat Bot 112:76–83
23. Minissale P, Molina JA, Sciandrello S (2017) *Pilularia minuta* Durieu (Marsileaceae) discovered in south-eastern-Sicily: new insights on its ecology, distribution and conservation status. Bot Lett 164:197–208
24. McMillan C, Jens CE, Adler WR, Blystone RV, Gillespie WH, Irwin JR, Janowsky RE, Kolle DO, McGlathery TR, Martin RR, Morey RW, Mynard CR, Patty TS (1968) Factors influencing the narrow restriction of *Pilularia americana* in Texas. Southwest Nat 13:117–127
25. Ganger M, Sturey T (2012) Antheridiogen concentration and spore size predict gametophyte size in *Ceratopteris richardii*. Botany 90:175–179
26. Ferreira AB, Piedade LR, Piedade MTF, Lopes A (2021) Biomass production of the aquatic macrophyte *Ceratopteris pteridoides* (Hook.) Hieron (Pteridaceae) in nutrient addition treatments. Acta Bot Bras 35:126–131
27. Chau MM, Reyes WR, Ranker TA (2013) Ecological factors influencing growth of the endangered Hawaiian fern *Marsilea villosa* (Marsileaceae) and implications for conservation management. Am J Bot 100:1532–1543
28. Gopal B (1969) Responses of some Indian species of *Marsilea* to different temperature treatments. Am Fern J 59:150–152
29. Wu T-C, Kao W-Y (2011) Ecophysiological traits of leaves of three *Marsilea* species distributed in different geographical regions. Taiwania 56:279–286
30. Warne TR, Lloyd RM (1980) The role of spore germination and gametophyte development in habitat selection: temperature responses in certain temperate and tropical ferns. Bull Torrey Bot Club 107:57–64
31. Goodnoe TT, Hill JP, Aho K (2016) Effects of variation in carbon, nitrogen, and phosphorus molarity and stoichiometry on sex determination in the fern *Ceratopteris richardii*. Botany 94: 249–259
32. Dong Y-H, Gituru RW, Chen J-M, Wang Q-F (2005) Effect of habitat modification on the distribution of the endangered aquatic fern *Ceratopteris thalictroides* (Parkeriaceae) in China. J Freshw Ecol 20:689–693
33. Hew C-S, Wong YS (1974) Photosynthesis and respiration of ferns in relation to their habitats. Am Fern J 64:40–48
34. Wu T-C, Lin B-L, Kao W-Y (2020) Active stomatal control of *Marsilea crenata*, an amphibious fern, in response to CO_2 and exogenous application of ABA. Taiwania 65:431–437
35. Kao W-Y, Lin B-L (2010) Phototropic leaf movements and photosynthetic performance in an amphibious fern, *Marsilea quadrifolia*. J Plant Res 123:645–653
36. Deans RM, Brodribb TJ, Busch FA, Farquhar GD (2019) Plant water-use strategy mediates stomatal effects on light induction of photosynthesis. New Phytol 222:382–395

37. Tseng M-H, Lin K-H, Huang Y-J, Chang Y-L, Huang S-C, Kuo L-Y, Huang Y-M (2017) Detection of chlorophylls in spores of seven ferns. J Plant Res 130:407–416
38. Lloyd RM, Klekowski EJ Jr (1970) Spore germination and viability in Pteridophyta: evolutionary significance of chlorophyllous spores. Biotropica 2:129–137
39. Raghavan V (1980) Cytology, physiology, and biochemistry of germination of fern spores. Int Rev Cytol 62:69–118
40. Suo J, Chen S, Zhao Q, Shi L, Dai S (2015) Fern spore germination in response to environmental factors. Front Biol 10:358–376
41. Pérez-García B, Mendoza-Ruiz A, Sánchez-Coronado ME, Orozco-Segovia A (2007) Effect of light and temperature on germination of spores of four tropical fern species. Acta Oecol 32: 172–179
42. Haufler CH, Welling CB (1994) Antheridiogen, dark spore germination, and outcrossing mechanisms in *Bommeria* (Adiantaceae). Am J Bot 81:616–621
43. Sugai M, Nakamura K, Yamane H, Sato Y, Takahashi N (1987) Effects of gibberellins and their methyl esters on dark germination and antheridium formation in *Lygodium japonicum* and *Anemia phyllitidis*. Plant Cell Physiol 28:199–202
44. Chia S-GE, Raghavan V (1982) Abscisic acid effects on spore germination and protonemal growth in the fern *Mohria caffrorum*. New Phytol 92:31–37
45. Ueguchi-Tanaka M, Nakajima M, Katoh E, Ohmiya H, Asano K, Saji S, Hongyu X, Ashikari M, Kitano H, Yamaguchi I, Matsuoka M (2007) Molecular interactions of a soluble gibberellin receptor, GID1, with a rice DELLA protein, SLR1, and gibberellin. Plant Cell 19: 2140–2155
46. Cooke TJ, Racusen RH (1988) The growth forms of developing fern gametophytes: a theoretical consideration of form and function relationships. Ann Bot 62:633–641
47. Raghavan V (2005) Developmental biology of fern gametophytes. Cambridge University Press, Cambridge
48. Banks JA (1999) Gametophyte development in ferns. Annu Rev Plant Physiol Plant Mol Biol 50:163–186
49. Döpp W (1950) Eine die Antheridienbildung bei Farnen fördernde Substanz in den Prothallien von *Pteridium aquilinum* (L.) Kuhn. Ber Deut Bot Ges 63:139–147
50. Näf U (1961) Mode of action of an antheridium-inducing substance in ferns. Nature 189: 900–903
51. Atallah NM, Banks JA (2015) Reproduction and the pheromonal regulation of sex type in fern gametophytes. Front Plant Sci 6:100
52. Warne TR, Hickok LG (1989) Evidence for a gibberellin biosynthetic origin of *Ceratopteris* antheridiogen. Plant Physiol 89:535–538
53. Mander LN (2003) Twenty years of gibberellin research. Nat Prod Rep 20:49–69
54. Romanenko KO, Babenko LM, Vasheka OV, Romanenko PO, Kosakivska IV (2020) In vitro phytohormonal regulation of fern gametophytes growth and development. Russ J Dev Biol 51: 71–83
55. Ohishi N, Hoshika N, Takeda M, Shibata K, Yamane H, Yokota T, Asahina M (2021) Involvement of auxin biosynthesis and transport in the antheridium and prothalli formation in *Lygodium japonicum*. Plan Theory 10:2709
56. Testo WL, Grasso MS, Barrington DS (2014) Beyond antheridiogens: chemical competition between gametophytes of *Polypodium appalachianum* and *Polypodium virginianum*. J Torrey Bot Soc 141:302–312
57. Haufler CH, Pryer KM, Schuettpelz E, Sessa EB, Farrar DR, Moran R, Schneller JJ, Watkins JE Jr, Windham MD (2016) Sex and the single gametophyte: revising the homosporous vascular plant life cycle in light of contemporary research. Bioscience 66:928–937
58. Oloyede FA, Olatunji TY, Ogbimi ER (2017) Spore germination, gametophyte development and studies of growth of *Adiantum capillus-veneris* L. Not Sci Biol 9:557–562
59. Sakamaki Y, Ino Y (1999) Contribution of fern gametophytes to the growth of produced sporophytes on the basis of carbon gain. Ecol Res 14:59–69

60. Klekowski EJ Jr (1967) Observations on pteridophyte life cycles: relative lengths under cultural conditions. Am Fern J 57:49–51
61. López-Pozo M, Fernández-Marín B, García-Plazaola JI, Ballesteros D (2018) Desiccation tolerance in ferns: from unicellular spore to multi-tissular sporophyte. In: Fernández (ed) Current advances in fern research. Springer International Publishing AG, Cham
62. Watkins JE Jr, Mack MC, Sinclair TR, Mulkey SS (2007) Ecological and evolutionary consequences of desiccation tolerance in tropical fern gametophytes. New Phytol 176: 708–717
63. Chambers SM, Watkins JE Jr, Sessa EB (2017) Differences in dehydration tolerance among populations of a gametophyte-only fern. Am J Bot 104:598–607
64. Nitta JH, Watkins JE Jr, Holbrook NM, Wang TW, Davis CC (2021) Ecophysiological differentiation between life stages in filmy ferns (Hymenophyllaceae). J Plant Res 134: 971–988
65. Reynolds TL, Bewley JD (1993) Characterization of protein synthetic changes in a desiccation-tolerant fern, *Polypodium virginianum*. Comparison of the effects of drying, rehydration and abscisic acid. J Exp Bot 44:921–928
66. Nadal M, Brodribb TJ, Fernández-Marín B, García-Plazaola JI, Arzac MI, López-Pozo M, Perera-Castro AV, Gulias J, Flexas J, Farrant JM (2021) Differences in biochemical, gas exchange and hydraulic response to water stress in desiccation tolerant and sensitive fronds of the fern *Anemia caffrorum*. New Phytol 231:1415–1430
67. Banupriya TG, Ramyashree C, Akash D, Yathish NS, Sharthchandra RG (2020) Studies on the mechanism of desiccation tolerance in the resurrection fern *Adiantum raddianum*. J Appl Biol Biotechnol 8:6–14
68. Zivkovic S, Popovic M, Dragisic-Maksimovic J, Momcilovic ID, Grubisic D (2010) Dehydration-related changes of peroxidase and polyphenol oxidase activity in fronds of the resurrection fern *Asplenium ceterach* L. Arch Biol Sci 62:1071–1081
69. Hõrak H, Kollist H, Merilo E (2017) Fern stomatal responses to ABA and CO_2 depend on species and growth conditions. Plant Physiol 174:672–679
70. Grantz DA, Linscheid BS, Grulke NE (2019) Differential responses of stomatal kinetics and steady-state conductance to abscisic acid in a fern: comparison with gymnosperm and an angiosperm. New Phytol 222:1883–1892
71. Gong L, Liu X-D, Zeng Y-Y, Tian X-Q, Li Y-L, Turner NC, Fang X-W (2021) Stomatal morphology and physiology explain varied sensitivity to abscisic acid across vascular plant lineages. Plant Physiol 186:782–797
72. Hasan MM, Gong L, Nie Z-F, Li F-P, Ahammed GJ, Fang X-W (2021) ABA-induced stomatal movements in vascular plants during dehydration and rehydration. Environ Exp Bot 186: 104436
73. Yang Y-J, Bi MH, Nie Z-F, Jiang H, Liu X-D, Fang X-W, Brodribb TJ (2021) Evolution of stomatal closure to optimize water-use efficiency in response to dehydration in ferns and seed plants. New Phytol 230:2001–2010
74. Cândido-Sobrinho SA, Lima VF, Friere FBS, de Souza LP, Gago J, Fernie AR, Daloso DM (2022) Metabolism-mediated mechanisms underpin the differential stomatal speediness regulation among ferns and angiosperms. Plant Cell Environ 45:296–311
75. Gago J, Coopman RE, Cabrera HM, Hermida C, Molins A, Conesa MÀ, Galmés J, Ribas-Carbó M, Flexas J (2013) Phoptosynthesis limitations in three fern species. Physiol Plant 149: 599–611
76. Nobel PS, Calkin HW, Gibson AC (1984) Influences of PAR, temperature and water vapor concentration on gas exchange by ferns. Physiol Plant 62:527–534
77. Nishida K, Kodama N, Yonemura S, Hanba YT (2015) Rapid response of leaf photosynthesis in two fern species *Pteridium aquilinum* and *Thelypteris dentata* to changes in CO_2 measured by tunable diode laser absorption spectroscopy. J Plant Res 128:777–789
78. Tosens T, Nishida K, Gago J, Coopman RE, Cabrera HM, Carriqui M, Laanisto L, Morales L, Nadal M, Rojas R, Talts E, Tomas M, Hanba Y, Niinemets Ü, Flexas J (2016) The

photosynthetic capacity in 35 ferns and fern allies: mesophyll CO_2 diffusion as a key trait. New Phytol 209:1576–1590
79. Saldaña A, Lusk CH, Gonzáles WL, Gianoli E (2007) Natural selection on ecophysiological traits of a fern species in a temperate rainforest. Evol Ecol 21:651–662
80. Anderson OR, Griffin KL (2021) Preliminary ecophysiological observations of the fern *Asplenium platyneuron* growing in two microenvironments varying in light intentsity at an urban location in New York City. J Torrey Bot Soc 148:308–317
81. Ludlow CJ, Wolf FT (1975) Photosynthesis and respiration rates of ferns. Am Fern J 65:43–48
82. Nasrulhaq-Boyce A, Haji Mohamed MA (1987) Photosynthetic and respiratory characteristics of Malayan sun and shade ferns. New Phytol 105:81–88
83. Hew C-S, Wong Y-S (1974) Photosynthesis and respiration of ferns in relation to their habitat. Am Fern J 64:40–48
84. Nishida K, Hanba YT (2017) Photosynthetic response of four fern species from different habitats to drought stress: relationship between morpho-anatomical and physiological traits. Photosynthetica 55:689–697
85. Dubuisson J-Y, Schneider H, Hennequin S (2009) Epiphytism in ferns: diversity and history. C R Biol 332:120–128
86. Pinson JB, Chambers SM, Nitta JH, Kuo L-Y, Sessa EB (2017) The separation of generations: biology and biogeography of long-lived sporophyteless fern gametophytes. Int J Plant Sci 178: 1–18
87. Camloh M (1993) Spore germination and early gametophyte development of *Platycerium bifurcatum*. Am Fern J 83:79–85
88. Nagmani R, Raghavan V (1983) Origin of the rhizoid and protonemal cell during germination of spores of *Drymoglossum*, *Platycerium* and *Pyrrosia* (Polypodiaceae). Bot Gaz 144:67–72
89. Camloh M, Ravnikar M, Zel J (1996) Jasmonic add promotes division of fern protoplasts, elongation of rhizoids and early development of gametophytes. Physiol Plant 97:659–664
90. Camloh M, Vilhar B, Zel J (2001) Jasmonic acid induces changes in growth and polypeptide composition of fern gametophytes. Acta Bot Croat 60:149–156
91. Kwa S-H, Wee Y-C, Lim T-M, Kumar PP (1995) IAA-induced apogamy in *Platycerium coronarium* (Koenig) Desv. gametophytes cultured *in vitro*. Plant Cell Rep 14:598–602
92. Garcés M, Ulloa M, Mirana A, Bravo LA (2018) Physiological and ultrastructural characterisation of a desiccation-tolerant filmy fern, *Hymenophyllum caudiculatum*: influence of translational regulation and ABA on recovery. Plant Biol 20:288–295
93. Minardi BD, Voytena APL, Santos M, Randi ÁM (2014) The epiphytic fern *Elaphoglossum luridum* (Fée) Christ. (Dryopteridaceae) from central and south America: morphological and physiological responses to water stress. Sci World J 2014:817892
94. John SP, Hasenstein KH (2018) Biochemical responses of the desiccation-tolerant fern *Pleopeltis polypodioides* to dehydration and rehydration. J Plant Physiol 228:12–18
95. Sillett SC, Bailey MG (2003) Effects of tree crown structure on biomass of the epiphytic fern *Polypodium scouleri* (Polypodiaceae) in redwood forests. Am J Bot 90:255–261
96. Klinghardt M, Zotz G (2021) Abundance and seasonal growth of epiphytic ferns at three sites along a rainfall gradient in Western Europe. Flora 274:151749
97. Testo WL, Watkins JE Jr (2012) Influence of plant size on the ecophysiology of the epiphytic fern *Asplenium auritum* (Aspleniaceae) from Costa Rica. Am J Bot 99:1840–1846
98. Flores-Bavestrello A, Król M, Ivanov AG, Hüner NPA, García-Plazaola JI, Corcuera LJ, Bravo LA (2016) Two Hymenophyllaceae species from contrasting natural environments exhibit a homoiochlorophyllous strategy in response to desiccation stress. J Plant Physiol 191:82–94
99. Hietz P, Briones O (1998) Correlation between water relations and within-canopy distribution of epiphytic ferns in a Mexican cloud forest. Oecologia 114:305–316
100. Ostria-Gallardo E, Larama G, Berrios G, Fallard A, Gutiérrez-Moraga A, Ensminger I, Manque P, Bascuñán-Godoy L, Bravo LA (2020) Decoding gene networks modules that explain *Hymenoglossum cruentum* Cav. after extreme desiccation. Front Plant Sci 11:574

101. Suhaimi N, Cicuzza D (2020) The effects of drought and the recovery phase of two tropical epiphytic ferns. J Trop Plant Phys 12:33–41
102. Sanger JC, Kirkpatrick JB (2017) The distribution of vascular epiphytes over gradients of light and humidity in north-east Australian rainforest. Austral Ecol 42:976–983
103. Zhang Q, Chen J-W, Li B-G, Cao K-F (2009) Epiphytes and hemiepiphytes have slower photosynthetic responses to lightflecks than terrestrial plants: evidence from, ferns and figs. J Trop Ecol 25:465–472
104. Winter K, Osmond CB, Hubick KT (1986) Crassulacean acid metabolism in the shade. Studies on an epiphytic fern *Pyrrosia longifolia*, and other rainforest species from Australia. Oecologia 68:224–230
105. Griffiths H, Ong BL, Avadhani PN, Goh CJ (1989) Recycling of respiratory CO_2 during Crassulacean acid metabolism: alleviation of photoinhibition in *Pyrrosia piloselloides*. Planta 179:115–122
106. Holtum JAM, Winter K (1999) Degrees of crassulacean acid metabolism in tropical epiphytic and lithophytic ferns. Aust J Plant Physiol 26:749–757
107. Ong BL, Kluge M, Friemert V (1986) Crassulacean metabolism in the epiphytic ferns *Drymoglossum piloselloides* and *Pyrrosia longifolia*: studies on responses to environmental signals. Plant Cell Environ 9:547–557
108. Minardi BD, Voytena APL, Santos M, Randi ÁM (2014) Water stress and abscisic acid treatments induce the CAM pathway in the epiphytic fern *Vittaria lineata* (L.) Smith. Photosynthetica 52:404–412
109. Quinnell R, Howell D, Ritchie RJ (2017) Photosynthesis of an epiphytic resurrection fern *Davallia angustata* (Wall. ex Hook. & Grev.). Aust J Bot 65:348–356
110. Tarakhovskaya ER, Maslov YI, Shishova MF (2007) Phytohormones in algae. Russ J Plant Physiol 54:163–170
111. Zhang J, Liu L, Shu J-P, Jin D-M, Shen H, Chen H-F, Zhang R, Yan Y-H (2019) Transcriptomic evidence of adaptive evolution of the epiphytic fern *Asplenium nidus*. Int J Genomics 2019:142931

Extracts and Composites of *Equisetum* for Bone Regeneration

26

Rosangela Maria Ferreira da Costa e Silva, Ivana Márcia Alves Diniz, and José Maria da Fonte Ferreira

Contents

Abstract

The present chapter reviews the different preparation and characterization of *Equisetum* extracts and composites, including their biological properties and uses in the regeneration of bone defects derived from traumas and diseases, such as osteoporosis and arthritis. The results have shown that the *Equisetum genus* contains high amounts of silicon, potassium, zinc, sulfur, carbon, and other organic compounds with potential use as therapeutic drugs. The experimental variations in the preparation of extracts showed that the maximum yield (~20% w/w) can be obtained for the concentration of 10–50% of ethanol in ethanol/water solutions. The maximum total phenolic and antioxidant contents can be obtained at concentrations of 50–90% ethanol. The *Equisetum* genus-derived composite

Rosangela Maria Ferreira da Costa e Silva, Ivana Márcia Alves Diniz and José Maria da Fonte Ferreira contributed equally with all other contributors.

R. M. F. da Costa e Silva (✉)
Universidade Estadual do Mato Grosso do Sul, Campus Dourados-MS, Mato Grosso do Sul, Brazil

I. M. Alves Diniz
Departamento de Odontologia Restauradora, Universidade Federal de Minas Gerais, Minas Gerais, Brazil

J. M. da Fonte Ferreira
CICECO -Aveiro Institute of Materials, Department of Materials and Ceramic Engineering, University of Aveiro, Aveiro, Portugal
e-mail: jmf@ua.pt

H. N. Murthy (ed.), *Bioactive Compounds in Bryophytes and Pteridophytes*, Reference Series in Phytochemistry, https://doi.org/10.1007/978-3-031-23243-5_31

materials and extracts are promising for bone regeneration applications. In vivo studies revealed that the trabecular bone thickness increased by 11% and 46% when treating SHAM and ovariectomized (OVX) rats, respectively, with a diet containing 60 mg · kg^{-1} *E. arvense (EA) extract*. The cortical bone thickness of the *EA* extract group has increased by 14% compared to the OVX group. Quantitative analysis of composite EA extract and poly (lactic acid) (PLA) revealed alkaline phosphatase (ALP) activity three times higher in comparison to the polystyrene (21 days) and an increase in mineralization of adipose tissue-derived mesenchymal stem cells (AT-MSCs) after 14 and 21 days. In vivo and in vitro studies have concluded that *E. arvense*, *E. hyemale*, *E. giganteum*, and other *Equisetum* species showed antioxidants and osteoconductive and osteoinductive properties, which are vital for bone regeneration.

Keywords

Arthritis · Bone lost · Bone regeneration · Equisetum · Osteoporosis

Abbreviations

AIA	Antigen-induced arthritis
ALP	Alkaline phosphatase
AT-MSC	Human adipose tissue-derived mesenchymal stem cell
AT-MSCs	Adipose tissue-derived mesenchymal stem cells
BHA	Butylated hydroxy anisole
BMD	Mandibular bone mineral density
BMP-2	*Bone morphogenetic protein 2* human
C. albicans	*Candida albicans*
C2C12	Mouse myoblast *cell* line
CD69	Cluster of differentiation 69
CFSE	Carboxyfluorescein diacetate succinimidyl ester
COL1	Collagen type 1
ConA	Concanavalin A
CsA	Cyclosporine A
DD	Daily diet
DMSO	Dimethyl sulfoxide
DPPH	1,1-diphenyl-2-picrylhydrazyl
E	*Equisetum*
E. coli	*Escherichia coli*
E. faecalis	*Enterococcus faecalis*
E1	Extraction fractions 1
E2	Extraction fractions 2
E3	Mixture of E1 and E2
EA	*E. arvense*
ECM	Calcium complexes
EH	*E. hyemale*
EM	*Equisetum* mixture (genus not identified)

EP	*E. pratense*
ER	*E. ramosissimum*
ES	*E. sylvaticum*
ESR	Electron spin resonance
Et	Ethanol
ET	*E. telmateia*
FRAP	Ferric reducing antioxidant power
Group N	OVX rats were fed with nutrients for body weight: calcium carbonate (104.16 $mg \cdot kg^{-1}$), zinc sulfate (1.56 $mg \cdot kg^{-1}$), L-ascorbic acid (93.75 $mg \cdot kg^{-1}$), L-lysine (93.75 $mg \cdot kg^{-1}$), L-proline (46.85 $mg \cdot kg^{-1}$), L-arginine (200 $mg \cdot kg^{-1}$), and Vitamin D (100 IU) + normal feed
HA	Hydroxyapatite phase $[Ca_{10}(PO_4)_6(OH)_2]$
hADSCs	Human adipose-derived stem cells
HDL	High-density lipoprotein
HPV	Human Papillomavirus
HUVECs	Human umbilical vein endothelial cells
IC_{50}	Half-maximal inhibitory concentration
IFN-γ	Human Interferon-gamma
IL-1	Interleukin 1
IL-2	Interleukin 2
IL-6	Interleukin 6
IL-10	Interleukin 10
K. pneumoniae	*Klebsiella pneumoniae*
L. monocytogenes	*Listeria monocytogenes*
LPS	*Lipopolysaccharide*
mBSA	Methylated bovine serum albumin
MC3T3-E1	Immortalized *murine calvarial cell line*
M-CSF	Macrophage-colony stimulating factor
MEK	Mitogen-activated protein kinase
MIC	Minimum inhibitory concentration
MMP-13	Matrix metalloproteases 13
MRSA-IC	Methicillin resistant *Staphylococcus aureus*
MTT	3-(4,5-dimethylthiazol-2-yl)-2,5-diphenyl-2H tetrazolium bromide
NIH3T3	Mouse fibroblast cell line
NF-kB	Nuclear factor-κB
NO	Nitric oxide
NSAID	Non-steroidal anti-inflammatory drugs
OA	Osteoarthritis
OPG	Osteoprotegerin
OPN	Osteopontin
OVX	Ovariectomized rat
PBMC	Peripheral blood mononuclear cells
PGE_2	Prostaglandin E_2

PHA-L	Phytohemagglutinin-L
PLA	Poly(lactic acid)
P. aeruginosa	*Pseudomonas aeruginosa*
RA	Rheumatoid arthritis
RANKL	Receptor activator of nuclear factor κB ligand
RANK	Receptor activator of nuclear factor Kappa-Beta
RAL	Raloxifene
RF	Rheumatoid factor
ROS	*Reactive oxygen species*
Runx2	*R*unt-related transcription factor 2
RT-PCR	*Real-time polymerase chain reaction*
S. aureus	*Staphylococcus aureus*
S. enteritidis	*Salmonella enteritidis*
S. epidermidis	*Staphylococcus epidermidis*
S. flexneri	*Shigella flexneri*
S. saprophyticus	*Staphylococcus saprophyticus*
SERMs	Selective estrogen receptor modulators
SHAM	Normal rats operated
S/L	Solid/liquid ratio
TEAC	Trolox equivalent antioxidant capacity
TNF	Tumor necrosis factor
TNFα	Tumor necrosis factor-α human
TRAP	Tartrate-resistant acid phosphatase
W	Water

1 Introduction

The *genus Equisetum* consists of about 30 species described in the flora of Asia, Europe, North America, Central America, Sul America, and Africa [1]. The word *Equisetum* derives from Latin *Equus* ("horse") + seta ("bristle"). The name "horse-tail" is used for all components of the *Equisetum* genus, because it resembles a horse's tail [2]. *Equisetum genus* prefers wet calcareous and sandy soils though some are semi-aquatic, and others are adapted to wet clay soils, and due to this are found near wet areas, such as marshes, streams, or rivers [2].

Equisetum genus is a herbaceous perennial plant without secondary tissues that is distinguished by jointed stems with a fused whorl of nodal leaves. The epidermis is thick because of encrusted silica [3]. *Equisetum* plants are divided into subgenera *Equisetum* and *Hippochaete*. The first has stomata flush with the epidermis and stems branched (e.g., *E. arvense – EA*), whereas the latter has stomata sunken below the surface of the epidermis and stems generally unbranched (e.g., *E. hyemale* (Fig. 1) [1].

Despite the existence of the species – *Equisetites arenaceus*, *E. arvense*, *E. bogotense*, *E. clarnoi*, *E. diffusum*, *E. dimorphum*, *E. fluviatoides*, *E. fluviatile*, *E. giganteum*, *E. haukeanum*, *E. hyemale*, *E. laevigatum*, *E. myriochaetum*, *E. thermale*, *E. laterale*, *Equisetites lyelli*, *E. pratense*, *E. palustre*, *E. ramosissimum*,

Fig. 1 *Equisetum hyemale*: (**a**) plants in situ and (**b**) strobilus

E. sylvaticum, *E. scirpoides*, *E. telmateia*, *E. thermale*, *E. vancouverense*, and *E. variegatum* [3] – that have variables uses (Table 1) described in folk and traditional medicine on almost all continents, studies confirming the action against the diseases were reported only for the species *E. arvense*, *E. pratense*, *E. sylvaticum*, *E. telmateia*, *E. ramosissimum*, *E. giganteum*, and *E. hyemale*.

2 *Equisetum* Genus Applications in Bone Regeneration

Equisetum genus and other plants offer advantages, such as high availability of raw material, synergic effects, low cost, and minimal side effects [4]. Several biological properties of these plant extracts have been attributed to their complex composition, including silicon, polyphenolic derivatives as flavonoids, and other components listed in Table 1. Francescato et al. identified in the *E. giganteum* the phenolic compounds: quercetin-tri-O-hexoxide, kaempferol-3-O-sophoroside-7-O-glucoside, 3-hydroxyhispidin-3, 4-di-O-glucoside, quercetin-3,7-di-O-glucoside, quercetin-3-O- (caffeoyl)-glucoside, kaempferol-3,7- di-O-glucoside, flavonol-di-O-hexoxide, equisetum pyrone, caffeic acid derivative, kaempferol-3-O- sophoroside, ferulic acid derivative, kaempferol-3-O-glucoside, and quantified isoquercitrin, quercetin, and kaempferol for species *E. giganteum* ($1.05 \pm 0.06\%$ w/w; 0.116 ± 0.005 mg $\cdot$ g^{-1}; 1.41 ± 0.06 mg $\cdot$ g^{-1}), *E. arvense* ($2.4 \pm 0.3\%$, w/w; 2.31 ± 0.18 mg $\cdot$ g^{-1}; 0.53 ± 0.04 mg $\cdot$ g^{-1}), *E. hyemale* ($1.01 \pm 0.04\%$, w/w; 0.083 ± 0.002 mg $\cdot$ g^{-1}; 1.38 ± 0.06 mg $\cdot$ g^{-1}), and *E. bogotense* ($1.74 \pm 0.05\%$, w/w; 0.033 ± 0.001 mg $\cdot$ g^{-1}; 2.36 ± 0.06 mg $\cdot$ g^{-1}), respectively. All species have shown quantities higher than established by the Pharmacopeia European (minimum of 0.3% of total flavonoids quantified as isoquercitrin for extract dried) [5, 6].

The biological capacity of the *Equisetum* genus to foster the regeneration of bone and cartilage – such as in osteoporosis and arthritis conditions – or to prevent bone loss density [4, 7, 8] has been documented in the following three main axes: (1) The first one includes studies providing evidence that the high availability of silicon increases the volume and density of bone, inhibits bone resorption, stimulates

Table 1 Indications, applications on the popular medicine and phytotherapy, and potential substance with pharmacological identified on the species

Specie Author/year	Indications and applications	Substances with potential pharmacological identified
E. arvense Grundemann, 2014 [6]	The use is described in several handbooks of phytotherapy and Anthroposophical Medicine, European Pharmacopoeia, and German Commission E Monograph (herbal substances) of the German Federal Institute for Drugs and Medical Devices as antioxidants, hepatoprotective, diuretic, antibacterial, anti-inflammatory, and arthritis	Silicon 5–8% w/w, potassium, calcium, flavonoids (0.2–0.9%), phytosterols, phenolic acids, oniti-9-O-glucoside, equisetumpyrone, triterpenoids Essential oil: cis-geranyl acetone, trans-phytol, thymol, hexahydrofarnesyl acetone
E. hyemale Alves, 2019 [18]	Popularly, this plant has been used as a diuretic, anti-anemic, anti-inflammatory, and digestive	
E. arvense Costa Rodrigues 2012 [10]	*E. arvense* has been described in the literature of use by ancient Greeks and Romans to treat wounds This has been utilized in folk remedies for arthritis, bleeding ulcers, hepatitis, jaundice, kidney diseases, and tuberculosis These have been described in several European pharmacopeias for use as antiseptic, anti-inflammatory, diuretic, antioxidant, hepatoprotective, vasorelaxant, and anti-nociceptive	
E. arvense Batir-Marin, 2021 [40]	*E. arvense* is described in the European Pharmacopoeia as a diuretic, for the treatment of urinary and respiratory tract infections, cardiovascular diseases, and skin diseases	Flavonoids, alkaloids, phenolic acids, tannins, phytosterols, and triterpenoids
E. arvense Nagai, 2005 [15]	*E. arvense* is utilized as drunk as a health drink Antipyretic, diuretic, anticough, and anti-inflammatory	Silicic acid, saponin, carotein, vitamin B1, B2, B6, nicotinic acid, folic acid, pantothenic acid, alkaloid, flavonoids as isoflavone, flavonol, flavone, and flavanol, vitamins K, E (4.9 mg 100 g^{-1}), and C (33 mg · 100 g^{-1}) Elements: Si, Na, K (640 mg · 100 g^{-1}), Ca, Mg, P, Fe, Zn (1.1 mg . 100 g^{-1}), Cu (0.22 mg · 100 g^{-1}), Mn, Sr, Ti
E. giganteum Farinon, 2013 [38]	There are indicated treatments, such as hemostatic, urinary disorders, diuretic, and several inflammatory and rheumatic diseases	Flavonoids, styrylpyrones glycosides, hydroxycinnamic acid derivatives, metals, silica, and some apolar compounds in

(continued)

Table 1 (continued)

Specie Author/year	Indications and applications	Substances with potential pharmacological identified
		their oleoresin, as alkanes, fatty acids, methyl esters, and steroid triterpenes
E. giganteum, *E. arvense*, *E. hyemale*, *E. bogotense* Francescato, 2012 [5]	*E. giganteum* is used in traditional medicine and herbal medicine of Argentina, Brazil, Guatemala, Venezuela, and Mexico as a diuretic, hemostatic, remineralizing, liver, and urinary diseases	*E. giganteum*: Phenolic compounds (quercetin-tri-O-hexoside, kaempferol-3-O-sophoroside-7-O-glucoside, 3-hydroxyhispidin-3, 40-di-O-glucoside, quercetin-3,7-di-O-glucoside, quercetin-3-O-(caffeoyl)-glucoside, kaempferol-3,7-di-O-glucoside, flavonol-di-O-hexoside, equisetumpyrone, caffeic acid derivative, kaempferol-3-O-sophoroside, ferulic acid derivative, kaempferol-3-O-glucoside) *E. giganteum* (isoquercitrin (%, w/w): 1.05 $\pm$ 0.06% w/w; quercetin (mg $\cdot$ g^{-1}): 0.116 $\pm$ 0.005; kaempferol (mg $\cdot$ g^{-1}): 1.409 $\pm$ 0.061) *E. arvense* (isoquercitrin (%, w/w): 2.42 $\pm$ 0.26; quercetin (mg $\cdot$ g^{-1}): 2.313 $\pm$ 0.176; kaempferol (mg $\cdot$ g^{-1}) 0.533 $\pm$ 0.036) *E. hyemale* (isoquercitrin (%, w/w) 1.01 $\pm$ 0.04; quercetin (mg $\cdot$ g^{-1}): 0.083 $\pm$ 0.002; kaempferol (mg $\cdot$ g^{-1}) 1.380 $\pm$ 0.055 *E. bogotense* (isoquercitrin (%, w/w) 1.74 $\pm$ 0.05; quercetin (mg $\cdot$ g^{-1}): 0.033 $\pm$ 0.001; kaempferol (mg $\cdot$ g^{-1}) 2.360 $\pm$ 0.063
E. hyemale Li, 2012 [42]	These are indicated for the treatment of acute stroke, hypertension, anti-inflammatory, bleeding, and neoplastic diseases	Phenolic and flavonoid compounds
E. hyemale Pandey 2017 [16]	These are indicated by kidney disorders, urinary infections, tuberculosis, burns, gastric hemorrhage, rheumatoid arthritis, and osteoporosis	Phenols (ferulic acid isomers, feruloyl, and caffeoyl glucosides), flavonoids, flavonol glucosides, and alkaloids (Fonsa et al. 2013 [45]; Li et al. 2012 [42]; Park and Tomohiko 2011 [46])

(continued)

Table 1 (continued)

Specie Author/year	Indications and applications	Substances with potential pharmacological identified
		Silicon and other several minerals and bioflavonoids
E. ramosissimum Yusuf, 2020 [43]	The folk medicine indicates gonorrhea, bone traumas, nasal polypus, tuberculosis, prevention and treatment of neoplastic diseases (breast, stomach, liver, intestine, kidneys, tongue), tuberculosis, several internal hemorrhages, HIV, diuretics, and rheumatic diseases	Flavonoids, phenolic proteins, alkaloids, saponins, triterpenoids, and phytosterols
E. arvense, *E. sylvaticum,* *Equiseta herba* Bye, 2010 [12]	The folk medicine indicates para use as a diuretic, disinfectant, and astringent	Saponins and flavones
E. arvense Saudelli, 2018 [35]	*E. arvense* is used in traditional East Asian medicine Mu Zei due to its capacity of stimulating bone and cartilage tissue regeneration	Silicon
	This is indicated by preventing osteoporosis, in the treatment of some neoplastic diseases, and papilloma virus (HPV) treatment	

osteoblast proliferation, and increases the synthesis and differentiation of collagen and glycosaminoglycans in vitro [9–11]; (2) the second axis is focused on the synergic action of flavonoids, steroids and phenol compounds, as well as several other components, except silicon [12]; and (3) the third axis gives special attention to the synergic association of silicon, flavonoids, steroids, and several phenol compounds in bone remodeling. The silicon increases calcium absorption and other compounds with antioxidant properties, such as quercetin, kaempferol, betulinic, oleanolic and ursolic acids, and apigenin and luteolin increases osteoblast, whereas inhibits osteoclast proliferation in bone tissue repair [4, 9, 13].

However, the diverse methodology used to prepare the extracts (Table 2) makes the comparisons difficult. The solvents can be aqueous, methanolic, or ethanolic, or their mixtures and sometimes the whole plants are used. Uslu et al. have studied the influence of the solvent [water (W) or ethanol (Et)], solid/liquid (S/L) ratio, temperature, and stirring speed, on the composition of *E. arvense* (EA) extracts. The maximum yield (18.67%) was obtained for 10–50% Et (20–30 °C); the maximum total phenolic content (123 $mg \cdot g^{-1}$) was obtained for 50–90% Et; the maximum total antioxidant content (1.608 μM TEAC (Trolox equivalent antioxidant capacity) mg^{-1}) was obtained for 70–90% Et; and the maximum silicic acid amount (0.0049 $mg \cdot mg^{-1}$) was registered for S/L = 1:20, after ~7 h at 25 °C [14]. Nagai et al. also obtained total variable contents of phenolic components (8–24 $mg \cdot g^{-1}$)

Table 2 Extract preparation, experimental summary, and main results of *Equisetum* genus

Specie Author/year Country	Extract preparation/experimental summary	Main results
E. arvense Uslu 2013 [14] Turquia	*E. arvense* extracts were prepared of leaves dried, grounded with a variation of ethanol/water relation, solid-liquid ratio, temperature, stirring speed, and extraction time. The liquid extracts were filtered, and the solvent was removed by using a rotary evaporator (40 °C) and lyophilization There are determined yield percentage, silicic acid content, phenolic content, and total antioxidant capacity for several conditions of extracts obtained Cytotoxic activity in NIH3T3 mouse fibroblast cell line and antibacterial effect and minimum inhibition concentrations (MIC) against the bacterial culture of *Escherichia coli* and *S. epidermidis* and fungal culture of *C. albicans*	Yield percent of 18,67%, 0.0049 mg of silicic acid, 123 mg · g^{-1} of phenol content, and 1608 μM · g^{-1} TEAC of total antioxidant capacity in the dry extract The maximum yield of extraction has been obtained for ethanol concentration between 10–50% (20–30 °C), and the maximum silicic acid amount was obtained for the 1:20 solid-liquid ratio at 25 °C (extraction time of 7 h) The maximum total phenolic and total antioxidant contents have been obtained for conditions: ethanol concentration between 50–90% and ethanol concentrations between 70–90%, respectively (with time longer than 12 h) Extracts obtained in conditions of 90% ethanol (12 h of extraction) showed higher cytotoxic activity Extracts obtained in conditions of 10% ethanol (2 h of extraction) have not decreased the viability of fibroblast cells
E. arvense Pallag, 2018 [17] Romania	*E. arvense* extract (*EA*) was prepared by a mixture of 10 g of the plant dried with 100 mL of ethanol/water (70/30 v/v) for 20 min. The solvent was evaporated, and the solids lyophilized Antimicrobial activity and MIC of *EA* were evaluated by disk diffusion method *S. aureus* (ATCC 29213), *S. pneumoniae* (ATCC 49619), *E. coli* (ATCC 25922), *P. aeruginosa* (ATCC 27853), and *C. albicans* (ATCC 90029) and human clinical isolates – *S. aureus*, *S. epidermidis*, *S. pyogenes*, *S. agalactiae*, and group G beta-hemolytic *streptococcus* Antioxidant, anti-inflammatory, and antiapoptotic activities of *EA* were evaluated on human umbilical vein endothelial cells (HUVECs cells)	*Equisetum arvense* L. extract exhibited an antibacterial effect on gram-positive bacteria MIC values were 25 mg · mL^{-1} of *EA* for *S. aureus* (ATCC 29213) and human isolate *S. aureus* and 12.5 mg · mL^{-1} for *S. pneumoniae* (ATCC 49619), *S. pyogenes* group A, and group G beta-hemolytic *Streptococcus* MIC values were 50 mg · mL^{-1} of *EA* for *staphylococci* and 25 mg · mL^{-1} for *streptococcal* *EA* (2.5 μg · mL^{-1}) showed an antioxidant effect in hypertonic conditions by decreasing lipid peroxidation

(continued)

Table 2 (continued)

Specie Author/year Country	Extract preparation/experimental summary	Main results
E. arvense Khakestani, 2017 [34] Iran	*E. arvense* extract (*EA*) was prepared for the addition of 1000 mL hydroethanolic solution (1:1) in 200 g of plant milled in a glass balloon for 3 days. The solvent was evaporated by a rotary evaporator at 50 °C Nanofibers of poly(lactic acid) (PLA) solution incorporated with *E. arvense* extract (10% w/w) were obtained by electrospinning Physical chemistry characterization, mechanical properties, porosity, and contact angle of the prepared webs and releasing behavior (buffer pH 7.2) Cell attachment, spreading, proliferation, capacity, and differentiation osteogenic (ALP activity and calcium content assay) was assessed (in vitro) using human adipose tissue-derived mesenchymal stem cell (AT-MSC)	The yield of extract was 17.5% (w/w) of dried plant, with 17.51 + 0.84 mg · g^{-1} routine (total flavonoids) and 564.4 mg · g^{-1} of silica contents Uniform and porous nanofibers (157 + 47 nm, porosity: 80% v/v) Increased tensile strength and Young's modulus Cellular adhesion with flattened normal morphology ALP activity and quantitative analysis for the mineralization of adipose tissue-derived mesenchymal stem cells (AT-MSCs) were higher as compared to the pattern (polystyrene, 3.0 times, 21 days) and PLA with nHA (1.8 times, 21 days) on times 14 and 21 days
E. arvense Marvi, 2021 [4] Iran	*E. arvense* extract (*EA*) was prepared with 25 mL cold ethanol: 5 g plant dried and milled, maintained in a shaker at 4 °C (24 h), and centrifugation at 10,000 rpm (30 min). The solvent was removed by freeze-dried *EA* extracted powder composition (major elements) Bioactive nanofibrous coating of polyvinyl alcohol sericin and *EA* powder were obtained by the electrospinning process Morphology, adhesive strength, and bioactivity of titanium plate coating by *EA* Bioactivity evaluation and *EA* loading efficiency and in vitro release hADSC cell adhesion, proliferation, and differentiation	Uniform and bioactive nanofibrous coating of polyvinyl alcohol sericin and *EA* (diameter: 239 ± 35 nm) show no variation in diameter due increase of *EA* Silicon quantified in *EA* was 8,7% w/w Adhesive strength had 100% higher (1.16 ± 0.07 MPa, 5 mg *EA*: 1 mL of nanofiber solution) for *EA* nanofibers in relation to the nanofibers pattern (0.54 ± 0.09 MPa) Pattern sample obtained lowest adhesion strength (0.54 ± 0.09 MPa) when compared to samples *EA* 1 mg · mL^{-1}: (0.98 ± 0.14 MPa), *EA* 3 mg · mL^{-1} (1.04 ± 0.08 MPa) and *EA* 5 mg · mL^{-1} (1.16 ± 0.07 MPa) There was increased attachment, proliferation, and differentiation of the human adipose-derived stem cells (hADSCs) verified by increased ALP activity and calcium

(continued)

Table 2 (continued)

Specie Author/year Country	Extract preparation/experimental summary	Main results
		deposition assays and mRNA expressions (genes: Runx2, COL1, and OPN) due to *EA* The viability of the hADSC cells at different times was higher than 90% viability The COL1, Runx2, and OPN expression of the seeded hADSCs under osteogenic conditions showed higher value in the presence of 5 $mg \cdot mL^{-1}$ of *EA* The formed calcium complexes (ECM) rose with increasing the *EA* content, and the highest value detected was sample *EA* (5 $mg \cdot mL^{-1}$)
E. arvense Bessa, 2012 [13] Portugal	*E. arvense* extract (*EA*) was prepared by extracted of dried aerial parts of the plant using a mixture of methanol and water (1:1), for 26 days. The solvent mixture of the extract was renewed every 3 days by distillation and separation of the extracted solids Three extraction fractions named E1 (4–10 days), E2 (11–26 days), and E3 (mixture of E1 and E2) were obtained according to the extraction period The fractions E1, E2, and E3 were tested (10–1000 $\mu g \cdot mL^{-1}$) in human osteoblastic cells for osteoblastic markers in cell cultures performed on top of HA substrates; cell viability/proliferation and alkaline phosphatase (ALP) activity of hBMC cultures and real-time polymerase chain reaction (RT-PCR) analysis for expression of COL1, ALP, BMP-2, RANKL, M-CSF, and OPG	The extracts showed the proprieties of bone healing and antibacterial activity E1 extract shown increased viability/ proliferation and ALP activity for concentrations $<100\ \mu g \cdot mL^{-1}$ E1 shown inhibition on *Staphylococcus aureus* for concentrations $>500\ \mu g \cdot mL^{-1}$
E. arvense Costa Rodrigues 2012 [10] Portugal	*E. arvense* extract (*EA*) was prepared by extraction of dried aerial parts of the plant using a mixture of methanol and water (1:1), for 10 days. The solvent mixture of the extract was renewed every 3 days by distillation and	The concentration of Si in the dimethyl sulfoxide (DMSO) solution was $11.1 \pm 0.8\ \mu g \cdot mL^{-1}$ *EA* has not modified cell increases at concentrations $<0.1\ mg \cdot mL^{-1}$ and has not affected spontaneous osteoclastogenesis

(continued)

Table 2 (continued)

Specie Author/year Country	Extract preparation/experimental summary	Main results
	separation of the extracted solids The solvent was evaporated in a rotating evaporator at 50 °C. The solid phase was dissolved in DMSO EA (0.00016–0.5 mg · mL^{-1}) was evaluated in relation to the activity and differentiation in human osteoblasts and osteoclast precursor cells in non-stimulated and the presence of osteoclastogenic factors (M-CSF and RANKL) MEK and NF-κB signaling pathways and PGE_2 synthesis were evaluated in osteoclastogenic	PBMC cultures (21 days) in the presence of *EA* had increase TRAP activity (15% for 0.004 mg · mL^{-1} and 54% for 0.5 mg · mL^{-1} of *EA*, respectively) and exposure to extract (0.004 mg · mL^{-1}) and caused dose-dependent inhibitory effect in enzyme activity for different osteoclast activation and differentiation markers (MEK and NF-κB signaling and PGE_2 production) *EA* has reduced function and development for osteoclast precursor cell cultures or in cultures of osteoclastic and osteoblastic cells
E. arvense Corleto, 1999 [36] Italia	*EA* was evaluated for the treatment of osteoporosis for women in menopause	Treatments proved effective in improving bone metabolism in a short time (80 days)
By Słupski, 2021 [44] Saudelli 2018 [35]	Patients' treatments: 30 women have been treated with *EA* (80 days) 31 women have been treated with a placebo (40 days) and sequentially have been treated with *EA* (40 days) 29 women have been not treated whatsoever 32 women have been treated with OSTEOSIL calcium (80 days). Women have been treated sequentially with *EA* and OSTEOSIL Calcium for 1 year	*EA* extract and OSTEOSIL treatments in the long term (1 year) have been efficient in the increase in the total body bone densitometry
E. arvense Arbabzadegan, 2019 [9] Iran	*E. arvense* extract (*EA*) dried in shade and powdered was mixed with 70% ethanol and agitated for 72 h. The solvent has been dried in the rotary evaporator *EA* has evaluated the daily diet (30 days) of male and female Wistar rats in accord with the methodology: Control group: Control group with daily diet (DD) Positive control group: DD + 200 mg · kg^{-1} calcium/vitamin D 60 *EA* group: DD + 60 mg · kg^{-1} *EA*	The compounds 2-furancarboxaldehyde (6.1%), 5-hydroxymethylfurfural (27%), 4-(ethoxymethyl) phenol (1.4%), ursolic acid (32.9%), 1-oxaspiro [2.5]oct-5-ene (2.5%), oleanolic acid (10.9%), betulinic acid (3.6%), borane (2.1%), phosphorus trifluoride (2.2%), and bis(2-ethylhexyl) phthalate (2.8%) were identified and quantified There was an increase higher than 5% of BMD mandibular for a daily diet with 90 or 120 mg · kg^{-1} and an increase higher than 6% of BMD

(continued)

Table 2 (continued)

Specie Author/year Country	Extract preparation/experimental summary	Main results
	90 *EA* group: DD + 90 mg · kg^{-1} *EA* 120 *EA* group: DD + 120 mg · kg^{-1} *EA* The parameters of the serum levels of calcium, phosphorus, vitamin D, and spongy bone density in the maxilla and mandible after 30 days are measured	maxillary for a daily diet with 60 or 120 mg · kg^{-1}
E. arvense Kotwal et al. 2016 [11] India	*E. arvense* extract (*EA*) was prepared with 95% ethanol and dried powder of aerial part plant using a Soxhlet apparatus. The solvent was vacuum dried *EA* was utilized in accord with the methodology: Wistar ovariectomized rats and SHAM-operated rats (after 15 days of recovery period) were divided into groups and fed for 3½ months in accord with the parameters: Group SHAM: normal feed Group OVX: normal feed Group RAL: OVX rats were fed with RAL (5.4 mg · kg^{-1}) + normal feed Group N: OVX rats were fed with nutrients for body weight: calcium carbonate (104.16 mg · kg^{-1}), zinc sulfate (1.56 mg · kg^{-1}), L-ascorbic acid (93.75 mg · kg^{-1}), L-lysine (93.75 mg · kg^{-1}), L-proline (46.85 mg · kg^{-1}), L-arginine (200 mg · kg^{-1}), and vitamin D (100 IU) + normal feed Group N + *EA*: N + *E. arvense* et extract [60 mg · kg^{-1}]) [9] + normal feed Serum parameters: Calcium, osteocalcin, ALP, TRAP, high-density lipoprotein (HDL) cholesterol, cholesterol, and triglycerides have been evaluated Femur bone volume, bone mineral content, and length of bone were measured	The trabecular bone thickness of group N + *EA* increased by 11% and 46% compared to the group SHAM and OVX, respectively The cortical bone thickness of group N + *EA* increases 14% compared to group OVX Group N + *EA* results in the same serum ALP and TRAP level (210 + 22, 2.6 + 0.6) UI · L^{-1} when compared to group OVX (201 + 13, 2.4 + 0.5) UI · L^{-1} No variation significant was observed in cholesterol HDL or LDL and triglycerides, but group N + *EA* recorded low-liver weight (no abnormality was observed by histological exam)

(continued)

Table 2 (continued)

Specie Author/year Country	Extract preparation/experimental summary	Main results
	The percentage content (%) of calcium, phosphorus, and zinc was quantified	
E. arvense Nagai, 2005 [15] Japan	*E. arvense* extracts (*EA*) were prepared by shaking a mixture of 5 g of plant with water or ethanol (5x) at 4 °C (24 h) and centrifugation (30 min). The solvents were evaporated, and the solids were diluted in 1 mL of the respective original solvent (water or ethanol) The total phenolic components and proteins were determined Autoxidation test, superoxide anion radical generated from the xanthine/xanthine oxidase system, DPPH free radical, and hydroxyl radical-scavenging tests were realized	The contents of total phenolic components (8–24 mg · g^{-1}) and protein (476–919 mg · g^{-1}) in the extracts showed variations due to solvent (water or ethanolic) and the part of the plant used Water extracts have shown quantities twice as high than ethanol extract for proteins, and ethanol extracts have shown quantities twice as high for extraction of phenolic components as compared to water extracts Ethanolic and water extracts (1% of extracts) have resulted in approximately the same antioxidative activities as 5 mM ascorbic acid Ethanolic extracts showed superoxide anion radical (54–93%) and hydroxyl radical-scavenging activities (87–100%) higher than 1 mM ascorbic acid and α-tocopherol The DPPH radical-scavenging activity of extracts was low
E. arvense Grundemann, 2014 [6] German	*E. arvense* extract (*EA*) was prepared by boiling a mixture of ethanol (36%v/v) and dried plant (9: 1) for 4 h. The solution was filtered sterile *EA* was characterized phytochemically Proliferation, viability, and activity of mitogen-activated human lymphocytes and intracellular levels of effector molecules (IL-2, IFN-γ, and TNF-α) were evaluated The immunomodulatory activity was evaluated by the effects on cell growth and performed proliferation assays with increasing concentrations of the Equi*setum* extract (0.05–0.8 μg · mL^{-1}) using CFSE-labeled mitogen (PHA-L)-	The immunomodulatory capacity of *EA* was evaluated on human primary lymphocyte function in vitro for application in inflammatory disorders The flow cytometry analysis demonstrated that the proliferation of CFSE + mitogen-activated lymphocytes was inhibited to levels of no detection in the presence of the positive control cyclosporine A (CsA) compared to stimulated cells alone The presence of *EA* concentration dependently (0.05 μg · mL^{-1}: 95% ± 3.2; 0.1 μg · mL^{-1}: 89% ± 6.6; 0.2 μg · mL^{-1}: 86% ± 5.1; 0.4 μg · mL^{-1}: 78% ± 9.8; 0.8 μg · mL^{-1}:

(continued)

Table 2 (continued)

Specie Author/year Country	Extract preparation/experimental summary	Main results
	activated lymphocytes The mechanism behind the reduced cell proliferation observed and quantified apoptotic and necrotic effects on activated lymphocytes in the presence of increasing concentrations (0.05–0.8 μg · mL^{-1}) and the influence on IL-2-receptor expression and IL-2 secretion was evaluated	54% ± 29.8) decreased the proliferative capacity of human immunocompetent cells as compared to stimulated cells alone (IC50 = 1.09 μg · mL^{-1}) *EA* has modified cell proliferation capacity by partial inhibition of lymphocyte activation and IL-2 *EA*-mediated inhibition of proliferation is not by apoptosis or necrosis induction (concentrations <0.4 ug · mL^{-1}), and *EA* inhibited effector functions shown by reduction of IFN-γ and TNF-α A reduction of the proliferative capacity of T-cell activation by the mediation of inhibition of CD69 and IL-2 surface receptor *EA* reduced the expression of CD69 (0.05 μg/mL: 95% ± 6.2; 0.1 μg · mL^{-1}: 91% ± 6; 0.2 μg · mL^{-1}: 86% ± 7.5; 0.4 μg · mL^{-1}: 78% ± 8.9; 0.8 μg · mL^{-1}: 71% ± 9.7) early compared to controls, and therefore, *EA* has antiproliferative effect through inhibition of cell activation
E. hyemale Costa e Silva, 2022 Brazil [32]	*E. hyemale* was dried and carbonized at 300 °C for 2 h. The EH ash was mixed in a 50:50 v/v ethanol (E)-deionized water (DI) solution and left under stirring for 2 h at 80 °C and filtered. The ash was dispersed again in the E:DI water solution for 15 min at 80 °C and filtered three times. The pH of the suspension was corrected to pH 7.0 and filtered, and the composite of EH ash dried (BG-Carb)	The composition of EH ash and BG carb was identified to resemble as the blood plasma that favors the HA formation MC3T3-E1 cell cultured with BG carb conditioned medium showed the formation of mineralized nodules similarly to the positive control and showed no toxicity
E. hyemale Pandey 2017 [16] India	*E. hyemale* extract (*EH*) has been prepared with dried aerial parts and ground of plants and solution ethanol-water (1:1) in three times for 6 h. The solvent has evaporated by a vacuum evaporator and lyophilization Phytochemical characterization and	Iron, zinc, manganese, copper, and chromium elements were identified. Trace elements of cobalt (0.513 ± 0.049 ppm), nickel (1.251 ± 0.112 ppm), selenium (1.512 ± 0.169 ppm), molybdenum (0.478 ± 0.025 ppm), cadmium (0.272 ± 0.045), lead

(continued)

Table 2 (continued)

Specie Author/year Country	Extract preparation/experimental summary	Main results
	quantification of macro and microelements have been evaluated Free radical scavenging activity, antioxidant activity, DPPH free radical scavenging, β-carotene-linoleic acid assay, and reducing power assay were evaluated	(1.760 ± 0.324), and arsenic (0.314 ± 0.033) were quantified The total phenol and total flavonoid content were 332.3 mg · g^{-1} and 38.5 mg · g^{-1} dry extract, respectively Routine (429.4 μg · g^{-1}), β-sitosterol (372.9 μg · g^{-1}), lupeol (259.4 μg · g^{-1}), diosgenin (163.1 μg · g^{-1}), and ellagic acid (27.5 μg · g^{-1}) were quantified The effective concentration of *EH* against DPPH (scavenging effect on free radical) was 3.52 mg · g^{-1} and 2.05 mg · g^{-1} for the pattern (ascorbic acid) The antiradical power was 28.38 and 48.75 for *EH* and ascorbic acid, respectively The antioxidant activity of *EH* (1.20 mg · mL^{-1}) and α-tocopherol was 48% and 61%, respectively
E. giganteum Farinon, 2013 [38] Brazil	*E. giganteum* extract (*EG*) was prepared by decoction (EG: water of 1:11.7, 95 °C for 15 min). The solvent was dry for spray drying The effect of *EH* in cell viability assay, lymphocyte proliferation assay, articular nociception, and in vivo total leucocyte migration has been evaluated in mice with induction of antigen-induced arthritis (AIA)	*EG* maintained cell viability similar to the control cells (no cytotoxicity on lymphocytes) Treatment with EG (600 mg · kg^{-1}) reduced inflammatory parameters of acute AIA for half in relation to the vehicle treatment and significantly reduced leukocyte recruitment (43.1%) to the site of inflammation (16.4 ± 6.5) × 10^4 leukocytes. $cavity^{-1}$, as compared with vehicle (38.1 ± 4.2) × 10^4 leukocytes. $cavity^{-1}$ *EG* has shown an anti-inflammatory potential in the acute model of inflammation, an immunomodulatory effect on both B and T lymphocytes, and inhibited lymphoproliferation induced by both mitogens ConA (23.6%) and LPS (31.8%)
E. arvense *E. ramosissimum* and *E. telmateia*	*E. arvense*, *E. ramosissimum*, and *E. telmateia* extracts were obtained by a mixture of ground fresh plant material in (1 g .5 mL^{-1}) of	*Equisetum arvense*, *Equisetum ramosissimum*, and *Equisetum telmateia* extracts showed higher antioxidant proprieties in the

(continued)

Table 2 (continued)

Specie Author/year Country	Extract preparation/experimental summary	Main results
Stajner, 2006 [39] Servia	0.1 mol · L^{-1} K_2HPO_4 (pH 7.0) centrifugated at 4 °C, 15,000 rpm by 10 min Inhibition of nitric oxide (NO)-radical production, radical scavenger capacity (RSC), total reducing power, and the scavenging activity were evaluated Influence of the phosphate buffer (pH 7) in the *Equisetum* extracts on hydroxyl radical (HO∗) formation	methods DPPH (radical scavenger capacity: 51, 63, and 91%) and ESR (scavenger effect: 74, 98, and 99%), respectively. *E. telmateia* showed relevant results for FRAP (44%) and NO inhibition (35%)
E. hyemale Alves, 2019 [18] Brazil	*E. hyemale* extract was prepared by a mixture of 70% ethanol: stems dried and a chopped maintained shakeup in the macerate for 4 weeks with the solvent renewed weekly. The ethanol was evaporated by a rotary evaporator that results in an aqueous extract Phenolic compounds were determined The antimicrobial activity of compounds (Ferulic acid, luteolin, coumarin, routine) was evaluated by microdilution in broth Cytotoxicity and genotoxicity were evaluated by MTT and through cellular damage, respectively	The substances luteolin, ferulic acid, and coumarin were identified The phenolic compounds showed the MIC for the microorganism *E. faecalis*, *S. epidermidis*, *S. saprophyticus*, *S. aureus*, MRSA-IC, *L. monocytogenes*, *E. coli*, *S. eneridis*, *S. flexineri*, *P. aeruginosa*, and *K. pneumoniae* at low concentrations: ferulic acid (MIC: 0.37–0.75 ug · mL^{-1}), luteolin (MIC: 0.18–0.75 ug · mL^{-1}), coumarin (MIC: 0.37–0.75 ug · mL^{-1}), routine (MIC: 0.75 ug · mL^{-1}) There is no cytotoxicity or genotoxicity at the concentrations studied
E. pratense, *E. sylvaticum,* *E. telmateia* Batir-Marin, 2021 [40] Romania	Extracts of *E. pratense* (*EP*), *E. sylvaticum* (*ES*), and *E. telmateia* (*ET*) have been obtained from 10 g of plants dried with methanol 70% in the water bath for 60 min. The solvents were evaporated using the rotary evaporator Extracts of *EP*, *ES*, and *ET* have been obtained from 10 g of dry plant material with ethanol 70% using an ultrasonic bath at 30 °C (30 min). The solvents were evaporated using the rotary evaporator Phytochemistry of *EP*, *ES*, and *ET* has been characterized Antioxidant activity (in vitro), lipoxygenase inhibition capacity, and scavenger capacity of	Different amounts of flavonoids (epicatechin, luteolin, quercetin-3-glucoside, luteolin-7-glucoside, apigenin-7-glucoside, quercetin, apigenin, kaempferol) and polyphenol carboxylic acids (neochlorogenic acid, chlorogenic acid, caffeic acid, and ferulic acid) were identified for the three investigated equisetum species The concentration of the compounds in the same specie varied due to the change in the solvent (methanol or ethanol) The chelating capacity of ferrous ions was higher than 50% for concentrations of 10 mg · mL^{-1} of three *equisetum genus*, and the *ES* (10 mg · mL^{-1}) has obtained 80%

(continued)

Table 2 (continued)

Specie Author/year Country	Extract preparation/experimental summary	Main results
	superoxide anion of extracts have been evaluated	of chelating capacity for methanolic extract *ES*, *EP*, and *ET* have shown lipoxygenase inhibition capacity and scavenger action of the hydroxyl radical of approximately 50% (10 mg · mL^{-1} of plants) *ES* (10 mg · mL^{-1}) has resulted in lipoxygenase inhibition capacity (methanolic and ethanolic extracts) and scavenger action of the hydroxyl radical of 100% *ET* and *ES* resulted in scavenger capacity of the superoxide anion of 80% (10 mg · mL^{-1} of plants in methanol)
Equisetum Jiang 2014 [41] China	The equisetum mixture (EM): 100 mL/bottle, produced by Shanghai Guanghua Hospital, was compared with methotrexate tablets (2.5 mg/tablet)/diclofenac sodium (25 mg/tablet) in applications for RA treatments (60 cases of patients) randomly divided into experimental group and control group (3 months): EM group: EM (15 mL): three times/day Methotrexate (7.5 mg): once a week; diclofenac sodium (25 mg): three times/day Control group: methotrexate (7.5 mg): once a week; diclofenac sodium (25 mg): three times/day The criteria of clinical symptoms are evaluated: time of joint morning stiffness, index of arthralgia, index of joint swelling, grip, and analysis of curative effect by laboratory index: blood sedimentation, C-reactive protein, RF, TNF-α, and IL-10	*EM* has shown a clinical effect on RA with a total effective rate of 80% against 17% of the control group Comparison of before/after treatments of morning stiffness time: (132. 7 ± 55.6/54.4 ± 32.2), hand grip (75.1 ± 18.6/ 99.9 ± 30.6), number of swelled joint (14.1 ± 5.9/5.6 ± 3.5), number of arthralgia (17.7 ± 3.6/ 7.9 ± 3.5); and grading of joint function (1.8 ± 0.7/0.8 ± 0.683) within-group was variable *EM* group: (IL-10 (ng · mL^{-1}): before/after treatments: 46.1 ± 69.0/88.1 ± 83. 9); (TNF-α (ng · mL^{-1}): before/after treatments: 44.8 ± 7.1/40.8 ± 7.6) Control group: (IL-10 (ng· mL^{-1}): before/after treatments: 52.2 ± 73.8/75.5 ± 73.3); (TNF-α (ng · mL^{-1}): before/after treatments: 44.5 ± 8.5/42.5 ± 7.9) The parameters number of arthralgia and the swelled joint, and grading of joint function had reduced higher than 50% for times before/after. The control group had a reduction of 13%
E. arvense Menghini, 2016	In vitro and in vivo (Sprague-Dawley rats), the protective effects of the formula (5–1000 μg · mL^{-1})	The formulations (5–1000 μg · mL^{-1}) have significantly inhibited the ROS and PGE_2 production and

(continued)

Table 2 (continued)

Specie Author/year Country	Extract preparation/experimental summary	Main results
[37] Italy	containing lactoferrin 12%, *E. arvensis* (EA) 54%, soy isoflavones 34%, and vitamin D3 0.002% have been evaluated, in PBMC and C2C12 cells: activities of several inflammations and bone homeostasis mediators (IL-6, TNFα, PGE2, osteoprotegerin, RANK, RANKL, and NF-κB) after 21 days	TNFα and IL-6 gene expression There is a significant increase in OPG/RANKL ratio in aged rats
E. arvense Bye 2010 [12] Norway	The silicon content of *E. arvense* was determined using polar and non-polar solvents (water, methanol, ethanol, potassium hydroxide solution, dichloromethane, hexane)	The silicon amount was approximately 5% and the maximum water-extractable silicon was 0.3%

and proteins (476–919 mg $\cdot$ g^{-1}) in the extracts of *EA* upon changing the solvent (W or Et) and using different parts of the plant, corroborating the findings of Uslu. The protein contents in W-based extracts were twice higher than those in Et-based extracts, with the opposite situation being registered towards phenolic components that have shown doubled quantities for Et-based extracts [15]. Pandey et al. characterized the chemical profiles of *E. hyemale* (*EH*) aqueous extracts aiming to understand its role in improving bone health. The researchers identified the presence of iron, zinc, manganese, copper, and chromium elements in the extracts. They also suggested that the role of the *EH* in bone healing can be justified by the total contents of phenol (332.3 mg $\cdot$ g^{-1}), flavonoid (38.5 mg $\cdot$ g^{-1}), rutin (429.4 μg $\cdot$ g^{-1}), β-sitosterol (372.9 μg $\cdot$ g^{-1}), lupeol (259.4 μg $\cdot$ g^{-1}), diosgenin (163.1 μg $\cdot$ g^{-1}), ellagic acid (27.5 μg $\cdot$ g^{-1}), and trace elements of cobalt (0.51 $\pm$ 0.05 ppm), selenium (1.51 $\pm$ 0.17 ppm), nickel (1.251 $\pm$ 0.112 ppm), molybdenum (0.478 $\pm$ 0.025 ppm), lead (1.76 $\pm$ 0.32) arsenic (0.31 $\pm$ 0.03), and cadmium (0.27 $\pm$ 0.05) [16].

An efficient bone regeneration process must involve the stimulation of bone resorption and deposition, collagen production, and using bone grafts with antibacterial proprieties to prevent microorganisms' contamination [10]. The antibacterial activity of *Equisetum* genus against *Staphylococcus aureus* was observed at concentrations of or higher than 500 μg $\cdot$ mL^{-1} of EA extract reported by Bessa et al. [13]. Pallag et al. concluded that the minimum inhibitory concentration (MIC) of *EA* extract against microorganism *S. aureus* (ATCC 29213) and human isolated *S. aureus* was 25 mg $\cdot$ mL^{-1} and decreased to 12.5 mg $\cdot$ mL^{-1} for other streptococcal strains: *S. pneumoniae* (ATCC 49619), *S. pyogenes* group *A*, and group G beta-hemolytic Streptococcus [17]. Alves et al. reported antimicrobial activity of isolated phenolic compounds from *EH* extracts against a series of microorganisms including *Enterococcus faecalis*, *Staphylococcus epidermidis*, *Staphylococcus saprophyticus*, *S. aureus*, Methicillinresistant *Staphylococcusaureus* (*MRSA-IC*), *Listeria monocytogenes*, *Escherichia coli*, *S. eneridis*, *S. flexineri*, *Pseudomonas aeruginosa*, and *Klebsiella pneumonia*. The MIC for those microorganisms varied in the interval concentration of the following compounds: ferulic acid (0.37–0.75 $\mu g \cdot mL^{-1}$), luteolin (0.18–0.75 $\mu g \cdot mL^{-1}$), coumarin (0.37–0.75 $\mu g \cdot mL^{-1}$), and routine (0.75 $\mu g \cdot mL^{-1}$) [18].

The need for bone regeneration can be triggered by bone loss due to traumatic accidents or diseases, such as osteoporosis and rheumatoid arthritis and yet for preventing lost bone density [7, 8]. The treatment of critical bone defects may include autografts, bone allografts, and metallic prostheses [19].

Autografts are gold-standard grafts, despite the limited bioavailability. Furthermore, harvesting in the donor site leads to surgical morbidity. Allografts from cadaveric donors are available in larger quantities but may present lower performance in terms of osteoconduction, need extra steps to be processed without the risks of disease transmission, and are still limited by rejection by host tissues. Metallic prostheses are expensive, exhibit poor cell adhesion and osteointegration, and sometimes implant loosening [4, 19]. Alternatively, synthetic materials, such as hydroxyapatite, bioactive glass, or ceramic materials with porous structures resembling the native bone tissues have been developed to improve specific properties of bone grafts like osteoconductivity, osteointegration, and osteoinductivity. Through the suitable control of their composition, shape, pore size, the porosity of scaffolds, or the nano-surface of composite coatings, synthetic bone grafts can also be endowed with antimicrobial activity [17].

Osteoinductive capacity can stimulate the native pluripotent cells to differentiate into bone-producing cells, and the osteoconductive capacity can accelerate osteointegration by scaffolding bone growth. The bioactive surface premises more efficient vascularization and ossification [4, 7, 19]. However, bone grafts are often prepared from high purity and costly raw materials through complex processes [19]. The composition of *Equisetum* genus partially charred and minimally processed can be a viable solution to mitigate the time-consuming processes and the high costs of synthetic materials while offering several additional components with antibacterial and osteoconductive activity.

Horsetail tissues contain up to 40% carbon, 25% amorphous silica, and high percentages of potassium, calcium, iron, magnesium, sodium, sulfur, phosphorus, and zinc elements [2, 20–26]. All these elements revealed benefits to humans [27]. Carbon composites are excellent for scaffold applications due to their low density, high specific surface area [4], and porosity to allow space for the growth of new bone tissue [28]. Through several mechanism, silicon contributes to the formation and regulation of collagen and other bone tissues. The hydroxyapatite phase [$Ca_{10}(PO_4)_6(OH)_2$ – HA] of bone is composed basically of calcium, phosphorus, and some magnesium. Zinc increases alkaline phosphatase activity and inhibits osteoclast differentiation. Almost all the other plant elements (Na, K, Mg, Ca, HCO_3^-, HPO_4^{2-}, and SO_4^{2-}) are part of the composition of blood plasma and contribute to the mineralization and formation of *HA* or are micro elements of the humans with several relevant activities (antibacterial, healing, and regulating bone metabolism) [27–31]. Costa e Silva et al. prepared a composite material produced by partially charring *E. hyemale* (*EH*) and evaluated its composition, physical chemistry properties, biomineralization capacity (the formation of a surface HA layer upon immersing in SBF solution), and cytotoxicity in immortalized murine calvarial cell line (MC3T3-E1) cell cultures for 14 days. The main identified components (in mg · kg^{-1}) were C (6.85 × 10^{4}), Ca (5.52 × 10^{4}, K (2.19 × 10^{4}, Si (1.18 × 10^{4}, Mg (1.73 × 10^{4}), P (7.20 × 10^{3}, Na (1.39 × 10^{3}) Fe (6.82 × 10^{2}), and Zn (1.72 × 10^{2}). The *EH* composite promoted cellular proliferation, leading to matrix mineralization, which is positively associated with osteoblast proliferation and differentiation [32, 33]. Therefore, the complete composition of composites and extracts of *EH* are potential candidates for being used as bone graft materials with synergic effects on the healing or regeneration of bone. In the same line, the proliferation and differentiation of human osteoclast precursor cells in the presence of *EA* extract (0.00016–0.5 mg · mL^{-1}) were evaluated in non-stimulated and in the presence of osteoclastogenic factors [macrophage-colony stimulating factor (M-CSF) and receptor activator of nuclear factor κB ligand (RANKL)]. In the presence of the *EA* extract, peripheral blood mononuclear cell (PBMC) cultures (21 days) showed increased tartrate-resistant acid phosphatase (TRAP) activity (15% for 0.004 mg · mL^{-1} and 54% for 0.5 mg · mL^{-1} of the EA extract), while the exposure to extract caused a dose-dependent reduction in the enzyme activity. *EA* extract reduced human osteoclast development and function of osteoclast precursor cell cultures, even in co-cultures of osteoclastic and osteoblastic cells [10].

Still, as an alternative to conventional grafts, *EA* extracts were utilized in studies by Khakestani et al. associated with nanofibers of poly(lactic acid) (PLA). The yield of extract was 17.5% (w/w) of dried plant, with 17.51 ± 0.84 mg · g^{-1} and 564.4 mg · g^{-1} of routine (total flavonoids) and silica contents, respectively. Quantitative analysis revealed higher values of alkaline phosphatase (ALP) activity and mineralization of adipose tissue-derived mesenchymal stem cells (AT-MSCs) seeded on the experimental surface after 21 days compared to the control (polystyrene, 3.0 times more) or PLA with nano HA (nHA) (1.8 times more) [34]. Marvi et al. obtained uniform bioactive nanofibrous coatings of polyvinyl alcohol-sericin and *EA* extract with a diameter of 239 ± 35 nm. Enhanced attachment, proliferation, and differentiation of human adipose-derived stem cells (hADSCs) in the presence of *EA* were confirmed by increased ALP activity and calcium deposition assays and upregulation of bone-related mRNA expressions (genes: runt-related transcription factor 2 (*Runx2*), collagen type I (*COL1*), and osteopontin (*OPN*)) [4].

Recurrent bone diseases, such as osteoporosis and arthritis appeal to suitable bone regeneration solutions. Osteoporosis affects more than 200 million people worldwide and is characterized by a progressive reduction of bone mass (osteopenia) and an increased risk for bone fractures [8, 35]. The reduction of bone mass is attributed to the microstructural deterioration of bone tissue and, consequently, to the reduction in bone density (osteoporosis) [8, 35].

The increase in bone porosity derives from the imbalance between bone formation (osteoblastic activity) and bone resorption (osteoclastic activity) processes, which can be recurrently associated with estrogen deficiency, aging, and oxidative stress [8]. The available osteoporosis treatments involve inhibition of osteoclast proliferation and induction of osteoclast apoptosis through the use of antiresorptive drugs, or by stimulating proliferation and differentiation of osteoblast cells responsible for bone formation. A balanced diet rich in complements, antioxidants, and calcium will help achieve this goal. Antiresorptive drugs include bisphosphonates, estrogen, selective estrogen receptor modulators (SERMs), and calcitonin. Anabolic steroids and PTH (teriparatide), abaloparatide, and romosozumabe are available sources for stimulating bone formation [8].

All types of treatments for osteoporosis might cause adverse effects, such as nausea, diarrhea, peptic ulceration, the possibility of osteomalacia, risk of pulmonary embolism and cancer, hypercalcemia, and hypercalciuria [8]. Garattini et al. conclude that for a 10-year treatment, about 2% of women experienced serious side effects [35]. Several in vitro and in vivo studies of extracts and composites of the *Equisetum* genus revealed their capacity to increase bone mass, accelerate fracture healing, and exert antioxidant activity, thus improving the bone microarchitecture and reducing or even reversing osteoporosis [8]. Corlleto has shown that the human use of *EA* extracts improved bone metabolism in short-term analysis (80 days). Two long-term (1 year) treatments also revealed the effectiveness of *EA* extracts in increasing the total body bone densitometry [35, 36]. Alimentary supplementations are also utilized to control osteoporosis. The *EA* extract used in the daily diet of male and female Wistar rats (at concentrations of 90 or 120 $mg \cdot kg^{-1}$ over 30 days) resulted in mandibular bone mineral density (BMD) increase of >5% and > 6%, respectively [9]. In the same context, Wistar ovariectomized rat groups (OVX) and healthy rats (SHAM) operated rat groups were examined before and after 3½ months of systematic incorporation of several nutrients in the feeding mixture for body weight: calcium carbonate (104.16 $mg \cdot kg^{-1}$), zinc sulfate (1.56 $mg \cdot kg^{-1}$), L-ascorbic acid (93.75 $mg \cdot kg^{-1}$), L-lysine (93.75 $mg \cdot kg^{-1}$), L-proline (46.85 $mg \cdot kg^{-1}$), L-arginine (200 $mg \cdot kg^{-1}$), vitamin D (100 IU), and *EA* extract [60 $mg \cdot kg^{-1}$]). The trabecular bone thickness increased by 11% and 46% for rat groups with the diet with 60 $mg \cdot kg^{-1}$ *EA* in comparison to the placebo SHAM and OVX groups, respectively. The cortical bone thickness of the *EA* group has increased by 14% compared to OVX groups [11]. In vitro and in vivo effects of *EA* formulation (*ES* 54%, lactoferrin 12%, soy isoflavones 34%, and vitamin D3 0.002%) were evaluated. In vivo, Sprague-Dawley rats treated for 21 days with the *EA* formulation (5 mg/rat) showed a significant reduction in inflammatory prostaglandin E_2 (PGE_2) and nuclear factor-κB (NF-κB) activity and an increase in osteoprotegerin (OPG): receptor activator of nuclear factor κB ligand (RANKL) ratio in aged rats compared to the respective control group [37]. In PBMC and mouse myoblast *cell* line (C2C12), cell evaluation of the several pro-inflammatory and bone homeostasis mediators [Interleukin 6 (IL-6), tumor necrosis factor-α human (TNFα), PGE_2, OPG, receptor activator of nuclear factor (RANK), RANKL, and NF-κB)] showed that the *EA* formulations (5–1000 $\mu g \cdot mL^{-1}$) were able to inhibit reactive oxygen species (ROS) and PGE_2 production and inhibited both TNFα and IL-6 gene expression [37].

Arthritis is a chronic joint inflammatory disorder that can be subclassified into osteoarthritis and rheumatoid arthritis (RA). Osteoarthritis (OA) is a musculoskeletal disorder affecting approximately 15% of the population [1] and is induced by pro-inflammatory cytokines, e.g., interleukin 1 (IL-1), IL-6, and TNF-α. Degradation of collagen type II by collagenase-1 and collagenase-3 (also called matrix metalloproteases 13 MMP-13) represents one of the biochemical hallmarks of osteoarthritis. The main factors that aggravate OA risk are aging, hormonal modification, overweight, genetics, diet, and physical activities with biomechanical stress. Rheumatoid arthritis (RA) is a chronic inflammatory joint disease characterized by inflammation of the synovium that leads to the destruction of cartilage and bone [38]. This is a chronic progressive systemic autoimmune disease associated with high levels of oxidative stress and inflammatory mediators affecting 1% of the world population. The disease evolution generates disability and increased risk for cardiovascular disease, lymphoma, and death. Treatments of RA are with a wide variety of medicines ranging from steroidal/nonsteroidal anti-inflammatory drugs (NSAID and pain killers) to potent biological agents targeting specific immune and inflammatory pathways, such as TNF-α inhibitors and IL-1 receptor antagonists. The adverse effects include gastrointestinal ulceration, hemorrhagic events, and nephrotoxicity induced by NSAID; infusion hypersensitivity reactions and auto-immune responses (e.g., lupus-like syndrome) triggered by TNF-α inhibitors; increased risk of severe infection, affecting mainly the respiratory tract, caused by biological drugs (anakinra, rituximab, or abatacept); fatal cytopenia induced by methotrexate; etc. [38].

Several studies converged on the beneficial action of the *Equisetum* genus against arthritis diseases. In vivo studies with *E. giganteum* (*EG*, 600 mg $\cdot$ kg^{-1}) have reduced inflammatory parameters of antigen-induced arthritis (AIA) for half and leukocyte recruitment in 43.1% [to the site of inflammation ($16.4 \pm 6.5 \times 10^4$ leukocytes/cavity)] compared to vehicle ($38.1 \pm 4.2 \times 10^4$ leukocytes/cavity). *EG* inhibited lymphoproliferation induced by both mitogens concanavalin A (ConA, 23.6%) and lipopolysaccharide (LPS, 31.8%), showing an immunomodulatory effect on both B and T lymphocytes and good cell viability [38].

Antioxidants as phenolic components of the *Equisetum* genus can realize the control of oxidative stress levels and inflammatory mediators. Nagai et al. evaluated the composition of extracts and the antioxidant proprieties of W-based or E-based extracts of *EA* extract. For proteins, W-based extracts have shown quantities twice as higher compared to E-based extracts, while the opposite situation was observed for the extraction of phenolic components. Ethanolic and water extracts (1% of extracts) have resulted in approximately the same antioxidative activities as 5 mM ascorbic acid. Ethanolic extracts showed a reduction of superoxide anion radical (54–93%) and hydroxyl radical-scavenging activities (87–100%) higher than 1 mM ascorbic acid and α-tocopherol [15]. The immunomodulatory capacity of *EA* extract was evaluated by Grundmann et al. on human primary lymphocyte function in vitro for application in inflammatory disorders. The flow cytometry analysis demonstrated that the proliferation of carboxyfluorescein diacetate succinimidyl ester (CFSE) + mitogen-activated lymphocytes was inhibited to levels of no detection in the presence of the positive control cyclosporine A (CsA) compared to stimulated cells alone. The presence of *EA* extract in increasing concentrations decreased the proliferative capacity of human immunocompetent cells compared to stimulated cells alone (IC_{50} = 1.09 μg · mL^{-1}): (0.05 μg · mL^{-1}: 95% ± 3.2; 0.1 μg · mL^{-1}: 89% ± 6.6; 0.2 μg · mL^{-1}: 86% ± 5.1; 0.4 μg · mL^{-1}: 78% ± 9.8; 0.8 μg · mL^{-1}: 54% ± 29.8). Accordingly, the *EA* extract has modified cell proliferation capacity by partial inhibition of lymphocyte activation. Furthermore, EA extract inhibited effector functions shown by the reduction of human interferon-γ (IFN-γ) and TNF-α concentrations. *EA* extract reduced the expression of CD69 (0.05 μg · mL^{-1}: 95% ± 6.2; 0.1 μg · mL^{-1}: 91% ± 6; 0.2 μg · mL^{-1}: 86% ± 7.5; 0.4 μg · mL^{-1}: 78% ± 8.9; 0.8 μg · mL^{-1}: 71% ± 9.7) at early stages compared to controls, thus confirming the antiproliferative role of *EA* extract through the inhibition of cell activation and anti-inflammatory effect [6].

The chemical profile and antioxidant potential of *E. hyemale* (*EH*) extract have been characterized in the search for support for its role in improving bone health. The effective concentration of *EH* extract against DPPH (1,1-diphenyl-2-picrylhydrazyl) scavenging effect on free radical was 3.52 mg · mg^{-1} and 2.05 mg · mg^{-1} compared to control (ascorbic acid). The antiradical power values were 28.38 and 48.75 for *EH* extract and ascorbic acid, respectively. The antioxidant activities of *EH* extract (1.20 mg · mL^{-1}) and α-tocopherol were 48% and 61%, respectively. For reducing power assay, the effective concentration at which the absorbance would be 0.5 was recorded as 2.31 ± 0.04 and 1.76 ± 0.03 for *EH* extract and butylated hydroxy anisole (BHA), respectively [16].

Stajner evaluated the ability of three *Equisetum* genus to control inflammatory effects. *Equisetum arvense (EA)*, *Equisetum ramosissimum* (*ER*), and *Equisetum telmateia (ET)* extract showed higher antioxidant proprieties in the DPPH assay (radical scavenger capacity: 51, 63, and 91%) and ESR (scavenger effect: 74, 98, and 99%), respectively. *E. telmateia* has shown relevant results for ferric reducing antioxidant power (FRAP: 44%) and nitric oxide (NO) inhibition (35%) [39]. Different amounts of flavonoids (epicatechin, luteolin, quercetin-3-glucoside, luteolin-7-glucoside, apigenin-7-glucoside, quercetin, apigenin, kaempferol) and polyphenol carboxylic acids (neochlorogenic acid, chlorogenic acid, caffeic acid, and ferulic acid) were identified for the three investigated *Equisetum* species. The concentrations of the compounds in the same species varied according to the solvent (methanol or ethanol). The chelating capacity of ferrous ions was higher than 50% for a concentration of 10 mg · mL^{-1} of each *Equisetum* genus in E-based extracts. Still, for an *ES* methanolic extract at the same concentration (10 mg · mL^{-1}), the chelating capacity increased to 80%. *ES, EP,* and *ET* have shown lipoxygenase inhibition capacity and scavenger action of the hydroxyl radical of approximately 50% (10 mg · mL^{-1} of plants). *ES* (10 mg · mL^{-1}) has resulted in lipoxygenase inhibition capacity (methanolic and ethanolic extracts) and scavenger action of the hydroxyl radical of 100%. *ET* and *ES* resulted in scavenger capacity of the superoxide anion of 80% (10 mg · mL^{-1} of plants in methanol) [40].

Equisetum mixture (genus not identified) (EM) extract has shown a clinical effect on RA in a trial with an effective rate of 80% against 17% of the control group.

Comparison of before/after treatments of morning stiffness time (132.67 ± 55.61/54.36 ± 32.21), hand grip (75.06 ± 18.61/99.88 ± 30.56), number of swelled joint (14.06 ± 5.93/5.64 ± 3.54), number of arthralgia (17.73 ± 3.63/7.93 ± 3.46), and grading of joint function (1.77 ± 0.68/0.77 ± 0.63) within group was variable. The parameters "number of arthralgia" and the "swelled joint" as wells as "grading of joint function" were reduced by more than 50% before/after. The control group had a reduction of 13%. The *EM* group in relation showed modifications in the following parameters: IL-10 ($ng \cdot mL^{-1}$) (before/after treatments: 46.06 ± 69.00/88.08 ± 83.89) and TNF-α ($ng \cdot mL^{-1}$) (before/after treatments: 44.75 ± 7.06/40.80 ± 7.58), whereas the control group showed IL-10($ng \cdot mL^{-1}$) (before/after treatments: 52.22 ± 73.81/75.54 ± 73.31) and TNF-α ($ng \cdot mL^{-1}$) (before/after treatments: 44.51 ± 8.53/42.52 ± 7.90) characterizing significant improvement in laboratory results for the *EM* group [41].

3 Conclusions

The synergic action of silicon and antioxidant compound derivatives of the *Equisetum* genus against bone loss is evidenced by its role in osteoporosis and arthritis diseases. Other species of *Equisetum* should be investigated for potential applications in bone regeneration. The *Equisetum* genus is a potential candidate drug for bone regeneration and yet for the prevention of bone loss. However, extensive standardized clinical studies for each species should be done to determine doses and possible adverse effects.

References

1. Kokten K, Ozel HB, Yazici H, Iskil R, Kaya Z (2020) Fatty acid compositions of different Equisetum species. Chem Nat Compd 56(6):1117–1119
2. Neumann M, Wagner S, Noske R, Tiersch B, Strauch P (2010) Morphology and structure of biomorphous silica isolated from Equisetum hyemale and *Equisetum telmateia*. Z Naturforsch B 65(9):1113–1120
3. Elgorriaga A, Escapa IH, Rothwell GW, Tomescu AM, Rubén Cúneo N (2018) Origin of *Equisetum*: evolution of horsetails (Equisetales) within the major euphyllophyte clade Sphenopsida. Am J Bot 105(8):1286–1303
4. Marvi MS, Nourmohammadi J, Ataie M, Negahdari B, Naderi M (2021) Surface modification of titanium implants via electrospinning of sericin and *Equisetum arvense* enhances the osteogenic differentiation of stem cells. Int J Polymeric Mat 71:1–12
5. Francescato LN, Debenedetti SL, Schwanz TG, Bassani VL, Henriques AT (2013) Identification of phenolic compounds in *Equisetum giganteum* by LC–ESI-MS/MS and a new approach to total flavonoid quantification. Talanta 105:192–203
6. Gründemann C, Lengen K, Sauer B, Garcia-Käufer M, Zehl M, Huber R (2014) *Equisetum arvense* (common horsetail) modulates the function of inflammatory immunocompetent cells. BMC Complement Altern Med 14(1):1–10
7. Xie C, Ye J, Liang R, Yao X, Wu X, Koh Y et al (2021) Advanced strategies of biomimetic tissue-engineered grafts for bone regeneration. Adv Healthc Mater 10(14):2100408

8. Putnam SE, Scutt AM, Bicknell K, Priestley CM, Williamson EM (2007) Natural products as alternative treatments for metabolic bone disorders and for maintenance of bone health. Phytother Res 21(2):99–112
9. Arbabzadegan N, Moghadamnia AA, Kazemi S, Nozari F, Moudi E, Haghanifar S (2019) Effect of *Equisetum arvense* extract on bone mineral density in Wistar rats via digital radiography. Caspian J Intern Med 10(2):176
10. Costa-Rodrigues J, Carmo S, Silva J, Fernandes M (2012) Inhibition of human in vitro osteoclastogenesis by *Equisetum arvense*. Cell Prolif 45(6):566–576
11. Kotwal SD, Badole SR (2016) Anabolic therapy with *Equisetum arvense* along with bone mineralising nutrients in ovariectomized rat model of osteoporosis. Ind J Pharmacol 48(3):312
12. Bye R, Thingstad SF, Paulsen BS (2010) Horsetail (*Equisetum* spp,) as a source of silicon supplement in human nutrition – a myth? Int J Geogr Inf Syst 16(2):119–125
13. Bessa Pereira C, Gomes P, Costa-Rodrigues J, Almeida Palmas R, Vieira L, Ferraz M et al (2012) *Equisetum arvense* hydromethanolic extracts in bone tissue regeneration: in vitro osteoblastic modulation and antibacterial activity. Cell Prolif 45(4):386–396
14. Uslu ME, Erdoğan İ, Bayraktar O, Ateş M (2013) Optimization of extraction conditions for active components in *Equisetum arvense* extract. Rom Biotechnol Lett 18(2):8115–8131
15. Nagai T, Myoda T, Nagashima T (2005) Antioxidative activities of water extract and ethanol extract from field horsetail (tsukushi) *Equisetum arvense* L. Food Chem 91(3):389–394
16. Pandey G, Khatoon S (2017) Evaluation of phytochemical profile and antioxidant activity of *Equisetum hyemale* L. World J Pharm Res 6(3):723–737
17. Pallag A, Filip GA, Olteanu D, Clichici S, Baldea I, Jurca T et al (2018) Equisetum arvense L. extract induces antibacterial activity and modulates oxidative stress, inflammation, and apoptosis in endothelial vascular cells exposed to hyperosmotic stress. Oxidative Med Cell Longev 2018:1–14
18. dos Santos Alves CF, Bonez PC, de Souza Ebling M, Casagrande C, Freitas L, Dolwitsch C et al (2019) Antimicrobial, cyto and genotoxic activities of *Equisetum hyemale*. Pharm J 11(6): 1563–1571
19. Li Y, Pan Q, Xu J, He X, Li HA, Oldridge DA et al (2021) Overview of methods for enhancing bone regeneration in distraction osteogenesis: potential roles of biometals. J Orthop Translat 27: 110–118
20. Permatasari N, Sucahya TN, Nandiyanto ABD (2016) Agricultural wastes as a source of silica material. Indonesian J Sci Technol 1(1):82–106
21. Perry CC, Keeling-Tucker T (2003) Model studies of colloidal silica precipitation using biosilica extracts from *Equisetum telmateia*. Colloid Polym Sci 281(7):652–664
22. Sapei L, Nöske R, Strauch P, Paris O (2008) Isolation of mesoporous biogenic silica from the perennial plant *Equisetum hyemale*. Chem Mater 20(5):2020–2025
23. Sola-Rabada A, Rinck J, Belton DJ, Powell AK, Perry CC (2016) Isolation of a wide range of minerals from a thermally treated plant: *Equisetum arvense*, a Mare's tale. J Biol Inorg Chem 21(1):101–112
24. Steinborn C, Potterat O, Meyer U, Trittler R, Stadlbauer S, Huber R et al (2018) In vitro anti-inflammatory effects of *Equisetum arvense* are not solely mediated by silica. Planta Med 50(08): 519–526
25. Tayebee R, Pejhan A, Ramshini H, Maleki B, Erfaninia N, Tabatabaie Z et al (2018) *Equisetum arvense* as an abundant source of silica nanoparticles. SiO2/H3PW12O40 nanohybrid material as an efficient and environmental benign catalyst in the synthesis of 2-amino-4H-chromenes under solvent-free conditions. Appl Organomet Chem 32(1):e3924
26. Zemnukhova L, Panasenko A, Fedorishcheva G, Ziatdinov A, Polyakova N, Kuryavyi V (2012) Properties of silicon prepared from plant raw materials. Inorg Mater 48(10):971–976
27. Shih S-J, Chou Y-J, Chien I-C (2012) One-step synthesis of bioactive glass by spray pyrolysis. J Nanopart Res 14(12):1–8

28. Onna D, Minaberry Y, Jobbágy M (2015) Hierarchical bioglass scaffolds: introducing the "milky way" for templated bioceramics. J Mater Chem B 3(15):2971–2977
29. Wu C, Zhou Y, Xu M, Han P, Chen L, Chang J et al (2013) Copper-containing mesoporous bioactive glass scaffolds with multifunctional properties of angiogenesis capacity, osteostimulation and antibacterial activity. Biomaterials 34(2):422–433
30. Karakuzu-Ikizler B, Terzioğlu P, Basaran-Elalmis Y, Tekerek BS, Yücel S (2020) Role of magnesium and aluminum substitution on the structural properties and bioactivity of bioglasses synthesized from biogenic silica. Bioactive Mater 5(1):66–73
31. Neščáková Z, Zheng K, Liverani L, Nawaz Q, Galusková D, Kaňková H et al (2019) Multifunctional zinc ion doped sol–gel derived mesoporous bioactive glass nanoparticles for biomedical applications. Bioactive Mater 4:312–321
32. RMC C e S, Diniz IM, Gomes N, Silva GJ, Ferreira J, Freitas Filho R et al (2022) *Equisetum Hyemale*-derived unprecedented bioactive composite for hard and soft tissues engineering. Sci Rep 12:13425
33. Santhiya D, Kumari Alajangi H, Anjum F, Murugavel S, Ganguli M (2013) Bio-inspired synthesis of microporous bioactive glass-ceramic using CT-DNA as a template. J Mater Chem B 1(45):6329–6338
34. Khakestani M, Jafari SH, Zahedi P, Bagheri R, Hajiaghaee R (2017) Physical, morphological, and biological studies on PLA/n HA composite nanofibrous webs containing *Equisetum arvense* herbal extract for bone tissue engineering. J Appl Polym Sci 134(39):45343
35. Saudelli G, Tinti L, Suffritti G (2018) A review on the treatment of osteoporosis with *Equisetum arvense*. Gen Med 6(313):2
36. Corletto F (1999) Female climacteric osteoporosis therapy with titrated horsetail (*Equisetum arvense*) extract plus calcium (osteosil calcium): randomized double blind study. Miner Ortoped Traumatol 50(5):201–206
37. Menghini L, Ferrante C, Leporini L, Recinella L, Chiavaroli A, Leone S et al (2016) A natural formula containing lactoferrin, *Equisetum arvensis,* soy isoflavones and vitamin D3 modulates bone remodeling and inflammatory markers in young and aged rats. J Biol Regul Homeost 30(4):985–996
38. Farinon M, Lora PS, Francescato LN, Bassani VL, Henriques AT, Xavier RM et al (2013) Effect of aqueous extract of giant horsetail (*Equisetum giganteum* L.) in antigen-induced arthritis. Open Rheumatol J 7:129
39. Štajner D, Popović B, Čanadanović-Brunet J, Boža P (2006) Free radical scavenging activity of three *Equisetum* species from Fruška gora mountain. Fitoterapia 77(7–8):601–604
40. Batir-Marin D, Boev M, Cioanca O, Mircea C, Burlec AF, Beppe GJ et al (2021) Neuroprotective and antioxidant enhancing properties of selective *Equisetum* extracts. Molecules 26(9):2565
41. Jiang X, Qu Q, Li M, Miao S, Li X, Cai W (2014) Horsetail mixture on rheumatoid arthritis and its regulation on TNF-α and IL-10. Pak J Pharm Sci 27(6 Suppl):2019–2023
42. Li H, Wang P, Liu Q, Cheng X, Zhou Y, Xiao Y (2012) Cell cycle arrest and cell apoptosis induced by *Equisetum hyemale* extract in murine leukemia L1210 cells. J Ethnopharmacol 144(2):322–327
43. Yusuf M, Shrivastav A, Porwal M, Khan NA (2020) A review on *Equisetum ramosissimum*. J Drug Deliv Ther 10(5):311–315
44. Słupski W, Jawień P, Nowak B (2021) Botanicals in postmenopausal osteoporosis. Nutrients 13(5):1609
45. Fons F, Froissard D, Bessière JM, Fruchier A, Buatois B, Rapior S (2013) Volatile composition of six horsetails: prospects and perspectives. Nat Product Communications 8(4): 1934578X1300800426
46. Park BJ, Tomohiko M (2011) Feruloyl, caffeoyl, and flavonol glucosides from Equisetum hyemale. Chemistry of Natural Compounds 47(3):363–365

Pteridophytes as Ecological Indicators in Legislation: A Case Study in Southern Brazil

27

Aline Possamai Della

Contents

Abstract

Pteridophytes are considered good ecological indicators since their geographic distribution is highly correlated with abiotic variables. They have already been used as indicators for various purposes, including to characterize vegetation types and successional stages. In Santa Catarina, a state located in Southern Brazil, there are three resolutions (04/1994 for forests, 261/1999 for Restinga, and 423/2010 for high-altitude fields) of the National Environment Council (Conama – Brazil), which establish the parameters to characterize the primary vegetation and the successional stages of vegetation formations. Among the aspects of the composition of taxa highlighted in these resolutions are the indicator species. In this chapter, I analyzed the pteridophytes species considered indicators of primary vegetation, and the initial, medium, and advanced

A. Possamai Della (✉)
Department of Botany, Institute of Biosciences, Laboratory of Systematics and Biogeography of Vascular Plants, University of São Paulo, São Paulo, Brazil

H. N. Murthy (ed.), *Bioactive Compounds in Bryophytes and Pteridophytes*,
Reference Series in Phytochemistry, https://doi.org/10.1007/978-3-031-23243-5_32

successional stages of regeneration of these formations. For this analysis, I established criteria that define good ecological indicators, and that characterize the species of these successional stages. Nine species were considered as indicators in the resolution of the high-altitude fields, five species in the Restinga, and one in the forests. Most of these species were not considered good indicators according to the established criteria. In addition, there are inconsistencies in these norms, such as names that do not exist and combinations that were not found in the literature. Revisions to these resolutions are needed, as well as greater integration between science and policy to make them more appropriate and informative.

Keywords

Bioindicators · Ferns · Forests · High-altitude fields · Lycophytes · Restinga

Abbreviations

Conama	The National Environment Council
SC	Santa Catarina state

1 Introduction

In this chapter, the use of pteridophytes as ecological indicators in laws that aim to promote the characterization of the successional stages of forests, high-altitude fields, and Restingas will be discussed. For this, Brazilian laws will be considered as an example. Initially, an overview of the study group will be presented, later a small bibliographic review of the central term of this chapter, ecological indicators, will be presented. Finally, the results of the analysis of the species considered as indicators in these laws will be presented and discussed.

2 Pteridophytes

Pteridophytes correspond to a group of seedless vascular plants, which present a pronounced alternation of heteromorphic and independent generations (gametophyte vs. sporophyte) [1]. Pteridophytes are popularly known as ferns, tree ferns, lycopods, and horsetails. This group is ancient in evolutionary terms, it is estimated that it appeared 400 million years ago [2]. Despite this, most of the current fern diversity (Leptosporangiates, which correspond to >80% of current fern species) has accumulated since the Cretaceous, after the emergence and radiation of angiosperms [3]. This recent diversification was possibly an opportunistic ecological response to the radiation of the flowering plants, which came to dominate the terrestrial ecosystems [3].

Pteridophytes comprise two evolutionary lineages, the lycophytes and ferns. Lycophytes diverged early in the phylogenetic tree of vascular plants, and encompass three families (Lycopodiaceae, Selaginellaceae, and Isoëtaceae). These families are part of the Lycopodiopsida, which comprises less than 1% of current tracheophytes. Ferns, Polypodiopsida, comprise 48 families [4]. The main difference between these groups is the presence of microphylls in lycophytes and megaphylls in ferns. This last group is part of a larger taxon, called euphyllophytes, which also includes spermatophytes. As can be seen from this paragraph, pteridophytes correspond to a paraphyletic group, however, the term pteridophyte is still used in scientific and colloquial literature, and is being used here.

It is estimated that there are around 12,000 species of pteridophytes worldwide, of which 10,578 are ferns and 1,338 are lycophytes [4]. This group of plants has a wide geographic distribution, from tundra to tropical forests, including deserts [5]. Despite this, some regions have greater species richness, such as the Andes and Southeastern and Southern Brazil [6, 7]. The tropical and subtropical mountain regions are home to a disproportionate amount of pteridophytes species in relation to the area they occupy, this can be driven by ecoclimatic factors, such as precipitation and edaphic specialization [8].

The microscopic spores of pteridophytes can disperse over great distances [6], which explains the wide geographic distribution of many species, as well as the existence of recent intercontinental disjunctions, and the absence of endemism on islands very close to the continents [6].

Despite this, the distribution of pteridophytes on a restricted spatial scale is not random. Soil characteristics (texture and fertility), atmospheric temperature and humidity, precipitation, and light intensity are determining factors for the distribution of these plants [9–12]. Furthermore, they are not dependent on pollens and seed vectors [13].

3 Ecological Indicators

Ecological indicators can be defined as species, or groups of species, that reflect and/or predict abiotic or biotic conditions in an environment [14, 15]. They can indicate changes in habitat, community, ecosystem, or even the diversity of other groups of organisms in an area [16]. Furthermore, they can also be used in extensive research, which studies very difficult or inconvenient attributes, or when time and budget are constrained [16, 17].

Ecological indicators can be characterized by different taxonomic levels, such as family, genus, and species, or even by ecological attributes, such as richness, abundance, and diversity [18, 19]. Nonetheless, indicator species are often used [12]. Indicator species can be defined as an organism that reflects a specific environmental condition, through changes in its presence, abundance, and frequency in a given location [18].

Nowadays, ecological indicators are mainly used to assess the condition of the environment (whether it is preserved, etc.), to warn about ecological problems (natural or caused by humans), and to propose remediation actions [20].

Ecological indicators can also be used as assessment tools in nature conservation norms/laws [21]. For this purpose, indicators should connect the quality of the environment with political measures, however, they are rarely used in this way [12]. The development of indicators for this purpose involves a process of selection and integration of parameters, which are not easy to obtain and establish ([21]; see Della & Falkenberg [12] for steps in selecting ecological indicators). An in-depth study of what is being indicated, what is correlated, and how the species (the taxon) fits into the ecosystem is necessary to employ the indicators accurately [18]. Furthermore, the ecological indicators should be positioned at the science-policy interface, and be able to connect these domains [21].

Many authors employ criteria for the selection of ecological indicators. Della and Falkenberg [12] presented a list of the most commonly used criteria. The use of criteria is important because it prevents bad ecological indicators from being proposed, thus reduces the waste of time and budget. According to Della and Falkenberg [12], the four criteria most used criteria are: (a) be relatively well known taxonomically; (b) have economic potential; (c) have well-known biology and natural history; and (d) be representative of related and unrelated taxa. Finally, it should be highlighted that the criteria for selecting indicators may vary depending on the objectives and/or purpose of the ecological indicators.

4 Pteridophytes as Ecological Indicators

Pteridophytes are good ecological indicators as dozen studies have already shown us (see review by Della & Falkenberg [12, 22]). These plants can provide a lot of information about the environment they occupy (such as integrity, fragmentation, etc.), due to their distribution having a high correlation with abiotic variables (such as soil characteristics, air humidity, etc.) [10, 23, 24]. Many species are sensitive to environmental changes (such as deforestation and fire), while other species may often be associated with environments that have been subjected to some level of disturbance (such as pollution, contamination, etc.) [25–27]; this contributes to their potential as ecological indicators.

Furthermore, pteridophytes are often small in size and easy to recognize in the field [12, 24]. And as mentioned in previous paragraphs, this group is rich in species and has a wide geographic distribution (occurs in most vegetation formations) [24].

Della and Falkenberg [12] from a review of 134 articles, which employed pteridophytes as ecological indicators, proposed seven types of ecological indications for which these plants are commonly used. The seven types of indications are: (a) classification of vegetation, soils, environments, and ecosystems; (b) integrity (or quality) of environments; (c) disturbance; (d) regeneration/restoration of environments; (e) climate changes; (f) contamination of the air, soil, and water; and (g) other groups of organisms. In this way, pteridophytes can be used to indicate several aspects of the environments in which they live, being useful in the analysis of environmental quality and the recovery of degraded areas [19]. Despite this potential indication, they are still little explored in norms that characterize successional stages of vegetation.

5 A Case Study in Southern Brazil

Santa Catarina (SC) is a Brazilian state located in the southern region (see Fig. 1), which has different vegetation formations. There are forests (dense and mixed ombrophilous forests, and semideciduous seasonal forest), high-altitude fields, Restingas (plant formations established in sandy soils, which occur in the coastal plain region of Brazil), and mangroves (adapted from Klein [28]). In this state, there are about 480 species of pteridophytes, of which only 52 are lycophytes [29–31]. The vast majority of these species occur in forest formations, about 420; for the high-altitude fields, 64 species were listed, Restinga 47 and the mangroves 7 [31].

The National Environment Council (in Portuguese Conama – Conselho Nacional do Meio Ambiente) is the consultative and deliberative agency of the Brazilian National Environment System, which creates norms and criteria for the development of the National Environmental Policy (http://www.mma.gov.br/port/conama/estr.cfm). Among these norms are the resolutions that establish the parameters to characterize the successional stages of vegetation formations. This characterization is important for environmental licensing because, from it, the viability of authorizing the suppression of vegetation by environmental agencies is verified. Thus, these resolutions have great practical relevance in the daily routine of many professionals in the environmental area, and the population in general.

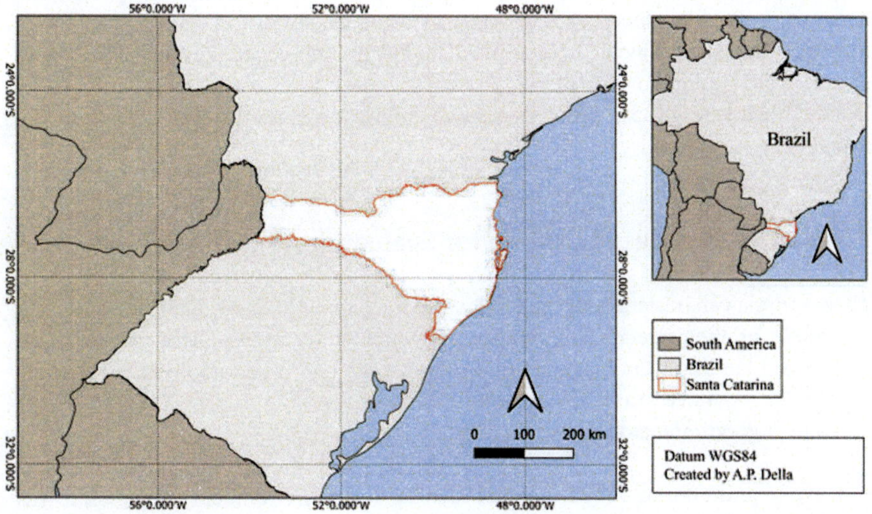

Fig. 1 Map of Brazil highlighting the state of Santa Catarina

For the state of Santa Catarina (SC), there are three Conama resolutions: 04/1994 [32], which establishes the parameters to characterize the successional stages of dense and mixed ombrophilous forests, and deciduous seasonal forest; 261/199 [33] for Restinga; and 423/2010 [34] for high-altitude fields in the south country. These norms characterize through the structure (characteristic phytophysiognomies, land cover, land use, etc.) and species composition (indicators, dominant and endemic species, height and diameter of the species, life forms, etc.), the primary or original vegetation (which can be defined as the vegetation of maximum local expression, with great biological diversity, and minimum human actions, which do not mainly affect its original characteristics of structure and species), and secondary vegetation with initial, medium, and advanced stages of regeneration (this vegetation is the result of natural processes of succession, after total or partial suppression of the primary vegetation by humans actions or natural causes) of these formations.

Among the aspects of taxa composition highlighted in these resolutions are the indicators species. These species directly reflect the abiotic and biotic characteristics of the stages in each formation. Many taxa, from different taxonomic levels, were cited as indicators in these resolutions, belonging mainly to angiosperms and pteridophytes.

Thus, in the following sections, I will present some results obtained by Della and Falkenberg [35], when examining the quality of indication of the pteridophytes considered Conama resolutions (04/1994, 261/1999, 423/2010) to the state of Santa Catarina.

Della and Falkenberg [35] elaborated two sets of criteria (presented in the next sections) to evaluate the quality of indication of pteridophytes species. The first set of criteria presents general aspects, which define a good indicator, and the second presents specific autecological characteristics for the species in each successional stage. From the examination of collection records (verified online on the *SpeciesLink* network [36]) of the species for the state of Santa Catarina, and from the review of floristic, phytosociological, or ecological studies in each vegetation formation, these criteria were applied to each pteridophyte (for more details on the methodology, see Della & Falkenberg [35]). Thus, how more criteria a species fits, the better indicator it will be.

5.1 The First Set of Criteria for Evaluation of the Indicators

These criteria can be defined as general, since they are used in the analysis of all taxa, regardless of the vegetation formation in which they occur. These criteria were established based on the bibliographic review [12, 19]. It should be noted that the species does not need to fit all the criteria.

To be a good indicator, the species needs to:

- Have a minimum of 50% collection records in the environment for which it was mentioned in the resolution
- Be often cited in studies about this environment

- Be easily identified in the field
- Be visible and/or prominent, in other words, have a more particular aspect and/or size that allows its identification in a simple visualization, without the need to "search" the plant
- Be exclusive of the environment and/or a successional stage for which it was appointed; if not exclusive, it should have a distribution with significant difference between the environments and/or stages
- Be frequent and/or abundant in the environment for which it was proposed

5.2 The Second Set of Criteria for Evaluation of the Indicators

These criteria, which evaluate the quality of indication for the successional stages, are unequal between the formations, due to their peculiarities. For resolutions 04/1994 and 261/1999, which encompass vegetation more similar to wood, a set of criteria was created; and another set was used to evaluate the resolution 423/2010, which deals with the vegetation that is more herbaceous. These criteria were also established based on a bibliographic review [19, 37–65].

5.2.1 Conama Resolutions 04/1994 and 261/1999

(A) In the initial successional stage, the species needs to:
  - Be able to survive in altered environments, in sometimes compacted soils
  - Have tolerance to intense light
  - Have tolerance to the reduction of humidity (of the soil and air)
  - Have tolerance to wide temperature variation
  - Have fast growth

(B) In the advanced successional stage, the species needs to:
  - Be able to survive in more preserved (or recovered) environments
  - Have tolerance to shading
  - Have tolerance to the increase of humidity (of the soil and air)
  - Have tolerance to the smallest temperature variation

(C) For the medium successional stage, the characteristics are not easily defined, because this stage presents both initial and advanced aspects. Thus, it is difficult to determine the characteristics of the species at this stage. Therefore, intermediate characteristics between the conditions considered above were considered.

5.2.2 Conama Resolution 423/2010

(A) In the initial successional stage, the species needs to:
  - Be able to survive in altered environments, which have suffered frequent fires, and/or intense exploitation by animals, and/or agricultural activities.

(B) To the medium and advanced successional stages, and primary vegetation, the species needs to:
  - Be able to survive in more preserved (or recovered) environments, which have not suffered recently the changes mentioned for the initial stage.

6 Analysis of the Pteridophytes Considered as Indicators in Conama Resolutions for SC

The analyzed resolutions considered 13 species of pteridophytes as indicators, see Table 1. The resolution 423/2010 considered nine indicator species, followed by 261/1999 (with five species; this resolution considered genera as indicators, these were not analyzed here; besides, for this analysis, the herbaceous and/or subshrubby Restinga was disregarded, because it does not have defined successional stages), and by resolution 04/1994 (one species). The smaller number of pteridophytes species in the forests resolution (04/1994) is incoherent because this group is more diverse in these vegetation formations (as previously mentioned). Nonetheless, in the three resolutions, pteridophytes represented a minimal percentage of the species (when compared with angiosperms), evidencing the little importance given to them in the elaboration of these norms. In the next sections, I will comment on the species considered as indicators in each resolution.

6.1 Conama Resolution 423/2010

Only *Lomariocycas schomburgkii* (treated as *Blechnum imperiale* [Fée & Glaz.] Christ in the resolution) and *Pteridium esculentum* subsp. *arachnoideum* (as *P. aquilinum* var. *arachnoideum* [Kaulf.] Brade) were considered good indicators. *L. schomburgkii* is an indicator of primary vegetation and medium and advanced stages, and *P. esculentum* subsp. *arachnoideum* of the initial stage. These two species present most of the characteristics of a good indicator (they are easily recognizable, quite visible and or prominent in their habitats, and frequent and abundant in this formation). *L. schomburgkii* has all the characteristics established here for more conserved stages, while *P. esculentum* subsp. *arachnoideum* presents all the characteristics for the initial successional stage.

Diphasiastrum thyoides (Willd.) Holub (*Lycopodiella thyoides* was the name used in the resolution; probably authors of the resolution wanted to refer to *Lycopodium thyoides* Humb. and Bonpl. ex Willd., but I did not find the combination of this species for *Lycopodiella*), *Parablechnum cordatum* (Desv.) Gasper and Salino (treated as *Blechnum regnellianum* [Kunze] C. Chr.), and *Selaginella microphylla* (Kunth) Spring need further studies to conclude if they are good indicators. There are few collection records of these three species in high-altitude fields in Santa Catarina (and few collection records showing the successional stages). However, they present some general criteria assumed for a good indicator (they are easily recognizable, quite visible, and/or prominent in their habitats, although the last one has a reduced size) and seem to survive in more preserved (or conserved) environments.

Lycopodiella alopecuroides and *Pseudolycopodiella caroliniana* (the last one as *Lycopodiella* in the resolution) should also be reviewed because they do not occur in Brazil, according to Øllgaard and Windisch [66]. In addition to this taxonomic problem, there are few collection records and references of these species in high-altitude fields in Santa Catarina. *Lycopodiella geometra* B. Øllg. and P.G. Windisch,

Table 1 Pteridophytes are considered as indicators of primary vegetation and successional stages of Conama (The National Environment Council – Brazil) resolutions, 04/1994 for forests, 261/1999 for Restinga, and 423/2010 for high-altitude fields. Species names have been updated to accepted ones. *This name was not found, so it has not been updated. Initial: refers to the initial stage of regeneration. Medium: medium stage of regeneration. Advanced: advanced stage of regeneration. Primary: primary or original vegetation

Conama Resolutions												
	04/1994	261/1999								423/2010		
	Forests	Shrubby Restinga				Arboreal Restinga				High-altitude fields		
	Initial	Primary	Initial	Medium	Advanced	Primary	Initial	Medium	Advanced	Initial	Medium	Advanced and primary
Diphasiastrum thyoides (Humb. & Bonpl. ex Willd.) Holub											X	X
*Gleichenia brasiliensis**											X	X
Gleichenella pectinata (Willd.) Ching											X	X
Lomariocycas schomburgkii (Klotzsch) Gasper & A.R. Sm.											X	X
Lycopodiella alopecuroides (L.) Cranfill											X	X
Parablechnum cordatum (Desv.) Gasper & Salino											X	X

(continued)

Table 1 (continued)

Conama Resolutions	04/1994	261/1999								423/2010		
	Forests	Shrubby Restinga				Arboreal Restinga				High-altitude fields		
	Initial	Primary	Initial	Medium	Advanced	Primary	Initial	Medium	Advanced	Initial	Medium	Advanced and primary
Pecluma robusta (Fée) M. Kessler & A.R. Sm.						X			X			
Pleopeltis lepidopteris (Langsd. & Fisch.) de la Sota		X		X	X							
Pseudolycopodiella caroliniana (L.) Holub											X	X
Pteridium esculentum subsp. *arachnoideum* (Kaulf.) Thomson	X		X				X			X		
Rumohra adiantiformis (G. Forst.) Ching		X	X		X		X	X	X			
Selaginella microphylla (Kunth) Spring											X	X
Telmatoblechnum serrulatum (Rich.) Perrie, D.J. Ohlsen & Brownsey		X		X	X							

Lycopodiella longipes (Grev. & Hooker) Holub, *Pseudolycopodiella carnosa* (Silveira) Holub, and *P. meridionalis* (recognized by Øllgaard & Windisch [66] as occurring in Brazil and previously identified with the names that begin this paragraph) survive in recovered environments and, especially, in the most conserved and humid ones. Despite this, these names have not yet been widely adopted in collection records and publications. Therefore, a review of Lycopodiaceae species is necessary for the next version of this resolution.

Gleichenella pectinata (Willd.) Ching was not considered a good indicator of primary vegetation and the medium and advanced stages of the high-altitude fields. There are no collection records or publications that mention it for these environments, showing that it is not frequent and abundant in these fields, as well as it does not survive in more recovered environments; on the contrary, it is more frequent in altered environments [38].

Gleichenia brasiliensis was cited by the resolution, but it was not found on the Flora do Brasil, Tropicos, and the International Plant Names Index (IPNI) websites. This name was also not found in the analyzed bibliography, so it was not discussed here. There is a species whose specific epithet is similar and which may have been confused in the elaboration of the resolution, *G. brasiliana* (Desv.) Spreng. This last name is appointed as a synonym of *G. pectinata* in the herbarium of the Botanical Garden of Edinburgh, in a supposed type of *G. brasiliana*.

In the resolution of high-altitude fields, the pteridophytes present secondary importance compared to the several hundred angiosperms included there. Few species were considered good indicators. In addition, there are several taxonomic problems with the species mentioned in the resolution, which determines the need for a broad revision of this norm.

6.2 Conama Resolution 261/1999

Of the five species considered as indicators, three stand out for the wide range of successional stages that were cited in the resolution: *Telmatoblechnum serrulatum* (treated as *Blechnum serrulatum* Rich. in the resolution), *Pleopeltis lepidopteris* (Langsd. & Fisch.) de la Sota (as *Polypodium lepidopteris* [Langsd. & Fisch.] Kunze), and *Rumohra adiantiformis* (G. Forst.) Ching. Despite the three species being easily identified, having high visibility, and being frequent and abundant in the Restinga, the wide range of habitats (that occur in this formation) reduces their potential to identify and, mainly, differentiate the stages in which they were cited in the resolution. Thus, it is important to verify in which type of Restinga (arboreal vs. shrub) each taxon is more abundant, to later define where they will be the best indicators. If they are more abundant in one vegetation than in the others, they can be considered good indicators; if not, their usefulness as indicators would be compromised, although they are really important elements of the Restinga flora.

It is worth mentioning that even though these plants are characteristic of the Restinga, they end up not being so useful in isolation for the characterization and distinction of the successional stages, due to the wide variety of environments in which they occur. Furthermore, for the selection of ecological indicators, exclusive species of a successional stage should be sought, because they will be more informative and more easily employed in the analysis.

Pteridium esculentum (G. Forst.) Cockayne subsp. *arachnoideum* (Kaulf.) Thomson (treated as *P. aquilinum* [L.] Kuhn) presents most of the general criteria considered for a good indicator species. This plant is easily recognized in the field, has great visibility, and a reasonable amount of collection records and studies mention it to the Restinga. In addition, it fulfills all the criteria assumed here for the initial stage and has a more restricted occurrence. Therefore, this species is considered a good indicator of the initial stage for arboreal and shrubby Restinga.

Pecluma robusta (Fée) M. Kessler and A.R. Sm. (treated as *Polypodium robustum* Fée) does not seem so frequent and abundant in the environments for which it was pointed, but it is abundant in several points of them. Furthermore, this species is not easily identified, because it is similar to *Pecluma paradiseae* (Langsd. & Fisch.) M.G. Price. There is a lack of information about this species, however, it presents characteristics established here for an advanced stage and/or for primary vegetation (personal observation). Despite some difficulties in identification, *Pecluma robusta* can be a reasonable indicator of these environments and stages of arboreal Restinga.

6.3 Conama Resolution 04/1994

In this resolution, *P. esculentum* subsp. *arachnoideum* (treated as *P. aquilinum*) was the only pteridophyte considered as an indicator (cited for the initial stage of the regeneration of the dense and mixed ombrophilous forests, and deciduous seasonal forest). This plant is a good indicator because it has a reasonable amount of collection records for forests (despite not having any explicit record for the deciduous seasonal forest of Santa Catarina) and many studies do not mention it for forest environments. Thus, it presents most of the general characteristics of a good indicator and has all the criteria assumed here for the initial stage of regeneration.

This resolution presented few indicator species, even of seeds plants. The numerical parameters (diameter, basal area, and height) also present in the resolution are more valued in its practical application than the species, which is an error frequently accepted by environmental licensing agencies, partly justified by the insufficiency of indicator species. Only one pteridophyte as an indicator is an insignificant number against more than four hundred species of this group existing in these formations [29, 31], in which many taxa have great potential for ecological indication [19]. One future revision of this resolution should incorporate new taxa.

6.4 Overview of the Three Resolutions

P. esculentum subsp. *arachnoideum* was the single species present in the three resolutions, being here considered a good indicator of the initial stage of the regeneration of all these formations. Of eight species considered in resolution 423/2010, only two are good indicators, in 216/1999 only two (of five), and in 04/1994 the only species mentioned was considered a good indicator.

Just over a dozen pteridophytes were treated as indicators in these three resolutions, which highlights the secondary importance of the group in such laws. There are inconsistencies in these norms, such as a great number of indicator pteridophytes species in formations where they are not so expressive (Restinga and high-altitude fields), and a single species where they are very prominent (forests), as well as the adoption of nonexistent or problematic taxa. Several taxa pointed out as indicators do not look good for this purpose. Therefore, it is necessary to revise these resolutions (mainly inserting more indicator pteridophytes species in the resolution of forests), aiming to make them more informative and appropriate.

7 Conclusions

Although pteridophytes are considered good indicators, as several studies have shown us, this group of plants is still little explored in the laws that characterize primary vegetation and the successional stages. As seen in the previous section, in the case study carried out with Brazillian laws (Conama), there are some difficulties, which should be resolved or further discussed, for the use of pteridophytes for this purpose, such as:

- Lack of ecological information, such as frequency, abundance, habitats, geographical coordinates, altitude, luminosity and humidity conditions, successional stages, etc. for many species. Of the analyzed registers by Della and Falkenberg [35] about half do not present information of this type.
- There are few ecological studies with pteridophytes, mainly in Restinga and high-altitude fields. Despite having a great number of studies with this group in the forests, few of them addressed the different successional stages. Therefore, the analysis realized by Della and Falkenberg [35] was limited. The ecological behavior of pteridophytes in these vegetational formations is little known. This information is important for the management of conserved areas, recuperation of altered areas, and establishment of ecological indicators.
- Insufficient scientific attention has been given to these laws, despite their enormous practical importance and the numerous political issues involved in them. The social pressure and the economic interest interfere directly with the objectives and the quality of these norms that regulate the area used for commercial, industrial, or residential projects. Professionals in the environmental area need to integrate the conservation of biodiversity with the most miscellaneous political interests. Unfortunately, nowadays, many environmental laws are elaborated by politicians who have little knowledge about flora and fauna, so the participation of scientists who are dedicated to environmental studies would be essential and could contribute to having fewer errors in these laws. In this way, we should seek to bring together and improve science and policy so we have more correct and

effective resolutions. Thus, the resolutions will be more comprehensive and employed by users and would allow the conservation of a significant part of the remaining natural areas.

I hope that in the future these difficulties will be solved, and the pteridophytes will be more explored/used in laws that deal with the characterization of vegetation and its successional stages, given the great potential that this group presents as ecological indicators.

References

1. Raven PH, Evert RF, Eichhorn SE (2007) Biologia Vegetal. Guanabara Koogan, Rio de Janeiro
2. Lehtonen S, Silvestro D, Karger DK, Scotese C, Tuomisto H, Kessler M, Peña C, Wahlberg N, Antonelli A (2017) Environmentally driven extinction and opportunistic origination explain fern diversification patterns. Sci Rep 7:4831. https://doi.org/10.1038/s41598-017-05263-7
3. Schneider H, Schuettpelz E, Pryer KM, Cranfill R, Magallon S, Lupia R (2004) Ferns diversified in the shadow of angiosperms. Nature 428:553–557
4. PPG I (2016) A community-derived classification for extant lycophytes and ferns. J Syst Evol 54:563–603
5. Moran RC (2008) Neotropical genera of ferns and lycophytes: a guide for students. New York Botanical Garden, New York
6. Tryon RM (1972) Endemic areas and geographic speciation in tropical American ferns. Biotropica 4:121–131
7. Suissa JS, Sundue MA (2020) Diversity patterns of neotropical ferns: revisiting Tryon's centers of richness and endemism. Am Fern J 110:211–232
8. Suissa JS, Sundue MA, Testo WL (2021) Mountains, climate and niche heterogeneity explain global patterns of fern diversity. J Biogeogr 48:1296–1308
9. Tuomisto H, Poulsen AD (2000) Pteridophyte diversity and species composition in four Amazonian rain forests. J Veg Sci 11:383–396
10. Nóbrega GA, Eisenlohr PV, Paciencia MLB, Prado J, Aidar MPM (2011) A composição florística e a diversidade de pteridófitas diferem entre a Floresta de Restinga e a Floresta Ombrófila Densa das Terras Baixas do Núcleo Picinguaba/PESM, Ubatuba/SP? Biota Neotrop 11:153–164
11. Patil S, Lavate R, Rawat V, Dongare M (2016) Diversity and distribution of pteridophytes from Satara District, Maharashtra (India). Plant Sci Today 3:149–156
12. Della AP, Falkenberg DB (2019) Pteridophytes as ecological indicators: an overview. Hoehnea 46:e522018
13. Abotsia KE, Boseb R, Adjossoua K, Deblauweb V, Rouhane G, Seglaa KN, Atsria KH, Kokou K (2020) Ecological drivers of pteridophyte diversity and distribution in Togo (West Africa). Ecol Indic 108:105741
14. McGeoch M (1998) The selection, testing and application of terrestrial insects as bioindicators. Biol Rev 73:181–201
15. Siddig AAH, Elisson AM, Ochs A, Villar-Leeman C, Lau MK (2016) How do ecologists select and use indicators species to monitor ecological change? Insights from 14 years of publication in Ecological Indicators. Ecol Indic 60:223–230
16. Landres PB, Verner J, Thomas JW (1988) Ecological uses of vertebrate indicator species: a critique. Conserv Biol 2:316–328

17. Oster M, Persson K, Eriksson O (2008) Validation of plant diversity indicators in semi-natural grasslands. Agric Ecosyst Environ 125:65–72
18. Jaffe D, McDonough C (2009) Indicator species. http://www.eoearth.org/view/article/153801. Accessed 11 Feb 2022
19. Della AP (2016) Pteridófitas como indicadores ecológicos: revisão geral e aplicações em Santa Catarina. Final work for the undergraduate, Federal University of Santa Catarina
20. Niemi GJ, McDonald ME (2012) Application of ecological indicators. Annu Rev Ecol Evol Syst 35:89–111
21. Turnhout E, Hisschemöller M, Eijsackers H (2007) Ecological indicators: between the two fires of science and policy. Ecol Indic 7:215–228
22. Della AP (2022) Ferns as ecological indicators. In: Marimuthu J, Fernández H, Kumar A, Thangaiah S (eds) Ferns. Springer, Singapore
23. Salovaara KJ, Cárdenas GG, Tuomisto H (2004) Forest classification in an Amazonian rainforest landscape using pteridophytes as indicator species. Ecography 27:689–700
24. Tuomisto H, Ruokolainen K (1993) Distribution of Pteridophyta and Melastomataceae along an edaphic gradient in an Amazonian rain forest. J Veg Sci 5:25–34
25. Chang JS, Yoon IH, Kim KW (2009) Heavy metal and arsenic accumulating fern species as potential ecological indicators in As-contaminated abandoned mines. Ecol Indic 9:1275–1279
26. Bergeron A, Pellerin S (2014) Pteridophytes as indicators of urban forest integrity. Ecol Indic 38:40–49
27. Heringer G, Almeida TE, Dittrich VAO, Salino A (2020) Assessing the effectiveness of protected areas for the conservation of ferns and lycophytes in the Brazilian state of Minas Gerais. J Nat Conserv 53:125775
28. Klein RM (1978) Mapa fitogeográfico do estado de Santa Catarina. Sudesul, FATMA & Herbário Barbosa Rodrigues, Itajaí
29. Gasper AL, Salino S, Vibrans AL, Sevegnani L, Verdi M, Korte A, Stival-Santos A, Dreveck S, Cadorin TJ, Schmitt L, Caglioni E (2012) Pteridófitas de Santa Catarina: um olhar sobre os dados do inventário florístico florestal de Santa Catarina, Brasil. Acta Bot Bras 26:421–434
30. Gasper AL, Salino A (2015) Samambaias e licófitas de Santa Catarina: composição, riqueza e espécies ameaçadas. Iheringia Ser Bot 70:321–342
31. Flora do Brasil 2020 (2022). http://floradobrasil.jbrj.gov.br/reflora/listaBrasil/ConsultaPublicaUC/ConsultaPublicaUC.do#CondicaoTaxonCP. Accessed 11 Feb 2022
32. Ministério do Meio Ambiente (2022) Resolução N° 04, de 4 de maio de 1994. http://www.ibama.gov.br/sophia/cnia/legislacao/MMA/RE0004-040594.PDF. Accessed 11 Feb 2022
33. Ministério do Meio Ambiente (2022) Resolução N° 261, de 30 de junho de 1999. http://www.ibama.gov.br/sophia/cnia/legislacao/MMA/RE0261-300699.PDF. Accessed 11 Feb 2022
34. Ministério do Meio Ambiente (2022) Resolução N° 423, de 12 de abril de 2010. https://cetesb.sp.gov.br/licenciamento/documentos/2010_Res_CONAMA_423.pdf. Accessed 11 Feb 2022
35. Della AP, Falkenberg DB (2019) Pteridófitas usadas na legislação como indicadoras de estágios sucessionais no Estado de Santa Catarina, Brasil. Hoehnea 46:e572018
36. SpeciesLink (2020). https://specieslink.net/. Accessed 08 Feb 2022
37. Sehnem A (1968) Blecnáceas. In: Reitz R (ed) Flora Ilustrada Catarinense. Herbário Barbosa Rodrigues, Itajaí
38. Sehnem A (1970a) Gleiqueniáceas. In: Reitz R (ed) Flora Ilustrada Catarinense. Herbário Barbosa Rodrigues, Itajaí
39. Sehnem A (1970b) Polipodiáceas. In: Reitz R (ed) Flora Ilustrada Catarinense. Herbário Barbosa Rodrigues, Itajaí
40. Sehnem A (1979) Salviniáceas. In: Reitz R (ed) Flora Ilustrada Catarinense. Herbário Barbosa Rodrigues, Itajaí
41. Sehnem A (1977) As filicíneas do sul do Brasil, sua distribuição geográfica, sua ecologia e suas rotas de migração. Pesquisas Bot 31:1–108

42. Athayde-Filho FP, Windisch PG (2006) Florística e aspectos ecológicos das pteridófitas em uma floresta de Restinga no estado do Rio Grande do Sul, Brasil. Iheringia Ser Bot 61:63–71
43. Behar L, Viégas GMF (1992) Pteridophyta da Restinga do Parque Estadual de Setiba, Espírito Santo. Bol Mus Biol Mello Leitão 1:39–59
44. Bresolin A (1979) Flora da Restinga da Ilha de Santa Catarina. Ínsula 10:1–54
45. Campestrini S (2014) Aspectos florísticos, parâmetros fitossociológicos e ecológicos nos Campos de Palmas, SC/PR, Brasil. Dissertation, Federal University of Santa Catarina
46. Daniel RB (2006) Florística e fitossociologia da Restinga herbáceo-arbustiva do Morro dos Conventos, Araranguá, SC. Disseretation, University of the Extreme South of Santa Catarina
47. Danilevicz E, Janke H, Pankowski LHS (1990) Florística e estrutura da comunidade herbácea e arbustiva da Praia do Ferrugem, Garopaba, SC. Acta Bot Bras 4:21–34
48. Gomes MAM (2009) Caracterização da vegetação de campos de altitude em unidades de paisagem na região do Campo dos Padres, Bom Retiro/Urubici, SC. Dissertation, Federal University of Santa Catarina
49. Gomes FS, Guedes MLS (2014) Flora vascular e formas de vida das formações de Restinga do litoral norte da Bahia, Brasil. Acta Biol Catarin 1:22–43
50. Gonzatti F (2015) Florística, fitogeografia e conservação das samambaias e licófitas da Região planície costeira do Rio Grande do Sul, Brasil. Dissertation, Federal University of Rio Grande do Sul
51. Gonzatti F, Valduga E, Wasum RA, Scur L (2014) Florística e aspectos ecológicos de samambaias e licófitas em um parque urbano do Rio Grande do Sul, Brasil. Rev Bras Biosci 12:90–97
52. Guimarães TB (2006) Florística e fenologia reprodutiva de plantas vasculares na Restinga do Parque Municipal das Dunas da Lagoa de Conceição, Florianópolis, SC. Dissertation, Federal University of Santa Catarina
53. Klein AS, Citadini-Zanette V, Santos R (2007) Florística e estrutura comunitária de Restinga herbácea no município de Araranguá, Santa Catarina. Biotemas 20:15–26
54. Korte A, Gasper AL, Kruger A, Sevegnani L (2013) Composição florística e estrutura das Restingas em Santa Catarina. In: Vibrans AC, Sevegnani L, Gasper AL, Lingner DV (eds) Inventário florístico floresal de Santa Catarina. Edifurb, Blumenau
55. Mania LF, Monteiro R (2010) Florística e ecologia de epífitas vasculares em um fragmento de floresta de Restinga, Ubatuba, SP, Brasil. Rodriguésia 61:705–713
56. Mauhs J (2002) Fitossociologia e regeneração natural de um fragmento de Floresta Ombrófila Mista exposto a perturbações antrópicas. Dissertation, University of Vale do Rio dos Sinos
57. Menezes-Silva S (1998) As formações vegetais da planície litorânea da Ilha do Mel, Paraná, Brasil: composição florística e principais características estruturais. Doctoral thesis, State University of Campinas
58. Paciencia MLB (2008) Diversidade de pteridófitas em gradientes de altitude na Mata Atlântica do estado do Paraná, Brasil. Doctoral thesis, University of São Paulo
59. Sampaio MB, Guarino ESG (2007) Efeitos do pastoreio de bovinos na estrutura populacional de plantas em fragmentos de Floresta Ombrófila Mista. Rev Árvore 31:1035–1046
60. Santos MG, Sylvestre LS, Araujo DSD (2004) Análise florística das pteridófitas do Parque Nacional da Restinga de Jurubatiba, Rio de Janeiro, Brasil. Acta Bot Bras 18:271–280
61. Simão C (2008) Caracterização florística e espectro biológico de refúgios vegetacionais altomontanos no Morro Anhangava, Parque Estadual Serra da Baitaca, Paraná. Dissertation, Federal University of Paraná
62. Sonehara JS (2005) Aspectos florísticos e fitossociológicos de um trecho de vegetação de Restinga no Parque Estadual do Rio da Onça – Matinhos. Dissertation, Federal University of Paraná

63. Souza MLDR, Falkenberg DB, Amaral LG, Fronza M, Araujo AC, Sá MR (1991) Vegetação do Pontal da Daniela, Florianópolis, SC, Brasil. I. Levantamento florístico e mapa fitogeográfico. Ínsula 21:87–117
64. Záchia RA, Waechter JL (2011) Diferenciação espacial de comunidades herbáceo-arbustivas em florestas costeiras do Parque Nacional da Lagoa do Peixe, Rio Grande do Sul. Pesquisas Bot 62:211–238
65. Zanin A, Longhi-Wagner HM, Souza MLDR, Rieper M (2009) Fitofisionomia das formações campestres do Campo dos Padres, Santa Catarina, Brasil. Ínsula 38:42–57
66. Øllgaard B, Windisch PG (2016) Lycopodiaceae in Brazil. Conspectus of the family II. The genera *Lycopodiella*, *Palhinhaea*, and *Pseudolycopodiella*. Rodriguésia 67:691–719

Index

H. N. Murthy (ed.), *Bioactive Compounds in Bryophytes and Pteridophytes*, Reference Series in Phytochemistry, https://doi.org/10.1007/978-3-031-23243-5

M

N

T